CW00537823

PLANTAE WILSONIANAE

PLANTAE WILSONIANAE

AN ENUMERATION OF THE WOODY PLANTS
COLLECTED IN WESTERN CHINA FOR THE
ARNOLD ARBORETUM OF HARVARD
UNIVERSITY DURING THE YEARS
1907, 1908, AND 1910
BY E. H. WILSON

EDITED BY

CHARLES SPRAGUE SARGENT

VOLUME III

Biosystematics, Floristic & Phylogeny Series,

Volume 3

Theodore R. Dudley, Ph.D., General Editor

DIOSCORIDES PRESS
Portland, Oregon

Volume III first published 1917 by Cambridge University Press

Reprinted 1988 by
Dioscorides Press
9999 S.W. Wilshire
Portland, Oregon 97225

ISBN 0-931146-01-1 (3 volume set)

Printed in Singapore

Library of Congress Cataloging-in-Publication Data

Wilson, Ernest Henry, 1876-1930.
    Plantae Wilsonianae : an enumeration of the woody plants collected
in western China for the Arnold Arboretum of Harvard University
during the years 1907, 1908, and 1910 / by E.H. Wilson ; edited by
Charles Sprague Sargent.
        p.   cm. -- (Biosystematics, floristic & phylogeny series ; v.
3)
    Reprint. Originally published: Cambridge : Cambridge University
Press, 1913, 1916-1917. (Publications of the Arnold Arboretum ; no.
4)
    Includes bibliographical references and indexes.
    ISBN 0-931146-01-1 (set) : $110.00
    1. Woody plants--China--Catalogs and collections--Massachusetts-
-Cambridge. 2. Arnold Arboretum--Catalogs.   I. Sargent, Charles
Sprague, 1841-1927. II. Title. III. Series: Publications of the
Arnold Arboretum ; no. 4.
QK355.W49 1988
582.1'5'0951--dc19   88-6946
                                                    CIP

EDITORIAL NOTE

THE publication of part III. of the third volume of the Plantae Wilsonianae finishes the enumeration of the trees and shrubs collected by Wilson during his two journeys in western China undertaken for the Arboretum. In this work, the first part of which appeared in July 1911, there are enumerated 2716 species and 640 varieties and forms of woody plants representing 429 genera in 100 natural families. Among these collections have been found 4 previously undescribed genera, 382 previously undescribed species, and 323 varieties and forms of species now first described. In addition to these, 18 new species and 11 new varieties collected by Wilson have been described in other works before the publication of the Plantae Wilsonianae. In earlier journeys in China, Mr. Wilson discovered 121 new species of plants and 22 undescribed varieties of species; of these 52 species and 13 varieties are first described in this work. His four journeys in China, therefore, have added to the ligneous flora of that country 4 genera, 521 species and 356 varieties and forms of species not previously known.

In addition to the descriptions of the new plants collected by Wilson there will be found in this work descriptions of 150 previously undescribed species, 51 new varieties, and 17 new forms found in China outside the region visited by Wilson and principally collected in Yunnan by Dr. Augustine Henry.

The study of the material in American and European herbaria necessary for the preparation of this work has made it possible to prepare partial synoptical accounts of about 30 important genera. A large part of these synoptical accounts in addition to those of the Chinese Empire include the species of Japan, and in several cases also those of the Himalayas. Among such synopses are those of such important genera as Salix, Populus, Ulmus, Carpinus, Corylus, Betula, Magnolia, Rhamnus, and the subgenus Cerasus of Prunus. Of Celtis, Sorbus, and Malus accounts are given of all the species which grow in China and Japan, and the treatment of Smilax includes the species of China and Korea. The synoptical accounts of Rosa, Dentzia, Hydrangea, and several other genera are restricted to the Chinese species.

The herbarium staff of the Arboretum, consisting of Messrs. Rehder, Wilson, Shaw and Schneider, have been assisted in the preparation of this work by a number of European specialists. Dr. E. Janczewski of Krakow has prepared the account of Ribes; Dr. E. Koehne of Berlin that of Prunus, Maddenia and Philadelphus; Dr. W. O. Focke of Bremen that of Rubus; Dr. T. Loesener of Berlin that of the Aquifoliaceae; Dr. F. Gagnepain of Paris that of the Vitaceae; Dr. A. B. Rendle of the British Museum has determined the Bambuseae collected by Wilson; Dr. J. S. Gamble, late of the Indian Forest Service, has elaborated the Lauraceae; Mr. W. G. Craib of the Royal Gardens, Kew, the Leguminosae; Mr. J. Hutchinson of the same institution the Euphorbiaceae and Rubiaceae; Dr. H. Harms of Berlin the Araliaceae; Dr. Henri Lecomte of Paris the Loranthaceae. Mr. J. B. Norton of the Department of Agriculture of the United States has prepared the account of Smilax and Heterosmilax. I take this opportunity to thank these men in the name of the Arboretum for the assistance they have given us in elaborating these difficult groups. Without it want of sufficient material in American herbaria would have long delayed the completion of this work.

For the arboriculturist and the gardener the results of Mr. Wilson's journeys are as important and interesting as they have been for the botanist. During his two journeys in China for the Arboretum, Wilson collected 1593 lots of seeds, and 168 lots of plants and cuttings representing 1193 species and varieties of woody plants. From these, plants of 918 species and varieties have been successfully raised at the Arboretum. Many of these plants, including most of the large collection of Rhododendrons and other broad-leaved evergreens, and the plants found only at low levels in central China, have, however, not proved hardy in Massachusetts; and of the plants raised here from the seeds and cuttings collected by Wilson 473 species and varieties are now established in the Arboretum. The seeds collected by Wilson were shared with the best cultivators in the United States and Europe; and plants of the 445 species and varieties raised here which cannot grow in Massachusetts have also been widely distributed and are now growing in the gardens of many temperate regions, especially in those of our Pacific coast states where the Arboretum has established large collections of the new Rhododendrons and other Chinese plants.

No man, perhaps, has discovered more new plants than Wilson, or made larger and more interesting contributions to gardens; and his specimens on which this work is based now found in the principal herbaria of the United States and Europe, and his plants now growing in many gardens are witnesses to his zeal, industry and success as a botanical traveler.

C. S. SARGENT.

DECEMBER, 1916.

TABLE OF CONTENTS

LILIACEAE.

Determined by JESSE BAKER NORTON.

SMILAX L.

Smilax herbacea Linnaeus, *Spec.* 1030 (1752). — A. De Candolle, *Monog. Phaner.* I. 50 (1878).

Coprosmanthus herbaceus Kunth, *Enum.* V. 265 (1850).
The type of the species is confined to North America.

Smilax herbacea, var. **acuminata** Wright in *Jour. Linn. Soc.* XXXVI. 97 (1903).

Western Hupeh: Fang Hsien, woodlands, alt. 1300–2000 m., June 2, 1907 (No. 3246, in part; climber 2–2.6 m., flowers greenish); Patung Hsien, woodland, alt. 1600 m., June 1907 (No. 3246, in part; climber 2.6–3 m., flowers greenish); Hsing-shan Hsien, woodland, alt. 1300–2000 m., July 1907 (No. 3246, in part; climber 2–2.6 m., flowers greenish).

Smilax herbacea, var. **oblonga** Wright in *Jour. Linn. Soc.* XXXVI. 98 (1903).

Kiangsi: Kuling, thickets, common, alt. 1300 m., July 28, 1907 (No. 1729; climber 1 m.). Western Hupeh: north and south of Ichang, woodland, alt. 1300–2000 m., October 1907 (No. 3245; sub-shrubby climber, fruit greenish). Szech'uan: E. Faber (" probably No. 274"). Mandshuria: Western Shengking, E. Faber.

A robust unbranched form with pulverulent pubescence on the underside of the leaves.

Smilax glabra Roxburgh, *Fl. Ind.* ed. 2, III. 792 (1832). — Wallich, *Cat.* No. 5114 B (1830). — Hooker & Arnott, *Bot. Voy. Beechey*, 218 (1841). — Seemann, *Bot. Voy. Herald*, 420, t. 100 (1852–57). — Bentham, *Fl. Hongk.* 369 (1861). — A. De Candolle, *Monog. Phaner.* I. 60 (1878). — Hooker f., *Fl. Brit. Ind.* VI. 302 (1892). — Wright in *Jour. Linn. Soc.* XXXVI. 97 (1903).

Smilax Hookeri Kunth, *Enum.* V. 162 (1850).

1

Kiangsi: Kuling, thickets, common, alt. 1300 m., July 28, 1907
(No. 1723; climber 3 m.). Chekiang: vicinity of Ningpo, 1908, *D. Macgregor*.

Smilax microphylla Wright in *Kew Bull. Misc. Inform.* 1895, 117; in
Jour. Linn. Soc. XXXVI. 99 (1903). — Warburg in *Bot. Jahrb.* XXIX.
259 (1900).

Western Hupeh: Fang Hsien, thickets, alt. 1600–2300 m., June
1907 (No. 3247, in part; climber 2–4 m., flowers yellowish); Patung
Hsien, ravine, alt. 1000–1600 m., June 1907 (No. 3247, in part; climber
2–4 m., flowers greenish yellow). Western Szech'uan: Mupin,
thickets, alt. 1000–1300 m., November 1908 (No. 1256; climber 2–4
m., fruit blue-black).

Smilax scobinicaulis Wright in *Kew Bull. Misc. Inform.* 1895, 117;
in *Jour. Linn. Soc.* XXXVI. 100 (1903). — Warburg in *Bot. Jahrb.*
XXIX. 259 (1900).

Smilax brevipes Warburg in *Bot. Jahrb.* XXIX, 258 (1900).

Western Hupeh: Wushan Hsien, thickets, alt. 1300–1600 m.,
October 1907 (No. 455; climber 2–3 m., fruit greenish black); north
and south of Ichang, alt. 600–1000 m., June and November 1907 (No.
627; climber 2–3 m., fruit black); Chang-yang Hsien, alt. 1300–2000
m., June and December 1907 (No. 671; climber 3–4 m., flowers yellowish, fruit greenish blue); Hsing shan Hsien, cliffs, alt. 1300–1600 m.,
December 1907 (No. 680, in part; climber 3 m., fruit greenish black);
same locality, June 1907 (No. 680, in part; flowers whitish); same locality, woodlands, alt. 1000–1300 m., May 26, 1907 (No. 3233; climber
2–4 m.).

Smilax pekingensis A. De Candolle, *Monog. Phaner.* I. 108 (1878). —
Wright in *Jour. Linn. Soc.* XXXVI. 100 (1903).

Western Szech'uan: Mupin, alt. 2000 m., October 1908 (No.
1231; 1 m., fruit blue-black); southeast of Tachien-lu, alt. 2000–2300
m., June and October 1908 (No. 1271; bush 1.3–2 m., flowers brownish,
fruits blue-black).

Smilax vaginata Decaisne in *Jaquemont, Voyage,* IV. Bot. 169, t.
169 (1844). — A. De Candolle, *Monog. Phaner.* I. 109 (1878).

Smilax stans Maximowicz in *Bull. Acad. St. Pétersbourg,* XVII. 170 (1872);
in *Mél. Biol.* VIII. 407 (1872). — A. De Candolle, *Monog. Phaner.* I. 49
(1878). — Wright in *Jour. Linn. Soc.* XXXVI. 101 (in part) (1903).

Western Szech'uan: west and near Wên-ch'uan Hsien, alt. 2000–2600 m., October 1908 (No. 1019; 1.3–2 m., fruits black); Mupin, thickets, alt. 1000–1600 m., June 1908 (No. 3249; climber 3 m., flowers brownish yellow). Shensi: La-hua-shan, shady places, alt. 1300 m., December 29, 1913, *F. N. Meyer* (No. 1390; shrub, deciduous, green branches, 1–1.6 m. in height, seeds under No. 2100ᵃ). Western Yunnan: Tɛli range, alt. 3300–3600 m., lat. 25° 40′ N., June 1906. *G. Forrest* (No. 4779).

Smilax trachypoda Norton, n. sp.

Smilax stans Wright in *Jour. Linn. Soc.* XXXVI. 101 (in part), not Maximowicz) (1903).

Frutex ramosus, erectus v. subscandens, 1–4-metralis; caules inermes, teretes, leviter striati, virides, in sicco flavescentes, internodiis 4–6 cm. longis; rami erecto-ascendentes, angulo 10–45° divergentes, leviter zig-zag, non penduli v. flexuosi. Folia disticha, ovata, cordata, acuminata, integra v. leviter crenulato-erosa, 7–10 cm. longa et 4–6 cm. lata, ad basin venarum subtus ut petioli papilloso-pubescentia pilis persistentibus rigidis 1 mm. longis v. brevioribus fulvescentibus v. albidis, venis 5 interdum in foliis majoribus pari marginali irregulari addito subtus prominentibus supra fere planis; petioli striato-costati, basi ⅖ vaginati et tumidi, apicem versus tenuiora, 1.5–3 cm. longi, ecirrhosi, vetusti persistentes, cicatricibus semiamplexicaulibus. Inflorescentiae axillares, plerumque una tantum ad basin rami cujusque; pedunculi initio graciles, in fructu validiusculi, saepe per annos persistantes, 3–5 cm. longi; receptaculum non manifeste evolutum; bracteae paucae, deciduae, anguste lanceolatae, rubescentes, circiter 3 mm. longae; pedicelli masculi 8–10, graciles, glauci, 10–15 mm. longi, feminei 3–10 mm. longi, in fructu crassiusculi; alabastra floris masculi obovata, 3 mm. longa; sepala gracilia, lanceolata, recurva, luteo-viridia, 4–5 mm. longa; stamina 6, ⅗ perianthii aequantia; filamenta antheris multo longiora; flos femineus masculo minor, sepalis 2–3 mm. longis; ovarium obovatum, glaucum; stigmata brevia, recurva. Baccae 1–6, globosae, 1–2-spermae, coeruleo-nigrae, pruinosae, 6–8 mm. diam.; semina aurantiaco-rubra, 3–4 mm. diam.

Western Hupeh: Fang Hsien, thickets, alt. 1600 m., May 1907 (No. 3242; semi-scandent bush 1.3–1.6 m., flowers greenish). Western Szech'uan: west and near Wên-ch'uan Hsien, thickets, alt. 2300 m., June and October 1908 (No. 1065, type; 2–3 m., fruits blue-black); same locality, alt. 2000 m., July 1908 (No. 3251, in part; 2 m.); same locality, alt. 1600–2000 m., October 1910 (No. 4182; 2–3 m., fruits black); same locality, alt. 2000 m., October 1908 (No. 1051; 2–3 m., fruits blue-black, seeds orange-red); same locality, alt. 2000 m., July 1908 (No. 3250; climber 2 m., flowers green); Wa-shan, thickets, alt. 1300–1600 m., June 1908 (No. 3251, in part; bush 1 m., flowers greenish).

Closely related to *S. vaginata* Decaisne from which it is easily distinguished by its larger size and by the most peculiar papillate pubescence of the petiole and leaf ribs. Some of the flowers on Wilson's No. 3242 are abnormal in possessing both stamens and pistil.

Smilax china Linnaeus, *Spec.* 1029 (1753). — Thunberg, *Fl. Jap.* 152 (1784). — Maximowicz in *Bull. Acad. Sci. St. Pétersbourg*, XVII. 171 (1874). — Kunth, *Enum. Pl.* V. 243 (1850). — A. De Candolle, *Monog. Phaner.* I. 46 (1878). — Warburg in *Bot. Jahrb.* XXIX. 255 (1900). — Wright in *Jour. Linn. Soc.* XXXVI. 96 (in part) (1903).

Smilax ferox Bentham, *Fl. Hongk.* 370 (not Wallich) (1861). — Maximowicz in *Bull. Acad. Sci. St. Pétersbourg*, XVII. 171 (1874).
Smilax japonica Gray in *Mem. Am. Acad.* n. s. V. 412 (1857).
Coprosmanthus japonicus Kunth, *Enum.* V. 268 (1850).
Smilax Taquetii Léveillé in Fedde *Rep. Spec. Nov.* X. 372 (1912).

Western Hupeh: Hsing-shan Hsien, alt. 1000–1300 m., May 8, 1907 (No. 3234; climber 3–5 m., flowers greenish yellow). Eastern Szech'uan: Wushan Hsien, thickets, alt. 1300 m., October 1907 (No. 448; climber 2–4 m., fruit red). Chekiang: vicinity of Ningpo, 1908, *D. Macgregor* (six sheets). Korea: Quelpaert, "in sylvis Hallaisan," June and July 1909, *Taquet* (No. 3306, co-type of *S. Taquetii* Léveillé); Hongno, May and August 1909, *Taquet* (Nos. 3307 and 3308).

Although this species is credited by most writers as the source of "China root," there is little doubt that some of its relatives are also used for this purpose. The Korean forms agree closely with the typical Japanese plants, having the same orbicular leaves and zigzag stems; the leaves of Wilson's No. 3234 from Hupeh also resemble closely the Japanese type, while the leaves of his No. 448 are like those of the Hongkong plant.

Smilax megalantha Wright in *Kew Bull. Misc. Inform.* 1895, 118; in *Jour. Linn. Soc.* XXXVI. 99 (1903).

Smilax polycolea Warburg in *Bot. Jahrb.* XXIX. 257 (1900).

Kiangsi: Kuling, thickets, abundant, alt. 1300 m., July 28, 1907 (No. 1719; climber 2 m.). Western Hupeh: north and south of

Ichang, ravines, etc., alt. 30–100 m., March 20 and December 1907 (No. 661, in part; climber 2.5–3 m., fruits coral red, " Tu fu ling "); same locality, March and December 1907 (No. 661, in part. Western Szech'uan: Wa-shan, thickets, alt. 1600 m., November 1908 (No. 3254; climber 3 m., fruits coral red); Ching-chi Hsien, Ta-hsiangling, alt. 1300–2000 m., May 1908 (No. 3253; climber 5 m., flowers greenish). Yunnan: Mengtsze, southwest mountains, alt. 2000 m., A. Henry (No. 9867; red fruit).

Smilax menispermoidea A. De Candolle, *Monog. Phaner.* I. 108 (1878). — Wright in *Jour. Linn. Soc.* XXXVI. 99 (1903).

Western Szech'uan: Tachien-lu, thickets, alt. 1300–2300 m., October 1908 (No. 1258; climber 1–2.5 m., fruit blue-black); southeast of Tachien-lu, thickets, alt. 2300–2600 m., June 1908 (No. 3248; climber 2–3 m., flowers greenish). Western Yunnan: dry rocky situations amongst scrub, in side valleys on the eastern flank of the Tali range, alt. 3000 m., lat. 25° 40′ N., July 1906, *G. Forrest* (No. 4793; shrub of 0.6–1.3 m., flowers dark green).

Smilax longipes Warburg in *Bot. Jahrb.* XXIX. 256 (1900).

Smilax china Wright in *Jour. Linn. Soc.* XXXVI. 96 (in part) (1903).

Western Hupeh: Patung Hsien, thickets, alt. 900–1500 m., April 1907 (No. 667, in part; climber 3–4 m., flowers yellowish); December 1907 (No. 667, in part; climber 3–4 m., fruits blue-black); north and south of Ichang, thickets, alt. 600–1200 m., April and October 1907 (No. 489, in part; climber 3–5 m., flowers greenish yellow); same locality, October 1907 (No. 489, in part; climber 3–5 m., fruits blue-black same locality, alt. 300–1200 m., April and October 1907 (No. 483; climber 3–4 m., flowers yellowish, fruits black); without precise locality, *A. Henry* (No. 1222).

Henry's No. 1222 is cited by Wright as *S. china*, but it certainly belongs here.

Smilax glauco-china Warburg in *Bot. Jahrb.* XXIX. 255 (1900).

Smilax china Wright in *Jour. Linn. Soc.* XXXVI. 96 (in part) (1903).

Kiangsi: Kuling, thickets, common, alt. 1300 m., July 29, 1907 (No. 1731; 2 m.). Western Hupeh: Hsing-shan Hsien, alt. 1000–1600 m., May and October 1907 (No. 3241; climber 3–4 m., flowers greenish, fruits blue-black.

The Chinese colloquial name of this species is "Chin-pa-tou."

Smilax discotis Warburg in *Bot. Jahrb.* XXIX. 256 (1900).

Smilax china Wright in *Jour. Linn. Soc.* XXXVI. 96 (in part) (1903).
Smilax ferox A. De Candolle, *Monog. Phaner.* I. 102 (in part) (1878).

Western Hupeh: Fang Hsien, woodlands, alt. 1000–1300 m., May 31, 1907 (No. 3231; large climber, flowers greenish; same locality, woods, alt. 1600 m., May 10, 1907 (No. 3235; large climber, flowers greenish); same locality, upland thickets, alt. 2300–2600 m., May 19, 1907 (No. 3236; climber 2–3 m., flowers greenish); same locality, thickets, alt. 1300–1600 m., June 2, 1907 (No. 3239; climber 4 m., flowers whitish); same locality, alt. 1300–1600 m., June 2, 1907 (No. 3240; climber 4–5 m., flowers greenish); north and south of Ichang, thickets, alt. 300–1000 m., May 28, 1907 (No. 3232; climber 2–4 m., flowers greenish yellow); Hsing-shan Hsien, alt. 1000–1300 m., May 11, 1907 (No. 3237; climber 2.6–4 m., flowers greenish yellow); same locality, roadsides, alt. 1000–1300 m., May 11, 1907 (No. 3238; climber 3–5 m., flowers greenish yellow). Western Szech'uan: Wa-shan, thickets, alt. 1200–2000 m., July 1908 (No. 3243; climber 3 m., flowers yellowish); west and near Wên-ch'uan Hsien, thickets, alt. 1600–2000 m., July and October 1908 (No. 3244; climber 3 m., fruit blue-black). Yunnan: Mengtsze, mountains north, alt. 2600 m., *A. Henry* (No. 10566; large climber); shady situations amongst scrub in side valley on the eastern flank of the Tali range, lat. 25° 40', alt. 2300–3000 m., July and August 1906, *G. Forrest* (No. 4796; spinous shrub of 1–2.5 m., flowers brownish green); Lichiang valley, lat. 27° 15', alt. 2800 m., May 1906, *G. Forrest* (No. 2180).

The Chinese colloquial name of this species is "Hsao-chin-pa-tou."

Smilax discotis, var. **concolor**, Norton, n. var.

A typo recedit habitu robustiore, foliis subtus non glaucis, auriculis stipularibus ⅔ petioli non excedentibus.

Western Hupeh: Hsing-shan Hsien, woodlands, alt. 1600 m., June 1907 (No. 193, in part, type; climber 3–5 m.); same locality, September 1907 (No. 193, in part; climber 3–5 m.); Changyang Hsien, roadside thickets, alt. 1000–1300 m., October 1907 (No. 420; climber 2.5 m., fruit black).

Smilax micropoda A. De Candolle, var. **reflexa** Norton, n. var.

Smilax flaccida Warburg in *Bot. Jahrb.* XXIX. 259 (non Wright) (1900).

Caules rugulosi, teretes, sed alis duobus angustis a basi petiolorum decurrentibus instructi; aculei parvi, recurvi; rami angulo obtuso, 100°–130°, divergentibus. Folia opaca et supra leviter scabrida, subtus pallidiora et plus minusve lucida. Ramuli floriferi valde reducti, ut videtur nulli; pedunculi breves, plerumque bracteis basalibus obtecti; bacca monosperma, nigra, 6 mm. diam.

Western Hupeh: without precise locality, *A. Henry* (No. 3327, type, in U. S. Nat. Mus.); Chang-yang Hsien, alt. 1300–1600 m., April and Dec. 1907 (No. 678; climber 1.6–2.6 m., flowers greenish yellow, fruits black).

Smilax cocculoides Warburg in *Bot. Jahrb.* XXIX. 257 (1900).

Smilax stans Wright in *Jour. Linn. Soc.* XXXVI. 101 (in part) (1903).
Smilax vaginata Wright in *Jour. Linn. Soc.* XXXVI. 101 (1903).

Western Hupeh: Ichang, ravines, alt. 30–300 m., March 20, 1907 (No. 3252; small climber, flowers dull yellow). Yunnan: Mengtsze, forests, alt. 2000 m., *A. Henry* (No. 11239; large climber, red fruit).

KEY TO THE CHINESE AND KOREAN SPECIES OF SMILAX.

Shoots strictly herbaceous, never spiny; leaves thin, mucronate-pointed; perianth-segments of the staminate flowers more than 3 mm. long, widely reflexed.
  Shoots short, usually erect; tendrils absent from leaves at flower bearing nodes; leaves ovate, thin . . . . 1. *S. Oldhamii.*
  Shoots usually more than one m. tall, scandent, with tendrils at flower-bearing nodes . . . . 2. *S. herbacea.*
    Leaves ovate-lanceolate.
      Leaves pulverulent beneath; peduncles usually 8–12 cm. long.
          var. *acuminata.*
      Leaves glabrous; peduncles usually 5–8 cm. long.
        Leaves cordate at base; perianth segments 5 or more mm. long.
          var. *angustata.*
        Leaves not cordate at base; perianth segments less than 5 mm. long.
          var. *nipponica.*
    Leaves oblong . . . . var. *oblonga.*
Shoots woody; leaves thicker, often coriaceous or evergreen.
  Flowers appearing with the leaves; peduncles single in the axil of a leaf, not arising from a cluster of scale-like bracts.
    Perianth segments of the staminate flowers incurved, usually less than 3 mm. long.
      Peduncle much longer than the petiole; receptacle not strongly developed and with few bracts . . . . 3. *S. riparia.*
      Peduncle shorter than the petiole, receptacle often nearly sessile, well developed and with numerous, persistent bracts.
        Leaves glaucous beneath, turning red-brown when wilted; fruit thick-walled, over 6 mm. in diam.
          Leaves ovate-lanceolate; ♂ flower buds normal . . 4. *S. hypoglauca.*

Leaves oblong-lanceolate; ♂ flower buds short, with a distinct carina, or ridge along outer segments . . . . 5. *S. glabra.*
Leaves scarcely glaucous, turning a blackish green when wilted.
Leaves ovate to ovate-hastate, up to 5 cm. or more long; receptacle with its scales often 5 mm. in diam.; a large climber with a few spines.
                                 6. *S. trigona.*
Leaves lanceolate-oblong, less than 5 cm. long; receptacle small, few-flowered; stems often quite spiny . . . . . . 7. *S. microphylla.*
Perianth segments of the staminate flowers not incurved, usually reflexed, and more than 3 mm. long; receptacle not strongly developed, with few deciduous bracts; peduncle approximating petiole in length.
Plant not at all glaucous; spines straight, acicular or bristle-like, usually abundant when present; berry greenish black.
Leaves ovate, usually cordate; peduncles longer than the petioles; spines stout, black . . . . 8. *S. Sieboldii.*
Leaves ovate-lanceolate, scarcely cordate; peduncles shorter than the petioles; spines when present bristle-like . . . . 9. *S. scobinicaulis.*
Plant glaucous, at least on pedicels and fruit; berry red or blue-black.
Stipular auricle not distinct from the petiole; petiole vaginate at base, without tendrils; erect, spineless, deciduous shrubs with blue-black berries.
Leaves ovate-lanceolate, narrowed at base, peduncle 1–3 cm. long, 1–2 flowered . . . . 10. *S. pekingensis.*
Leaves ovate, broad at base, usually cordate, up to 5 or more cm. long.
Petioles glabrous, striate . . . . 11. *S. vaginata.*
Petioles scabrous, papillate pubescent . . . . . 12. *S. trachypoda.*
Stipular auricle distinct, ending in tendrils on the larger shoots; climbing vines usually with stout, recurved spines.
Berries red.
Leaves ovate to orbicular, not long-pointed, not glaucous beneath.
                                 13. *S. china.*
Leaves oblong lanceolate, apiculate, glaucous beneath.
                               14. *S. megalantha.*
Berries blue-black; leaves usually glaucous below.
Petiole 1–2 cm. long.
Tendrils near apex of the petiole; leaves ovate; branches strongly zigzag . . . . 15. *S. menispermoidea.*
Tendrils below the middle of the petiole; leaves oval; branches scarcely zigzag. . . . . 16. *S. longipes.*
Petiole rarely 1 cm. long.
Stipules not large; leaves oval-lanceolate, narrowed at the base, glaucous beneath . . . . 17. *S. glauco-china.*
Stipules developed into a semicircular auricle; leaves broad at the base, usually cordate.
Auricle extending entire length of petiole; leaves glaucous beneath.   18. *S. discotis.*
Auricle extending about half way to blade, leaves not glaucous beneath . . . . var. *concolor.*
Flowers produced on special, leafless branches, sometimes much reduced but indicated by a group of bud scales in the axil of the mature leaf where the flowers appear.
Flower branches simple, normally with one umbel.

Erect shrub without tendrils; leaves ovate, small, 1–3 cm. long; umbels few-flowered; perianth segments less than 3 mm. long . . . . 19. *S. rigida.*
Climbing vines, with tendrils.
Umbels nearly or quite sessile, peduncle hidden by basal bracts.
Stem smooth . . . . 20. *S. micropoda.*
Stem scabrous; winged . . . . var. *reflexa.*
Umbels distinctly peduncled.
Flowering branch and peduncle combined longer than the petiole; leaves thick, opaque . . . . 21. *S. opaca.*
Flowering branch and peduncle combined shorter than the petiole.
Leaves thin, green, somewhat shining . . . . . . 22. *S. lanceaefolia.*
Leaves thick, light colored, opaque.
Leaves oval-lanceolate . . . . 23. *S. cocculoides.*
Leaves long-lanceolate . . . . var. *lanceolata.*
Flower branches normally with more than one umbel; large climbing vines.
Stipules developed into a large spreading auricle encircling the stem.
24. *S. ocreata.*
Stipules not abnormally developed.
Stem scabrous, leaves oval . . . . 25. *S. stenopetala.*
Stem smooth; leaves ovate.
Exerted stamens longer than the perianth . . . . . . 26. *S. ovalifolia.*
Exerted stamens shorter than the perianth . . . . . . 27. *S. indica.*

ENUMERATION OF THE CHINESE AND KOREAN SPECIES.

Sect. I. NEMEXIA. A. De C.

The section *Nemexia* has been used by De Candolle and later writers as a depository for a large number of species that have no immediate relationship to the American members of the section. *Smilax china* Linnaeus and its relatives were placed in both *Nemexia* and *Eusmilax* because of the variable number of ovules. De Candolle recognized the apparent relationship of *S. ferox* Wallich to the *S. china* group and the absolute similarity between *S. stans* Maximowicz and *S. vaginata* Decaisne, but still placed them in different sections. Neither is it possible to allow *S. lanceaefolia* Roxburgh and its relatives to remain in this section, as the habit of the inflorescence in this genus is plainly a better indicator of relationship than the presence of one or two ovules, which is often a variable character even in the same ovary.

1. **Smilax Oldhamii** Miquel in *Versl. Med. Kon. Acad. Weten.* ser. 2, II. 86 (1868); in *Ann. Mus. Lugd.-Bat.* III. 150 (1868). — A. De Candolle, *Monog. Phaner.* I. 53 (1878).

Korea: Quelpaert, north in "sylvis Hallaisan," alt. 800 m., May 15, 1910, *Taquet* (Nos. 4059, 4060).

2. **Smilax herbacea** Linnaeus. See p. 1.

Smilax herbacea, var. **acuminata** Wright. See p. 1.

* **Smilax herbacea,** var. **angustata** Wright in *Jour. Linn. Soc.* XXXVI. 97 (1903).

Hupeh: Nanto, *A. Henry* (No. 5600 F; co-type) Yunnan: Mengtsze, alt. 1600 m., *A. Henry* (No. 13649; large climber).

Smilax herbacea, var. **nipponica** Maximowicz in *Bull. Acad. Sci. St. Péters-bourg,* XVII. 174 (1872). — A. De Candolle, *Monog. Phaner.* I. 52 (1878). — Wright in *Jour. Linn. Soc.* XXXVI. 98 (1903).

Smilax nipponica Miquel in *Versl. Med. Kon. Akad. Weten.* ser. 2, II. 87 (1868); in *Ann. Mus. Lugd.-Bat.* III. 150 (1868).

Korea: Quelpaert, " in sepibus Hongno," July 1908, *Taquet* (No. 1598); " in sylvis Nokan," alt. 600 m., September 7, 1908, *Taquet* (No. 1601).

Smilax herbacea, var. **oblonga** Wright. See p. 1.

Smilax herbacea with its many forms extending over temperate North America and Eastern Asia needs a careful revision, but the number of good specimens in the larger herbaria in this country is still very inadequate. The open land and the forest forms of the same variety often look so different that the collector assumes them to be entirely different species.

Sect. II.   COILANTHUS A. De C.

3. **Smilax riparia** A. De Candolle, *Monog. Phaner.* I. 55 (1878). — Wright in *Jour. Linn. Soc.* XXXVI. 100 (1903).

Central Fokien: Dunn's Exped., April–June 1905 (Hongk. Herb. Nos. 3570, 3576).

De Candolle described this species as herbaceous; the above numbers were partly identified on the original label as *S. herbacea*, and a similar specimen by Wright in the U. S. Nat. Mus. is labeled *S. consanguinea*, a supposed synonym of *S. herbacea*; yet all the specimens seen have perennial shoots from which the apparently herbaceous growth arises. The flowers in this Wright specimen reveal the *Coilanthus* affinities of this species.

4. **Smilax hypoglauca** Bentham, *Fl. Hongk.* 369 (1861). — A. De Candolle, *Monog. Phaner.* I. 61 (1878). — Wright in *Jour. Linn. Soc.* XXXVI. 98 (1903).

Yunnan: Szemao, mountains south, alt. 1500 m., *A. Henry* (Nos. 12115, 12115[a]; climber; black fruit). Hongkong: January 19, 1893, *C. Ford.*

5. **Smilax glabra** Roxburgh. See p. 1.

6. **Smilax trigona** Warburg in *Bot. Jahrb.* XXIX. 258 (1900).

Yunnan: Mengtsze, exposed mountains, alt. 1800 m., *A. Henry* (No. 9330; large climber); same locality, rocky mountains, alt. 1800 m., *A. Henry* (No. 9330[a]; large climber, reddish flowers).

7. **Smilax microphylla** Wright. See p. 2.

Sect. III.   EUSMILAX A. De C.

8. **Smilax Sieboldii** Miquel in *Versl. Med. Kon. Akad. Weten.* ser. 2, II. 87 (1868); in *Ann. Mus. Lugd.-Bat.* III. 150 (1868). — Maximowicz in *Bull. Acad. Sci. St. Pétersbourg,* XVII. 169 (1872); in *Mél. Biol.* VIII. 406 (1872). — A. De Candolle, *Monog. Phaner.* I. 48 (1878). — Wright in *Jour. Linn. Soc.* XXXVI. 100 (1903).

Korea: Seoul, September 21–24 1905, *J. G. Jack.* Korean Archipelago: Quelpaert, " in sepibus Gylungen (?)," June 11, 1910, *Taquet* (No. 4061; same island, " in sylvis Nokayi," July 1908, *Taquet* (Nos. 1602, 1603); same island, " in sepibus Hallaisan," June 1909, *Taquet* (Nos. 3303, 3304, 3305).

Smilax Sieboldii and *S. scobinicaulis* Wright are very closely related to *Smilax hispida* Muhlenberg and have been removed from Nemexia to their proper place in *Eusmilax.*

9. **Smilax scobinicaulis** Wright.　See p. 2.

10. **Smilax pekingensis** A. De Candolle.　See p. 2.

11. **Smilax vaginata** Decaisne.　See p. 2.

12. **Smilax trachypoda** Norton.　See p. 3.

13. **Smilax china** Linnaeus.　See p. 4.

14. **Smilax megalantha** Wright.　See p. 4.

15. **Smilax menispermoidea** A. De Candolle.　See p. 5.

16. **Smilax longipes** Warburg.　See p. 5.

17. **Smilax glauco-china** Warburg.　See p. 5.

18. **Smilax discotis** Warburg.　See p. 6.

Smilax discotis, var. **concolor** Norton.　See p. 6.

19. **Smilax rigida** Wallich, *Cat.* No. 5120 (1830). — Kunth, *Enum.* V. 164 (1850). — A. De Candolle, *Monog. Phaner.* I. 105 (1878).

Western Yunnan: dry situations amongst scrub, in side valleys on the eastern flank of the Tali range, alt. 3000–3300 m., lat. 25° 40′ N., June–July 1906, *G. Forrest* (No. 4785; shrub of 0.3–1.3 m., flowers brown).

20. **Smilax micropoda** A. De Candolle, *Monog. Phaner.* I. 58 (1878).
The type of this species occurs in Assam and Burma.

Smilax micropoda, var. **reflexa** Norton.　See p. 6.

21. **Smilax opaca** Norton, n. comb.

Smilax lanceaefolia, var. *opaca* A. De Candolle, *Monog. Phaner.* I. 57 (1878).— Wright in *Jour. Linn. Soc.* XXXVI. 99 (1903).

Smilax lanceaefolia Seemann, *Bot. Voy. Herald*, 420, t. 99 (non Roxburgh) (1852–57). — Bentham, *Fl. Hongk.* 370 (1861). — Franchet in *Bull. Soc. Bot. France*, XLVI. 214 (1901).

Hongkong: *C. Ford.*
This plant is plainly distinct from the *S. lanceaefolia* Roxburgh of India and southwest China.

22. **Smilax lanceaefolia** Roxburgh, *Hort. Beng.* 72 (1814); *Fl. Ind.* ed. 2, III. 792 (1832). — Wallich, *Cat.* No. 5132 (1830). — A. De Candolle, *Monog. Phaner.* I. 57 (1878).

Yunnan: Szemao, forests, alt. 1600 m., *A. Henry* (No. 12902; large climber, red flowers; No. 12902ᵃ; large climber; No. 12902ᵇ; large climber, white flowers; No. 12799; climber, white flowers); Mengtsze, forests to southeast, alt. 1600 m., *A. Henry* (No. 11489); same locality, mountain forests to southeast, alt. 2000 m., *A. Henry* (No. 10397); same locality, forests to southwest, alt. 1600 m., *A. Henry* (No. 10911; large climber).

23. **Smilax cocculoides** Warburg.　See p. 7.

Smilax cocculoides, var. **lanceolata** Norton, n. var.
A typo recedet foliis anguste lanceolatis, 15–20 cm. longis 3–5 cm. latis subtus vix glaucis.
Yunnan: Szemao, mountains east, alt. 1600 m., *A. Henry* (No. 12577).

24. **Smilax ocreata** A. De Candolle, *Monog. Phaner.* I. 191 (1878).
Yunnan: Szemao, forests, alt. 1600 m., *A. Henry* (No. 12903; large climber, whitish flowers; No. 12903ᵃ; large climber, reddish fruits); same locality, southern

forests, alt. 1300 m., *A. Henry* (No. 12902^c; large climber); Mengtsze, forests, alt. 1800 m., *A. Henry* (No. 11237; large climber, green flowers). **Kwangtung**: Hainan, *C. Ford* (in part, and partly *Smilax china* Linnaeus).

25. **Smilax stenopetala** Gray in *Mem. Amer. Acad.* n. s. VI. 412 (1859). — Maximowicz in *Bull. Acad. Sci. St. Pétersbourg,* XVI. 169 (1872); in *Mél. Biol.* VIII. 405 (1872). — A. De Candolle, *Monog. Phaner.* I. 189 (1878). — Henry in *Trans. As. Soc. Jap.* XXIV. suppl. 96 (1896). — Wright in *Jour. Linn. Soc.* XXXVI. 101 (1903).

Formosa: Bankinsing, *A. Henry* (Nos. 52, 55, 115, 144); South Cape, *A. Henry* (No. 284).

26. **Smilax ovalifolia** Roxburgh, *Hort. Beng.* 72 (1814); *Fl. Ind.* ed. 2, III. 794 (1832). — Seemann, *Bot. Voy. Herald,* 421 (1852–57). — Bentham, *Fl. Hongk.* 370 (1861). — A. De Candolle, *Monog. Phaner.* I. 199 (1878). — Wright in *Jour. Linn. Soc.* XXXVI. 100 (1903).

Smilax prolifera Roxburgh, *Fl. Ind.* ed. 2, III. 795 (1832). — Kunth, *Enum.* V. 247 (pro parte) (1850).

Yunnan: Mengtsze, woods, alt. 1800 m., *A. Henry* (No. 9415^a; flowers green); same locality, mountains southeast, alt. 2300 m., *A. Henry* (No. 9225; green fruit, very large climber); Szemao, forests, alt. 1500 m., *A. Henry* (No. 12719; large climber, white flowers.)

With plenty of material this species may prove a complex of several forms.

27. **Smilax indica** Vitman, *Summa Pl.* V. 422, (1879). — A. De Candolle, *Monog. Phaner.* I. 187 (1878).

Smilax Hohenackeri Kunth, *Enum.* V. 240 (1850).

Yunnan: Mengtsze, alt. 1500 m., *A. Henry* (No. 11238; large climber, green flowers).

DOUBTFUL SPECIES.

The following species reported from China have not been included in this list for various reasons, mostly because of insufficient evidence of their validity.

S. Davidiana A. De Candolle, *Monog. Phaner.* I. 104 (1878). Not seen.

S. flaccida Wright in *Kew Bull. Misc. Inform.* 1895, 118. This species is placed by Wright in Eusmilax, but all the material I have seen of Henry's No. 3630, on which the species is based, is strictly herbaceous and closely related to *S. herbacea,* var. *angustata.*

S. laevis, var. *ophirensis* A. De Candolle fide Wright in *Jour. Linn. Soc.* XXXVI. 99 (1903). A doubtful identification.

S. cyclophylla Warburg in *Bot. Jahrb.* XXIX. 257 (1900). This may be *S. menispermoidea* which is not included by Warburg in his list.

S. cinerea Warburg in *Bot. Jahrb.* XXIX. 258 (1900). Probably not distinct from *S. ocreata* De Candolle.

S. Bockii Warburg in *Bot. Jahrb.* XXIX. 259 (1900).

S. Nebelii Gilg in *Bot. Jahrb.* XXXIV., *Beibl.* LXXV. 26 (1904). Probably *S. Sieboldii* Miquel.

S. Lyi Léveillé in *Fedde Rep. Spec. Nov.* VIII. 171 (1910). Probably *S. ocreata.*

S. perulata Léveillé in *Fedde Rep. Spec. Nov.* IX. 78 (1910). Probably *S. ovalifolia* Roxburgh.

Léveillé and Vaniot in Léveillé, *Liliac. etc. Chine,* 26–28 (1905) have published several new names under this genus, but without sufficient description to allow one to even guess at the synonymy. Their *S. ocreata* is preoccupied by the *S.*

ocreata of A. De Candolle. *S. gracillima* (p. 26) is described as a monoecious plant and is probably not a *Smilax*.

Hayata in *Jour. Coll. Sci. Tokio*, XXX. 1–471, *Mat. Fl. Formosa* (1911) publishes several new species, but until more Formosan material is available for study little can be said as to the value of these names.

HETEROSMILAX Kunth

Heterosmilax Gaudichaudiana Maximowicz in *Bull. Acad. Sci. St. Pétersbourg*, XVII. 176 (1872); in *Mél. Biol.* VIII. 415 (1872). — A. De Candolle, *Monog. Phaner.* I. 44 (1878). — Wright in *Jour. Linn. Soc.* XXXVI. 95 (in part) (1903).

Smilax Gaudichaudiana Kunth, *Enum.* V. 252 (1850).
Oligosmilax Gaudichaudiana Seemann in *Jour. Bot.* VI. 258, t. 83 (1868).

Western Hupeh: Fang Hsien, thickets, alt. 1300 m., June 1907 (No. **674**, in part; climber 3–4 m.); north and south of Ichang, ravines, common, alt. 300–1150 m., June and December 1907 (No. **674**, in part; climber 2–4 m., fruits blue-black.)

To complete the enumeration of the Chinese *Smilaceae* two other *Heterosmilax* native to eastern China and Formosa may be added here.

Heterosmilax Gaudichaudiana, var. **hongkongensis** A. De Candolle, *Monog. Phaner.* I. 45 (1878).

Heterosmilax Gaudichaudiana Bentham, *Fl. Hongk.* 370 (1861). — Wright in *Jour. Linn. Soc.* XXXVI. 95 (in part), (1903).
Smilax hongkongensis Seemann, *Bot. Voy. Herald*, 420 (1852–57).
Hongkong: *C. Ford*.

Heterosmilax japonica Kunth, *Enum. Pl.* V. 270 (1850). — Maximowicz in *Bull. Acad. Sci. St. Pétersbourg*, XVII. 175 (1872); in *Mél. Biol.* VIII. 415 (1872).— A. De Candolle, *Monog. Phaner.* I. 43 (1878). — Franchet & Savatier, *Enum. Pl. Jap.* II. 50 (1879). — Wright in *Jour. Linn. Soc.* XXXVI. 96 (1903).
Formosa: Bankinsing, *A. Henry* (No. 74; fruit); Tchow, *A. Henry*; South Cape, *A. Henry* (No. 1302; fruit).

DIOSCOREACEAE.

Determined by ALFRED REHDER.

DIOSCOREA L.

Dioscorea acerifolia Uline apud Diels in *Bot. Jahrb.* XXIX. 261 (nomen nudum) (1900). — Prain & Burkill in *Jour. As. Soc. Bengal*, LXXIII. pt. 2, Suppl. 7 (1904); in Prain, *Contrib. Ind. Bot.* 419 (1906).

Dioscorea quinqueloba C. H. Wright in *Jour. Linn. Soc.* XXXVI. 92 (1903), synon. exclud. — Pampanini in *Nuov. Giorn. Bot. Ital.* n. ser. XVII. 243 (1910); XVIII. 110 (1911).

Western Hupeh: north of Ichang, thickets, common, alt. 1200–1500 m., July 1907 (No. 2923; climber 3–5 m.); without precise locality, *A. Henry* (Nos. 150, 5870, 5870[b], 7358).

Dioscorea zingiberensis C. H. Wright in *Jour. Linn. Soc.* XXXVI. 93 (1903). — Pampanini in *Nuov. Giorn. Bot. Ital.* n. ser. XVII. 243 (1910); XVIII. 165 (1911).

Western Hupeh: north and south of Ichang, thickets, alt. 1200–1500 m., July 1907 (No. 2921; climber 2–3 m., rootstock edible); without precise locality, *A. Henry* (No. 1621).

Dioscorea japonica Thunberg, *Fl. Jap.* 151 (1784). — Miquel in *Ann. Bot. Lugd.-Bat.* III. 159 (1867), exclud. synon.; *Prol. Fl. Jap.* 323 (1867), exclud. synon. — Franchet & Savatier, *Enum. Fl. Jap.* II. 47 (1879). — Makino, *Ill. Fl. Jap.* I. t. 22 (1890). — Diels in *Bot. Jahrb.* XXIX. 261 (1900). — C. H. Wright in *Jour. Linn. Soc. Bot.* XXXVI. 92 (1903).

Western Hupeh: Patung Hsien, thickets, alt. 900 m., July 1907 (No. 2922; climber, 3 m.; staminate flowers).

14

CHLORANTHACEAE.

Determined by ALFRED REHDER.

CHLORANTHUS Sw.

Chloranthus brachystachyus Blume, *Fl. Jav.* fasc. VIII. 13, t. 2 (1828–51). — Miquel, *Fl. Ind. Bat.* I. pt. 1, 801 (1855); in *Ann. Mus. Lugd.-Bat.* III. 129 (1867); *Prol. Fl. Jap.* 293 (1867). — Bentham, *Fl. Hongk.* 334 (1861). — Solms in De Candolle, *Prodr.* XVI. pt. 1, 475 (1869). — Franchet & Savatier, *Enum. Pl. Jap.* I. 444 (1875). — Maximowicz in *Bull. Soc. Nat. Mosc.* LIV. 56 (1879). — Hooker f., *Fl. Brit. Ind.* V. 100 (1886). — Hemsley in *Jour. Linn. Soc.* XXVI. 367 (1891). — Diels in *Bot. Jahrb.* XXIX. 272 (1900). — Dunn & Tutcher in *Kew Bull. Misc. Inform.* add ser. X. 221 (*Fl. Kwangtung & Hongk.*) (1912). — Léveillé, *Fl. Kouy-Tchéou,* 74 (1914).

? Chloranthus monander R. Brown in *Bot. Mag.* XLVIII. t. 2190, in nota (1821).
Ascania serrata Blume, *Enum. Pl. Jap.* I. 80 (non Chloranthus serratus Roemer & Schultes) (1827).
Sacandra chloranthoides Gardner in *Calcutta Jour. Nat. Hist.* VI. 348 (1846). — Wight, *Icon. Pl. Ind. Or.* VI. t. 1946 (1853) — Le Maout & Decaisne, *Traité Gén. Bot.* 516, fig. (1876).
Chloranthus ceylanica Miquel, *Fl. Ind. Bat.* I. pt. 1, 802 (1855).
Chloranthus denticulatus Cordemoy in *Adansonia,* III. 296 (1863).
Chloranthus ilicifolius Blume in herb. ex Miquel in *Ann. Mus. Lugd.-Bat.* III. 129 (pro synon.) (1867); *Prol. Fl. Jap.* 293 (1867).
Chloranthus montanus Siebold ex Miquel, l. c. (pro synon.) (1867).

Western Szech'uan: Wa-shan, alt. 11–1300 m., October 1908 (No. 1124; suffruticose, 0.6 m., fruit scarlet). Yunnan: Mengtsze, mountains to southwest, alt. 1200 m., *A. Henry* (No. 9361ᵃ; shrub 0.5 m.); same locality, mountains to southeast, alt. 1500 m., *A. Henry* (No. 9361ᶜ; shrub 1.6 m.); Szemao, alt. 1300–1500 m., *A. Henry* (Nos. 12341, 12341ᵇ, 12341ᶜ; shrub 1.6–2 m., fruit red). Hongkong: *C. Wilford, C. Wright* (No. 469), *H. F. Hance* (No. 379); Tsin-tso-ngam, Jan. 22, 1907 (ex Herb. Hongk. No. 3337).

Henry's No. 12341 and 12341ᵇ differ from the typical form in their larger inflorescence, which measures, including the peduncle, from 6 to 13 cm. in length.

SALICACEAE.

Determined by CAMILLO SCHNEIDER.

POPULUS L.

Populus Wilsonii Schneider, n. sp.

Arbor late pyramidalis, 5–24 m. alta, trunco circuitu 0.5–1.5 m., primo aspectu *Idesiae polycarpae* non absimilis; ramuli crassi, annotini cylindrici, purpurascentes v. olivaceo-brunnei, saepe subglauci, glabri; vetustiores cinereo-brunnei, deinde cinerascentes; gemmae satis magnae, ut videtur obtuse ovatae, rubro- v. flavo-brunneae, glabrae sed (an semper?) leviter viscosae, perulis glabris margine ciliatis obtectae. Folia papyracea, initio rubescentia, deinde supra obscure glauco-viridia, ut videtur semper glabra, subtus discoloria, cinereo-viridia, opaca, initio floccoso-tomentosa, cito subglabra, ad costam et nervos primarios sparse pubescentia v. pleraque tantum basi in axillis nervorum barbata utrinque laxe reticulata, late ovata v. subrotunda v. late ovato-oblonga, basi plus minusve cordata v. rotundo-truncata, apice obtusa, margine crenato-dentata dentibus 3 5 mm. distantibus distincte callosis, mınora circiter 8 cm. longa et 7 cm. lata, majora ad 18 cm. longa at ad 15 cm. lata v. interdum ad 20 cm. longa et ad 16 cm. lata; glandulae ima basi foliorum duae (v. interdum nullae), subrotundae, leviter excavatae; petioli purpurascentes, supra leviter caniculati, (4–)6–11 cm. longi. Inflorescentiae tantum femineae cognitae, iis *P. lasiocarpae* simillimae, pubescentes, juveniles c. 7 cm., fructiferae ad 15 cm. longae; flores breviter pedicellati. Fructus subglabri (an semper?), ovati.

Western Hupeh: Hsing-shan Hsien, side of streams, very rare, alt. 1800 m., May 20 and August 1907 (No. **706ᵃ**, type; pyramidal bushy tree 8 m. tall, 0.5 m. girth); Changyang Hsien, alt. 1300–2000 m., May 1907 (No. **706ᵇ**; tree 20–25 m. tall, girth 1.5 m.); Fang Hsien, woodlands, alt. 1800–2600 m., June 16, 1910 (No. **4450**; tree 5–15 m. tall, girth 0.75–1.25 m., usually very bushy; sterile); north and south of Ichang, high mountains, 1901 (Veitch Exped. No. 384, in part; tree 7–13 m. tall; flowering branchlets). Western Szech'uan:

northeast of Sungpan Ting, upland thickets, alt. 3000–3300 m., August 1910 (No. 4450ᵃ; bushy tree, 6–7.5 m. tall; sterile).

In the herbarium it is not easy to distinguish this species from the nearly related *P. lasiocarpa* Oliver, but according to Mr.Wilson's description, and judging by the young living plants in the Arnold Arboretum *P. Wilsonii* seems to be a very distinct species. It is readily distinguished from *P. lasiocarpa* by its habit, by the glabrous and purple, not tomentose and yellowish branchlets and by the color of the leaves. These are bluish green above and rather whitish gray beneath, while the leaves of *P. lasiocarpa* are clear-green above and greenish and a little shining beneath. The shape of the leaves of *P. Wilsonii* is round-ovate, and not so deeply cordate at the base as the more broadly oblong leaves of *P. lasiocarpa*. The nearest relative to *P. Wilsonii* seems to be *P. glauca* Haines (see p. 30).

This is a rare tree in western Hupeh and in Szech'uan, where it grows in forested country between 1300 and 3300 m., especially near watercourses. In the forests it is pyramidal in outline with rather thin ascending-spreading branches and dense foliage; in the open the outline is more or less columnar and the branches are short and spreading. The dark bluish-green leaves are strikingly handsome, and in habit and general appearance it reminds one of *Idesia polycarpa* Maximowicz. Pictures of this Poplar will be found under Nos. 090 and 099 of the collection of my photographs.   E. H. W.

Populus lasiocarpa Oliver in *Hooker's Icon.* XX. t. 1943 (1890). — Burkill in *Jour. Linn. Soc.* XXVI. 536 (1899). — *J. H. Veitch* in *Jour. Hort. Soc. Lond.* XXVII. 65, fig. 27 (1903). — Schneider, *Ill. Handb. Laubholzk.* I. 17 (1904). — Dode in *Mém. Soc. Hist. Nat. Autun*, XVIII. (*Extr. Monog. Inéd. Populus*, 66) (1905). — Ascherson & Graebner, *Syn. Mitteleur. Fl.* IV. 51 (1908). — Gombocz in *Math. Termesz. Közl.* XXX. 120 (*Monog. Gen. Populi*) (1908).—Henry in Elwes & Henry, *Trees Gr. Brit. & Irel.* VII. 1846, t. 408, fig. 9 (1913). — Bean, *Trees & Shrubs Brit. Isl.* II. 215 (1914). — Skan in *Bot. Mag.* CXLI. t. 8625 (1915).

Populus Fargesii Franchet in *Bull. Mus. Hist. Nat. Paris*, 11. 280 (1896).

Western Hupeh: North and south of Ichang, woods, alt. 1300–2300 m., May 1907 (No. 706, in part; trcc 16–24 m. tall, girth 1.25–1.65 m.); same locality, June 1900 and 1901 (Veitch Exped. No. 384, in part; tree 7–13 m. tall; ripe fruits); Hsing-shan Hsien, woodlands, alt. 2000 m., July 1907 (No. 706, in part; tree 20 m. tall, girth 1.25 m.); Fang Hsien, woodlands, common, 1800–2300 m., June 1910 (No. 706, in part); without locality, May 1904, mountains, 2500 m. (Veitch Exped. No. 4538; tree, 20 m. tall); district of Chienshih, alt. 1300–2000 m., common in mountains, *A. Henry* (No. 5423ᵃ, type; good timber tree); without locality, *A. Henry* (No. 5423; male catkins, No. 4013, fruits).

Possibly *P. Fargesii* may prove to be the preceding species, but according to Burkill, l. c., Franchet himself pointed out that his *P. Fargesii* should be united with *P. lasiocarpa*. Unfortunately 1 have not seen Franchet's type specimen which came from Tchen-keou-tin, Szech'uan, and his description is too insufficient. The pubescence of the leaves and branchlets of *P. lasiocarpa* is variable, and cultivated plants usually are much more glabrous than the wild trees.

In moist woods this is a very common tree in western Hupeh and eastern Szech'uan, but I do not remember meeting with it in western Szech'uan. It is a tree of medium size with a straight and relatively thin trunk clothed with dark gray fissured bark. The crown is pyramidal or rounded at the top and rather sparingly branched and the branchlets are thick. Colloquially it is known as the Tai-erh-po. Pictures of this Poplar will be found under Nos. 561, 563, 572, 0127 and 0130 of the collection of my photographs and also in my *Vegetation of Western China*, Nos. 401, 402.

The specimen figured in the *Botanical Magazine* (l. c.) has polygamous flowers, but this is not normal in this or any other species of Poplar, though it has been detected occasionally in many species and probably occurs from time to time in them all. In the *Botanical Magazine* (l. c.) it is stated that the introduction of this Poplar was effected by means of "a living plant." This should read "plants," since the original consignment which 1 sent to Messrs. Veitch comprised about a dozen small specimens of this tree, and probably normal flowers of the two sexes will appear when these trees flower.   E. H. W.

Populus suaveolens Fischer in *Allg. Gartenzeit.* IX. 404 (1841); in *Bull. Acad. Sci. St. Pétersbourg*, IX. 348 (1842).—Ledebour, *Fl. Ross.* III. 2, 629 (1850). — Maximowicz in *Bull. Soc. Nat. Mosc.* LIV. 51 (1879), tantum forma a pro parte.—Koehne, *Deutsche Dendr.* 84 (pro parte) (1893). — Schneider, *Ill Handb. Laubholzk.* I. 14 (pro parte) (1904).—Dode in *Mém. Soc. Nat. Hist. Autun*, XVIII. (*Extr. Monog. Inéd. Populus*, 61) vix v. tantum pro parte (1905). — Ascherson & Graebner, *Syn. Mitteleur. Fl.* I. 48 (pro parte) (1908).— Gombocz in *Math. Termesz. Közl.* XXX. 110 (1908), exclud. var. *β.* — Henry in *Gard. Chron.* ser. 3, LIII. 198, fig. 88 (1913); in Elwes & Henry, *Trees Gr. Brit. & Irel.* VII. 1841, t. 410, fig. 25 (1913).

Populus balsamifera Pallas, *Fl. Ross.* I. 1, 67, t. 41 (excl. fig. B) (non Linnaeus) (1784).
Populus balsamifera, var. *γ intermedia* Loudon, *Arb. Brit.* III. 1674 (1838).
Populus balsamifera, var. *ζ suaveolens* Loudon, *Arb. Brit.* III. 1674 (1838). — Wesmael in De Candolle, *Prodr.* XVI. 2, 330 (1868); in *Mém. Soc. Sci. Hainaut*, III. 246 (*Monog. Populus*, 66) (pro parte) (1869). — Burkill in *Jour. Linn. Soc.* XXVI. 535 (1899), pro parte, includ. var. *laurifolia.* — Matsumura, *Ind. Pl. Jap.* II. pt. 2, 7 (1912).
Populus pseudobalsamifera Turczaninow in *Bull. Soc. Nat. Mosc.* I. 101 (nomen nudum, non Fischer) (1838) secundum specim. originale, fide Henry.

Populus suaveolens, var. *macrocarpa* Schrenk[1] in Fischer & Meyer, *Enum. Pl. Nov. Schrenk.* II. 16 (1842).
Populus suaveolens, var. *angustifolia* Regel in *Mém. Acad. Sci. St. Pétersbourg*, ser. 7, IV. 132 (*Tent. Fl. Ussur.*) (1862). — Gombocz in *Math. Termesz. Közl.* XXX. 111 (1908).

Western Szech'uan: Min River valley, Sungpan to Mao-chow, common, alt. 2000–3000 m., August 27, 1910 (No. 4577; tree 24–32 m. tall, girth 1.25–3 m., sterile branches with elliptical acuminate leaves with acute base); Monkong Ting, common in dry valleys, alt. 2300–3000 m., June 28, 1908 (No. 2162, tree 7–32 m. tall, girth 0.75–3 m.; fruiting branches, inflorescences to 15 cm. long, leaves rather ovate and shortly acuminate); near Mon-kong Ting, ascent of Hsiao Chin Ho, alt. 2300–2600 m., June 29, 1908 (No. 2164; tree 12–24 m. tall, girth 1.5–2.5 m.; fruiting branches similar to No. 2162, but leaves more elliptic-ovate, base a little cordate-lobate); Ching-chi Hsien, Fu-yueh-ling, alt. 2300–3000 m., May 1908 (No. 1432; tree 10 m. tall, girth 1.25 m.; fruiting branches, leaves mostly broad-ovate, acuminate, base slightly cordate, to 10.5 cm. long and 7 cm. broad); same locality, October 1908 (No. 1431; bark from young and old tree); Tachien-lu, planted around temples, July 1903 (Veitch Exped. No. 4539, tree 24–32 m. tall; fruits and young twig, leaves broad-ovate or broad-elliptical, base ovate or slightly acute, twigs somewhat angulate). Kansu: "intra prov. Kansu, inter fl. Hoangho et murum magnum, in Liang-tschou et Shan-dan-siang viis publicis, in arenosis Han-tschou," July, August 1875, *P. J. Piasezki* (ex Maximowicz) Chili: Hsiao Wu-tai-shan, alt. 1800 m., August 23 and 28, 1913, *F. N. Meyer* (No. 1311); Jehol, 1910, *W. Purdom* (No. 2).

TIBET. "Ad Hoangho superiorem ejusque affluentes nec non ad lacum Kukunor," 1872 and 1880, *N. M. Przewalski* (type, ex Maximowicz).
NORTHEASTERN ASIA. Transbaikalia: "Dahuria," *Fritsch;* southern Mongolia: "Mont Muniula," 1872, *N. M. Przewalski* (flowering and sterile branches, old female flowers glabrous, seems to be mixed with young female flowers of *Populus tremula*, var. *Davidiana;* in herb. Kew). Mandshuria: Harbin, August 16, 1903, *C. S. Sargent* (tree with pale bark, deeply furrowed near its base). Korea: Ping-yang, September 18, 1905, *J. G. Jack* (in the shape of the leaves this specimen agrees with typical *P. suaveolens*, but in texture and color they are more like those of *P. Maximowiczii*.
SONGARIA: "an den Quellen des Aksu," July 19, 1840, *A. Schrenk* (sterile). Alatau, August 27, 1843, *A. Schrenk* (No. 118).
TURKESTAN: "ad flumen Sairam," July 19, 1878, *Fetissow.*

---

[1] In the pubescence of the branchlets, petioles and leaves this form resembles *P. Maximowiczii* Henry, but in the shape and texture of the leaves it agrees with *P. suaveolens;* and I am not quite sure to what species it really belongs.

Fischer founded this species, as Henry has already pointed out (l. c.) on the type of Pallas's figure, of which Fischer had the original specimen in his collection. Later authors, like Maximowicz and Komarov (see under *P. Maximowiczii*) took the eastern plant for the type (at least partly); this has been separated with good reason by Henry as a different species, *P. Maximowiczii*. The type is also represented for the first time as *P. foliis ovatis acutis serratis* by Gmelin's plate (*Fl. Sibir.* 152, t. 33 [1747]). To this typical *P. suaveolens* may belong Maximowicz's *P. suaveolens*, var. a (in *Bull. Soc. Nat. Mosc.* LIV. 51 [1879]), at least concerning the specimens of Tatarinow and Bretschneider if they do not belong to *P. Simonii*. The Kansu specimen of Piasezki I refer to var. *Przewalskii* (see p. 32). It is often very difficult to determine herbarium specimens and to keep apart the forms of *P. suaveolens*, *P. Maximowiczii*, *P. Simonii* and *P. laurifolia*. Turczaninow (in *Bull. Soc. Nat. Mosc.* XXVII. 398 (1854); *Fl. Baical'-Dahur.* II. 124 [1856]) describes the variability of the leaves of *P. suaveolens*, mentioning three different forms. Gombocz does not separate *S. Maximowiczii* from *P. suaveolens*.

In western Szech'uan from 1000–3300 m. altitude, this is very common on the banks of rivers and in the forest and has been very extensively planted as a shade tree in the warm semi-arid valleys. It is a favorite with Tibetan monks, and their lamasaries and dwellings are generally shaded by groves of this Poplar. It is a large tree, often 30 m. tall, with massive wide-spreading branches. The trunk is thick and the bark is smooth and pale gray on young trees but becomes dark and fissured with age. The leaves vary in size considerably and the branchlets are always glabrous. Pictures of this tree will be found under Nos. 223 and 0334 of the collection of my photographs and also in my *Vegetation of Western China*, No. 396.   E. H. W.

Populus szechuanica Schneider, n. sp.

Arbor ad 40 m. alta, trunco circuitu ad 4 m.; ramuli juveniles (turiones) distincte angulati, olivacei v. purpurascentes, glabri, vetustiores teretes, flavo-brunnei, deinde cinerascentes; gemmae purpurascentes, glabrae, viscosae, acutissimae. Folia initio rubescentia, papyracea, supra intense viridia, glabra v. basim versus puberula, subtus albescentia, non v. vix puberula, nervis primariis et ex parte secundariis distinctis, turionum saepissime ovato-elongata, basi leviter v. distincte cordata v. rotunda, apicem versus sensim acuta v. breviter acuminata, margine anguste glanduloso-crenato-dentata, 11–20 cm. longa et 5–11 cm. lata v. usque ad 28 cm. longa et ad 16 cm. lata, ramulorum fructiferorum late ovata v. ovato-rotunda, basi rotunda v. plus minusve cordata; minima ovato-lanceolata basi leviter acuta, apice saepissime subito breviter acuminata, margine plus minusve distincte glandulosodentata, initio ciliolata, 8–18 cm. longa et 5–15 cm. lata; petioli foliorum turionalium satis breves, 2–4 cm. longi, foliorum ramulorum satis elongati, 2.5–7 cm. longi, semper glabri. Amenta tantum fructifera vidi, ad 16 cm. longa, glabra; capsulae subsessiles, glabrae, 3–4-valvae, ut videtur, ovato-globosae, 7–9 cm. longae, basi perianthio glabro irregulariter lobato cinctae.

Western Szech'uan: west of Kuan Hsien, forests of Pan-lan-shan, alt. 2300–3700 m., July 6, 1908 (No. 2163, type; tree 32 m. tall, girth 4 m.); same locality, woodlands, alt. 2600–3300 m., October 1910 (Nos. 4346, 4348; tree 24–32 m. tall, girth 4–4.5 m.); northeast of Tachien-lu, forest of Ta-p'ao-shan, alt. 2600–3700 m., July 3, 1908 (No. 1413; tree 7.5–40 m. tall, girth 1–3 m.); same locality, alt. 2300–4600 m., July 6, 1908 (No. 2165; tree 20–40 m. tall, girth 1.25–4 m.); south-east of Tachien-lu, woods, alt. 2300–2800 m., October 1908 (No. 1434; tree 20–24 m. tall, girth 1.5–2 m., bark gray); same locality, forests, alt. 2600–3300 m., October 1910 (No. 4355, tree 32–36 m. tall, girth 4–4.5 m.); Lungan Fu, Tu-ti-liang-shan, alt. 2300–3000 m., August 1910 (No. 4578; tree 20–32 m. tall, girth 1.25–2 m.); Mupin, alt. 2400–2600 m., October 1910 (No. 4361; tree 17–20 m. tall, girth 2.8–3.4 m.; sterile).

Populus szechuanica is a very handsome species, judging by the young plants (from Nos. 4355 & 4361) in the Arnold Arboretum, which resemble in habit those of *P. lasiocarpa* Oliver.

In the moist forests of western Szech'uan between 2200 and 3300 m. altitude this is a very common tree. It grows to a large size, and though usually associated with Firs, Spruces and Larches it often forms pure stands. In habit and general appearance it resembles *P. suaveolens* Fischer, but the branches are more massive and the branchlets are stouter. Even on old trees the leaves are large, approximating in size those of *P. lasiocarpa* Oliver.

Pictures of this tree will be found under Nos. 123, 159, 183, 184 and 186 of the collection of my photographs, and also in my *Vegetation of Western China*, Nos. 392, 393, 394, 395.
E. H. W.

Populus Simonii Carrière in *Rev. Hort.* 1867, 360. — Wesmael in De Candolle, *Prodr.* XVI. 2, 330 (1868); in *Mém. Soc. Sci. Hainaut*, III. 247 (*Monog. Populus*, 67) (1869). — Maximowicz in *Bull. Soc. Nat. Mosc.* LIV. 52 (1879). — Dippel, *Handb. Laubholzk.* II. 211, fig. 105 (1892). — Koehne, *Deutsche Dendr.* 85 (1893). — Mouillefert, *Traité Arb. & Arbriss.* II. 1114 (1897). — Komarov in *Act. Hort. Petrop.* XXII. 746 (1904). — Schneider, *Ill. Handb. Laubholzk.* I. 16, fig. 6 o–q (1904). — Dode in *Mém. Soc. Hist. Nat. Autun*, XVIII (*Extr. Monog. Inéd. Populus*, 57) (1905). — Ascherson & Graebner, *Syn. Mitteleur. Fl.* IV. 47 (1908). — Gombocz in *Math. Termesz. Közl.* XXX. 105 (*Monog. Gen. Populi*) (1908). — Henry in Elwes & Henry, *Trees Gr. Brit. & Irel.* VII. 1839, t. 410, fig. 28 (1813).

Populus laurifolia, var. *Simonii* Regel, *Русс. Дендр.* ed. 2, 152 (1883).
Populus balsamifera, var. *Simonii* Wesmael in *Bull. Soc. Bot. Belg.* XXVI. 378 (1887). — Burkill in *Jour. Linn. Soc.* XXVI. 536 (1899).
Populus balsamifera, var. *laurifolia* Burkill, l. c. 535 (non Wesmael) (1899), quoad specim. citata.

Populus brevifolia Carrière ex Schneider, *Ill. Handb. Laubholzk.* I. 16 (pro synon.) (1904).

Western Hupeh: Hsing-shan Hsien, roadside, alt. 1300 m., very rare, May 8, 1909 (No. 1454; tree 12 m. tall, girth 1.50 m.). Western Szech'uan: Wên-chuan Hsien, Min River valley, alt. 1300–2300 m., abundant, October 1908 (No. 1420; tree 5–12 m.). Chili: near Santun-ying, on sandy places along mountain streams, May 31, 1913, *F. N. Meyer* (Nos. 974, 975; attains often great size); Nankow, October 6, 1905, *J. G. Jack;* Siwantze, *G. E. Simon* (No. 151, type, ex Burkill).

NORTHEASTERN ASIA. Mandshuria: upper Amur, August 13, 1891, *S. Korshinsky;* common in country of Harbin, August 29, 1903, *C. S. Sargent* (branches slightly angular, leaves broadly obovate, a somewhat doubtful specimen).

This species, which has been so often confounded with *P. suaveolens*, appears sometimes to grow with that species. Its nearest relative may be *P. laurifolia* Ledebour.

The shape of the leaves is variable, and in some forms the leaves are very small. On young vigorous shoots and also near the ends of the branches of older trees the leaves are round-obovate with a long acute base and are rounded and only shortly cuspidate at the apex, and the petioles are very short (2–20 mm.) and usually dark red. Such leaves measure from 3–13 cm. in length and from 2–10 cm. in width. The long-petiolate leaves of younger plants often greatly resemble those of *P. suaveolens*, but are nearly always obovate-lanceolate. Unfortunately I have not yet seen any female flowers or fruits. The whole plant seems to be glabrous.

In Hupeh I met with only three trees of this Poplar, and these were growing by a house half a day's march from the city of Hsing-shan Hsien. In the valley of the Min River in western Szech'uan it is common between 1300 and 2300 m. altitude and is usually planted round houses and in villages. It is a more slender tree than *P. suaveolens* Fischer, with shorter and thinner branches, which form a rounded or oval crown. The Hupeh trees had ascending-spreading branches, but in Szech'uan the branches were spreading and the branchlets pendulous. Young trees raised from cuttings from both sources and growing in the Arnold Arboretum exhibit no apparent difference in branching habit and resemble the Szech'uan type. A picture of this tree will be found under No. 0337 of the collection of my photographs.

E. H. W.

Here may be added two forms cultivated in the Arnold Arboretum.

Populus Simonii, f. pendula Schneider, n. forma.
A typo recedit ramis distincte pendulis, ramulis satis angulatis v. subalatis.
To this form belongs a living specimen in the Arnold Arboretum.

Populus Simonii, f. fastigiata Schneider, n. forma.
A typo recedit habitu late pyramidali, ramis angulo acuto a trunco divergentibus, ramulis subangulatis v. subrotundis.
This interesting form was discovered and introduced by F. N. Meyer through the Department of Agriculture of the United States. See his photographs in the collection of the Department of Agriculture Nos. 5288, 5335 and 5412.

Populus adenopoda Maximowicz in *Bull. Soc. Nat. Mosc.* LIV. 50 (1879). — Schneider, *Ill. Handb. Laubholzk.* I. 20, in adnot. (1904). — Dode in *Mém. Soc. Nat. Hist. Autun,* XVIII. (*Extr. Monog. Inéd. Populus,* 34) (1905). — Gombocz in *Math. Termesz. Közl.* XXX. 133 (Monog. Gen. Populi) (1908). — Léveillé, *Fl. Kouy-Tchéou,* 380 (1915).

Populus tremula, var. *adenopoda* Burkill in *Jour. Linn. Soc.* XXVI. 537 (1899).
Populus Silvestrii Pampanini in *Nuov. Giorn. Bot. Ital.* n. ser. XVII. 247, fig. 2 (1910).

Western Hupeh: north and south of Ichang, common, alt. 1300 m., March 11 and October 1907, March 23 and April 1908, March 1909 (No. 724; tree 6–25 m. tall, girth 0.5–1.8 m.); Changyang Hsien, woodlands, alt. 1000–2300 m., May and June 1907 (Nos. 1400, 1437; tree 6–20 m. tall, girth 0.75–1.5 m.); Hsing-shan Hsien, uplands, alt. 1600–2500 m. (Plant No. 1460; tree up to 20 m. tall; young plants in the Arnold Arboretum); Fang Hsien, woods, alt. 2500–3400 m. (Plant No. 4440; tree 13–20 m. tall, girth 0.9–1.5 m.; young plants in the Arnold Arboretum); without exact locality, April 1900 (Veitch Exped. No. 7); without locality, *A. Henry* (Nos. 5281, 5281ᵃ). Northern Hupeh: " Kao-kien-scian," alt. 800 m., May–June 1907, *C. Silvestri* (Nos. 296ᵃ and 296, ex Pampanini); "Sce-kio-ho," alt. 600, April 15–20, 1906, *C. Silvestri* (No. 378, ex Pampanini, type of *P. Silvestri*); " Monte Triora," alt. 1950 m., July 3, 1907, *P. C. Silvestrii* (No. 297, ex Pampanini); " ab urbe Hankan boream versus, in Yun-yan Fu," March, *P. J. Piasezki* (staminate flowers, ex Maximowicz). Western Szech'uan: southeast of Tachien-lu, alt. 2300 m., October 1908 (No. 1430; tree 5–10 m. tall, girth 0.75–1.25 m.); Mupin, alt. 1300–2000 m., woodlands, October 1907 (No. 1400ᵃ; tree 12–18 m. tall, girth 0.75–1.25 m.). Kweichou: "Cy-Po," February 1900, *J. Cavalerie* (ex Léveillé). Kiangsi: Kiukiang, alt. 300 m., foothill, not common, August 2, 1907 (No. 1738; tree 10 m. tall). Shensi: "ad fl. Han," April, 1875, *P. J. Piasezki* (type, ex Maximowicz).

This is a very distinct species, easily recognized by the long-acuminate closely crenate leaves of old mature trees. On vigorous young plants the leaves are not yet distinctly acuminate and may resemble those of shoots of *P. tremula* Linnaeus or of *P. Sieboldii* Miquel. The leaves are greenish beneath, while those of *P. tremula,* var. *Davidiana* Schneider are distinctly glaucous.

This is the common low-level Poplar of Hupeh and Szech'uan and the more eastern parts of the Yangtsze valley. In western Hupeh it is very abundant in open country and woods from river-level up to 1500 m. altitude. It is a rather slender shapely tree, 20–25 m. or more tall, with a straight trunk clear of branches for 10–15 m. and clothed with smooth pale gray bark which on old trees and near the ground becomes dark and slightly fissured. The branches are thin, ascending-spreading and form an oval crown. The leaves vary considerably in degree of pubescence, but on old trees they are glabrous at maturity. On young trees and on adventitious shoots they are densely pubescent.

Pictures of this tree will be found under Nos. 523, 531, 540, 543, 564 and 0181 of the collection of my photographs and also in my *Vegetation of Western China*, Nos. 397, 398, 399 and 400.   E. H. W.

Populus tremula Linnaeus, var. Davidiana Schneider, var. nov.

Populus tremula Maximowicz in *Mém. Sav. Étr. Acad. Sci. St. Pétersbourg*, IX. 245 (*Prim. Fl. Amur.*) (1859). — Burkill in *Jour. Linn. Soc.* XXVI. 557 (pro parte▪ maxima) (1899). — Komarov in *Act. Hort. Petrop.* XXII. 14 (*Fl. Mansh.* II.) 14 (1904). — Nakai in *Jour. Coll. Sci. Tokyo*, XXXI. 211 (*Fl. Kor.* II.) (1911).

Populus Davidiana Dode in *Mém. Soc. Nat. Hist. Autun*, XVIII. (*Extr. Monog. Inéd. Populus*, 31, t. 11, fig. 31) (1905).

?Populus pellostachya Dode, l. c. t. 11, fig. 32 (1905).

?Populus wutaica Mayr, *Fremdl. Wald- u. Parkbäume*, 494 (1906).

Western Hupeh: Hsing-shan Hsien, upland thickets, alt. 1600–2500 m., October 1907 (No. **722**, in part; tree 8–24 m. tall); same locality (Plant No. 1459; young plants in the Arnold Arboretum); Fang Hsien, alt. 1600–2000 m., October 15, 1907 (No. **722**, in part; tree 8–24 m. tall, girth 0.5–1.5 m.); mountains north of Ichang, May 1901 (Veitch Exped. No. 1886; tree 8 m. tall; female flowers). Western Szech'uan: west of Kuan Hsien, Pan-lan-shan, woodlands, alt. 3000 m., October 1910 (No. **4347**); north of Tachien-lu, woodlands, alt. 2500 m., October 1910 (No. **4358**; tree 6–16 m. tall, bark gray, smooth). Chili: Jehol, *A. David* (No. 1687, type of the variety); Weichang, 1910, *W. Purdom* (No. 11); Hsiao-Wu-tai-shan, alt. 2000 m., Aug. 5, 1913, *F. N. Meyer* (Nos. 43, 1113).

NORTHEASTERN ASIA. Mandshuria: Mukden, May 29, 1906, *F. N. Meyer* (No. 101). Amur: "ad fl. Amur," 1855, *R. Maack* (Nos. 10, 11); Amur, *C. Maximowicz*. Ussuri: Khabarovsk, August 23, 1903, *C. S. Sargent*. Korea: Wanson, September 5, 1903, *C. S. Sargent*.

It is only on account of their geographical distribution that I refer these northeastern Asiatic and Chinese forms to a distinct variety. There are living plants of some of Wilson's numbers in the Arnold Arboretum. It is possible that these plants when fully grown may present some characters to distinguish the eastern forms from those of the Altai and Europe. Even to separate var. *Davidiana* from the following species is a difficult task.

Léveillé, *Fl. Kouy-Tchéou*, 380 (1915), mentions Bodinier's No. 2101 from Kouyyang, February 1898, as *P. tremula*. I have not seen this specimen, which may belong to our variety.

This is a common tree on the higher mountains of Hupeh and Szech'uan, where it covers extensive areas and often forms pure stands. It is always a slender tree, rarely exceeding 20 m. in height with thin branches forming a loose and rounded to oval crown. The bark is smooth greenish to pale gray except on very old trees where it is dark gray and fissured on the lower part of the trunk. The young leaves are reddish-purple and very beautiful as they unfold, and the habit of the tree is light and attractive. It produces suckers very freely, like its Japanese relative *P. Sieboldii* Miquel. Pictures of this tree will be found under Nos. 087, 088 of my collection of photographs.   E. H. W.

Populus tremula, var. Davidiana, f. tomentella Schneider, n. forma.

A typo glabro v. tantum initio sparse pubescente recedit ramulis hornotinis foliisque adultis plus minusve distincte pubescentibus v. tomentellis.

Western Szech'uan: Mupin, forests, alt. 2300–3000 m., October 1910 (No. **4359**; trees 20–24 m. tall, trunk 1.25–2 m. in girth).

This pubescent form of var. *Davidiana* seems to be analogous to f. *pubescens* of var. *typica* of *P. tremula*, or it may even have the same systematic value as var. *villosa* Lang (see Schneider, *Ill. Handb. Laubholzk.* I. 19 [1904]).

This pubescent form is common in western Szech'uan, but is very rare in western Hupeh. The leaves are usually larger than those of the type.   E. H. W.

Populus rotundifolia Griffith, var. Duclouxiana Gombocz in *Math. Termesz. Közl.* XXX. 130 (*Monog. Gen. Populi*) (1908).

Populus macranthela Léveillé & Vaniot in *Bull. Soc. Bot. France*, LII. 142 (March 1905); in *Monde Plant.* XII. 9 (1910); in Fedde, *Rep. Spec. Nov.* VIII. 446 (1910). — Gombocz in *Math. Termesz. Közl.* XXX. 157 (*Monog. Gen. Populi*) (1908).

Populus Duclouxiana Dode in *Mem. Soc. Nat. Hist. Autun*, XVIII. (*Extr. Monog. Inéd. Populus*, 32, t. 11, fig. 34ª) (1905).

Populus rotundifolia Griffith, var. *macranthela* Gombocz in *Bot. Közl.* X. 25 et (7) (1910). — Léveillé, *Fl. Kouy-Tchéou*, 380 (1915).

Western Szech'uan: Ching-chi Hsien, Fei-yueh-ling, woods, alt. 2–3000 m., May 1908 (No. **1432**; tree 12–20 m. tall, girth 0.75–1.25 m.). Kweichau: Pin-Fa, April 1, 1903, *J. Cavalerie* (No. 974, type of *P. macranthela*, ex Léveillé, 1005). Yunnan: without locality, *F. Ducloux* (ex Dode, type of *P. Duclouxiana*).

I have seen neither the type of Léveillé's nor of Dode's species, but Gombocz says, l. c. (1911) of *P. macranthela*: " D'après les échantillons originaux elle est très voisine à la *P. rotundifolia* Griff. et ne diffère de la dernière que par ses châtons plus longs (15–24 cm.) et par ses capsules à pedicelles plus longs. Elle peut donc être identifié avec la *P. Duclouxiana* (distinguée par Dode), qui a d'ailleurs la même répartition: China, Yunnan." I am not quite sure if Wilson's specimens really represent the same form. They may be described as follows:

Ramuli initio tomentelli sed cito glabrescentes v. glabri, rubro-brunnei, vetustiores cinerascentes; gemmae satis evolutae ovato-acutae, rubro-brunneae, paulo puberulae, perulis ciliatis. Folia turionum (apice ramorum fructiferorum) late ovata, basi leviter cordata, apice acuta v. subacuminata, margine sinuato-crenata, utrinque glabra v. subtus initio puberula, supra viridia, subtus glaucescentia, 8–11 cm. longa et 6.5–10 cm. lata; minora (ramulorum abbreviatorum) simillima, sed basi truncata; petioli foliorum majorum 4.5–6.5 cm., foliorum minorum 2–4 cm. longi, initio tomentelli v. plerique glabri, compressi. Amenta fructifera ad 15–16 cm. longa, rhachi puberula breviter pedunculata. Capsulae ovato-oblongae, apice acutae, glabrae, bivalvatae, circiter 5 mm. longae et 3 mm. crassae, pedicellis subglabris 2–3 mm. longis.

This tree is common in western Szech'uan in the district of Ching-chi Hsien and in habit and general appearance is very similar to *P. tremula*, var. *Davidiana* Schneider.   E. H. W.

CONSPECTUS SECTIONUM SPECIERUMQUE ASIAE ORIENTALIS NEC NON HIMALAYAE.

Disci florum decidui; antherae longae et apiculatae. Folia valde polymorpha, utrinque cinerea et stomatibus numerosis praedita et distincte elevato-reticulata . . . . Sect. 1. TURANGA.
    Folia polymorpha dentata v. lanceolata et integerrima . . . 1. *P. euphratica.*
    Folia ovato-elliptica v. reniformia, integra, nunquam lanceolata 2. *P. pruinosa.*
Disci florum persistentes. Folia discoloria, haud distincte elevato-reticulata.
    Antherae longae, apiculatae; styli satis elongati. Folia ramorum vetustiorum turionumque vix diversa, semper longe petiolata.   Sect. 2. LEUCOIDES.
        Ramuli hornotini annotinique brunnescentes v. purpurascentes. Folia supra opaca, coeruleo-viridia, subtus in sicco albescentia v. glauca, matura tantum ad basim parce villosula, circuitu pleraque late ovata v. ovato-rotunda, breviter cordata, subcordata v. truncato-rotunda; petioli saepissime laminam aequilongi v. ⅓ breviores.
            Ramuli ut videtur semper glabri, vix sulcati. Folia subtus in sicco albescentia, basi saepe breviter anguste cordata. Fructus ovati, longiores quam lati . . . . 3. *P. Wilsonii.*
            Ramuli ut videtur in sulcis plus minusve initio villosuli. Folia subtus glauca, basi ut videtur nunquam anguste cordata. Fructus subglobosi . . . . 4. *P. glauca.*
        Ramuli viridi- v. flavo-cinerei, vix brunnescentes, hornotini subangulati, pubescentes. Folia circuitu pleraque late ovato-oblonga, distincte basi cordata, supra claro-viridia, subopaca, subtus pallidiora, sed viridescentia et leviter nitentia (in planta viva); petioli dimidio laminae vix longiores.
                                5. *P. lasiocarpa.*
    Antherae oblongae v. subrotundae, nunquam apiculatae; styli plerique breves v. brevissimi. Folia saepissime inter se plus minusve variabilia.

* Folia subtus distincte (in vivo) discoloria, albescentia, turionum et ramulorum inter se pleraque satis dimorpha;[1] petioli basi tereti, apice vix v. non compressi supra canaliculati; gemmae satis viscosae, balsamicae. Bracteae florum valde fimbriatae sed non v. vix ciliatae.

Sect. 3. TACAMAHACA.

Flores feminei fructusque distincte pedicellati, florum masculorum antherae 3–3½-plo longiorae quam latae. Turiones [teste Hooker et Dode] angulatae. Folia turionum ramulorumque diversorum, ut videtur, haud valde dissimilia, nunc ovato-acuminata, basi rotunda v. cordata v. raro subtruncata, in specim. visis 7:4.5–17:12 cm. magna, nunc late ovato-cordata, brevius acuminata v. acuta, c. 8:8–15:12 cm. magna, omnia longe petiolata, pedicellis subteretis, glabris v. versus apicem sparse puberulis, 3–12 cm. longis, crenato-glanduloso-dentata, fere semper distincte ciliata, supra intense v. claro-viridia, subtus albescentia, utrinque (an semper?) glabra. Amenta glabra, mascula ad 10 cm. longa, fructifera ad 30 cm. longa, fructibus satis distantibus, glabris, ovato-globosis, in apicem productis . . . . 6. *P. ciliata*.

Flores feminei fructusque sessiles v. subsessiles; antherae (a me visae) vix plus quam duplo longiores quam latae.

Ramuli juveniles cylindrici, interdum apice ut turiones angulati v. striati, biennes semper teretes (confer etiam 9. *P. szechuanicam*, var. *tibeticam*).

Ramuli initio olivacei vix v. non brunnescentes, deinde flavescentes v. ochroleuci, glabri v. supra nodos versus gemmas puberuli. Folia membranacea v. papyracea, supra plus minusve intense viridia, subnitentia et vix opaca, glabra v. basim versus sparse pilosa, subtus discoloria, opaca, albescentia v. in sicco flavescentia, glabra (sed vide var. *macrocarpam* ut p. 19 indicata!); margine glanduloso-crenulata,

---

[1] It is helpful to keep the following in mind in regard to the variability of the leaves of *Populus* (of this section and also of the others). L. A. Dode was the first who clearly described the variation of the leaves according to the age of the plants, the position of the branches, and the climatic and other conditions under which they were produced. He distinguishes many different kinds of shoots (pousses) and of leaves. Unfortunately Dode's descriptions in his "Extraits" are so incomplete that it is impossible to form an opinion of the real value of all his different species.

I think it sufficient if we distinguish 3 different kinds of leaves: 1. the leaves of the offshoots or suckers (folia turionum radicalium); 2. the leaves of the normal but not the vigorous shoots of the young plants (folia ramorum [sed non turionum] plantae juvenilis); and 3. the leaves of the old trees (folia arboris adultae fructiferae). The leaves of the third kind we usually find at the ends of vigorous branches of young plants when they are growing well and becoming mature. If we compare leaves of different species it is necessary to be very careful only to compare the same kind of leaves!

Referring to the pubescence of branchlets, leaves and petioles, many of the species of *Populus* seem to produce a glabrous and a pubescent variety, both closely connected by intermediate forms, but sometimes quite distinct in the extreme forms.

There is another character in *Populus* which seems to be very often overestimated, that is whether the young shoots are angular or cylindric. As far as my observations go vigorous shoots (and offshoots) of nearly every species with cylindrical branchlets may be more or less angular, at least at their apex or when the branch is very young.

saepe ciliata, radicalia (teste Fischer) saepe lanceolata, utrinque acuminata, plantae juvenilis elliptica, basi acuta, apice plus minusve
acuminata, v. ovato-elliptica, basi subrotunda, 5 : 2.5–9 : 4.5 cm. magna,
plantae maturae pleraque ovata v. late ovata, basi rotunda, raro
subcordata, apice acuminata v. acuta, 5.5 : 3.5–1 : 6.5 cm. magna,
in plantis cultis etiam majora, cordato-ovata, ad 18 : 14 cm. magna;
petioli semper satis longi (turionum paullo breviores), 1.5–5 cm. longi.
Amenta mascula 4–5 cm. longa (teste Maximowicz), fructifera ad
13 cm. longa, parce puberula v. glabra, capsulis satis approximatis,
sessilibus v. subsessilibus, ovoideis, glabris (in var. *Przewalskii* pubescentibus maturis glabratis), trivalvibus nondum apertis circiter
7 : 5 mm. magnis . . . . **7. *P. suaveolens*.**
Ramuli initio ut videtur plus minusve rubescentes, deinde olivacei,
    vetustiores cinereo-flavi, v. cinerei, plus minusve dense puberuli v.
    pubescentes sed interdum satis glabri. Folia papyracea v. saepissime
    subcoriacea, supra intense v. livide viridia, opaca, tota facie v. tan-
    tum nervis puberula, raro glabrescentia, rete nervorum angusta,
    distincta, subtus valde discoloria, albescentia v. saepissime rubello-
    flavescentia, margine glanduloso-crenulata, plus minusve ciliata,
    radicalia non visa, cetera inter se satis conformia, pleraque elliptica,
    v. ovato-elliptica, basi subcordata, apice breviter oblique acuminata
    v. acuta, interdum elliptico-rotunda, apice obtusa, 6.5 : 3–12 : 9 cm.
    magna; petioli 1–4 cm. longi, valde (v. raro vix) puberuli. Amenta
    mascula 5–10 cm. longa, antheris 30–40; feminea initio 3.5–5 cm.,
    fructifera ad 18(–25) cm. longa, breviter tomentella, puberula v. plus
    minusve glabra, ovariis glabris, capsulis sessilibus v. subsessilibus
    subglobosis, nondum apertis circ. 8–9 : 7–8 mm. magnis, 3–4-valvibus.
                                        8. *P. Maximowiczii*.
Ramuli juveniles turionesque distincte angulati v. biennes striati.
Folia turionum (et pleraque ramulorum) infra medium latissima, lance-
    olata, elliptica, ovata v. cordata (haud distincte obovata) (confer
    etiam 10. *P. yunnanensem* et 11. *P. Gamblei*).
Ramuli floriferi tereti vel subtereti (haud distincte angulati), olivaceo-
    brunnei v. ochroleuci (haud pallido-flavi v. grisei).
                                        9. *P. szechuanica*.
Ramuli etiam floriferi satis angulati, pallido-flavi v. grisei.
                                        12. *P. laurifolia*.
Folia turionum supra medium latissima, saepissime distincte obovalia,
    basi plus minusve cuneata v. rotunda, apice obtusa sed pleraque in
    acumen plus minusve longum producta . . . . . **13. *P. Simonii*.**
** Folia subtus concoloria viridescentia v. discoloria et glaucescentia, non albes-
    centia sed interdum tomento albo obtecta.
  † Folia subtus plus minusve concoloria, viridescentia, utrinque stomatibus
    numerosis praedita, turionum ramulorumque vetustiorum vix v. paullo
    diversa, deltoidea, basi truncata, apice plus minusve acuminata. Gem-
    mae viscosae . . . . Sect. 4. AIGEIROS.
Folia parva obovata v. ovato-deltoidea, basi plus minusve longe cuneata,
    apice breviter acuminata, supra medium crenato-denticulata v. sub-
    serrata, 1.4–4 cm. longa et 1.2–2 cm. lata, initio margine ciliolata et
    subtus parce puberula; petioli graciles, ut videtur non compressi, initio
    puberuli, 0.3–2 cm. longi. . . . . 14. *P. afghanica*.
Folia majora, turionum late subcordato-triangularia, apice subito in

acumen producta, toto circuitu glanduloso-crenata, glabra, ad 9 cm. longa et 10 cm. lata; petioli 4–5 cm. longi, plus minusve compressi.

15. *P. nigra*, var. *italica*.

††Folia subtus concoloria v. discoloria, superne stomatibus haud praedita, turionum et ramorum plus minusve diversa, forma variabilia. Gemmae non viscosae, glabrae v. pubescentes . . . . . . Sect. 5. LEUCE.

‡Folia turionum subtus tomento albo persistente obtecta, margine lobata v. irregulariter lobato-denticulata, ramorum fructiferorum saepe glabrescentia et viridescentia, crasse papyracea. Gemmae plus minusve pubescentes . . . . Subsect. ALBIDAE.

Folia turionum plus minusve palmato-lobata, ramulorum fructiferorum subtus vix glabrescentia, plus minusve cinerea, satis parva.

16. *P. alba*.

Folia turionum irregulariter lobato-duplicato-dentata, ramulorum fructiferorum saepe valde glabrescentia, viridescentia, satis magna.

17. *P. tomentosa*.

‡‡Folia turionum subtus cinerea, pubescentia v. deinde glabrescentia, margine breviter denticulata, ramulorum fructiferorum sericea v. glabra, saepe glaucescentia, membranacea tenuiter papyracea. Gemmae saepissime glabrae . . . . Subsect. TREPIDAE.

Folia ramulorum fructiferorum distincte longe acuminata.

18. *P. adenopoda*.

Folia ramulorum fructiferorum obtusa, acuta v. breviter acuminata.

Folia ramulorum fructiferorum late ovata v. ovato-rotunda, basi saepissime subito in petiolum brevem producta, saepissime glandulifera, apice breviter acuminata (v. acuta), margine satis anguste et regulariter glanduloso-serrulata (dentibus pro 1 cm. circiter 3).

19. *P. Sieboldii*.

Folia ramulorum fructiferorum saepissime suborbicularia, basi rotundata v. pleraque truncata v. plus minusve cordata, rariter glandulifera, apice obtusa v. acuta v. breviter acuminata, margine sinuato-glanduloso-dentata v. eroso-crenulata, dentibus satis remotis circiter 1–2 pro 1 cm.

Folia ramulorum fructiferorum basi nunquam cordata, rariter rotunda, sed saepissime truncata, apice obtusa v. acuta. Amenta fructifera vix plus quam 12 cm. longa . . . 20. *P. tremula*.

Folia ramulorum fructiferorum basi saepe plus minusve cordata et apice acuta v. breviter acuminata. Amenta saepe plus quam 12 cm. longa . . . . 21. *P. rotundifolia*.

ENUMERATIO SECTIONUM SPECIERUMQUE ASIAE ORIENTALIS NEC NON HIMALAYAE.

Sect. 1. TURANGA Bunge in *Mém. Sav. Étr. Acad. Sci. St. Pétersbourg*, VII. 498 (*A. Lehmann. Rel. Bot.* 322) (1851). — Gombocz in *Math. Termesz. Közl.* XXX. 67 (*Monog. Gen. Populi*) (1908).

Populus, subgen. *Turanga* Dode in *Mém. Soc. Hist. Nat. Autun*, XVIII. (*Extr. Monog. Inéd. Populus* 13) (1905).

Folia polymorpha, utrinque cinereo-glauca et stomatibus numerosis praedita et distincte reticulata, satis coriacea, basi glandulis vix v. haud prominentibus munita; petioli plus minusve rotundi. Gemmae haud v. paullo viscosae. Disci florum decidui, distincte lobati; stigmata 3, stylosa; stamina, circiter 12, antheris longis, apice apiculatis. Capsulae elongatae, pedicellatae.

1. **Populus euphratica** Olivier, *Voy. Emp. Othoman.* III. fig. 45–46 (1807). — Wesmael in *Mém. Soc. Sci. Hainaut,* III. (*Monog. Populus,* 54, t. 10–13) (1869). — Brandis, *Forest Fl. Brit. Ind.* 474, t. 63 (1874). — Hooker f., *Fl. Brit. Ind.* V. 638 (1888). — Burkill in *Jour. Linn. Soc.* XXVI. 536 (1839). — Dode in *Mém. Soc. Hist. Nat. Autun,* XVIII. (*Extr. Monog. Inéd. Populus,* 16, t. 11, fig. 1, 1ᵃ, 1ᶜ, 1ᵉ) (1905). — Gombocz in *Math. Termesz. Közl.* XXX. 68 (*Monog. Gen. Populi*) (1908).

Populus diversifolia Schrenk in *Bull. Acad. Sci. St. Pétersbourg,* X. 253 (1842); in Fischer & Meyer, *Enum. Alt. Pl. Nov. Schrenk.* 15 (1842). — Trautvetter, *Pl. Imag. Fl. Russ.* 23, t. 16 (1844). — Dode in *Mém. Soc. Hist. Nat. Autun,* XVIII. (*Extr. Monog. Inéd. Populus,* 15) (1905).

Balsamiflua deltoides Griffith, *Icon. Pl. Asiat.* IV. t. 526 (1854).

Populus Ariana Dode in *Mém. Soc. Hist. Nat. Autun,* XVIII. (*Extr. Monog. Inéd. Populus,* 16) (1905).

Populus Litwinowiana Dode, l. c. 17 (1905).

CHINA. Chili, Peking, *R. Alcock* (fide Burkill); Kanou, beyond the Great Wall, *P. J. Piasezkei* (fide Maximowicz). Also in northern Africa and western and central Asia.

I doubt, however, if Alcock's specimen from Peking belongs to this species.

2. **Populus pruinosa** Schrenk in *Bull. Acad. Sci. St. Pétersbourg,* III. 210 (1845). Wesmael in *Mém. Soc. Sci. Hainaut,* III. (*Monog. Populus,* 56, t. 14) (1869). — Dode in *Mém. Soc. Hist. Nat. Autun,* XVIII. (*Extr. Monog. Inéd. Populus,* 18, t. 11, fig. 2ᵃ) (1905). — Gombocz in *Math. Termesz. Közl.* XXX. 72 (*Monog. Gen. Populi*) (1908).

Populus glaucicomans Dode in *Mém. Soc. Hist. Nat. Autun,* XVIII. (*Extr. Monog. Inéd. Populus,* 18, t. 11, fig. 2) (1905).

TURKESTAN, SOUTHWESTERN SIBERIA.
According to the photographs and notes taken by the well-known explorer F. N. Meyer on February 19, 1911 in Chinese Turkestan for the Department of Agriculture of the United States, *P. pruinosa* may be distinguished from *P. euphratica* by the bark of old trunks. Meyer's photograph No. 5685 shows a trunk of *P. pruinosa* with a distinctly and deeply grooved bark as in Ash and Elm trees, while the photograph No. 5689 represents *P. euphratica* with a more or less shaggy bark.

Sect. 2. **LEUCOIDES** Spach, *Hist. Vég.* X. 385 (1841). — Gombocz in *Math. Termesz. Közl.* XXX. 118 (*Monog. Gen. Populi*) (1908).

Populus, subgen. *Eupopulus* Dode, sect. *Leucoideae* Dode in *Mém. Soc. Hist. Nat. Autun,* XVIII. (*Extr. Monog. Inéd. Populus,* 14 et 35) (1905).

Folia ramulorum turionumque vix diversa subtus plus minusve discoloria, basi glandulifera, superne stomatibus non praedita; petioli basi cylindrica, apice paulo compressi, semper satis longi. Gemmae vix viscosae. Disci florum persistentes, satis evoluti, fere ad basim lobati; stigmata 2–3, stylis elongatis; stamina 30–40, antheris longis, apice apiculatis. Capsulae crassae, pedicellatae (an semper?), villosae.

3. **Populus Wilsonii** C. Schneider. See p. 16.

4. **Populus glauca** Haines in *Jour. Linn. Soc.* XXXVII. 408, fig. (1906). — Gombocz in *Math. Termesz. Közl.* XXX. 157 (*Monog. Gen. Populi*) (1908).

? *Populus Jacquemontiana* Dode in *Mém. Soc. Hist. Nat. Autun*, XVIII. (*Extr. Monog. Inéd. Populus*, 66, t. 12, fig. 108) (1905).

INDIA. Bhutan.

Through the kindness of the Keeper of the Herbarium of the Royal Gardens, Kew three sheets of Haines' plant bearing the No. 826 have been loaned to the Arnold Arboretum: one with fruits, collected June 1904, Labah ridge, alt. 2200 m.; another with leaves, collected August 1904, Bhotan, without precise locality, alt. 2400 m.; and the third with flowers only, collected April 1905, Pankasari ridge, alt. 2700 m. In their shape and in the color of the upper surface the leaves resemble those of *P. Wilsonii* Schneider, and are up to 20 cm. long and 18.5 cm. broad, the petioles being up to 13.5 cm. long. The color of the under surface of the leaves is glaucous, not whitish as in the dried leaves of *P. Wilsonii*, but the scarce pubescence is almost the same in both, while in *P. lasiocarpa* Oliver even the mature leaves of cultivated plants are hairy along the midrib and the main veins beneath. The flowers of *P. glauca* are mostly bi-sexual, as indicated by Haines, otherwise the perianth and the ovaries scarcely differ from those of *P. Wilsonii*, the perianth of which is mostly not so largely developed and so deeply lobed as in *P. glauca*. In this species the stigmas seem to be a little shorter-stalked, but as far as I can judge from the material before me all these characters of the flowers are too variable to represent good distinguishing features. The fruits of *P. glauca* are roundish and apparently a little more pubescent or villose than those of *P. Wilsonii*. Both species need further study.

5. **Populus lasiocarpa** Oliver. See p. 17.

Sect. 3. TACAMAHACA Spach, *Hist. Vég.* X. 392 (1841).

Populus, subgen. *Eupopulus* Dode, sect. *Tacamahacae* Dode in *Mém. Soc. Hist. Nat. Autun*, XVIII (*Extr. Monog. Inéd. Populus*, 14 et 34) (1905).

Folia ramulorum turionumque inter se pleraque variabilia, subtus distincte discoloria, albescentia, superne stomatibus plus minusve numerosis [1] v. nullis praedita, basi glandulifera; petioli longitudine variabiles. Disci florum persistentes, margine lobati v. crenati; stigmata 2–4, stylis brevibus v. nullis; stamina 18–60, antheris oblongis v. subglobosis, haud apiculatis. Capsulae variabiles.

6. **Populus ciliata** Wallich apud Royle, *Ill. Bot. Himal.* I. 346; II. t. 84ᵃ or 98, fig. 1 (1839). — Wesmael in De Candolle, *Prodr.* XVI. 2, 329 (1868); in *Mém. Soc. Sci. Hainaut*, III. 63 (*Monog. Populus*) (1869). — Brandis, *Forest Fl. Brit. Ind.* 475 (pro parte) (1874); *Ind. Trees*, 640 (pro parte) (1906). — Stewart, *Punjab Pl.* 204 (1869). — Hooker f., *Fl. Brit. Ind.* V. 638 (1888). — Collett, *Fl. Siml.* 481 (1902). — Dode in *Mém. Soc. Nat. Hist. Autun*, XVIII. (*Extr. Monog. Inéd. Populus*, 65) (1905). — Haines in *Jour. Linn. Soc.* XXXVII. 407 (1906). — Gombocz in *Math. Termesz. Közl.* XXX. 116 (*Monog. Gen. Populi*) (1908). — Henry in Elwes & Henry, *Trees Gr. Brit. & Irel.* VII. 1840, in adnot. (1913).

INDIA. Western Himalaya to Bhutan: Kumaon, northern face of Choor, *R. Blinkworth* (No. 2796 of Wallich's Cat., type; with ripe fruits); Srinaghur, *Kamrup* (No. 2796 B of Wallich's Cat.); Nain-tál, alt. 2100 m. *Strachey & Winterbottom* (No. 1); northwest Himalaya, without locality (Lachen?), 2000–3000 m., *T. Thomson;* Sikkim, without locality, 2–3000 m., *J. D. Hooker;* Chakrata, alt. 2300 m., June 26, 1912, *K. Narayana Jyengar* (a lofty tree with gray, smooth bark; common in blanks in Deodar forests); same locality, May 5 and June 25, 1912,

---

[1] According to Gombocz the plants of this section have no stomata on the upper surface of the leaves.

P. K. Subramaina Jyer (No. 86); Bhotan, without locality, *W. Griffith* (Nos. 957, 2559, ex Haines).

This species may represent a distinct group of sect. *Tacamahaca* according to its rather long stalked fruits, its densely ciliate leaves, its long anthers, etc. Dode refers it to his "groupe *Ciliata*" together with *P. Jacquemontiana* Dode (in *Mém. Soc. Hist. Nat. Autun,* XVIII. [*Extr. Monog. Inéd. Populus,* 66, t. 12, fig. 108] [1905]) from the same geographical region without referring to any specimens. The only difference I can detect in his description is in the pubescent fruit. Having seen no specimen I cannot judge from Dode's insufficient indications if his species is distinct or only a variety of *P. ciliata.*

7. **Populus suaveolens** Fischer. See p. 18.

Populus suaveolens, var. Przewalskii Schneider, n. var.

Populus suaveolens, var. a Maximowicz in *Bull. Soc. Nat. Mosc.* LIV. 51 (1879) quoad specim. Piasezki.

Populus Przewalskii Maximowicz in *Bull. Acad. Sci. St. Pétersbourg,* XXVII. 540 (1882); in *Mél. Biol.* XI. 321 (1881). — Dode in *Mém. Soc. Hist. Nat. Autun,* XVIII. (*Extr. Monog. Inéd. Populus,* 55) (1905), exclud. synon.! — Gombocz in *Math. Termesz. Közl.* XXX. 101 (*Monog. Gen. Populi*) (1908).

Populus Przewalskii, f. *microphylla* Gombocz, l. c. (1908).

TIBET. Kukunor: " ad Hoangho superiorem ejusque affluentes nec non ad lacum Kukunor," 1872 and 1880, *N. M. Przewalski* (type of *P. Przewalskii,* ex Maximowicz).

CHINA. Kansu: " inter fl. Hoangho et murum magnum, in Liang-tshouc et Shan-dan-siang viis publicis, in arenosis Han-tschou," July, August 1875, *P. J. Piasezki* (co-type of *P. Przewalskii,* ex Maximowicz).

MONGOLIA: " circa lacus Ubsa," 1879, *G. N. Potanin* (with not quite ripe fruits).

According to Maximowicz his *P. Przewalskii* is distinguished by " capsulis parvis subsessilibus ovoideis obtusis pubescenti-pilosis maturis glabratis, stigmatibus 3." I have not seen a type specimen, but by the kindness of the Keeper of the Kew Herbarium I have been able to examine Potanin's specimen mentioned above. It agrees well with Maximowicz's description. The leaves are ovate-oblong and somewhat acuminate at the apex and orbicular-cuneate at the base, both surfaces being more or less sparsely and finely puberulent; the petioles are loosely villose. The perianth and the ovaries of the male flowers are more or less pubescent; but further observations are needed to determine if the pubescence of the floral parts of *Populus* is a good character or not. The shape, the texture, and the pubescence of *P. suaveolens* Fischer seem to be very variable. See also p. 20; some of the sterile specimens here mentioned may belong to var. *Przewalskii.*

8. **Populus Maximowiczii** A. Henry in *Gard. Chron.* ser. 3, LIII. 198, fig. 89 (1913); in Elwes & Henry, *Trees Gr. Brit. & Irel.* VII. 1838, t. 410, fig. 24 (1913).

Populus suaveolens, var. *latifolia* Regel in *Mém. Acad. Sci. St. Pétersbourg,* ser. 7, IV. 133 (*Tent. Fl. Ussur.*) (1862). — Gombocz in *Math. Termesz. Közl.* XXX. 112 (*Monog. Gen. Populi*) (1908).

Populus suaveolens Maximowicz in *Bull. Soc. Nat. Mosc.* LIV. 52 (formae sub b et c indicatae) (non Fischer) (1879). — Sargent in *Gard. & Forest,* VI. 404 (1893); *Forest Fl. Jap.* 71 (1894). — Komarov in *Act. Hort. Petrop.* XXII. 17 (*Fl. Mansh.* II.) (pro parte maxima) (1904). — Dode in *Mém. Soc. Hist. Nat. Autun,* XVIII. (*Extr. Monog. Inéd. Populus,* 61) (pro parte) (1905). — Nakai in *Jour. Coll. Sci. Tokyo,* XXXI. 211 (*Fl. Kor.* II.) (1911).

Populus balsamifera, var. *suaveolens* Burkill in *Jour. Linn. Soc.* XXVI. 536 (pro parte) (non Loudon) (1899). — Shirasawa, *Icon. Ess. For. Jap.* I. 37, t. 18, fig. 11–24 (1900).

NORTHEASTERN ASIA. Mandshuria: ad fl. Amur., Sept. 1, 1855, *R. Maack;* ad fl. Li-Fudin, 1860, *C. Maximowicz;* prov. Sheng-king, June 25, 1906, *F. N. Meyer* (No. 22; young shoots slightly angular ¹); Kabarovka, August 23, 1903, *C. S. Sargent;* banks of Chita River, Chita, August 12, 1903, *C. S. Sargent.* Saghalien: without locality, *Fr. Schmidt* (ex Herb. Petrop.); Korsakof, July and September 1908, *U. Faurie* (Nos. 280, 280ᵇⁱˢ). Kamtschatka: Petropavlovski, 1853–56, *C. Wright.*

JAPAN. Hokkaido: Sapporo, Prov. Ishikari, May 1884, *K. Miyabe;* same locality, August 21, 1905, *J. G. Jack;* " in arena rivorum Kamikawa," July 1905, *U. Faurie* (No. 6642). Hondo: prov. Shimotsuke, Lake Chuzenji, August 11, 1905, *J. G. Jack.*

This magnificent Poplar grows to a larger size than any other species of eastern Asia, and ranks with the largest trees which grow there. It is abundant in Hokkaido, rather less so in southern Saghalien, and is rare in Hondo, where it occurs sparingly in the Nikko region and northward. At its best it is a tree from 27 to 30 m. tall with a trunk 5 or 6 m. in girth. The branches are massive and widespreading and form a flattened rounded and somewhat oval crown. The bark is gray and deeply fissured on the trunk of old trees but smooth and yellowish-gray on the branches and also on the trunks of young trees. It is seen at its best in the moist bottom lands of river-valleys. Pictures of this tree will be found under Nos. x237, x238, x239, x244, x367, x507 of my collection of Japanese photographs.

E. H. W.

9. **Populus szechuanica** C. Schneider. See p. 20.

Populus szechuanica, var. **tibetica** Schneider, n. var.

Populus balsamifera Thomson, *W. Himalaya & Tibet*, 180 (1852). — Stewart, *Punjab Pl.* 204 (1869). — Hooker f., *Fl. Brit. Ind.* V. 638 (in part, not Linnaeus) (1888). — Brandis, *Ind. Trees*, 640 (1906).

Populus tristis Koehne, *Deutsche Dendr.* 83 (non Fischer) (1893), quoad specimina citata No. 2–4. — Henry in Elwes & Henry, *Trees Gr. Brit. & Irel.* VII. 1840, in adnot., 1841 (1913).

A typo recedit gemmis, ramulis leviter angulatis ramisque et petiolis breviter pubescentibus, foliis initio utrinque deinde tantum nervis plus minusve puberulis v. subglabrescentibus etiam axi amentorum foemineorum hirtella.

Arbor ad 28 m. alta. Folia turionum ut videtur elliptico-lanceolata, basi subrotundata v. late cuneata, apicem versus sensim acuta, visa ad 12 cm. longa et ad 5.5 cm. lata, margine glanduloso-crenato-dentata, ramulorum fructiferorum ovata, basi plus minusve rotunda, apice acuta v. subacuminata v. late ovata v. ovato-rotunda, basi leviter v. distinctius cordata, apice plus minusve acuminata, 10–14 cm. longa et 6–8.5 cm. lata, margine satis crebre glanduloso dentata, pleraque ciliata; petioli 2–6 cm. longi. Flores non vidi. Capsulae subsessiles, glabrae, ovoideoglobosae, 3–4-valvae.

CHINA. Western Szech'uan: without precise locality, ravines, alt. 3000–3600 m., June 1904 (Veitch Exped. No. 4527; tree 16 m. tall, with smooth bark).

INDIA. Kashmir: "Northwestern Tibet," cultivated, alt. 2000–4500 m., *T. Thomson* (ripe open fruits, type); Ladak, Khárbu Koma to Sháksi, southwest

¹ This specimen may represent the same form as *P. suaveolens*, var. *macrocarpa* Schrenk, see p. 19. I am not quite sure whether this is a form of *P. suaveolens* or *P. Maximowiczii.*

of Dah, July 3, 1856, *Schlagintweit* (No. 5327, sterile branch); Upshi to Gulab-Gárh, left side of Indus valley, June 28, 1856, *Schlagintweit* (No. 1563, old sterile branch); right shore of the Indus near Leh, July 4–9, 1856, *Schlagintweit* (No. 1162, small part of young shoot); prov. Dras, Múlbe to Dras, October 8–11, 1856, *Schlagintweit* (No. 4976, old branch, distributed sub nom. *P. laurifolia*).

There is another Himalayan specimen of Thomson's from Zanskar in Tibet with young leaves and rather old female catkins. The ovaries or young fruits are deeply furrowed and glabrous, but the pedicels are rather longer and hairy. I am not quite sure about the value of the furrows and believe it is only an effect of drying.

It seems very difficult to distinguish the Chinese *P. szechuanica* from this Tibetan form, but nevertheless I should have treated the latter as a distinct species on account of its geographical distribution had I not seen Wilson's No. 4527, which shows the same pubescence and differs only in the more deeply cordate and roundish leaves. I cannot refer the Chinese nor the Tibetan form to any of Dode's species of this section [1] or to a form described by previous authors. *Populus tristis* of Fischer, which Koehne and Henry believed to be probably identical with these Tibetan specimens, seems to me quite distinct. Fischer says that he secured dried specimens from Sitka. I believe that *P. tristis* is very near to *P. Maximowiczii* or to *P. trichocarpa* Hooker of western North America.

10. **Populus yunnanensis** Dode in *Mém. Soc. Nat. Autun*, XVIII. (*Extr. Monog. Inéd. Populus*, 63, t. 11, fig. 103 a) (1905).

CHINA. Yunnan: without locality, *F. Ducloux* (Dode received a living plant).

I have seen only young vigorous plants of this species raised from cuttings from Dode's type specimen. Dode's description is insufficient; and I can only say that this species seems to be nearly related to *P. szechuanica* and needs further investigation.

11. **Populus Gamblei** Dode in *Mém. Soc. Nat. Autun*, XVIII. (*Extr. Monog. Inéd. Populus*, 63, t. 12, fig. 103) (1905). — Haines in *Jour. Linn. Soc.* XXXVII. 407, fig. (1906).

INDIA. Brit. Bhutan: district Darjeeling, Kalimpung, alt. 1300 m., March 1875, *J. S. Gamble* (No. 2646[a]; type, with old fruits; No. 6707; male co-type).

This species was first mentioned by Gamble (*Man. Ind. Timb.* ed. 2, 690 [1902]), without description. Dode bases his description upon male and female specimens collected by Gamble without citing any number, while Haines founded it upon Gamble's No. 2646[a] and 7607, of which I have before me excellent photographs and some male flowers and fruits kindly sent to the Arnold Arboretum by the Keeper of the herbarium of Kew Gardens. Haines says that it is extremely doubtful if Dode's imperfect description refers to his *P. Gamblei*, but I do not think that there can be the least doubt that both descriptions refer to the same plant. Dode's figure of the leaves agrees well with those of Haines' type specimen before me, he also mentions the puberulent perianth and the narrow bracts. I have not seen young female flowers, and unfortunately I have no leaves. Haines describes a dimorphism of the leaves and shoots which, he believes, is possibly due to fungous agency. But, as far as I can judge from what he says, the appearance of large pubescent cordate leaves with two large glands just above the petioles on some branches while the normal leaves are quite glabrous, not cordate and without glands, may not be unusual in *Populus*, as the same variation occurs in *Populus*

---

[1] The only species which might prove to be the same as var. *tibetica* is *P. Jacquemontiana* Dode, which I mentioned under *P. ciliata* Wallich, but Dode says: " ses capsules sont pubescentes."

adenopoda Maximowicz and in other species. With regard to it and to the flowers I would be inclined to refer *P. Gamblei* to sect. *Leuce,* but Dode says that it is most nearly related to *P. yunnanensis* Dode, of which I know living plants, and which certainly belongs to sect. *Tacamahaca.* Gombocz (in *Math. Termesz. Közl.* XXX. 156 [1908]) knew these species only from the descriptions of the authors. That of *P. Gamblei* being insufficient, I add a more complete one.

Arbor, ut videtur satis alta; ramuli annotini ut videtur glabri sed interdum villosuli; turiones (ex Dode) 5-angulati. Folia ramulorum fructiferorum ovato-lanceolata, basi subcordata v. truncata v. subtrapezoidea, apice breviter v. longius acuminata, margine distanter sinuoso-dentata, subtus (ex Dode) albescentia, glabra, maxima ad 15.5 cm. longa et 9.5 cm. lata, v. (ex Haines) subtus valde pubescentia villosave, basi cordata, ad 33 cm. longa et 25 cm. lata; petioli foliorum glabrorum ⅛–¼ laminis longi, ut videtur glabri, foliorum pubescentium (ex Haines) breviores, crassiores, pubescentes. Amenta mascula ad 8 cm. longa, rhachi pilosa; bracteae lanceolatae, basi attenuatae, apice acutae, saepissime plusminusve incisae, brunneae, margine satis dense longe ciliatae, facie glabrae v. pilis paucis longis praeditae, ad 7 mm. longae; flores pedicello distincto piloso suffuliti; discus oblongus, margine crenulato-lobulatus, ciliatus et extus pilosus, intus glaber; stamina 12–20, filamentis tenuibus glabris quam antheras purpureas subglobosas vix v. paulo longioribus; amenta feminea tantum fructifera visa, circiter 10 cm. longa, rhachi pilosa; bracteae mihi ignotae; ovaria subsessilia, matura ad 11 mm. longa, e basi ellipsoidea oblongo-linearia, glabra, 2-carpellaria (ex Dode 2–4-valvata), basi perianthio piloso irregulariter inciso cincta; stigmata non vidi.

12. **Populus laurifolia** Ledebour, *Fl. Alt.* IV. 297 (1833) excl. syn. Gmelinii; *Icon. Fl. Ross.* V. t. 479 (1834); *Fl. Ross.* III. pt. 2, 629 (1850). — Fischer in *Allg. Gartenzeit.* IX. 404 (1841); in *Bull. Acad. Sci. St. Pétersbourg,* IX. 347 (1842). — Dippel, *Handb. Laubholzk.* II. 208, fig. 103 (1892). — Koehne, *Deutsche Dendr.* 85 (1893). — Schneider, *Ill. Handb. Laubholzk.* I. 16, fig. 5 m–n, 6 v–y (1904). — Dode in *Mém. Soc. Nat. Hist. Autun,* XVIII. (*Extr. Monog. Inéd. Populus,* 59, t. 12, fig. 94) (1905). — Ascherson & Graebner, *Syn. Mitteleur. Fl.* IV. 47 (1908). — Gombocz in *Math. Termesz. Közl.* XXX. 102 (*Monog. Gen. Populi*) (1911). — Henry in Elwes & Henry, *Trees Gr. Brit. & Irel.* VII. 1842, t. 410, fig. 30 (1913).

> *Populus balsamifera* Pallas, *Fl. Ross.* t. 41, fig. B (non Linnaeus) (1784).
> *Populus balsamifera,* var. 2, *viminalis* Loudon, *Arb. Brit.* III. 1673 (1838), excl. syn. Fischer. — Wesmael in De Candolle, *Prodr.* XVI. pt. 2, 330 (1868).
> *Populus balsamifera,* var. *laurifolia* Wesmael, l. c.; in *Mém. Soc. Sci. Hainaut,* III. 246 (*Monog. Populus,* 66) (1869).
> *Populus Lindleyana* Carrière in *Rev. Hort.* XXXIX. 380 (non Booth) (1867). — Dode in *Mém. Soc. Nat. Hist. Autun,* XVIII. (*Extr. Monog. Inéd. Populus,* 59) (1905).
> *Populus laurifolia,* var. *viminalis* Dippel, *Handb. Laubholzk.* II. 209 (pro parte) (1892).
> *Populus laurifolia,* var. *Lindleyana* Ascherson & Graebner, *Syn. Mitteleur. Fl.* IV. 47 (1908).

WESTERN SIBERIA. Prov. Tomsk: Altai Mts., near Alexandrofka, June 6, 1911, *C. A. Meyer* (No. 783).

As far as I know this species has been collected only on the Altai Mts. The leaves of the young shoots and offshoots are rather narrow-oblong with a roundish or obtuse, rarely subcordate base, gradually tapering into an acute or shortly acuminate apex. These long leaves measure between 10: 3 and 15: 6 cm., have a narrow glandular serrature and are borne on short petioles from 8–15 mm. long. The shape of the long-petioled leaves of the older plant is rather variable, being lanceolate-elliptic or elliptic-ovate with an acute base, or ovate, even round-ovate, with an obtuse, round or slightly cordate base and an acute or shortly acuminate apex. The largest measure over 13:8 cm. The color is rather dull green above and whitish beneath. The branchlets, petioles and the midribs of the leaves are commonly more or less pubescent. According to Henry the fruits are slightly pubescent. 1 have not yet seen female flowers or fruits. Dode distinguishes the typical *P. laurifolia* by its glabrous fruits from *P. Lindleyana* with pubescent capsules.

13. **Populus Simonii** Carrière. See p. 21.

Populus Simonii, f. pendula Schneider. See p. 22.

Populus Simonii, f. fastigiata Schneider. See p. 22.

Sect. 4. AIGEIROS Duby in De Candolle *Bot. Gall.* ed. 2, I. 427 (1828). — Gombocz in *Math. Termesz. Közl.* XXX. 75 (*Monog. Gen. Populi*) (1908).

Populus, subgen. *Eupopulus* Dode, sect. *Aegiri* Dode in *Mém. Soc. Hist. Nat. Autun*, XVIII. (*Extr. Monog. Inéd. Populus*, 14, 34) (1905).

Folia ramulorum turionumque satis similia, subtus vix v. paullo discoloria, viridescentia, utraque pagina stomatifera, basi glandulifera; petioli plus minusve compressi, semper satis longi; gemmae plus minusve viscosae. Disci florum persistentes, sinuati; stigmata 2–4, plus minusve sessilia; stamina pleraque 15–30, antheris subglobosis, non apiculatis. Capsulae variabiles.

14. **Populus afghanica** Schneider, n. sp.

Populus nigra, var. *afghanica* Aitchison & Hemsley in *Jour. Linn. Soc.* XVIII. 96 (*Fl. Kuram Vall.*) (1880). — Gombocz in *Math. Termesz. Közl.* XXX. 91 (*Monog. Gen. Populi*) (1908).

AFGHANISTAN : Kuram valley, 1879, *J. E. T. Aitchison* (No. 161, sterile). The authors' description is quite sufficient. They say: " possibly it may prove a distinct species; but in the absence of very complete material we have not ventured to give it that rank. A large tree, fully 100 feet in height and 8 feet in girth; quite wild, also cultivated in the vicinity of Shálizán, but only at one shrine. In the Hariáb district it is common, cultivated, and apparently wild also; in fruit May."
The extremely small leaves show on the upper side the same numerous stomata as the typical *P. nigra* Linnaeus. The branches are exceedingly slender. Dode seems to have overlooked the description of this peculiar species.

15. **Populus nigra** Linnaeus, var. **italica** Du Roi, *Harbk. Baumz.* II. 141 (1772).

Populus nigra, var. *pyramidalis* Spach in *Ann. Sci. Nat.* sér. 2, XV. 31 (1841). — Gombocz in *Math. Termesz. Közl.* XXX. 89 (*Monog. Gen. Populi*) (1908).
? *Populus nigra*, var. *sinensis* Carrière in *Rev. Hort.* 1867, 340. — Wesmael in De Candolle *Prodr.* XVI. 2, 327 (1868); in *Mém. Soc. Sci. Hainaut*, III. 258 (*Monog. Populus*, 59) (1869). — Maximowicz in *Bull. Soc. Nat. Mosc.* LIV. 50 (1879). — Burkill in *Jour. Linn. Soc.* XXVI. 536 (1899). — Gombocz in *Math. Termesz. Közl.* XXX. 91 (*Mong. Gen. Populi*) (1908).
Populus sinensis Dode in *Mém. Soc. Nat. Autun*, XVIII. (*Extr. Monog. Inéd. Populus*, 50, t. 12, fig. 72) (1905).

CHINA. Northern China: without locality, *G. E. Simon* (fide Carrière). Southern Kansu: *P. J. Piasezki* (fide Maximowicz). Shensi: 1875, *P. J. Piasezki* (fide Maximowicz).

According to Hooker f., Stewart and others the Lombardy Poplar is cultivated here and there in the northwestern Himalaya and western Tibet. To this form belongs a specimen in the Gray Herbarium with the label: *Populus balsamifera*, Herb. Falconer No. 957.

The typical *P. nigra* Linnaeus *Spec.* 1034 (1753) (*P. nigra*, var. *typica* Beck, *Fl. Nied.-Oestr.* 303 [1890]). — *P. europaea* Dode in *Mém. Soc. Nat. Hist. Autun*, XVIII [*Extr. Monog. Inéd. Populus*, 51, t. 12, fig. 78] [1905] [non *P. nigra* Dode]) appears to find the eastern limits of its range in the Altai region.

Of the very doubtful *P. nigra*, var. *sinensis* Carrière, I have seen only a cultivated plant distributed by Dode and supposed to have come from the type plant. I do not know how to distinguish this cultivated plant from the Lombardy Poplar.

The Lombardy Poplar has been introduced into China by foreigners and is cultivated especially in many of the treaty ports, including that of Ichang. It has also been introduced and is now cultivated in Tokyo and other parts of Japan. If the obscure *P. nigra*, var. *sinensis* Carrière, really came from China, and is really referable to *P. nigra* Linnaeus, the specimen was probably taken from a cultivated plant of the Lombardy Poplar. I do not believe that *P. nigra* Linnaeus or any of its forms grows spontaneously in China.   E. H. W.

Sect. 5. LEUCE Duby in De Candolle, *Bot. Gall.* I. 427 (1828). — Gombocz in *Math. Termesz. Közl.* XXX. 137 (*Monog. Gen. Populi*) (1908).

Populus, subgen. *Leuce* Dode, *Mém. Soc. Hist. Nat. Autun*, XVIII. (*Extr. Monog. Inéd. Populus*, 13) (1905).

Folia ramulorum turionumque saepe satis diversa, subtus pleraque discoloria, superne haud stomatifera, basi saepe v. haud glaudulifera; petioli plus minusve compressi v. foliorum turionalium plus minusve cylindrici, semper satis longi. Gemmae haud v. vix viscosae. Disci florum persistentes, sinuati; stigmata pleraque 2, sessilia; stamina 5–20, subglobosa v. elongata, non apiculata. Capsulae satis parvae, elongatae.

16. **Populus alba** Linnaeus, *Spec.* 1034 (1753). — Ledebour, *Fl. Alt.* IV. 295 (1833). — Stewart, *Punjab Pl.* 203 (1869). — Hooker f., *Fl. Brit. Ind.* V. 638 (1888). — Brandis, *Ind. Trees* 640 (1906). — Gombocz in *Math. Termesz. Közl.* XXX. 141 (*Monog. Gen. Populi*) (1908).

Populus triloba Dode in *Mém. Soc. Hist. Nat. Autun*, XVIII. (*Extr. Monog. Inéd. Populus*, 21, t. 11, fig. 5) (1905).
Populus Morisetiana Dode, l. c. 22, t. 11, fig. 11 (1905).

The species seems to be wild (and often planted) in the northwestern Himalaya, Tibet and Altai. I have seen no specimens from those regions, and I do not know if Dode's species may really indicate any distinct form.

17. **Populus tomentosa** Carrière, in *Rev. Hort.* X. 340 (1867). — Wesmael in De Candolle, *Prodr.* XVI. 2, 325 (1868); in *Mém. Soc. Sci. Hainaut*, III. 228 (*Monog. Populus*, 48, t. 17) (1869). — Schneider, *Ill. Handb. Laubholzk.* 1. 21, fig. 5 f–g², 7 t–u (1904). — Dode in *Mém. Soc. Nat. Hist. Autun*, XVIII. (*Extr. Monog. Inéd. Populus*, 25, t. 11, fig. 21) (1905). — Ascherson & Graebner, *Syn. Mitteleur. Fl.* IV. 24 (1908). — Gombocz in *Math. Termesz. Közl.* XXX. 140 (*Monog. Gen. Populi*) (1908). — Henry in Elwes & Henry, *Trees Gr. Brit. & Irel.* VII. 1786, t. 408, fig. 2 (1913).

Populus alba, var. *denudata* Maximowicz in *Bull. Soc. Nat. Mosc.* LIV. 48 (non Hartig) (1879).

Populus alba, var. *tomentosa* Wesmael in *Bull. Soc. Bot. Belg.* XXVI. 373 (1887).— Burkill in *Jour. Linn. Soc.* XXVI. 535 (1899).
Populus alba Burkill, 1. c. (non Linnaeus) (1899). — ?Nakai in *Jour. Coll. Sci. Tokyo*, XXXI. 211 (*Fl. Kor.* II.) (1911).
Populus pekinensis L. Henry, *Rev. Hort.* 1903, 355, fig. 142.
Populus alba, var. *seminuda* Komarov in *Act. Hort. Petrop.* XXII. 20 (1903).
Populus glabrata Dode in *Mém. Soc. Hist. Nat. Autun*, XVIII. (*Extr. Monog. Inéd. Populus*, 27, t. 11, fig. 25ª) (1905).

CHINA. Chili: Siwantze, *G. E. Simon* (ex Carrière, type); near Peking, *Skatschkow* (fide Maximowicz); same locality, *E. Bretschneider;* same locality, *S. W. Williams* ("common over northern China, 50 feet high, timber not durable, flowers in April"); Yellow Temple, near Peking, Sept. 16, 1903, *C. S. Sargent* ("leaves very lustrous"); Peking, October 7, 1905, *J. G. Jack.* Chekiang: Shanghai, cultivated in Public Garden, 1908, *D. Macgregor.*

A picture of a cultivated specimen of this tree will be found under Nos. 650 and 0321 of the collection of Wilson photographs and also in his *Vegetation of Western China*, No. 403.

18. **Populus adenopoda** Maximowicz. See p. 23.

19. **Populus Sieboldii** Miquel in *Ann. Mus. Lugd.-Bat.* III. 29 (1867), exclud. plantis masculis. — Wesmael in De Candolle, *Prodr.* XVI. pt. 2, 327 (1868); in *Mém. Soc. Sci. Hainaut*, III. (*Monog. Populus*, 57) (1869). — Dippel, *Handb. Laubholzk.* II. 192, fig. 91 (1892). — Koehne, *Deutsche Dendr.* 80 (1893). — Schneider, *Ill. Handb. Laubholzk.* I. 17, fig. 3 h–i, 7 h–l (1904). — Dode in *Mém. Nat. Hist. Autun*, XVIII. (*Extr. Monog. Inéd. Populus*, 32, t. 11, fig. 33) (1905). — Ascherson & Graebner, *Syn. Mitteleur. Fl.* IV. 29 (1908). — Gombocz in *Math. Termesz. Közl.* XXX. 131 (*Monog. Gen. Populi*) (1908). — Henry in Elwes & Henry, *Trees Gr. Brit. & Irel.* 1794, t. 408, fig. 6 (1913).

Populus tremula, var. *villosa* Franchet & Savatier, *Enum. Pl. Jap.* I. 463 (non Lang) (1875). — Maximowicz in *Bull. Soc. Nat. Mosc.* LIV. 49 (1879). — Shirasawa, *Icon. Ess. For. Jap.* I. 37, tab. 18, figs. 1–10 (1900). — Matsumura *Ind. Pl. Jap.* II. 7 (1912).

NORTHEASTERN ASIA. Saghalien: Korsakof, rare, June 1908, *U. Faurie* (No. 279).

JAPAN. Hokkaido: Hakodate 1861, *C. Maximowicz;* Shibetsu, July 12, 1884, *K. Miyabe;* (near Mori, September 26, 1892, *C. S. Sargent.* Hondo: without locality, *P. von Siebold* (type, ex Herb. Lugd. Bat.); prov. Mutsu, prope Aomori, in sylvis, May 1902, *U. Faurie* (Nos. 5085, 5087, 5088); same locality, June 1905, *U. Faurie* (No. 6644); prov. Shimotsuke, Lake Chuzenji, August 11, 1905, *J. G. Jack;* Nikko, April 25, 1904, *N. Mochizuki;* prov. Shinano, above Narai, alt. 1200 m., September 3, 1905, *J. G. Jack.*

This Japanese species seems to be distinct in the rather close dentation of its leaves.

This tree is widely distributed in Japan from about central Hondo northward through Hokkaido to southern Saghalien. In Hondo it is not plentiful, but in Hokkaido it was formerly very abundant; but as its wood makes the best match-splints in Japan it has been promiscuously felled for this purpose. In the government forests in the more inaccessible parts of Hokkaido it still remains in fair quantity. It is a tree of medium size with smooth grayish-green bark, and is very like the common Aspen (*P. tremula* Linnaeus) in habit and general appearance, and like that species it produces suckers freely. E. H. W.

20. **Populus tremula** Linnaeus, *Spec.* 1043 (1753).

Populus tremula, var. a *typica* Schneider, *Ill. Handb. Laubholzk.* I. 19
(1904). — Ascherson & Graebner, *Syn. Mitteleur. Fl.* IV. 25 (1908). — Gombocz in *Math. Termesz. Közl.* XXX. 126 (*Monog. Gen. Populi*) (1908).
WESTERN SIBERIA.

Populus tremula, var. **Davidiana** Schneider. See p. 24.

Populus tremula, var. **Davidiana**, f. **tomentella** Schneider. See p. 25.

21. **Populus rotundifolia** Griffith, *Notul. Pl. As.* IV. 382 (1854); *Icon. Pl. As.*
IV. t. 546 (1854). — Dode in *Mém. Soc. Nat. Autun,* XVIII. (*Extr. Monog.
Inéd. Populus,* sub Errata et Addenda, Espèce 34) (1905). — Gombocz
in *Math. Termesz. Közl.* XXX. 130 (*Monog. Gen. Populi*) (1908).
Populi spec. Griffith, *Itin. Notes,* II. No. 881 (1848).
Populus microcarpa Hooker f., *Fl. Brit. Ind.* V. 639 (1888). — Dode in *Mém.
Soc. Nat. Hist. Autun,* XVII. (*Extr. Monog. Inéd. Populus,* 32, t. 11, fig. 34)
(1905).

INDIA. Bhutan: *W. Griffith* (No. 4495). This specimen may be the type of
Hooker's *P. microcarpa:* Bhutan, about Panga and towards Chupeh, alt. 2500 m.,
Griffith.
Having seen only one fruiting specimen of the typical *P. rotundifolia* by the
kindness of the Keeper of the Kew Herbarium, I cannot decide whether var.
Duclouxiana Gombocz to which I have referred some of Wilson's plants from
Szechuan is sufficiently distinct. The shape of the leaves is variable, but the
fruits of the type are very small, and the pedicels are glabrous.

Populus rotundifolia, var. **Duclouxiana** Gombocz. See p. 25.

SPECIES INCERTAE.

Populus Bonati Léveillé in *Monde Plant.* XII. 9 (1910); in Fedde, *Rep. Spec.
Nov.* VIII. 445 (1910). — Gombocz in *Bot. Közl.* X. 25, fig. (7) (1911).
CHINA. Yunnan.
Populus Bonati has been collected in Yunnan: Pa-ta-ouan, près Pin Tchouan,
March 3, 1907, *Jean Py* (No. 665 (type ex Léveillé), and according to Gombocz
also by Ducloux at Tchong-chan. Gombocz considers *P. Bonati* a distinct species
on account of its trilobed or trifid stigmas. His figures, l. c. p. 26, represent rather
old female flowers of *P. tremula, P. Bonati* and *P. adenopoda,* and I have never seen
such narrow and round stigmas on young flowers. The shape of the lobes of the
stigmas seems variable even in the same species. The male flowers of *P. Bonati* are
described with 12–15 stamens. In the flowers of typical *P. adenopoda* I have never
found more than 7–9 stamens. I suggest that *P. Bonati* may be rather a form
of *P. rotundifolia,* var. *Duclouxiana* than a form of *P. adenopoda.*

Populus alaschanica Komarov in Fedde, *Rep. Spec. Nov.* XIII. 233 (1914).
SOUTHERN MONGOLIA. Prov. Alashan: "ad ripas canalium irriga-
torium et lacuum circa oppidum Dyn-juanj-in," March 27, April 15, June 4, 1908,
Tshetyrkin (male and female types, ex Komarov).
According to the author this species belongs to Sect. *Leuce,* but I do not quite
understand Komarov's remark which follows: "An origine hybrida inter *P. tremula* L. et *P. Przewalskii* Max. (sect. Tacamahaca?) *Populus tremula* in vicinitate
deest autem perfecte." The description is very short and rather insufficient.

SALIX L.

Salix Wilsonii Seemen in *Bot. Jahrb.* XXXVI. Beibl. LXXXII. 28 (1905). — Léveillé in *Bull. Soc. Bot. France,* LVI. 301 (1909).

Salix Mesnyi Burkill in *Jour. Linn. Soc.* XXVI. 530 (pro parte, non Hance) (1899).
Salix Argyi Léveillé in Fedde, *Rep. Spec. Nov.* X. 473 (1912).

Western Hupeh: Ichang, side of streams, alt. 300 m., April and May 1907 (No. 2121; tree 6.5 m. tall); Hsing-shan Hsien, woodlands, alt. 1300–1600 m., woodlands, May 1907 (No. 2140; tree 13 m. tall, girth 1.8 m.); same locality, *A. Henry* (No. 3442); without locality, April 1900 (Veitch Exped. No. 334, in part [juvenile flowers only in Herb. Arnold Arboretum]); 415ᵃ, ex von Seemen; 334 and 415ᵃ are cotypes); without locality, *A. Henry* (No. 246, No. 1277, type, ex von Seemen, No. 3538). Kweichou: without locality, *E. Faber* (No. 116; tall tree; fruits). Kiangsu: without locality, *d'Argy* ("vulgo Se Me Jam Zu; on fait du thé avec les feuilles"; type of *S. Argyi,* ♂ and ♀); Nanking, *E. Faber* (No. 901; ♂). Chekiang: vicinity of Ningpo, 1908, *D. Macgregor* (fruits).

This species is certainly very nearly related to the Japanese *S. glandulosa* Seemen, which is said by von Seemen to occur in Shensi, but it seems to differ in the shape of the old leaves and of the stipules. The latter are wanting or small in *S. Wilsonii,* which has no glands on the petioles, while in *S. glandulosa* the semicordate stipules are mostly distinctly developed, and the petioles bear glands which often are somewhat leaf-like. Von Seemen did not compare *S. Wilsonii* with *S. glandulosa* and with *S. Rosthornii* Seemen (in *Bot. Jahrb.* XXIX. 276, t. 2, E–H [1900]), which was collected by von Rosthorn's men (No. 1512) in southwestern Szech'uan, Nanch'uan. I have a strong belief that it is the same as *S. Wilsonii,* although I have seen only a leaf of it. *S. Rosthornii* Seemen would be the older name.

This is the common Willow of the mountains of western Hupeh and eastern Szech'uan, where it is abundant on the banks of streams and mountain torrents between 1200 and 2000 m. altitude. It is a rather low tree with a short, very thick trunk and very numerous and thick ascending spreading or spreading branches and slender branchlets. Pictures of this tree will be found under Nos. 512, 515, and 520 of the collection of my photographs and in my *Vegetation of Western China,* Nos. 456, 457, and 458.   E. H. W.

Salix paraplesia Schneider, n. sp.

Arbor 6–7 m. alta. Planta ♂ glabra, ramulis foliis juvenilibus pedunculisque iis plantae ♀ simillimis, vide inferius. Amenta coetanea, densiflora, pulchra, pedunculis 1–1.5 cm. longis foliatis suffulta, 3.5–6 cm. longa, circiter 1–1.2 cm. crassa; flores ♂ 5–7andri; filamenta basi pilosa, bracteis subduplo longiora, antheris flavis ovalibus coronata; glandulae duae, dorsalis lata, ovata, apice obtusa, ventralis haud minor, apice saepe bifida; bracteae oblongae, obtusae, flavae, utrinque villosae, v. pleraeque extus basi excepta glabrae. Planta ♀ (fructifera!); ramuli glabri, sublucidi, purpurascentes, deinde cinereo-olivacei v. cinerei. Folia obovato-elliptica, ovata v. rarius ovato-lanceolata, basi acuta v. obtusa, apice obtusa et subito breviter acuminata v. plus minusve sensim acuta, supra intense viridia, subtus pallide viridia v. in eodem ramulo albescentia, margine densissime subtiliter glanduloso-serrata, minora circiter 3.5 cm. longa et 1.8 cm. lata, maxima usque ad 6 cm. longa et 3.2 cm. lata v. ad 7 cm. longa et 2.5 cm. lata; petioli 5–10 mm. longi, superne ad apicem glanduliferi. Amenta pedunculis 2–3 cm. longis glabris foliiferis (foliis 3–5 ceteris simillimis) suffulta, 2.5–4 cm. longa et 1.3 cm. crassa, rhachi villosa, densiflora; fructus glabri, breviter pedicellati, pedicello glandulam unam dorsalem latam bifidam paullo longiore; ovaria 7–8 mm. longa, ovato-conica, apice stigmatibus ut videtur brevibus subsessilibus coronata.

Western Szech'uan: mountains west of Tachien-lu, alt. 2600–3800 m., June 1904 (No. **4518**, type; tree 6–7 m. tall; ♂); same locality, September 1904 (No. **4518ᵃ**; with fruits).

This species seems most nearly related to *S. pentandra* Linnaeus and may represent only a variety of that species. The ♂ specimen differs from those of *S. pentandra* collected in the Altai Mountains only in the broader glands, the shape of which is very variable in both the species. The fruiting branch bears on the upper part leaves which are very whitish underneath. I have not seen or found a description of a form of *S. pentandra* Linnaeus with such leaves. The ♀ flowers may present some better differences, but the fruits are very much alike.

There is a sterile specimen from western Hupeh: Changyang Hsien, alt. 800 m., November 1907 (No. **1441**; tree 18 m. tall, girth 2 m.); the leaves of which are glabrous and white tinged with red beneath. They are elliptic-lanceolate, from 6–10 cm. long and from 2–3.5 cm. broad (others are deformed); the petioles are from 8–10 mm. long and bear two glands; the stipules are semicordate, half as long as the petioles and glandular-serrate. From this very tree Mr. Wilson sent cuttings to the Arnold Arboretum, from which a young plant is growing very freely. The leaves are ovate- or elliptic-lanceolate, acute at the base and acuminate at the apex, measuring from 8 to 11 cm. in length and from 2.5 to 3.6 cm. in width. They show the same white color underneath. I suppose that No. 1441 may belong to this new species.

The specific name is derived from παραπλήσιος, resembling.

The specimens of my No. 1441 came from a tree which is common in one locality in Changyang Hsien growing by the side of watercourses between altitudes of from 700 to 800 m. This tree grows tall (20–25 m.) and has a straight, rather slender trunk and dark gray deeply fissured bark. The branches are rather sparse, short, spreading, rather slender, and form a loose oval crown. Pictures of this tree will be found under Nos. 584 and 586 of the collection of my photographs and in my *Vegetation of Western China*, Nos. 459 and 460. E. H. W.

Salix babylonica Linnaeus, *Spec.* 1017 (1753). — Brandis, *Forest Fl. Brit. Ind.* 465, t. 59 (1874); *Ind. Trees*, 637 (1908). — Hooker f., *Fl. Brit. Ind.* V. 629 (1888). — Burkill in *Jour. Linn. Soc.* XXVI. 525 (pro parte) (1899). — Wolf in *Izv. S.-Peterburg. Liesn. Inst.* IV. 28, t. 6, fig. 9–13, t. 8, fig. 4 (*Мат. Изуч. Нов Европ. Росс.*) (1900); in *Act. Hort. Petrop.* XXIII. 192 (1903). — Collett, *Fl. Siml.* 479 (1902). — Seemen, *Salic. Jap.* 29, t. 3, fig. A–E (1903); in Ascherson & Graebner, *Syn. Mitteleur. Fl.* IV. 82 (1908). — Schneider, *Ill. Handb. Laubholzk.* I. 36, fig. 11 m–n, fig. 15 b (1904). — Henry in Elwes & Henry, *Trees Gr. Brit. & Irel.* VII. 1749 (1913).

> *? Salix chinensis* Burman, *Fl. Ind.* 211 (errore typogr. 311) (1768).
> *? Salix cantoniensis* Hance in *Jour. Bot.* VI. 49 (1868).
> *Salix alba* F. P. Smith, *Chin. Mat. Med.* 232 (non Linnaeus) (1871). — Debeaux in *Act. Soc. Linn. Bordeaux*, XXX. 109 (*Fl. Shangh.* 57) (1875). — Fauvel in *Mém. Soc. Sci. Nat. Cherbourg*, XXII. 354 (1879). — Burkill in *Jour. Linn. Soc.* XXVI. 526 (1899).

Western Hupeh: Patung Hsien, roadside, alt. 650 m., April 1907 (No. **2122**; tree 8 m. tall; ♂; No. **2122ª**; ♀); Ichang, side of streams, etc., alt. 300–800 m., April 1907 (No. **1435**, tree 5–12 m. tall; ♂; No. **1435ª**; ♀); same locality, *A. Henry* (No. 1328; with fruits; No. 3355; ♂); Nanto, side of streams, March 1900 (Veitch Exped. No. 475; tree 4 m. tall; ♂); Changlo Hsien, *A. Henry* (No. 6325; ♂). Chekiang: without precise locality, *Barchet* (♂ and ♀); Ningpo, without precise locality, 1908, *D. Macgregor* (♂). Kwangtung: " ad rivulorum margines in delta fl. Cantoniensis, certe spontanea," February 1867, *Th. Sampson* (Herb. Hance No. 13757, type of *S. cantoniensis;* ♂). Kiangsu: Shanghai, *W. W. Perry* (fruits).

JAPAN. Kyushu: " in locis depressis principatus Fizen," *Pierot* (ex von Seemen; I have not seen a wild specimen from Japan).

Wilson's specimens agree very well with the description and figure by von Seemen. The female plant of No. 1435 differs a little in the somewhat denser silky pubescence, shorter bracts, a more distinct style and in its rather longer stigmas. In some of Barchet's specimens the ovaries are subsessile.

Hance, l. c., says of his *S. cantoniensis:* " this is the only willow really found wild in southern China." I cannot separate the specimen of Hance's species in the Gray Herbarium from *S. babylonica* Linnaeus. Burkill, l. c., says: " The lowest flowers on the catkins have 5–7 stamens, while the uppermost have 3 or 2. This Hance failed to observe." With regard to this statement I suspect that Hance may have collected different things and perhaps a hybrid between *S. babylonica* Linnaeus (or *S. cantoniensis* Hance) and *S. Mesnyi* Hance (see p. 95). Leaves of such a hybrid might " pass for *S. fragilis*," as Burkill says. Judging by the ovaries, which are hairy at the base, *S. elegantissima* K. Koch rather than *S. babylonica* seems to be represented in Brandis's plate.

Salix babylonica is cultivated in all the warmer parts of the world, although it does not stand very well our northern climate. In northern countries often hybrids of this species with *S. alba* Linnaeus and *S. fragilis* Linnaeus or pendent forms of these species are planted as *S. babylonica*. See also *S. Matsudana* Koidzumi (on p. 107).

This is the common tree Willow of the region bordering the Yangtsze River from its mouth westward for nearly two thousand miles. On the alluvial soils from the region of the Tungting Lake eastward to the sea it is most abundant, and it has been very generally planted in the neighborhood of houses and villages. It is plentiful near Shanghai and is doubtless widely spread throughout the mild and warmer parts of China. Near Ichang it does not ascend above 400 m. altitude, but in the warm valleys of western Szech'uan it grows up to 1300 m. altitude, and sometimes magnificent trees may be seen. From Ichang westward it is by no means common, whereas from Ichang eastward on the alluvial plains it is the most common tree.

The habit is variable, and all forms are met with from the typical form with pendent branchlets and arching branches to one in which the branches are ascending spreading. The most common form is intermediate between these extremes, and this probably represents the phylogenetic type.

Salix babylonica is a favorite tree with the Chinese and is commonly depicted in Chinese drawings, pictures on porcelains, in wood-carvings and other works of art. The fruits are occasionally used for stuffing cushions and the like, and by peasants the leaves are locally used as " tea."

This Willow has been introduced into Japan, where it is now widely cultivated, being a favorite with Buddhist priests. In Tokyo and other Japanese cities it is a common street tree, and each year these trees are severely pruned and only the slender pendent branchlets are allowed to develop freely.

Pictures of this tree will be found under Nos. 86, 398, 470, 594, 630, 715 and 0167 of the collection of my photographs and in my *Vegetation of Western China*, Nos. 455, 461, 462, 463, 464. In this series of photographs the various forms of this Willow are shown.   E. H. W.

Salix atopantha Schneider, n. sp.

Frutex parvus, erectus, ramosus, 0.6–2.5 m. altus; ramuli hornotini laxe puberuli v. glabri, vetustiores glabri, omnes intense purpurascentes; gemmae 6–7 mm. longae, oblongae, adpressae, glabrae, purpureobrunneae. Folia nondum satis evoluta oblongo-elliptica, utrinque acuta v. obtusa, supra viridia, initio puberula, deinde costa paulo impressa excepta glabrescentia, nervis tenuiter incisis, subtus pallida v. glauca, initio tantum ad costam prominulam sparse puberula v. glaberrima, nervis lateralibus utrinque circiter 6 vix visibilibus v. paulo prominulis, margine breviter glanduloso-denticulata v. rarius subintegra, minora circiter 1 cm. longa et 0.4 cm. lata, maxima ad 2.5 cm. longa et 1 cm. lata. Amenta coetanea, pedunculis foliis 3–6 parvis normalibus instructis puberulis suffulta, rhachi villosa, densiflora; amenta ♂ pedunculis 0.5–1 cm. longis exceptis 1–2 cm. longa et circiter 0.5–0.6 cm. crassa; flores diandri, filamentis liberis basi pilosis, antheris subglobosis flavis instructi; glandulae 2, ventralis latior, 2–3-partita v. 3-lobata, dorsalis simplex, minor, tenuior, omnes interdum quasi pseudodiscum 4-partitum formantes; bracteae obovatae, interdum truncatae, filamentis subduplo breviorcs, apice fuscescentes, utrinque breviter villosae v. ad apicem extus subglabrae; amenta ♀ pedunculis ad 1.5 cm. longis exclusis 0.8–2 cm. longa et circiter 0.6 cm. crassa; ovaria sessilia, ovata, sericeo-tomentella, stylo brevi sed distincto usque basim bipartito ramis divergentibus coronata; stigmata brevia, biloba; bracteae ovariis longiores v. subaequilongae glandulaeque duae iis florum ♂ simillimae.

Western Szech'uan: west of Kuan Hsien, Pan-lan-shan, alt. 3700–4000 m., June 24, 1908 (No. **2137**; shrub 1.2–2.4 m., ♀ ; No. **2137**ª; ♂); same locality, summit of Niu-tou-shan, alt. 3100 m., June 20, 1908 (No. **2134**, type; shrub 0.6–1.2 m.; ♀ ; No. **2134**ª; ♂).

In 'its foliage this species very much resembles forms of sect. *Longiflorae*, but it is well distinguished by the peculiar glands and by the hairy ovaries with the two separated arms of the style. *S. atopantha* may represent the type of a new section. The specific name is derived from ἄτοπος, strange, and ἄνθος, flower.

Salix magnifica Hemsley in *Kew Bull. Misc. Inform.* 1906, 163. — Léveillé in *Bull. Soc. Bot. France,* LVI. 299 (1909). — Bean, *Trees & Shrubs Brit. Isl.* II. 484 (1914).

Western Szech'uan: Mupin, mountains, alt. 3000 m., July 1903 (Veitch Exped. No. 4526, in part, type number; shrub 7.5 m., only two seen; fruits); same locality, June 1904 (Veitch Exped. No. 4526, in part, type number; ♂); west and near Wên-ch'uan Hsien, thickets, alt. 2600–3000 m., July 1908 (No. **1401**ª; bush 0.9–6 m.); Chiu-ting-shan, thickets, alt. 2000–2500 m., May 23, 1908 (No. **1401**; bush 1.8–3 m.); west of Kuan-Hsien, Niu-tou-shan, alt. 2600–3000 m., thickets, June 20, 1908 (No. **1401**ᵇ; bush 1.5–6 m.); Mupin, thickets, 2600–3000 m., June 1908 (No. **1401**ᶜ); same locality, October 1910 (No. **4363**; sterile; bush 1.5–6 m.).

An extremely well-marked species. See also the following species and the description of sect. *Magnificae,* p. 113.

Salix ulotricha Schneider, n. sp.

Frutex *S. magnificae* Hemsley signis omnibus valde similis, ab ea praecipue differt: ovariis crassioribus, subsessilibus ovato-rotundis, crispato-tomentellis, subito in stylum crassum contractis, stigmatibus paulo longioribus, distinctius revolutis, bracteis latioribus, dimidio ovario aequilongis v. subbrevioribus.

Western Szech'uan: Mupin, thickets, alt. 2300–3000 m., June 1908 (No. 1401ᵈ; ♀).

In the branches, the size and shape of the large leaves, the glabrous peduncle and rhachis and in the glabrous bracts there seems to be no real difference between this species and *S. magnifica* Hemsley. But the typical *S. magnifica* has: ovaria longius pedicellata, oblonga v. ovata, glabra, apice attenuata stylo bifido stigmatibus brevioribus bifidis coronata; bracteae ovato-oblongae. The length of the bracts and of the pedicels differs very much according to the age of the flowers, the uppermost in the catkins being the youngest. See also the description of the sect. *Magnificae* on p. 113.

The specific name is derived from οὐλότριχος, with curly hair.

Salix pella Schneider, n. sp.

Frutex 4–6 m. altus; ramuli elongati, crassi, in sicco nigro-purpurascentes, glabri. Folia oblongo-elliptica v. elliptica, apice acuta v. subobtusa, basi rotunda v. obtusa, supra viridia, laevia, costa flavescente subimpressa, subtus distincte pallida, ad costam pilis longis sericeis obtecta v. glabra, costa nervisque prominulis, nervis lateralibus fere modo *S. magnificae* Hemsley ante marginem evanescentibus, 1 pro 1 cm., reticulo ex parte satis distincto, margine plus minusve glanduloso-serrato-dentata v. crenulata v. ex parte integra, minora 3.5–8 cm. longa et 2–2.5 cm. lata, maxima usque ad 13 cm. longa et 5.5 cm. lata; petioli satis breves, 6–15 mm. longi, glabri v. pilis sericeis sparse obtecti, purpurascentes v. flavescentes. Amenta fructifera pedunculo 1–2 cm. longo sericeo suffulta, 4–9 cm. longa, circiter 1.2 cm. crassa, rhachi sericea. Fructus maturi circiter 5 mm. longi, glabri, brevipedicellati, pedicello glandulam unam ovato-truncatam subaequilongo; styli ut videtur mediocres, apice bifidi, stigmatibus bifidis; bracteae glabrae, flavo-brunneae, ovatae, apice truncatae, capsulis 3plo breviores.

Western Szech'uan: west of Kuan Hsien, Niu-tou-shan, woodlands, alt. 2600–3000 m., October 1910 (No. 4350, type; bush 4–6 m.; with fruits).

This species differs from *S. magnifica* Hemsley in its shorter petioles, smaller and narrower, more distinctly serrate leaves, its much shorter catkins and in its truncate bracts. The nervation of the leaves is somewhat intermediate between that of sections *Magnificae* and *Eriostachyae*. At first I thought it might be a hybrid between *S. magnifica* and *S. moupinensis* Franchet which grow in the same region and has the same kind of silky pubescence on the under side of the midrib of the leaves, but the rhachis of the catkins is more glabrous in Franchet's species. The color of the under side of the leaves is the same as in *S. magnifica*, while the leaves of *S. moupinensis* Franchet have a greenish shining lower surface. *S. pella* may be nearly related to *S. plocotricha* Schneider, which is readily distinguished, however, by its hairy ovaries. As long as the male flowers are unknown it is always difficult to determine the real relationship of a Willow.

The name is derived from πελλός, dark colored.

Salix apatela Schneider, n. sp.

Frutex 1.8–3 m. altus; ramuli glabri, purpurascentes; gemmae late ovatae, obtusae, purpureae, subpruinosae, 7–8 mm. longae. Folia obovato-oblonga, basim versus angustata, acuta v. subrotundata, apice obtusa v. breviter acuta, supra saturate viridia, glabra, costa flava impressa, subtus valde pallida, glabra, costa nervisque flavis elevatis et subtiliter reticulatis, nervis angulo lato a costa orientibus parallelibus fere ad marginem currentibus 2 pro 1 cm., margine integra, minora circiter 5 cm. longa et 3 cm. lata, maxima ad 12 cm. longa et ad 5 cm. lata; petioli glabri, purpurei, 9–15 mm. longi, supra late sulcati. Amenta fructifera ramulis glabris 3–5 cm. longis folia 3–5 normalia gerentibus suffulta, glabra, densiflora, 13–16 cm. longa, circiter 1.3 cm. crassa; fructus maturi ovato-oblongi, subsessiles, apice stylo bipartito stigmatibus bifidis coronato; bracteae ovatae, apice truncatae, flavo-brunneae, glabrae; glandula una, lata, rectangularis, pedicello aequi-longa.

Western Szech'uan: Lungan Fu, Tu-ti-liang-shan, woodlands, alt. 2600–3000 m., August 1910 (No. **4575**; shrub 1.8–3 m.; fruits).

This species looks somewhat like a form of *S. magnifica* Hemsley with smaller obovate-oblong leaves, but judging by the nervation it seems more nearly related to *S. Fargesii* Franchet, from which it differs chiefly in its entire leaves only very finely reticulate beneath, and in the glabrous catkins and bracts; ♂ flowers are unknown.

The specific name is derived from ἀπατηλός, misleading.

Salix moupinensis Franchet in *Nouv. Arch. Mus. Paris*, sér. 2, X. 82 (*Pl. David.* II. 120) (1887). — Burkill in *Jour. Linn. Soc.* XXVI. 531 (1899). — Seemen in *Bot. Jahrb.* XXIX. 277 (1900). — Léveillé in *Bull. Soc. Bot. France*, LVI. 300 (1909).

Western Szech'uan: Mupin, thickets, alt. 2000–2300 m., June 1908 (No. **1412**; shrub 2–3 m.; ♀ flowers and fruits); same locality, 2600–3000 m., October 1910 (No. **4370**; bush 3–4.5 m.; sterile); same locality, in silvis, March 1869, *A. David* (type; arbor ad 30 m. alta, ex Franchet); west and near Wên-ch'uan Hsien, woodlands, alt. 2300–3000 m., July 1908 (No. **1417**; shrub 1.5–6 m.; with fruits); west of Kuan Hsien, Niu-tou-shan, alt. 3000 m., June 20, 1908 (No. **1417ª**; shrub 5–6 m.; with fruits); Wa-shan, thickets, alt. 2300–2700 m., June 1908 (No. **2126**; bush 3–5 m.; with fruits); same locality, alt. 3000 m., July 1903 (Veitch Exped. No. 4522; shrub 2.5–5 m.; ♀); same locality, *A. Henry* (No. 8891; fruits); Tachien-lu, *A. E. Pratt* (No. 335; ex Burkill).

This fine species may be distinguished by its serrate acute leaves greenish yellow, glabrous and more or less reticulate beneath, by its thin slender flowering aments and by its small glabrous fruits. These are about 4 or 5 mm. long without the 1–2 mm. long pedicels. The almost verticillate arrangement of the flowers on the rachis is the same as in *S. magnifica* Hemsley, and is more prominent in these two species than in the other species of the sections *Magnificae* and *Eriostachyae*.

Salix Fargesii Burkill in *Jour. Linn. Soc.* XXVI. 528 (1899). — Seemen in *Bot. Jahrb.* XXIX. 277, t. 3, fig. A–F (1900); l. c. XXXVI. Beibl. LXXXII. 30 (1905). — Léveillé in *Bull. Soc. Bot. France*, LVI. 299 (1909).

Western Hupeh: Hsing-shan Hsien, thickets, alt. 1600–2300 m., June 1 and 5, July and October 1907 (No. **720**; shrub 0.6–3 m. tall; with fruits; No. **720ᵃ**; ♀ ; No. **720ᵇ**; ♂ ; No. **720ᶜ**; sterile); Changlo Hsien, thickets, alt. 600–2000 m., May 1907 (No. **2156**; shrub 0.6–2.20 m.; ♂); Fang Hsien, woodlands, alt. 1800–2300 m., November 1910 (No. **4439**; shrub 0.5–1.5 m., often prostrate or procumbent; sterile); same locality May 1900 and later (Veitch Exped. No. 1820; ♂ and ♀ flowers and fruits); Changyang Hsien (Veitch Exped. No. 679; ex von Seemen). Eastern Szech'uan: Wushan Hsien, *A. Henry* (No. 5678; co-type; with fruits): Tchen-keou-tin, *P. Farges* (No. 795, type; ♂ ; ex Burkill).

This handsome species is easily distinguished from *S. moupinensis* Franchet by its mostly larger leaves with a pale under surface, its thicker flowering aments and by the long silky hairs of the bracts. The ovaries mostly bear a longer style only cleft at the apex. Its nearest relative may be *S. Ernesti* Schneider which, like *S. moupinensis* Franchet, inhabits a different geographical region.

Salix Ernesti Schneider, n. sp.

Frutex breviter ramosus, erectus, 1.2–6.5 m. altus; ramuli intense purpurascentes, tantum initio sericeo-villosi, mox glabri, vetustiores purpureo-brunnei. Folia elliptica, obovato-elliptica v. late oblanceolata, basi obtusa v. acuta (v. in No. 2152 ex parte rotunda v. subaurita) apice subobtusa v. pleraque brevius v. longius acuta, supra sordide viridia, plus minusve, costa nervisque distinctius, puberula v. tomentella, subtus cinerea, sericea v. sericeo-villosa v. lanata et submicantia, deinde pleraque plus minusve glabrescentia, costa nervisque prominentibus, sed vix reticulata, nervis parallelis circiter 2 pro 1 cm., margine integra v. raro satis indistincte undulato-glanduloso-crenulata, minora latiora obtusiora 3–6 cm. longa, 1.5–3 cm. lata, maxima elliptica v. obovato-lanceolata acuta, usque 8–10 cm. longa et 2.2–4 cm. lata v. ad 13 cm. longa et 5 cm. lata; petioli 5–10 mm. longi, undique v. tantum in sulco

superne sericeo-villosi; stipulae parvae, semi-ovatae, acutae v. nullae.
Amenta ♂ pedunculis 5–10 mm. longis folia 2–3 parva sericea gerentibus suffulta, 2–6 cm. longa, 0.8–1 cm. crassa, rhachi sericeo-villosa;
flores diandri, filamentis fere ad apicem villosis, antheris ovato-oblongis
flavis; bracteae late obovatae, intus plus minusve glabrae, filamentis
fere duplo breviores; glandulae duae, ventralis ovata v. ovato-rectangularis, dorsalis minor, saepe angustior.   Amenta ♀ pedunculis villosis
foliis 3–4 normalibus instructis 1–2.5 cm. longis suffulta, 3–11 cm.
longa, ad 1 cm. (fructifera ad 1.5 cm.) crassa, rhachi villosa, densiflora.
Flores sessiles v. subsessiles; ovaria ovato-oblonga, breviter villosa,
stylis satis longis fere ad basim partitis ramis divergentibus purpureis coronatis; stigmata stylis breviora, angusta, bifida, divergentia;
bracteae obovato-oblongae, apice rotundae, ovariis juvenilibus aequilongae, flavo-brunneae, utrinque sericeo-villosae v. pleraeque intus
glabrae; glandulae 1–2, ventralis late ovata v. ovato-elongata, apice
saepe truncata v. lobulata, ovario 3–4plo brevior, dorsalis minima,
satis difficile recognoscenda v. nulla.   Fructus laxe villosi, circiter 6 mm.
longi, ovariis similes, sed breviter pedicellati, pedicello glandula ventrali
breviora; bracteae fructibus paullo v. duplo breviores.

Western Szech'uan: west of Kuan Hsien, near summit of Niu-
tou-shan, alt. 3000–3300 m., June 20, 1908 (No. 2151, type; ♂; No.
2151ᵃ, ♀; No. 2153; old ♀ flowers; shrub 2.5–6 m.; No. 2132; ♂;
No. 2132ᵃ; ♀). west of Tachien-lu, thickets, alt. 4000 m., July 25, 1908
(No. 2149; with fruits; No. 2149ᵃ; ♂); southwest of Tachien-lu, alt.
3300–4200 m., July 1903 (Veitch Exped. No. 4520; shrub 1.2–1.8 m.);
Tachien-lu, alt. 2600–3000 m., June 1908 (No. 2152; shrub 1.5–3.6 m.,
♀; forma foliis intensius molliter sericeo-villosis, basi saepe sub-
auritis); Wa-shan, 2600–3000 m., June 1908 (No. 2159; shrub 1.5–3
m.; ♂; forma glabriuscula floribus abnormalibus staminibus 3–4, 1–2
minoribus, instructis); Mupin, thickets, alt. 2600–3000 m., June 1908,
No. 1406; shrub 2.4–3 m.; ♀); same locality, woodlands, alt. 2600–
3300 m., October 1910 (No. 4371; shrub 2.4–3.6 m.; sterile); without locality, side of streams, alt. 3300 m., June 1904 (Veitch Exped.
No. 4521; shrub 3 m.; ♀).

A very distinct species apparently nearly related to *S. Fargesii* Franchet (see
above) but well distinguished from it by its nearly entire, more pubescent leaves,
its hairy ovaries and fruits, its larger bracts and by its more pilose filaments. The
dorsal gland does not always seem to be present, and it would be a very artificial
arrangement to place this species in another section on account only of the development of a second gland.

The species is named in compliment to Mr. Ernest H. Wilson.

Salix argyrophegga Schneider, n. sp.

Frutex 1.5–3.6 m. altus, breviter ramosus; ramuli initio crispo-villosi, deinde glabrescentes, flavo-brunnei. Folia elliptica, ovato-elliptica v. elliptico-oblonga, basi pleraque rotunda, apice breviter acuta v. obtusa, supra sordide viridia, initio plus minusve sericeo-villosula, deinde costa tomentella excepta glabrescentia, subtus dense v. laxius molliter sericeo-villosa, albescentia (v. subglabra, pallide viridia, tenuiter reticulata), costa nervisque 2 pro 1 cm. elevatis, margine distincte argute serrata, dentibus callosis apicem versus productis, sinuosa, 5–8 cm. longa, 3.2–3.8 cm. lata, maxima usque 13 cm. longa et 5.2 cm. lata; petioli 10–15 mm. longi, sericeo-villosi; stipulae ovatae v. semiovatae, acutae, petiolis fere triplo breviores, ad apicem glanduloso-serratae, pilosae. Amenta fructifera ramulis subvillosis 2.5–4 cm. longis folia 3–4 normalia gerentibus suffulta, 9–14 cm. longa, circiter 1.5 cm. crassa, rhachi sericeo-villosa; flores (inter fructus remanentes) sessiles; ovaria ovato-elliptica, argenteo-sericeo-villosa, stylis bifidis quam stigmata oblonga bifida paulo longioribus coronata; bracteae late ovato-rotundae, brunnescentes, extus sericeae, intus (et interdum apice) glabrae, ovariis duplo breviores; glandula una ventralis oblonga, bracteis duplo brevior. Fructus circiter 6 mm. longi, subsessiles, ad apicem paulo attenuati, ut ovaria villosi.

Western Szech'uan: west of Kuan Hsien, Niu-tou-shan, alt. 2600–3000 m., June 1908 (No. 2154, type; bush 1.5–3.6 m.; with fruits).

This species seems most nearly related to *S. Ernesti* Schneider, but differs clearly in its serrate leaves, the smaller bracts, the narrower gland and in the yellowish brown twigs. Unfortunately there are no ♂ flowers.

The specific name is derived from ἀργυροφεγγής, shining like silver.

Salix plocotricha Schneider, n. sp.

Frutex 3–7 m. altus, ramosus; ramuli tantum juveniles sericei, mox glabri, annotini nigro-purpurascentes glabri. Folia (nondum satis evoluta) oblongo-elliptica v. obovato-lanceolata, basi acuta, obtusa v. rotunda, apice obtusa v. breviter acuta, supra viridia, initio sparse puberula, deinde costa excepta glabra, subtus pallidiora, initio sericeo-tomentella saepe pilis fulvis intermixtis, deinde tantum ad costam paulo elevatam plus minusve sericea, nervis vix prominulis, rete nervorum tenuissimo, margine brevissime et distanter crenulato-serrata v. subintegra, minora (ramulorum florif.) 4–5.5 cm. longa et 1.8–2 cm. lata, maxima usque 10 cm. longa et 3 cm. lata; petioli initio sericei, deinde tantum sulco superne tomentelli, purpurascentes, 5–15 mm.

longi. Stipulae ut videtur minimae v. nullae. Amenta ♂ et ♀ similia ramulis 3–5.5 cm. longis laxe sericeis folia 3–4 normalia gerentibus suffulta, rhachi villosa, 5–9 cm. longa, circiter 1 cm. crassa, densiflora; flores ♂ diandri, filamentis basi villosis, antheris flavis ovalibus, glandulis 2 separatis, ventrali late ovata obtusa crassiuscula, dorsali minore, angustiore; bracteae ovatae, obtusae, filamentis 3plo breviores, utrinque intus laxius crispo-sericeae, brunneae; ovaria sessilia, ovata, crispo-tomentella, stylo purpureo fere aequilongo, apice bifido, stigmatibus brevibus bifidis coronata; glandula una, lata, apice truncata v. emarginata; bracteae ovario circiter duplo breviores ceterum ut in floribus masculis. Fructus ignoti.

Western Szech'uan: west of Kuan Hsien, near summit of Niutou-shan, 3000 m., June 20, 1908 (No. **2147**, type; shrub 3–7 m.; ♀; No. **2147ª**, co-type; ♂).

Almost identical with those of this species seem to be specimens from Mupin, thickets, alt. 2600–3300 m., May and October 1908 (No. **1407**; bush 1.5–3 m., ♂; No. **1407ª**; sterile). The mature leaves are elliptic, mostly acute at the ends, finely glandular-serrate, grayish and glabrous beneath, with a distinctly prominent rib and nerves and a very fine reticulation. They measure from 6–10 cm. in length and from 2.5–3.5 in width. There are two other specimens from Mupin, alt. 2600–3000 m., May 1908 (No. **2157**; shrub 1.8–2 m., and **2158**; shrub 3 m.); No. 2158 has abnormal ♀ flowers with hairy ovaries and may very well belong to *S. plocotricha* Schneider, but No. 2157 with its nearly glabrous ovaries looks somewhat like a hybrid between this species and *S. moupinensis* Franchet. See also *S. spathulifolia* Seemen, p. 114.

The specific name is derived from πλόκος, curl, and θρίξ, hair.

Salix phanera Schneider, n. sp.

Frutex altissimus v. arbor 7–12 m. alta; ramuli crassi, ut videtur olivacei, sed in sicco nigrescentes, initio tomentelli, sed mox glabri. Folia ovato-lanceolata v. late elliptico-lanceolata, basi obtusa v. rotundata (v. subaurita) ad apicem sensim breviter acuminata v. acuta, supra sordide viridia, sparse v. vix costa impressa excepta puberula, satis distincte inciso-reticulata, subtus initio densissime molliter sericeo-tomentosa, deinde paullo glabrescentia, cinerea v. cinereo-alba, nervis ut costa glabriusculis flavidis elevatis numerosis parallelis circiter 2 pro 1 cm., distincte elevato-reticulata, margine breviter crenato-denticulata, minora 6–11 cm. longa et 3.5–4 cm. lata, maxima 15–22 cm. longa et 5.5 cm. lata; petioli 1–2 cm. longi, superne sulco aperto tomentelli; stipulae parvae v. satis evolutae, semicordatae, glanduloso-dentatae, ut folia tomentellae et reticulatae. Amenta (♂ tantum visa) ramulis circiter 3 cm. longis folia 3 normalia gerentibus suffulta, 10–12 cm. longa, vix 1 cm. crassa, rhachi villosa; flores diandri, fila-

mentis liberis ⅔ villosis, antheris flavis ovato-oblongis; bracteae late
ovatae, apice rotundae, glanduliferae, flavo-brunneae, utrinque sericeo-
villosae, filamentis duplo brevioribus; glandula ventralis oblonga, satis
crassa, bracteis vix duplo brevior, saepe incisa, dorsalis paullo brevior,
saepe bifida.

Western Szech'uan: west of Kuan Hsien, Pan-lan-shan, thickets,
2300–3000 m., June 1908 (No. 2155, type; very large bush or tree, 7–12
m.; ♂); same locality, October 1910 (No. 4352; bush 2.6–4 m.;
sterile).

A striking species, exceedingly well marked by its very long and distinctly reticu-
late leaves and by its long slender male aments. It seems most nearly related to
S. phaidima Schneider, which has the same long and narrow aments.
The specific name is derived from φανερός, distinct.

Salix phaidima Schneider, n. sp.

Frutex 1.8–3 m. altus; ramuli initio sparse sericeo-villosi, mox glabri,
purpureo-nigri; gemmae ovatae, purpureae, divaricantes, petiolis subduplo breviores. Folia lanceolata v. minora ovato-lanceolata, basi
obtusa v. rotunda, rarius acuta, apice acuta v. acuminata, supra ut
videtur intense viridia, initio sparse, costa distinctius, sericeo-villosula,
deinde fere glabra, costa paulo impressa, nervis vix visibilibus, subtus
etiam adulta dense molliter sericea, alba, micantia, tantum costa elevata glabriuscula, margine integra, interdum leviter undulata, minora
3–7 cm. longa, 1.2–2 cm. lata, maxima sensim acuminata 9–15 cm.
longa et 1.3–2 cm. lata; petioli 5–12 mm. longi, initio sericeo-villosuli,
deinde glabri, leviter sulcati; stipulae pleraeque nullae v. minimae,
sericeae. Amenta ramulis 1.5–3 cm. longis subsericeo-villosis folia
3–5 parva v. mediocra normalia gerentibus suffulta, elongata, gracilia,
saepe curvata, rhachi villosa, ♂ 8–10 cm. longa, circiter 5 mm. crassa,
♀ 7–12 cm. longa, circiter 8 mm. crassa, fructifera nondum visa; flores
♂ diandri, filamentis liberis ad apicem pilosis, antheris flavis ovato-
elongatis; bracteae flavo-brunneae, oblongae, apice obtusae, utrinque
sericeo-villosae; glandula ventralis oblonga, apice saepe bifida, bracteis
duplo brevior, dorsalis subaequalis, saepe bipartita, interdum cum
ventrali quasi pseudodiscum 3–4-partitum formans; flores ♀: ovaria
sessilia, ovato-oblonga, breviter sericea, stylo mediocri apice bifido
stigmatibus angustis bifidis coronata; bracteae ut in floribus ♂; glandula una ventralis elongata v. filiformis, fere medium ovarii attingens.

Western Szech'uan: Mupin, thickets, alt. 1700–2300 m., May
and November, 1908 (No. 1409, type; ♀ flowers and old leaves; No.
1409ᵃ, ♂ co-type).

A very graceful species well distinguished by its long narrow catkins and leaves. In the silvery pubescence of the leaves it resembles *S. psilostigma* Andersson and *S. viminalis* Linnaeus. By its catkins it seems most nearly related to *S. phanera* Schneider. In *S. psilostigma* Andersson the catkins are considerably shorter, otherwise, so far as I can judge from the material before me, the Hupeh form of this species comes very near my new *S. phaidima*, which has longer petioles and quite entire leaves.

The specific name is derived from φαίδιμος, illustrious.

Salix dissa Schneider, n. sp.

Frutex ramosus, 0.6–1.5 (–2.4) m. altus, erectus; ramuli hornotini plus minusve dense breviter albo- (v. interdum fulvo-) tomentelli, annotini parce tomentelli v. glabrescentes, cinereo-brunnei, deinde cinerascentes; gemmae glabrae v. sparse pilosae, ovatae, acutae, adpressae, 3–3.5 mm. longae. Folia (nondum matura) oblongo-elliptica, utrinque obtusa v. basi subaurita, (in 2145 apice breviter acuta), supra viridia, glabra, sed ad costam elevatam flavidam puberula, subtus glauca v. initio pruinosa, glaberrima, costa elevata, nervis lateralibus vix v. paulo supra interdum distinctius prominulis, margine integerrima, minora 1–1.5 cm. longa, 5–6 mm. lata, majora usque 2.8–3 cm. longa et 0.8–1 cm. lata; petioli 1–3 mm. longi, glabri. Amenta ramulis folia 3–6 normalia gerentibus ut rami pilosis 0.5–3.5 cm. longis, rhachi villosa, pedunculata, ♂ 2–5.5 cm. longa, circiter 0.5–0.8 cm. crassa, subdensiflora brevius (0.5–1 cm.) pedunculata, ♀ 3–5 cm. longa, circiter 0.6–0.7 mm. crassa, pedunculis usque 3.5 cm. longis; flores ♂ diandri, filamentis liberis ad medium villosis, antheris flavis subglobosis parvis; glandula ventralis anguste ovata, dorsali simili paulo major, bracteam late obovatam parvam brunnescentem glabram subaequans; ovaria ovato-oblonga, sessilia, glabra, stylo brevissimo stigmatibus brevibus saepe bifidis subbreviore coronata. Fructus maturi nondum visi.

Western Szech'uan: west of Kuan Hsien, summit of Niu-toushan, alt. 3300 m., June 20, 1908 (No. **2136**; type; bush 0.9–1.5 m.; ♀ ; No. **2136ᵃ**, ♀ ; No. **2160**; bush 0.6 m.; ♂); west and near Wên-ch'uan Hsien, thickets, alt. 2300–3300 m., July 1908 (No. **2145**; bush 1.5–2.4 m.; ♂ ; No. **2145ᵃ**, ♀ ; leaves of ♀ twig acute at the apex, otherwise I can see no difference).

This species very much resembles *S. cathayana* Diels, but on account of the dorsal gland in the ♂ flowers I place it in a different section. The region from which *S. dissa* Schneider comes is extremely rich in endemic Willows. The male flowers are very much like those of *S. dyscrita* Schneider, the dorsal gland of which seems to be very small.

The specific name is derived from δισσός, ambiguous.

Salix dyscrita Schneider, n. sp.

Frutex 1.5–3 m. altus, ramosus, erectus; ramuli initio sericeo-tomentelli, deinde glabri v. glabrescentes, olivaceo-brunnei v. plerumque cinerascentes. Folia (juvenilia tantum visa) elliptica, utrinque obtusa, supra viridia, tantum costa subincisa villosula v. glabra, subtus cinerea v. paulo pruinosa, initio distincte sericea v. sericeo-tomentella, mox subglabra, tantum ad costam sericea, nervis utrinque satis difficile visibilibus, margine integra, minora circiter 1 cm. longa et 0.6 cm. lata, majora usque ad 2.5 cm. longa et 1 cm. lata; petioli 1–3 mm. longi, sericei. Amenta ♂ pedunculis sericeo-villosis vix 1 cm. longis folia 3–4 normalia gerentibus suffulta, rhachi villosa, 3–4.5 cm. longa, 0.6–0.9 cm. crassa, densiflora; flores ♂ diandri, filamentis 2 liberis ad medium villosis, antheris ovalibus flavis; glandula ventralis ovato-oblonga, bracteis late ovatis in eodem amento viridibus v. fere nigro-brunneis intus glabris extus sericeis v. subglabris satis longe sericeo-ciliatis subduplo brevior, dorsalis minima.

W e s t e r n   S z e c h ' u a n : Ching-chi Hsien, Ta-hsiang-ling, alt. 2300–3000 m., May 1908 (No. **1429**[bis], type; bush 1.5–3 m.; ♂).

This species resembles *S. dissa* Schneider, but differs from it in the somewhat silky under surface of the leaves, the larger and denser aments, the larger anthers, the very small dorsal gland and in the slightly larger bracts which are partly greenish. *S. dyscrita* may be of hybrid origin. At first I thought that Wilson's No. 2129 (see *S. driophila* Schneider, p. 59) from the same place might represent the ♀ plant of *S. dyscrita*, but the leaves have a denser and different pubescence.

The specific name is derived from δύσκριτος, ambiguous.

Salix hypoleuca Seemen in *Bot. Jahrb.* XXXVI. Beibl. LXXXII. 31 (1905). — Léveillé in *Bull. Soc. Bot. France*, LVI. 299 (1909).

W e s t e r n   S z e c h ' u a n : Mupin, thickets, alt. 2300–2600 m. (No. **1411**; shrub 1.5–2.4 m.; ♀; No. **1411**[a]; ♂); same locality, November 1909 (No. **1404**; shrub 0.75–1.5 m., sterile; gemmae ovato-oblongae, subacutae, purpureae, petiolis aequilongae); same locality, woodlands, alt. 2600–3000 m., October 1910 (No. **4336**; shrub 2–4 m.; sterile; foliis utrinque acutis); same place (No. **4367**; bush 1.5–2.4 m.; living plants in the Arboretum). W e s t e r n   H u p e h : without locality, April 1900 (Veitch Exped. No. 334, in part; mixed with *S. glandulosa* Seemen; with fruits; folia ovata v. ovato-elliptica); Fang Hsien, alt. 2000–3000 m. (Nos. **4434, 4437**; living plants in the Arnold Arboretum). S h e n s i : "In kia pu, Lao-y-san," March, *G. Giraldi* (No. 5362; type, ex Seemen); "Lao-y-san," June 4, 1897, *G. Giraldi* (with fruits; co-type).

The description of von Seemen being based on rather incomplete fruiting material, I add a fuller description based on Wilson's specimens, especially on No. 1411.

Frutex 0.9–3.6 m. altus, ramosus; ramuli tantum initio sparse pilosi, mox glabri, purpureo- v. nigro-brunnei. Folia elliptica, elliptico-oblonga, lanceolata v. rarius ovata, basi in petiolum breviter attenuata v. saepissime obtusa v. rotundata, apice acuta v. subacuta, supra viridia, initio saepe tota facie breviter pilosula pilis fulvis et albis mixtis v. tantum costa albo-tomentella, mox glabra, subtus glauca, semper glabra v. tantum costa initio pilosula, margine integra, minora 2–3.5 cm. longa et 1–1.5 cm. lata, majora lanceolata usque ad 5 cm. longa et 2 cm. lata v. latiora ad 4.5 cm. longa et 2.3 cm. lata, utrinque costa nervisque tenuiter prominentibus, vix reticulata; petioli graciles, mox glabri, (3–)5–10 mm. longi. Amenta coetanea, pedunculo 3–10 mm. longo foliis minimis v. parvis saepe caducis instructo suffulta, rhachi glabra v. sparse pilosa, ♂ laxiflora, 2.5–4.5 cm. longa et 0.5–0.6 mm. crassa, ♀ densiflora, 2.5–5 cm. longa et 0.5 (v. fructifera 0.7) cm. crassa; flores ♂ diandri, filamentis liberis fere ad medium villosis, antheris parvis subglobosis flavis (v. dein brunneis); glandula una ventralis ovata v. apice incisa, satis crassa, bractea duplo brevior; bracteae obovatae, obtusae, brunnescentes, glabrae; flores ♀ ut a cl. von Seemen descripti.

Salix hypoleuca, var. platyphylla Schneider, n. var.

A typo differt foliis latioribus late ellipticis basi rotundis apice brevissime apiculatis, 2.3–4.7 cm. longis et 1.5–3 cm. latis, amentis fructiferis longioribus, ad 7.5 cm. longis, ramulis annotinis flavidis.

Western Szech'uan: west and near Wên-ch'uan Hsien, thickets, alt. 2000–2600 m., July 1908 (No. 1418ᵃ, type; shrub 3–6 m.; with fruits); same locality, October 1908 (No. 1418; sterile; ramulis annotinis nigro-purpurascentibus).

Giraldi's specimen of the type agrees very well with Wilson's No. 1411. The Hupeh plant is intermediate between the type and var. *platyphylla*, which otherwise agrees with this species. *S. hypoleuca* may be distinguished from *S. longiflora* Andersson by its glabrousness, by the somewhat longer petioles and by the almost sessile stigmas.

Of No. 1418 there are living plants in by the Arnold Arboretum, but they have not yet flowered.

Salix rhoophila Schneider, n. sp.

Frutex 1.2–2.4 m. altus, valde ramosus; ramuli juveniles tantum villosuli, interdum pilis albis et fulvis mixtis, hornotini dein subglabri, annotini vetustioresque glabri, purpurascentes, subpruinosi. Folia valde juvenilia tantum primaeva subtus v. utrinque albo-tomentella (pilis fulvis intermixtis), mox glabra, elliptica v. elliptico-oblonga, integra, ceterum nondum satis evoluta. Amenta ♂ subsessilia, basi foliolis minimis 2 v. squamiformibus saepe caducis instructa, 1–1.5 cm. longa, circiter 0.6 cm. crassa, rhachi villosula; flores diandri, filamentis libris basi villosis, antheris parvis flavis ovato-globosis, glandula una ventrali plus minusve latiore quam longa, bracteis ovalibus apice rotundis v. subtruncatis flavo-brunneis glabris filamentis duplo brevioribus. Amenta ♀ pedunculis circiter 5 mm. longis pilis albis et fulvis praeditis foliola minima 2–3 gerentibus instructa, rhachi sparse villosa, divaricata, tenuissima, 2–3 cm. longa, 3(–4) mm. crassa; ovaria brevia, ovata, sessilia, glabra, stigmatibus brevibus subintegris; bracteae late ovatae, obtusae, ovariis subaequilongae, brunneae, glabrae; glandula una ventralis ovata v. subrectangularis, ovario triplo brevior. — Haud sine dubio addeo ramulos steriles No. 2117[bis]: ramuli hornotini plus minusve laxe villosuli, annotini glabri v. subglabri, purpureocinerei, deinde cinerascentes; gemmae ut videtur parvae, ovato-oblongae, subacutae purpureae, subglabrae, petiolis 3plo breviores. Folia minora elliptica v. elliptico-lanceolata, utrinque acuta, 2–4.5 cm. longa, 1–1.5 cm. lata, majora oblanceolata, basim versus angustata, acuta, apice acuta v. sensim breviter acuminata, ad 4.5:2 cm. magna v. pleraque angustiora 5–8 cm. longa et 1.5–1.8 cm. lata, supra viridia, tantum costa puberula, nervis paulo prominulis, subtus cinereo-glauca, tota facie pilis brevibus sparsis albis et fulvis obtecta et costa sericeopuberula v. in toto glabra; petioli 4–10 mm. longi, superne puberuli.

Western Hupeh: Fang Hsien, side of streams, alt. 2300–2600 m., May 16, 1909 (No. **2117**, type; bush 1.2–2.4 m., much branched; ♀, No. **2117ᵃ**, co-type; ♂); same locality, in the fall (No. **2117**[bis]; sterile).

This species seems nearly related to *S. hypoleuca* Seemen by its small ovaries with sessile stigmas, but the male catkins are different, so far as I can judge from the material before me. If the No. 2117[bis] with mature leaves really belongs to the same species, the shape of the leaves is somewhat different from those of *S. hypoleuca*, and the petioles and young twigs are pubescent. See also the following species.

The specific name is derived from ῥόος, river, and φιλέω, I love.

Salix polyclona Schneider, n. sp.

Frutex 2.4–3.6 m. altus, valde ramosus; ramuli initio laxe sericei, annotini glabri, purpureo-brunnei, vetustiores cinereo-brunnei. Folia (juvenilia tantum visa) elliptico-oblonga (v. in 2116[bis] elliptica) supra costa puberula excepta mox glabra, subtus plus minusve, saepe dense, albo-sericea (interdum pilis fulvis intermixtis), colore incerta, basi obtusa, apice acuta, integra, 1.5–3 cm. longa, 0.6–1 cm. lata (in No. 2116[bis] supra initio saepe fulvo-pilosa, mox costa puberula excepta glabra, subtus ad costam dense sericeo-villosa, mox glabra, ut videtur pallidiora, 2–3.5 cm. longa, 1–1.7 cm. lata); petioli 1–2 cm. longi, sericei (in No. 2116bis 2–6 mm. longi, tomentelli).   Amenta ♀ florifera pedunculo vix 3–5 mm. longo sericeo foliola minima normalia 2–3 gerente suffulta, 2–3 cm. longa, circiter 5 mm. crassa, rhachi sericea, densiflora; ovaria sessilia v. subsessilia, glabra, ovato-oblonga, stylo mediocri apice subbifido stigmatibus bifidis brevibus sublongiore coronata, glandula una ventralis, oblonga, bracteis subrotundis brunnescentibus intus glabris extus et margine satis longe sericeis fere aequilonga (in No. 2116bis amenta pedunculo brevi sæpissime defoliolato, bracteis glabris).

Western Hupeh: Fang Hsien, alt. 2300–2600 m., May 16, 1909 (No. 2116, type; ♀ ; bush 2.4–3.6 m., much branched); same place, side of streams, alt. 1700–2300 m., May 16, 1909 (No. 2116bis; bush 1.8–2.4 m.; ♀ flowers a little older).

This species comes from the same place as *S. rhoophila* Schneider, but it differs from that species in the distinct styles and the silky bracts. No. 2116bis looks somewhat intermediate with its glabrous bracts and distinct styles. I do not know if No. 2116bis really belongs to *S. polyclona*, but it differs also from *S. rhoophila* Schneider. Both these species may be distinguished from the forms of *S. longiflora* Andersson or *S. cathayana* Diels by their leafless or shorter peduncles and by their different geographical distribution. It is extremely difficult to interpret the Chinese Willows of this section, and I think it better to separate at present the different forms rather than to unite them without sufficient reason. As I have stated above, central China has an uncommonly rich development of the genus *Salix*, and anybody who has been earnestly interested in the study of this genus knows the difficulty in obtaining sufficient material of the flowers of the two sexes and of mature leaves of both. Mr. Wilson, however, has been more successful in securing complete material than other collectors of Chinese Willows.

The specific name is derived from πολύκλωνος, with many branches.

Salix mictotricha Schneider, n. sp.

Frutex 4.5–6 m. altus, valde ramosus; ramuli initio laxe sericeo-villosuli, annotini glabri v. subglabri, purpureo-brunnei v. in sicco nigrescentes, vetustiores cinerascentes.   Folia (juvenilia) elliptica v. late elliptica, basi rotunda, apice rotunda usque subacuta, supra ut videtur laete viridia, initio sparse sericea v. sericeo-tomentella, mox glabra v. tantum costa puberula, subtus cinerea v. glauca, initio pilis longis sericeis obtecta v. tantum costa villosula, mox glabrescentia, margine integra, minora 1.5–2 cm. longa et 0.9–1.5 cm. lata, majora (plantae ♂) ad 2.3:1.2 cm. v. (plantae ♀) ad 2.6:1.7 cm. magna; petioli sericei v. villosuli, 3–5 mm. longi, graciles.   Amenta ♂ subsessilia, basi foliolis minimis sericeis 1–2 suffulta, 2 cm. longa, vix 4 mm. crassa, densiflora, rhachi sericea; flores diandri, filamentis liberis basi pilosis quam bracteae 2½plo longioribus, antheris ovalibus flavis;

glandula una oblonga, bracteis obovato-rotundis utrinque fulvo-sericeis (pilis albis intermixtis) duplo brevior. Amenta ♀ brevissime pedunculata, basi foliolis parvis 2 normalibus instructa, 2–2.5 cm. longa, densiflora, rhachi pilosa; ovaria ovato-elliptica, sessilia, glabra, stylo brevi subbifido stigmatibus brevissimis bifidis coronata; bracteae late obovatae, flavo-brunnescentes, tantum extus basi pilosae, ovariis juvenilibus subaequilongae (in fructibus valde breviores); glandula una, obovata, obtusa, ovariis circiter 3plo brevior.

Western Hupeh: Hsing-shan Hsien, thickets, 1300–1700 m., May 14, 1907 (No. 2118, type; bush 4.5–6 m.; ♂; No. 2118ᵃ; ♀ co-type).

This species may be easily distinguished by the brownish silky bracts of the ♂ plant. In the ♀ plant the leaves are broad and roundish at the apex, somewhat resembling those of *S. hypoleuca* Seemen, but are pubescent like the petioles and branchlets while young. The shape of the leaves is different from that of those of *S. longiflora* Andersson or *S. cathayana* Diels, which have a little larger fruits and well developed leaves on the peduncles.

The specific name is derived from μικτός, mixed, and θρίξ, hair.

Salix cathayana Diels in *Not. Bot. Gard. Edinburgh*, V. 281 (*Pl. Chin. Forrest.*) (1912).

Salix longiflora Burkill in *Jour. Linn. Soc.* XXVI. 530 (pro parte, non Andersson) (1899).

Western Szech'uan: Mupin, thickets, alt. 1600–2000 m., June 1908 (No. 1402; bush 2.4–3.6 m.; with fruits); same locality, alt. 2000–2300 m., June 1908 (No. 1408; bush 1.8–3 m.; with fruits; No. 1408ᵃ; ♂; No. 2131; with fruits); same locality, woodlands, alt. 2300–2600 m., October 1910 (No. 4365; bush 4.5–6 m.; sterile [foliis utrinque acutis subtus distincte sericeis]); same locality, thickets, alt. 2600 m., October 1910 (No. 4368; bush 3–6.6 m.; sterile [forma satis parvifolia fere glaberrima]); Ching-chi Hsien, Ta-hsiang-ling, thickets, alt. 2000–3000 m., May 1908 (No. 1428; bush 1.2–2.4 m.; with immature fruits; [amentis ad 5 cm. longis]); Chiu-ting-shan, thickets, alt. 2300–2600 m., May 23, 1908 (No. 2127; bush 1.8–2.4 m.; ♂); west of Kuan Hsien, summit of Niu-tou-shan, alt. 3300 m., June 20, 1908 (No. 2135; bush 0.6–1.2 m.; ♀; No. 2135ᵃ, ♂); Wa-shan, thickets, alt. 3300 m., June 1908 (No. 1405; bush 1.8–3 m; ♀ flowers and fruits; No. 1405ᵃ, sterile, collected in the fall); same locality, alt. 2600–3000 m., June 1908 (No. 2130; bush 3–4.5 m.; ♂); without exact locality, mountain tops, May 1904 (Veitch Exped. No. 4516; bush 0.9–1.2 m.; ♂); without locality, side of streams, 3000–3300 m., June 1904 (Veitch Exped. No. 4517; bush 1.8–3.6 m.; ♂). Yunnan:

Tali Range, alt. 2600–3000 m., June 1906, *G. Forrest* (No. 4600, type; ♀); Mosoyn, April 4, 1887, *J. Delavay* (ex Diels).

The description by Diels is rather incomplete, and I think it best to add a full description based on the good material collected by Wilson.

Frutex 0.6–4.5 m. altus, valde ramosus; ramuli initio sericeo-villosuli v. interdum tomentelli (imo apice tomentosi), vetustiores subglabri v. glabri, flavo-brunnescentes v. saepissime cinerascentes; gemmae breviter ovatae, plus minusve obtusae, pilosae v. in eodem ramulo glabrae, petiolis breviores v. subaequilongae. Folia elliptica v. elliptico-lanceolata, utrinque obtusa v. apice plus minusve acuta (No. 1405, 1408, 4368), rarius basi etiam acuta (No. 4365), supra satis viridia, initio saepe plus minusve, deinde costa tantum puberula v. sericea (adulta in No. 4368 interdum tota glabra), subtus glauca v. cinerea, initio sericea (rarius tomentella), interdum pilis fulvis intermixtis, deinde pleraque plus minusve glabra (in 1405, 4365 etiam adulta tota sericea), costa nervisque subtus distincte prominulis, margine integra, minora (pedunculorum) 1–3 cm. longa et 0.6–1 cm. lata, majora 3–5 cm. longa et 1.2–1.5 cm. lata (v. in No. 1408 folia nondum matura ad 5 cm. longa et ad 2 cm. lata); petioli 2–5 mm. longi, plus minusve pilosi. Amenta ♂ pedunculis 0.5–1.5 cm. longis sericeis v. villosis foliola 3–6 normalia gerentibus suffulta, 2–3(–3.5) cm. longa, 0.6–0.8 cm. crassa, densa v. interdum sublaxiflora; flores diandri, filamentis liberis ad medium pilosis bracteis 2–3plo longioribus, antheris satis parvis ovalibus v. subglobosis flavis; glandula una ovato-oblonga, satis crassa, bracteis late ovatis v. obovatis apice rotundis intus glabris extus margineque sericeis flavo-brunneis paulo v. vix brevior. Amenta ♀ ut a cl. Diels descripta, sed interdum ad 5 cm. longa, pedunculis (etiam in typo ad 1 cm.) ad 1.8 cm. longis; styli breves sed distincti, apice bifidi, stigmatibus brevibus bifidis. Fructus 3–4 mm. longi, ovato-oblongi, sessiles v. subsessiles, glabri; bracteae interdum ovato-oblongae.

This is a rather variable species, very near to *S. longiflora* Andersson (see p. 121). Of No. 1428 Mr. Wilson sent plants to the Arboretum of which the young branchlets are slightly hairy. The old leaves are glabrous, except the midrib on both sides, deep green above, glaucous-gray beneath, elliptical, nearly roundish at the base, shortly acute at the apex, about 1–2 cm. long, 2 cm. broad. The petioles are 3–4 mm. long, of about the same length as the buds.

The specimen collected by *E. Faber* on Mount Omei (No. 103; tree 3 m. tall; with young fruits) which is referred to *S. longiflora* Andersson by Burkill, may represent a different form. The branchlets, the petioles and the midrib of the leaves bear a brownish tomentum. The leaves are narrow-elliptical, more or less acute at both ends and up to 4.5 cm. long and 1.6 cm. broad. The petioles of the larger leaves are 5–8 mm. long. The ♀ flowers or young fruits agree with the description given above, but the ovaries are slightly silky at the base, and their bracts are more silky than in the other specimens before me. Mount Omei being so rich in endemic species, this form needs further investigation; it seems to be indeed very nearly related to *S. longiflora* Andersson, of which I have not seen sufficient material.

Salix macroblasta Schneider, n. sp.

Frutex ramosus, 2–6 m. altus; ramuli annotini plus minusve breviter pilosi, flavescentes v. flavo-brunnei, vetustiores glabri, cinereo-brunnei; gemmae satis evolutae oblongae, acutae, flavo-rubrae, glabrae v. minute puberulae, petiolis circiter duplo longiores. Folia adulta an-

guste lanceolata (in No. 4356 latiora, elliptico-lanceolata), basi rotunda
v. subcordata (aurita), apice acuta, supra intense viridia, subnitida,
costa minute puberula, ceterum glabra, costa nervisque subprominulis,
subtus albescentia v. glaucescentia, glabra (v. sub lente pilis sparsis
obtecta), costa flava nervisque plus minusve prominulis vix reticulata,
integra, 1.4–3 cm. longa et 0.5–0.9 cm. lata (in No. 4356 2.5–4 cm. longa
et ad 1–1.2 cm. lata v. ad 5 cm. longa et ad 2 cm. lata); petioli brevissimi,
1–3 mm. longi, pilosuli, supra late sulcati. Amenta (tantum fructifera
visa) pedunculo ad 1.5 cm. longo folia circiter 4 normalia gerente
piloso suffulta, 2.5–3 cm. longa, valde deflorata circiter 0.6 mm. crassa,
rhachi pilosa, densiflora. Flores inter fructus remanentes subsessiles;
ovaria ovato-oblonga, glabra, stylo brevissimo, stigmatibus brevibus
emarginatis coronata; glandula una, ovato-oblonga, bracteis obovato-
oblongis brunnescentibus subglabris duplo brevior. Fructus oblongi,
circiter 4 mm. longi.

Western Szech'uan: west of Kuan Hsien, Pan-lan-shan, upland
thickets, alt. 3300 m., October 1910 (No. 4354, type; bush 2–3.6 m.;
with fruits; No. 4356; sterile); Mupin, alt. 1600–2000 m., June 1908,
(No. 1402; bush 2.4–3.6 m.; with fruits).

Mr. Wilson sent cuttings of both No. 4354 and 4356 to the Arnold Arboretum.
Plants raised from these agree exactly; the twigs get black in drying, and the leaves,
which do not differ from the larger ones in No. 4356 and measure up to 5 cm. in
length and 1.8–2.5 cm. in width, are nearly sessile. No. 1402 is a little more pubescent, but has the same kind of long buds. The leaves are nearly as narrow as
those of the type specimen and measure up to 4 cm. in length and to 1.2 in width.
In its leaves and buds the species looks very distinct, but is certainly nearly related to S. cathayana Diels.

The specific name is derived from μακρός, long, and βλαστός, bud.

Salix driophila Schneider, n. sp.

Frutex ramosus, 1.5–3 m. altus; ramuli novelli sericeo-villosuli v.
tomentelli, biennes plus minusve glabrescentes, nigrescentes et cinerascentes, vetustiores glabri; gemmae, ut videtur, ovatae, adpressae,
paullo pilosae, flavo-brunneae. Folia (nondum matura) ovata, elliptica v. obovato-oblonga, basi rotunda v. obtusa, apice obtusa et apiculata v. acuta, supra initio tomentella, mox costa excepta glabrescentia,
viridia, subtus cinerea v. subglaucescentia, initio densius sericea v.
sericeo-tomentosa, vix v. paulo glabrescentia (adulta ignota!), nervis
utrinque paulo prominulis, margine integra, minora 2–2.5 cm. longa,
1–1.5 cm. lata, majora ad 4 cm. longa et ad 2 cm. lata v. ad 5 cm. longa
et ad 1.8 cm. lata; petioli 3–6 mm. longi, sericei v. tomentelli, superne
sulcati. Amenta ♂ et ♀ pedunculis 1.5–2.5 cm. longis sericeo-villo-

sulis folia 3–4 normalia gerentibus suffulta, rhachi sericeo-villosa, 4–6 cm. longa, 0.8(–0.9) cm. crassa, densiflora; flores ♂ diandri, filamentis liberis fere ad medium villosis quam bracteae duplo longioribus, antheris flavis ovalibus; glandula una ventralis, oblonga, satis crassa, bracteis ovato-oblongis obtusis intus glabris extus (apice interdum excepta) sericeis flavo-brunneis subtriplo brevior; ovaria sericea, sessilia, juvenilia crasse ovata, dein ovato-oblonga, stylo satis brevi apice bifido stigmatibus breviter oblongis partitis divaricatis coronata; bracteae late ovato-rotundae, intus glabrae, extus sericeae et ciliatae, ad apicem glabrescentes, flavo-brunneae, ovariis initio vix duplo breviores; glandula una, bracteis subaequilonga, oblonga, crassa. Fructus ignoti.

Western Szech'uan: Mupin, thickets, bush 1.5–2.4 m., alt. 2300–2600 m., June 1908 (No. 2150; co-type, ♂); Chiu-ting-shan, thickets, alt. 2300–2600 m., May 23, 1908 (No. 2123, type; bush 3 m.; ♀).

The ♂ and ♀ plants look very much alike, but the leaves of the males are a little more glabrescent, and the ♂ plant resembles somewhat S. dyscrita Schneider, but that species has a small dorsal and a ventral gland. The ♀ specimens are not unlike S. plocotricha Schneider, which is much more glabrous and has longer petioles. The fruiting state may be represented by No. 2129 from Ching-chi Hsien, thickets, alt. 2000–2600 m., June 1908, bush 3 m. The following differences may only be due to the difference in age between No. 2123 and 2129: folia majora, supra tenuiter sparse adpresse pilosa, costa puberula subtus cinerea, tota facie molliter sericeo-tomentella, nervis satis prominentibus circiter 3–4 pro 1 cm., minutissime reticulata, elliptica, utrinque obtusa v. apice subacuta, minora 2.2–4 cm. longa et 1.2–1.8 cm. lata, maxima ad 7 cm. longa et ad 2.3 cm. lata. Amenta fructifera eodem modo pedunculata, 5.5–7.5 cm. longa et 1 cm. crassa. Fructus stylo excluso 4 mm. longa, ovata, apice angustata (flores inter fructus remanentes iis S. driophilae Schneider vix diversa).

The specific name is derived from δρίος, thicket, and φιλέω, I love.

Salix amphibola Schneider, n. sp.

Frutex 1.5 m. altus; ramuli hornotini plus minusve sericeo-villosi, annotini subglabri, purpurascentes, vetustiores cinerascentes. Folia ovato-elliptica v. late elliptica, basi obtusa v. rotunda, apice subrotunda usque subacuta, supra intense viridia (initio ut videtur etiam facie laxe villosula), costa pilis fulvis et albis puberula excepta glabra, subtus cinerea, plus minusve sericea v. costa elevata flava excepta glabra, nervis circiter 2 pro 1 cm. prominulis, tenuissime v. vix reticulata, margine integra, minora 2.5–4 cm. longa et 1.5–2 cm. lata, maxima ad 5 cm. longa et 3.3 cm. lata v. ad 5.5 cm. longa et 3 cm. lata; petioli (3–)5–11 mm. longi, villosuli. Amenta ♀ pedunculo villosulo circiter cm. longo folia 3–4 normalia gerente suffulta, 3.5 cm. longa et tantum 5 mm. crassa, densiflora, rhachi villosa; ovaria sessilia, ovato-elliptica, sericeo-villosa, sed ad apicem glabra, stylo brevi stigmatibus bilobis brevibus divaricatis coronata; glandula una ventralis, ovato-oblonga, crassiuscula, ovario vix duplo brevior; bracteae late ovatae, obtusae, brunnescentes, utrinque glabrescentes, margine sericeae, ovario tertia parte breviores.

Western Szech'uan: Chiu-ting-shan, thickets, alt. 2300 m., May 23, 1908 (No. 2124, type; ♀).

I am very doubtful if this species belongs to the section *Longiflorae*. In the rather large leaves it resembles some forms of sect. *Eriostachyae*, but the catkins are exceedingly small and thin; and I suspect a hybrid origin between a species with glabrous and another with hairy ovaries.

The specific name is derived from ἀμφίβολος, doubtful.

Salix heterochroma Seemen in *Bot. Jahrb*. XXI. Beibl. LIII. 56 (1896); XXXVI. Beibl. LXXXII. 30 (1905).

> *Salix Henryi* Burkill in *Jour. Linn. Soc*. XXVI. 530 (1899). — Léveillé in *Bull. Soc. Bot. France* LVI. 299 (1909).

Western Hupeh: Hsing-shan Hsien, woods, alt. 1700–2000 m., May 1909 (No. 2119; bush or tree, 3–15 m. tall; ♀; No. 2119ᵃ, ♂); same place, fall 1909 (No. 2119ᵇ; sterile); same place, mountain thickets, alt. 2000–2300 m. (No. 1461; bush 1–1.8 m.; cultivated in Arnold Arboretum, sterile); Changyang Hsien, May 1900 (Veitch Exped. No. 685; tree 3.5 m.; ♀); Chienshi Hsien, A. *Henry* (No. 5843; fruiting co-type of *S. heterochroma* and *S. Henryi*); same place, A. *Henry* (No. 5349, type of *S. Henryi*; ♀). Eastern Szech'uan: Wushan Hsien, A. *Henry* (No. 5671; fruiting type of *S. heterochroma*, co-type of *S. Henryi*); Tchen-keou-tin, P. *Farges* (ex Burkill). Western Szech'uan: Mt. Omei, May 1904 (Veitch Exped. No. 5794; with fruits).

Von Seemen and Burkill knew only the ♀ plant. I add the description of the ♂: Amenta ♂ subsessilia, basi foliolis 2 oblongis squamiformibus subtus dense sericeis instructa, 3–5.5 cm. longa, circiter 0.8–1.1 cm. crassa, densissima, flavescentia, rhachi sericea; flores diandri, filamentis liberis tantum basi pilosis bracteis 2–2½plo longioribus, antheris ovato-oblongis flavis; bracteae oblongae, obtusae, flavo-brunneae, utrinque dense sericeae, longe ciliatae; glandula una obovata v. claviformis, crassa, bracteis subtriplo brevior. Ceterum ut a cl. von Seemen v. Burkill descripta.

This handsome species has lanceolate entire leaves which are whitish beneath; the fruits are about 5 mm. long, the pedicels being 2–4 times longer than the broad gland. The shape of the glands is similar to that of those of *S. tetrasperma* Roxburgh, and von Seemen first believed that *S. heterochroma* might belong to sect.

Tetraspermae. Burkill placed his *S. Henryi* in the sect. *Hastatae,* but so far as I can judge it belongs with two other species to a new section (see p. 121). The catkins of both the sexes are long-cylindrical.

Salix microphyta Franchet in *Nouv. Arch. Mus. Paris,* sér. 2, X. 83 (*Pl. David.* II. 121) (1887). — Burkill in *Jour. Linn. Soc.* XXVI. 531 (1899). — Léveillé in *Bull. Soc. Bot. France,* LVI. 299 (1909).

Western Szech'uan: Wa-shan, on rocks, alt. 2300–3000 m., June 1908 (No. 2141; bush 15–20 cm.; ♀ flowers and fruits); same place, alt. 3300 m., July 1903 (Veitch Exped., No. 4513; ♀); west of Kuan Hsien, Pan-lan-shan, moist banks, alt. 2300–3700 m., June 1908 (No. 2142; bush 15–20 cm.; ♂; No. 2142ª; with young fruits); Mupin, "in humidis regionis altissimae," June 1869, *A. David* (type, female, ex Franchet).

Franchet's description is rather incomplete, and it seems best to give here a fuller one:

Frutex minimus, trunco subterraneo radicante, ramulis gracilibus erectis fastigiato-flabellatis, 10–30 cm. altus; ramuli novelli villosuli, mox glabri, flavo-virides v. in sicco nigrescentes, vetustiores brunnescentes v. cinerascentes. Folia ovata, obovata, oblonga v. fere spathulata, basi subrotunda v. pleraque acuta, apice obtusa v. subito breviter acuta, utrinque glabra (v. interdum initio costa pilis paucis conspersa), viridia, saepe paulo nitentia, elevato-reticulata, textura papyracea, margine crenato-glanduloso-denticulata v. serrata, minora 1–1.5 cm. longa et 0.8–1.3 cm. lata v. ad 1.5 cm. longa et 0.7 cm. lata, maxima 2–2.7 cm. longa et 1.2–1.7 cm. lata; petioli glabri v. sulco superne sparse pilosi, (3–)4–9 mm. longi. Amenta ♀ et ♂ ramulos elongatos foliatos sparse villosos 3–5 cm. longos terminantia, subdensiflora, anguste cylindrica, 2–4 cm. longa, florifera circiter 5 mm. crassa, fructifera vix longiora, sed crassiora, rhachi sparse villosula; flores ♂ diandri, filamentis basi ⅓ villosulis bracteas duplo superantibus, antheris parvis ovato-globosis rubescentibus; glandula una ventralis, lata, ovata, latere compressa, bracteis fere 3plo brevior; bracteae flavo-brunneae, obovatae, concavae, apice subtruncatae, erosae, glabrae; ovaria sessilia v. subsessilia, ovato-oblonga, glabra, stylo triplo breviore integro v. apice bifido stigmatibus angustis bifidis. coronata; glandula una ovato-oblonga, ½ ovarii aequans; bracteae ut in floribus masculis sed margine ciliatae, ovariis initio subaequilongae. Fructus maturi conici, 5 mm. longi, bracteis glandulisque brevioribus.

I believe the species is best placed in sect. *Myrsinites,* see p. 138.

Salix Souliei Seemen in Fedde, *Rep. Spec. Nov.* III. 23 (1906).

Western Szech'uan: Tachien-lu, alt. 3600–4000 m., July 1908 (No. 2143; prostrate on rocks; ♀); same locality, July 1903 (Veitch Exped. No. 4512; prostrate shrub); same locality, June and July 1894, *Soulié* (No. 2289, type; ♂ ex von Seemen).

Von Seemen described only a male specimen, and I add the following description of the female plant, chiefly based on Wilson's No. 2143:

Frutex pygmaeus, trunco subterraneo, ramis procumbentibus radicantibus; ramuli prostrati, flavo-brunnei v. rubescentes, glabri v. sparse sericei. Folia parva, firma, lanceolato-elliptica v. obovato-elliptica, basim versus angustata, acuta v. subobtusa, apice acuta v. subobtusa, supra laete virentia, sublucida, costa vix impressa saepe villosula, ceterum glabra (v. subglabra), fere enervia, subtus glauca v. albescentia, costa elevata et interdum etiam facie sparse villoso-sericea, nervis paulo prominulis vix reticulata, margine integra, 0.9–2 cm. longa, 0.3–1 cm. lata; petioli 3–10 mm. longi, supra saepe sericeo-villosuli. Amenta ♀ ramulos foliatos 0.5–1 cm. longos terminantia, ovato-capitata, pluriflora, circiter 1 cm. longa, fructifera ad 1 cm. crassa, rhachi sparse villosa v. fere glabra; ovaria oblonga, glabra, pedicellata, stylo distincto brevi bifido stigmatibus bipartitis brevibus coronata; glandula una ventralis, oblongo-rectangularis, satis crassa, saepe incisa v. partita, pedicello longior v. subbrevior, interdum glandula minima dorsalis adest; bracteae oblongae, obtusae, flavae, nervatae, pleraeque tantum breviter ciliatae, ceterum glabrae, ½ ovarii aequantes. Capsulae nondum maturae circiter 4 mm. longae, apice attenuatae.

The typical *S. Souliei* Seemen seems to differ from Wilson's plant only in its shorter petioles (2 mm.) and in its somewhat smaller leaves (up to 1.2–0.6 cm.). See also *S. brachista* Schneider and *S. Lindleyana* Andersson. Burkill (in *Jour. Linn. Soc.* XXVI. 531 [1899]) places Soulie's No. 543 from Tachien-lu under *S. oreophila* Hooker f. This species however differs from our plant in the different serration and nervation of the smaller, narrower leaves; the glandular teeth are incurved. Delavay's specimens from Yunnan cited by Burkill, l. c., under *S. oreophila* I have not seen.

Salix opsimantha Schneider, n. sp.

Frutex 1–3 m. altus, breviter ramosus; ramuli satis crassi, novelli in sicco nigrescentes, basi sericei v. glabri, vetustiores purpurascentes v. brunnescentes. Folia obovato-elliptica v. elliptica, apice obtusa v. subrotunda v. breviter acuta, basi obtusa v. rotunda, saepe angustata, supra intense viridia, costa puberula, ceterum glabra, nervis paulo v. vix prominulis, subtus opaca v. glauca, initio ut videtur subsericea, mox glabra, laevia, sed reticulo nervorum visibili, margine plus minusve glanduloso-crenulata v. subintegra, minora 1.5–3 cm. longa, 1–2 cm. lata, maxima ad 5 cm. longa et ad 2.7 cm. lata; petioli 3–12 mm. longa, superne pilosula. Amenta ramulos puberulos normaliter foliatos terminantia, cylindrica, densiflora, ♂ 3–3.5 cm. longa, circiter 1 cm. crassa, pedunculis circiter 1 cm. longis, rhachi puberula, ♀ usque 3 cm. pedunculata, 4–7 cm. longa, fructifera circiter 1.3 crassa, rhachi fere glabra; flores ♂ diandri, filamentis liberis ½ pilosis bracteis fere duplo longioribus, antheris ovalibus flavis; glandulae duae, ventralis late obovata, apice saepe crenulata, latere compressa, ⅓ bracteae aequans, dorsalis minor, similis v. bipartita v. lobulata; bracteae brunnescentes, oblongae, satis angustae, apice obtusae, utrinque laxe pilosae ciliataeque; ovaria sessilia, glabra, ovato-oblonga, stylis distinctis apice bifidis, stigmatibus bifidis oblongis divaricatis; glandula una ventralis, e basi latiore oblonga, obtusa, dimidium ovarii juvenilis aequans; bracteae iis florum ♂ similes, ovariis subaequilongae, glabriusculae. Fructus circiter 6 mm. longi, basi late ovati, apice subito attenuati, subsessiles, bracteis brevioribus.

Western Szech'uan: southeast of Tachien-lu, thickets, alt. 3700 m., July 1908 (No. **2139**, type; ♂; No. **2139ᵃ**; bush 1–3 m.; ♀); without locality, alt. 3700–4000 m., July 1903 (Veitch Exped. Nos. 4519, 4519ᵃ; bush 1.2 m.; ♀ flowers and fruits).

This species seems to be nearly related to *S. floccosa* Burkill, of which I have not seen any material. According to Burkill's description his species differs in having somewhat longer and narrower leaves and slightly hairy ovaries. Both species need further investigation.

The specific name is derived from ὄψιμος, late, and ἄνθος, flower.

Salix Wallichiana Andersson in *Svensk. Vetensk. Akad. Handl.* 1850, 477 (1851); in *Jour. Linn. Soc.* IV. 50 (1860); in *Svensk. Vetensk. Akad. Handl.* VI. 80, tab. 5, fig. 46 (*Monog. Salic.*) (1867); in De Candolle, *Prodr.* XVI. pt. 2, 223 (1868). — Brandis, *Forest Fl. Brit. Ind.* 468, t. 61 (1874). — Hooker f., *Fl. Brit. Ind.* V. 628 (1888). — Collett, *Fl. Siml.* 478 (1902).

> *Salix grisea* Wallich, *Cat.* No. 3700 (nomen nudum) (1829).
> *Salix Wallichiana*, var. *grisea* Andersson in *Svensk. Vetensk. Akad. Handl.* VI. 80 (*Monog. Salic.*) (1867); in De Candolle, *Prodr.* XVI. pt. 2, 223 (1868). — Burkill in *Jour. Linn. Soc.* XXVI. 534 (1899). — Seemen in *Bot. Jahrb.* XXIX. 277 (1900).
> *Salix Mairei* Léveillé in Fedde, *Rep. Spec. Nov.*, XIII. 342 (1914).
> *Salix Caprea* Léveillé, *Fl. Kouy-Tchéou*, 381 (non Linnaeus) (1915).

Western Hupeh: Changlo Hsien, side of streams, alt. 1000–2600 m., May 1907 (No. **2111**; bush 3 m.; fruits; No. **2111ᵃ**, ♂); north and south of Ichang, alt. 1000–2600 m., April 1907 (No. **2113**, ♀; bush or small tree 1.8–4.5 m. tall, abundant, No. **2113ᵃ**, ♂); same locality, May 1907 (No. **2113ᵇ**; sterile); Changyang Hsien, woodlands, 1000–2600 m., May 1907 (No. **2114**; bush 2.4–3.6 m. tall; fruits); same locality, April 1907 (No. **2114ᵃ**, ♀; No. **2114ᵇ**, ♂); Hsing-shan Hsien, thickets, alt. 1000–2600 m., April 1907 (No. **2115**; bush 1.8–4.5 m. tall, ♀; No. **2115ᵃ**, ♂; No. **2115ᵇ**; with fruits); Fang Hsien, upland thickets, 1800–2800 m. (No. **4447**; bush 1.8–2.4 m. tall; young plants in the Arnold Arboretum; also from Nos. **4432** and **4436** from the same place, which differ from 4447 in their leaves being quite glabrous sometimes); without locality, April 1900 (Veitch Exped. No. 13; ♀); without locality, A. Henry (No. 1466, ♂; Nos. 7653, 3399, ♀); "Kao-kien-scian," alt. circiter 800 m., May–June 1907, C. Silvestri (No. 299, ♀). Western Szech'uan: Wa-shan, thickets, alt. 2700 m., June 1908 (No.

2148; bush 4.5 m. tall; fruits); Ching-chi Hsien, thickets, alt. 1600–2700 m., May 1908 (No. 2128; bush 1.8–3.6 m. tall; ♀); west of Kuan Hsien, Pan-lan-shan, woodlands, alt. 2700–3100 m., October 1910 (No. 4353, tree 5–10 m. tall, girth 0.9–1.8 m., sterile; a somewhat uncertain form); same locality (No. 4352; bush 2.4–4 m. tall; young plants in the Arboretum). Kwei-chou: "Kouy-yang, mont du Collège," February and March 1898, *E. Bodinier* (Nos. 2070, 2102; ex Léveillé, sub *S. Caprea*). Yunnan: "vallée de Kiao-mé-ti, bords du torrent," alt. 3000 m., April 1913, *E. E. Maire* (small tree; type of *S. Mairei* Léveillé); "montagnes à Sen-tchai-tse," alt. 2800 m., April 1913, *E. E. Maire* (♂).

Whether the other specimens, cited by Burkill, l. c., really belong to *S. Wallichiana* or not, I cannot tell. Henry's No. 5296 (in Herb. Gray) from Hupeh, south of Patung, is a ♂ specimen, the flower of which seems to have only a single stamen with a hairy filament. Burkill mentions this number in part under *S. longiflora* Andersson (see p. 121). Henry's specimen resembles *S. gracilistyla* Miquel, but the gland is short, rectangular or oval, somewhat like that of *S. Wallichiana*. Andersson describes the filaments as hairy, but I have found them always glabrous, as on Brandis' plate (l. c.). The Chinese form seems to agree well with the Indian. I have seen the following specimens from India:

Kashmir: "reg. temp. alt. 7–8000 ped.," *T. Thomson* (with fruits); "Western Thibet, reg. temp. 6–11000 ped.," *T. Thomson* (with fruits). Kumaon: "Gagar pass, alt. 6000–6500 feet," *Strachey & Winterbottom* (No. 3; with fruits; No. 5; ♂ and ♀ flowers); "Kalimat, 6400 feet," *Strachey & Winterbottom* (No. 11; ♀). Eastern Afghanistan: Kurrum valley, 1879, *J. E. T. Aitchison* (No. 389); ♀; without locality (No. 4501, Herb. Griffith).

Andersson distinguished 3 varieties of *S. Wallichiana*: 1. var. *a. grisea*, in *Svensk. Vetensk. Akad. Handl.* VI. 80 (*Monog. Salic.*) (1867); in De Candolle, *Prodr.* XVI. pt. 2, 223 (1868), the type of which was collected by *Wallich* sub No. 3700 in Nepal and Kumaon; 2. var. *b. julacea*, l. c. 81 (1867), l. c. 224 (1868) (*S. julacea* Andersson in *Svensk. Vetensk. Handl.* 1850, 476 [1851]; in *Jour. Linn. Soc.* IV. 50 [1860]), of which the type was collected by *Jacquemont*, "in sylvis excelsis supra Hayderabad," alt. 2600–2700 m., May 3, 1831; 3. var. *c. sericea*, l. c. 81 (1867) et l. c. 224 (1868), collected by *Hooker* f. & *T. Thomson* "ad Banahal, reg. temp. 6000–9000 ped." I have seen no specimen of this var. All these forms need a careful study.

In the type the leaves according to Andersson are "juniora valde tenuia utrinque pubescentia subtus cano-villosa, adulta glaberrima, nitentia, venis prominulis striata, laete viridia, plana, rigida." Of the cultivated plants from China in the Arboretum there seem to be two forms, one with rather dull green leaves which are softly pubescent beneath, and another with more shining leaves, which are nearly or wholly glabrous beneath.

From *S. Caprea* Linnaeus and *S. cinerea* Linnaeus *S. Wallichiana* may be chiefly distinguished by its thinner, much less reticulate leaves, which are mostly ovate-oblong and more acuminate. The length of the pedicels seems rather variable; the style is usually very short.

Salix dolia Schneider, n. sp.

Frutex valde et breviter ramosus, 0.6–0.9 m. altus; ramuli novelli villosuli, annotini glabri v. subglabri, purpurascentes, vetustiores cinerascentes; gemmae crasse ovatae, obtusae, basi sericeae, purpureo-brunneae. Folia (nondum satis evoluta) lanceolata, oblanceolata v. elliptico-lanceolata, apice acuta v. subobtusa, basin versus pleraque attenuata, acuta, supra viridia, plus minusve v. tantum costa impressa puberula, nervis saepe paulo incisis, subtus ut videtur vix pallidiora, adpresse sericea, costa (et interdum etiam nervis) elevata, margine integra; petioli nondum satis evoluti, sericei. Amenta sessilia, basi tantum squamis parvis suffulta, ovato-cylindrica, densiflora, rhachi sericeo-villosa, 1–2 cm. longa, ♂ usque 1 cm. crassa, ♀ paulo tenuiora, fructifera ignota; flores ♂ diandri, filamentis liberis ad ⅓ pilosis, antheris ovalibus aurantiacis, glandula una ventrali oblonga obtusa ⅓ bracteae aequante; bracteae oblongae, subacutae v. obtusae, ½ filamentorum aequantes, fuscae, utrinque sericeae; flores ♀ glandula una ventrali ut in flore masculo instructi; ovaria sessilia v. subsessilia, breviter sericea, ovata, stylo satis longo (½ ovarii aequante) apice bifido stigmatibus brevibus plus minusve partitis coronata; bracteae ovatae, obtusae, ½ ovarii aequantes, glandulam duplo superantes, utrinque sericeae.

Western Szech'uan: west of Kuan Hsien, summit of Niu-tou-shan, alt. 3300 m., June 20, 1908 (No. 2133, type; bush 0.6–0.9m. tall; No. 2133ᵃ, ♂).

In habit this species resembles some species of the sect. *Longiflorae*, but on account of the short sessile catkins and the silky ovaries I have placed it in the sect. *Argenteae*. The style is rather long, but the stigmas are short.

The specific name is derived from δόλιος, misleading.

Salix Rehderiana Schneider, n. sp.

Frutex elatus, 2–3.6 m. altus, v. arbor usque 9 m. alta, ramis elongatis, habitu ut videtur *S. viminali* non absimilis; ramuli novelli sparse villosuli v. glabri, fusco-brunnei (v. in plantis cultis olivacei), deinde cinereo-brunnei; gemmae ovato-oblongae, compressae, flaves-centes v. brunneae, petiolis breviores v. subaequilongae, v. floriferae majores, crassiores, ovatae, purpurascentes, interdum subpruinosae, glabrae v. sparse puberulae. Folia lanceolata (v. interdum infima oblanceolata), basi obtusa v. interdum acuta, apice sensim breviter acuminata, supra intense viridia, juvenilia sparse breviter pilosa v. costa saepe impressa puberula excepta glabra, subtus albo-cinerea v. glauca, plus minusve modo *S. viminalis* sericea (v. in plantis cultis subglabra v. glabra), costa flava valde prominente, nervis prominulis et plus minusve reticulata, margine irregulariter glanduloso-crenulata, rarius subintegra, 5–11.5 cm. longa, 1.2–2.5 cm. lata; petioli 2–8 mm.

longi, supra canaliculati et plerique puberuli. Amenta (plantae cultae) praecocia v. subpraecocia, sessilia v. basi foliolis parvis 2–3 lanceolatis subtus sericeis integris suffulta, cylindrica, densiflora, rhachi sericeo-villosa, ♂ ad 2.5 longa et ad 1 cm. crassa, ♀ 2–3 cm. longa, circiter 0.8 cm. crassa, fructifera ignota; flores ♂ glandula una ventrali oblonga angusta obtusa bracteis circiter ⅓ aequante, filamentis glabris v. basi pilosis liberis v. interdum basi connatis, antheris ovalibus initio purpureis deinde flavis; bracteae oblongae, obtusae, brunnescentes, utrinque sparse sericeo-villosae; ovaria subsessilia, glabra (v. sparse sericeo-villosula) ovato-oblonga, glandula una ut in flore ♂, stylo distincto (ovarium juvenilem aequante) duplo breviore apice interdum bifido, stigmatibus brevibus emarginatis coronata; bracteae iis floris ♂ similes v. obovato-oblongae, pleraeque longius densiusque sericeae, ¾ ovarii aequantes.

Western Szech'uan: Mupin, thickets, alt. 2000 m., November 1908 (No. 1403, co-type; bush 2–3 m. tall; sterile and specimens from cuttings cultivated under the same number in the Arnold Arboretum, coll. April 24, 1912, type, ♀, and April 13, 1910, co-type, ♂); west and near Wên-ch'uan Hsien, alt. 1600–2000 m., side of streams, November 1909 (No. 1421; bush 2–3.6 m. tall; sterile; ♂ specimens from cuttings cultivated under the same number in the Arnold Arboretum, coll. April 1910, flowers apparently touched by frost, bracts short, roundish); Mupin, thickets in river-bottoms, alt. 2000–2500 m., October 1910 (No. 4364; tree 6–9 m. tall, girth 0.9–1.8 m.; sterile); same place (No. 4368; tree 5–8 m. tall, girth 1–1.5 m.; living plants in Arboretum); west of Kuan Hsien, Pan-lan-shan, woodlands, alt. 2600–3000 m., October 1910 (No. 4357; bush or small tree, 4.5–9 m. tall; sterile); Wa-shan (No. 1424; living plants from cuttings, ♂ and ♀, in the Arnold Arboretum).

The type of this species differs from the other species of this section by its mostly glabrous ovaries. No. 2125 from Wa-shan, thickets, alt. 2300–2600 m., June 1908, bush 2.4 m. tall, with glabrous fruits and not yet fully developed leaves is certainly very near this species. It differs, as far as I can see, only in the oval bracts which are often denticulate at the apex, in the often deeply cleft styles and in the sessile fruits. The leaves are up to 5.5 cm. long and to 2 cm. broad and are not different from those of the type.

The name is given in compliment to Mr. Alfred Rehder.

Salix Rehderiana, var. brevisericea Schneider, n. var.

Frutex 1.5–3 m. altus; ramuli novelli sericeo-villosi, annotini glabri, fusci, dein brunneo-cinerei. Folia juvenilia lanceolata, oblanceolata v.

ovato-lanceolata, utrinque acuta, initio subtus intensius sericea v. sericeo-villosula (supra interdum pilis brunneis mixtis), dein glabrescentia, subtus paulo pallidiora, integra, ad 4 cm. longa et ad 1.4 cm. lata, nondum distincte nervosa, breviter petiolata, petiolis villosulis. Amenta ♀ subsessilia, basi foliolis parvis lanceolatis subtus sericeis suffulta, 3–5 cm. longa, circiter 1–1.2 cm. crassa, densa, rhachi villosa; ovaria sessilia v. subsessilia, laxe breviter sericea v. subglabra, circiter 3 mm. longa, ovato-oblonga, in stylum longum ovarii ½–⅔ aequantem apice subbifidum attenuata, stigmatibus brevibus subbifidis subdivaricatis; glandula una ventralis, late ovato-oblonga, obtusa, ⅓ bracteae aequans; bracteae ovato-acutae, fuscae, utrinque sericeae, ovario subaequilongae.

Western Szech'uan: Ching-chi Hsien, Ta-hsiang-ling, alt. 2300–3000 m., May 1908 (No. **1429**, type; bush 1.5–3 m. tall; with young fruits).

This variety, which needs further investigation, seems to differ from the type principally in its more acute bracts and in its short silky pubescent young fruits. There may be some relationship between it and *S. myrtillacea* Andersson (see p. 71).

Salix hylonoma Schneider, n. sp.

Frutex 3–5 v. arbor parva ad 6 m. alta; ramuli etiam novelli parce pilosi, mox glaberrimi, in sicco fusci, vetustiores rubro-brunnei. Folia ovata, ovato-elliptica v. ovato-lanceolata, rarius obovato-elliptica et subcordata, basi rotunda v. obtusa, saepe paulo obliqua, apice subito v. sensim acuminata, supra saturate viridia, initio tantum sparse sericeo-villosula, dein glabra, interdum costa paulo impressa pilis sparsis praedita, indistincte reticulata, subtus albo-cinerea v. cerea, laevia, novella plus minusve dense sericea, micantia, adultiora tantum versus basim ad costam subsericea, tenuissime nervata, margine distanter et tenuiter glanduloso-serrata v. subintegra, minora latiora obtusiora 2.5–5.5 cm. longa et 1.5–2.3 cm. lata, majora angustiora acutiora 6–8.5 cm. longa et 2–3 cm. lata; petioli satis graciles, subpilosi, 3–6 mm. longi. Amenta coetanea, brevipedunculata, pedunculo 5–13 mm. longo foliis 2–3 parvis saepissime deinde caducis instructo, rhachi villosa, anguste cylindrica, densa v. subdensiflora, ♂ circiter 3 cm. longa, vix 6 mm. crassa, ♀ 5–7 cm. longa, circiter 7 mm. crassa, fructifera ad 8 cm. longa, vix 1 cm. crassa; flores ♂ staminibus 2 instructa, filamentis basi villoso plus minusve coalitis v. tantum stamen unicum adest; antheris ovalibus flavis; glandula satis longa, angusta, sublinearis, bracteas obovatas apice rotundas brunnescentes utrinque longe sericeas ⅓ superans; ovaria sessilia, breviter sericea, crasse ovata, stylo breviter subbifido divaricato stigmatibus oblongis brevibus bifidis coronata; glandula ut in flore ♂, bracteis iis floris ♂ similibus aequilonga, medium ovarii subaequans. Fructus brevissime pedicellati, sparse sericei, circiter 3 mm. longi.

Western Szech'uan: Wa-shan, thickets, alt. 2300–2600 m., June 1908 (No. 2138, type; bush 3–5.4 m. tall; ♀ ; No. 2138ᵃ, ♂); Mupin, woodlands, alt. 2300–3300 m., June 1908 (No. 2144; bush or small tree, 4.5–6 m. tall; ♀ flowers and fruits).

At the first sight this species is not unlike *S. heterochroma* Seemen in the shape of the leaves and of the catkins. But *S. hylonoma* may be easily distinguished from it by its serrate leaves, which are smooth and somewhat waxy, not reticulate beneath, and by the long narrow gland of the flowers. See also *S. isochroma* Schneider (p. 122), which differs in the greenish and reticulate under surface of the leaves. A specimen from Hupeh (Henry's No. 5296) was referred by Burkill to *S. longiflora* (see my remarks under *S. Wallichiana* Andersson, p. 65), but it seems to have only a single stamen. The nearly sessile silky catkins are 2.5 long and 1 cm. thick, and are not unlike those of *S. Wallichiana* and of *S. heterochroma*. The bracts are narrow and acute, somewhat like those of *S. gracilistyla* Miquel. The gland is short (⅓ the length of the bracts), rather broad and rectangular. I have seen no other specimen from central China which I could place with Henry's number. If the presence of only one stamen is a normal condition this number may represent an undescribed species of this section.

The specific name is derived from ὑλονόμος, living in woods.

Salix cheilophila Schneider, n. sp.

Frutex elatus, 3–5.4 m. altus, ramis elongatis; ramuli novelli adpresse sericei, dein glabri, purpurascentes; gemmae acutae, sericeo-villosae, petiolis ⅓–½ longiores. Folia linearia v. lineari-oblanceolata, basi acuta, rarius subobtusa, apice acuta v. breviter acuminata mucronata, supra initio adpresse sericeo-villosa, dein glabriuscula, sordide viridescentia, costa elevata nervis subprominulis, subtus distinctius sericea, cinerea, costa flava elevata, nervis angulo acuto a costa ad apicem divergentibus prominulis, interdum tenuiter reticulata, margine saepe revoluta, ad apicem distinctius glanduloso-serrata, basin versus subintegra, 1.5–5 cm. longa, 3–7 mm. lata; pctioli 1–3 mm. longi, sericeo villosi. Amenta coetanea, rhachi villosa, ♂ pedunculis foliiferis ad 1 cm. longis suffulta, gracillima, densiflora, 1.5–2.3 cm. longa, 3–4 mm. crassa; fructifera pedunculis vix ad 5 mm. longis foliis parvis 3–4 praeditis suffulta, vix 1.5 cm. longa, 1 cm. crassa, densa; flores ♂ glandula una ventrali oblonga angusta interdum apice bifida bracteis subduplo breviore, filamentis usque ad apicem coalitis glabris, antheris 2 liberis subglobosis aurantiacis; bracteae obovato-oblongae, apice obtusae v. leviter emarginatae, versus apicem saepissime rubescentes v. fuscae, glabrae, basin versus sericeae; ovaria florum inter fructus remanentium sessilia, dense breviter sericea, ovata v. ovato-oblonga, stylo nullo v. brevissimo purpureo, stigmatibus minimis; glandula ut in flore ♂; bracteae iis floris ♂ similes, apice saepe truncatae, ovariis ½–⅔ breviores. Fructus circiter 3 mm. longi, subacuti.

Western Szech'uan: near Monkong Ting, alt. 2600–3000 m., side of river, June 29, 1908 (No. 2146, type; bush 3–4.5 m. tall; ♀; No. 2146ᵃ, ♂); round Sungpan Ting, bed and banks of Min River, alt. 2300–3200 m., August 1910 (No. 4576; bush 3–5 m. tall; sterile).

At first sight this species much resembles *S. Wilhelmsiana* M. von Bieberstein (see p. 169), but so far as I can judge from the material before me it differs from that species chiefly in the very small stigmas and in the mostly two-colored bracts. The very thin ♂ catkins are borne on rather long peduncles, and the fruiting aments are very short. The value of the color of the bracts needs further investigation, because they are partly of one color in Wilson's specimen. *S. Wilhelmsiana*, *S. microstachya* Turczaninow and their relatives seem to form a rather distinct group of the sect. *Purpureae*, but all the species are very nearly related, and there may occur forms like *S. cheilophila*, which seem to connect these species with *S. purpurea* Linnaeus and its allies.

The specific name is derived from χεῖλος, bank of a river, and φιλέω, I love.

Salix variegata Franchet in *Nouv. Arch. Mus. Paris*, sér. 2, X. 82 (*Pl. David.* II. 120) (1887). — Burkill in *Jour. Linn. Soc. Bot.* XXVI. 534 (1899). — Seemen in *Bot. Jahrb.* XXIX. 278 (1900). — Léveillé, in *Bull. Soc. Bot. France*, LVI. 301 (1909).

Salix densifoliata Seemen in *Bot. Jahrb.* XXI. Beibl. LIII. 57 (1896), XIX. 278 (1900). — Léveillé in *Bull. Soc. Bot. France*, LVI. 289 (1909).

Western Hupeh: Ichang gorge, alt. 30–300 m., rocks, December 1907 (No. 2120; bush 0.3–0.9 m. tall; ♂; No. 2120ᵃ; ♂); same place, *A. Henry* (No. 7175; ♂; type of *S. densifoliata*, ex von Seemen; the specimen ex Herb. Gray is a fruiting one); Ichang and Nanto, *A. Henry* (No. 48; ♂; Nos. 3974, 7182; ♀; also Nos. 46 and 957, ex Burkill); Mitan Gorge, October, *G. Niederlein* (No. 98, ex von Seemen). Eastern Szech'uan: "in valle fluminis Yang-tze kiang ad rupes immersas" (between Ichang and the end of the gorges), December 1868, *A. David* (type, ♂ and fruiting, ex Franchet).

A comparison of the original descriptions of Franchet and von Seemen leaves little doubt that the two species are the same. Even the "ramulos saepius elongatos" of the ♀ catkins measure only up to 1.5 cm. according to Franchet, while in Henry's No. 7175 several peduncles are 2 cm. long. The bracts of the ♀ flowers are described as "lanceolatae acutiusculae" by Franchet. The arrangement of the ♂ catkins in *S. densifoliata* Seemen is precisely the same as described by Franchet. See the keys on p. 79 and p. 90.

This pretty low-growing and creeping Willow is abundant on the rocks and on the sandy foreshores of the Yangtsze River from Ichang westward through the gorges for some 300 miles, but scarcely reaches Chungking. During the summer floods this plant is submerged for weeks together, but in the winter it is very conspicuous with its vivid green leaves, the plants forming broad mats of verdure. The flowers open in December and January and are borne at the ends of the previous season's branchlets as is usual in most species of Willow, whereas in *S. Bockii* Seemen, which in herbaria looks similar, the flowers are borne on the shoots of the current season and while these shoots are still growing.   E. H. W.

Salix Bockii Seemen in *Bot. Jahrb.* XXIX. 278, t. 3, fig. G–M (1900).—Léveillé in *Bull. Soc. Bot. France*, LVI. 297 (1909). — Bean, *Trees & Shrubs Brit. Isl.* II. 477 (1914).

Western Szech'uan: near Wa-shan, side of streams, common, alt. 600–1500 m., July 1908 (No. 1414; bush 0.9–1.5 m. tall; flowers white, conspicuous; ♂ and ♀); Yachou Fu, side of streams, abundant, alt. 600–1500 m., August 1908 (No. 1414ᵃ; ♀ flowers and fruits); Mupin, alt. 2600 m., September and October 1910 (No. 4369; ♀ flowers and fruits); generally shingly beds of streams, alt. 1000–3000 m., September 1910 (No. 4351; bush 0.9–3 m. tall; ♂); banks of Yangtsze River, June 1903 (Veitch Exped. No. 4510; bush 0.3–0.6 m. tall; ♂ and ♀). Eastern Szech'uan: Chungking, *E. Faber* (No. 3; small shrub; ♂); Mt. Omei, 1600 ft., *E. Faber* (No. 74; ♂); Kiating Fu, *E. Faber* (No. 211; ♂). Southeastern Szech'uan: Nanch'uan, *A. v. Rosthorn* (No. 1509, type; ♀).

In western Szech'uan at altitudes between 300 and 3000 m. this Willow is abundant on the sandy foreshores and in the gravelly and stony beds of rivers and mountain-torrents. It also occurs sparingly on the banks of the Yangtsze River west of Chungking and near Sui Fu and on the lower reaches of the Min River in the neighborhood of Kiating Fu. It grows from 1 to 3 m. tall and often covers large areas. The leaves are gray-green and in late summer and autumn it is a decidedly ornamental flowering shrub. The catkins are freely borne on the current season's and actively growing shoots, and in this character this Willow is unique among the Chinese species.   E. H. W.

Salix myrtillacea Andersson in *Jour. Linn. Soc.* IV. 51 (1860). — Hooker f., *Fl. Brit. Ind.* V. 637 (1888).

Salix subpycnostachya Burkill in *Jour. Linn. Soc.* XXVI. 532 (1899). — Léveillé in *Bull. Soc. Bot. France*, LVI. 301 (1909).

Western Szech'uan: west of Kuan Hsien, Pan-lan-shan, alt. 3300–3600 m., June 1908 (No. 2161; bush 0.6–1.2 m. tall; with young fruits); Tachien-lu, alt. 3000–4300 m., *E. Pratt* (No. 835, type of *S.*

subpycnostachya, ♀ ; and No. 751, ex Burkill); valley of Jerikhou, *J.A. Soulié* (No. 509, ex Burkill). Sikkim : Lachen, "alt. 12000 ped.," *J. D. Hooker* (type of *S. myrtillacea;* with fruits).

There is no marked difference between Pratt's No. 835 and Wilson's No. 2161. Both closely resemble the co-type in Herb. Gray of Andersson's *S. myrtillacea;* the descriptions, too, of both authors agree very well. The ovaries are very short-stalked or subsessile in all the three specimens before me, the pedicels being about half as long as the gland. The styles are more or less elongated and are as long or longer than the oblong stigmas. The bracts are acutish, more or less silky, and often glabrous at the apex on the outside. The leaves are glabrous or very soon become glabrous and are oblanceolate or elliptical, mostly acutish at the ends and entire. I have not seen mature leaves; in Andersson's type they measure up to 3 cm. in length and 1.5 in width, the petioles being very short (1–2 mm.). The flat appressed acutish leaf-buds are alike in both the types; the purplish-black color of the somewhat pruinose branchlets is the same in all the specimens.

Salix allochroa Schneider, n. sp.

Frutex 1.2–3 m. altus; ramuli satis crassi, novelli tantum basi sericei, in sicco fusco-nigri, biennes rubro-brunnei, glabri, deinde cinereo-fusci; gemmae ovato-oblongae, obtusae, subdivaricatae, circiter 8 mm. longae, ramulis concolores. Folia satis crassa, late elliptica v. obovato-elliptica, basi subrotunda usque subacuta, apice rotunda v. obtusa, supra saturate sed sordide viridia, glabra, distincte subinciso-reticulata (iis *S. reticulatae* haud absimilia), subtus tantum novella sericea, mox glabra, cinerea, nervis lateralibus pro 1 cm. 2, angulo satis lato a costa exeuntibus sed satis subito apicem versus divergentibus, distincte elevato-reticulata, margine brevissime glanduloso-serrulata v. subintegra, minora saepe subrotunda 3–6 cm. longa, 2.5–4 cm. lata, majora usque 8 cm. longa et 4 cm. lata; petioli satis longi, glanduliferi (glandulis 2–4 conspersis), subrotunda, mox glabra, 1–1.5 cm. longa. Amenta (fructifera tantum visa) ramulis folia 3–4 normalia gerentibus subglabris 2–3 cm. longis suffulta, longa, densa, cylindrica, 9–11 cm. longa, circiter 1.3 cm. crassa, rhachi villosula; flores (inter fructus remanentes) subsessiles; ovaria villosa, ovata, stylo mediocri ad basim fisso divaricato stigmatibus oblongis bifidis divaricatis coronata; glandula una ventralis, latior quam alta, obtusa, subamplectens, apice interdum pilosa, pedicello brevissimi subaequilonga; bracteae ovatae, obtusae, ½ ovarii aequantes, brunneae, extus sparse villosae, v. ad apicem glabrae, intus tantum basi pilosae. Fructus villosi, rubescentes, ovato-oblongi, circiter 5 mm. longae.

Western Szech'uan: Mupin, thickets, alt. 2300–2600 m., June 1908 (No. **1410**, type; bush 1.2–3 m. tall; with fruits).

In the large leaves, the long aments on long leafy peduncles and in the distinct deeply cleft styles this species resembles some forms of sect. *Psilostigmatae,* but it differs from them in the reticulation of both the surfaces of the leaves. As long as the ♂ plant is unknown, it seems difficult to determine the systematic position of *S. allochroa.* In *S. moupinensis* Franchet we find similar glands on the petioles.

The specific name is derived from ἀλλόχροος, strange-looking.

Salix etosia Schneider, n. sp.

Frutex 0.9–1.2 m. altus, ramulis ut videtur virgatis; ramuli floriferi satis crassi, sericeo-tomentelli, grisei. Folia nondum visa. Amenta ♀ praecocia (v. coetanea?) pedunculos 8–10 mm. longos sericeo-villosos foliolis 3–4 parvis angusto-ellipticis integris basi obtusis apice acutis supra glabris subtus sericeis pallidioribus 10–13 mm. longis instructos terminantia, cylindrica, subdensiflora, saepe curvata, rhachi laxe sericeo-villosa, 5–8 cm. longa, circiter 1 cm. crassa; ovaria e basi ovata oblonga, glabra, pedicello glabro quam glandula una ventrali lineari-oblonga duplo breviore suffulta, stylo pedicellum subaequante v. paulo longiore stigmatibus parvis oblongis subbifidis ½–⅔ styli aequantibus coronata; bracteae ovato-obtusae, flavo-brunneae, ⅓ ovarii aequantes, glandulam fere duplo superantes, sparse longo-sericeae v. extus glabriusculae. Fructus maturi non vidi.

Western Hupeh: Patung Hsien, alt. 1300–2000 m., side of streams, May 1907 (No. **2112**, type; bush 0.9–1.2 m. tall; ♀).

There are only flowering branchlets without leaves. The catkins look very much like those of *S. Wallichiana* Andersson with which they were partly mixed, but the flowers are totally different and resemble those of *S. japonica* Thunberg, which, however, has a different short gland and does not occur in China. The flowers also are not unlike those of *S. Fargesii* Franchet, which is a Hupeh plant but differs considerably in its glabrous branchlets and in the large leaves at the base of the catkins.

The specific name is derived from ἐτώσιος, useless.

CONSPECTUS ANALYTICUS SALICUM ASIAE ORIENTALIS HIMALAYAEQUE.

I. CLAVIS SPECIERUM SECUNDUM SPECIMINA MASCULA.

(Specimina feminea, see p. 83.)

A. **Flores glandulis duobus (ventrali et dorsali** [1]**) instructi** (B, see p. 78).

a. *Stamina 3 v. plura* (b, see p. 75).

Amenta (pedunculo excluso) valde elongata, gracilia, cylindrica, pleraque (7–)8–10 cm. longa. Folia adulta ramulorum floriferorum ovato-lanceolata v. late lanceolata, plus minusve longe acuminata v. caudata, majora 8–15 cm. longa (pe-

[1] Glandula ventralis inter florem et rhachim sita, nunquam deest. Confer p. 94.

tiolis exclusis), nervis secundariis utraque facie distincte prominulis (confer etiam 20. *S. Urbanianam*).

Stamina 8–9 . . . . 1. *S. tetrasperma*.
Stamina 3–6.
Stamina 5–6 (ovaria floris ♀ glabra) (confer etiam 13. *S. Kusanoi*).
    4. *S. araeostachya*.
Stamina 3–4 (ovaria floris ♀ tomentella) . . . . 6. *S. ichnostachya*.
Amenta (pedunculo excluso) vix ad 7 cm. longa, v. foliorum forma diversa.
* Folia matura late-ovata v. elliptica, 1½–2½plo longiora quam lata v. elliptico-lanceolata, basi saepe subcordata et petiolis 1.3–2.5 cm. longis instructa, utraque facie plus minusve distincte nervosa v. reticulata. Bracteae apice subrotundae v. fere truncatae.
† Folia glabra, etiam juvenilia tantum sparse sericeo-pilosiuscula, subtus concoloria v. pallidiora, vix distincte glaucescentia, saepe tantum 1½–2plo longiora quam lata.
    Folia 3–9 cm. longa, apice breviter acuminata. Stamina 5(–7), (ovaria glabra satis breviter pedicellata) . . . . 21. *S. cardiophylla*.
    Folia pleraque longiora, ad 12 cm. longa, apice acuminato-cuspidata. Stamina 6–10.
        Ovaria floris ♀ glaberrima, satis longe pedicellata . . . . 3. *S. Mesnyi*.
        Ovaria floris ♀ tomentella, subsessilia . . . . 23. *S. populifolia*.
†† Folia juvenilia subtus distincte tomentella, deinde glabra, glaucescentia, pleraque 2–2½plo longiora quam lata. Amenta 5–7 cm. (pedunc. excluso) longa et circiter 1–1.2 cm. crassa; stamina 5 (ovaria tomentella, sessilia).
    20. *S. Urbaniana*.
** Folia matura lanceolata v. elliptico-lanceolata, (3–)3½–5plo longiora quam lata, basi nunquam subcordata (sed in 7. *S. tonkinensis* aurita).
† Folia late lanceolata, apice longe acuminata, basi saepe rotundata, nervis utrinque satis prominentibus. Stamina 4–6.
    Folia basi aurita. Bracteae stamina involucrantes (flores ♀ glandula una ventrali instructa) . . . . 7. *S. tonkinensis*.
    Folia basi non aurita. Bracteae staminibus breviores. (Flores ♀ glandulis 2 instructi) . . . . 13. *S. Kusanoi*.
†† Folia non longe acuminata v. anguste lanceolata, nervis utrinque vix v. haud prominulis, v. stamina tantum 3.
    ‡ Glandula ventralis et dorsalis plus minusve aequalis et pleraque incisa, lobulata v. emarginata, interdum dorsalis paullo altior quam ventralem; stamina (3–)4–12.
        ‖ Folia late-lanceolata v. elliptico-lanceolata, haud apice longe-acuminata, margine anguste distincteque glanduloso-serrata.
            Stamina pleraque (5–)6–12; inflorescentiae plus minusve densiflorae, ad 1.5 cm. crassae.
                Amenta pleraque tantum 3–4 cm. (pedunculo saepe longo excluso) longa, stamina 5–12 (stylus floris ♀ distinctus).
                    15. *S. pentandra*.
                Amenta pleraque 5–7 cm. longa (pedunculo excluso), stamina 5–9(–12), (stigmata floris ♀ sessilia v. subsessilia).
                    Glandulae quasi discum 6-lobatum formantes; stamina 8–12 (ovaria floris ♀ satis longe pedicellata; glandulae in flore ♀ ut videtur 2) (confer etiam *S. anisandram*, p. 102) . . . . 18. *S. Cavaleriei*.
                    Glandula dorsalis et ventralis separata, tantum interdum lobulata; stamina 5–7 (ovaria florum ♀ subsessilia; glandulam dorsalem non vidi) . . . . 16. *S. paraplesia*.

Stamina tantum 3–5; inflorescentiae satis laxiflorae, circiter 0.8 cm. crassae.

Petioli plerique glandulosi . . . . 11. *S. glandulosa.*

Petioli eglandulosi (confer etiam 17. *S. dodecandram*).

10. *S. Wilsonii.*

‖‖ Folia anguste-lanceolata, apice acuminata, margine subintegra v. vix anguste serrata.

Folia vix supra 4.5 cm. longa. Bracteae latae, stamina circiter 5 quasi involucrantes . . . . 19. *S. macrolepis.*

Folia pleraque 5–6–12 cm. longa. Bracteae staminibus 6–8 breviores.

24. *S. dealbata.*

‡‡ Glandula ventralis quam dorsalem crassior v. major; stamina 3. Folia lanceolata v. ovato-lanceolata, distincte serrata, subtus distincte pallida.

25. *S. amygdalina,* var. *nipponica.*

b. *Stamina tantum 2.*

Folia lanceolata v. ovato-lanceolata, apice plus minusve acuminata, (5–)6–15 cm. longa.

* Amenta (pedunculis exclusis) satis parva v. vix plus quam 5(–6) cm. longa.

† Folia (et ramuli) tantum initio plus minusve sericeo-pubescentia, deinde valde glabrescentia, v. semper glabra (in formis diversis 35. *S. albae* folia adulta sericea, sed ovaria glabra et bracteae deciduae sunt).

Amenta, brevia pleraque 1.5–2.5 cm. longa, 4–6 mm. crassa. Folia anguste lanceolata (majora circiter ad 12 cm. longa et 1–1.6 cm. lata), tantum juvenilia parce sericeo-pilosa (ovaria florum ♀ glaberrima).

Ramuli olivacei v. flavi. Bracteae late ovatae, obtusae. 26. *S. Matsudana.*

Ramuli plus minusve brunnei v. rubescentes. Bracteae oblongae, subacutae . . . . 34. *S. babylonica.*

Amenta plus minusve longiora et crassiora, v. folia magis sericea (v. ovaria florum ♀ pilosa).

Folia plus minusve distincte serrulata (species sequentes sine speciminibus femineis accurate discernere mihi valde difficile videtur).

Flores ♀ glandulis duobus praediti.

Ovaria glabra, pedicellata, stylo brevi instructa. Folia glaberrima, ad 16 : 4 cm. magna . . . . 25. *S. fragilis.*

Ovaria tota v. ad medium sericeo-pubescentia, sessilia, stylo satis longo instructa. Folia initio sericeo-pilosa, minora.

28. *S. eriocarpa.*

Flores ♀ glandula una ventrali praediti.

Ovaria plus minusve sericea v. tomentella, sessilia.

Styli ovariorum brevissimi.

Bracteae ellipticae, apice rotundatae v. subemarginatae.

31. *S. jessoensis.*

Bracteae ovato-oblongae, subacutae . . . . 32. *S. lasiogyne.*

Styli ovariorum distincti.

Ramuli hornotini velutino-tomentosi. Folia lanceolata, acuminata, rarius anguste oblonga, acuta. Bracteae late ellipticae, apice rotundatae, interdum subemarginatae.

29. *S. hondoensis.*

Ramuli hornotini glabri. Folia lanceolata, utrinque attenuata.

Bracteae ovatae, obtusae . . . . 33. *S. koreensis.*

Ovaria glaberrima, brevipedicellata. Folia etiam adulta subtus plus
minus v. utrinque sericea . . . . 35. *S. alba.*
Folia integra v. vix et indistincte denticulata; ovaria sericeo-tomentella,
sessilia, stylis brevibus instructa. . . . . 36. *S. sericocarpa.*
†† Folia (et ramuli annotini) etiam adulta subtus intense sericea, micantia.
Amenta gracilia, densa sed satis tenuia, circiter 5 mm. crassa.
55. *S. psilostigma.*
**Amenta (pedunculis exclusis) longissima, densa 4–8 mm. crassa (7–)8–11 cm.
longa.
Folia integerrima, maxima ad 12 cm. longa et 3 cm. lata, etiam adulta subtus
dense sericea, micantia, nervis invisibilibus, apice sensim acuminato-
caudata . . . . 54. *S. phaidima.*
Folia irregulariter crenulata, perlonga, maxima ad 22:6 cm. magna, apice sensim
breviter acuminata, subtus initio villoso-tomentella, deinde plus minusve
glabrescentia, albescentia, rete nervorum distincte elevato . 53. *S. phanera.*
Folia elliptica v. ovata, apice obtusa, v. parva et vix ad 4–5 cm. longa.
* Folia majora, plus minusve elliptica, 8–20 cm. longa et semper 3–12 cm. lata (confer etiam 52. *S. Daltonianam*).
Amenta (pedunculis exclusis) (8–)9–12 cm. longa, circiter 11–15 mm. crassa.
Folia permagna, ad 20 cm. longa et 12 cm. lata, utrinque obtusa v. rotundata,
apice apiculata, basi interdum subcordata, glabra, nervis secundariis inter se
valde distantia, margine integra . . . . 40. *S. magnifica.*
Amenta (pedunculis exclusis) brevia, vix 8 cm. longa. Folia minora v. utrinque
acuta v. dentata, nervis secundariis angustioribus, numerosioribus, distinctius a costa fere ad marginem percurrentibus.
Inflorescentiae (bracteaeque) sparse v. vix pilosae, gracillimae, vix 8 mm.
crassae. Folia ovata, acuta v. elliptica utrinque acuta, adulta subtus
flavo-viridia, tantum ad costam tenuiter (v. etiam paulo ad nervos) sericea, margine anguste distincteque glanduloso-serrata, circiter ad 13 : 5.5
cm. magna . . . . 44. *S. moupinensis.*
Inflorescentiae bracteaeque distinctius pilosae v. sericeae, pleraeque crassiores v. folia subtus pallida, v. margine integra.
Filamenta staminum subglabra v. vix ad medium pilosa; bracteae oblongae.
Folia fere semper satis distincte serrulata, apice acuta (ovaria floris ♀
glabra) . . . . 45. *S. Fargesii.*
Filamenta staminum distincte ½–⅘ pilosa; bracteae late ovatae v. obovatae. Folia integra v. ex parte indistincte denticulata, v. apice obtusa
(ovaria florum ♀ puberula) (confer etiam 48. *S. plocotricham*).
46. *S. Ernesti.*
**Folia minora, vix plus quam 7(–8) cm. longa, v. parva.
† Folia maxima circiter 2–8 cm. longa, elliptica, ovata v. obovata. Frutices
erectae v. in 104. *S. anglorum* et 106. *S. altaica* saepe prostratae. (Confer
etiam 80. *S. formosam* et 82. *S. kolymensem*).
‡ Folia margine graciliter denticulata, interdum subintegerrima (confer etiam
65. *S. fruticulosam* et 126. *S. floccosam*).
‖ Folia juniora subtus dense sericea, etiam adulta plus minusve sericeo-pilosa.
Bracteae apice rotundo v. truncato, plus minusve denticulatae.
Folia elliptica, v. lanceolato-elliptica, apice acuta, majora ad 6–8 cm.
longa et 2–2.8 cm. lata, subtus, praecipue juniora, ferrugineo-sericea.
Bracteae oblongae, apice truncatae. Frutex altus v. arbuscula.
52. *S. Daltoniana.*
Folia oblanceolata, vix ad 2.5(–3) cm. longa et 6–10 mm. supra medium

lata, subtus albo-sericea et glauca. Bracteae ovato-rotundatae.
Frutex parvus. . . . . 56. *S. Thomsoniana.*

|||| Folia etiam juvenilia vix v. parce pilosa, adulta glabra, subtus glauca.
Bracteae apice obtusae, raro emarginatae (confer etiam 180. *S. algistam*).
Folia obovata, supra medium latissima, basi pleraque acuta, apice obtusa, saepe apiculata, ad 5 : 2.5 cm. magna . . 125. *S. opsimantha.*
Folia elliptica v. elliptico-oblonga, raro obovato-elliptica, apice acuta
v. subacuta.
Amenta brevia, pedunculo excluso vix 1 cm. longa; glandulae ventralis
dorsalis floris ♂ (et ♀) pseudodiscum pluripartitum formantes.
Folia adulta ignota. . . . . 39. *S. atopantha.*
Amenta longiora; glandulae floris ♂ duae, separatae (in flore ♀
tantum ventralis adest). Folia adulta ad 5.5 : 2 cm. magna.
58. *S. denticulata.*

‡‡ Folia margine integra (in 65. *S. fruticulosa* et in 126. *S. floccosa* interdum ex
parte dentata).
Frutices erecti. Amenta ♂ subsessilia v. pedunculi foliis quam alia plus
minusve minoribus praediti.
Folia obovato-oblonga v. elliptico-oblonga, ad 7 : 2.8 cm. magna, juniora
dense longe sericea, micantia.
Ovaria florum ♀ sessilia . . . . 126. *S. floccosa.*
Ovaria breviter sed distincte pedicellata. . . . . . 124. *S. glauca.*
Folia valde minora v. anguste elliptico-lanceolata, etiam juniora subglabra v. sericeo-tomentella.
Inflorescentiae ♂ (et ♀) elongatae, cylindricae, pedunculo excluso
3–5 cm. longae.
Bracteae obovato-rotundae v. late ovatae obtusae.
Filamenta basi distincte pilosa . . . . 59. *S. dissa.*
60. *S. dyscrita.*
Filamenta glabra . . . . 63. *S. Camusii.*
Bracteae oblongae, obtusae; filamenta basi sparse pilosa.
62. *S. erioclada.*
Inflorescentiae breves, pedunculo excluso usque 2.5 cm. longae,
v. subsessiles, 1–2 cm. longae.
Amenta subsessilia, brevia; ramuli hornotini et saepe annotini plus
minusve pubescentes.
Bracteae apice denticulatae, plus minusve glabrae, antherae ut
videtur violaceae. . . . . 65. *S. fruticulosa.*
Bracteae apice integrae, extus plus minusve pubescentes; antherae flavae.
Folia juniora utraque facie villosula . . . 37. *S. sclerophylla.*
Folia juniora ad costam tantum tomentella . 04. *S. luctuosa.*
Amenta pedunculata; ramuli glaberrimi . . . 61. *S. Biondiana.*
Frutices plus minusve procumbentes. Amenta ♂ in apice ramulorum brevium folia normalia gerentium provenientia. Folia integerrima.
Folia obovata v. elliptico-rotunda, basi obtusa v. breviter cuneata,
subtus plus minusve sericeo-villosa, distinctius reticulata, glauca
(confer etiam 117. *S. reticulatam*) . . . . 103. *S. arctica.*
Folia obovato-oblonga v. elliptico-obovata, basi sensim in petiolum attenuata, subtus subglabra, glauca, vix reticulata (confer etiam 104. *S.
anglorum*). . . . . 106. *S. altaica.*

†† Folia etiam maxima minora, raro ad 2.5 cm. longa (confer etiam 37. *S. sclerophyllam*, 59. *S. denticulatam* et 65. *S. fruticulosam*). Frutices prostrati ramulis radicantibus (confer etiam 117. *S. reticulatam*).

‡ Glandulae fere semper separatae, simplices.

|| Folia rotunda, late ovata v. obovata, vix duplo longiora quam lata v. distincte arguteque serrata crenatave, vix ad 2.5 : 2 cm. magna, subtus nervis lateralibus distincte elevatis v. graciliter reticulata.

° Folia distincte crenata v. irregulariter serrata.

Folia argute denseque serrata. Bracteae sericeae, apice atrae. 113. *S. berberifolia.*

Folia crenato-serrata. Bracteae glabrae v. sparse pilosae, virides v. in sicco brunneae.

Folia subtus viridia, concoloria . . . . 115. *S. herbacea.*

Folia subtus glaucescentia, discoloria.

Amenta subcylindrica, pleraque 1–2 cm. longa. 114. *S. flabellaris.*

Amenta subovata, vix 1 cm. longa . . . . 122. *S. calyculata.*

°° Folia integra, rotundata, basi leviter cordata v. obtusa (v. folia obovato-elliptica) (confer etiam 109. *S. polarem*, 111 *S. ovalifoliam* et 112 *S. phlebophyllam*). . . . . 116. *S. rotundifolia.*

|||| Folia lanceolata, elliptica, oblanceolata v. spathulata, pleraque 2–4plo longiora quam lata, subtus tantum costa distincte elevata nervis lateralibus vix visibilibus (confer etiam 104. *S. anglorum*).

Folia subtus distincte pallida v. glauca, integra.

Amenta apice ramulorum brevium vix 1 cm. longorum, brevia subcapitata, 0.5 cm. longa, pauciflora . . . . . 118. *S. Lindleyana.* 120. *S. Souliei.*

Amenta apice ramulorum 2–3 cm. longorum, subcylindrica, 1 cm. longa, pluriflora; stamina pilosa . . . . 119. *S. brachista.*

Folia subtus concoloria, supra medium utrinque 2–3-dentata v. lobulato-incisa integris intermixtis. Amenta pauciflora, subcapitata. 121. *S. oreophila.*

‡‡ Glandulae fere semper partitae v. distincte lobatae, pseudodiscum 4- v. plurilobatum formantes; filamenta basi villosa . . . 123. *S. Serpyllum.*

B. **Flores glandula unica (ventrali) instructi.**

a. *Filamenta staminum tota longitudine v. ⅓–⅘ connata v. stamen unicum tantum adest* (b, see p. 80).

Stamen unicum.

Bracteae ovatae, obtusae (confer etiam 152. *S. Saidaeanam* et 153. *S. Harmsianam*). 151. *S. Buergeriana.*

Bracteae lanceolatae acutae (confer observationes sub 129. *S. Wallichiana*).

Stamina dua filamentis ⅓–⅘ v. tota connatis, v. interdum in singulis floribus stamen tantum unicum adest.

* Flores stamine unico et staminibus 2 filamentis saepe liberis in eodem amento mixti; bracteae ovatae, apice saepissime obtusae.

Amenta praecocia, densa, circiter 0.7–0.8 cm. crassa; filamenta ima basi parce pilosa . . . . 150. *S. Sieboldiana.*

Amenta coetanea (foliis ramulorum sterilium jam satis evolutis), tenuia, vix 0.5–0.6 cm. crassa; filamenta ⅓ pubescentia . . . . . 155. *S. hylonoma.*

**Flores semper staminibus duobus instructi; filamenta ⅓ ad ⅘ v. tota connata (si inflorescentiae longae, crassae, aureo-micantes et folia utrinque sericeo-tomentosa confer 139. *S. lanatam*).

† Bracteae ovato-lanceolatae, sensim acuminatae; glandula anguste linearis, elongata; amenta praecocia, ad 1.5 cm. crassa, stylus floris ♀ valde elongatus, gracillimus) . . . . 156. *S. gracilistyla*.

†† Bracteae plus minusve obtusae v. amenta v. glandula (et stylus) diversa.

‡ Folia adulta 6–16 cm. longa, late v. anguste lanceolata, acuminata, toto margine plus minusve serrata v. dentata.

  * Amenta ♂ evoluta pleraque 4–5(–6) cm. longa (pedunculis exclusis), densiflora, 1–1.5 cm. crassa. Folia adulta glabra, lanceolata.
    Folia subintegra. Filamenta ad medium tantum coalita.

172. S. oxycarpa.

    Folia serrata (dentibus circiter 3 pro 1 cm.).

    Folia utrinque serrata; stipulae circiter 2 cm. longae, petiolum superantes . . . . 158. *S. Miyabeana*.

    Folia apice acuminata; stipulae circiter 1 cm. longae, petiolo breviores

160. S. lepidostachys.

  ** Amenta ♂ evoluta breviora v. tenuiora.

  ° Folia late lanceolata, ad 14:2.4 cm. magna, basi subito attenuata, pleraque obtusa, apicem versus longe acuminata, anguste serrata (dentibus 5–6 pro 1 cm.) . . . . 157. *S. Pierotii*.

  °° Folia anguste v. lineari-lanceolata, apice acuta (si acuminata vix plus quam 7:0.8 cm. magna).

    Folia apice plus minusve acuminata (ovaria floris ♀ stylo et pedicello brevi instructa) . . . . 162. *S. Gilgiana*.

    Folia apice plus minusve acuta (ovaria floris ♀ sessilia; stylus nullus v. brevissimus) . . . . . . 161. *S. purpurea* ssp. *eupurpurea*.

‡‡ Folia adulta vix plus quam 4 cm. longa v. margine integra v. ut in forma sequente.

  ‖ Folia plus minusve opposita v. alterna, basi subcordata, sessilia, interdum petiolata, apice denticulata, ad 6:2.5 cm. magna, raro longiora.

161. S. purpurea ssp. amplexicaulis.

  ‖‖ Folia omnia alterna, petiolata (?167. *S. Duclouxii*), breviora v. angustiora.

    Stamina pleraque tantum basi, raro tota longitudine connata, interdum libera. Folia late ovata v. obovato-oblonga, integra, ad 4:2 cm. magna, brevissima petiolata . . . . 175. *S. minutiflora*.

    Stamina omnia plus minusve ad apicem connata. Folia elliptica, lanceolata v. linearia.

    Ramuli dense sericei v. pubescentes. Folia etiam adulta sericea v. pilosa, raro glabra, plus minusve remote denticulata, apice rotunda apiculata, subtus glauca.

    Folia linearia, apice subacuminata, minute denticulato-serrata.

    Bracteae saepe parte superiore atrae (confer etiam 165. *S. microstachyam*) . . . . 163. *S. cheilophila*.

    Folia anguste elliptica v. oblongo-lanceolata, apice satis obtusa v. subacuta. Bracteae pleraeque totae flavo-brunneae.

    Amenta ad ramulos hornotinos crescentes aestate apparentia (confer etiam 167. *S. Duclouxii*) . . . . . 166. *S. Bockii*.

    Amenta ad ramulos maturos (annotinos) autumno v. hieme apparentia.

    Antherae glabrae (confer etiam 170. *S. kouytchensem*).

168. S. variegata.

Antherae apice saepissime pilosae . . . 169. *S. andropogon.*
Ramuli ab initio glabri v. sparse pilosi. Folia adulta glaberrima, integra.
Folia apice plus minus obtusa v. rotundata, tantum apiculata.
Bracteae subacutae (ovaria floris ♀ breviter hirta, sessilia; stylus·
satis longus). Folia subtus pallide viridia.
                                          174. *S. myrtillacea.*
Bracteae obtusae (ovaria floris ♀ stylo brevi coronata).
Folia subtus pallide viridia (ovaria glabra, sessilia).
                                          171. *S. pycnostachya.*
Folia subtus glauca (ovaria tomentella, subsessilia).
                                          173. *S. myricaefolia.*
Folia apice acuta, linearia, ad 6:0.5 cm. v. longiora. Bracteae apice
obtusae v. truncatae (ovaria floris ♀ sessilia, breviter sericea;
styli breves) . . . . 164. *S. Wilhelmsiana.*

b. *Filamenta staminum 2 libera, raro ima basi v. ⅓ coalita (species sequentes sine
    specimine femineo accurate discernere saepe impossibile videtur).*

Pubescentia bractearum et inflorescentiae axis fulva v. rufo-ferruginea interdum
    pilis albis mixta; amenta densiflora, cylindrica, ad 5 cm. longa et 0.6–0.7 cm.
    crassa (confer etiam 69. *S. mictotricham* et 88. *S. Reinii).* . . . 92. *S. vulpina.*
Pubescentia inflorescentiarum bractearumque alba v. grisea v. aureo-micans, con-
    color.
  * Amenta densissime sericea, aureo-micantia, 4–6 cm. longa et 1.5 cm. crassa.
       Folia integra, utrinque plus minusve tomentello-sericea . . 139. *S. lanata.*
  ** Amenta nunquam aureo-micantia, breviora v. tenuiora v. folia diversa.
       † Inflorescentiae laxiflorae, ad 10 cm. longae, vix 0.6 cm. crassae; bracteae parvae,
          ovatae, sparse pilosae. Folia ovato-lanceolata, apice caudata v. acumi-
          nata, subtus opaca v. glauca, margine satis distanter argute serrata, adulta
          glabra . . . . 94. *S. japonica.*
      †† Inflorescentiae densiflorae crassioresque v. breves v. bracteae foliave diversa.
        ‡ Amenta ♂ plus minusve distincte praecocia v. subcoetanea, satis (1–2 cm.)
           crassa, sessilia v. subsessilia, pedicello brevi tantum foliis parvis v.
           squamiformibus instructa. Arbores v. frutices alti, ramis elongatis
           satis crassis.
          ‖ Folia late v. anguste lanceolata, apicem versus sensim acuminata, majora
             4–8plo longiora quam lata (confer etiam 129. *S. Wallichianam* et 132.
             S. insignem).
            ° Folia mox perfecte glabrescentia v. ab initio glabra; ramuli floriferi
               glabri (confer etiam 147. *S. Rehderianam).*
               Folia distincte angusteque glanduloso-serrata; ramuli saepe pruinosi;
                 amenta circiter 1.5 cm. crassa, breviter v. ovato-cylindrica (confer
                 etiam 132. *S. insignem)* . . . . 138. *S. rorida.*
               Folia subintegra v. satis irregulariter et distanter crenato-denticulata;
                 ramuli semper epruinosi; amenta cylindrica, vix plus quam 1 cm.
                 crassa . . . . 143. *S. sachalinensis.*
            °° Folia etiam adulta subtus dense sericea v. plus minusve pubescentia,
               margine integra v. satis indistincte crenulato-denticulata (confer
               etiam 135. *S. sibiricam).*
               Folia anguste v. lineari-lanceolata, 6–12-plo longiora quam lata, mar-
                 gine integra, basi acuta, subtus intense sericea (saepissime micantia).
                                                              142. *S. viminalis.*

Folia lanceolata v. elliptico-lanceolata, 4–5plo longiora quam lata, margine saepe crenulato-denticulata, basi obtusa, subtus pubescentia v. fere glabra . . . . 147. *S. Rehderiana.*

|||| Folia saepissime latiora, 1½–3½(–4)plo longiora quam lata, integra v. irregulariter crenato-denticulata.

  ° Folia etiam juvenilia glabra v. mox glabra (ramulis glabris), supra saturate viridia, subtus glauca (confer etiam 86. *S. chlorostachyam*).

Folia late lanceolata v. oblongo- v. rotundato-obovata, basi cuneata, ad 6: 3.5 cm. magna. Amenta cylindrica, ad 2 cm. longa et 1 cm. crassa (ovaria floris ♀ breviter pilosa) (confer etiam 79. *S. phylicifoliam*) . . . . 131. *S. Starkeana.*

Folia ovato- v. oblongo-elliptica, interdum fere orbicularia, ad 7: 5.5 cm. magna. Amenta ad 4.5 cm. longa et 1.5 cm. crassa (ovaria glaberrima) . . . . 96. *S. pyrolaefolia.*

  °° Folia etiam adulta subtus sericea v. pubescentia.

Ramuli floriferi glabri v. tantum in axillis amentorum puberuli.

Folia adulta subtus griseo-tomentello-villosa, oblonga v. late-ovata, apice breviter acuta, basi acuta v. obtusa, ad 11: 5.5 cm. magna, textura crasse-papyracea. Amenta ovata, ad 3 cm. longa et 2 cm. crassa . . . . 128. *S. Caprea.*

Folia adulta subtus sericea v. subglabra, oblongo-lanceolata v. ovato-lanceolata, apice acuta v. subacuminata, basi pleraque obtusa, ad 10: 3–4 cm. magna, textura tenuiora. Amenta 2–4 cm. longa et ad 1.8 cm. crassa . . . . . . 129. *S. Wallichiana.*

Ramuli floriferi tomentelli v. puberuli; folia etiam supra vix omnino glabrescentia.

Folia adulta subtus distincte reticulata, oblanceolata v. obovato-lanceolata v. plus minusve ovalia, ad 10: 4.5 cm. magna, textura crasse papyracea. Amenta ovalia v. cylindrica, ad 5 cm. longa et 2 cm. crassa . . . . 127. *S. cinerea.*

Folia adulta subtus vix reticulata, cetera ut supra sub No. 131 indicata . . . . 131. *S. Starkeana*, var. *cinerascens.*

‡‡ Amenta ♂ distincte coetanea v. longe cylindrica, tenuiora v. perparva. Plerique frutices ramis saepissime brevibus tenuioribusque.

|| Amenta cylindrica, gracilia, circiter 2–6 cm. longa et tantum 0.4–7 cm. crassa. Frutices arbusculaeque erectae, raro arbores (confer etiam 135. *S. sibiricam*).

  ° Bracteae inflorescentiaeque pilis rufis et albis mixtis sericeae; amenta tantum (an semper?) 2 cm. longa et 0.4 cm. lata. Folia parva, ovata, integra, subtus glauca, parce sericea (confer etiam 88. *S. Reinii*).

                                 69. *S. mictotricha.*

  °° Bracteae inflorescentiaque pilis concoloribus tantum albis sericeae v. villosae (in 76. *S. heterochroma* pili paulo subflavescentes).

  § Folia integerrima, satis magna et apice plus minusve acuminata v. minora et elliptica, apice breviter acuta v. rotunda (confer etiam 80. *S. formosam*).

Folia late lanceolata v. elliptica, juniora dense sericea, basi obtusa, apice acuminata, ad 12: 3.8 cm. magna. Amenta subsessilia, sericea, 4–5.5 cm. longa et ad 1 cm. crassa. 76. *S. heterochroma.*

Folia elliptica, obovato-elliptica v. orbicularia, apice acuta v. obtusa v. rotunda (confer etiam 74. *S. driophilam*).

Folia elliptica, obovato-elliptica v. orbicularia, apice rotunda v.

oblique obtusiuscula, ad 4.5: 3.5 cm. magna; petioli ad 1.5 cm. longi. Amenta 2–4 cm. longa, brevipedunculata.

101. *S. Nakamurana.*

Folia elliptica, apice acuta, ad 5: 2 cm. magna, petioli ad 10 mm. longi. Amenta pedicellata, foliis 2–4 parvis distinctis suffulta, 2.5–6 cm. longa et 0.5–0.7 cm. crassa (pedunculo excluso) (confer etiam 67. *S. rhoophilam* et 71. *S. cathayanam*).

66. *S. hypoleuca.*

§§ Folia circumcirca v. ex parte crenato-serrata v. denticulata, subtus glauca.

Folia etiam adulta subtus adpresse pubescenti-tomentella; petioli pubescentes.

Folia oblonga v. elliptica, rarius obovato-elliptica, basi pleraque obtusa usque rotundata, margine crenato-serrata, nervis lateralibus 5–9 (ovaria subsessilia, sericeo-tomentosa).

141. *S. vulpinoides.*

Folia elliptica, breviter lateque acuminata, basi acuta usque obtusissima, margine medio tantum remote serrulata, nervis lateralibus 11–16 (ovaria longe pedicellata, glabra).

98. *S. kenoensis.*

Folia adulta subtus glabra v. subglabra; petioli glabri v. parce pilosi.

Folia ovata v. ovato-lanceolata, basi rotunda v. pleraque cordata, margine argute serrata, juvenilia dense et longe sericea.

Bracteae ovato-rotundae, longe sericeae (ovaria glabra, longe-pedicellata) . . . . 99. *S. Shiraii.*

Folia nunquam basi cordata, margine crenata v. crenato-dentata v. subintegra.

Bracteae plus minusve glabrae, tantum basi pilosae et ciliatae.

Ramuli novelli ut folia juvenilia glabra v. saepe parce fulvo-pilosi. Folia majora obovata v. obovato-oblonga, basi angustata sed pleraque rotundata . . 88. *S. Reinii.*

Bracteae distincte et longe sericeae v. crispato-sericeae.

Folia (ramulique) etiam juvenilia glabra. Bracteae sericeae (ovaria tomentella).

Folio obovato-lanceolata v. elliptica, mediocra.

79. *S. phylicifolia.*

Folia lanceolata, longa . . . . 145. *S. opaca.*

Folia (ramulique) initio pubescentia v. sericea, pleraque ovalia v. elliptica. Bracteae crispato-sericeae (ovaria glabra).

97. *S. hastata.*

‖‖ Amenta brevia, 1–2.5 cm. longa et satis crassa. Frutices parvi v. minimi v. prostrati.

Frutices parvi sed ramis brevibus v. longioribus (ad 0.5 m.) erecti, trunco solum procumbente et radicante. Folia majora 2.5–7 cm. longa (confer etiam 136. *S. doliam*).

Folia lanceolata v. lineari-lanceolata v. obovato-lanceolata, pubescentia, sericea. Amenta ♂ 0.8–1.8 cm. longa. 135. *S. sibirica.*

Folia orbiculata usque anguste elliptica, ovata, v. raro lanceolata, ad 5: 3 cm. magna, glanduloso-denticulata v. serrata v. integra, adulta glabra.

Folia adulta utrinque viridia, nitentia, pleraque glanduloso-serrata v. denticulata, ad 5: 3 cm. magna. Amenta ♂ ad 3: 1.2 cm. magna . . . . 108. *S. Myrsinites.*

Folia subtus glauca, pleraque integra, ad 3.5:1.8 cm. magna.
Amenta ♂ ad 1.5:0.8 cm. magna . . . . 133. *S. myrtilloides*.
Frutices prostrati ramulis saepe radicantibus .
Folia crenato-serrata. Amenta elongata, cylindrica.
110. *S. microphyta*.
Folia integra v. pro parte minute denticulata. Amenta ovato-rotunda,
pauciflora . . . . 109. *S. polaris*.

II. CLAVIS SPECIERUM SECUNDUM SPECIMINA FEMINEA.

A. **Flores glandulis duobus (dorsali et ventrali) instructi; glandula dorsalis**[1] **interdum valde reducta difficile recognoscenda (B, see p. 85).**

Arbores v. frutices alti. Folia 4–12 cm. longa, distincte serrata.
* Ovaria glabra, pedicellata v. rarius sessilia; stylus nullus v. brevis, quam stigmata
haud v. vix longior. (Confer etiam 3. *S. Mesnyi* and 1. *S. tetraspermam*).
† Stigmata satis angusta, bifida, stylo distincto bifido aequilonga. Folia glabra
v. mox glaberrima.
Amenta fructifera 10–14 cm. longa (pedunculis 2–3 cm. longis inclusis). Folia
e basi ovata v. subacuta ovato-lanceolata, longe acuminata. Ovaria satis
longe graciliter pedicellata . . . . 14. *S. Maximowiczii*.
Amenta fructifera (pedunculis exclusis) 4–5 cm. longa. Folia basi ovata v.
elliptica, cordata v. rotundata, apice acuminata. Ovaria brevipedicellata.
21. *S. cardiophylla*.
†† Stigmata brevia, satis crassa; stylus distinctus, crassus v. nullus. Folia novella
saepe distincte sericea.
‡ Pedicelli ovariorum v. fructuum nectariis vix v. haud longiores; stylus brevis,
plerumque apice bifidus; bracteae oblongae ovariis florum subaequilongae (confer etiam 26. *S. fragilem* et 27. *S. hamatidentem*).
Foliola amentorum anguste glanduloso-serrata. Stylus distinctus.
15. *S. pentandra*.
Foliola amentorum integra v. subintegra. Stylus brevissimus.
26. *S. Matsudana*.
‡‡ Pedicelli ovariorum v. fructuum glandulis 2–4plo longiores; stigmata sessilia;
bracteae latiores v. ovariis florum dimidio breviores.
Folia juvenilia (ramulique) distincte ferrugineo-sericea, etiam adulta vix
glabra. Capsulae brevi-pedicellatae, pedicellis quam glandulae circiter
2plo longioribus, quam capsulae triplo brevioribus . 13. *S. Kusanoi*.
Folia juvenilia (ramulique) glabra v. pube albescente pilosa. Capsulae
longi-pedicellatae, pedicelli medium capsulae aequantes v. longiores (confer etiam 9. *S. dictyoneuram*).
Petioli plerique glanduliferi; stipulae saepissime evolutae, semicordatae . . . . 11. *S. glandulosa*.
Petioli eglandulosi; stipulae deficientes v. minimae.
Glandula dorsalis parva, separata v. interdum obsoleta.
10. *S. Wilsonii*.
Glandula dorsalis lata, quam ventralis altior 18. *S. Cavaleriei*.
**Ovaria tomentella v. pilosa, sessilia v. subsessilia; stylus distinctus stigmatibus
elongatis aequilongus v. paullo brevior.

---

[1] Glandula dorsalis inter bracteam et ovarium sita est, glandula ventralis inter
ovarium et rachim.

Amenta brevia, pedunculo excluso etiam fructifera vix 3 cm. longa; glandula
ventralis quam dorsalis minor et tenuior, saepe valde reducta. Folia lanceolata sensim acuminata v. utrinque acuta (confer etiam 33. *S. koreensem*).
28. *S. eriocarpa.*
Amenta pedunculo excluso 8–14 cm. longa.
Glandulae aequiformes, satis laterales; stigmata stylis sublongiora. Folia
late oblonga, ovalia v. ovato-elliptica, crenato-serrata, adulta subtus
reticulata (confer etiam 23. *S. populifoliam*) . . . . 20. *S. Urbaniana.*
Glandulae diversae, dorsalis minima. Folia integra v. indistincte glanduloso-
crenulata. Stigmata stylis breviora . . . . 46. *S. Ernesti.*
Frutices parvi, breviramei, v. minimi, prostrati, ramulis radicantibus; folia vix
usque 5 cm. longa v. integra.
* Frutices erecti. Amenta (pedunculo foliifero excluso) pleraque 1–2, tantum in
126. *S. floccosa* 4–7 cm. longa. Folia ovata, obovata, elliptica v. late lanceolata, integra.
Folia satis magna, majora 5–8 cm. longa et usque 3 cm. lata; Amenta ad
7 : 1.2 cm. magna; ovaria subsessilia; styli distincti, bifidi stigmatibus
angustis bifidis; glandula dorsalis minima . . . . 126. *S. floccosa.*
Folia minora, ad 2.5 cm. longa (adulta ignota). Amenta tantum ad 2.5 cm.
longa.
Styli ad basim divisi, divergentes. Ramuli floriferi glabri. Folia lanceolato-
elliptica. Amenta 0.5–2 cm. longa, circiter 0.7 cm. crassa.
39. *S. atopantha.*
Styli tantum apice bifidi. Ramuli floriferi pubescentes. Folia elliptica v.
obovato-elliptica, basi distinctius rotundata.
Glandulae variabiles, quasi pseudodiscum variabiliter lobulatum formantes.
38. *S. oritrepha.*
Glandulae duae distinctae, graciles, ventralis dorsali minor.
37. *S. sclerophylla.*
**Frutices parvi v. minimi ramis v. ramulis plus minusve prostratis saepissime
radicantibus. Amenta pleraque apice ramulorum normalium.
† Ovaria pubescentia; amenta anguste cylindrica ramulo foliato excluso usque
4 cm. longa; glandulae pseudodiscum plurilobulatum formantes. Folia
ovata, late elliptica, obovata v. rotunda, apice rotunda, ad 5.5 cm. longa, satis
crassa, subtus albescentia, distincte reticulata, integra . 117. *S. reticulata.*
†† Ovaria glabra, folia minora, tenuiora.
‡ Folia crenato-serrata v. dentata. Amenta fructifera 2–4.5 cm. longa (ramulo
foliato excluso).
Stylus mediocris v. subnullus, simplex. Folia utrinque viridia, obovato-
rotunda v. oblonga (confer etiam 115. *S. herbaceam*) . 114. *S. flabellaris.*
Stylus longus, gracillimus, bipartitus. Folia subtus pallida, obovato- v.
elliptico-lanceolata . . . . 123. *S. Serpyllum.*
‡‡ Folia integerrima. Amenta etiam fructifera vix usque 2 cm. longa.
Folia utrinque viridia, vix v. paullo longiora quam lata, apice rotunda v.
saepe emarginata, basi rotunda v. saepe cordata. Amenta parva,
3–6 flora. Bracteae obovatae . . . . . 116. *S. rotundifolia.*
Folia subtus glauca, 2–3plo longiora quam lata, apice acuta v. obtusa,
basi saepissime acuta. Amenta 6–15-flora.
Folia subtus reticulata. Ovaria sessilia, stylo brevissimo coronata;
bracteae atrae . . . . 111. *S. ovalifolia.*
Folia subtus laevia, haud reticulata. Ovaria subsessilia, stylo satis
distincto apice bifido coronata; bracteae viridescentes.
120. *S. Souliei.*

B. Flores glandula una ventrali instructi (glandula pedicellum ovarii interdum subamplectens).

a. *Ovaria plus minusve distincte pedicellati; pedicelli glandulam 2plo usque 6plo superantes.*[1] (b, see p. 86).

1. Ovaria et pedicelli glaberrima. Stigmata sessilia v. styli breves stigmatibus vix v. paulo longiores (tantum in 96. *S. pyrolaefolia* distincte longiores). (2, see p. 86).

Folia adulta lanceolata v. ovato-lanceolata, plus minusve acuminata, 3–6plo longiora quam lata (confer etiam 12. *S. ochrophyllam*).

* Folia anguste lanceolata, margine integra v. vix anguste serrata.

Stigmata bifida lineariloba; bracteae late obovatae, ovaria fere tota involucrantes; pedicelli glandula saepe vix duplo longiores . . 19. *S. macrolepis.*

Stigmata lati- et breviloba; bracteae breviores angustioresque; stylus brevis; pedicellus glandula 3–4plo longior . . . . 24. *S. dealbata.*

**Folia saepissime late lanceolata v. margine distincte serrata v. glanduloso-denticulata.

Glandula vix altiora quam lata, pedicellum basi subamplectens; pedicelli glandula 2–5plo longiores.

Folia plus minusve serrato-dentata. Styli plerumque brevissimi (confer etiam 12. *S. eucalyptoidem*).

Bracteae rhachisque amentorum fructiferorum sericeo-tomentellae.
4. *S. araeostachya.*

Bracteae rhachisque amentorum fructiferorum subglabrescentes (confer etiam 5. *S. Balansaei*) . . . . 1. *S. tetrasperma.*

Folia integra, vix denticulata. Stigmata sessilia; bracteae rhachisque sericeo-villosae (confer etiam 1. *S. tetraspermae* varietates) . . 2. *S. pyrina.*

Glandula ovalis v. subrectangularis, fere duplo altiora quam lata, basim pedicelli 2–3plo longioris non amplectens . . . . . . 25. *S. amygdalina.*

Folia adulta ovata, obovata, elliptica v. oblonga 1½–2½plo longiora quam lata, sive lanceolato-elliptica, apice haud acuminata (si amenta longissima confer etiam Sect. *Magnificae*, p. 113).

* Bracteae rhachisque amentorum pilis ferrugineis interdum cum albis mixtis sericeae; pedicelli 2–2½plo longiores quam glandula; styli breves sed distincti stigmatibus emarginatis paullo longiores (confer etiam 88. *S. Reinii*, stylis longioribus, pedicellis plerisque pilosis, glandulis longioribus differt).
92. *S. vulpina.*

**Bracteae et rhachis pilis tantum albis, v. griseis indutae (confer etiam 94. *S. japonicam* et 122. *S. calyculatam*).

Pedicelli glandulam 3–5plo superantes. Folia lanceolato-elliptica v. integra.

Folia lanceolato- v. oblongo-elliptica, 2½–3½plo longiora quam lata, plus minusve breviter glanduloso-denticulata. Fructus anguste ovato-elliptici, pedicello paullo longiores; stigmata ut videtur brevia, sessilia.
8. *S. Dunnii.*

Folia rotundato- usque angusto-elliptica, vix 2½plo longiora quam lata, integra. Fructus elongati, pedicello plerique duplo longiores (confer etiam 134. *S. fuscescentem*) . . . . 133. *S. myrtilloides.*

Pedicelli glandulam vix triplo superantes. Folia latiora, dentata.

Stylus distinctus, stigmatibus bifidis longior, interdum medium ovarii aequi-

---

[1] Si amenta longissima adsunt confer etiam 44. *S. moupinensem*, 45. *S. Fargesii* et 78. *S. omeiensem.*

longus; pedicellus glandula 2–4plo longior. Folia orbiculata v. elliptica, satis obtusa, irregulariter glanduloso-serrata . . . . 96. *S. pyrolaefolia.*
Stylus brevior; pedicellus glandula 2plo longior. Folia ovata v. ovato-lanceolata serrata, acuta, basi saepe cordata (confer etiam 97. *S. hastatam).*
99. *S. Shiraii.*

2. Ovaria (et v. interdum tantum pedicelli) pilosi (confer etiam sub b, 2).
Styli brevissimi v. breves, stigmatibus vix v. tantum paullo longiores.
* Glandula latior quam alta, basim pedicelli subamplectens. Folia ovato-lanceolata, acuminata, margine glanduloso-serrata.
Folia basi acuta v. obtusa. Amenta (pedunculo satis longo excluso) usque 7: 1 cm. magna . . . . 6. *S. ichnostachya.*
Folia basi aurita; amenta usque 3: 1 cm. magna, pedunculo brevi.
7. *S. tonkinensis.*
**Glandula pleraque altior quam lata, haud subamplectens.
† Folia lanceolata, glanduloso-serrata. Bracteae nigro-brunneae, oblongae, ovariis subaequilongae, dorso a medio ad apicem glabrae; stylus subnullus.
132. *S. insignis.*
†† Folia latiora v. integerrima (confer etiam 79. *S. phylicifoliam).*
‡ Folia late lanceolata, elliptica, obovata v. subrotunda; pedicelli glandula 3–6plo longiores. Amenta fructifera plus quam 2.5 cm. longa (pedunculo excluso).
Amenta fructifera 7–12 cm. longa, vix usque 1.5 cm. crassa. Folia ovata v. ovato-elliptica, majora apice saepissime satis longe acuminata v. caudata, subtus albescentia, juniora sericea, integra; stylus distinctus, apice bifidus. Ramuli novelli mox glabrescentes, valde tenues.
76. *S. heterochroma.*
Amenta fructifera crassiora v. breviora v. folia diversa (confer etiam 92. *S. vulpinam,* var. *tomentosam,* p. 131).
Folia pro parte maxima haud plus quam 5: 2.5 cm. magna (confer p. 81).
131. *S. Starkeana.*
Folia pro parte maxima 6 cm. longa v. majora; vide 128. *S. Capream,* 127. *S. cineream,* et 129. *S. Wallichianam* (confer etiam p. 81).
‡‡ Folia lanceolata, oblanceolata, spathuliformia v. obovato-lanceolata; pedicelli glandulam 2–4plo superantis. Amenta fructifera vix 2 cm. longa (pedunculis exclusis) v. vix 0.6 cm. crassa.
Stylus quam stigmata brevior; pedicelli glandula 2–3plo longiores; folia 2–7 cm. longa. Folia lanceolata, lineari-lanceolata v. anguste obovato-lanceolata, pubescentia, sericea . . . . 135. *S. sibirica.*
Stylus quam stigmata distincte longior; pedicelli glandula circiter 4plo longiores. Folia obovato-lingulata, vix 2.5 cm. longa.
87. *S. leptoclados.*
Styli elongati, quam stigmata distincte longiores, ovario interdum aequilongi.
Pedicelli tantum pilosi; ovaria glabra . . . . 183 S. *Leveilleana.*
Pedicelli ut ovaria pilosi . . . . 146. *S. Sziuzevii.*

b. *Ovaria sessilia v. pedicelli glandulam saepe longam non v. paulo (haud duplo) superantes.*

1. Ovaria et pedicelli glaberrima (2, see p. 89).

Amenta florifera fructiferaque longissima (8–27 cm.), apice ramulorum brevium folia magna 2–4 gerentium. Folia permagna, majora 9–20 cm. longa, 4–10 cm. lata. Styli breves v. mediocres, tenues, apice bifidi, stigmatibus brevibus.

* Folia integra v. subintegra apice rotundata, interdum apiculata, basi obtusa v. aurita, subtus glauca, nervis vix prominentibus. Stylus satis brevis, stigmatibus bifidis vix v. paullo longior; pedicelli glandulam interdum 3plo superantes (confer etiam 42. *S. pellam* et 43. *S. apatelam*) . . . . . 40. *S. magnifica.*

**Folia plus minusve distincte serrata, apice basique pleraque acuta, nunquam aurita. Ovaria subsessilia v. pedicelli glandulam ½–3plo superantes.

Ovaria apice sensim in stylum satis longum attenuata. Folia subtus pallida.

Folia distincte glanduloso-serrata . . . . 45. *S. Fargesii.*

Folia subintegra v. tenuissime serrulata . . . . 78. *S. omeiensis.*

Ovaria apice satis subito in stylum paullo breviorem contracta. Folia subtus pallide viridia, subnitentia . . . . 44. *S. moupinensis.*

Amenta breviora v. basi foliis parvis squamiformibusve instructa. Folia minora v. angustiora.

* Arbores v. frutices erecti, ramulis nunquam radicantibus.

† Ovaria (fructusque) distincte sessilia.

‡ Styli subnulli v. breves, stigmatibus v. ½ ovarii haud longiores.

‖ Bracteae plus minusve acutae; amenta brevia, vix 2 cm. longa (pedunculo 2–3-foliolato excluso). Folia angusta acuminata, serrata v. juniora subintegra.

Folia lanceolata vel anguste lanceolata, 5–17 cm. longa.

Ramuli brunnei v. purpurascentes. Folia longe acuminata, ad 17 cm. longa . . . . 34. *S. babylonica.*

Ramuli viridi-brunnescentes v. plus minusve flavescentes. Folia breviter acuminata, ad 10 cm. longa . . . . . 26. *S. Matsudana.*

Folia linearia, 1.5–4.5 cm. longa . . . . . . 165. *S. microstachya.*

‖‖ Bracteae late ovatae, obtusae, subrotundae v. oblongae. Folia apice acuta v. obtusa, margine integra v. obscure crenulata.

° Folia subrotunda usque elliptica v. lanceolato-elliptica, apice pleraque acuta, nervis secundariis subtus plus minusve visibilibus. Ramuli haud glaucescentes.

§ Folia late elliptica, obovata v. subrotunda, 1.8–2.5 cm. lata, obscure crenata. Bracteae oblongae, obtusae; styli distincti, apice bifidi, stigmatibus bifidis satis elongatis paullo longiores.

125. *S. opsimantha.*

§§ Folia angustiora, integra. Bracteae late ovatae v. subrotundae.

Amenta pedunculo distincto foliis normalibus instructo suffulta.

Folia, amenta et bracteae glabra v. subglabra. Styli nulli, stigmata sessilia. Petioli satis longi . . . 66. *S. hypoleuca.*

Folia, amenta et bracteae plus minusve sericeo-villosa.

Petioli 1–3 mm. longi, gemmis elongatis breviores; styli brevissimi . . . . 72. *S. macroblasta.*

Petioli ad 5 mm. longi, gemmis plus minusve longiores; styli breves . . . . 70. *S. longiflora.*

71. *S. cathayana.*

Amenta subsessilia, basi nuda v. foliolis perparvis suffulta.

Styli nulli; stigmata sessilia. Folia utrinque plus minusve acuta. Amenta valde cylindrica, tenuia . . . . . 67. *S. rhoophila.*

Styli plus minusve distincti et bifidi .

Ovaria crasse ovata, styli brevissimi; bracteae tantum basi extus pilosae. Folia utrinque subrotunda.

69. *S. mictotricha.*

Ovaria elongata, styli satis distincti; bracteae extus et apice sericeae.

Amenta valde cylindrica, ad 3.5 cm. longa, circiter 0.5 cm.
crassa. Ramuli annotini glabri, hornotini albo-tomentelli.
68. *S. polyclona*.
Amenta circiter 7 mm. crassa, vix 3 cm. longa. Ramuli
annotini puberuli, hornotini flavo-tomentelli.
64. *S. luctuosa*.
°° Folia lineari-lanceolata, apice obtusa, fere enervosa. Ramuli plerique
glaucescentes. Stylus distinctus, brevis, haud bifidus, stigmatibus
brevibus, bifidis . . . . 171. *S. pycnostachya*.
‡‡ Styli elongati, stigmatibus et ½ ovarii longiores.
Amenta magna, crassa (ad 8:2 cm.), densissime sericea, leviter flavescentia; styli gracillimi, stigmatibus elongatis haud bifidis. Folia utrinque
sericeo-tomentella, integra, late ovata v. elliptica . . 139. *S. lanata*.
Amenta minora, haud dense sericea. Folia diversa.
Folia late elliptica v. rotundata, ad 4.5 cm. longa; petioli ad 1.5 cm.
longi. Stigmata bifida, anguste linearia . . 101. *S. Nakamurana*.
Folia lanceolata v. lanceolato-elliptica, apice acuta, ad 11.5 cm. longa;
petioli 2-8 mm. longi. Stigmata brevia . . . . 147. *S. Rehderiana*.
†† Ovaria (fructusque) breviter v. distincte pedicellata (confer etiam 70. *S. longiflo-
ram* et 71. *S. cathayanam*).
‡ Styli subnulli v. breves, stigmatibus haud longiores v. stigmata valde brevia
(confer etiam 132. *S. etosiam*).
‖ Folia integerrima v. anguste lanceolata et interdum ad apicem distanter
denticulata.
Folia anguste lanceolata, ad 4.5 cm. longa, brevipetiolata. Stylus
quam stigmata bifida linearia satis longa subbrevior.
19. *S. macrolepis*.
Folia elliptica v. obovata, ad 4.5:2.3 cm. magna; petioli ad 1.5 cm.
longi; stylus stigmatibus linearibus integris sublongior.
101. *S. Nakamurana*.
‖‖ Folia distincte serrata v. denticulata v. haud anguste lanceolata.
° Petioli glanduliferi; folia ovato- v. elliptico-lanceolata, breviter v. longius acuminata (v. interdum obovato-lanceolata, acuta). Pedicelli
breves. . . . . 16. *S. paraplesia*.
°° Petioli eglandulosi.
§ Folia late v. anguste lanceolata, apice acuminata.
Folia anguste lanceolata, apice sensim acuminata, margine satis
anguste glanduloso-serrata, etiam adulta subtus saepe sericea.
Amenta fructifera (pedunculo excluso) circiter 3 cm. longa; stylus
subnullus . . . . .35. *S. alba*.
Folia late v. anguste lanceolata, apice subintegra subito caudato-
acuminata, margine satis distanter grossius serrata. Amenta
fructifera (pedunculo excluso) ad 10 cm. longa; stylus distinctus
(confer etiam 95. *S. Fauriei*) . . . . .94. *S. japonica*.
§§ Folia latiora, apice acuta v. obtusa (confer etiam 98. *S. kenoensem*).
Bracteae (praesertim apice) longe crispo-sericeae; stylus pedicello
et stigmatibusque distincte longior (confer etiam 86. *S. chlorostachyam*) . . . . 97 *S. hastata*.
Bracteae subglabrae, apice haud crispo-sericeae; stylus pedicello
stigmatibus subaequilongus.
Folia elliptica, utrinque acuta v. subacuta, ad 5:2 cm. magna,
margine breviter denticulato-serrata . . . 58. *S. denticulata*.
Folia variabilia, elliptico-lanceolata v. -obovata, basi acuta v. ro-

tundata, apice pleraque acuta v. subacuminata, margine saepissime satis grosse serrata v. crenato-serrata.

              94. *S. japonica.*

‡‡ Styli elongati v. stigmata satis longa et linearia.

Bracteae basi saepe glanduloso-denticulatae, longe sericeae; amenta crassa, ad 6 : 1.5 cm. magna, sessilia. Folia lanceolata, longa, anguste serrata, glabra . . . . 138. *S. rorida.*

Bracteae integrae.

Amenta sessilia, crassa . . . . 137. *S. daphnoides,* var. *indica.*

Amenta pedunculata, tenuia. . . . . 57. *S. radinostachya.*

**Frutices parvi, saepissime prostrati ramulis radicantibus. Folia parva v. minima (confer etiam 122. *S. calyculatam*).

 † Folia integerrima (confer etiam 116. *S. rotundifoliam*) v. minutissime sparse denticulata.

Folia obovalia v. late oblonga, integra, utrinque viridia, nervis secundariis prominentibus. Amenta (sine pedunculo) cylindrica, ad 2 cm. longa; stylus simplex, quam stigmata bifida oblonga longior; pedicellus distinctus, glandulae aequilongus . . . . 112. *S. phlebophylla.*

Folia elliptica, utrinque acuta, integra v. minutissime denticulata, subtus glauca, costa exclusa enervia v. nervis indistinctis. Amenta brevia, ovato-rotunda (si nervi subtus sunt distincti et folia ad 2 cm. longa, confer etiam 120. *S. Souliei*) . . . . 118. *S. Lindleyana.*

‡‡ Folia serrata v. inciso-dentata.

Amenta cylindrica (fructifera ad 4 cm. longa).

Folia circumcirca crenato-serrata v. irregulariter acuto-serrata, utrinque viridia, nitentia, reticulata. Styli breves, apice bifidi, stigmatibus brevibus bifidis.

Ovaria sessilia. Folia crenato-serrata. . . . . . 110. *S. microphyta.*

Ovaria pedicellis glandulam vix usque duplo superantibus suffulta.

Folia irregulariter acuto-serrata . . . . 113. *S. berberifolia.*

Folia crenato-denticulata . . . . 115. *S. herbacea.*

Folia a medio ad apicem tantum inciso-dentata, oblanceolata, subtus glauca, vix reticulata. Styli distincti, bipartiti; stigmata bifida.

           123. *S. Serpyllum.*

Amenta ovato-capitata, pauciflora. Folia saepe tantum versus apicem inciso-adpresso-serrata v. lobulata, subtus pallidiora, vix nervosa. Ovaria subsessilia; styli perbreves, stigmatibus brevibus integris.

           121. *S. oreophila.*

  2. Ovaria, rarissime tantum pedicelli v. basis ovarii, pilosa.

a Ovaria distincte sessilia v. interdum pedicelli brevissimi glandula 3–5plo breviores (confer etiam sub *β*). (*β,* see p. 92).

Stigmata sessilia; stylus deficiens v. brevissimus.

 * Folia plus minusve opposita v. alterna, basi subcordata, sessilia, interdum petiolata, usque 6 : 2.5 cm. magna v. longiora, praesertim versus apicem denticulato-serrata . . . . 161. *S. purpurea* ssp. *amplexicaulis.*

**Folia omnia alterna, plus minusve petiolata (si folia magna, late ovato-acuminata confer 23. *S. populifoliam*).

 † Bracteae concolores, viridescentes (v. siccae flavo-brunneae).

Bracteae ovato-ellipticae, apice truncatae, emarginatae v. erosae; stigmata elongata; amenta (pedunculo excluso) 2.5–4 cm. longa . 31. *S. jessoensis.*

Bracteae ovatae, acutae (v. obtusae?); stigmata obovata; amenta (pedunculo excluso) vix usque 2.5 cm. longa . . . . 32. *S. lasiogyne.*

†† Bracteae bicolores, apice v. pro parte maxima nigro-brunneae, basi viridescentes v. in 168. *S. variegata* omnino flavo-brunneae.

‡ Folia adulta pleraque 6–12 cm. longa.

Amenta satis brevia, pedunculo brevissimo excluso vix 2.5 cm. longa. Folia ad 11: 1.5 cm. magna, anguste v. lineari-lanceolata, acute serrata.
161 *S. purpurea.*

Amenta 3.5–8 cm. longa, saepe curvata.

Folia ad 16: 2.5 cm. magna, plus minusve crenulato-serrata.
158. *S. Miyabeana.*

Folia ad 6: 1 cm. magna, subintegra . . . . 172. *S. oxycarpa.*

‡‡ Folia adulta breviora v. tantum in ramulis robustis ad 6 cm. longa.

Amenta etiam fructifera vix 1 cm. longa, subsessilia; bracteae parte superiore atrae, valde obtusae; stigmata subsessilia brevia, integra. Folia lineari-lanceolata, minutissime denticulata, subacuminata.
163. *S. cheilophila.*

Amenta 1.5–3 cm. longa; bracteae tantum imo apice atrae v. flavo-brunneae; stigmata subsessilia, bifida, separata. Folia obtusa v. breviter acuta.

Amenta aestate ad ramulos hornotinos crescentes apparentia (confer etiam 167. *S. Duclouxii*) . . . . 166. *S. Bockii.*

Amenta autumno v. hieme ad ramulos maturos (annotinos) saepe efoliosos apparentia (confer etiam 169. *S. andropogon* et 170. *S. kouytchensem*) . . . . 168. *S. variegata.*

Stigmata stylo brevi v. longiore suffulta.

* Stylus brevis stigmatibus haud v. paullo longior (136. *S. dolia* excepta).

† Amenta satis praecocia, sessilia v. breviter pedunculata, basi tantum foliis parvis v. squamiformibus instructa (confer etiam 52. *S. Daltonianam*).

‡ Folia juvenilia et adulta subtus (v. utrinque) dense sericea v. sericeo-tomentella, lanceolata, integra v. subintegra (si folia cinereo-tomentella, oblonga ellipticave confer etiam 141. *S. vulpinoidem*).

Amenta fructifera ad 8 cm. longa; styli ad basim partiti. Folia subtus sericea, micantia . . . . 55. *S. psilostigma.*

Amenta fructifera circiter 3.5 cm. longa; styli integri, stigmatibus angustis adpressis apice bifidis aequilongi. Folia utrinque sericeo-tomentella . . . . 178. *S. polia.*

‡‡ Folia etiam juvenilia glabra v. tantum sparse sericea v. amenta vix 2 cm. longa.

Folia lineari-lanceolata, integra, semper glabra. Stylus apice tantum divisus, stigmatibus bifidis satis brevibus; amenta circiter 1.5 cm. longa, crassa . . . . 164. *S. Wilhelmsiana.*

Folia non lineari-lanceolata v. serrata.

Amenta gracilia, pedunculo brevi excluso 5–8 cm. longa, fructifera vix 0.9 cm. crassa; glandula angusta, dimidio ovarii aequilonga. Folia ovato-lanceolata, acuminata, tenuissime glanduloso-denticulata.

Folia subtus albescentia, vix reticulata . . . . . 155. *S. hylonoma.*

Folia subtus viridescentia, reticulata . . . . . . 77. *S. isochroma.*

Amenta breviora, ad 3 cm. longa, crassiora; glandula brevior, latior.

Folia longa, lanceolata, acuminata, serrulata. Amenta 1–3 cm. longa; stigmata oblonga, emarginata (confer etiam 30. *S. Makinoanam* et 29. *S. hondoensem*) . . . . 33. *S. koreensis.*

Folia (juvenilia) elliptica, integra. Amenta circiter 1.5 cm. longa;
stylus apice bifidus, stigmatibus bifidis brevibus (confer etiam 173.
S. myricaefoliam) . . . . 136. *S. dolia.*

†† Amenta tardiflora, coetanea, apice ramulorum foliiferorum (foliis satis magnis
ceteriis conformibus).

‡ Folia mediocra v. maxima, 4–20 cm. longa.

‖ Amenta fructifera 20 cm. longa et 1.5 cm. crassa; ovaria globosa, crispo-
tomentella; styli breves crassi; stigmata bifida. Folia maxima, glabra,
integra . . . . 41. *S. ulotricha.*

‖‖ Amenta breviora et saepissime tenuiora. Folia minora, saepe sericea et
plus minusve serrata.

° Pubescentia foliorum juvenilium ramulorumque alba v. grisea v. nulla
(in 51. *S. sikkimensi* interdum flavescens, sed non cuprea).

§ Folia lanceolata, acuminata, integra, ad 14: 2.5 cm. magna, subtus
dense sericea, micantia. Amenta 9–11 cm. longa, florifera vix 0.8
cm. crassa; glandula longa, linearis . . . . . 54. *S. phaidima.*

§§ Folia late v. ovato-elliptica, obovato-oblonga v. elliptica, acuta v.
obtusa.

‡ Amenta distincte pedunculata.

Amenta (pedunculo excluso) 5.5–14 cm. longa (confer etiam 126.
S. floccosam et 177. *S. allochroam*).

Folia distincte glanduloso-serrata, subtus tota facie molliter
pubescentia.[1] Amenta 9–14 cm. longa; styli stigmatibus
subaequilongi . . . . 47. *S. argyrophegga.*

Folia integra v. indistincte serrulata, subtus mox glabra v.
parce sericea. Amenta vix usque 10 cm. longa.

Petioli tantum 3–6 mm. longi. Styli stigmatibus aequilongi
v. subduplo longiores . . . . 74. *S. driophila.*

Petioli 7–12 mm. longi. Styli stigmatibus brevibus 3–4plo
longiores (confer etiam 49. *S. spathulifoliam*).
48. *S. plocotricha.*

Amenta (pedunculo excluso) 2.5–5 cm. longa (confer etiam 61. *S.
Biondianam* et 80. *S. formosam*).

Amenta (florifera) ad 3.5 cm. longa, vix 0.5 cm. crassa. Folia
late elliptica v. obovato-elliptica, utrinque obtusa v. subrotunda. . . . . 75. *S. amphibola.*

Amenta ad 5.5 cm. longa, crassiora. Folia utrinque plus
minusve acuta, obovato-oblonga . . . . 50. *S. eriostachya.*

‡‡ Amenta subsessilia, ad 4 cm. longa et circiter 1.5 cm. lata.
51. *S. sikkimensis.*

°° Pubescentia foliorum juvenilium ramulorumque cupreo-sericea pilis
griseis mixta; folia lanceolato-elliptica, ad 8: 2.8 cm. magna. Amenta
florifera ad 12 cm. longa, vix 0.6 cm. crassa . . .52. *S. Daltoniana.*

‡‡ Folia (visa) tantum ad 2.5 cm. longa. Bracteae apice truncatae; styli distincti, apice bifidi; stigmata bifida v. emarginata (si bracteae sunt apice
subacutae, confer etiam 179. *S. algistam*).

Frutices parvi v. erecti. Amenta 3–9 cm. longa.

Stigmata brevia, emarginata. Folia subtus glauca, integra, obtusa,
amentorum ad 2.5 : 1.3 cm. magna . . . . 73. *S. resecta.*

---

[1] Si folia integra confer etiam 46. *S. Ernesti*, p. 47, quia glandula dorsalis saepe
deest.

Stigmata bifida. Folia subtus cinerea, lanceolata, acuta, denticulata,
   ad 2.5 : 1 cm. magna . . . . 56. *S. Thomsoniana.*
   Frutex humilis, ramis prostratis radicantibus. Amenta brevia, 1–2 cm.
   longa; stigmata bifida. Folia subtus pallida ad 2.5 : 1.7 cm.
   magna . . . . 107. *S. oreinoma.*
** Stylus elongatus (⅓–)½ ovarii aequilongus v. longior; amenta precocia (si stylus est valde partitus confer etiam supra sub "Amenta tardiflora, coetanea").
   † Bracteae ovato-obtusae v. obovatae, apice rotundatae.
      Styli conspicue furcati; stigmata bifida. Folia subintegra, obovato-elliptica,
      utrinque plus minusve obtusa; amenta subcoetanea; frutex parva.
                                                        126. *S. floccosa.*
      Styli vix apice bifidi; stigmata incisa. Folia lanceolata, utrinque acuta v.
      apice acuminata. Arbores v. frutices alti (confer etiam 147. *S. Rehderia-*
      nam et 149. *S. obscuram*).
      Folia subtus sericea v. albescentia.
         Stylus gracillimus; glandula cylindrica, angusta. Folia subtus dense
         sericea micantia, integra . . . . 142. *S. viminalis.*
         Stylus crassiusculus; glandula brevior, latior. Folia adulta subtus
         glabra, albescentia, serrata . . . . 157. *S. Pierotii.*
      Folia subtus viridescentia . . . . 143. *S. sachalinensis.*
   †† Bracteae ovatae v. lanceolatae, acutae (confer 140. *S. taimyrensem*).
      Bracteae acutissimae; glandula valde elongata, anguste linearis, dimidio
      ovarii sublongiora. Folia lanceolata v. obovato-lanceolata, tenuiter glan-
      duloso-serrata, subtus cinerea . . . . 156. *S. gracilistyla.*
      Bracteae acutae; glandula brevior, latior.
         Folia integra, apice obtusa. Ovaria brevissime pedicellata.
                                                        174. *S. myrtillacea.*
         Folia serrata v. apice subacuminata. Ovaria sessilia.
            Folia distincte serrata (confer etiam 82. *S. kolymensem*).
                                                        83. *S. boganidensis.*
            Folia plus minusve integra . . 147. *S. Rehderiana*, var. *brevisericea.*

β. Ovaria pedicellis brevibus v. longioribus stipitata (pedicelli glandula saepe satis
      longa breviores v. vix fere duplo longiores).

Stylus ½ ovarii aequilongus v. longior; glandula linearis, elongata, pedicello sub-
   aequilonga v. longior; stigmata brevia v. integra. Folia lanceolata v. late
   oblanceolata, utrinque acuta, subintegra (confer etiam 148. *S. Blinii*).
                                                        143. *S. sachalinensis.*
Stylus brevior stigmatibus vix v. usque duplo longior v. glandula latior et brevior.
   * Amenta 2.5–8 cm. (in 177. *S. allochroa* 9–11 cm.) longa (valde juvenilia interdum
      breviora).
   † Folia anguste lanceolata. Pedicelli glandula distincte v. subbreviores (confer
      etiam 79. *S. phylicifoliam*, 82. *S. kolymensem* et 135. *S. sibiricam*).
      Folia integra. Bracteae concolores . . . . 36. *S. sericocarpa.*
      Folia serrata. Bracteae apice nigrescentes . . . . . 162. *S. Gilgiana.*
   †† Folia late lanceolata, oblongo-lanceolata, elliptica, ovata v. obovata.
      ‡ Folia plus minusve distincte serrata, crenata v. denticulata (confer etiam
         79. *S. phylicifoliam*).
      ‖ Folia late lanceolata, basi acuta, apice acuminata, serrata, ad 12 : 2.5 cm.
         magna. Amenta praecocia (confer etiam 145. *S. opacam*).
                                                        160. *S. lepidostachys.*
      ‖‖ Folia diversa. Amenta saepe coetanea.

Folia subtus cinerea v. glauca (confer etiam 145. *S. opacam*).
Ovaria (v. fructus) parce pilosa v. glabra et tantum pedicelli breves
    pilosi.
Folia obovalia v. obovali-elliptica, apice breviter producta, satis
    grosse crenato-dentata, mox glabra (confer etiam 93. *S. Matsu-*
    muraei) . . . . 88. *S. Reinii.*
Folia lanceolata, apice saepe breviter acuminata breviter et satis
    argute glanduloso-serrata . . . . 81. *S. characta.*
Ovaria distincte pilosa v. sericeo-villosa (confer etiam 151. *S. Buergeri-*
    anam, 153. *S. Harmsianam* speciesque iis valde affines).
                           150. *S. Sieboldiana.*
Folia utrinque concoloria, viridia, subnitida, ovata v. obovata v. plus
    minusve lanceolata. Amenta coetana v. fructifera apice ramulorum
    brevium foliiferorum; pedicelli glandula breviores v. aequilongi.
                           108. *S. Myrsinites.*
‡‡ Folia integra (v. in 79. *S. phylicifolia* pro parte serrata).
 ‖ Folia 1–3plo longiora quam lata, vix lanceolata.
 ° Folia 2–3plo longiora quam lata, utrinque pleraque plus minusve acuta
    (confer etiam 91. *S. amnicolam* et 180. *S. ampheristam*).
Folia glaberrima v. mox glabra, interdum serrata. 79. *S. phylicifolia.*
Folia subtus etiam adulta sericea v. villosa . . . . 124. *S. glauca.*
8 Folia vix 2plo longiora quam lata v. apice obtusa (confer etiam 104. *S.*
    anglorum, 106. *S. altaicam* et 177. *S. allochroam*) . . 103. *S. arctica.*
‖‖ Folia 3½–5plo longiora quam lata, lanceolata v. oblanceolata. Amenta
    sessilia, cylindrica, circiter 7 cm. longa . . . . . . 181. *S. melea.*
**Amenta etiam fructifera vix 2.5 cm. longa v. minora (confer etiam 135. *S. sibi-*
    ricam). Folia parva, elliptica v. oblonga, integra, subtus pallida v. glauca.
† Frutices erecti, ramis haud prostratis radicantibusve.
‡ Folia apice acuta, v. satis lata.
Folia 0.8–3 cm. longa. Amenta parva; stigmata brevia; pedicelli glan-
    dula plerumque breviores . . . . 175. *S. minutiflora.*
Folia 3–7 cm. longa. Amenta majora; stigmata longiora; pedicelli glan-
    dulam aequantes v. duplo superantes . . . . . . 176. *S. Kochiana.*
‡‡ Folia apice obtusa, interdum apiculata (confer etiam 65. *S. fruticulosam*).
                        173. *S. myricaefolia.*
†† Frutex humilis prostratus, ramis radicantibus . . . . . . 109. *S. polaris.*

ENUMERATIO SECTIONUM SPECIERUMQUE.
ASIAE ORIENTALIS ET HIMALAYAE.[1]

Sect. 1. TETRASPERMAE Andersson in *Svensk. Vetensk. Akad. Handl.* VI. 1
(*Monog. Salic.*) (1867); in De Candolle, *Prodr.* XVI. pt. 2, 192 (pro parte maxima)
(1868).

    Arbores v frutices alti. Folia pleraque magna, late lanceolata, saepissime acu-
minata, adulta satis crassa. Amenta ♂ brevi-pedunculata, satis longa, cylindrica,
saepe laxiflora, floribus pleiandris, glandulis duobus saepe pseudodiscum plurilobatum formantibus; amenta ♀ longius pedunculata, saepe satis elongata, plus
minusve laxiflora; ovaria distincte pedicellata, glabra v. pilosa; styli nulli v. bre-

---

[1] In our area I include eastern Afghanistan and that part of Tibet which polit-
ically belongs to Kashmir, and Siberia east of a line running from the eastern
shores of Lake Baikal along the Lena River to the Arctic Sea.

vissimi; stigmata satis brevia, crassa, emarginata v. bifida; glandula una, ventralis,[1] basin pedicelli plus minusve subamplectens.

1. **Salix tetrasperma** Roxburgh, *Pl. Corom.* I. 66, t. 97 (1795); *Fl. Ind. ed.* 2, III. 753 (1832). — Forbes, *Salicet. Woburn.* 61, t. 61 (pro parte) (1829). — Andersson in *Svensk. Vetensk. Akad. Handl.*, 1850, 484 (*Ost-Ind. Pilarter*) (1851); in *Jour. Linn. Soc.* IV. 41 (pro parte) (1860); in *Svensk. Vetensk. Akad. Handl.* VI. 1 (*Monog. Salic.*) (pro parte) (1867); in De Candolle, *Prodr.* XVI. pt. 2, 192 (pro parte) (1868). — Brandis, *Forest Fl. Brit. Ind.* 462, t. 58 (pro parte) (1874); *Ind. Trees*, 636 (pro parte) (1906). — Hooker f., *Fl. Brit. Ind.* V. 626 (pro parte) (1888). — Cooke, *Fl. Pres. Bombay*, II. pt. 4, 661 (1907). — Talbot, *Forest Fl. Bombay*, II. 557, cum icone mala (1911).

INDIA. Madras Presidency: "banks of rivulets and moist places among the Circar Mountains," W. Roxburgh (type, ex Roxburgh). East Bengal: without locality (Herb. Griffith No. 4503); Mungut River, Jainteas, alt. 900 m., October 20, 1867, *C. B. Clarke* (No. 5911ᵃ, ♀; 5911ᵇ, ♀); "Bengala inferior," 1815, *N. Wallich* (No. 3707; sterile). Assam: Khasia, Nurtung(?), alt. 1200 m., December 1871, *C. B. Clarke* (No. 14580; with ripe fruits). United Provinces: Gorakhpur Division, Domariaganj (Dumuniyagunj), (Wallich Cat. Nos. 3707ᵇ and 3707ᶜ; sterile). Doubtful are the two following specimens representing probably different forms: Eastern Bengal: Dacca, April 12, 1868, *C. B. Clarke* (No. 6780; sterile); Noatilee, alt. 700 m., November 12, 1883, *C. B. Clarke* (No. 34176ᵃ; with fruits, branchlets and petioles finely puberulent).

* Without having seen Roxburgh's type specimen it is difficult to describe the typical *S. tetrasperma*, a name which has been applied to very different forms by different authors. As far as I can judge from Roxburgh's description and plate the ♂ flowers have mostly more than 6 stamens, and the ♀ flowers have long-stalked ovaries with a more or less distinct style, although not so long as the capsule, as Roxburgh described it; the leaves are serrate, and the petioles are mostly less than 1 cm. long. The specimens mentioned above may represent the typical *S. tetrasperma*, which seems to be widespread in the subtropical and tropical parts of India and probably also in the Malayan peninsula, but does not, I believe, occur in the Himalaya, expecially not between eastern Nepal and Kashmir, nor in China.

Besides the following variety there is mentioned a var. *β pubescens* Lindley sub No. 3707ᵈ of Wallich's Catalogue. By the kindness of the Keeper of the Herbarium of the Royal Garden, Kew, Dr. O. Stapf, to whom I take this occasion to express my very best thanks for his kind help, the Arnold Arboretum has received a photograph of Wallich's No. 3707, showing very poor sterile specimens which were referred to our species by Andersson in 1851.

Salix tetrasperma, var. **suaveolens** Andersson in *Jour. Linn. Soc.* IV. 41 (1860); in *Svensk. Vetensk. Akad. Handl.* VI. 2 (*Monog. Salic.*) (1867); in De Candolle, *Prodr.* XVI. pt. 2, 193 (1868). — Hooker f., *Fl. Brit. Ind.* V. 627 (1888).

Salix suaveolens Andersson in *Svensk. Vetensk. Akad. Handl.* 1850, 491 (1851).

---

[1] Different authors have given different names to these glands. A. Kerner in *Verh. Zool.-Bot. Ges. Wien.* X. 37 (*Niederöstr. Weid.*) (1860) uses the word "innere (interna)" for the gland between the flower and the rachis of the catkin. By von Seemen and other authors this gland, which is always developed, is named "postica." The gland between the flower and the bract, which is present only in some groups of species, has been named "externa" or "antica," but Lundström and other authors use the terms antica and postica in the opposite sense. I propose the names ventralis for the gland between rachis and flower, and dorsalis for the gland between flower and bract.

INDIA. Rajputana: near Adjmir (Ajmere), March 9–13, 1832, *V. Jacquemont* (No. 96, type, ♂, ex Andersson).

According to the description this variety is a quite glabrous tree, only the rhachis of the aments, the bracts and filaments being densely hairy. Not having seen the type or Huegel's No. 526 from the northwestern Himalaya, cited by Andersson in 1860, I do not know whether it is a good variety or only a form of *S. tetrasperma*. Huegel's specimen may belong to a different variety or to another species from a different geographical region.

Wight (*Icon. Pl. Ind. Or.* VI. 6, t. 1954 [1853]) figures a form of *S. tetrasperma* with the bract of the ♂ and ♀ flowers slightly dentate at the apex. The same character is found on plate 302, fig. 13–14 in Beddome's *Fl. Sylv. S. Ind.* VI. 302 (1874). The bracts of the ♂ flowers, fig. 11–12 of this plate, are entire. Beddome's figures 1–10 represent *S. ichnostachya* Lindley, and only fig. 13–17 belong to *S. tetrasperma*.

2. **Salix pyrina** Wallich apud Andersson in *Svensk. Vetensk. Akad. Handl.* 1850, 486 (1851); VI. 4 (*Monog. Salic.*) (1867); in De Candolle, *Prodr.* XVI. pt. 2, 192 (1868).

? *Salix disperma* D. Don, *Prodr. Fl. Nepal.* 58 (1825).

Salix tetrasperma, var. *pyrina* Andersson in *Jour. Linn. Soc.* IV. 41 (1860). — Hooker f., *Fl. Brit. Ind.* V. 627 (pro parte) (1888). — Brandis, *Ind. Trees*, 636 (1906).

INDIA. Nepal: without precise locality, 1821, *N. Wallich* (No. 3705, type of *S. pyrina*, ♂); without precise locality, *G. S. Perrotet* (apparently ♂ co-type, ex Andersson); without precise locality, *F. Hamilton* (type of *S. disperma*, ex Don).

This species differs from the typical *S. tetrasperma* Roxburgh in its more tomentose young branchlets, leaves and petioles which are glabrous on the fruiting branchlets of *S. tetrasperma*, and in its entire or nearly entire leaves, which seem to be grayish and not very glaucous beneath. In the shape of the bracts and of the capsules there seems to be no real difference. The ♂ plant is unknown to me, as I have seen only Wallich's type by the kindness of the Keeper of the Herbarium of the Royal Gardens, Kew. The fruiting aments have short leafy peduncles and are from 6 to 8 cm. long; the leaves are up to 11 cm. long and to 3.3 cm. wide; the petioles are from 5 to 9 mm. in length.

According to the description I believe that *S. disperma* D. Don is the same species. Don's types were collected in Nepal by Hamilton and Wallich.

3. **Salix Mesnyi** Hance in *Jour. Bot.* XX. 38 (1882). — Burkill in *Jour. Linn. Soc.* XXVI. 531 (pro parte) (1899). — Dunn & Tutcher in *Kew Bull. Misc. Inform.* add. ser. X. 255 (*Fl. Kwangtung & Hongk.*) (1913).

CHINA. Kwangtung: "ad ripas limosas fl. Cantonensis," January 1870, *T. Sampson* (♂ type); Kwangsi: "juxta fl. Liang-fung," June 1879, *W. Mesny* (No. 16446, Herb. Hance; ♀ type).

By the kindness of Dr. Rendle, I have been able to see flowers and a photograph of both the types from the collection in the British Museum. The ♂ flowers have 6 stamens which are finely hairy at the base. There are two rectangular separate glands two-thirds shorter than the somewhat obovate obtuse bract which is nearly glabrous on the outer surface, short-villose within and ciliate on the margins. The fruit does not show any remnants of a style or stigmas. The oblong ovary is glabrous, on a pedicel of about one-third the length of the ovary and about twice longer than the gland. There is only one ventral gland which seems to be of the same shape as that in sect. *Tetraspermae*; the bracts appear to be deciduous. This species has large leaves. It seems to be absent from central China. I do not know if any of the specimens referred to it by Burkill belong to this species except the types.

4. **Salix araeostachya** Schneider, n. sp.

Salix tetrasperma Burkill in *Jour. Linn. Soc.* XXVI. 533 (pro parte, non Roxburgh) (1899).

Frutex 2.5 m. altus v. arbor usque ad 6.5 m. alta; ramuli hornotini plus minusve tomentelli, annotini glabri, purpureo-brunnei. Folia juvenilia (plantae floriferae) ovato- v. obovato-lanceolata, basi acuta v. obtusa, apice sensim v. pleraque subito breviter v. longius acuminata v. breviter caudata, margine breviter glanduloso-serrata, rarius subintegra, supra facie sparse, costa densius villosula, cito glabrescentia, subtus initio satis dense breviter sericea, mox glabriuscula, concoloria v. glaucescentia, adulta (plantae fructiferae) glaberrima, crasse papyracea, laete viridia, utrinque concoloria (v. glaucescentia, fide Henry), nitidula, supra distinctius quam subtus nervosa, tenuissime reticulata, nervis lateralibus angulo 45–70° a costa divergentibus satis parallelis circiter (2–)3 pro 1 cm., distincte v. obscure crenato-serrata, minora circiter 6 cm. longa et 2.5 cm. lata, majora usque ad 12 cm. longa et ad 3.5–4 cm. lata; petioli initio tomentelli, mox glabri, purpurascentes, 1–1.5 cm. longi, supra sulcati, sulco saepissime puberulo, eglandulosi; stipulae satis parvae, semicordatae, puberulae, vix glanduloso-denticulatae, caducae. Amenta coetanea, florifera pedunculis 1–2 cm. longis foliolis 2–3 parvis instructis suffulta; ♂ cylindrica, laxiflora, 4–8 cm. longa, circiter 0.8 cm. crassa, breviter villosa; flores subverticillati; stamina 5(–6), bracteis aequilonga v. duplo longiora, basi pilosa, antheris ovato-rotundis flavis; bracteae late ovato-rotundae, viridi-flavae, intus margineque villosae, extus plus minusve glabrae; glandulae duae, similes, obovatae, apice incisae v. interdum pluripartitae, forma irregularia. Amenta ♀ cylindrica, tenuia, laxiflora, rhachi villosa, 5–8 cm. longa, circiter 0.8 cm. crassa; flores glabri pedicellati; ovaria basi ovata, apicem versus attenuata pedicello paullo v. ⅓ longiora; pedicelli glandulam unam latam apice obtusam v. rotundam v. incisam 3–4-plo superantes; stylus subnullus; stigmata brevia subbifida, obovata; bracteae pedicello sub- v. aequilongae, interdum breviores, late obovatae v. iis florum ♂ simillimae, sed saepissime utrinque villosae, nervosae. Amenta fructifera longius pedunculata (ad 3.5 cm.), foliis majoribus iis ramulorum sterilium similibus, ad 10 cm. longa et 1.7 cm. crassa, minus pilosa; fructus 7–9 mm. longi pedicellis inclusis glandula marcida subamplectente circiter 4–5-plo et bracteis remanentibus interdum duplo longioribus.

CHINA. Yunnan: Mengtsze, alt. 1800 m., *A. Henry* (No. 9338, type ♂; tree 6.5 m. tall); same locality, alt. 1600 m., near water, *A. Henry* (No. 9338ᶜ, ♀ co-type, 9338ᵈ; tree 1.5–3.5 m. tall; with fruits); same locality, southeastern mountain forests, alt. 1800 m., *A. Henry* (No. 11250; tree 3 m. tall; ♀).

This species certainly is very nearly related to *S. tetrasperma* Roxburgh, but the ♂ flowers of that species have 8–9 stamens. As long as the Indian forms are not clearly limited, it seems better to keep those from China distinct.

The ♂ flowers of *S. araeostachya* Schneider do scarcely differ from those of *S. Kusanoi* Schneider, but the leaves of the Formosan plant have a different pubescence, shape, serration and reticulation.

The specific name is derived from ἀραιός, slender, and στάχυς, spike.

5. **Salix Balansaei** Seemen in *Bot. Jahrb.* XXIII. Beibl. LVII. 44 (1897).

INDO-CHINA. Tongking: Hanoi, near the lake, December 1890, *B. Balansa* (No. 4999, type, ♀; tree 4–5 m., ex von Seemen); same locality, August 1891, *B. Balansa* (No. 4753; sterile, ex von Seemen).

According to the description this species seems most nearly related to *S. tetra-*

sperma Roxburgh as understood by me, as that species may be planted through-out the tropical regions of that part of the world. Unfortunately I have not seen any of the above-mentioned specimens. The ♂ plant is still unknown.

6. **Salix ichnostachya** Lindley apud Andersson in *Svensk. Vetensk. Akad. Handl.* 1850, 488 (1851). — Wight, *Icon. Pl. Ind. Or.* VI. 6, t. 1953 (1853). — Hooker f., *Fl. Brit. Ind.* V. 628 (1888). — Talbot, *Trees Bombay*, ed. 2, 336 (1902); *Forest Fl. Bombay*, II. 539 (1911). — Cooke, *Fl. Pres. Bombay*, II. pt. 4, 662 (1907).

    Salix tetrasperma, var. *ichnostachya* Andersson in *Jour. Linn. Soc.* IV. 41 (1860); in *Svensk. Vetensk. Akad. Handl.* VI. 3 (*Monog. Salic.*) (1867); in De Candolle, *Prodr.* XVI. pt. 2, 193 (1868).

    Salix tetrasperma Beddome, *Fl. Sylv. S. Ind.* II. 302, t. 302 (exclud. fig. 11–17) (non Roxburgh) (1874).

INDIA. Pondicherry: near Karikal, 1836, *G. S. Perrotet* (type, ex Andersson); Mysore: Shevary Hills (ex Wight); Madras Presidency: Madras, near Salem (ex Wight; probably the same as Wallich, Cat. No. 3702, ex Herb. Wight, of which I have seen a photograph); Assam: Khasi, "reg. trop. 2–4000 ped.," *J. D. Hooker & T. Thomson* (♂ and ♀ specimens ex Herb. Ind. Or., distributed as *S. tetrasperma*).

The plant from Khasia which I take for this species agrees very well with Wight's description and plate. The ovaries are hairy and the pedicels shorter than the bracts. There are only 3 or 4 stamens in the ♂ flowers, while Wight's plate shows 5 stamens and Andersson says: "stamina octo."

There may occur in India different species of *Salix* of this section with pubescent ovaries, and a critical study could only be made with a large amount of material. The tree Willows in the tropical and subtropical parts of India and eastern Asia seem to be widely spread by planting, and it is therefore difficult to fix the limits of their geographical distribution.

7. **Salix tonkinensis** Seemen, in *Bot. Jahrb.* XXI. Beibl. LIII. 53 (1896).

INDO–CHINA. Tongking: near Tu Fap, in swamps, December 1887, *A. Balansa* (No. 3787; tree 2–5 m. tall, ex von Seemen); Notre Dame (Black River), behind the rock, *A. Balansa* (No. 3788, ex von Seemen).

In the grayish brown pubescence of the twigs and leaves this species seems to resemble *S. Kusanoi* Schneider, but the capsules are loosely hairy and there is only one gland as in *S. tetrasperma* Roxburgh. For other characters see the keys on p. 74 and p. 86.

8. **Salix Dunnii** Schneider, n. sp.

    ? *Salix tetrasperma* Dunn & Tutcher in *Kew Bull. Misc. Inform.* add. ser. X. 255 (*Fl. Kwangtung & Hongk.*) (non Roxburgh) (1912).

Frutex ?; ramuli hornotini sparse villosuli, ut annotini purpurascentes, vetustiores glabri. Folia pro sectione parva, elliptico-lanceolata v. oblanceolata, utrinque acuta v. subobtusa, apice saepe apiculata, supra viridia, glabra v. basim versus ad costam parce puberula, subtus cinerea, sparse pilosa v. subglabra, nervibus secundariis angulo satis acuto a costa divergentibus parallelis 3–4 pro 1 cm., costa nervisque utrinque prominulis reticulata, margine distanter breviter glanduloso-serrata v. basim versus integra, minora circiter 2 cm. longa et 0.8 cm. lata, maxima usque ad 5 cm. longa et ad 1.5 cm. lata; petioli 3–5 mm. longi, supra in sulco pilosi. Amenta (tantum fructifera vidi) apice ramulorum, circiter 2 cm. longorum, foliiferorum (foliis circiter 4–5 ceteris simillimis) ad 5 cm. longa et 1 cm. crassa, subdensiflora, rhachi laxe villosa. Fructus glaberrimi, longe pedicellati; pedicelli capsulis angustis 3–4 mm. longis subbreviores, glandulam unam ventralem latam subamplectentem 4–5plo superantes; stigmata parva, subsessilia; bracteae late ovatae, subacutae, laxe villosulae, pedicello vix aequilongae v. deciduae.

CHINA. Fokien: without locality, Dunn's Exped. to central Fokien, April to June 1905 (Hongk. Herb. No. 3509; with fruits).

This specimen is named *S. tetrasperma* Roxburgh in the Hongkong Herbarium. It seems to belong to this section, but differs widely from all the other species in its small leaves. Without having seen young ♀ and also ♂ flowers the real relationship of the new species cannot be established. *S. tetrasperma* Dunn & Tutcher from Kwangtung, "muddy banks of the Canton River; Shekmun near Canton," may be our new species.

Sect. 2. PENTANDRAE Dumortier, in *Bijdr. Natuurk. Wetensch.* I. 58 (*Verh. Geslacht Wilgen*, 17) (1825). — Borrer in Hooker, *Brit. Fl.* 416 (1830); in Loudon, *Arb. Frut. Brit.* III. 1503 (1838). — Andersson in *Svensk. Vetensk. Akad. Handl.* VI. 30 (*Monog. Salic.*) (1867); in De Candolle, *Prodr.* XVI. 2 (1868). — Seemen, *Salic. Jap.* 15 (1903).

> *Salix*, sect. *Lucidae* Andersson in *Svensk. Vetensk. Akad. Handl.* VI. 30 (*Monog. Salic.*) (1867). — Seemen in Ascherson & Graebner, *Syn. Mitteleur. Fl.* IV. 56 (1908).

Arbores v. frutices alti. Folia pleraque magna, lanceolata v. ovalia, saepissime acuminata, iis sectionis 1 similia, juvenilia saepe viscosa; petioli saepe glanduliferi. Amenta ♂ subpraecocia v. coetanea, satis densiflora, floribus pleiandris,[1] glandulis duobus ut in sect. 1, bracteis concoloribus; amenta ♀ floresque ut in sect. 1; glandulae duae, distinctae.[1]

9. **Salix dictyoneura** Seemen in *Bot. Jahrb.* XXIX. 275, t. 6, II. fig. A–D (1900).

CHINA. Southeastern Szech'uan: Nan-ch'uan, *A. von Rosthorn* (No. 1511, type; with fruits).

I have seen only a photograph and one leaf of the type specimen of this species. According to the author *S. dictyoneura* belongs to sect. *Safsaf;* he describes the glands as follows: " glandula brevi stipitem 6-plo longiorem amplectente bipartita parte postica lata integra parte antica tripartita." To me it seems always very difficult to form an opinion from withered glands of fruiting specimens. It would be a very peculiar fact if an African species like *Salix safsaf* Forskal had a relative in central China. I suppose that *S. dictyoneura* stands very near *S. Wilsonii* Seemen.

10. **Salix Wilsonii** Seemen. See p. 40.

11. **Salix glandulosa** Seemen in *Bot. Jahrb.* XXI. Beibl. LIII. p. 55 (1896); XXIX. 276 (1900); *Salic. Jap.* 22, t. 1, fig. A–F (1903). — Léveillé in *Bull. Acad. Intern. Géogr. Bot.* XIV. 208 (1904). — Pavolini in *Nuov. Giorn. Bot. Ital.* n. ser. XV. 439 (1908). — Nakai in *Jour. Coll. Sci. Tokyo,* XXXI. 214 (*Fl. Kor.* II) (1911). — Koidzumi in *Tokyo Bot. Mag.* XXVII. 87 (1913).

> *Salix triandra,* var. *vulgaris* Seemen in *Bot. Jahrb.* XXIX. 276 (1900).

CHINA. Shensi: Ko lu pa, June 1890–96, *G. Giraldi* (No. 1158; ex von Seemen).

---

[1] The typical *S. pentandra* Linnaeus has a ventral gland which is higher than broad and does not embrace the pedicel, while in *S. glandulosa* Seemen and other species this gland is very similar to the ventral gland of the first section. These species may probably be better referred to a separate section.

NORTHEASTERN ASIA. Korea: Chung-shan, June 1887, *O. Warburg* (No. 6522; ex von Seemen); "Kyoeng-san, circa Tai-ku, Na-Tong, Chang-chyoeng," *T. Uchiyama* (ex Nakai).

See my remarks under *S. Wilsonii* Seemen and also under *S. eucalyptoides* Meyer. The following variety probably represents a distinct species.

Salix glandulosa, var. Warburgii Koidzumi in *Tokyo Bot. Mag.* XXVII. 88 (1913).

Salix Warburgii Seemen in *Bot. Jahrb.* XXIII. Beibl. LVI. 43 (1892). — Burkill in *Jour. Linn. Soc.* XXVI. 534 (1899). — Matsumura & Hayata in *Jour. Coll. Sci. Tokyo*, XXII. 394 (*Enum. Pl. Formos.*) (1906).

Salix Mesnyi Hayata in *Jour. Coll. Sci. Tokyo*, XXX. art. 1, 306 (*Mat. Fl. Formos.*) (non Hance) (1911).

CHINA. Northern Formosa: Sintiam, January 1888, *O. Warburg* (No. 10201, type; ♀; No. 10203; ♂ type, ex von Seemen); Tai-peh Fu, January 1888, *Warburg* (No. 10202; ♀, ex von Seemen); same locality, secus aquas Taipeh, April 30, 1903, *U. Faurie* (No. 519); Kelung, January 1888, *O. Warburg* (ex von Seemen); without locality, 1864 (*R. Oldham* No. 509).

I have seen only Oldham's and Faurie's specimens which represent old fruiting branches on which I could find no indication of dorsal glands.

12. **Salix eucalyptoides** F. N. Meyer in litt., n. sp.

Arbor, ut videtur satis magna, pulchra, trunco cortice albo tecto; ramuli elongati, glabri, fusco-brunnei, vetustiores purpurascentes, subpruinosi. Folia adulta (juvenilia nondum visa) lanceolata v. elliptico-lanceolata, utrinque acuta, supra viridia, subtus glaucescentia v. albescentia, interdum in sicco utrinque pallida, glaberrima, costa utrinque distincte flava v. brunnea, nervis lateralibus utrinque paulo distinctis tenuibus sub 45–60° a costa divergentibus 2–3 pro 1 cm., margine integra v. versus apicem breviter denticulato-serrata, 4–6.5 cm. longa, 1.2–2 cm. lata, v. maxima ad 7 cm. longa et ad 2.3 cm. lata; petioli glabri, 4–6 mm. longi. Amenta (tantum fructifera visa) pedunculos ad 2 cm. longos glabros foliis parvis normalibus praeditos terminantia, glabra, 4–4.5 cm. longa, vix 1 cm. crassa, laxiflora; capsulae pedicellis inclusis circiter 5 mm. longae, glabrae, oblongo-ellipticae, angustae, pedicellis circiter 4–5-plo brevioribus glandulam unam ventralem latam subamplectentem 3–4-plo superantibus suffultae, stigmatibus ut videtur sessilibus brevibusque coronatae; bracteae non visae, deciduae.

NORTHEASTERN ASIA. Mandshuria: by streams, mountains 12 hours east of Harbin by railroad, August 31, 1903, *C. S. Sargent* (slender tree; sterile). Korea: without locality, valleys, August 14, 1906, *F. N. Meyer* (type; very fine tree, white bark; with fruits).

In a note accompanying his photograph (No. 5225 of the collection of the Department of Agriculture) from North Korea, August 16, 1906, which shows a tree of this Willow, Mr. Meyer says: "A remarkably tall willow, growing from 80–100 feet tall. Has, when young, a bark covered with white bloom; when older, this disappears and the bark becomes shaggy like that of an *Eucalyptus*. If not yet named, *Salix eucalyptoides* would be very appropriate." It may be the same as *S. glandulosa* of von Seemen and Nakai from Korea and seems to be a well marked species. It chiefly differs from *S. glandulosa* Seemen in its narrower lanceolate often entire leaves which are acute at the base and apparently bear no glands on the petioles. The fruits are narrow-elliptical (not ovate), and the pedicel is comparatively short. The only gland I have seen has the same shape as those of *S. tetrasperma* Roxburgh, and it may belong to sect. *Tetraspermae* near *S. Dunnii* Schneider. The specimen collected by Professor Sargent is a sterile one, but it is very like Meyer's plant.

13. **Salix Kusanoi** Schneider, n. sp.

Salix Oldhamiana Henry in *Trans. As. Soc. Jap.* XXIV. suppl. 90 (non Miquel) (1896).

Salix tetrasperma Burkill in *Jour. Linn. Soc.* XXVI. 530 (pro parte, non Roxburgh) (1899). — Matsumura & Hayata in *Jour. Coll. Sci. Tokyo,* XXII. 395 (*Enum. Pl. Formos.*) (1906).

Salix tetrasperma, var. *Kusanoi* Hayata in *Jour. Coll. Sci. Tokyo,* XXX. art. 1, 307 (*Mat. Fl. Formos.*) (1911).

Arbor 3–6 m. alta; ramuli hornotini praecipue apice cinereo- et brunneo-sericeo-villosi, annotini vetustioresque glabri, purpurascentes; gemmae ovato-conicae, subglabrae, purpureae. Folia ovata v. ovato-elliptica, basi rotunda v. obtusa, rarius subaurita, apice acuta v. saepius satis subito longe acuminata, caudata, supra saturate viridia, ad costam fulvo-sericea, subtus pallidiora, initio dense fulvo- et griseo-sericea, deinde tantum ad costam pilosa, utrinque distincte nervosa, sed vix reticulata, nervis angulo 45–60° a costa divergentibus angustis parallelis 3–4 pro 1 cm., margine grossius dense glanduloso-crenato-serrata, minora 4–6 cm. longa, 2–3 cm. lata, maxima usque ad 10 cm. longa et ad 3.8 cm. lata; petioli 5–9 mm. longi, superne in sulco pilosi, eglandulosi; stipulae parvae, ovato-lanceolatae, caducissimae. Amenta subpraecocia, ♂ pedunculis 5–10 mm. longis foliolis parvis lanceolatis 2–3 instructis suffulta, 5–6 cm. longa et 0.6 cm. crassa, sublaxiflora, villosa; flores 5–6-andri; filamenta basi pilosa bracteis vix v. duplo longiora, antheris ovalibus flavis; bracteae late ovato-rotundae, intus villosae, ciliatae, extus plus minusve glabrae; glandula ventralis late ovata, apice rotunda, saepe lobulata, dorsalis vix minor, saepissime distinctius lobata. Amenta ♀ (fructifera) pedunculis ♂ simillimis vix longioribus suffulta, circiter 3 cm. longa et 1 cm. lata, rhachi villosa; flores ignoti; fructus glabri, brevipedicellati, densi; pedicelli glandulis circiter duplo longiores ovariis circiter 5-plo breviores; ovaria ovato-cylindrica, 3–4 mm. longa, apice stigmatibus brevibus subsessilibus coronata; glandula ventralis lata, superne subemarginata; dorsalis paulo minor, sublobulata; bracteae pedicellis duplo longiores, iis florum ♂ similes, extus distinctius pilosae.

CHINA. Formosa: Tamsui, *A. Henry* (No. 1404; ♂; No. 1473, sterile), creek, February 24, *A. Henry* (No. 1068ª; tree 3–5 m. tall; with fruits), May 8, *A. Henry* (No. 1068ᵇ; with old fruits); Ta kao, bank of creek far in (3 miles), July 15, *A. Henry* (No. 1068; tree 3–5 m. tall; with fruits); Banchoryo, 1909, *S. Kusano* (type of var. *Kusanoi,* ex Hayata).

This species is readily distinguished by the brownish pubescence of its young parts, the coarsely serrate multi-nerved leaves, the short fruiting catkins, and by the nervation of its old leaves. It appears to be nearly related to *S. tonkinensis* Seemen or to some of those forms of *S. tetrasperma* Roxburgh of which I have not yet been able to see sufficient material.

14. **Salix Maximowiczii** Komarov in *Act. Hort. Petrop.* XVIII. 442 (1901); XXII. 25, t. 1 (*Fl. Mansh.* II.) (1903); XXV. 813 (1907). — Nakai in *Jour. Coll. Sci. Tokyo,* XXXI. (*Fl. Kor.* II.) 214 (1911).

NORTHEASTERN ASIA. Korea: "in montibus Zatan-ien secus rivulos in regione fl. Jalu," June 28, 1897 (type, ex Komarov); "in monte des Diamants," June 24, 1906, *U. Faurie* (No. 177).

According to von Seemen (*Salic. Jap.* 25 [1903]) this species is the same as *S. cardiophylla* Trautvetter, see p. 103, which is not mentioned either by Komarov or by Nakai, although Komarov says (1907) that *S. Maximowiczii* is a very different species. The plant collected by Faurie agrees very well with Komarov's description and plate. The ovate-lanceolate, long-acuminate leaves measure up to

12 cm. in length and to 3.3 cm. in width. The fruiting aments of the type are without the leafy peduncle and are from 10–14 cm. long and 1 cm. thick, while Faurie's specimen shows only a part of one catkin. The fruits possess a distinct deeply bifid ventral gland; the dorsal gland is very small or wanting. The pedicel is nearly three times as long as the gland and about a quarter the length of the narrow ovate ovary. Komarov does not mention the style or stigma. I have seen only the remnant of a short but distinct, deeply bifid style with narrowish bifid stigmas. The leaves become black in drying, and in their nervation resemble more those of *S. Kusanoi* Schneider than those of *S. pentandra* Linnaeus.

15. **Salix pentandra** Linnaeus, *Spec.* 1016 (1753). — Wolf in *Act. Hort. Petrop.* XXI. 179 (1903). — Komarov in *Act. Hort. Petrop.* XXII. 27 (*Fl. Mansh.* II.) (1904). — Schneider, *Ill. Handb. Laubholzk.* I. 30, fig. 12 f–f¹, 13 d–h (1903). — Henry in Elwes & Henry, *Trees Gr. Brit. Irel.* VII. 1745 (1913). — Moss, *Cambridge Brit. Fl.* II. 15, t. 18 (1914).

For additional synonyms see von Seemen in Ascherson & Graebner, *Syn. Mitteleur. Fl.* IV. 62 (1908).

NORTHEASTERN ASIA. Amur: Amur River, 1855, *R. Maack* (No. 207). I have seen only this one specimen from northeastern Asia. Komarov, l. c., says: " A clar. Korshinskio in provincia Amurensi ad pagum Tambovka solum adhuc collecta." Wolf, l. c., mentions " Kamtschatka (*Stubendorff; Lubarski; Peters*)." Von Seemen keeps " *S. bracteosa* Turczaninow in Herb. Hort. Petrop," a synonym of *S. pentandra*, while Wolf refers this form to *S. macrolepis* Trautvetter. *S. pentandra* has very smooth leaves and densely flowered ♂ catkins. The dorsal gland of the ♀ flowers is often very small or even wanting. The pedicel of the fruit is not much longer than the ventral gland, and the style is rather long and often bifid at the apex. The stigmas of the type are rather short and broad. See also Wolf, in *Izv. S.-Peterburg. Liesn. Inst.* IV. 17, t. 2, fig. 12–13, t. 3, fig. 1–10 (*Мат. Изуч. Нъ Европ. Росс.*) (1900).

In Maack's specimen the bracts of the female flowers show several small glandular teeth at the end. The ventral glands of both the sexes are mostly deeply cleft.

16. **Salix paraplesia** Schneider. See p. 40.

17. **Salix dodecandra** Léveillé in *Bull. Soc. Bot. France*, LII. 141 (1905); *Fl. Kouy-Tchéou*, 181 (1915).

CHINA. Kweichau: Pin-fa, March 22, 1902, *J. Cavalérie* (No. 1317, type; ♂).

Through the kindness of the author, the Arboretum has received some ♂ and ♀ flowers of this species. The ♂ flowers have only 5 stamens (not 12 as Léveillé says) which are hairy at the base. The bracts are ovate-oblong, somewhat obtuse, and in one ament glabrous with the exception of some hairs near the base within, while in a second ament they are shortly pubescent on both surfaces. There are two different glands forming a nearly 4-lobed disc, the lateral lobes being not very distinct. The ♀ flowers are nearly the same as those of *S. Wilsonii* Seemen, but the stigmas are raised on a rather short but distinct style somewhat like that of *S. pentandra* Linnaeus. There are also two glands, of which the dorsal one is a little 3-lobed. This species needs further investigation.

18. **Salix Cavaleriei** Léveillé in *Bull. Soc. Bot. France*, LVI. 298 (1909).

Salix polyandra Léveillé in *Bull. Soc. Agric. Sci. Sarthe*, LIX. (sér. 2, XXXI) 325 (non Weigel, nec Gleditsch ¹) (1904); in Fedde, *Rep. Spec. Nov.* VI. 377 (1909); in *Bull. Soc. Bot. France*, LVI. 300 (1909).

¹ A. Toepffer (*Sched. Salic. Exsicc. Fasc.* III. 70 [1908]) gives an exact account of the earlier *S. polyandra*. This name was first used by Wilcke (1765) apparently

Salix Pyi Léveillé in *Bull. Soc. Bot. France,* LVI. 300 (1909).
Salix yunnanensis Léveillé, l.c. 301 (1909).
Salix tetrasperma Diels in *Not. Bot. Gart. Edinburgh,* VII. 91 (*Pl. Chin. Forrest.*) (non Roxburgh) (1912).

CHINA. Yunnan: " environs de Yunnan-sen, bord des canaux dans la plaine," February 15, 1897, *E. Bodinier* (No. 65; type of *S. polyandra;* tree; ♂ and ♀); same locality, " pagode de He-long-tan," March 19, 1905, April 11, 1906, *F. Ducloux* (Nos. 658, 669; ♀ type of *S. Cavaleriei*); same locality, " vallons de Tchong-Chan," February 21, 1906, *F. Ducloux* (♂ type of *S. Pyi*); same locality, " plaines," April 4, 1906, *F. Ducloux* (No. 653; ♂ type of *S. yunnanensis*); Li chiang fu, plain, alt. 2500 m., lat. 26° 50′, May 1906, *G. Forrest* (No. 2033; ♀).

By the kindness of Mgr. Léveillé I have flowers of his different types except those of *S. yunnanensis,* of which I have seen only one leaf. There seems to be no real difference in the ♀ and ♂ flowers of the different types. The ♂ flowers have from 8–12 stamens and a somewhat five-lobed disc. The bracts are ovate-deltoid, rather acute, distinctly pubescent on the inner surface and on the margin and often glabrous on the outer surface. The ♀ flowers have sessile stigmas, glabrous [1] ovaries, and are distinctly stalked with pedicels from one-half to three-fifths as long as the ovaries. The bracts are like those of the ♂ flowers, only in *S. Cavaleriei* Léveillé they are obtuse or even truncate at the apex. There are two glands: the ventral is broad and low, embracing the base of the pedicel; the dorsal is 2- or 3-lobed in *S. Cavaleriei* Léveillé and *S. polyandra* Léveillé, but it is simple and rather large in Forrest's plant which might represent a variety.

An apparently nearly related species is *S. anisandra* Léveillé et Vaniot (in Fedde, *Rep. Spec. Nov.* III. 22 [1906]. — Léveillé, *Fl. Kouy-Tchéou,* 381 [1915]), the type of which was collected by *J. Esquirol* in Kweichou at Pia Fong, March 1905 [No. 362; ♂]. It has from 10 to 12 stamens and orbicular-ovate, obtuse and nearly glabrous bracts. The two glands seem to be more distinctly separated. This species needs further observation.

19. **Salix macrolepis** Turczaninow in *Bull. Soc. Nat. Mosc.* XXVII. 371 (1854); *Fl. Baical.-Dahur.* II. 98 (1856). — Andersson in *Svensk. Vetensk. Akad. Handl.* VI. 52, t. 3, fig. 33 (*Monog. Salic.*) (1867); in De Candolle, *Prodr.* XVI. pt. 2, 213 (1868), ut videtur pro parte. — Glehn in *Beitr. Kennt. Russ. Reich.* XXV. 211 (1868); in *Act. Hort. Petrop.* IV. 81 (?) (1875). — Fr. Schmidt in *Mém. Acad. Sci. St. Pétersbourg,* sér. 7, XII. 172 (*Reis. Amur. Sachal.*) (1868). — Herder in *Act. Hort. Petrop.* XI. 400 (pro parte) (1891). — Wolf in *Act. Hort. Petrop.* XXI. 48 (1903).

Salix bracteosa Turczaninow, *Pl. Exsicc.,* sed ut videtur non apud Trautvetter in Middendorff, *Reise Sibir.* 1. pt. 2, Bot. abt. 2, 77 (*Fl. Ochot.*) (1856), quoad specim. prope Udskoi lectum.

NORTHEASTERN ASIA. Transbaikalia: "in virgultis subalpinis Dahuriae," *N. Turczaninow* (type, ex Turczaninow); " ad flumen Selenga prope Werchne-Udinsk," *N. Turczaninow* (ex Andersson et Wolf). Amur: Stanovoi

as a nomen nudum, while Weigel (*Fl. Pomerano-Rugica,* 80 [1769]) first describes it. This *S. polyandra* is the same as *S. polyandra* Bray (in *Denkschr. Bot. Ges. Regensb.* I. pt. 2, 41, t. 1 [1818]), namely, the hybrid between *S. pentandra* Linnaeus and *S. fragilis* Linnaeus. *S. polyandra* Gleditsch (*Syst. Einl. Forstwissensch.* II. 4 [1775]) is only a synonym of *S. pentandra* Linnaeus.

[1] In Fedde, l. c. (1909), the description is: " ovariis fusiformibus, pubescentibus, vix pedicellatis." This must be a mistake.

Mountains, *Fr. Schmidt* (ex Andersson). Saghalien: "im Arkaithal," May 27, 1891, *P. von Glehn* (ex Schmidt et fide Wolf); without locality, *Fr. Schmidt* (in Herb. Gray).

This is a very interesting species with rather small and slender catkins and narrow leaves which are nearly entire in Schmidt's plant from Saghalien. Neither the ♂ nor the ♀ flowers of this specimen have any glands. Further information is given in the key, pp. 75 and 85.

Sect. 3. URBANIANAE Seemen, *Salic. Jap.* 15 (emend.) (1903).

A sectione praecedente praecipue differt foliis tamen majoribus, saepe basi subcordatis, ovariis (capsulisque) sessilibus v. brevipedicellatis, glabris v. tomentosis, stylis distinctis bipartitis (in *S. populifolia* Andersson nullis), stigmatibus bifidis elongatis angustis.

20. **Salix Urbaniana** Seemen, *Salic. Jap.* 24, t. 1, G–L (1903). — Léveillé in *Bull. Acad. Intern. Géogr.-Bot.* ser. 3, XIV. 208 (1904); XVI. 151 (1906).

JAPAN. Hondo: prov. Shimotsuke, west end of Lake Chuzenji, forming woods, June 1, 1914, *E. H. Wilson* (No. 6775; tree 8–16 m. tall, girth 0.3–1.5 m., bark gray, fissured, branches erect, spreading; ♂); Nikko, June 3, 1894, *M. Shirai* (No. 33, ex von Seemen); "prov. Nambu, in subalpinis ad rivulos," 1865, *Tschonoski* (No. 4, Maximowicz, Iter Sec.; type; ♀); prov. Mutsu, Aomori, "in petrosis torrentum," June 1902, *U. Faurie* (No. 507, tree 7–8 m.; with fruits); same prov., July 29, 1910 (ex Herb. Sakurai; with fruits). Hokkaido: prov. Oshima, Hakodate, 1861, *C. Maximowicz* (No. 2, fruiting co-type); prov. Ishikari, Sapporo, May 14 and 18, 1890 and 1891, October 5 and 15, 1890, *Y. Tokubuchi* (♂ and ♀); same locality, August 21, 1905, *J. G. Jack* (with fruits); same locality "secus torrentes," June 14, 1908, *U. Faurie* (No. 251, ♂; 252, ♀); prov. Kushiro, July 21, 1910 (with fruits).

This is a striking Willow with its large leaves bluish white beneath, its long catkins and its silvery fruits. Sterile branchlets often resemble those of *S. Caprea* Linnaeus, but the adult leaves of *S. Urbaniana* are nearly or quite glabrous, their serration is rather regular, the petioles are comparatively longer, the stipules are more obtuse and the buds are acute and compressed, not obtuse and thick as in *S. Caprea*. For further remarks see *S. cardiophylla* Trautvetter & Meyer.

This and the closely allied *S. cardiophylla* grow to a greater size than any other Willows of eastern Asia. Both grow from 20 to 25 m. tall and have a thick trunk from 6 to 8 m. in girth with deeply fissured bark and massive spreading branches, and in general appearance suggest *Populus Maximowiczii* Henry. *S. Urbaniana* is common on the shores of Lake Chuzenji, especially at the western end. Elsewhere in Japan from Mt. Ontake in Shinano province in central Hondo and northward I found it growing sparingly in mountain valleys. In Hokkaido it is more abundant, and some fine specimens are growing in the park in the city of Sapporo.

Pictures of this Willow will be found under Nos. x303, x372, x373 of the collection of my Japanese photographs. E. H. W.

21. **Salix cardiophylla** Trautvetter & Meyer in Middendorff, *Reise Sibir.* I. pt. 2, Bot. abt. 2, 77, t. 19, 20 a–i (*Fl. Ochot.*) (1856). — Regel & Tiling in *Nouv. Mém. Soc. Nat. Mosc.* XI. 117 (*Fl. Ajan.*) (1858). — Andersson in *Svensk. Vetensk. Akad. Handl.* VI. 37, t. 3, fig. 25 (*Monog. Salic.*) (1867); in De Candolle, *Prodr.* XVI. pt. 2, 207 (1868). — Schmidt in *Mém. Acad. Sci. St. Pétersbourg*, sér. 7, XII. 172 (*Reis. Amur. Sachal.*) (1868). — Herder in *Act. Hort. Petrop.* XI. 399 (1891). — Wolf in *Act. Hort. Petrop.* XXI. 47 (1903). — Shirai in *Tokyo Bot. Mag.* XVIII. 161, t. 2 (1904). — Koidzumi in *Tokyo Bot. Mag.* XXVII. 97 (1913).

NORTHEASTERN ASIA. Maritime prov.: "ad fl. Polowinnaja, prope oppid. Udskoi," June 7, 1844, *A. T. von Middendorff* (type, ♂, ex Trautvetter & Meyer); "ad sinum Ujakon," August 23, September 1, 1844, *A. T. von Middendorff* (co-type, ex Trautvetter & Meyer; with fruits); near Ajan, *H. Tiling* (ex Regel & Tiling). Amur: Stanowoi Mountains, July 1860, *Fr. Schmidt* (ex Andersson et Wolf). Saghalien: Konuma, moors, not common, August 4, 1914, *E. H. Wilson* (No. 7332; tree 13–17 m. tall, girth 2.1–3 m., bark gray, deeply furrowed, branches thick, spreading; sterile); "im Thal des Baches von Adugiwo," July 5, 1860 (with fruits; ex Schmidt).

JAPAN. Hokkaido: prov. Oshima, Tokatsi (ex Koidzumi); Hondo: prov. Echigo, Simizutoge (ex Koidzumi); prov. Shimotsuke, Nikko (ex Koidzumi).

According to Shirai's plate this species must be most closely related to *S. Urbaniana* Seemen, which may represent only a variety with sessile hairy ovaries. I have not been able yet to compare any type specimens, and the figures of Trautvetter & Meyer as well as Wolf's description are somewhat different from these Japanese forms, of which I have not seen the specimens mentioned by Koidzumi. The following specimens may represent the true *S. cardiophylla* or a variety of *S. Urbaniana* Seemen with short-stalked, more or less or wholly glabrous ovaries and fruits.

Hokkaido: prov. Ishikari, Sapporo, "secus torrentes," June 14, 1908, *U. Faurie* (No. 253); "in arena rivorum Kamikawa," July 1905, *U. Faurie* (No. 6643); without precise locality, August 21, 1905, *J. G. Jack;* Sapporo, May 20, 1885 (Agric. College). Hondo: prov. Shimotsuke, Nikko, road to Lake Chuzenji, Sept. 3, 1892, *C. S. Sargent* (tree 6–38 m.).

This Willow, which in general appearance and size is similar to *S. Urbaniana* Seemen, I met with only in Saghalien in the neighborhood of Toyohara. The bark on the branches is smooth and pale gray and on the trunk it is gray and deeply fissured.   E. H. W.

22. **Salix angiolepis** Léveillé in Fedde, *Rep. Spec. Nov.* III. 22 (1906); in *Bull. Soc. Bot. France*, LVI. 297 (1909); *Fl. Kouy-Tchéou*, 381 (1915).

CHINA. Kweichau: Pin-fa, 1903, *J. Cavalerie* (No. 2069, type; ♂).

By the kindness of Mgr. Léveillé, I have received a small branchlet of the type of this very interesting species. Unfortunately the material was collected when the flowers were very young and not yet fully developed; and it seems best to give the following description of what I have seen:

Ramuli novelli dense albo-tomentosi, vetustiores subglabri, nigrescentes. Folia valde juvenilia supra plus minusve subtus dense albo-sericeo-tomentosa, ut videtur ovato-oblonga basi subacuta, apice breviter acuminata, margine integra (v. serraturis nondum evolutis), nervis nondum visibilibus, ad 2 cm. longa et 0.5–0.8 cm. lata; petioli breves, sed distincti, circiter 3 mm. longi, cano-villosi. Amenta ♂ valde juvenilia, pedunculos breves foliis 2–3 parvis lanceolatis instructos villosos terminantia, cylindrica, densiflora, usque 3 cm. longa et 0.3–0.5 cm. crassa, rhachi tomentosa; flores ut videtur nondum satis evoluti, staminibus 5–6 (–8, ex Léveillé) filamentis brevissimis (an basi villosis?), antheris ovalibus flavis (interdum nondum satis evolutis); glandulae 2, ut videtur pseudodiscum lobulatum formantes; bracteae late ovatae, apice subtruncatae, interdum denticulatae, in sicco nigrescentes, nervatae, concolores, basi utrinque distincte albo-villosae, ceterum sparse pilosae (haud glabrae) et ciliatae, stamina circiter duplo superantes.

As far as I can judge from the young flowers and leaves, this plant apparently belongs in the sect. *Urbanianae*. It certainly is an interesting Willow, but needs further investigation. The flora of Kweichau seems to contain a number of distinct species of *Salix*, closely related to those of Hupeh and Szech'uan, although different in several respects.

23. **Salix populifolia** Andersson in *Svensk. Vetensk. Akad. Handl.* 1850, 494 (*Ost-Ind. Pilarter*) (1851); in *Jour. Linn. Soc.* IV. 48 (1860); in *Svensk. Vetensk. Akad. Handl.* VI. 6, t. 1, fig. 5 (*Monog. Salic.*) (1867); in De Candolle, *Prodr.* XVI. pt. 2, 194 (1868).

INDIA. Without precise locality, *G. S. Perrotet & V. Jacquemont* (ex Andersson).

This seems to be a curious species which was not mentioned by Hooker in his *Flora of British India.* According to the author the large ovate-acuminate leaves are from 5 to 10 cm. long and from 2.5 to 5 cm. broad, with petioles up to 2.5 cm. in length. They are scarcely paler beneath and show a distinct reticulation. The catkins and the capsules are almost sessile. There is no style and the stigmas are deeply divided.

Sect. 4. ACMOPHYLLAE Andersson in *Svensk. Vetensk. Akad. Handl.* VI. 7 (*Monog. Salic.*) (1867); in De Candolle, *Prodr.* XVI. pt. 2, 195 (1868).

Salix, sect. *Orientales* Andersson, l. c. (nomen solum) (1867 et 1868).

Arbores v. frutices elati. Folia anguste lanceolata, acutissima, integra v. obsolete remote glanduloso-denticulata, subtus glaucescentia, crassiuscula, laevia. Amenta iis sect. *Tetraspermae* similia; flores ♂ pleiandri, glandulis 2; ovaria v. capsulae distincte pedicellata, glabra, stylo brevi v. nullo, stigmatibus brevibus emarginatis; glandula una, lata v. partita.

24. **Salix dealbata** Andersson in *Svensk. Vetensk. Akad. Handl.* 1850, 472 (1851); in *Jour. Linn. Soc.* IV. 43 (1860); in *Svensk. Vetensk. Akad. Handl.* VI. 8 (*Monog. Salic.*) (1867).

? *Salix glaucophylla* Andersson in *Svensk. Vetensk. Akad. Handl.* 1850, 474 (1851); in *Jour. Linn. Soc.* IV. 43 (1860); in *Svensk. Vetensk. Akad. Handl.* VI. 8 (*Monog. Salic.*) (1867).

? *Salix urophylla* Lindley apud Andersson in *Svensk. Vetensk. Akad. Handl.* 1850, 487 (1851). — Andersson in *Svensk. Vetensk. Akad. Handl.* VI. 5 (*Monog. Salic.*) (1867); in De Candolle, *Prodr.* XVI. pt. 2, 194 (1868). — Hooker f., *Fl. Brit. Ind.* V. 637 (1888).

? *Salix tetrasperma,*** *urophylla* Andersson in *Jour. Linn. Soc.* IV. 41 (1860).

? *Salix acmophylla* Hooker f., *Fl. Brit. Ind.* V. 628 (pro parte) (1888).

INDIA. Northwest Provinces: "inter Saharupora (Saharunpur) et pedem montium Sulir Nanka (Nankah) et Mohur (Mohun), in planitie secus torrentem ad Ghautka roure, d. 11 Aug. 182– "; *V. Jacquemont* (No. 308[1], type of *S. dealbata;* ♀, ex Andersson); without precise locality and without date, *V. Jacquemont* (No. 146, type of *S. suaveolens;* ♀, ex Andersson); Oude: Kootukonnaut Forest, March 5, 1825, *N. Wallich* (No. 3708, type of *S. urophylla*, ♀); Kashmir: Balti, Saling, July 13–15, 1856, *Schlagintweit* (No. 5458; sterile).

AFGHANISTAN. Without locality (No. 4505, Herb. Griffith; with fruits).

S. dealbata Andersson and his two species referred to it as being apparently the same plant or nearly related forms seem to belong to a group of species hitherto very little known. Without having seen Jacquemont's specimens it is impossible

---

[1] This Willow is apparently the one mentioned by Jacquemont, *Voy.* II. 7 (1841), as: " Une espèce de Saule (B. 208), la première que je vis dans l'Inde, est commune sur les grèves du torrent. C'est un arbre qui ressemble beaucoup, par son port, au *Salix alba* d'Europe." Jacquemont was at Saharunpur from April second to ninth,1830, and camped at Mohun April 11, not in August.

to interpret correctly Andersson's species. Of Wallich's type specimen of *S. uro-phylla* the Arnold Arboretum has received a photograph and some fruits by the kindness of the Keeper of the Kew Herbarium, but the material is very poor and, as already said by Hooker f., " too incomplete to found a species upon." According to Hooker it came " from a plant no doubt cultivated at Oude," while Dr. Stapf, in a letter, says that No. 3708 is "from the Kootukonnaut Forest," of which he has not been able to trace the name, " but from the fact that Wallich collected 3708 on the 5th March and was at Rampur on the 27th February it is quite clear that the locality must be in the extreme northwest of Oude." The species of sect. *Acmophyllae* need a very careful study.

Sect. 5. TRIANDRAE Dumortier in *Bijdr. Natuurk. Wetensch.* I. 58 (*Verh. Geslacht Wilgen*, 17) (1825). — Borrer apud Hooker, *Brit. Fl.* 414 (1830); apud Loudon, *Arb. Frut. Brit.* III. 1496 (1838). — Seemen, *Salic. Jap.* 16 (1903); in Ascherson & Graebner, *Syn. Mitteleur. Fl.* IV. 57 (1908).

Salix, sect. *Amygdalinae* W. D. Koch, *Salic. Europ. Comment.* 17 (pro parte) (1828). — L. Reichenbach, in *Moessler, Handb.* ed. 2, III. 1753 (1829). — Andersson, in *Svensk. Vetensk. Akad. Handl.* VI. 19 (*Monog. Salic.*) (1867); in De Candolle, *Prodr.* XVI. pt. 2, 200 (1868).

Arbores v. frutices, ramis elongatis flexilibus. Folia lanceolata, acuta, serrata, glabra v. mox glabrescentia. Amenta coetanea; flores ♂ glandulis duobus liberis, staminibus 3 v. pluribus; ovaria longe pedicellata, glabra, stylis nullis v. brevissimis, stigmatibus brevibus, glandula una.

25. **Salix amygdalina** Linnaeus, *Spec.* 1016 (1753). — W. D. Koch, *Salic. Eur.* 18 (1828). — Seemen in Ascherson & Graebner, *Syn. Mitteleur. Fl.* IV. 74 (1908).

Salix triandra Linnaeus, *Spec.* 1016 (1753). — Herder in *Act. Hort. Petrop.* XI. 395 (1891). — Wolf in *Act. Hort. Petrop.* XXI. 52 (1903). — Komarov in *Act. Hort. Petrop.* XXII. 30 (*Fl. Mansh.* II) (1903). — Moss, *Cambridge Brit. Fl.* II. 22, t. 25–26 (1914).
For further synonyms and literature see Seemen, l. c., Herder, l. c., and Komarov, l. c.

NORTHEASTERN ASIA. Transbaikalia to Maritime prov.: (see Herder, l. c., Komarov, l. c., and Wolf, l. c.).
Besides the type there is the following variety:

Salix amygdalina, var. **nipponica** Schneider, n. comb.
Salix nipponica Franchet & Savatier, *Enum. Pl. Jap.* I. 495 (1875); II. 502 (1879).
Salix triandra, var. *nipponica* Seemen, *Salic. Jap.* 27. t. 2, fig. E–J (1903). — Shirasawa, *Icon. Ess. For. Jap.* II. t. 9, fig. 11–23 (1908).
Salix Kinashii Léveillé in *Bull. Soc. Bot. France*, LII. 141 (1905), fide Koidzumi.
Salix amygdalina Koidzumi in *Tokyo Bot. Mag.* XXVII. 94 (1913).

CHINA. Shantung: Kiao-chao, Iltis Mountain, April 30, 1899, *Nebel* (ex Seemen).
NORTHEASTERN ASIA. Korea: Chinnampo, September 20, 1905, *J. G. Jack* (sterile).

JAPAN. Hokkaido: prov. Ishikari, Sapporo, May 23, 1892, *Y. Tokubuchi* (♂); same locality, 1903, *S. Arimoto* (♂ and with fruits). Hondo: prov. Mutsu, " secus aquas Hirosaki," May 1904, *U. Faurie* (No. 5760; ♂); Aomori " secus aquas," May 1904, *U. Faurie* (No. 5766; with fruits); same locality, May 1902, *Kinashi* (no. 11, type of *S. Kinashii*, ex Léveillé); prov. Echigo, Niigata, *L. Savatier* (No. 2717, co-type, ex Franchet & Savatier); prov. Shimotsuke, Nikko, April 21, 1900, *H. Shirasawa* (♂); prov. Musashi, Toda, April 12, 1891, *K. Watanabe* (♂); same prov., " prov. Nambu, in silvis subalpinis ad rivulos ad ipsam aquam," 1865, *Tschonoski* (♂; distributed sub nom. *S. Oldhamiana*); prov. Suruga, Fuji-san, October 1909, *M. Koyama* (sterile); prov. Sagami, "circa Yokoska," *L. Savatier* (No. 1139, type; ex Franchet & Savatier).

According to von Seemen this variety differs from the typical *S. amygdalina* in the more copious pubescence of the young twigs and young leaves, the almost entire leaves of the peduncles and the denser flowered catkins.

Sect. 6. FRAGILES W. D. Koch, *De Salic. Europ. Comment.* 13 (pro parte) (1828). — E. Fries in *Syll. Pl. Nov.* II. 36 (pro parte) (1828). — Borrer in Hooker, *Brit. Fl.* 417 (1830); in Loudon, *Arb. Brit.* III. 1507 (pro parte) (1838). — Seemen in Ascherson & Graebner, *Syn. Mitteleur. Fl.* IV. 70 (1908).

Salix, sect. *Subfragiles* Seemen, *Salic. Jap.* 15 (1903).

Arbores v. frutices, ramis elongatis, saepe ad insertionem fragilibus. Folia lanceolata, acuta, serrata; petioli saepe glanduliferi. Amenta coetanea. Flores ♂ diandri, glandulis duobus separatis; ♀ bracteis concoloribus, flavis v. flavo-brunneis, saepe deciduis; ovaria brevipedicellata v. sessilia, glabra v. pilosa, stylis brevibus v. longioribus, stigmatibus ovatis v. angustis, glandulis duobus, dorsali interdum obsoleta.

25. **Salix fragilis** Linné, *Spec.* 1017 (1753). — Brandis, *Forest Fl. Brit. Ind.* 466 (1874). — Hooker f., *Fl. Brit. Ind.* V. 630 (1888). — Wolf in *Izv. S.-Peterburg. Liesn. Inst.* IV. 21, t. 1, fig. 8–9, t. 5, fig. 1–4 (*Мам. Изуч. Нѣ Еороп. Росс.*) (1900). — Seemen in Ascherson & Graebner, *Syn. Mitteleur. Fl.* IV. 70 (1908). — Moss, *Cambridge Brit. Fl.* II. 17, t. 20–21 (1914).

For further information see the keys and von Seemen, l. c., and Wolf, l. c.

So far as I know, this species does not occur in a wild state within the limits of our area, but may be cultivated in northern Kashmir (see Hooker f., l. c., and Brandis, l. c.).

26. **Salix Matsudana** Koidzumi in *Tokyo Bot. Mag.* XXIX. 312 (1915).

? Salix babylonica Franchet in *Nouv. Arch. Mus. Paris*, sér. 2, VII. 92 (*Pl. David.* I. 282) (not Linnaeus) (1884). — Burkill in *Jour. Linn. Soc.* XXVI. 526 (1899), quoad specim. Chinæ septentrionalis.

Arbor 3.5–13 m. alta; ramuli fragiles, erecti v. pendentes, initio puberuli, mox glabri (tantum ad gemmas pilosi) olivacei v. satis flavi, vetustiores cinereo-brunnei. Folia anguste lanceolata, basi obtusa v. subrotunda, rarius acuta, apice sensim longe acuminata, margine distincte satis anguste argute glanduloso-serrata, supra laete viridia, mox glaberrima v. interdum basim versus ad costam tomentella, subtus glaucescentia v. albescentia, tantum initio laxe sericeo-villosa, mox glabra, utrinque (supra distinctius) tenuiter nervosa et tenuissime reticulata, minora acuta, circiter 5–6 cm. longa et 1.2–1.5 cm. lata, maxima acuminata, ad 8 cm. longa et ad 1.5 lata v. ad 10 cm. longa et ad 1.1 lata; petioli breves, 2–8 mm. longi, in sulco supra pilosuli; stipulae lanceolatae, breves, glanduloso-serratae, saepissime nullae. Amenta ♂ (a cl. Wilson collecta) praecocia, breviter cylindrica, 1–1.5 cm. longa et circiter 0.6 cm. lata, pedunculis 2–3 mm. longis foliola 2 lanceolata obtusa integra subtus paullo sericea v. glabra gerentibus, rhachi villosa; flores diandri, conferti, glandulis 2 separatis ovatis obtusis, filamentis basi pilosis, antheris ovalibus flavis, bracteis ovatis obtusis flavo-viridibus, plerisque tantum extus basi sericeis. Amenta ♀ (plantae in Arboreto Arnoldiano cultae) coetanea, perparva, circiter 12 mm. longa, 0.4 mm. lata, pedunculis 3–5-foliatis 5 mm. longis, pilosis suffulta, rhachi villosa; folia linearia, basi acuta, apice fere acuminata, integra, sparse sericea, sed cito glabra, ad 2.5 cm. longa et 0.3 cm. lata; ovaria sessilia, oblonga, glabra, stylo nullo v. brevissimo stigmatibus separatis brevibus ovalibus sublobatis coronata; glandulae 2, ventralis ovato-oblonga, apice angustiora, satis crassa, dorsalis minima.

CHINA. Chili: Tientsin, near race course (also planted everywhere between Tientsin and Peking), April 18, 1909, *E. H. Wilson* (♂); Peking, September 16, 1902, *C. S. Sargent* (large tree with pendulous branches, "the common Peking Willow"; sterile); Peking-Nankow road, October 6, 1906, *J. G. Jack* ("common Willow of Peking plain," large tree; sterile). Kansu: Ranshiu, April 17, 1907, *E. Umemura* (No. 17, type, ex Koidzumi).

NORTHEASTERN ASIA. Transbaikalia: Chita, bottom lands of Chita River, August 12, 1903, *C. S. Sargent* (round-headed tree, 10 m. tall, bark very deeply furrowed; sterile, doubtful form with broadly ovate stipules as long as the petioles). Ussuri: Tien-shan mountains, June 9, 1906, *F. N. Meyer* (No. 54; weeping Willow, rare; with fruits). Amur: south of Harbin, common, August 24, 1903, *C. S. Sargent* (large handsome tree; sterile, leaves long-acuminate, up to 12:2 cm.). Korea: Chinnampo, September 20, 1905, *J. G. Jack* (tree, probably introduced; sterile).

Jack sent cuttings to the Arboretum, and these produced a small tree with upright branches; and this tree (No. 5737 of the Arboretum, ♀ type) bore the ♀ flowers described above. There are other cultivated ♂ plants with more pendulous branchlets in the Arboretum, originated from cuttings sent by the Department of Agriculture under No. 22450 and collected in northern China, prov. Chili, near Paoting Fu, probably by *F. N. Meyer* on January 30, 1908 (No. 250; the ordinary Willow which grows excellently everywhere on the dry lands in northern China; needs no water supply beyond a scanty summer rainfall). There are also plants raised from cuttings sent by *W. Purdom* (No. 281) from northern China. After having described the specimens from Chili as a new species, No. 348 of the *Tokyo Botanical Magazine* has been received with the description of Koidzumi's new species which exactly agrees with the plant from Chili.

This species seems to be the "*Salix babylonica*" of northern China, as mentioned by Burkill and others (see *S. babylonica*, p. 42). *S. Matsudana* is easily distinguished by its greenish or yellowish twigs, while those of *S. babylonica* are reddish brown or purplish.

27. **Salix hamatidens** Léveillé in *Bull. Soc. Bot. France*, LVI. 301 (1909).

JAPAN. Hokkaido: prov. Ishikari, Sapporo, "secus aquas," June 13, 1908, *U. Faurie* (No. 263, type; ♀).

The ♀ flowers of this species very much resemble those of the true *S. fragilis* Linnaeus. They have a small dorsal gland, a pedicel of about the same length as the ventral gland, a short, slightly bifid style and ovate stigmas. The bracts are long. This curious form looks somewhat like a hybrid between *S. eriocarpa* Franchet & Savatier or related forms and *S. amygdalina*, var. *nipponica* Schneider. The ovaries are glabrous. It needs further observation.

28. **Salix eriocarpa** Franchet & Savatier, *Enum. Pl. Jap.* I. 459 (1875); II. 503 (1879). — Koidzumi in *Tokyo Bot. Mag.* XXVII. 88 (pro parte) (1913).

Salix dolichostyla Seemen in *Bot. Jahrb.* XXX. Beibl. LXVII. 39 (1901); *Salic. Jap.* 26, t. 2, fig. A–D (1903). — Léveillé in *Bull. Int. Acad. Géogr. Bot.* XIV. 208 (1904); XVI. 144 (1906).

JAPAN. Hondo: prov. Sagami, " circa Hami prope Yokoska in locis inundatis," *L. Savatier* (No. 2718, type of *S. eriocarpa*, ex Franchet & Savatier); Kamakura, April 6, 1880 (No. 11, ex Bot. Inst. Coll. Sci. Tokyo, type of *S. dolichostyla*, ex von Seemen); prov. Mutsu, Aomori, in river valleys, May 6, 1898, *U. Faurie* (Nos. 1259, 1260, co-types, ex von Seemen); "in plateis Aomori," May 1905, *U. Faurie* (No. 6623; tree 7–8 m. tall; ♂). Shikoku: prov. Tosa, Nanokawa, April 4, 1890, *K. Watanabe* (♀).

According to the description of the authors this species certainly is the same as *S. dolichostyla* Seemen, the type of which came from about the same region. It must be very nearly related to *S. jessoensis* Seemen, and there is a ♀ specimen before me from Sapporo, May 1892 (No. 51 ex Herb. Bot. Gard. Tokyo), which agrees with this species in every respect but has two glands like *S. eriocarpa*.

According to Lackschewitz (in *Schedae Herb. Fl. Ross.* VII. 107, No. 2255 [1911]). — Toepffer, *Salicol. Mitt.* No. 5, 236 [1912]), *S. mixta* Korshinsky (in *Act. Hort. Petrop.* XII. 391 [1892]), is the same as *S. dolichostyla* Seemen. Having before me ♂ and ♀ specimens collected by Korshinsky " ad ostium Ussuri prope Polowinnaja," May 16, 1891 (♀) and on the middle Amur, May 5, 1891 (♂), I agree with Lackschewitz that *S. mixta* is a very similar plant, although I have not found a dorsal gland in the ♀ flowers which according to Lackschewitz is not always present; Korshinsky does not say anything about the glands. As it is the case with other species, the dorsal gland of the ♀ flowers may be sometimes wanting. Further investigation of good specimens of all these species is needed to prove whether they are connected with *S. jessoensis* Seemen and whether the sections *Fragiles* and *Albae* can be separated by the glands in the ♀ flowers or not. The ventral glands of the ♂ flowers of *S. mixta* Korshinsky are variously dentate or lobulate in the specimen before me, and the dorsal gland is smaller and often entire, while Lackschewitz says: "nectarium duplex, internum oblongum, externum lineare." According to Komarov in *Act. Hort. Petrop.* XXII. 28 (*Fl. Mansh.*) (1903) *S. mixta* is a hybrid between *S. purpurea* Linnaeus and *S. viminalis* Linnaeus, but this is certainly a mistake. Nakai (in *Jour. Coll. Sci. Tokyo*, XXXI. 215 [*Fl. Kor.* II.] [1911]) mentions *S. mixta* from Korea.

There is another species of this section, *Salix elegantissima* K. Koch (in *Wochenschr. Ver. Beförd. Gartenb. Preuss.* XIV. 380 [1871]; *Dendr.* II. pt. 2, 505 [1872]), which is supposed to have been introduced into Europe from Japan. It may represent a hybrid between *S. babylonica* Linnaeus and *S. hondoensis* Koidzumi. Von Seemen (*Salic. Jap.* 77 [1903]) believes *S. elegantissima* a hybrid between *S. babylonica* and *S. fragilis* Linnaeus, like *S. blanda* Andersson (in *Svensk. Vetensk. Akad. Handl.* VI. 50 [*Monog. Salic.*] [1867]); and Wolf (1900) and I (1904) reached the same conclusion. But in Ascherson & Graebner (*Syn. Mitteleur. Fl.* IV. 73 [1908]) Seemen places *S. elegantissima* in the sect. *Fragiles* without any other remarks than: " Vielleicht aus Japan stammend." See also *S. lasiogyne* Seemen, p. 111.

Sect. 7. ALBAE Borrer in Hooker, *Brit. Fl.* 418 (1830); in Loudon, *Arb. Brit.* III. 1522 (1838). — Kerner in *Verh. Zool.-Bot. Ges. Wien.* X. 185 (1860). — Seemen in Ascherson & Graebner, *Syn. Mitteleur. Fl.* IV. 78 (1908).

Salix, sect. *Subalbae* Koidzumi in *Tokyo Bot. Mag.* XXVII. 88 (1913), in adnot.

Vide sect. 6, ab ea praecipue differt petiolis semper eglandulosis, floribus femineis glandula una ventrali instructis.

29. **Salix hondoensis** Koidzumi in *Tokyo Bot. Mag.* XXVII. 88 (1913).

Salix dolichostyla, var. *hirosakensis* Léveillé & Vaniot in Fedde, *Rep. Spec. Nov.* III. 22 (1906).

Salix hirosakensis Koidzumi in *Tokyo Bot. Mag.* XXVII. 264 (1913).

JAPAN. Hondo: prov. Mutsu, Hirosaki, "secus aquas," May 1905, *U. Faurie* (No. 6602, type of var. *hirosakensis,* ♀); Aomori, in streets, May 1905, *U. Faurie* (No. 6622; ♀); " basi montis Iwagisan in pago Dake," May 1905, *U. Faurie* (No. 6601; ♂); Hokkaido: prov. Ishikari, "in planitie Sapporo," June 14, 1908, *U. Faurie* (No. 256; ♀).

Koidzumi founded *S. hondoensis* on a specimen from Sapporo without mentioning a precise type. Taking up the name *hirosakensis* he says: floribus femineis saepe biglandulosis. In Faurie's type the ♀ flowers have only one gland, but otherwise the specimen looks very much like *S. eriocarpa* Franchet & Savatier. The ♂ flowers in No. 6601 have rather small glands sometimes appearing like a single ventral gland almost surrounding the pedicel. Our species certainly is very nearly related to *S. eriocarpa* and *S. jessoensis* Seemen. Whether they are hybrids or varieties of one species has not yet been determined.

30. **Salix Makinoana** Seemen in Fedde, *Rep. Spec. Nov.* I. 173 (1905).

Salix gymnolepis Léveillé in Fedde, *Rep. Spec. Nov.* III. 22 (1906).

Salix purpurea, subspec. *gymnolepis* Koidzumi in *Tokyo Bot. Mag.* XXVII. 267 (1913).

JAPAN. Hondo: prov. Mutsu, "in sylvis Aomori," May 1904, *U. Faurie* (No. 5761, type, ♀, ex Seemen), "in plateis Aomori," May 1904, *U. Faurie* (No. 2769; co-type, ♀, ex Seemen); Hirosaki, "secus rivos," May 1905, *U. Faurie* (No. 6615, type of *S. gymnolepis;* ♀).

This species has been very differently interpreted by Koidzumi, l. c., but unfortunately I have not seen one of von Seemen's types. Koidzumi says, that No. 5761 partly belongs to *S. stipularis* Smith (in which he includes *S. opaca* Andersson and *S. sachalinensis* Schmidt, see p. 143), and partly to *S. Miyabeana* Seemen, see p. 166. No. 2769 is referred by Koidzumi to his subspec. *gymnolepis.* Léveillé's type agrees very well with von Seemen's good description of *S. Makinoana.* Otherwise *S. gymnolepis* Léveillé much resembles *S. hondoensis* Koidzumi (*S. dolichostyla,* var. *hirosakensis* Koidzumi). The bracts are not perfectly glabrous, as Léveillé says, but hairy at the base. The ovaries have very short pedicels, as indicated by von Seemen for *S. Makinoana,* and von Seemen's whole description of the ♀ flowers and fruits fits exactly the specimen of *S. gymnolepis* Léveillé before me. I believe that *S. Makinoana* Seemen and *S. gymnolepis* Léveillé are nearly related to *S. hondoensis* Koidzumi and have no connection with *S. purpurea* Linnaeus.

31. **Salix jessoensis** Seemen, *Salic. Jap.* 31, t. 3, fig. F–L (1903).

JAPAN. Hokkaido: prov. Ishikari, Sapporo, Park, August 23, 1914, *E. H. Wilson* (No. 7414; tree 15–17 m. tall, girth 2.4–3.9 m., bark grayish, shallowly fissured; sterile); same locality, May 18, 1891, *Y. Tokubuchi* (type, ex Herb. Sapporo Coll., ♀ and ♂); same locality, May 1890, *Y. Tokubuchi;* same locality, May 11, 1892, *Y. Tokubuchi* (♂); same locality, July 1, 1891, *Y. Tokubuchi* (sterile).

The sterile specimen of Tokubuchi has very hairy yellowish brown branchlets (suckers?), and the leaves are up to 13.5 cm. in length and 3 cm. in width and grayish and loosely silky beneath; the petioles are about 1 cm. long, and the stipules are broadly ovate at the base, very acuminate at the apex and ¾ the length of the petioles. This species differs from *S. eriocarpa* Franchet & Savatier and from *S. hondoensis* Koidzumi in the larger ♀ catkins, the shorter style and the shorter, truncate bracts, but, as I have said before, it is very doubtful whether these three species are specifically distinct or mere varieties of one species.

This Willow is common round Sapporo and in the park there are some notable trees. It is a tree from 20–25 m. tall with a trunk from 4 to 5 m. in girth, gray shallowly fissured bark, large ascending-spreading and spreading branches and gray-green leaves. The habit is singularly like that of the English Oak.

Pictures of this tree will be found under Nos. x452 and x453 of the collection of my Japanese photographs.   E. H. W.

32. **Salix lasiogyne** Seemen, *Salic. Jap.* 32, t. 4, fig. A–C (1903).

NORTHEASTERN ASIA. Korea: Soeul, May 22, 1901, *U. Faurie* (No. 632, co-type ex Seemen).

JAPAN. Hondo: Yamakita, May 2, 1899, *U. Faurie* (No. 3702, type, ♀, ex Seemen); prov. Suruga, Numadzu, April 1895 (No. 63, ex Herb. Bot. Gard. Tokyo).

This species, of which I have not seen the type, is distinguished by its acute bracts, its nearly sessile stigmas and by the glabrous upper part of the ovaries. A specimen, collected by *U. Faurie*, in Hokkaido, Otara, culta, June 12, 1908 (No. 259), agrees well with von Seemen's description and figures, but the ovaries are wholly glabrous. As von Seemen points out, the species looks very much like *S. babylonica* Linnaeus. I am inclined to believe that it may have a parentage similar to that of *S. elegantissima*, Koch, p. 109. The specimen from Korea may belong to *S. koreensis* Andersson.

33. **Salix koreensis** Andersson in De Candolle, *Prodr.* XVI. pt. 2, 271 (1868). — Nakai in *Jour. Coll. Sci. Tokyo*, XXXI. 215 (*Fl. Kor.* II.) (1911). — Koidzumi in *Tokyo Bot. Mag.* XXVII. 89 (1913).

Salix pogonandra Léveillé in Fedde, *Rep. Spec. Nov.* X. 436 (1912).
Salix pseudo-Gilgiana Léveillé, l. c. (1912).
Salix pseudo-lasiogyne Léveillé, l. c. (1912).
Salix pseudo-jessoensis Léveillé, l. c. (1912).
Salix Feddei Léveillé, l. c. (1912).

NORTHEASTERN ASIA. Korea: " in ora boreali," *A. Schlippenbach* (type, ex Andersson); Quelpaert: " Piento Tchimpat," 400 m., April 14, 1908, *U. Faurie* (No. 4706, type of *S. pogonandra*); " in pago Polmongi," April 8, 1908, *U. Faurie* (No. 4707); same locality, end of April 1908, *U. Faurie* (No. 1447, type of *S. pseudo-jessoensis*); " in pago Hiotan(?)," April 14, 1908, *U. Faurie* (No. 4708); without locality, April 1909, *U. Faurie* (No. 3240, type of *S. pseudo-Gilgiana*); " in humidis vulcani," April 1909, *U. Faurie* (No. 3241), " in sepibus Setchimin," May 1909, *U. Faurie* (No. 3242, type of *S. Feddei*); in Chemulpo, May 1909, *U. Faurie* (No. 3243, type of *S. pseudo-lasiogyne;* No. 3244); " in jugo Pomasa," May 21, 1906, *U. Faurie* (No. 175); Fusan, "secus rivulos," May 1906, *U. Faurie* (Nos. 176, 181).

I have not seen Andersson's type, but all of Faurie's specimens agree well with Andersson's and Koidzumi's descriptions. There are really no differences at all between these forms described by Léveillé as different species. The pubescence of the ovaries is variable, as is the shape of the gland and the stigmas and the length of the style and of the ♀ catkins. At the time of flowering the ♀ catkins are short (1 cm.) and oval; the fruiting aments are up to 2.5 cm. long and 0.9 cm. broad. The catkins are not really sessile, although they are precocious, often losing early the small leaves of the short peduncle. The ♂ catkins are from

2 to 3.5 cm. long without the peduncle which is from 0.5 to 1 cm. in length. The smaller leaves are ovate or ovate-lanceolate and acute, the largest broadly lanceolate, very acuminate, and up to 9 cm. long and to 2.5–3 cm. wide.

34. **Salix babylonica** Linnaeus. See p. 42.

35. **Salix alba** Linnaeus, *Spec.* 1021 (1753). — Brandis, *Forest Fl. Brit. Ind.* 466 (1874); *Ind. Trees* 637 (1906). — Hooker f., *Fl. Brit. Ind.* V. 629 (1888). — Seemen in Ascherson & Graebner, *Syn. Mitteleur. Fl.* IV. 78 (1908). — Henry in Elwes & Henry, *Trees Gr. Brit. & Irel.* VII. 1759 (1913).

For further synonyms see Seemen, l. c.

INDIA.   Kashmir: Mulbe to Dras, October 8–11, 1856, *Schlagintweit* (No. 4970; sterile).

AFGHANISTAN.   Without precise locality, 1884–5, *J. E. T. Aitchison* (No. 1118; sterile).

The above specimens are referred not without doubt to this species, which may be planted in the northwestern Himalaya and Tibet according to Brandis and Hooker f.   The Willow of Kiangsu and Chusan Archipelago referred by Debeaux, Burkill and others to *S. alba* belongs to *S. babylonica* Linnaeus (see p. 00).

36. **Salix sericocarpa** Andersson in *Jour. Linn. Soc.* IV. 43 (1860). — Hooker f., *Fl. Brit. Ind.* V. 637 (1888).

? Salix lenta Fries, *Novit. Fl. Suec. Mant.* I. 78 (1832).
Salix alba, var. *eriocarpa* Hooker & Thomson, in *Herb. Ind. Or.*

INDIA.   Kashmir: " reg. temp., 600 ped." *T. Thomson* (types; ♂ and ♀).

AFGHANISTAN.   Without locality, 1884–5, *J. E. T. Aitchison* (No. 1116; ♀); Kurrum valley, *J. E. T. Aitchison* (No. 1207; ♀).

In the types the leaves are more or less denticulate; the ♂ flowers have two glands, of which the dorsal gland is only a little smaller; the ovaries are nearly sessile and bear a short style with narrow shortly cleft stigmas. The specimen No. 1116 from Afghanistan has entire leaves which are more or less silky on both sides or are nearly glabrous above. The ♀ flower and fruit show a short pedicel nearly as long as the single gland. This interesting species needs further investigation.

S. lenta Fries is a very uncertain species, found in Nepal by Wallich, according to the author. The ♂ flowers are described as " diandri."

Sect. 8. SCLEROPHYLLAE Schneider, n. sect.

Frutices mediocres v. parvi, erecti, breviter ramosi. Folia elliptica v. elliptico-oblonga v. ovata. Amenta coetanea, pedunculata v. ♂ sessilia; flores ♂ diandri, glandulis 2, filamentis sub medio pilosis; ♀ glandulis 2 simplicibus v. interdum plurilobatis, ovariis sessilibus v. subsessilibus pubescentibus, stylis distinctis bifidis v. bipartitis, stigmatibus bifidis.

It is with a good deal of hesitation that I unite in this new section the following species. The value of the presence or absence of the second (dorsal) gland in the ♀ flower in taxonomic classification needs further investigation. It seems certainly not sufficient to justify the large groups made by von Seemen (as *Didymadeniae*, *Heteradeniae* and *Monadeniae*) based on the glands of the ♀ flowers.

37. **Salix sclerophylla** Andersson in *Jour. Linn. Soc.* IV. 52 (1860); in *Svensk. Vetensk. Akad. Handl.* VI. 148, t. 8, fig. 82 (*Monog. Salic.*) (1867); in De Candolle, *Prodr.* XVI. pt. 2, 248 (1868). — Brandis, *Ind. Trees*, 638 (1906).

INDIA.   Kashmir: " Laptal, 15000 ped.," *Strachey & Winterbottom* (No. 8, in part, type; ♂); " Rimkim, 13500 ped.," *Strachey & Winterbottom* (No. 8, in part; ♂); without locality, " 15000 ped.," *Strachey & Winterbottom* (No. 10; ♀); " Dras, 10000 ped.," *T. Thomson* (type of var. *pubescens;* sterile).

The male and female flowers of Nos. 8 and 10 have two narrow glands of which the dorsal is often very small, and apparently was overlooked by Andersson. In his first description he says: " filamentis glabris," but in the second " filamentis liberis inferne pilosis." The style is short, a little cleft with short bifid stigmas. According to Brandis, this species is an inhabitant of " inner arid valleys."

Andersson, l. c. (1868), describes two varieties: var. *a glabra* " foliis utrinque glabris laete viridibus," which seems to be a mistake, none of his specimens having glabrous leaves, and var. *β pubescens*, the leaves of which are " densius albovillosa " beneath. So far as I see, all the specimens have pubescent leaves especially Thomson's plant from Dras, in which the under surface of the leaves is very pubescent; they are broad, elliptical and often subcordate at the base. The nearest relative to this species seems to be the following:

38. **Salix oritrepha** Schneider, n. sp.

Frutex parvus, erectus, breviter ramosus, 0.6–1.2 m. altus; ramuli annotini vetustioresque griseo-tomentelli, deinde glabri, cinerei. Folia elliptica v. ovata, basi rotunda v. obtusa, apice obtusa v. acuta, apiculata, supra sordide viridia, initio pubescentia, deinde costa excepta interdum pilis fuscis instructa glabra, tenuiter inciso-reticulata, subtus pallida v. glauco-cinerea, initio sericeo-pilosa, mox glabra, nervis visibilibus sed vix prominentibus reticulata, margine integra, textura crassiuscula, minora 1 cm. longa et 0.8 cm. lata, maxima usque ad 2.4 cm. longa et ad 1.5 cm. lata; petioli 5–8 mm. longi, purpurei, pubescentes v. subtus glabri. Amenta tantum ♀ cognita, coetanea, pedunculo brevi piloso 3–7 mm. longo foliis parvis 2–3 normalibus instructa, 1–1.5 cm. longa, circiter 1 cm. crassa, densa; ovaria satis matura sessilia, crasse ovata, apice breviter attenuata, breviter crispo-villosa, pilis albis et ferrugineis mixtis, stylis brevibus integris quam stigmata angusta bifida subaequilongis coronata; bracteae late obovatae, ovario subduplo breviores, purpurascentes, nervosae, utrinque albo-sericeae; glandulae duae, pleraeque irregulariter anguste lobatae, pseudodiscum quasi formantes.

CHINA. W e s t e r n S z e c h ' u a n : southwest of Tachien-lu, mountains, alt. 3600–4000 m., July 1903 (Veitch Exped. No. 4525, type; ♀).

This interesting species looks very much like the Himalayan *S. sclerophylla* Andersson, but *S. oritrepha* may be easily distinguished from that species by the glands, the brownish and white pubescence of the ovaries, and by the pubescence of the older branchlets.

The specific name is derived from ὀριτρεφής, dwelling on the mountains.

39. **Salix atopantha** Schneider. See p. 43.

Sect. 9. MAGNIFICAE Schneider, n. sect.

Frutices alti ramulis elongatis satis crassis purpureis saepe leviter glaucis. Folia maxima, papyracea, laevia, late elliptica, obovata v. obovato-rotunda, obtusa, subtus glauca v. albescentia, nervis lateralibus paulo elevatis distantibus 1 (rarius 2) pro 1 cm., angulo fere recto a costa divergentibus et 5–10 mm. ante marginem evanescentibus, tenuissime reticulata, integra v. versus apicem nonnihil breviter denticulata; petioli longi, crassi. Amenta apice ramulorum folia normalia magna gerentium 18–30 cm. longa et usque 1.5–2 cm. crassa; flores ♂ diandri, glandulis 2, ventrali lata et crassa saepe emarginata bipartita v. triloba, dorsali minore, crasse obovata v. minima interdum ut videtur obsoleta; flores ♀ ovariis subsessilibus v. in fructu brevipedicellatis glabris v. tomentellis, stylis mediocribus crassis bifidis, stigmatibus crassis brevibus v. oblongis bilobis, glandula una ventrali crassa late ovata obtusa.

The living plants of *S. magnifica* Hemsley look extremely distinct, scarcely resembling any other known Willow.

40. **Salix magnifica** Hemsley. See p. 44.

41. **Salix ulotricha** Schneider. See p. 44.

42. **Salix pella** Schneider. See p. 45.

Sect. 10. ERIOSTACHYAE Schneider, n. sect.
Frutices mediocres v. alti, ramosi. Folia satis magna, elliptica, ovata, obovata v. obovato-oblonga pleraque dentata v. serrata, textura satis firma, nervis parallelibus subtus satis prominulis et etiam reticulata; petioli distincti, interdum glanduliferi. Amenta ramulis plus minusve longis folia normalia gerentibus suffulta, 5–13 cm. longa; flores ♂ diandri, glandulis 2 separatis; flores ♀ ovariis sessilibus subsessilibus v. in fructus satis distincte pedicellatis glabris v. tomentellis, stylis saepe satis longis plerisque bipartitis v. apice bifidis, stigmatibus brevibus v. oblongis bifidis, glandula una ventrali v. interdum etiam glandula parva dorsali praediti.

S. eriostachya Andersson, of which the ♂ plant is yet unknown, agrees so well with the other species of this section that I do not hesitate to make this, which is the oldest species of the group, the type of my new section.

43. **Salix apatela** Schneider. See p. 46.

44. **Salix moupinensis** Franchet. See p. 46.

45. **Salix Fargesii** Burkill. See p. 47.

46. **Salix Ernesti** Schneider. See p. 47.

47. **Salix argyrophegga** Schneider. See p. 49.

48. **Salix plocotricha** Schneider. See p. 49.

49. **Salix spathulifolia** Seemen in *Bot. Jahrb.* XXXVI. Beibl. LXXXII. 31 (1905). — Léveillé in *Bull. Soc. Bot. France*, LVI. 304 (1909).

CHINA.   Shensi: Huon-tou shan, June 18, *G. Giraldi* (No. 5359, type, ♀, ex Seemen).

As far as I can judge from the description this species belongs to this (or the following) section and resembles *S. plocotricha* Schneider in many respects. The bracts are "eroso-denticulata" at the apex, the leaves are up to 8 cm. in length and to 2.5 cm. in width, and are described as "spathulata (v. inferiora oblonga)." Not having seen the type specimen, I cannot decide whether *S. plocotricha* Schneider is only a form of this species or not.

50. **Salix eriostachya** Wallich apud Andersson in *Svensk. Vetensk. Akad. Handl.* 1850, 493 (1851). — Andersson in *Jour. Linn. Soc.* IV. 46 (1860); in De Candolle, *Prodr.* XVI. pt. 2, 270 (1868). — Hooker f., *Fl. Brit. Ind.* V. 633 (1888). — Brandis, *Ind. Trees,* 637 (1906).

INDIA.   Nepal: "ad Gossain Than," 1821, *N. Wallich* (No. 3704, type; ♀).
By the kindness of the Keeper of the Herbarium of the Kew Gardens, I have been able to examine Wallich's type specimen. It shows the same long foliaceous peduncles, the deeply cleft styles and the broad, obovate rather truncate bracts, nearly enveloping the young ovaries as they do in most of the species of this section. The leaves are very indistinctly glandular denticulate or nearly entire, acute at the base and hairy on both sides, especially beneath; the petioles are from 3 to 9 mm. long and the leaves are up to 6.5 cm. in length and 2.3 cm. in width; with 2 or 3 lateral nerves in each cm. of length.

51. **Salix sikkimensis** Andersson in De Candolle, *Prodr.* XVI. pt. 2, 268 (1868).— Hooker f., *Fl. Brit. Ind.* V. 632 (1888). — Brandis, *Ind. Trees*, 638 (1906).

INDIA. Sikkim: " Lachen valley, alt. 13000 ped." June 13, 1849, *J. D. Hooker* (type; ♀).

I have seen a branchlet of the type specimen which agrees very well with Andersson's description, only the pubescence of the leaves and catkins is not really " cupreo-micans," but more yellowish white. The species differs from the others of this section in the nearly sessile catkins, but I have only seen very young aments appearing with the leaves. The ♂ plant mentioned by Hooker f. I do not know, but the ♀ flowers are very much like those of *S. Ernesti* Schneider or of other species of the *Eriostachyae*. The ovaries are not sessile, but very short-stalked, the pedicel being only one third the length of the rather long gland. There is no small dorsal gland as in *S. Ernesti*. The bracts of the young flowers are very large, broadly obovate, round at the apex and somewhat crenulate, and nearly envelop the whole flower. The deeply cleft styles are about one-half as long as the pubescent ovaries, the stigmas are rather short, obovate, emarginate or bifide. The stout flowering branch resembles that of *S. Caprea* Linnaeus or of *S. daphnoides* Villars and is apparently somewhat glaucous. The mature leaves are not yet known.

Burkill (in *Jour. Linn. Soc.* XXVI. 532 [1899]) mentions *S. sikkimensis* from Yunnan (Delavay No. 2792), but I have not seen the specimen.

Sect. 11. PSILOSTIGMATAE Schneider, n. sect.

Arbores v. frutices alti, rarius parvi. Folia elliptica v. pleraque ovato-lanceolata v. lanceolata, satis magna v. mediocra, rarius parva, subtus pleraque sericea v. sericeo-tomentella, plus minusve nervosa sed rarius reticulata. Amenta brevius v. longius pedunculata, cylindrica, densiflora, ♀ pleraque 3–12 cm. longa; flores ♂ diandri glandulis 2 separatis; flores ♀ v. fructus sessiles; ovaria sericea v. tomentella (in specie dubiae affinitatis glabra), stylo brevi v. satis longo, bipartito v. bifido, stigmatibus bifidis coronata; glandula una ventralis.

I have some doubt whether the species, united by me in this new section, form a natural group. *S. Daltoniana* Andersson may perhaps be better placed in the preceding section, while *S. Thomsoniana* Andersson differs from the others in being a very low shrub with small leaves.

52. **Salix Daltoniana** Andersson in *Jour. Linn. Soc.* IV. 49 (1860); in De Candolle, *Prodr.* XVI. pt. 2, 269 (1868). — Hooker f., *Fl. Brit. Ind.* V. 632 (1888). — Brandis, *Ind. Trees*, 637 (1908).

INDIA. Sikkim: " reg. temp., alt. 9000–14500 ped." (Andersson says 1868: 14000), *J. D. Hooker* (type, of which I have seen a co-type). Bhutan: without locality, *W. Griffith* (No. 4498, ex Hooker).

Burkill (in *Jour. Linn. Soc.* XXVI. 528 [1899]), refers to *S. Daltoniana* some forms collected by Delavay in Yunnan, of which I have not seen specimens. He also describes a var. *Franchetiana*, the types of which were collected by David in Mupin and by Mussot (No. 348) at Tachien-lu. In this variety the gland appears, according to Burkill, to be single, while two are present in the type. Without having seen this specimen, I cannot tell anything about this variety or about Delavay's No. 988 from Tali in which Burkill recognized " a hybrid between *S. Daltoniana* and *S. elegans* " (= *S. denticulata* Andersson). The last species, so far as I know, is a native of the northwestern Himalaya and Afghanistan and does not occur in China.

S. Daltoniana, var. *crassijulis* Andersson (in De Candolle, *Prodr.* XVI. pt. 2, 279 [1868]), the type of which was collected by *J. D. Hooker* in Sikkim between " 9000–14000 ped.," I know only from Andersson's description, but I think it belongs to a different species.

The typical *S. Daltoniana* is easily distinguished by the brownish pubescence mixed with white of the lower surface of the elliptic-lanceolate indistinctly crenulate-dentate leaves, the short ♂ catkins, the bracts of which are oblong and truncate and denticulate at the apex, and by the long female catkins. In the ♀ flowers with their large bracts and the long deeply cleft style *S. Daltoniana* much resembles *S. Ernesti* Schneider and other species of sect. *Eriostachyae.*

53. **Salix phanera** Schneider. See p. 50.

54. **Salix phaidima** Schneider. See p. 51.

55. **Salix psilostigma** Andersson in *Svensk. Vetensk. Akad. Handl.* 1850, 496 (1851).

  Salix eriophylla Andersson in *Jour. Linn. Soc.* IV. 48 (1860); in De Candolle, *Prodr.* XVI. pt. 2, 270 (1868). — Hooker f., *Fl. Brit. Ind.* V. 633 (1888). — Burkill in *Jour. Linn. Soc.* XXVI. 528 (1899). — Brandis, *Ind. Trees*, 638 (1908). — Diels in *Not. Bot. Gard. Edinburgh*, VII. 286 (*Pl. Chin. Forrest.*) (1912).
  ? Salix Smithiana Andersson in *Jour. Linn. Soc.* IV. 48 (non Wildenow) (1860).
  ? Salix viminalis, var. *Smithiana* Hooker f., *Fl. Brit. Ind.* V. 632 (1888).

CHINA. Hupeh: Changlo Hsien, *A. Henry* (No. 6274; ♂). Yunnan: Mengtsze, alt. 3500 m., *A. Henry* (No. 9338[b]; tree, 4 m. tall; ♂); same locality, grass mountains, alt. 2000 m., *A. Henry* (No. 10493; shrub 1.5 m. tall; ♂ and ♀); same locality, mountains north, alt. 2000 m., *A. Henry* (No. 14493[a]; tree 3–4.5 m. tall; ♂); Ta-lei-shan, July 23, *A. Henry* (No. 10209; with fruits and old leaves); Tali valley, alt. 2100 m., May 1906, *G. Forrest* (No. 4967; shrub 0.6–1.8 m. tall; ♂).

INDIA. Assam: " Khasi Hills, 4–5000 ped.," *J. D. Hooker & Thomson* (type of *S. eriophylla*, of which I have seen only a sterile branch; the type of *S. psilostigma* was collected by *Jacquemont* in India). Sikkim: "reg. temp. 5–8000 ped.," *J. D. Hooker* (type of *S. viminalis*, var. *Smithiana*, a doubtful sterile form). Eastern Bengal: without locality, *W. Griffith* (No. 4500; ♂).

As stated above, p. 51, the specimen from Hupeh resembles *S. phaidima* Schneider in its pubescence, but agrees with *S. psilostigma* Andersson in its short petioles and short aments. The typical *S. psilostigma* has a dense silky and woolly pubescence; the ♂ catkins are very short-stalked and from 3 to 6 cm. long; the fruiting aments are up to 8 cm. in length and 1 cm. thick; the style is deeply cleft and hidden at the base by the hairs of the ovary. The stigmas are bifid, oblong and often a little recurved.

56. **Salix Thomsoniana** Andersson in *Jour. Linn. Soc.* IV. 54 (1860); in De Candolle, *Prodr.* XVI. pt. 2, 297 (1868).—Hooker f., *Fl. Brit. Ind.* V. 635 (1888).—Brandis, *Ind. Trees*, 638 (1908).

INDIA. Sikkim: "reg. temp. alt. 10000 ped.," *J. D. Hooker* (types; ♂ and ♀); Lachoong, *R. Pantling* (ex Hooker).

On account of the two glands of the ♂ flowers I put this species in this section. It differs from the other species, however, in being a very small shrub with small lanceolate-elliptic leaves. The ovaries and styles are much the same as in *S. psilostigma* Andersson; the solitary gland is about half as long as the ovary; the broad-ovate or roundish bracts are glabrous within, and in the male plant lobulate-denticulate.

57. **Salix radinostachya** Schneider, n. sp.

Frutex?; ramuli initio sparse sericei, ut videtur mox glabri, purpureo-brunnei; gemmae ovato-oblongae, obtusae, purpureo-brunneae, glabrae, subdivaricantes.

Folia (tantum juvenilia visa) oblanceolata, basin versus angustata, acuta, apice brevius longiusve acutata, minora elliptica, utrinque acuta, supra viridia, tantum costa sericea, subtus glauca, initio satis, deinde tantum ad costam albo-sericea, nervis angustis parallelis paullo prominulis circiter 3 pro 1 cm., margine subintegra v. brevissime glanduloso-undulato-denticulata, minora ramulorum floriferorum 1.5–3.5 cm. longa, 1.8 cm. lata, maxima lanceolata ad 9 cm. longa et supra medium 2.5 cm. lata; petioli subsericei, 5–9 mm. longi. Amenta ♀ gracillima, curvata, ramulis circiter 1 cm. longis folia 3–4 parva normalia gerentibus pedunculata, 5–7 cm. longa, circiter 0.5 cm. crassa, rhachi sparse sericea; ovaria glabra, breviter pedicellata, pedicellis glandula una oblonga apice saepe erosa ventrali subbrevioribus, stylis subaequilongis vix bifidis stigmatibus satis longis bifidis recurvatis coronata; bracteae ovato-oblongae, apice plus minusve truncatae, brunnescentes, tantum ad basim pilosae, ovariis subaequilongae.

INDIA. Sikkim: "reg. temp. alt. 9000 ped.," *F. D. Hooker* (type, ♀; sub nomine *S. elegans* Wall. γ, ex Herb. Ind. Or. in Herb. Gray).

This interesting species differs from all the species of this section in its glabrous ovaries. In the long thin aments it resembles *S. Daltoniana* Andersson, in other respects somewhat *S. longiflora* Andersson, from which it differs in its short pedicels and long styles. As far as I can judge from the poor specimen before me, it seems near *S. Daltoniana* Andersson, but needs further investigation.

The specific name is derived from ῥαδινός, slender, and σταχύς, spike.

Sect. 12. DENTICULATAE Schneider, n. sect.

Frutices breviramosi v. interdum arbores parvae. Folia satis parva, elliptica, breviter denticulata v. integra, pleraque tantum initio pilosa, subtus glauca. Amenta coetanea pedunculis foliolatis instructa, ♂ breviora, saepissime satis laxiflora, ♀ satis longa et densiflora; flores ♂ diandri, glandula dorsali et ventrali instructa, staminibus liberis basi v. ad medium villosis; bracteae obovatae, obtusae, v. obovato-spathulatae; ovaria breviter pedicellata v. sessilia, glabra v. pilosa, stylis subnullis v. distinctis apice breviter bifidis, stigmatibus bifidis; glandula una, ventralis, oblonga; bracteae obovatae.

I unite the following species in this section on account of the two glands of the ♂ flowers.

58. **Salix denticulata** Andersson in *Svensk. Vetensk. Akad. Handl.* 1850, 481 (1851). — Klotzsch & Garcke, *Bot. Ergeb. Reise Prinz. Waldemar,* 119, t. 89 (1862).

Salix kamaunensis Lindley in Wallich, *Cat.* No. 3701 (nomen nudum) (1829).
Salix elegans Wallich, *Cat.* No. 3699 (nomen nudum) (non Besser, nec Host) (1829) et apud Andersson in *Jour. Linn. Soc.* IV. 51 (1860); in *Svensk. Vetensk. Akad. Handl.* VI. 168, t. 9, fig. 99 (*Monog. Salic.*) (1867); in De Candolle, *Prodr.* XVI. pt. 2, 256 (1868). — Brandis, *Forest Fl. Brit. Ind.* 466 (1874). *Ind. Trees,* 637 (1908). — Hooker, *Fl. Brit. Ind.* V. 630 (1888). — Collett, *Fl. Siml.* 479, fig. 156 (1902).

INDIA. Kashmir: "Rajhoti, alt. 15000 feet," *Strachey & Winterbottom* (No. 9; ♂). Garhwal: "Niti, 11500 feet," *Strachey & Winterbottom* (No. 7; with fruits); Badyah, alt. 2600 m., June 1894, *C. G. Rogers* (with fruits, androgyn); "Chakrata, 7000 ft.," May 14 and June 24, 1912, *K. N. Jyengar* (♂ and with fruits); without precise locality, "reg. temp. 6–9000 ped.," *T. Thomson* (♂ and ♀); without precise locality, *H. Falconer* (No. 961; ♀); Kumaon: "Dugli, 13000 ft.," *Strachey & Winterbottom* (No. 6; with fruits); Baltol, August 28, 1831, *V. Jacquemont* (Nos. 565, 966, 1035; co-types of *S. denticulata,* ex Andersson); without precise locality, *R. Blinkworth* (No. 3701 Herb. Wallich, type of *S. denticulata,* ex Andersson), Nepal: without precise locality, 1821, *N. Wallich* (No. 3699ᵃ, type of *S. elegans;* with fruits); Punjab: Sirmore, *G. Govan & Kamrup* (No. 3699ᵇ, in part, Herb. Wallich).

AFGHANISTAN. Kurrum valley, *J. E. T. Aitchison* (No. 413; ♂ and ♀).

So far as I know this species does not occur in the eastern Himalaya nor in China. Burkill (in *Jour. Linn. Soc.* XXVI. 528 [1899]), confused it with another species. For further information see the keys on pp. 77 and 78.

No. 9 of Strachey & Winterbottom seems to be a smaller high-alpine form. The bracts are glabrous. It looks somewhat intermediate between *S. denticulata* and *S. flabellaris* Andersson.

Andersson described the following variety, of which I have not seen a specimen:

Salix denticulata, var. **himalensis** Andersson in *Svensk. Vetensk. Akad. Handl.* 1850, 482 (1851).

Salix elegans, β *Govaniana* Wallich, *Cat.* No. 3699ᶜ (nomen nudum) (1829).
Salix himalensis Klotzsch in *herb.* (ex Andersson).
Salix elegans, var. *himalensis* Andersson in *Jour. Linn. Soc.* IV. 51 (1860); in *Svensk. Vetensk. Akad. Handl.* VI. 168 (*Monog. Salic.*) (1867).
Salix elegans, var. *Govaniana* Andersson in De Candolle, *Prodr.* XVI. pt. 2, 257 (1868). — Hooker f., *Fl. Brit. Ind.* V. 630 (1888).

INDIA. Kashmir: without precise locality, *W. Hofmeister* (ex Andersson); "Serinagur, 8000 feet," *G. Govan* (No. 3699ᵇ, Herb. Wallich, in part); Punjab: Sirmore, *G. Govan & Kamrup* (No. 3699ᶜ, Herb. Wallich, in part, ex Wallich).

"Taller, leaves larger, almost lanceolate, more sharply serrulate, more glaucous beneath," fide Andersson and Hooker.

59. **Salix dissa** Schneider. See p. 52.

60. **Salix dyscrita** Schneider. See p. 53.

61. **Salix Biondiana** Seemen in *Bot. Jahrb.* XXXVI. Beibl. LXXXII. 32 (1905). — Léveillé in *Bull. Soc. Bot. France*, LVI. 297 (1909).

CHINA. Western Hupeh: Fang Hsien, top of mountains, June 1900 (Veitch Exped. No. 2045, co-type; bush 1.2–1.5 m.; ♂). Shensi: "Pao-ki-scen, Miao-wang-shan," July, *Hugh Scallan* (No. 5361, type; Herb. Giraldi; ♀, ex Seemen).

I have seen only a ♂ co-type of this species (Herb. N. York Bot. Gard.), the flowers of which have a rather long ventral gland two-fifths the length of the bracts, and a very small dorsal gland from one-third to one-fourth as long as the ventral gland. The bracts are oblong, rounded at the apex, while von Seemen says: "bracteis late ovatis irregulariter dentatis." According to von Seemen the dorsal gland seems to be somewhat larger in his type, and the ♂ flowers have sometimes 3 stamens, a fact I did not observe in the co-type. Among the plants collected by Wilson for the Arboretum I have not found a specimen agreeing with the co-type or with Seemen's description. The fruits are described as "sitzend, kurz dünn grau behaart," the style as short and thick, the stigmas as "oval, tief ausgerandet, aufrecht gabelig," the gland as "a basi lata truncata capsulae ¼ aequante." The species of this section are very difficult to distinguish, and a careful field study of them is needed.

62. **Salix erioclada** Léveillé in Fedde, *Rep. Spec. Nov.* III. 22 (1906); in *Bull. Soc. Bot. France*, LVI. 299 (1909); *Fl. Kouy-Tchéou*, 381 (1915).

CHINA. Kweichau: "Montée de Pia-Fong à Sa-Jang," March 4, 1905, *J. Esquirol* (No. 567, type; ♂).

I have only seen a ♂ catkin and two small leaves of the type which were kindly sent to me by Mgr. Léveillé. The flowers have two glands and the species very much resembles *S. dyscrita* Schneider. The bracts of *S. erioclada*, however, are more oblong, the filaments are very slightly hairy at the base, and the young leaves have a different nervation, the lateral nerves being more numerous, about 10 in a leaf of 1.8 cm. in length, while a leaf of the same size of *S. dyscrita* Schneider has only about 5 or 6. As long as it is impossible to compare better material with fully grown leaves, I prefer not to unite these two forms, which are geographically well separated.

Léveillé (in litt. 1915) says that under the two names *S. erioclada* and *S. pachyclada* (see p. 150) " se cachent certainement plusieurs espèces."

63. **Salix Camusii** Léveillé in *Bull.Soc. Agr. Sci. Sarthe*, sér. 2, XXXI. 326 (1904); in *Bull. Soc. Bot. France*, LVI. 297 (1906); *Fl. Kouy-Tchéou*, 381 (1915).

CHINA. Kweichau: " Kouy-yang, Gan-pin, etc., mont du Collège," April 1898, *E. Bodinier* (No. 2134, types).

Through the kindness of the author, I have received a ♂ and a ♀ catkin of this species the original description of which I have not been able to consult. The following is the description of what I have seen:

Ramuli?. Folia juvenilia elliptica, utrinque subobtusa, in sicco nigrescentia, supra costa albo-villosa subincisa excepta glabra, venis vix visibilibus, subtus glaucescentia, costa elevata facieque sparse v. distinctius adpresse sericea v. costa sericeo-villosula, nervis paullo elevatis pro 2.5 cm. circiter 8, margine minute et indistincte adpresse glanduloso-denticulata, ad 2.5 cm. longa et 1.1 cm. lata; petioli circiter 2–4 mm. longi, dense albo-cani. Amenta ut videtur subcoetanea, subsessilia, basi foliolis 1–3 parvis subtus sericeis suffulta, cylindrica, rhachi villosula, ♂ ad 5.5 cm. longa, circiter 0.7 cm. crassa, ♀ (florifera) ad 3.5 cm. longa, circiter 0.6 cm. crassa; flores ♂ diandri, filamentis glabris liberis quam bracteæ 4–5-plo longioribus, antheris parvis flavis subglobosis; glandulae 2, ventralis oblonga, apice incisa, bractea paullo brevior, dorsalis parva, tenuia, ventrali minor; bracteae parvae, ovato-rotundae, concavae, flavo-brunnescentes, margine et basi villosae, ceterum plus minusve glabrae; ovaria sessilia v. subsessilia, ovata, stylo parvo sed distincto satis crasso apice bifido stigmatibus minimis emarginatis coronata; glandula una ventralis, bractea fere duplo brevior, oblonga, angusta; bracteae late ovatae, obtusae, iis florum ♂ non absimiles, pleraeque tantum basi utrinque villosae.

This species seems nearly related to *S. erioclada* Léveillé, p. 118, which chiefly differs in its longer oblong bracts. The ♂ catkins and young leaves look very much alike, but the latter seem to be entire in *S. erioclada*. *S. dyscrita* Schneider also much resembles the ♂ plant of *S. Camusii*, but in *S. dyscrita* the filaments are very hairy at the base, and the young entire leaves show fewer pairs of lateral nerves. Unfortunately, the ♀ sex neither of *S. dyscrita* Schneider nor of *S. erioclada* Léveillé is known.

64. **Salix luctuosa** Léveillé in Fedde, *Rep. Spec. Nov.* XIII. 342 (1914).

CHINA. Eastern Yunnan: " brousse des montes à Kiao-Mê-ti, alt. 3200 m.," May 1913, *E. E. Maire* (type; ♂ and ♀).

According to the two glands in the ♂ flower, this species belongs to sect. *Denticulatae*. The ♀ flowers have a rather long style bifid at the apex with very small stigmas. Old leaves are unknown. See the key, pp. 77 and 78.

65. **Salix fruticulosa** Andersson in *Jour. Linn. Soc.* IV. 53 (exclud. planta ♀) (1860). — Hooker f., *Fl. Brit. Ind.* V. 637 (1888).

Salix furcata Andersson in De Candolle, *Prodr.* XVI. pt. 2, 291 (pro parte) (1868).

INDIA. Kumaon: "Pindari, alt. 12000 feet," *Strachey & Winterbottom* (No. 13; ♂ and ♀).

The types of *S. fruticulosa* and of *S. furcata* Andersson both came from Pindari. Andersson, in 1868, made apparently a mistake in confounding these two specimens. But there are some other points to clear up. In 1860 the ♀ plant is described: "capsula sessilis . . . dense sericea, . . .; stylus vix conspicuus; stigmata brevissima," as it is the case, except that there is a very short pedicel in Anderson's ♀ co-type from Zanzkar I have before me. There is, however, a different ♀ plant from Pindari with: "capsulis sessilibus . . . glabris . . . stylo bipartito, laciniis erectis, stigmatibus integris capitatis," as described by Andersson in 1868. This ♀ specimen has the same reddish brown branchlets as the ♂ co-type, while the Zanskar plant has branchlets with "cortice glabro fusco-nitente vel testaceo interdum glaucescente." After all the ♀ plant of the type belongs to *S. myricaefolia* Andersson, see p. 172. The real *S. fruticulosa* Andersson (1860), not Kerner (1864), the systematic position of which is yet uncertain, may be described as follows:

Fruticulus ut videtur parvus, breviramosus; ramuli satis crassi, rubro-brunnei, deinde cinerascentes, novelli villosuli. Folia novella paulo evoluta lanceolata v. oblanceolata, basi acuta, apice obtusiora, superne viridia, glabra, nervis paulo impressis, subtus glaucescentia, costa pilis sericeis obtecta v. mox glabra, costa nervisque vix prominulis margine tenuiter inciso-denticulata, rarius subintegra, vix 10 mm. longa et 4 mm. lata, brevissime petiolata. Amenta lateralia, brevipedunculata, basi foliis parvis suffulta, densiflora, rhachi laxe villosa; ♂ 1–1.5 cm. longa, circiter 0.9 cm. crassa; ♀ ovato-oblonga, 1 cm. longa, sed nondum satis evoluta; flores ♂ diandri, filamentis liberis basi pilosis bracteas paulo superantibus, antheris ovato-globosis, violaceis, glandula ventralis anguste oblonga, ⅓ bracteae aequans; dorsalis subaequilonga, tenuior; bracteae obovato-oblongae, apice plus minusve truncatae, denticulatae, flavescentes v. violascentes, nervatae, glabrae; ovaria sessilia, glabra, juvenilia bracteis breviora, stylo bipartito subaequilongo stigmatibus brevibus emarginatis capitatis coronata; glandula una ventralis, late ovata, obtusa v. lobulata, compressa, ovario juvenili ½ brevior; bracteae iis florum ♂ similes, dorso sericeae.

Sect. 13. LONGIFLORAE Schneider, n. sect.

Frutices breviter ramosi. Folia parva v. mediocra, elliptica, elliptico-rotunda, elliptico-lanceolata v. rarius oblanceolata, integra (v. indistincte denticulata), subtus pallidiora, cinerea v. glauca, pleraque tantum initio distincte pilosa. Amenta coetanea pedunculis foliatis instructa, v. rarius subpraecocia, subsessilia, cylindrica, densi- v. sublaxiflora; flores ♂ diandri, glandula una ventrali instructi, staminibus liberis basi pilosis, bracteis obovatis obtusis; ovaria sessilia v. subsessilia, glabra v. pilosa, stylis nullis v. plus minusve distinctis, stigmatibus sessilibus brevibus bifidis angustioribus, glandula una ventrali, bracteis obovatis v. oblongis obtusis.

This section differs from the *Denticulatae* chiefly in the single gland of the ♂ flowers. The species, of which only the ♀ flowers are known, are doubtfully included here, but the co-type of *S. longiflora* Andersson is so much like *S. cathayana* Diels that I use the name *Longiflorae* for this section.

66. **Salix hypoleuca** v. Seemen. See p. 53.

  Salix hypoglauca, var. **platyphylla** Schneider. See p. 53.

67. **Salix rhoophila** Schneider. See p. 54.

68. **Salix polyclona** Schneider. See p. 55.

69. **Salix mictotricha** Schneider. See p. 56.

70. **Salix longiflora** Andersson in *Jour. Linn. Soc.* IV. 50 (1860); in De Candolle, *Prodr.* XVI. pt. 2, 271 (1868). — Hooker f., *Fl. Brit. Ind.* V. 633 (1888). — Brandis, *Ind. Trees*, 637 (1908).

INDIA. Sikkim: " Lachen, alt. 9000 ped.," *J. D. Hooker* (type; ♀, of which I have seen a co-type ex Herb. Gray).

Judging by the co-type of this species I was at first inclined to unite it and *S. cathayana* Diels (see p. 57), of which Diels says: " The differences between this species and *Salix longiflora* And., of the Sikkim Himalaya, are obvious." Unfortunately they are not so in the specimens before me. According to the descriptions given by Andersson and Hooker f., *S. longiflora* may be distinguished by its long ♀ catkins ("2–4 pollicaria" [Andersson] "2–5 in." [Hooker]) and by the apparently longer leaves ("adulta 1½–2½ pollices longa" [A.]; " larger 3–6 in." [H.]). In the co-type the largest leaves are up to 5 cm. long and 1.5 cm. broad. Of the pubescence, there seem to be at least two forms: one with the leaves only slightly silky beneath when young and glabrous later in the season; and another with tomentose young leaves, the pubescence being thickish and somewhat brownish. It needs a careful examination of many specimens before it is possible to decide, if the Chinese and Sikkim forms are distinct or not. See *S. cathayana* Diels, p. 57.

Burkill describes a variety *S. longiflora*, var. *albescens* (in *Jour. Linn. Soc.* XXVI. 530 [1899]), the type of which was collected by *Mussot* (No. 347) at Ta-chien-lu, Szech'uan. Not having seen the specimen, I cannot tell from the short description to what species this plant may really belong.

71. **Salix cathayana** Diels. See p. 57.

72. **Salix macroblasta** Schneider. See p. 58.

73. **Salix resecta** Diels in *Not. Bot. Gard. Edinburgh*, vol. VII. 281 (*Pl. Chin. Forrest.*) (1912).

CHINA. Yunnan: Moist situations on the margins of thickets on the eastern flank of the Tali Range, alt. 3000 m., July 1906, *G. Forrest* (No. 4602, type; with fruits).

In the co-type before me the catkins are at the end of branchlets from 2 to 3 cm. long, with normal leaves and are not subsessile, as described by Diels; they are from 5 to 9 cm. long and about 1 cm. thick. The nearly sessile old flowers or fruits are loosely silky and have a rather long style, mostly cleft at the apex, with short bifid stigmas. The bracts are truncate at the apex. Without having seen any ♂ specimen or good sterile material it seems impossible to judge the relationship of this species.

74. **Salix driophila** Schneider. See p. 59.

75. **Salix amphibola** Schneider. See p. 60.

Sect. 14. HETEROCHROMAE Schneider, n. sect.

Arbores v. frutices. Folia satis magna, lanceolata, ovato-lanceolata v. late ellip-tica, integra v. breviter serrata, subtus pallida v. concoloria, textura tenui, sed firma, initio sericea. Amenta pedunculata, satis longa, cylindrica, densiflora; flores ♂ diandri, filamentis liberis pilosis, glandula una ventrali, bracteis oblongis obtusis; ovaria breviter v. longius pedicellata, sericea v. glabra, stylis distinctis apice bifidis stigmatibus bifidis oblongis, glandula una ventrali pedicellis breviore v. duplo longiore, bracteis ovatis obtusis.

This section differs from sect. *Phylicifoliae* and *Hastatae* in the longer catkins, in this resembling sect. *Eriostachyae*. The leaves are thin but firm and show a fine but distinct reticulation beneath.

76. **Salix heterochroma** Seemen. See p. 61.

77. **Salix isochroma** Schneider, n. sp.

Frutex 1.2 m. altus; ramuli annotini glabri, olivaceo-fusci, deinde purpureo-brunnescentes. Folia ovato-oblonga v. oblanceolata, basi rotunda, apicem versus plus minusve subito acuminata v. acuta, supra viridia, costa pilis fulvis et albis obdita excepta glabra, subtus fere concoloria, viridescentia, pilis sericeis saepe fulvis plus minusve obtecta, interdum paullo glabrescentia, costa nervisque circiter 3 pro 1 cm. prominulis et tenuiter reticulata, margine satis indistincte et distanter glanduloso-serrata, ex parte integra, minora 2–4.5 cm. longa, 1–1.5 cm. lata, maxima ad 7 cm. longa et ad 2.3 cm. lata; petioli 4–6 mm. longi, graciles, ut costa pilosuli, supra sulcati. Amenta (fructifera tantum vidi) pedunculo circiter 5 mm. longo piloso foliola minima lanceolata 1–2 gerente suffulta, 4.5–5.5 cm. longa, circiter 1 cm. crassa, laxe sericea; flores ♀ (inter fructus remanentes) brevipedicellati; ovaria ovato-oblonga, apicem versus attenuata, sericea, bracteas fere triplo superantia, stylo distincto ad medium bifido stigmatibus partitis oblongis divaricatis coronata; glandula anguste-oblonga, pedicellum duplo superans, bracteis late-ovatis brunnescentibus extus sericeis intus subglabris aequilonga. Fructus anguste oblongi, circiter 4 mm. longi, breviter pedicellati, subsericei.

CHINA. Western Szech'uan: Wa-shan, alt. 3300–3800 m., July 1903 (Veitch Exped. No. 4524, type; with fruits).

This species differs from *S. heterochroma* Seemen in the greenish under surface of the leaves, which are a little serrulate, and in the glands being longer than the short pedicels. It seems to be only a low shrub. The gland is very like that of *S. hylonoma* Schneider, and *S. isochroma* may be most nearly related to that species! The specific name is derived from ἴσος, equal, and χρῶμα, color.

78. **Salix omeiensis** Schneider, n. sp.

Arbor ad 10 m. alta; ramuli annotini vetustioresque glabri, purpureo-nigri; gemmae purpureae, ovato-oblongae, adpressae, glabrae v. pilis sparsis praeditae, petiolis subtriplo breviores. Folia satis magna, late ovata v. paulo obovata, basi rotunda, apice subito acutata v. sensim acuta, supra saturate viridia, glabra, subtus valde pallida, saepissime tantum ad costam flavam sericea (juniora tota sericea?), utrinque costa nervisque circiter 1–2 pro 1 cm. prominentibus, et (subtus distinctius) reticulata, margine breviter et distanter glanduloso-serrata v. integra, minora (pedunculorum) 3–6 cm. longa, 1.8–3.5 cm. lata, maxima usque 11 cm. longa et 5.3 cm. lata; petioli 7–15 mm. longi, glabri v. initio sericei. Amenta (fructifera tantum vidi) pedunculo 1–1.5 cm. longo piloso folia 1–2 normalia gerente suffulta, 10–12.5 cm. longa, circiter 1.1 cm. crassa, rhachi sparse pilosa. Fructus brevipedicellati, ovato-oblongi, apice angustati, circiter 3–3.5 mm. longi, glabri (stylis marcescentibus distinctis apice bifidis, stigmatibus oblongis bifidis); bracteae ovatae, subacutae, brunneae, extus sericeae, intus glabratae, pedicellis 2–3plo longiores; glandula una crassiuscula, saepe plus minusve latior quam alta, pedicello subduplo brevior v. subaequilonga, oblonga.

CHINA. Western Szech'uan: Mt. Omei, May 1904 (Veitch Exped., No. 5193, type; tree 10 m., with fruits).

This peculiar species seems to be related to *S. heterochroma* Seemen, but differs widely from it in its broad leaves and in its longer petioles. The long fruiting aments somewhat resemble those of *S. Fargesii* Franchet and its relatives, but the reticulation of the leaves is more like that of the former species.

Sect. 15. **PHYLICIFOLIAE** Dumortier, *Fl. Belg. Prodr.* 12 (1827).—E. Fries in *Syllog. Pl. Nov.* II. 36 (pro parte) (1828). — Andersson in *Svensk. Vetensk.*

Akad. Handl. VI. 125 (*Monog. Salic.*) (1867); in De Candolle, *Prodr.* XVI. pt. 2, 240 (1868). — Seemen, *Salic. Jap.* 18 (1903).

Salix, sect. *Nigricantes* Borrer in Hooker, *Fl. Brit.* 426 (1830); in Loudon, *Arb. Brit.* III. 1563 (1838).

Salix, sect. *Bicolores* Borrer in Hooker, *Fl. Brit.* 428 (1830); in Loudon, *Arb. Brit.* III. 1577 (1838).

Salix, sect. *Virescentes* Andersson in *Svensk. Vetensk. Akad. Handl.* VI. 125 (*Monog. Salic.*) (1867); in De Candolle, *Prodr.* XVI. pt. 2, 240 (1868). — Seemen in Ascherson & Graebner, *Syn. Mitteleur. Fl.* IV. 130 (1909).

Frutices alti v. minores, breviter ramosi. Folia pleraque ovato-lanceolata usque obovata, saepissime glabra, interdum pubescentia, subtus viridia v. pallida, margine remote inflexo-serrata, marcescentia saepe nigrescentia. Amenta ovali-cylindrica, subsessilia v. breviter pedunculata; ovaria pleraque pilosa, pedicellata; glandula semper una ventralis.

This section [1] seems to include rather different forms. Most of the species of northeastern Asia are only imperfectly known, and I have not been able to see much material. I am following Andersson and von Seemen, but I cannot agree with them in their systematic arrangement. The forms of this and the following section specially need careful study.

79. **Salix phylicifolia** [2] Linnaeus, *Spec.* 1016 (exclud. var. β) (1753). —Andersson in *Svensk. Vetensk. Akad. Handl.* VI. 131, t. 7, fig. 70 (*Monog. Salic.*) (1867); De Candolle, *Prodr.* XVI. pt. 2, 241 (1868). — ? Franchet in *Nouv. Arch. Mus. Paris*, sér. 2, VII. 93 (*Pl. David.* I. 283) (1884). — Burkill in *Jour. Linn. Soc.* XXVI. 532 (pro parte ?) (1899). — Herder in *Act. Hort. Petrop.* XI. 414 (pro parte) (1891). — Komarov in *Act. Hort. Petrop.* XXII. 28 (*Fl. Mansh.* II.) (1903). — Seemen in Ascherson & Graebner, *Syn. Mitteleur. Fl.* IV. 140 (1909). — Nakai in *Jour. Coll. Sci. Tokyo* XXXI. 42 (*Fl. Kor.* II.) (1911). — Moss, *Cambridge Brit. Fl.* II. 44, t. 45 (1914).

[1] According to Ledebour (*Fl. Ross.* III. 609 [1850]), *S. nigricans* Smith (in *Trans. Linn. Soc.* VI. 120 [1802]), was found in Kamtchatka (by *Chamisso* and by *Beechey* ex Hooker & Arnott). I do not find any mention of *S. nigricans* in Hooker & Arnott, *Bot. Voy. Beechey.* According to Siuzev (in *Trav. Mus. Bot. Acad. Sci. St. Pétersbourg*, IX. [*Contr. Fl. Mansh.*] [1912]) ex Toepffer, *Salicol. Mitt.* No. 5, 248 (1912), *S. nigricans* has been recently found in Mandshuria. I only mention this fact with the remark that according to the Vienna rules the oldest name for this species apparently is *S. myrsinifolia* Salisbury, *Prodr.* 394 (1796), founded on *S. myrsinites* Hoffmann, *Hist. Salic.* I. 71, t. 17–19 and 24, fig. 2 (non Linnaeus) (1787). The type of *S. nigricans* Smith is *S. phylicifolia*, var. β Linnaeus, *Spec.* 1016 (1753). For further synonyms and literature see von Seemen in Ascherson & Graebner, *Syn. Mitteleur. Fl.* IV. 132 (1909).

[2] According to the Vienna rules there is no reason why this name should be changed. It is founded on " Salix foliis serratis glabris lanceolatis, crenis undulatis " in *Fl. Lappon.* 283, No. 351 (misprint 358 in *Spec.*), t. 8, fig. d (1737). This is Linnaeus' type according to Andersson, *Salic. Lappon.* 42 (1845), Wimmer, *Salic. Europ.* 79 (1866), Enander, *Salic. Linn. Herb.* 96 (1907) and other authors. I believe Toepffer, *Salicet. Exsiccat.* No. 31, made a mistake in saying that Linnaeus' No. 351 of the Flora Lapponica represents *S. nigricans* Smith and that No. 350 (*Salix phylicifolia*, var. β Linné, *Spec. Pl.* 1018 [1753]) is *S. bicolor* Ehrhart. According to Smith the type of *S. nigricans* is " *S. phylicifolia* β Linn. Sp. Pl. 1442, Fl. Lapp. ed. 2, 291, t. 8, f. c. n. 350." See *S. myrsinifolia* Salisbury above.

Salix bicolor Ehrhart, *Beitr.* V. 162 (1790). — Schneider, *Ill. Handb. Laubholzk.*
I. 55. fig. 19 b, 24 k–l (1904).
Salix Weigeliana Willdenow, *Spec.* IV. pt. 2, 678 (1805).
Salix arbuscula, var. *phylicifolia* Wolf in *Izv. S.-Peterburg. Liesn. Inst.* IV. 93,
t. 36, fig. 8–13, t. 43, fig. 8–12 (*Мат. Изуч. Ив Европ. Росс.*) (1900).

For further synonyms and literature see Herder, l. c., and Seemen, l. c.
CHINA.   Chili: Hsiao Wu-tai-shan, alt. 2300 m., August 12, 1913, *F. N.
Meyer* (No. 1204; shrub; sterile).
NORTHEASTERN ASIA.   Maritime prov.: Plover Bay, Lat. 64° N.,
1865–6, *W. H. Dall* (with fruits).   Kamtchatka: Petropavlovski, 1853–6, *C.
Wright* (with fruits).

Franchet, l. c., mentions specimens from Chili collected by David which I have
not seen and which probably do not belong to our species, and may be the same as
S. mongolica Siuzev (see p. 178).   Burkill cites a specimen of Bretschneider from
the Po-hua-shan, and the species is also found in Mandshuria and Korea according to Komarov and Nakai.   The specimens before me collected by Dall and
Wright differ somewhat from the type. The ovaries are rather short-stalked, and
the glands are longer than the pedicels. The peduncles of the fruiting aments of
Dall's plant are up to 2 cm. long. They may represent *S. oblongifolia* Trautvetter
& Meyer, see p. 126.

All these eastern forms of this section need a very careful study.   Meyer's sterile
specimen agrees rather well with some European forms of this variable species.

80.   **Salix formosa** Willdenow, *Berl. Baumz.* 452 (1796); *Spec.* IV. 680 (1805).

Salix arbuscula [1] Linnaeus, *Spec.* 1018 (tantum var. γ) (1753).   Herder in
   Act. Hort. Petrop. XI. 417 (1891). — Wolf in *Izv. S.-Peterburg. Liesn. Inst.*
   IV. 92, t. 35, fig. 12–18, t. 36, fig. 1 (*Мат. Изуч. Ив Европ. Росс.* I.)
   (tantum var. *typica*) (1900). — Schneider, *Ill. Handb. Laubholzk.* I. 55, fig.
   19 p, 24 q–u (1904). — Seemen in Ascherson & Graebner, *Syn. Mitteleur.
   Fl.* IV. 146 (1909). — Moss, *Cambridge Brit. Fl.* II. 39, t. 41 (1914).

[1] According to the Vienna rules this name cannot stand.   The type of Linnaeus
S. arbuscula is founded on "Salix foliis subserratis glabris subdiaphanis subtus
glaucis, caule suffruticoso" in *Fl. Suec.* 291, No. 798 (1745), var. *a,* and the type of
var. *a* is "Salix foliis serratis glabris verticaliter ovatis" in *Fl. Lappon.* 284, No.
352, tab. 8, fig. e (1737).   This No. 352 is the same as *S. livida* Wahlenberg (vide
Enander, *Salic. Linn. Herb.* 97 [1907]).   Linnaeus' herbarium specimen may belong
to *S. phylicifolia* Linnaeus or to a hybrid of *S. nigricans* Smith with *S. phylicifolia*
(see Enander, l. c.).   Linnaeus' *S. arbuscula,* var. *β, Spec.* 1018 (1753) is founded
on "Salix foliis integris glabris ovatis confertis pellucidis" in *Fl. Lapp.* 287, No.
356 (1757).   This No. 356 represents *S. livida* Wahlenberg (see Enander, l. c.).

Only *S. arbuscula,* var. γ represents the true *S. arbuscula* Auctorum plurim. (see
Enander, l. c. 149).   So far as I can see the oldest name for this form is *S. formosa*
Willdenow.   Andersson and von Seemen cite a *Salix coruscans* Jacquin, *Fl.
Austr.* V. t. 408 (non Willdenow) (1778).   But Jacquin describes, l. c., under No.
408 *Salix arbuscula.*   It was Willdenow, *Spec.* IV. 681 (1805), who founded a *S.
coruscans* on *S. arbuscula* Jacquin.   *S. formosa* Willdenow was published 1796, and
this name is older than *S. glaucescens* Moench, who merely changed Willdenow's
name.   *S. alpina* Scopoli, *Fl. Carn.* ed. 2, II. 255 (1772) is a mixture of *S. arbuscula*
and *S. myrsinites,* var. *Jacquiniana* Koch.

Salix cinerea Willdenow, *Berl. Baumz.* 350 (non Linnaeus, nec Willdenow, 1805) (1796).

Salix glaucescens Moench, *Meth.* Suppl. 116 (1802).

For further synonyms and literature see Herder, l. c., and Seemen, l. c.

NORTHEASTERN ASIA. Maritime prov.: Tschukschen country, river Anadyr, June 7 and 14, 1869, *K. Maydell* (ex Herder).

Not having seen a specimen of this species from our area, I do not know if it is really found so far east. According to Camus, *Classif. Saul. Europe*, I. 123 (1904), it has stomata on the upper surfaces of the leaves.

81. **Salix characta** Schneider, n. sp.

Frutex ?; ramuli initio puberuli, dein glabri, flavo- v. rubro-brunnei, vetustiores cinerascentes; gemmae visae ovatae, obtusae, subadpressae, subpilosae glabraeve, flavo-rubrae, petiolis ⅓ v. vix breviores. Folia satis adulta lanceolata, basi acuta, rarius subobtusa, apice acuta v. breviter acuminata, supra laete viridia, costa impressa sparse puberula excepta glabra, rarius tota facie plus minusve sparse sericea nervis subdistinctis vix incisis, subtus glauca, tantum costa flava elevata sparse sericea v. glabra, nervis 3–4 pro 1 cm. prominulis et tenuissime reticulata, margine circumcirca breviter sed satis argute glanduloso-serrata dentibus circiter 2 pro 1 mm., minora 1.5–3 cm. longa, 0.5–0.9 cm. lata, maxima usque 4.5 cm. longa et 1.2 cm. lata; petioli breves, 1–3 mm. longi, superne sulco puberuli. Amenta (tantum valde matura fructifera visa) pedunculo brevi sericeo foliato suffulta, ut videtur circiter 2.5 cm. longa et densiflora, rhachi villosa; flores ♀ inter fructus remanentes pedicellati; ovaria oblonga, tantum basi et pedicello glandulam unicam oblongam apice incisam subaequante pilosa ceterum glabra, stylo distincto ⅔ ovarii aequante ut videtur integro stigmatibus ovato-oblongis brevibus (? bifidis) coronata; bracteae ovatae, acutae, brunneae, sparse sericeae, pedicello duplo longiores. Fructus circiter 4 mm. longi, oblongi, ut ovaria pilosi et pedicellati.

CHINA. Chili: Hsiao Wutai-shan, mountain slopes, alt. 1700–3000 m., August 8, 1913, *F. N. Meyer* (No. 1160, type; with fruits).

This species much resembles some forms of *S. formosa* Willdenow from the Altai, but it can be distinguished from that species by its very closely serrate leaves, by its short petioles and by its rather acute bracts. I have seen only one old fruiting catkin. It may be nearly related to *S. kolymensis* Seemen, of which I have not seen a specimen.

The specific name is derived from χαρακτός, with teeth.

82. **Salix kolymensis** Seemen in Fedde, *Rep. Spec. Nov.* V. 18 (1908).

Salix boganidensis, var. *angustifolia* Herder in *Act. Hort. Petrop.* XI. 434 (pro parte) (1891).

NORTHEASTERN ASIA. Yakutsk: district Kolyma, on the river Kolyma, June 10, 14 and 30, 1875, *F. M. Augustinowicz* (ex Seemen); banks of the Alasei, June 8 [21] and 11[24] 1905, *Roznovski* (ex von Seemen).

The leaves of this species are described as sometimes entire and darker green above; the length of the petioles is not given by the author who saw only young leaves. The ovaries are wholly pubescent and seem to have longer pedicels.

83. **Salix boganidensis** Trautvetter in Middendorff, *Reise Sibir.* I. 2, Bot., Abt. 1, 154, t. 2, 3 (*Fl. Boganid.*) (1847). — Ledebour, *Fl. Ross.* III. 616 (1850). — Andersson in De Candolle, *Prodr.* XVI. pt. 2, 277 (1868). — Herder in *Act. Hort. Petrop.* XI. 434 (pro parte) (1891).

NORTHEASTERN ASIA. Maritime prov.: Boganida River, June 7 and 18, August 3, 1843, *A. T. von Middendorff* (♂ and ♀ types, ex Trautvetter).

Salix boganidensis, var. **latifolia** Trautvetter in *Act. Hort. Petrop.* VI. 34 (*Fl. Terr. Tschuktsch.*) (1879). — Herder in *Act. Hort. Petrop.* XI. 435 (1891).

Salix taimyrensis Trautvetter in *Act. Hort. Petrop.* V. pt. 2, 557 (*Fl. Rip. Kolym.* 63) (non Trautvetter) (1878).

NORTHEASTERN ASIA. Maritime prov.: Anadyr River, May 23, 26, 30, June 1, 3, 7, 12, 14, 15, 20, 1869, *K. Maydell* (ex Trautvetter); Kolyma River, May 30, June 24, July 6, 1875 and June 14, 1876, *F. M. Augustinowicz* (ex Trautvetter); river Jana, June 27, 1885, *A. Bunge f.* (ex Herder).

According to Trautvetter (1847) the type is "inter *S. phylicifoliam* L. et *S. arbusculam* L. quasi media." The figure shows a nearly sessile ovary with a long style, not unlike that of *S. viminalis* Linnaeus, but von Seemen says that in *S. kolymensis* the styles seem to be rather longer than in *S. boganidensis,* and he places *S. kolymensis* in the group "*Meiostylae.*" Ledebour and Andersson place *S. boganidensis* in the same group with *S. lanata* Linnaeus, but I cannot decide to which section it really belongs.

84. **Salix oblongifolia** Trautvetter & Meyer in Middendorff, *Reise Sibir.* I. 2, Bot., Abt. 2, 81 (*Fl. Ochot.*) (1856). — Andersson in De Candolle, *Prodr.* XVI. pt. 2, 248 (1868). — Herder in *Act. Hort. Petrop.* XI. 419 (1891).

NORTHEASTERN ASIA. Maritime prov.: "ad sinum Ujakon," August 23–September 1, *A. T. von Middendorff* (type, ex Trautvetter & Meyer; with fruits). Kamtchatka: Petropavlovsk, *Kastalsky* (ex Herder).

This very little known species may represent *S. phylicifolia* Linnaeus from eastern Asia (see p. 124). According to the description it is near *S. characta* Schneider, p. 125, but differs from it chiefly in the remotely serrulate leaves and in the silky ovaries. It has been omitted in the key.

85. **Salix udensis** Trautvetter & Meyer, in Middendorff, *Reise Sibir.* I. pt. 2, Bot., Abt. 2, 81 (*Fl. Ochot.*) (1856). — Andersson in De Candolle, *Prodr.* XVI. pt. 2, 248 (1868). — Herder in *Act. Hort. Petrop.* XI. 419 (1891).

NORTHEASTERN ASIA. Maritime prov.: near Udskoi, June 15, 1844, *A. T. von Middendorff* (♀ type, ex Trautvetter & Meyer).

This is an uncertain species, which is said to be nearly related to *S. arbuscula* (*S. formosa* Willdenow), but has the leaves greenish on both surfaces. It has been omitted in the keys.

86. **Salix chlorostachya** Turczaninow in *Bull. Soc. Nat. Mosc.* XXVII. 373 (1854). — Andersson in *Svensk. Vetensk. Akad. Handl.* VI. 143, t. 7, fig. 78 (*Monog. Salic.*) (1867); in De Candolle, *Prodr.* XVI. pt. 2, 246 (1868). — Herder in *Act. Hort. Petrop.* XI. 416 (1891).

NORTHEASTERN ASIA. Transbaikalia: "in insulis fluvii Angarae, ad torrentem Chalagum prope Turan, in transbaicalensibus et cet.," *N. Turczaninow* (♀ type, ex Turczaninow).

According to Turczaninow the type of this species was apparently collected in Gouv. Irkutsk, but the species also occurs in Transbaikalia on the western limits of our area. Andersson, l. c. (1867), describes these forms, and he says: "Est species sat singularis amentis *S. retusae,* foliis nunc (integerrimis) *S. myrtilloidi* nunc (serrulatis) *S. arbusculae* similis, sed habitu quasi inter *S. arbusculam* et *S. phylicifoliam* intermedia." Turczaninow says: "pedicellis nectarium aequantibus," while Andersson says: "pedicello nectarium duplo superante." The ovary is glabrous. I have not seen any specimen, and I doubt if Andersson's different forms belong to the same species.

Herder, l. c. 417, cites a *Salix submyrtilloides* Andersson, which seems to be a herbarium name, and according to Trautvetter is the same as *S. chlorostachya* Turczaninow.

87. **Salix leptoclados** Andersson in *Svensk. Vetensk. Akad. Handl.* VI. 144, t. 7, fig. 79 (*Monog. Salic.*) (1867); in De Candolle, *Prodr.* XVI. pt. 2, 247 (1868). — Herder in *Act. Hort. Petrop.* XI. 417 (1891).

NORTHEASTERN ASIA. Transbaikalia: on the river Selenga near Verchne Udinsk, 1832, *N. Turczaninow* (ex Andersson).

This is a doubtful species. The hairy ovary has a pedicel which is 3 or 4 times longer than the gland. The style is usually bifid.

88. **Salix Reinii** Franchet & Savatier, *Enum. Pl. Jap.* I. 459 (nomen nudum) (1875); apud Seemen, *Salic. Jap.* 41. t. 6, fig. A–E (1903). — Léveillé in *Bull. Acad. Int. Geogr. Bot.* XIV. 208 (1904); XVI. 143 (1906). — Koidzumi in *Tokyo Bot. Mag.* XXVII. 91 (1913).

Salix glabra Franchet & Savatier, *Enum. Pl. Jap.* II. 503 (1879). — Kawakami in *Tokyo Bot. Mag.* X. 50 (1896).

JAPAN. Hondo: prov. Suruga, slopes of Fuji-san, common, woodlands, above alt. 800 m., May 8, 1914, *E. H. Wilson* (No. 6642; bush 0.3–1.5 m. tall; ♂; No. 6642ᵃ; ♀); same locality, August 1906, *M. Koyama* (with fruits); same locality (ex Herb. Bot. Gard. Tokyo; ♂); same locality, July 29, 1891, *K. Watanabe* (♂); prov. Shimotsuke, Yumoto, open country, alt. 2000–2800 m., June 23, 1914, *E. H. Wilson* (No. 6844; bush 0.6–1.5 m. tall; ♂; No. 6844ᵃ, ♀); prov. Mutsu, Hakkoda-yama, abundant on upper slopes and summit, July 4–6, 1914, *E. H. Wilson* (No. 7080; bush 0.3–0.9 m. tall; ♂); Iwaki san, May and August 1908 (ex Herb. Yokohama Nursery Co., ♀ flowers and fruits); leaves large, ovate-elliptical, up to 11 cm. long and 3.8 cm. broad); Mt. Iwate, July 1903, *S. Arimoto* (with fruits); on Gassan, July 23, 1887 (No. 66 ex Herb. Bot. Gard. Tokyo; with fruits); prov. ? "in monte Hak'san, in fruticetis humidis," *J. Rein* (Savatier, Nos. 2923, 2924; types, ex Franchet & Savatier).

The fruits of typical *S. Reinii* are mostly somewhat hairy, at least on the pedicels, but are sometimes wholly glabrous. The ♂ catkins are often rather small; the bracts are oblong, hairy or nearly glabrous. The glands of both the sexes are rather long compared with those of *S. vulpina* Andersson or forms of *S. japonica* Thunberg. The pedicel of the ovary is about as long as the gland or up to twice its length in the fruit. The pubescence of the bracts may sometimes contain some brownish hairs like those of *S. vulpina* Andersson, which differs in its short more orbicular-ovate bracts, shorter styles and short glands. There are certainly hybrids between it and *S. vulpina* Andersson and also between it and *S. japonica*, var. *padifolia* Seemen (and var. *Oldhamiana* Franchet & Savatier?). I suppose that *S. Reinii* would be better referred to sect. *Hastatae*.

It seems to be a variable species the forms of which need a careful study. Von Seemen (*Salic. Jap.* 42 [1903]) described 4 different forms, the types of which I have not seen, namely:

f. *typica* Seemen: folia obovata, apice brevi-acuta, saepe plicata, subtus cinereo-glauca (type collected by *J. Rein*).

f. *nikkoensis* Seemen: reptans; folia magna, ad 6.5 cm. longa, 3.5 cm. lata, oblonga: Nikko, on moist ground; *M. Shirai* (No. 20; ex von Seemen).

f. *riishiriensis* Seemen: folia oblonga usque lanceolata, ad 4.5 cm. longa, 2 cm. lata, in apicem acutam attenuata, subtus paullo pallidiora, nervatura utrinque distinctius prominula: Riishiri, *U. Faurie* (No. 3711; ex Seemen).

f. *aomorensis* Seemen: folia late ovata, ad 4 cm. longa et 3.5 cm. lata, breviter acuta (iis *Populi tremulae* similia): Aomori, *U. Faurie* (No. 808, in part; ex Seemen). To this form may belong *C. S. Sargent's* specimen from Mt. Hakkoda, near Aomori, from foothills up to 1600 m., Oct. 2, 1892 (shrub 0.6–1.2 m. tall; with fruits).

> **Salix Reinii, var. cyclophylloides** Koidzumi in *Tokyo Bot. Mag.* XXVII. 91 (1913).

JAPAN. Hondo: "Rikuchu, Mt. Iwate and Hayachine-san" (fruiting types, ex Koidzumi); prov. Rikuchu, Mt. Iwate, July 15, 1903, *S. Arimoto* (with fruits; may be the same as the type!).

In this variety the leaves are described as mostly obovate or orbicular, and often brownish pubescent on both surfaces when young; the catkins are only 15 mm. long.

There are three specimens collected by *Faurie* in Hondo " in pago Sambongi," May 8, 1908 (No. 6613, ♂, No. 6614, 1 m. tall, ♀) and " in planitie Sambongi," May 1908 (No. 6626, 5–6 m. tall, ♂) which are similar to *S. Reinii*, but I am not quite sure whether or not they really belong to this species. The young leaves are somewhat blackish, and not paler on the lower surface.

89. **Salix kakista** Schneider, n. hybr.

Frutex 0.6–1.8 m. altus, valde ramosus; ramuli fructiferi glabri, fusci v. fusco-brunnei, dense cinereo-brunnei. Folia obovata v. obovato-oblonga, basi rotunda, interdum paulo angustata, apice pleraque satis subito breviter acuminata, ultra medium latissima, supra intense viridia, glaberrima v. costa subimpressa sparse pilosula, nervis paulo v. vix prominulis, subtus albescentia v. pallide viridia, glaberrima (juvenilia tantum ad costam pilosa), costa nervisque prominulis, tenuiter subreticulata, nervis sub 70–60° a costa divergentibus, margine satis distincte glanduloso-crenato-serrata v. versus basim apicemque integra, minora 3–4.5 cm. longa, 1.8–2.5 cm. lata, majora 5–7 cm. longa, 3–3.5 cm. lata; petioli superne laxe pilosi, 7–10 mm. longi. Stipulae raro evolutae, vix 2 mm. longae, semiovatae, subacutae, glanduloso-denticulatae. Amenta (fructifera tantum visa) pedunculos normaliter foliatos 1.2–3 cm. longos terminantia, cylindrica, 4–5 cm. longa, 1 cm. crassa, rhachi sparse albo-villosula, pilis paucis brunneis intermixtis; flores inter fructus remanentes glabri, pedicellis ovario circiter 3–3½-plo brevioribus glandulam ⅓ superantibus, stylo satis brevi apice bifido stigmatibus parvis emarginatis circiter 2–3-plo longiore coronata; glandula una ventralis, oblongo-rectangularis, ⅔ pedicelli aequans; bracteae oblongae, obtusiusculae, sparse albo-sericea et facie pilis paucis brunneis obditae, pedicello 2-plo longiores. Capsulae saepe longius pedicellatae, valvis valde recurvatis, 3–4 mm. longae.

JAPAN. Hondo: prov. Mutsu, Hakkoda-yama, common, lower and middle slopes, July 5, 1914, *E. H. Wilson* (No. 7103, type; bush 0.9–1.8 m. tall, much branched; with fruits); prov. Ugo, " Akita, in sylvis," June 18, 1905, *U. Faurie* (No. 5752; with fruits); prov. Eshigo, Mt. Myoko, July 27, 1914 (ex Herb. Bot. Gard. Tokyo; with fruits); prov. Kozuke, Mt. Asama, alt. 2200 m., August 4, 1910 (ex Herb. Sakurai; with fruits); prov. Suruga, Fuji-san, alt. 2600 m., August 2, 1907 (ex Herb. Sakurai; with fruits); prov. Izumi, Nanakoshi, May 3, 1896 (No. 106, ex Herb. Bot. Gard. Tokyo; ♀).

The different specimens referred by me to this hybrid are, of course, not quite alike. The type resembles more *S. Reinii* Franchet & Savatier than *S. vulpina* Andersson, while others of these forms are nearer to the last named parent. All show a certain mixture of the principal characters of the two parents. According to the author, *S. vulpina*, var. *discolor* Seemen (*Salic. Jap.* 40 [1903]) may represent the

same hybrid. The type was collected in Hondo, near Mt. Indonosan, July 22, 1887 (No. 21; ex Herb. Bot. Gard. Tokyo; ex von Seemen); see p. 131.
A rather distinct form seems to be *Wilson's* No. 7104 from the same locality as No. 7103. Here the fruits with their short bracts and their short broad gland much more resemble *S. vulpina* Andersson. Most of the leaves are distinctly whitish beneath, and the young ones bear a rusty tomentum on the lower surface. The catkins are very long, measuring from 6 to 8 cm. in length, nearly sessile, and resemble those of *S. japonica* Thunberg, especially var. *Oldhamiana*, which can, however, be distinguished at once by its longer bracts with whitish (not rusty) pubescence.
The specific name is derived from κάκιστος, useless.

90. **Salix daiseniensis** Seemen, *Salic. Jap.* 65, t. 15, fig. A–D (1903). — Léveillé in *Bull. Acad. Int. Géogr. Bot.* XIV. 210 (1904); XVI. 145, 146 (1906). — Koidzumi in *Tokyo Bot. Mag.* XXVII, 266 (1913).

Salix vulpina, var. *daiseniensis* Koidzumi in *Tokyo Bot. Mag.* XXVII. 90 (1913).

JAPAN. Hondo: prov. Hoki, Mt. Daisen, May 26, 1899, *U. Faurie* (Nos. 3708, 3709, types; ♂ and ♀, ex Seemen); prov. Itachi, Tsukuba, April 14, 1909 (ex Herb. Sakurai); prov. Rikuchu, Amibari, August 1909 (ex Herb. Yokohama Nursery Co.; with fruits); same locality, April 1, 1911 (ex Herb. Sakurai). Hokkaido: Rebunziri, mountains, August 1, 1899, *U. Faurie* (No. 3712; with ? ex von Seemen).
A doubtful species. According to von Seemen the catkins of *S. daiseniensis* resemble those of *S. Buergeriana* Miquel, but they appear with the leaves, and the ♂ flowers have one or two stamens, while the flowers of *S. Buergeriana* are precocious and the ♂ flowers have only one stamen. Koidzumi first made *S. daiseniensis* a variety of *S. vulpina* Andersson, but later kept it distinct and placed it in the sect. *Phylicifoliae*, saying: "Species habitu *S. Sieboldianae* Blume sat similis sed foliis tenuioribus; filamentis liberis fere glabris diagnoscenda." The specimen ex Herb. Sakurai from the type locality has glabrous ovaries. So far as I can judge, *S. daiseniensis* may be a hybrid between one of the forms of *S. vulpina* Andersson and *S. Sieboldiana* Blume. I am not, however, quite sure if all the specimens above represent such a hybrid. See also *S. ampherista* Schneider.

91. **Salix amnicola** Wolf in *Act. Hort. Petrop.* XXVIII. 31 (1911).
NORTHEASTERN ASIA. Maritime prov.: distr. Khabarovsk, " in alveo amnis Amur," May 19, 1910, *N. Kuznetzov* (No. 38, type; ♀ ex Wolf).
The leaves of this species are, according to Wolf, lanceolate or elliptic-lanceolate, entire, from 2.5 to 5.4 cm. long and about ⅓ as broad, somewhat pubescent on both sides and glaucous beneath. The female catkins are sessile and appear with the leaves; they are cylindric and 3 to 4 cm. long, the bracts are lingulate and acute or obtusish; the ovaries are shortly stipitate, silky, and bear a style being a little longer (¼–⅕ of the ovary) than the pedicel; the stigmas are bilobate or bipartite and about as long as the style; the gland is as long as the pedicel, which in the fruit becomes ⅓ longer than the gland.

Sect. 16. **HASTATAE** Borrer in Hooker, *Brit. Fl.* 433 (1830); in Loudon, *Arb. Brit.* 1592 (pro parte) (1838). — Andersson in *Svensk. Vetensk. Akad. Handl.* VI. 157 (*Monog. Salic.*) (1867); in De Candolle, *Prodr.* XVI. pt. 2, 251 (pro parte) (1868). — Seemen, *Salic. Jap.* 18 (1903). — Koidzumi in *Tokyo Bot. Mag.* XXVII. 89 (1913).

Salix, sect. *Rigidae* Andersson in *Svensk. Vetensk. Akad. Handl.* VI. 157 (*Monog. Salic.*) (1867); in De Candolle, *Prodr.* XVI. pt. 2, 251 (1868). — Seemen in Ascherson & Graebner, *Syn. Mitteleur. Fl.* IV. 152 (1909). *Salix*, sect. *Vulpinae* Seemen, *Salic. Jap.* 17 (1903).

Species iis sect. *Phylicifoliae* saepe valde similes, sed praecipue differentes: foliis saepe lanceolatis, amentis interdum rufescenti-pilosis, ovariis fere semper glabris.[1]

I follow Koidzumi in uniting sect. *Vulpinae* (*S. vulpina*) again with sect. *Hastatae*, and I believe that this section is most closely related to sect. *Phylicifoliae*. Both groups, as now limited, seem to be more artificial than natural. It is an interesting fact that apparently there is no species of these groups in the interior of China.

92. **Salix vulpina** Andersson in *Mem. Am. Acad.* VI. 452 (Gray, *Bot. Jap.*) (1859). — Franchet & Savatier, *Enum. Pl. Jap.* I. 461 (1875). — Kawakami in *Tokyo Bot. Mag.* X. 50 (1896). — Seemen, *Salic. Jap.* 37, t. 5, fig. F–I (1903). — Léveillé in *Bull. Acad. Int. Géogr. Bot.* XIV. 208 (1904); XVI. 144 (1906). — Koidzumi in *Tokyo Bot. Mag.* XXVII. 89, 264 (1913).

Salix Miquelii Andersson in *Svensk. Vetensk. Akad. Handl.* VI. 166, t. 9, fig. 98 (*Monog. Salic.*) (1867); in De Candolle, *Prodr.* XVI. pt. 2, 256 (1868). *Salix Miquelii*, var. *vulpina* Andersson, l. c. 256 (1868). *Salix Shiraii*, var. *vulcaniana* Léveillé & Vaniot in *Bull. Acad. Int. Géogr. Bot.* XIV. 209 (1904). *Salix ignicoma* Léveillé & Vaniot, l. c. XVI. 143 (1906). *Salix vulpina*, var. *typica* Koidzumi in *Tokyo Bot. Mag.* XXVII. 90 (1913).

JAPAN. Hokkaido: prov. Oshima, Hakodate, 1861, *C. Maximowicz* (♂); same locality, 1861, *M. Albrecht* (♀); prov. Shiribeshi, around Otaru, June 12,1908, *U. Faurie* (Nos. 258, 258 bis; with fruits); without precise locality, July 1905, *U. Faurie* (No. 6608; with fruits); "volcan de Tarumai," June 17, 1893, *U. Faurie* (Nos. 10026, 10027, ♀; type of var. *vulcaniana*, ex Léveillé). Hondo: prov. Bitchu, Takaya, in forests, June 24, 1904, *U. Faurie* (No. 5751; with fruits); prov. Izumi, Shinoda, April 24, 1897 (No. 109, ex Herb. Bot. Gard. Tokyo; ♀); prov. Musashi, Yokohama, *S. W. Williams & J. Morrow* (♂ type of *S. vulpina*, ex Andersson); prov. Hitachi, Mt. Tsukuba, April 14, 1895 (No. 33 ex Herb. Bot. Gard. Tokyo; ♂); prov. Rikuchu, Mt. Iwate, April 20, 1903 (ex Herb. Bot. Gard. Tokyo, ♂); prov. Mutsu, Aomori, in forests, May 1904, *U. Faurie* (No. 5763, type of *S. ignicoma*, ex Léveillé; "in basi montis Iwagi," May 6, 1905, *U. Faurie* (No. 6624; ♂); "in basi montis Hayachine," June 6, 1905, *U. Faurie* (No. 6605, with fruits; No. 6606; ♂); same locality, June 1905, *U. Faurie* (No. 6621; with fruits); "in sylvis Ubayu," July 1, 1905, *U. Faurie* (No. 5757; with

---

[1] Andersson also places in this section *S. rhamnifolia* Pallas (*Fl. Ross.* I. pt. 2, 84 [1788]) which was founded by Pallas upon Gmelin's No. 13 in his *Fl. Sibir.* I. 159, t. 35, fig. I (1747), and which came from western Siberia, river Tunguska. Ledebour (*Fl. Ross.* III. pt. 2, 612 [1850]) mixed this species with *S. rhamnifolia* Hooker & Arnott, see *S. fuscescens* Anderson, p. 153. Andersson (in *Svensk. Vetensk. Akad. Handl.* VI. 169, t. 9, fig. 100 [*Monog. Salic.*] [1867]) describes one of Ledebour's plants " ex Sibiria baicalensi." This species is a very doubtful plant, and I do not know to which form really belong *C. Wright's* specimens from Arakamtchatchene Island, cited by Herder (in *Act. Hort. Petrop.* XI. 420 [1891]) under *S. rhamnifolia* Pallas.

fruits); "in planitie Sambongi," May 8, 1905, *U. Faurie* (No. 6595; ♂).
Kyushu: prov. Hizen, Nagasaki, 1862, *R. Oldham* [1] (No. 551, type of *S. Miquelii*, ex Andersson).

The typical *S. vulpina* to which the above specimens are referred seems to be distinguished by the pale-green under surface of its leaves, its glabrous ovaries and by the reddish brown or gray and brown hairs of its bracts. There are also described the following varieties:

Salix vulpina, var. discolor Seemen, *Salic. Jap.* 40 (1903).

Folia subtus discoloria, cinerea v. rubescenti-cinerea.
NORTHEASTERN ASIA. Ochots Sea: without precise locality, *J. Small* (in Collect. Wright).
JAPAN. Hondo: near Mount Indonosan, July 22, 1887 (No. 21, type; ex Herb. Tokyo, ex Seemen; ♀). Hokkaido: Oshima, July 10, 1890, *K. Miyabe & Y. Tokubuchi* (with fruits).
Small's specimen seems to be a co-type of Wright's No. 4 mentioned by Andersson (1867), and *S. Miquelii* sensu stricto seems to belong to this variety, which, I suppose, is not identical with the hybrid *S. kakista* Schneider, p. 128.

Salix vulpina, var. nikkoensis Koidzumi in *Tokyo Bot. Mag.* XXVII. 90 (1913).

"Ovaria subsessilia, glabra; stigmate obscuriter bilobulato; bracteis albo-tomentosis, florum ♀ suborbicularibus, ♂ obovato-ellipticis."
JAPAN. Hondo: prov. Shimotsuke, Nikko (ex Koidzumi).
Not having seen a specimen, I cannot judge the value of this form. It may belong to *S. kakista* Schneider, see p. 128.

Salix vulpina, var. pubescens Koidzumi, l. c. (1913).

"Folia adulta utrinque secus costas pubescentia."
JAPAN. Southern Hondo (ex Koidzumi).
I have not seen this form.

Salix vulpina, var. coriacea Koidzumi, l. c. (1913).

"Folia rigide coriacea."
JAPAN. Hokkaido: prov. Oshima; Hondo: prov. Mutsu (ex Koidzumi).
This seems to me a very doubtful form.
As far as I can see, *S. vulpina* Andersson may produce hybrids with several species. I have already mentioned *S. daiseniensis* Seemen, p. 129, and *S. kakista* Schneider, p. 128. Another hybrid may be *S. Matsumuraei* Seemen, see below. An interesting form seems to be *S. vulpina*, var. *tomentosa* Koidzumi, l. c. 265 (1913), from Hondo, prov. Shinano, Togakusimura, and prov. Kaga, Kanazawa. This has very hairy leaves and fruits. Koidzumi's description is insufficient, and I have not seen a specimen.

93. **Salix Matsumuraei** Seemen, *Salic. Jap.* 71, t. 78, A–B[1] (1903).

Salix vulpina, var. *Matsumuraei* Koidzumi in *Tokyo Bot. Mag.* XXVII. 90 (1913).

[1] The co-types seen by von Seemen and myself belong to *S. Harmsiana* Seemen, see p. 163. Koidzumi does not mention any specimens of *S. vulpina* from Kyushu.

JAPAN. Hondo: prov. Idzu, Shimoda, *C. Wright* (type; Herb. N. Pacif. Exp. Ringgold & Rodgers 1853-6; with fruits). Hokkaido: prov. Oshima, Hakodate, July 10, 1890, *K. Miyabe & Y. Tokubuchi* (with fruits; doubtful).

Judging by a co-type in Herb. Gray, this may be a hybrid between *S. japonica*, var. *padifolia* Seemen and *S. vulpina* Andersson, but the material is insufficient, and the co-type seems to differ in the shape of the longer, more acuminate leaves and in the partly brownish pubescence of some of the catkins which are put on the same sheet separately from the sterile branch.

Léveillé (in *Bull. Acad. Int. Géogr. Bot.* XVI. 144 [1906]) mentions *U. Faurie's* No. 1, from Hondo, "Ubayu, dans les forêts," July 1, 1904, as belonging to *S. Matsumuraei*, but I have not seen this specimen. The same author (in *Bull. Soc. Bot. France*, LVI. 141 [1905]), describes the ♂ form of this species after a specimen from Hondo, prov. Mutsu, Aomori, May 1902, collected by *Kinashi* (sub No. 10). Not having seen this specimen, I cannot tell if this ♂ plant really belongs to *S. Matsumuraei*, the hybrid origin of which has already been suggested by von Seemen.

94. **Salix japonica** Thunberg, *Fl. Jap.* 24 (1784); *Icon. Pl. Jap.* Dec. IV. t. 1 (1802). — Andersson in *Mem. Am. Acad.* VI. 450 (Gray, *Bot. Jap.*) (1859). — Miquel in *Ann. Mus. Lugd.-Bat.* III. 24 (1867). — Franchet & Savatier, *Enum. Pl. Jap.* 459 (1875). — Seemen, *Salic. Jap.* 43, t. 7, fig. A–E (1903). — Schneider, *Ill. Handb. Laubholzk.* I. 50, fig. 20 z, 23 b (1904).

Salix babylonica, var. *japonica* Andersson in *Svensk. Vetensk. Akad. Handl.* VI. 51 (*Monog. Salic.*) (1867); in De Candolle, *Prodr.* XVI. pt. 2, 213 (1868).

Salix japonica, var. *nipponensis* Léveillé in *Bull. Acad. Int. Géogr.-Bot.* IV. 209 (1904).

Salix japonica, f. *typica* Koidzumi in *Tokyo Bot. Mag.* XXVII. 91 (1913).

JAPAN. Hondo: prov. Musashi, Yokohama, Bluff, common, April 11, 1914, *E. H. Wilson* (No. 6402; bush 0.9-1.2 m. tall; ♂ and ♀); prov. Sagami, Hakone mountains, wet rocks, April 16, 1914, *E. H. Wilson* (No. 6447; bush 0.9-1.8 m. tall, common, wet rocks); same locality, April 14, 1904 (♂), April 21, 1907 (♀, ex Herb. Sakurai); Mt. Oyama, May 1910 (ex Herb. Yokohama Nursery Co.; ♀); prov. Shimotsuke, Nikko, June 1, 1901, *J. Matsumura & Yabe* (No. 70, ex Herb. Bot. Gard. Tokyo); prov. Idzu, April 26, 1893, *H. Shirasawa* (♂). Hokkaido: prov. Oshima, Hakodate, *S. W. Williams & J. Morrow* (ex Andersson).

The type may be distinguished as follows: Folia lanceolata, caudato-acuminata, argute serrata serraturis productis incisive, apice saepe integra. Glandula florum ♀ altior quam lata.

I refer to this species the following form:

Salix japonica, var. **typica**, f. **pygmaea** Schneider, n. forma.

Salix japonica, var. *pygmaea* Franchet & Savatier, *Enum. Pl. Jap.* I. 459 (1875); II. 503 (1879).

JAPAN. Hondo: prov. Sagami, "circa Yokoska" (type, ex Franchet & Savatier); Hakone, March 27, 1911, *Tonosawa* (ex Herb. Tokyo).

Frutex parvus, circiter 30 cm. altus. Amenta satis brevia, erecta. Ceterum ut videtur var. *typicae* simillima.

Besides this we have the following varieties:

Salix japonica, var. **Oldhamiana** Franchet & Savatier, *Enum. Pl. Jap.* I. 459 (1875); II. 503 (1879). — Seemen, *Salic. Jap.* 45, t. 7, fig. F–H (1903).

Salix Oldhamiana Miquel in *Ann. Mus. Lugd.-Bat.* III. 25 (1867).

Salix japonica, f. *Oldhami* Koidzumi in *Tokyo Bot. Mag.* XXVII. 91 (1913).

JAPAN. Hondo: prov. Sagami, near Kamakura, roadsides, May 4, 1914, *E. H. Wilson* (No. 6602; bush 0.9 m. tall; leaves shining; with fruits); prov. near Yokohama, 1863, *R. Oldham* (No. 720, type; with fruits); Oyama, April 14, 1911 (ex Herb. Sakurai; with fruits); prov. Suruga, Fuji-san, October 1910, *M. Koyama* (sterile); without precise locality, *P. von Siebold* (? ♂ co-type; ex Miquel); prov. Musashi, April 13, 1895 (ex Herb. Bot. Gard. Tokyo; ♀).

Folia ovato-oblonga, interdum acuminata, v. elliptico-oblonga, aut obverse oblonga et tum basi acutiuscula, satis subito brevius v. longius acuta v. subacuminata, fere ut in var. *typica* serrata. Amenta ♂ nondum cognita; glandula florum ♀ fere ut in var. *padifolia*.

Salix japonica, var. padifolia Seemen, *Salic. Jap.* 45, t. 8, fig. A–B (1903).

Salix padifolia Andersson in *Mem. Am. Acad.* VI. 451 (Gray, *Bot. Jap.*) (1859); in *Svensk. Vetensk. Akad. Handl.* VI. 165. t. 8, fig. 67 (*Monog. Salic.*) (1867); in De Candolle, *Prodr.* XVI. pt. 2, 255 (exclud. var. β) (1868).
Salix japonica, f. *padifolia* Koidzumi in *Tokyo Bot. Mag.* XXVII. 91 (1913).

Folia ovato-lanceolata, apice in cuspidem obliquum abrupte producta, margine plus minusve sed non profunde serrulata. Glandula florum ♀, saepissime latior quam alta.

JAPAN. Hondo: " ad Simoda," *S. W. Williams & J. Morrow* (♀ type ex Andersson); prov. Echigo, Mt. Miyokozan, July 17, 1897 (No. 114, ex Herb. Tokyo; with fruits). Hokkaido: prov. Oshima, Hakodate, 1890, *K. Miyabe* (ex Seemen); see *S. vulpina*, var. *discolor* on p. 131.

Unfortunately there are no types of Andersson's Willows collected by *Williams & Morrow* in the Gray Herbarium. Seemen has seen the type of *S. padifolia* Andersson from the Riks Herbarium at Stockholm. Most of the specimens I have seen which are referred to *S. padifolia* by Japanese botanists seem to belong to *S. vulpina*, var. *discolor* Seemen or to *S. kakista* Schneider. Even No. 114 from Echigo mentioned above may not represent the true var. *padifolia* even if it represents a variety of *S. japonica* Thunberg. There is, however, a Japanese specimen at the Arboretum from the Herb. Yokohama Nursery Company, April 1903, without locality, consisting of ♂ flowers and also old leaves which may be the ♂ *S. padifolia* Andersson, as the catkins and flowers resemble those of typical *S. japonica* Thunberg. It may be described as follows:

Ramuli foliiferi glabri v. tantum juxta gemmas puberuli, flavo-cinereo-brunnei. Folia late ovato-lanceolata v. paulo obovato-lanceolata, basi rotunda, interdum subito brevissime in petiolum producta, apice satis subito breviter acuminata, supra ut videtur intense viridia, subtus glauca, utrinque glabra, costa nervisque prominulis graciliter reticulata, nervis sub 80° a costa divergentibus 1–2 pro 1 cm., margine satis regulariter et distanter glanduloso-serrata (dentibus circiter 3 pro 1 cm.), 7–10 cm. longa, 2.2–4 cm. lata; petioli glabri, tantum juxta gemmas superne pilosi, 8–10 mm. longi; stipulae parvae, anguste lanceolatae, circiter 2–3 mm. longae, integrae v. sparse glanduloso-denticulatae; ramuli floriferi glabri foliiferis simillima. Amenta precocia, breviter pedunculata, foliolis parvis lanceolatis subtus sericeis 2–3 suffulta, gracilia, anguste cylindrica, usque 5 cm. longa (haud perfecte evoluta), vix 6–7 mm. crassa, subdensiflora, rhachi subvillosa; flores ♂ diandri filamentis glabris ima basi coalitis bracteis 2–2½-plo longioribus, antheris parvis flavis ovalibus; glandula una ventralis, satis crassa, ovato-rectangularis, apice truncato-obtusa, bracteis oblongis obtusis viridi-brunneis fere tantum versus basim albo-sericeo-villosulis 2–3-plo brevior.

Certainly *S. japonica*, var. *padifolia* Seemen needs further observation. The rather broad and thick gland seems to be a very characteristic feature in all the forms belonging to *S. japonica* Thunberg.

95. **Salix Fauriei** Seemen in *Bot. Jahrb.* XXX. Beibl. LXVII. 40 (1901); *Salic. Jap.* 49, t. 8, fig. C–E (1903).—? Léveillé in Fedde, *Rep. Spec. Nov.* III. 351 (1907). JAPAN. Hondo: Yamakita, on rocks, May 8, 1899, *U. Faurie* (No. 3700; ♀ type, ex von Seemen).

This interesting plant has, according to von Seemen, thin, very reddish-brown branchlets, narrow and acuminate leaves and very short petioles. The ovaries are glabrous and have short pedicels of about the length of the oval gland. I have not yet seen a specimen of the ♀ plant nor of what is called the male *S. Fauriei* by Léveillé, l. c. (Hakone, rocks, July 2, 1892, *U. Faurie*, No. 7691). Léveillé says: "rhachide rufo-villosa; stamina 3."

96. **Salix pyrolaefolia** Ledebour, *Fl. Alt.* IV. 270 (1833); *Icon. Pl. Fl. Ross.* V. 22, t. 476 (1834); *Fl. Ross.* III. pt. 2, 613 (1850). — Turczaninow in *Bull. Soc. Nat. Mosc.* XXVII. 385 (1854); *Fl. Baical.-Dahur.* II. 112 (1856). — Pokorny, *Oesterr. Holzpfl.* 105, t. 21, fig. 289–294 (1864). — Andersson in *Svensk. Vetensk. Akad. Handl.* VI. 169, t. 9, fig. 101 (*Monog. Salic.*) (1867); in De Candolle, *Prodr.* XVI. pt. 2, 257 (1868). — Herder in *Act. Hort. Petrop.* XI. 420 (1891). — Wolf in *Izv. S.-Peterburg Liesn. Inst.* V. 106, t. 38, fig. 1–5, t. 44, fig. 14, t. 45 (*Mam. Изур. Нов Европ. Росс.*) (1900). — Schneider, *Ill. Handb. Laubholzk.* I, 15, fig. 20 z¹, 23 n (1904). — Krylov, *Фл. Алтая* 1221 (1909).

Salix alnoides [1] Schangin ex Sievers in Pallas, *Neu. Nord. Beitr.* VII. 347, 349 (1797?), fide Ledebour. — Georgi, *Beschreib. Russ. Reich.* pt. III. vol. IV. 1340 (nomen nudum) (1800).

Salix sabulosa Turczaninow, *Pl. Exsicc.* a. 1830, ex Ledebour, *Fl. Ross.* III. pt. 2, 613 (pro synon.) (1850).

Salix corylifolia Turczaninow, l. c.

NORTHEASTERN ASIA. Transbaikalia to Amur and Yakutsk (ex Herder).

I have not seen a specimen from our area, and the species is not mentioned by Komarov (*Fl. Mansh.*).

Ledebour (*Fl. Alt.* IV. 270 [1833]) describes three varieties: var. *ovata*, var. *orbiculata*, and var. *cordata*, based on the shape of the leaves. Andersson followed him in 1867, but in 1868 he called the var. *ovata* var. *alnoides* and var. *cordata* var. *sabulosa*. As Turczaninow (l. c.) says, we can find leaves of all the three kinds on the same plant. The species is very near *S. hastata* Linnaeus, but differs from it chiefly in the longer petioles and pedicels and in the straight hairs of the silky pubescence of the catkins which are more matted in *S. hastata* Linnaeus.

97. **Salix hastata** Linnaeus, *Spec.* 1017 (1753).—Brandis, *Forest Fl. Brit. Ind.* 467 (1874); *Ind. Trees*, 673 (ut videtur tantum pro parte) (1906). — Hooker f., *Fl. Brit. Ind.* V. 630 (pro parte) (1888). — Herder in *Act. Hort. Petrop.* XI. 421 (1891). — Schneider, *Ill. Handb. Laubholzk.* I. 51, fig. 230, 24 m–n (1904). — Seemen in Ascherson & Graebner, *Syn. Mitteleur. Fl.* IV. 152 (1909).

For further literature see Seemen, l. c., and Herder, l. c.

NORTHEASTERN ASIA. Yakutsk and Maritime prov. (ex Herder).

I have only seen specimens of the following variety.

Salix hastata, var. **himalayensis** Andersson in *Svensk. Vetensk. Akad. Handl.* VI. 173 (*Monog. Salic.*) (1867).

---

[1] If this is a valid name it would be the oldest.

? *Salix hirta* Royle, *Ill. Bot. Himal.* I. 343 (nomen nudum) (1839).
Salix Roylei Klotzsch apud Andersson in *Svensk. Vetensk. Akad. Handl.* 1850, 479 (1851); *Bot. Ergeb. Reise Prinz. Waldemar*, 120 (nomen nudum) (1862).
Salix hastata, f. *rotundifolia*, f. *oblongifolia* Andersson in *Svensk. Vetensk. Akad. Handl.* 1850, 479 (1851); in *Jour. Linn. Soc.* IV. 51 (1860).

INDIA. Kashmir: " Hazara distr., Kogan valley, Safr Haluk Sar, 18000 ft.," June 5, 1910 (♂ and ♀); "alt. 6–9000 ped.," *T. Thomson* (ex Andersson); "Western Tibet, alt. 11–15000 ped.," *T. Thomson* (type, ♀); " Ladak, alt. 15000 ped.," *T. Thomson* (sterile); same district, " Kharbu Koma to Shaksi, southwest of Dah," July 3, 1886, *Schlagintweit* (No. 5329; ♀); " Balti, Tsogosbang above Barol Brok, right side of the Sos pór glacier," July 17, 1856, *Schlagintweit* (No. 6096; sterile); " Hasóra, Sángu Sár, on the right side of the Tsunger glacier," September 12, 1856, *Schlagintweit* (No. 6592; sterile).

The type of this variety has a long style, but No. 5392 is a rather typical *S. hastata* Linnaeus. Perhaps this Himalayan form is the same as *S. hastata*, var. *alpestris* Andersson, l. c. in Herb. Ind. Or. *S. hastata*, var. *himalensis* is often mixed with *S. sclerophylla* Andersson. See p. 112.

Although I have not seen the Sikkim specimens mentioned by Hooker f., l. c. 631, I doubt if they belong to *S. hastata* Linnaeus.

98. **Salix kenoensis** Koidzumi in *Tokyo Bot. Mag.* XXVII. 265 (1913).
JAPAN. Hondo: prov. Musashi, Chitsibu; prov. Kozuke, Akagisan (ex Koidzumi).
" Species *S. Shiraii* Seemen affinis, sed florum glandulis oblongis; filamentis basi distincte connatis pilosis; foliis basi obtusis non cordatis subtus secus costas adpresse villoso-tomentoso differt."
I have not seen a specimen of this plant.

99. **Salix Shiraii** Seemen in *Bot. Jahrb.* XXX. Beibl. LXVII. 40 (1901); *Salic. Jap.* 42, t. 6, fig. F–K (1903).—Léveillé in *Bull. Acad. Int. Géogr. Bot.* IV. 208 (1904).
JAPAN. Hondo: prov. Shimotsuke, Nikko, on rocks, alt. 2000 m., May 1898, *M. Shirai* (Nos. 42, 43, types, ♂, ex Seemen); same locality, May 27, 1898, *U. Faurie* (No. 2142, co-type; ♀, ex Seemen); same locality, April 1898, *H. Shirasawa* (young ♀ flowers); prov. Mutsu, Aomori, "in monte Hakkoda usque ad 2000 m.," June 8, 1894, *U. Faurie* (Nos. 13103, 13105; ♀, ex Léveillé).
I have seen only Shirasawa's specimen, the flowers of which well agree with von Seemen's figures. But there is a plant collected by *E. H. Wilson*, at Nikko, a bush from 0.6–1.2 m. tall, roadsides, etc., common, alt. 700–1100 m., May 14, 1914 (No. 6686; ♂ and ♀), which very much resembles *S. Shiraii* Seemen, except that the glands are shorter and broader, and that the pedicels of the ovaries are scarcely longer than the gland; the filaments seem to be somewhat united at the very base.

100. **Salix mezereoides** Wolf in *Act. Hort. Petrop.* XXVIII. 529 (1911).
NORTHEASTERN ASIA. Ussuri: "distr. Chabarowsk, mons Janykan," July 16, 1910, *N. Kuznetzov* (No. 187, ♂ type, ex Wolf).
Wolf doubtfully refers this not fully known species to sect. *Hastatae*. According to the description the leaves seem to resemble those of *S. characta* Schneider (see p. 125), but they are "basi in petiolum longum sensim attenuata." The ♂ catkins appear with the leaves and are cylindrical, from 12 to 22 mm. long, the filaments are glabrous or nearly so and the bracts are ligulate.

101. **Salix Nakamurana** Koidzumi in *Tokyo Bot. Mag.* XXVII. 96 (1913); in Matsumura, *Icon. Pl. Koisikav.* I. 149 t. 75 (1913).

Salix cyclophylla Seemen in *Bot. Jahrb.* XXX. Beibl. LXVII. (non Rydberg 1899) (1902); *Salix. Jap.* 69, t. 16, fig. E–F (1903). — Léveillé in *Bull. Acad. Int. Géogr. Bot.* IV. 210 (1904).

JAPAN. Hokkaido: summit of Mt. Riishiri, July 25, 1899, *U. Faurie* (No. 3713; ♀ type of *S. cyclophylla* ex von Seemen); same locality, "repens in caveis ubi nix fusa fuerit," June 27, 1891, *U. Faurie* (No. 3713; ♀, ex Léveillé). Hondo: prov. Shinano, on Yatsugatake, very rare, alt. 2800 m., August 1, 1910 (ex Herb. Sakurai; ♀); same locality, August 1907 (ex Herb. Yokohama Nursery Company; ♀); same prov., on Siroumatake, August 8, 1908 (ex Herb. Tokyo sub nom. *S. arctica*); Dailenyezan, August 1912, *Nakamura* (types of *S. Nakamurana* ex Koidzumi).

The leaves of this species are rather variable in shape, but there is otherwise no real difference between Koidzumi's species and von Seemen's plant according to Matsumura's plate, and I cannot separate these two forms. Von Seemen's name having been used earlier by Rydberg for another species must be changed. The specimens before me have almost entire or not deeply cleft stigmas, and the pedicel of the ovaries is more or less distinct, while in von Seemen's plate the stigmas are represented as deeply cleft and the ovaries almost sessile.

Sect. DIPLODICTYAE Schneider, n. nom.

Salix, sect. *Arcticae* [1] Rydberg, in *Bull. N. York Bot. Gard.* I. 263 (pro parte) (1899).

Frutices prostrati ramulis ramisque radicantibus v. nani ramulis plus minusve erectis. Folia integra, superne ut videtur stomatibus paucis instructa, subtus pallida, satis firma, saepissime distincte petiolata (petiolis saepe $\frac{1}{8}$–$\frac{1}{4}$ laminae longis). Amenta apice ramulorum foliiferorum multiflora, densiflora; flores ♂ glandulis duobus (an semper?) instructi, glandula dorsali saepe minima, antheris (an semper?) rubescentibus; flores ♀ glandula una ventrali; ovaria subsessilia v. brevipedicellata, plus minusve pilosa v. tomentosa, stylis distinctis apice saepe bifidis, stigmatibus bifidis elongatis (? v. brevibus).

Rydberg includes in this section (with some doubt, I believe) *S. polaris* Wahlenburg, see p. 139, and also *S. glauca* Linnaeus. Von Seemen (in Ascherson & Graebner, *Syn. Mitteleur. Fl.* IV. 162 [1909]) says that *S. arctica* Pallas, *S. Brownei* Lundström and *S. ovalifolia* Trautvetter belong to sect. *Myrtosalix*. It is difficult to classify the species No. 103–106 and their relatives. It is unfortunate that Rydberg did not apparently pay any attention to the glands of the ♂ and ♀ flowers.

103. **Salix arctica** Pallas, *Fl. Ross.* I. pt. 2, 86 (1788). — Georgi, *Beschreib. Russ. Reich.* pt. III. vol. IV. 1339 (1800). — Lundström in *Nov. Act. Soc. Sci. Upsal.* 1877, 31, t., fig. 1, 1–3 (*Weid. Now. Semljas*) (1877). — Bebb in *Bot. Gaz.* XIV. 115 (pro parte) (1889). — Rydberg in *Bull. N. York Bot. Gard.* I. 265 (1899). — Schneider, *Ill. Handb. Laubholzk.* I. 41, fig. 20 n–o, 26 d (1904).—Koidzumi in *Tokyo Bot. Mag.* XXVII. 97 (1913). — See also Herder in *Act. Hort. Petrop.* XI. 438 (1891).

Salix Pallasii, var. *diplodictya* Andersson in De Candolle, *Prodr.* XVI. pt. 2, 285 (pro parte) (1868).

[1] The name *Arcticae* (*Arctica*) was first used by Andersson (in *Öfvers. K. Vet.-Akad. Förh.* 1858, 119 and in *Proc. Am. Acad.* IV. 59 [*Salic. Bor.-Am.* 13] [1858]) for a section containing *S. Hookeriana* Barrat, *S. speciosa* Hooker & Arnott, etc. which Andersson (in De Candolle, *Prodr.* XVI. pt. 2, 274, 275 [1868]) referred to sect. *Niveae*.

NORTHEASTERN ASIA. Maritime Prov.: Plover Bay, 1865–6, *W. H. Dall* (ex Rydberg); Berings Island, 1891, *J. M. Macoun* (No. 18884³, ex Rydberg). Also Arctic zone of Russia, Asia, North America and Japan: Kurile Islands (fide Koidzumi).

The type of Pallas's species was collected by Sujef on the Gulf of Obi near the Arctic Ocean. Andersson and other botanists mixed *S. arctica* with several different species which Lundström first distinguished. According to Rydberg the type has thick, broad, obovate or obcordate, strongly reticulate, obtuse leaves which are from 2.5–5 cm. long; the catkins are thick and from 2.5–8 cm. long. It is a shrub. See also the following species.

104. **Salix anglorum** Chamisso[1] in *Linnaea* VI. 541 (1831). — Rydberg in *Bull. N. York Bot. Gard.* I. 266 (1899).

    Salix arctica R. Brown in Ross, *Voy. Expl. Baffin's Bay*, app. cxliii. and ed. 2, II. 194 (nomen nudum; haud Pallas) (1819); *Capt. Parry's Voy.* App. Suppl. p. cclxxxvi (*Chloris Melvill.*) (1823); in Nees von Esenbeck, *R. Brown's Verm. Bot. Schr.* I. 405 (1825); in Bennett, *Misc. Bot. Works R. Brown*, I. 215 (1866). — Trautvetter in *Nouv. Mém. Soc. Nat. Mosc.* II. 307, t. 6 (fide Lundström) (1832). — Herder in *Act. Hort. Petrop.* XI. 438 (pro parte) (1891).

    Salix Brownei Lundström in *Nov. Act. Soc. Sci. Upsal.* 1877, 37 (*Weid. Now. Semljas*) (1877). — Bebb in *Bot. Gaz.* XIV. 115 (1889).

    Salix arctica, var. *Brownei* Andersson in De Candolle, *Prodr.* XVI. pt. 2, 286 (exclud. f. 3) (1868).

NORTHEASTERN ASIA. Yakutsk to Kamtchatka.

Brown's type came from the Baffin Bay region. I have not seen a Siberian specimen.

This species differs according to Lundström and Rydberg from the true *S. arctica* Pallas in being a prostrate shrub, with thin obovate or elliptic-lanceolate, not strongly reticulate, more glabrous and often acutish leaves. The catkins are exceedingly large, and the large conical capsules are a little less hairy. The ♀ specimens from Alaska which I have seen, have a very long style, and the ♂ a large ventral and a small dorsal gland.

105. **Salix diplodictya** Trautvetter in *Nouv. Mém. Soc. Nat. Mosc.* II. 307, t. 14 (1832), fide Rydberg in *Bull. N. York Bot. Gard.* I. 264 (1899).

    Salix Pallasii, var. *diplodictya* Andersson[2] in De Candolle, *Prodr.* XVI. pt. 2 285 (1868).

NORTHEASTERN ASIA. Maritime Prov. (ex Rydberg).

This plant differs from *S. arctica* Pallas and *S. anglorum* Chamisso (according to Rydberg) in the smaller and more rounded leaves, from 1 to 3 cm. long, which are rather crowded and short-stalked, and in the shorter catkins from 1 to 3 cm. in length. The stem is less creeping than in *S. arctica* and the branches are shorter. As

---

[1] Chamisso also cites some specimens which according to Hooker (*Fl. Bor.-Am.* II. 153 [1839]) belong to *S. retusa* Hooker, l. c., which is the same as *S. phlebophylla* Andersson, hence Hooker and Andersson add the name *S. anglorum* Chamisso as a synonym to *S. phlebophylla* Andersson, and Trautvetter (in *Act. Hort. Petrop.* VI. 37 [1879]) accepted Chamisso's name for Andersson's species.

[2] According to Andersson his variety includes the type of Pallas's *S. arctica*. I have not seen Trautvetter's plate.

far as I can judge this seems to be a rather doubtful species, at least I am not sure that Rydberg's specimens are identical with Trautvetter's form. It is omitted in the keys.

106. **Salix altaica** Lundström in *Nov. Act. Soc. Sci. Upsal.* 1877, 36 (*Weid. Now. Semljas*) (1877).

> *Salix crassijulis* Trautvetter in *Nouv. Mém. Soc. Nat. Mosc.* II. 308, t. 15 (1832), fide Lundström.
> *Salix torulosa* Trautvetter, l. c. 309 (1832), fide Lundström.
> *Salix arctica* Trautvetter in Ledebour, *Fl. Alt.* IV. 283 (pro parte, non Pallas) (1833). — Ledebour, *Icon. Pl. Fl. Ross.* V. 18, t. 460 (1834).
> *Salix Pallasii*, var. *crassijulis* Andersson in De Candolle, *Prodr.* XVI. pt. 2, 285 (an tantum pro parte?) (1868).
> *Salix arctica*, f. *altaica* Wolf in Krylov, Фл. Алтая, 1229 (1909).

NORTHEASTERN ASIA. Altai. ? Kamtchatka.
The type of Lundström species is Ledebour's plate of the Altai form. Andersson cites specimens from Kamtchatka. So far as I can judge by the plate and by the specimens from the Altai, the leaves are elliptic and acute at both ends or obovate-oblong and gradually narrowed into the rather short petioles. They are rather smooth on the lower surface and only slightly reticulate.

It would need copious and good material clearly to fix the limits of the different species of this group which, according to Lundström, seem connected with *S. glauca* Linnaeus and other arctic species by so many intermediate forms. There may be several hybrids among them.

107. **Salix oreinoma** Schneider, n. sp.
Frutex humilis, ramis prostratis radicantibus; ramuli initio sericeo-villosi, sed mox glabri, flavo-brunnei v. purpurei, vetustiores nigrescentes. Folia minora rotunda v. obovata, utrinque obtusa v. rotunda, majora elliptica v. obovato-elliptica, basi rotunda v. obtusa, apice saepe subacuta, novella ut videtur utrinque sparse sericeo-villosa, dein mox glabra, supra satis viridia, tenuiter reticulata, subtus pallida, tenuissime albo-punctata, costa nervisque flavescentibus elevatis, sed vix reticulata, margine minute distanter glanduloso-crenato-serrata, minora 1–1.7 cm. longa, 0.7–1.3 cm. lata, maxima usque ad 2.5 cm. longa et ad 1.5–1.7 cm. lata; petioli 4–6 mm. longi, superne sulcati, glabri. Amenta (♀ tantum visa) apice ramulorum lateralium 2–4 cm. longorum folia plura normalia gerentium 1.3–2 cm. longa, circiter 0.9 cm. crassa, pluri- et densiflora, rhachi villosula; ovaria sessilia, sericeo-villosa, ovato-oblonga, stylo satis longo apice bifido, stigmatibus partitis linearibus divaricatis coronata; glandula una ventralis, pleraque ovato-oblonga, sed interdum lata et variabiliter incisa, circiter ⅓ ovarii aequans; bracteae flavo-brunnescentes, ovato-ellipticae, apice truncatae, tantum extus basi villosae, ⅔–¾ ovarii aequantes.

CHINA. Western Szech'uan: southwest of Tachien-lu, bed of torrents, alt. 4000 m., July 1904 (Veitch Exped. No. 4515, type, ♀).
This interesting species most resembles *S. diplodictya* Trautvetter, but the leaves are not entire. As long as the ♂ flowers are unknown I place it with some doubt in this section.
The specific name is derived from ὀρεινόμος, living on the mountains.

Sect. 17. MYRSINITES Borrer in Hooker, *Brit. Fl.* 431 (1830); in Loudon, *Arb. Brit.* III. 1587 (pro parte) (1838).

Salix, sect. *Myrtosalix* Kerner in *Verh. Zool.-Bot. Ges. Wien.* X. 203 (*Niederoest. Weid.*) (1860). — Seemen in Ascherson & Graebner, *Syn. Mitteleur. Fl.* IV. 161 (1909).

A sect. *Diplodictyae* praecipue recedit foliis utrinque viridibus et reticulatis, saepe serratis, floribus ♂ glandula tantum ventrali instructis.
In my *Ill. Handb. Laubholzk.* (I. 41 [1904]) I referred *S. arctica* Pallas (see p. 136), which is placed by Koehne (*Deutsche Dendr.* 95 [1903]) and by von Seemen (1909) in the sect. *Myrtosalix*, to the section where *S. glauca* Linnaeus belongs (see p. 147). In *S. arctica* Pallas I have found stomata in the upper surface of the leaves, as did Camus (*Classif. Saul. d'Europe*, II. 54 [1905]). Camus keeps *S. arctica* Pallas with *S. Myrsinites* Linnaeus in the sect. *Myrtosalix*, but I cannot accept his rather artificial classification based on anatomical characters. Only a comparative study of all the existing Willows with regard to every character presented by the sexual and vegetative organs can lead us to a natural arrangement. If there are only a few stomata in the upper surface of the leaves, this character does not seem to be very useful. I add *S. polaris* Wahlenberg to this section because this species agrees very well with *S. Myrsinites* Linnaeus in the structure of the ♂ and ♀ flowers and in the color and reticulation of the leaves. The position of *S. microphyta* Franchet is still doubtful.

108. **Salix Myrsinites** Linnaeus, *Spec.* II. 1018 (1753). — Pokorny, *Oester. Holzpflz.* 81, t. 17, fig. 200–205 (1864). — Herder in *Act. Hort. Petrop.* XI. 442 (1891). — Wolf in *Izv. S.-Peterburg. Liesn. Inst.* V. 108, t. 28, fig. 6–74 (*Mam. Изур. Иѕ Ѕѕроп. Росс.*) (1900). — Schneider, *Ill. Handb. Laubholzk.* I. 42, fig. 19 o, 21 v–x (1904). — Seemen in Ascherson & Graebner, *Syn. Mitteleur. Fl.* IV. 162 (1909). — Moss, *Cambridge Brit. Fl.* II. 31, t. 33 (1914).
For further literature and synonyms see Herder, l. c., and von Seemen, l. c.
NORTHEASTERN ASIA.   Maritime Prov. and Kamtchatka (ex Herder).
I have not seen any specimen of this species from our area. It is a very variable species, and a prostrate or somewhat erect much branched low shrub. The leaves are shining green, and vary from orbicular-oval to narrow-lanceolate. The petioles are only from 1 to 5 mm. long. The catkins resemble those of *S. arctica* Pallas and are very silky when they first appear.

109. **Salix polaris** Wahlenberg, *Fl. Lapp.* 261, t. 13, fig. 1 (1812). — Andersson, *Salic. Lapp.* 86, fig. 28 (1845); in De Candolle, *Prodr.* XVI. pt. 2, 299 (1868). — Lundström in *Nov. Act. Soc. Sci. Upsal.* 1877, 29 (*Weid. Now. Semljas*) (1877). — Herder in *Act. Hort. Petrop.* XI. 448 (1891). — Wolf in *Izv. S.-Peterburg. Liesn. Inst.* IV. 113, t. 39, fig. 5–7, t. 46, fig. 10 (*Mam. Изуѕ. Иѕ Ѕѕроп. Росс.*) (1900); Wolf & Palibin, *Опред. Деѕѕ. Куѕт. Ѕѕроп. Росс.* 78, fig. (1904). — Schneider, *Illustr. Handb. Laubholzk.* I. 39, fig. 27 d–e (1904). — Camus, *Classif. Saules d'Europe*, II. 48 (1905).
NORTHERN ASIA.   Yakutsk: Arakam Island, 1853–6, *C. Wright.*
The proper position of this species is uncertain. Most authors include it in the sect. *Herbaceae*, and Andersson even says, l. c. (1868): "*S. herbaceae* ita similis, ut ab ea vix distinguatur." In my *Laubholzkunde*, l. c., I stated, that I had never seen a ♂ flower with a dorsal gland which seems to be always present in *S. herbacea* Linnaeus. In habit *S. polaris* is different from *S. Myrsinites* Linnaeus, but otherwise, I do not see any difficulty in placing it in the same section. In Wright's specimen the ♀ flowers sometimes have a small dorsal gland and the length of the pedicel and style is very variable.

It is with some doubt that I add also the following species with glabrous ovaries to this section, but I have been able to find only one gland in the ♂ flowers.

110. **Salix microphyta** Franchet. See p. 62.

Sect. 18. OVALIFOLIAE [1] Rydberg in *Bull. N. York Bot. Gard.* I. 274 (1899). Frutices humiles, plerique prostrati, truncis v. ramis radicantibus. Folia integra, 1–2 cm. longa, subtus pallida, reticulata. Amenta pluriflora, ramulos foliiferos terminantia; flores ♂ diandri, glandula ventrali et dorsali, antheris flavis?; flores ♀ ovariis pubescentibus v. glabris, stylis brevibus sed distinctis, stigmatibus bifidis, glandula una ventrali v. interdum ut videtur etiam dorsali.

I doubt if the species united by Rydberg in this section really belong in the same group.

111. **Salix ovalifolia** Trautvetter in *Nouv. Mém. Soc. Nat. Mosc.* II. 306 (1832), fide Ledebour, *Fl. Ross.* III. pt. 2, 620 (1850). — Andersson in De Candolle, *Prodr.* XVI. pt. 2, 291 (pro parte) (1868). — Lundström in *Nov. Act. Soc. Sci. Upsal.* ser. 3, 1877, 40, t., fig. 2 (*Weid. Now. Semljas*) (1877), ut videtur tantum var. *typica.* — Herder in *Act. Hort. Petrop.* XI. 444 (1891). — Rydberg in *Bull. N. York Bot. Gard.* I. 275 (1899).

NORTHEASTERN ASIA. Yakutsk: "in terra Tschuktschorum," *A. von Chamisso & J. F. Eschscholtz* (ex Ledebour); Maritime Prov.: "ad fl. Ajan," *H. Tiling* (ex Herder); Kamtchatka: "in alpibus," *K. H. Mertens* (in herb. Trautvetter).

According to Lundström (l. c. 16), *S. ovalifolia* is nearly related to *S. reptans* Ruprecht, but it certainly is very difficult to limit the typical forms. They are apparently very variable and are connected by intermediate forms with *S. arctica* Pallas, *S. glauca* Linnaeus and other species. The typical *S. ovalifolia* Trautvetter has glabrous ovaries, the pedicels and styles are sometimes short, but distinct and sometimes are wanting or nearly obsolete.

According to the description, I believe that *S. erythrocarpa* Komarov (in Fedde, *Rep. Spec. Nov.* XIII. 165 [1914]) is the same as or very closely related to *S. ovalifolia.* The type was collected by Komarov "in alpibus peninsulae Kamtschatkae circa lacus Natshika at Kronotzkoe, et ad fontes fl. Kamtshatka, annis 1908–09." Or it may possibly represent a new species of sect. *Lindleyanae* and may be closely related to *S. Souliei* Seemen and *S. brachista* Schneider.

112. **Salix phlebophylla** Andersson in *Öfvers. Vetensk.-Akad. Förh.* XV. 132 (1858); in *Proc. Am. Acad.* IV. 27 (*Salic. Bor.-Am.*) (1858); in De Candolle, *Prodr.* XVI. pt. 2, 290 (pro parte) (1868).—Rydberg in *Bull. N. York Bot. Gard.* I. 275 (1899).

Salix retusa Hooker & Arnott, *Bot. Voy. Beechey,* 130 (non Linnaeus) (1832).— Hooker, *Fl. Bor.-Am.* II. 153 (1839).—Seemann, *Bot. Voy. Herald,* 40 (1852). *Salix Uva-ursi* Herder in *Act. Hort. Petrop.* XI. 444 (non Pursh) (1891).

NORTHEASTERN ASIA. Maritime Prov.: Behring Straits, Arakam Island, 1853–6, *C. Wright* (ex Rydberg).

Rydberg, l. c., says that the type is "No. 96 of Herb. Hook., Barratt & Torr. from the Arctic Coast." Hooker, l. c. (1839), cites Hooker and Arnott, l. c. (1832), where the *S. retusa* is said to come from Kotzebue Sound. Seemann quotes the same specimen and also a specimen from Pelly Island collected by *Pullen.* Seemann's

---

[1] Toepffer (*Salicol. Mitteil.* No. 4, 186 [1911]) cites a sect. *Ovalifoliae* Pokorny (*Oester. Holzpfl.* 58 [1864]), but there is no such section. The name *Ovalifoliae* is only a paragraph in a key, otherwise Pokorny has adopted the systematic arrangement of Kerner.

specimen is the type of *S. phlebophylla* Andersson, f. *major*. The second form, f. *media*, is based upon "Cap Mulgrave: *Beechey*" and "Arctic coast: Dr. *Richardsson*." The third form, f. *minor*, is described without citing a specimen. In all these 3 forms the capsules are described as "glaberrimae," but in 1868 Andersson mixed these forms with others saying: "capsulis . . . tenuiter puberulis sessilibus." Rydberg says: "and the capsule somewhat puberulent when young." He quotes *C. Wright's* specimen mentioned above. I suppose this specimen is the same as that quoted by Herder, l. c., under *S. Uva-ursi* Pursh, which is a species of northeastern America. I have seen no specimen of Wright which belongs to *S. phlebophylla*.

Sect. 19. BERBERIFOLIAE Schneider, n. sect.
Frutices humiles trunco crasso prostrato saepe radicante, ramulis parvis erectis v. procumbentibus. Folia ut in sect. *Myrsinites*, utrinque viridia et reticulata, argute, saepe inaequaliter inciso-serrata v. crenato-serrata. Amenta ramulos foliatos plus minusve longos terminantia, breviter v. longius cylindrica, densiflora, satis crassa; flores ♂ diandri, biglandulosi, glandula dorsali saepe minima, antheris initio ut videtur flavis deinde nigrescentibus; flores ♀ glandula una ventrali v. interdum etiam glandula dorsali minima, ovariis glabris plus minusve pedicellatis, stylis brevibus v. mediocribus, stigmatibus bipartitis oblongis.
This section differs from sect. *Myrsinites* especially in the biglandular ♂ flowers, and from the sect. *Herbaceae* in its longer many-flowered catkins and more distinct styles.

113. **Salix berberifolia** Pallas, *Fl. Ross.* I. pt. 2, 84 (excl. tabula)[1] (1788). — Georgi, *Beschreib. Russ. Reich.* pt. III. vol. IV. 1339 (1800). — Forbes, *Salic. Woburn.* 276, fig. 140 (1829). — Ledebour, *Icon. Pl. Fl. Ross.* V. 15, t. 449, fig. g–k (1834). — Trautvetter in Ledebour, *Fl. Ross.* III. 621 (1850).

Salix Brayi, var. *berberifolia* Andersson in De Candolle, *Prodr.* XVI. pt. 2, 293 (1868). — Herder in *Act. Hort. Petrop.* XI. 445 (1891).
Salix berberifolia, var. *genuina* Glehn in *Act. Hort. Petrop.* IV. 81 (1876).
Salix berberifolia, var. *leiocarpa* Trautvetter in *Act. Hort. Petrop.* VI. 35 (*Fl. Terr. Tschuktschor.*) (1879).

NORTHEASTERN ASIA. Transbaikalia: Barguin River, rocks, *G. W. Steller* (type of Gmelin); "Davuria, in rupibus calvis montium excelsissimorum, juxta nives," *Sujeff* (Pallas's type). Kamtchatka: *G. W. Steller* (Pallas's co-type).
I have not seen a type specimen. According to Trautvetter the var. *genuina* Glehn, l. c., may be distinguished by "foliis profunde inciso-serratis, serraturis patentissimis, immo deorsum erectis," while the following var. *Brayi* differs by "foliis minute serratis, serraturis sursum spectantibus." There is a var. *eriocarpa* Trautvetter (in *Act. Hort. Petrop.* VI. 35 [1879]) with pubescent ovaries; the type of it was found by Turczaninow "in alpe Czekondo tractus baicalensis," together with var. *genuina*.
I suppose the following variety is very closely connected with the typical form.

Salix berberifolia, var. **Brayi** Trautvetter ex Herder in *Act. Hort. Petrop.* XI. 445 (1891).

---

[1] This species was first figured by Gmelin as *Salix pumila foliis dense congestis ovalibus cristatis* (*Fl. Sibir.* I. 161, t. 35, fig. 3 [1747]), and later by Pallas as *Salix foliis sessilibus ovatis dentato-serratis venosis nitentibus* (*Reise*, III. App. 759, t. K.k., fig. 7 [1776]).

Salix Brayi Ledebour, *Fl. Alt.* IV. 289 (1833); *Icon. Pl. Fl. Ross.* V. 15, t. 449 (excl. fig. g–k) (1834). — Trautvetter, *Fl. Ross.* III. pt. 2, 621 (1850). — Andersson in De Candolle, *Prodr.* XVI. pt. 2, 293 (excl. var. γ) (1868). — Krylov, *Фл. Алтая*, 1231 (1909).

Salix berberifolia Pallas, *Fl. Ross.* I. pt. 2, t. 82 (1788), exclud. descriptione fide Trautvetter.

NORTHEASTERN ASIA. Altai: type collected by *A. Bunge* (ex Trautvetter); "Verchov. r. dscholo, pod Lednikom," July 4, 1901, *P. Krylov* (♂ and ♀ flowers), without locality (ex Herb. Mus. Bot. Tomsk, fruiting specimen). Kamtchatka: *Kastalsky & Rieder* (ex Herder).

In Krylov's specimen the ♂ flowers have two glands, and the bracts are covered with a fine silky pubescence somewhat like those of *S. Myrsinites* Linnaeus. The fruits are ovate, acute with a pedicel twice longer than the gland and about ⅔ shorter than the capsule. The serration of the leaves is very sharp and irregular.

114. Salix flabellaris Andersson in *Svensk. Vetensk. Akad. Handl.* 1850, 497 (1851); in *Jour. Linn. Soc.* IV. 54 (1860); in De Candolle, *Prodr.* XVI. pt. 2, 295 (1868). — Klotzsch & Garcke, *Bot. Ergeb. Reise Prinz. Waldemar*, 120, t. 90 (1862). — Hooker f., *Fl. Brit. Ind.* V. 634 (1888). — Brandis, *Ind. Trees*, 638 (1906).

Salix obovata Wallich, *Cat.* No. 3698 (nomen nudum) (1829).
Salix lucida Herb. Jacquemont No. 1160, ex Andersson in *Svensk. Vetensk. Akad. Handl.* 1850, 497 (pro synon.) (1851).
Salix rotundifolia Herb. Royle ex Andersson, l. c. (pro synon.) (1851).

INDIA. Kashmir: " in humidis herbosis sub jugis versus Soognum," alt. 4000 m., August 1, 1829, *V. Jacquemont* (type of *S. flabellaris*, ex Andersson); Kumaon: *N. Wallich* (No. 3698; co-type ex Andersson); " in alpibus himalensibus," *W. Hoffmeister* (ex Andersson); Punjab: Kunawur, " reg. alp., 15000 ped.," *T. Thomson* (with fruits).

Andersson describes several forms. I have only seen Thomson's specimen and a photograph of Wallich's No. 3698. Judging from Klotzsch and Garcke's plate this species is nearly related to *S. berberifolia* Pallas, but the serration of the leaves is much more like that of *S. herbacea* Linnaeus, the flowering forms of which from the Altai I mention on p. 143 seem to be not unlike some forms of *S. flabellaris*.

Sect. 20. HERBACEAE Borrer in Hooker, *Brit. Fl.* 432 (1830); in Loudon, *Arb. Brit.* III. 1590 (pro parte) (1838). — Rydberg in *Bull. N. York Bot. Gard.* I. 277 (1899). — Wolf in *Izv. S.-Peterburg. Liesn. Inst.* V. 110 (*Mam. Изуч. Ив. Евроn. Росс.*) (pro parte) (1900). — Seemen in Ascherson & Graebner, *Syn. Mitteleur. Fl.* IV. 64 (pro parte) (1908).

Salix, sect. *Chamaetia* Dumortier in *Bijdr. Natuurk. Wetensch.* 1825 (*Verh. Gesl. Wilgen* 15) (1825), exclud. *S. reticulata*.
Salix, sect. *Glaciales* W. D. Koch, *De Salic. Europ. Comm.* 61 (pro parte) (1828).
Salix, sect. *Retusae* Kerner in *Verh. Zool.-Bot. Ges. Wien* X. 195 (*Niederöstr. Weid.*) (1860). — Rydberg in *Bull. N. Y. Bot. Gard.* I. 277 (1899). — Seemen in Ascherson & Graebner, *Syn. Mitteleur. Fl.* IV. 84 (1908).
Salix, sect. *Nitidulae* (sive *Glaciales*), subsect. b. *Retusae* Andersson in De Candolle, *Prodr.* XVI. pt. 2, 293 (pro parte) (1868) et subsect. d. *Herbaceae* Andersson, l. c. 297 (pro parte).

Fruticuli pygmaei, trunco decumbente subterraneo repente, ramulis prostratis, saepe radicantibus. Folia glabra, rotundata, obovata, elliptica v. rarius lanceolata, utrinque viridia et reticulata, crasse papyracea, integra v. plus minusve denticulata. Amenta ramulos foliosos terminantia, pleraque brevia et pauciflora; flores ♂ biglandulosi; ♀ plerique tantum glandula una ventrali instructa, sed interdum biglandulosi; ovaria saepissime glabra, brevipedicellata v. subsessilia, stylo brevi stigmatibus bipartitis coronata.

It seems to me also impossible to separate the sect. *Retusae* from the sect. *Herbaceae* on account of the serrate or entire leaves or the presence of a second gland in the♀ flowers. As I find and as it is stated by Toepffer (*Salicet. Exsiccat.* sub No. 27), the ♀ flowers of *S. herbacea* Linnaeus usually have only a ventral gland. Even in the ♂ flowers the gland very rarely seems to be " ringartig, oberwärts unregelmässig eingeschnitten," as described by von Seemen.

115. **Salix herbacea** Linnaeus, *Spec.* 1018 (1753). — Pokorny, *Oesterr. Holzpfl.* 80, t. 17, fig. 197–199 (1864). — Herder in *Act. Hort. Petrop.* XI. 447 (1891). — Hempel & Wilhelm, *Bäume & Sträucher*, II. 107, fig. 192 G (1896). — Wolf in *Izv. S.-Peterburg. Liesn. Inst.* V. 112, t. 38, fig. 21–22 a, t. 39, fig. 1–4 (*Мам. Изуч. Иɜъ Еɜроп. Росс*). (1900); Wolf & Palibin, *Опред. Дереɛ. Куɛт. Еɜроп. Росс.* 76, fig. (1904). — Schneider, *Ill. Handb. Laubholzk.* I. 39, fig. 19 f, 20 r–r² (1904). — Seemen in Ascherson & Graebner, *Syn. Mitteleur. Fl.* IV. 64 (1908). — Moss, *Cambridge Brit. Fl.* II. 27, t. 30(1914).

For further information see Herder, l. c., and von Seemen, l. c.

NORTHEASTERN ASIA. Transbaikalia to the Maritime Prov. (ex Herder).

Specimens collected by *F. N. Meyer*, Altai, Eemonskai, June 20, 1911 (No. 858; ♂) and Altaisk, June 10, 1911 (No. 820; ♀) have rather long, narrow, cylindrical catkins. There is a very small dorsal gland in both the sexes. I have not seen a specimen from the northeastern parts of Siberia. This species may have been confused with *S. polaris* Wahlenberg and *S. rotundifolia* Trautvetter, both of which have usually entire leaves. See also *S. flabellaris* Andersson, above.

116. **Salix rotundifolia** Trautvetter in *Nouv. Mém. Soc. Nat. Mosc.* II. 304, t. 11 (*Salic. frigid.*) (1832), fide Lundström. — Andersson in De Candolle, *Prodr.* XVI. pt. 2, 299 (1868). — Lundström in *Nov. Act. Soc. Sci. Upsal.* 1877, t. 30, fig. 3 (*Weid. Now. Semljas*) (1877). — Rydberg in *Bull. N. York Bot. Gard.* I. 276 (1899). — Wolf in *Izv. S.-Peterburg. Liesn. Inst.* V. 112, t. 38, fig. 15–20, t. 46, fig. 7–9 (*Мам. Изуч. Иɜъ Еɜроп. Росс.*) (1900).

Salix retusa, var. *rotundifolia* Treviranus ex Trautvetter in *Nouv. Mém. Soc. Nat. Mosc.* II. 305 (pro synon.) (*Salic. Frig.*) (1832), fide Lundström. — Bunge in *Mém. Sav. Étr. Acad. Sci. St. Pétersbourg*, II. 607 (*Verz. Altai-Geb. Pfl.* 85) (1835); *Verz. Altai-Geb. Pfl.* ed. 8°, 114 (1836). — Ledebour, *Fl. Ross.* III. 624 (1850). — Herder in *Act. Hort. Petrop.* XI. 446 (1891).

Salix nummularia Andersson in De Candolle, *Prodr.* XVI. pt. 2, 298 (1868). — Krylov, *Фл. Алтая*, 1263 (1909).

For further information see Lundström, l. c.

NORTHEASTERN ASIA. Arctic Siberia: Behring Straits, Arakam Island, *C. Wright*, 1853–6.

The types of *S. rotundifolia* Trautvetter and *S. nummularia* Andersson were the same: *A. T. von Middendorff*, Maritime Prov., Boganida River, and *A. Bunge*, Altai, " in summis alpibus ad fl. Tschuja." Whether Andersson's *S. nummularia*, var. *subretusa* (in De Candolle, *Prodr.* XVI. pt. 2, 298 [1868]) represents a distinct variety I do not know. *S. rotundifolia* Trautvetter is very near *S. retusa* Linnaeus, but the leaves are rounded or cordate at the base and not attenuated and acute.

Sect. 21. RETICULATAE E. Fries, *Sylloge Pl. Nov.* II. 38 (pro parte) (1828). — Borrer in Hooker, *Brit. Fl.* 422 (1830); in Loudon, *Arb. Brit.* III. 1542 (1838). — Rydberg in *Bull. N. York Bot. Gard.* I. 259 (1899). — Seemen in Ascherson & Graebner, *Syn. Mitteleur. Fl.* IV. 67 (1908). — Ball in Coulter & Nelson, *New Man. Bot. Rocky Mts.* 138 (1909).

Salix, sect. *Nitidulae* s. *Glaciales,* d. *Reticulatae* Andersson in De Candolle, *Prodr.* XVI. pt. 2, 300 (1868).

Fruticuli pygmaei habitu iis sect. *Herbaceae* similes, interdum frutices nani. Folia satis crassa, supra nervis saepissime impressis subtus elevatis reticulata, discoloria, integra, longe petiolata. Amenta multos foliatos terminantia, anguste cylindrica, pleraque multiflora; flores ♂ et ♀ glandulis duobus saepe pseudodiscum lobulatum formantibus instructi; ovaria sessilia, plus minusve tomentosa, stylis nullis v. brevissimis, stigmatibus bipartitis.

117. **Salix reticulata** Linnaeus, *Spec.* 1018 (1753). — Andersson in De Candolle, *Prodr.* XVI. pt. 2, 301 (excl. var. β) (1868). — Lundström in *Nov. Act. Soc. Sci. Upsal.* 1877, 31 (*Weid. Now. Semljas*) (1877). — Herder in *Act. Hort. Petrop.* XI. 450 (1891). — Wolf in *Izv. S.-Peterburg. Liesn. Inst.* V. 114, t. 39, fig. 8–10, t. 46, fig. 4–6 (*Mam. Изуч. Нов. Европ. Росс.*) (1900). — Hempel & Wilhelm, *Bäume & Sträucher,* II. 107, fig. 192 b–f (1892). — Schneider, *Ill. Handb. Laubholzk.* I. 40, fig. 14 g–h, 18 a–d (1904). — Moss, *Cambridge Brit. Fl.* II. 26, t. 29 (1914).

Salix reticulata, var. *glabra* Ledebour, *Fl. Ross.* III. 623 (1850).
? *Salix reticulata,* var. *villosa* Ledebour, l. c. (1850). — Toepffer, *Schedae Salic. Exsicc. Fasc.* III. 71 (1908).
Chamitea reticulata Kerner in *Verh. Zool.-Bot. Ges. Wien.* X. 277 (*Niederoestr. Weid.*) (1860). — Pokorny, *Oesterr. Holzpfl.* 126, t. 24, fig. 364–365 (1864).

For further information see Herder, l. c., and von Seemen, l. c.
NORTHEASTERN ASIA. Transbaikalia to Kamtchatka.
According to Rydberg (in *Bull. N. York Bot. Gard.* I. 260 [1899]), the true *S. reticulata* is absent from North America, and *S. vestita* Pursh (*Fl. Am. Sept.* II. 610 [1814]) does not occur in Asia. The specimens mentioned by Herder, l. c. 451, as *S. reticulata,* var. *sericea* Andersson belong to *S. reticulata* Linnaeus and not to *S. vestita* Pursh, which differs in its shorter-stalked leaves with a persistent silky pubescence on the lower surface. The petioles are only about as long as the buds.

Sect. 22. LINDLEYANAE Schneider, n. sect.

Salix, sect. *Nitidulae,* a. *Myrtosalix* Andersson in De Candolle, *Prodr.* XVI. pt. 2, 285 (1868) et b. *Retusae* Andersson, l. c. 293 (pro parte).

Fruticuli pygmaei habitu iis sect. *Herbaceae.* Folia parva v. minima, pleraque lanceolata, oblanceolata v. ovato-lanceolata, integra v. serrata v. lobulato-incisa, supra viridia, nervis plus minusve impressis v. enervia, subtus discoloria, pallida v. glauca v. pallide viridia, costa nervisque elevatis v. enervia. Amenta ramulos foliolatos terminantia, pleraque brevia et pauciflora; flores ♂ diandri, antheris (an semper) flavis, glandulis duobus; flores ♀ glandula pleraque unica ventrali; ovaria glabra, brevipedicellata, stylo brevi v. mediocri apice bifido, stigmatibus bipartitis.

This section is apparently nearly related to sect. *Herbaceae,* but differs in the leaves, which are paler below and have a different venation. The geographical distribution of the species is also entirely different; the Himalaya and the high mountains of western China being regions where no species of sect. *Herbaceae* occurs.

118. **Salix Lindleyana** Wallich apud Andersson in *Svensk. Vetensk. Akad. Handl.* 1850, 499 (1851); in *Jour. Linn. Soc.* IV. 56 (1860); in De Candolle, *Prodr.* XVI. pt. 2, 296 (1868). — Hooker f., *Fl. Brit. Ind.* V. 634 (1888). — Brandis, *Ind. Trees,* 638 (1906).

Salix furcata Andersson in De Candolle, *Prodr.* XVI. pt. 2, 291 (pro parte) (1868).

INDIA. Kumaon: " Pindari, alt. 12000 feet," *Strachey & Winterbottom* (No. 12, types of *S. furcata,* ♂ and ♀); " distr. Johar, Milum glacier, alt. 11200–12000 feet," June 25, 1865, *Schlagintweit* (No. 9626; ♂); Nepal: " ad Gossain Than," *R. Blinkworth* (Herb. Wallich No. 3697, type of *S. Lindleyana,* ♀, ex Andersson); Sikkim: Kin chin jhow, " 16000 ped.," September 14, 1849, *J. D. Hooker* (♀).

Andersson describes two forms: var. *latifolia* (l. c. 300 [1851] et l. c. 56 [1860]): foliis 3–4 lin. longis, 1–2 lin. latis, apicem versus subserrulatis, and var. *microphylla* (l. c. [1851] and l. c. 296 [1868]): foliis 1–2 lin. longis, lineam latis, margine subintegro revolutis. To the last form belong the specimens cited above. Hooker f. (l. c. 635 [1888]) mentions a " forma *major* Andersson in Herb. Wall." (*S. clavata* Wallich, Cat. 3698," by *R. Blinkworth*, Kumaon): " Leaves 1–1½ in. rounded obovate narrowed into a petiole ½ in., and capsules nearly ¼ in. long." It seems to be a different species. According to Andersson the leaves are " subtus opaco-glaucescentia costa flava nitente eximie prominente plana," which is the case in the specimens I have seen.

The leaves of the type of *S. furcata* Andersson are denticulate, and the nerves are more or less prominent beneath; the petioles are very short compared with blade. The broad ventral glands of the flowers are the same in both sexes as in the typical *S. Lindleyana;* the stamens are glabrous. I cannot see a real difference between it and *S. furcata* Andersson. Andersson apparently mixed *S. furcata* with his *S. fruticulosa* which he omitted from the Prodromus, only citing *S. arbuscula* Andersson (1860) as a synonym of *S. furcata*. The *S. arbuscula* is apparently a misprint for *S. fruticulosa*, which belongs to a different section (see p. 119).

119. **Salix brachista** Schneider, n. sp.

Frutex pygmaeus facie *S. Lindleyanae* Andersson; rami prostrati, radicantes, flavo-brunnei; ramuli procumbentes (v. erectiusculi?) graciles, glabri, rubescentes; gemmae parvae, obtusae, glabrae. Folia perparva, firma, elliptica, utrinque acuta, v. elliptico-spathulata, apice obtusa, supra viridia, glabra, costa impressa, nervis interdum subprominulis, subtus pallida, sed non glauca, praesertim ad costam pilis longis sericeis conspersa v. glabra, costa nervisque distincte prominulis interdum subreticulata, margine versus apicem minute denticulata v. rarius subintegra, 4–9(–10) mm. longa, 1.5–5 mm. lata, basi in petiolos 1–3 mm. longos saepe sericeos angustata. Amenta ♂ tantum visa, ramulos densifoliatos 1.5–2.5 cm. longos interdum laxe sericeos terminantia, ovato-cylindrica, 7–10 mm. longa, circiter 5 mm. lata, pluriflora, rhachi sparse villosa, flores ♂ diandri, filamentis basi villosis bracteas vix duplo superantibus, antheris parvis ovalibus flavis instructi; glandulae duae, ventralis anguste oblonga, basi subcrassior, bracteis duplo brevior, dorsalis minor, angustior; bracteae obovato-oblongae, flavae, glabrae, apice interdum truncatae et denticulatae.

CHINA. Western Szech'uan: Ching-chi Hsien, Ta-hsiang-ling, mountains, alt. 2800 m., May 1904 (Veitch Exped. No. 4511, type; ♂).

It is with some hesitation that I describe this plant as a new species. It may be only a form of *S. Souliei* Seemen, the ♂ of which according to von Seemen chiefly differs in its entire more glaucous leaves, its shorter catkins and in its glabrous filaments. Unfortunately I have not been able to examine the type of *S. Souliei* (see

p. 62). *S. brachista* also is nearly related to *S. Lindleyana* Andersson, which differs in its more glaucous leaves, in the glabrous filaments and in the much broader ventral gland.

The specific name is derived from βράχιστος, very small.

120. **Salix Souliei** Seemen. See p. 62.

121. **Salix oreophila** Hooker f. apud Andersson in *Jour. Linn. Soc.* IV. 57 (1860); in De Candolle, *Prodr.* XVI. pt. 2, 296 (1868). — Hooker f., *Fl. Brit. Ind.* V. 635 (1888).

INDIA. Sikkim: "reg. alp. alt. 15–16000 ped.," *J. D. Hooker* (type; ♂ flowers and fruits).

This is a very small, flabellately branched, prostrate shrub. The leaves are greenish beneath; their nerves are impressed above, and do not show on the lower surface; the serration is rather deep, and the teeth are incurved, or the leaves are rather trilobate at the apex. The bracts, at least of the ♂ plant, are slightly lobulate; the filaments are glabrous.

Salix .oreophila, var. **secta** Andersson in De Candolle, *Prodr.* XVI. pt. 2, 297 (1868). — Hooker f., *Fl. Brit. Ind.* V. 635 (1888).

Salix secta Hooker f. apud Andersson in *Jour. Linn. Soc.* IV. 57 (1860).

INDIA. Sikkim: "reg. alp., alt. 14000 ped.," *J. D. Hooker* (type, ♂ and also ♀, ex Andersson, who says: "17000 ped.").

S. secta Hooker f. seems to be only a very small form of *S. oreophila* Hooker f., differing in its smaller, somewhat broader leaves, which are lobulate at the apex (foliis cuneatis apice 3–5-fidis). The bracts in the male co-type before me are obtuse at the apex and the filaments are glabrous as in the type of *S. oreophila.* It resembles the European *S. serpyllifolia* Scopoli. These are, so far as I know, the smallest Willows known.

122. **Salix calyculata** Hooker f. apud Andersson in *Jour. Linn. Soc.* IV. 55 (1860); in De Candolle, *Prodr.* XVI. pt. 2, 296 (1868). — Hooker f., *Fl. Brit. Ind.* V. 635 (1888).

INDIA. Sikkim: "reg. alp., alt. 12–15000 ped.," *J. D. Hooker* (type; ♂ and ♀).

According to Hooker, this is "a very small gnarled shrub, with ascending branchlets." I have seen the types the leaves of which apparently are not yet fully developed. The ♂ flowers have two very long and narrow glands, the dorsal being scarcely smaller than the ventral. They are about half as long as the oblong glabrous bracts which are ciliate and somewhat emarginate at the apex. The free glabrous filaments are only ⅓ longer than the bracts. The ♀ flowers apparently are not in a normal condition, the ovaries being narrowly oblong with short sessile stigmas. Andersson says: "capsulis breviter pedicellatis ovatis glabris, stylo mediocri, stigmatibus brevibus." The gland is rather broad, ovate-rectangular, truncate or emarginate at the apex; the bracts are glabrous, broadly obovate, crenulate at the apex, nearly enveloping the ovaries. The younger branchlets and leaves are loosely covered with long silky hairs.

A ♂ specimen, collected by *G. Forrest*, Yunnan, Tali Range, alt. 3300–3800 m., August 1906 (No. 4603), which was named *S. microphyta* Franchet by Diels (in *Not. Bot. Gard. Edinburgh*, VII. 252 [1912]) looks very much like *S. calyculata.* The flowers have glabrous filaments and two rather large and broad glands. The Yunnan plant, however, differs widely from *S. microphyta* Franchet (see p. 62) and may represent a new species; certainly it needs further study.

Sect. 23. SERPYLLA Schneider, n. sect.

A sectione *Lindleyanae* praecipue differt glandulis duobus florum ♂ bipartitis v. bilobis pseudodiscum 4-partitum formantibus, stylis florum ♀ gracillimis bipartitis ramis divaricantibus, stigmatibus minimis emarginatis.

According to these characters the following species seems to represent a distinct group.

123. **Salix Serpyllum** Andersson in *Jour. Linn. Soc.* IV. 55 (1860); in De Candolle, *Prodr.* XVI. pt. 2, 292 (1868). — Hooker f., *Fl. Brit. Ind.* V. 634 (1888). — Brandis, *Ind. Trees*, 638 (1906).

Salix longipes Hooker & Thomson in *Herb. Ind. Or.* apud Andersson in *Jour. Linn. Soc.* IV. 56 (pro synon.) (1860).

INDIA. Sikkim: " reg. alp., alt. 10–14000 [non 17000] ped.," *J. D. Hooker* (♂ and ♀ types).

Having seen the types from the Kew Herbarium [Lachen, 10000′, June 3, 49], I agree with Andersson, who says: " Distinctissima est species, crescendi modo (fere ut in *Thymo serpyllo*) ramis et foliis omnibus diversa." The fruiting aments are up to 4.5 cm. long; while the young ♀ catkins are only from 1–1.5 cm. in length. Andersson describes the rachis of the ♂ catkins as " aureopilosa." I find only a grayish pubescence, otherwise his description is correct.

Sect. 24. GLAUCAE E. Fries,[1] *Syllog. Pl. Nov.* II. 36 (pro parte) (1828). — Borrer in Hooker, *Brit. Fl.* 422 (pro parte) (1830); in Loudon, *Arb. Brit.* III. 1543 (pro parte) (1838).

Salix, sect. *Frigidae* W. D. Koch, *De Salic. Comm.* 53 (pro parte) (1828).

Salix, sect. *Niveae* s. *Glaucae*, c. *Sericeae* Andersson in De Candolle, *Prodr.* XVI. pt. 2, 280 (1868).

Salix, sect. *Sericeae* Koehne, *Deutsche Dendr.* 93 (1903). — Seemen in Ascherson & Graebner, *Syn. Mitteleur. Fl.* IV. 57 (1908).

Frutices erecti, plerique parvi. Folia elliptico-lanceolata v. elliptica usque obovata, integra, mollia. Amenta plus minusve foliato-pedunculata; flores ♂ glandulis duobus instructi, antheris (an semper?) rubescentibus; flores ♀ glandula una ventrali (v. interdum etiam dorsali) instructi, ovariis sericeo-lanatis pedicellatis, stylis distinctis saepe apice bifidis, stigmatibus bifidis oblongis.

This section shows a relationship with several other groups, and it is with some doubt that I refer to it *S. floccosa* Burkill and *S. opsimantha* Schneider.

124. **Salix glauca** Linnaeus, *Spec.* 1019 (1753). — Andersson in De Candolle, *Prodr.* XVI. pt. 2, 280 (1868).—Lundström in *Nov. Act. Soc. Sci. Upsal.* 1877, 38 (*Weid. Now. Semljas*) (1877). — Herder in *Act. Hort. Petrop.* XI. 435 (1891). — Wolf in *Izv. S.-Peterburg. Liesn. Inst.* V. 24, t. 24, fig. 5, t. 25, fig. 1–10 (*Мат. Изуч. Ив. Европ. Pocc.*) (1900). — Schneider, *Ill. Handb. Laubholzk.* I. 41, fig. 21 e–g, 25 k (1904). — Seemen in Ascherson & Graebner, *Syn. Mitteleur. Fl.* IV. 58 (1909).

For further information see Herder, l.c., and von Seemen, l. c.

NORTHEASTERN ASIA. Transbaikalia to Kamtchatka.

A variable species which seems to be connected by intermediate (or hybrid?) forms with *S. arctica* Pallas, *S. anglorum* Chamisso, *S. lanata* Linnaeus, etc., see also Lundström, l. c.

125. **S. opsimantha** Schneider. See p. 63.

---

[1] Fries's name is older than sect. *Frigidae* W. Koch.

126. **Salix floccosa** Burkill in *Jour. Linn. Soc.* XXVI. 529 (1899). — Diels in *Not. Bot. Gard. Edinburgh*, VII. 119 (*Pl. Chin. Forrest.*) (1912).

CHINA. Yunnan: " Maeul chan, and rocks of Lao long tong above Yen-tze-hay at 11000 ft.," also " Li chiang at 12500 ft., always on a calcareous formation," *J. Delavay* (Nos. 2200, 3105, 4323, 4678, ex Burkill); " summit of exposed cliffs forming a side valley on the eastern flank of the Lichiang range, alt. 10000–10500 ft., June 1906," *G. Forrest* (No. 2318; shrub 0.6–0.9 m. tall; ex Diels).

I have not seen a specimen of this Willow, which seems to be most nearly related to *S. opsimantha* Schneider (p. 63). Burkill placed it in sect. *Viminales*, but according to his description it may be better placed in sect. *Glaucae*, and certainly it does not belong to sect. *Viminales*. It also resembles some species of sect. *Eriostachyae*.

Sect. 25. CAPREAE Bluff & Fingerhuth,[1] *Comp. Fl. Germ.* II. 565 (pro parte) (1825).—Dumortier, *Fl. Belg. Prodr.* 11 (*Capraeae*) (1827).—W. D. Koch, *De Salic. Comm.* 31 (pro parte) (1828). — Andersson in *Svensk. Vetensk. Akad. Handl.* VI. 57 (1867); in De Candolle, *Prodr.* XVI. pt. 2, 215 (pro parte) (1868). — Schneider, *Ill. Handb. Laubholzk.* I. 57 (1904). — Seemen, *Salic. Jap.* 17 (1903); in Ascherson & Graebner, *Syn. Mitteleur. Fl.* IV. 58 (1908).

> *Salix*, sect. *Cinereae* E. Fries in *Syllog. Pl. Nov.* II. 37 (1828). — Borrer in Hooker, *Brit. Fl.* 424 (1830); in Loudon, *Arb. Brit.* III. 1553 (1838).
>
> *Salix*, sect. *Capreae*, subsect. γ *Rugosae* Reichenbach, *Fl. Germ. Excurs.* I. 169 (1830).
>
> *Salix*, sect. *Rugosae* Kerner in *Verh. Zool.-Bot. Ges. Wien*, X. 242 (1860).

Arbores mediocres v. frutices alti parvive. Folia ovata, elliptica, obovata v. obovato-lanceolata, supra saepe venulis impressis rugosa, subtus pleraque elevato-reticulata, novella v. etiam adulta plus minusve pubescentia. Amenta pleraque praecocia et sessilia, ovata v. cylindrica, satis crassa; bracteae pleraeque bicolores; flores ♂ diandri, filamentis plerisque liberis, antheris flavis, glandula una ventrali; flores ♀ glandula una, ovariis plerisque pilosis longe pedicellatis, stylo nullo v. brevissimo, stigmatibus brevibus v. oblongis.

The species of this section are distinguished by the long pedicels and by the short styles or sessile stigmas.

127. **Salix cinerea** Linnaeus, *Spec.* 1021 (1753). — Pokorny, *Oesterr. Holzpfl.* 112, t. 23, fig. 316–317 (1864). — Herder in *Act. Hort. Petrop.* XI. 400 (1891). — Hempel & Wilhelm, *Bäume & Sträucher*, II. 116, fig. 199 F–G (1897).—Wolf in *Izv. S.-Peterburg. Liesn. Inst.* V. I. 72, t. 36, fig. 6–16 (*Мат. Изуч. Нов. Европ. Росс.*) (1900); XIV. 189, 195, t. (*Азиам. Нов*, II.) (1906). — Komarov in *Act. Hort. Petrop.* XXII. 22 (1903). — Schneider, *Ill. Handb. Laubholzk.* I. 58, fig. 24 a–a¹, 25 f–g (1904). — Seemen in Ascherson & Graebner, *Syn. Mitteleur. Fl.* IV. 93 (1909); apud Siuzev in *Trav. Mus. Bot. Acad. Sci. St. Pétersbourg*, IX. (1912), ex Toepffer, *Salicol. Mitt.* No. 5, 248 (1912).—Nakai in *Jour. Coll. Sci. Tokyo*, XXXI. 213 (*Fl. Kor.* II.) (1911). — Moss, *Cambridge Brit. Fl.* II. 54, t. 52–53 (1914).

For further synonyms and literature see Herder, l. c., and von Seemen, l. c.

NORTHEASTERN ASIA. Amur: (fide Siuzev). Korea: (fide Nakai).

According to Komarov the *S. cinerea* of his *Flora Manshuriae* is *S. gracilistyla* Miquel (see p. 164), but von Seemen apud Siuzev mentions two varieties of this variable species and he also mentions *S. Thunbergiana* Blume, which is only a synonym of Miquel's species. Unfortunately I have not seen any specimens of either from

---

[1] These authors apparently did not use the term *Capreae* to designate a natural section.

Mandshuria except a sterile specimen collected by C. S. Sargent, Khingan Mountains, Yulo Pass, alt. 1000 m., August 15, 1913, which probably belongs to *S. cinerea*.

128. **Salix Caprea** Linnaeus, *Spec.* 1020 (exclud. var. γ et δ) (1753). — Pokorny, *Oesterr. Holzpfl.* 110, t. 22, fig. 302–305 (1864). — Herder in *Act. Hort. Petrop.* XI. 402 (1891). — Hempel & Wilhelm, *Bäume & Sträucher*, II. 114, fig. 198 A-B, t. 32 (1897). — Wolf in *Izv. S.-Peterburg. Liesn. Inst.* V. 66, t. 31, fig. 16–17, t. 32, fig. 1–10 (*Mam. Изуч. Изъ Европ. Росс.*) (1900). — Seemen, *Salic. Jap.* 33, t. 4 D-H (1903); in Ascherson & Graebner, *Syn. Mitteleur. Fl.* IV. 98 (1909); apud Siuzev in *Trav. Mus. Bot. Acad. Sci. St. Pétersbourg*, IX. (1912) ex Toepffer, *Salicol. Mitt.* No. 5, 248 (1912). — Komarov, in *Act. Hort. Petrop.* XXII. 21 (*Fl. Mansh.* II.) (1903). — Schneider, *Ill. Handb. Laubholzk.* I. 60, fig. 11 v–v², 15 l (1904). — Léveillé in *Bull. Acad. Int. Géogr. Bot.* XVI. 150 (1906). — Shirasawa, *Icon. Ess. For. Jap.* II. t. 8, fig. 12–22 (1908). — Henry in Elwes & Henry, *Trees Gr. Brit. & Irel.* VII. 1745 (1913). — Moss, *Cambridge Brit. Fl.* II. 52, t. 49 (1914).

    Salix brachystachys Franchet & Savatier, *Enum. Pl. Jap.* I. 460 (1875); II. 506 (1879).

    For further synonyms and literature see Herder, l. c., and von Seemen, l. c.
    CHINA. Chili: Hsiao Wu-tai-shan, well-wooded mountain slopes, alt. 1600–2600 m., August 20, 1903, *F. N. Meyer* (No. 1248, sterile; a doubtful form [1]). Shensi: Tsingling, *A. David* (fide von Seemen); Khin-lin-san, *G. Giraldi* (No. 1160, ex von Seemen).
    NORTHEASTERN ASIA. Ussuri: Khabarovsk, in upland oakwoods, August 3, 1903, *C. S. Sargent* (sterile); near Vladivostok, August 18, 1903, *C. S. Sargent* (a small tree; sterile). Korea: Chinampo, September 20, 1905, *J. G. Jack* (sterile); Quelpaert, "secus rivulos montium," alt. 1800 m., May 17, 1907, *U. Faurie* (No. 1503; 3–4 m. tall; ♀). Kamtchatka: Petropaulovski, 1853–6, *C. Wright* (with fruits). Saghalien: Sakaihama, moist woodlands, seashore, August 5, 1914, *E. H. Wilson* (No. 7366; bush 1.5–1.8 m. tall; sterile); Toyo-hara, side of streams, abundant, August 3, 1914, *E. H. Wilson* (No. 7342; bush 1.8–4.5 m. tall; sterile); Korsakoff, July 1908, *U. Faurie* (Nos. 272, 272ᵇⁱˢ in part; " arbor lata et frequentissima"; with fruits); same locality, October 1908, *U. Faurie* (No. 272ᵇⁱˢ in part; with old leaves; No. 273ᵗᵉʳ; sterile); without precise locality, *Fr. Schmidt* (♀).
    JAPAN. Kurile Island: Etorofu, Furebetsu, July 28, 1884, *K. Miyabe* (with fruits). Hokkaido: prov. Oshima, near Mosi, September 26, 1892, *C. S. Sargent* (sterile); Hakodate, in forests, June 1, 1904, *U. Faurie* (No. 5759; with fruits); Jirafu, in forests, June 27, 1905, *U. Faurie* (No. 6611; sterile); Shiribeshi-san, common around mountain base, alt. 330 m., July 27, 1914, *E. H. Wilson* (No. 7267; small tree 6 m. tall, girth 0.3 m., branches spreading; sterile); prov. Ishikari, Sapporo, April 23–24, 1890, *Y. Tokubuchi* (♂ and ♀); same locality, April 1891, *Y. Tokubuchi* (♂ and ♀); same locality, May 21, 1903, *S. Arimoto* (♀); same

----

[1] There is another sterile specimen of Meyer's from the mountains in Sheng-king, June 24, 1906 (No. 20), with entire shortly acuminate ovate leaves, covered below with a thick soft pubescence. The leaves are up to 8 cm. long and 3.8 cm. broad, and the petioles are from 4–9 mm. in length. Meyer names this form " broad leaved willow." There is also a specimen of Purdom's from northern China, Weichang, 1910 (No. 79). The leaves of this are very broad-elliptic, entire, from 6–16 cm. long and 3.5–8 cm. broad and thickly grayish pubescent beneath; the apex is roundish and shortly apiculate; the petioles are from 1–2 cm. long. All these forms need further study.

locality, July 1904, *H. Shirasawa* (sterile); same locality, June 14, 1908, *U. Faurie*
(No. 254; ♀); same locality, April 29, 1906, *K. Kondo* (♂ and ♀). Hondo:
prov. Mutsu, Mt. Hakkoda, near Aomori, alt. 1000 m., October 23, 1892, *C. S.
Sargent* (small tree; sterile) " in basi montis Iwagisan," May 13, 1905, *U. Faurie*
(♂ and ♀); "in sylvis Hayashima," June 6, 1905, *U. Faurie* (No. 6612; with
fruits); prov. Kozuke, Karuizawa, September 1, 1905, *J. G. Jack* (sterile); prov.
Shimotsuke, shores of Lake Chuzenji, September 3, 1892, *C. S. Sargent* (tree 7–10 m.
tall; sterile); same locality, woods, common, alt. 1200 m., May 24, 1914; *E. H.
Wilson* (No. 6732; bush 3–5 m. tall; with fruits); same locality, June 1, 1914, *E.
H. Wilson* (No. 6779; tree 6–8 m. tall; girth 0.3–0.9 m.; with fruits); Nikko,
1904, *N. Mochizuki* (♂); same locality, April 31, 1908, *H. Shirasawa* (♀); prov.
Shinano, on Tsubakura-dake, side of torrents, common, alt. 1600 m., September
13, 1914, *E. H. Wilson* (No. 7500; bush or small tree, 2–7 m. tall; sterile); prov.
Musashi, Mt. Buko, May 1901 (No. 11 ex Herb. Tokyo; with fruits); Tokyo,
Meguro, April 1912, *H. Shirasawa* (♀); prov. Sagami, around Hakone, woodlands,
common, April 17, 1914, *E. H. Wilson* (No. 6423; bush or small tree, 3–4.5 m.
tall; ♀).

So far as I can judge from the material before me these forms of *S. Caprea* from
eastern Asia may represent a somewhat different form or include some hybrids.
Von Seemen (in Ascherson & Graebner, *Syn. Mitteleur. Fl.* IV. 101 [1909]), describes *S. Caprea,* var. *villosa* with roundish, entire, tomentose leaves and tomentose branchlets from eastern Asia, collected by Siuzev. Nakai does not mention
S. Caprea from Korea; he cites *S. cinerea* Linnaeus (see above), but Léveillé (in
Fedde, *Rep. Spec. Nov.* X. 435 [1912]) describes a *S. hallaisanensis* with a var.
nervosa from Korea, Quelpaert, " in sylvis Hallaisan," circ. 1200 m., June 9, 1908,
Taquet (No. 1442; almost creeping shrub, 0.6–0.7 m. tall; with fruits; No. 1443);
same locality, alt. 1200 m., August 12, 1908, *Taquet* (No. 1444, type of var.
nervosa, sterile); same place, alt. 1200 m., June 1909, *Taquet* (Nos. 3251, 3252,
3253, 3255, 3256, 3258, 3259, 3260, ♀ co-types, 3257, ♂ co-type); " in Monte des
Diamants," June 20, 1906, *U. Faurie* (No. 174; ♀), "in aridis montium Quelpaert,
supra 1200 m.," July 1907, *U. Faurie* (No. 1502; 1 m. alta; ♀); Hallaisan, alt. 1300-
m., October, 1907, *U. Faurie* (No. 324; sterile). All these specimens have orbicular-
ovate, entire or crenate leaves, prominently reticulate beneath. There seems to be
no real difference between these and *S. Caprea,* the older branchlets being glabrous,
not hairy as in *S. cinerea.* See also *S. Wallichiana* Andersson, p. 64.

In the northwestern Himalaya *S. Caprea* is only cultivated. See Brandis (*Forest
Fl. Brit. Ind.* 467, t. 60 [1874]); the ♀ flower on the plate shows a very short
pedicel.

According to von Seemen (apud Siuzev in *Trav. Mus. Bot. Acad. Sci. St. Pétersbourg,* IX. No. 2 [1912] ex Toepffer, *Salicol. Mitt.* No. 5, 248 [1912]), also *S. aurita*
Linnaeus has been found in Mandshuria. All the species of sect. *Capreae* are very
closely related, and they need a special and careful study.

129. **Salix Wallichiana** Andersson. See p. 64.

130. **Salix pachyclada** Léveillé in Fedde, *Rep. Spec. Nov.* III. 22 (1906); in
Bull. Soc. Bot. France, LVI. 300 (1909); *Fl. Kouy-Tchéou,* 381 (1915).

CHINA. Kweichau: " Montée de Sa-Yang à Pia-Fong," March 4, 1905,
J. Esquirol (No. 368, type; ♂).

The author has kindly sent me a male catkin of this species. This is 2.2 cm. long
and about 1.3 cm. thick, subsessile, the peduncle being only about 5 mm. long,
bearing two small leaves, which are glabrous above and a little silky beneath. The
flowers have a short and rather broad flat ventral gland, 4 or 5 times shorter than

the very narrow lanceolate acuminate dark bracts, which are densely silky on the inner side and on the margins and are nearly glabrous on the outside. The two free stamens are somewhat pubescent at the very base and about twice as long as the bracts; the anthers are oblong and yellow. The young leaves of which I have seen only a small one (2 cm. long, 1 cm. broad) are ovate-elliptic, acute at the ends, silky above and more densely silky-villose beneath. I suspect *S. pachyclada* is nearly related to *S. Wallichiana* Andersson and that it may represent only a form of this variable species. All these forms with very precocious flowers are difficult to understand from the lack of good old leaves from the same plants of the two sexes from which the flowers were gathered.

131. **Salix Starkeana** Willdenow, *Spec.* IV. 677 (1805). — Trautvetter in Ledebour, *Fl. Alt.* IV. 274 (1833).

Salix livida Wahlenberg, *Fl. Lapp.* 272 (1812). — Koehne, *Deutsche Dendr.* 100 (1893). — Hempel & Wilhelm, *Bäume & Sträucher*, II. 119, fig. 201 (1897). — Wolf in *Izv. S.-Peterburg. Liesn. Inst.* V. 85. t. 34, fig. 21–22, t. 35, fig. 1–11 (*Мат. Изуч. Иѣ Европ. Росс.*) (1900). — Schneider, *Ill. Handb. Laubholzk.* I. 61, fig. 15 k, 24 b–c (1904). — Toepffer, *Salicol. Mitt.* No. 4, 213–215 (1911).

Salix depressa Fries, *Nov. Fl. Suec. Mant.* I. 56 (pro parte, haud Linnaeus [1]) (1832). — Pokorny, *Oesterr. Holzpfl.* 115, t. 33, fig. 323–326 (1864). — Seemen in Ascherson & Graebner, *Syn. Mitteleur. Fl.* IV. 115 (1909). — Krylov, *Фл. Алтая*, 1217 (1909).

Salix vagans Andersson in *Öfs. Vetensk. Akad. Förh.* (1858) 121; in *Proc. Am. Acad.* IV. 61 (*Salic. Bor. Am.* 15) (1858); in *Svensk. Vetensk. Akad. Handl.* VI. 86 (1867); in De Candolle, *Prodr.* XVI. pt. 2, 226 (1868). — Herder in *Act. Hort. Petrop.* XI. 404 (1891). — Komarov in *Act. Hort. Petrop.* XXII. 31 (*Fl. Mansh.* II.) (1904). — Nakai in *Jour. Coll. Sci. Tokyo*, XXXI. 212 (*Fl. Kor.* II.) (1911). — Seemen apud Siuzev in *Trav. Bot. Mus. Acad. Sci. St. Pétersbourg*, IX. (1912) ex Toepffer, *Salicol. Mitt.* No.5, 248 (1912).

Salix livida, var. *Starkeana* Toepffer, *Salicol. Mitt.* No. 4, 215 (1911).

The typical form (represented by *S. Starkeana* Willdenow and ,by typical *S. livida* Wahlenberg) has glabrous twigs and leaves, which are only sometimes hairy when young. As far as I can see, this form does not occur in eastern Asia or at least is much rarer than the following variety.

Salix Starkeana, var. cinerascens Schneider, n. comb.

Salix livida, var. *cinerascens* Wahlenberg, *Fl. Lapp.* 275 (1812).
Salix depressa, var. *cinerascens* Fries, *Nov. Fl. Suec. Mant.* I. 57 (1832).
Salix depressa, var. *velutina* W. D. Koch, *Syn. Fl. Germ. Helv.* 653 (1857).
Salix vagans, subspec. *cinerascens*, var. *Linnaeana* Andersson in *Proc. Am. Acad.* IV. 61 (*Salic. Bor. Am.* 16) (1858).
Salix vagans, var. *a cinerascens* Andersson in De Candolle, *Prodr.* XVI. pt. 2, 226 (1868).

NORTHEASTERN ASIA. Transbaikalia to Kamtchatka and Korea. I have seen only a specimen from the Amur River, 1865, R. *Maack* (No. 126).

---

[1] *Salix depressa* Linnaeus (*Fl. Suec.* ed. 2, 352 [1755]) is, according to Enander (*Stud. Salic. Linné's Herb.* 102 [1907]), a mixture of *S. Caprea* and some hybrid forms. The name *S. foliolosa* Afzel is only cited as a synonym by Smith (in Linnaeus *Fl. Lapp.* ed. 2, 295 [1792]). *S. Starkeana* Willdenow seems to be according to the citations of Andersson, von Seemen and Toepffer the oldest name of any form belonging to *S. livida* Wahlenberg or *S. vagans* Andersson.

Probably *U. Faurie's* No. 178 from Korea, June 26, 1906, "in monte des diamants" belongs to this variety.

In var. *cinerascens* the young branchlets are pubescent, as are the leaves on both surfaces. To var. *cinerascens* may belong *S. vagans*, f. *manshurica* Siuzev (in *Trav. Bot. Mus. Acad. Sci. St. Pétersbourg*, IX. 88 [1912]; in Fedde, *Rep. Spec. Nov.* XIII. 328 [1914]. — Toepffer, *Salicol. Mitt.* No. 5, 248 [1912]), the leaves of which are glabrous on the upper side.

The species as a whole is extremely variable, and may be distinguished, from *S. Caprea* Linnaeus and the other related species by its short petioles, hardly more than 5 mm. in length, and by its mostly obovate-lanceolate leaves, which are more or less attenuate at the base and pointed at the apex. The flowers are much like those of *S. Caprea* Linnaeus.

132. **Salix insignis** Andersson in *Jour. Linn. Soc.* IV. 47 (1860); in De Candolle, *Prodr.* XVI. pt. 2, 262 (1868). — Brandis, *Forest Fl. Brit. Ind.* 470 (1874); *Ind. Trees*, 637 (1906). — Hooker f., *Fl. Brit. Ind.* V. 631 (1888).

INDIA.   Kashmir: " reg. temp. alt. 6–8000 ped.," *T. Thomson* (type, ♂ and ♀); "Piti, 11000 feet," September 1847, *T. Thomson* (ex Brandis [1906]); "Pangi, 12000 feet," August 1899, *J. H. Lace* (ex Brandis [1906]); " W. Tibet, reg. temp., alt. 6–8000 ped.," *T. Thomson* (type of the form with narrower more pubescent leaves, ex Andersson).

AFGHANISTAN.   Kurrum Valley, 1879, *J. E. T. Aitchison* (No. 574; fruits). This is a doubtful species of which I have seen branchlets with ♀ and ♂ flowers of the type. According to Andersson's description and my own observations *S. insignis* looks somewhat intermediate between sect. *Capreae* (*Salix Wallichiana*) and sect. *Daphnoideae* (*Salix daphnoides*), but on account of the ♀ flowers I place it in sect. *Capreae*. In Aitchison's No. 574 the flowers and fruits are very similar to the type, but the young twigs and leaves are covered with a somewhat yellowish tomentum. The old leaves are shining green above, only puberulous on the midribs, and glaucous beneath with only scattered hairs; they are ovate, somewhat acute at the base and apex, and entire or somewhat crenulate-dentate. They measure up to 10 cm. in length and 4.2 cm. in width; the petioles are about 1 cm. long; the fruiting catkins are 7 cm. long and 1.2 cm. thick. The old branches become purplish.

Sect. 25. MYRTILLOIDES Borrer in Loudon, *Arb. Brit.* III. 1587 (1838). — Koehne, *Deutsche Dendr.* 102 (1893).

Salix, sect. *Roseae* (sive *Myrtilloides*) Andersson in *Svensk. Akad. Handl.* VI. 94 (*Monog. Salic.*) (1867); in De Candolle, *Prodr.* XVI. pt. 2, 229 (1868). — Seemen in Ascherson & Graebner, *Syn. Mitteleur. Fl.* IV. 58 (1908).
Salix, sect. *Argenteae*, subsect. *Myrtilloides* Schneider, *Ill. Handb. Laubholzk.* I. 63 (1904).

Frutices humiles, ramulis gracilibus, trunco repente subterraneo. Folia tenuia, elliptica v. lanceolata, subtus glaucescentia, pleraque glabra. Amenta foliato-pedunculata, sublaxiflora. Bracteae apice parum infuscatae. Flores ♂ diandri, glandula una ventrali, filamentis liberis (v. pro parte coalitis?), antheris flavis ovalibus. Flores ♀ glandula una ventrali, ovariis saepissime glabris pedicellatis, stylis brevibus, stigmatibus satis crassis subbilobis.

133. **Salix myrtilloides** Linnaeus, *Spec.* 1019 (1753). — Pokorny, *Oesterr. Holzpfl* 123, t. 24, fig. 350–352 (1864). — Herder in *Act. Hort. Petrop.* XI. 407 (1891). — Hempel & Wilhelm, *Bäume & Sträucher*, 124, fig. 209 (1897). — Wolf in *Izv. S.-Peterburg. Liesn. Inst.* V. 99, t. 37 fig. 1–7 (*Мам. Изуч. Нѣи Европ. Росс.*) (1900). —

Komarov, in *Act. Hort. Petrop.* XXII. 26 (*Fl. Mansh.* II.) (1904). — Seemen in Ascherson & Graebner, *Syn. Mitteleur. Fl.* IV. 120 (1909). — Nakai in *Jour. Coll. Sci. Tokyo,* XXXI. 214 (*Fl. Kor.* II.) (1911).
For further synonyms and literature see Herder, l. c. and von Seemen, l. c.
NORTHEASTERN ASIA. Transbaikalia to Kamtchatka and Korea. Of this species I have only seen a specimen collected by V. Komarov, prov. Amur, fluvium Sutár, June 13, 1895, with ♂ flowers and with fruits). The plant which Maximowicz (in *Mém. Sav. Étr. Acad. Sci. St. Pétersbourg,* IX. 244 [*Prim. Fl. Amur.*] [1859]) collected and distributed under the name *S. myrtilloides,* var. *nmarkica* Trautvetter & Meyer (in Middendorff, *Reise Sibir.* I. pt. 2, Bot. abt. 1, 80 [*Fl. Ochot.*] [1847]), is mentioned by Andersson (in *Svensk. Vetensk. Akad. Handl.* VI. 98 [1867]) under *S. rugulosa,* subspec. *finmarkica.* According to Andersson and von Seemen, l. c. 226 (1909), *S. rugulosa* is a hybrid between *S. aurita* Linnaeus and *S. myrtilloides* Linnaeus, the oldest name of which would be *S. onusta* Besser (*Enum. Pl. Volhyn.* 78 [1821]). See also the keys on p. 83 and p. 85.

134. **Salix fuscescens** Andersson in *Svensk. Vetensk. Akad. Handl.* VI. 97 (*Monog. Salic.*) (1867); in De Candolle, *Prodr.* XVI. pt. 2, 230 (1868). — Herder in *Act. Hort. Petrop.* XI. 401 (1891).

Salix rhamnifolia Hooker & Arnott, *Bot. Voy. Beechey,* 117, t. 26 (non Pallas) (1832). — Trautvetter & Meyer in Middendorff, *Reise Sibir.* I. pt. 2, Bot. abt. 1, 80 (*Fl. Ochot.*) (1847).

NORTHEASTERN ASIA. Kamtchatka: Avatschka Bay, *Beechey & K. H. Mertens* (type, ex Andersson).
I know this species only from Andersson's description and from Hooker & Arnott's plate. Its taxonomic position seems rather uncertain. Very doubtful are the forms *S. fuscescens,* β? *minor* Andersson, l. c. 230 (1868) (*S. myrtilloides* f. 1 et 2, Chamisso in *Linnaea,* VI. 539 [1831]), and *S. fuscescens,* var. *dasycarpa* Trautvetter (ex Herder in *Act. Hort. Petrop.* XI. 409 [1891]), types: "Providenzbucht," end of July 1881, *Dobrotworsky,* and "Beringsinsel," 1879, *Dybowsky.*

Sect. 27. INCUBACEAE Dumortier, *Fl. Belg. Prodr.* 12 (1827). — Fries, *Nov. Fl. Suec. Mant.* I. 64 (pro parte) (1832).

Salix, sect. *Argenteae* W. D. Koch, *De Salic. Comm.* 46 (pro parte) (1828). — Andersson in *Svensk. Vetensk. Akad. Handl.* VI. 106 (*Monog. Salic.*) (1867); in De Candolle, *Prodr.* XVI. pt. 2, 233 (pro parte) (1868). — Seemen in Ascherson & Graebner, *Syn. Mitteleur. Fl.* IV. 123 (1909).
Salix, sect. *Rosmarinifoliae* Borrer in Hooker, *Brit. Fl.* 419 (1830).
Salix, sect. *Fuscae* Borrer, l. c. 420 (1830).
Salix, sect. *Repentes* Wimmer, *Fl. Schles.* ed. 2, 335 (pro parte) (1841). — Pax in Engler & Prantl, *Nat. Pflanzfam.* III. abt. 1, 37 (1889). — Seemen, *Salic. Jap.* 17 (1903).

Frutices humiles v. mediocres, trunco saepe subterraneo repente. Folia ovalia usque linearia. Amenta praecocia v. coetanea, breviter cylindrica; flores ♂ glandula una ventrali, filamentis liberis v. pro parte coalitis, antheris flavis v. rubescentibus; flores ♀ glandula una ventrali, ovariis glabris v. sericeo-pubescentibus, plus minusve pedicellatis, stylis brevibus, stigmatibus oblongis.
Of this section there seems to be in eastern Asia only the following species.[1]

---

[1] Herder, in *Acta Hort. Petrop.* XI. 410 (1891) mentions sterile specimens of *Salix tristis* Aiton (Hort. Kew. III. 393 [1789]) from Kamtchatka, *Kastalsky* (in Herb. Petrop.). The identification of this material may be wrong.

135. **Salix sibirica** Pallas, *Fl. Ross.* I. pt. 2, 72, t. 81, fig. 3 (1788). — Wolf in *Izv.*
S.-Peterburg. Liesn. Inst. XIV. 193, 197 (*Азiam. Ивъ*, II.) (1900). — Krylov, *Фл.*
Алмая 1223 (1909).

Salix repens Auct., quoad plantam asiat.; vide Herder in *Acta Hort. Petrop.*
XI. 410 (1891). — Komarov in *Act. Hort. Petrop.* XXII. 29 (*Fl. Mansh.* II.)
(1903). — Nakai in *Jour. Coll. Sci. Tokyo*, XXXI. 214 (*Fl. Kor.* II.) (1911).
Salix repens, subspec. *rosmarinifolia*, var. *flavicans* Andersson in *Svensk. Vetensk.*
Acad. Handl. VI. 116 (*Monog. Salic.*) (1867).
Salix repens, var. γ *flavicans* Andersson in De Candolle, *Prodr.* XVI. pt. 2,
238 (1868). — Seemen in Ascherson & Graebner, *Syn. Mitteleur. Fl.* IV. 128
(1909).

NORTHEASTERN ASIA.   Transbaikalia to Kamtchatka.
Wolf, l. c., believes that the true *Salix repens* Linnaeus (see von Seemen, l. c. 123)
does not occur in Asia.   I am not yet convinced that it is possible to keep separate
S. sibirica Pallas, but Wolf knows the Russian and Siberian Willows so well that
I decide to accept Pallas's name for the forms in question.   Whether the following
variety does occur on the mainland and in Korea, or whether it represents a good
form or not, I am not able to decide without having seen much more material from
eastern Asia:

Salix sibirica, var. **subopposita** Schneider, n. comb.
Salix subopposita Miquel in *Ann. Mus. Lugd.-Bat.* III. 28 (1867); *Prol. Fl. Jap.*
216 (1867). — Franchet & Savatier, *Enum. Pl. Jap.* I. 461 (1875).
Salix repens, var. *subopposita* Seemen, *Salic. Jap.* 35, tab. 5 A–E[1] (1903); in
Ascherson & Graebner, *Syn. Mitteleur. Fl.* IV. 128 (1909).

NORTHEASTERN ASIA.   Korea: Quelpaert: "in herbidis Hallaisan,"
1700 m., June 1909, *U. Faurie* (No. 3247; with fruits; No. 3246; sterile).
JAPAN.   Shikoku: prov. Tosa, April 1888 (No. 53, ex Herb. Tokyo; ♀).
Hondo: prov. Sagami, "in montibus Hakone," *Pierot* (type, ex Miquel; a fruiting branchlet in Herb. Gray seems to be a co-type). See also von Seemen, l. c.
This variety differs chiefly in the richer pubescence of the juvenile parts, the
stronger nervation and reticulation of the old leaves and in the more distinct
stipules.

136. **Salix dolia** Schneider. See p. 65.

Sect. 28.   DAPHNOIDEAE Dumortier, *Fl. Belg. Prodr.* 12 (1827).
Salix, sect. *Pruinosae* W. D. Koch, *De Salic. Comm.* 22 (1828). — Andersson in
De Candolle, *Prodr.* XVI. pt. 2, 261 (1868). — Seemen, *Salic. Jap.* 19 (1903);
in Ascherson & Graebner, *Syn. Mitteleur. Fl.* IV. 59 (1908).
Salix, sect. *Acutifoliae* Borrer in Loudon, *Arb. Brit.* III. 1494 (1838).
Salix, ser. *Daphnoides* Moss, *Cambridge Brit. Fl.* II. 58 (1914).
Arbores v. frutices elati, ramis elongatis saepe glauco-pruinosis.   Folia longa,
lanceolata, acuminata, subtus glauca.   Amenta sessilia, crasse cylindrica, aureo- v.
cinereo-sericea; flores ♂ glandula una, staminibus 2 liberis, antheris flavis; flores
♀ glandula una, ovariis subsessilibus v. pedicellatis plerisque glabris, stylo longo,
stigmatibus elongatis integris; bracteae basi saepe glanduloso-denticulatae.

137. **Salix daphnoides** Villars, var. **indica** Andersson in *Svensk. Vetensk. Akad.*
Handl. 1850, 475 (1851); in *Jour. Linn. Soc.* IV. 46 (1860).

? *Salix japonica* D. Don, *Prodr. Fl. Nepal.* 59 (haud Thunberg) (1825).
Salix oxycarpa, var. *serratifolia* Andersson in *Jour. Linn. Soc.* IV. 46 (1860).

Salix oxycarpa, var. *serrata* Andersson in De Candolle, *Prodr.* XVI. pt. 2, 310 (1868).
Salix daphnoides Brandis, *Forest Fl. Brit. Ind.* 409, t. 62 (non Villars) (1874); *Ind. Trees*, 637 (1906). — Hooker f., *Fl. Brit. Ind.* V. 631 (1888). — Collett, *Fl. Siml.* 480 (1902).

INDIA. Kashmir: "in summa valle Jumnath supra fontem thermalem a 2500 ad 3300 m. alt. d. 16 Maji 1829," [1] *V. Jacquemont* (type; ♀, ex Andersson); "Western Tibet, reg. temp. alt. 6–8000 ped.," *T. Thomson* (sterile; sub nom. *S. acutifolia* Willd.). Punjab: Simla, "reg. temp. 8000 ped.," *T. Thomson* (co-type of *S. oxycarpa*, var. *serratifolia;* with fruits).

The specimen from Simla agrees well with Brandis's plate. The ♀ bracts are short and without any glandular dentation, the styles are hardly longer than the stigmas, and the fruiting aments are up to 9 cm. long and 1.5 cm. thick. The stipules are small and lanceolate-semicordate, acuminate, and more like those of *S. acutifolia* Willdenow. The shape of the leaves seems to vary considerably. These Indian forms need a careful study. See also *S. insignis* Andersson, p. 152. The type of D. Don's *S. japonica* was collected in Nepal by Hamilton near Narainhetty.

138. **Salix rorida** Lackschewitz in *Schedae Herb. Fl. Ross.* VII. 131 (1911). — Toepffer, *Salicol. Mitt.* No. 5, 238 (1912).

Salix coerulescens Turczaninow, *Pl. Exsicc. ann.* 1828 (non Doell), fide Lackschewitz.
Salix acutifolia Ledebour, *Fl. Ross.* III. pt. 2, 601 (pro parte) (non Willdenow) (1850). — Turczaninow in *Bull. Soc. Nat. Mosc.* XXVII. 374 (1854); *Fl. Baical. Dahur.* II. 101 (1856). — Franchet & Savatier, *Enum. Pl. Jap.* I. 461 (1875). — Herder in *Act. Hort. Petrop.* XI. 424 (pro parte) (1891). — Komarov in *Act. Hort. Petrop.* XXII. 23 (*Fl. Mansh.* II.) (1903). — Nakai in *Jour. Coll. Sci. Tokyo*, XXXI. 215 (*Fl. Kor.* II.) (1911).
Salix daphnoides Ledebour, *Fl. Ross.* III. pt. 2, 602 (pro parte, non Villars) (1850). — Herder in *Act. Hort. Petrop.* XI. 423 (pro parte) (1891). — Seemen, *Salic. Jap.* 49, t. 9, fig. A–E (1903). — Shirasawa, *Icon. Ess. For. Jap.* II. t. 10, fig. 13–22 (1908). — Nakai in *Tokyo Bot. Mag.* XXVI. 168 (1912).
Salix praecox Trautvetter & Meyer in Middendorff, *Reise Sibir.* I. pt. 2, Bot. abt. 2, 78 (*Fl. Ochot.*) (non Hoppe) (1856). — Trautvetter in *Mém. Sav. Etr. Acad. Sci. St. Pétersbourg*, IX. 242 (Maximowicz, *Prim. Fl. Amur.*) (1859).

NORTHEASTERN ASIA. Amur: without precise locality, *C. Maximowicz* (♀); same region, 1855, *R. Maack* (Nos. 5–6; ♂ and ♀). Korea: Song Chang, September 4, 1903, *C. S. Sargent* (large tree; sterile); Monsan, September 5, 1903, *C. S. Sargent* (large tree; sterile; a doubtful form with rather long petioles). Saghalien: Toyo-hara, side of streams, common, August 3, 1914, *E. H. Wilson* (No. 7343; bush or tree, 1.8–12 m. tall, girth 2.1 m.; sterile); without locality, "secus aquas communis," June 1908, *U. Faurie* (No. 278; with fruits); without locality, *F. Schmidt* (♀).
JAPAN. Hokkaido: prov. Kitami, Rubeshibe, side of stream Okelo, August 11, 1914, *E. H. Wilson* (No. 7403; bush 1.5 m. tall; sterile); prov. Ishikari, Sapporo, April 15, 1890 (♀), April 23, 1890 (♂), May 1890 (with fruits and sterile),

---

[1] The date 1829 must be a misprint. According to his Journal (*Voy.* II. 84, 95 [1841]) Jacquemont visited "les sources de la Jumna" and "Jumnoutri," where "la source thermale" is, on May 16, 1830. He mentions a "*Salix incerta*" which apparently is the same as our variety.

October 15 and 21, 1890 (sterile), April 1891 (♂), April 15, 1891 (♀), *Y. Tokubuchi;* same locality, common, September 18, 1892, *C. S. Sargent* (small tree; sterile). According to Lackschewitz this species differs from *S. daphnoides* Villars: "ramulis tenuioribus, flaccidis, amentis paullo minoribus angustioribusque, squamis in amentis ♀ ad basin glanduloso-crenulatis, germinibus vix compressis, stylo longiore, stipulis ovatis v. remiformibus (non semicordatis)," and from *S. acutifolia* Willdenow " amentis approximatis, ramulis conferte assidentibus, squamis in floribus ♀ glanduloso-crenulatis, foliis latioribus in acumen minus longum productis nec non forma stipularum (non lanceolatis longe acutatis)."

There is a specimen of Wilson's from Hondo: Lake Towada, alt. 300–600 m., October 5, 1914 (No. 7616; tree 16–20 m. tall; girth 1.8–2.4 m.; sterile) which has short-tomentose branchlets and remains doubtful.

Sect. 29.   CHRYSANTHEAE W. D. Koch, *De Salic. Comm.* 52 (1828).

Salix, sect. *Niveae* s. *Glaucae,* subsect. A. *Lanatae* Andersson in De Candolle, *Prodr.* XVI. pt. 2, 273 (1868).

Salix, sect. *Lanatae* E. Fries in *Tidskr. Landtm.- och Kommun.-Ekon.* 1859, 21 (*Anmärk. Sverige Pilart.*) (pro parte) (1859).—Koehne, *Deutsche Dendr.* 93 (pro parte) (1893).

Salix, sect. *Lanatae,* subsect. a *Chrysantheae* Schneider, *Ill. Handb. Laubholzk.* I. 47 (1904).

Frutices parvi v. satis alti, ramulis crassis. Folia ovali-rotundata v. late lanceolata, subtus reticulata. Amenta sessilia v. subsessilia, versus apicem ramulorum conferta, magna, crassa, densiflora, saepe aureo-sericea; flores ♂ glandula una, filamentis liberis glabris, antheris flavis; flores ♀ glandula una, ovariis sessilibus v. pedicellatis glabris v. tomentellis, stylo longo apice saepe bifido, stigmatibus angustis bifidis; bracteae atrofuscae.

139.   **Salix lanata** Linnaeus, *Spec.* 1019 (1753). — Andersson in De Candolle, *Prodr.* XVI. pt. 2, 273 (1868). — Lundström in *Nov. Act. Soc. Sci. Upsal.* 1877, 42 (*Weid. Now. Semljas*) (1877). — Herder in *Act. Hort. Petrop.* XI. 429 (1891). — Wolf in *Izv. S.-Peterburg. Liesn. Inst.* V. 18, t. 23, fig. 6–11, t. 14, fig. 1 (*Мам. Изуч. Нѣъ Европ. Росс.*) (1900). — Schneider, *Ill. Handb. Laubholzk.* I. 48, fig. 21 a–b, 25 i (1904). — Moss, *Cambridge Brit. Fl.* II. 29, t. 31–32 (1914).

Salix lanuginosa Pallas, *Fl. Ross.* I. pt. 2, 83 (fide Wolf) (1788).
Salix chrysanthos Vahl, *Fl. Dan.* VI. fasc. XVIII. 6, t. 157 (1792).

For further literature see Herder, l. c.

NORTHEASTERN ASIA. Yakutsk and Maritime Province.

This species is distinguished by its densely white tomentose leaves and thick silky hairy catkins.

It is with a good deal of doubt that I include in this section the two following species.

140.   **Salix taimyrensis** Trautvetter in Middendorf, *Reise Sibir.* I. pt. 2, Bot. abt. 1, 27, t. 5–6 (*Fl. Taimyr.*) (1847).—Lundström in *Nov. Act. Soc. Sci. Upsal.* 1877, 17, 41 (*Weid. Now. Semljas*) (1877).

Salix arctica, var. *taimyrensis* Andersson in De Candolle, *Prodr.* XVI. pt. 2, 287 (1868). — Herder in *Act. Hort. Petrop.* XI. 440 (1891).

NORTHEASTERN ASIA. Yakutsk: Taimyr River, June–July 1843, *A. T. von Middendorff* (types, ex Trautvetter).

This interesting Willow has glabrous obovate or oblanceolate leaves, which are glaucous beneath and sometimes denticulate, and silky ovaries. It needs further investigation to fix its taxonomic position.

141. **Salix vulpinoides** Koidzumi in *Tokyo Bot. Mag.* XXVII. 94 (1913). JAPAN. Hondo. (The author does not cite a type specimen.) I have seen no specimen and place this species in the sect. *Chrysantheae* on the authority of the author.

Sect. 30. VIMINALES Bluff & Fingerhuth, *Comp. Fl. Germ.* II. 562 (1825). — W. D. Koch, *De Salic. Comm.* 27 (1828). — Borrer in Hooker, *Brit. Fl.* 423 (1830). — Seemen, *Salic. Jap.* 19 (1903); in Ascherson & Graebner, *Syn. Mitteleur. Fl.* IV. 60 (pro parte) (1908).

Salix, sect. *Micantes* (seu *Viminales*) Andersson in De Candolle, *Prodr.* XVI. pt. 2, 264 (pro parte) (1868).

Arbores v. frutices alti, ramis vimineis flexilibus. Folia pleraque anguste v. latius elongato-lanceolata. Amenta pleraque praecocia, sessilia v. brevipedunculata, elongato-cylindrica, densiflora; flores ♂ diandri, filamentis liberis, antheris aureis, glandula una; flores ♀ glandula una, ovariis sessilibus v. pedicellatis satis dense pilosis, stylis elongatis, stigmatibus integris bifidisve.

The section is closely connected by hybrids with the sections *Capreae*, *Daphnoideae*, *Helix* and probably also with *Sieboldianae* and *Gracilistylae*.

142. **Salix viminalis** Linnaeus, *Spec.* 1021 (1753). — Pokorny, *Oesterr. Holzpfl.* 85, t. 18, fig. 224–226 (1864). — Andersson in De Candolle, *Prodr.* XVI. pt. 2, 264 (1868). — Brandis, *Forest Fl. Brit. Ind.* 470 (1874); *Ind. Trees*, 638 (1906). — Hooker f., *Fl. Brit. Ind.* V. 631 (1888). — Herder in *Act. Hort. Petrop.* XI. 425 (1891). — Burkill in *Jour. Linn. Soc.* XXVI. 534 (1899). — Hempel & Wilhelm, *Bäume & Sträucher*, II. 112, fig. 197 and t. 31 (1896). — Wolf in *Izv. S.-Peterburg. Liesn. Inst.* IV. 70, t. 19, fig. 1–8 (*Мат. Изуч. Ист Европ. Росс.*) (1900); l. c. XIII. 45, 54, t. 1–2 (*Азіат. Ист*, I.) (1905). — Seemen, *Salic. Jap.* 50, t. 9 ʀ-ᴋ (1903); in Ascherson & Graebner, *Syn. Mitteleur. Fl.* IV. 173 (1909). — Komarov in *Act. Hort. Petrop.* XXII. 32 (*Fl. Mansh.* II.) (pro parte) (1903). — Schneider, *Ill. Handb. Laubholzk.* 1. 45, fig. 12 o, 20 i–k (1904). — Nakai in *Jour. Coll. Sci. Tokyo*, XXXI. 214 (*Fl. Kor.* II.) (1911). — Moss, *Cambridge Brit. Fl.* II. 60, t. 59–61 (1914).

For additional synonyms and references see Herder, Wolf and von Seemen, l. c.

NORTHEASTERN ASIA. Transbaikalia: banks of streams near Sryechinsk, August 13, 1903, *C. S. Sargent* (tree 7 m. tall, girth 2.4 m.; sterile). Amur and Ussuri: Amur River, Kwashima, May 8, 1891, *S. Korshinsky* (♀); Khabarovska, May 18, 1891, *S. Korshinsky* (with fruits); Bidshana, July 11, 1891, *S. Korshinsky* (sterile); Tschernajaewa, August 13, 1891, *S. Korshinsky* (sterile); station grounds, Nikolks, August 12, 1903, *C. S. Sargent* (sterile); without precise locality, *C. Maximowicz* (♀ and ♂); without precise locality, 1855, *R. Maack* (with fruits). Saghalien: Toyo-hara, side of streams, abundant, August 3, 1914, *E. H. Wilson* (No. 7341; bush 1.8–4.5 m. tall; sterile); " secus aquas communis," end of June, *U. Faurie* (No. 273; with fruits); without locality, *Fr. Schmidt* (♀).

INDIA. Punjab: distr. Simla, Bashahr state, bank of the Baspa River near Sangla, May 1908 (with fruits). Kashmir: " reg. temp. 9–11000 ped.," *T. Thomson* (sterile).

This is the typical European form varying in the shape of the leaves, which always are covered below with a dense silky pubescence, the hairs lying parallel to the lateral nerves. The ovaries are sessile; the stigmas are long and narrow, and a little shorter or as long as the long styles. For further specimens see Herder, l. c., and Komarov, l. c., although I have some doubt if all these specimens belong to the typical form.

There are two other varieties:

Salix viminalis, var. Gmelinii Andersson in De Candolle, *Prodr.* XVI. pt. 2, 266 (1868), sensu Seemen in Ascherson & Graebner, *Syn. Mitteleur. Fl.* IV. 174 (1908).

Salix Gmelini Pallas, *Fl. Ross.* I. 2, 77 (1788), — Ledebour, *Fl. Ross.* III. 606 (pro parte) (1850). — Syreishtschikof, *Фл. Моск. Губ.* II. 33, fig. (1907).

For additional synonyms see von Seemen, l. c.

A typo recedit ovariis sessilibus obtusioribus stigmatibus valde elongatis angustis stylo breviore plus minusve superantibus coronatis.

This seems the *S. viminalis* of most parts of Russia and western Siberia, but is apparently wanting in eastern Siberia.

Salix viminalis, var. yezoensis Schneider, n. var.

A typo recedit ovariis sessilibus stigmatibus brevibus angustis stylo longo multo brevioribus coronatis.

JAPAN.   Hokkaido: prov. Ishikari, Sapporo, 1878 ($\male$ and $\female$); same place, May 26 and Oct. 15, 1890, *Y. Tokubuchi* (sterile); same place, April 17, 1891, *Y. Tokubuchi* (No. 17, type; $\female$); same place, April 17, 1893, *Y. Tokubuchi* ($\female$); same place, April 29, 1906, *K. Kondo* ($\male$); same place, " secus aquas," June 14, 1908, *U. Faurie* (No. 255; with fruits); same place, April 28 and May 5, 1913 (ex Herb. Sakurai ; $\male$ and sterile). Hondo: prov. Rikuchu, Mt. Iwate, May 5, 1911 (co-type, $\male$, ex Herb. Sakurai); prov. Musashi, Kichijoji, April 12, 1914, *E. H. Wilson* (No. 6419; small bushy tree; branches erect spreading; $\male$).

See von Seemen, l. c. (1903). The short stigmas seem to be the only real difference compared with certain forms of var. *typica*. Sometimes var. *yezoensis* may be difficult to distinguish from forms of *S. opaca* Andersson (see p. 159), and it needs further investigation to decide whether certain forms are to be regarded as hybrids or not.

143. **Salix sachalinensis** Fr. Schmidt in *Mém. Acad. Sci. St. Pétersbourg*, sér. 7, XII. 173 (*Reis. Amurl. Sachal.*) (1868). — Herder in *Act. Hort. Petrop.* XI. 429 (1891). — Seemen, *Salic. Jap.* 53, t. 10 G-L (1903).

Salix korsakoviensis Léveillé in *Bull. Soc. Bot. France*, LVI. 302 (1909).
Salix stipularis, var. *sachalinensis* Koidzumi in *Tokyo Bot. Mag.* XXVII. 94 (1913).

NORTHEASTERN ASIA.   Saghalien: "beim Posten Dui am Bach Anfang Juni 1860," *Fr. Schmidt* (sub nom. *S. sachalinensis;* $\female$); same place? *Fr. Schmidt* (sub nom. *S. stipularis;* $\male$ flowers and old leaves); without locality "secus aquas," June and July 1908, *U. Faurie* (No. 274, $\female$ type, and Nos. 275, 276, leaves of *S. korsakoviensis.*

JAPAN.   Hokkaido: prov. Oshima, Hakodate, 1861, *C. Maximowicz* (No. 1; $\female$); prov. Ishikari, near Sapporo, June 12–13, 1908, *U. Faurie* (No. 265, 270; with fruits; stigmas short).   Northern Hondo: prov. Mutsu, Iwagi-san, May 1905, *U. Faurie* (No. 6609; 1–2 m. tall; $\male$; No. 6610; $\female$).

All these specimens have short-stalked ovaries, the pedicel being shorter than the gland, and styles about as long as the ovaries with comparatively short stigmas ⅓ the length of the style. The $\female$ catkins appear with the young leaves, which are silky beneath but soon become glabrous and pale green, not glaucous. The old leaves of Faurie's No. 276 are up to 17 cm. long and 3.5 cm. broad, acute at the base, and coarsely crenate with veins prominent beneath.

Von Seemen supposes that this species may be of hybrid origin, *S. viminalis* Linnaeus being one of the parents. I am inclined to believe that *S. sachalinensis* represents a good species, but there may be some hybrids between this species and *S. viminalis* or other species. See also under *S. Miyabeana* Seemen, p. 166.

Salix sachalinensis, var. Pilgeriana Schneider, n. var.

Salix Pilgeriana Seemen in Fedde, *Rep. Spec. Nov.* VII. 134 (1909).

A typo recedit amentis ♀ longioribus ad 16 cm. longis, glandula apice bifida v. emarginata.

JAPAN.   Hokkaido: "in sylvis Jirafu," June 26, 1905, *U. Faurie* (No. 6600, type; ♀).

The leaves are not yet fully developed and I cannot find any other difference between this form and *S. sachalinensis*. I suppose Faurie's Nos. 265 and 270 from Hokkaido are best placed under this variety.

144. **Salix stipularis** Smith, *Engl. Bot.* XVII. t. 1214 (1803); *Fl. Brit.* III. 1069 (1804).

Salix dasyclados, subspec. *stipularis* Seemen in Ascherson & Graebner, *Syn. Mitteleur. Fl.* IV. 180 (1909).
Salix dasyclados Seemen apud Siuzev in *Trav. Mus. Bot. Acad. Sci. St. Pétersbourg*, IX. No. 2 (1912) ex Toepffer, *Salicol. Mitt.* No. 5, 248 (1912).

This species is said by von Seemen to be found in Mandshuria, but the interpretations of it by different authors differ widely. Smith's type was well figured and described in 1803: "Nectary long cylindrical, obtuse. Germen on a short stalk, ovate, downy. Style a little elongated. Stigmas remarkably long, awl-shaped, recurved, undivided." Seemen, l. c., says that in *S. stipularis* the style is longer than in the type, which is *S. dasyclados* Wimmer (in *Flora*, XXXII. 35 [1849]); of this von Seemen, l. c. 178 (1909), says that the style is as long as the ovary. According to von Seemen, *S. dasyclados* was collected by Maximowicz in Amurland. The specimen before me (ex Herb. Gray) shows ovaries with short pedicels of about the length of the style and of the long stigmas, and agrees with *S. opaca* sensu von Seemen. After all, I am not convinced that either *S. stipularis* Smith or *S. dasyclados* Wimmer occurs in eastern Asia. There may be hybrids between *S. viminalis* Linnaeus and any species of sect. *Capreae*, but it is always a most difficult matter to refer such hybrids to any form already described. This species has been omitted in the keys.

145. **Salix opaca** Andersson apud Seemen, *Salic. Jap.* 51, t. 10, fig. A–F (1903). — Shirasawa, *Icon. Ess. For. Jap.* II. t. 9, fig. 1–10 (1908).

Salix stipularis Trautvetter in *Mém. Sav. Etr. Acad. Sci. St. Pétersbourg*, IX. 243 (Maximowicz, *Prim. Fl. Amur.*) (non Smith) (1859). — Schmidt in *Mém. Acad. Sci. St. Pétersbourg*, sér. 7, XII. 172 (*Reis. Amur. Sachal.*) (1868). — Nakai in *Tokyo Bot. Mag.* XVI. 168 (1012).
Salix viminalis Komarov in *Act. Hort. Petrop.* XXII. 32 (*Fl. Mansh.* II.) (pro parte) (1903).

So far as I can see, this species was never described by Andersson. Schmidt, l. c., mentions " *S. opaca* Andersson in sched. hort. bot. Petr." as a synonym, and he never described *S. stipularis* f. *amurensis*, cited by von Seemen. This last name he only mentions under *S. sachalinensis*, saying: ". . . foliis concoloribus ab affinibus *S. stipulari amurensi* (*opaca* And.) et *udensi* Trautv. differt." Herder, in *Act. Hort. Petrop.* XI. 428 (1891) only cites specimens under the name *S. opaca* Andersson in Herb.

I unite under *S. opaca* the following specimens without making an attempt to decide whether this species is a hybrid between *S. viminalis* Linnaeus and *S. Caprea* Linnaeus or whether some of the specimens mentioned represent hybrids of a different origin.

NORTHEASTERN ASIA.   A m u r : without locality, *C. Maximowicz* (♂ and ♀); S a g h a l i e n : Toyohara, moorlands, side of streams, abundant, August 3 ,1914, *E. H. Wilson* (No. 7334; bush 1.8–3 m. tall; sterile); same locality and date, *E. H. Wilson* (No. 7344; bush 1.8–6 m. tall; sterile, looks like *S. Caprea* × *S. sachalinensis*); Dui, along the river, beginning of June, 1860, *P. von Glehn* (type of *S. opaca*, with fruits; ex Schmidt).

JAPAN.   H o k k a i d o : prov. Ishikari, Sapporo, April and September, 1877, *Watase* (♂, ♀ and sterile); same locality, April 21, 1890, May 26, 1890, *Y. Tokubuchi* (♂ and with fruits; partly as *S. sachalinensis*, partly as *S. stipularis*); same locality, October 15, 1890, April 21 and 26, 1891, May 2, 1891, *Y. Tokubuchi* (♂, ♀ and sterile; partly as *S. stipularis*); same locality, May 27 and August 11, 1891, *Y. Tokubuchi* (with fruits and sterile; as *S. sachalinensis*); same locality, April and July 1893, *Y. Tokubuchi* (♀ and with fruit; as *S. sachalinensis*); same locality, September 17, 1892, *C. S. Sargent* (tree 7–10 m. tall, girth 0.3 m.; sterile); same locality, "secus aquas," June 12, 1908, *U. Faurie* (No. 260; with fruits); same locality, May 31, 1903, *S. Arimoto* (♀; as *S. sachalinensis*); same locality, *H. Shirasawa* (sterile); Maruyama, May 7, 1892, *Miura* (♂ and ♀; as *S. sachalinensis*); prov. Oshima, near Mori, September 13, 1892, *C. S. Sargent* (small tree; sterile).   H o n d o : prov. Rikuzen, Sendai, August 27, 1905, *J. G. Jack* (sterile); prov. Shimotsuke, Nikko to Lake Chuzenji, roadside, October 25, 1905, *J. G. Jack* (sterile); Lake Chuzenji to Lake Yumoto, August 11, 1905, *J. G. Jack* (sterile); shores of Lake Yumoto, September 6, 1892, *C. S. Sargent* (sterile); Nikko, June and October 2 (Nos. 88, 89 ex Herb. Tokyo; sterile); prov. Shinano, on Tsubakura-dake, side of torrents, common, alt. 1600 m., *E. H. Wilson* (No. 7499; small tree 4.5–7 m. tall; sterile, looks like *S. viminalis* × *S. Caprea*); Kanogawa, 1862, *C. Maximowicz* (No. 5; ♀ type of von Seemen); prov. Musashi, Toda, April 14, July 22, 1913 (♀ and sterile; looks like *S. viminalis* × *S. purpurea*); Tokyo, Meguro, April 1910, *H. Shirasawa* (♀ and sterile); prov. Suruga, near Gotemba, side of streams, April 14, 1914, *E. H. Wilson* (No. 6456; bush 1.8–2.4 m. tall; ♀, looks like a hybrid with *S. purpurea*).   S h i k o k u : prov. Tosa, Nanokawa, April 3 and May 20, 1888, *K. Watanabe* (♀ and sterile).

Very nearly related to *S. opaca* seems also *S. aequitriens* Seemen (*Salic. Jap.* 70, t. 15, fig. E–G [1903]), from Hokkaido, prov. Ishikari, Sapporo, May 1892, *Y. Tokubuchi* (type; ♀), which is only known from young flowering branchlets. The pedicel and the style are both as long as the ovary; the stigmas are linear and bifid; the gland is oval and only ⅓ the length of the pedicel.

146.   **Salix Siuzevii** Seemen in Fedde, *Rep. Spec. Nov.* V. 17 (1908), apud Siuzev in *Trav. Mus. Bot. Acad. Sci. St. Pétersbourg,* IX. No. 2 (1912) ex Toepffer, *Salicol. Mitt.* No. 5, 248 (1912). — Wolf in *Act. Hort. Petrop.* XXVIII. 527 (1911).

NORTHEASTERN ASIA.   U s s u r i : Nikolsk Ussuriiski, April 29, 1905, *P. V. Siuzev* (No. 34, type, ♀, ex von Seemen).   M a r i t i m e   P r o v . : Sachoba river, 1906, *Paltchevsky* (♂ and ♀, ex Wolf).

Von Seemen states that this Willow belongs to sect. *Viminales* and is most nearly related to *S. opaca* Andersson, differing from it especially by the shorter and narrower gland and by the shorter entire style.   According to the description the pedicel is as long or even longer than the capsule, the style only ⅓ the length; the stigmas are short and thin; the lanceolate leaves are grayish green beneath and are up to 13 cm. long and 2 cm. broad.

147. **Salix Rehderiana** Schneider. See p. 66.

Salix Rehderiana, var. **brevisericea** Schneider.   See p. 67.

148. **Salix Blinii** Léveillé in Fedde, *Rep. Spec. Nov.* X. 435 (1912).
Salix Taquetii Léveillé, l. c. 436 (1912).

Frutex ut videtur breviramosus; ramuli novelli sparse sericeo-villosuli, deinde glabri, purpurascentes, nitidi. Folia obovata v. obovato-elliptica, apice acuta v. obtusa, basin versus attenuata, acuta, in sicco brunnescentia, subconcoloria v. subtus cinerea, initio utrinque, subtus intensius, sericeo-villosa, demum plus minusve glabrescentia, nervis costaque subtus prominulis et paulo reticulata, margine crenato-serrata v. basi subintegra, minora 2–4 cm. longa, 1–2 cm. lata, maxima usque 5.5 cm. longa et 2.2 cm. lata; petioli pilosi, 2–6 mm. longi. Amenta (tantum matura fructifera vidi) subsessilia, basi foliis 2–3 parvis suffulta, ad 4.5 cm. longa, circiter 1 cm. crassa, densa, rhachi villosa; flores ♀ (inter fructus remanentes) glandula una ventrali elongata obtusa ⅓ ovarii aequante instructa; ovaria subsessilia, breviter sericea, ovata, stylo ½ ovarii aequilongo integro stigmatibus angustis parallelis integris coronata; bracteae oblongae, subacutae, brunneae, utrinque sericeae, ¾ ovarii aequantes.   Fructus ovato-oblongi, subacuti, sericei, circiter 5 mm. longi, pedicellis brevibus glandulae aequilongis suffulti.

NORTHEASTERN ASIA.   Korea: Quelpaert, "in sylvis torrentium Hallaisan," alt. 1200 m., June 1909, *U. Faurie* (Nos. 3248, type, 3249, co-type, both with fruits; 3245, type of *S. Taquetii*).

On account of the rather long style, the narrow stigmas and gland I place this species in this section. In the leaves it resembles *S. Starkeana,* var. *cinerascens* Schneider (see p. 151) and also forms of *S. Caprea* Linnaeus (see p. 149).

149. **Salix obscura** Andersson in De Candolle, *Prodr.* XVI. pt. 2, 269 (1868). — Hooker f., *Fl. Brit. Ind.* V. 632 (1888). — Brandis, *Ind. Trees,* 638 (1906).

INDIA.   Sikkim: "Lachen valley, alt. 9000 ped.," June 3, 1849, *J. D. Hooker* (♀ type).

Through the kindness of the Keeper of the Kew Herbarium I have seen part of the type specimen, and I must say, as Andersson did, that this species "mihi quod ad affinitatem obscura manet." It resembles somewhat *S. Rehderiana,* var. *brevisericea* Schneider, but that has a longer style and acute bracts. In *S. obscura* the style is about ⅓ as long as the ovary and in young flowers only half as long; it is very shortly bifid at the apex, with narrow emarginate stigmas. The bracts are obovate-oblong, very obtuse, and in the young flowers nearly as long as the sessile ovary. The leaves are apparently oblong-lanceolate, with a short acute apex, and are gradually narrowed toward the base. The tomentum of the young branchlets and leaves is somewhat brownish and gray. The leaves apparently become soon rather glabrous, only the midribs being puberulent; they are entire or obscurely denticulate. The almost sessile fruiting catkins are 4 cm. long and about 1 cm. thick.

Burkill (in *Jour. Linn. Soc.* XXVI. 531 [1899]) refers to *S. obscura* No. 22 of *Soulié* from Tachien-lu. Among the copious material collected by Wilson in this part of Szech'uan I do not find any form which seems to me identical with *S. obscura* Andersson.

Sect. 31. SIEBOLDIANAE Seemen, *Salic. Jap.* 21 (1903). — Schneider, *Ill. Handb. Laubholzk.* I. 69 (1904).

Frutices v. arbores parvae. Folia ovata, obovato-oblonga v. elliptica, integra v. crenulata, subtus cinerea v. glauca, juniora v. etiam adulta sericea. Amenta praecocia sessilia v. subcoetanea, brevi-pedunculata, cylindrica, densa; flores glandula una ventrali plus minusve ovato-elongata; ♂ staminibus 2 liberis v. basi coalitis v. stamine uno, antheris flavis; ♀ ovariis pedicellatis v. sessilibus sericeis, stylis brevibus v. mediocribus, stigmatibus brevibus v. oblongis.

This is an interesting section on account of the 1–2 stamens of the ♂ flowers. The following species need careful comparison; they are certainly very closely related, but without having seen many complete specimens of the different forms it is impossible to obtain a clear idea of their taxonomic value.

150. **Salix Sieboldiana** Blume, *Bijdr. Fl. Ned. Ind.* X. 517 (1825). — Andersson in *Mem. Am. Acad.*, n. ser. VI. 451 (Gray, *Bot. Jap.*) (1859); in *Svensk. Vetensk. Akad. Handl.* VI. 81, t. 5, fig. 47 (*Monog. Salic.*) (1867); in De Candolle, *Prodr.* XVI. pt. 2, 224 (1868). — Miquel in *Ann. Mus. Lugd.-Bat.* III. 28 (1867); *Prol. Fl. Jap.* 216 (1867). — Franchet & Savatier, *Enum. Pl. Jap.* I. 460 (1875). — Seemen, *Salic. Jap.* 63, t. 14, fig. ғ–к (1903). — Léveillé in *Bull. Acad. Int. Géogr. Bot.* XIV. 210 (1904). — Koidzumi in *Tokyo Bot. Mag.* XXVII. 93 (tantum var. *typica*) (1913).

? Salix caloptera Miquel in *Ann. Mus. Lugd.-Bat.* III. 26 (1867); *Prol. Fl. Jap.* 214 (1867).

? Salix brachylepis Franchet & Savatier, *Enum. Pl. Jap.* II. 503 (1879).

JAPAN. Kyushu: near Nagasaki, open places, common, March 18, 1914, *E. H. Wilson* (No. 6301; bush 0.6–1.8 m. tall, erect, spreading; ♂; No. 6301ᵃ, ♀); Kagoshima, roadside, common, March 3, 1914, *E. H. Wilson* (No. 6165; bush 0.6–1.5 m. tall; ♂; No. 6165ᵃ, ♀); same place, Fogo, roadsides, abundant, March 14, 1914, *E. H. Wilson* (No. 6277; bush 0.6–1.5 m. tall; ♂; No. 6277ᵃ, ♀; *S. Buergeriana?*); Mt. Kirishima, roadside thickets, March 6, 1914, *E. H. Wilson* (No. 6179; bush 0.6–1.8 m. tall; ♂; No. 6179ᵃ, ♀). Shikoku: prov. Tosa, April 11, 1888, *T. Makino* (No. 37ᵃ; ♂); same prov., Nanokawa, April 10, 1888, *K. Watanabe* (♀). Hondo: prov. Musashi, Renkoji-mura, April 13, 1895 (ex Herb. Tokyo; ♀).

This may be distinguished from the following species, according to von Seemen, by the broadly rounded or cordate base of the leaves, and by the male flowers with 1–2 stamens; but not having seen mature leaves of the two species, I do not know whether these are forms of one species (Koidzumi) or of two distinct species (Seemen). The type was sent to Java by P. von Siebold under the name *S. alba*. See Blume, l. c., and Seemen, l. c.

151. **Salix Buergeriana** Miquel in *Ann. Mus. Lugd.-Bat.* III. 28 (1867); *Prol. Fl. Jap.* 216 (1867). — Franchet & Savatier, *Enum. Pl. Jap.* I. 460 (1875). — Seemen, *Salic. Jap.* 66, t. 16, fig. ᴀ–ᴅ¹ (1903). — Léveillé in *Bull. Acad. Intern. Géogr. Bot.* XIV. 210 (1904); XVI. 143 (1906).

Salix Sieboldiana, var. *Buergeriana*, f. 1. *genuina* Koidzumi in *Tokyo Bot. Mag.* XXVII. 93 (1913).

JAPAN. Kyushu: base of Nishi-Kirishima, roadsides, March 10, 1914, *E. H. Wilson* (No. 6248; bush 0.5–1.5 m. tall; ♂; No. 6248ᵃ, ♀); Nagasaki, March 21, April 2, 1862, *C. Maximowicz* (type, ♂; mixed with ♀ *S. vulpina* Andersson in Herb. Gray); same locality, March 20, April 10, 1863, *C. Maximowicz* (♀; both specimens as *S. Miqueli*); Ehiko-san, May 3, 1907, *U. Faurie* (Nos. 10, 11; ♂; No. 12, with fruits).

This species is very similar to *S. Sieboldiana* Blume, from which it differs, according to von Seemen, who saw the types, in its leaves being attenuate, obtusish or acute at the base and in its ♂ flowers with only one stamen. In both the species the pedicels measure from ½ to ¾ and the styles from ⅓ to ½ of the length of the ovaries; the glands are mostly shorter than the pedicels. The aments appear before the leaves and are almost sessile. In *S. Sieboldiana* Blume the ♀ bracts seem to be somewhat longer (ovate-lanceolate) and the stigmas shorter than in *S. Buergeriana*.

152. **Salix Saidaeana** von Seemen, *Salic. Jap.* 68, t. 17, fig. A–E (1903).

Salix Sieboldiana, β *Buergeriana,* f. 2. *Saidaeana* Koidzumi in *Tokyo Bot. Mag.* XXVII. 93 (1913).

JAPAN. Kyushu: April 1875, *J. Rein* (co-type, ex von Seemen); Goto Island, May 1901, *U. Faurie* (No. 4995, type, ex von Seemen); Mt. Aso, June 20, 1899, *U. Faurie* (No. 3694; co-type, ex von Seemen). Hondo: Daisen, May 26 and 28, 1899, *U. Faurie* (Nos. 3705, 3706, 3708, ex von Seemen).

This species seems most closely related to *S. Buergeriana* Miquel, from which according to von Seemen it chiefly differs in its shorter female catkins, which appear with the leaves on short peduncles, and in its somewhat shorter pedicels and styles. I have not seen any specimen I could positively refer to it.

153. **Salix Harmsiana** Seemen, *Salic. Jap.* 73 (♀) and 80 (♂), t. 18, fig. C–E (1903).

Salix Sieboldiana, β *Buergeriana,* f. 3. *Harmsiana* Koidzumi in *Tokyo Bot. Mag.* XXVII. 93 (1913).

JAPAN. Kyushu: Nagasaki, 1862, *R. Oldham* (No. 551, in part, type, ♂, ex von Seemen; No. 551, in part; ♀, co-type in Herb. Gray); Hondo: prov. Shinano, July 1894 (No. 113 ex Herb. Tokyo; doubtful form with fruits).

This is a doubtful species very near *S. Buergeriana* Miquel, but, according to von Seemen, differing in its nearly sessile short ovaries with very short stigmas resembling those of *S. purpurea* Linnaeus. The ♀ co-type which I have seen shows distinct pedicels and styles, the flowers resembling those of *S. Buergeriana* Miquel. The specimen from Shinano has nearly sessile ovaries but a distinct style and remains doubtful. It may be a hybrid between *S. purpurea* Linnaeus and *S. Buergeriana* Miquel.

154. **Salix propitia** Koidzumi in *Tokyo Bot. Mag.* XXVII. 266 (1913).

Salix Sieboldiana, γ *sikokiana* Koidzumi, l. c. 93 (1913).

JAPAN. Sikoku: prov. Awa, Tsurugi-san (ex Koidzumi).

Koidzumi says: " Amentis floribusque iis *S. Matsumuraei* Seemen simillimis; foliis iis *S. Sieboldianae* Blume valde affinis, sed tenuioribus subtus secus costas adpresse sericeo-villosis." This is a doubtful form, of which only the ♀ plant is known. It has been omitted from the key.

155. **Salix hylonoma** Schneider. See p. 68.

Sect. 31. GRACILISTYLAE Schneider, n. sect.

Salix, sect. *Subviminales* Seemen, *Salic. Jap.* 20 (pro parte) (1903); in Ascherson & Graebner, *Syn. Mitteleur. Fl.* IV. 60 (1908). — Schneider, *Ill. Handb. Laubholzk.* I. 65 (pro parte) (1904).

Frutices elati v. arbores parvae. Folia lanceolata, elliptica v. obovato-lanceolata, acuta v. breviter acuminata, rarius obtusa, subtus cinerea, dein glabra, nervata et graciliter reticulata. Amenta saepissime praecocia, subsessilia, cylindrica, densa; bracteae oblongae, acutae, apice nigrescentes; flores ♂ diandri, filamentis totis coalitis, antheris liberis ovalibus flavis (? aurantiacis), glandula una ventrali longa angusta; ovaria sessilia v. pedicellata, dense sericea, stylo gracillimo longo integro, stigmatibus oblongis emarginatis.

Von Seemen in his sect. *Subviminales* names first *S. Pierotii* Miquel and adds *S. gracilistyla* (or *S. Thunbergiana* Blume). According to my opinion *S. Pierotii* is much more closely related to the species of sect. *Purpureae* than to *S. gracilistyla* Miquel, which widely differs in its long thin styles, in the long narrow gland and in the narrow acute bracts. A taxonomic arrangement of sections or even groups of higher value with regard to the length of the style seems to me an artificial one.

156. **Salix gracilistyla** Miquel in *Ann. Mus. Lugd.-Bat.* III. 26 (1867); *Prol. Fl. Jap.* 214 (1867). — Franchet & Savatier, *Enum. Pl. Jap.* I. 461 (1875). — Schneider, *Ill. Handb. Laubholzk.* I. 65, fig. 26 f, 27 i–k (1904). — Koidzumi in *Tokyo Bot. Mag.* XXVII. 92 (1913).

Salix Thunbergiana Blume apud Andersson in De Candolle, *Prodr.* XVI. pt. 2, 271 (1868). — Burkill in *Jour. Linn. Soc.* XXVI. 533 (1899). — Komarov in *Act. Hort. Petrop.* XXII. 30 (*Fl. Mansh.* II.) (1903). — Seemen, *Salic. Jap.* 61, t. 14, fig. A–E (1903). — Léveillé in *Bull. Acad. Intern. Géogr. Bot.* XIV. 210 (1904); XVI. 148 (1906). — Shirasawa, *Icon. Ess. For. Jap.* II. t. 7, fig. 1–9 (1908). — Nakai in *Jour. Coll. Sci. Tokyo,* XXXI. 213 (*Fl. Kor.* II.) (1911).
Salix cinerea Komarov in *Act. Hort. Petrop.* XXII. 22 (1903); XXV. 813 (1907).

NORTHEASTERN ASIA. Maritime Prov.: Shingking, *J. Ross* (Nos. 16, 175; fide von Seemen); Korea: Seoul, Puk Han, September 25, 1905, *J. G. Jack* (sterile); see also von Seemen. JAPAN. Hondo: prov. Mutsu, Aomori, rivers, May 1904, *U. Faurie* (No. 5767; bush 1 m. tall; with fruits); prov. Shinano, Nojiri, September 5, 1905, *J. G. Jack* (sterile); upper Kisogawa, alt. 1000 m., September 3, 1905, *J. G. Jack* (sterile); prov. Musashi, Mt. Tokao, April 16, 1911 (ex Herb. Sakurai; ♀); Mt. Buko, May 1901 (No. 103 ex Herb. Tokyo; sterile); without precise locality, July 19, 1891 (sterile); prov. Kai, Kusakabe, side of torrents near Jawata village, May 28, 1914, *E. H. Wilson* (No. 6331; bush 0.5–1 m. tall; with fruits); on the Yatsugatake, side of torrents, September 17, 1904, *E. H. Wilson* (No. 7528; bush 0.9–1.8 m. tall; sterile); prov. Sagami, Hakone, April (No. 74 ex Herb. Bot. Gard. Tokyo; ♀); prov. Izu, Awasima, March 20, 1897 (♂); Shuzenzi, March 27, 1905, April 10, 1912 (ex Herb. Sakurai; ♀ and ♂); without precise locality, April 2, 1895, *J. Matsumura* (No. 54; ♀); prov. Izumi, March 22, 1896 (No. 73 ex Herb. Bot. Gard. Tokyo; ♂); Otsu, March 6, 1896 (No. 108 ex Herb. Bot. Gard. Tokyo; ♀); prov. Mimasaka, April 10, 1904, *S. Arimoto* (with fruits). Kyushu: base of Kirishima, side of stream Miya, not common, March 7, 1914, *E. H. Wilson* (No. 6215; bush 1.5–3 m. tall; ♀); Nagasaki, March 10–24, 1862, *C. Maximowicz* (♂); same locality, 1863, *C. Maximowicz* (with fruits); same locality, 1862 (No. 527; ♂ and ♀; No. 719, sterile). See remarks in the section above.

Salix gracilistyla, var. **melanostachys** Schneider, n. comb.

Salix Thunbergiana, subsp. *melanostachys* Makino in *Tokyo Bot. Mag.* XVIII. 141 (1904).
Salix melanostachys Makino, l. c. (pro synon.) (1904).
Salix nigrolepis Shirai ex Makino, l. c. (pro synon.) (1904).

JAPAN. Known only in cultivation. According to the specimens I have seen ex Hort. Bot. Tokyo, March 18, 1890, ex Hort. Sakurai, March 19, 1911, October 27, 1912, this plant differs from the type only in the more glabrous branchlets, leaves and bracts, which are a little broader toward the base. The gland is the same in both. Only the ♂ sex is known. There may be other forms of *S. gracilistyla* Miquel or hybrids of it with different species. I mention a ♀ specimen of *Wilson's* from Hondo: slopes of Fuji-san, abundant, above 800 m., May 8, 1914, bush 1.8–3.6 m. tall, in which the ovaries have a distinct pedicel nearly as long as the gland; the style is only half as long as the ovary, and the bracts are very acute and black in the upper part. The leaves are only half grown and are entire, with a pubescence very much like that of the typical form.

Sect. 33. HELIX Dumortier in *Bijdr. Natuurk. Wetensch.* 1825 (*Verh. Gesl. Wilgen*, 15) (1825).

Salix, sect. *Purpureae* E. Fries in *Syllog. Pl. Nov.* II. 37 (*Conspec. Disp. Salic. Suec.*) (1828). — W. D. Koch, *De Salic. Comm.* 24 (1828). — Andersson in De Candolle, *Prodr.* XVI. pt. 2, 306 (1868). — Seemen, *Salic. Jap.* 20 (1903); in Ascherson & Graebner, *Syn. Mitteleur. Fl.* IV. 60 (1908), exclud. *S. caesia.* — Schneider, *Ill. Handb. Laubholzk.* I. 68 (1904).
Salix, sect. *Monandrae* Borrer in Hooker, *Brit. Fl.* 413 (1830). — Wolf in *Act. Hort. Petrop.* XXI. 135 (1903).
Salix, sect. *Subviminales* Seemen, *Salic. Jap.* 20 (pro parte) (1903).

Arbores et frutices elati v. frutices satis parvi, ramis plerisque elongatis flexilibus. Folia saepissime late usque lineari-lanceolata, 4–12-plo longiora quam lata et pleraque acuta, rarius elliptica v. oblanceolata et breviora latioraque, margine integra v. serrato-denticulata. Amenta praecocia v. subcoetanea sessilia v. subsessilia, angusto-cylindrica v. cylindrica, densa; flores ♂ glandula una, staminibus 2 filamentis plus minusve v. totis coalitis, interdum etiam antheris flavis coalitis; flores ♀ ovariis sericeis v. glabris sessilibus v. subsessilibus v. breviter pedicellatis, stylis nullis v. brevibus v. elongatis, stigmatibus satis brevibus.

I place *S. Pierotii* Miquel in this section and keep separate sect. *Caesiae* (see p. 173). Probably also such species as *S. myrtillacea* Andersson and *S. Bockii* Seemen and their relatives should be referred to distinct sections.

157. **Salix Pierotii** Miquel in *Ann. Mus. Lugd.-Bat.* III. 27 (1867); *Prol. Fl. Jap.* 215 (1867). — Franchet & Savatier, *Enum. Pl. Jap.* I. 461 (1875); II. 506 (1879). — Seemen, *Salic. Jap.* 60, t. 13, fig. E–F (1903); apud Siuzev in *Trav. Mus. Bot. Acad. Sci. St. Pétersbourg*, IX. No. 2 (1912) ex Toepffer, *Salicol. Mitt.* No. 5, 248 (1912). — Schneider, *Ill. Handb. Laubholzk.* I. 66, fig. 23 a, 27 l–m (1904). — Léveillé in *Bull. Intern. Acad. Géogr. Bot.* XIV. 210 (1904); XVI. 145 (1906).

Salix japonica Dippel, *Handb. Laubholzk.* II. 221 (non Thunberg) (1892).

NORTHEASTERN ASIA. Amur and Ussuri: fide von Seemen (1912).
JAPAN. Hondo: "Iwajama, ubi frequens in jugis et in vallibus," April, *P. von Siebold* (type, ex Miquel). Shikoku: prov. Tosa, Nanokawa, June 20, 1889, *K. Watanabe* (sterile). Kyushu: "in promontorio Nomo Saki," *Pierot* (co-type, ex Miquel).
So far as I can judge mainly from plants cultivated in the Arnold Arboretum it is a good species, easily distinguished from *S. purpurea* Linnaeus and *S. Miyabeana* Seemen. The mature leaves are broadest below the middle, and their serration is closer (5–6 teeth in 1 cm.) than in *S. Miyabeana* Seemen, where the leaves are broadest at or above the middle and have a coarser serration (about 3 teeth in 1 cm.). The young twigs of *S. Pierotii* are pubescent, and often glabrous and yellow in *S. Miyabeana*, which has longer stipules.

158. **Salix Miyabeana** Seemen in *Bot. Jahrb.* XXI. Beibl. LIII. 50 (1896); *Salic. Jap.* 57, t. 12, fig. A–E (1903). — Tokubuchi [1] in *Tokyo Bot. Mag.* X. 69, t. 6 (1896). — Léveillé in *Bull. Acad. Intern. Géogr. Bot.* XIV. 210 (1904); XVI. 144 (1906). — Koidzumi in *Tokyo Bot. Mag.* XXVII. 264 (1913).

JAPAN. Hokkaido: prov. Ishikari, Sapporo, May 15, 1890 (♂), October 15 and 21, 1890 (sterile), April 23 and 30, 1891 (♀), May 2, 1891 (♀), *Y. Tokubuchi* (types and co-types).

There is a Chinese ♀ specimen collected by Wilson in Chili, near Tientsin, April 18, 1909, which has very yellow glabrous branchlets with precocious aments the flowers of which much resemble those of *S. Miyabeana*, but it may be *S. purpurea* Linnaeus with yellowish bark. For further information, see under *S. Pierotii* Miquel and in the keys (p. 79 and p. 90).

There seems to be a variety of this species with tomentose branchlets, judging by a ♂ specimen collected by *Mochizuki*, 1904, at Nikko (in collect. Jack) and a sterile specimen collected by *C. S. Sargent*, Oct. 1, 1892, at Aomori, banks of river, tree, 10 m. tall. The catkins resemble those of *S. lepidostachys* Seemen, being about 5 cm. long and 12 mm. thick.

159. **Salix sapporoensis** Léveillé in *Bull. Soc. Bot. France*, LVI. 302 (1909).

JAPAN. Hokkaido: prov. Ishikari, Sapporo, " secus aquas," June 13, 1908, *U. Faurie* (Nos. 266, 268; fruiting types).

Léveillé also mentions Faurie's No. 262 from the same locality, but the specimen of this number which I have seen belongs to another species. There is No. 267, named *S. Miyabeana* Seem.? which is not different from 266 or 268; it consists of a fruiting branchlet and a sterile one, the long lanceolate leaves of which are up to 10 cm. long and 1.8 cm. broad and hairy beneath while young. They have linear-lanceolate dentate stipules ⅔ as long as the petioles, which are about 1 cm. in length. The leaves closely resemble those of *S. Miyabeana* Seemen, but are apparently greenish beneath and not unlike those of *S. sachalinensis*, var. *Pilgeriana* Schneider, which, however, are more entire. The types of *S. sapporoensis* have sessile or subsessile ovaries with very short styles, resembling those of *S. purpurea* Linnaeus. Faurie's No. 257 from the same place and date is a little different; it is named *S. daphnoides*. In this specimen even the young very serrate leaves are glabrous, the stipules are short, ovate and dentate, the fruits are nearly glabrous with a distinct pedicel. Without fully grown leaves and young flowers it seems impossible to decide whether *S. sapporoensis* represents a new species or may be of hybrid origin. In the neighborhood of Sapporo there are many different species of *Salix*, wild and cultivated, and hybrids of different origin may occur there.

160. **Salix lepidostachys** Seemen in *Bot. Jahrb.* XXI. Beibl. LIII. 51 (1896); *Salic. Jap.* 58, t. 12, fig. F–K (1903); apud Siuzev in *Trav. Mus. Bot. Acad. Sci. St. Pétersbourg*, IX. No. 2 (1912) ex Toepffer, *Salicol. Mitt.* No. 5, 248 (1912). — ? Léveillé in *Bull. Acad. Intern. Géogr. Bot.* XIV. 210 (1904); XVI. 145 (1906).

NORTHEASTERN ASIA. Amur and Ussuri: fide von Seemen (1912).

JAPAN. Hokkaido: prov. Ishikari, May and October 1891, *Y. Tokubuchi* (No. 31 ex Herb. Sapporo, co-type; ♀); prov. Oshima, Hakodate, 1861, *C. Maximowicz* (No. 6, co-type; ♂). Hondo: prov. Mutsu, Hirosaki, June 1897, *U. Faurie* (No. 806, ex von Seemen); prov. Izumi, Ishitzu, April 3, 1896 (No. 105 ex Herb. Bot. Gard. Tokyo; ♀). Shikoku: prov. Tosa, April 11, 1888 (No. 37[b] ex Herb. Bot. Gard. Tokyo; ♀).

Tokubuchi's specimens from the type locality of May 2, 1891, which I have seen

---

[1] This author has published, l. c., p. 120, an article in Japanese on the Willows of Hokkaido.

in the herbarium of the Arnold Arboretum agree well with the author's description, except that the bracts are shorter and a little more acute, and the pedicels twice as long as the gland. The following branchlets are glabrous, and the long stigmas are a little longer than the style. On Maximowicz's No. 6 in the Gray Herbarium the ♂ flowers agree with von Seemen's figure, the branchlets being covered with the remnants of a tomentum near their apex. The ♀ flowers, however, are very different and almost like those of *S. gracilistyla* Miquel; the bracts are more obtuse, the ovaries short-stalked and the branchlets almost tomentose. U. Faurie's No. 806 I have not seen; it was collected at Hirosaki, where Faurie found also several different forms (see *S. hondoensis* Koidzumi, p. 110). The specimens from Ishitzu and prov. Tosa are doubtfully referred to our species. Certainly *S. lepidostachys* and its relatives need further observation. It is almost impossible to interpret correctly flowering branchlets of these forms without having seen mature leaves from the same plant.

161. **Salix purpurea** Linnaeus, *Spec.* 1017 (1753). — Pokorny, *Oesterr. Holzpfl.* 124, t. 24, fig. 362–363 (1864). — Andersson in *Mem. Am. Acad.* VI. 451 (*Gray, Bot. Jap.*) (1859); in De Candolle, *Prodr.* XVI. pt. 2, 306 (1868). — Franchet & Savatier, *Enum. Pl. Jap.* I. 462 (1875). — Herder in *Act. Hort. Petrop.* XI. 452 (1891). — Hempel & Wilhelm, *Bäume & Sträucher,* II. 107, fig. 30 (1897). — Burkill in *Jour. Linn. Soc.* XXVI. 532 (1899). — Wolf in *Izv. S.-Peterburg. Liesn. Inst.* IV. 42, t. 13, fig. 1–11 (*Мат. Изуч. Пез Европ. Росс.*) (1900); in *Act. Hort. Petrop.* XXI. 148 (1903). — Seemen, *Salic. Jap.* 54, t. 11, fig. A–E (1903); in Ascherson & Graebner, *Syn. Mitteleur. Fl.* IV. 192 (1909); apud Siuzev in *Trav. Mus. Bot. Acad. Sci. St. Pétersbourg,* IX. No. 2 (1912) ex Toepffer, *Salicol. Mitt.* No. 5, 248 (1912). — Komarov in *Act. Hort. Petrop.* XXII. 27 (*Fl. Mansh. II.*) (1904). — Schneider, *Ill. Handb. Laubholzk.* I. 68, fig. 20 v–v¹, 23 m (1904). — Shirasawa, *Icon. Ess. For. Jap.* II. t. 7, fig. 10–18 (1908). — Nakai in *Jour. Coll. Sci. Tokyo,* XXXI. 215 (*Fl. Kor. II.*) (1911). — Koidzumi in *Tokyo Bot. Mag.* XXVII. 92 (1913). — Moss, *Cambridge Brit. Fl.* II. 65, t. 65–66 (1914).

Salix purpurea, var. *typica* Beck, *Fl. Nied.-Oestr.* 288 (1890).
Salix purpurea, subspec. *eupurpurea,* var. *typica* Schneider, *Ill. Handb. Laubholzk.* I. 68 (1904).

CHINA. See Burkill and von Seemen.
NORTHEASTERN ASIA. Transbaikalia to Maritime Prov.: probably together with var. *sericea.* Korea: Chinampo, " secus aquas," August 1906, *U. Faurie* (No. 180; sterile); " in monte des diamants," June 21, 1906, *U. Faurie* (No. 179; with fruits).
JAPAN. Hokkaido: prov. Ishikari, Sapporo, May 19, 1889 (No. 34 ex Herb. Bot. Gard. Tokyo; ♀). Hondo: prov. Shimotsuke, abundant in swamps between Chuzenji and Yumoto, May 26, 1914, *E. H. Wilson* (No. 6765; shrub 0.3–0.9 m. tall; ♂; No. 6765ᵃ, ♀); prov. Musashi, March 1895 (No. 70 ex Herb. Bot. Gard. Tokyo; ♂); Nambu, 1865, *Tschonoski* (with fruits); Azusawa near Tokyo, April (No. 81 ex Herb. Bot. Gard. Tokyo; ♀); prov. Tamba, April 4, 1910 (ex Herb. Sakurai; ♀); prov. Tajima, March 17, 1907 (ex Herb. Sakurai; ♀). Kyushu: Mt. Kirishima, roadside thickets, rare, March 6, 1914, *E. H. Wilson* (No. 6180; bush 1.5–2.4 m. tall; ♀ ; No. 6180ᵃ, ♂).
This is the typical form with glabrous or slightly pilose young branchlets. The leaves are mostly alternate with distinct petioles, and are glabrous even when young. There are also the following varieties:

Salix purpurea, subspec. **eupurpurea,** var. **sericea** Schneider, *Ill. Handb. Laubholzk.* I. 69 (1904). — Koidzumi in *Tokyo Bot. Mag.* XXVII. 92 (1913).

Salix purpurea, var. *sericea* W. D. Koch, *Syn. Deutsch. Schweiz. Fl.* 647 (1837).— Seemen, *Salic. Jap.* 56 (1903).

Ramuli novelli plus minusve sericeo-tomentelli. Folia pleraque alterna, petiolata, subtus ut petioli initio sericeo-tomentella, dein plus minusve glabrescentia. NORTHEASTERN ASIA. Transbaikalia to Korea: probably together with the type. JAPAN. Hokkaido: prov. Ishikari, Sapporo, low ground, September 11, 1892, *C. S. Sargent* (tree 10 m. tall, girth 0.3 m.; sterile). Hondo: prov. Shinano, April 1904, August 1905, *H. Shirasawa* (♂ and sterile); prov. Musashi, Tokyo, probably cultivated (many specimens).

Salix purpurea, subspec. amplexicaulis, var. multinervis Schneider, n. comb.

Salix amplexicaulis Chaubard in Bory, *Exp. Sci. Morée*, III. 2, Bot. 277; *Atlas*, sér. 4, Bot. t. 34 (1832); in Bory, *Nouv. Fl. Pélop.* 64, t. 36 (1838).
Salix multinervis Franchet & Savatier, *Enum. Pl. Jap.* II. 504 (1879).
Salix purpurea, var. *multinervis* Matsumura, *Shokubutsu Mei-J*. 261 (1895). — Seemen, *Salic. Jap.* 56, t. 11, fig. F-K (1903). — Shirasawa, *Icon. Ess. For. Jap.* II. t. 8, fig. 7 (folia tantum) (1908).
? *Salix purpurea*, subspec. *amplexicaulis*, var. *petiolata* Koidzumi in *Tokyo Bot. Mag.* XXVII. 92 (1913); from Mandshuria.

Ramuli glabri. Folia glabra, sessilia v. petiolata, oblonga v. anguste elliptica, marginibus plus minusve parallelis, basi rotunda v. cordata, apice pleraque obtusa, margine serrata, basin versus integra, ad 6.5 longa et ad 2.5 cm. lata. JAPAN. Hokkaido: prov. Oshima, Hakodate, 1861, *C. Maximowicz* (♀); prov. Iburi, Yubutsu, June 12, 1884, *K. Miyabe* (♀); Chitose, June 11, 1890, *Y. Tokubuchi* (♀). Hondo: prov. Shimotsuke, around Lake Chuzenji, lava beds, common, alt. 1800 m., May 27, 1914, *E. H. Wilson* (No. 6749; bush 1.2–1.8 m. tall; with fruits); Nikko, August 8, 1905, *J. G. Jack* (sterile); prov. Shinano, Narai, September 2, 1905, *J.G. Jack* (sterile); Kituadzumi, August 1903, *H. Shirasawa* (sterile); without precise locality, July 6, 1893 (ex Herb. Bot. Gard. Tokyo; sterile); prov. Musashi, Komaba, March 24, 1911 (♀); without precise locality, April 11, 1891, April 15, 1894 (with fruits); prov. Kai, Kusakabe, Yawata village, side of streams, March 28, 1914, *E. H. Wilson* (No. 6330; bush 0.6–1.2 m. tall; ♂; No. 6330ᵃ, ♀); prov. Suruga, slopes of Fuji-san, roadsides, etc., abundant, May 7, 1914, *E. H. Wilson* (No. 6639; bush 0.3–1.8 m. tall; with fruits); prov. Sagami, "in locis humidis," *L. Savatier* (No. 1145, type, ex Franchet & Savatier; with fruits); Tatteyama, near Yokoska, *L. Savatier* (co-type; ♂, ex Franchet & Savatier).

I regard the subspec. *amplexicaulis*, var. *multinervis* distinct from var. *genuina* Schneider, n. comb. (*S. amplexicaulis* Chaubard, l. c.; *S. purpurea*, δ *amplexicaulis* Boissier, *Fl. Orient.* IV. 1187 [1879]): Folia in comparatione angustiora, acutiora, distinctius v. angustius serrata, usque ad 6 cm. longa et ad 1.5 cm. lata.

S. purpurea, subspec. *amplexicaulis*, var. *petiolata*, subvar. *angustifolia* Koidzumi, l. c. 92, from Hokkaido seems to be only a form of subspec. *eupurpurea*, var. *typica* with opposite leaves.

Shirasawa, on his plate (8, l. c.), figures ♀ flowers, which may belong to *S. Gilgiana* Seemen, but the color of the bracts of both his ♀ and ♂ flowers is yellowish brown.

The typical *S. purpurea* Linnaeus has sessile ovaries and sessile or subsessile short stigmas; the color of the young anthers is purple.

There is also the following doubtful variety:

Salix purpurea, var. stipularis Franchet in *Nouv. Arch. Mus. Paris*, sér. 2, VII. 91 (*Pl. David.* I. 284) (1884). — Burkill in *Jour. Linn. Soc.* XXVI. 532 (1899).

CHINA. Chili: " Jéhol, bords des ruisseaux," May and March, *A. David* (No. 1681, type, ex Franchet).
I have not seen the type of this variety, and the description is too short to judge whether it is distinct or not.

162. **Salix Gilgiana** Seemen, *Salic. Jap.* 59, t. 13, fig. A–D (1903). — Léveillé in *Bull. Acad. Intern. Géogr. Bot.* XVI. 145 (1906).
JAPAN. Hondo: prov. Musashi, Tokyo (Jedo), March 15, April 23, June 8, 1874, *Hilgendorf* (♂ and ♀ types ex Seemen). I doubtfully include here the following specimens: prov. Shimotsuke, Chuzenji, lava beds, common, alt. 1800 m., May 27, 1914, *E. H. Wilson* (No. 6748; bush 1.5–2.4 m. tall; ♀); prov. Musashi, Toda, April 12, 1891, *K. Watanabe* (♀); Tokyo, March 1895 (No. 46 Herb. Bot. Gard. Tokyo; ♀); same place, April 22, 1893 (No. 112; Herb. Bot. Gard. Tokyo; with fruits); same place, April 3, 1901 (No. 123; Herb. Bot. Gard. Tokyo; ♀); Hokkaido: prov Ishikari, Sapporo, cultivated, May 3, 1906, *J. Hanzawa* (♂ and ♀).
These specimens agree with one another and with von Seemen's description, but the leaves are not so distinctly and sharply serrate as they are shown in Seemen's figure. *S. Gilgiana* needs further study.

163. **Salix cheilophila** Schneider. See p. 69.

164. **Salix Wilhelmsiana** Marschall v. Bieberstein, *Fl. Taur.-Cauc.* III. Suppl. 627 (1819). — Trautvetter in *Mém. Sav. Etr. Acad. Sci. St. Pétersbourg,* III. 627, t. 3 *(Salicet.)* (1837). — Regel in *Act. Hort. Petrop.* VI. 464 (1880). — Herder in *Act. Hort. Petrop.* XI. 457 (pro parte) (1891).—Wolf in *Act. Hort. Petrop.* XXIII. 153 (1903); Wolf & Palibin, *Опред. Дерев. Куст. Европ. Росс.* 89, fig. (1904).

Salix angustifolia Willdenow, *Spec.* IV. 699 (non Wulfen)[1] (1805). — Marschall von Bieberstein, *Fl. Taur.-Cauc.* II. 414 (1808). — Andersson in De Candolle, *Prodr.* XVI. pt. 2, 315 (exclud. var. β) (1868).—Brandis, *Forest Fl. Brit. Ind.* 471 (1874). — Hooker f., *Fl. Brit. Ind.* V. 636 (1888).—Wolf in *Izv. S.-Peterburg. Liesn. Inst.* V. 33, t. 16, fig. 10–17, t. 29, fig. 1–2 *(Мат. Изуч. Ивъ Европ. Росс.)* (1900). — Schneider, *Ill. Handb. Laubholzk.* I. 68, fig. 20 x–y (1904).

Salix angustifolia, var. *eriocarpa* Ledebour, *Fl. Ross.* III. pt. 2, 604 (1850).

AFGHANISTAN. Kurrum Valley, 1879, *J. E. T. Aitchison* (No. 111; ♂ and ♀).
INDIA. Kashmir: " reg. temp., alt. 12000 ped.," *T. Thomson* (with fruits).
This species is very nearly related to the following, which chiefly differs in its glabrous ovaries and seems to have a different geographical distribution. See Wolf, l. c. (1903). The length of the short style, which is sometimes wanting, varies in both species; the bracts are yellowish and not two-colored, as in *S. purpurea* and its nearest relatives. The species with bracts of one color may perhaps represent a different section.

---

[1] *Salix angustifolia* Wulfen (in Jacquin, *Collect.* III. 48 [1789]) is often regarded as a synonym of *S. rosmarinifolia* Linnaeus, but according to several authors Linnaeus's plant may represent a hybrid between *S. repens* Linnaeus and *S. viminalis* Linnaeus, while Wulfen's plant is the same as *S. repens,* var. *rosmarinifolia* Wimmer & Grabowski (*Fl. Siles.* II. 380 [1829].—*S. repens* var. *angustifolia* Neilreich, *Fl. Wien,* 79[1846]).

170 WILSON EXPEDITION TO CHINA

165. **Salix microstachya** Turczaninow apud Trautvetter in *Mém. Sav. Étr. Acad. Sci. St. Pétersbourg*, III. 628, t. 4 (*Salicetum*) (1837); in *Bull. Soc. Nat. Mosc.* XXVII. 377 (1854); *Fl. Baical.-Dahur.* 104 (1856). — Regel in *Act. Hort. Petrop.* VI. 466 (1880). — Wolf in *Act. Hort. Petrop.* XXIII. 156 (1903).

Salix angustifolia, β leiocarpa Ledebour, *Fl. Ross.* III. pt. 2, 604 (1850).

Salix angustifolia, β microstachya Andersson in De Candolle, *Prodr.* XVI. pt. 2, 315 (1868).

? *Salix cyanolimnea* Hance in *Jour. Bot.* XX. 294 (1882).

? *Salix Wilhelmsiana* Franchet in *Nouv. Arch. Mus. Paris*, sér. 2, VII. 93 (*Pl. David.* I. 283) (non Marshall von Bieberstein) (1884).

Salix Wilhelmsiana, β microstachya Herder in *Act. Hort. Petrop.* XI. 456 (1891).

The type was collected by Turczaninow in western Siberia on the Irkut River, and according to Hance's description I have little doubt that the following specimen belongs to the same species:

TIBET. Kukunor: "ad lacum Ko-ko-nor," 1881, *W. Mesny* (Herb. Hance No. 22009; ex Hance).

MONGOLIA: Ordos "dans les plaines humides," June, *A. David* (No. 2719; with fruits, ex Franchet); "Ou la chan, bords des ruisseaux des montagnes," *A. David* (No. 2676; ex Franchet).

David's specimens may belong to *S. cheilophila* Schneider (see p. 69), which is very similar, although it may be distinguished by its mostly two-colored bracts and minute stigmas.

166. **Salix Bockii** Seemen. See p. 71.

167. **Salix Duclouxii** Léveillé in *Bull. Soc. Bot. France*, LVI. 298 (1909).

CHINA. Yunnan: Yunnan Fu, valleys, July 23, 1905, *F. Ducloux* (No. 670; ♂ and ♀ types).

Through the kindness of Mgr. Léveillé I have received ♂ and ♀ catkins and a leaf of his type. The flowers appear in summer like those of *S. Bockii* Seemen, and I suppose both species are very closely related. The ♂ have the same long gland as the ♀ flowers, the bracts are acute and somewhat like those of *S. variegata* Franchet. It needs more material in order to decide whether our plant is a good species or only a form of *S. Bockii*.

168. **Salix variegata** Franchet. See p. 70.

169. **Salix andropogon** Léveillé in Fedde, *Rep. Spec. Nov.* III. 21 (1906); in *Bull. Soc. Bot. France*, LVI. 297 (1909); *Fl. Kouy-Tchéou*, 381 (1915).

CHINA. Kweichau: "Lit du fleuve, submergé aux grandes eaux," December 15, 1904, *J. Esquirol* (No. 327; ♂ and ♀ types).

Mgr. Léveillé has kindly sent to the Arnold Arboretum a ♂ and a ♀ catkin of this Willow. The species is very closely related to *S. variegata* Franchet, and I cannot distinguish the ♀ flowers from those of that species. The pubescence of the catkins is sometimes somewhat yellowish, but I would not say: "rhachi fulvo-villoso," as Léveillé does. The ♂ catkins differ in several respects. The young anthers bear some hairs at the end of each connective, something I have never seen in any other species of *Salix*. These hairs seem to fall off from the older anthers toward the base of the catkin. The filaments are pubescent at the base, while in *S. variegata* Franchet I have found them always glabrous, but von Seemen describes the base of the filaments of the type of his *S. densifoliata* as "dicht weissgrau behaart," a character which is variable in several species. The bracts of *S. andropogon* are villose and ciliate. Léveillé's indication "squama . . . albide villosa, ebarbata" seems to be a misprint for "et barbata." I have seen only very young small leaves which are nearly glabrous, the very short petioles being tomentose.

170. **Salix kouytchensis** Schneider, n. sp.

Salix Duclouxii, var. *kouytchensis* Léveillé in *Bull. Soc. Bot. France*, XVI. 298 (1909).

Salix Duclouxii Léveillé, *Fl. Kouy-Tchéou*, 381 (non Léveillé, 1909) (1915).

CHINA. Kweichau: "bord de la rivière de Ouen-Tsen-Kiao (Kouy-Tin)," Nov. 23, 1902, *J. Cavalerie* (No. 728; fruiting type).

I have received from the author a small piece of a branchlet with a fruiting catkin which I describe as follows:

Ramuli satis dense breviter adpresse albo-sericei, substriati. Folia lanceolata, utrinque acuta v. apice obtusiora, supra sparse sericea, in sicco brunnescentia, costa nervisque subprominulis, subtus distincte pallidiora, distinctius, initio ut videtur dense, sericea, glabrescentia, nervis costaque subprominulis, minutissime reticulata, margine satis dense glanduloso-serrato-dentata (dentibus 5–7 pro 1 cm.), circiter 1.8 cm. longa, 0.5 cm. lata; petioli 2–3 mm. longi, plus minusve sericeo-villosuli. Amenta breviter pedunculata, basi foliis 1–2 suffulta, sericeo-villosa, circiter 2.5 cm. longa et 1 cm. crassa. Fructus sessiles, ovato-oblongi, subacuti, sericei, stigmatibus oblongis bifidis sessilibus coronati. Glandula 2–3-fida, partibus linearibus, bracteis triplo brevior. Bracteae oblongae v. obovato-oblongae, obtusae, margine sericeo-villosae, faciebus subglabrescentes.

The deeply cleft gland and the rather obtuse bracts seem to distinguish this species from *S. Duclouxii* Léveillé. In the shape of the bracts it resembles *S. Bockii* Seemen, but the bracts of that species are often rather acutish and the leaves are much more silky and mostly obtuse at the ends. *S. kouytchensis* seems to flower in the autumn, and it needs further study to decide whether it is a good species or only a variety of *S. Bockii* Seemen or of *S. variegata* Franchet.

171. **Salix pycnostachya** Andersson in *Jour. Linn. Soc.* IV. 44 (1860); in De Candolle, *Prodr.* XVI. pt. 2, 309 (1868).

INDIA. Kashmir: Zanskar (Gauskar, misprint in Andersson), "reg. alp., alt. 13000 ped.," *T. Thomson* (♂ and ♀ types).

This species is described with glabrous ovaries, while Hooker f. (*Fl. Brit. Ind.* V. 636 [1888]), and Brandis (*Forest Fl. Brit. Ind.* 470 [1874]) say: "capsules . . . silky." They also say that the catkins stand "on leaf-bearing peduncles," although they are nearly sessile in the type. I suppose there may be some other forms mixed up with our species, or it may be connected with the doubtful *S. oxycarpa* Andersson by intermediate or hybrid forms. *S. pycnostachya*, so far as I know, has sessile glabrous ovaries with short styles, nearly entire obtusish leaves, obovate bracts, and entirely united filaments, the var. *alpina* being a little more hairy on the young shoots and leaves. See also *S. oxycarpa* Andersson and the keys on p. 80 and p. 88.

Salix pycnostachya, var. **alpina** Hooker f. apud Andersson in De Candolle, *Prodr.* XVI. pt. 2, 309 (1868). — Hooker f., *Fl. Brit. Ind.* V. 636 (1888).

INDIA. Kashmir: Zanskar "reg. alp. alt. 13–14000 ped.," *T. Thomson* (♂ type); "Tibet, Ladak, Timti La pass," July 2 and 3, 1856, *Schlagintweit* (No. 6531; sterile).

This variety seems to be only a dwarf alpine form with smaller leaves and catkins. It may belong to *S. oxycarpa* Andersson because the leaves are somewhat denticulate.

172. **Salix oxycarpa** Andersson in *Jour. Linn. Soc.* IV. 45 (1860), exclud. var. b *serratifolia*; in De Candolle, *Prodr.* XVI. pt. 2, 310 (1868), exclud. var. β *serrata*. — Brandis, *Forest Fl. Brit. Ind.* 471 (1874); *Ind. Trees*, 638 (1906). — Hooker f., *Fl. Brit. Ind.* V. 636 (1888). — Collett, *Fl. Siml.* 480 (pro parte) (1902).

INDIA. Kashmir: Kishtwar "alt. 6–11000 ped.," *T. Thomson* (♂ and ♀ types, ex Andersson); "Tibet, Ladak, Yuru Kiom via Kandzi up the Timti La pass," July 2, 1856, *Schlagintweit* (No. 5261; with fruits); "Tsanskar, Sulle to Padum," June 22–24, 1856, *Schlagintweit* (No. 6744; sterile).

This seems to be a very uncertain species. I have received from the Kew Herbarium a specimen which is supposed to be Andersson's type. But there are three different things upon this sheet. First a fruiting branch to which apparently belongs the small label " Ind. Valley, 7/7/48." These fruits agree very well with those of the co-type of var. *breviuscula*, and may belong to Andersson's form a. *angustifolia* (1860). The ovaries are subsessile, the stigmas rather narrow and as long as the short style. The bracts are ovate-oblong, rather obtusish, and light brown. There are also two ♂ specimens. To one seems to belong a small label, " Kishtwar 8–9000, 19/6/48." Of this specimen the filaments are totally united, only the anthers being free, and are hairy in the lower part. The gland is rather long and narrow, and half as long as the obovate-oblong short silky dark-tipped bract. The old branchlets of both these ♂ and ♀ specimens have the same purplish color and are somewhat pruinose; the young twigs are distinctly tomentose; the short stipules of the ♀ branch are ovate, acute, or semi-cordate and glandular-serrate; the ♂ catkins have short leafy peduncles and are 5–6 cm. long and 1 cm. thick. This ♂ form does not agree with Andersson's description. To the second ♂ specimen apparently belongs the main label: " Kishtwar, 6–11000 ped., coll. *Thomson*," which would be Andersson's type. But Andersson says in his description: " Rami . . . in speciminibus masculis a me visis molliter incano-pubescentes . . ." and " Stamina 2, filamentis usque ad medium connatis glabris . . ." In the specimen before me only the very young twigs are tomentose, the " rami " being glabrous, purplish and somewhat pruinose as in the other forms. The filaments are united only at the very base and are very hairy on the lower third. The gland is shorter and somewhat broader than in the other ♂ specimen, the obovate-oblong light-brown bracts are very silky on the inner surface and nearly glabrous except at the very base on the outer surface. The young leaves show no difference in shape, color, etc. from those of the other specimens.

There is perhaps another ♂ form mixed with those mentioned above, representing Andersson's ♂ type with pubescent branches and glabrous filaments united for half of their length.

Salix oxycarpa, var. **breviuscula** Andersson in *Jour. Linn. Soc.* IV. 46 (1860); in De Candolle, *Prodr.* XVI. pt. 2, 310 (1868).

INDIA. Kashmir: " Zanskar, alt. 10–14000 ped.," *T. Thomson* (fruiting type).

According to the co-type this variety agrees well with the ♀ form mentioned above. Schlagintweit's No. 5261 differs slightly in the very sessile stigmas. If the ♂ specimen with the completely united stamens does belong to the same form, this variety could be kept as a distinct species much like *S. pycnostachya* Andersson, but differing in its hairy ovaries. Further study of good material is needed to prove whether this ♀ plant belongs to the typical *S. oxycarpa* or to a different species.

173. **Salix myricaefolia** Andersson in *Svensk. Vetensk. Akad. Handl.* 1850, 483 (1851); in *Jour. Linn. Soc.* IV. 53 (1860).

Salix fruticulosa Andersson in *Jour. Linn. Soc.* IV. 53 (1860), quoad plant. ♀.

Salix divergens Andersson in De Candolle, *Prodr.* XVI. pt. 2, 316 (1868). — Hooker f., *Fl. Brit. Ind.* V. 637 (1888). — Brandis, *Ind. Trees*, 639 (1906).

INDIA. Kashmir: " Kishtwar, 12000 ped.," *T. Thomson* (♂ and ♀ types of *S. divergens*); " Zanskar, 15000 ped.," *T. Thomson* (♀ type of *S. fruticulosa* of

1860); " in India orientali superiori," *V. Jacquemont & G. S. Perrotet* (♀ type of *S. myricaefolia*, ex Andersson).

See also my remarks under *S. fruticulosa* Andersson (p. 119).

S. myricaefolia certainly is closely related to *S. oxycarpa* Andersson, which has the same silky ovaries, and to *S. pycnostachya* Andersson, which has the same obtusish and entire leaves, but it differs from both in its much shorter catkins. Of the ♂ flowers I have not seen sufficient material.

174. **Salix myrtillacea** Andersson. See p. 71.

Sect. 33. CAESIAE Kerner in *Verh. Zool. Bot. Ges. Wien*, X. 205 (*Niederöstr. Weid.*) (1860). — Schneider, *Ill. Handb. Laubholzk.* I. 67 (1904).

Fruticuli v. frutices glabri plerique valde et breviter ramosi. Folia ovata, elliptica, lanceolata v. oblanceolata, discoloria, in sicco saepe nigrescentia, plus minusve tenuiter nervata et reticulata, margine integra. Amenta coetanea, subsessilia, ovato-cylindrica v. breviter cylindrica, densa; flores ♂ glandula una dorsali oblonga, filamentis ex parte v. totis coalitis; ovaria subsessilia v. breviter pedicellata, sericea, stylis brevibus v. mediocribus, stigmatibus brevibus oblongis.

The type of this section, *Salix caesia* Villars (*Pl. Dauph.* III. 768 [1789]) (see Seemen in Ascherson & Graebner, *Syn. Mitteleur. Fl.* IV. 198 [1908]) does not occur in Asia. The two following species are kept distinct by Wolf.

175. **Salix minutiflora** Turczaninow apud Wolf in *Act. Hort. Petrop.* XXIII. 142 (1903). — Krylov, *Фл. Алтая*, 1208 (1909).

Salix sibirica, a *glabra* Ledebour, *Fl. Ross.* III. pt. 2, 622 (1850), quoad syn.
    S. minutiflora Turczaninow, Pl. Exsicc.
Salix caesia, a *glabra* Turczaninow in *Bull. Soc. Nat. Mosc.* XXVII. 394 (1854); *Fl. Baical.-Dahur.* II. 121 (1856).
Salix caesia, var. *minutiflora* Andersson in De Candolle, *Prodr.* XVI. pt. 2, 317 (1868). — Herder in *Act. Hort. Petrop.* XI. 458 (1891).
Salix sibirica Herder in *Act. Hort. Petrop.* XI. 457 (non Pallas) (1891).

NORTHEASTERN ASIA. Transbaikalia to Yakutsk (see Herder, l. c., and Wolf, l. c.).

S. minutiflora is most closely related to *S. Kochiana* Trautvetter (see the differences in the key, p. 79 and p. 93).

Salix minutiflora, γ **pubescens** Wolf in *Act. Hort. Petrop.* XXIII. 143 (1903).
Salix caesia, β *pubescens* Turczaninow in *Bull. Soc. Nat. Mosc.* XXVII. 394 (1854); *Fl. Baical.-Dahur.* II. 121 (1856).
Salix sibirica Schneider, *Ill. Handb. Laubholzk.* I. 67, fig. 20 w¹, 20 c (non Pallas) (1904).

This form does not occur in our area, so far as I know; I have seen specimens from the Altai, *P. Krylov*, July 5, 1901.

176. **Salix Kochiana** Trautvetter in *Mém. Sav. Étr. Acad. Sci. St. Pétersbourg,* III. 632, t. 1 (*Salicet.*) (1837). — Ledebour, *Fl. Ross.* III. pt. 2, 602 (1850). — Turczaninow in *Bull. Soc. Nat. Mosc.* XXVII. 375 (1854); *Fl. Baical.-Dahur.* II. 102 (1856). — Andersson in De Candolle, *Prodr.* XVI. pt. 2, 314 (1868). — Herder in *Act. Hort. Petrop.* XI. 455 (1891). — Wolf in *Act. Hort. Petrop.* XXIII. 143 (1903). — Krylov, *Фл. Алтая*, 1209 (1909).

Salix Pontederana Trautvetter in Ledebour, *Fl. Alt.* III. 263 (exclud. synon., non Willdenow) (1833).
Salix loniceraefolia Turczaninow, Pl. Exsicc., ex Turczaninow in *Bull. Soc. Nat. Mosc.* XXVII. 375 (pro synon.) (1850).

NORTHEASTERN ASIA. Transbaikalia (see Herder, l. c., and Wolf, l. c.).
I have seen specimens of this species only from the Altai, collected by Sapo-
schnikof, June 10, 1897. It is very closely related to *S. minutiflora* Turczaninow
(see the keys, p. 79 and p. 93).

SPECIES SECTIONIS INCERTAE.

177. **Salix allochroa** Schneider. See p. 72.

178. **Salix polia** Schneider, n. sp.
Frutex erectus, ut videtur satis longe ramosus; ramuli novelli annotinique dense
albo-griseo-tomentelli, deinde glabrescentes, flavi, vetustiores brunneo-cineras-
centes; gemmae ovato-acutae, subappressae, tomentellae, rubro-brunneae. Folia
juvenilia lanceolata v. oblanceolata, apice breviter acuta v. subobtusa, basin versus
sensim attenuata, in petiolum 1–4 mm. longum villosum producta, supra dense
sericeo-villosa, paulo v. vix viridescentia, subtus densius albo-sericeo-villosa, nervis
nondum visibilibus, margine dense villosula, ut videtur interdum minute glanduloso-denticulata (glandulis inter pubescentiam occultis), 2–4 cm. longa, 0.4–0.8
cm. lata; stipulae petiolis breviores, ovato- v. angusto-lanceolatae, tomentellae.
Amenta (♀ tantum visa) subcoetanea, subsessilia, basi foliis parvis lanceolatis
2–3 suffulta, erecta, densa, cylindrica, 3–4 cm. longa, 1 cm. crassa, rhachi villoso
tomentella; ovaria sessilia, juvenilia crasse ovata, dein crasse ovato-oblonga, dense
sericeo-villosa, stylo mediocri satis tenui integro stigmatibus linearibus satis
longis apice bifidis subaequilongo coronata; bracteae late obovatae; ovaria juvenilia plus minusve amplectentes, dein ovariis triplo breviores, flavo-brunneae,
concolores, utrinque sericeae v. sericeo-villosae et ciliatae; glandula una ventralis
plana, oblongo-rectangularis, apice truncata v. emarginata, bracteis circiter 2–3-plo
brevior.   Capsulae griseo-sericeo-villosae, circiter 6 mm. longae, sessiles (v. subsessiles).
MONGOLIA: near Saisansk, along a watercourse in sandy soil, May 21, 1911,
F. N. Meyer (No. 733, type; ♀).
This species resembles *S. Lapponum* Linnaeus (*Spec.* 1019 [1753]) and especially
S. Krylovii Wolf (in *Act. Hort. Petrop.* XXVIII. 537 [1911].—*S. pseudolapponum*
Wolf apud Krylov, *Фл. Алтая*, 1226 [non Seemen] [1909]), to which I refer No. 836
of *F. N. Meyer*, Altai, Birel, June 16, 1911, and No. 793, Marka Kul, June 7, 1911.
In *S. Lapponum* Linnaeus, however, and in *S. Krylovii* Wolf, the ovaries are distinctly stalked, while they are sessile in our new species. *S. Lapponum* Linnaeus
differs also in the more or less acute bracts, and the mostly longer style as well as
in the glabrous brown branchlets. *S. Krylovii* Wolf has the same brown, somewhat
shining branchlets of the second year as *S. Lapponum* Linnaeus and broader, elliptical or oblong-obovate leaves, while in *S. polia* the leaves are lanceolate or oblanceolate and the branchlets are distinctly yellowish. As long as I do not know the
♂ flowers I am not sure to which section *S. Krylovii* Wolf belongs. *S. Lapponum* Linnaeus is placed by von Seemen (in Ascherson & Graebner, *Syn. Mitteleur.
Flora*, IV. 182 [1908]) in sect. *Viminales*, but I think it would be better to keep
this species in a distinct section. Andersson (in De Candolle, *Prodr.* XVI. 275
[1868]) placed *S. Lapponum* Linnaeus under sect. *Niveae* subsect. *Villosae* Andersson, the first species of which is *S. speciosa* Hooker & Arnott. There is no
S. villosa under the different forms united by Andersson in his subsection, which
to my opinion does not form a natural group.
The specific name is derived from πολιός, white of hair.

179. **Salix algista** Schneider, n. sp. v. hybr.
Frutex **v.** arbuscula, 0.9–4.5 m. alta, ut videtur satis breviter ramosa; ramuli
novelli sparse sericei, biennes glabri, purpureo-rubri, vetustiores cinerascentes.

Folia valde juvenilia lanceolata, elliptico-lanceolata, elliptica v. obovato-lanceo-lata, basi pleraque ut videtur subacuta, apice acuta v. subobtusa, supra viridia v. subnigrescentia, glabra v. basim versus ad costam sericea, subtus-pallidiora, ad costam sparse sericea v. subglabra, margine satis distincte crenato-serrata, nervis visibilibus angulo 45° a costa divergentibus, 1–2 cm. longa, 0.5–1 cm. lata; petioli nondum satis evoluti brevissime sericei. Amenta coetanea, brevipeduncu-lata, basi foliis parvis ceteris similibus 2–4 suffulta, cylindrica, densiflora, rhachi sericeo-villosa, ♂ 1.5–2.5 cm. (pedunculo excluso) longa, circiter 0.6–0.7 cm. crassa, ♀ tantum 1–1.5 cm. longa, circiter 5 mm. crassa; flores ♂ diandri, filamentis liberis glabris bracteis fere duplo longioribus, antheris ovalibus flavis; glandula in floribus juvenilibus una ventralis, ovato-rectangularis, obtusa, in floribus evolutis etiam glandula dorsalis multo minor adest; bracteae ovato-oblongae, subobtusae, glandula ventrali adulta 2½–3-plo longiores, initio apice fuscae, dein fere concolores, flavo-brunneae, sparse sericeae v. facie utrinque subglabrae; ovaria subsessilia, ovato-elongata, laxe sericea sed versus apicem glabra, in stylum distinctum apice bifidum attenuata, stigmatibus brevibus emarginatis; glandula una ventralis late ovata, apice truncata, saepe incisa, bracteis ½–⅔ brevior; bracteae ovato-oblongae, apice subacutae v. acutae, ovario sub-aequilongae, utrinque breviter sericeae, flavo-brunneae.

JAPAN. Hondo: prov. Suruga, slopes of Fuji-san, abundant above alt. 300 m., May 7, 1914, *E. H. Wilson* (No. 6637; type; bush or small tree, 0.9–4.5 m. tall, ♀; No. 6637ᵃ, ♂ co-type).

This is a curious form, looking somewhat like a hybrid between *S. Reinii* Franchet & Savatier and a species of sect. *Albae.* The old ♂ flowers have a very small dorsal gland. Unfortunately there are no mature leaves. The habit seems to be different from that of all the species of sect. *Albae* resembling *S. Reinii,* which is apparently a common plant at the type locality.

The specific name is derived from ἄλγιστος, very difficult.

180. **Salix ampherista** Schneider, n. sp.

Frutex?; ramuli hornotini tenuiter albo-sericeo-villosuli, vetustiores glabri, purpurascentes, leviter nitentes, satis tenues. Folia tenuia (ut videtur nondum satis evoluta) late elliptica, obovata v. obovato-elliptica, basi acuta et saepe angustata, apice breviter acuta, supra viridia, sparse villosula, subtus fere con-coloria v. paullo pallidiora, initio tomentella, mox subglabrescentia, sparse ad-presse breviter sericea pilis fulvis intermixtis, costa prominula distinctius pilosula, nervis lateralibus utrinque vix prominulis satis angustis sub angulo 70–50° a costa divergentibus, 3–5 pro 1 cm., margine integerrima, minora 2–3 cm. longa, 1–2 cm. lata, majora usque 4.2 cm. longa et 2.5 cm. lata v. ad 4.5 cm. longa et 2 cm. lata; petioli 2–4 mm. longi, villosuli. Amenta (tantum fructifera visa) subsessilia (ut videtur praecocia et initio squamis paucis suffulta) 3–6 cm. longa, circiter 1.2 cm. crassa, satis densa, rhachi sericeo-villosa; ovaria ovato-oblonga, breviter sericea, breviter pedicellata pedicellis glandula una ventrali ovato-rectangulari apice truncata compressa initio duplo brevioribus sed in fructu in-terdum aequilongis v. sublongioribus, stylo distincto pedicello subaequilongo integro, stigmatibus oblongis ut videtur integris stylo brevioribus coronata; brac-teae ovatae, obtusae, concavae, fuscae, glandulis junioribus subaequilongae dis-tincte et satis longe subtus sparsius pilis albis v. rarius paullo flavescentibus sericeae et ciliatae. Fructus pedicello incluso circiter 5 mm. longi, sparsius quam ovaria juniora sericei.

JAPAN. Hokkaido: prov. Oshima, Hakodate, June 1, 1904, *U. Faurie* (No. 5758 type; ♀; sub nom. *S. daisenensis* Seemen).

This species looks somewhat like *S. phylicifolia* Linnaeus and resembles also *S.*

daisenensis Seemen, although the latter, according to von Seemen's figure, has longer catkins which are more distinctly pedunculate; the leaves are rather coarsely denticulate, and the bracts and stigmas seem to be different.

The specific name is derived from ἀμφήριστος, doubtful.

181. **Salix melea** Schneider, n. sp.

Frutex?; ramuli novelli sparse villosuli, vetustiores purpureo-fusci, satis tenues, glabri. Folia (nondum satis evoluta) lanceolata v. pleraque oblanceolata, basim versus sensim attenuata, acuta, apice breviter acuta v. obtusiuscula, supra satis obscure viridia, costa tenuissime pilosula ceterum glabra, subtus cinerea, costa elevata flavescente plus minusve laxe sericea, rarius etiam facie sparse pilosa, ceterum glabra, nervis vix prominulis angulo 75–50° a costa divergentibus, circiter 3–4 pro 1 cm., margine integra, 1.5–5 cm. longa, 0.5–1.2 cm. lata; petioli breves, 2–5 mm. longi, sparse sericeo-villosi; stipulae minimae, lineares, caducissimae. Amenta (fructifera tantum vidi) fere sessilia, ut videtur initio foliolis parvis v. squamis paucis suffulta et praecocia, longe cylindrica, 5.5–7 cm. longa, circiter 1 cm. crassa, rhachi sericeo-villosa, densissima; flores inter fructus remanentes subsessiles; ovaria ovata, puberula, stylo pedicello subaequilongo integro, stigmatibus stylo subaequilongis oblongis ut videtur integris coronata; glandula una ventralis, pedicello paullo longior, oblonga obtusa; bracteae ½–⅔ ovarii aequantes, obovatae, brunneae, plus minusve albo-sericeae. Fructus tantum tenuissime pilosuli, circiter 4 mm. longi, oblongi, pedicello glandula vix aequilongo suffulti, bracteis pedicello duplo longioribus.

CHINA. Without locality or date, *W. Purdom* (fruiting type).

With its long fruiting aments and its entire oblanceolate leaves this species looks distinct, but somewhat resembles forms of *S. opaca* Andersson (see p. 159). The Willows of northern China are still very badly known and need further study.

The specific name is derived from μέλεος, useless.

182. **Salix etosia** Schneider. See p. 73.

183. **Salix Léveilléana** Schneider, n. sp.

Frutex v. arbor?; ramuli floriferi breviter sericeo-villosuli, olivaceo-brunnei, novelli ignoti; gemmae oblongae, adpressae, breviter pilosae. Folia ignota. Amenta (♀ tantum nota) praecocia, sessilia, basi squamis paucis extus sericeis suffulta, ad 5 cm. longa (an satis evoluta?) et circiter 1 cm. crassa, paulo curvata, sericea, densiflora, rhachi villosa; ovaria glabra ovata, in stylo subaequilongo satis crasso sensim producta, stigmatibus linearibus stylo 2–3-plo brevioribus adpressis bifidis coronata, pedicello distincto piloso glandulam duplo superante suffulta; glandula una ventralis, obtuso-rectangularis; bracteae oblongae, subobtusae, satis angustae, brunnescentes, longe sericeae, sed basi intus et apice extus interdum glabrae, pedicello duplo longiores. Fructus ignoti.

CHINA. Yunnan: "vallée de Kiao-mé-ti, ruisseau," alt. 3000 m., April 1913, *E. E. Maire* (type, ♀).

I have received a part of a branchlet of this interesting Willow from Mgr. Léveillé in honor of whom I have named this species. Not having seen the leaves and ♂ flowers, I cannot refer it to a section, but it is certainly different from all the numerous Willows I have seen from China and the other parts of our area.

SPECIES VARIETATESQUE SECUNDUM PLANTAS IN DITIONE NOSTRA COLLECTAS DESCRIPTAE SED DUBIAE.

Salix apiculata Andersson in *Svensk. Vetensk. Akad. Handl.* 1850, 470 (1851); *Jour. Linn. Soc.* IV. 42 (1860).

Salix cuspidata Wallich, *Cat.* No. 3703 A-B (nomen nudum) (non Schultz, nec D. Don) (1829). — Hooker f., *Fl. Brit. Ind.* V. 627 (1888).

INDIA. Madras presidency: Nilgherry, Dimhutty, *D. Noton* (Wallich's No. 3703[b]; sterile); without locality (Wallich's No. 3703[a], ex Herb. Wright, apparently ♀ type of *S. apiculata*); East Bengal?, Sonog?, December 1827, *Dr. Gomez* (Wallich's No. 9106; ex Andersson).

This is a very doubtful species and probably the same as *S. tetrasperma* Roxburgh. I have seen only a photograph of the type specimen and a drawing of a single leaf of No. 3703[b]. According to Hooker f. No. 3703[a] "is *S. tetrasperma* with long bracts and glabrous capsules," and it is from this specimen that Andersson must have drawn up his description. He cites as a synonym *S. cuspidata* D. Don (*Prodr. Fl. Nepal.* 59 [1825]), which was founded on a sterile specimen collected by Kamrup at Sirinagur (Kashmir). I doubt if the true *S. tetrasperma* occurs in this region.

Salix calophylla Wallich apud Andersson in *Svensk. Vetensk. Akad. Handl.* 1850, 502 (1857). — Hooker f., *Flor. Brit. Ind.* V. 628 (1888).

INDIA. Burma: banks of Attran, March 1827, *N. Wallich* (Cat. No. 9102; sterile type, ex Andersson).

This seems to be a large-leafed form of *S. tetrasperma* Roxburgh.

Salix calostachya Andersson in *Svensk. Vetensk. Akad. Handl.* 1850, 490 (1851); in *Jour. Linn. Soc.* IV. 42 (1860); in *Svensk. Vetensk. Akad. Handl.* VI. 5 (*Monogr. Salic.*) (1867); in De Candolle, *Prodr.* XVI. pt. 2, 194 (1868). — Hooker f., *Fl. Brit. Ind.* V. 637 (1888).

INDIA. Kashmir: "a Kahouta ad Mahabad, in sylvis humidis, alt. 2000 m." *V. Jacquemont* (No. 250; type ♀ ex Andersson).

This seems to be a form of *S. Wallichiana* Andersson with glabrescent ovaries. According to his journal (Voy. III. 168 [1841]) Jacquemont left Kohouta and camped at Ilahabad on May 2, 1831. Hooker writes the names Kahvata and Mahadeb.

Salix densa Wallich apud Andersson in *Svensk. Vetensk. Akad. Handl.* 1850, 502 (non Fries 1832) (1851). — Hooker f., *Fl. Brit. Ind.* V. 628 (1888).

INDIA. Burma: Martaban, March 1827, *N. Wallich* (Cat. No. 9103; sterile type ex Andersson).

This is a form with very long linear-oblong leaves. According to Hooker f., it is also probably referable to *S. tetrasperma* Roxburgh.

Salix fulcrata, var. **subphylicifolia** Andersson in De Candolle, *Prodr.* XVI. pt. 2, 244 (1868).

NORTHEASTERN ASIA. Without locality, *Stubendorff* (type, with fruits, ex Andersson).

This is a very doubtful form, not even mentioned by Herder (1891).

Salix futura Seemen, *Salic. Jap.* 71, t. 17, fig. F-G[1] (1903).

JAPAN. Without locality, *M. Shirai* (No. 47, type, ♀, ex von Seemen).

This species is described from a ♀ flowering branch only; the leaves, the ♂ flowers, and the original locality are not known.

Salix macilenta Andersson in *Svensk. Vetensk. Acad. Handl.* VI. 141, t. 7, fig. 75 (*Monog. Salic*) (1867); in De Candolle, *Prodr.* XVI. pt. 2, 245 (1868).

Salix phylicifolia Chamisso in *Linnaea*, VI. 542 (1831).

NORTHEASTERN ASIA. Kamtchatka: Redowski-land, *A. Chamisso* (♀ type ex Andersson).

Andersson says: "Est sine ullo dubio *S. phylicifoliae* v. *S. arbusculae* sat affinis." The fruits are glabrous, and the pedicels are scarcely as long as the gland. This little-known species is omitted by Herder (1891).

Salix mongolica Siuzev in *Trav. Mus. Bot. Acad. Sci. St. Pétersbourg*, IX. 90 (1912); in Fedde, *Rep. Spec. Nov.* XIII. 328 (1914). — Toepffer, *Salicol. Mitt.* No. 5, 248 (1912).

Salix mongolica, f. gracilior Siuzev in *Trav. Mus. Bot. Acad. Sci. St. Pétersbourg*, IX. 90 cum icone; in Fedde, *Rep. Spec. Nov.* XIII. 328 (1914). — Toepffer, *Salicol. Mitt.* No. 5, 248 (1912).

NORTHEASTERN ASIA. "Prov. Kirinensis, inter st. Jao-mên et p. U-dsiu-dsja, Sept. 1905 (fol.)."

By *S. mongolica* (Franch.) I suppose Siuzev means *S. phylicifolia* Franchet in *Nouv. Arch. Mus. Paris*, sér. 2. VII. 93 (Pl. David. I. 283) (1884), which is a very uncertain plant. Not having seen the original publication or any of the specimens, I can only mention this species.

Salix nobilis Fries, *Nov. Fl. Suec. Mant.* I. 78 (1832). — Andersson in *Svensk. Vetensk. Akad. Handl.* 1850, 492 (1851); VI. 4 (in textu sub *S. pyrina*) (1867). — Hooker f., *Fl. Brit. Ind.* V. 628 (1888).

Salix tetrasperma, var. *nobilis* Andersson in *Jour. Linn. Soc.* IV. 42 (1860).

INDIA. Nepal: without locality, *N. Wallich* (in Herb. Hornemann, ex Fries).

So far as I can judge from the description this species may be a mixture of two forms, the ♀ plant belonging to a species like *S. tetrasperma* Roxburgh and the ♀ plant to *S. daphnoides*, var. *indica* Andersson. The style is described as very long, but not as bifid, as Hooker f. says, and the stigmas as linear and cleft.

Salix (phylicaefolia?) macrocarpa Andersson in *Svensk. Vetensk. Akad. Handl.* 1850, 479 (1851).

INDIA. Garhwal, "in frigidis umbrosis et fertilibus a Bari ad Kounass, d. 14 Maji 1829," *V. Jacquemont* (No. 70*, type; ♀, ex Andersson).

Andersson never mentioned this doubtful Willow again, which apparently was collected by Jacquemont on May 14, 1830, between Bâri and Bounasse (see *Voy. Jour.* II. 79 [1841]). In addition to his description Andersson refers to "*S. glabrescens* Lindley in Oude et Rohilcund 1825; Wallich (*Cat.* n. 3706)," of which I have a photograph and some fruits and leaves before me. Wallich collected this number in the Kootuhonnaut Forest, March 6, 1825, and in the Rampoor Forest, February 27, 1825. According to the shape of the gland of the glabrous long-pedicelled capsules this Willow seems to belong to sect. *Tetraspermae*. But I have to confess, as Andersson did, that I cannot "certum de iis judicium pronuntiare."

Salix phylicoides, var. attenuata Andersson in *Svensk. Vetensk. Akad. Handl.* VI. 141 (Monog. Salic.) (1867); in De Candolle, *Prodr.* XII. pt. 2, 245 (1868). — Herder, in *Act. Hort. Petrop.* XI. 416 (1891).

NORTHEASTERN ASIA. Kamtchatka: Paratan, July 31, 1849, *Stubendorff* (No. 223, sterile type, ex Andersson and Herder); Petropavlowsk, *Kastalsky* (ex Herder).

Herder says that according to C. A. Meyer the type specimen belongs to *S. boganidensis* Trautvetter & Meyer (see p. 125). Andersson (1867) describes the leaves as "anguste lineari-lanceolatis 3–4 (in 1868, 3–7!) pollices longis vix ½ poll. latis." This is a very doubtful form.

Salix sitchensis Sanson, var. **ajanensis** Andersson in De Candolle, *Prodr.* XVI. pt. 2, 233 (1868).

Salix sitchensis Regel & Tiling in *Nouv. Mém. Soc. Nat. Mosc.* XI. 117) (*Fl. Ajan.* (non Sanson) (1858).

NORTHEASTERN ASIA. Yakutsk: Ajan, *H. Tiling* (No. 252; with fruits; ex Andersson).

See also Herder (in *Act. Hort. Petrop.* XI. 409 [1891]). The true *S. sitchensis* Sanson apud Bongard in *Mém. Acad. Sci. St. Pétersbourg*, sér. 6, II. 162 (*Végét. de Sitcha* 44) (1833) does not occur in Asia.

Salix speciosa Hooker & Arnott, var. **ajanensis** Andersson in De Candolle, *Prodr.* XVI. pt. 2, 275 (1868).

Salix Lapponum Regel & Tiling in *Nouv. Mém. Soc. Nat. Mosc.* XI. (*Fl. Ajan.* 118) (non Linnaeus) (1858).

NORTHEASTERN ASIA. Yakutsk: near Ajan, *H. Tiling* (with fruits).

Salix speciosa, var. **Trautvetteriana** Andersson in De Candolle, *Prodr.* XVI. pt. 2, 276 (1868). — Herder in *Act. Hort. Petrop.* XI. 431 (1891).

NORTHEASTERN ASIA. Transbaikalia: "in Dahuria," *N. Turczani now* (see Herder, l. c.; ♀ type).

Salix speciosa, var. **chrysea** Andersson in De Candolle, *Prodr.* XVI. pt. 2, 276 (1868).

NORTHEASTERN ASIA. Yakutsk: without locality, *Stubendorff* (♂ and ♀ types, ex Andersson).

It seems extremely doubtful if any of these forms belong to *S. speciosa* Hooker & Arnott (*Bot. Voy. Beechey*, 130 [non Host, 1828] [1832]), the correct name of which is *S. alaxensis* Coville (in *Proc. Wash. Acad. Sci.* II. 280 [1900]). — *S. speciosa*, *β alaxensis* Andersson in De Candolle, *Prodr.* XVI. pt. 2, 275 [1868]. — *S. longistylis* Rydberg in *Bull. N. York Bot. Gard.* II. 163 [1901]).

Salix subfragilis Andersson in *Mem. Am. Acad.* n. ser. VI. 450 (Gray, *Bot. Jap.*) (1859). — Franchet & Savatier, *Enum. Pl. Jap.* II. 502 (1879). — Seemen, *Salic. Jap.* 74 (1903).

JAPAN. Hokkaido: prov. Oshima, Hakodate, *S. W. Williams & J. Morrow* (♀ type, ex Andersson).

Unfortunately I have not been able to see the type specimen of this plant, and so far as I can judge from the rather short description it may be the same as *S. jessoensis* Seemen or *S. hondoensis* Koidzumi (see p. 110).

Salix viridula Andersson *Mem. Am. Acad.* n. ser. VI. 451 (Gray, *Bot. Jap.*) (1859). — Seemen, *Salic. Jap.* 74 (1903).

Salix padifolia, var. *viridula* Andersson in De Candolle, *Prodr.* XVI. pt. 2, 256 (1868).

JAPAN. Hondo: Yokohama, *S. W. Williams & J. Morrow* (♂ type, ex Andersson).

I have not seen the type specimen of this species, which according to the description may be *S. Reinii* Franchet & Savatier (see p. 127).

Salix Yoshinoi Koidzumi in *Tokyo Bot. Mag.* XXIX. 314 (1915).

JAPAN. Hondo: "prov. Bittsiu, Kawakamigori, Abe" (♀ type, ex Koidzumi).

Without having seen a specimen I am not able to determine the relationship of this Willow, which may belong to sect. *Albae* or to sect. *Helix*. Koidzumi says that the catkins resemble those of *S. hirosakensis* Koidzumi, which is the same as *S. hondoensis* Koidzumi (see p. 110).

JUGLANDACEAE.

Determined by ALFRED REHDER and E. H. WILSON.[1]

PLATYCARYA Sieb. and Zucc.

Platycarya strobilacea Siebold & Zuccarini in *Abh. Akad. Münch.* III. 742, t. 5, fig. 1 (1843). — Maximowicz in *Bull. Acad. Sci. St. Pétersbourg*, sér. 3, XVIII. 64 (1873); in *Mél. Biol.* VIII. 640 (1873). — Franchet in *Nouv. Arch. Mus. Paris*, sér. 2, VII. 92 (*Pl. David. I.* 282) (1884). — Skan in *Jour. Linn. Soc.* XXVI. 495 (1899). — Pritzel in *Bot. Jahrb.* XXIX. 273 (1900). — Pampanini in *Nuov. Giorn. Bot. Ital.* n. ser. XVII. 31 (1910). — Dunn & Tutcher in *Kew Bull. Misc. Inform.* add. ser. X. 251 (*Fl. Kwangtung & Hongk.*) (1912). — Léveillé, *Fl. Kouy-Tchéou*, 203 (1914).

Fortunaea chinensis Lindley in *Jour. Hort. Soc. Lond.* I. 150 (1846).

Kiangsi: Kuling, thickets, alt. 1300 m., July 31, 1907 (No. **1665**; bush 2.5–4 m.). Western Hupeh: north and south of Ichang, thickets and woods, alt. 30–1300 m., June and October 1907 (No. **495**; bush or small tree 3–12 m. tall, 0.3–1 m. girth); Ichang, glens, May 1900, Veitch Exped. (No. 451); Chienshi Hsien, *A. Henry* (No. 6014); "Monte di Ku-tchen," alt. 600 m., July 1906, *C. Silvestri* (No. 309); "lungo il fiume Han-Kiang," alt. 700 m., June 1907, *C. Silvestri* (No. 311); "Monte Si-ho," July 1909, *C. Silvestri* (No. 2923). Yunnan: Milê district, *A. Henry* (No. 9937). Chekiang: vicinity of Ningpo, 1908, *D. Macgregor*, near Hanchou, June 1907, *F. N. Meyer* (No. 427).

This is a common bush or small tree, more rarely a tree from 12–15 m. tall with a trunk 1 m. in girth but occasionally 20 m. tall and 2.5 m. in girth. The bark is thick, dark and deeply furrowed; the branches are moderately thick and spreading and form a rounded or flattened head.

In Hupeh the colloquial name is Huan-hsiang-shu and the fruit is used to prepare a black dye for cotton goods. Photographs of this tree will be found under Nos. 495, 0179, 0154 of the collection of Wilson's photographs and in his *Vegetation of Western China*, No. 389.

[1] Carya by C. S. Sargent.

PTEROCARYA Kunth.

Sect. EUPTEROCARYA Rehder and Wilson, n. sect.
Gemmae nudae, plerumque plures superpositae; amenta mascula
e gemmis nudis axillaribus infra apicem ramuli anni praeteriti orientia.

Pterocarya stenoptera C. De Candolle in *Ann. Sci. Nat.* sér. 4,
XVIII. 34 (1862); *Prodr.* XVI. pt. 2, 140 (1864). — Hance in *Jour.
Bot.* XI. 376 (1873). — Maximowicz in *Bull. Acad. Sci. St. Péters-
bourg*, sér. 3, XVIII. 64 (1873); (in *Mél. Biol.* VIII. 639 (1873). —
Lavallée, *Icon. Arb. Segrez.* 65, t. 19 (1885). — Skan in *Jour. Linn.
Soc.* XXVI. 494 (1899). — Pritzel in *Bot. Jahrb.* XXIX. 274 (1900).
— Dunn & Tutcher in *Kew Bull. Misc. Inform.* add. ser. X. 250 (*Fl.
Kwangtung & Hongk.*) (1912).

Pterocarya laevigata Hort. ex Lavallée, *Icon. Arb. Segrez.* 65 (pro synon.)
(1885).
Pterocarya chinensis Hort. ex Lavallée, l. c. (pro synon.) (1885).
Pterocarya japonica Hort. apud Dippel, *Handb. Laubholzk.* II. 329, fig. 151
(1892).
Pterocarya stenoptera, a typica Franchet in *Jour. de Bot.* XII. 317 (1898). —
Pampanini in *Nuov. Giorn. Bot. Ital.* n. ser. XVII. 249 (1910).
Pterocarya stenoptera, β kouitchensis Franchet in *Jour. de Bot.* XII. 318
(1898). — Léveillé, *Fl. Kouy-Tchéou*, 203 (1914).

Kiangsi: Kiukiang, side of streams, alt. 300 m., August 2, 1907
(No. 1660; tree 10–25 m. tall, girth 1–5 m.). Western Hupeh:
Ichang, 30–1000 m., April, June and October 1910 (No. 3214); same
locality, April 1900 (Veitch Exped. No. 117); Changyang Hsien, alt.
1000 m., December 1907 (No. 788; tree 30 m. tall, girth 6 m.); without locality, *A. Henry* (No. 1332); " On-kia-ki," April to May, *C.
Silvestri* (No. 316). Yunnan: Lunan, river-side, *A. Henry* (No.
10573). Chekiang: Ningpo, 1861, *R. Oldham*. Fokien: Dunn's
Exped., April to June 1905 (Hongkong Herb. No. 3863). Kwangtung: Canton, cultivated, 1872 (Herb. H. F. Hance, No. 17623).

This is one of the commonest trees on river-banks and on the stony and sandy
beds of summer torrents in Hupeh and Szech'uan, up to 1000 m. altitude. On the
Yangtsze River and its main tributaries it is the first tree to appear on newly
formed islands. It is a quick-growing tree, attaining the height of from 25–30 m.
and a girth of from 4–6 m. with massive spreading branches and thick, deeply
fissured gray bark. Young plants spring from the roots and frequently form
thickets on dry stony or sandy river-beds. The wood is soft, brittle and of no value
except for fuel. In Shanghai, Hankow and other cities it is commonly planted as
a street tree and by foreigners is called " Chinese Ash." Near Ichang, where this
tree is particularly abundant, it is colloquially known as the Liu-shu.

Pictures of this tree will be found under Nos. 430, 482, 487, 496, 497, 498, 598, 646 of the collection of Wilson's photographs and in his *Vegetation of Western China*, Nos. 413, 414, 415, 416, 417, 418.

The characters on which Franchet (l. c.) based his varieties are inconstant and of no taxonomic value, being dependent upon situation, vigor and the season of the year.

Pterocarya hupehensis Skan in *Jour. Linn. Soc.* XXVI. 493 (1899). — Pampanini in *Nuov. Giorn. Bot. Ital.* n. ser. XVII. 31 (1910).

Western Hupeh: north and south of Ichang, moist woods, alt. 1300–2000 m., June and October 1907 (No. 404; tree 5–16 m. tall, girth 0.6–2 m.); Fang Hsien, woods, alt. 1800–2000 m., November 1907, October 1910 (Nos. 639, 4429; tree 13–20 m., girth 1–2 m.); Changyang Hsien, woods, June 1901 (Veitch Exped. No. 905); same locality, A. *Henry* (No. 6158). Western Szech'uan: southeast of Tachien-lu, woodlands, alt. 2300–3000 m., October 1910 (No. 4141; tree 16–20 m. tall, 2–2.5 m. in girth).

This is a common tree in moist woods and by the side of streams in western Hupeh between 1200–2000 m. altitude. The bark is fibrous, smooth and pale gray, becoming deeply fissured on the trunk of old trees. The number of leaflets varies from 7 to 13.

In Hupeh a colloquial name for this tree is Shan-liu-shu.

The specimen from Szech'uan (No. 4141) is fragmentary and may possibly belong to *P. Delavayi* Franchet, which we have not seen.

Pterocarya Paliurus Batalin in *Act. Hort. Petrop.* XIII. 101 (1893). — Franchet in *Jour. de Bot.* XII. 318 (1898). — Skan in *Jour. Linn. Soc.* XXVI. 494 (1899). — Pritzel in *Bot. Jahrb.* XXIX. 274 (1900). — J. H. Veitch in *Jour. Hort. Soc. Lond.* XXVIII. 65, fig. 26 (1903).

Descriptioni adde:

Amenta mascula geminata v. terna, e gemmis lateralibus nudis 1–2 infra apicem rami anni praeteriti orta, 8–13 cm. longa, pendula, satis laxa; axis angulata, breviter villosula; bracteae oblongae, obtusae, 1–2 mm. longae, lepidotae, cito caducae; flores breviter pedicellati; perianthium erectus lepidotum; stamina 12–24 in seriebus 2–4 disposita.

Western Hupeh: north and south of Ichang, woods, alt. 1000–1600 m., June and October 1907 (No. 452; tree 13–20 m. tall, 1.5–2.5 m. girth); same locality, June 1900 (Veitch Exped. No. 901; flowers); Kui-chou, August 1900 (Veitch Exped. No. 901; fruit); Changyang Hsien, A. *Henry* (No. 7694). Eastern Szech'uan: "Tchen-kéou-tin," *R. P. Farges.* Western Szech'uan: Chiu-ting-shan, road-side, alt. 2150 m., May 23, October 1908 (No. 3213; tree 20 m. tall, 2 m. girth, only one seen).

This tree is fairly common scattered through the moist woods in western Hupeh, but is very rare in western Szech'uan. The bark is thick, gray, fibrous and deeply fissured on old trees. The leaflets vary from 7–9 and are subopposite; the fruit on some of the specimens is 7.5 cm. across.

In Hupeh the colloquial name for this tree is Chin-chien-li.

Franchet's section *Cycloptera* based on this species can hardly be maintained as a distinct section. The only difference is in the great development of the wings of the fruit, which become confluent so as to form one circular wing. Its origin from two wings, however, is apparent in many fruits which are distinctly contracted in the middle of the upper and of the lower margins. The other more important character attributed by Franchet to his section *Cycloptera* concerning the position of the staminate catkins does not exist in *P. Paliurus,* which has the staminate catkins arranged exactly like those in the two preceding species, as flowering specimens before us show. The flowering specimen on which Franchet based his description of the flowers of *Cycloptera* apparently belongs to the following species.

Sect. Chlaenopterocarya Rehder and Wilson, n. sect.

Gemmae perulis magnis imbricatis circiter 3 caducis praeditae, solitariae, manifeste stipitatae; amenta mascula ex axillis foliorum novellorum infra amentum femineum orientia.

Typus sectionis: *P. rhoifolia* Siebold & Zuccarini.

Pterocarya insignis Rehder & Wilson, n. spec.

Pterocarya Paliurus Franchet in *Jour. de Bot.* XII. 318 (non Batalin) (1898) quoad descriptionem florum et specimen floribus masculis e Tchen-kéou-tin.

Arbor 16–25-metralis, trunco 30–60 cm. diam.; trunci cortex cinereus, laevis demum fissus; ramuli lenticellati; gemmae perulis 3, oblongo-lanceolatis 2.5–3 cm. longis convolutis extus lepidotis et sparse pubescentibus praeditae. Folia 7–11-foliolata, petiolo incluso 20–45 cm. longa; foliola basin versus decrescentia, lateralia opposita v. subopposita, breviter petiolulata v. subsessilia, oblonga, acuta, basi oblique rotundata, 8–20 cm. longa et 3–6.5 cm. lata, terminale petiolulatum petiolo 1.5–2.5 cm. longo, oblongo-lanceolatum v. ovato-lanceolatum, acuminatum, basi attenuatum, 15–20 cm. longum et 5–8 cm. latum, omnia serrulata, supra glabrescentia, subtus initio floccoso-tomentosa glandulis minutis sparsis intermixtis, mox fere glabra costa nervisque fulvo-tomentosis exceptis, nervis utrinsecus circiter 20 subtus elevatis; petioli validi, 4.5–6 cm. longi, supra plani, basi dilatati; rhachis leviter marginata ut petioli petiolulique cinereo-tomentosi. Flores non visi. Racemi fructiferi foliis longiores, axi-glabri leviter angulata; fructus alis orbiculari-ovatis apice rotundatis 2 cm. longis et 2–2.5 cm. latis glandulis fuscis nitentibus conspersis.

Western Szech'uan: Wa-shan, alt. 1600 m., September 18, 1908 (No. 3212).

This is a very distinct species, differing in its covered winter-buds from all others known except the Japanese *P. rhoifolia* Siebold & Zuccarini. We have seen no staminate flowers, but we are convinced that the staminate aments described by Franchet under *P. Paliurus* belong here, as they agree in regard to their position exactly with those of *P. rhoifolia*, which is closely related to *P. insignis*. The Japanese species is well distinguished by its smaller more numerous leaflets markedly lepidote-glandular below, its sparsely pubescent axis of the fruit-catkin and by its much smaller wings (1.2 cm. long, 1.6 cm. wide) to the fruit.

This new *Pterocarya* is a rare tree and known to us from only one locality near the base of Mt. Wa. A picture of this tree will be found under No. 355 of the collection of Wilson's photographs and in his *Vegetation of Western China*, No. 412.

JUGLANS L.

Juglans regia Linnaeus, *Spec.* 997 (1753). — Loureiro, *Fl. Cochin.* 573 (1790). — Bunge in *Mém. Sav. Étr. Acad. Sci. St. Pétersbourg*, II. 136 (*Enum. Pl. Chin. Bor.* 62) (1835). — C. De Candolle in De Candolle, *Prodr.* XVI. pt. 2, 135 (1864). — Kurz in *Jour. Bot.* XI. 193 (1873). — Brandis, *Forest Fl. Brit. Ind.* 497 (1874); *Ind. Trees*, 619 (1906). — Hooker f., *Fl. Brit. Ind.* V. 595 (1888). — Skan in *Jour. Linn. Soc.* XXVI. 493 (1899).

> *Juglans regia*, var. *sinensis* C. De Candolle in *Ann. Sci. Nat.* sér. 4, XVIII. (1862); in De Candolle, *Prodr.* XVI. pt. 2, 135 (1864). — Maximowicz in *Bull. Acad. Sci. St. Pétersbourg*, ser. 3, XVIII. 57; in *Mél. Biol.* VIII. 630 (1873). — Franchet & Savatier, *Enum. Pl. Jap.* I. 455 (1875). — Debeaux in *Act. Soc. Linn. Bordeaux*, XXXIII. 63 (*Fl. Tientsien*, 40) (1879). — Skan in *Jour. Linn. Soc.* XXVI. 493 (1899). — Pritzel in *Bot. Jahrb.* XXIX. 274 (1900). — Shirasawa, *Icon. Ess. For. Jap.* II. t. 5, fig. 1–15 (1908). — Matsumura, *Ind. Pl. Jap.* II. pt. 2, 4 (1912).
>
> *Juglans japonica* Siebold apud Miquel in *Ann. Mus. Bot. Lugd.-Bat.* III. 103 (1867).
>
> *Juglans Duclouxiana* Dode in *Bull. Soc. Dendr. France*, 1906, 81, fig.; in Fedde, *Rep. Spec. Nov.* VII. 90 (1909).
>
> *Juglans Orientis* Dode l. c. 1906, 91, fig.; l. c. 92, (1909).
>
> *Juglans sinensis* Dode l. c. 1906, 92, fig.; l. c. 92 (1909).
>
> *Juglans sigillata* Dode l. c. 1906, 94, fig.; l. c. 93 (1909).

Western Hupeh: north and south of Ichang, cultivated, alt. 300–2000 m., May 7, August and October 1907 (No. 203; tree 10–20 m. tall, 1–3 m. girth); Changyang Hsien, cultivated, alt. 1300–1600 m., May and October 1907 (Nos. 653; tree 20 m. tall, girth 3 m.); same locality (Nos. 654, 654ᵃ; fruit only); Hsing-shan Hsien, cultivated, alt. 1300-2100 m., April, May, August and September 1907 (Nos. 389, 206; tree 10–16 m. tall, girth 1–2.5 m.); Fang Hsien, cultivated, alt. 1300–2000 m., May, October 1907 (No. 390; tree 13–20 m. tall, girth 2–3 m.); without locality, April 1900 (Veitch Exped. No. 330). Western Szech'uan: west and near Wên-ch'uan Hsien, cultivated, alt. 1300–2600 m., October 1908 (Nos. 1144, 1145; tree 13–25 m. tall, girth 2–5 m.); Tachien-lu, cultivated, alt. 1300–2600 m., October 1908 (No. 1145ᵃ; tree 13–26 m. tall, girth 2–5 m.). Yunnan: Mengtsze, cultivated, alt. 1600 m., A. Henry (No. 10507). Chili: near Changli, October 1905, F. N. Meyer (No. 198); north and west of Peking, mountains, 1913, F. N. Meyer (Nos. 35611, 35612, 35613, nuts only); purchased in market, Peking, September 16, 1903, C. S. Sargent (nuts only); Ming Tombs, north of Peking, October 6, 1905, J. G. Jack; hills near Great Wall, Peking-Kalgan Road, October 5, 1905, J. G. Jack.

The common Walnut is generally cultivated in Hupeh and Szech'uan; its nuts are valued for food, and the wood is used for making rifle-stocks. In Japan it is only sparingly cultivated. Wilson nowhere saw trees that could be declared spontaneous, and considers it highly improbable that Juglans regia is indigenous to China.

In addition to the specimens enumerated we have before us commercial samples of nuts from other parts of China, Mandshuria, and Gyantse in Thibet and others from central Asia, Persia and Europe. No two samples are exactly alike; they differ greatly in size, shape, sculpturing, and thickness of shell. The nuts purchased in the market at Peking by C. S. Sargent are nearly globose, 2–2.3 cm. in diameter and have a virtually smooth, thick shell; they agree very well with C. De Candolle's description of his var. sinensis. In Wilson's No. 1144, from western Szech'uan the nuts are 4–4.5 cm. in diameter and have a thin deeply pitted shell. Between these extremes there is every conceivable gradation and we find it utterly impossible to distinguish botanically any variety or form. The leaflets vary in number from 5 to 9 and are rounded, obtuse, acute or acuminate, but in their general appearance and character and in the characters of the shoot and fruit there can be no mistaking the common Walnut for any other species.

The characters on which Dode (l. c.) relies to distinguish many of his species of Juglans may be of interest to pomologists, but they have absolutely no specific value.

Pictures of this tree will be found under Nos. 146, 536, 539, 546 of the collection of Wilson's photographs and also in his Vegetation of Western China, Nos. 260, 261, 262.

Juglans cathayensis Dode in Bull. Soc. Dendr. France, 1909, 47, fig.; in Fedde, Rep. Spec. Nov. X. 298 (1911). — Wilson in Gard. Chron. ser 3, L. 189, fig. 88, t. (1911). — Bean, Trees & Shrubs Brit. Isl., I. 664 (1914).

Juglans mandshurica Skan in Jour. Linn. Soc. XXVI. 493 (pro parte, non Maximowicz) (1899). — Pritzel in Bot. Jahrb. XXIX. 274 (1900).

Juglans Sieboldiana Pritzel in *Bot. Jahrb.* XXIX. 274 (non Maximowicz) (1900). — Pampanini in *Nuov. Giorn. Bot. Ital.* n. ser. XVII. 249 (1910). *Juglans Draconis* Dode in *Bull. Soc. Dendr. France*, 1909, 49, fig.; in Fedde, *Rep. Spec. Nov.*, X. 298 (1911).

Western Hupeh: Hsing-shan Hsien, thickets and woods, alt. 600–2000 m., May and October 1907 (No. 371; bush or tree 2–12 m. tall); same locality, June 3, 1907 (No. 371ᵃ, bush 1–6 m. tall); Fang Hsien, woods, alt. 2000 m., May 20, 1907 (No. 371ᵇ; tree 12 m. tall, girth 1 m.); without locality, *A. Henry* (Nos. 3834, 5233ᵃ). Western Szech'uan: Wa-shan, woodlands, alt. 1300–2000 m., June and September 1908 (No. 371ᶜ; bush or tree 2–13 m. tall); Hungya Hsien, Wa-wu-shan, thickets, alt. 1300–2000 m., September 14, 1908 (No. 371ᵈ; bush or tree 2–13 m. tall); Mupin, woodlands, alt. 1000–2000 m., June 1908 (No. 371ᵉ); west and near Wên-ch'uan Hsien, woodlands, alt. 1600–2000 m., May 1908, October 1910 (Nos. 371ᶠ, 371ᵍ; tree 16 m. tall, girth 2 m.); Mt. Omei, alt. 600–2000 m., June and October 1903, Veitch Exped. (No. 4500). Yunnan: Mengtsze, woods, alt. 1800–2000 m., *A. Henry* (Nos. 10498, 10498ᵃ). Kiangsu: mountains near Nanking, June 6, 1915, *F. N. Meyer* (No. 1453).

This is the Chinese Butternut and it is abundant throughout Hupeh and Szech'uan. Usually it is a bush, but in moist woods it forms a tree from 12–15 m. tall, with a trunk from 1–1.5 m. in girth. The bark is smooth and pale gray, becoming fissured with age. The leaves vary in length from 45–90 cm.; the male catkins are from 20–30 cm. long and the fruit is very glandular.

We have not seen *Juglans formosana* Hayata (in *Jour. Coll. Sci. Tokyo*, XXX. art. 1, 283 [1911]), but from the description we suspect that it belongs here.

Pictures of this tree will be found under Nos. 344, 370 of the collection of Wilson's photographs and in his *Vegetation of Western China*, Nos. 258, 259.

Here may be added notes on two genera not collected during the Arnold Arboretum expeditions.

ENGELHARDTIA Leschen.

Engelhardtia chrysolepis Hance in *Ann. Sci. Nat.* sér. 4, XV. 227 (1861); in *Jour. Linn. Soc.* XIII. 124 (1873).

Engelhardtia Wallichiana Lindley in Wallich, *Cat.* No. 4942 (nomen nudum) (1830). — C. De Candolle, *Prodr.* XVI. pt. 2, 141 (1864). — Skan in *Jour. Linn. Soc.* XXVI. 495 (1899). — Dunn & Tutcher in *Kew Bull. Misc. Inform.* add. ser. X. 251 (*Fl. Kwangtung & Hongk.*) (1912). *Engelhardtia pterococca* Roxburgh, *Fl. Ind.* III. 631 (pro parte) (1832). *Engelhardtia Wallichiana*, var. *chrysolepis* C. De Candolle, *Prodr.* XVI. pt. 2, 142 (1864). — Skan in *Jour. Linn. Soc.* XXVI. 495 (1899).

Western Szech'uan: Kiating Fu, alt. 400 m., August 1903 (Veitch Exped. No. 4487). Yunnan: Szemao, fruits, alt. 1300 m., *A. Henry* (No. 13049). A very rare tree in western Szech'uan.

CARYA.

Determined by C. S. Sargent.

Carya cathayensis Sargent, n. sp.

Arbor 12–20-metralis, trunco 30–60 cm. diam.; cortex laevigatus, griseus; ramuli tenues, novelli squamis luteo-aurantiacis obtecti, demum puberuli et sparse lepidoti, annotini purpureo-grisei, versus apicem puberuli, lenticellis parvis instructi. Folia 5–7-foliolata, petiolo incluso 20–30 cm. longa; petioli et rhaches juniora puberula, sparse squamis obtecta; foliola breviter petiolulata, lanceolata v. obovato-lanceolata, interdum leviter falcata, acuminata, basi attenuata, cuneata v. rotundata, simpliciter serrata, margine ciliata, 10–14 cm. longa, 3.5–5 cm. lata, juniora subtus dense squamis luteo-aurantiacis vestita, demum sparse lepidota, costa media pilosa excepta glabra, pallide luteo-aurantiaca. Flores ignoti. Fructus juvenilis obovoideus, basi constrictus, 4-costatus costis alatis, dense squamis luteo-aurantiacis obtectus; exocarpium maturum 2.5–3 mm. crassum, sparse lepidotum; nux ovoidea v. ovalis, subcylindrica, leviter compressa, basi rotundata, apice acuta, mucronulata, obscure 4-angulata, 2–2.5 cm. longa, 1.5–2 cm. diam.; endocarpium 2.5–3 mm. crassum; cotyledones apice profunde lobatae.

Chekiang: mountains round Changhua Hsien, 70 miles west of Hangchou, at an elevation of about 130–400 m. above the sea-level, July 8–12, 1915 (with old nuts picked up from the ground), (near Yi-tsun, No. 1521) F. N. Meyer.

Without buds it is impossible to be sure of the section of the genus to which this species belongs, but in spite of the few leaflets, the comparatively thick husk of the fruit, the thick shell of the nut and the deeply lobed cotyledons, I believe that it belongs to the *Apocarya* C. De Candolle which is distinguished by the valvate scales of the bud, and that it is most closely related to *C. myristiciformis* Nuttall. This species, unlike the other American species of this section, has usually 7 to 9 but occasionally 5 leaflets. Like *C. cathayensis*, it has obovoid conspicuously winged young fruit covered with yellow scales, a thick-shelled elliptic nut and deeply lobed cotyledons. That is, in these two species are found characters which connect *Eucarya* and *Apocarya*, which without these intermediate forms might be considered distinct genera.

Since the finding in China of a species of *Liriodendron* and of *Sassafras*, previously believed to be monotypic genera of eastern North America, no addition to our knowledge of the distribution of the trees of the northern hemisphere is so important and interesting as Mr. Meyer's discovery of a representative of the genus *Carya* in Asia, for it has always been supposed that this genus was confined to eastern America where it is represented by several widely distributed and common trees.[1]

Meyer's attention was first drawn to this tree by finding the nuts offered for sale on a fruit-stand in the city of Hanchou. In writing of his discovery he says, "The upper surface of the leaves is of a soft green color, while the under surface is rusty brown. When the wind blows a group of these Hickories presents a remarkably reddish brown blotch of color in the midst of the ordinary green vegetation. The nuts are collected for sale; they are eaten as a sweetmeat, and a fine clear yellow oil is extracted from them and used in fancy pastry. The wood is tough and strong and is used for tool-handles. The trees are more or less pro-

---

[1] *Carya sinensis* Dode (in *Bull. Soc. Dendr. France*, 1912, 39, fig.) is certainly not a *Carya* and, judging by the figure of the fruit, is probably *Aleurites triloba* Forster, the Candle-nut tree of southern China, often cultivated in tropical countries for the oil contained in the cotyledons.

tected; that is, when firewood is needed they are generally spared. In a few places mountaineers have even made small plantations near their houses of the Hickory and of a red-fruited Cornus. The Hickory trees thrive best at the foot of the mountains and in narrow moist valleys where they grow in deep rich humus. They love shelter and when exposed to much wind they become crippled. Apparently they cannot stand much frost. They grow with *Liquidambar formosana, Castanea mollissima, Diospyros kaki, Pistacia chinensis, Albizzia* species, *Aleurites Fordii* and *Pinus Massoniana.* The colloquial name for this tree is *Shan-gho-to.*"

In China there are many endemic trees, but in eastern North America, now that *Carya* has been found in China, *Sabal, Platanus* (found also in Europe and western Asia), *Robinia, Oxydendrum, Arbutus* (found also in Europe and western Asia), *Kalmia, Pinckneya, Taxodium* and *Chamaecyparis* are the only genera of eastern American extratropical trees which are not also represented in eastern continental Asia. In China, however, *Taxodium* appears in the elosely related *Glyptostrobus* which some botanists consider a *Taxodium,* and *Chamaecyparis,* although it has not yet been found in China, occurs in Japan and Formosa.

MYRICACEAE.

Determined by E. H. WILSON.

MYRICA L.

Myrica rubra Siebold & Zuccarini in *Abh. Akad. Münch.* IV. pt. 3, 230 (*Fl. Jap. Fam. Nat.* II. 106) (1846). — Bentham in *Hooker's Jour. Bot. & Kew Gard. Misc.* VI. 115 (1854); *Fl. Hongk.* 322 (1861). — Franchet & Savatier, *Enum. Pl. Jap.* I. 454 (1875). — Skan in *Jour. Linn. Soc.* XXVI. 496 (pro parte) (1899). — Shirasawa, *Icon. Ess. For. Jap.* II. t. 6, fig. 12–23 (1908). — Dunn & Tutcher in *Kew Bull. Misc. Inform.* add. ser. X. 251 (*Fl. Kwangtung & Hongk.*) (1912). — Léveillé, *Fl. Kouy-Tchéou*, 281 (1914).

> *Morella rubra* Loureiro, *Fl. Cochin.* 548 (1790).
> *Myrica Nagi* C. De Candolle, *Prodr.* XVI. pt. 2, 151 (non Thunberg [1]) (1864). — Miquel in *Ann. Mus. Lugd.-Bat.* III. 129 (1867); *Prol. Fl. Jap.* 293 (1867). — Hooker f. in *Bot. Mag.* XCIV. t. 5725 (1868). — Hance in *Jour. Linn. Soc.* XIII. 124 (1873). — Chevalier in *Mém. Soc. Sci. Nat. Cherbourg*, XXXII. 198 (*Monog. Myric.* 114) (1901).

Kiangsi: Kiukiang, foothills, alt. 300 m., August 1, 1907 (No. 1652; bush 3 m. tall). Chekiang: "hills near Tangsi," March 1906, *F. N. Meyer* (No. 228). Fokien: Dunn's Exped., April to June 1905 (Hongk. Herb. 3462). Korean Archipelago: Quelpaert, July 1907, *U. Faurie* (No. 1541); same locality, April 1908, 1911, May 1908, July 1910, Taquet (Nos. 4703, 4704, 1402, 4821, 4426).

This is a rare plant in Kiangsi; it is not recorded from farther west. On the island of Yakushima off the extreme south of Japan it is an extremely common fluviatile bush or small tree. In eastern China where it is much cultivated it is known as the Yangmae. In Japan it is known as Yama-momo and under this name it is figured in Banks, *Icon. Select. Kaempf.* t. 37 (1791). In southeastern China the genus is represented by *Myrica esculenta* Hamilton (apud D. Don, *Prodr. Fl. Nepal.* 56 [1825]. — Chevalier in *Mém. Soc. Sci. Nat. Cherbourg*, XXXII. 204 [*Monog. Myric.* 120] [1901–02]). Yunnan: Mengtsze, mountains, alt. 1600–2300 m., *A. Henry* (Nos. 9015, 9015ᵃ, 9015ᶜ, 9015ᵈ; tree 3–5 m. tall); Szemao, mountains, alt. 1600 m., *A. Henry* (No. 9015ᵉ). The villose shoots and petioles, the thinner leaves with prominent veins and the much larger and panicled male inflorescence readily distinguish this species from *M. rubra*.

[1] The *Myrica Nagi* Thunberg (*Fl. Jap.* 76 [1784]) is *Podocarpus nagi* Makino.

FAGACEAE.

Determined by ALFRED REHDER and E. H. WILSON.

FAGUS L.

Fagus longipetiolata Seemen in *Bot. Jahrb.* XXIII. Beibl. No. 57, 56 (1897).

Fagus sylvatica, var. *longipes* Oliver in *Hooker's Icon.* XX. t. 1936, in textu (1890). — Franchet in *Jour. de Bot.* 201 (1899). — Skan in *Jour. Linn. Soc.* XXVI. 525 (1899), pro parte, exclud. specim. Henryi Nos. 6793, 6797.
Fagus sinensis Oliver in *Hooker's Icon.* t. 1936 (in tabula tantum) (1890). — Diels in *Bot. Jahrb.* XXIX. 284, fig. j–k (1900). — Koidzumi in *Tokyo Bot. Mag.* XXX. 95 (1916).
Fagus longipes Léveillé, *Fl. Kouy-tchéou,* 126 (1914).

Western Hupeh: Patung Hsien, *A. Henry* (Nos. 5334, 5334ᵃ, co-types, 7444); same locality, woods, alt. 1000–1600 m., April, June and November 1907 (Nos. **493, 700ᵃ**; tree 10–16 m. tall, girth 0.6–1 m.); same locality, April 1900 (Veitch Exped. No. 537); Changyang Hsien, woods, alt. 1600–2100 m., June 1907 (No. **700ᵇ**; tree 6–16 m. tall, girth 0.6–1.3 m.); same locality, June and October 1900 (Veitch Exped. No. 608).   Eastern Szech'uan: Wushan Hsien, south of Yangtsze River, woods, alt. 1000–1600 m., August and October (Nos. **700, 460**; tree 10–16 m. tall, girth 1.3–2 m.).   Western Szech'uan: Mupin, woods, alt. 1300–2000 m., October 1908 (No. **2791**; tree 26 m. tall, girth 4 m.); same locality, July 1903 (Veitch Exped. No. 4499); west of Kuan Hsien ascent of Niu-tou-shan, alt. 1600 m., June 19, 1908 (No. **2791ᵃ**; tree 13 m. tall, girth 1.6 m.).   Yunnan: Mengtsze, mountains to southeast, alt. 2600 m., *A. Henry* (No. 9027; tree 16 m. tall).

Although nowhere very abundant, this is the common Beech of central and western China.   In southwestern Hupeh it occasionally forms pure woods of no great extent, but more usually it grows in association with Oaks, Maples and other deciduous leaved trees.   In western Szech'uan, Wilson noted the largest trees, but observes that it is rare.   Usually it is rather small, but at its best it is a stately tree which in habit and appearance resembles the Japanese *F. Sieboldii* Endlicher.   The trunk is single or very rarely divided near the base and is covered with smooth, very pale gray bark.   The leaves equal in size those of the American Beech (*F. grandifolia*

Ehrhart), are slightly glaucous on the under side and are irregularly, often coarsely toothed, and the base is usually abruptly and broadly cuneate; the petioles and peduncles vary somewhat in length, but are relatively long.

A picture of this tree will be found under No. 514 of the collection of Wilson's photographs.

Fagus Engleriana Seemen apud Diels in *Bot. Jahrb.* XXIX. 285, fig. a–d (1900).

> *Fagus sylvatica*, var. bracteolis involucri exterioribus anguste spatulatim dilatatis Oliver in *Hooker's Icon.* XX. sub. t. 1936 (1890).
>
> *Fagus silvatica*, var. *chinensis* Franchet in *Jour. de Bot.* XIII. 201 (1899).
>
> *Fagus silvatica* Léveillé, *Fl. Kouy-tchéou*, 126 (non Linnaeus) (1914).

Western Hupeh: Fang Hsien, woods, alt. 1500–2300 m., August 1907 (No. **704**; tree 6–16 m. tall, girth 1–2 m.); same locality, *A. Henry* (No. 6797, co-type); Hsing-shan Hsien, forming pure woods, alt. 1600–2500 m., May 31, 1907 (No. **703**; tree 10–23 m. tall, girth 0.3–2 m.); Chienshi Hsien, woods, April and May 1900 (Veitch Exped. Nos. 256, 447). Eastern Szech'uan, Wushan Hsien, south of Yangtsze River, woods 1600–2000 m., May 1900 (Veitch Exped. No. 747); Nan-ch'uan, *A. von Rosthorn* (No. 1525, co-type).

This Beech is common on the higher mountains of northwestern Hupeh and eastern Szech'uan, where it often forms pure woods. It is much more rare in southwestern Hupeh, and Wilson did not meet with it in western Szech'uan. It is a tree of moderate height and almost invariably the trunk divides at the very base into many stems which diverge somewhat from one another and never attain any great thickness. In the Japanese *F. japonica* Maximowicz the trunk divides in a similar manner, but into fewer stems, which grow to a greater size. In both species the leaves are sinuous on the margins, dark green above and somewhat glaucous on the under side, where the venation is prominently netted. The fruits are totally different, but these two species are undoubtedly closely related.

Fagus lucida Rehder & Wilson, n. sp.

Arbor 6–25-metralis, trunco 1–3 m. in circuitu, semper indiviso, cortice laevi obscure cinereo obtecto, ramis validis patentibus; ramuli arborum juniorum tenues, penduli, initio adpresse sericeo-pilosi, mox glabri. Folia breviter petiolata, ovata v. ovato-lanceolata, acuminata, basi rotundata v. saepe subcordata, rarius late cuneata, margine leviter, interdum obsolete sinuata, in sinubus dentibus brevibus v. brevissimis triangularibus instructo inter dentibus convexo, 5–10 cm. longa et 2–4.5 cm. lata, glabra pilis longis sericeis subtus ad costam et venas exceptis, utrinque lucida, supra intense viridia, subtus pallidiora et tenuiter reticulata, nervis utrinsecus circiter 10 rectis in dentes exeuntibus subtus ut costa manifeste elevatis; petioli 3–12, plerique 5–8 mm. longi, adpresse sericeo-pilosi. Flores et fructus ignoti.

Western Hupeh: Hsing-shan Hsien, woods, alt. 1600–2600 m.,
May 31, 1907 (No. **715**, type); Fang Hsien, *A. Henry* (No. 6793).

It is unfortunate that our material of this interesting tree is so incomplete, yet
it is so distinct from all known species that we do not hesitate to give it a name.
The leaves are shining green on both surfaces, even when quite young, and the dentation is remarkable. The margin is sinuate (sometimes obscurely so) and is
convex between the secondary veins, which are usually projected from the base of
the sinus and form short triangular teeth of irregular size often reduced to a small
and thickened mucro. Wilson paid considerable attention to the Chinese Beeches
and states that this new species is easily distinguished in the woods by its thick
and always single trunk and by its bark, which is a much duller gray than that
of the other two species. The branches are thick and spread to form a broad
flattened or rounded crown. In northwestern Hupeh and in northeastern Szech'-
uan this new species is fairly common in mixed forests and with *F. Engleriana*
Seemen often forms pure woods. Henry's No. 6793 is by Skan (in *Jour. Linn. Soc.*
XXVI. 527 [1899]) doubtfully referred to Oliver's *F. sylvatica*, var., which is the
same as *F. Engleriana* Seemen.

Plants of this and the other two Chinese species brought home by Wilson in 1911
are growing in the Arnold Arboretum.

A picture of this new Beech will be found under No. 0132 of the collection of
Wilson's photographs.

CASTANEA Mill.

Castanea mollissima Blume, *Mus. Bot. Lugd.-Bat.* I. 286 (1850). —
Seemen in *Bot. Jahrb.* XXIX. 288 (1900). — Rehder in Bailey, *Stand.
Cycl. Hort.* II. 682 (1914). — Nakai in *Toyko Bot. Mag.* XXIX. 54
(1915).

Castanea vesca Bunge in *Mém. Sav. Étr. Acad. Sci. St. Pétersbourg*, II. 137
(*Enum. Pl. Chin. Bor.* 62) (non Gaertner) (1833).
Castanea Bungeana Blume, *Mus. Bot. Lugd.-Bat.* I. 284 (1850). — Nakai in
Tokyo Bot. Mag. XXIX. 54 (1915).
Castanea vulgaris Hance in *Jour. Bot.* X. 69 (non Lamarck) (1872). — Debeaux
in *Act. Soc. Linn. Bordeaux*, XXXI. 363 (*Fl. Tché-foû*, 130) (1876), quoad
plantam Chinensem; l. c. XXXIII. 64 (*Fl. Tien-tsin*, 41) (1879), quoad plan-
tam Chinensem.
Castanea vulgaris, var. *yunnanensis* Franchet in *Jour. de Bot.* XIII. 196 (1899).
Castanea sativa Skan in *Jour. Linn. Soc.* XXVI. 525 (pro parte, non Miller)
(1899). — Léveillé, *Fl. Kouy-tchéou*, 125 (1914).
Castanea sativa, a *typica* Seemen in *Bot. Jahrb.* XXIX. 287 (1900).
Castanea Duclouxii Dode in *Bull. Soc. Dendr. France*, 1908, 150; in Fedde,
Rep. Spec. Nov. X. 239 (1911). — Schneider, *Ill. Handb. Laubholzk.* II. 899
(1912). — Koidzumi in *Tokyo Bot. Mag.* XXX. 99 (1916).
Castanea hupehensis Dode in *Bull. Soc. Dendr. France*, 1908, 151, fig. (1908);
in Fedde, *Rep. Spec. Nov.* X. 240 (1911). — Schneider, *Ill. Handb. Laubholzk.* II. 899, fig. 563 c–d (1912). — Koidzumi in *Tokyo Bot. Mag.* XXX.
99 (1916).

Castanea crenata Henry in Elwes & Henry, *Trees Great Brit. & Irel.*, IV. 854
(pro parte, non Siebold & Zuccarini) (1909).
Castanea sativa, var. *mollissima* Pampanini in *Nuov. Giorn. Bot. Ital.* n. ser.
XVII. 250 (1910).

Western Hupeh: north and south of Ichang, woods, May, June
and October 1907 (Nos. 550, 507ᵃ; bush or tree 0.6–16 m. tall, girth
1–2.3 m.); same locality, May 1900 (Veitch Exped. No. 812); same
locality, *A. Henry* (No. 1570); Changyang Hsien, woods, alt. 1300–
1600 m., May and October 1907 (No. 549; tree 6–16 m. tall, girth 1–2.5
m.); Patung Hsien, woods, alt. 1000–2000 m., June, July and October
1907 (Nos. 344ᵃ, 3663, 548; tree 6–16 m. tall, girth 1–2.5 m.); Hsing-
shan Hsien, woods, alt. 600–2500 m., June, July, October 1907 (Nos.
344, 3615, 3662, 3661, 3668; tree 6–20 m. tall, girth 0.6–2.5 m.); Fang
Hsien, woods, alt. 1300–1600 m., August and October 1907 (No. 547;
tree 6–13 m. tall). Western Szech'uan: southeast of Tachien-lu,
woods, alt. 1300 m., October 1908 (No. 1141ᵃ, tree 6–12 m. tall);
Chienshi Hsien, Fei-yüeh-ling, woods, alt. 2600–3000 m., October
1910 (No. 4581; tree 6–10 m. tall); Wa-shan woods, alt. 1000–1600 m.,
October 1908 (No. 1141; tree 6–10 m. tall); Nan-ch'uan, *A. von Rosthorn* (No. 1513). Yunnan: Mengtsze, mountains south, alt. 1600
m., *A. Henry* (No. 10701; tree 5 m. tall, cultivated?). Chekiang:
vicinity of Ningpo, 1908, *D. Macgregor;* near Hanchou, June 26, 1915,
F. N. Meyer (No. 1465; tree 10–26 m. tall); Mokan-shan, alt. 550 m.,
July 21, 1915, *F. N. Meyer* (No. 1589; shrub 0.6–1.3 m.). Kiangsu:
Shanghai, 1915, *D. Macgregor* (No. 44). Kianghuai: Nanking, June
1, 1915, *F. N. Meyer* (No. 1409; tree 13 m. tall). Chili: Peking,
1903, *C. S. Sargent* (seeds only); "near San-tun-ying," May 30, 1913,
F. N. Meyer; "Ying-tan-ho," cultivated, September 12, 1913, *F. N.
Meyer* (No. 1288). Shantung: Tsing-tau, 1901, *R. Zimmermann*
(No. 352). Shensi: "Ya-tze-ko, southwest of Sian Fu," cultivated,
September 2, 1914, *F. N. Meyer* (No. 1400ᵃ). Fokien: Dunn's
Exped., April to June 1905 (Hongkong Herb., No. 3501).

INDIA. Sikkim: Mungpoo, cultivated, May 1895, *J. A. Gammie* (as
"Yunnan Chestnut ").

This is the common Chestnut of China and is distributed from the neighborhood
of Peking in the northeast to the extreme limits of Szech'uan and Yunnan in the
west and southwest. Near Peking it is cultivated and we do not know if it is wild
in that region; the specimens collected in Sikkim by Gammie are also from cultivated trees. In western Hupeh and in Szech'uan it is an exceedingly common tree
and is unquestionably indigenous. Near villages and towns where the woody vegetation is continually cut down to furnish fuel this Chestnut is met with as a bush
or a low scrub, but in the thinly populated areas it is a tree from 15 to 20 m. tall,

with a trunk from 1.5 to 2 m. in girth and is an important constituent of the woods and forests. It is readily distinguished from the other old world species of the section *Eucastanon* by the complete absence of lepidote glands on the under surface of the leaves, which on the same tree are glabrous or nearly so or are densely covered with a nearly white felt of stellate hairs. In its typical form the branchlet is sparsely or densely pilose, and this is especially the case on vigorous shoots and on young trees, but with a little searching such hairs can be found on every tree. As a rule, however, on adult trees the branchlets are densely or sparsely clothed with a short, gray velutinous pubescence only. The winter-buds are short and broadly ovoid and pubescent. The leaves are very variable in size, the dentation is very irregular, and the teeth are broad and triangular and terminate in a long aristate point or they are reduced to a short mucro. The male aments are variable in length, but are usually shorter than the leaves; the fruit is also variable in size and there are forms in which the nut is as large as that of the best forms of the European Chestnut (*C. sativa* Miller). The involucral spines at maturity are pale straw color and are densely clothed with appressed villose pubescence.

This Chinese Chestnut has been confused with *C. sativa* Miller, which grows in the Mediterranean region and eastward to the Caucasus and in northern Persia, and is distinguished by the presence in fewer or greater numbers of minute lepidote glands scattered over the under surface of the leaves and occasionally confined to the primary and secondary veins and tissues nearby; the shoot too is different, being nearly glabrous, and clothed only with a scurvy puberulous indumentum; the male aments are often longer than the leaves, though this character has little or no significance.

Castanea mollissima was first introduced to cultivation at this Arboretum by Professor Sargent, who in 1903 sent seeds which he purchased in a market at Peking. The plants raised from these seeds have proved hardy here and have grown well. They are now quite established and have produced fruit. They have shown no signs of the dreaded chestnut-bark disease. But in China this species is subject to this disease, for in a note on his No. 1409 from Nanking, Meyer says, " the tree was attacked by the bark disease *Diaportha parasitica* Murrill."

In regard to *C. Bungeana* Blume there can be no question that it is identical with his *C. mollissima*. Seemen first took up this latter name, and as priority of position has no standing under the international rules, *C. Bungeana* Blume becomes a synonym. Hayata's *C. sativa*, var. *formosana* (in *Jour. Coll. Sci. Tokyo*, XXX. 304 [1911]), of which we have seen no specimen, in all probability is referable to *C. mollissima* Blume.

The Chinese name for this Chestnut is Pan-li, and the nuts are a valued article of food.

Pictures of this tree will be found under Nos. 506, 540 of the collection of Wilson's photographs, and also in his *Vegetation of Western China*, Nos. 146, 147.

Castanea Seguinii Dode in *Bull. Soc. Dendr. France*, 1908, 152, fig.; in Fedde, *Rep. Spec. Nov.* X. 240 (1911). — Schneider, *Ill. Handb. Laubholzk.* II. 899 (1912). — Wilson, *Naturalist West. China*, II. 32 (1913). — Koidzumi in *Tokyo Bot. Mag.* XXX. 100 (1916).

> *Castanea vulgaris*, var. *Japonica* Hance in *Jour. Bot.* XII. 262 (non A. De Candolle) (1874). — Franchet in *Nouv. Arch. Mus. Paris*, sér. 2, VII. 87 (*Pl. David.* I. 277) (1884).
>
> *Castanea sativa*, var. *japonica* Seemen in *Bot. Jahrb.* XXIX. 287 (1900).
>
> *Castanea Davidii* Dode in *Bull. Soc. Dendr. France*, 1908, 153, fig.; in Fedde,

Rep. Spec. Nov. X. 241 (1911). — Schneider, *Ill. Handb. Laubholzk.* II. 899, fig. e–f (1912).
Castanea crenata Henry in Elwes & Henry, *Trees Great Brit. & Irel.*, IV. 854 (pro parte, non Siebold & Zuccarini) (1909).
Castanea sativa, var. *Bungeana* Pampanini in *Nuov. Giorn. Bot. Ital.* n. ser. XVII. 250 (1910).

Kiangsi: Kuling, thickets, alt. 1300–1500 m., July 29, 1907 (Nos. 1529, 1542; bush or small tree, 1–5 m. tall). Western Hupeh: Ichang, hillsides, 30–1100 m., May, October 1907 (Nos. 507, 374; bush 0.6–2.5 m.); Changyang Hsien, hillsides and woods, alt. 1000–1600 m., July, August and September (Nos 374ª, 3667); Changlo Hsien, woods, alt. 1300–1600 m., July 1907 (No. 3665; tree 5–10 m. tall, girth 0.6–1.5 m.). Patung Hsien, woods, alt. 600–1200 m., June and October 1907 (No. 3618; tree 6–13 m. tall, girth 0.6–2 m.); Hsing-shan Hsien, woods and thickets, alt. 1300–2000 m., June and July 1907 (Nos. 3660, 3666, 3659; bush or small tree, 1.5–13 m. tall); Fang Hsien, woods and thickets, alt. 1000–1600 m., June 8, 1907 (No. 3664; bush or small tree, 2–8 m. tall); without locality, *A. Henry* (Nos. 2867, 5800ª, 6046). Shensi: " mountains near Tze-wu Hsien, south of Sian Fu," September 1, 1914, *F. N. Meyer.* Chekiang: vicinity of Ningpo, 1908, *D. Macgregor;* near Hanchou, June 1907, June 26, 29, 1915, *F. N. Meyer* (Nos. 441, 1479, 1486).

This Chestnut is very abundant on the hills and mountainsides throughout the Yangtsze Valley from the neighborhood of Ningpo in the east to eastern Szech'uan in the west. It also grows in the provinces of Shensi and Kweichou, but Wilson does not remember meeting with it in western Szech'uan. Usually it is a bush or a low bushy tree, but under very favorable conditions it forms a shapely tree from 12 to 15 m. tall, with spreading branches and a trunk from 1 to 1.5 m. in girth. The species is well distinguished by the under side of the leaves, which is more or less densely lepidote and glabrous or nearly so, except on the primary and secondary veins, which are villose. The branchlets in their first year are usually more or less densely clothed with a short velvety pubescence, but occasionally they are nearly glabrous. The winter-buds are small, conical to subglobose and pubescent. The leaves vary much in size, shape and dentation; the upper surface is deep green and the under side pale green and often more or less glaucescent, more especially when young. The ripe fruit is variable in size and contains from three to six small nuts, which have a peculiarly sweet and pleasant flavor. In the ripe involucre the spines are pale straw color, often tinged with purple and are sparsely villose.

It has been known for a long time that two distinct species of Chestnut grew in China. Abel (*Narr. Journ. China*, 165 [1818]) was the first European to write of a shrubby Chestnut with small fruits, and he observed it near the city of Tatung on the lower Yangtsze River. Fortune (*Residence among Chinese*, 51 and 144 [1857]) speaks of two kinds of Chestnuts which grow near Ningpo and mentions that he sent seeds of both to India. Later (in *Gard. Chron.* 1860, 170) Fortune tells of having introduced both Chestnuts to England, but apparently they have been lost to cultivation there.

Hance and others have confounded this Chestnut with *C. crenata* Siebold & Zuccarini, which is confined to Japan and Korea and has the under side of the leaves lepidote and densely felted with short stellate hairs; only occasionally is it subglabrous, usually on leaves growing in the shade. The branchlet is sparsely velutinous or glabrous. When allowed to grow unchecked the Japanese Chestnut makes a large tree from 20 to 25 m. tall and has thick widespreading branches and a trunk from 3 to 5 m. in girth.

In western Hupeh the vernacular name for *C. Seguinii* is Mao Pan-li and the fruits are collected and eaten by the peasantry.

Castanea Henryi Rehder & Wilson, n. comb.

> *Castanopsis Henryi* Skan in *Jour. Linn. Soc.* XXVI. 523 (1899). — Seemen in *Bot. Jahrb.* XXIX. 287 (1900). — Koidzumi in *Tokyo Bot. Mag.* XXX. 101 (1916).
> *Castanea sativa* Skan in *Jour. Linn. Soc.* XXVI. 525 (pro parte, non Miller) (1899).
> *Castanea sativa,* γ var. *acuminatissima* Seemen in *Bot. Jahrb.* XXIX. 287 (1900). — Pampanini in *Nuov. Giorn. Bot. Ital.* n. ser. XVII. 250 (1910).
> *Castanea Vilmoriniana* Dode in *Bull. Soc. Dendr. France,* 1908, 156, fig.; in Fedde, *Rep. Spec. Nov.* X. 242 (1911). — Schneider, *Ill. Handb. Laubholzk.* II. 899, fig. 563ʰ (1912). — Rehder in Bailey, *Stand. Cycl. Hort.* II. 682 (1914). — Koidzumi in *Tokyo Bot. Mag.* XXX. 100 (1916).
> *Castanea Fargesii* Dode in *Bull. Soc. Dendr. France,* 1908, 158, fig.; in Fedde, *Rep. Spec. Nov.* X. 242 (1911). — Schneider, *Ill. Handb. Laubholzk.* II. 899, fig. 563 i–k (1912). — Koidzumi in *Tokyo Bot. Mag.* XXX. 99 (1916).
> *Castanea crenata* Henry in Elwes & Henry, *Trees Great Brit. & Irel.,* IV. 854 (pro parte, non Siebold and Zuccarini) (1909).

Kiangsi: Kuling, thickets, alt. 1300 m., July 31, 1907 (No. **1521**; tree 10 m. tall, girth 0.6 m.). Western Hupeh: north and south of Ichang, woods, alt. 600–1800 m., June and October, 1907 (No. **551**; tree 8–26 m. tall, girth 1–3 m.); same locality, May 1900 (Veitch Exped. No. 759); Changyang Hsien, woods, October 1907 (No. **3617**; tree 8–13 m. tall, girth 1–2 m.); same locality, October 1900 (Veitch Exped. Seed No. 777); Changlo Hsien, woods, alt. 1300–1600 m., June 1907 (No. **551ᵃ**; tree 8–20 m. tall, girth 1–2 m.); Hsing-shan Hsien, woods, alt. 1300–2000 m., June 2, October, November 1907, June 11, 1910 (Nos. **551ᵇ**, **352**, **552**, **4582**; tree 5–26 m. tall, girth 1–2 m.); Fang Hsien, woods, alt. 1300–2000 m., July and November 1907 (No. **551ᶜ**, **3616**; tree 6–13 m. tall, girth 1–2 m.); Patung Hsien, *A. Henry* (No. 2878, co-type of *Castanopsis Henryi* Skan); without locality, *A. Henry* (Nos. 5793, 5793ᵃ, 5800, co-type of *Castanea sativa* var. *acuminatissima* Seemen); "Ma-pan-scian," alt. 1000 m., May 1907, *C. Silvestri* (No. 331). Western Szech'uan: Mt. Omei, June 1904 (Veitch

Exped. No. 5186; tree 8 m. tall). Chekiang: vicinity of Ningpo, 1908, *D. Macgregor.*

This very distinct species is distributed from the neighborhood of Ningpo through the valley of the Yangtsze River as far west as Mt. Omei. On the mountains of western Hupeh and of eastern Szech'uan it is common in woods. This Chestnut grows to a larger size than any other Chinese species and trees from 20 to 25 m. tall with trunks from 1 to 3 m. are common. Occasionally trees 30 m. tall and 5 m. in girth of trunk are met with. The leaves are green on both surfaces and entirely glabrous except for a few appressed hairs on the under side of the primary and secondary veins. The leaves are without lepidote glands except on the upper surface of the very young leaves, from which they disappear very early. Although variable in size the leaves are very characteristic; they are always caudate-acuminate and broadest below or at the middle, and the secondary veins are projected in long aristate points. The shoots are dark-colored and quite glabrous and the winterbuds are brownish, short, broadly ovoid, obtuse or subacute and are glabrous or nearly so. The styles vary in number from 6 to 9, and the fruit may be solitary or two or three on a short spike. The spines of the ripe involucre are sparsely villose. All the fruits we have seen contain a solitary nut, but it is probable that occasionally two occur, as they do in the American *C. pumila* Miller. Dode's *C. Fargesii*, as his figure shows, is apparently founded on such an abnormal fruit, for the description of the leaves, shoots and buds can apply to no other Chinese species.

How distinct this species is is well shown by the fact that Skan describes it as a *Castanopsis.* Certainly it suggests this genus more than that of Castanea, but the leaves are deciduous. The specimens with male flowers Skan referred to *C. sativa* Mill.

A picture of this tree will be found under No. 0126 of the collection of Wilson's photographs. In Hupeh this Chestnut is known as the Ch'in Pan-li.

The Asiatic species of Castanea have been little understood. Some botanists have referred all to the European *C. sativa* Miller and others have divided them into numerous species and varieties. With the great mass of material before us it is evident that there are four well-marked species in eastern Asia and these may be readily distinguished by their leaves, not to mention other characters, as follows:

Castanea Henryi Rehder & Wilson: Leaves not lepidote except on the upper surface when very young and quite glabrous save for a few appressed hairs on the primary and secondary veins on the under side before the leaves unfold.

Castanea mollissima Blume: Leaves not lepidote, normally densely clothed on the under side with a pale felt of stellate hairs, only occasionally quite glabrous.

Castanea Seguinii Dode: Leaves densely lepidote on the under side, otherwise normally quite glabrous except the primary and secondary veins on the under side, which are villose, very rarely obscurely pubescent.

Castanea crenata Siebold & Zuccarini: Leaves lepidote on the under side, usually more or less clothed with a short felt of stellate hairs, occasionally quite glabrous.

The last species is confined to Japan and to Korea and the other three to China, where they are widely distributed in the temperate parts of the Empire.

CASTANOPSIS Spach.

Castanopsis hystrix A. De Candolle in *Jour. Bot.* I. 182 (1863); *Prodr.* XVI. pt. 2, 111 (1864). — Miquel in *Ann. Mus. Lugd.-Bat.* I.

119 (1863–64). — Hooker f., *Fl. Brit. Ind.* V. 620 (1888). — King
in *Ann. Bot. Gard. Calcutta*, II. 95, t. 84 (1889). — Skan in *Jour. Linn.
Soc.* XXVI. 524 (1899). — Dunn in *Jour. Linn. Soc.* XXXVIII. 367
(1908).

> *Castanea hystrix* Hooker f. & Thomson mss. ex A. De Candolle, *Prodr.*
> XVI. pt. 2, 111 (pro synon.) (1864).
> *Quercus rufescens* Hooker f. & Thomson ex Hooker f., *Fl. Brit. Ind.* V. 620
> (pro synon.) (1888).

Western Szech'uan: Kiating Fu, Pagoda Hill, alt. 300–500 m.,
May 1908 (No. 3623; tree 10 m. tall, trunk 1.3 m. girth; Mt. Omei,
woods, alt. 1000–1600 m., June 1904 (Veitch Exped. Nos. 4502, 5187;
tree 13 m. tall). Yunnan: Mengtsze, forests, alt. 1600 m., *A.
Henry* (No. 11313); Szemao, woods, alt. 1500 m., *A. Henry* (Nos.
11738, 11738ª, 11738ᵇ). Fokien: without locality, Dunn's Exped.
April to June 1905 (Hongkong Herb. No. 3498).

This species is fairly common at low altitudes on Mt. Omei and in the surrounding country. It is a low tree of no great size, but with widespreading branches. Like all species of *Castanopsis* it is evergreen, and the contrast between the rufous brown unfolding young leaves and the shining green of the upper surface of adult leaves is striking.

Castanopsis Fargesii Franchet in *Jour. de Bot.* XIII. 195 (1899). —
Skan in *Jour. Linn. Soc.* XXVI. 523 (1899). — Seemen in *Bot. Jahrb.*
XXIX. 287 (1900).

Western Hupeh: Patung Hsien, woods, alt. 1300 m., November
1907 (No. 686; tree 11–16 m. tall, trunk 1–2 m. girth); Changyang
Hsien, woodlands, alt. 1300 m., May 1907 (No. 3641; tree 15–20 m.
tall, trunk 2–3 m. girth); without locality, *A. Henry* (No. 5551) in
Herb. Gray ex Herb. Kew as " *Quercus glauca*, var. ? "). Western
Szech'uan: Hungya Hsien near Wa-wu-shan, woods, alt. 1000 m.,
September 8, 1908 (No. 3620; tree 15–33 m. tall, trunk 1–4 m. girth).
Yunnan: south of the Red River from Manmei, alt. 2000 m., *A.
Henry* (No. 9202ª; tree 6 m. tall).

This tree is common at low altitudes in western Hupeh and in eastern Szech'uan, and is the only species from that region which has spinescent fruit. It is also fairly common in western Szech'uan, where it grows in company with other species. It is a handsome evergreen tree from 15 to 33 m. tall, with a trunk from 1 to 4 m. in girth and widespreading branches. The leaves vary in degree of dentation and the under side of the leaf is pale grayish or rusty brown. The male aments which have not been described are solitary in the axils of minute, caducous scales or occasionally in the axils of young foliage leaves. Often the weak lateral floriferous branchlets are very short and quite leafless and give the impression of a panicled inflorescence. The perianth segments are from 1 to 1.5 mm. long, broadly ovate-rotundate to orbicular, ciliolate and nearly or quite glabrous without; the stamens are about 6 mm. long and at their base there is a conspicuous tuft of white villose hairs. The male flowers are white and the aments are produced in great quantities and when it is in flower the tree is quite conspicuous.

This species is closely related to *C. tribuloides* A. De Candolle and especially to the var. *longispina* King, which may be distinguished by its pubescent branchlets and large leaves. In western Hupeh *C. Fargesii* is colloquially known as the Tsze-li-tzu-shu.

A picture will be found under No. 317 of the collection of Wilson's photographs and also in his *Vegetation of Western China*, No. 148.

Castanopsis ceratacantha Rehder & Wilson, n. sp.

Arbor 6–25-metralis, trunco in circuitu 1–2.5 m. metiente; ramuli dense tomento fulvo villoso obtecti, anno secundo v. tertio glabrescentes; gemmae ovoideae, compressae, obtusae v. acutiusculae, adpresse pubescentes, perulis biserialibus scariosis. Folia persistentia, coriacea, elliptico-oblonga v. oblongo-lanceolata, rarius oblanceolato-oblonga, subito v. rarius sensim acuminata, basi late cuneata v. rotundata, integra v. apicem versus dentibus paucis parvis instructa, 8–14 cm. longa et 3–6 cm. lata, matura supra lucida, glabra, subtus dense tomento fulvo villoso anno secundo saepe evanescente obtecta, costa media supra impressa subtus elevata, nervis utrinsecus 14–17 angulo circiter 45° divergentibus leviter curvatis supra obsoletis subtus manifeste elevatis trabeculis numerosis parallelis leviter elevatis conjunctis; petioli dense villosi, 0.5–1 cm. longi. Flores ignoti. Fructus biennes, sessiles, spicati, in rhachi axillari valida pubescente 15–22 cm. longa congesti; involucra matura subglobosa v. depresso-globosa et interdum leviter lobata, 1.5–2.5 cm. diam., undique spinis validis 0.6–1.2 cm. longis circa medium in ramos plures patentes pungentes divisis obtecta, extus spinis inclusis dense breviter fulvo-villosa, intus pilosa, maturitate fere ad basin irregulariter dehiscentia; glandes pleraeque 2, rarius 3 v. 1, late conico-ovoideae, mucronatae, 1–1.4 cm. altae et fere ac crassae, pallide brunneae, tomentulae.

Western Szech'uan: Wa-shan, woodlands, alt. 1000–1500 m., October 1908 (No. 1142, type); Hungya Hsien, near Wa-wu-shan, alt. 900–1400 m., September 8, 1908 (No. 3619); on Wa-wu-shan, alt. 1000–2000 m., September 12, 1908 (No. 3621); Pingling-shih, alt. 1000 m., September 8, 1908 (No. 3621ª). Yunnan: Szemao, forests, alt. 1500 m., A. *Henry* (No. 12831).

This is a very distinct and remarkable species characterized by the dense and persisting yellow-brown villous tomentum on the shoots, on the under side of the leaves and on the ripe fruits, by the long and massive fruiting spikes and by the numerous long antler-like spines covering the ripe involucre. It is perhaps most closely related to *Castanopsis tribuloides*, var. *longispina* King, which is a very much more glabrous plant, with leaves inclined to be broadest below the middle and with fewer, less prominent secondary nerves; the spines on the involucre are straighter and branch from near the base, and the ripe gland is glabrous. The whole appearance of specimens of the two plants is very different. The typical *C. tribuloides* De Candolle and its other varieties are much farther removed from this new species. Henry's specimen differs in the leaves which are from 15 to 18 cm. long, cuneate at the base and slightly more prominently toothed, with petioles from 0.8 to 1.5 cm. long, but we have no hesitation in referring it to our new species.

C. ceratacantha is a handsome evergreen tree and is fairly common in woods at low altitude in the little known region between Mt. Omei, Mt. Wa and Mt. Wa-wu in western Szech'uan.

Pictures of this tree will be found under Nos. 314, 322 of the collection of Wilson's photographs and also in his *Vegetation of Western China*, No. 150.

Castanopsis platyacantha Rehder & Wilson, n. sp.

Arbor 6–25-metralis, trunco 1–3 m. in circuitu metiente, cortice cinereo in lamellas tenues irregulares fisso; rami validi patentes; ramuli glabri, atropurpurei; gemmae ovato-oblongae, rarius ovatae, compressae, acutiusculi, 0.5–1.3 cm. longae perulis biserialibus scariosis late ovatis v. ovatis, pallide olivaceis nitidulis glabris ciliolatis. Folia persistentia, coriacea, elliptico-lanceolata v. elliptico-oblonga, rarius elliptica, subito acuminata, basi obliqua, late cuneata v. fere rotundata, supra medium remote paucidentata, rarius integra v. fere integra, 8–16 cm. longa et 2.5–6, rarius ad 7 cm. lata, supra intense viridia, glabra, nitida, subtus tomento crustaceo fulvescente demum albo-cinereo costa media glabra excepta dense obtecta, costa media supra leviter elevata v. plana, subtus elevata, nervis utrinsecus 8–10 supra obsoletis subtus leviter elevatis plerumque ante marginem anastomosantibus; petioli crassiusculi, 8–10 mm. longi, glabri. Amenta mascula in axillis foliorum v. bractearum ad basin innovationum solitaria rhachis glabra stricta, angulata, 7–16 cm. longa; flores fasciculati; perianthii lobi anguste ovata v. obovata, concava, 1.5–2 mm. longi, extus glabri, intus villosi; stamina exserta, 4.5–5 mm. longa, filamentis glabris filiformibus, antheris globosis; ovarii rudimentum dense villosum. Spicae fructiferae axillares in parte superiore ramulorum, 5–6 cm. longae; fructus biennes, sessiles, subglobosi v. late ovoidei; involucra dense echinata, spinis fasciculatis 3–5 mm. longis in seriebus concentricis dispositis triangularibus compressis rarius fere subulatis apice pungente glabro, ceterum ut involucrum cinereo-tomentosis; glandes 1–3, plerumque 1, late conico-ovoideae apiculata, 1.2–1.5 cm. altae, 1.2–1.8 cm. diam., pallide brunneae, puberulae, maculo basali latissimo, interdum fere dimidiam glandem obtegente.

Western Szech'uan: Mupin, woods, alt. 1000–1300 m., October 1908 (No. **1096**, type; fruits); Wa-shan, woods, alt. 1300–2400 m., June and October 1908 (Nos. **3622**, co-type ♂, **1096ª**); Hungya Hsien, Wa-wu-shan, woods, alt. 1000–2000 m., September 12, 1908 (No. **1096ᵇ**); Mt. Omei, alt. 2000–2300 m., October 1903, May 1904 (Veitch Exped. Nos. 4508, 5188).   Yunnan: Mengtsze, forests, alt. 2300 m., *A. Henry* (No. 10610).

This is a very distinct species, well characterized by its winter-buds, its leaves, its solitary, axillary male aments, its short fruiting spike, its zonate-ridged strongly echinate fruit and by its pubescent, pyramidate nut, with an unusually broad basal scar. It may be contrasted with *C. orthacantha* Franchet, which we have not seen, but which, according to the description, has a somewhat similar fruit, but has differently shaped and thicker leaves, glabrous and glaucescent on both surfaces, and fasciculate male aments.

This new species is common in woods on the lower slopes of Mt. Omei and the regions to the immediate west and south, and is a handsome, umbrageous tree.

A picture will be found under No. 323 of the collection of Wilson's photographs and also in his *Vegetation of Western China*, No. 149.

Castanopsis caudata Franchet in *Nouv. Arch. Mus. Paris*, sér. 3, VII. 87 (*Pl. David.* I. 277) (1884); in *Jour. de Bot.* XIII. 196 (1883).— Skan in *Jour. Linn. Soc.* XXVI. 522 (1899).—Seemen in *Bot. Jahrb.* XXIX. 287 (1900).

Kiangsi: Kuling, thickets, alt. 1300 m., July 31, 1907 (No. **1522**; bush 4 m., densely branched).

Our specimen is from the type locality and agrees well with Franchet's description, except that the leaves can scarcely be said to be glaucescent.   This plant is not very common on the Lushan mountains and is usually a large, much branched bush or small tree.   The leaves are coriaceous and shining and all parts of the plant are glabrous.

Castanopsis sclerophylla Schottky in *Bot. Jahrb.* XLVII. 638 (1912).

Quercus chinensis Abel, *Narr. Journ. China*, 165, t. 363 (1818). — Forbes in *Jour. Bot.* XX. 81 (1884). — Franchet in *Nuov. Arch. Mus. Paris*, sér. 2, VII. 85 (*Pl. David.* I. 275) (1884). — Skan in *Jour. Linn. Soc.* XXVI. 509 (1899).

Quercus sclerophylla Lindley in Lindley & Paxton, *Flower Gard.* I. 59, fig. 37 (1850–51). — Walpers, *Ann. Bot.* III. 384 (1852–53). — Masters in *Gard. Chron.* n. ser. I. 632 (1874). — Hance in *Jour. Bot.* XII. 242 (1874); XIII. 366 (1875); XX. 294 (1882). — Forbes in *Jour. Bot.* XXII. 86 (1884). — Skan in *Jour. Linn. Soc.* XXVI. 520 (1899). — Seemen in *Bot. Jahrb.* XXIX. 295 (1900). — Dunn & Tutcher in *Kew Bull. Misc. Inform.* add. ser. X. 252 (*Fl. Kwangtung & Hongk.*) (1912).

Quercus cuspidata, var. *sinensis* A. De Candolle, *Prodr.* XVI. pt. 2, 103 (1864). — Hance in *Jour. Bot.* XIII. 366 (1875).

Castanopsis chinensis Schottky in *Bot. Jahrb.* XLVII. 687 (1912).

Synaedrys sclerophylla Koidzumi in *Tokyo Bot. Mag.* XXX. 187 (1916).

Western Hupeh: Changyang Hsien, woods, alt. 1300–1500 m., May 8 and October 1907 (No. 542; tree 8–20 m. tall, girth 1–3 m.); Patung Hsien, woods, alt. 1000–1200 m., May 1900 (Veitch Exped. No. 236); north and south of Ichang, woods, alt. 600–1300 m., May 1907 (No. 3642; tree 10–20 m. tall, girth 1–3 m.); same locality, October 1900 (Veitch Exped. No. 575, fruiting specimens only); same locality, A. Henry (Nos. 3218, 3218ᵃ); without locality, A. Henry (Nos. 7596, 3870, 7707).   Chekiang: vicinity of Ningpo, 1908, D. Macgregor; "Tangsi," March 1906, F. N. Meyer (No. 227).   Fokien: without locality, Dunn's Exped. April to June 1905 (Hongk. Herb. No. 3491).   Hongkong: cultivated November 1862 (Herb. Hance, No. 367; in Herb. Gray).

This is a common evergreen tree in the woods of western Hupeh from river-level to 1500 m. altitude, and this region apparently is the western limit of the species.   It is a handsome tree with a nearly smooth dark gray bark and a very densely branched flattened round crown.   The leaves are usually toothed above the middle only as depicted in Abel's figure, but in the specimen from Fokien some of the leaves are toothed nearly to the base, as shown in Lindley's figure of his *Q. sclerophylla.*   The fructification is annual and the cupule almost completely encloses the small conical nut and splits at maturity to liberate it.

In Hupeh this tree is known as the Chu-li and the fruit is gathered and crushed and converted into a paste known as tou-fu.   In appearance this paste resembles bean-curd and it is an article of food among the peasants.   In flavor the nut is like that of the Chinquapin (*Castanea pumila* Miller).

This *Castanopsis* is distinct from all others.   Abel's figure is very good and it is strange that its identity with Lindley's plant should have been so completely overlooked.   By A. De Candolle and others it has been considered closely related to *C. cuspidata* Schottky (*Quercus cuspidata* Thunberg), but the relationship is remote. In Thunberg's plant the leaves are smaller, with indistinct venation, the male flowers are in a shorter and differently arranged inflorescence, the fruit is biennial, densely clustered, the cup is ovoid and acute and the nut is also ovoid.   The specimens from Chinkiang and Kiukiang collected by C. Maries and referred by Skan (in *Jour. Linn. Soc.* XXVI. 510 [1899]) to *Q. cuspidata* probably belong to *C. sclerophylla.*

We are not sure that *C. sclerophylla* grows wild in Hongkong, as asserted by Dunn and Tutcher, for such specimens as we have received from them named *Q. sclerophylla* belong to *C. cuspidata* Schottky.

The oldest specific name cannot be used for this species as there exists already a *C. chinensis* described by Hance in 1868.

Here may be added notes on some Chinese species of *Castanopsis* not collected during the Arnold Arboretum Expeditions.

Castanopsis indica A. De Candolle in *Jour. Bot.* I. 182 (1863); *Prodr.* XVI. pt. 2, 109 (1864). — Miquel in *Ann. Mus. Lugd.-Bat.* I. 119 (1863–64). — Brandis, *Forest Fl. Ind.* 490 (1874); *Ind. Trees,* 634, fig. 196 (1906). — Hooker f., *Fl. Brit. Ind.* V. 620 (1888). — King in *Ann. Bot. Gard. Calcutta,* II. 94, t. 83 (1889).

   Quercus dubia Lindley in Wallich, *Cat.* No. 2786 (nomen nudum) (1828).
   Quercus serrata Roxburgh, *Fl. Ind.* ed. 2, III. 641 (non Thunberg) (1832).

Castanea indica Roxburgh, *Fl. Ind.* ed. 2, III. 643 (1832). — Wight, *Icon.* II.
t. 417 (1843). — Blume, *Mus. Bot. Lugd.-Bat.* I. 284 (1850). — Kurz, *Forest
Fl. Brit. Burma*, II. 478 (1877).
Yunnan: near Menglieh, alt. 1300 m., *A. Henry* (No. 12834; tree 10 m.
tall).

Castanopsis tribuloides A. De Candolle, var. **ferox** King apud Hooker f., *Fl.
Brit. Ind.* V. 623 (1888); in *Ann. Bot. Gard. Calcutta*, II. 102, t. 95, fig. 1–2
(1889). — Brandis, *Ind. Trees*, 635 (1906).
Quercus ferox Roxburgh, *Fl. Ind.* ed. 2, III. (639) (1832). — Wight, *Icon. Fl.
Ind. Orient.* I. t. 218 (1840).
Castanopsis ferox Spach, *Hist. Vég.* XI. 185 (1842). — Miquel in *Ann. Mus.
Lugd.-Bat.* I. 119 (pro parte, exclud synon.) (1863–64).
Yunnan: Szemao, forests, alt. 1800 m., *A. Henry* (No. 12831ª).
The specimen bears female aments and the young shoots are densely velutinous,
but year-old branchlets are quite glabrous.

Castanopsis tribuloides, var. **echidnocarpa** King apud Hooker f., *Fl. Brit. Ind.*
V. 623 (1888); in *Ann. Bot. Gard. Calcutta*, II. 103, t. 96, fig. 1–3 (1889). — Brandis, *Ind. Trees*, 635 (1906).
Quercus ? caudata Lindley in Wallich, *Cat.* No. 2787 (nomen nudum) (1828).
Castanopsis echidnocarpa A. De Candolle in *Jour. Bot.* I. 182 (1863); *Prodr.*
XVI. pt. 2, 112 (1864). — Miquel in *Ann. Mus. Lugd.-Bat.* I. 119 (1863–64).
Castanea echidnocarpa Hooker f. & Thomson mss. ex A. De Candolle *Prodr.*
XVI. pt. 2, 112 (pro synon.) (1864).
Yunnan: valley of Red River, near Manpan, alt. 1150 m., *A. Henry* (No.
10847); Szemao, woods, alt. 1500–1600 m., *A. Henry* (Nos. 11565, 11565ª,
11565ᶜ, 11565ᵈ, 12328ª).

Castanopsis concolor Rehder & Wilson, n. sp.
Arbor circiter 10-metralis (ex Henry); ramuli glabri, dense minute lenticellati;
gemmae ovoideae, compressae, obtusae, perulis biserialibus puberulis. Folia coriacea, ovato-elliptica v. oblongo-lanceolata, acuminata, basi late cuneata saepe obliqua, vix rotundata, supra medium dentibus paucis remotis parvis mucronulatis
instructa, raro integra, 5–11.5 cm. longa et 2–4.5 cm. lata, utrinque glaberrima et
fere concoloria, supra nitida, costa utrinque leviter elevata, nervis utrinsecus 8–9
leviter curvatis et ante marginem dissolutis supra obsolete prominulis subtus
distinctius sed leviter tantum elevatis, subtus venulis leviter reticulatis; petioli 8–15
mm. longi, glabri. Flores non visi. Spica fructifera circiter 6 cm. longa, rhachi
valida glabra; fructus biennes subglobosi irregulariter et obsolete 2–3-lobati, 2.5–
3 cm. alti et crassi; involucrum crassum, dehiscens, extus tomentosulum, echinatum,
aculeis crassis conicis 2–4 mm. longis fasciculatis et in seriebus concentricis irregulariter curvatis dispositis; glans 1 (semper?), puberula.
Yunnan: Yuan-chiang, alt. 2300 m., *A. Henry* (No. 13342, type); Milê,
A. Henry (No. 10369).
The fruit of this new species is similar to that of *C. armata* Spach, which is
probably the species nearest to it, but in that species the branchlets are pubescent,
the leaves thinly coriaceous and entire, with prominent veins below.

Castanopsis fissa Rehder & Wilson, n. comb.
Quercus fissa Champion apud Bentham in *Hooker's Jour. Bot. & Kew Misc.*
VI. 114 (1854). — Bentham, *Fl. Hongk.* 320 (1861). — Seemann, *Bot. Voy.
Herald*, 415, t. 92 (1857). — Hance in *Jour. Bot.* I. 175 (1863); XIII. 37
(1875). — A. De Candolle, *Prodr.* XVI. pt. 2, 104 (1864). — Wenzig in *Jahrb.
Bot. Gart. Berlin*, IV. 238 (1886). — Skan in *Jour. Linn. Soc.* XXVI. 512

(1899). — Dunn & Tutcher in *Kew Bull. Misc. Inform.* add. ser. X. 252 (*Fl. Kwangtung & Hongk.*) (1912).

Castanea regia Hance in *Ann. Sci. Nat.* sér. 4, XVIII. 231 (1862).

Pasania fissa Oersted in *Kjoebenh. Vidensk. Meddel.* XVIII. 76 (1866).

Synaedrys fissa Koidzumi in *Tokyo Bot. Mag.* XXX. 187 (1916).

Hongkong: Happy Valley woods, November 5, 1903, *C. S. Sargent;* same locality, April 19, 1895, *C. Ford;* without precise locality, *C. Wright* (No. 467; in Herb. Gray). Fokien: Dunn's Exped., April to June 1905, Hongk. Herb., No. 3500).

Castanopsis calathiformis Rehder & Wilson n. comb.

Quercus calathiformis Skan in *Jour. Linn. Soc.* XXVI. 508 (1899).

Synaedrys calathiformis Koidzumi in *Toyko Bot. Mag.* XXX. 188 (1916).

Yunnan: Mengtsze, mountains, southwest, alt. 1500 m., *A. Henry* (Nos. 9070ᵃ, 9070ᵇ, co-types, 13682); Szemao, mountains, alt. 1600 m., *A. Henry* (Nos. 11696ᵃ, 11696ᵇ, 11696ᶜ; shrub 2–3 m. high, flowers with disagreeable odor).

This species is very near *C. fissa* Rehder & Wilson, and resembles it in the leaves and in the flowers so closely that flowering specimens of the two species are almost indistinguishable. The fruits, however, are very different; in *C. fissa* the cupule is ovoid and encloses nearly the entire nut, leaving only a very small opening at the apex, while in this species the cupule is cup-shaped and covers only one-half or two-thirds of the nut and its scales are much more distinct. *C. calathiformis* forms thus a transition to *Pasania* as regards the shape of the cupule, but in all other characters it shows a closer relation to *Castanopsis.*

Castanopsis cuspidata Schottky in *Bot. Jahrb.* XLVII. 625 (1912).

Quercus cuspidata Thunberg, *Fl. Jap.* 176 (1784), *Icon. Pl. Jap.* dec. V. t. 7 (1805). — Siebold & Zuccarini, *Fl. Jap.* I. 8, t. 2 (1835). — Blume, *Mus. Bot. Lugd.-Bat.* I. 288 (1850). — Miquel in *Ann. Mus. Lugd.-Bat.* I. 117 (1863–64); III. 194 (1867); *Prol. Fl. Jap.* 358 (1867). — Franchet & Savatier, *Enum. Pl. Jap.* I. 449 (1875). — Masters in *Gard. Chron.* n. ser. XII. 232, fig. 38 (1879). — Wenzig in *Jahrb. Bot. Gart. Berlin,* IV. 237 (1886). — Hance in *Jour. Bot.* XXV. 13 (1887). — Skan in *Jour. Linn. Soc.* XXVI. 510 (1899). — Dunn & Tutcher in *Kew Bull. Misc. Inform.* add. ser. X. 252 (*Fl. Kwangtung & Hongk.*) (1912).

Quercus cuspidata, β var. *pusilla* Blume, *Mus. Bot. Lugd.-Bat.* I. 288 (1850).

Pasania cuspidata Oersted in *Kjoebenh. Vidensk. Meddel.* XVIII. 76 (1866). — Prantl in Engler & Prantl, *Pflanzenfam.* III. Abt. 1, 55, fig. 38 (1894). — Shirasawa, *Icon. Ess. For. Jap.* I. t. 34, fig. 1–13 (1900). — Matsumura & Hayata in *Jour. Coll. Sci. Tokyo,* XXII. 392 (*Enum. Pl. Formos.*) (1906). — Nakai in *Jour. Coll. Sci. Tokyo,* XXXI. 207 (*Fl. Kor.* II.) (1911). — Matsumura, *Ind. Pl. Jap.* II. pt. 2, 24 (1912).

Pasania cuspidata, α Thunbergii Makino in *Tokyo Bot. Mag.* XXIII. 141 (1909).

Pasania cuspidata, β Sieboldii Makino, l. c. (1909).

Pasania cuspidata, β Sieboldii, forma *pusilla* Makino, l. c. 142 (1909).

Pasania Sieboldii Makino in *Tokyo Bot. Mag.* XXIV. 232 (1910).

Pasania Sieboldii, var. *pusilla* Makino, l. c. 233 (1910).

Lithocarpus cuspidata Nakai in *Tokyo Bot. Mag.* XXIX. 55 (1915).

Lithocarpus Sieboldii Nakai, l. c. (1915).

Synaedrys cuspidata Koidzumi in *Tokyo Bot. Mag.* XXX. 186 (1916).

Synaedrys Sieboldii Koidzumi, l. c. 187 (1816).

Hongkong: ravine south of Wongneicheong Gap, March 4, 1912, *W. J. Tutcher* (ex Herb. Bot. Gard. Hongkong, No. 10202, as *Quercus sclerophylla*). Formosa: South Cape, *A. Henry* (No. 1313).

Korea: Quelpaert, woods, October 1906 and June 1907, *U. Faurie* (Nos. 184, 1524); same locality, October 1907, May and August 1908, May and October 1909, *Taquet* (Nos. 329, 1428, 1429, 1438, 2554, 2556, 2557, 2560, 2562, 2563).

This is apparently a rare tree in China, where it was not met with by Wilson. In Japan it is exceedingly common from Tokyo southward, especially near the sea, and is also commonly cultivated. It grows from 10 to 25 m. tall and has massive widespreading branches. The leaves are variable in size and shape and the under side varies from brownish to nearly white. We have before us a co-type of Blume's var. *pusilla*, consisting of a slender branchlet with small leaves, and a similar specimen can be found among other material before us. The characters on which Makino establishes his *P. Sieboldii* are both trivial and inconstant and, moreover, as Thunberg's figure proves, his statement that *Q. cuspidata* Thunberg has a globular nut is erroneous. When young the fruit is round and flattened, but at maturity it is always oblong-ovoid to ovoid.

Pictures of this tree will be found under Nos. x53, x65, x66, x67, x109, x159 x294, x573, x578 of the collection of Wilson's Japanese photographs.

This plant is figured by Banks (*Icon. Kaempf.* t. 38 [1791]) under the native Japanese name of Ssi-no-ki.

Castanopsis cuspidata, f. latifolia Rehder & Wilson, n. comb.

Quercus cuspidata latifolia Nicholson, *Dict. Gard.* III. 264 (1887).
Pasania cuspidata, β Sieboldii, forma *rotundifolia* Makino in *Tokyo Bot. Mag.* XXIII. 142 (1909).
Pasania Sieboldii, var. *rotundifolia* Makino in *Tokyo Bot. Mag.* XXIV. 233 (1910).

Cultivated: Arboretum, Royal Gardens, Kew, March 10, 1880, *G. Nicholson* (No. 1677; in Herb. Arnold Arboretum).

The leaves of this cultivated plant are usually broadly elliptic-ovate and abruptly broad cuneate at the base.

Castanopsis cuspidata, f. variegata Rehder & Wilson n. comb.

Quercus cuspidata variegata Nicholson, *Dict. Gard.* III. 264 (1887).

Cultivated: Arboretum, Royal Gardens, Kew, October 13, 1880, *G. Nicholson* (No. 2291; in Herb. Arnold Arboretum).

In the *Gardeners' Chronicle*, n. ser. XII. 233, fig. 38 in part, there is a figure of this plant.

LITHOCARPUS Bl.[1]

Lithocarpus cleistocarpa Rehder & Wilson, n. comb.

Quercus cleistocarpa Seemen in *Bot. Jahrb.* XXIII. Beibl. No. 57, 52 (1897); XXIX. 295 (1900). — Skan in *Jour. Linn. Soc.* XXVI. 510 (1899). — Franchet in *Jour. de Bot.* XIII. 157 (1899).

[1] **Lithocarpus** Blume, *Bijdr.* 526 (1825); *Fl. Jav.* I. Cupulif. 34, t. 20 (1828). — Endlicher, *Gen.* 275 (1846). — Miquel, *Fl. Ind. Bot.* I. 865 (1855). — Nakai in *Tokyo Bot. Mag.* XXIX. 55 (excl. sect. *Chlamydobalanus*) (1915).
Synaedrys Lindley, *Nat. Syst. Bot.* ed. 2, 441 (1836). — Koidzumi in *Tokyo Bot. Mag.* XXX. 185 (1916).
Cyclobalanus Oersted in *Kjoebenh. Vidensk. Meddel.* XVIII. 80 (*Bidr. Egesl. Syst.* 72) (1866).

Quercus fragifera Franchet in *Jour. de Bot.* XIII. 157 (1899).
Quercus Wilsonii Seemen in Fedde, *Rep. Nov. Spec.* III. 53 (1906).
Pasania Wilsonii Schottky in *Bot. Jahrb.* XLVII. 660 (1912).
Pasania cleistocarpa Schottky in *Bot. Jahrb.* XLVII. 660 (1912). — Koidzumi
    in *Icon. Pl. Koisikav.* I. 113, t. 57 (1912).
Synaedrys Wilsonii Koidzumi in *Tokyo Bot. Mag.* XXX. 187 (1916).
Synaedrys cleistocarpa Koidzumi, l. c. 188 (1916).

Western Hupeh: Changyang Hsien, woods, alt. 1300–1600 m.,
September 1907 (No. 3636; tree 6–16 m. tall, girth 1–2 m.); same locality, June and December 1900 (Veitch Exped. No. 1521, type of *Q.
Wilsonii* Seemen). Patung Hsien, woods, alt. 1000–1600 m., June 1907
(No. 3636ª; tree 10–13 m. tall, girth 1–1.5 m.); same locality, November 1901 (Veitch Exped., Seed No. 1204); Changlo Hsien, woods, alt.
1000–1500 m., July 1907 (No. 3636ᵇ; tree 10–13 m. tall); Hsing-
shan Hsien, woods, alt. 1000–1800 m., June 2 and July 1907 (Nos.
3636ᵈ, 3635, 3635ª; tree 6–13 m. tall); Fang Hsien, woods, alt. 1500
m., July 1907 (No. 3636ᶜ; tree 13 m. tall, girth 2 m.); without locality, *A. Henry* (Nos. 6002, 6524, 6538, 6715). Western Szech'uan:
Wa-shan, woods, alt. 2000–2600 m., July 1908 (No. 3624; tree 10 m.
tall, girth 1 m.).

We can find no valid differences between von Seemen's two species and it would
appear that he must have overlooked his *Q. cleistocarpa* when describing *Q. Wilsonii*.
The summit of the acorn is flat or even depressed in imperfectly developed
fruits, but usually it is slightly raised. Sometimes the upper part of the cup does not
properly develop, and in that case a greater or lesser part of the acorn is exposed and
the appearance of the fruit is changed. The species is obviously very closely re-

---

Pasania Oersted in *Kjoebenh. Vidensk. Meddel.* XVIII. 81 (*Bidr. Egesl. Syst.* 73)
(excl. sect. *Chlamydobalanus*) (1866). — Prantl in Engler & Prantl, *Nat.
Pflanzenfam.* III. abt. 1, 55 (excl. sect. *Chlamydobalanus*) (1889). —
Schottky in *Bot. Jahrb.* XLVII. 620 (1912).

There can be little doubt that the four genera enumerated above are so closely
related that they cannot be considered distinct, and as *Lithocarpus* is the oldest of
these generic names, it becomes, as already pointed out by Nakai, the name of the
genus, the type of which is *L. javensis* Blume. The reason why Koidzumi rejected
Lithocarpus and uses the name *Synaedrys* is apparently the supposed existence of an
earlier homonym which he quotes as " genus Styracaceorum in Cat. Buitenzorg.
1823." We are, however, unable to find in Blume's *Catalogus van . . . Gewassen te
vinden in 's Lands plantentuin te Buitenzorg* of 1823 any mention of a genus *Lithocarpus*. The first place where we find this name mentioned is in Endlicher's *Genera
plantarum*, 743 (1836–40) under *Styrax* a. *Eustyrax* as " Lithocarpus Blume *Hort.
Buitenz. non Fl. Jav.* St. Benzoin Dryand.," and Miquel refers to it in 1855 in his
Flora Indiae Batavae, I. pt. 1, 865, adding after *Lithocarpus* Bl. *Bijdr.* p. 527 " non
homonymum ejusdem genus Styracem Benzoin spectans." Apparently *Lithocarpus*
Blume Hort. Buitenzorg is an unpublished name and being, moreover, a synonym
of *Styrax*, it cannot, even if it should have been published earlier, invalidate Blume's
Lithocarpus of 1825.

lated to **L. truncata** Rehder & Wilson, n. comb. (*Q. truncata* King). On the mountains of western Hupeh and of eastern Szech'uan, *L. cleistocarpa* is a common inhabitant of the woods and forests between 800 and 1500 m. altitude. In western Szech'uan it is rare. It forms a tree of moderate size, rarely exceeding 30 m. in height, and has a much branched widespreading and flattened crown. The fructification is biennial. Colloquially this tree is known as the Chou-li or Chou-ko-li and it is valued for its wood, which is hard and tough.

Lithocarpus spicata Rehder & Wilson, n. comb.

Quercus spicata Smith in Rees, *Cyclop.* XXIX. No. 12 (1819). — D. Don, *Prodr. Fl. Nepal.* 56 (1824). — Wallich, *Pl. As. Rar.* I. 40, t. 46 (1830). — Miquel, *Fl. Ind. Batav.* II. 848 (1855); in *Ann. Bot. Lugd.-Bat.* I. 106 (1863–64). — A. De Candolle, *Prodr.* XVI. pt. 2, 85 (1864). — Oudemans in *Verh. Akad. Wetensch. Amsterdam,* XI. No. 3, 2, t. 1, fig. 1–3 (*Annot. Cupul. Jav.*) (1865). — Brandis, *Forest Fl. Ind.* 489 (1874); *Ind. Trees,* 629, fig. 194 (1906). — Kurz, *Forest Fl. Brit. Burma,* II. 486 (1877). — Wenzig in *Jahrb. Bot. Gart. Berlin,* IV. 224 (1886). — Hooker f., *Fl. Brit. Ind.* V. 609 (1888). — King in *Ann. Bot. Gard. Calcutta,* II. 47, t. 41, fig. 1–2 (1889). — Franchet in *Jour. de Bot.* XIII. 155 (1899), quoad specimen e Yunnan. — Skan in *Jour. Linn. Soc.* XXVI. 521 (1899), quoad specimen Delavayi. — Koorders, *Exkursionsfl. Java,* II. 65 (1912).

Quercus elegans Blume in *Batav. Verhand.* IX. 208 (1825); *Bijdr.* 518 (1825); *Fl. Java,* Cupulif. 21, t. 10 (1828–29). — Oudemans in *Verh. Akad. Wetensch. Amsterdam,* XI. No. 3, 2, t. 3, fig. 1–4 (*Annot. Cupul. Java*) (1865).

Quercus Arcaula Hamilton ex D. Don, *Prodr. Fl. Nepal.* 56 (pro synon.) (1825); apud Sprengel, *Syst.* III. 857 (1826). — Blume, *Ann. Mus. Bot. Lugd.-Bat.* I. 290 (1850).

Quercus squamata Roxburgh, *Fl. Ind.* III. 638 (1832). — Wight, *Icon. Pl. Ind. Orient.* I. t. 213 (1840).

Pasania spicata Oersted in *Kjoebenh. Vidensk. Meddel.* XVIII. 83 (1866).

Synaedrys spicata Koidzumi in *Tokyo Bot. Mag.* XXX. 198 (1916).

Yunnan: Yuan-chiang, alt. 1150 m., *A. Henry* (No. 13238; tree 10–13 m. tall).

This variable species, which ranges from Yunnan through the Malay Peninsula to Java, is represented in western Szech'uan by the following variety.

Lithocarpus spicata, var. mupinensis Rehder & Wilson, n. var.

A typo praecipue differt gemmis subglobosis, foliis minoribus et angustioribus, cupula parva breviore imam tantum basin glandis conicae minoris amplectente.

Arbor 13–20-metralis; gemmae subglobosae, leviter pubescentes. Folia tenuiter coricaea, lanceolata v. oblongo-lanceolata, subito acuminata, basi attenuata, integra, 5–9 cm. longa et 2–3 cm. lata; petioli circiter 0.5 cm., rarius 1 cm. longi. Fructus biennes, in spicis subterminalibus 9–10 cm. longis; cupula 10–13 mm. diam., circiter 5 mm. alta, bracteis leviter incrassatis arcte adpressis; glans conico-ovoidea, apiculata, circiter 1.5 cm. altus et 1.3–1.5 cm. diam.

Western Szech'uan: Mupin, woodlands, alt. 1000–1600 m., October 1908 (No. **3629**).

This variety is well distinguished by its subglobose, slightly pubescent winter-buds, by its small and narrow somewhat chartaceous leaves and by its thin, shallow cup embracing only the base of the conical acorn, which is slightly over half an inch in diameter. It is probable that the specimens from Khasia, referred by King (in *Ann. Bot. Gard. Calcutta*, II. 48, t. 43, fig. 9–11) to A. De Candolle's var. *microcalyx*, belong here, although they are described as having rather broader leaves and an acorn " less than half an inch in diameter." A. De Candolle's var. *microcalyx* is founded upon *Q. microcalyx* Korthals (in *Verh. Nat. Gesch. Bot.* 206 [1839–42]), which is a native of Borneo and has hemispherical acorns which are often depressed at the summit. That a tree native of Borneo should also grow in western China is highly improbable and is contrary to our experience.

This new variety is not uncommon in moist woods at low altitudes in the petty state of Mupin and contiguous country to the southeast. It is a shapely evergreen tree growing from 15 to 20 m. tall. The wood is hard and tough.

Here may be recorded another variety of this polymorphous species collected by Henry in Yunnan.

Lithocarpus spicata, var. **brevipetiolata** Rehder & Wilson, n. comb.

Quercus grandifolia D. Don in Lambert, *Pinus*, II. 27, t. 8 (1824); ed. 2, II. Append. 3, t. 2 (1828). — D. Don, *Prodr. Fl. Nepal.* 57 (1824).
Quercus spicata, δ *brevipetiolata* A. De Candolle, *Prodr.* XVI. pt. 2, 86 (1864).
Quercus spicata, var. *brevipetiolata* King apud Hooker f., *Fl. Brit. Ind.* V. 610 (1888). — King in *Ann. Bot. Gard. Calcutta*, II. 48, t. 41, fig. 3 (1889).

Yunnan: Szemao, mountains to east and southwest, alt. 1300–1600 m., *A. Henry* (Nos. 11614, 11614ᵃ, 11614ᵇ, 11614ᶜ, 11614ᵈ; tree 6–16 m. tall).

Lithocarpus megalophylla Rehder & Wilson, n. sp.

Arbor 16-metralis, trunco 0.3 m. diam.; ramuli validi, striati, glabrati. Folia coriacea, oblongo-elliptica, subito breviter acuminata, basi oblique, late cuneata v. fere rotundata, integra, margine revoluta, 15–27 cm. longa et 7.5–12 cm. lata, glabra, nitidula, concoloria, costa supra impressa subtus elevata, nervis utrinsecus circiter 10, curvatis, secus marginem adscendentibus et anastomosantibus supra leviter impressis subtus valide elevatis trabeculis elevatis subparallelis conjunctis, utrinque in sicco saltem leviter reticulata, plus minusve bullata; petioli validi, 3–4.5 cm. longi, supra anguste canaliculati, glabri. Fructus biennes, sessiles v. subsessiles saepissime plures basi confluentes, in spicis 12 cm. longis v. ultra, rhachi valida lenticellata pubescente; cupula lignea, scutelliformis, 2–2.5 cm. diam. et vix 1 cm. alta, basin tantum glandis amplectens, intus cinereo-tomentella margine glabrescente excepta, cicatrice glandis elevata, bracteis late triangulari-ovatis acuminatis carinatis tomentosis, marginalibus arcte adpressis indistinctis; glans fere globularis, apiculata, basi concava, brunnea, nitidula, 2–2.5 cm. diam., pericarpio osseo 2 mm. crasso.

Western Szech'uan: Hungya Hsien, near Wa-wu-shan, woods, alt. 1000 m., September 1908 (No. 3631).

This new species is most closely related to *L. spicata* Rehder & Wilson, but in that species and its numerous varieties the leaves, though varying greatly in size, are nearly always broadest above the middle and are narrowly cuneate or attenuate at the base, the petioles never exceed one inch and are usually only about half an inch in length, and the cup scales are thickened, usually very much so, tuberculate and indistinct one from another and the acorn is smaller.

Wilson met with this new *Lithocarpus* in only one locality, so it would appear to be a rare tree. The wood is exceedingly tough.

Lithocarpus sp.

Western Hupeh: Changyang Hsien, alt. 1300–1600 m., June 1907 (No. **3638**).

Apparently closely related to *L. megalophylla* Rehder & Wilson and *L. Henryi* Rehder & Wilson, but distinguished by a rufous gray fascicled pubescence on the young shoots; the leaves are also smaller. Our specimen consists of a leafy branch with unopen flowers and is too incomplete for determination, although it is unlike any other we have seen.

Lithocarpus Henryi Rehder & Wilson, n. comb.

Quercus Henryi Seemen in *Bot. Jahrb.* XXIII. Beibl. No. 57, 50 (1897); XXIX. 294 (1900).
Quercus spicata Franchet in *Jour. de Bot.* XIII. 155 (non Smith) (1899), quoad specimen e Szech'uan orientali. — Skan in *Jour. Linn. Soc.* XXVI. 521 (1899), quoad specimina e Hupeh et Szech'uan.
Pasania Henryi Schottky in *Bot. Jahrb.* XLVII. 665 (1912).

Western Hupeh: south of Ichang, woods, October 1907 (No. **543**; tree 6–16 m. tall, girth 1–2.3 m.); Changlo Hsien, woods, alt. 1000–1500 m., June 1907 (No. **543ᵃ**; tree 6–16 m. tall); Patung Hsien, woods, alt. 1100–1300 m., May, June, July and September 1907 (No. **543ᵇ, 543ᶜ, 3637**; tree 8–20 m. tall, girth 1–3 m.); same locality, July 1901 (Veitch Exped. No. 2228, in part, Seed No. 775); Fang Hsien, woods, alt. 1300 m., June and July 1907 (Nos. **543ᵃ, 3637ᵃ**; tree 5–13 m. tall, girth 1–2 m.); Paokang Hsien, June and July 1901 (Veitch Exped. No. 2228, in part, tree 16 m. tall); without locality, *A. Henry* (Nos. 5805, 5805ᵃ, 6023, co-type). **Eastern Szech'uan:** Wushan Hsien, *A. Henry* (Nos. 7030, 7030ᵃ).

This is a very common evergreen tree in the woods of western Hupeh and eastern Szech'uan between 600 and 1500 m. altitude. It seldom exceeds 16 m. in height, and has ascending-spreading branches which form a neat oval or rounded crown. The fructification is biennial and subterminal on the season's growth and the acorn may be flattened-round on the summit or slightly raised and pointed; the cup is always very shallow and thin and the scales are keeled on the back. Franchet and Skan both refer this plant to the variable *L. spicata* Rehder & Wilson (*Quercus spicata* Smith), but if this is correct, it is certainly a very distinct variety. Von Seemen has pointed out certain differences between his species and Smith's, and we may add that in all the specimens of *L. Henryi* before us the petioles are always longer and the leaves narrower; the cup is never thickened and tubercled and the bracts subtending the branches of the male inflorescence are ovate-lanceolate acuminate and three times the size of those on *L. spicata* Rehder & Wilson. *Lithocarpus Henryi* occupies a distinct geographical region, and until our knowledge of the Chinese Oaks is more complete we think it is best to maintain it as a distinct species.

Pictures of this tree will be found under Nos. 499, 538, 650 of the collection of Wilson's photographs and also in his *Vegetation of Western China*, Nos. 425, 427, 430.

Lithocarpus viridis Rehder & Wilson, n. comb.

Quercus polystachya Skan in *Jour. Linn. Soc.* XXVI. 519 (non Wallich) (1899), quoad specimina e Yunnan, No. 9636 excepto.
Pasania viridis Schottky in *Bot. Jahrb.* XLVII. 668 (1912).
Synaedrys viridis Koidzumi in *Tokyo Bot. Mag.* XXX. 198 (1916).

Western Szech'uan: Yachou Fu, woods, alt. 1000–1500 m., October 1908 (No. 3630; tree 20 m. tall, girth 2.5 m.). Yunnan: Mengtsze, woods, alt. 1600–1800 m., *A. Henry* (Nos. 10520, 11434, 11434[a]); Szemao, mountains, alt. 1500–1600 m., *A. Henry* (Nos. 12329, 12329[a], 12329[b], 12329[c], 12329[d], 12329[e]).

Schottky also includes Henry's No. 9636, but that has leaves pubescent on the under side and the male inflorescence is much shorter and the flowers are different. In the neighborhood of Mt. Omei *L. viridis* is a common evergreen tree. The fruit is biennial.

QUERCUS L.

Quercus dentata Thunberg, *Fl. Jap.* 177 (1784); *Icon. Pl. Jap.* pt. 5, t. 6 (1805). — Blume, *Mus. Bot. Lugd.-Bat.* I. 297 (1849–51). — A. De Candolle, *Prodr.* XVI. pt. 2, 13 (1864). — Hance in *Ann. Sci. Nat.* sér. 5, V. 243 (1866); in *Jour. Linn. Soc.* X. 488 (1869); XIII. 7 (1873); in *Jour. Bot.* XI. 172 (1873). — Franchet & Savatier, *Enum. Pl. Jap.* I. 445 (1875). — Bretschneider, *Chinese Silkworm Trees*, 5 (1881). — Forbes in *Jour. Bot.* XXII. 86 (1884). — Franchet in *Nouv. Arch. Mus. Paris*, sér. 3, VII. 82 (*Pl. David.* I. 272) (1884). — Wenzig in *Jahrb. Bot. Gart. Berlin*, IV. 218 (1886). — Sargent in *Garden and Forest*, VI. 383, fig. 59 (1893); *Forest Fl. Japan*, 67, t. 23 (1894). — Shirai in *Tokyo Bot. Mag.* IX. 407, t. 7 fig. 9 (1895). — Skan in *Jour. Linn. Soc.* XXVI. 511 (1899). — Shirasawa, *Icon. Ess. For. Jap.* I. 52, t. 27, fig. 1–15 (1900). — Komarov in *Act. Hort. Petrop.* XXII. 76 (*Fl. Mandsh.*) (1904). — Matsumura & Hayata in *Jour. Coll. Sci. Tokyo*, XXII. 393 (*Enum. Pl. Formos.*) (1906). — Pampanini in *Nuov. Giorn. Bot. Ital.* n. ser. XVII. 251 (1910). — Nakai in *Jour. Coll. Sci. Tokyo*, XXXI. 209 (*Fl. Kor.* II.) (1911). — Matsumura, *Ind. Pl. Jap.* II. pt. 2, 26 (1913). — Léveillé, *Fl. Kouy-Tchéou*, 127 (1914). — Nakai in *Tokyo Bot. Mag.* XXIX. 60 (1915).

> *Quercus obovata* Bunge in *Mém. Sav. Étr. Acad. Sci. St. Pétersbourg*, II. 136 (*Enum. Pl. Chin. Bor.* 62) (1833). — Carruthers in *Jour. Linn. Soc.* VI. 32 (1862).
> *Quercus dentata*, β *Wrightii* A. De Candolle, *Prodr.* XVI. pt. 2, 13 (1864). — Shirai in *Tokyo Bot. Mag.* IX. 408, t. 7, fig. 6 (1895).
> *Quercus Daimio* Hort. ex K. Koch, *Dendr.* II. pt. 2, 45 (pro synon.) (1873).
> *Quercus yunnanensis* Franchet in *Jour. de Bot.* XIII. 146 (1899).
> *Quercus dentata*, var. *grandifolia* Koidzumi in *Tokyo Bot. Mag.* XXVI. 161 (1912).
> *Quercus nipponica* Koidzumi in *Tokyo Bot. Mag.* XXVI. 161 (1912). — Nakai in *Tokyo Bot. Mag.* XXIX. 61 (1915).

Western Hupeh: Fang Hsien, woods, alt. 1000–1300 m., June 1910 (No. 4584; tree 6–16 m. tall, girth 0.6–2 m.); " Ou-tan-scian," alt. 2090 m., July 1907, *C. Silvestri* (No. 344). Western Szech'uan: southeast of Tachien-lu, woods, alt. 1300 m., October 1908 (No. 1297; tree 10 m. tall, girth 1.5 m.). Yunnan: Mengtsze, mountains to northward, alt. 2000 m., *A. Henry* (No. 9201). Shantung: mountains near Chefoo, *E. Faber* (No. 101; in Herb. Gray). Chili: north of Peking, Ming tombs, October 6, 1905, *J. G. Jack;* same locality, October 1905, *F. N. Meyer* (No. 225); near " San-tun-ying," May 31, 1913, *F. N. Meyer* (Nos. 38, 976); hot springs, " Tang-shan," October 1905, *F. N. Meyer* (No. 214).

NORTHEASTERN ASIA. Eastern Mongolia: Jehol, *A. David* (No. 1704; in Herb. Gray); same locality, 1910, *W. Purdom* (Nos. 273, 274). Korea: Wonsan, September 5, 1903, *C. S. Sargent;* Fusan, May 21, 1906, *U. Faurie* (Nos. 187, 188); " Ouen-san," July 1906, *U. Faurie* (No. 193); Chemulpo, September 1906, *U. Faurie* (No. 186); Quelpaert, September 1908 and June 1909, *Taquet* (Nos. 1434, 2549); Port Hamilton, 1859, *C. Wilford* (No. 2; in Herb. Gray).

This Oak is widely dispersed, but is not common in western Hupeh or in Szech'uan and large trees are rare. Neither can it be said to be a common tree in southern or central Japan, but in Hokkaido, and especially in the north and west, it forms pure woods in open valleys and on low mountain slopes. At its best it is a tree from 20 to 25 m. tall with ascending and spreading branches forming an oval or flattened crown and a trunk from 2 to 3 m. in girth, covered with handsome, deeply furrowed, firm and dark gray bark. The wood is very hard and tough and of little commercial value, but the bark is very valuable and is largely employed in Japan in tanning hides. In Japan it is known as the Ko-gashiwa. For so widely distributed a tree it is remarkably constant in its characters. A slight variation in length of the bristly cup scales and of the degree of pubescence on the leaves and shoots is all that is noticeable. The fruit ripens in one season and on plants a foot high it is often as large and as perfect as on full-grown trees.

Pictures of this tree will be found under Nos. x437, x438, x439 of the collection of Wilson's Japanese photographs.

Quercus glandulifera Blume, *Mus. Bot. Lugd.-Bat.* I. 295 (1850). — Miquel in *Ann. Mus. Lugd.-Bat.* I. 104 (1863–64). — A. De Candolle, *Prodr.* XVI. pt. 2, 40 (1864). — Moore in *Jour. Bot.* XIII. 231 (1875). — Franchet & Savatier, *Enum. Pl. Jap.* I. 447 (1875). — Forbes in *Jour. Bot.* XXII. 86 (1884). — Franchet in *Nouv. Arch. Mus. Paris*, sér. 2, VII. 84 (*Pl. David.* I. 274) (1884). — Wenzig in *Jahrb. Bot. Gart. Berlin*, IV. 220 (1886). — Sargent in *Garden and Forest*, VI. 385 (1893); *Forest Fl. Jap.* 68 (1894). — Shirai in *Tokyo Bot. Mag.* IX. 410, t. 7, fig. 1 (1895). — Skan in *Jour. Linn. Soc.* XXVI. 514 (1899). — Seemen in *Bot. Jahrb.* XXIX. 290 (1900). — Shirasawa, *Icon. Ess. For. Jap.* I. 50, t. 26, fig. 13–24 (1900). — Pampanini in *Nuov. Giorn. Bot. Ital.* n. ser. XVII. 251 (1910). — Nakai in *Jour. Coll. Sci. Tokyo*, XXXI. 207 (*Pl. Kor.* II.) (1911); in *Tokyo Bot. Mag.* XXIX. 59 (1915). — Matsumura, *Ind. Pl. Jap.* II. pt. 2, 27 (1913).

> *Quercus glandulifera, β* var. *subattenuata* Blume, *Mus. Bot. Lugd.-Bat.* I. 295 (1850).
> *Quercus glandulifera, γ* var. *inaequali-serrata* Blume, l. c. (1850).
> *Quercus glandulifera, δ* var. *subcrenata* Blume, l. c. (1850).
> *Quercus glandulifera, ε* var. *polymorpha* Blume, l. c. (1850).
> *Quercus urticaefolia* Blume, l. c. 296 (1850).
> *Quercus canescens* Blume, l. c. (1850). — A. De Candolle, *Prodr.* XVI. pt. 2, 15 (1864). — Shirai in *Tokyo Bot. Mag.* IX. 411, t. 7, fig. 2 (1895).
> *Quercus canescens, β urticaefolia* Miquel in *Ann. Mus. Lugd.-Bat.* I. 105 (1863–64).
> *Quercus urticaefolia, β brevipetiolata* A. De Candolle, *Prodr.* XVI. pt. 2, 16 (1864).
> *Quercus Griffithii*, var. *glanduligera* Franchet in *Jour. de Bot.* XIII. 149 (1899).
> *Quercus coreana* Léveillé in litt. ex Nakai in *Tokyo Bot. Mag.* XXIX. 59 (pro synon., non Q. *koreana* Nakai) (1915).
> *Quercus grosseserrata* Léveillé in litt. ex Nakai, l. c. (pro synon.; non Blume) (1915).
> *Quercus serrata* Léveillé in litt. ex Nakai, l. c. (pro synon.; non Thunberg) (1915).

Kiangsi: Kuling, forming scrub, alt. 1300 m., July 29, 1907 (No. 1560; bush 1–2 m.).   Western Hupeh: Ichang, alt. 30–1000 m., April and May 1907 (Nos. 3652, 3654, 3656; tree 5–15 m. tall, girth 0.3–2 m.); north and south of Ichang, forming scrub, alt. 30–1300 m., November 1907 (No. 528; bush 1–3 m.); Changyang Hsien, woodlands, alt. 300–1600 m., October, November 1907 (Nos. 521, 785, 526, 520; tree 3–13 m. tall, girth 0.3–2 m.); Patung Hsien, woods, alt. 600–

1500 m., October 1907 (Nos. 518, 519; tree 3–12 m. tall, girth 0.2–1 m.);
Hsing-shan Hsien, woods, alt. 600–1500 m., May, October 1907 (Nos.
3650, 525, 530; tree 13–16 m. tall, girth 1–2.5 m.); Fang Hsien, woods,
alt. 1000–1600 m., October 1907 (Nos. 350, 524, 546; tree 6–13 m. tall,
girth 0.6–2 m.); without locality, September 1900 (Veitch Exped.
No. 1688); without locality, A. Henry (Nos. 157, 2760, 3906);
" Ou-tan-scian," alt. 2090 m., September 1907, C. Silvestri (No. 350);
" Kao-kien-scian," alt. 800 m., May 1907, C. Silvestri (No. 348).
Szech'uan: banks of Yangtsze River, E. Faber (No. 215). Western
Szech'uan: Kiating Fu, sandstone hills, alt. 300–600 m., September
1908 (No. 3649; bush or small tree 3–6 m.); Wa-shan, woodlands, alt.
1000–1600 m., October 1908 (No. 1143; tree 13 m. tall, girth 2.5 m.);
Mupin, woods, alt. 1300–1600 m., October 1908 (No. 1294; tree 10–20
m. tall, girth 1.5–2.5 m.); west and near Wên-ch'uan Hsien, forming
scrub, alt. 1000–3000 m., July and September 1908 (Nos. 3627, 1095ᵃ;
bush 1.5–3 m.). Shantung: "Lau-shan," August 1907, F. N.
Meyer (No. 321); without locality, September 1862, A. Maingay (No.
192; in Herb. Gray).

NORTHEASTERN ASIA. Korea: Seoul, September 24, 1905, J. G. Jack;
Fusan, temple gardens, September 6, 1903, C. S. Sargent; same locality, September
11, 1905, J. G. Jack; Chemulpo, September 1908, U. Faurie (No. 198); Quelpaert,
May 15, 1907, U. Faurie (No. 1527); same locality, woods, May 1908, July and
August 1910, Taquet (Nos. 2433, 2551, 4444, 4443, 4442).

This is a very common tree throughout the Yangtsze Valley from river-level to
altitudes of 1600 m. On low hills and near villages and towns it occurs as low scrub
or coppice, but when allowed to develop it forms a tree from 20 to 25 m. tall with a
trunk from 2 to 3 m. in girth. In Japan it is also abundant both as a forest tree and
as low scrub or coppice growth. It exhibits considerable variation in the size and
shape of leaves, in the length of petioles and in the degree of pubescence on the shoots
and on the underside of the leaves, but we find it quite impossible to distinguish
any variety or form. Normally the leaves on the under side have a short fascicled, almost stellate pubescence heavily masked by long straight appressed hairs;
sometimes one or other of these forms of pubescence is absent or nearly so and commonly one or other has the ascendency. Wilson's No. 3649 from Kiating and western
Szech'uan has very short petioles and leaves glabrous on the under side with the
exception of a few long appressed hairs on the principal veins. No. 1294 from
Mupin is similarly glabrous, but the petioles are long and slender. No. 520 has
only a very sparse fasciculate pubescence. A specimen before us from a tree cultivated in the Botanic Garden, Tokyo, has the shoots and under side of the leaves
densely clothed with a silky pubescence. Meyer's specimen from Shantung has
the leaves deeply toothed, subsessile, slightly pointed at the base and clothed
with appressed hairs on the under side. Maingay's specimen from the same province has subsessile to short-petiolate leaves, glabrous on the under side and with
few very shallow teeth. Between these extremes specimens before us show every
gradation. The teeth are always gland-tipped and point forward and are usually
slightly incurved. The fruit ripens in one season and the shallow cup embraces

one-third or less of the oblong-cylindric acorn; the scales of the cup are closely appressed and are densely villose. In autumn the leaves change to orange-red and crimson. In Hupeh this Oak is known as the Peh-fan-li.

A picture of this tree will be found under No. 568 of the collection of Wilson's photographs, in his *Vegetation of Western China*, No. 426, and also in the collection of his Japanese photographs, Nos. x90, x613.

Quercus aliena Blume, *Mus. Bot. Lugd.-Bat.* I. 298 (1850). — A. De Candolle, *Prodr.* XVI. pt. 2, 14 (1864). — Hance in *Jour. Bot.* XIII. 361 (1875). — Franchet & Savatier, *Enum. Pl. Jap.* I. 445 (1875). — Bretschneider, *Chinese Silkworm Trees*, 5 (1881). — Forbes in *Jour. Bot.* XXII. 85 (1884). — Wenzig in *Jahrb. Bot. Gart. Berlin*, IV. 219 (1886). — Skan in *Jour. Linn. Soc.* XXVI. 505 (exclud. synon.) (1899). — Seemen in *Bot. Jahrb.* XXIX. 288 (1900). — Shirasawa, *Icon. Ess. For. Jap.* I. 55, t. 28, fig. 12–22 (1900). — Palibin in *Act. Hort. Petrop.* XVIII. 196 (*Consp. Fl. Kor.*) (1900). — Komarov in *Act. Hort. Petrop.* XXII. 75 (*Fl. Mandsh.* II.) (1903). — Pavolini in *Nuov. Giorn. Bot. Ital.* n. ser. XV. 439 (1908). — Pampanini in *Nuov. Giorn. Bot. Ital.* n. ser. XVII. 251 (1910). — Nakai in *Jour. Coll. Sci. Tokyo*, XXXI. 209 (*Fl. Kor.* II.) (1911). — Matsumura, *Ind. Pl. Jap.* II. pt. 2, 25 (1912). — Koidzumi in *Tokyo Bot. Mag.* XXVI. 163 (sphalmate " *alieana* ") (1912). — Nakai in *Tokyo Bot. Mag.* XXIX. 59 (1915), exclud. syn. *Q. Griffithii* Hooker f. & Thomson.

Quercus hirsutula Blume in *Mus. Bot. Lugd.-Bat.* I. 298 (1850).

Western Hupeh: Changyang Hsien, woods, alt. 1300–1600 m., November 1907 (No. **782**; tree 6–13 m. tall, girth 1–1.5 m.); Ichang, *Dr. Aldridge;* without locality, *A. Henry* (No. 3080); Patung Hsien, woods, alt. 1000–1600 m., May 1907 (No. **527**[b]); bank of Yangtsze River, April 1903 (Veitch Exped. No. 4501, tree 8 m. tall). Western Szech'uan: southeast of Tachien-lu, near Moshi-mien, alt. 1600 m., October 1910 (No. **4282**; small tree 5–8 m. tall). Chekiang: " near Tangsi," June 1907, *F. N. Meyer* (No. 442).

NORTHEASTERN ASIA. Korea: Fusan, temple gardens, September 6, 1903, *C. S. Sargent;* same locality, May 1906, *U. Faurie* (No. 197); Seoul, east Palace park, September 24, 1905, *J. G. Jack.*

This form, with broad, rounded, obtuse teeth separated by shallow sinuses, which Blume makes the type of the species, is very rare in China; in Japan Wilson did not meet with it. In habit and other features it is identical with the following variety, which unquestionably represents the phylogenetic type. The different dentation, however, makes the leaves look very distinct.

A picture of *Q. aliena* will be found under No. 615 of the collection of Wilson's photographs and also in his *Vegetation of Western China*, No. 432.

Quercus aliena, var. acuteserrata Maximowicz apud Wenzig in *Jahrb. Bot. Gart. Berlin*, IV. 219 (1886).

Quercus aliena, γ acutedentata Maximowicz in litt. apud Franchet & Savatier, *Enum. Fl. Jap.* I. 445 (nomen nudum) (1875). — Shirai in *Tokyo Bot. Mag.* IX. 409 (*acutidentata*) (1895). — Matsumura, *Ind. Fl. Jap.* II. pt. 2, 25 (1912).

Western Hupeh: Patung Hsien, woods, alt. 1000–1600 m., October 1907 (No. 527; tree 8–16 m. tall, girth 1–2 m.); same locality, May 1900 (Veitch Exped. No. 562); same locality, *A. Henry* (No. 68ᵃ); Ichang, *A. Henry* (No. 2293); Hsing-shan Hsien, woods, alt. 1000–2000 m., May and November 1907 (Nos. 517, 529, 3658; tree 6–20 m. tall, girth 1.5–2.5 m.); Fang Hsien, woods, alt. 1000–1600 m., October 1907 (No. 516; tree 8–16 m. tall, girth 1–2 m.); " Ou-tanscian," August 1909, *C. Silvestri* (No. 2931); without locality, *A. Henry* (No. 2293). Hunan: Shihmen Hsien, *A. Henry* (No. 7945). Western Szech'uan: west and near Wên-ch'uan Hsien, woods, alt. 1000–2000 m., November 1908 (Nos. 1094, 1095; bush or small tree 2–13 m. tall). Yunnan: Mengtsze, alt. 1600 m., *A. Henry* (Nos. 9394, 9394ᵃ, 11298).

NORTHEASTERN ASIA. Mandshuria: Sheng-king, June 28, 1906, *F. N. Meyer* (No. 38). Korea: Chinnampo, September 17, 1905, *J. G. Jack;* same locality, August 1906, *U. Faurie* (No. 192); " Ouen-san," July 1906, *U. Faurie* (No. 195).

JAPAN. Hondo: Rikuchu province, Ishidoriya, rare, September 26, 1914, *E. H. Wilson* (No. 7562; tree 11–20 m. tall, girth 0.3–1.5 m.); Uzen province, without locality, June 21, 1902, *K. Sakurai*. Kyushu: Hizen province, " Simshara," 1863, *C. Maximowicz* (co-type; in Herb. Gray).

In western Hupeh and in Szech'uan this is a very common Oak in mixed woods on the mountains from near river-level up to 2000 m. altitude. In Japan it is rare, and Wilson met with it there in only one locality. It is a tree from 20 to 25 m. tall, with massive widespreading branches, and the bark is gray and fissured. The leaves vary much in size and the petioles in length, but the under side of the leaves is always more or less densely clothed with a matted felt of short pale gray hairs; the shoots are always glabrous at maturity except occasionally immediately below the clustered terminal winter-buds, where often a sparse puberulous indumentum can be detected. In Hupeh this Oak is known as the Hu-li. Pictures of this tree will be found under Nos. 501, 507 of the collection of Wilson's photographs and also in his *Vegetation of Western China*, Nos. 428, 429.

Quercus aliena, var. acuteserrata, f. calvescens Rehder & Wilson, n. forma.

A typo varietatis recedit foliis subtus non tomentosis sed tantum pilis brevibus fasciculatis v. simplicibus plus minusve dense conspersis, interdum fere glabrescentibus.

Western Hupeh: Fang Hsien, woods, alt. 1000–1600 m., May 31 and October 1907 (No. 515, type; tree 7–20 m. tall, girth 1–2.6 m.); Hsing-shan Hsien, woods, alt. 1200–1800 m., June and October 1907 (No. 3658ᵃ; tree 10 m. tall, girth 1.2 m.); north and south of Ichang, woods, alt. 1000–1200 m., June and December 1907 (No. 531; tree 14 m. tall, girth 1.6 m.), "monte Triora," alt. 1950 m., July 3, 1907, *C. Silvestri* (No. 339). Shensi: "Lao-y-san," 1897, *G. Giraldi*. Chili: near San-tun-ying, among Pine trees on a sheltered mountain slope, May 29, 1913, *F. N. Meyer* (No. 955).

NORTHEASTERN ASIA. Korea: Quelpaert, June 8, 1908, and July 1909 (Nos. 1436, 2546).

This seems to be only a glabrescent form of *Q. aliena*, var. *acuteserrata*, which occurs occasionally in the whole region of this variety. Sometimes the pubescence is rather dense and the under side of the leaf is soft to the touch, as in Meyer's No. 37 and in Silvestri's No. 339, though the fascicled hairs are separate and not matted as in the type, or the hairs are rather scattered and partly simple, appressed and very short, as in Wilson's Nos. 515 and 3658, while young seedlings raised from seed of No. 515, as well as those raised from seed of No. 517, have the leaves nearly glabrous. The lobes of the leaves of the glabrescent form are acute or acutish; in Giraldi's specimen, however, and partly in Meyer's No. 37 they are more or less rounded, but the leaves agree in size and shape with those of var. *acuteserrata*.

Quercus Fabri Hance in *Jour. Linn. Soc.* X. 202 (1869); in *Jour. Bot.* XIII. 362 (1875); XX. 362 (1882). — Forbes in *Jour. Bot.* XXII. 86 (1884). — Franchet in *Nouv. Arch. Mus. Paris*, sér. 2, VII. 84 (*Pl. David.* I. 274) (1884); in *Jour. de Bot.* XIII. 160 (1899). — Skan in *Jour. Linn. Soc.* XXVI. 512 (1899). — Koidzumi in *Tokyo Bot. Mag.* XXVI. 162 (1912).

Western Hupeh: Ichang, alt. 30–1000 m., common, April, May and October 1907 (Nos. 3653, 3655, 3655ᵃ, 3657; tree 6–13 m. tall, girth 0.3–1.5 m.); north and south of Ichang woodlands, alt. 600–1300 m., May and October 1907 (No. 523; slender tree 6–12 m. tall); same locality, *A. Henry* (Nos. 2294, 2636); Changyang Hsien, woods, alt. 600–1300 m., October and November 1907 (Nos. 522, 783; tree 8–13 m. tall, girth 1–1.5 m.); Patung Hsien, woodlands, common, alt. 1000–1600 m., May 1907 (No. 527ᵃ; tree 8–16 m. tall, girth 1–2 m.); Hsing-shan Hsien, woodlands, alt. 1000–1300 m., May 1907 (No. 3651; tree 10–16 m. tall, girth 1–2 m.); Fang Hsien, alt. 1000–1600 m., May 1907 (No. 516ᵃ). North-central Szech'uan: Pa-chou, alt. 1200 m., July 13, 1910 (No. 4585; tree 25 m. tall, girth 4 m.). Western Szech'uan: Wên-ch'uan Hsien, woods, alt. 1300–2000 m., October 1908 (No. 1093; tree 6–16 m. tall, girth 1–2 m.). Chekiang: vicinity of Ningpo, 1908, *D. Macgregor;* Pan-shan, near Hanchou, alt. 300 m., June 29, 1915, *F. N. Meyer* (No. 1485). Kiangsu: Shanghai, 1915, *D. Macgregor.* Kwangtung: without locality, *C. Ford* (No. 357).

In western Hupeh and in Szech'uan this tree is common from river-level to 1500 m. altitude and usually as scrub or coppice growth; large trees are very rare. In the type as described by Hance the leaves have prominent reticulate veins on the under side. This character is very variable; it is prominent in Henry's No. 2294, less so in Wilson's No. 522 and in Wilson's No. 523 it is scarcely noticeable; between the extremes there is every gradation. The branchlets are always furrowed and pubescent and the fruit ripens in one season.

Komarov (in *Act. Hort. Petrop.* XXII. 74 [1904]) and others have reported this species from Korea, but we have seen no specimens from there and are very doubtful if it grows so far north.

At Ichang this Oak is colloquially known as the Hsiao Peh-fan-li.

A picture of this tree will be found under No. 0211 of the collection of Wilson's photographs.

Quercus serrata Thunberg, *Fl. Jap.* 176 (1784). — Siebold & Zuccarini in *Abh. Akad. Münch.* IV. pt. III. 226 (*Fl. Jap. Fam. Nat.* II. 102) (1846). — Blume, *Mus. Bot. Lugd.-Bat.* I. 296 (1849–51). — A. De Candolle, *Prodr.* XVI. pt. 2, 50 (1864). — Hance in *Jour. Linn. Soc.* XIII. 8 (1871). — Brandis, *Forest Fl. Brit. Ind.* 486 (1874); *Ind. Trees,* 626, fig. 192 (1906). — Franchet & Savatier, *Enum. Pl. Jap.* I. 447 (1875). — Debeaux in *Act. Soc. Linn. Bordeaux,* XXXI. 362 (*Fl. Tché-foû,* 128) (1876). — Bretschneider, *Chinese Silkworm Trees,* 4 (1881). — Forbes in *Jour. Bot.* XXII. 86 (1884). — Franchet in *Nouv. Arch. Mus. Paris,* sér. 2, VII. 85 (*Pl. David.* I. 275) (1884). — Wenzig in *Jahrb. Bot. Gart. Berlin,* IV. 221 (1886). — Hooker f., *Fl. Brit. Ind.* V. 601 (1888). — Kanitz in *Növen. Gyüjt. Ered. Szechenyi,* II. 842 (*Pl. Enum.* 57) (1891); in *Wiss. Ergeb. Reise Szechenyi,* II. 731 (1891). — Shirai in *Tokyo Bot. Mag.* IX. 412, t. 7, fig. 9 (1895). — Skan in *Jour. Linn. Soc.* XXVI. 520 (1899). — Shirasawa, *Icon. Ess. For. Jap.* I. t. 26, fig. 1–12 (1900). — Komarov in *Act. Hort. Petrop.* XXII. 74 (*Fl. Mansh.* II.) (1903). — Pampanini in *Nuov. Giorn. Bot. Ital.* n. ser. XVII. 252 (1910). — Nakai in *Jour. Coll. Sci. Tokyo,* XXXI. 208 (*Fl. Kor.* II.) (1911). — Dunn & Tutcher in *Kew Bull. Misc. Inform.* add. ser. X. 253 (*Fl. Kwangtung & Hongk.*) (1912). — Matsumura, *Ind. Pl. Jap.* II. pt. 2, 29 (1913). — Léveillé, *Fl. Kouy-Tchéou,* 128 (1914).

Quercus polyantha Lindley apud Wallich, *Cat.* No. 2771 (nomen nudum) (1829).
Quercus serrata, β var. *tanbakuri* Blume, *Mus. Bot. Lugd.-Bat.* I. 297 (1850).
Quercus serrata, γ var. *attenuata* Blume, l. c. (1850).

Quercus serrata, ♂ var. *nana* Blume, l. c. (1850).
Quercus serrata, ε var. *obtusata* Blume, l. c. (1850).
Quercus acutissima Carruthers in *Jour. Linn. Soc.* VI. 33 (1862). — Schottky in *Bot. Jahrb.* XLVII. 638 (1912). — Nakai in *Tokyo Bot. Mag.* XXIX. 57 (1915). — Koidzumi in *Tokyo Bot. Mag.* XXX. 203 (1916).
Quercus serrata, var. *Roxburghii* A. De Candolle, *Prodr.* XVI. pt. 2, 51 (1864), excl. synon. "*Q. serrata* Roxb." — King in *Ann. Bot. Gard. Calcutta*, II. 22, t. 16 (1889).
Quercus Bombyx Hort. Leroy ex Koch, *Dendr.* II. pt. 2, 72 (pro synon.) (1873).
Quercus acutissima, var. *Roxburghii* Schottky in *Bot. Jahrb.* XLVII. 638 (1912).
Quercus Ushiyamana Nakai in Fedde, *Rep. Spec. Nov.* XIII. 250 (1914).

Western Hupeh: Patung Hsien, woods, alt. 1000 m., May 1907 (No. 536ᵃ; tree 10–20 m. tall, girth 1–2.5 m.); north and south of Ichang, alt. 30–1000 m., abundant, May 1907 (No. 3645; tree 10–20 m. tall, girth 1–2 m.). Western Szech'uan: Hokiang Hsien, *E. Faber* (No. 216); Wa-shan, woodlands, alt. 1300–2000 m., November 1908 (No. 1332; tree 8–12 m. tall, girth 1–3 m.); Mt. Omei, June 1904 (Veitch Exped. No. 5189). Chekiang: vicinity of Ningpo, 1908, *D. Macgregor;* Changhua Hsien, alt. 300 m., July 12, 1915, *F. N. Meyer* (No. 1546). Chili: near "Santun-ying," May 1913, *F. N. Meyer* (Nos. 39, 977); near " Changli," October 1905, *F. N. Meyer* (No. 197); Peking, cemetery, October 10, 1905, *J. G. Jack.* Shantung: Chifu, September 22, 1903, *C. S. Sargent.*

NORTHEASTERN ASIA. Korea: Seoul, September 25, 1905, *J. G. Jack;* Quelpaert, " Hongno," August 1907, *U. Faurie* (No. 1525); Hallaisan, September 4, 1908, *Taquet* (No. 1435); in woods, July 1909, June, July 1910, *Taquet* (Nos. 2544, 2548, 4446).

This is the common low-level Oak of the Yangtsze Valley and up to 1000 m. altitude is abundant everywhere in western Hupeh and in Szech'uan. It often forms pure woods or grows mixed with Pines, and as low coppice growth or scrub it covers low steep hills. In Japan it is common in Kyushu, Shikoku and in the warmer parts of Hondo and is cultivated in Tokyo and other places to the northward. When allowed to develop this Oak makes a handsome tree 25 m. tall, with stout widespreading branches and a trunk from 3 to 4 m. in girth. The leaves vary greatly in shape, in length of petiole and in length of the setae; when young the leaves are densely hairy, but at maturity they are glabrous or nearly so and always green on the under surface. The bark is dark gray, rough and fissured, but never corky. The fruit matures the second season. Its wood is valued for boat building and for general construction purposes, and that of coppice growth is largely used as fuel and for making charcoal; the cups are used for dyeing silk-yarn black. In Hupeh it is known as the Hwa-li and occasionally Hung (red) Hwa-li, but in the prefecture of Paoning in north-central Szech'uan, where a silkworm (*Antheraea Pernyi*) feeds on the leaves, it is known as Ching-kang. In the province of Kweichou this silkworm feeds on this Oak.

We do not think that the Indian plant can be maintained as a distinct variety, for the specimens before us from Korea exhibit as much variation as specimens from Japan and India do. Meyers No. 1546 agrees perfectly with that of *Q. acutissima* Carruthers.

Pictures of this tree will be found under Nos. 1, 65, 429, 438 and 0175 of the collection of Wilson's photographs and also in his *Vegetation of Western China*, Nos. 437, 438, 439 and 440.

Quercus variabilis Blume, *Mus. Bot. Lugd.-Bat.* I. 297 (1850). — A. De Candolle, *Prodr.* XVI. pt. 2, 50 (1864). — Sargent in *Garden & Forest*, VI. 385 (1893); *Forest Fl. Japan*, 68 (1894). — Shirai in *Tokyo Bot. Mag.* IX. 413, t. 7, fig. 3 (1895). — Shirasawa, *Icon. Ess. For. Jap.* I. 54, t. 28, fig. 1–11 (1900). — Matsumura & Hayata in *Jour. Coll. Sci. Tokyo*, XXII. 394 (*Enum. Pl. Formos.*) (1906). — Nakai in *Jour. Coll. Sci. Tokyo*, XXXI. 208 (*Fl. Kor.* II.) (1911). — Léveillé, *Fl. Kouy-Tchéou*, 129 (1914).

> *Quercus chinensis* Bunge in *Mém. Sav. Étr. Acad. Sci. St. Pétersbourg*, II. 135 (*Enum. Pl. Chin. Bor.* 61) (non Abel) (1835). — Bretschneider, *Chinese Silkworm Trees*, 5 (1881).
> ? *Quercus Moulei* Hance in *Jour. Bot.* XIII. 563 (1875). — Skan in *Jour. Linn. Soc.* XXVI. 519 (1899).
> *Quercus Bungeana* Forbes in *Jour. Bot.* XXII. 83 (1884). — Franchet in *Nouv. Arch. Mus. Paris*, sér. 2, VII. 85 (*Pl. David.* I. 275) (1884); in *Jour. de Bot.* XIII. 154 (1899). — Kanitz in *Növen. Gyüjt. Ered. Szechenyi*, II. 841 (*Pl. Enum.* 56) (1891); in *Wiss. Ergeb. Reise Szechenyi*, II. 731 (1891). — Skan in *Jour. Linn. Soc.* XXVI. 508 (1899). — Seemen in *Bot. Jahrb.* XXIX. 291 (1900). — Léveillé, *Fl. Kouy-Tchéou*, 127 (1914).
> *Quercus serrata*, var. *a chinensis* Wenzig in *Jahrb. Bot. Gart. Berlin*, IV. 221 (1886). — Matsumura, *Ind. Pl. Jap.* II. pt. 2, 29 (1913).
> *Quercus Bombyx tomentosa* Hort. ex *Kew Handlist*, 201 (pro synon. *Q. serratae* Thunberg) (1894).
> *Quercus serrata* Carruthers in *Jour. Linn. Soc.* VI. 32 (non Thunberg) (1862). — Schottky in *Bot. Jahrb.* XLVII. 638 (1912). — Nakai in *Tokyo Bot. Mag.* XXIX. 57 (1915). — Koidzumi in *Tokyo Bot. Mag.* XXX. 205 (1916).

W e s t e r n   H u p e h : Patung Hsien, woods, alt. 1000–1500 m., October 1907 (Nos. **532, 536**; tree 8–20 m. tall, girth 1–2.5 m.); north and south of Ichang, woods, alt. 300–1500 m., May 6 and October 1907 (Nos. **3648, 533, 534, 535**; tree 10–25 m. tall, girth 1–2 m., bark thick and cork-like); same locality, *A. Henry* (Nos. 2291, 2291ª); Hsingshan Hsien, woods, alt. 300–1500 m., September and October 1907 (Nos. **3647, 539**; tree 13–25 m. tall, girth 1–2 m.); Fang Hsien, woodlands, alt. 1000–1500 m., May and October 1907 (Nos. **538, 537**; tree 10–25 m. tall, girth 1–2 m.). W e s t e r n   S z e c h ' u a n : southeast of Tachien-lu, woods, alt. 1300 m., September 1908 (Nos. **1295, 1296**; tree 15–20 m. tall, girth 1.5–2.3 m.); without locality, *A. von Rosthorn* (No. 168). Y u n n a n : Mengtsze, woods, alt. 1600 m., *A. Henry*

(Nos. 9913, 9913ᵃ); Yuan-chiang, alt. 1600 m., *A. Henry* (No. 13247). Chili: Peking, western hills, April 30, 1912, *W. Purdom;* Peking, August 1865, July 1876, *S. W. Williams* (Herb. Hance, No. 11416; in Herb. Gray); same locality, *Skatschkoff* (in Herb. Gray); "Chang-li," November 1905, *F. N. Meyer* (No. 326). Shantung: Lau-shan, August 1907, *F. N. Meyer* (No. 307).

NORTHEASTERN ASIA. Korea: "Taikou," May 1906, *U. Faurie* (No. 194).

On the mountains of western Hupeh and in eastern Szech'uan between altitudes of from 800 to 1600 m. this is a very common tree in mixed woods and often forms pure stands; in western Szech'uan it is much less common. In Japan it is sometimes cultivated, but Sargent and Wilson did not see it there as a wild tree. At its best it is a tree 25 m. tall with a straight trunk 3 m. in girth and clothed with handsome pale gray deeply furrowed corky bark. The fruit ripens the second season, and is usually solitary, but occasionally in pairs. The shoots are quite glabrous, slightly pubescent or tomentose, but the under side of the leaves is always clothed with a dense pale gray felt. On young plants — seedlings or coppice growth — the petioles are very short and the leaves may be described as subsessile; the base is rounded or subauriculate and the secondary veins are more numerous. The cup and the acorn are variable in size. In central and western China this tree is known as the Hwa-k'o-li and its wood is valued for boat building and general construction purposes; the cups are employed in dyeing silk-yarn black; the bark is used by the peasants for roofing their houses. On the trunks of felled saplings of this Oak an edible fungus (*Hirneola polytricha*) is cultivated by the Chinese in western Hupeh.

Pictures of this tree will be found under Nos. 537, 544, 555, 0142 of the collection of Wilson's photographs and also in his *Vegetation of Western China*, Nos. 441, 442, 443.

We have not seen Hance's *Q. Moulei*, but believe that it belongs here. The vernacular name given, Ma-lieh, is doubtless a transliteration of Hwa-li, the Chinese name for *Q. serrata* Thunberg, and often applied also to *Q. variabilis* Blume.

Quercus Engleriana Seemen in *Bot. Jahrb.* XXIII. Beibl. No. 57, 47 (1897); XXIX. 291 (1900). — Skan in *Jour. Linn. Soc.* XXVI. 512 (1899). — Koidzumi in *Icon. Pl. Koisikav.* I. 111, t. 56 (1912).

Quercus obscura Seemen in *Bot. Jahrb.* XXIII. Beibl. No. 57, 49 (1897). — Skan in *Jour. Linn. Soc.* XXVI. 519 (1899).
Quercus sutchuenensis Franchet in *Jour. de Bot.* XIII. 512 (1899).

Western Hupeh: north and south of Ichang, woods, alt. 1300–2000 m., May 31, 1907 (No. **3633**; tree 6–10 m. tall); Changyang Hsien, woods, alt. 1600 m., May and November 1900 (Veitch Exped. No. 678, Seed No. 1223); without locality, *A. Henry* (No. 6167, cotype of *Q. obscura* Seemen). Eastern Szech'uan: Wushan Hsien, *A. Henry* (Nos. 5682, 5682ᵃ, co-type). Western Szech'uan: Wa-shan, woods, alt. 1600–2500 m., June 1908 (No. **3632**; tree 10 m. tall, girth 1.5 m.).

This species is common in rocky places in western Hupeh and in Szech'uan between 1000 and 2000 m. altitude and is always a small tree. The leaves persist through the winter and fall as the new ones unfold. They vary much in degree of dentation and are sometimes quite entire; the fruit ripens in one season. The specimen from Wa-shan is much less hairy than those from Hupeh and may possibly represent a distinct variety. Seemen's *Q. obscura* is founded on a flowering specimen with young leaves only, and some of the leaves are quite entire, but it is without doubt specifically identical with his *Q. Engleriana*.

Quercus semicarpifolia Smith in Rees, *Cyclop.* XXIX. No. 20 (1819). — Wallich, *Pl. As. Rar.* II. 56, t. 174 (1831). — Miquel in *Ann. Mus. Lugd.-Bat.* I. 110 (1863–64). — A. De Candolle, *Prodr.* XVI. pt. 2, 15 (1864). — Brandis, *Forest Fl. Ind.* 479, t. 64 (1874). — Wenzig in *Jahrb. Bot. Gart. Berlin*, IV. 219 (1886). — Hooker f., *Fl. Brit. Ind.* V. 601 (1888). — King in *Ann. Bot. Gard. Calcutta*, II. 21, t. 15, fig. 1–3 (1889). — Kanitz in *Növen. Gyüjt. Ered. Szechenyi*, II. 841 (*Pl. Enum.* 56) (1891); in *Wiss. Ergeb. Reise Szechenyi*, II. 731 (1891). — Franchet in *Jour. de Bot.* XIII. 150 (1899). — Skan in *Jour. Linn. Soc.* XXVI. 520 (1899).

> *Quercus obtusifolia* Roxburgh, *Prodr. R. Nepal.* 56 (1825).
> *Quercus Cassura* Roxburgh, l. c. 57 (1825).

Western Szech'uan: northwest of Tachien-lu, Ta-p'ao-shan, forests, alt. 3000–4100 m., October 1908 (No. **4579**; small tree 6–15 m. tall, girth 1–2 m.); west of Tachien-lu, uplands, alt. 3300–4600 m., October 1910 (No. **4281**ᵃ; bush 0.6–5 m.); same locality, July 1903 (Veitch Exped. Nos. 4503, 4503ᵃ).

On the uplands round Tachien-lu this is a common gregarious shrub, and with allied species, *Juniperus squamata* Lambert and certain species of Rhododendrons with small leaves, ascends to the altitudinal limits of woody vegetation. The size of the bushes and of the leaves depends upon situation and altitude, and the more exposed and austere the conditions the smaller the leaves and the dwarfer the bush. In the ravines and wooded areas between 3000 and 4000 m. it forms a tree, from 12 to 15 m. tall, with a trunk 2 m. in girth, covered with dark gray bark which is fissured into thin, elongated scaly flakes. On the trees the leaves are much longer and frequently quite entire. The fruit is solitary or more rarely in pairs and ripens in one season, and the nearly globular acorn, seated in a shallow cup with a fringe of loose scales, is most distinct. The leaves persist through the winter to the summer and a few until the autumn and the plant is entitled to be termed evergreen. The wood is very hard and makes the best of charcoal.

A picture of this Oak will be found under No. 186 of the collection of Wilson's photographs.

Quercus aquifolioides Rehder & Wilson, n. sp.

Frutex v. arbor parva, 1–10-metralis, ramosissima; ramuli hornotini fusci, fasciculato-pilosi, annotini glabrescentes. Folia persistentia per tres annos, coriacea, subsessilia, ovalia v. ovata v. elliptica, rarius obovato-elliptica, apice rotundata saepe spinoso-mucronata, basi subcordata v. auriculata, sinuoso-spinoso-dentata v. integra, margine leviter revoluta, 3–7.5 cm. pleraque 4–5.5 cm. longa et 2–5.5 cm. lata, supra maturitate fere glabra v. leviter scabrella, nitidula, subtus dense tomento fusco-aureo obtecta costa nervisque glabrescentibus exceptis, nervis utrinsecus 5–7 ante marginem plerisque furcatis supra impressis subtus manifeste elevatis, costa media supra leviter subtus manifeste elevata; petioli 1–5 mm. longi, fasciculato-pilosi, demum glabrescentes, stipulis ad secundum annum persistentibus, lineari-lanceolatis 5–8 mm. longis membranaceis fasciculato-pubescentibus. Flores feminei plures in spicis axillaribus erectis pedunculatis 1–2 cm. longis fasciculato-pubescentibus; bracteae triangulares acutae villosae; styli plerumque 3, recurvati. Amenta mascula fasciculato-pubescentia, floribus remotis; perianthium rotatum, membranaceum, 2 mm. longum, plerumque lobis 5 brevibus imbricatis ciliatis v. irregulariter eroso-denticulatis, intus villosum, extus sparse pubescens; stamina 5 filamentis perianthium aequantibus glabris, antheris ovoideis 1.25 mm. longis. Fructus biennes, 1–5 in spicis erectis; cupula cupuliformis, 8–10 mm. diam., 4–5 mm. alta, intus adpresse sericeo-pilosa; bracteae liberae, lineari-lanceolatae, laxe adpressae, infra dense villosae, in parte superiore ciliatae ceterum fere glabrae, brunneae, superiores cupulam superantes et marginem brevem fimbriatum formantes; glans ovoidea, 1.2–1.5 cm. longa et 1–1.1 cm. diam., apiculata.

Western Szech'uan: west of Kuan Hsien, Pan-lan-shan, uplands, alt. 3500–4500 m., June 24, 1908 and October 1910 (Nos. 3632ª, 4372ª, 4580, type).

This new species is perhaps most closely related to *Q. semicarpifolia* Smith, but that species has an annual fructification and a larger, nearly globular acorn. In western Szech'uan these two species grow in the same regions at similar altitudes and superficially very strongly resemble each other. Both are gregarious plants and on the windswept uplands cover large areas in the form of scrub; in sheltered places they develop into small trees. In both the under surface of the leaves is a rich golden brown, but in *Q. aquifolioides* they persist at least until the end of the third season. *Quercus spinosa* David agrees with our new species in having biennial fruit, but in David's species the leaves at maturity are glabrous, except the midrib in its lower half, and the tomentum on the young leaves is gray, never golden brown; the veins are more spreading and the leaf strongly bullate; the scales of the cup are very short, firmly appressed and quite distinct one from another. In

Hupeh *Q. spinosa* grows always on limestone, whereas the region of the Chino-Thibetan borderland, where *Q. aquifolioides* grows, is composed of granite, gneiss and mud shales. No. 4372 differs slightly from the type in having the scales of the cup more uniformly linear-lanceolate, more loosely appressed and more decidedly villose.

A picture of this Oak will be found under No. 141 of the collection of Wilson's photographs and also in his *Vegetation of Western China*, No. 435.

We have preferred to coin a new name for this species instead of elevating the following variety to specific rank, as there exists an older, though not valid, homonym, namely, *Q. rufescens*, Hooker f. & Thomson, a synonym of *Castanopsis hystrix* A. De Candolle.

Quercus aquifolioides, var. rufescens Rehder & Wilson, n. comb.

Quercus Ilex, var. *rufescens* Franchet in *Jour. de Bot.* XIII. 151 (1899). — Skan in *Jour. Linn. Soc.* XXVI. 516 (1899). — Koidzumi in *Tokyo Bot. Mag.* XXVI. 161 (1912).

Quercus semicarpifolia, var. *rufescens* Schottky in *Bot. Jahrb.* XLVII. 642 (1912).

Western Szech'uan: Tachien-lu, alpine moorlands, alt. 2600–4600 m., June and September 1908 (Nos. **3626, 4372**, seeds only; shrub 0.3–3 m., forming scrub).

This variety differs from the type in its pubescence, which is pale gray and loosely matted on the young leaves and yellowish or brownish gray and loose on the old leaves, from which it often disappears entirely; the cup scales are also rather thicker, not so long and more densely villose than in the type species. The name *rufescens* is more applicable to the type than to this variety, but Franchet's description is clearly of this plant. We have not seen any of the specimens referred by Franchet to this variety, but ours agree exactly with his description and are from the same locality as several of his specimens.

This is a common Scrub Oak on the uplands and moors around Tachien-lu and regions to the north and west. When mixed with allied species it covers vast areas.

A picture of this Oak will be found under No. 225 of the collection of Wilson's photographs and also in his *Vegetation of Western China*, No. 436.

Quercus Gilliana Rehder & Wilson, n. sp.

Arbor parva, 5–8-metralis v. frutex ramosissimus 0.25–1.5 m. altus; ramuli purpureo-brunnei, fasciculato-pilosi, annotini tarde glabrescentes. Folia coriacea, per duos annos persistentia, subsessilia, ovalia v. obovata, apice rotundata spinoso-mucronata, basi subcordata v. leviter auriculata, sinuoso-spinoso-dentata, 3–5 cm. longa et 2–3.5 cm. lata, sub maturitate utrinque glabra, costa media utrinque basin versus interdum fasciculato-pilosa excepta, concoloria v. subtus fuscescentia, nervis utrinsecus 5–6 ante marginem furcatis supra leviter subtus manifeste elevatis ut costa; petioli 1–3 mm. longi, glabrescentes; stipulae erectae, subulatae, 4–5 mm. longae, persistentes. Fructus annui in axillis foliorum superiorum 2–4 aggregati, breviter pedunculati pedunculo crasso fasciculato-piloso; cupula fere hemisphaerica 5–6 mm. alta et 9–10 mm. diam., intus villosa, margine non fimbriata; bracteae adpressae, triangulari-ovatae, basi leviter tumidae, acutiusculae, apice brunneæ glabrescentes, ceterum tomento cinereo-flavascente velutino obtectae; glans ovoidea, circiter 1.2 cm. alta, 1 cm. diam., glabra, stigmatibus divaricatis coronata.

Western Szech'uan: southeast of Tachien-lu, alt. 2000–2600 m., October 1910 (Nos. **4583**, type **4281**, seeds only).

A well-marked species distinguished by its annual fructification from all Chinese Oaks with spinescent leaves except *Q. semicarpifolia* Smith. The latter has a more or less persistent felt of yellow-brown or gray-brown hairs on the under side of the leaves, a shallow saucer-shaped cup with fewer, triangular-lanceolate to lanceolate scales and is loosely fringed round the rim; the acorn is more or less globular. The cup of our new species somewhat resembles that of *Q. Ilex* Linnaeus, but is much smaller in every way. In arid parts of the Tung River valley and in those of its tributaries in western Szech'uan between 1300 and 2600 m. altitude this new Oak is very common and covers large areas in the form of scrub from 0.3 to 1 m. high. Under more favorable conditions and on the margins of woods it is a small tree with rather slender branches which form an oval or flattened, dense crown. It is named for the late Captain W. J. Gill, who in 1877 made an extended tour through western China. He was one of the first travellers to visit Tachien-lu and in his book (*The River of Golden Sand*) frequent references to the "Holly-leaved Oak" of western Szech'uan are made.

Quercus spinosa David apud Franchet in *Nouv. Arch. Mus. Paris*, sér. 2, VII. 84 (*Pl. David*. I. 274) (1884).

Quercus bullata Seemen in *Bot. Jahrb.* XXIII. Beibl. No. 57, 48 (1897); XXIX. 289 (1900).
Quercus ilicioides David apud Bretschneider, *Hist. Europ. Bot. Discov. China*, 850 (nomen seminudum) (1898).
Quercus Ilex, var. *spinosa* Franchet in *Jour. de Bot.* XIII. 152 (1899). — Skan in *Jour. Linn. Soc.* XXVI. 516 (1899). — Seemen in *Bot. Jahrb.* XXIX. 290 (1900). — Koidzumi in *Tokyo Bot. Mag.* XXVI. 160 (1912).
Quercus Ilex, var. *bullata* Franchet in *Jour. de Bot.* XIII. 152 (1899). — Seemen in *Bot. Jahrb.* XXIX. 290 (1900).
Quercus semicarpifolia, var. *glabra* Skan in *Jour. Linn. Soc.* XXVI. 520 (1899), quoad specimen e Mengtsze; an Franchet?
Quercus semicarpifolia, var. *spinosa* Schottky in *Bot. Jahrb.* XLVII. 642 (1912).

Western Hupeh: Hsing-shan Hsien, exposed cliffs, alt. 2000–2600 m., July 1907 (No. **3643**; bush or small tree, 2–8 m. tall); Changyang Hsien, woods and rocky places, alt. 1500–2200 m., July 1901 (Veitch Exped. No. 1493; tree 2–6 m. tall); without locality, *A. Henry* (Nos. 5981ᶜ, 5981ᵃ, type of *Q. bullata*). Yunnan: Mengtsze, "Ta-hei-shan," summit, *A. Henry* (No. 10217; tree 6 m.).

This Oak is common as a shrub on exposed limestone cliffs and in rocky places generally throughout western Hupeh and eastern Szech'uan between altitudes of from 800 to 2500 m. and is colloquially known as the " Chi-kang shu." The bushes grow 1–5 m. tall and are very densely and intricately branched. Under favorable conditions it develops into a small tree 10 m. tall with a short, thick trunk, sharply ascending-spreading branches and slender pendent branchlets, the whole forming a shapely oval crown.

The young leaves are usually more or less densely clothed on the under side with a loose felt of yellow-brown to yellowish gray hairs which disappears early, but occasionally they are practically glabrous even when not fully unfolded. This variation in degree of pubescence on the young leaves and its early disappearance reconciles the diverse statements of Franchet and von Seemen. The basal half of the midrib on the under surface of the leaf is always clothed with a dense gray felt of villose hairs. The leaves are dark green, strongly bullate and normally quite entire, though they are often coarsely dentate and spiny, and persist until the summer of the second season. The fructification is biennial and the fruit is pedunculate in the axils of the lower leaves of its season's growth. Henry's No. 10217, referred by Skan to *Q. semicarpifolia* var. *glabra*, differs only in the leaves being slightly less bullate and rather larger than in the Hupeh form, which may be owing to different ecological conditions. The fructification is clearly biennial, so it cannot belong where Skan placed it, since in that species the fructification is annual. Where the specimens belong on which Franchet based the variety cannot be stated without seeing them.

Pictures of this Oak will be found under Nos. 566 and 571 of the collection of Wilson's photographs and also in his *Vegetation of Western China*, No. 434.

Quercus acrodonta Seemen in *Bot. Jahrb.* XXIII. Beibl. No. 57, 48 (1897); XXIX. 290 (1900).

Quercus phylliræoides Franchet in *Nouv. Arch. Mus. Paris*, sér. 2, VII. 85 (*Pl. David.* I. 275) (non Gray) (1884).
Quercus Ilex, var. *phyllireoides* Franchet in *Jour. de Bot.* XIII. 152 (pro parte) (1899).
Quercus Ilex, var. *acrodonta* Skan ın *Jour. Linn. Soc.* XXVI. 516 (1899). — Koidzumi in *Tokyo Bot. Mag.* XXVI. 160 (1912).

Western Hupeh: Hsing-shan Hsien, ravine, alt. 1300 m., June 8, 1907 (No. **3644**; tree 5 m. tall, girth 0.3, one only seen); Ichang and neighborhood, *A. Henry* (Nos. 2954, 2954ᵇ, 3425); without locality, *A. Henry* (No. 7619).

This Oak grows on the limestone cliffs in western Hupeh, but is rare. It is a small, densely branched evergreen tree or large bush. The leaves are shining green above and persist for two seasons, and the fructification is biennial, subsessile and subterminal in the axils of the uppermost leaves of its current season's growth. It is very closely related to *Q. phillyraeoides* Gray, and further knowledge may prove it to be only a variety of that species. The close-matted, interwoven felt of pale gray hairs on the under side of the leaf and the midrib slightly impressed near the base of the leaf serve to distinguish it, and since it has a name we think it is best, at present, to consider it distinct.

Quercus spathulata Seemen in *Bot. Jahrb.* XXIII. Beibl. No. 57, 49 (1897). — Skan in *Jour. Linn. Soc.* XXVI. 521 (1899). — Pampanini in *Nuov. Giorn. Bot. Ital.* n. ser. XVII. 252 (1910).

Western Hupeh: Hsing-shan Hsien, rocky places, alt. 1300 m., May and November 1907 (No. 540; tree 6–12 m. tall, girth 1–2 m.); Changyang Hsien, woods, alt. 1300 m., May 1907 (No. 3640; tree 5–8 m. tall, girth 1.5–2 m.); without locality, *A. Henry* (No. 6359). Western Szech'uan: Chien-chi Hsien, summit of Fei-yueh-ling, alt. 3000 m., May 1908 (No. 3625; bush 2.5 m. high); without locality, alt. 2300 m., July 1903 (Veitch Exped. No. 4504; tree 10 m. tall).

This is not a common tree in western Hupeh or Szech'uan, but is scattered over a wide area. It never attains a large size. The crown is dense and flattened-oval in outline. The leaves are coriaceous and shining green above and persist for two or three seasons. They vary slightly in size and are frequently quite entire and the under side of mature leaves is either quite glabrous or clothed with a short, curled gray pubescence. The fruit is biennial and short-stalked in the axils of the leaves immediately below the winter-bud of its current season's growth. The scales of the cup are narrow and recurved and the cup resembles that of *Q. cerris* Linnaeus, but is smaller.

Quercus Baronii Skan in *Jour. Linn. Soc.* XXVI. 507 (1899).

Quercus Dielsiana Seemen in *Bot. Jahrb.* XXIX. 291, fig. 2 (1900).

Western Szech'uan: Wên-ch'uan Hsien, valley of Min River, thickets, alt. 1500 m., May 26, 1908 (No. 3628; bush 2 m.); "Tsa-ku-lao", *A. von Rosthorn* (No. 2533, type of *Q. Dielsiana*). Shensi: "Lao-y-san" and "Kin-qua-san," 1897, *G. Giraldi;* "Ta-hua-shan," December 29, 1913, *F. N. Meyer;* south of Sian Fu, mountains near "Nan-tu-tchu," January 21, 1914, *F. N. Meyer*(No. 1692).

In warm semi-arid regions of the Min River valley this Oak is a common shrub, but Wilson did not meet with it elsewhere. Meyer observes that it grows on sterile mountain slopes in Shensi and is semi-evergreen, the leaves persisting through the winter. The fruit is biennial; the scales of the cup are recurved and the cup resembles that of *Q. spathulata* Seemen. The leaves are membranous with spinescent teeth.

Quercus glauca Thunberg, *Fl. Jap.* 175 (1784). — Banks, *Icon. Kaempfer.* t. 17 (1791). — Siebold & Zuccarini in *Abh. Akad. Münch.* IV. pt. 3, 226 (*Fl. Jap. Fam. Nat.* II. 102) (1846). — Blume, *Mus. Bot. Lugd.-Bat.* I. 302 (1850). — Miquel in *Ann. Mus. Lugd.-Bat.* I. 115 (1863–64). — A. De Candolle, *Prodr.* XVI. pt. 2, 100 (1864). — Masters in *Gard. Chron.* n. ser. I. 632 (1874). — Hance in *Jour. Bot.* XIII.

363 (1874); XX. 294 (1882). — Franchet & Savatier, *Enum. Pl. Jap.*
I. 448 (1875). — Forbes in *Jour. Bot.* XXII. 86 (1884). — Franchet in
Nouv. Arch. Mus. Paris, sér. 2, VII. 86 (*Pl. David.* I. 276) (1884); in
Jour. de Bot. XIII. 159 (1899). — Wenzig in *Jahrb. Bot. Gart. Berlin,*
IV. 233 (1886). — Hooker f., *Fl. Brit. Ind.* V. 604 (1888). — King in
Ann. Bot. Gard. Calcutta, II. 29, t. 23 (1889). — Skan in *Jour. Linn.
Soc.* XXVI. 515 (1889). — Seemen in *Bot. Jahrb.* XXIX. 293 (1900). —
Shirasawa, *Icon. Ess. For. Jap.* I. 56, t. 30, fig. 13–24 (1900). — Mat-
sumura & Hayata in *Jour. Coll. Sci. Tokyo,* XXII. 392 (*Enum. Pl.
Formos.*) (1906). — Matsumura, *Ind. Pl. Jap.* II. pt. 2, 27 (1912). —
Léveillé in *Fl. Kouy-Tchéou* 128 (1914). — Nakai in *Tokyo Bot. Mag.*
XXIX. 61 (1915).

> *Quercus annulata* Smith in Rees, *Cyclop.* XXIX. No. 22 (1819). — Miquel
> in *Ann. Mus. Lugd.-Bat.* I. 114 (1863–64). — A. De Candolle, *Prodr.* XVI.
> pt. 2, 100 (1864). — Brandis, *Forest Fl. Ind.* 487, t. 65 (1874), exclud.
> synon. *Q. semiserrata.*
> *Quercus Phullata* Hamilton apud D. Don, *Prodr. Fl. Nepal.* 57 (1825).
> *Quercus laxiflora* Lindley in Wallich, *Cat.* No. 2774 (nomen nudum) (1829). —
> A. De Candolle, *Prodr.* XVI. pt. 2, 108 (1864).
> *Quercus dentosa* Lindley in Wallich, *Cat.* No. 2775 (nomen nudum) (1829).
> *Quercus glauca,* var. *caesia* Blume, *Mus. Bot. Lugd.-Bat.* I. 303 (1850).
> *Cyclobalanopsis glauca* Oersted in *Kjoebenh. Vidensk. Meddel.* XVIII. 70
> (1866). — Schottky in *Bot. Jahrb.* XLVII. 655 (1912).
> *Cyclobalanopsis annulata* Oersted in *Kjoebenh. Vidensk. Meddel.* XVIII. 70
> (1866).

Kiangsi: Kuling, thickets, alt. 1300–1500 m., July 29 and 31,
1907 (Nos. **1558, 1559**; dense bush, 3–6 m.). Western Hupeh:
Changyang Hsien, woods, alt. 1000–1500 m., June and October 1907,
January 1909 (Nos. **545, 542ª**; tree 10–20 m. tall, girth 1.5–3 m.);
Changlo Hsien, woods, alt. 1300–1600 m., May 1907 (Nos. **3634,
3639**; tree 5–15 m. tall, girth 1–2 m.); Patung Hsien, woods, alt.
1300 m., October 1907 (No. **541**; tree 6–12 m. tall); Hsing-shan
Hsien, woods, alt. 600–1000 m., May and June 1907 (Nos. **3646, 541ª**;
tree 6–20 m. tall, girth 1–2.5 m.); neighborhood of Ichang, *A. Henry*
(Nos. 3098, 1561, 3426); without locality, *A. Henry* (Nos. 5277ᵇ, 5632ᵉ;
in fruiting specimens only in Herb. New York). Western Szech'uan:
southeast of Tachien-lu, woodlands, alt. 1600–2000 m., October 1908
(No. **1298**; tree 10 m. tall). Yunnan: Mengtsze, woods, alt. 1800
m., *A. Henry* (No. 9299ª; in Herb. Arnold Arboretum). Chekiang:
vicinity of Ningpo, 1908, *D. Macgregor.* Fokien: without locality,
Dunn's Exped., April to June 1905 (Hongkong Herb. No. 3494).
Formosa: Bankinsing, *A. Henry* (No. 428). Korea: Quaelpaert,

woods, October 1906, *U. Faurie* (No. 185); same locality, May and July 1908, *Taquet* (Nos. 1427, 1432).

This is the most common evergreen Oak in western Hupeh and in eastern Szech'-uan, where it grows from river-level to 1600 m. above the sea. On the Lushan mountains in Kiangsi it is abundant, but in western Szech'uan it is not very common. It is a handsome tree with a bushy flattened-round head of widespreading branches. In Hupeh it is known as the Tieh-chou-li and the wood, which is hard and tough, is highly valued for general construction purposes.

The numerous specimens before us show considerable variation in degree of pubescence on the young shoots and of glaucescence on the under side of the leaves. The leaves also vary in size, but in all of them the basal part of the leaf is entire and the teeth, with their gland-tipped mucro, are pointed forward and are never spreading. The fruit is annual and the concentric rings of the cup are very regular and are quite entire or more or less crenulate on the same cluster. The acorns are much longer than broad, narrowly cylindric or ovoid-conic. Our No. 1298 has the venation rather reticulate on the under side of the leaf and the cup is slightly villose, and in these respects approaches *Q. glauca*, var. *villosa* Skan, of which we have seen a co-type. Skan (in *Jour. Linn. Soc.* XXVI. 515 [1899]) cites Henry's No. 9299[a] under his var. *villosa*. We suspect the specimens under this number have been mixed, for in this herbarium the number must be referred to typical *Q. glauca* Thunberg.

A picture of this tree will be found under No. 585 of the collection of Wilson's photographs and also in his *Vegetation of Western China*, No. 431.

Quercus glauca, f. gracilis Rehder & Wilson, n. forma.

A typo recedit foliis minoribus angustioribusque 4.5–8 cm. longis et 1.5–2.5 cm. latis ovato-oblongis v. oblongis v. lanceolato-oblongis infra v. circa medium latissimis sensim acuminatis basi late cuneatis v. fere rotundatis minus profunde serratis.

> *Quercus Vibrayeana* Seemen in *Bot. Jahrb.* XXIX. 293 (non Franchet & Savatier) (1900).

Western Hupeh: Patung Hsien, woods, alt. 1500 m., June and November 1907 (No. **687,** type; tree 12 m. tall, girth 2 m.); neighborhood of Ichang, *A. Henry* (No. 3088). **Eastern Szech'uan:** Wushan Hsien, *A. Henry* (No. 7076).

In its small leaves this variety suggests *Q. myrsinaefolia* Blume, but the leaves are pubescent on the under side, not glabrous, as in that species. In western Hupeh this variety is not uncommon and grows with the type species.

Quercus oxyodon Miquel in *Ann. Mus. Lugd.-Bat.* I. 114 (1863–64).— A. De Candolle, *Prodr.* XVI. pt. 2, 98 (1864).

> *Cyclobalanopsis oxyodon* Oersted in *Kjoebenh. Vidensk. Meddel.* XVIII. 71 (1866).
> *Quercus lineata*, var. β *oxyodon* Wenzig in *Jahrb. Bot. Gart. Berlin*, IV. 232 (1886). — Hooker f., *Fl. Brit. Ind.* V. 605 (1888). — King in *Ann. Bot. Gard.*

Calcutta, II. 33, t. 26, fig. 3 (1889). — Skan in *Jour. Linn. Soc.* XXVI. 517 (1899). — Seemen in *Bot. Jahrb.* XXIX. 293 (pro parte) (1900).
Quercus glauca, var. *lineata* Franchet in *Jour. de Bot.* XIII. 159 (1899).
Quercus lineata, var. *grandifolia* Skan in *Jour. Linn. Soc.* XXVI. 517 (1899).
Quercus lineata, var. *macrophylla* Seemen in *Bot. Jahrb.* XXIX. 294 (1900).
Cyclobalanopsis lineata, var. *oxyodon* Schottky in *Bot. Jahrb.* XLVII. 654 (1912).
Cyclobalanopsis lineata, var. *grandiflora* Schottky, l. c. (1912).

Western Hupeh: north and south of Ichang, woods, alt. 1100 m., October 1907 (No. **544**; tree 6–10 m. tall); same locality, May and October 1900 (Veitch Exped. No. 647, Seed No. 800); without locality, *A. Henry* (No. 5632ᵈ). Eastern Szech'uan: Wushan Hsien, *A. Henry* (Nos. 5632, 5632ᵃ, 5632ᵈ, 5632 , 5277ᵃ).

INDIA.   Assam: Khasia, alt. 5000 ft., *Hooker f. & Thomson* (in Herb. Gray, co-type).

In western Hupeh and in eastern Szech'uan this is a common tree in woods between 600 and 1200 m. above the sea level. It is a low-growing tree with wide-spreading branches which form a broad, flattened crown. The leaves persist through the winter, but fall as the new leaves unfold in the spring. The winter-buds are angular and are composed of many pale gray chaff-like, acute or mucronate scales. The fruit is annual, spicate and clustered near the ends of the branchlets. The cup is sessile and saucer-shaped and the gray concentric rings are somewhat irregular, not closely appressed, and are crenulate. The acorn is subglobose to ovoid and is broad or broader than high. On the co-type the blade of the leaf is from 10 to 12 cm. long and the petiole is from 2 to 2.5 cm. in length; and on the numerous Chinese specimens the blade varies in length from 12 to 21 cm. and the petiole from 2 to 5 cm. The lateral veins are deeply or only slightly impressed on the upper surface of the leaf.

Wenzig united this species with *Q. lineata* Blume, and his view has been generally accepted; to us, however, it is untenable. We have before us a co-type of Blume's species and consider it distinct from Hooker f. & Thomson's Khasia plant, with which the Chinese plant agrees exactly. In the Java plant the leaves are narrower and uniformly smaller, quite entire or with a few shallow, minutely pointed teeth near the apex of the leaf; the winter-buds are small, ovoid and covered with few brown obtuse scales. The cup is nearly urn-shaped, distinctly higher than broad, and the concentric rings are very regular, quite entire or slightly crenulate, and the whole cup is covered with a yellowish villose tomentum; the fruit is subterminal and, as the co-type clearly shows, ripens in the second season.

It is probable that the specimens collected by von Rosthorn which von Seemen (in *Bot. Jahrb.* XXIX. 294 [1900]) refers to *Q. lamellosa* Smith belong to *Q. oxyodon*.

Quercus oxyodon, var. Fargesii Rehder & Wilson, n. comb.

Quercus Fargesii Franchet in *Jour. de Bot.* XIII. 158 (1899).
Quercus lineata, var. *Fargesii* Skan in *Jour. Linn. Soc.* XXVI. 517 (1899).
Quercus lineata, var. *oxyodon* Seemen in *Bot. Jahrb.* XXIX. 293 (pro parte) (non Wenzig) (1900).
Cyclobalanopsis lineata, var. *Fargesii* Schottky in *Bot. Jahrb.* XLVII. 654 (1912).

Western Szech'uan: Wa-shan, alt. 1300–2300 m., October 1908
(No. 1218; tree 6–10 m. tall); Hungya Hsien, Wa-wu-shan, woods,
alt. 1300–2000 m., September 14, 1908 (No. 1218a; tree 6 m. tall);
Mt. Omei, June 1904 (Veitch Exped. No. 5190; tree 6 m. tall).

This is a common tree in the moist woods round the base of Mt. Omei. It is
distinguished from the type by the concentric rings of the cup being entire or only
slightly crenulate. The pubescence on the cupule and on the winter-buds is also
more yellowish than in the type.

Here may be added notes on some species not collected during the Arnold
Arboretum Expeditions.

Quercus mongolica Fischer apud Turczaninow in *Bull. Soc. Nat. Mosc.* I.
101 (1838); XXVII. pt. 1, 409 (1854). — Ledebour, *Fl. Ross.* III. pt. 2, 589
(1850). — Maximowicz in *Bull. Phys.-Math. Acad. Sci. St. Pétersbourg*, XV.
137 (1857); in *Mém. Sav. Étr. Acad. Sci. St. Pétersbourg*, IX. 241 (*Prim. Fl.
Amur.*) (1859). — Regel in *Mém. Acad. Sci. St. Pétersbourg*, sér. 7, IV. No. 4,
130 (*Tent. Fl. Ussur.*) (1861). — Carruthers in *Jour. Linn. Soc.* VI. 32 (1862). —
A. De Candolle, *Prodr.* XVI. pt. 2, 14, 487 (1864). — Fr. Schmidt in *Mém.
Acad. Sci. St. Pétersbourg*, sér. 7, XII. No. 2, 171 (*Reis. Amur. Sachal.*) (1868). —
Hance in *Jour. Linn. Soc.* X. 487 (1869); XIII. 7 (1873). — Bretschneider,
Chinese Silkworm Trees, 5 (1881). — Forbes in *Jour. Bot.* XXII. 86 (1884). —
Wenzig in *Jahrb. Bot. Gart. Berlin*, IV. 218 (1886). — Herder in *Act. Hort. Petrop.*
XI. 365 (1892). — Korshinsky in *Act. Hort. Petrop.* XII. 388 (1892). — Skan in
Jour. Linn. Soc. XXVI. 518 (1899). — Komarov in *Act. Hort. Petrop.* XXII. 68
(*Fl. Mandsh.*) (1904). — Nakai in *Jour. Coll. Sci. Tokyo*, XXXI. 208 (*Fl. Kor.*
II.) (1911). — Koidzumi in *Tokyo Bot. Mag.* XXVI. 165 (1912). — Nakai in *Tokyo
Bot. Mag.* XXIX. 57 (1915). — Miyabe & Miyake, *Fl. Saghal.* 420, t. 11, fig. 3
(1915).

Quercus Robur Pallas, *Fl. Ross.* II. 3 (pro parte) (non Linnaeus) (1788).

Quercus sessiliflora, var. *mongolica* Franchet in *Nouv. Arch. Mus. Paris*, sér.
2, VII. 83 (*Pl. David.* I. 273) (1884).

Quercus crispula, var. *sachalinensis* Koidzumi in *Tokyo Bot. Mag.* XXVI. 164
(1912).

Quercus crispula, var. *manshurica* Koidzumi, l. c. (1912).

Quercus mongolica, γ *manshurica* Nakai in *Toyko Bot. Mag.* XXIX. 58 (1915).

CHINA. Chili: west Weichang, 1910, *W. Purdom* (Nos. 105, 114); Weichang, 1910, *W. Purdom* (Nos. 42, 43, 97, 95).

NORTHEASTERN ASIA. Mandshuria: east slope Khingan Mts., Yolo
Pass, August 15, 1903, *C. S. Sargent*. Amur: valley of Amur River, May and June
1891, *S. Korshinsky;* same locality, 1855, *L. Schrenk;* same locality, *C. Maximowicz*. Ussuri: Khabarovsk, August 18, 23, 1903, *C. S. Sargent;* near Vladivostock, August 18, 1903, *C. S. Sargent*. Saghalien: Sakhai-hama, seacoast,
August 5, 1914, *E. H. Wilson* (No. 7372). Korea: Song-chong, September 4,
1903, *C. S. Sargent;* Chinnampo, September 20, 1905, *J. G. Jack;* same locality,
August 1906, *U. Faurie* (No. 191); Seoul, September 25, 1905, *J. G. Jack;* " Ouen-
san," July 1906, *U. Faurie* (No. 190); Wan-san, September 5, 1903, *C. S.
Sargent;* " Kan-ouen-to," July 3, 1901, *U. Faurie* (No. 608). " Hoang-hai-to,"
August 1906, *U. Faurie* (No. 189); Quelpaert, woods, July 1910, *Taquet* (Nos.
4447, 4448); Quelpaert, Hallaisan, alt. 1000–1700 m., August 1907, June 1909,
Taquet (Nos. 326, 2547, 2550).

JAPAN. Hokkaido: Kitami province, Rubeshihe, August 16, 1914, *E. H.*

Wilson (No. 7384; tree 15 m. tall, bark deeply furrowed); Ishikari province, Sapporo, September 20, *K. Miyabe;* Oshima province, Hakodate, woods, August 1903, *U. Faurie* (No. 5421).

This is the common Oak of northeast continental Asia. It is apparently common in Mandshuria, Korea and the Amur region and occurs sparingly in Saghalien and Hokkaido. In Japan it is abundantly represented by its variety *grosseserrata*, which is doubtfully distinct. We have much material before us, but unfortunately have little information of taxonomic importance. Wilson met with it as scrub only in Saghalien, but in the Hokkaido he notes that the very few trees he saw had dark and deeply furrowed bark. Whether this character is constant or not we do not know, and Wilson states that old trees of the var. *grosseserrata* are occasionally found with dark and furrowed bark. In herbaria the type may be recognized by the more thickened scales of the cup, which is usually slightly fringed, and by its broader and larger acorns. The dentation of the leaf is variable and the teeth are broad and rounded or narrow and acute. The quality of timber depends very much upon the habitat of the tree, but it is stated (W. S. in *Timber Trades Journal*, July 31, 1915) that " that of Mandshurian *Q. mongolica* is indistinguishable from that of *Q. grosseserrata*." The wood of the continental *Q. mongolica* is not yet so well known commercially as the wood of the var. *grosseserrata* of Hokkaido, but it would appear that the continental area is a timber field of much promise. It is vastly more extensive than that of Hokkaido, but we have no statistics giving the size and abundance of *Q. mongolica*. However, such information as we have points to its being a very common tree.

We have not seen Mayr's *Q. wutaishanica* (*Fremdl. Wald- & Parkbäume*, 504, fig. 224 [1906]), but from his brief description and figure we strongly suspect that it is *Q. mongolica* Fischer with small leaves.

A picture of this Oak will be found under No. x425 of the collection of Wilson's Japanese photographs.

Quercus mongolica, var. grosseserrata Rehder & Wilson.

  Quercus crispula Blume, *Mus. Bot. Lugd.-Bat.* I. 298 (1850). — Miquel in
    Ann. Mus. Lugd.-Bat. I. 104 (1863–64). — Franchet & Savatier *Enum. Pl.
    Jap.* I. 446 (1875). — Sargent in *Garden & Forest*, VI. 385 (1893); *Forest
    Fl. Japan*, 67 (1894). — Shirai in *Tokyo Bot. Mag.* IX. 410, t. 7, fig. 5
    (1895). — Matsumura, *Ind. Pl. Jap.* II. pt. 2, 26 (1912). — Koidzumi in
    Tokyo Bot. Mag. XXVI. 164 (1912).
  Quercus grosseserrata Blume, *Mus. Bot. Lugd.-Bat.* I. 306 (1850). — Miyabe in
    Boston Soc. Nat. Hist. IV. 7, 259 (*Fl. Kurile Isl.*) (1890). — Sargent in *Gar-
    den & Forest*, VI. 385 (1893); *Forest Fl. Japan*, 67 (1894). — Shirai in
    Tokyo Bot. Mag. IX. 410, t. 7, fig. 4 (1895). — Shirasawa, *Icon. Ess. For.
    Jap.* I. 53, t. 27, fig. 16–28 (1900). — Komarov in *Act. Hort. Petrop.*
    XXII. 74 (*Fl. Mansh.* II.) (1904). — Nakai in *Jour. Coll. Sci. Tokyo*,
    XXXI. 209 (*Fl. Kor.* II.) (1911). — Matsumura, *Ind. Pl. Jap.* II. pt. 2,
    27 (1912). — Miyabe and Miyake, *Fl. Saghal.* 421 (1915).
  Quercus crispula, β *grosseserrata* Miquel in *Ann. Mus. Lugd.-Bat.* I. 104 (1863–
    64). — Franchet & Savatier, *Enum. Pl. Jap.* I. 446 (1875).

NORTHEASTERN ASIA. S a g h a l i e n : without locality, 1861, *P. von Glehn.*

JAPAN. H o k k a i d o : Ishikari province, around Sapporo, August 23, 1914, *E. H. Wilson* (No. 7416); same locality, September 1892, *C. S. Sargent;* same locality, August 25, 1885, *K. Miyabe;* Iburi province, Muroran, September 24, 1892, *C. S. Sargent;* Oshima province, near Mori, September 26, 1892, *C. S. Sargent;* Hakodate, woods, *C. Wright* and *C. Maximowicz* (in Herb. Gray). H o n d o : Rikuchu province, slopes of Hayachine-san, September 27, 1914, *E. H. Wilson*

(No. 7580; tree 20–30 m., girth 3–4 m.); Ugo province, slopes of Chokai-san, October 1914, *E. H. Wilson* (No. 7177); Uzen province, June 7, 1904, *K. Sakurai;* Shimotsuke province, Nikko region, around Lake Chuzenji, September 3, 1892, *C. S. Sargent;* same locality, August 11, 1905, *J. G. Jack;* Shinano province, slopes of Tsubakura-dake, alt. 1000–1500 m., September 13, 1914, *E. H. Wilson* (No. 7448).   Kyushu: Hizen province, Nagasaki, 1863, *C. Maximowicz* (in Herb. Gray); slopes of Kirishima, alt. 500–1000 m., March 10, 1914, *E. H. Wilson* (No. 6256); without locality, 1886, *H. Mayr;* without locality or collector ex Herb. Lugd.-Bat. in Herb. Gray as *Q. crispula.*

With abundant material before us we find it impossible to separate this Oak specifically from *Q. mongolica* Fischer, and, indeed, we have doubts whether it is entitled to rank even as a variety. The less thickened scales of the cup, which is usually not thickened at the rim, and the usually more narrow acorn seem to be the principal differences. In extreme forms the few veins and the broad rounded teeth of one and the numerous veins and the narrow acute teeth of another look very distinct, but there is every gradation between these extremes. In Japan Wilson paid much attention to this Oak, but was unable to discover any valid difference between *Q. crispula* Blume and *Q. grosseserrata* Blume. The habit of the trees and the character of the bark are the same whether the trees are growing in Kyushu, Hondo or Hokkaido. Blume's principal difference appears to be in the number of the veins of the leaves, but this has no significance since it breaks down on almost every tree. Miyabe separates them on differences in the fruit, but such differences cannot be read into Blume's descriptions, since the fruits were unknown to him. Moreover, although such differences in the size of the cup as pointed out by Miyabe do exist, they are not constant, and Wilson found that the cups often varied on the same individual tree. Nor is there any taxonomic value in the fact that the cup falls with the acorn or sheds it. On the same individual both of these conditions occur, but most usually the acorn is shed first and the cup falls later.

This Oak grows through the length and breadth of Japan, but it is most abundant in the colder parts from central Hondo northward. It is especially plentiful in Hokkaido, where it is being rapidly felled, and it is the most important hardwood timber tree in northern Japan. The Japanese name is Nara, and the wood, though very durable, is for a hardwood fairly easily worked, and for making furniture it approximates in value the best European and American Oak timber, ranking commercially after Austrian or Hungarian Oak (*Q. robur* Linnaeus, *Q. sessiliflora* Salisbury and *Q. conferta* Kitaibel) and Indiana Oak (*Q. alba* Linnaeus, *Q. bicolor* Wildenow, *Q. macrocarpa* Michaux). In recent years a large and increasing trade in Nara lumber has developed, and every year it is exported from Hokkaido to America and Europe for furniture making, flooring, panelling, etc., and to China, India, Australia, Egypt and other parts of Africa for railway ties. The demand annually exceeds the supply, and in about a decade more the tree will be, commercially speaking, non-existent. Unfortunately it is not being replanted by the Japanese government foresters and the trees are mostly too old for a successful coppice growth to develop from the old stools.

The Nara grows from 25 to 30 m. tall, with a trunk from 2 to 4 m. in girth, and massive spreading branches forming a flattened crown. The bark is pale gray and scaly, but occasional old trees are found on which the bark is dark and furrowed very much as in *Q. mongolica* Fischer.

Pictures of this tree will be found under Nos. x234, x259, x261, x264, x265, x279, x300, x301, x302, x354, x407, x451, x458, x540 of the collection of Wilson's Japanese photographs. An interesting account of wood of this Oak is to be found in the *Timber Trades Journal*, July 24, 1915.

Quercus liaotungensis Koidzumi in *Tokyo Bot. Mag.* XXVI. 166 (1912); in Matsumura, *Icon. Pl. Koisikav.* I. 109, t. 55 (1912).

Quercus mongolica, β *liaotungensis* Nakai in *Tokyo Bot. Mag.* XXIX. 58 (1915).

Quercus funebris Léveillé in litt. ex Nakai, l. c. (pro synon.) (1915).

Quercus undulatifolia Léveillé in litt. ex Nakai, l. c. (pro synon.) (1915).

Chili: Hsiao Wu-tai-shan, alt. 2000 m., August 13, 1913, *F. N. Meyer* (Nos. 1211, 40; low scrubby growth).

It is possible that this is merely a form or state of *Q. mongolica* Fischer with very small leaves and fruits. The cup resembles that of *Q. mongolica,* var. *grosseserrata* Rehder & Wilson, but is smaller. Unfortunately our material is insufficient to decide the question.

Quercus Griffithii Hooker f. & Thomson apud Miquel in *Ann. Mus. Lugd.-Bat.* I. 104 (1863–64). — De Candolle, *Prodr.* XVI. pt. 2, 14 (1864). — Wenzig in *Jahrb. Bot. Gart. Berlin,* IV. 218 (1886). — Hooker f., *Fl. Brit. Ind.* V. 602 (1888). — King in *Ann. Bot. Gard. Calcutta,* II. 21, t. 18 (1889). — Franchet in *Jour. de Bot.* XIII. 147 (1899), exclud. speciminibus e Tibet et Sutchuen. — Koidzumi in *Tokyo Bot. Mag.* XXVI. 164 (1912).

Quercus aliena Skan in *Jour. Linn. Soc.* XXVI. 505 (non Blume) (1899), quoad synon. *Q. Griffithii.*

Quercus aliena, var. *Griffithii* Schottky in *Bot. Jahrb.* XLVII. 635 (1912).

CHINA.   Yunnan: Szemao, forests, alt. 1600–1800 m., *A. Henry* (Nos. 11687, 12435, 12435[a]; tree 6 m. tall).

Skan reduces this species to *Q. aliena* Blume, from which, however, the pubescent branchlets and the fringed rim of the cup of the fruit readily distinguish it.

Quercus phillyraeoides Gray in *Mem. Am. Acad.* n. ser. VI. 406 (Bot. Jap.) (1859). — Miquel in *Ann. Mus. Lugd.-Bat.* I. 104 (1863–64). — A. De Candolle, *Prodr.* XVI. pt. 2, 39 (1864). — Masters in *Gard. Chron.* n. ser. I. 632 (1874). — Franchet & Savatier, *Enum. Pl. Jap.* I. 446 (1875). — Wenzig in *Jahrb. Bot. Gart. Berlin,* IV. 220 (1886). — Shirasawa, *Icon. Ess. For. Jap.* I. 58, t. 31, fig. 1–12 (1900). — Seemen in *Bot. Jahrb.* XXIX. 290 (1900).

Quercus Ilex, var *phillyraeoides* Franchet in *Jour. de Bot.* XIII. 152 (sphalmate " phyllireoides ") (pro parte) (1899). — Skan in *Jour. Linn. Soc.* XXVI. 516 (1899). — Matsumura, *Ind. Fl. Jap.* II. pt. 2, 28 (sphalmate " phylliraeoides ") (1912). — Koidzumi in *Tokyo Bot. Mag.* XXVI. 160 (sphalmate " phyllireoides ") (1912).

Quercus phillyreoides, var. *sinensis* Schottky in *Bot. Jahrb.* XLVII. 643 (1912).

CHINA.   Western Hupeh: Patung Hsien, May 1900 (Veitch Exped. No. 363, small tree 6 m. tall). Szech'uan: Nanch'uan, August 25, 1891, *A. von Rosthorn* (No. 621). Fokien: Dunn's Exped., April to June 1905 (Hongkong Herb. No. 3482).

In central and western China this Oak, as far as is known, is rare. In Hupeh it grows on the cliffs of the gorges and glens at a low altitude, but in exposed situations. In Japan it is found from the neighborhood of Yokohama southward to Yakushima and in the warmer parts is common, especially on the seacoast. Usually it is a dense bush from 1 to 3 m. high, but occasionally it forms a small bushy tree from 6 to 8 m. tall. The leaves are coriaceous, dark green, quite glabrous, or with slight floccose tomentum, usually finely serrate, sometimes quite entire or occasionally dentate in the apical part and slightly undulate on the margin. The fruit is biennial, very short-peduncled and subterminal in axils of the upper leaves of the season's growth; the cup is shallow with short closely appressed scales.

A specimen in the Gray Herbarium collected in the Liukiu Islands by *C. Wright* (No. 306) differs markedly from the type in the under surface of the leaves, which are clothed with a loose floccose gray tomentum.

Franchet reduces this species to a variety of *Q. Ilex* Linnaeus and makes other Chinese Oaks varieties of Linnaeus's species. In this he is followed by Skan, but both have overlooked the important character of the fruit. In *Q. Ilex* it matures in one season, but in all the Asiatic Oaks referred to it as varieties by Franchet and Skan the fructification is biennial. Von Seemen (in *Bot. Jahrb.* XXIX. 289 (1900)) refers specimens collected by *A. von Rosthorn* at Nanch'uan, southeastern Szech'uan, to *Q. Ilex* and remarks that they resemble Kotschy's var. *lanceolata*. His brief description suggests to us a species very closely related to *Q. Franchetii* Skan.

We have seen no specimen of *Quercus* from China or any other part of eastern Asia that is in any sense referable to *Q. Ilex* Linnaeus, and we are strongly of the opinion that neither the species nor any of its varieties or forms grow there. Franchet and other authors evidently attached too much importance to the general appearance of the leaves and overlooked important characters in the fruit. In China there are a number of different species of *Quercus* in which the leaves exhibit the same variations as are found on those of *Q. Ilex*. They are spiny or entire, pubescent or glabrous, according to the different ecological conditions under which they grow, and these same variations occur on different species of Oaks native of western North America.

Quercus salicina Blume, *Mus. Bot. Lugd.-Bat.* IV. 305 (1850). — A. De Candolle, *Prodr.* XVI. pt. 2, 100 (pro parte) (1864). — Franchet & Savatier, *Enum. Pl. Jap.* I. 449 (1875). — Matsumura, *Ind. Pl. Jap.* II. pt. 2, 28 (1912) exclud. synon.

    Quercus glauca, δ var. *stenophylla* Blume, *Mus. Bot. Lugd.-Bat.* IV. 303 (1850). — Franchet & Savatier, *Enum. Pl. Jap.* I. 448 (in nota) (1875). — Matsumura, *Ind. Pl. Jap.* II. pt. 2, 27 (1912).

    Cyclobalanops salicina Oersted in *Kjoeb. Vedensk. Meddel.* XVIII. 70 (1866).

    Quercus myrsinaefolia Shirasawa, *Icon. Ess. For. Jap.* I. 59, t. 31, fig. 13–24 (non Blume) (1900).

    Quercus stenophylla Makino in *Tokyo Bot. Mag.* XXIV. 17 (1910). — Nakai in *Tokyo Bot. Mag.* XXIX. 62 (1915).

    Quercus stenophylla, var. *salicina* Makino in *Tokyo Bot. Mag.* XXIV. 54 (1910).

    Cyclobalanopsis stenophylla Schottky in *Bot. Jahrb.* XLVII. 657 (1912).

    Quercus longinux Hayata in *Jour. Coll. Sci. Tokyo,* XXX. art. 1, 292 (*Mat. Fl. Formosa*) (1911), fide Nakai.

    Quercus pseudo-myrsinaefolia, Hayata, l. c. 295 (1911), fide Nakai.

    Cyclobalanopsis stenophylla, var. *salicina* Schottky in *Bot. Jahrb.* XLVII. 657 (1912).

    Quercus angustissima Makino in *Tokyo Bot. Mag.* XXVII. 114 (1913).

NORTHEASTERN ASIA. Korea: Quelpaert, Hongno, August 1907, *U. Faurie* (No. 1529); same locality, May and October 1909, *Taquet* (Nos. 2558, 2559, 2561).

JAPAN. Kyushu: prov. Osumi, island of Yakushima, forests, alt. 300–1000 m., February 20, 1914, *E. H. Wilson* (No. 6033; tree 10–20 m. tall, girth 1.5–2.3 m.); Mt. Kirishima, woods, October 1914, *T. Miyoshi* (No. 7868 ex *E. H. Wilson*); without locality, ex Herb. Bot. Gard. Tokyo. Shikoku: prov. Tosa, Nanokawa April 25, 1888, *K. Watanabe*. Hondo: prov. Musashi, Takao-san, September 24, 1914, *E. H. Wilson;* without locality, ex Herb. Lugd.-Bat. (co-type of *Quercus salicina* Blume; in Herb. Gray); without locality, ex Herb. Lugd.-Bat. (co-type of *Q. glauca,* δ var. *stenophylla* Blume; in Herb. Gray).

Blume founded this species on a fragment the leaves of which are unusually narrow, mostly entire, and more or less rounded at the base, and in consequence it has remained obscure. His material on which he founded his *Quercus glauca*, δ var. *stenophylla* is much more typical of the species. It consists of leafy branches with oblong-lanceolate acuminate leaves, attenuate at the base, from 1.2 to 2.7 cm. wide, and sharply toothed above the middle; the petioles are from 1 to 1.5 cm. long On specimens before us the leaves vary in width from 0.8 to 3 cm., are rounded or attenuate at the base, and the petioles vary from 0.5 to 2.5 cm. in length. The leaves are whiter on the undersurface than those of the very closely related *Q. glauca* Thunberg, and the branchlets are always pale gray and not dark and purplish as in Thunberg's species. The fruit ripens in one season. In the Gray Herbarium there is a specimen from the Liukiu Islands, collected by *C. Wright* (No. 307) and named *Q. salicina*, but it is *Q. Miyagii* Koidzumi (in *Tokyo Bot. Mag.* XXVI. 167 [1912]), a very distinct species.

Pictures of *Q. salicina* will be found under Nos. x575, x578 of the collection of Wilson's Japanese photographs.

Quercus bambusifolia Hance in *Jour. Bot.* XIII. 364 (non Masters, nec *Q. bambusaefolia* Fortune) (1875).

> *Quercus salicina* Seemann, *Bot. Voy. Herald*, 415 (non Blume) (1852–57). — Bentham, *Fl. Hongk.* 321 (1861). — Miquel in *Ann. Mus. Lugd.-Bat.* I. 116 (1863–64). — A. De Candolle, *Prodr.* XVI. pt. 2, 100 (1864), quoad plantam chinensem. — Wenzig in *Jahrb. Bot. Gart. Berlin*, IV. 231 (1886). — Skan in *Jour. Linn. Soc.* XXVI. 519 (1899). — Dunn & Tutcher in *Kew Bull. Misc. Inform.* add. ser. X. 253 (*Fl. Kwangtung & Hongk.*) (1912).

> *Quercus bambusaefolia* Hance apud Seemann, *Bot. Voy. Herald*, 415, t. 91 (pro synon.) (1852–57). — Miquel in *Ann. Mus. Lugd.-Bat.* I. 116 (pro synon.) (1863–64).

> *Cyclobalanopsis neglecta* Schottky in *Bot. Jahrb.* XLVII. 650 (1912).

> *Quercus neglecta* Koidzumi in *Tokyo Bot. Mag.* XXX. 201 (1916).

CHINA. Hongkong: without precise locality, common, November 5, 1903, *C. S. Sargent;* Bowen Road, January 14, 1904 (ex Herb. Hongk. Bot. Gard. No. 1067); without precise locality, *C. Wilford; C. Wright* (No. 466); *H. F. Hance* (No. 787).

We have before us a co-type of Blume's *Q. salicina*, and it is obvious that Hance and Schottky are correct in considering the Hongkong plant distinct from the Japanese tree. The confusion was primarily caused by Blume basing his species on fragmentary material, but even the leafy shoots are quite different. In Blume's co-type of his species the year-old shoot is pale gray, densely lenticellate and channelled; the leaves are slightly oblique and rounded or very abruptly narrowed and broadly cuneate at the base, the apex is acuminate and mucronate and the margin is entire or sparingly toothed above the middle; the petiole is from 5 to 12 mm. long. The Hongkong plant has dark purplish year-old branchlets, leaves attenuate at the base, rounded or more rarely obtuse at the apex, usually entire, but occasionally with one or two obscure rounded teeth, and petioles from 2 to 5 mm. long. With complete material the differences are very great, but it is only necessary to mention here that in the Hongkong plant the fruit is very much larger and does not ripen until the second season, while the Japanese plant ripens its fruit in one season. Moreover, there is no species of *Quercus* of the section *Cyclobalanops* known from Japan in which the fruit does not ripen until the second year. There has been considerable confusion about Hance's name, but there never has been any doubt as to the plant he applied it to. Since Fortune's *Q. bambusaefolia* is a nomen nudum, and Master's *Q. bambusifolia* is a synonym of *Q. myrsinaefolia* Blume, Hance's name remains valid under the International rules.

Quercus myrsinaefolia Blume, *Mus. Bot. Lugd.-Bat.* I. 305 (1850). — Miquel in *Ann. Mus. Lugd.-Bat.* I. 117 (1863–64). — Franchet & Savatier, *Enum. Pl. Jap.* I. 449 (1875). — Matsumura, *Ind. Pl. Jap.* II. pt. 2, 28 (1912).
Quercus bambusaefolia Fortune in *Gard. Chron.* 1860, 170 (nomen nudum, non *Q. bambusifolia* Hance).
Quercus bambusifolia Masters in *Gard. Chron.* n. ser. I. 632 (non Hance) (1874).
Quercus Vibrayeana Franchet & Savatier, *Enum. Pl. Jap.* I. 449 (1875); II. 498 (1879). — Forbes in *Jour. Bot.* XXII. 85 (1884). — Skan in *Jour. Linn. Soc.* XXVI. 522 (1899). — Shirasawa, *Icon. Ess. For. Jap.* I. 55, t. 29, fig. 16–31 (1900). — Matsumura, *Ind. Pl. Jap.* II. pt. 2, 30 (1912).
Quercus acuta, var. *bambusaefolia* Nicholson in *Kew Handlist Trees & Shrubs,* pt. 2, 181 (1896).
Quercus glauca Léveillé in litt. ex Nakai in *Tokyo Bot. Mag.* XXIX. 62 (pro synon.; non Thunberg) (1915).
Quercus Taquetii Léveillé, nov. hybr. ex Nakai; l. c. (pro synon.) (1915).
Cyclobalanopsis myrsinifolia Schottky in *Bot. Jahrb.* XLVII. 656 (1912).
Cyclobalanopsis Vibrayeana Schottky in *Bot. Jahrb.* XLVII. 656 (1912).

CHINA. Western Hupeh: Patung Hsien, October 1900 (Veitch Exped. No. 651; tree 6 m. tall). Eastern Szech'uan: Wushan Hsien, woods, 1000 m. alt., May 1900 (Veitch Exped. No. 573; tree 10 m. tall). Yunnan: Szemao, forests, alt. 1500 m., *A. Henry* (No. 11698; tree 8 m. tall). Hongkong: without locality, November 5, 1903, *C. S. Sargent;* Repulse Bay, March 29, 1900 (ex Herb. Hongkong, No. 1704 in part); Little Hongkong woods, February 1905 (ex Hongkong, No. 1704 in part); without locality or date, *C. Ford.*

From Tokyo southward this is a common tree in Japan, and it is often planted for ornament or used for making tall hedges. In central China it is very rare. The specimens from Hongkong are all labelled *Q. glauca,* but we have seen no material of the real *Q. glauca* Thunberg from that locality. In this herbarium there is a specimen from J. Veitch & Sons of *Q. bambusifolia* Masters, sent by the late Mr. George Nicholson with the date October 13, 1880. This is identical with Blume's *Q. myrsinaefolia* and there can no longer be any question that these plants are the same.

Pictures of this tree will be found under Nos. x36, x42, x47 of the collection of Wilson's Japanese photographs.

Quercus vestita Rehder & Wilson, n. nom.
Quercus velutina Lindley apud Wallich, *Pl. As. Rar.* II. 41, t. 150 (non Lamarck) (1831). — Miquel in *Ann. Mus. Lugd.-Bat.* I. 115 (1863–64). — A. De Candolle, *Prodr.* XVI. pt. 2, 99 (1864). — Kurz, *Forest Fl. Brit. Burma,* II. 487 (pro parte) (1877). — Wenzig in *Jahrb. Bot. Gart. Berlin,* IV. 236 (1886). — Hooker f., *Fl. Brit. Ind.* V. 606 (1888). — King in *Ann. Bot. Gard. Calcutta,* II. 35, t. 29 a (1889). — Koidzumi in *Tokyo Bot. Mag.* XXX. 202 (1916).
Cyclobalanopsis velutina Oersted in *Kjoebenh. Vidensk. Meddel.* XVIII. 71 (1866). — Schottky in *Bot. Jahrb.* XLVII. 651 (1912).
CHINA. Yunnan: Szemao, forests, alt. 1500–1600 m., *A. Henry* (Nos. 11675, 11675[b], 11675[c], 11675[d], 11675[e]; tree 3–6 m. tall).
This is a very interesting addition to the Chinese flora.

Quercus Delavayi Franchet in *Jour. de Bot.* XIII. 158 (1899). — Skan in *Jour. Linn. Soc.* XXVI. 511 (1899). — Koidzumi in *Tokyo Bot. Mag.* XXX. 200 (1916).

Quercus gilva Skan in *Jour. Linn. Soc.* XXVI. 514 (non Blume) (1899), quoad specimina Henryana.

Cyclobalanopsis Delavayi Schottky in *Bot. Jahrb.* XLVII. 657 (1912).

Synaedrys Delavayi Koidzumi in *Tokyo Bot. Mag.* XXX. 191 (1916).

CHINA. Yunnan: Mengtsze woods, alt. 1800 m., *A. Henry* (Nos. 10504, 10504ᵃ; tree 6 m. tall).

As Schottky points out, this species is easily distinguished from *Q. gilva* Blume by its biennial fruit. In addition to Henry's specimens Skan refers other material to Blume's species, but we have seen no material from China referable to this Japanese tree, and we doubt if it grows there.

Quercus Schottkyana Rehder & Wilson, n. nom.

Cyclobalanopsis glaucoides Schottky in *Bot. Jahrb.* XLVII. 657 (1912).

Quercus glaucoides Koidzumi in *Tokyo Bot. Mag.* XXX. 200 (non Martens & Galeotti) (1916).

CHINA. Yunnan: Szemao, woods, alt. 1600 m., *A. Henry* (Nos. 11770, 11770ᵃ, 11698ᵃ, 11698ᵇ; bush or tree 2–6 m. tall).

This species is very distinct from its allies and is characterized by its long female aments and by its very small acorn and cup. Schottky suggests that *Q. glauca*, var. *villosa* Skan may belong here, but to us this seems highly improbable, and we are doubtful if Skan's plant is entitled to any rank to distinguish it from the type.

ULMACEAE.

Determined by CAMILLO SCHNEIDER.

ULMUS L.

Ulmus Wilsoniana Schneider, *Ill. Handb. Laubholzk.* II. 904, fig. 565 e, 566 c–d (1912); in *Oester. Bot. Zeitschr.* LXVI. (1916).

Ad descriptionem addenda: Arbor ad 17 m. alta; ramuli novelli hornotinique griseo-villosuli, annotini plus minusve glabrescentes, rubro-brunnei, lenticellis flavis sparse obtecti, vetustiores cinerascentes, paulo rimosi (interdum suberosi); gemmae ovatae, plus minusve acutae, perulis pluribus bicoloribus versus basim flavo-brunneis versus marginem purpurascentibus dorso vix pilosis margine tenuiter ciliatis obtectae. Folia (plantae maturae) ovata, late ovata v. ovato-elliptica, rarius obovato-oblonga, basi plus minusve rotunda v. semicordata satis asymmetrica, apice satis subito breviter acuminata, supra intense viridia, tantum in costa pilosa sed pleraque plus minusve scabra, interdum sublaevia, subtus in nervis elevatis lateris longioris (imo apice excluso) 16–18–22 et in costa griseo-villosula, in facie sparsius pilosa, barbulata (barbulis saepe valde indistinctis), praeterea saepe plus minusve scabra, margine dupliciter crenato-serrata serraturis primariis saepe incurvatis, textura satis firma; petioli breves, 3–9 mm. longi, villosuli, subcrassi. Inflorescentiae fasciculato-cymosae; flores nondum vidi. Samarae late obovatae, circiter 16–19 mm. longae (pedicello excluso) et 13–14 mm. latae, undique (emarginatura excepta) glabrae, brevissime stipitatae; semina juxta emarginaturam sita; perigonia sub fructu remanentia glabra, lobis 4–5 brevibus latis griseo- et fulvo-ciliatis instructa, sensim in partem glabram pedicelli transientia; pars inferior pedicelli pilosa parte superiore subaequilonga v. paulo longior; filamenta marcida staminum perigonio multo longiora; antherae non visae.

Western Hupeh: Changyang Hsien, side of streams, alt. 700 m., November 1907 (Nos. **745, 746**; tree 10–17 m. tall, girth 1.2–2.4 m.; sterile); same locality, woods, alt. 1200–1600 m., May 1907 (No. **2806**; with ripe fruits and not yet fully grown leaves); same locality, mountains, May 1900 (Veitch Exped. No. 677 in part, type; tree 10 m. tall, with ripe fruits and almost fully grown leaves [the sterile branchlets of this No. belong to *U. castaneifolia*]); Fang Hsien, woods, 2000–2500 m., November 1910 (No. **4442**; living material).

This species seems to be closely related to *U. japonica* Sargent. As far as I can judge from the material before me the color of the one-year-old branchlets and of the well-developed buds (as indicated in the key) seem to afford good characters to distinguish these two Elms. But there are glabrous forms of both which it is much more difficult to separate and which need further observations in the field.

To the typical form of *U. Wilsoniana* may belong *Ulmus* sp. nov.? Hemsley in *Jour. Linn. Soc.* XXVI. 448 (1894); E. Pritzel in *Bot. Jahrb.* XXIX. 296 (1900), the type of which is a sterile specimen collected by *A. Henry* in Wushan Hsien, eastern Szech'uan (No. 5537; ex Hemsley), and to which E. Pritzel refers with some doubt *A. von Rosthorn's* No. 678 from southeastern Szech'uan " Nanch'uan, Ch'ien ts'un kou, an Flussufern," August 1891 (tree 12–15 m. tall; sterile). The leaves are described as lanceolate and minutely serrulate, and without having seen a specimen I cannot refer it to any of the Elms I have seen from central China.

On the margins of moist woods and in thickets on the mountains of western Hupeh this Elm occurs, but is nowhere common. It is a tree of moderate size, growing from 15 to 25 m. tall, with a trunk from 1.5 to 3 cm. in girth and dark gray shallowly fissured bark. Usually the branches are rather slender and ascending-spreading, forming a narrow crown, but on trees growing in more open country the branches are fairly stout and spreading. Quite commonly the branchlets are corky-winged. The species is in cultivation in this Arboretum from graftwood sent from Fang Hsien in 1910.

A picture of this tree will be found under No. 573 of the collection of my photographs and in my *Vegetation of Western China*, No. 493.   E. H. W.

Ulmus Wilsoniana, var. psilophylla Schneider, n. var.

Arbor ad 23 m. alta, trunco cortice rugoso longitudinaliter fisso obtecto; ramuli etiam novelli glabri, annotini olivaceo-brunnescentes, vetustiores cinerascentes, rimosi, interdum suberosi (in No. 2802); gemmae adhuc ignotae. Folia tantum juvenilia visa, obovata v. obovato-oblonga, basi angustata, subrotunda, plus minusve v. vix asymmetrica, apice subito in acumen distinctum contracta, supra glabra v. costa parcissime pilosa, subscabriuscula, subtus pallidiora, tantum axillis nervorum lateralium lateris longioris 17–20 distincte griseo-barbata, laevia v. vix scabriuscula, margine argute dupliciter serrata serraturis primariis subincurvis, maxima ad 6.5 cm. longa et 3.2 cm. lata; petioli satis graciles, ad 8 mm. longi, supra minute villosuli; stipulae anguste lanceolatae, petiolis longiores, mox caducae. Inflorescentiae fasciculato-cymosae; flores ignoti. Samarae (an satis maturae?) vix a typo diversae.

Western Hupeh: Fang Hsien, side of streams, alt. 1200–1600 m., fairly common, May 21, 1907 (No. **2803**, type; tree 13–23 m. tall, girth

1.8–3 m., bark rough and splitting longitudinally; with fruits and not yet fully grown leaves); same locality, woods, alt. 1800 m., May 28, 1907 (No. 2802; tree 13 m. tall, girth 2.4 m.; sterile with young leaves, older parts of the branches somewhat corky).

By the fruits this variety cannot be separated from the type, but the shape of the leaves is more obovate-oblong. Unfortunately I have seen no mature leaves, but there is a sterile branch collected by *C. Silvestri* in northern Hupeh " Fan-sien, circa 800 m. s. m.," May 20 and June 3, 1906 (No. 360), which may belong to var. *psilophylla*. It can be described as follows: Ramuli foliiferi flavo-brunnei, basi minutissime puberuli, ceterum glabri, lenticellis sparsis obtecti, vetustiores cinerascentes, laeves. Folia pleraque ovato-elliptica, basi rotundo-truncata v. subcordata paulo asymmetrica, apice subito breviter acuminata, supra scabra, tantum costa nervisque pro parte tenuissime pilosula, subtus in sicco vix pallidiora, plus minusve scabra, tantum axillis nervorum lateris longioris (imo apice excluso) 15–18 elevatorum barbulata, margine dupliciter serrata serraturis primariis subincurvis; petioli satis tenues, superne pilosuli, 3–6 mm. longi.

For further remarks see *U. japonica* Sargent.

Ulmus Bergmanniana Schneider, *Ill. Handb. Laubholzk.* II. 902, fig. 565 a–b, 566 a–b (1912); in *Oester. Bot. Zeitschr.* LXVI. (1916).

> *Ulmus montana* Hemsley in *Jour. Linn. Soc.* XXVI. 448 (non Withering) (1894), quoad plantam Henryi. — E. Pritzel in *Bot. Jahrb.* XXIX. 296 (1900).

Ad descriptionem addenda: Arbor 10–23 m. alta; ramuli etiam novelli glaberrimi, annotini olivaceo-brunnei v. rubro-brunnei, lenticellis flavis sparsis conspersi, deinde cinerascentes, laeves, ut videtur vix rimosi; gemmae satis evolutae, ovato-conicae, acutae, perulis dorso plerisque glabris intense purpurascentibus versus basim plus minusve discoloribus margine tenuissime ciliatis cinctae, petiolis aequilongae v. sublongiores. Folia obovata, obovato-oblonga, elliptica v. rarius ovata, basim versus pleraque sensim paulo attenuata, basi rotunda sed distincte asymmetrica parte longiore deducta saepe petiolum et ramulum obtegente, apice subito fere caudato-acuminata, supra intense viridia, tantum in costa impressa sparse pilosa, ceterum glabra, sed plus minusve scabra, subtus pallidiora, tantum axillis nervorum lateralium elevatorum lateris longioris (imo apice excepto) (15–)17–23 plus minusve albo-barbulata ceterum tantum distincte scabra, margine dupliciter serrata serraturis primariis acutis saepissime incurvatis, minora 7–10 cm. longa, 3.5–5 cm. lata, majora ad 12:6.5 cm. v. ramulorum sterilium ad 16:7 cm. magna; petioli satis tenues, glabri v. superne pilosuli, 2–7 mm. longi. Inflorescentiae pilosae, fasciculato- v. fructiferae paulo elongato-cymosae, circiter 12-florae; perigonia glabra, lobis oblongis ad medium v. paulo ultra incisis apice ciliatis intensius coloratis, 4–5 instructa; stamina 4–5, filamentis satis elongatis, antheris oblongis; pars superior pedicellorum parte inferiore pilosa vix v. paulo brevior. Samarae late obovatae v. obovato-rotundae, basi in stipitem perigonio breviorem subito contractae, apice interdum leviter emarginatae, fere clausae, emarginatura pilosa brevi instructae, ceterum glaberrimae, circiter 1.3–1.8 cm. longae et 1.3–1.5 cm. latae; semina in centro samarae sita, ab emarginatura distincte remota.

Western Hupeh: Patung Hsien, side of streams, alt. 800 m., April 1907 (No. 743; tree 7–17 m. tall, girth 0.6–1.8 m.; with flowers, without leaves); same locality, April 1900 (Veitch Exped. No. 320, co-type; with fruits and leaves); same locality, alt. 1800 m. (Veitch Exped. Seed No. 305; young plants in cultivation); Hsing-shan Hsien, woods, alt. 1600 m., May 1907 (No. 2804; tree 8–13 m. tall, girth 1.2 m.; with young fruits and leaves); same locality, July 1907 (No. 2804ª; sterile); Hsing-shan Hsien, woods, rare, alt. 1600–1800 m., June 3, 1907 (No. 2805; thin tree 12 m. tall, girth 0.6 m., with fruits and leaves); same locality, woods, alt. 1600 m., May 31, 1907 (No. 2805ª; small tree, 10 m. tall, with fruits and leaves); same locality, mountains, rare, May 1901 (Veitch Exped. No. 1855, type; tree 7 m. tall, with fruits and leaves); Fang Hsien, in ravine, alt. 1600–2400 m., June 16, 1910 (No. 4442ª; tree 23 m. tall, girth 6 m., very large head; sterile). Eastern Szech'uan: Wushan Hsien, A. Henry (No. 5690; with fruits and leaves).

A well-marked species which certainly seems to be more closely related to the Himalayan *U. Brandisiana* Schneider than to *U. laciniata* Mayr from northeastern Asia or to *U. glabra* Hudson from Europe and western Asia. The following is a hairy variety, the leaves of which sometimes resemble those of typical *U. japonica* Sargent, but may be easily distinguished by their very short petioles.

This Elm is fairly common above 800 m. alt. in moist woods throughout western Hupeh and eastern Szech'uan. As usually seen it is a low tree with a slender trunk and thin, spreading branches, but trees from 25 to 28 m. tall and from 4 to 6 m. in girth of trunk, with massive widespreading branches, are occasionally met with. The bark is dark gray with shallow fissures and is rather scaly. In Hupeh the colloquial name for this and the other species of Elm is Lang shu. This species is in cultivation in this Arboretum and in Kew from seeds I sent to Messrs. Veitch in 1900.

A picture of this tree will be found under No. 091 of the collection of my photographs.   E. H. W.

Ulmus Bergmanniana, var. lasiophylla Schneider, n. var.

A typo praecipue recedit foliis etiam adultis subtus in costa nervisque facieque villosis v. interdum autumno in facie glabrescentibus, petiolis distinctius pilosis, fructibus stipitibus quam perigonium subduplo longioribus suffultis.

Western Szech'uan: Wa-shan, woods, alt. 2200 m., July 1908 (No. 820, type; tree 17 m. tall, girth 1.8 m.; with ripe fruits and good leaves); near Wa-shan, roadside, alt. 1600 m., September 9, 1908 (No. 1423; tree 27 m. tall, girth 2.4 m.; sterile, leaves very rough on both sides, up to 15.5 cm. long and 7 cm. broad); southeast of Tachien-lu, alt. 2400 m., woods, October 1910 (No. 4360; tree 8–10 m. tall, girth 0.6 m.; sterile, leaves very rough, but a little more glabrous beneath).

This seems to be a distinct variety, and it may even represent a species. The young twigs are glabrous, and the winter-buds are glabrous or slightly hairy. The shape of the leaves is scarcely different from that of the type, the compound serration being very similar. The main teeth usually possess 3–4 small, rather sharp secondary teeth. The most distinct character is found in the fruit, which, on account of its long stipe, much resembles that of *U. Brandisiana* Schneider.

This tree is not common, but is found scattered over a wide area in western Szech'uan. In habit and in general appearance it differs in no way from the type species. It is in cultivation in this Arboretum from living material I sent in 1908 under No. 1423.

A picture will be found under No. 349 of the collection of my photographs and also in my *Vegetation of Western China*, No. 492.   E. H. W.

Ulmus pumila Linnaeus, *Spec.* 226 (1753), exclud. synon. Plukenet.—Pallas, *Fl. Ross.* I. 76, t. 48, fig. A, C, E (1784), exclud. plantas e Rossia et forma suberosa e Sibiria.[1] — Aiton, *Hort. Kew*, I. 320 (1789). — Willdenow, *Berlin Baumz.* 395 (1796); *Spec.* I. pt. 2, 1326 (1798), exclud. synon. pro parte. — Persoon, *Syn.* I. 291 (1805). — Planchon in *Ann. Sci. Nat.* sér. 3, X. 271 (1848); in De Candolle, *Prodr.* XVII. 159 (1873). — Turczaninow in *Bull. Soc. Nat. Mosc.* XXVII. 368 (1854), exclud. var. β; *Fl. Baical-Dahur.* 95 (1856), exclud. var. β. — Maack & Ruprecht in *Bull. Phys.-Math. Acad. Sci. St. Pétersbourg*, XV. 375 (1857); in *Mél. Biol.* II. 556 (1858). — Trautvetter in *Mém. Sav. Étr. Acad. Sci. St. Pétersbourg*, IX. 248 (Maximowicz, *Prim. Fl. Amur.*) (1859). — Franchet in *Nouv. Arch. Mus. Paris*, sér. 2, VII. 78 (*Pl. David.* I. 268) (1884). — Dippel, *Handb. Laubholzk.* II. 27 (1892). — Koehne, *Deutsche Dendr.* 135 (1893). — Palibin in *Act. Hort. Petrop.* XIV. 139 (1895). — Mouillefert, *Traité Arb. Arbriss.* II.

---

[1] Pallas includes three different forms. His "A. *Ulmus pumila* Rossiae australis," t. 48, fig. D, E is *U. foliacea* Gilibert; "B. *Ulmus pumila* transbaicalensis" represents the true *U. pumila* with exception of the branch with corky wings (fig. B), which apparently belongs to *U. japonica*, var. *levigata* Schneider. Pallas did not describe a variety *transbaicalensis*, as Henry, l. c., cites, for "transbaicalensis" is not printed in italics and is used only as a geographical term.

1200 (1897). — Komarov in *Act. Hort. Petrop.* XXII. 85 (*Fl. Mansh.* II.) (1903). — Schneider, *Ill. Handb. Laubholzk.* I. 221, fig. 137 m–p, 139 i–k (1904); in *Oester. Bot. Zeitschr.* LXVI. (1916). — Ascherson & Graebner, *Syn. Mitteleur. Fl.* IV. 551 (1911). — Nakai in *Jour. Coll. Sci. Tokyo,* XXXI. 189 (*Fl. Kor.* II.) (1911). — Henry [1] in Elwes & Henry, *Trees Great Brit. & Irel.* VII. 1926, t. 411, fig. 1 (1913). — Bean, *Trees & Shrubs Brit. Isl.* II. 619 (1914).

Ulmus humilis Gmelin apud Amman, *Stirp. Rar. Ruth.* 180 (1739). — Gmelin, *Fl. Sibir.* III. 105 (1768). — Lamarck, *Encycl. Méth.* IV. 611 (1797), exclud. synon. Plukeneti.

Ulmus pumila *microphylla Persoon, *Syn.* 291 (1805).

Ulmus campestris, var. *parvifolia* Loudon, *Arb. Brit.* III. 1377, fig. 1230 (1838), pro parte, non *U. parvifolia* Jacquin! — Kirchner in Petzold & Kirchner, *Arb. Musc.* 555 (1864).

Ulmus microphylla Persoon [2] ex Loudon, *Arb. Brit.* III. 1337 (pro synon.) (1838).

Ulmus campestris, β *suberosa,* b. *pumila* Ledebour, *Fl. Ross.* III. pt. 2, 647 (1850), tantum pro parte! — Herder in *Act. Hort. Petrop.* XII. 44 (*Pl. Radd.*) (pro parte) (1892).

Ulmus campestris, a *vulgaris,* lusus d. *pumila* Regel in *Mém. Acad. Sci. St. Pétersbourg,* sér. 7, IV. 134 (*Tent. Fl. Ussur.*) (1861), exclud. synon. pro parte! — Korshinsky in *Act. Hort. Petrop.* XII. 387 (1892).

Ulmus campestris, δ *pumila* Maximowicz in *Bull. Acad. Sci. St. Pétersbourg,* XVIII. 290 (1873); in *Mél. Biol.* IX. 23 (1873). — E. Pritzel in *Bot. Jahrb.* XXIX. 296 (1900).

Ulmus campestris Hemsley in *Jour. Linn. Soc.* XXVI. 446 (pro parte, non Linnaeus) (1894).

Kiangsi: Kuling, roadsides, abundant, alt. 1700 m., July 31, 1907 (No. **1565**; tree 5–8 m. tall; sterile, young shoots pubescent). Western Szech'uan: Tung-chuan Fu, cultivated, alt. 1600 m., July 1910 (No. **4612**; tree 10–13 m. tall, girth 0.9–1.5 m., bark deeply corrugated; sterile). Shantung: Chifu, September 27, 1903, *C. S. Sargent* (sterile; young shoots hairy). Chili: Hsiao Wu-tai-shan, alt. 1600 m., August 30, 1913, *F. N. Meyer* (No. 1385; leaves small, narrow-lanceolate, rough on both sides, petioles short, shoots pubescent, a somewhat doubtful form); Weichang, 1910, *W. Purdom* (No. 96; sterile); "Cal-ceen-wong," 1910, *W. Purdom* (No. 61; sterile); Peking, Temple of Agriculture, September 15, 1903, *C. S. Sargent* (with mature leaves and flowers); same locality, park of Temple of Heaven,

[1] See note on page 242.

[2] Persoon did not describe a species *U. microphylla,* as usually stated by different authors. He only says in his remarks on *U. pumila:* "Sequens quae in horto Celsii colitur diversa species videtur." According to the quotation "*U. pumila,* β *transbaicalensis.* Pall. l. c. fig. A. B. C. E." his variety is to be regarded as typical *U. pumila,* but he may include, as Pallas did, some other shrubby forms.

April 19, 1909, *E. H. Wilson* (tree 8–13 m. tall, girth 1.8–3 m., bark rough, deeply furrowed; with flowers, leaflets); same place, May 1910 (No. **4000**, seeds only); same locality, on city wall, April 18, 1913, *F. N. Meyer* (No. 923; with very young leaves and fruits); same locality, May 20, 1913, *F. N. Meyer* (No. 928; sterile, "var. fol. variegatis "; the hairy dark-colored branchlets somewhat resemble *U. parvifolia* Jacquin); near Peking, "common Elm of the Peking plain," October 7, 1905, *J. G. Jack* (sterile and with flower-buds); same locality, October 9, 1905, *J. G. Jack* (one-year-old plants with 5–6 pairs of opposite leaves); Peking plain, " common Elm," *W. Purdom* (with ripe fruits and young leaves and sterile; young branchlets hairy); Nankow, October 6, 1905, *J. G. Jack* (with mature leaves and flower-buds).

NORTHEASTERN ASIA: Mandshuria: south of Harbin, August 29, 1903, *C. S. Sargent* (sterile, leaves up to 8.5 cm. long and 3.8 cm. broad); on plain east of the Khingan mountains, very common, September 15, 1903, *C. S. Sargent* (tree 10–13 m. tall, pale bark, round head of pendulous branches; sterile). Transbaicalia: Nertchinsk, 1849, *Sentinoff* (with fruits); bank of stream near Sryetensk, August 13, 1903, *C. S. Sargent* (sterile); in montibus transbaicalensibus, 1829, *N. Turczaninow* (with young fruits; there is on the sheet in Herb. Gray a branch with mature leaves which are very rough, roundish-ovate, up to 7 cm. long and 5 cm. broad and with a coarse dentation, which may belong to *U. japonica* Sargent). Amur: " ad. fl. Amur," 1855, *R. Maack* (No. 103, with very young fruits and leaves).

I have not seen a specimen from Korea, where it was found by Komarov and Nakai. *U. pumila* is not recorded from Japan nor from the Altai, but it is apparently common in Turkestan, where it is represented by the very closely related var. *Arborea* Litwinow (see p. 262). Some of the specimens enumerated above may belong to *U. glaucescens* Franchet (see p. 263), which seems to be a very similar species.

Ulmus pumila is common in the form of a bush or low bushy tree on the Lushan mountains in Kiangsi, but I never met with it in Hupeh and only with two or three doubtfully wild trees in Szech'uan. It does not grow in Japan, but is common in Mandshuria and Transbaical. In northern China it is widely distributed, having been reported from the shores of the gulf of Chili westward to Chinese Turkestan. In and around Peking this Elm is abundant, and in the park surrounding the Temple of Heaven many fine old trees may be seen. It is a rather low tree, growing from 10 to 16 m. tall, with a short trunk from 1.3 to 2.6 m. in girth; the bark is rough and deeply corrugated, and the branches are spreading and ascending-spreading and form a bushy crown. Under cultivation in this Arboretum it grows rapidly and makes a neat and shapely tree.   E. H. W.

Ulmus parvifolia Jacquin, *Pl. Rar. Hort. Schoenbr.* III. 6, t. 262 (1798). — Willdenow, *Hort. Berol.* I. 295 (1809), quoad synon. Jacquin; *Berlin. Baumz.* ed. 2, 521 (1811), quoad synon. Jacquin. — Poiret, *Encycl. Méth.* Suppl. IV. 189 (1816). — Roemer in Roemer & Schultes,

Syst. Veg. VI. 302 (1820). — Planchon in *Ann. Sci. Nat.* sér. 3, X. 280 (1848); in De Candolle, *Prodr.* XVII. 161 (1873). — K. Koch, *Dendr.* II. pt. 1, 423 (1872). — Maximowicz in *Bull. Acad. Sci. St. Pétersbourg,* XVIII. 292 (1873); in *Mél. Biol.* IX. 25 (1873). — Franchet & Savatier, *Enum. Pl. Jap.* I. 431 (1875). — Franchet in *Mém. Soc. Nat. Cherbourg,* XXIV. 253 (1882), ex Hemsley. — Dippel, *Handb. Laubholzk.* II. 34, fig. 11 (1892). — Koehne, *Deutsche Dendr.* 134 (1893). — Hemsley in *Jour. Linn. Soc.* XXVI. 448 (1894). — Henry in *Trans. As. Soc. Jap.* XXIV. Suppl. 85 (*List Pl. Formosa*) (1896). — Mouillefert, *Traité Arb. Arbriss.* II. 1203 (1898). — E. Pritzel in *Bot. Jahrb.* XXIX. 296 (1900). — Shirasawa, *Icon. Ess. For. Jap.* I. 68, t. 37, fig. 1–9 (1900). — Schneider, *Ill. Handb. Laubholzk.* I. 222, fig. 137 g–s, 139 b–d (1904); in *Oester. Bot. Zeitschr.* LXVI. (1916). — Ph. de Vilmorin, *Hort. Vilm.* 52 (1906). — Matsumura & Hayata in *Jour. Coll. Sci. Tokyo,* XXII. 369 (*Enum. Pl. Formos.*) (1906). — Mayr, *Fremdl. Wald- & Parkbäume,* 524 (1906) exclud. figura.[1] — Mottet in *Rev. Hort.* 1909, 397, fig. 167–168. — Nakai in *Jour. Coll. Sci. Tokyo,* XXXI. 189 (*Fl. Kor.* II.) (1911). — Matsumura, *Ind. Pl. Jap.* II. pt. 2, 33 (1912). — Dunn & Tutcher in *Kew Bull. Misc. Inform.* add. ser. X. 242 (*Fl. Kwangtung & Hongk.*) (1912). — Henry in Elwes & Henry, *Trees Great Brit. & Irel.* VII. 1928, t. 411, fig. 5 (1913). — Daveau in *Bull. Soc. Dendr. France,* 1914, 26, fig. 1, e, e', fig. A–A". — Bean, *Trees & Shrubs Brit. Isl.* II. 619 (1914).

Ulmus chinensis Persoon, *Syn.* I. 291 (1805). — Kirchner in Petzold & Kirchner, *Arb. Muscav.* 558 (1864).
Planera parvifolia Sweet, *Hort. Brit.* ed. 2, 464 (1830).
Ulmus virgata Roxburgh, *Fl. Ind.* ed. 2, II. 67 (1832).
Ulmus campestris, var. *parvifolia* Loudon, *Arb. Brit.* III. 1377 (1838), quoad synonyma pro parte.
Ulmus campestris, var. *chinensis* Loudon, l. c. fig. 1231 (1838).
Microptelea parvifolia Spach in *Ann. Sci. Nat.* sér. 2, XV. 358 (1841); *Hist. Vég.* XI. 113 (1842). — Siebold & Zuccarini in *Abh. Akad. Münch.* IV. Abt. 3, 224 (*Fl. Jap. Fam. Nat.* II. 100) (1846). — Miquel in *Ann. Mus. Lugd.-Bat.* III. 65 (1867); *Prol. Fl. Jap.* 253 (1867).
Ulmus Shirasawana Daveau in *Bull. Soc. Dendr. France,* 1914, 27, fig. 1, b, c, c'.

Western Hupeh: around Ichang, common, alt. 300–600 m., October 1907 (Nos. **2801, 726,** living material; tree 8–20 m. tall, girth 0.3–1.5 m.; leafy branchlets with unripe fruits); same locality, *A.*

[1] Mayr's fig. 244 shows a fruiting branchlet of which the samaras are very much like those of *U. levis* Pallas, and the leaves are not those of *U. parvifolia.* I suppose Mayr made a mistake in figuring a piece of the European *U. levis,* which does not occur in eastern Asia.

Henry (No. 3219; old leafy branchlets with ripe fruits); Paokang Hsien, mountains, October 1901 (Veitch Exped. No. 2650; tree 7 m. tall; leafy branches with flowers); without precise locality, *A. Henry* (No. 7518; leafy branchlets with fruits). **North-central China:** Kusan, *Hugh Scallan* (sterile). **Northern Shensi:** "In-kia-po," October 1897, *G. Giraldi* (leafy branchlets with fruits). **Chili:** Peking, Temple of Heaven grounds, September 1905, *F. N. Meyer* (No. 187; a tall, spreading Elm, with many small branches bearing small leaves and flowering in fall, in winter the bark peels off in curiously formed pieces; leafy branchlets with flowers); Hsiao Wu-tai-shan, alt. 1800 m., August 30, 1903, *F. N. Meyer* (No. 1385; sterile); Shimen, in crevices of decomposed slate rock, August 2, 1913, *F. N. Meyer* (No. 1069; sterile). **Kiangsu:** Shanghai, cultivated, October 15, 1905, *J. G. Jack* (leafy branchlets with fruits). **Fokien:** Amoy, *H. F. Hance* (No. 1414; leafy branchlets with fruits). **Formosa:** Bankinsing, *A. Henry* (No. 1529; tree 10 m. tall; leafy branchlets with young fruits).

NORTHEASTERN ASIA. **Korea:** Fusan, temple grounds, September 6, 1903, *C. S. Sargent* (sterile); Quelpaert, " circa Hongno, rara," July 1907, *U. Faurie* (No. 2008; sterile, leaves small, narrow, acuminate); " in sepibus Hongno," June 1909, *Taquet* (No. 3268; with a few old fruits, similar to the foregoing specimen); " in sylvis Sampargtan," October 1908, *Taquet* (No. 1334; leafy branchlets with young fruits, typical).

JAPAN. **Hondo:** prov. Musashi, Tokyo, Garden Agricultural College, October 9, 1892, *C. S. Sargent* (leafy branchlets with flowers); Itabashi, September 6, 1907 (ex Herb. Sakurai; with young fruits); Hanno, September 1908 (ex Herb. Yokohama Nursery Co.; with flowers); Tokyo, November 1909 (ex Herb. Yokohama Nursery Co.; leafy branchlets with fruits); Meguro, December 1914, *H. Shirasawa* (" bark: lenticels clear, cracked, narrow, scales small and thin "); leafless branchlets with ripe fruits); same place and date, *H. Shirasawa* (" bark coarse, lenticels not clear, scales little larger "); leafless branchlets with fruits); Momoyama, near the emperor Meiji's tomb, December 5, 1914, *E. H. Wilson* (No. 7856; small bushy tree, 8 m. tall, girth 0.9 m., leaves said to fall early in the spring; leafy branchlets with fruits); prov. Yamashiro, Nara, temple grounds, 1892, *C. S. Sargent* (small tree, 7 m. tall; leafy branchlets with fruits). **Kyushu:** prov. Hizen, Nagasaki, 1863, *C. Maximowicz* (leafy branchlets with fruits); without locality, 1842, *P. de Siebold* (ex Herb. Zuccarini as *Planera;* sterile); without locality (ex Herb. Mus. Lugd.-Bat. as *Microptelea;* leafy branchlets with fruits).

Daveau (in *Bull. Soc. Dendr. France*, 1914, 21) describes besides *U. Shirasawana*, mentioned above, *U. Sieboldii* (l. c. 26, fig. 1, d–d′ and B–B″), which is said to represent the true *U. japonica* Siebold in *Verh. Batav. Gen.* XII. 28 (nomen nudum, non Sargent) (1830). I have failed to detect sufficient differences between the three species proposed by Daveau even to keep them as distinct forms. The characters on which he has founded his species are too variable. Of *U. Sieboldii* the samaras are said to be larger, 15 mm. long and 8 mm. broad, the leaves are deciduous, and the bark is persistent not " s'exfoliant annuellement par plaques." I have not seen any fruits longer than 12 mm., although often as broad as 8–9 mm., their shape and the length of the stipe being very variable, as are also the dentation and shape of the leaves. Daveau, in his key, says that the typical *U. parvifolia* has 16–20 pairs of lateral nerves and the perigones are more or less reddish or red, while in *U. Shirasawana* the number of the lateral nerves is only 8–10, the perigones being more or less greenish or yellowish. I have failed to see a specimen with more than 16 pairs of lateral nerves, and I do not believe that the color of the perigone is of any importance in distinguishing even a form. *U. Sieboldii* Daveau may represent a variety if the large fruits and the different bark prove to be constant; otherwise I cannot accept these new species.

At low altitudes this species is common in western Hupeh, especially in the neighborhood of the city of Ichang. On rich soil and near water-courses it grows fully 25 m. tall and has a trunk from 1.5 to 2.5 m. in girth and clear of branches for half its height. The bark is gray and smooth, and on old trees peels off in roundish patches showing the brown inner bark; on young trees it is merely scaly. The branches are numerous and slender and form a rather broad, rounded crown. In open country it is a low broad-topped tree with a trunk seldom more than 1 m. in girth. It also grows on the cliffs of the glens and gorges in western Hupeh and in eastern Szech'uan, where it is usually a bush. It flowers in September and the fruit ripens in October. The leaves are retained late into the autumn and often assume rich shades of red and purple; in localities where the climate is warm and moist the leaves on individual trees sometimes remain quite green through the winter and fall as the young leaves unfold in the spring.

Japanese botanists consider this Elm indigenous in Japan, but I saw there only planted trees and not very many of them. At Meguro, near Tokyo, I saw the tree from which came the material on which *U. Shirasawana* Daveau is based, but I could detect no valid difference between it and the typical *U. parvifolia* Jacquin, trees of which grew near by.

Pictures of *U. parvifolia* Jacquin will be found under Nos. 474, 635 and 654 of the collection of my photographs and also in my *Vegetation of Western China*, Nos. 494 and 495.   E. H. W.

CONSPECTUS SECTIONUM SPECIERUMQUE ASIAE ORIENTALIS NEC NON HIMALAYAE.[1]

Perigonia circiter ⅓ v. ad medium (rarius ultra medium) incisa, lobis late oblongis
  v. fere rectangulis haud v. vix 2–2½-plo longioribus quam latis. Folia fere
  semper decidua. Flores vere apparentes . . .   Sect. I. **MADOCARPUS.**
Semina ab emarginatura samarae distincte remota, plus minusve in centro
  samarae sita (confer etiam 14. *U. pumilam*) . . Subsect. 1. GLABRAE.
  Samarae undique pilosae et ciliatae . . . . Ser. a. Wallichianae.
  Pedicellorum[2] pars superior nuda perigonio et parte inferiore pilosa circiter
  2–3-plo longior. Samarae tenuiter pilosae et satis sparse ciliatae.
                                              1. *U. Wallichiana.*

---

[1] To present the distinguishing characters more clearly the two European species, *U. foliacea* Gilibert and *U. glabra* Hudson, which have been constantly confused with certain species of eastern Asia, have been included in the key.

[2] The pedicels of the flowers are articulated, the part above the joint being mostly glabrous and more or less gradually passing into the perigone, while the part below the joint is mostly pilose, like the peduncle of the inflorescence. The "perigone" forms that part of the flowers which surrounds the stamens and the ovaries, which are inserted on the top of the pedicel at the very base of the perigone.

Pedicellorum pars superior parte inferiore plus minusve aequilonga v. brevior.
Inflorescentia fasciculata, 10–15-flora; perigonia in facie partesque superiores pedicellorum fere glabra. Samarae maturae lanceolatae, circiter 1–1.2 cm. longae. Folia (fide Brandis) ovato-oblonga, 4–11 cm. longa, glabra . . . . 2. *U. villosa.*
Inflorescentia fasciculata (fide Hance) 5–9-flora; perigonia in facie et pedicellorum partes superiores plus minusve distincte puberulae. Samarae maturae circiter 2.5 cm. longae. Folia ramulorum fertilium rhomboideo-subrotunda, ad 5 cm. longa, subtus pilosa et barbata.
3. *U. macrocarpa.*

Samarae maturae undique glabrae . . . . Ser. b. EUGLABRAE.
Inflorescentia plus minusve elongato-cymosa; pedicellorum pars superior parte inferiore pilosa v. perigonio saepe ad 2–2¼-plo longior v. samararum maturarum stipites distincti perigonio circiter duplo longiores. Folia satis lanceolato-elliptica, circiter 2½-plo longiora quam lata; petioli circiter 1 cm. longi . . . . 4. *U. Brandisiana.*
Inflorescentia fasciculata (v. in *U. Bergmanniana* et *U. Uyematsui* paulo elongata); pedicellorum pars superior parte inferiore pilosa v. pergionio vix longior v. subbrevior. Samararum maturarum stipites perigonio breviores v. rarius in *U. Bergmanniana* var. *lasiophylla* longiores. Petioli plerique breviores.
Pedicelli fructiferi toti 5–7 mm. longi . . . . . . 5. *U. Uyematsui.*
Pedicelli fructiferi toti 2–4 mm. longi.
Samarae late obovatae v. obovato-rotundae. Ramuli annotini brunnei v. rubro-brunnei, etiam hornotini glabri v. perulae gemmarum dorso glabrae. Folia obovato-oblonga, obovato-elliptica v. elliptica.
6. *U. Bergmanniana.*
Samarae elliptico- v. obovato-oblongae v. elliptico-rhomboideae. Ramuli hornotini annotinique fere semper pilosi, saepe scabri v. gemmarum perulae in dorso pilosae v. folia distinctius obovata v. apice 3(–5)-lobata.
Ramuli novelli plus minusve hirsuti v. annotini distincte brunnescentes. Folia ramulorum fructiferorum nunquam apice trilobata.
7. *U. glabra.*
Ramuli novelli vix hirsuti v. cito glabri, annotini plerique grisei v. flavescentes. Folia etiam ramulorum fructiferorum pro parte apice 3(–5)-lobata . . . . 8. *U. laciniata.*
Semina apice samarae juxta emarginaturam sita v. tantum paulo ab ea remota (et tantum in fructibus satis brevibus et latis *U. pumilae* in centro samarae sita) . . . . Subsect. 2. FOLIACEAE.
Pedicellorum pars inferior pilosa parte superiore subaequilonga v. brevior, vix v. rarius ea v. perigonio paulo longior (v. folia non coriacea lucidaque et non simpliciter obtuse serrata).
Samarae pleraeque obovatae v. satis anguste ellipticae, interdum in disco seminitego pilosae, v. folia satis magna, nervis utrinsecus 16 v. ultra instructa, basi distincte inaequalia, margine dupliciter serrato-dentata.
Ser. a. NITENTES.
Samarae undique glabrae (v. in *U. japonica* tantum juveniles in disco parce pilosae).
Samarae oblongo-ellipticae, circiter duplo longiores quam latae. Folia lanceolata v. elliptico-lanceolata, subtus tantum in axillis nervorum lateralium (lateris longioris 18–24, apice satis sensim producto incluso) barbata, coriacea. Gemmae apice ramulorum obovato-

oblongae, subacutae, perulis concoloribus purpurascentibus ciliatis et
in facie pilosis . . . . 9. *U. castaneifolia*.
Samarae obovato-oblongae v. obovatae, haud duplo longiores quam
    latae v. folia subobovata ovato-oblongave, satis subito apiculata et
    latere longiore (imo apice excluso) nervis paucioribus instructa.
    Folia majora in latere longiore nervis 18 v. ultra instructa. Gemmae
    perulis bicoloribus in parte superiore intensius coloratis obtectae,
    ramuli annotini brunnescentes v. purpurascentes. Samarae late
    obovatae . . . . 10. *U. Wilsoniana*.
    Folia majora in latere longiore nervis 8–14(–16) instructa v. gemmae
    perulis concoloribus obtectae et ramuli annotini plus minusve
    cinereo-brunnei v. grisei. Samarae plus minusve oblongo-obo-
    vatae v. late ellipticae.
    Ramuli hornotini autumno v. etiam annotini floriferique plus
    minusve pilosi, saepe scabri, flavescentes v. cinereo-brunnescentes
    v. glabrescentes et distinctius rubescentes. Folia subtus villosula
    v. glabra et tantum barbulata, obovata et basim versus pleraque
    satis angustata (sensim cuneata) v. circuitu elliptica et ima basi
    vix v. minus distincte quam in formis speciei sequentis asym-
    metrica. (Samarae interdum in disco parce pilosae.)
                                  12. *U. japonica*.
    Ramuli hornotini et annotini sparse v. non pilosi, plerique plus
    minusve distincte rubescentes v. flavo-rubri v. fere purpurei.
    Folia matura subtus pleraque tantum barbata, forma variabilia
    sed basi saepissime distincte asymmetrica [1] . . 11. *U. foliacea*.
Samarae maturae in disco seminitego distincte pilosae et etiam in facie
    plus minusve sparse pilosae sed versus marginem glabrae. Folia ma-
    tura ut videtur iis *U. japonicae* subsimilia . . . . 13. *U. Davidiana*.
Samarae elliptico-rotundae v. late ellipticae, glaberrimae. Folia satis
    parva, maxima vix ad 6 cm. longa et 3 cm. lata, nervis utrinsecus 6–14,
    serraturis obtusioribus brevioribusque saepe subsimplicibus.
                            Ser. b. Pumilae.
    Samarae vix plus quam 15 mm. longae . . . . 14. *U. pumila*.
    Samarae (fide Franchet) 2–2.5 cm. longae . . . . . 15. *U. glaucescens*.
Pedicellorum pars inferior pilosa fere semper valde elongata, parte superiore
    glabra perigonioque multoties longior. Samarae magnae, circiter 2–2.5 cm.
    longae, distincte stipitatae; semina inter emarginaturam et centrum v.
    juxta emarginaturam samarae sita. Folia oblongo-elliptica, acuminata,
    subpersistentia, coriacea, lucida, simpliciter obtuse serrata.
                            Ser. c. Lanceaefoliae.
                           16. *U. lanceaefolia*.
Perigonia fere ad basim fissa, lobis lanceolatis, saepe 3–4-plo longioribus quam latis,
    sub fructu pleraque caduca; inflorescentiae fasciculato-cymosae; pedicellorum
    pars inferior pilosa brevissima v. parte superiore glabra haud plus quam duplo
    longior. Samarae glabrae; semina in centro sita. Folia saepe subpersistentia.
    Flores autumno hyemeve apparentes . . . . . Sect. II. **MICROPTELEA**.
                            17. *U. parvifolia*.

---

[1] The description of the leaves in the key always refers to leaves of mature
plants, leaves of young sterile plants being often very different and much alike in
different species.

ENUMERATIO SPECIERUM ASIAE ORIENTALIS
NEC NON HIMALAYAE.[1]

Sect. I. **MADOCARPUS** Dumortier, *Prodr. Fl. Belg.* 25 (1827). — Henry in Elwes & Henry, *Trees Great Brit. & Irel.* VII. 1848 (1913). — Schneider in *Oester. Bot. Zeitschr.* LXVI. (1916).

Ulmus, sect. *Dryoptelea* Spach in *Ann. Sci. Nat.* sér. 2, XV. 361 (1841).

Ulmus, subgen. *Dryoptelea* Planchon in *Ann. Sci. Nat.* sér. 3, X. 260 (1848); in De Candolle, *Prodr.* XVII. 156 (1873). — Engler in Engler & Prantl, *Nat. Pflanzenfam.* III. Abt. 1, 62 (1888). — Sargent, *N. Am. Silva*, VII. 40 (1895).

Ulmus, subgen. *Euulmus* K. Koch, *Dendr.* II. pt. 1, 405 (1872).

Ulmus, subgen. *Euulmus*, sect. *Dryoptelea* Dippel, *Handb. Laubholzk.* II. 22 (1892). — Koehne, *Deutsche Dendr.* 135 (1893). — Schneider, *Ill. Handb. Laubholzk.* I. 216 (1904).

Ulmus, sect. A. *Euulmus*, II. *Madocarpus* Ascherson & Graebner, *Syn. Mitteleur. Fl.* IV. 550 (1911).

This section contains the largest number of species, but it is represented in the New World only by *U. fulva* Michaux.

Subsect. a. GLABRAE Schneider in *Oester. Bot. Zeitschr.* LXVI. (1916).

Ulmus, series *Glabrae* Moss, *Cambridge Brit. Fl.* II. 89 (emend.) (1914).

This subsection is well marked by the position of the seeds in the samara, as indicated in the key, p. 247.

Series a. WALLICHIANAE Schneider in *Oester. Bot. Zeitschr.* LXVI. (1916).

The species of this group, which are confined to India and eastern Asia, need further observation.

1. **Ulmus Wallichiana** Planchon in *Ann. Sci. Nat.* sér. 3, X. 277 (1848); in De Candolle, *Prodr.* XVII. 158 (1873). — Brandis, *For. Fl. Ind.* 432, t. 51 (1874); *Ind. Trees*, 594 (ut videtur pro parte) (1906). — Hooker f., *Fl. Brit. Ind.* V. 480 (pro parte) (1888). — Collett, *Fl. Siml.* 455 (probabiliter tantum pro parte) (1902). — Gamble, *Man. Ind. Timb.* ed. 2, 627, t. 13, fig. 1 (pro parte?) (1902). — Schneider, *Ill. Handb. Laubholzk.* I. 216, in adnot. (1904), II. 902, fig. 565 g (1912); in *Oester. Bot. Zeitschr.* LXVI. (1916).

Ulmus erosa Wallich, *Cat.* No. 3546 (nomen nudum) (non Roth) (1831).

INDIA. Kashmir: Kishtwar, "reg. temp. alt. 8–10000 ped.," *T. Thomson* (ex. Herb. Ind. Or. sub nom. *U. campestris macrophylla;* with fruits). Kumaon: without precise locality, *R. Blinkworth* (type, ex Herb. Wallich No. 3546; fide Planchon).

According to Planchon's description the fruits of *U. Wallichiana* are hairy only upon the disk containing the seed, while the wing and the margins are glabrous. Unfortunately, I did not see Wallich's specimen No. 3546. Brandis (1874) describes the samara as " pubescent," and the plant before me agrees well with his plate and also with Planchon's description of the leaves and other parts. Hooker f. (1888) says: "Samara . . . glabrous or disk puberulous." I have not seen any Elm from India with fruits hairy only upon the disk. There may be some mistake in Planchon's and Hooker's statements; otherwise such a variability in the pubescence of the fruit would be a strange fact in the genus. The Elms of British India are still imperfectly known and need careful observation in the field.

---

[1] *Ulmus Cavaleriei* Léveillé is *Pteroceltis Tatarinowii* Maximowicz; see under *Pteroceltis.*

2. **Ulmus villosa** Brandis in *Ind. Forester*, XXV. 230 (1899); *Ind. Trees*, 594, fig. 185 (1906). — Gamble, *Man. Ind. Timb.* ed. 2, 628 (1902). — Schneider in *Oester. Bot. Zeitschr.* LXVI. (1916).
Ulmus campestris Brandis, *Forest Fl. Ind.* 433 (pro parte, non Linnaeus vel Auct. Al.) (1874).
Ulmus campestris Smith, var., Aitchison in *Jour. Linn. Soc.* XVIII. 93 (1880).
Ulmus Wallichiana Hooker f., *Fl. Brit. Ind.* V. 480 (pro parte, non Planchon) (1888).

INDIA. Kashmir: Kulu (or Kullu), *D. Brandis* (with unripe fruits and without leaves; co-type); same locality, October 1876, *D. Brandis* (" a large tree," ex Brandis); Bussahir, " at 6500 ft. in May 1881," *D. Brandis* (ex Brandis); Kunawar, " near Tranda at 7000 ft.," *D. Brandis;* Thelum River, "between 5000 and 7000 ft.," *J. E. T. Aitchison* (" a good sized tree "); ex Brandis). Punjab: Pabar Valley, " at an elevation of 4000 ft.," May 1881, *D. Brandis* (ex Brandis).
AFGHANISTAN. Kurrum Valley, " in the woods at 7000–9000 ft., not common," 1879, *J. E. T. Aitchison* (No. 677; sterile).
I am not quite sure whether all the specimens mentioned by Brandis belong to this species or partly to *U. Brandisiana* Schneider. Aitchison's No. 677 seems to be *U. villosa*, the leaves of which I know only from Brandis's description and figure. His description being somewhat incomplete, I think it best to describe No. 677 as follows: Arbor magna; ramuli hornotini flavo-rubri, tenuissime scabri, vetustiores laevi, cinerascentes, deinde nigrescentes, rimosi; gemmae nondum satis evolutae parvae, ovatae, puberulae, in perularum margine fulvo-ciliatae. Folia obovata v. obovato-oblonga, basi cuneata v. rotundata, paulo v. vix asymmetrica, apice satis subito in acumen brevem contracta, supra-viridia, plus minusve scabra, tantum in costa incisa sparse pilosa, subtus in sicco vix pallidiora, scabra, axillis nervorum 11–14 (imo apice excluso) albo-barbata, minora obovata v. elliptica, 4–7 cm. longa et 3–4.5 cm. lata, maxima obovato-oblonga, 8–12 cm. longa et 5–6.5 cm. lata, margine lobulato-serrata serraturis primariis dorso dentibus 2–4 instructis; petioli 3–7 mm. longi, satis tenues, praesertim in sulco ventrali pilosa.
See also my remarks on No. 403 of Aitchison under *Ulmus* spec. on p. 265. It is extremely difficult to determine an *Ulmus* from sterile specimens alone.

3. **Ulmus macrocarpa** Hance in *Jour. Bot.* VI. 332 (November 1, 1868). — Planchon in *Compt. Rend. Acad. Sci. Paris*, LXXIV. pt. 1, 1498 (1872); in De Candolle, *Prodr.* XVII. 162 (1873). — Maximowicz in *Bull. Acad. Sci. St. Pétersbourg*, XVIII. 289 (1873); in *Mél. Biol.* IX. 22 (1873). — Franchet in *Nouv. Arch. Mus. Paris*, sér. 2, VII. 78, t. 8, fig. c (*Pl. David.* I. 268) (1884). — Hemsley in *Jour. Linn. Soc.* XXVI. 447 (1894). — Komarov in *Act. Hort. Petrop.* XXII. 81 (*Fl. Mansh.* II.) (1903). — Nakai in *Jour. Coll. Sci. Tokyo*, XXXI. 190 (*Fl. Kor.* II.) (1911). — Schneider, *Ill. Handb. Laubholzk.* II. 904, fig. 566 h (1912).

Ulmus sp. novae 2 Maximowicz in *Mém. Sav. Étr. Acad. Sci. St. Pétersbourg*, IX. 477 (*Prim. Fl. Amur. Ind. Fl. Pekin.*) (nomen nudum) (1859), fide Maximowicz.
?Ulmus rotundifolia Carrière in *Rev. Hort.* 1868, 374, fig. 40 (October 1, 1868).

CHINA. Chili: " in montosis prope Jehol," *A. David* (No. 1718, ex Planchon, and Herb. Hance, No. 14538, ex Hance, type); Nankou Pass, November 1905, *F. N. Meyer* (No. 208; sterile branch with corky wings); Nankou, rocky cliffs, July 27, 1913, *F. N. Meyer* (without No.; of shrubby growth; branches winged as in No. 208, leaves very small); near Shimen, rocky crevices and loess cliffs, August 3, 1913, *F. N. Meyer* (No. 1088; apparently the same as the preceding specimen, but the leaves partly larger); Shiling, January 25, 1908, *F. N. Meyer* (No.

231; according to Meyer's note in *Bull. U. S. Dept. Agric. Bur. Pl. Industr.*, No. 137, 48, No. 22364 [1909], " a shrubby Elm, often having irregular corky wings along its branches. Grows on very dry and rocky mountain slopes, growing a couple of feet up to 20 or 30 feet high. Chinese name Shan yushu. Seems to be very variable in habitus." Distributed as *U. macrocarpa* Hance? Young plants are growing in the Arnold Arboretum). Shansi: Yento, March 1, 1908, *F. N. Meyer* (No. 275; according to Meyer's note in *Bull. U. S. Dept. Agric. Bur. Pl. Industr.*, No. 142, 24, No. 22678 [1909], " a densely branched Elm of shrubby growth, occasionally growing into a small tree; found growing on a sunny rocky mountain slope at about 4000 feet altitude." Cultivated plants in the Arboretum look much like those of the preceding number).

Of this species I have seen only one fruit from David's No. 1718, kindly sent to the Arnold Arboretum by the Director of the botanical department of the Muséum in Paris. According to the characters given in the key on p. 248 this species belongs to subsect. *Glabrae* and has no relationship whatever with *U. americana* Linnaeus, *U. pedunculata* Fougeraux or *U. alata* Michaux mentioned by Hance as " closely allied " to his species. Unfortunately I have not seen mature leaves and I do not know how to distinguish without fruits *U. Davidiana* Planchon (see p. 261) and *U. japonica* Sargent from *U. macrocarpa*. According to Komarov (1903) this Elm seems to be a common plant in northern China and Mandshuria, and it may show the same variations as appear in the European *U. foliacea* Gilibert. Judging by the sterile specimens collected by F. N. Meyer and W. Purdom, mentioned under the different species, and also by young cultivated plants I believe that *U. macrocarpa* can be distinguished by the rather flat corky wings of the branches, while in *U. japonica* the corky excrescences are not wing-like, but are more irregular in shape and grow almost round the whole branch. The specimens I am inclined to take for *U. Davidiana* show no corky wings and are much more glabrous (branchlets, petioles, leaves), the leaves being large, broadly or even roundish obovate, with many parallel nerves and a somewhat longer petiole. See p. 261. Some specimens of Meyer's mentioned under *U. pumila* Linnaeus (p. 242) bear some resemblance to shrubby forms of either *U. macrocarpa* or *U. japonica*.

The description given by Carrière of his *U. rotundifolia* is insufficient and the figure shows only a part of a sterile branch with 3 leaves. In the shape the leaves somewhat resemble those of the specimens of Meyer and Purdom mentioned under *U. Davidiana*, but Carrière says " feuilles . . . très regulièrement dentées " and in the figure the dentation is simple, not compound, as in all the specimens before me. According to Planchon Carrière's species is the same as *U. macrocarpa;* it was raised from seeds collected by David. Without having seen original specimens of Carrière's species, it cannot be determined if his name is the oldest for either *U. macrocarpa* or *U. Davidiana*. Both came from Jehol, but they are not the same, as Hemsley believes, the latter being most nearly related to *U. japonica* Sargent.

Series b. EUGLABRAE Schneider in *Oester. Bot. Zeitschr.* LXVI. (1916).

Without fruits it is sometimes very difficult to distinguish the species of this series from those of the *Nitentes*, and we need further investigation on the variability of each species according to its age and the situation where it is found. Plants grown in shady moist places always have the leaves very different from those of plants growing in dry sunny places, and the leaves of young well-growing cultivated plants usually look different from those of old trees. There seem to be no forms with corky branches of the species of this series so far as we now know.

4. **Ulmus Brandisiana** Schneider in *Oester. Bot. Zeitschr.* LXVI. (1916).

?*Ulmus effusa* Brandis, *For. Fl. Ind.* 432 (non Willdenow) (1874), quoad specimen Thomsonii.

?*Ulmus Wallichiana* Hooker f., *Fl. Brit. Ind.* V. 480 (pro parte) (1888).

Ad descriptionem addenda et emendanda: Arbor?; ramuli novelli annotinique glabri, flavo-brunnescentes, lenticellis elongatis flavescentibus satis numerosis obtecti, vetustiores cinerascentes, rimosuli; gemmae satis evolutae non visae. Folia lanceolato-elliptica, basi breviter attenuata v. rotundata v. fere cordata, plus minusve asymmetrica, apice satis sensim in acumen distinctum producta, supra ut videtur intense viridia, satis levia, tantum in costa incisa sparsissime pilosa, subtus pallidiora, tantum in axillis nervorum lateralium elevatorum lateris longioris (imo apice excepto) 14–19 albo-barbulata, ceterum glabra et vix scabriuscula, margine dupliciter serrata, serraturis primariis haud (v. vix) incurvis, minora 5–8 cm. longa, 2–3.2 cm. lata, maxima ad 10–11 cm. longa et 4–4.3 cm. lata; petioli satis tenues, tantum superne pilosuli, 0.6–1 cm. longi; stipulae late lanceolatae, ciliatae, petiolis juvenilibus plus quam duplo longiores, mox caducae. Inflorescentiae plus minusve elongato-cymosae, ad 12 mm. longae, pilosae; flores juveniles non vidi; perigonia in basi fructum 5–6-loba, lobis oblongis ad medium v. paulo ultra medium incisis ciliatis, ceterum glabra, pars superior pedicellorum parte inferiore pilosa brevior (in typo) v. 2–2¼-plo longior (in specimine Thomsoniano). Samarae maturae obovatae v. obovato-rectangulares, basi subito in stipitem perigonio duplo longiorem contractae, apice leviter emarginatae emarginatura brevi pilosa instructae ceterum glaberrimae, ad 18 mm. longae et 12 mm. latae; semina in centro samarae sita, ab emarginatura distincte remota.

INDIA. Kumaon: "Dwali, 8500 feet," *R. Strachey & J. E. Winterbottom* (No. 3, type in Herb. Gray; with ripe fruits and leaves). Kashmir: "Banahal, reg. temp., alt. 7500 ped.," *T. Thomson* (ex Herb. Ind. Or. sub nom. *U. campestris, a glabra* in Herb. Gray; with young fruits and leaves). Kashmir: "reg. temp. 5000 ped.," *T. Thomson* (a very poor sterile specimen distributed in Herb. Ind. Or. as *U. pedunculata* Foug.?; an uncertain form).

This seems to be a very interesting species, to which apparently belongs Thomson's specimen from Kashmir, at Ganderbal, at 5000 feet, April 1848, which was doubtfully referred to *U. effusa* by Brandis. Whether or not Thomson's specimen from Banahal represents a distinct variety on account of the elongated upper part of the pedicels needs further investigation, as I have seen only a poor specimen with unripe fruits. The flowers seem to be very similar to those of the European *U. glabra* Hudson, but otherwise the Himalayan plant may be more closely related to *U. Bergmanniana* Schneider from central China, the inflorescence of which sometimes is a little elongated. *U. Brandisiana*, however, differs from all the other species of this series in the long stipe of its fruit (see also *U. Bergmanniana*, var.). The species is named in honor of the late Sir Dietrich Brandis.

5. **Ulmus Uyematsui** Hayata, *Icon. Pl. Formos.* III. 174, t. 32 (1913).

CHINA. Formosa: Mt. Arisan, March 1913, *K. Uyematsu* (type, ex Hayata).

According to the description and plate this species seems closely related to *U. Bergmanniana*, var. *lasiophylla* Schneider. The leaves are described as elliptic or oblong, up to 10 cm. long and to 4.5 cm. broad, and densely pubescent beneath. The petioles are 4 mm. long. The inflorescence is elongate-cymose, the pedicels are about 6 mm. (in the drawing up to 8 mm.) long and articulated in the middle. The ripe fruits are yet unknown, and according to the plate the species belongs to the ser. *Glabrae*, the seeds being in the centre of the samara. Besides the rather long pedicels I cannot detect much difference between the Formosan Elm and the pubescent variety of the species from central China. In the drawing the stipe of the young samara is not longer than the 5-lobed perigone. The filaments of the 5 stamens seem to be only as long as the perigone.

6. **Ulmus Bergmanniana** Schneider. See p. 240.

Ulmus Bergmanniana, var. **lasiophylla** Schneider. See p. 241.

7. **Ulmus glabra** Hudson, *Fl. Angl.* 95 (1762), exclud. var. *β*. — Rehder in *Mitt. Deutsch. Dendr. Ges.* 1908, 57 (1909). — Moss in *Gard. Chron.* ser. 3, LI. 217 (1912); *Cambridge Brit. Fl.* II. 95, t. 104–105 (1914).

Ulmus campestris Linnaeus, *Spec.* 225 (pro parte et in herb.) (1753). — Schmidt *Oester.-Baumz.* IV. 43, t. 226 (1822). — Kerner in *Oester. Bot. Zeitschr.* XXVI. 52 (1876).

Ulmus scabra (*scabris!*) Miller, *Gard. Dict.* ed. 8, No. 2 (1768). — K. Koch, *Dendr.* II. pt. 1, 412 (pro parte) (1872). — Dippel, *Handb. Laubholzk.* II. 27 (pro parte) (1892). — Koehne, *Deutsche Dendr.* 135, fig. 27 E (1893). — Schneider, *Dendr. Winterstud.* 161, fig. 153 f–h (1903); *Ill. Handb. Laubholzk.* I. 216, fig. 136 u–v, 137 h–l (pro parte) (1904). — Ascherson & Graebner, *Syn. Mitteleur. Fl.* IV. 560 (pro parte) (1911). — Nakai in *Jour. Coll. Sci. Tokyo,* XXXI. 190 (*Fl. Kor.* II.) (1911), exclud. var. *major.*

Ulmus montana Stokes in Withering, *Bot. Arr. Veg. Gr. Brit.* ed. 2, I. 259 (1787). — Loudon, *Arb. Brit.* III. 1398 (1838). — Planchon in *Ann. Sci. Nat.* sér. 3, X. 274 (1848); in De Candolle, *Prodr.* XVII. 159 (pro parte) (1873). — Trautvetter in *Bull. Phys.-Math. Acad. Sci. St. Pétersbourg,* XV. 352 (1857). — Hempel & Wilhelm, *Bäume & Sträucher,* III. 7, fig. 225 B, 230–233, t. 38 (1897). — Henry in Elwes & Henry, *Trees Great Brit. & Irel.* VII. 1864, t. 400, 411, fig. 13 (1913). — Bean, *Trees & Shrubs Brit. Isl.* II. 616 (1914).

Ulmus campestris, var. *latifolia* Aiton, *Hort. Kew.* I. 319 (1789).

Ulmus nuda Ehrhart, *Beytr. Naturk.* VI. 86 (1791).

Ulmus latifolia Salisbury, *Prodr.* 391 (1796), an Persoon?

Ulmus excelsa Borkhausen, *Handb. Forstbot.* I. 839 (1800).

Ulmus suberosa Michaux, *N. Am. Sylva,* III. 98 (t. 129, fig. sub nom. *U. latifolia!*) (non Moench et Auct. Al.) (1819).

Ulmus corylacea Dumortier, *Fl. Belg. Prodr.* 25 (1827).

Ulmus major Reichenbach, *Icon. Fl. German.* XII. 13, t. 665, fig. 1335 (ncn Smith) (1850), exclud. synon.

Ulmus campestris, var. *scabra* Neilreich, *Fl. Wien,* 165 (1846).

Ulmus campestris, var. *montana* Hartig, *Vollst. Naturg. Forstl. Kulturpfl.* 458 (1850), forma satis dubia. — Ascherson, *Fl. Prov. Brandenbg.* I. 614 (exclud. forma) (1864).

EUROPE (with exception of Portugal); NORTHERN AFRICA; WESTERN ASIA.

I include this well-known species because *U. laciniata* Mayr has been hitherto referred to it. As the many synonyms show, the nomenclature is complicated. The leaves of sterile shoots or of branches growing in the shade sometimes are tricuspidate at the apex and cannot be distinguished from leaves of the same shape of *U. laciniata,* where such trilobate leaves prevail even on mature trees. This form of *U. glabra* has many different names, and the synonymy may be given as follows:

Ulmus glabra, f. **grandidentata** Moss, *Cambridge Brit. Fl.* II. 96 (1914).

Ulmus corylacea, var. *grandidentata* Dumortier, *Prodr. Fl. Belg.* 25 (1827).

Ulmus tridens Hartig, *Vollst. Naturg. Forstl. Kulturpfl.* 460 (1850).

Ulmus sublaciniatus Mathieu, *Fl. Gén. Belg.* I. 480 (in textu pro forma) (1853).

Ulmus expansa Rota in *Bot. Zeit.* XIII. 469 (1855).

Ulmus tricuspidata Besser ex Trautvetter in *Bull. Phys.-Math. Acad. Sci. St. Pétersbourg,* XV. 352 (1857), in textu.

Ulmus montana, var. *tridens* Lange, *Haandb. Danske Fl.* 267 (1886–8), fide Moss.

Ulmus montana, subspec. *major*, f. *tricuspis* Dippel, *Handb. Laubholzk.* II.
29 (1892).
Ulmus montana triserrata Mouillefert, *Traité Arb. Arbriss.* II. 1201 (1898).
Ulmus montana, f. *lobata* Waisbecker in *Oester. Bot. Zeitschr.* XLIX. 67 (1899).
Ulmus montana, var. *corylifolia* Zapałowicz, *Consp. Fl. Galic. Crit.* II. 98
(1908), fide Ascherson & Graebner.
Ulmus glabra, f. *tricuspis* Rehder in *Mitt. Deutsch. Dendr. Ges.* XXIII. (1915).

8. **Ulmus laciniata** Mayr, *Fremdl. Wald- u. Parkbäume*, 523, fig. 243 (1906)
exclud. descript. pro parte maxima.

Ulmus major, var. *heterophylla* Maximowicz & Ruprecht in *Bull. Phys.-Math.
Acad. Sci. St. Pétersbourg*, XV. 139 (1856); in *Mél. Biol.* II. 434 (1857).
Ulmus montana Maack & Ruprecht, in *Bull. Phys.-Math. Acad. Sci. St.
Pétersbourg*, XV. 376 (non Stokes) (1857); in *Mél. Biol.* II. 557 (1858). —
Maximowicz in *Bull. Acad. Sci. St. Pétersbourg*, XXVII. 291 (1873); in
Mél. Biol. IX. 25 (1873). — Planchon in De Candolle, *Prodr.* XVII. 159
(1873), quoad plantas Transbaical. — Franchet & Savatier, *Enum. Pl.
Jap.* I. 431 (1875). — Herder in *Act. Hort. Petrop.* XII. 47 (*Pl. Radd.*)
(1892), quoad plantas Transbaical. — Hemsley in *Jour. Linn. Soc.* XXVI.
448 (1894), quoad plantas Japonicas. — Komarov in *Act. Hort. Petrop.*
XXII. 88 (*Fl. Mansh.* II.) (1903).
Ulmus montana, var. *laciniata* Trautvetter in *Mém. Sav. Étr. Acad. Sci. St.
Pétersbourg*, IX. 246 (Maximowicz, *Prim. Fl. Amur.*) (1859). — Schmidt in
Mém. Acad. Sci. St. Pétersbourg, sér. 7, XII. No. 2, 174 (*Reis. Amur. Sachal.*)
(1868). — Miquel in *Ann. Mus. Lugd.-Bat.* III. 65 (1867); *Prol. Fl. Jap.*
253 (1867). — Shirasawa, in *Bull. Agric. Coll. Tokyo*, II. 266, t. 13, fig. 7
(*Jap. Laubh. Winter.* t. 9, fig. 7) (1895); *Icon. Ess. For. Jap.* II. t. 15, fig.
1–9 (1908). — Henry in Elwes & Henry, *Trees Great Brit. & Irel.* VII. 1865
(1913). — Miyabe & Miyake, *Fl. Saghal.* 404 (1915).
Ulmus scabra, var. *typica*, f. *heterophylla* Schneider, *Ill. Handb. Laubholzk.*
I. 218 (pro parte) (1904). — Ascherson & Graebner, *Syn. Mitteleur. Fl.*
IV. 564 (pro parte) (1911).

CHINA. Chili: Weichang, 1910, *W. Purdom* (No. 87; with fruits and good
leaves).
NORTHEASTERN ASIA. Mandshuria: mountains 12 hours east of Har-
bin by railroad, August 31, 1903, *C. S. Sargent* (very large tree; sterile); Shengking,
Tien-shan, June 8, 1906, *F. N. Meyer* (No. 124; sterile). Amur: "Amur," *C.
Maximowicz* (fruiting branchlet with leaves); "Amur med.," May 27, 1891, *S.
Korshinsky* (sterile; distributed sub nom. *Corylus mandshurica*). Ussuri: near
Vladivostock, cultivated, August 20, 1903, *C. S. Sargent* (sterile). Saghalien:
without locality, *Fr. Schmidt* (very young fruits and young leaves).
JAPAN. Hokkoido: prov. Ishikari, Sapporo, June 5, 1891, *Y. Tokubuchi*
(with fruits and leaves); hill near Sapporo, September 17, 1892, *C. S. Sargent* (tree
7–10 m. tall, girth 0.3 m., pale bark; sterile).
The mature plants of this species are well distinguished from those of *U. glabra*
Hudson, but young vigorous plants are much more alike. The pale color of the
branchlets in the autumn of their first year and in the second year, being somewhat
brownish only on those parts which have been much exposed to the sun, seems to be
the best character to distinguish *U. laciniata* from *U. glabra*. The perigones are
often hairy besides being ciliated, as are the upper part of the pedicels. The fruits
seem to be always quite glabrous, while, according to Trautvetter (1857), some
Russian forms of *U. glabra* have a distinctly hairy disk, at least when young. The

winter-buds of Japanese plants of *U. laciniata* are spindle-shaped and rather acute, resembling those of *U. laevis* Pallas, but in plants from Chili grown in the Arnold Arboretum the buds are more ovate and the deep red-brown scales are very hairy on the margin, while in Japanese plants which were brought by J. G. Jack from Chuzenji the scales are only very finely ciliate and of a lighter color in the lower part. Whether or not the plants from northern China may represent a distinct variety needs further investigation.

This species occurs sparingly in moist woods near Lake Yumoto, in the Nikko region, and in slightly increasing numbers northward, but it is nowhere plentiful in Hondo. In Hokkaido it is fairly common, more especially in the central parts of this island, and always is associating with mixed trees. As I saw it this Elm is in Japan always a small, slender and unimportant tree, not more than 12 m. tall, with a trunk scarcely 1 m. in girth and covered with gray, fairly smooth, fibrous bark. My observations as to the size of this tree agree with those of Maximowicz and Sargent. Mayr, who says it grows 30 m. tall and is an important constituent of the forests in central Hokkaido, confuses it with *U. japonica* Sargent. On the mainland of eastern Asia it would appear to be a larger tree than in Japan, for Sargent noted it as a " very large tree on the mountains twelve hours east of Harbin by railroad."

A picture will be found under No. x365 of the collection of my Japanese photographs.   E. H. W.

Subsect. 2.   FOLIACEAE Schneider in *Oester. Bot. Zeitschr.* LXVI. (1916).

This subsection contains the most difficult forms, especially those of the following series.

Series a.   NITENTES Moss, *Cambridge Brit. Fl.* II. 89 (1914). — Schneider in *Oester. Bot. Zeitschr.* LXVI. (1916).

The species of this series are very closely related, and it is difficult to distinguish some of them without carefully collected material with flowers, fruits and mature leaves. In Europe they are connected with the species of ser. GLABRAE by many natural hybrids, but we do not yet know if there are any such hybrids in eastern Asia or in India.

9.   **Ulmus castaneifolia** Hemsley in *Jour. Linn. Soc.* XXVI. 446, t. 10 (1894). — E. Pritzel in *Bot. Jahrb.* XXIX. 296 (1900). — Schneider, *Ill. Handb. Laubholzk.* II. 904, fig. 565 c–d, 566 e–f (1912).

CHINA. Hupeh: Changyang Hsien, *A. Henry* (No. 7780, type; with ripe fruits on leafless branchlets); same locality, mountains, summer 1900, *E. H. Wilson* (Veitch Exped. No. 677 in part; sterile); Chienshi Hsien, May 1900, *E. H. Wilson* (Veitch Exped. No. 545; with ripe fruits and very young leaves). Eastern Szech'uan: Wushan Hsien, *A. Henry* (No. 5498, co-type; sterile, young and old leaves).

The young flowers of this species which seems to have a very limited distribution in central China are not yet known. The species is well characterized by the shape of its leaves and fruits, as described in the key. The largest leaf I have seen is about 14 cm. long and 5 cm. broad, without the rather thick petiole, which measures 10 mm. in length. The surfaces of the leaves are more or less rough, but otherwise glabrous with the exception of a few hairs on the impressed rib of the upper surface and the clusters of gray hairs in the axils of the prominent lateral veins beneath.

10.   **Ulmus Wilsoniana** Schneider.   See p. 238.

Ulmus Wilsoniana Schneider, var. **psilophylla** Schneider.   See p. 239.

Ulmus Wilsoniana, var. **subhirsuta** Schneider, n. var.

A typo praecipue recedit ramis annotinis dense pilis griseo-brunneis obtectis fere hirsutis.

CHINA. Southern Szech'uan: prope Hohsi versus Te-li-pu, May 7, 1914, *C. Schneider* (No. 1113; frutex squarrosus, circiter 2–3 m. altus; with unripe fruits). I have seen this Elm only in this one locality, and the young fruits and leaves agree well with those of the type. The branchlets of the preceding year bear a very dense grayish and brownish, almost hirsute, pubescence.

11. **Ulmus foliacea** Gilibert, *Exercit. Phyt.* II. 395 (*foliaceus*) (1792). — Sargent in *Bull. Pop. Inform. Arn. Arb.* No. 11 (1911); in *Gard. Chron.* ser. 3, L. 202 (1911). — Schneider in *Oester. Bot. Zeitschr.* LXVI. (1916).

> *Ulmus campestris* Linnaeus, *Spec.* 225 (pro parte) (1753). — Planchon in De Candolle, *Prodr.* XVII. 156 (pro parte) (1873). — Hempel & Wilhelm, *Bäume & Sträucher* II. 1, fig. 225 A, 226–229, t. 37 (1897). — Ascherson & Graebner, *Syn. Mitteleur. Fl.* IV. 551 (pro parte) (1911).
> *Ulmus glabra* Miller, *Gard. Dict.* ed. 8, No. 4 (*glabris*) (non Hudson) (1768). — Smith, *Engl. Bot.* XXXII. t. 2248 (1811). — Reichenbach, *Icon. Fl. German.* XII. 13, t. 664, fig. 1334 (1850). — Koehne, *Deutsche Dendr.* 135 (1893). — Schneider, *Ill. Handb. Laubholzk.* I. 219, fig. 136 g–h, 137 a–g, n. 138 i–o (pro parte) (1904). — Ley in *Jour. Bot.* XLVIII. 69 (1910).
> *Ulmus sativa* Duroi, *Harbk. Baumz.* II. 502 (non Miller) (1772).
> *Ulmus vulgaris* Pallas, *Iter*, III. 314 (nomen nudum) (1776). — Gueldenstedt, *Iter*, II. 24 (nomen nudum) (1787).
> *Ulmus carpinifolia* Gleditsch, *Pflanzenverz. Lust- u. Baumgärt.* 334 (1773).
> *Ulmus nitens* Moench, *Meth. Pl.* 333 (1794). — Rehder in *Mitt. Deutsch. Dendr. Ges.* 1908, 157 (1909). — Moss in *Gard. Chron.* ser. 3, LI. 199 and 217 (1912); *Cambridge Brit. Fl.* II. 89 (1914). — Henry in Elwes & Henry, *Trees Great Brit. & Irel.* VII. 1887, t. 402, 412, fig. 23 (1913). — Bean, *Trees & Shrubs Brit. Isl.* II. 618 (pro parte) (1914).
> *Ulmus campestris*, var. *glabra* Aiton, *Hort. Kew.* I. 319 (1789). — Pokorny, *Oester. Holzpfl.* 46, t. 12, fig. 140–141 (1864).
> *Ulmus campestris*, var. *laevis* Spach in *Ann. Sci. Nat.* sér. 3, XV. 362 (1841). — Trautvetter in *Bull. Phys.-Math. Acad. Sci. St. Pétersbourg*, XV. 351 (1857).
> *Ulmus campestris*, var. *vulgaris* Ledebour, *Fl. Ross.* III. pt. 2, 646 (pro parte) (1850).
> *Ulmus campestris*, a *genuina* Ascherson, *Fl. Prov. Brandenb.* I. 614 (1864).
> *Ulmus eu-campestris* Ascherson & Graebner, *Fl. Nordostdeutsch. Flachl.* 259 (1898).
> *Ulmus vulgaris*, a *campestris* Rouy, *Fl. France*, XII. 266 (1910).
> *Ulmus vulgaris*, γ *carpinifolia* Rouy, l. c. (1910).

This European species does not occur in eastern Asia, where it is represented by *U. japonica* Sargent, from which it differs by the characters given in the key on p. 249. It is mostly known as *U. campestris* or *U. glabra*, but neither of these names can be properly applied to it. Linnaeus's *U. campestris* includes all the European Elms and cannot be used for any one of them. By different authors mentioned above the name *U. nitens* Moench has been adopted, but it is antedated by *U. foliacea* of Gilibert, who gave a much better description of this tree than Moench did. In the *Oester. Bot. Zeitschr.* I have pointed out that Gilibert's name undoubtedly refers to our species.

12. **Ulmus japonica** Sargent, *Trees & Shrubs*, II. 1, t. 101 (1907). — Schneider, *Ill. Handb. Laubholzk.* II. 906 (1912). — Henry in Elwes & Henry, *Trees Great Brit. & Irel.* VII. 1923, t. 411, fig. 4 (1913). — Bean, *Trees & Shrubs Brit. Isl.* II. 675 (1914).

Ulmus pumila, β suberosa Turczaninow in *Bull. Soc. Nat. Mosc.* XXVII. 369 (an pro parte?) (1854); *Fl. Baic.-Dahur.* II. 96 (1856).
Ulmus glabra Maximowicz & Ruprecht in *Bull. Phys.-Math. Acad. Sci. St. Pétersbourg*, XV. 138 (non Miller) (1856); in *Mél. Biol.* II. 434 (1857). — Maack & Ruprecht in *Bull. Phys.-Math. Acad. Sci. St. Pétersbourg*, XV. 375 (1857); in *Mél. Biol.* l. c. 557 (1858). — Dippel, *Handb. Laubholzk.* II. 25 (pro parte) (1892). — Koehne, *Deutsche Dendr.* 135 (pro parte) (1893). — Schneider, *Ill. Handb. Laubholzk.* I. 220 (1904), quoad var. *typicam* pro parte.
Ulmus suberosa Maack & Ruprecht in *Bull. Phys.-Math. Acad. Sci. St. Pétersbourg*, XV. 375 (non Moench) (1857); in *Mél. Biol.* II. 557 (1858).
Ulmus campestris, var. *laevis* Trautvetter in *Mém. Sav. Étr. Acad. Sci. St. Pétersbourg*, IX. 247 (Maximowicz, *Prim. Fl. Amur.*) (non Spach et Auct. Al.) (1859). — Schmidt in *Mém. Acad. Sci. St. Pétersbourg*, sér. 7, XII. No. 2, 174 (*Reis. Amur. Sachal.*) (1868). — Maximowicz in *Bull. Acad. Sci. St. Pétersbourg*, XVIII. 290 (1873); in *Mél. Biol.* IX. 23 (1873). — Franchet & Savatier, *Enum. Pl. Jap.* I. 431 (an pro parte?) (1875). — Herder in *Act. Hort. Petrop.* XII. 45 (*Pl. Radd.*) (1892), quoad specim. Transbaicalensia. — Shirasawa in *Bull. Agric. Coll. Tokyo*, II. 266, t. 13, fig. 8–9 (*Jap. Laubh. Winter.* t. 9, fig. 8–9) (1895). — Matsumura, *Ind. Pl. Jap.* II. pt. 2, 33 (an pro parte?) (1912).
Ulmus campestris, var. *major* Trautvetter in *Mém. Sav. Étr. Acad. Sci. St. Pétersbourg*, IX. 248 (Maximowicz, *Prim. Fl. Am.*) (non Walpers) (1859). — Maximowicz in *Bull. Acad. Sci. St. Pétersbourg*, XVIII. 290 (1873); in *Mél. Biol.* IX. 23 (1872). — Franchet & Savatier, *Enum. Pl. Jap.* I. 431 (1875). — Herder in *Act. Hort. Petrop.* XII. 45 (*Pl. Radd.*) (1892), quoad specim. Transbaicalensia. — Matsumura, *Ind. Pl. Jap.* II. pt. 2, 32 (1912).
Ulmus campestris, a vulgaris Regel in *Mém. Acad. Sci. St. Pétersbourg*, sér. 7, IV. 133 (*Tent. Fl. Ussur.*) (non Spach) (1861), exclud. lusus d. *pumila.* — Maximowicz in *Bull. Acad. Sci. St. Pétersbourg*, XVIII. 290 an tantum pro parte? (1873); in *Mél. Biol.* IX. 22 (1873). — Franchet & Savatier, *Enum. Pl. Jap.* I. 430 (1875). — Korshinsky in *Act. Hort. Petrop.* XII. 386 (1892), exclud. 3, f. *pumila.* — Shirasawa, *Icon. Ess. For. Jap.* II. t. 15, fig. 10–21 (1908). — Matsumura, *Ind. Pl. Jap.* II. pt. 2, 32 (1912).
Ulmus campestris, β suberosa Regel in *Mém. Acad. Sci. St. Pétersbourg*, sér. 7, IV. 134 (*Tent. Fl. Ussur.*) (non Wahlenberg) (1861). — Herder in *Act. Hort. Petrop.* XII. 44 (*Pl. Radd.*) (pro parte) (1892). — Korshinsky in *Act. Hort. Petrop.* XII. 387 (1892). — Nakai in *Jour. Coll. Sci. Tokyo*, XXXI. 191 (*Fl. Kor.* II.) (1911).
? *Ulmus sibirica* Kirchner in Petzold & Kirchner, *Arb. Musc.* 566 (1864).
Ulmus campestris, var. *major*, f. *suberosa* Miquel in *Ann. Mus. Lugd.-Bat.* III. 65 (1867); *Prol. Fl. Jap.* 253 (1867).
Ulmus campestris Miyabe in *Mém. Bost. Soc. Nat. Hist.* IV. 258 (*Fl. Kurile Isl.*) (non Linnaeus) (1890). — Sargent in *Garden & Forest*, VI. 323, fig. 50 (1893); *For. Fl. Jap.* 57, t. 18 (1894). — Hemsley in *Jour. Linn. Soc.* XXVI. 446 (pro parte) (1894). — Komarov in *Act. Hort. Petrop.* XXII. 82 (*Fl. Mansh.* II.) (1903). — Seemen & Loesener in *Bot. Jahrb.* XXXIV. Beibl. LXXV. 29 (1904). — Nakai in *Jour. Coll. Sci. Tokyo*, XXXI. 190 (*Fl. Kor.* II.) (1911).

Ulmus campestris, var. *Japonica* Rehder in Bailey, *Cycl. Am. Hort.* IV. 1882 (1902). — Miyabe & Miyake, *Fl. Saghal.* 403 (1915). *Ulmus montana*, var. *major* Nakai in *Jour. Coll. Sci. Tokyo*, XXXI. 190 (*Fl. Kor.* II.) (1911).

CHINA. Chili: Hsiao Wu-tai-shan, alt. 1800 m., August 23, 1913, *F. N. Meyer* (No. 1328; sterile branches with corky excrescences); Shi-feng-ko, in rocks and at the edges of precipices, June 3, 1913, *F. N. Meyer* (No. 993; of shrubby growth; sterile). Chekiang: near Yü-hang, in earth banks, alt. 100 m., July 6, 1915, *F. N. Meyer* (No. 1508; small tree; sterile, needs further observation). Shangtung: " Lauschan-Gebirge bei Lauting," May 1900, *Nebel* (ex Seemen & Loesener sub *U. campestris*).

JAPAN. Hokkaido: prov. Ishikari, Sapporo, May 1884, *K. Miyabe* (with flowers and fruits, no leaves); same locality, common, September 18, 1892, *C. S. Sargent* (tree 23–27 m. tall, girth 1.2–1.5 m., habit of *U. americana;* sterile co-type of *U. japonica*); same locality, August 19, 1905, *J. G. Jack* (sterile co-type of *U. japonica*); same locality, April 15 and September 3, 1903, *S. Arimoto* (with flowers and sterile); same locality, April 1907 and summer (ex Herb. Yokohama Nursery Co., with flowers and sterile); same locality, May 2, 1909, September 11, 1912 (ex Herb. Sakurai; with flowers and sterile); prov. Tokachi, May 12, 1907 and later (ex Herb. Sakurai; with flowers and sterile); prov. Iburi, Shira-oi, August 24, 1905, *J. G. Jack* (sterile); prov. Oshima, " ad pedem vulcani Kuma-ga-take, in silvis," 1861, *C. Maximowicz* (type of *U. japonica* in Herb. Gray; with ripe fruits and young leaves); same locality, 1861–63, *M. Albrecht* (with some fruits and young leaves); same locality, June 1, 1904, *U. Faurie* (No. 5877; with fruits and young leaves, young branchlets almost glabrous). Hondo: prov. Shimotsuke, forming woods west end of Lake Chuzenji, June 1, 1914, *E. H. Wilson* (No. 6777, tree 13–27 m. tall, girth 0.9–3.6 m., bark gray, fairly smooth, wood pale brown; with fruits and young leaves, disk of fruits with fine hairs); same locality, August 13, 1905, *J. G. Jack* (sterile; pubescence of branchlets and petioles brownish and gray); same locality, October 26, 1905, *J. G. Jack* (like the foregoing specimen); same locality, not common, September 3, 1892, *C. S. Sargent* (small tree; sterile branches with corky wings, pubescence as in Jack's specimens[1]); prov. Suruga, Fuji shrine, alt. 900 m., May 8, 1914, *E. H. Wilson* (sterile). Kyushu: prov. Osumi, Higashi-Kirishima, woods, common, March 6, 1914, *E. H. Wilson* (No. 6195; tree 13–17 m. tall, girth 1.5–1.8 m., flat topped, flowers reddish; with flowers and very young fruits, without leaves); prov. Satsuma, Togo, thickets, not common, March 14, 1914, *E. H. Wilson* (No. 6279; small tree 8–10 m. tall, trunk 0.3–0.6 in girth).

NORTHEASTERN ASIA. Saghalien: without locality, *Fr. Schmidt* (sterile); moist pastures and moors, common, August 4, 1914, *E. H. Wilson* (No. 7329, tree 13–25 m. tall, girth 1.8–3.6 m., bark light gray, fissured, branches thick, spreading; sterile). Korea: Sang-chang, cliffs, September 4, 1003, *C. S. Sargent* (large tree; sterile, shoots with corky wings); Ping-yang, September 18, 1905, *J. G. Jack* (sterile). Ussuri province: Khabarovsk, August 27, 1903, *C. S. Sargent* (sterile); Nikolsk, station grounds, August 22, 1903, *C. S. Sargent* (young tree; sterile). Mandshuria: Shengking, Tsienshan Mountains, June 9, 1906, *F. N.*

---

[1] The same brownish and gray, somewhat hirsute pubescence is found on cultivated plants in the Arnold Arboretum, raised from seed sent by Dr. Mayr in 1889 from Tokyo as *U. parvifolia* and in 1895 from Sapporo by the Agricultural College as *U. campestris* var. The leaves, etc., of Sargent's specimen agree well with Shirasawa's drawing, l. c. t. 15, fig. 15.

Meyer (No. 109; sterile). Transbaikal: " e Sibiria," July 1825, *Fischer* (corky winged branchlet with young leaves in Herb. Gray).

The specimens referred above to the typical *U. japonica* have more or less hairy branchlets and petioles and the color of the branchlets in the autumn of the first year or in the second year is light yellow or pale brown or tawny. The shape of the fruits is somewhat variable, and they are not always entirely glabrous. Especially the fruits of Wilson's No. 6777 are more or less minutely hairy on the disk, and it needs further observation to decide if this pubescence affords any good character for distinguishing a variety. The texture of the leaves is very firm, and they are more or less softly pubescent beneath when young and rather rough on both sides later in the season, while in the following variety the leaves usually are thinner and smooth. With its glabrous young shoots Faurie's No. 5877 is somewhat intermediate between the type and var. *levigata*, which seems to be absent from Hondo and even from Hokkaido. See also the note under *U. macrocarpa* Hance (p. 252).

In Japan this Elm is found from extreme southern Kyushu northward through Hondo and Hokkaido to Saghalien. South of the Nikko region in Hondo, it is a rare tree, and it is nowhere really common south of Hokkaido, although at the western end of Lake Chuzenji it forms nearly pure woods of no great size. In moist valleys in Hokkaido, and especially in the central and northern parts of this island, it is abundant and is one of the most prominent and important constituents of the forest. In Japanese Saghalien it is also common and grows to a larger size than any other deciduous tree. At its best the Japanese Elm is a handsome and lofty tree, often 30 m. or more tall, with a trunk 6 m. in girth. In the forests the trunk is clear of branches for from 10 to 20 m. from the ground. In the open, however, the trunk usually divides into several ascending-spreading stems at about 5 or 6 m. from the ground, and such trees in habit resemble the American Elm (*U. americana* Linnaeus) more than they do that of any other Asiatic or any European species. The bark is light gray, slightly fissured and fibrous; the branches are stout and spread to form a broad flattened or rounded crown.

Vernacular names for this tree are Nire and Aka-damo. By timber surveyors of foreign governments this tree is known as Red Ash, and under this trade name the lumber is exported. The wood warps badly and is of comparatively little value though it lasts well in water. To the Ainu or aboriginal people of Japan this Elm is of importance, though formerly much more so than now. The inner layers of the bark after maceration, brought about by soaking in warm stagnant water for some ten days, are woven into cloth which formerly was in general use for making wearing apparel. The bark is always largely used by these people for roofing and covering the sides of their huts, and the dried roots were formerly used to generate fire by friction. Naturally such an important and useful tree figures largely in Ainu mythology. A great many of these people believe it was the first of all trees, and that it was sent direct from Heaven already grown. Now, however, in spite of its usefulness it is looked upon as a tree of ill omen and is supposed to harbor evil spirits.

Pictures of this Elm will be found under Nos. x244, x245, x250, x298, x375, x411, x427, x432 of the collection of my Japanese photographs.   E. H. W.

Ulmus japonica, var. levigata Schneider, n. var.

> *Ulmus pumila* Pallas, *Fl. Ross.* I. pt. 1, 77, t. 48, fig. B (non Linnaeus) (1784), quoad formam ramulis suberosis.
>
> *? Ulmus campestris*, var. *major* Miyabe and Miyake, *Fl. Saghal.* 403 (non Walpers) (1915). — Some of the synonyms quoted above may, at least partly, belong to this variety.

A typo praecipue recedit ramulis etiam novellis glabris, foliis superne plus minusve nitentibus glabris levibus subtus tantum in axillis nervorum pro parte barbulatis basi saepe distinctius asymmetricis tenuioribus, petiolis minus puberulis, fructibus glaberrimis.

NORTHEASTERN ASIA. Northern Korea: " fluvium Jalu super. vallis Samsumuris," July 6, 1897, V. *Komarov* (type; with some small fruits and almost fully grown leaves). Saghalien: Konuma, moorlands, not common, August 4, 1914, *E. H. Wilson* (No. 7325; tree 13 m. tall, girth 3 m., round-headed; sterile). Ussuri: near Vladivostok, by streams in low moist ground, August 18, 1903, *C. S. Sargent* (sterile); Khabarovsk, August 23 and 27, 1903, *C. S. Sargent* (small tree, leaves very lustrous; sterile, folia maxima obovato-oblonga ad 15:7 cm. magna). Amur: " ad fl. Amur", 1855, *R. Maack* (Nos. 113, with flowers; 379, sterile branch with corky wings); " Amur superior et medius," August 26, 1891, *S. Korshinsky* (sterile branch with corky wings). Mandshuria: mountains, 12 hours east of Harbin, August 31, 1903, *C. S. Sargent* (sterile, branches partly corky).

This is a well-marked variety which like the type has a form with corky winged branches. On dry sunny places the color of the branches seems to be more distinctly reddish. It is difficult to determine any shrubby form of *Ulmus* growing in dry stony situations; the leaves of plants in such positions become mostly much smaller and very rough on both surfaces. See also my remarks under *U. macrocarpa* Hance (p. 252) and the following species. Such shrubby small-leaved forms very often are taken for *U. pumila* Linnaeus.

Here may be mentioned a sterile specimen collected by *F. N. Meyer* in Shansi, near Tsin-tse, May 1907 (No. 416), of which the small leaves in shape and pubescence resemble those of *U. japonica*, but the length of the petioles is about one third of the length of the whole leaf except the elongated tip. This curious form needs further observation.

In Saghalien *U. japonica*, var. *levigata* Schneider is a rather low and broad-topped tree and it is not very common. A picture will be found under No. x365 of the collection of my Japanese photographs. E. H. W.

13. **Ulmus Davidiana** Planchon in *Compt. Rend. Acad. Sci. Paris*, LXXIV. pt. 1, 1498 (nomen nudum) (1872); in De Candolle, *Prodr.* XVII. 158 (1873). — Maximowicz in *Bull. Acad. Sci. St. Pétersbourg*, XVIII. 291 (1873); in *Mél. Biol.* IX. 24 (1873). — Franchet in *Nouv. Arch. Mus. Paris*, sér. 2, VII. 76, t. 8, fig. B (*Pl. David.* I. 266) (1884). — Hemsley in *Jour. Linn. Soc.* XXVI. 447 (1894). — Schneider, *Ill. Handb. Laubholzk.* II. 904, fig. 565 i, 566 i (1912).

CHINA. Chili: Jehol, *A. David* (No. 1716, type ex Planchon); same locality, December 11, 1907, *F. N. Meyer* (No. 201; according to Meyer's note in *Bull. U. S. Dept. Agric. Bur. Pl. Indust.* No. 137, 25, no. 21932 [1909], " an Elm growing to be a medium-sized tree with a round, spread-out head; when young has two corky wings along its young branches; is not a common tree at all. Grows in very dry and exposed localities." Plants distributed by the Department of Agriculture as *U. Davidiana*, cultivated in the Arnold Arboretum under No. 5927); south of Weichang, 1910, *W. Purdom* (No. 262; apparently the same as a plant growing in the Arnold Arboretum received as Purdom No. 261 from Hort. Veitch in 1912).

This is a very little known species of which I have seen only a fruit and a young leaf of the type kindly sent to the Arboretum by the Director of the botanical department of the Muséum in Paris. *U. Davidiana* is certainly nearly related to the preceding species. The leaves of the sterile specimens mentioned above are more glabrous and more broadly or roundish obovate than those of *U. japonica*, of which the leaves are mostly more cuneate towards the base. The length of the petiole and the number of the lateral nerves seem to be variable, and the color of the branchlets and buds is not unlike in both the species as far as I can judge by the material before me. *U. Davidiana* certainly needs further observation. See also my remarks under *U. macrocarpa* Hance (p. 252).

Series d. PUMILAE Schneider in *Oester. Bot. Zeitschr.* LXVI. (1916).

This series could probably be raised to the rank of a subsection equal to subsect. *Foliaceae*, but there are some forms, especially of *U. foliacea* Gilibert in Europe and western Asia, as also of other species in India and northeastern Asia, which are not yet sufficiently known to decide whether or not *U. pumila* Linnaeus is really closely allied to any species of ser. *Nitentes*.

15. **Ulmus pumila** Linnaeus. See p. 242.

> **Ulmus pumila, var. arborea** Litwinow in *Sched. Herb. Fl. Ross.* VI. 166, t. 3 (1908).
> *Ulmus campestris,* γ *pumila* Regel in *Act. Hort. Petrop.* VI. 478 (1879).
> *Ulmus parvifolia* Lauche, *Deutsche Dendr.* 348 (non Jacquin) (1880), exclud. synon. — ?Hooker f., *Fl. Brit. Ind.* V. 481 (1881).
> *? Ulmus turcestanica*[1] Regel in *Gartenflora,* XXXIII. 28 (nomen nudum) (1884).
> *Ulmus pinnato-ramosa* Dieck ex Koehne in Fedde, *Rep. Spec. Nov.* VIII. 74 (1910); in *Mitt. Deutsch. Ges.* 1910, 92 (1911).
> *Ulmus pumila, var. pinnato-ramosa* Henry in Elwes & Henry, *Trees Great Brit. & Irel.* VII. 1926 (1913).

TURKESTAN.

INDIA. ?Kashmir.

As far as I can judge by the cultivated specimens before me and by those collected by *P. Sintenis,* "regio transcaspica," Krasnowodok, March 22, 1900 (No. 29ᵃ; with flowers), and Kisil-Arwat, Karakala, July 5, 1901 (No. 2016; sterile), and distributed as *U. glabra* Miller, this variety is most closely related to the typical *U. pumila.* The characters indicated by Koehne as distinguishing his *U. pinnato-ramosa* can also be seen in specimens from northeastern Asia.

There is an Elm from Kashmir: "Western Tibet, Nubra, alt. 10,000 ft." collected by *T. Thomson,* which was distributed in Herb. Ind. Or. as *U. pumila,* and later was referred by Hooker f., to *U. parvifolia* Jacquin. But Hooker says: "The Tibetan plant may be a small-leaved form of *U. Wallichiana.*" Unfortunately the specimen is only a sterile one, and it may belong to *U. pumila,* var. *pinnato-ramosa* as far as I can judge by the leaves and branchlets.

In many respects very similar is also the type of *U. campestris,* var. Aitchison in *Trans. Linn. Soc.* ser. 2, III. 108 (1880) from Khorasan, July 11, 1885 (No. 711; a cultivated tree; sterile), which may be described as follows: Arbor; ramuli hornotini graciles, glabri, flavescentes, deinde cinerascentes; gemmae (nondum satis evolutae) parvae, obtuso-ovatae, perulis paucis fere nigrescentibus extus vix v. paulo puberulis sed margine distincte longe fulvo-fimbriatis obtectae. Folia parva, ovata v. ovato-elliptica, basi rotundata, valde asymmetrica, apice sensim acuta, supra laevia, tantum costa parcissime pilosa, ut videtur laete

---

[1] According to Litwinow in *Sched. Herb. Fl. Ross.* VI. 164, t. 1–2 (1908), *U. turcestanica* Regel is a synonym of *U. densa* Litwinow, l. c. 163 (1908). Unfortunately I have not seen any material, but according to Litwinow the sterile branchlets must be very similar to those of *U. pumila* Linnaeus, while the fruits are up to 2 cm. long and to 1.2 cm. broad, and like those of *U. foliacea* Gilibert. Litwinow adds as a synonym *U. campestris,* var. *laevis* Regel in *Act. Hort. Petrop.* VI. 477 (1879), quoad plant. Turcestanicam.

viridia, subtus paulo pallidiora, glabra v. in axillis nervorum lateralium lateris longioris circiter 10–15 paulo barbulata, margine subsimpliciter distincte dentata, minora 3–4.5 cm. longa et 1.5–2.2 cm. lata, majora 5–7 cm. longa, ad 3 cm. lata; petioli satis longi, graciles, flavescentes, superne tantum sparse villosuli, 6–10 mm. longi. Flores fructusque ignoti.

Except the very unequal base of the leaves this interesting form is very much like *U. pumila* Linnaeus.

16. **Ulmus glaucescens** Franchet in *Nouv. Arch. Mus. Paris*, sér. 2, VII. 76, t. 8, fig. A (*Pl. David.* I. 266) (1884). — Schneider, *Ill. Handb. Laubholzk.* II. 904, fig. 565 h, 566 g (1912).

CHINA. Shansi: "Toumet, Sartchy," *A. David* (No. 2634, type, ex Franchet). According to the description this species must be very closely related to *U. pumila* Linnaeus. Franchet says that the leaves are only up to 4 cm. long and 2.5 cm. broad, but on the plate the leaves are represented up to 5 cm. long and very similar in shape to those of specimens of *U. pumila* before me. But I have never seen such large samaras as *U. glaucescens* is said to have (" samara 2–2½ cm. longa "). Hemsley does not mention this species in the *Index Florae Sinensis* and I have not seen any specimen of *Ulmus* from eastern Shansi.

Series c. LANCEAEFOLIAE Schneider in *Oester. Bot. Zeitschr.* LXVI. (1916).

Ulmus, subgen. *Microptelea* Planchon in *Ann. Sci. Nat.* sér. 3, X. 281 (1848), quoad *U. lancifoliam;* in De Candolle, *Prodr.* XVII. 162 (1873), quoad *U. Hookerianam.* — Engler in Engler & Prantl, *Nat. Pflanzenfam.* III. Abt. 1, 62 (1888), quoad *U. Hookerianam.*

I suppose that this series may be better regarded as a subsection or even as a section, and it needs further investigation to fix the real relationship of the following species, which hitherto has been referred to sect. *Microptelea*.

15. **Ulmus lanceaefolia** Roxburgh apud Wallich, *Icon. Pl. As. Rar.* II. 86, t. 200 (1831).

Ulmus lancifolia Roxburgh, *Fl. Ind.* ed. 2, II. 66 (1832). — Planchon in *Ann. Sci. Nat.* sér. 3, X. 281 (1848); in De Candolle, *Prodr.* XVII. 162 (1873). — Kurz, *Forest Fl. Brit. Burma*, III. 473 (1877). — Hooker f., *Fl. Brit. Ind.* V. 480 (1888). — Hemsley in *Jour. Linn. Soc.* XXVI. 447 (1894). — Gamble, *Man. Ind. Timber*, ed. 2, 628 (1902). — Prain, *Bengal Pl.* 958 (1903). — Schneider, *Ill. Handb. Laubholzk.* 222, in adnot. (1904); II. 904, fig. 565 f (1912). — Brandis, *Ind. Trees.* 594 (1906).

Ulmus Hookeriana Planchon in De Candolle, *Prodr.* XVII. 162 (1873). — Engler in Engler & Prantl, *Nat. Pflanzenfam.* III. Abt. 1, 62 (1888).

CHINA. Yunnan: Linan district, March 1, *A. Henry* (No. 10571; tree 3 m. tall; with ripe fruits and leaves); Szemao, Yulo forest, April 9, *A. Henry* (No. 12861; tree 7 m. tall; with ripe fruits and leaves).

INDIA. Assam: "Dimapur, 400 ft.," March 23, 1896, *Dr. King's Collector* (No. 100; with fruits and leaves). Burma: without locality (No. 4678, Herb. Griffith; with fruits and leaves). Eastern Bengal: "Khasia, reg. trop., alt. 0.–3000 ped.," *J. D. Hooker* (sterile): Chittagong, hilly parts, *W. Roxburgh* (type ex Roxburgh). Sikkim: "reg. trop., alt. 4–5000 ped.," *J. D. Hooker* (with fruits and leaves; type of *U. Hookeriana*). Sikkim: "Environs of Darjiling, 6000 to 8000 feet," July 1855, *Schlagintweit* (No. 12319; young plant).

The leaves of this interesting species are subdeciduous, and according to Brandis the tree is generally leafless during part of the hot season when it flowers. The pubescence of the branchlets, especially of the strong shoots, is usually more or less fulvous and often somewhat scabrous. The lobes of the perigone are of the same shape as in the other species of this section and not deeply cleft as in sect. *Microptelea.* In Hooker's specimen from Sikkim, the type of Planchon's *U. Hookeriana,* the fruits (apparently not yet fully ripe) are smaller, and the lower pilose part of the pedicels is mostly shorter than in the other specimens and only a little longer than the upper part from the joint to the perigone. It needs further investigation before it is possible to know whether this fact is of any taxonomic value in distinguishing a separate form.

Sect. II. **MICROPTELEA** Bentham & Hooker, *Gen. Pl.* 352 (1883). — Henry in Elwes & Henry, *Trees Great Brit. & Irel.* VII. 1848 (1913). — Daveau in *Bull. Soc. Dendr. France,* 1914, 21 (pro parte).

> *Microptelea* Spach in *Ann. Sci. Nat.* sér. 2, XV. 358 (1841); *Hist. Vég.* XI. 113 (1842).
>
> *Ulmus,* subgen. *Microptelea* Planchon in *Ann. Sci. Nat.* sér. 3, X. 279 (pro parte) (1848); in De Candolle, *Prodr.* XVII. 161 (1873), exclud. *U. Hookeriana.* — K. Koch, *Dendr.* II. pt. 1, 422 (1872). — Engler in Engler & Prantl, *Nat. Pflanzenfam.* III. Abt. 1, 62 (1888), exclud. *U. Hookeriana.* — Dippel, *Handb. Laubholzk.* II. 34 (1892). — Koehne, *Deutsche Dendr.* 134 (1893). — Schneider, *Ill. Handb. Laubholzk.* I. 221 (1904).

To this section also belongs the North American *U. crassifolia* Nuttall, which represents the series CRASSIFOLIAE Schneider in *Oester. Bot. Zeitschr.* LXVI. (1916) " fructibus undique pilosis, dense ciliatis."

16. **Ulmus parvifolia** Jacquin. See p. 244.

There may be mentioned another Elm probably representing a new species the taxonomic position of which remains doubtful as long as the fruits are unknown.

17. **Ulmus, spec. nova?**
Arbor?; ramuli floriferi plus minusve pubescentes (ut videtur novelli villosuli) flavo-brunnei; gemmae ovatae, perulis extus pilosis et distincte ciliatis brunneis. Inflorescentiae fasciculatae, circiter 10-florae; flores breviter pedicellati, parte pedicellorum superiore glabra quam pars inferior pilosa vix longiore; perigonia 4-lobata, glabra, lobis paulo ultra medium incisis obtusis fusco-ciliatis; stamina 4, filamentis perigonio plusquam duplo longioribus, antheris nondum visis; ovarium glabrum, sessile. Samarae ignotae.

INDIA. "Suring, Sarju Valley, alt. 5000 feet," *R. Strachey & J. E. Winterbottom* (No. 2 in Herb. Gray; leafless flowering branchlets and also the sterile branchlets described beneath).

The specimen also contains sterile branchlets with apparently not yet fully grown leaves, and I am not quite sure whether or not they belong to the same plant. They may be described as follows: Ramuli novelli annotinique dense griseo- et fulvo-hirsuti. Folia ovata, ovato-lanceolata v. late ovata, basi plus minusve rotunda, paullo asymmetrica, apice sensim v. subito acuminata, supra pilis scabris obtecta, subtus pallidiora, ad costam nervosque utrinsecus circiter 12 (imo apice excepto) praecipue in axillis et paulo ad faciem hirtella, margine dupliciter serrato-dentata, minora 4-5.5 cm. longa, 1.5-2 cm. lata, maxima ad 7:3 cm. v. 8:5 cm. magna, petioli 3-7 mm. longi, hirsuti, stipulis late lanceolatis sparse sericeis breviores.

The branchlets described above apparently are from young plants which in several species develop shoots with such a brownish hirsute pubescence. The same Elm may be represented by two sterile specimens from Afghanistan collected by

J. E. T. Aitchison, 1879, " Kurrum valley " (No. 403), which is the same as *Ulmus*
sp. Aitchison in *Jour. Linn. Soc.* XVIII. 93 (*Fl. Kuram Valley*) (1880), and " Hari-
rud valley," June 5, 1885 (No. 1112), which is the type of *U. montana* Aitchison in
Trans. Linn. Soc. ser. 2, III. 108 (non Withering) (1888). In both specimens the
young branchlets are hirsute, becoming more or less glabrous and of a yellowish
brown or a very pale color. The shape of the leaves, which are more or less rough on
both sides, vary from ovate-elliptic with a very unequal base to obovate with an
almost symmetrical base. In No. 1112 the leaves are up to 10 cm. long and 5–6 cm.
broad, with a very coarse compound serration, while in No. 403 the leaves are
6–8.5 cm. long and 3–4 cm. broad, with a little sharper serration. No. 711 is said
by Aitchison to be "a more common, but smaller tree than 677," which number
I referred to *U. Brandisiana* Schneider, see p. 253, while No. 1112 is " a cultivated
tree, 30 feet high and 9 feet in circumference." As far as I can judge by this material
it might represent a new species the taxonomic position of which remains uncertain
as long as the fruits are unknown.

CELTIS L.

Celtis Julianae Schneider, n. sp.

Arbor ad 27 m. alta; ramuli hornotini dense flavo-tomentelli v.
subhirsuti, annotini paulo glabrescentes, distinctius striato-angulati,
dein nigrescentes, lenticellis plus minusve numerosis rubiginosis obtecti; gemmae floriferae ovato-rotundae, obtusae, perulis extus rufo-
hirsutis cinctae. Folia magna crassa, oblique late ovata, obovata v.
obovato-elliptica, basi plus minusve inaequali obtusa v. acuta, apice
satis subito breviter acuminata v. caudata, supra laete viridia, subscabra, in facie et costa nervisque flavescentibus paulo impressis laxe
hirtella, subtus flavo-viridia v. distincte (in sicco) flavescentia, costa nervisque distincte elevatis reticulato-venosa (nervis lateralibus a costa
exeuntibus utrinsecus basalibus 2 inclusis 3–5), omnino molliter villosula v. pubescentia, in facie sub microscopio reticulato-papillosa, margine praecipue ultra medium crenato-dentata dentibus apiculatis et
junioribus saepe incurvis, interdum subintegra, 7–14 cm. longa et
5–8.5 cm. lata v. elliptica ad 11 cm. longa et ad 5.5 cm. lata; petioli
satis crassi, supra sulcati, tomentosi, 1–1.3 cm. longi. Flores ♂ basi
ramulorum novellorum cymoso-aggregati, plerique 5-meri; perigonii
lobi liberi 5, ovato-lanceolati, ut videtur 2 exteriores 3 interioribus
sublongiores, omnes ad apicem extus fusco-villosi v. comosi, ceterum
glabri v. margine ciliati, extus purpureo-maculati; stamina juvenilia
5 (–6), lobis breviora, glabra, normalia, filamentis disco villoso inserta;
rudimenta gynaecei parva; pedicelli breves, hirsuti. Flores ♀ apice
ramulorum floriferum singuli axillares, plerique 4-meri; perigonia staminaque vix ab iis florum ♂ diversa sed lobi etiam basin versus extus

sparse pilosi; ovaria glabra stigmatibus 2 pro genere normalibus elongatis coronata; pedicelli elongati, villosi. Fructus aurantiaci, ovato-globosi, glabri, ad 13 mm. longi et ad 11 mm. lati, obtusi; putamina ovato-globosa, hilo acuto-dentata et distincte foveolata, ceterum facie tantum indistincte v. vix foveolata et costa mediana instructa, circiter 10 mm. longa et 8 mm. crassa; pedicelli fructiferi ut ramuli tomentosi, satis crassi, 1.5–2.2 cm. longi.

Western Hupeh: Patung Hsien, woodlands, alt. 900–1100 m., November 1907 (No. 635, type; tree 12–27 m. tall, girth 1.2–3 m., fruit orange; with ripe fruits); same locality, mountainside, alt. 1200 m., April 1907 (No. 635ᵃ, co-type; tree 27 m. tall, girth 3 m.; with ♀ flowers); same locality and date (No. 635ᵇ, co-type; with ♂ flowers); same locality, mountainsides, alt. 900–1100 m., May 1907 (No. 635ᶜ; tree 12–27 m. tall, girth 1.2–3 m.; with very young fruits); same locality, alt. 900–1200 m., June 1907 (No. 635ᵈ; with leaves and unripe fruits); same locality, June 1901 (Veitch Exped. No. 1941; tree 13 m. tall; sterile); Changyang Hsien, side of streams, alt. 900 m., May 1907 (No. 635ᵉ; tree 12–13 m. tall, girth 1.2–1.8 m.; with very young fruits).

This is a most distinct species, the nearest relative of which seems to be *C. Vandervoetiana* Schneider. According to Planchon's description *C. mollis* Wallich apud Planchon (in *Ann. Sci. Nat.* sér. 3, X. 297 [1848]) has a similar pubescence, but it has much smaller entire leaves and smaller fruits. The type of *C. mollis* was collected by *N. Wallich* in Upper Burma " ad ripas fluminis Irrawadi ad Avam," and distributed sub No. 7203. There are young plants of *C. Julianae* growing in the Arnold Arboretum. The species is named in compliment to my wife.

This Celtis is fairly common in open country and on the margins of woods throughout western Hupeh between 600 and 1300 m. altitude. It forms a handsome tree often 26 m. tall, with a clean straight trunk 3 m. in girth. The bark is smooth and very pale gray; the branches are stout and spread to form a round-topped crown. In winter and early spring the large red-brown flower-buds which densely stud the branchlets make the tree quite conspicuous. In autumn the large globose orange-colored fruits are very attractive.

A picture of this tree will be found under No. 521 of the collection of my photographs and also in my *Vegetation of Western China*, No. 158.   E. H. W.

The following variety may be distinguished.

Celtis Julianae, var. calvescens Schneider, n. var.

A typo recedit ramulis pedicellisque fructiferis fere v. omnino calvescentibus distinctius lenticellosis et tenuissime verruculosis.

Western Szech'uan: Chingchi Hsien, mountainsides, alt. 1200 m., October 1908 (No. 1305, type; tree 30 m. tall, girth 3 m., fruits orange; with ripe fruits).

This tree is not uncommon in the warm dry valleys of the Yachou prefecture in western Szech'uan and here and there fine specimens are to be met with. A picture of one of these will be found under No. 277 of the collection of my photographs and in my *Vegetation of Western China*, No. 156.

Celtis Vandervoetiana Schneider, n. sp.

Arbor ad 20 m. alta; ramuli etiam ut videtur novelli glaberrimi, hornotini rubro-brunnei, angulati, lenticellis concoloribus sparsis obtecti, annotini obscurius colorati; gemmae nondum satis evolutae purpureae, ovatae, plus minusve fusco-hirsutae. Folia magna, firma, ovato-elliptica, basi plus minusve inaequali subrotunda v. acuta, apice subsensim acuminata, supra intense v. dilute viridia, ut videtur leviter nitentia, glabra, costa nervis nervillisque planis flavescentibus, subtus plus minusve viridi-albescentia, costa nervisque elevatis, sed vix distincte reticulata, nervis lateralibus a costa exeuntibus utrinsecus basalibus 2 inclusis 3–5, in facie sub microscopio dense papillosa, tantum in axillis paulo barbata, margine praecipue ultra medium utrinque regulariter crenato-serrato-dentata, minimis exceptis 8–13 cm. longa, 4.5–7 cm. lata; petioli 1.4–1.9 cm. longi, superne anguste sulcati, glabri, flavo-brunnei. Flores adhuc ignoti. Fructus axillares, singuli, pedicellis 3–3.5 cm. longis glabris suffulti, aurantiaci, glabri, ovato-elliptici, apiculati (in sicco), circiter 1.5 cm. longi et 1.1 cm. crassi; putamina ovato-oblonga v. ovata, hilo dentato et excavato excepto plus minusve leviter foveolata et costata.

Western Szech'uan: Yungyang Hsien, roadsides, rare, alt. 1200 m., August 8, 1908 (No. **2320**, type; tree 20 m. tall, girth 2.4 m.; with ripe fruits); without exact locality, ravine, alt. 600 m., May 1904 (Veitch Exped. No. 4467; tree 10 m. tall; with unripe fruits).

This is a well-marked species, apparently most closely related to the preceding, but easily distinguished by its glabrousness and by its smooth leaves.

The name is given in compliment to Mr. Christian van der Voet, Superintendent of the Arnold Arboretum.

A picture of this tree will be found under No. 282 of the collection of Wilson's Chinese photographs and also in his *Vegetation of Western China*, No. 157.

Celtis labilis Schneider, n. sp.

Celtis sinensis Hemsley in *Jour. Linn. Soc.* XXVI. 450 (non Persoon) (1894), quoad Henry's Nos. 3404 et 7866.

Arbor ad 17 m. alta; ramuli hornotini paullo angulati, flavo-tomentosi, breves fructiferi autumno v. hieme decidui, annotini glabrescentes vetustioresque cinereo-rubiginosi, teretes, lenticellis paullo discoloribus parvis obtecti; folia adulta satis crassa, pleraque ovato- v. elliptico-oblonga, basi inaequali semirotunda v. obtusa, apice satis sensim breviter acuminata, supra saturate viridia, in costa nervisque paulo impressis sparse pilosa, in facie pilis adpressis sparse obtecta, scabriuscula, subtus in costa nervisque elevatis satis hirtella (nervis lateralibus a costa exeuntibus utrinsecus basalibus 2 inclusis 3–5 inter se parallelibus), vix barbata, nervillis leviter prominulis sparse hirtellis reticulata, etiam in facie pilis minimis difficile recognoscendis saepe brunnescentibus glandulosis plus minusve instructa, discoloria, in sicco cinereo-brunnescentia, margine tantum ultra medium distincte regulariter subcrenato-serrata, 6–10 cm. longa, 2.5–5 cm. lata (juniora in No. 2318 tenuia, subtus flavescentia, aeque pilosa, nondum reticulata, pro parte late ovato-rhomboidea et tantum 8 cm. longa et 5 cm. lata); petioli 5–8 mm. longi, crassi, tomentelli, superne sulcati. Flores ignoti (sed vide infra!). Fructus maturi plerique bini v. terni, pedicellis 6–8 mm. longis tomentellis suffulti, aurantiaci, ut videtur globosi, circiter 7–8 mm. crassi, glabri; putamina subglobosa, hilo lato obtuse rectangulari distincte foveolata, ceterum exigue v. vix foveolata et costata, circiter 5 mm. crassa.

Western Hupeh: Changyang Hsien, alt. 900–1300 m., October 1907 (No. **444**, type; fruits orange, branchlets articulating; tree 12–17 m. tall, girth 0.9–1.8 m.; with ripe fruits); Patung Hsien, alt. 900–1300 m., October 1907 (No. **444**[a]; same as No. **444**); Ichang, not common, alt. 1–300 m., April 1907 (No. **2318**; tree 7 m. tall, girth 0.9 m.; with well-developed young leaves and very young fruits); same locality, *A. Henry* (No. 3404; with young leaves and fruits, very similar to No. 2318, but all leaves oblong-elliptic).

This is a very interesting species on account of its throwing off the fruit-bearing branchlets. It seems to be very closely related to *C. cercidifolia* Schneider (p. 276), and it needs further investigation to decide if that species may not represent only a variety of *C. labilis*. The types, however, look very different, but there are forms like Wilson's Nos. 2318 and 1761[a] which seem to connect the two species. There are young plants growing in the Arnold Arboretum raised from seed, sent by Wilson under No. 444, the branchlets of which are covered with a very dense yellowish and rough pubescence. The leaves are more hairy than in the fruiting type and have a more copious and more crenate dentation, but they are ovate-oblong like those of old plants.

There is a flowering specimen of Henry's (No. 7866 in the herbarium of the Arnold Arboretum from Hupeh without precise locality), which may belong to *C. labilis*, but the flowers are very much alike in the different species of this genus. Of No. 7866 I add the following description: Ramuli annotini vetustioresque ut in forma typica, novelli floriferi 3–4 cm. longi, flavo-tomentosi, foliis valde juvenilibus 1–1.5 cm. longis oblongis acuminatis supra laxius subtus densius ut ramuli pilosis praediti. Inflorescentiae ♂ basi ramulorum axillares, fasciculato-cymosae; flores plerique 4-meri; perigonii lobi ovato-oblongi, margine ciliati, ceterum glabri, extus leviter rubescentes; stamina normalia, juvenilia lobis aequilonga, disco piloso in-

serta; rudimenta gynaecei parva; pedicelli breves, flavo-tomentelli; flores ♀ (v. ☿)
ad apicem ramulorum axillares, plerique bini v. terni, a ♂ vix diversi sed gynaeceo
distincto normali ultra medium flavo-piloso instructi.

Here may be mentioned also a specimen collected by myself in Yunnan, between Yungning and Yung-peh-ting, near Lan-ti-cho, June 27, 1914 (No. 1667).
In the pubescence of the branchlets it resembles *C. labilis*, but the leaves are more
glabrous and show a somewhat different nervation. Unfortunately my material
comes from a sterile plant which had been nearly cut down.

Celtis labilis is common on the margins of woods and in open country throughout
western Hupeh and eastern Szech'uan from river-level up to 1300 m. altitude.
It is a tree from 12 to 16 m. tall, with a straight trunk from 1 to 2 m. in girth, pale
gray smooth bark, and relatively thin branches which form a neat, more or less
oval crown. The species is easily recognized by the small fruit-bearing branchlets
which drop off when the globose orange-colored fruits are ripe. It is in cultivation,
but has not proved very hardy in this Arboretum.

A picture of this tree will be found under No. 588 of the collection of my photographs and also in my *Vegetation of Western China*, No. 159. E. H. W.

Celtis Bungeana Blume, *Mus. Bot. Lugd.-Bat.* II. 71 (1852). —
Planchon in De Candolle, *Prodr.* XVII. 171 (1873). — Franchet
in *Nouv. Arch. Mus. Paris*, sér. 2, VII. 79 (*Pl. David.* I. 269)
(1884). — Dippel, *Handb. Laubholzk.* II. 47 (1892). — Koehne,
Deutsche Dendr. 137 (1893). — Hemsley in *Jour. Linn. Soc.* XXVI.
449 (1894), exclud. No. 5735 Henryi. — E. Pritzel in *Bot. Jahrb.* XXIX.
296 (1900). — Komarov in *Act. Hort. Petrop.* XXII. 90 (*Fl. Mansh.*
II.) (1903). — Schneider, *Ill. Handb. Laubholzk.* I. 229, fig. 147 q,
148 q (1904). — Diels in *Bot. Jahrb.* XXXVI. Beibl. LXXXII. 33
(1905). — Pampanini in *Nuov. Giorn. Bot. Ital.* n. ser. XVII. 253
(1910). — Nakai in *Jour. Coll. Sci. Tokyo*, XXXI. 192 (*Fl. Kor.* II.)
(1911).

Celtis chinensis Bunge in *Mém. Sav. Étr. Acad. Sci. St. Pétersbourg*, II. 135
(*Enum. Pl. Chin. Bor.* 61) (non Persoon) (1833).
Celtis sinensis Planchon in *Ann. Sci. Nat.* sér. 3, X. 286 (non Persoon)
(1848). — Maximowicz in *Bull. Acad. Sci. St. Pétersbourg*, XVIII. 293 (1873),
quoad specim. e China boreali; in *Mél. Biol.* IX. 27 (1873). — Baker &
Moore in *Jour. Linn. Soc.* XVII. 386 (1879). — Hemsley in *Jour. Linn.
Soc.* XXVI. 450 (1894), quoad specim. Henryi, No. 4214. — Diels in *Bot.
Jahrb.* XXXVI. Beibl. LXXXII. 33 (1905).
Celtis Davidiana Carrière in *Rev. Hort.* 1868, 300. — Planchon in De Candolle,
Prodr. XVII. 172 (1873). — André in *Rev. Hort.* 1894, 97. — Mouillefert,
Traité Arb. Arbriss. II. 1210 (1898). — Schneider, *Ill. Handb. Laubholzk.*
I. 228, fig. 147 k (1904). — Henry in Elwes & Henry, *Trees Great Brit. &
Irel.* IV. 929, t. 267, fig. 11 (1909). — Bean, *Trees & Shrubs Brit. Isl.* I.
328 (1914).
Celtis Mairei Léveillé in Fedde, *Rep. Spec. Nov.* XIII. 264 (1914).

Western Hupeh: Fang Hsien, side of streams, rare, alt. 1200
m., November 1907 (No. **595**; tree 7–10 m. tall, girth 0.6–1.2 m., fruit

black); same locality, etc., May 23, 1907 (No. 595[a]; tree 10 m. tall, girth 1.2 m.; with young fruits; forma foliis satis late ovato-rhomboideis, pedicellis quam petioli vix duplo longioribus); Hsing-shan Hsien, roadsides, etc., alt. 50–200 m., October 1907 (No. 343; tree 10 m. tall, girth 0.9 m., fruits black; with ripe fruits and bark); same locality, alt. 600–900 m., roadside, June 1907 (No. 2319; tree 10 m. tall, girth 0.9 m.; with young fruits); Patung Hsien, thickets, alt. 900 m., November 1907 (No. 2317; bush 3 m. tall, fruits black); without precise locality, *A. Henry* (Nos. 4214, 6483; with not yet ripe fruits). Northern Hupeh: "Niang-Niang, monte 1890 m.," July 1907, *C. Silvestri* (No. 359; with unripe fruits); " Catena di Ou-tan-scian," November 1909, *C. Silvestri* (No. 2939; with ripe fruits). Southeastern Szech'uan: "Huang pet'ang, Wald," September 1891, *A. v. Rosthorn* (No. 804; ex Pritzel). Yunnan: Mengtsze, alt. 1500 m., *A. Henry* (No. 9323 in part in Herb. Arbor. Arnoldiano; tree 3–7 m. tall; with ripe fruits), " plaine près Long-Tan à Tong-Tchouan," alt. 2500 m., August 1912, *E. Maire* (type of *C. Mairei;* with fruits); " rives des canaux à Tong-Tchouan," alt. 2000 m., April 1913, *E. Maire* ("arbre peu élevé"; with flowers). Northern Shensi: "presso Ta-sce-tsuen," September 15, 1897, *G. Giraldi* (with ripe fruits). Shantung: Lung-tung, rocky situations, September 1907, *F. N. Meyer* (No. 272; attains only a small size when growing wild, if planted and cared for, however, seems to grow much larger; with ripe fruits). Chili: near San-tun-ying, in rocky mountain ravines, May 29, 1913, *F. N. Meyer* (No. 965; with young fruits); Weichang, *W. Purdom* (sterile with young leaves and finely pubescent branchlets).

NORTHEASTERN ASIA. Korea: Song-chang, September 4, 1903, *C. S. Sargent* (with almost ripe but not yet black fruits); Wan-san, September 5, 1903, *C. S. Sargent* (similar to the preceding specimen); Chinnampo, September 17, 1905, *J. G. Jack* (with ripe black fruits). Mandshuria: Shengking, Tsien-shan, June 9, 1906, *F. N. Meyer* (No. 94; sterile, uncertain form with rather hairy branchlets and petioles, dentation of leaves rather coarse).

A well-marked, widely spread species with rather firm leaves which are greenish on both sides and somewhat glossy above. The black fruits have white and very smooth, almost globular stones, sometimes even broader than high. On young plants and vigorous shoots the leaves are much more hairy on the veins and more or less roughish, and also the branchlets are densely pubescent. The leaves are sometimes narrow-lanceolate or lyrate as described in *C. sinensis* Persoon, p. 277, the dentation being very coarse and obtuse. In Sargent's specimens from Korea the stones are very indistinctly reticulate, and the branchlets are partly very gibbous on account of the numerous distinct lenticels. Henry's specimen from Mengtsze agrees well with the forms from central and northern China. So does the type of *C. Mairei* Léveillé of which I have received some fragments through the kindness of the author.

This species occurs in open country, in thickets and by the side of streams from river-level up to 1600 m. altitude throughout western Hupeh and eastern Szech'uan, but is nowhere common. It is a tree of medium size, growing from 10 to 15 m. tall, with a trunk from 1 to 2 m. in girth. The bark is smooth and light gray and the rather thick branches spread to form a broad flattened or rounded crown.

A picture of this tree will be found under No. 625 of the collection of my photographs and in my *Vegetation of Western China*, No. 160.   E. H. W.

Celtis cerasifera Schneider, n. sp.

Arbor ad 10 m. alta; ramuli (novelli ignoti) hornotini angulati, glabri v. versus basim internodiorum supra gemmas plus minusve puberuli, flavo-brunnescentes, lenticellis numerosis subconcoloribus obtecti, annotini cinereo-rubiginosi, distinctius lenticellosi; gemmac acuto-ovatae, adpressae, perulis brunneis margine intensius coloratis et tenuissime ciliatis in dorso plus minusve pilosis cinctae. Folia satis magna, firma, late lanceolata, late ovata v. ovato-elliptica, basi inaequali rotunda v. fere subcordata sed nervis basalibus paulo in petiolum productis, apice pleraque subsensim breviter acuminata, supra intense viridia, in costa nervisque leviter impressis sparse pilosa, ceterum glabra, levia, subtus cinereo-albescentia sub microscopio dense reticulato-papillosa, in costa nervisque elevatis sparse hirtella (nervis lateralibus a costa exeuntibus utrinsecus basalibus 2 inclusis plerisque 3), in nervillis vix v. haud prominulis facieque glabra v. pilis paucis difficile recognoscendis instructa, margine satis grosse crenato-serrata, dentibus acutis 2–4 pro 1 cm., minora ovato-lanceolata 6–8 cm. longa et 3–4 cm. lata, majora latiora ad 9.5 cm. longa et ad 6.2 cm. lata; petioli (8–)10–15 mm. longi, purpurascentes, glabri v. in sulco superne pilosuli. Flores adhuc ignoti. Fructus maturi, nigri, glauco-pruinosi, globosi, circiter 1 cm. crassi, glabri, pedicellis glabris 2–2.5 cm. longis suffulti; putamina flavo-brunnescentia, ovato-elliptica, hilo dentata, costa mediana distincta instructa, ceterum in facie satis irregulariter leviter v. distinctius foveolata et tuberculata, circiter 8 mm. longa et 6 mm. crassa.

Western Hupeh: Fang Hsien, thickets, alt. 1100 m., October 1907 (No. **593**, type; tree 7–10 m. tall, girth 0.6–1.2 m., fruits black; with ripe fruits); Wushan Hsien, thickets, alt. 1100 m., September 1907 (No. **442**; 3 m. tall, fruits black; with almost ripe fruits. Young plants are growing in the Arnold Arboretum).

This species seems to be well characterized by its papillose leaves and by its rather large bluish-black fruits, not unlike small cherries, and by its irregularly pitted and gibbous stones. It may be most nearly related to *C. jessoensis* Koidzumi, which has thinner, more closely serrate, mostly narrower and longer-pointed leaves, and smaller fruits which are apparently not pruinose but shining. See also my remarks under *C. jessoensis*, p. 281.

Not uncommon in open country, by the side of streams and on the mountains of western Hupeh and eastern Szech'uan. It is a slender tree from 10 to 23 m. tall, with a trunk from 0.6 to 1.5 m. in girth; the bark is smooth and very pale gray (almost white); the branches are thin and rather sparse, forming a narrow crown. This species is in cultivation and has proved quite hardy in this Arboretum.

A picture of this tree will be found under No. 0174 of the collection of my photographs. E. H. W.

Celtis Biondii Pampanini in *Nuov. Giorn. Bot. Ital.* n. ser. XVII. 252, fig. 3 (1910).

> *Celtis sinensis* Hemsley in *Jour. Linn. Soc.* XXVI. 450 (non Persoon) (1894), quoad specim. Henryi No. 5276 et probabiliter alia. — E. Pritzel in *Bot. Jahrb.* XXIX. 296 (1900).
> *Celtis Bungeana* Hemsley in *Jour. Linn. Soc.* XXVI. 449 (non Blume) (1894), quoad specimen Henryi No. 5735.
> *?Celtis* spec. E. Pritzel, l. c. (1900).

Western Hupeh: north and south of Ichang, roadside, etc., alt. 100–950 m., April–May 1907 (No. **2321**; tree 8–13 m. tall, girth 0.9–1.5 m.; with young leaves and old ♀ flowers); same locality, cliffs, glens in gorges, May 1900 and later (Veitch Exped. No. 249; shrub 0.8 m. tall; the number consists of young flowering branchlets and branchlets with young leaves and young fruits and also with old leaves and ripe fruits); same locality, etc., April 1900 (Veitch Exped. No. 1761; shrub 1.2 m. tall; with flowers); same locality, *A. Henry* (No. 3100; with old leaves and ripe fruits); same locality?, *A. Henry* (Nos. 3150, 5276; same as No. 3100); " Sian-men-kou, alt. circ. 900 m.," May 1, December 10, 1906, *C. Silvestri* (No. 357, type; ex Pampanini); " Monte Triora, alt. 1950 m.," July 3, 1907, *C. Silvestri* (Nos. 358, 358ᵃ, co-types; ex Pampanini). **Eastern Szech'uan**: Wushan Hsien, *A. Henry* (No. 5735; with young fruits, leaves up to 10 cm. long and 3.5 cm. broad); " Shan-tzu-p'ing," August 1891, *A. von Rosthorn* (No. 301; sterile); district Cheng-kou-ting, *P. Farges* (Hort. Vilmorin, No. 681; living plants in Arnold Arboretum). **Kiangsi**: Kuling, thickets, not common, alt. 1200 m., July 31, 1907 (No. **1520**; tree 8 m. tall; fruits not yet ripe). **Kiangsu**: Spirit Valley near Nanking, in woods and on rocky places, June 4, 1915, *F. N. Meyer* (No. 1425; tree; with unripe fruits).

Unfortunately I have not seen a type specimen of Pampanini's, but the specimens cited above agree with the author's description, except that the leaves are not so broad as those of Pampanini's figure, which in their shape more closely resemble the leaves of the following variety. A part of Wilson's No. 249 with young fruits has broad young leaves, while in Henry's No. 5735 the leaves are large but narrow. The old leaves are yellowish brown beneath and the finer veins are more or less impressed and not prominent. The pubescence and other characters are the same as described in var. *heterophylla* Schneider on p. 282. *C. Biondii* seems to be a widely distributed and variable species. The two varieties look rather distinct in their typical forms and have also a different geographical distribution, but the type seems to be connected with var. *heterophylla* by the specimens from Kiangsi, the stones of which are globose and more distinctly pitted and gibbous; on the other hand var. *Cavaleriei* cannot be separated as a species.

The type may be briefly characterized as follows: Folia pleraque ovato-oblonga, 2–2½-plo longiora quam lata, margine plus minusve integra v. ultra medium satis breviter et acute dentata, adulta utrinque pilis sparsis adpressis instructa v. in facie fere glabra. Putamina (an semper?) ovato-elliptica v. ovato-oblonga. Cetera ut in var. *heterophylla* descripta.

Celtis Biondii is a common tree at low altitudes in Hupeh and in Szech'uan and is known in the vernacular as the P'o shu. In Chinese Herbals and other books it is referred to as the Ch'ing-t'an shu.   E. H. W.

Celtis Biondii, var. Cavaleriei Schneider, n. var.

Celtis Cavaleriei Léveillé in Fedde, *Rep. Spec. Nov.* X. 440 (1912).

A typo praecipue differt foliis adultis plerisque late ovatis v. ovato-rotundatis paullo v. vix duplo longioribus quam latis 4–9 cm. longis et 3–5.5 cm. latis ultra medium satis grosse et obtuse dentatis; putaminibus plus minusve globosis tenuiter et anguste sed satis distincte foveolato-gibboso-reticulatis.

Western Szech'uan: west and near Wên-ch'uan Hsien, valleys, alt. 900–1600 m., September 1908 (No. 972; tree 10–17 m. tall, girth 1.2–2.4 m., fruits orange; with ripe fruits); near Wa-shan, Tung River valley, rare, alt. 550–650 m., September 21, 1908 (No. 933; small tree 7 m. tall, girth 0.6 m., fruits orange; with ripe fruits). Kweichou: Pin-fa, September 4, 1902, *J. Cavalerie* (No. 394, type of *C. Cavaleriei*; with ripe fruits). Yunnan: Milê district, Teng-tien, old woods, November 1, *A. Henry* (No. 9938; tree 7 m. tall, fruit red; with ripe fruits); Puerh Fu, alt. 1300 m., *A. Henry* (No. 13509; tree 7 m. tall; sterile).

Through the kindness of Mgr. Léveillé I have received some leaves and fruits of his type. It agrees well with Wilson's specimens. According to Henry the fruit is "red," but it seems to have the same yellowish or brownish red color as in the type the fruits of which are said by Wilson to be "orange." Otherwise there is hardly any difference between the specimens from Yunnan and those collected by Wilson. In No. 9938 the old leaves are distinctly reticulate beneath, although the finest veinlets are not prominent but almost impressed. The fruits are mostly 2(–3) on the same pedicel, but I have seen neither flowers, young leaves or branchlets. The stones are hardly distinguishable from those of var. *heterophylla* Schneider or from those of Wilson's specimen from Kiangsi.

CONSPECTUS SPECIERUM CHINAE JAPONIAEQUE.

Inflorescentiae ♀ uniflorae v. fasciculato-cymosae, 2–3 florae, pedicellis separa-
tis v. ima basi coalitis v. interdum singulis bifloris . . . Sect. 1. EUCELTIS.
Fructus maturi maximi, aurantiaci; putamina circiter 10 mm. longa et 8 mm.
    crassa. Folia adulta magna, 8–13 cm. longa et 5–9 cm. lata, crassa, supra
    flavo-viridia, subtus (in sicco) distincte flavescentia.
Folia adulta supra plus minusve aspera, subtus molliter villosa, elevato-
    reticulata, in facie sub microscopio reticulato-papillosa. Ramuli, petioli
    pedicellique fructiferi crassi dense tomentelli v. in var. *calvescente* partim
    glabrescentes. Fructus ovato-globosi; putamina hilo dentato et excavato
    excepto vix v. indistincte foveolata et costata . . . . . . 1. *C. Julianae.*
Folia adulta tantum in axillis nervorum subtus paulo barbata, ceterum glaber-
    rima, supra nitentia, subtus vix reticulata, densissime sub lente papillosa.
    Ramuli, petioli pedicellique fructiferi glabri. Fructus ovato-elliptici;
    putamina ovata, plus minusve leviter foveolata et costata.
                                    2. *C. Vandervoetiana.*
Fructus maturi putaminaque minora v. folia non distincte flavescentia v.
    minora et tenuiora.
Pedicelli fructiferi satis breves, petiolis paulo v. vix subduplo longiores v.
    interdum breviores.
Folia subtus distincte reticulata v. flavescentia cinnamomeave (non glauco-
    albescentia et sub microscopio dense papillosa).
Ramuli pedicellique fructiferi glabri. Folia subcoriacea, subtus in sicco
    brunnescentia, in costa nervisque primariis elevatis sparse pilosa, in
    axillis barbata, ceterum glabra, rete nervorum vix v. parum prominulo.
Folia satis sensim acuminata, vix 6 cm. longa. Putamina hilo acùto
    excepto laevissima; pedicelli vix ultra 12 mm. longi.
                                    7. *C. amphibola.*
Folia apice abrupte et longe acuminata v. caudata, maxima ad 11 cm.
    longa. Putamina in tota facie leviter foveolata et indistincte costata,
    hilo obtusiora; pedicelli saepe ad 15 mm. longi.
                                   8. *C. yunnanensis.*
Ramuli pedicellique fructiferi puberuli. Folia in facie subtus pilosula v.
    distinctius elevato-reticulata.
Folia adulta subtus facie glabra, elevato-reticulata, vix discoloria (non
    brunnescentia), satis parva, pleraque 2–4 cm. longa, apice plus
    minusve obtusa, margine vix v. sparse dentata. Putamina ovato-
    elliptica, fere laevia . . . . 9. *C. nervosa.*
Folia adulta subtus facie tenuissime pilosula (pilis saepe minimis glandu-
    losis), pleraque distincte discoloria et majora v. putamina distincte
    foveolata.
Folia magna, membranacea, integerrima, maxima ramulorum fructi-
    ferorum 11–12 cm. longa et 5–6 cm. lata. Fructus aurantiaci;
    putamina ovata, obtusa, tenuissime punctato-reticulata.
                                    3. *C. Bodinieri.*
Folia minora v. plus minusve distincte dentata serratave.
Ramuli breves fructiferi autumno v. hieme toti (an omnes?) decidui,
    plus minusve dense flavo-tomentelli. Folia ovato- v. obovato-
    elliptica (v. rhomboidea?). Putamina subglobosa, hilo obtuso
    distincte ceterum exigue v. vix foveolata et costata.
                                    4. *C. labilis.*

Ramuli fructiferi non decidui, minus dense pilosi v. putamina eximie foveolata et costata v. levissima.
Putamina levia, hilo distincte dentata. Folia ovato- v. rhomboideo-rotunda, paulo longiora quam lata, basi inaequali rotunda, truncata v. subcordata . . . . 5. *C. cercidifolia.*
Putamina in tota facie foveolata et costata. Folia saepissime
ovata v. ovato-rhomboidea v. ovato-lanceolata, majora, 2–
2½-plo longiora quam lata, basi inaequali pleraque plus minusve
acuta . . . . 6. *C. sinensis.*
Folia subtus distincte glauco-albescentia, sub microscopio dense papillifera,
oblongo-ovata, basi plus minusve acuta, apice acuminata. Ramuli, petioli
pedicellique fructiferi glabri. Putamina ovata, tantum versus hilum
satis obtusum exigue foveolata, ceterum levia, indistincte costata.
10. *C. formosana.*
Pedicelli fructiferi (saltem longiores) petiolos plerique 2–4-plo superantes.
Putamina facie levia (rarius valde indistincte reticulata), hilo valde obtuso
excepto plus minusve irregulariter globosa. Ramuli, petioli (sulco
saepe excepto) pedicellique fructiferi glabri. Folia satis crassa, subtus
vix discoloria, pleraque barbulis exceptis glabra . . . 11. *C. Bungeana.*
Putamina plus minusve distincte et anguste (punctiforme) foveolata v.
gibbosa et costata. Ramuli, petioli pedicellique fere semper plus
minusve pilosi v. folia subtus eximie discoloria.
Folia ramulorum fructiferorum subtus satis distincte viridi- v. albo-
cinerascentia, sub microscopio dense papilloso-reticulata, pleraque
late ovata, basi excepta circumcirca serrato-dentata; petioli 0.8–
1.5 cm. longi. Fructus singuli, nigri; putamina irregulariter gibbosa
foveolatave et costata.
Fructus magni, glauco-pruinosi, circiter 1 cm. crassi. Folia chartacea,
satis grosse crenato-serrata, dentibus 2–4 pro 1 cm.
12. *C. cerasifera.*
Fructus minores, nigri (ut videtur nitentes). Folia satis membranacea,
plus minusve angustius argutiusque serrata, dentibus acutioribus
4–6 pro 1 cm. . . . . 13. *C. jessoensis.*
Folia ramulorum fructiferorum subtus in sicco cinnamomea v. flavescentia,
haud dense papilloso-reticulata rate nervillarum paullo impresso;
petioli vix ad 8 mm. longi. Fructus aurantiaci, saepe bini v. terni;
putamina anguste punctiforme foveolata . . . . . 14. *C. Biondii.*
Inflorescentiae ♀ elongato-cymosae, pluri-florae. Fructus saepe singuli sed pedunculo cicatricibus pedicellorum aliorum delapsorum instructo (confer etiam 16.
C. foveolatam) . . . . Sect. 2. SPONIOCELTIS.
15. *C. Salvatiana.*

ENUMERATIO SPECIERUM JAPONIAE CHINAEQUE.[1]

Sect. 1. EUCELTIS Planchon in *Ann. Sci. Nat.* sér 3, X. 263 (1848); in De
Candolle, *Prodr.* XVII. 169 (1873).

1. **Celtis Julianae** Schneider. See p. 265.

   Celtis Julianae, var. **calvescens** Schneider. See p. 266.

[1] *Celtis polycarpa* Léveillé in Fedde, *Rep. Spec. Nov.* XI. 296 (1912), of which I
have seen a specimen, is *Bischofia javanica* Blume.

2. **Celtis Vandervoetiana** Schneider. See p. 267.

3. **Celtis Bodinieri** Léveillé in Fedde, *Rep. Spec. Nov.* XIII. 265 (1914); *Fl. Kouy-Tchéou,* 424 (1915).

Ad descriptionem valde incompletam auctoris addenda v. emendenda:

Arbor ad 10 m. alta; ramuli hornotini (initio ut videtur densius) flavescenti-villosuli, satis tenues, subangulati, annotini obscure rubro-brunnescentes, valde v. omnino glabrescentes, lenticellis paullo discoloribus obtecti, teretes. Folia magna, firma sed tenuia, late ovato-oblonga, basi inaequali plus minusve rotunda, apicem versus subsensim acuminata, interdum fere caudata, supra intense viridia (in vivo nitidula?), tantum in costa nervisque vix elevatis v. subimpressis sparse pilosula, subtus plus minusve discoloria, in sicco flavo-brunnescentia, in costa nervisque elevatis sparse pilosa, tenuiter elevato-reticulata, etiam in facie pilis minimis saepe fuscis et glandulosis conspersa, in axillis barbulata, sub microscopio valde indistincte reticulato-papillosa, margine integerrima, minimis exceptis 9-12 cm. longa, 4.5-6 cm. lata; petioli superne sulcati, tenuiter pilosi, 5-10 mm. longi. Inflorescentiae et flores normales, ♀ axillares singuli ad terni; ovaria glabra. Fructus (an satis maturi?) aurantiaci, ovato-globosi, circiter ad 7 mm. longi et 5 mm. crassi, glabri; putamina ovata, ad hilum obtusa et non foveolata, in facie tenuissime punctato-reticulata, sutura mediana tenui incisa, circiter 5 mm. longa et 4 mm. lata; pedicelli tenuiter pilosi, 8-12 mm. longi.

CHINA. Kwei-chou: "environs de Kouy-chang, bois de la pagode Lan-Yo-Chan, 10 avril 1899," *E. Bodinier* (No. 1633, co-type; with flowers); "jardin du Pe-Tang et environs de la ville," June 10, 1897, *E. Bodinier* (No. 2587, type; with not yet ripe fruits and leaves). Yunnan: Szemao, Yulo, mountains to the south, alt. 1600 m., *A. Henry* (No. 12881; with young fruits).

This is apparently a well-marked species. Mgr. Léveillé has kindly sent me fruits and flowers and a leaf of Bodinier's type-numbers. I have little doubt that Henry's specimen represents the same species. The leaf of the type has the same fine brownish glandular hairs on the lower surface as I have described above from Henry's plant, and it is not " glabra " as stated by the author. Léveillé calls Bodinier's plant " maxima arbor." The lobes of the perigone are glabrous on the inner side and not " intus tomentelli," a fact I have never observed in any Celtis.

4. **Celtis labilis** Schneider. See p. 267.

5. **Celtis cercidifolia** Schneider, n. sp.

Celtis sinensis Hemsley in *Jour. Linn. Soc.* XXVI. 450 (non Persoon) (1894), quoad specimen Henryi, No. 2262.

Frutex v. arbor?; ramuli hornotini fructiferi distincte v. tantum partim flavescenti-tomentelli, satis angulati, rubiginosi, lenticellis flavescentibus sparsis conspersi, annotini glabrescentes vetustioresque plus minusve nigrescentes. Folia adulta crasse papyracea, ovato- v. rhomboideo-rotunda, basi inaequali rotunda, truncata v. subcordata, apice acutata v. satis subito breviter acuminata, tantum infra acuminem margine breviter serrata, 4-6 cm. longa, 2.5-4.5 cm. lata (in No. 1761[a] ad 7.5 cm. longa et 5 cm. lata), supra intense viridia (an subnitentia?), levia, in costa nervisque impressis et in nervillis subimpressis plus minusve sparse v. vix pilosa, subtus discoloria, in sicco plus minusve brunnescentia, in costa nervisque elevatis et nervillis paulo prominulis sparse pilosa, in facie tenuissime sparse glanduloso-pilosa, nervis lateralibus a costa exeuntibus utrinsecus basalibus 2 inclusis 3-4, interdum basi sub 5-nervia; petioli 5-8 mm. longi, superne sulcati, tomentelli. Flores adhuc ignoti. Fructus maturi ut videtur aurantiaci, subglobosi, circiter 6-7 mm. crassi, glabri; putamina globosa, hilo distincte dentato excepto levia, circiter 5 mm. crassa (in No. 1761[a] leviter costata et exigue v. vix striato-foveolata).

Western Hupeh: Ichang, *A. Henry* (No. 2262, type; with ripe fruits); same locality, glens in gorges, August 1901 (Veitch Exped. No. 1761ᵃ in Herb. New York Bot. Gard.; with almost ripe fruits).

This species looks very distinct on account of its leaves, which in shape resemble those of a Cercis, but, as I have stated under *C. labilis* Schneider, on p. 267, Wilson's No. 1761ᵃ seems to be somewhat intermediate between these two species. According to Wilson *C. labilis* is a tree, while No. 1761ᵃ forms only a low shrub. The stone of the only fruit of the type I have seen is very smooth, like that of *C. Bungeana* Blume, while in the typical *C. labilis* the stones are more or less finely pitted and ribbed; but the stones of the almost ripe fruits of No. 1761ᵃ are somewhat intermediate between these two conditions.

6. **Celtis sinensis** Persoon, *Syn.* I. 292 (1805), descriptio valde manca, sed fide Blume.[1] — Willdenow, *Berl. Baumz.* ed. 2, 81 (1811); *Enum. Pl. Hort. Berol. Suppl.* 68 (1813). — Schultes in Roemer & Schultes, *Syst.* VI. 306 (1820). — Loudon, *Arb. Brit.* III. 1416 (an pro parte?) (1838). — Spach in *Ann. Sci. Nat. sér.* 2, XV. 37 (1841); *Hist. Vég.* XI. 126 (1842). — Blume, *Mus. Bot. Lugd.-Bat.* II. 70 (1852). — Bentham, *Fl. Hongk.* 324 (1861). — Miquel in *Ann. Mus. Lugd-Bat.* II. 197 (1865–6); *Prol. Fl. Jap.* 129 (1866–7). — K. Koch, *Dendr.* II. pt. 1, 431 (1872). — Maximowicz in *Bull. Acad. Sci. St. Pétersbourg,* XVIII. 293 (1873); in *Mél. Biol.* IX. 27 (1873), exclud. specim. e China boreali. — Planchon in De Candolle, *Prodr.* XVII. 172 (1873). — Franchet & Savatier, *Enum. Pl. Jap.* I. 431 (1875). — Dippel, *Handb. Laubholzk.* II. 46 (1892), exclud. icone. — Hemsley in *Jour. Linn. Soc.* XXVI. 450 (pro parte) (1894). — Henry in *Trans. As. Soc. Jap.* XXIV. Suppl. 85 (*List Pl. Formosa*) (1896), exclud. No. 1616. — Mouillefert, *Traité Arb. Arbriss.* II. 1209 (1898). — Palibin in *Act. Hort. Petrop.* XVIII. 190 (*Consp. Fl. Kor.* II.) (1900). — Schneider, *Ill. Handb. Laubholzk.* I. 229, fig. 147 r–r², 148 r (1904). — Matsumura & Hayata in *Jour. Coll. Sci. Tokyo,* XXII, 370 (*Enum. Pl. Formos.*) (1906). — Nakai in *Jour. Coll. Sci. Tokyo,* XXXI. 192 (*Fl. Kor.* II.) (1911); *Icon. Pl. Koisik.* I. 3, t. 2, fig. II. (1911). — Matsumura, *Ind. Pl. Jap.* II. pt. 2, 32 (1912). — Dunn & Tutcher in *Kew Bull. Misc. Inform.* add. ser. X. 243 (*Fl. Kwangtung & Hongk.*) (1912).

Celtis orientalis Thunberg, *Fl. Jap.* 114 (non Linnaeus) (1784).

Celtis Willdenoviana Schultes in Roemer & Schultes, *Syst.* VI. 306 (1820). — Loudon, *Arb. Brit.* III. 1416 (1838). — Siebold & Zuccarini in *Abh. Akad. Münch.* IV. pt. 3, 222 (*Fl. Jap. Fam. Nat.* II. 98) (1846). — Planchon in *Ann. Sci. Nat.* sér. 3, X. 287 (1848).

Celtis japonica Planchon in De Candolle, *Prodr.* XVII. 172 (1873).

CHINA. Kwangtung: Soo-kun-po, March 8, 1893, *C. Ford* (flowering branchlets). Hongkong: 1853–6, *C. Wright* (No. 457; with very young fruits); same locality (No. 1068 Herb. Hance; sub nom. *C. serotina* Planch. distributa; vix indigena; with young fruits); same locality, planted, *C. Wilford* (with flowers and with very young fruits). Kiangsu: "S'un Sɐ'on Chi," temple of the God of War, June 2, 1913, *J. Bailie* (No. 6; with unripe fruits). Formosa: without precise locality, 1864, *R. Oldham* (Nos. 512, 513; both with unripe fruits).

[1] The description runs: " fol. lato-ovatis crenatis glaberrimis, venis prominentibus. Hab. in Sina. Ex. hort. Celsii. Fol. majuscula, obtusa." According to the phrase " fol. glaberrimis," one might be inclined to refer Persoon's name to what is now called *C. Bungeana*, but Blume had seen Persoon's type, and says: "Specimen genuinum Herbarii Person exacte congruit cum aliis arborum juniorum ex Japonia." It is on the authority of Blume's statements that I accept Persoon's name for this species.

NORTHEASTERN ASIA. Korea: Port Hamilton, 1859, *C. Wilford* (sine num. in Herb. Gray; with very young fruits); Fusan, temple grounds, September 5, 1903, *C. S. Sargent* (large tree; leaves more or less obovate, above the middle very coarsely crenate-dentate); same locality, September 6, 1903, *C. S. Sargent* (sterile; leaves large, ovate); Chinnampo, September 17, 1905, *J. G. Jack* (sterile, apparently from a young plant, leaves similar to those of the preceding specimen); Seoul, July 1906, *U. Faurie* (No. 904; with young fruits); Quelpaert, "in pagis," October 1907, *Taquet* (No. 344; with ripe fruits); same locality, "in pagis," June 1909, *Taquet* (No. 3211; with young fruits); same locality, "in pagis Taipyang (?)," July 1909, *Taquet* (No. 3212; sterile); same locality, "in pago Hongno," July 1910, *Taquet* (No. 4417; with unripe fruits); same place, April 10, 1908, *Taquet* (No. 4702; with flowers).

JAPAN. Hondo: prov. Rikuzen, Sendai, cultivated, August 27, 1905, *J. G. Jack* (with unripe fruits); prov. Iwashiro, Fukura, September 13, 1892, *J. H. Veitch* (with fruits); prov. Musashi, Tokyo, April 18, 1905 and September 17, 1911 (ex Herb. Sakurai; flowers and with fruits); same locality, April 19, 1882, *K. Miyabe* (with flowers); same locality, May 23, 1888, *K. Miyabe* (with very young fruits); same locality, May 1908 (ex Herb. Yokohama Nursery Co.; with ♂ flowers); same locality, June 1908 (ex Herb. Yokohama Nursery Co.; with ♀ flowers); same locality, Arakawa, common, April 20, 1914, *E. H. Wilson* (No. 6477; tree 10–20 m. tall, girth 0.9–2.4 m., bark gray, smooth; with old flowers); Yokohama, 1862, *C. Maximowicz* (with flowers and with ripe fruits); prov. Sagami, road from Atami to Odawara, August 25, 1892, *C. S. Sargent* (large tree, pale bark; with fruits); prov. Idzu, "Simodah," May 17, 1855, *C. Wright* (a spreading tree; with young fruits); "Simoda," *S. W. Williams & J. Morrow* (with young fruits); prov. Mimasaka, August 6, 1903, *S. Arimoto* (with fruits). Kyushu: prov. Hizen, Nagasaki, 1862, *R. Oldham* (No. 722; with unripe fruits; without No., flowering branchlets); same locality, common, June 5, 1899, *U. Faurie* (No. 3653; with young fruits); near Naja near Nagasaki, November 12, 1903, *C. S. Sargent* (with ripe fruits); prov. Buzen, Noji, September 10, 1905, *J. G. Jack* (with ripe fruits); without locality (ex Herb. Lugd.-Bat., distributed as *C. Willdenowiana* Schultes; with fruits).

This species is well distinguished by the characters given in the key on p. 275. The leaves of vigorous plants usually are large and somewhat resemble in their shape those of *C. jessoensis* Koidzumi. On vigorous shoots the pubescence of the leaves is not only very rough on both sides, but also the shape of the leaves is often variable, as they are often abruptly contracted below their middle into a more or less distinctly caudate, mostly entire narrow-lanceolate apex; there may be every transition from this lyrate form to the normal shape. Such a change of shape occurs in the leaves of young plants in different (probably in most of the) species of Celtis.

The stones of *C. sinensis* are more or less irregularly pitted and ribbed or sometimes rather smooth (see Taquet's No. 344).

To this species seems closely related, according to the author's description, *C. boninensis* Koidzumi (in *Tokyo Bot. Mag.* XXVII. 183 [1913]), from the Ryukyu (Liukiu) Islands, Okinawa. The young branchlets are "ferrugineo-tomentosi mox glabri," the leaves "membranacea, nascentia utrinque rufo-pubescentia cito glabra . . . usque 14 cm. longa, 6 cm. lata; petiolis cito glabris 10–5 mm. longis." The fruit is "carnea 6–7 mm. in diametro; pedicellis 10–14 mm. longis solitariis v. geminis, endocarpio laeve." The main differences from *C. sinensis* seem to be the large leaves and smooth stones. The only specimen before me which somewhat agrees with the description is Jack's specimen from Chinnampo, but neither the branchlets nor the petioles are glabrous, and there are no fruits.

A picture of this tree will be found under No. 668 of the collection of Wilson's photographs and also in his *Vegetation of Western China*, No. 161.

7. Celtis amphibola Schneider, n. sp.

Arbor ad 7 m. alta; ramuli (novelli ignoti) hornotini fructiferi satis tenues, subteretes, rubro-brunnescentes, glabri, lenticellis discoloribus conspersi, vetustiores cinereo-rubiginosi. Folia adulta firma, levia, ovato-rhomboidea v. ovato-oblonga, basi paulo v. vix obliqua plus minusve rotunda, apice sensim v. subito in acumen integrum producta, interdum subcaudata, margine triente inferiore excepto satis regulariter acute dentata, 3.5–5.5 cm. longa, ad 2.5 cm. lata v. latiora ad 3.8 cm. lata, nervis lateralibus utrinsecus basalibus 2 inclusis 3–4, supra saturate viridia, in costa nervisque subimpressis sparse pilosa, ceterum glabra, subinciso-reticulata, subtus in sicco plus minusve brunnescentia, in costa nervisque elevatis sparsissime pilosa pleraque barbulata, ceterum glabra, rete nervorum vix v.paullo prominente; petioli 5–7 mm. longi, superne sulco pilosuli. Fructus maturi aurantiaci (v. nigrescentes?), subglobosi, circiter 7 mm. crassi, glabri; putamina subglobosa, levia, sed hilo obtuso dentata et foveolata, circiter 6 mm. longa et 5 mm. crassa; pedicelli glabri v. ima basi pilosi, 7–9 mm. longi, singuli.

CHINA. Yunnan: Mengtsze, great gully, wood, October 18, *A. Henry* (No. 9323 in part, type, in Herb. New York Bot. Garden; tree 7 m. tall, black and yellow fruits; with ripe fruits).

In many respects this species is very similar to the following species, of which it may represent only a variety, but it can be distinguished from it by its smaller leaves and its smooth stones. Henry's No. 9323 contains different species; the specimen in the herbarium of the Arnold Arboretum belongs to *C. Bungeana* Blume.

The specific name is derived from ἀμφίβολος, misleading.

8. Celtis yunnanensis Schneider, n. sp.

Frutex v. arbor?; ramuli (novelli ignoti) hornotini fructiferi subangulati, pro parte pilosuli, purpurascentes, lenticellis sparsis discoloribus obtecti, vetustiores rubiginosi, distinctius lenticellosi; gemmae ovatae, adpressae, perulis pluribus fimbriatis extus puberulis obtectae. Folia adulta firma, levia, basi inaequali rotundata, ovata v. obovata sed apice abrupte in acuminem 3–4 cm. longum integrum caudatum producta, in margine unius lateris a medio ad basim acuminis irregulariter dentata, alteri lateris integra (an semper?), ovato-lanceolata subsensim acuminata, 4–8.5 cm. longa et 2.3–3.5 cm. lata, latiora abrupte caudata 6–8.5 cm. longa et 3.5–4.5 cm. lata, nervis lateralibus utrinsecus basalibus 2 inclusis 3–4, supra saturate viridia, in costa nervisque subimpressis sparse (v. ad basim intensius) pilosa, ceterum glabra, subinciso-reticulata, subtus in sicco brunnescentia v. rubiginosa, in costa nervisque elevatis sparse pilosa, vix barbulata, in nervillis vix prominulis facieque glabra; petioli 5–9 mm. longi, satis crassi, superne pilosi, sulcati. Flores adhuc ignoti. Fructus maturi singuli v. bini, globosi, ut videtur aurantiaci, circiter 8 mm. crassi, glabri; putamina globosa, hilo paulo producto obtusa, in tota facie leviter foveolata, vix distincte costulata, circiter 6 mm. crassa; pedicelli tantum basi parce pilosi, 10–15 mm. longi.

CHINA. Yunnan: Mi-lê, *A. Henry* (No. 9323 A, type, in Herb. New York Bot. Garden; with ripe fruits).

As I have stated above, this species is apparently closely related to the preceding, but as long as we do not know the flowers and young leaves it seems better to keep the two distinct.

Here may be mentioned a Celtis which I collected in Yunnan in the Tali region not far from Teng-chuan, September 27, 1914 (No. 2709). The ripe fruits are bluish black, about 8 mm. thick, and the stones much resemble those of *C. yunnanensis.* The leaves are ovate with a more or less distinctly long-pointed apex or somewhat rhomboid-ovate, and they measure up to 11 cm. in length and 5.8 cm. in width. I do not know the color of the ripe fruits of *C. yunnanensis,* and I am not quite sure if my specimen represents a form of it or a new species. There is also some resemblance to *C. Salvatiana* (see p. 283), of which the fruits and the mature leaves are still unknown. The shape of the leaves seems to be rather variable in some species, those of my number are somewhat intermediate in shape between *C. amphibola* and *C. yunnanensis,* but the stones of these two species look very different.

9. **Celtis nervosa** Hemsley in *Jour. Linn. Soc.* XXVI. 450 (1894). — Henry in *Trans. As. Soc. Jap.* XXIV. Suppl. 85 (*List Pl. Formosa*) (1896). — Matsumura & Hayata in *Jour. Coll. Sci. Tokyo,* XXII. 370 (*Enum. Pl. Formos.*) (1906). — Matsumura, *Ind. Pl. Jap.* II. pt. 2, 32 (1912).

CHINA. Formosa: Ape's Hill, *Playfair* (No. 458, type; ex Hemsley); same locality, summit, *A. Henry* (shrub 0.6 m. tall; with ripe fruits); Takow, December 11, *A. Henry* (No. 2035, in Herb. New York Bot. Garden; tree 7 m. or more tall; with old leaves, fruits wanting).

The leaves of No. 2035 are up to 5.5 cm. long and 2.5 cm. wide, and are somewhat rough on both surfaces, the midrib and nerves being not quite so glabrous as in the other specimen. The fruits are described as " glauca," but Henry says nothing about the color. The stones may be described as: ovato-elliptica, hilo paullo producta vix apiculata et foveolata, lateraliter compressa, facie exigue foveolata v. pro parte maxima levia, tenuiter costulata, circiter 5 mm. longa et 3.5 mm. crassa. The flowers are unknown, and the relationship of this well-marked species remains uncertain.

10. **Celtis formosana** Hayata in *Jour. Coll. Sci. Tokyo,* XXX. Art. 1, 272 (*Mat. Fl. Formos.*) (1911).

Celtis sinensis Henry in *Trans. As. Soc. Jap.* XXIV. Suppl. 85 (*List Pl. Formos.*) (non Persoon) (1896), quoad No. 1616.

Celtis philippinensis Matsumura & Hayata in *Jour. Coll. Sci. Tokyo,* XXII. 369 (*Enum. Pl. Formos.*) (non Blanco) (1906).

CHINA. Formosa: Kelung and Pachina, 1896, *T. Makino* (type ex Hayata); Naibun, February 1907, *G. Nakahara* (co-type, ex Hayata); Bankinsing, *A. Henry* (No. 1616, in Herb. Arnold Arb.; tree 17 m. tall; with ripe fruits).

I have not seen a type specimen, but Henry's No. 1616 agrees well with Hayata's description, only the petioles and pedicels are a little longer, measuring about 12–14 mm. each, while according to Hayata the petioles are 8 mm. long and the pedicels 10 mm. long. The dried leaves are whitish beneath and densely papillose under the microscope. The species is totally different from *C. philippinensis* Blanco which belongs to sect. *Solenostigma.* Hayata says it is " near to *C. australis* Linn. and *C. tetrandra* Roxb.," but the former differs widely in its densely cuspidate-serrate leaves and in other respects, while the latter, as far as I know, has distinctly furrowed and ribbed stones, which are almost smooth in Henry's specimen and may be described as: ovato-oblonga, versus hilum truncatum paullo productum leviter foveolata, ceterum facie levia, vix costulata, lateraliter vix v. paulo compressa, circiter 7 mm. longa et 5 mm. crassa. This species may be most nearly related to some of the hitherto very little known subtropical or tropical species of this section from southeastern Asia.

11. **Celtis Bungeana** Blume. See p. 269.

12. **Celtis cerasifera** Schneider. See p. 271.

13. **Celtis jessoensis** Koidzumi in *Tokyo Bot. Mag.* XXVII. 183 (1913).

Celtis Bungeana Nakai, *Icon. Pl. Koisik.* I. 3, t. 2, fig. 1 (non Blume) (1911). — Matsumura, *Ind. Pl. Jap.* II. pt. 2, 31 (1912).

Ad descriptionem addenda v. emendanda: Arbor ad 23 m. alta.; ramuli novelli (v. surculorum) hirtello-pubescentes v. glabri, hornotini annotinique fructiferi plerique glaberrimi, subangulati, intense brunnescentes v. subfusci, ut vetustiores cinerascentes lenticellis numerosis discoloribus obtecti; gemmae oblongo-ovatae, acutae, adpressae, perulis brunnescentes ad marginem intense coloratis et tenuissime ciliatis in dorso subhirtellis cinctae. Folia ramulorum fructiferorum late ovata v. ovato-oblonga, basi inaequali rotundata v. plus minusve acuta, nervis basalibus paulo in petiolum productis, apice subsensim acuminata v. fere caudata, margine ima basi excepta satis anguste arguteque serrata serraturis 4–6 pro 1 cm., minora angustiora 5–8 cm. longa, 2.5–4 cm. lata, majora latiora ad 10 cm. longa et 5–5.5 cm. lata, supra satis viridia, in costa nervisque impressis sparse pilosa, ceterum glabra, levia, subtus discoloria, albo-viridia v. cinerascentia, in facie sub microscopio plus minusve reticulato-papillosa, in costa nervisque elevatis sparse pilosa, vix barbata, rete nervorum paulo prominulo, nervis lateralibus a costa exeuntibus utrinsecus basalibus 2 inclusis 3, rarius 4, plantarum juvenilium v. surculorum utrinque plus minusve scabra, subtus vix discoloria, margine grossius crenato-serrata v. crenata, ad 11 cm. longa et ad 6 cm. lata; petioli 6–11 mm. longi, glabri v. in sulco superne pilosi v. in plantis juvenilibus surculisve hirtelli. Flores adhuc ignoti. Fructus etiam juveniles glabri, maturi nigri, ut videtur subnitentes, globosi?, circiter 8 mm. crassi; putamina brunnescentia, ovata, hilo lato obtusa v. plus minusve tridentata et foveolata, ceterum in facie plus minusve irregulariter foveolata et costata v. gibbosa, costa mediana distincte elevata, circiter 7 mm. longa et 5–6 mm. crassa; pedicelli glabri, 1.8–2.3 cm. (v. in Taquet's No. 1376 tantum 1.3–1.5 cm.) longi.

NORTHEASTERN ASIA. Korea: Quelpaert "in pago Kang-kyeng"?, July 1908, *Taquet* (No. 1376; "fruit gros"; fruit not yet ripe, pedicels rather short); same locality, "in sylvis," alt. 1600 m., October 1, 1910, *Taquet* (No. 4418; with ripe fruits, stones very similar to those of Sargent's specimen from Moiwa-yama).

JAPAN. Hokkaido: prov. Ishikari, Moiwa-yama (sterile co-type, ex Koidzumi); same locality, July 30, 1914, *E. H. Wilson* (sterile); same locality, "hill near Sapporo," deep moist soil, September 18, 1892, *C. S. Sargent* (tree 17–20 m. tall, girth 5.4 m., with almost ripe fruits); probably same locality, Sapporo, wild, September 1885, *K. Miyabe* (sterile, named *C. sinensis*); Sapporo, Garden of the Forest Department, September 23, 1892, *C. S. Sargent* (shoot from young plant); same locality, Botanical Garden, May 1914, *K. Miyabe* (with young leaves and very young fruits, named *C. Bungeana*); prov. Kitami, isl. Okujiri (sterile type; ex Koidzumi). Hondo: prov. Shinano, bank of the Kiso-gawa near Fukushima, October 27, 1892, *C. S. Sargent* (tree 20–23 m. tall; fruits apparently shining black); prov. Mino, Nakatsugawa, September 6, 1905, *J. G. Jack* (with almost ripe fruits).

According to Koidzumi this species has also been found in the province Uzen and Iwashiro in Hondo and in Shikoku, prov. Awa, Tsurugi-san. I have not seen a type specimen, but Wilson's sterile specimen from the Moiwa-yama agrees well with the author's description. In Sargent's fruiting specimen from the same locality the stones are distinctly gibbous and ribbed on the surface and sharply dentate and pitted at the narrow end, while the stones of the specimens from Kiso-gawa and Nakatsu-gawa have a very obtuse apex and are more or less distinctly pitted and gibbous. The specimens from Quelpaert certainly belong to this species. There are young plants in the Arnold Arboretum raised from seeds sent from the Agricultural Department at Sapporo.

According to the author's description *C. koraiensis* Nakai in *Tokyo Bot. Mag.* XXVII. 191 (1909); in *Jour. Coll. Sci. Tokyo*, XXXI. 191 (*Fl. Kor.* II.) (1911) from Korea " Kyöngsan: Changdo et monte Phalchoryöng," *T. Uchiyama* (type ex Nakai) and " Ham-gyöng: circa Somui et Quensan," *T. Nakai* (co-type ex Nakai), has larger black fruits ("vulgo 1.3 cm. longa"), the pedicels of which are 2.5–3 cm. long. The leaves are described as " 5–11 cm. longis, 2–7 cm. latis . . . praeter basi argute serratis, serratulis apice incurvis." Unfortunately Nakai does not clearly describe the stone. His phrase " endocarpio osseo integro " seems to indicate that the stones are smooth. In that case *C. koraiensis* would be closely related to *C. Bungeana* Blume, but on account of the serration of the leaves I am inclined to believe that Nakai's species may be the same as *C. jessoensis*. Not having seen a specimen, I cannot decide the question. Nakai compares his species with *C. occidentalis* Linnaeus from North America, which differs in many respects, while Koidzumi says of his *C. jessoensis:* " Species *Celtis Tournefortii* Lam. affinis." Lamarck's species from southern Europe and western Asia differs widely in its bluish green leaves, its yellowish red fruits, in its almost smooth stones, etc. The nearest relative to either *C. jessoensis* or *C. koraiensis* is *C. cerasifera* Schneider see p. 271.

14. **Celtis Biondii** Pampanini. See p. 272.

Celtis Biondii, var. **Cavaleriei** Schneider. See p. 273.

Celtis Biondii, var. **heterophylla** Schneider, n. comb.

?*Celtis sinensis* Nakai in *Jour. Coll. Sci. Tokyo*, XXXI. 192 (*Fl. Kor.* II.) (pro parte, non Persoon) (1911).

Celtis Bungeana, var. *heterophylla* Léveillé in Fedde, *Rep. Spec. Nov.* X. 476 (descriptione valde incompleta) (1912).

Frutex v. arbor?; ramuli hornotini (novelli intensius) flave hirtello-villosi, fructiferi interdum subglabri, rubiginosi, dense lenticellosi, saepe tenuissime rimulosi, subteretes, annotini cinereo-purpurascentes, glabri, lenticellis parvis numerosis discoloribus obtecti, vetustiores intensius cinerascentes; gemmae ovatae, paulo divaricatae, plus minusve dense adpresse flavo-hirtellae. Folia crasse membranacea, obovata v. obovato-subrotunda, basi inaequali rotunda sed plus minusve in petiolum producta v. acuta, apice rotunda v. fere truncata sed abrupte caudato-acuminata, margine tantum ultra medium ad basim acuminis satis grosse serrato-dentata serraturis saepe acutis et leviter incurvis, supra ut videtur obscure viridia, in costa nervisque impressis et etiam in facie sparse adpresse pilosa et levia v. distincte hirta, scabra, subtus discoloria, cinerascentia v. in sicco brunnescentia, in costa nervisque et in facie sparse adpresse pilosa, paulo aspera v. (in Nos. 1375 et 2542) copiosius hirtella et distincte scabra, barbata, pilis saepissime flavescentibus, nervis lateralibus a costa orientibus utrinsecus basalibus 2 inclusis 2–3, basi interdum fere 5-nervia, minora 3–4 cm. longa, 1–2 cm. lata, majora ramulorum fructiferorum ad 5.5 cm. (acumine 1–1.5 cm. longo incluso) longa et ad 2.7 cm. (versus apicem) lata, v. ramulorum sterilium (No. 1375) latiora, ad 6.5 cm. (acumine fere ad 2 cm. longo incluso) longa et ad 4.5 cm. lata; petioli 3–8 mm. longi, subhirsuti; stipulae lineares, pilosae et ciliatae, petiolis sublongiores. Flores ♂ ad basim ramulorum novellorum in No. 2542 remanentes normales, 4 meri; perigonii lobi ovato-lanceolati, ciliati, extus rubelli; ♀ singuli v. plerique bini; perigonia et stamina ut in ♂, ovaria (ut pedicelli longi) dense flavo-hirsuta, ceterum normalia. Fructus maturi ut videtur aurantiaci, glabri, globosi, 5–6 mm. crassi, pedicellis flavo-hirtellis v.

subglabris 1–1.5 cm. longis suffulti; putamina globosa, hilo valde obtuso vix foveolata, ceterum in facie subregulariter exigue foveolata, tenuiter costata, 4–5 mm. longa et 3–4 mm. crassa.

CHINA. Fokien: without precise locality, Dunn's Exped. April–June 1905, (Hongkong Herb. No. 3433, as *C. australis;* with very young hairy fruits). NORTHEASTERN ASIA. Southern Korea: Mokhpo, May 1909, *Taquet* (No. 2542; with old flowers); Quelpaert, " in pago Tschangmani "? (" Tupyangenapi " ex Léveillé), August 1909, *Taquet* (No. 3213, type-number; fruits almost ripe); "in sylvis Sampangsan," October 1908, *Taquet* (No. 1375; with ripe fruits and sterile).

This variety is well characterized by the peculiar shape of its leaves, which mostly are broadest at the upper end below the caudate apex. The specimen from Fokien agrees well with those from Korea. Nakai, in his *Flora Koreana*, apparently confused it with *C. sinensis* Persoon, which is easily distinguished by the shape of its leaves, their different pubescence and serration, and also by its glabrous ovaries and the more irregularly and more deeply pitted stones. Our plant has no close relation to *C. australis* Linnaeus from southern Europe and western Asia, but it is most nearly related to the typical *C. Biondii* Pampanini, see p. 272, which has the same nervation and pubescence of the leaves and the same fruits and stones. Wilson's specimens from Kuling, mentioned under the type, look somewhat intermediate between it and this variety, but the shape of the leaves is more oblong as it is in the type and not obovate.

Sect. 2. SPONIOCELTIS Planchon in *Ann. Sci. Nat.* sér. 3, X. 263 (1848); in De Candolle, *Prodr.* XVII. 180 (1873).

The main difference of this section from sect. *Euceltis* is the elongated-cymose female inflorescence. But, as far as I can judge by the material before me, the species of sect. *Sponioceltis* are so closely connected with those of *Euceltis* that it would be a more natural taxonomic arrangement to unite both these sections and to arrange the species in certain series according to their affinity. But as long as I cannot study all these species I think it best not to make any attempt to distinguish different series. According to Planchon's disposition the following new species belongs to sect. *Sponioceltis*.

15. **Celtis Salvatiana** Schneider, n. sp.

Arbor ad 5 m. alta; ramuli novelli laxe flavescenti-villosi, hornotini glabrescentes, olivacei, subangulati, annotini glabri, rubescentes, lenticellis sparsis elongatis discoloribus, vetustiores cinereo-rubiginosi, tenuissime rimosuli. Folia juvenilia ramulorum floriferorum membranacea, ovato- v. rhomboideo-lanceolata, basi inaequali subrotunda v. acuta, apice acuminata, supra tantum in costa nervisque sparse pilosa, subtus valde initio plus minusve villosula sed mox glabra et tantum barbata, a medio ad acumen utrinsecus plus minusve serrato-dentata, 4–6.5 cm. longa, 1.5–3 cm. lata, petiolis 2–5 mm. longis flavo-pilosis suffulta, ramulorum fructiferorum sed nondum matura circiter duplo majora, late ovato-oblonga, rarius paulo obovata, basi paulo inaequali plus minusve rotunda, apice subsensim late acuminata, textura pubescentiaque ut in juvenilibus, nervatione normali, subtus paulo discoloria, levia, margine paulo infra medium ad basim acuminis satis regulariter subcrenato-serrata, 7–10.5 cm. longa, 3.8–5 cm. lata v. tantum 10 cm. longa et 5.5 cm. lata; petioli 3–6 mm. longi, flavo-pilosi. Inflorescentiae ♂ ad basim ramulorum floriferorum breviter elongato-cymosae, pluriflorae, pleraeque delapsae, ♀ in axillis foliorum superiorum elongatae, pleraeque triflorae; flores ♂ normales, 4-meri; perigonii lobi ovato-oblongi, ad marginem ciliati, ceterum glabri, extus rubescentes; stamina juvenilia incurvata, post anthesin lobis paullo longiora, antheris flavis in sicco basi thecarum albidis; discus villosus; flores ♀ a ♂ vix diversi, lobis perigonii paulo angustioribus ad apicem extus pilosis; ovaria normalia apice excepto glabra, stigmatibus normalibus longis linearibus coronata; pedicelli (et pedunculi) initio flavo-villosuli, deinde glabrescentes; inflorescentiae fructiferae petiolis 2–3-plo longiores. Fructus maturi ignoti.

CHINA. Yunnan: Man-pan, red river bank, October 10, *A. Henry* (No. 10848, type in Herb. New York Bot. Garden; tree 5 m. tall; with flowers and very young leaves and also with young fruits and older leaves; same No. in Herb. Arnold Arboretum, with flowers and young leaves only).

This species seems to be well distinguished from all the other species of this and the first section. According to Henry's note the flowering branches were collected in October and he says "♂ and ♀ flowers and young leaves (some old leaves)." It may be closely related to *C. cinnamomea* Lindley (in Wallich, *Cat.* No. 3696 [nomen nudum] [1829]; apud Planchon in *Ann. Sci. Nat.* sér. 3, X. 303 [1848]), the type of which has been collected by *N. Wallich* in Assam, Silhet (No. 3696). To this species seem to belong a flowering specimen collected by *Lane* from a tree cultivated in Hort. Bot. Calcutta, April 26, 1899, and a specimen from Sikkim, "reg. trop. alt. 3000 ped.," *J. D. Hooker*, with very young fruits. In both specimens the ♀ inflorescences are longer, bearing 3–5 flowers, the pedicels of which are very short. The leaves of Hooker's specimen are "subtus siccitate cinnamomeis," as said by Planchon. It is often confounded with *Celtis tetrandra* Roxburgh from the Himalaya and *C. Roxburghii* Planchon (*C. trinervia* Roxburgh, non Lamarck) from Burma. The Indian species and also those of southeastern Asia need a careful study.

The species is named in honor of Père Auguste Salvat, a distinguished French missionary, who was in charge of the French Catholic Mission at Tali Fu, in appreciation of valued service rendered to me while I was staying in that town during the autumn of 1914.

PTEROCELTIS Max.

Pteroceltis Tatarinowii Maximowicz in *Bull. Acad. Sci. St. Pétersbourg*, XVIII. 293, fig. (1873); in *Mél. Biol.* IX. 27, t. (1873); in *Bull. Soc. Nat. Mosc.* LIV. 53 (*Fl. As. Or. Fragm.*) (1879). — Hemsley in *Jour. Linn. Soc.* XXVI. 451 (1894). — E. Pritzel in *Bot. Jahrb.* XXIX. 297 (1900). — Schneider, *Ill. Handb. Laubholzk.* I. 227, fig. 146 (1904). — Vilmorin & Bois, *Frutic. Vilmorin.* 205, fig. A-C (1905). — Bean, *Trees & Shrubs Brit. Isl.* II. 264 (1914).

> *Ulmus Cavaleriei* Léveillé in Fedde, *Rep. Spec. Nov.* XI. 296 (1912); *Fl. Kouy-Tchéou*, 436 (1915).

Western Hupeh: Ichang, cliffs, alt. 35 m., September 1907 (No. **255**; much branched tree 17 m. tall, girth 4.5 m.; with ripe fruits); same locality, March 23, 1907 (No. **255**ᵃ; with ♂ and ♀ flowers); same locality, April 24, 1907 (No. **255**ᵇ; leafy branchlets with young fruits); same locality, April 1900 (Veitch Exped. No. 136; tree 10 m. tall; with ♂ flowers only); Hsing-shan Hsien, roadside, alt. 1100 m., October 1907 (No. **268**; tree 13 m. tall, branching from near base;

with ripe fruits); same locality, May 1907 (No. 268[a]; with young fruits); same locality, mountains, riversides, very rare, May 1901 (Veitch Exped. No. 1875; tree 10 m. tall; with young fruits). Eastern Szech'uan: " Tchen-kéou-tin," *P. Farges* (with ♂ flowers and also with young and ripe fruits). Western Szech'uan: near Wa-shan, Tung River valley, roadside, alt. 600 m., September 21, 1908 (No. 3215; tree 13 m. tall, girth 1.8 m.; with ripe fruits). Northern Shensi: " Lao-y-san (Zu-lu)," September 6, 1887, *G. Giraldi* (with ripe fruits). Chili: Peking, " in horto ecclesiae rossicae Pekini institutae," June 1847, *A. Tatarinow* (type, ex Maximowicz). Shantung: Su-yung, September 25, 1907, *F. N. Meyer* (large tree having a scaly whitish bark and small leaves); Lung-tung, September 25, 1907, *F. N. Meyer* (No. 247; sterile, leaves coarsely serrate, up to 12 cm. long and 7 cm. wide). Kweichou: " Gan-Chouen, bois," May 1910, *J. Cavalerie* (No. 3784, type of *Ulmus Cavaleriei;* with fruits).

A very distinct genus of which Maximowicz described only the ♂ flowers and fruits. I add the following description of the ♀ flowers from Wilson's No. 255[a]: flores ♀ in axillis foliorum ramulorum brevium novellorum mense Martii v. Aprili singuli, pedicello distincto sparse piloso suffulti; perigonia lobis 4 liberis lanceolatis viridescentibus plus minusve (praesertim apice) pilosis ovario (stigmatibus exclusis) vix longioribus; ovaria sessilia, elliptico-rotunda, latere compressa, in facie sparse pilosa, stigmatibus 2 lanceolatis plus minusve divaricatis paulo brevioribus coronata.

The ripe fruits of the type of *Ulmus Cavaleriei* Léveillé are distinctly pilose upon the seed, while they are wholly glabrous in the other specimens before me. Further investigation is needed to decide if the form from Kweichou represents a distinct variety.

As far as I have been able to observe this is everywhere a rare tree, but it is to be found here and there at low altitudes in western Hupeh and throughout Szech'uan. It is usually found near streams and is partial to rocky places in warm valleys. Some 5 miles above Ichang, by the side of the Yangtsze River and near the entrance to San-yu-tung Glen there is a fine old specimen about 16 m. tall, with a short gnarled trunk 5 m. in girth and much branched to form a wide-spreading crown. As it usually appears it is a tree from 12 to 15 cm. tall with a short trunk from 1.5 to 2.5 m. in girth, and divided near the base into several ascending and spreading stems which branch to form a wide-spreading flattened-round head. The bark is pale gray and peels off in elongated flakes of very irregular shape, and is very characteristic.

Pictures of this tree will be found under Nos. 34, 368 and 469 of the collection of my photographs and also in my *Vegetation of Western China*, Nos. 419, 420, 421.

E. H. W.

ZELKOVA Spach.

Zelkova sinica Schneider, n. sp.

Planera japonica Hemsley in *Jour. Bot.* XIV. 209 (non Miquel) (1876).
Zelkova acuminata Hemsley in *Jour. Linn. Soc.* XXVI. 449 (pro parte, non
Planchon) (1894). — Diels in *Bot. Jahrb.* XXXVI. Beibl. LXXXII. 33
(1905).

Arbor ad 17 m. alta, trunco cortice levi pro genere normali obtecto;
ramuli hornotini dense breviter villosuli v. pro parte glabri, brunnescentes, lenticellis satis magnis discoloribus conspersi, annotini cinereo-
rubiginosi, glabri, vetustiores cinerascentes. Folia satis parva, adulta
firma, ovato-oblonga, basi vix v. paulo inaequali subrotunda v. late
cuneata, apicem versus satis sensim acuminata, interdum fere breviter caudata, supra satis viridia, tantum in costa impressa nervisque
plus minusve sparse pilosula, subtus in sicco cinerascentia (v. paulo
brunnescentia), saepissime tantum in axillis nervorum lateralium utrinsecus 7–9 (imo apice excepto) paulo prominentium albo-barbata,
ceterum glabra v. pilis sparsis difficile recognoscendis praedita, margine regulariter et breviter crenato-serrata serraturis non v. vix
apiculatis, 1.5–4.8 cm. longa et 0.8–2.3 cm. lata; petioli 1–4 mm.
longi, villosuli. Flores ignoti. Fructus maturi singuli, axillares,
irregulariter obovato-rhomboidei, glabri (v. sparse pilosuli), circiter
5 mm. longi et crassi, basi perigonio 5 lobato persistente suffulti, apice
stigmatibus 2 incurvis coronati; pedicelli brevissimi, pilosi.

Western Hupeh: Hsing-shan Hsien, roadside, alt. 900 m., October 1907 (No. **2699**, type; tree 7–17 m. tall, girth 0.35–1.8 m.; with
ripe fruits and bark); same locality, etc., July 1907 (No. **2699**[a]; with
unripe fruits). Northern Shensi: "Kian-san," August 8, 1897,
J. Giraldi (with fruits); "Lao-y-san," 1897, *J. Giraldi* (sterile; leaves
up to 6.3 cm. long and 2.5 cm. wide, petioles up to 8 mm. long).
Chekiang: Ningpo, *E. Faber* (mature leaves as in the type, but
loosely pubescent beneath). Kiangsu; Shanghai, *E. Faber* (younger
leaves with a soft pubescence beneath, young fruits hairy).

The specimens enumerated above agree well with each other in their small leaves
with a short crenate serration. The pubescence of Faber's specimens is, so far as I
can see, due to a younger condition of the leaves. In F. N. Meyer's specimens
mentioned below the leaves are mostly larger and have a distinctly apiculate serration, their texture and nervation being somewhat different too. There is a
sterile specimen collected by him in Kiangsu, Spirit Valley, near Nanking, June
4, 1915 (No. 1444; at edges of forest in rich soil), the leaves of which are up to
9.5 cm. long and 4.5 cm. wide. They are rough above and bear a distinct pu-

bescence, especially on the veins, beneath; the serration is somewhat pointed as in *Z. serrata*. Meyer also collected what is apparently the same form near Soochow in Kiangsu, April 26, 1908 (No. 342), of which he sent seeds to the Department of Agriculture in Washington. From these seeds are growing young plants under No. 22985 (in part, mixed with *Ulmus parvifolia* Jacquin) in Chico, California. Sterile specimens of these plants before me show the same firm leaves with a more or less dense grayish pubescence beneath. The serration is the same as in Meyer's No. 1444 or even sharper. All these plants apparently represent a distinct new species which may be distinguished by its firmer leaves, which are more or less pubescent beneath at least when young and have a more incumbent shorter serration.

This is a rare tree in the parts of China where I have collected, and is only known to me from one locality near the town of Hsing-shan Hsien, where there are several trees from 6 to 16 m. tall, with trunks from 0.5 to 2 m. in girth. The bark is smooth and pale gray, and exfoliates in small, thin roundish flakes leaving behind brown scars.   E. H. W.

Here may be added the other species from eastern Asia, with which *Z. sinica* has been confused by previous authors.

Zelkova serrata Makino in *Tokyo Bot. Mag.* XVII. 13 (1903).
Corchorus hirtus Thunberg, *Fl. Jap.* 228 (non Linnaeus) (1784).
Corchorus serratus Thunberg in *Trans. Linn. Soc.* II. 335 (1794).
Ulmus Keaki Siebold in *Verh. Bat. Genoot.* XII. 28 (*Syn. Pl. Oec. Jap.*) (1830).
Planera acuminata Lindley in *Gard. Chron.* 1862, 428. — Regel in *Gartenfl.*
  XII. 56 (1863).
Planera Kaki Kirchner in Petzold & Kirchner, *Arb. Musc.* 567 (1864).
Planera japonica Miquel in *Ann. Mus. Lugd.-Bat.* III. 66 (1867); *Prol. Fl.
  Jap.* 254 (1867).
Planera Keaki K. Koch, *Dendr.* II. pt. 1, 427 (1872). — Graebener in *Gartenfl.*
  XXXVII. 21, fig. 6–8 (1888).
Zelkova acuminata Planchon in *Compt. Rend. Acad. Paris,* LXXIV. 1496
  (1872), fide Maximowicz; in De Candolle, *Prodr.* XVII. 166 (1873). —
  Hemsley in *Jour. Linn. Soc.* XXVI. 449 (1894). — Shirasawa in *Bull.
  Agric. Coll. Tokyo,* II. 267, t. 13, fig. 14 (*Jap. Laubh. Winter.* t. 9, fig. 14)
  (1895); *Icon. Ess. For. Jap.* I. 65, t. 36, fig. 1–17 (1900). — Mouillefert,
  Traité Arb. Arbriss. II. 1206 (1898). — Palibin in *Act. Hort. Petrop.* XVIII.
  190 (*Consp. Fl. Kor.* II. 44) (1900). — Henry in Elwes & Henry, *Trees Great
  Brit. & Irel.* IV. 920, t. 267, fig. 7 (1909). — Bean, *Trees & Shrubs Brit. Isl.*
  II. 693 (1914).
Zelkova Keaki Maximowicz in *Bull. Acad. Sci. St. Pétersbourg,* XVIII. 288
  (1873); in *Mél. Biol.* IX. 21 (1873). — Franchet & Savatier, *Enum. Pl.
  Jap.* I. 430 (1875). — Mayr, *Aus Wald. Jap.* 32, 48 (1891); *Fremdl. Wald-
  und Parkbäume,* 525, fig. 246–249 (1006). — Dippel, *Handb. Laubholzk.* II.
  40, fig. 15 (1892).— Sargent in *Garden & Forest,* VI. 323, fig. 49 (1893);
  Forest Fl. Jap. 58, t. 19 (1894). — Koehne, *Deutsche Dendr.* 137 (1893). —
  Nakai in *Jour. Coll. Sci. Tokyo,* XXXI. 188 (*Fl. Kor.* II.) (1911).
Abelicea acuminata O. Kuntze, *Rev. Gen. Pl.* II. 621 (1891).
Abelicea Keaki Schneider, *Dendr. Winterst.* 238, fig. 166 e–g (1903).
Abelicea hirta Schneider, *Ill. Handb. Laubholzk.* I. 226, fig. 143–144 (1904).
Zelkova hirta Schneider, l. c. 806 (1906).

NORTHEASTERN ASIA. Mandshuria: Mukden, March 11, 1913, *F. N.
Meyer* (No. 1796ª; very useful timber; cultivated in Yarrow Plant. Intro. Field

Sta., whence I saw a sterile branchlet under No. 35301). Korea: Seoul, common, May 24, 1901, *U. Faurie* (No. 583; with young fruits); same locality, East Palace Park, September 24, 1905, *J. G. Jack* (with fruits); same locality, May 1909, *Taquet* (No. 2537; with flowers); Quelpaert, "in sylvis Hallaisan," alt. 1300 m., August 1909, *Taquet* (No. 3215; sterile); same locality, alt. 1200 m., July 1909, *Taquet* (No. 3267; sterile).

JAPAN. Hondo: prov. Mutsu, "pagis circa Aomori," May 10, 1904, *U. Faurie* (No. 5879; with flowers); prov. Kozuke, Asama, July 1904, *U. Faurie* (No. 5878; with fruits); prov. Shinano, Nagasendo, banks of Kiso-gawa near Agematsu, October 25, 1892, *C. S. Sargent* (sterile, leaves very large, measuring up to 16 cm. in length and 8.5 cm. in width); Nagano, September 1, 1905, *J. G. Jack* (with fruits); prov. Musashi, Tokyo, April 27, 1882, *K. Miyabe* (with flowers); same locality, May 1907, June 1909 (ex Herb. Yokohama Nursery Co.; with flowers and young fruits); same locality, April 12, 1913 (ex Herb. Sakurai; "riesengrosser Baum"; with flowers); same locality, common, April 19, 1914, *E. H. Wilson* (No. 6377; tree 13–33 m. tall, girth 0.9–6 m., bark light gray; with flowers).

The species is not mentioned by Komarov in his *Flora Manshuriae*, and I believe Meyer's specimen from Mukden came from a cultivated plant.

Pictures of this tree will be found under Nos. x211, x213, x247, x610, x668 of the collection of Wilson's Japanese photographs.

Here may be added the following genus not yet collected in central China.

HEMIPTELEA Planch.

Hemiptelea Davidii Planchon in *Compt. Rend. Acad. Paris*, LXXIV. 132, 1496 (1872), fide Planchon in De Candolle, *Prodr.* XVII. 165 (1873). — Maximowicz in *Bull. Acad. Sci. St. Pétersbourg*, 289 (1873); in *Mél. Biol.* IX. 22 (1873). — Franchet in *Nouv. Arch. Mus. Paris*, sér. 2, VII, 78, t. 9 (*Pl. David.* I. 268 (1884). — Palibin in *Act. Hort. Petrop.* XVIII. 190 (*Consp. Fl. Kor.* II. 44) (1900). — Schneider, *Ill. Handb. Laubholzk.* I. 224, fig. 141 a–b (1904).

Planera Davidii Hance in *Jour. Bot.* VI. 333 (1868).
Hemiptelea Davideana Priemer in *Bot. Jahrb.* XVII. 455 (1893).
Zelkova Davidii Hemsley in *Jour. Linn. Soc.* XXVI. 449 (1894). — Komarov in *Act. Hort. Petrop.* XXII. 90 (*Fl. Mansh.* II.) (1903). — Nakai in *Jour. Coll. Sci. Tokyo*, XXXI. 188 (*Fl. Kor.* II.) (1911).
Zelkova Davidiana Bean,[1] *Trees & Shrubs Brit. Isl.* II. 694 (1914).

CHINA. Chekiang: Zah-kou near Hangchou, in earth banks, July 3, 1915, *F. N. Meyer* (No. 1501; scrubby, growing into a tree, as hedge material here and there; sterile). Kianghuai: Nanking, in banks of soil and debris, May 31, 1915, *F. N. Meyer* (No. 1407; shrub or small tree; with young fruits). Chili: near San-tun-ying, in loess banks, June 2, 1913, *F. N. Meyer* (No. 990; mostly as a shrub; sterile); Wei-chang, *W. Purdom* (sterile); north of Jehol, 1910, *W. Purdom* (No. 296; sterile).

NORTHEASTERN ASIA. Mandshuria: "circa oppidum Mukden in silva Fulim," September 27, 1897, *V. Komarov* (with ripe seeds). Korea: Seoul, September 24, 1905, *J. G. Jack* (with fruits); same locality, May 1909, *Taquet* (No.

---

[1] Bean cites Franchet as author, but I do not find that Franchet has published such a combination. Hemsley seems to be the author of the combination *Zelkova Davidii*, because Bentham & Hooker, *Gen. Pl.* III. 353 (1883), cited by Hemsley, do not mention any specific name.

2536; with flowers); same locality, in Namsan, July 1906, *U. Faurie* (No. 903; with fruits); Ping-yang, September 18, 1905, *J. G. Jack* (with fruits).

This species is easily distinguished from every *Zelkova* by its spinescent branchlets and winged fruits. It represents a distinct genus which, according to Priemer, also differs widely from *Zelkova* in its anatomical characters.

TREMA Lour.

Trema virgata Blume, *Mus. Bot. Lugd.-Bat.* II. 59 (1852).

Celtis virgata Roxburgh apud Wallich, *Cat.* No. 3694 (nomen nudum) (1828), fide Planchon.

Sponia virgata Planchon in *Ann. Sci. Nat.* sér. 3, X. 316 (1848); in De Candolle, *Prodr.* XVII. 195 (exclud. var.) (1873).

Sponia timorensis Kurz in *Flora, LV.* 447 (1872), ut videtur tantum pro parte; an Decaisne?. — Maximowicz in *Bull. Acad. Sci. St. Pétersbourg*, XVIII. 295 (1873); in *Mél. Biol.* IX. 29 (1873).

Trema timorensis Hemsley in *Jour. Linn. Soc.* XXVI. 452 (pro parte; an Planchon?) (1894). — E. Pritzel in *Bot. Jahrb.* XXIX. 297 (1900).

Western Szech'uan: near Wa-shan, thickets, alt. 900–1200 m., July 1908 (No. **2812**; bush 1.5 m. tall, flowers greenish). Western Hupeh: Changyang Hsien, *A. Henry* (No. 7170). Hainan: without precise locality, *A. Henry* (No. 8559). Yunnan: Mengtsze, alt. 1300 m., *A. Henry* (No. 10011; tree 10 m. tall).

This species seems to be well characterized by its small narrow lanceolate leaves. Most authors refer it to *Trema timorensis* Blume, *Mus. Bot. Lugd.-Bat.* II. 60 (1852. — *Sponia timorensis* Decaisne in *Nouv. Ann. Mus. Paris*, III. 498 [*Herb. Timor. Descr.*] [1834]), but as I have been unable to see more material I think it best to keep *T. virgata* separate.

Trema spec. (verisim. spec. nova).

Trema timorensis Hemsley in *Jour. Linn. Soc.* XXVI. 452 (non Blume) (1894), quoad specimen Henryii, No. 6210.

Kiangsi: Kuling, roadsides, common, August 1, 1907, alt. 900 m. (No. **1586**; bush 0.6–1.2 m. tall). Hupeh: Patung, *A. Henry* (No. 6210).

This is the form mentioned by Hemsley with "thicker, rougher, pubescent leaves." Both Wilson's and Henry's specimens are much alike, but in No. 6210 the petioles are somewhat longer. Without having seen mature leaves and fruits, it is very difficult to decide whether these plants represent a new species or may be referred to one of the many species already known. It is certainly very different from *T. virgata*, the leaves measuring from 7 to 11 cm. in length and from 2.5 to 4.5 cm. in width.

APHANANTHE Planch.

Aphananthe aspera Planchon in De Candolle, *Prodr.* XVII. 208 (1873). — Franchet & Savatier, *Enum. Pl. Jap.* I. 432 (1875). — Hemsley in *Jour. Linn. Soc.* XXVI. 452 (1894). — Shirasawa in *Bull. Agric. Coll. Tokyo,* II. 265, t. 12, fig. 10 (*Jap. Laubh. Winter* t. 8, fig. 10) (1895); *Icon. Ess. For. Jap.* I. 67, t. 37, fig. 10–20 (1900). — E. Pritzel in *Bot. Jahrb.* XXIX. 297 (1900). — Schneider, *Ill. Handb. Laubholzk.* I. 227, fig. 145 (1904). — Matsumura & Hayata in *Jour. Coll. Sci. Tokyo,* XXII. 370 (*Enum. Pl.' Formos.*) (1906). — Nakai in *Jour. Coll. Sci. Tokyo,* XXXI. 192 (*Fl. Kor.* II.) (1911). — Matsumura, *Ind. Pl. Jap.* II. pt. 2, 31 (1912). — Bean, *Trees & Shrubs Brit. Isl.* I. 196 (1914).

> *Prunus aspera* Thunberg, *Fl. Jap.* 201 (1784). — Willdenow, *Spec.* II. pt. 2, 993 (1799).
>
> *Celtis Muku* Siebold in *Verh. Bat. Genoot.* XII. 28 (*Syn. Pl. Oec. Jap.*) (1830). — Siebold & Zuccarini in *Abh. Akad. Münch.* IV. Abt. 3, 223 (*Fl. Jap. Fam. Nat.* II. 99) (1846). — Planchon in *Ann. Sci. Nat.* sér. 3, X. 315 (1848).
>
> *Sponia nudiflora* Siebold & Zuccarini in *Abh. Akad. Münch.* IV. Abt. 3, 223 (*Fl. Jap. Fam. Nat.* II. 99) (1846). — Planchon in *Ann. Sci. Nat.* sér. 3, X. 337 (1848).
>
> *Homoiceltis aspera* Blume, *Mus. Bot. Lugd.-Bat.* II. 64 (1852). — Miquel in *Ann. Mus. Lugd.-Bat.* II. 197 (1867); *Prol. Fl. Jap.* 129 (1867). — Maximowicz in *Bull. Acad. Sci. St. Pétersbourg,* XVIII. 295 (1873); in *Mél. Biol.* IX. 28 (1873).
>
> *Celtis sinensis* Dunn & Tutcher in *Kew Bull. Misc. Inform.* add. ser. X. 243 (*Fl. Kwangtung & Hongk.*) (pro parte) (1912).
>
> *Homoceltis japonica* Hort. ex Rehder in Bailey, *Stand. Cycl. Hort.* I. 308 (pro synon.) (1914).

CHINA. Kiangsi: Kiukiang plain, rare, alt. 100 m., July 27, 1907 (No. **1523**; tree 17 m. tall, girth 1.8 m., bark rough; with ripe fruits). Chekiang: Tientai Mountain, *E. Faber* (No. 199; ex Hemsley). Kwangtung: Lo-fan Mountains, September 1893, *C. Ford* (No. 327; with ripe fruits).

NORTHEASTERN ASIA. Korea: Quelpaert, "in pago Hogno," May 14, 1908, *Taquet* (No. 1387; with flowers); same locality, July 1910, *Taquet* (No. 4419; with almost ripe fruits): same locality, October 1907, *Taquet* (No. 346). JAPAN. Hondo: prov. Sagami, Atami, August 27, 1892, *C. S. Sargent;* Kanazawa, May 4, 1914, *E. H. Wilson* (No. 6597; tree 8–13 m. tall, girth 0.6–1.5 m.); prov. Musashi, Tokyo, October 6, 1892, *N. Okada* (sterile); same locality, May 15, 1906 (ex Herb. Sakurai; with flowers); same locality, Uyeno Park, August 4, 1905, *J. G. Jack* (with fruits); Yokohama, 1862, *C. Maximowicz* (with flowers and with ripe fruits); without locality (ex Herb. Lugd.-Bat. sub *Homoiceltis aspera* and *Celtis Muku*). Shikoku: prov. Tosa, Nanokawa, July 7, 1889, *K. Watanabe* (with fruits). Kyushu: Nagasaki, common, June 1899, *U. Faurie*

(No. 3654; with young fruits); same locality, 1862, *R. Oldham* (No. 723; with unripe fruits; No. 724; with ripe fruits). Goto Islands: common, May 1901, *U. Faurie* (No. 4841; with ♂ flowers and young leaves).

A well-marked species easily distinguished from any *Celtis* by its leaves with the straight parallel lateral veins, each ending in a tooth at the margin. The fruits are hairy, the styles are more or less persistent, and the dark-colored stones show a very fine granulation.

A picture of this tree will be found under No. x88 of the collection of Wilson's Japanese photographs.

MORACEAE.

Determined by CAMILLO SCHNEIDER.

MORUS L.

Morus cathayana Hemsley in *Jour. Linn. Soc.* XXVI. 456 (1894). — E. Pritzel in *Bot. Jahrb.* XXIX. 298 (1900). — Henry in Elwes & Henry, *Trees Great Brit. & Irel.* VII. 1606 (1912). — Bean, *Trees and Shrubs Brit. Isl.* II. 85 (1914).

Western Hupeh: Fang Hsien, side of streams, alt. 1300 m., May 6, 1907 (No. 10; bush 4.5 m. tall; ♀, leaves more or less lobed); same locality, woods, alt. 900–1300 m., May 10, 1907 (No. 10ª; tree 13 m. tall, girth 1.5 m.; ♂); north and south of Ichang, side of streams, alt. 900–1300 m., May 1907 (No. 10ᵇ; bush or tree, 3–8 m. tall; ♀, leaves mostly lobed); same locality, side of streams, woodlands, alt. 900–1300 m., June 1907 (No. 10ᶜ; with young fruits); Changlo Hsien, woodlands, alt. 600–900 m., May 1907 (No. 10ᵈ; bush 2.1–3 m. tall; ♂); Patung Hsien, alt. 600 m., May 1907 (No. 10ᵉ; tree 13 m. tall, girth 1.5 m.; ♂); same locality, cliffs, April 1900 (Veitch Exped. No. 365; bush 4 m. tall; ♂); same locality, *A. Henry* (No. 5860, co-type, ex Hemsley); Hsing-shan Hsien, side of streams, alt. 1300 m., October 1907 (No. 10ᶠ; tree 8 m. tall, girth 0.6 m.; sterile and bark, leaves lobed, up to 16 cm. long and 15 cm. wide); same locality, alt. 900 m., November 1907 (No. **749**; " Hu-lu-sang "; sterile); Nanto, *A. Henry* (No. 6378, co-type; ♀); Chienshih Hsien, *A. Henry* (No. 5543, type; ♀, ex Hemsley); without precise locality, *A. Henry* (Nos. 1409, 5548; ♂, in Herb. Gray). Eastern Szech'uan: without precise locality, *A. Henry* (No. 5487; ♂; in Herb. Gray); without precise locality, *A. von Rosthorn* (No. wanting; ♀). Western Szech'uan: Chiuting-shan, near Mao-chou, thickets, alt. 1300 m., May 22, 1908 (No. **3310**; bush 3 m. tall; ♀); without precise locality, alt. 600 m., May 1904 (Veitch Exped. No. 4468; tree 10–13 m. tall, fruits white). Chekiang: Mokan-shan, rocky slopes, alt. 600 m., July 21, 1915; *F. N. Meyer* (No. 1588; tree 5–7 m. tall; sterile). Kianghuai: Purple Mt. near Nanking, on rocky mountain slopes, June 3, 1915, *F. N. Meyer* (No. 1419; small tree; sterile).

This is a well-marked species with its hairy leaves, shoots and petioles and with the cylindrical white fruiting catkins. The style is very short, the two stigmas being long and short-pubescent or papillose. The hirsute pubescence of the ovate-oblong perigone lobes of the ♂ flowers is often somewhat yellowish. The lobes of the perigone of the ♀ flowers are round-obovate or the smaller ones ovate, the margins are more or less ciliate or hirsute at the apex, while otherwise there are only a few hairs on the outer surface. The large leaves show the usual Morus variation, being very often more or less lobed or deeply trilobate. The serration is often rather coarse and obtuse. In Wilson's Nos. 10ᵃ and 10ᵉ the dentation of the rather young leaves is closer and more acute, and the apex is more distinctly pointed. The ripe fruits are up to 3 cm. long and about 0.7 cm. thick. The nearest relatives are the following species and *M. laevigata* Wallich (see p. 301).

This species is common by the sides of streams, in thickets and in moist woods on the mountains of western Hupeh and of Szech'uan between 900 and 1600 m. alt. As usually seen it is a small tree from 3 to 6 m. tall, but occasional specimens are met with 15 m. tall with a trunk 1.5 m. in girth. The bark is gray, smooth or nearly so; the branches are numerous and widespreading and form a flattened round crown. The leaves are either entire or much dissected and the fruit is white, red or black when ripe. A colloquial name for this tree in Hupeh is Hu-lu-sang; the leaves are not used for feeding the silkworm. E. H. W.

Morus notabilis Schneider, n. sp.

Arbor ad 8 m. alta; ramuli hornotini autumno glabri, purpureo-brunnescentes, lenticellis distinctis discoloribus conspersi, annotini vetustioresque vix diversi; gemmae ovato-acutae, circiter 8 mm. longae, perulis 4–5 satis laxis inaequalibus obtusis glabris eciliatis partim flavescentibus partim purpurescentibus cinctae. Folia adulta rotundata, basi cordata, apice breviter acutata, margine satis anguste crenulato-serrato-dentata, dentibus saepissime paullo incurvis, supra satis viridia, glabra sed subaspera, subtus in sicco flavescentia, 7–13 cm. longa et 7–12 cm. lata, costa nervisque lateralibus elevatis utrinsecus 4–6 sparse v. vix pilosulis, facie glabra sed tactu plus minusve asperula, axillis vix v. haud barbulata, textura crasse membranacea; petioli 2–3 cm. longi, latere ut videtur compressi, glabri v. pilis sparsis conspersi. Inflorescentiae tantum fructiferae (syncarpia) visae, cylindricae, albae, circi 3 cm. longae et 0.7 cm. crassae, glabrae, pedunculis fere glabris ad 3.5 cm. longis suffultae; fructus (ovaria) glabri, tepalis 4 late obovato-rotundis subcucullatis glabris (tantum margine sub lente minutissime et indistincte denticulatis) involuti, ovato-oblongi, angulati, apicem versus angustati et stylo distincto coronati, stigmatibus 2 satis marcidis glandulosis.

Western Szech'uan: Hungya Hsien, round Wa-wu-shan, forests, alt. 1800 m., September 14, 1908 (No. **919**, type; fruits white, tree 8 m. tall, girth 0.9 m.).

This species looks like a very glabrous variety of *M. cathayana* Hemsley, but I think it better to make it a species on account of the distinctly longer style and the long peduncles. The style is of about the same length as in *M. acidosa* Griffith, but in its narrow cylindric fruiting aments, in the shape of its leaves and in other characters *M. notabilis* belongs to the same group as *M. cathayana* and *M. laevigata* Wallich, which seem to represent a distinct section.

Morus alba Linnaeus, *Spec.* 986 (1753). — Roxburgh, *Fl. Ind.* ed. 2, III. 594 (1832). — Loudon, *Arb. Brit.* III. 1348 (1838). — Moretti in *Giorn. Ist. Lombardo,* I. 180 (1841); *Prodr. Monog. Gen. Morus,* 19 (1842). — Spach, *Hist. Vég.* XI. 42 (1842). — Seringe, *Descr. Cult. Muriers,* 191, t. 1–18 (1855). — Kirchner in Petzold & Kirchner, *Arb. Musc.* 543 (1864). — Miquel in *Ann. Mus. Lugd.- Bat.* II. 197 (1866); *Prol. Fl. Jap.* 129 (1866). — Brandis, *Forest Fl. Ind.* 407, t. 47 (an pro parte?) (1874); *Ind. Trees,* 612 (1906). — Franchet & Savatier, *Enum. Pl. Jap.* I. 432 (1875). — Hooker f., *Fl. Brit. Ind.* V. 492 (1888). — Hemsley in *Jour. Linn. Soc.* XXVI. 455 (pro parte) (1894). — E. Pritzel in *Bot. Jahrb.* XXIX. 297 (exclud. var. plur.) (1900). — Collett, *Fl. Siml.* 457 (1902). — Komarov in *Act. Hort. Petrop.* XXII. 91 (pro parte minima) (1903). — Schneider, *Ill. Handb. Laubholzk.* I. 236, fig. 151 b–c, 152, 153 h–p (1904). — Nakai in *Jour. Coll. Sci. Tokyo,* XXXI. 193 (*Fl. Kor.* II.) (pro parte) (1911). — Ascherson & Graebner, *Syn. Mitteleur. Fl.* IV. 578 (1911). — Henry in Elwes & Henry, *Trees Great Brit. & Irel.* VI. 1609 (exclud. var. 1–2) (1912).

Morus indica [1] Linnaeus, *Spec.* ed. 2, 1398 (pro parte) (1763). — Burman in Rumphius, *Herb. Amboin. Auctuar.* 8, t. 5 (1755). — Poiret, *Enc. Méth.*

[1] *M. indica* Linnaeus, *Spec.* 986 (1753) is founded on " Fl. zeyl. 337 " and on " Tinda-parua Rheed. mal. I. p. 87, t. 49 " (recte 48). Linnaeus (*Fl. Zeyl.* 160, No. 337 [1747]) in the first place cites " Betulae species, conis oblongis villosis, foliis oblongis serratis. Herm. zeyl. 33. Burm. zeyl. 47." I have not seen Hermann's book. Burman (*Thes. Zeyl.* 47 [1737]) only quotes the same phrase from Hermann (*Mus. Zeyl.* 33 [1717]) that Linnaeus does. In the second place Linnaeus quotes " Arbor malabarica baccifera cortice albicante, glomerato flore. Comm. mal. 29." Commelin (*Fl. Malab.* 29 [1700]) quotes the same phrase and adds " Tinda Parua Mal.," which tree is quoted also by Linnaeus in *Flora Zeylanica*. The " Tinda Parua " is no *Morus*, but the same as *Streblus asper* Loureiro (*Fl. Cochin.* II. 615 [1790]). I do not know what this " Betulae species " of Hermann really is, but there is no indication of a *Morus* in any Flora of Ceylon.

In the second edition of the *Species plantarum* Linnaeus added as a synonym " Rumph. Amb. VII. t. 5," which apparently is nothing else than *M. alba* and not the same as *M. indica* Roxburgh (see under *M. acidosa*). Seringe (*Descr. Cult. Muriers,* 229 [1855]) also believes that Rumphius's plate represents *M. alba*, and he reproduces parts of Rumphius's and Rheede's drawings in his Atlas, t. 21, but he regards the " Tinda-Parua " as a good species of *Morus* under the name *M. indica*.

IV. 378 (pro parte) (1797). — Willdenow, *Spec.* IV. 370 (pro parte) (1805). — Matsumura in *Tokyo Bot. Mag.* V. 167 (1891).
Morus tatarica Linnaeus, *Spec.* 986 (1753). — Pallas, *Fl. Ross.* I. pt. 2, 9, t. 2 (1784). — Loudon, *Arb. Brit.* III. 1358, fig. 1225 (1838).
?Morus rubra Loureiro, *Fl. Cochin.* II. 555 (pro parte, non Linnaeus) (1790); ed. 2, II. 679 (1793).
Morus multicaulis Perrottet in *Mém. Soc. Linn. Paris*, II. 129 (1824), ex Perrotet in *Ann. Fromont*, I. 336, t. 3 (1829); III. 338, 341 (1831); in *Arch. Bot.* I. 228 (1833). — Spach, *Hist. Vég.* XI. 46 (1842). — Seringe, *Descr. Cult. Muriers*, 213 (1855). — Miquel in *Ann. Mus. Lugd.-Bat.* II. 198 (1866); *Prol. Fl. Jap.* 130 (1866).
?Morus atropurpurea Roxburgh, *Fl. Ind.* ed. 2, III. 595 (1832). — Wight, *Icon. Pl. Ind. Or.* II. t. 677 (1843).
Morus alba, var. *multicaulis* Loudon, *Arb. Brit.* III. 1348 (1838).
Morus alba, var. *tatarica* Seringe,[1] *Descr. Cult. Muriers*, 202, Atl. t. 12 (1855).
Morus alba, var. *Lhou* Seringe, l. c. 208 (1855).
Morus Tók-wa Siebold ex Petzold & Kirchner, *Arb. Musc.* 547 (pro synon.) (1864).
Morus alba, var. *vulgaris* Bureau in De Candolle, *Prodr.* XVII. 238 (1873).
?Morus alba, var. *Bungeana* Bureau, l. c. 241 (1873).
Morus alba, var. *indica* Bureau, l. c. 243 (pro parte) (1873). — Matsumura, *Ind. Pl. Jap.* II. pt. 2, 40 (1912).
Morus alba, var. *atropurpurea* Bureau, l. c. 244 (pro parte) (1873).
Morus alba, var. *latifolia* Bureau, l. c. 244 (pro parte) (1873).

For further references and synonyms see Seringe, l. c., Bureau, l. c., and Ascherson & Graebner, l. c.

Western Hupeh: Ichang, alt. 30–800 m., May 1907 (No. **3300**; tree 3–8 m. tall, cultivated everywhere for feeding the silkworm; ♂); same locality, May 1907 (No. **3308**; tree 7 m. tall, cultivated; ♀); same locality, A. *Henry* (No. 3498; ♀); Fang Hsien, alt. 900 m., May 8, 1907 (No. **3304**; tree 5 m. tall, probably cultivated, this form rare; ♂); Hsing-shan Hsien, side of streams, one only, alt. 900 m., May 6, 1907 (No. **3303**; small much branched tree 5 m. tall, " Yeh-sang-shu "; probably wild, ♀; No. **3303**ᵃ, ♂); without precise locality, A. *Henry* (No. 1339; with fruits). Kiangsi: Kuling, mountainsides, alt. 1200 m., July 31, 1907 (No. **1653**; 0.9 m. tall; sterile with lobed leaves). Yunnan: without precise locality, 1912, *E. E. Maire* (with fruits).

JAPAN. Hokkaido: Kakkumi Pass, Volcano Bay, September 30, 1892, *C. S. Sargent* (sterile). Kyushu: Nagasaki, 1862, *R. Oldham* (No. 778; with fruits).

---

[1] Seringe is the first author who makes this combination. The quotation " *M. alba*, var. *tatarica* Marschall a Bieberstein" is wrong because the author writes only " *M. alba*, var. *β*." Loudon also is often quoted as the author of this combination, but he cites as his fourth species " *M.* (*A.*) *tatarica* Pal." and seems to regard it as a kind of subspecies, saying in the text: "Very closely akin to *M. alba* L., and, perhaps, originally produced from that species."

INDIA. Kashmir: " Himal. Bor. Occ. Reg. temp. 4–6000 ped.," *T. Thomson* (sub nom. *M. tatarica* L.; ♂). Pondicherry: without precise locality, *G. S. Perrottet* (No. 454; with fruits).

I do not know where *M. alba* really comes from, but its native country seems to be China. The only specimen from China which may represent the wild form is Wilson's No. 3303. The young ♀ flowers have sessile stigmas which are not at all hairy on the inner surface, but are only covered with very minute papillae. The perigone lobes are broadly obovate and glabrous, with the exception of a very fine ciliation on the margins. The sepals of the ♂ flowers are ovate-oblong, almost acute at the apex and minutely pubescent on the outer surface. In the other specimens the stigmas are somewhat more distinctly papillose-hairy and always sessile or subsessile. It is impossible to mention here all the forms of this species, which has been cultivated for many centuries in the warmer and tropical parts of the Old World. I have quoted above the most important synonyms which, in my opinion, should be referred to *M. alba* sensu lato.

Except in the cold northern parts and on the higher mountains this tree is cultivated everywhere in China for its leaves, which are used for feeding the silk-worm (*Bombyx mori*). It is also common by waysides as an escape, and it is quite impossible to state where in China it is genuinely wild, although it is certainly indigenous somewhere in that country. It is also largely cultivated in Japan, and there too it is naturalized by waysides. This tree is too well known to need any description, but I may state that the fruit when ripe is usually dark red or black and only occasionally white, though often fruit of all three colors may be seen on the same branch. The Chinese name for this tree is Sang-shu.

E. H. W.

Morus mongolica Schneider, n. sp.

Morus alba, var. *mongolica* Bureau in De Candolle, *Prodr.* XVII. 241 (1874), ut videtur tantum ex parte. — Franchet in *Nouv. Arch. Mus. Paris*, sér. 2, VII. 80 (*Pl. David.* I. 270) (1884). — E. Pritzel in *Bot. Jahrb.* XXIX. 297 (1900). — Henry in Elwes & Henry, *Trees Great Brit. & Irel.* VII. 1609 (1912).

Arbor ad 8 m. alta v. frutex pauciramosus; ramuli novelli parce pilosuli, saepissime cito glabri, hornotini brunneo-rubri v. purpurascentes, lenticellis sparsis discoloribus obtecti, annotini plerique plus minusve flavescentes, vetustiores ut videtur cinerascentes; gemmae ovato-acutae, circiter 7 mm. longae, perulis 4–5 saepe laxis obtusis v. subacutis dorso glabris v. parce puberulis margine satis flavescenti-ciliatis flavo-brunneis versus marginem purpurascentibus cinctae. Folia membranacea, in plantis visis indivisa, late ovata v. ovato-oblonga, basi satis cordata, apice longe acuminato-caudata, supra viridia, tantum costa (et interdum nervis) minute pilosula, ceterum glabra et laevia, subtus ut videtur pallide viridia, glabra v. basim versus in costa nervisque lateralibus utrinsecus 5–7 flavescentibus v. rubicundis sparse pilosa, laevia, margine late dentata dentibus magnis triangularibus longe cuspidato-subulatis (aristis circiter 3

mm. longis), 8–17 cm. (acumine integro 1–3 cm.˙ longo incluso) longa
et 4–8(–9) cm. lata; petioli pilosuli v. plerique glabri, 2.5–4 cm. longi;
stipulae late lineares, acutae, extus flavo-sericeae, petiolis subduplo
breviores, caducae. Inflorescentiae ♂ cylindricae, laxiflorae, circiter
3–4 cm. longae et 0.7 cm. crassae, pedunculis gracilibus subpilosis ad
1.5 cm. longis suffultae, rhachi laxe hirtellae; flores sessiles, 4-meri,
tepalis ovato-oblongis subacutis extus satis flavo-hirtellis, staminibus
4, filamentis adultis quam tepala fere duplo longioribus basim versus
sensim sed paulo dilatatis, gynaecei rudimento minimo. Inflorescentiae ♀ breviter cylindricae, fructiferae ad 2 cm. longae, circiter 0.7 cm.
crassae, rhachi laxe v. densius flavo-villosae, pedunculis ad 1.5 cm.
longis laxe pilosis suffultae; flores tepalis 4 late ellipticis apice obtusis
facie glabris margine praecipue apice minute v. distinctius ciliatis;
ovaria glabra, in stylum distinctum ½–⅔ ovarii aequantem producta,
stigmatibus 2 stylo longioribus papillosis. Fructus nigri v. pallide rubri.

Western Hupeh: Changlo Hsien, cliffs, alt. 900 m., June 1907
(No. 8; thin tree 3 m. tall, fruits black); Lung-wang-tung, near
Ichang, cliffs, alt. 300 m., April 20, 1907 (No. 8ᵇ; bush 2.1–3.6 m.
tall; ♂); north and south of Ichang, cliffs, alt. 600–1200 m., May
1907 (No. 8ᶜ; thin tree, 3–4.5 m. tall; ♂); Hsing-shan Hsien, ravine,
alt. 600 m., May 24, 1907 (No. 8ᵈ; bush 8 m. tall, little branched, fruits
pale red, sweet, palatable); same locality, cliffs, 900–1200 m., July 1907
(No. 8ᵉ; thin tree, 3 m. tall; sterile); same locality, etc. (No. 8ᶠ; ♀).
Szech'uan: without precise locality, A. von Rosthorn (without No.;
fruiting specimen). Yunnan: "Brousse des rochers à Kiao-
Mé-ti," alt. 3100 m., May 1911, E. E. Maire (" grand arbuste, fruits
roses "); "Rochers des coteaux à Pan-Pien-Kai," alt. 2500 m.,
May 1912, E. E. Maire (" petit arbre dont l'écorce donne le papier
chinois "). Chili: "ad Gehol, Toumet, Sart-chy," A. David (Nos.
1804, 2635, types ex Bureau).

NORTHEASTERN ASIA. Korea: Chinnampo, September 20, 1905, J. G.
Jack (sterile).

This is a very distinct species in the peculiar long aristate teeth of the leaves,
which I have not found in any other species of the genus. I have not seen any
type specimen, and according to Bureau's description I believe he mixed his variety with hairy specimens of M. acidosa Griffith. In its leaves M. mongolica
somewhat resembles M. serrata Roxburgh (see p. 302).

Morus acidosa Griffith [1] *Not. Pl. As.* IV. 388 (*acidosus*) (1854).

[1] Griffith's species has been overlooked by Bureau and other authors, and
it is only mentioned by Hooker f. with a " ? ". Griffith's description is good and,

Morus cuspidata Wallich, *Cat.* No. 4646 (nomen nudum) (1830), fide Bureau.
Morus indica Roxburgh, *Fl. Ind.* ed. 2, III. 596 (non Linnaeus sensu stricto)
(1832). — Wight, *Icon. Pl. Ind. Orient.* II. t. 674 (1843). — Miquel in *Ann.
Mus. Lugd.-Bat.* II. 198 (1866); *Prol. Fl. Jap.* 130 (1866). — Brandis,
Forest Fl. Ind. 408 (1874); *Ind. Trees,* 612 (1906). — Kurz, *Forest Fl.
Brit. Burma,* II. 468 (1877). — Hooker f., *Fl. Brit. Ind.* V. 492 (1888). —
Collett, *Fl. Siml.* 457 (1902). — Gamble, *Man. Ind. Timb.* ed. 2, 635 (1902),
an tantum pro parte?
Morus stylosa Seringe,[1] *Descr. Cult. Muriers,* 225 (Atl. 15, t. 22 sub nom.
M. longistyla) (1855).
Morus alba, var. *nigriformis* Bureau in De Candolle, *Prodr.* XVII. 242 (1873).
Morus alba, var. *cuspidata* Bureau, l. c. 243 (1873).
Morus alba, var. *indica* Bureau, l. c. 243 (1873), quoad citat. Seringe. —
Palibin in *Act. Hort. Petrop.* XVIII. 191 (*Consp. Fl. Korea,* II. 45) (1900).
Morus alba, var. *stylosa* Bureau in De Candolle, *Prodr.* XVII. 243 (1873).—
Shirasawa, *Icon. Ess. For. Jap.* II. t. 6, fig. 1–11 (1908). — Matsumura,
Ind. Pl. Jap. II. pt. 2, 40 (1912). — Henry in Elwes & Henry, *Trees Great
Brit. & Irel.* VI. 1610 (1912). — Miyabe & Miyake, *Fl. Saghal.* 407 (1915).
Morus alba Hemsley in *Jour. Linn. Soc.* XXVI. 455 (pro parte, non Linnaeus)
(1894). — E. Pritzel in *Bot. Jahrb.* XXIX. 297 (1900). — Komarov in *Act.
Hort. Petrop.* XXII. 91 (*Fl. Mansh.* II) (pro parte) (1903). — Matsumura
& Hayata in *Jour. Coll. Sci. Tokyo,* XXII. 373 (*Enum. Pl. Formos.*) (pro
parte) (1906). — Nakai in *Jour. Coll. Sci. Tokyo,* XXXI. 193 (*Fl. Kor.* II.)
(an pro parte?) (1911).
Morus Cavaleriei Léveillé in Fedde, *Rep. Spec. Nov.* X. 146 (1911); *Fl. Kouy-
Tchéou,* 434 (1915).
Morus longistylus Diels in *Not. Bot. Gard. Edinburgh,* V. 293 (*Pl. Chin. Forrest.*)
(1912).
Morus inusitata Léveillé in Fedde, *Rep. Spec. Nov.* XIII. 265 (1914).
Morus bombycis Koidzumi in *Tokyo Bot. Mag.* XXIX. 313 (1915).
Morus japonica Bailey,[2] *Stand. Cycl. Hort.* IV. 2070 (pro parte maxima,
ut videtur non Siebold) (1916).

Western Hupeh: north and south of Ichang, cliffs, etc., alt. 300–
1200 m., July 1907 (No. **33**; bush 1.8–3 m., fruits dark red, leaves
variable, "Ai-sang "); same locality, etc., May 15, 1907 (No. **33**[b], ♀;
No. **33**[c], ♂); same locality, etc., May 1907 (No. **3306**; bush 1.8–2.4
m. tall; ♂); same locality, etc., June 1907 (No. **56**; bush 3–7 m. tall,

according to the statement that the fruits are "stylis stigmatibusque persistentibus
quasi echinata," undoubtedly refers to our species.

[1] Seringe quotes, p. 225, "*Morus stylosa* Sering. pl. XXII. (1854)," and in the
Atlas, p. 15, he gives the name as *M. longistyla.* According to his quotation it
seems that the Atlas appeared before the description, but the books before me both
bear the date 1855. On p. 229, under *Morus indica,* Seringe again refers to a
species " que je décris sous le nom de *Morus longistyla.*"

[2] Bailey quotes *M. japonica* Audibert ex Seringe 226, but Seringe mentions only
a " *M. japonica* jard. Audib." as a synonym of his *Morus stylosa* (see under *M.
acidosa* Griffith). Koidzumi quotes the same name and also a *M. japonica*
Noisette, but the last name is also a garden name for a plant in Hort. Noisette.
Koidzumi's new name, *M. bombycis,* is not appropriate at all because this Mulberry
is not (or very rarely?) used to feed silkworms.

fruits red); Fang Hsien, side of streams, alt. 600–1200 m., May 11, 1907 (No. 33^d; bush 1.5–2.4 m. tall; ♀); same locality, roadsides, alt. 900 m., May 15, 1907 (No. 33^e; bush 1.8 m. tall; ♂); same locality, thickets, alt. 900 m., July 1907 (No. 33^f; fruits dark red); same locality, thickets, alt. 900–1200 m., August 1907 (No. 176; small tree 3–6 m. tall, fruits dark red); Hsing-shan Hsien, cliffs, alt. 600–1200 m., July 1907 (No. 8^a; bush 1.2–3 m. tall, fruits dark red); same locality, roadside thickets, alt. 600–900 m., May 15, 1907 (No. 176; bush 1.8–3 m. tall; ♀); same locality, cliffs, alt. 600–900 m., May 1907 (No. 3302; bush 1.8 m.; ♀; No. 3305; bush 1.2–3 m. tall; ♂; Nos. 3309, ♀, 3309^a, ♂; bush 2.1–3 m. tall); Patung Hsien, cliffs, alt. 900 m., July 1907 (No. 33^a; bush 3 m. tall, fruits blackish red); same locality, etc., June 1907 (No. 3301; bush 1.8–3 m. tall, fruits variable in color); same locality, etc., May 1907 (No. 3301^a; ♂); Changyang Hsien, cliffs, etc., alt. 900–1200 m., July 1907 (No. 33^g; bush 3 m. tall, fruits dark red, "Hi-sang"); same locality, etc., May 1907 (No. 33^h; ♀); same locality, roadside thickets, alt. 1200–1600 m., July 1907 (No. 3299; bush 3 m. tall, fruits red); without precise locality, A. Henry (Nos. 5453, ♂, 6094, ♀, 6249^a; with ripe fruits). Eastern Szech'uan: Wushan Hsien, A. Henry (Nos. 5668, ♂, 5669, ♀, 5741, ♀, 5745, ♀). Western Szech'uan: west of Kuan Hsien, Niu-tou-shan, thickets, alt. 1200–1600 m., June 19, 1908 (No. 3307; bush 3 m. tall, fruits white; an forma distincta?); Chiuting-shan, near Mao-chou, cliffs, alt. 1200 m., May 22, 1908 (No. 3307^a; ♀); without precise locality, roadsides, alt. 1100 m. (Veitch Exped. No. 4467; shrub 0.9 m. tall, fruits white). Yunnan: "Tali Range, Lat. 25° 40′ N., alt. 7000–8000 ft.," August 1906, G. Forrest (No. 4672, type of M. longistylus Diels); Mengtsze, mountain forests, alt. 1900 m., A. Henry (No. 10535; tree 3 m. tall; ♂); same locality, northern mountains, alt. 1900 m., A. Henry (No. 10535^B; shrub 2.4 m. tall; ♀, forma valde hirsuta); Szemao, southwestern mountains, alt. 1600 m., A. Henry (No. 12980; tree 7 m. tall; ♀); "Brousserochers, pied des montagnes à Tong Tchouan, alt. 2550 m.," May 1912, E. E. Maire (♀); "coteaux arides, rocailles à La-Kou," alt. 2400 m., March 1912, E. E. Maire (type of M. inusitata, ♀ and ♂). Kweichou: "Ma-Jo," April 20, 1908, J. Cavalerie (No. 3283, type of M. Cavaleriei; ♀ and ♂). Chili: Peking, on the city wall, April 30, 1913, F. N. Meyer (No. 1382; ♂). Formosa: Tamsui, 1864, R. Oldham (No. 530; distributed sub nom. M. indica; ♀ and ♂); Bankinsing, A. Henry (No. 100; sterile); same locality, A. Henry

(No. 134; the inner bark of the root of shrubs [4'–10'] used by savages for making cloth; ♂); Takow, *A. Henry* (No. 744; shrub 0.6 m. tall; ♀); same locality, *A. Henry* (No. 1784; shrub 1.2 m. tall, white to red fruits; ♀). Kwangtung: Macao, *J. Calléry* (type of var. *stylosa*, ex Bureau); same locality, *J. Calléry* (No. 157, type of var. *nigriformis*, ex Bureau).

NORTHEASTERN ASIA. Korea: Quelpaert, "secus torrentes Hongno," end of April 1908, *Taquet* (Nos. 1392, 1393, 1394; ♀); same locality, "in sylvis Nokatji," May 8, 1908, *Taquet* (No. 1396; ♂); "in sepibus Typyengei"?, June 1909, *Taquet* (No. 3216, ♀); "in sylvis Taitpjeng"?, May 1909, *Taquet* (No. 3217; folia lobata, ♀ ; No. 3218, folia indivisa, ♀); "in sylvis secus torrentes Hallaisan," alt. 600 m., June 5, 1910, *Taquet* (No. 4421; ♀); "in dumosis littoris," April 8, 1908, *Taquet* (No. 4701; ♂); Tsu-sima Island, 1859, *C. Wilford* (♀).

JAPAN. Hokkaido: prov. Ishikari, Sapporo, June 1878 (ex Herb. Sapporo Agr. Coll.; ♀ and ♂); same locality, June 11, 1885 (same herbarium; ♂); same locality, June 1899, *J. Tokubuchi* (♂ and ♀); same locality, May 27, 1903, *S. Arimoto* (♂); prov. Oshima, Hakodate, 1861, *C. Maximowicz* (♀ and ♂); same locality, August 5, 1888, *J. Tokubuchi* (♀). Hondo: prov. Mutsu, Aomori, in forests, May 1904, *U. Faurie* (No. 5880; ♂); prov. Musashi, Ookan-yama, April 19, 1891, *K. Watanabe* (♂); same prov., Tokyo, April 17, 1911, ♂, May 22, 1912, ♀ (ex Herb. Sakurai); prov. ?, "Volcano Mt. Hakodadi, *J. Small* (ex collect. C. Wright 1853–6; ♀); "Simoda & Hakodadi," shady hillsides, 1855, *C. Wright* (a spreading bush, berries red or purple; ♀ ; mixed with *Broussonetia* in Herb. Gray). Kyushu: Nagasaki, 1862, *R. Oldham* (♀).

This widely dispersed species is very common in Hupeh and Szech'uan, more especially in rocky places. Usually it is a broad shrub from 1 to 5 m. high, but often it forms a small and bushy tree from 6 to 8 m. tall. As in other species of *Morus* the leaves are very variable in size and shape; the fruits when ripe are shining black and palatable. A colloquial name for this plant is Ai-Sang (Cliff Mulberry). The leaves are not used for feeding silkworms.   E. H. W.

CONSPECTUS ANALYTICUS
SPECIERUM ASIAE ORIENTALIS INDIAEQUE.

Ovaria stigmatibus sessilibus v. subsessilibus coronata.
  Syncarpia valde elongata, anguste cylindrica, ad 12 cm. longa . 1. *M. laevigata.*
  Syncarpia cylindrica v. elliptico-cylindrica, vix ultra 3 cm. longa (confer etiam
    3. *M. tiliaefoliam*).
    Folia margine plus minusve aequaliter dentata v. serrato-dentata (dentibus
      nunquam late triangularibus valde inaequalibus et apice breviter aristatis).
      Syncarpia anguste cylindrica, vix 7 mm. crassa. Folia etiam adulta subtus
        plus minusve molliter pubescentia, supra aspera . . . 2. *M. cathayana.*
      Syncarpia elliptico-cylindrica v. ovato-oblonga v. elliptica, crassiora.
        Folia etiam juvenilia pleraque vix pilosa v. cito subglaberrima 5. *M. alba.*
    Folia margine grosse irregulariter et acute dentato-serrata, dentibus saepe
      triangularibus, breviter v. vix aristata . . . . 6. *M. serrata.*
Ovaria stylo distincto brevi v. quam stigmata vix breviori coronata.
  Folia margine grosse serrato-dentata, dentibus magnis triangularibus apice
    longe aristatis, basi cordata, apice longe caudato-acuminata . 7. *M. mongolica.*

Folia margine satis anguste serrata v. dentata, dentibus nunquam apice longe aristatis.
Syncarpia glabra, distincti cylindrica, longe pedunculata. Stylus distinctus sed stigmatibus ut videtur brevior. Folia magna, fere rotunda, basi cordata, apice breviter acuta. Planta (an etiam juvenilis) fere glaberrima.
4. *M. notabilis.*
Syncarpia in rhachi plus minusve villosa, breviter elliptica. Stylus longus stigmatibus saepe vix brevior. Folia magnitudine et forma valde variabilia. Planta etiam adulta fere semper plus minusve pilosa . . . 8. *M. acidosa.*

ENUMERATIO SPECIERUM ASIAE ORIENTALIS INDIAEQUE.[1]

1. **Morus laevigata** Wallich, *Cat.* No. 4649 (nomen nudum) (1830). — Moretti, *Prodr. Gen. Morus,* 21 (nomen nudum) (1842). — Brandis, *Forest Fl. Ind.* 409 (1874); *Ind. Trees,* 613 (1906). — Kurz, *Forest Fl. Brit. Burma,* II. 467 (1877). — Hooker f., *Fl. Brit. Ind.* V. 492 (1888). — Gamble, *Man. Ind. Timb.* ed. 2, 636 (1902). — Cooke, *Fl. Pres. Bomb.* II. 658 (1907).

Morus viridis Hamilton in Wallich, *Cat.* No. 4650 (nomen nudum) (1830).
Morus alba, var. *laevigata* Bureau in De Candolle, *Prodr.* XVII. 245 (1873).
Morus alba, var. *laevigata,* subvar. *viridis* Bureau, l. c. (1873).
Morus laevigata, var. *viridis* Hooker f., *Fl. Brit. Ind.* V. 493 (1888).
Morus glabrata Wallich apud Hooker f., l. c. (1888), quam synon.

CHINA. Yunnan: Szemao, mountains to south, alt. 1200 m., *A. Henry* (No. 11975; tree 7 m. tall; ♀); same locality, Yulo forest to the south, *A. Henry* (No. 12019ᴬ; ♂).
INDIA. Kashmir: "N. W. India," cultivated, February 1881, *D. Brandis* (♀). Kumaon: "Kota-Dún, 2000 feet," *R. Strachey & J. E. Winterbottom* (No. 1; ♀). Nepal: without precise locality, 1821, *N. Wallich* (No. 4649ᵃ, type, ex Bureau). United Provinces: "Plan. Gangst. Sup. Regio trop., cult.," *T. Thomson* (♀); Saharanpur, 1825, *N. Wallich* (No. 4649ᴮ, co-type, ex Bureau). Sikkim: "Regio trop. 1–4000 ped.," *J. D. Hooker* (♀). Bengal: "Behar. Regio trop.," *J. D. Hooker* (♀); Patna, *A. Hamilton* (No. 4650 Cat. Wallich, type of *M. viridis,* ex Bureau). Burma: without locality (No. 4650, Herb. Griffith; ♀).
This species is easily distinguished by its very long fruiting catkins, the color of which, according to Hooker f., is yellowish white. Its nearest relative seems to be *M. macroura* Miquel (*Pl. Jungh.* 42 [1850]. — Bureau in De Candolle, *Prodr.* XVII. 247 [1873], sub species dubiae. — Koorders, *Exkursionsfl. Java,* II. 84 [1912]), the type of which was collected by Junghuhn in Java. According to Koorders this is a real *Morus,* forming a high tree (30 m.) in the forests. Judging by the description it must be very similar to *M. laevigata,* but unfortunately I have not seen a specimen.

2. **Morus cathayana** Hemsley. See p. 292.

---

[1] The following three species described by Léveillé, of which I have seen the types, do not belong to *Morus* or even to the *Moraceae.* The first, *Morus calva* Léveillé in Fedde, *Rep. Spec. Nov.* XIII. 265 (1914), is *Coriaria sinica* Maximowicz, while the second, *M. Mairei* Léveillé, l. c. (1914), is *Acalypha szechuanensis* Hutchinson in Sargent, *Pl. Wilson,* II. 524 (1916), which therefore has to be named **Acalypha Mairei** Schneider, n. nom. Also the third, *M. integrifolia* Léveillé & Vaniot, in *Bull. Acad. Int. Géogr. Bot.* XVII. No. 210–211, p. iii (1907), does not belong to the *Moraceae.*

3. **Morus tiliaefolia** Makino in *Tokyo Bot. Mag.* XXIII. 88 (1909).

Morus nigra Matsumura in *Tokyo Bot. Mag.* XVI. 18 (non Linnaeus) (1902).
?Morus japonica Siebold in *Verh. Bat. Genoot.* XII. 27 (*Syn. Pl. Oec. Jap.*)
(nomen nudum) (1830).
Morus rubra, var. *japonica* Makino in *Tokyo Bot. Mag.* XIX. 134 (1905). —
Matsumura, *Ind. Pl. Jap.* II. pt. 2, 40 (1912).
Morus japonica Bailey, *Stand. Cycl. Hort.* IV. 2071 (ut videtur pro parte)
(1916), quoad notam sub *M. rubra*.

JAPAN. Hondo: prov. Nagato, "tractu Otsugori, pago Misumi," September 22, 1895, *J. Nikai* (type ex Matsumura [1902]).

Not having seen a specimen I am not sure if this is a good species or only a form of *M. cathayana* Hemsley. There seems to be no Morus in Japan which is closely related to the North American *M. rubra* Linnaeus as assumed by Bailey, his *Morus japonica* being *M. acidosa* Griffith (see p. 297).

4. **Morus notabilis** Schneider. See p. 293.

5. **Morus alba** Linnaeus. See p. 294.

6. **Morus serrata** Roxburgh, *Fl. Ind.* ed. 2, III. 596 (1832). — Brandis, *Forest Fl. Ind.* 409 (1874); *Ind. Trees,* 612 (1906). — Hooker f., *Fl. Brit. Ind.* V. 492 (1888). — Collett, *Fl. Siml.* 457 (1902). — Gamble, *Man. Ind. Timb.* ed. 2, 635 (1902).

Morus pabularia Decaisne in Jacquemont, *Voy.* IV. 149, Atl. II. t. 151 (1854).
Morus vicorum Jacquemont ex Decaisne, l. c. (pro synon.) (1854).
Morus alba, var. *serrata* Bureau in De Candolle, *Prodr.* XVII. 242 (1873).

INDIA. Kashmir: forests about Dosa near Shrenagur, *Hardwicke* (type ex Roxburgh); "in sylvis circa Ilahabad, alt. 2122 m.," *V. Jacquemont* (type, ex Decaisne; Nos. 248, 686, ex Bureau); Hazara district, Kagan Valley, Balakot, cultivated, alt. 800 m., May 1910 (with fruits and sterile); same locality, Malkandi Reserve, alt. 1400 m., May 1910 (♂); "Tibet, Ladak," July 1856, *Schlagintweit* (No. 1358; ♀); without exact locality, *J. F. Royle* (♀); "Him. Bor. occ. reg. temp. 6000 ped.," *T. Thomson* (♀). Kumaon: without precise locality, *R. Strachey & J. E. Winterbottom* (No. 4; ♂). Punjab: Dehra Dun, Chakrata, alt. 2200 m., May 11, 1912, *Sulakhan Singh* (♀ and ♂).

This distinct species seems to be confined to the northwestern Himalaya. It is easily distinguished by the very coarse irregular dentation of its long-pointed leaves. The styles are sessile, and the sepals of the ♀ flowers are obovate-oblong, glabrous and only ciliate at the apex. The sepals of the ♂ flowers are oblong and rather densely hirsute-pubescent on the outer surface. The shape and serration of the leaves are similar to those of *M. mongolica* Schneider, but in that species the ovaries bear a distinct style.

7. **Morus mongolica** Schneider. See p. 296.

8. **Morus acidosa** Griffith. See p. 297.

There is also the following species which needs further investigation.

Morus australis Poiret, *Encycl. Méth.* IV. 380 (1797). — Willdenow, *Spec.* IV. 371 (1805).

?Morus indica Loureiro, *Fl. Cochin.* II. 555 (pro parte, non Linnaeus) (1790);
ed. 2, II. 679 (1793). — Poiret, *Encycl. Méth.* IV. 378 (1797).
Morus latifolia Poiret, *Encycl. Méth.* IV. 381 (1797).

Morus intermedia Perrottet in *Arch. Bot.* I. 234, t. 7 (1833); in *Ann. Sci. Nat.* sér. 2, XIII. 315 (1840).
Morus indica Perrottet in *Ann. Sci. Nat.* sér. 2, XIII. 315 (non Linnaeus)
. (1840). — Spach, *Hist. Vég.* XI. 47 (pro parte) (1842).
Morus alba, var. *latifolia* Bureau in De Candolle, *Prodr.* XVII. 244 (pro parte) (1873).

According to Perrottet and Spach this represents the true *M. indica* of Linnaeus, but, as I have shown on p. 294, this is a mistake. Poiret's type came from the island of Bourbon, where, according to Perrottet (1840), "le *Morus indica* ou *australis* est sauvage et cultivé." In his first description of *M. intermedia*, Perrottet says that the ovary is "terminé par deux stigmates sessiles," and his rather bad plate does not show a distinct style at the apex of the ovary. But in 1840 Perrottet states that the stigma "est toujours pédicellé, bipartite, recourbé en crosse"; and he says that his *M. intermedia* "était évidemment le *M. indica* de Linné, de Roxburgh et des auteurs en général." Without having seen Poiret's and Perrottet's types I am not able to decide whether *M. australis* sensu Perrottet represents a distinct species or is a variety of *M. alba* Linnaeus or a mixture of forms belonging to different species.

BROUSSONETIA L'Hér.

Broussonetia papyrifera L'Héritier in Ventenat, *Tabl. Règn. Vég.* III. 547 (1799). — Loiseleur-Deslongchamps in *Nouv. Duhamel,* II. 26, t. 7 (circa 1804). — Sims in *Bot. Mag.* L. t. 2358 (1823). — Andrews, *Bot. Repos.* VII. t. 488 (1807). — Blume, *Bijdr.* X. 487 (1825). — Siebold in *Verh. Bat. Genoot.* XII. 28 (*Syn. Pl. Oecon. Jap.*) (1830). — Loudon, *Arb. Brit.* III. 1361 (1838). — Siebold & Zuccarini in *Abh. Akad. Münch.* IV. Abt. 3, 220 (*Fl. Jap. Fam. Nat.* II. 96) (1846). — Blume, *Mus. Bot. Lugd.-Bat.* II. 85 (1852). — Seringe, *Descr. Cult. Muriers,* 236, Atl. 11, t. 26 (1855). — Miquel in *Ann. Mus. Lugd.-Bat.* II. 198 (1865); *Prol. Fl. Jap.* 130 (1865). — K. Koch, *Dendr.* II. pt. 1, 439 (1872). — Bureau in De Candolle, *Prodr.* XVII. 224 (1873). — Franchet & Savatier, *Enum. Pl. Jap.* I. 433 (1875). — Franchet in *Nouv. Arch. Mus. Paris,* sér. 2, VII. 79 (*Pl. David.* I. 269) (1884). — Hooker f., *Fl. Brit. Ind.* V. 490 (1888). — Dippel, *Handb. Laubholzk.* II. 16 (1892). — Koehne, *Deutsche Dendr.* 139 (1893). — Hemsley in *Jour. Linn. Soc.* XXVI. 455 (1894). — Mouillefert, *Traité Arb. & Arbriss.* II. 1217 (1898). — E. Pritzel in *Bot. Jahrb.* XXIX. 298 (1900). — Shirasawa, *Icon. Ess. For. Jap.* I. t. 38 (1900). — Schneider, *Dendr. Winterstud.* 100, fig. 112 i–o (1903); *Ill. Handb. Laubholzk.* I. 241, fig. 151 e–g, 155 a–c, g–m, 156 i–o (1904). — Matsumura & Hayata in *Jour. Coll. Sci. Tokyo,* XXII. 373 (*Enum. Pl. Formos.*) (1906). — Nakai in *Jour. Coll. Sci. Tokyo,* XXXI. 193 (*Fl. Kor.* II.) (1911). — Ascherson & Graebner, *Syn. Mit-*

teleur. Fl. IV. 583 (1911). — Matsumura, *Ind. Pl. Jap.* II. pt. 2, 35 (1912). — Bean, *Trees & Shrubs Brit. Isl.* I. 267, fig (1914).

Morus papyrifera Linnaeus, *Spec.* 986 (1753). — Thunberg, *Fl. Jap.* 72 (1784).
Streblus cordatus Loureiro, *Fl. Cochin.* 615 (1790), fide Bureau.
Papyrius Lamarck, *Rec. Pl. Encycl.* IV. t. 762 (1798).
Papyrius japonica Lamarck apud Poiret, *Encycl. Méth.* V. 3 (1804).
Stenochasma ancolanum Miquel, *Pl. Jungh.* 45 (1850).

Western Hupeh: north and south of Ichang, abundant, alt. 30–900 m., April 1907 (No. **3401**; tree 5–13 m. tall, girth 0.3–1.8 m., fibre used for making twine; ♂); same locality, etc., May 1907 (No. **3401**ᵃ; ♀); without exact locality, *A. Henry* (No. 3666; ♀). Southeastern Szech'uan: "Wang-shan-tsin," *A. von Rosthorn* (No. 374; ♀); Nanch'uan, *A. von Rosthorn* (No. 2224; sterile). Yunnan: Mi-lê district, *A. Henry* (No. 10569; tree 7 m. tall; ♂); Mengtsze, northern mountains, alt. 1600 m., *A. Henry* (No. 10569ᴬ; tree 3 m. tall; ♀); Szemao, eastern forests, alt. 1900 m., *A. Henry* (No. 10569ᴮ; tree; ♀). Chili: Peking, on city wall, May 20, 1913, *F. N. Meyer* (No. 928; ♀). Formosa: Ape's Hill, May 1889, *G. M. H. Playfair* (No. 391; tree; ♀); Takow, August 2, 1889, *G. M. H. Playfair* (♀).

NORTHEASTERN ASIA. Korea: Quelpaert, in Hongno, May 14, 1908, *Taquet* (No. 1398; ♂).
JAPAN. Hondo: prov. Musashi, Tokyo, September 1888 (ex Herb. Sci. Coll. Imp. Univ.; ♀); same locality, April 26, 1913 (ex Herb. Sakurai; ♂ and ♀); same locality, Botanic Garden, August 30, 1892, *C. S. Sargent* (♀); same locality, October 7, 1892, *C. S. Sargent* (small spreading tree; ♀).
The leaves of this species are very variable, and there are several varieties (see Seringe and Bureau, l. c.).
This tree is common everywhere in the warmer parts of China, and in Hupeh and Szech'uan it is abundant. In Japan I saw it only as a planted tree in one or two localities and I do not think it is genuinely wild anywhere in Japan. When allowed to develop it forms a tree 16 m. tall with a trunk 2 m. or more in girth; the bark is gray and smooth; the branches are widespreading and form a broad rounded crown. Birds are very fond of the fruits, which when ripe are bright brick-red and conspicuous. The Chinese name for this tree is Kou-shu and the bark is used for making string and also for making the famous leather paper (P'i-chih) of western China. Some account of these industries will be found in my *A Naturalist in Western China,* II. 74 (1913).
Pictures of this Broussonetia will be found under Nos. 297, 591 of the collection of my photographs and also in my *Vegetation of Western China,* Nos. 6 and 142.

E. H. W.

Broussonetia Kaempferi Siebold in *Verh. Bat. Genoot.* XII. 28 *(Syn. Pl. Oecon. Jap.)* (1830). — Siebold & Zuccarini in *Abh. Akad. Münch.* IV. pt. 3, 221 *(Fl. Jap. Fam. Nat.* II. 97) (1846). — Blume,

Mus. Bot. Lugd.-Bat. II. 87, t. 40 (1852). — Miquel in *Ann. Mus. Lugd.-Bat.* II. 198 (1865); *Prol. Fl. Jap.* 130 (1865). — Bureau in De Candolle, *Prodr.* XVII. 226 (1873). — Kurz in *Jour. Bot.* XI. 193 (1873). — Franchet & Savatier, *Enum. Pl. Jap.* I. 433 (1875). — Hemsley in *Jour. Linn. Soc.* XXVI. 455 (1894). — E. Pritzel in *Bot. Jahrb.* XXIX. 298 (1900). — Schneider, *Ill. Handb. Laubholzk.* I. 242 in adnot., fig. 151 h (1904). — Matsumura, *Ind. Pl. Jap.* II. pt. 2, 34 (1912). — Léveillé, *Fl. Kouy-Tchéou,* 424 (1915).

> *Morus nigra* Thunberg, *Fl. Jap.* 71 (non Linnaeus) (1784), fide Blume.
> *Broussonetia monoica* Hance in *Jour. Bot.* XX. 294 (1882).
> *Morus alba* Hemsley in *Jour. Linn. Soc.* XXVI. 455 (non Linnaeus) (1894), quoad Henry No. 1586.
> *?Broussonetia Kazinoki* E. Pritzel in *Bot. Jahrb.* XXIX. 298 (1900).

Western Hupeh: Patung Hsien, thickets, alt. 300–700 m., April 1907 (No. **34**ª; bush 0.6–1.5 m.; ♀); same locality, etc., June 1907 (No. **34**; bush 1.8 m. tall, fruits red); Hsing-shan Hsien, thickets, ravines, etc., alt. 30–900 m., April 1907 (No. **3400**; bush 1.2–1.8 m.; ♀ and ♂); same locality, etc., June 1907 (No. **3400**ª; ♀); Ichang, *A. Henry* (Nos. 1586, 3668ª; with fruits); Chienshih Hsien, *A. Henry* (No. 5972; ♀). Yunnan: Mengtsze, southeastern mountain forests, alt. 1800 m., *A. Henry* (No. 10959; large climber, 3–5 m. tall; ♀ ; No. 10959ª; ♂); same locality, alt. 1700 m., *A. Henry* (No. 13644; shrub 3 m. tall; ♀); Szemao, eastern mountains, alt. 1500 m., *A. Henry* (No. 12019; shrub 3 m. tall, red fruits). Chekiang: Mo-kan-shan, on rocky slopes, alt. 550 m., July 21, 1915, *F. N. Meyer* (No. 1587; tall shrub, fruit red, small, insipid; sterile).

NORTHEASTERN ASIA. Korea: Quelpaert, "in sepibus Typyengai"?, July 1909, *Taquet* (No. 3221; with fruits).

JAPAN. Hondo: prov. Musashi, near Tokyo, April 22, 1913 (ex Herb. Takurai; ♀, sub nom. *B. Kasinoki*); on "Jizogadake," July 1903, *U. Faurie* (No. 5535; ♀).

This is a distinct species with small leaves and very short petioles. It has the habit of a *Rubus* with often creeping or even climbing branches. The leaves are rarely deeply lobed (Henry's No. 5972). The pubescence seems very variable and the plant is apparently monoecious as well as dioecious.

Here may be added the synonyms of the following species, which is closely related to *B. papyrifera,* but apparently absent from China. It differs from it in its thinner more caudate and more glabrous leaves, in its darker colored branchlets, which are pubescent only when very young, and in its mostly shorter glabrescent petioles.

Broussonetia kazinoki Siebold in *Verh. Bat. Genoot.* XII. 28 (*Syn. Pl. Oec. Jap.*) (nomen nudum) (1830). — Siebold & Zuccarini in *Abh. Akad. Münch.* IV. pt. 3, 221 (*Pl. Jap. Fam. Nat.* II. 97) (1846). — Miquel in *Ann. Mus. Lugd.-Bat.* II. 198 (1865); *Prol. Fl. Jap.* 130 (1865). — K. Koch, *Dendr.* II. pt. 1, 441 (1872). — Bureau in De Candolle, *Prodr.* XVII. 226 (1873). — Franchet & Savatier, *Enum. Pl.*

Jap. I. 433 (*Kasinoki*) (1875). — Engler in Engler & Prantl, *Nat. Pflanzenfam.* III. Abt. 1, 76, fig. 54–55 (1888). — Dippel, *Handb. Laubholzk.* II. 17, fig. 6 (1892). — Koehne, *Deutsche Dendr.* 139 (1893). — De Wildeman, *Icon. Hort. Then.* II. 17, t. 45 (1900). — Schneider, *Ill. Handb. Laubholzk.* I. 241, fig. 151 i–k, 155 d–f (1904). — Matsumura, *Ind. Pl. Jap.* II. pt. 2, 35 (*Kasinoki*) (1912). — Bean, *Trees & Shrubs Brit. Isl.* I. 266 (1914).

> *Broussonetia Kaempferi* Siebold in *Jaarb. Kon. Nederl. Maatsch. Tuinb.* 25 (*Kruidk. Naaml.*) (nomen nudum, non Siebold 1830) (1844). — Mouille-fert, *Traité Arb. Arbriss.* II. 1218 (1898), exclud. synon.
> *Broussonetia Sieboldii* Blume, *Mus. Bot. Lugd.-Bat.* II. 86 (1852).
> *Morus Kaempferi* Seringe, *Descr. Cult. Muriers,* 228, Atl. 11, t. 23 (1855).
> *Broussonetia papyrifera,* var. *Kaempferi* Hort. ex Lavallée, *Arb. Segrez.* 242 (pro synon.) (1877).

NORTHEASTERN ASIA. Korea: Quelpaert " in sepibus Typyengai "?, August 1909, *Taquet* (No. 3220; sterile).

This species is common in Japan from Tokyo southward, and in parts of Shikoku and Kyushu it is abundant and is largely used for making a tough kind of paper.

E. H. W.

CUDRANIA Trécul.

Cudrania tricuspidata Bureau in Lavallée, *Arb. Segrez.* 243 (1877). — Dippel, *Handb. Laubholzk.* II. 19 (1892). — Koehne, *Deutsche Dendr.* 139 (1893). — Mouillefert, *Traité Arb. Arbriss.* II. 1220 (1898). — Schneider, *Dendr. Winterstud.* 96, fig. 112 a–e (1903); *Ill. Handb. Laubholzk.* I. 242, fig. 151 l, 156 a–e, 157 (1904). — Ascherson & Graebner, *Syn. Mitteleur. Fl.* IV. 585 (1911).

> *Maclura tricuspidata* Carrière in *Rev. Hort.* 1864, 390, fig. 37; 1872, 55, 138,\ fig. 6–7. — Nicholson, *Ill. Dict. Gard.* II. 312, fig. 496–7 (1887).
> *Cudranus trilobus* Hance in *Jour. Bot.* VI. 49 (1868).
> *Cudrania triloba* Forbes in *Jour. Bot.* XXI. 145 (1883). — Franchet in *Nouv. Arch. Mus. Paris,* sér. 2, VII. 80 (*Pl. David.* I. 270) (1884). — Oliver in *Hooker's Icon. Pl.* XVIII. t. 1792 (1888). — Hosie, *Three Years in Western China,* 21 (1889). — *Kew Bull. Misc. Inform.* 1888, 291, t. — Hemsley in *Jour. Linn. Soc.* XXVI. 470 (1894). — E. Pritzel in *Bot. Jahrb.* XXIX. 298 (1900). — André in *Rev. Hort.* 1905, 363. — Nakai in *Jour. Coll. Sci. Tokyo,* XXXI. 199 (1911). — Matsumura, *Ind. Pl. Jap.* II. pt. 2, 35 (1912). — Bean, *Trees & Shrubs Brit. Isl.* I. 440, fig. (1914).

Western Hupeh: Ichang, etc., alt. 30–900 m., October 1907 (No. **252**; tree 3–10 m. tall, fruits strawberry-red); same locality, etc., May 1907 (No. **252ᵃ**; bush or tree, leaves used for feeding the silkworms; ♂); same locality, *A. Henry* (No. 2322ᵃ; ♀). Yunnan: Mengtsze, alt. 1400 m., *A. Henry* (No. 9987; tree 3–7 m. tall, red fruit eaten by children; ♀). Chekiang: Ningpo, 1908, *D. Macgregor* (sterile). Central China: " Kur-san," *Hugh Scallan* (No. 119; ♂). Kiangsu: Purple Mount near Nanking, on rocky slopes, June

3, 1915, *F. N. Meyer* (No. 1422; tall shrub, fruit red, sweetish, at times leaves used as a food for silkworms; ♀). Shantung: Chifu, ocean bluff, September 22, 1903, *C. S. Sargent* (small shrub; ♀).

NORTHEASTERN ASIA. Korea: Quelpaert, " in pago Hongno," August 5, 1908, *Taquet* (No. 1509; ♀); same locality, June 15 and 25, 1908, *Taquet* (Nos. 1510, 1511; ♂); same locality, May 4, 1908, *Taquet* (Nos. 1512, 1513; with young flowers); same locality, " in sepibus Htepseng "?, July 20, 1911, *Taquet* (No. 4420; ♀, leaves orbicular-elliptic, about 9 cm. long).

JAPAN. Hondo: prov. Musashi, Tokyo, Botanic Garden, 1892, *C. S. Sargent* (♂); same locality, June 5, 1882, *K. Miyabe* (♂).

This is a variable species in the shape and the size of the leaves and in the length of the peduncle of the fruits. The lateral veins of the leaves are not so numerous as in the species mentioned below, but they are much more prominent on the under surface.

In western Hupeh and in Szech'uan this is a common wayside shrub or tree; in western Szech'uan, and especially in the prefecture of Kiating Fu, it is planted in quantity and its leaves are used with those of *Morus alba* Linnaeus for feeding the silkworm (*Bombyx mori*). Usually it is a bush or bushy tree from 3 to 8 m. high, but under exceptional conditions it forms a tree 18 m. tall with a trunk 2 m. in girth. The bark is pale gray and exfoliates in thin flakes of irregular shape; the branches are slender rather dense and form usually a rounded or flattened crown and the branchlets are spiny. The fruit is bright red and edible, but is not very palatable. The colloquial name for this tree is Tsa(or Cha)-shu.

Pictures will be found under Nos. 392, 576 of the collection of my photographs and also in my *Vegetation of Western China*, Nos. 175, 176.

E. H. W.

Here may be added notes on the other species known from China.

Cudrania fruticosa Wight ex Kurz, *Forest Fl. Brit. Burma*, II. 434 (1877). — Hooker f., *Fl. Brit. Ind.* V. 539 (1888). — Prain, *Bengal Pl.* 970 (1903). — Brandis, *Ind. Trees*, 614 (1906),

?Batis fruticosa Roxburgh, *Fl. Ind.* ed. 2, III. 763 (1832).

CHINA. Yunnan: Szemao, southeastern forests, alt. 1600 m., *A. Henry* (No. 13156; climber, flowers yellowish; ♂).

INDIA. Eastern Bengal: " Khasia Hills, alt. 4000 ft.," *J. D. Hooker & T. Thomson* (ex Hooker f.). Upper Burma: Ava (type, ex Kurz).

Henry's specimen agrees well with Kurz's description. This species seems to be most closely related to *C. tricuspidata* Bureau in its membranous leaves with few strong and prominent lateral nerves. The leaves are elliptic-oblong, from 7 to 11 cm. long and from 3.5 to 6 cm. wide, with a rounded base and a rather abruptly and slender acuminate apex. The petioles are about 9 mm. long and somewhat hairy in Henry's specimen. The grayish branchlets are glabrous or loosely hairy. I do not know tho ♀ flowers or the fruits. In the specimen before me the flowering branchlets are unarmed, but Kurz says " armed with curved sharp spines."

Cudrania pubescens Trécul in *Ann. Sci. Nat.* sér. 3, VIII. 125 (1847). — Kurz, *Forest Fl. Brit. Burma*, II. 435 (1877). — Hooker f., *Fl. Brit. Ind.* V. 539 (1888). — Brandis, *Ind. Trees*, 614 (1906). — Koorders, *Exkursionsfl. Java*, II. 91 (1912).

Cudranus pubescens Miquel, *Fl. Ind. Bat.* I. pt. 2, 290 (1859).

CHINA. Yunnan: Szemao, mountains to the west, alt. 1600 m., *A. Henry* (No. 11906; large climber; with young flowers); same locality, mountains to the east, alt. 1400 m., *A. Henry* (11906ª; large climber; ♂); same locality, mountains to the north, alt. 1600 m., *A. Henry* (No. 11906ᵇ; large climber; with fruits).

INDIA. Burma: " not unfrequent in the tropical forests of the eastern slopes of the Pegu Yomah and Martaban, up to 3000 ft. elevation " (ex Kurz). JAVA. Without precise locality, *Leschenault* (type, ex Trécul).

Henry's specimens agree with Trécul's description. The largest leaves measure up to 18 cm. in length and 8.5 cm. in width; like the branchlets they bear underneath a soft yellowish pubescence, especially on the prominent rib and on the equally prominent lateral veins, of which there are from 7 to 11 on each side of the rib. At the base the leaves are rounded or acute, and they are more or less abruptly pointed at the apex.

Cudrania javanensis Trécul in *Ann. Sci. Nat.* sér. 3, VIII. 123, t. 3, fig. 76–85 (1847).[1] — Bentham, *Fl. Austral.* VI. 179 (1873). — Brandis, *Forest Fl. Ind.* 425 (1874); *Ind. Trees*, 614 (1906). — Hooker f., *Fl. Brit. Ind.* V. 538 (1888); in Trimen, *Handb. Fl. Ceyl.* IV. 98 (1898). — Hemsley in *Jour. Linn. Soc.* XXVI. 469 (1894). — Gamble, *Man. Ind. Timb.* ed. 2, 651 (1902). — Matsumura & Hayata in *Jour. Coll. Sci. Tokyo*, XXII. 380 (*Enum. Pl. Formos.*) (1906).

Trophis spinosa Willdenow,[2] *Spec.* IV. 734 (pro parte) (1805), quoad citat. Rumphii. — Blume, *Bijdr.* X. 489 (1825).

Cudrania obovata Trécul in *Ann. Sci. Nat.* sér. 3, VIII. 126 (1847). — Hemsley in *Jour. Linn. Soc.* XXVI. 459 (1894).

Maclura javanica Blume, *Mus. Bot. Lugd.-Bat.* II. 83, t. 31 (1852). — Miquel, *Fl. Ind. Bat.* I. pt. 2, 280 (1859).

Maclura timorensis Blume, *Mus. Bot. Lugd.-Bat.* II. 84 (1852).

Maclura amboinensis Blume, l. c. (1852).

Plecospermum cuneifolium Thwaites in *Hooker's Jour. Bot. & Kew Gard. Misc.* VI. 303 (1854).

Cudrania acuminata Miquel apud Zollinger, *Syst. Verz. Ind. Archip.* 90, 96 (1854).

Cudranus amboinensis Miquel, *Fl. Ind. Bat.* I. pt. 2, 290 (1859).

Cudranus acuminatus Miquel, l. c. 291 (1859).

Cudranus Rumphii Thwaites, *Enum. Pl. Zeyl.* 262 (1864). — Beddome, *Fl. Sylv. South Ind.* II. For. Man. Bot. p. ccxx. t. 27, fig. 1 (1874).

Cudrania rectispina Hance in *Jour. Bot.* XIV. 365 (1876). — Hemsley in *Jour. Linn. Soc.* XXVI. 470 (1894). — Henry in *Trans. As. Soc. Jap.*

---

[1] Trécul's quotation " *Cudranus javanensis* Rumph." ·is not correct, nor is Willdenow's " *Cudranus javensis* Rumph." It should read " *Cudranus Amboinensis silvestris* Burmann in Rumphius, *Herb. Amb.* V. 25, t. 15–16 (1747)."

The Latin descriptions in Rumphius's book are written by Burmann, who is the editor. In the explanation of the figures he gives the name cited above, while on p. 22 in the text he mentions two forms, " *Cudranus Bimanus* sive *Javanus* " and " *Cudranus Amboinicus*."

[2] According to the description Willdenow's plant is *Plecospermum spinosum* Trécul. Willdenow cites as a synonym " *Trophis spinosa* Roxburgh," which must be a manuscript name. Roxburgh (*Fl. Ind.* ed. 2, III. 762 [1832]) changed Willdenow's *Trophis spinosa* to *Batis spinosa* Roxburgh, which also is *Plecospermum* (except the synonym " *Cudranus* Rumph."), while the real *Trophis spinosa* Roxburgh, l. c., is *Phyllochlamys spinosa* Bureau, the correct name of which seems to be *Ph. taxoides* (Heyne) Koorders.

XXIV. suppl. 88 (*List Pl. Formosa*) (1896). — Matsumura & Hayata in *Jour. Coll. Sci. Tokyo*, XXII. 380 (*Enum. Pl. Formos.*) (1906). *Cudrania amboinensis* Kurz, *Forest Fl. Brit. Burma*, II. 434 (1877). *Cudrania spinosa* Koorders, *Exkursionsfl. Java*, II. 90 (1912).

CHINA. Yunnan: Mengtsze, mountains to southeast, forests, alt. 1600 m., *A. Henry* (No. 9987[a]; tree 3 m. tall; ♂); same locality, rocky ravine, alt. 1600 m., *A. Henry* (No. 10821; bushy shrub 1.8 m. tall, thorns few on old branches, flowers white; ♂); Szemao, mountain forests, alt. 1400 m., *A. Henry* (No. 11937; large climbing shrub, yellowish flowers; ♀; No. 11937[a]; with fruits); same locality, mountains to southwest, alt. 1600 m., *A. Henry* (No. 11937[b]; climbing shrub; ♂); same locality, forests, alt. 1400 m., *A. Henry* (No. 12385; tree 7 m. tall; with fruits, leaves broad-obovate); same locality, *A. Henry* (No. 12385[b]; large cluster orange-red; with fruits, leaves narrow obovate-oblong). Fokien: Dunn's Exped., April to June 1905 (Hongkong Herb. No. 3479; ♂, branchlets very spiny, leaves small). Hongkong: "in valle Wongneichung," April 1874, *J. Lamont* (Herb. Hance No. 19216, type of *C. rectispina*, ex Hance). Formosa: Takow, *A. Henry* (No. 720; shrub 1.2–1.8 m.; ♀); Bankinsing, *A. Henry* (No. 135; shrub; ♀).

INDIA. Punjab: Dehra Dun, Lachiwala, alt. 650 m., November 12, 1912, *W. S. Fernandes* (No. 60; ♂, ♀); same district, *E. King* (♂). Assam: "Teock Ghat near Tingali Bam.," October 1898, *Prain's Collector* (No. 274; with fruits).

PHILIPPINE ISLANDS. Luzon: "circa Manillam," *Hugh Cuming* (No. 1017, type of *C. obovata*, ex Trécul).

This is a widely distributed and very variable species. Forms described by Trécul as *C. obovata* with rather broadly obovate leaves seem to be as frequent as the small-leaved forms Hance had in mind in describing *C. rectispina*. The young branches and petioles (or even the ribs of the leaves) are often more or less yellowish hairy, but sometimes the whole plant, except the peduncles and flower parts, appears to be quite glabrous. The variation seems to be due to climatic conditions.

Cudrania crenata C. H. Wright in *Jour. Linn. Soc.* XXVI. 439 (1894).

CHINA. Hainan: without precise locality, *A. Henry* (No. 8389, type, ♂ ex Wright).

According to the author this species is distinguished by its crenately serrate leaves, which are obovate and up to 10 cm. long and 5 cm. broad, with very short petioles. This, perhaps, is not a Cudrania at all.

Cudrania Bodinieri Léveillé in Fedde, *Rep. Spec. Nov.* XIII. 265 (1914).

CHINA. Hongkong: "Torrent de la Baie du Télégraphe près Béthanie," January 14, 1896, *E. Ḇodinier* (No. 1413, type).

According to the author this species is related to *C. rectispina* Hance. I have seen only two leaves which Mgr. Léveillé kindly sent to the Arnold Arboretum. They are elliptic and somewhat hairy on the midrib on both sides. I am not quite sure if this is a Cudrania.

FICUS [1] L.

Sect. UROSTIGMA (Gasparini) King.

Ficus lacor Hamilton in *Trans. Linn. Soc. Lond.* XV. 150 (1827), fide King.

[1] The *Ficus* of China need a thorough investigation. Being unable to make a special study of this difficult genus, I can give only approximate determinations according to King's monograph and Hemsley's enumeration.

Ficus infectoria Roxburgh, *Fl. Ind.* ed. 2, III. 551 (non Willdenow) (1832), exclud. synon. Rheede. — King in *Ann. Bot. Gard. Calcutta*, I. 60, t. 75 (1888). — Hemsley in *Jour. Linn. Soc.* XXVI. 463 (1899). — Warburg in *Bot. Jahrb.* XXIX. 299 (1900).

Szech'uan: Yangtsze Valley, alt. 100–500 m., common, April 1907 (No. 2796; tree 10–26 m. tall, girth 1.8–13 m., " Huang kuo shu ").

According to the international rules Hamilton's name has to be used instead of that of Roxburgh's.

This is a common tree at low altitudes in the prefecture of Kiating Fu in western Szech'uan and also in the neighborhood of the Yangtsze River from the city of Sui Fu eastward to that of Patung in western Hupeh. It grows to a very large size and the finest trees I met with occur near the foot of the sacred Mt. Omei and are about 26 m. tall, the composite trunks measuring from 15.5 to 16 m. in circumference. The branches are massive and spread to form a broad rounded or flattened crown; the bark is dark gray and smooth and the leaves fall late in the winter or are retained until spring, when they fall as the season's growth commences. In Szech'uan the colloquial name for this tree is Huang-kou-shu and wayside shrines are commonly associated with it.

Pictures of this Fig will be found under Nos. 12, 29, 30, 39, 48, 305, 379 and 380 of the collection of my photographs and also in my *Vegetation of Western China*, Nos. 217, 218, 219, 220, 221, 222 and 223.   E. H. W.

Sect. SYCIDIUM King.

Ficus clavata Wallich, *Cat.* No. 4495 (nomen nudum) (1828); apud Miquel in *London Jour. Bot.* VII. 431 (1848). — King in *Ann. Bot. Gard. Calcutta*, I. 87, t. 111 (1888). — Hemsley in *Jour. Linn. Soc.* XXVI. 458 (1899). — Warburg in *Bot. Jahrb.* XXIX. 299 (1900).

Western Szech'uan: Mien-chu Hsien, roadsides, common, alt. 800 m., May 21, 1908 (No. 2799; bush 1.8 m. tall).

As far as I know this species from King's description and other specimens Wilson's number seems to represent *F. clavata*. The perigone of the female flowers consists of (3–)5 lanceolate lobes which are minutely ciliate. But there is another specimen of Wilson's from western China: roadside, alt. 600–1100 m., July 1903 (Veitch Exped. No. 4472; tree 10–13 m. tall), the leaves and fruits of which well agree with those of *F. clavata*, but the female flowers I have seen seem to have only 2 perigone lobes, which are not ciliate. I cannot refer this specimen to any other species hitherto mentioned from Szech'uan or Hupeh.

Ficus foveolata Wallich, var. **Henryi** King apud Oliver in *Hooker's Icon.* XIX. t. 1824 (1889).

Ficus foveolata Hemsley in *Jour. Linn. Soc.* XXVI. 460 (pro parte, non Wallich) (1899). — Warburg in *Bot. Jahrb.* XXIX. 299 (pro parte) (1900).

Western Hupeh: Fang Hsien, climbing over rocks, alt. 300–900 m., July 1907 (No. 2797); Ichang, *A. Henry* (No. 3552[a], type, ex Oliver).

Ficus heteromorpha Hemsley in *Hooker's Icon.* XXVI. t. 2533, 2534 (1897); in *Jour. Linn. Soc.* XXVI. 461 (1899). — Warburg in *Bot. Jahrb.* XXIX. 300 (1900).

Western Hupeh: Fang Hsien, woods, alt. 1300 m., September 1907 (No. 311; bush 1.2–1.8 m. tall); Ichang, alt. 300–900 m., thickets, June 1907 (No. 2795; bush 1.2–1.8 m. tall); Hsing-shan Hsien, woods, alt. 900–1300 m., May 6, 1907 (No. 2793; bush 1.8 m. tall); same locality, thickets, August 1907 (No. 2794; bush 2.1–2.4 m. tall). Kiangsi: near Kuling, thickets, common, alt. 900 m., August 1, 1907 (No. 1588; thin shrub 0.9–1.8 m. tall).

Ficus impressa Champion in *Hooker's Jour. Bot. & Kew Gard. Misc.* VI. 76 (1854). — Hemsley in *Jour. Linn. Soc.* XXVI. 463 (1899). — Warburg in *Bot. Jahrb.* XXIX. 300 (1900).

Ficus foveolata, var. *impressa* King in *Ann. Bot. Gard. Calcutta,* I. 134, t. 167 F (1888).

Western Hupeh: Changyang Hsien, alt. 900 m., September 1907 (No. 131; bush 0.5–0.9 m. tall). Kiangsi: near Kuling, roadsides, alt. 600 m., August 8, 1907 (No. 1587; bush 0.9–1.2 m. tall).

I am following Hemsley and Warburg in keeping this species separate from *F. foveolata.* Wilson's No. 131 agrees well with Henry's No. 6506 and A. von Rosthorn's No. 408 cited by these authors, while in No. 1587 the petioles are longer and the leaves more distinctly cuspidate.

Ficus pumila Linnaeus, *Spec.* 1060 (1753). — Thunberg, *Ficus,* 9 (1786). — King in *Ann. Bot. Gard. Calcutta,* I. 124, t. 158 (1888). — Hemsley in *Jour. Linn. Soc.* XXVI. 465 (1899). — Warburg in *Bot. Jahrb.* XXIX. 299 (1900).

Ficus stipulata Thunberg, *Ficus,* 8 (1786).

Western Hupeh: around Ichang, a common epiphyte on trees, alt. 300 m., February 18, 1908 (No. 2798). Kiangsi: Kiu-kiang plain, a common epiphyte on trees and buildings, August 2, 1907 (No. 1596).

URTICACEAE.

Determined by Camillo Schneider.

BOEHMERIA Jacq.

Boehmeria nivea Gaudichaud[1] in Freycinet, *Voy. Bot.* 499 (1826). — Hooker & Arnott, *Bot. Voy. Beechey*, 214 (1836). — Hooker in *Hooker's Jour. Bot. & Kew Gard. Misc.* III. 315, t. 8 (1851). — Blume, *Mus. Bot. Lugd.-Bat.* II. 210 (1856). — Weddel in *Ann. Sci. Nat.* sér. 4, I. 200 (1854); in *Arch. Mus. Paris*, IX. 380, t. 11, fig. 10–17 (*Monog. Urticac.*) (1856); in De Candolle, *Prodr.* XVI. pt. 1, 206 (1859). — Bentham, *Fl. Hongk.* 331 (1861). — Miquel in *Ann. Mus. Lugd.-Bat.* III. 131 (1867); *Prol. Fl. Jap.* 295 (1867). — Brandis, *Forest Fl. Ind.* 402 (1874). — Maximowicz in *Bull. Acad. Sci. St. Pétersbourg*, XXII. 249 (1876); in *Mél. Biol.* IX. 638 (1877). — Hooker f., *Fl. Brit. Ind.* V. 576 (1888). — Dippel, *Handb. Laubholzk.* II. 4 (1892). — Koehne, *Deutsche Dendr.* 140 (1893). — Wright in *Jour. Linn. Soc.* XXVI. 486 (1894). — Diels in *Bot. Jahrb.* XXIX. 303 (1900). — Schneider, *Ill. Handb. Laubholzk.* I. 246, fig. 159 a–h (1904). — Matsumura & Hayata in *Jour. Coll. Sci. Tokyo*, XXII. 385 (*Enum. Pl. Formos.*) (1906). — Koorders, *Exkursionsfl. Java*, II. 143 (1912). — Diels in *Not. Bot. Gard. Edinburgh*, VII. 269 (1912), 342 (*Pl. Chin. Forrest.*) (1913).

> *Urtica nivea* Linnaeus, *Spec.* 985 (1753). — Jacquin, *Hort. Vindob.* II. 78, t. 166 (1772).

Western Hupeh: wild and abundant around Ichang, 300–900 m., July 1907 (No. **3759**). Kiangsi: Kuling, rocky places, abundant, alt. 1200 m., August 1, 1907 (No. **1518**; bush 0.9–1.8 m. tall). Yunnan: "Tali Range, Lat. 25° 40′ N., alt. 8000–9000 ft.," June 11, 1896, *G. Forrest* (No. 4782).

This plant is much cultivated in Szech'uan for its fibre, which is an important article of commerce in the west and in other warm parts of China. Some account of the industry will be found in my *A Naturalist in Western China*, II. 82 (1913) and a picture of the plant under No. 0173 of the collection of my photographs.

E. H. W.

[1] Gaudichaud writes: *Boehmeria (Procris) nivea;* hence he is the first author of the combination.

DEBREGEASIA Gaudich.

Debregeasia longifolia Weddell in De Candolle, *Prodr.* XVI. pt. 1, 235²⁴ (1869). — Brandis, *Forest Fl. Brit. Ind.* 405 (1874). — Koorders, *Exkursionsfl. Java,* II. 149 (1912). — Rehder in Bailey, *Stand. Cycl. Hort.* II. 973 (1914).

Urtica longifolia Burmann, *Fl. Ind.* 197 (sphalm. 297) (1768).
Urtica muricata Heyne in *Wallich,* Cat. No. 4612ᵃ, ᶜ, ᵈ (nomen nudum) (1828), fide Weddell.
Urtica angustata Blume, *Bijdr.* 499 (1825), fide Weddell.
Debregeasia velutina Gaudichaud, *Voy. Bonite* Atl. t. 90 (1844–9?) et *Expl. Descr. Planch.* par Ch. d'Alleizette 158 (1866). — Weddell in *Arch. Mus. Paris,* IX. 460, t. 15ᵃ (1856). — Hooker f., *Fl. Brit. Ind.* V. 590 (1888). — André in *Rev. Hort.* 1896, 321, fig. 118. — *Gard. Chron.* ser. 3, XXXIX. 232, t. suppl. (1906). — Brandis, *Ind. Trees,* 618 (1906).
Conocephalus niveus Wight, *Icon. Pl. Ind. Or.* VI. t. 1959 (1853).
Missiessya velutina Weddell in *Ann. Sci. Nat.* sér. 4, I. 195 (1854).
Morocarpus longifolius Blume, *Mus. Bot. Lugd.-Bat.* II. 156 (1855). — Beddome, *Fl. Sylv. S. Ind.* II. Forest Man. Bot. p. ccxxvi. t. 26, fig. 5 (1874). — Kurz, *Forest Fl. Brit. Burma,* II. 428 (1877).
Morocarpus velutinus Blume, *Mus. Bot. Lugd.-Bat.* II. 156 (1855).
Debregeasia edulis Wright in *Jour. Linn. Soc.* XXVI. 492 (non Weddell) (1894), quoad plantas ex Hupeh et Yunnan. — E. Pritzel in *Bot. Jahrb.* XXIX. 305 (1900).

Western Hupeh: north and south of Ichang, alt. 700–1100 m., June 1907 (No. 21; bush 0.9–1.8 m. tall, much branched, fruits orange-red, edible); same locality, May 1907 (No. 21ᵃ; ♂). **Yunnan**: Mengtsze, forests, alt. 1800 m., *A. Henry* (Nos. 10536, 10536ᵇ; tree 3 m. tall); same locality, mountain forests to southeast, alt. 1600 m., *A. Henry* (No. 10702; shrub 2.1 m. tall); Szemao, forests, alt. 1300 m., *A. Henry* (No. 12387; tree 3.6 m. tall, flowers white); same locality, mountains to east, alt. 2400 m., *A. Henry* (No. 12387ᵃ; tree 6 m. tall).

INDIA. Assam: " Kohima, 4500 ft.," March 1896, *King's Collector* (No. 173).

Wilson's plant, which has been in cultivation in the Arnold Arboretum, is somewhat intermediate between *D. longifolia* and *D. edulis* Weddell, of which the synonyms are given below. It is the same as Henry's 10536, of which the leaves are also rather or quite smooth on the upper side, while the pubescence of the young branchlets and petioles is not so hirsute and spreading as in the typical form and not so short and appressed as in the true *D. edulis.* I am not quite sure if it is possible to distinguish a distinct variety intermediate between the typical *longifolia* and what is now called *D. edulis* from Japan and probably eastern China. The other numbers of Henry's from Yunnan agree well with the Indian forms. They have a spreading villous or hirsute pubescence on the branchlets and petioles and the upper side of the leaves is more or less scabrid-hairy. It needs further investigation to decide if the flowers of these forms afford any good characters, but with the material before me I am inclined to believe that the Japanese plant represents only a variety of *D. longifolia.*

Debregeasia edulis Weddell in *Arch. Mus. Paris,* IX. 462 (1856); in De Candolle, *Prodr.* XVI. pt. 1, 235[24] (1869). — Maximowicz in *Bull. Acad. Sci. St. Pétersbourg,* XXII. 256 (1876); in *Mél. Biol.* IX. 649 (1877). — ?Hance in *Jour. Bot.* XX. 38 (1882). — Franchet & Savatier, *Enum. Pl. Jap.* I. 442 (1875). — Wright in *Jour. Linn. Soc.* XXVI. 492 (pro parte) (1894). — Matsumura & Hayata in *Jour. Coll. Sci. Tokyo,* XXII. 390 (*Enum. Pl. Formos.*) (1906). — Matsumura, *Ind. Pl. Jap.* II. pt. 2, 43 (1912). — Rehder in Bailey, *Stand. Cycl. Hort.* II. 973 (1914).

Morocarpus edulis Siebold & Zuccarini in *Abh. Akad. Münch.* IV. pt. 3, 218 (*Fl. Jap. Fam. Nat.* II. 94) (1846), exclud. planta ♂. — Blume, *Mus. Bot. Lugd.-Bat.* II. 155, t. 16ᵃ (1855). — Miquel in *Ann. Mus. Lugd.-Bat.* III. 130 (1867); *Prol. Fl. Jap.* 294 (1867).

Missiessya parvifolia Weddell in *Ann. Sci. Nat.* sér. 4, I. 195 (1854), fide Weddell.

JAPAN. Hondo: prov. Musashi, Tokyo, Botanic Garden, June 18, 1871 (ex Herb. Univ. Tokyo).

In the typical *D. edulis* the branchlets are described by Weddell as "pube brevi adpressa vestitis," as it is the case in the specimen before me. See my remarks above.

LORANTHACEAE.

Determined by HENRI LECOMTE.

LORANTHUS L.

A. Flores tetrameri, gamopetali.

Loranthus Balfourianus Diels in *Not. Bot. Gard. Edinburgh*, V. 250 (*Pl. Chin. Forrest.*) (1912).

> *Phyllodesmis Delavayi* Van Tieghem in *Bull. Soc. Bot. France*, XLII. 241 (1895).

Western Szech'uan: southeast of Tachien-lu, parasitic on Viburnum and various Willows, alt. 2800–3000 m., June 1908 (No. 3526; flowers orange and gold); without precise locality, parasitic on Willows, alt. 2500 m. (Veitch Exped. No. 4485; flowers scarlet).

A picture of this plant will be found under No. 630 of the collection of Wilson's photographs.

Loranthus caloreas Diels in *Not. Bot. Gard. Edinburgh*, V. 251 (1912).

Yunnan: Lichiang Range, alt. 3000–4000 m., July 1906, *G. Forrest* (No. 2600).

Loranthus caloreas, var. **oblongifolius** Lecomte, *Not. Syst.* III. 49 (1916).

Western Hupeh: parasitic on *Pterocarya stenoptera* De Candolle, alt. 300–1000 m., November 1907 (No. 3523).

The colloquial Chinese name of this plant is Sang-ch'e-sung.
Here may be added a variety not collected during the Arnold Arboretum Expeditions.

Loranthus caloreas, var. **Fargesii** Lecomte, *Not. Syst.* III. 49 (1916).
Western Szech'uan: parasitic on *Abies Deissneriana* Rehder & Wilson, alt. 300 m. (Veitch Exped. No. 4441).

Loranthus yadoriki Siebold & Zuccarini, var. **hupehanus** Lecomte, n. var.

A typo recedit foliis subtus cinereo-tomentosis.

Western Hupeh: north and south of Ichang, a pest on Oaks and many other trees, alt. 300–1200 m., May 1907 (No. 3521). Szech'uan: Nanch'uan, *A. von Rosthorn* (No. 2443).

The Chinese colloquial name of this plant is Yui-pao.

A picture of this plant will be found under No. 474 of the collection of Wilson's photographs.

Loranthus sutchuenensis Lecomte, *Not. Syst.* III. 167 (1916).

Western Hupeh: Ichang, parasitic on many trees, on deciduous Oaks in particular, alt. 30–600 m., April 1907 (No. **3522**).

B. Flores hexameri dialypetali.

Loranthus Delavayi Van Tieghem in *Bull. Soc. Bot. France*, XLI. 535 (nomen nudum) (1894); Lecomte, *Not. Syst.* III. 168 (1916).

?Loranthus Hemsleyanus King in *Jour. Linn. Soc.* XXVIII. 120 (1890).
Loranthus odoratus Hemsley in *Jour. Linn. Soc.* XXVI. 406 (non Wallich) (1894).
Loranthus odoratus Wallich, f. *Hemsleyanus* E. Pritzel in *Bot. Jahrb.* XXIX. 305 (1900).

Western Hupeh: Patung Hsien, parasitic on *Quercus variabilis* Blume, alt. 300–1000 m., December 1907 (No. **3524**); Ichang, *A. Henry* (No. 7849). Yunnan: parasitic on Oaks, alt. 2000 m., *A. Henry* (No. 9112); Chao-chio Valley, alt. 2300 m., March 1905, *G. Forrest* (No. 540).

Loranthus Delavayi, var. latifolius Van Tieghem in *Bull. Soc. Bot. France*, XLI. 535 (1894).

Yunnan: alt. 2700 m., *A. Henry* (No. 12892).

Here may be added notes on some species not collected during the Arnold Arboretum Expeditions. The first five species belong to the group with tetramerous gamopetalous flowers; the last species belongs to a group with pentamerous dialypetalous flowers.

Loranthus chinensis De Candolle, *Coll. Mém.* VI. 28, t. 7 (1830); *Prodr.* IV. 301 (1830). — Bentham, *Fl. Hongk.* 141 (1861). — Maximowicz in *Bull. Acad. Sci. St. Pétersbourg*, XXII. 230 (1876); in *Mél. Biol.* IX. 611 (1876). — Hemsley in *Jour. Linn. Soc.* XXVI. 405 (1894).

Western Szech'uan: without precise locality, parasitic on various trees (Veitch Exped. No. 4486).

Loranthus chinensis, var. formosanus Lecomte, n. var.

A typo recedit foliis basi attenuatis, fructibus basi paullo stipitatis.

Formosa: South Cape, *A. Henry* (No. 979).

Loranthus estipitatus Stapf in *Trans. Linn. Soc.* ser. 2, IV. 221 (1894). — Hemsley in *Jour. Linn. Soc.* XXVI. 405 (1894).

Loranthus estipitatus, var. longiflorus Lecomte, n. var.

A typo recedit floribus usque 3.5 cm. longis, antheris locellatis 5 mm. longis.

Yunnan: *A. Henry* (No. 10057; flowers scarlet).

Loranthus ferrugineus Roxburgh, *Fl. Ind.* II. 207 (1824). — De Candolle, *Prodr.* IV. 299 (1830). — Hooker f., *Fl. Brit. Ind.* V. 210 (1886).

Dendrophthoe ferruginea G. Don, *Gen. Syst.* III. 420 (1834). — Miquel, *Fl. Ind. Bat.* I. pt. 1, 812 (1855).
Ciclanthus ferrugineus Van Tieghem in *Bull. Soc. Bot. France,* XLII. 253 (1895).

Yunnan: Mi-lê district, on Oaks, *A. Henry* (No. 9947).

Loranthus philippensis Chamisso & Schlechtendal in *Linnaea,* III. 204 (1821).
Yunnan: Mengtsze, alt. 1500 m., on *Ehretia corylifolia* C. H. Wright, *A. Henry* (No. 9991).

Loranthus sootepensis Craib in *Kew Bull. Misc. Inform.* 1911, 454. — Lecomte, *Fl. Gén. Indo-Chine,* V. 192 (1916).
Yunnan: Mengtsze, alt. 1400–2000 m., *A. Henry* (No. 10315).

Loranthus adpressus Lecomte, *Not. Syst.* III. 53 (1916).

Leucobotrys adpressa Van Tieghem in *Bull. Soc. Bot. France,* 1894, 503 (nomen nudum).

Yunnan: *A. Henry* (No. 11664).

ELYTRANTHE G. Don.

Elytranthe ampullacea G. Don, *Gen. Syst.* III. 425 (1834).

? *Loranthus cochinchinensis* Loureiro, *Fl. Cochin.* 195 (1790).
Loranthus ampullaceus Roxburgh, *Fl. Ind.* II. 189 (1824). — De Candolle, *Prodr.* IV. 216 (1830). — Hooker f., *Fl. Brit. Ind.* V. 220 (1886). — Kurz, *Forest Fl. Brit. Burma,* II. 316 (1877). — Hemsley in *Jour. Linn. Soc.* XXVI. 405 (1894).

Yunnan: Mengtsze, *A. Henry* (No. 11473).

Elytranthe ampullacea, var. **tonkinensis** Lecomte, *Not. Syst.* III. 99 (1916).

Western Szech'uan: near Sui Fu, parasitic on *Aleurites Fordii* Hemsley, alt. 3000 m., April 1908 (No. **3525**; flowers orange-scarlet). Yunnan: Szemao, on trees and shrubs, alt. 2000 m., *A. Henry* (Nos. 11755, 11755[b]; flowers yellow).

Here may be added notes on two species collected by A. Henry in Yunnan.

Elytranthe bibracteolata Lecomte, n. comb.
Loranthus bibracteolatus Hance in *Jour. Bot.* XVIII. 301 (1880). — Hemsley in *Jour. Linn. Soc.* XXVI. 405 (1894).

Elytranthe bibracteolata, var. **sinensis** Lecomte, n. var.
A typo recedit foliis usque 12 cm. longis et 5 cm. latis, petiolis alatis 7–8 mm. longis.
Yunnan: alt. 1600 m., *A. Henry* (No. 12631; flowers pink).

Elytranthe Henryi Lecomte, n. sp.
Rami teretes, longitudinaliter bicarinati; ramuli compressiusculi, glabri. Folia opposita, subcoriacea, glabra; petiolus 2.25–2.75˙ cm. longus, basi articulatus,

versus apicem paulatim leviter alatus; limbus 12–14 cm. longus, 4.5–5.5 cm. latus, ovalis v. oblongus, basi inaequaliter rotundatus, paulum decurrens, apice acuminatus, acumine obtuso; costa subtus valde prominens; nervi 10–12-jugi, irregulares, subtus vix prominentes. Pedunculi crassi, 1 cm. longi, solitarii, axillares, basi bracteis 2 parvis adpressis instructi; flores geminati ad basim eminentia intermedia duriuscula compressaque 1 cm. alta separati; flos unusquisque 3-bracteatus; bractea externa 1 (antica) ovata, apice acuta, crassa, duriuscula; bracteolae internae 2, laterales, imbricatae, oblongae, intus concavae, apice obtusae, duriusculae, dorso carinatae; calyx bracteis lateralibus duplo brevis, limbo alto ore integerrimo, basi ovario adnatus; corollae tubus ruber, subclavatus, usque 3.5–4 cm. longus, glaber, paulum supra calycem constrictus, tunc gradatim ampliatus, valde curvatus; corollae segmenta 6, angusta, apice acuta, tandem reflexa, 1.5–1.8 cm. longa; stamina 6, filamentis supra medium liberis, ima basi incrassatis, antheris 4.5 mm. longis, terminalibus, angustis, apice subacutis; pollen stellatus; ovarium basi adhaerens, apice liberum, extus 6-costatum; stylus subulatus stamina superans stigmate capitato instructus. Fructus ignotus.

Yunnan: Szemao, alt. 1600 m., on trees, *A. Henry* (No. 11604; flowers red).

This new species resembles *Elytranthe albida* Blume figured in his *Flora Javae*, III. Lorantheae, t. 22, but it differs: 1° in its larger oblong, not oval leaves, 2° in its longer petiole always exceeding 2 cm. in length, 3° in the prominent blade separating the two flowers at the base, which becomes narrower toward the apex, while in Blume's figure it becomes narrower toward the base, 4° in the flowers which are red according to the collector's note and not white, 5° in the calyx which bears below its upper margin a slight circular excrescence. This plant cannot be confounded with *Loranthus leucosiphon* Griffith (*Notul.* IV. 623 [1854]; *Icon. Pl. Asiat.* t. 619, 621, 622, 623, [1854]), which has been referred to *E. albida* by different authors, particularly by Gamble, but which differs markedly in the crest separating the flowers.

VISCUM, L.

Viscum album Linnaeus, *Spec.* 1023 (1753). — De Candolle, *Prodr.* IV. 278 (1830). — Boissier, *Fl. Orient.* IV. 1068 (1879). — Brandis, *Forest Fl. Brit. Ind.* 392 (1874). — Kurz, *Forest Fl. Brit. Burma*, II. 323 (1877). — Hemsley in *Jour. Linn. Soc.* XXVI. 407 (1894).

Western Hupeh: Ichang, alt. 30–150 m., parasitic on *Pterocarya stenoptera* De Candolle, Dec. 23, 1907 (No. **796**; fruits golden yellow); without precise locality, *A. Henry* (No. 7883).

A picture of this plant will be found under No. 720 of the collection of Wilson's photographs.

Viscum articulatum Burman f., *Fl. Ind.* 311 (1768). — De Candolle, *Prodr.* IV. 284 (1830). — Hooker f., *Fl. Brit. Ind.* V. 226 (1886). — Hemsley in *Jour. Linn. Soc.* XXVI. 407 (1894).

Viscum moniliforme Blume, *Bijdr.* 667 (1825); *Fl. Jav.* III. Loranth. t. 25[b] (1851).

Western Hupeh: Ichang, growing on *Dalbergia hupeana* Hance, March 1907 (No. **3262**); Patung Hsien, alt. 800 m.; on *Diospyrus kaki*

Linnaeus, April 1907 (No. 3263); without precise locality, *A. Henry*
(No. 3206). Szech'uan: Yangtsze River banks, alt. 30–300 m.,
parasitic on *Aleurites Fordii* Hemsley, April 1908 (No. 2709); same
locality and same host, May 1903 (Veitch Exped. No. 4482); without precise locality, cliffs, July 1903 (Veitch Exped. No. 4484).
Yunnan: Mi-lê district, on Zanthoxylum, *A. Henry* (No. 9942).

A picture of this plant will be found under No. 0164 of the collection of Wilson's
photographs.
Here may be added the description of a new species collected by *A. Henry*.

Viscum stipitatum Lecomte, n. sp.
Frutex parasiticus; rami teretes, dichotomi; ramuli tenuiter longitudinaliterque
striati; internodia basi articulata, usque 5–6 cm. longa. Folia opposita, lanceolata,
utrinque attenuata, saepe falciformia, membranacea v. subcoriacea, basi articulata, apice subacuta, 6–7.5 cm. longa, 1.3–2 cm. lata, costis 5–7 utrinque prominentibus. Flores monoici; triades axillares, terminalesque; pedunculus communis
3–5 mm. longus, apice 2-bracteatus, bracteis triangularibus parvis; flos medius
semper ♀; flores laterales 2 saepe ♀ interdum ♂; flos ♂ ovoideus, sessilis, 0.75
mm. longus, lobis 4 triangularibus instructus; antherae petalis adnatae, transversaliter loculosae; flos ♀ ovoideus, primo sessilis lobis 4 triangularibus parvis
instructus; stamina 0; ovarium inferum, stigmate parvo cylindrico instructum;
perianthii lobi decidui. Fructus baccatus, ovoideus, demum longe stipitatus,
usque 9 mm. latus, sed maturus claviformis, 1-spermus. Semen ovoideum;
albumen carnosum; embryo verticalis, teres.
Yunnan: Szemao, alt. 1600 m., on trees, *A. Henry* (No. 12758).
This species is well characterized by its long-stipitate fruits and slender and
often falcate leaves.

SANTALACEAE.

Determined by Camillo Schneider.

OSYRIS L.

Osyris Wightiana Wallich, *Cat.* No. 4036 (nomen nudum) (1829); apud Wight, *Icon. Pl. Ind. Orient.* V. 17, t. 1853 (1852). — Diels in *Not. Bot. Gard. Edinburgh*, VII. 40, 265 (*Pl. Chin. Forrest.*) (1912).

> *Osyris arborea* Wallich, *Cat.* No. 4035 (nomen nudum) (1829); apud A. De Candolle, *Prodr.* XIV. 633 (1857). — Hooker f., *Fl. Brit. Ind.* V. 232 (1886).

Western Szech'uan: valley of Tung River near Tachien-lu, alt. 1000–1300 m., October 1908 (No. **4138**; bush 0.9–1.2 m. tall, fruits orange); same locality, October 1910 (No. **4138ᵃ**; bush 0.9–1.2 m. tall, fruits red); Monkong Ting, arid regions, alt. 1200–1700 m., June 30, 1908 (No. **3207**; bush 1.2–1.8 m. tall, flowers green, fruits red); without precise locality, cliffs, June 1902 (Veitch Exped. No. 4443; bush 0.5–0.75 m., flowers yellow). Southern Szech'uan: near Lo-ma-pu, dry mountainsides, alt. about 2200 m., May 9, 1914, *C. Schneider* (No. 1148; bush 1 m. tall). Yunnan: Tali Range, alt. 1800–2500 m., July 1906, *G. Forrest* (No. 4745; shrub 7–18 m. tall, flowers brownish green); same locality, October 1914, *C. Schneider* (No. 2597; shrub about 1 m. tall); dry open situations in thickets in most valleys between Tali Fu and Yunnan Fu, alt. 1800–2300 m., January 1905, *G. Forrest* (No. 522); Mengtsze, barren hills, alt. 1600 m., *A. Henry* (Nos. 9906, 9906ᵃ; shrub 0.75–1.5 m., flowers yellow).

OLACACEAE.

Determined by CAMILLO SCHNEIDER.

SCHOEPFIA Schreb.

Schoepfia jasminodora Siebold & Zuccarini in *Abh. Akad. Münch.* IV. pt. 3, 135 (*Fl. Jap. Fam. Nat.* II. 11) (1846). — Hemsley in *Jour. Linn. Soc.* XXVI. 114 (1886). — Diels in *Bot. Jahrb.* XXIX. 306 (1900); in *Not. Bot. Gard. Edinburgh*, VII. 30, 92 (*Pl. Chin. Forrest.*) (1912).

Schoepfiopsis jasminodora Miers in *Jour. Linn. Soc.* XVII. 77 (1878).

Western Hupeh: north and south of Ichang, thickets, alt. 100–900 m., June 1907 (No. **58**; small tree 4.5 m. tall, fruits orange); same locality, May 1907 (No. **58ᵃ**; bush or small tree, 1.8–6 m. tall, flowers white, fragrant); Hsing-shan Hsien, alt. 700 m., May 10, 1907 (No. **58ᵇ**; erect bush 7 m. tall, flowers greenish yellow, fragrant); without precise locality, June 1901 (Veitch Exped. No. 1929); without precise locality, *A. Henry* (No. 5975). Szech'uan: without precise locality, cliffs, alt. 1300 m., October 1904 (Veitch Exped. No. 4450; shrub 1.2–1.8 m. tall, fruits orange); Wushan Hsien, *A. Henry* (Nos. 5597, 5597ᵇ). Chekiang: vicinity of Ningpo, 1908, *D. Macgregor;* near Hanchou, "Shi-bun-shan," June 26, 1915, *F. N. Meyer* (No. 1478; shrub 1.5 m. tall). Yunnan: open situations in pine forests on the southern foothills of the Lichiang Range, alt. 2700–3000 m., May 1906, *G. Forrest* (No. 2054; shrub or tree 4.5–12 m. tall, flowers green, powerfully fragrant, appearing before foliage is developed, fruit an ovoid dark crimson drupe); same locality, August 1914, *C. Schneider* (No. 2173; fruits orange red); Mi-lê district, *A. Henry* (No. 10605; shrub 2.4 m. tall, flowers yellowish).

This species is distinguished by the sessile or almost sessile flowers. *Schoepfia chinensis* Gardner & Champion, of which I have been able to see but one specimen, is similar, but the flowers are almost twice as large and the leaves seem to be narrower and longer than those of *S. jasminodora*.

Here may be added the following species not collected during the Arnold Arboretum Expeditions.

Schoepfia fragrans Wallich in Roxburgh, *Fl. Ind.* II. 188 (1824); *Tent. Fl. Nepal.* I. 18, t. 9 (1824). — A. P. De Candolle, *Prodr.* IV. 320 (1830). — Hooker f., *Fl. Brit. Ind.* I. 580 (1875).

Schoepfiopsis fragrans Miers in *Jour. Linn. Soc.* XVII. 76 (1878).

CHINA. Yunnan: Szemao, eastern forests, alt. 1600 m., *A. Henry* (No. 12660; tree 7 m. tall); same locality, hills, alt. 1600 m., *A. Henry* (No. 12274; shrub 1.8 m. tall, white fragrant flowers); same locality, mountains to south, alt. 1400 m., *A. Henry* (Nos. 12274[a], 12274[b]; shrub 3–4.5 m. tall, flowers white, fragrant); same locality, mountains, alt. 1600 m., *A. Henry* (No. 12274[c]; shrub 1.8 m. tall, pale yellow fragrant flowers).

This species, which was first found in the eastern Himalaya (Nepal, Upper Bengal), is easily distinguished by the rather long pedicels of its flowers and by the somewhat longer leaves which are pointed towards the ends.

ARISTOLOCHIACEAE.

Determined by ALFRED REHDER and E. H. WILSON.

ARISTOLOCHIA L.

Aristolochia debilis Siebold & Zuccarini in *Abh. Akad. Münch.* IV. pt. 3, 197 (*Fl. Jap. Fam. Nat.* II. 73) (1846). — Duchartre in De Candolle, *Prodr.* XV. pt. 1, 483 (1864). — Miquel in *Ann. Mus. Lugd.-Bat.* II. 135 (1865–66); *Prol. Fl. Jap.* 67 (1866). — Franchet & Savatier, *Enum. Pl. Jap.* I. 420 (1875). — Hemsley in *Jour. Linn. Soc.* XXVI. 361 (1891). — Diels in *Bot. Jahrb.* XXIX. 310 (1900). — Pampanini in *Nuov. Giorn. Bot. Ital.* n. ser. XVII. 258 (1910).

> *Aristolochia longa* Thunberg, *Fl. Jap.* 144 (non Linnaeus) (1784).
> *Aristolochia Sinarum* Lindley in *Gard. Chron.* 1859, 708. — Duchartre in De Candolle, *Prodr.* XV. pt. 1, 497 (1864). — Hemsley in *Jour. Linn. Soc.* XXVI. 363 (1891).
> *Aristolochia recurvilabra* Hance in *Jour. Bot.* XI. 75, fig. (1873); XVIII. 301 (1880). — Franchet in *Nouv. Arch. Mus. Paris*, sér. 2, VII. 67 (*recurvilabris*) (*Pl. David.* I. 258) (1884).

Kiangsi: Kiukiang, roadside thickets, alt. 100 m., July 29, 1907 (No. **1507**; climber 3 m., flowers dark). Western Hupeh: Ichang, grassy hills, alt. 30–300 m., September 1907 (No. **1986**; flowers dark).

Very common at low altitudes round Kiukiang, rare at Ichang.

Aristolochia thibetica Franchet in *Jour. de Bot.* XII. 313 (1898).

Western Szech'uan: Tachien-lu, thickets, alt. 2300 m., August 1908 (No. **1985**; climber 2 m., flowers with chrome-yellow lip and bronze-colored throat).

On our specimen most of the leaves are obovate-oblong to panduriform; otherwise it agrees well with Franchet's description. The fruit is smaller and has more prominent and wider wings than that of *A. heterophylla* Hemsley, and the short petioles and different-shaped corolla also serve to distinguish it from Hemsley's species. The plant is rather rare.

Aristolochia heterophylla Hemsley in *Jour. Linn. Soc.* XXVI. 361 (1891). — Diels in *Bot. Jahrb.* XXIX. 310 (1900). — Bean, *Trees & Shrubs Brit. Isl.* I. 207 (1914).

> *Aristolochia setchuenensis* Franchet in *Jour. de Bot.* XII. 312 (1898). — Diels in *Bot. Jahrb.* XXIX. 310 (1900).

Western Hupeh: Hsing-shan Hsien, thickets, alt. 1300–1600 m., June and October 1907 (No. 367; climber 2–2.5 m., flowers dark); Chienshi Hsien, thickets, alt. 1300–1600 m., June and October, 1900 (Veitch Exped. No. 925, Seed No. 538); without locality, *A. Henry* (No. 6417). Eastern Szech'uan: Taning Hsien, thickets, alt. 1600 m., June 1910 (No. 4564; climber 2–3 m., flowers chocolate and gold); Wushan Hsien, *A. Henry* (No. 5620).

This species is common in western Hupeh and eastern Szech'uan in thickets on mountain slopes, but has not yet been reported from other parts of China. The leaves are extraordinarily variable in size and shape, but the nature and degree of pubescence are very constant and the species is a well-marked one. The capsule is conspicuously 6-costate, as Henry's No. 6417 clearly shows, and we suppose Hemsley's description of the capsule as " inconspicue 6-costata " is a slip of the pen.

Aristolochia moupinensis Franchet in *Nouv. Arch. Mus. Paris,* sér. 2, X. 79 (*Pl. David.* II. 117) (1887); in *Jour. de Bot.* XII. 314 (1898). — Diels in *Bot. Jahrb.* XXIX. 310 (1900); in *Not. Bot. Gard. Edinburgh,* VII. 251 (*Pl. Chin. Forrest.*) (1912). — Wright in *Bot. Mag.* CXXXVI. t. 8325 (1910). — Bean, *Trees & Shrubs Brit. Isl.* I. 207 (1914).

Western Szech'uan: Mupin, thickets, alt. 2000–2500 m., July 1908 (No. 1982; climber 3–4 m., flowers brownish); Chiuting-shan, alt. 1600 m., May 22, 1908 (No. 1983; climber 2.5–3 m., flowers brownish); Wa-shan, thickets, alt. 2000–2600 m., July and October 1903 (Veitch Exped. No. 4481; Seed No. 1423); same locality, *A. E. Pratt* (No. 3).

This is a very common climber in thickets and on the margins of woods in extreme western Szech'uan. Wright suggests that this and *A. Kaempferi* Willdenow may be the same, but the flowers and fruit of the latter species are very much smaller and the two plants are dissimilar in appearance.

Aristolochia mollissima Hance in *Jour. Bot.* XVII. 300 (1879). — Hemsley in *Jour. Linn. Soc.* XXVI. 362 (1891). — Diels in *Bot. Jahrb.* XXIX. 310 (1900). — Pampanini in *Nuov. Giorn. Bot. Ital.* n. ser. XVII. 258 (1910).

Aristolochia sechtuenensis [sic], var. *holotricha* Pavolini in *Nuov. Giorn. Bot. Ital.* n. ser. XV. 437 (non Diels) (1908).

Western Hupeh: Ichang, grassy hills, alt. 30–300 m., April 1907 (No. 1984; slender twiner 0.5–1 m., flowers yellow); same locality, *A. Henry* (No. 4144[a]).

Very common at low altitudes around Ichang.

POLYGONACEAE.

Determined by CAMILLO SCHNEIDER.

POLYGONUM L.

Polygonum multiflorum Thunberg, *Fl. Jap.* 169 (1784). — Meisner, *Mon. Gen. Polygon.* 64, t. 4, fig. Q (1826); in De Candolle, *Prodr.* XIV. 136 (1856). — Hemsley in *Jour. Linn. Soc.* XXVI. 342 (1891). — Dammer in *Bot. Jahrb.* XXIX. 314 (1900). — Schneider, *Ill. Handb. Laubholzk.* I. 259, fig. 169 i–m (1904).

Western Hupeh: Ichang, alt. 100–900 m., September and October 1907 (No. **437**; climber 7–10 m. tall, flowers white); same locality, *A. Henry* (No. 2343); without precise locality, *A. Henry* (Nos. 82, 2488). Formosa: 1864, *R. Oldham* (No. 436).

This species differs from the shrubby species of western Szech'uan, *P. Aubertii* L. Henry (in *Rev. Hort.* 1907, 82, fig. 23, 24), which was raised from seeds collected by Père Aubert near Tachien-lu, in its much more branched inflorescence the axes of which bear a very dense fine glandular pubescence at the edges, in the glabrous filaments, the smaller flowers and in the broader-winged fruits. *P. Aubertii*, of which I know only cultivated specimens, is more closely related to *P. baldschuanicum* Regel (in *Act. Hort. Petrop.* VIII. 684, t. 10 [1883]), but in the last species the axes of the inflorescence are wholly glabrous, while in *P. Aubertii* there is a fine glandular pubescence somewhat like that of *P. multiflorum* but not so dense. The filaments of both the other species are finely pubescent toward their base and the wings of the larger fruits are narrower compared with their size.

Here may be mentioned a *Rumex* not collected during the Arnold Arboretum Expeditions which is a very common woody plant in southern Szech'uan and northern Yunnan.

Rumex hastatus D. Don, *Prodr. Fl. Nepal.* 74 (1825), fide Meisner in Wallich, *Pl. As. Rar.* III. 04 (1832); in De Candolle, *Prodr.* XIV. 72 (1856). — Hooker f., *Fl. Brit. Ind.* V. 60 (1886). — Collett, *Fl. Siml.* 428, fig. 136 (1902).

Northern Yunnan: on the walls of the city of Hoching, September 25, 1914, *C. Schneider* (No. 2482; densely branched erect or procumbent shrub 0.3–0.8 m. tall); dry stony situations on the eastern flank of the Tali Range and on the walls of the city of Tali Fu, October 8, 1914, *C. Schneider* (No. 2880; fruits reddish, showy).

This species is a common plant in the Yangtsze Valley of southern Szech'uan; I collected it in flower, in May 1914, between Yunnan Fu and Ningyüan Fu, where it seems to reach the northern limit of its range.

PITTOSPORACEAE.

Determined by ALFRED REHDER and E. H. WILSON.

PITTOSPORUM Banks.

Pittosporum daphniphylloides Hayata in *Jour. Coll. Sci. Tokyo,* XXX. art. 1, 34 (1911).

Western Szech'uan: Wa-shan, cliffs, alt. 1000 m., October 1908 (No. 1219; bush 2 m. high); Chiuting-shan, cliffs, alt. 1500 m., May 22, 1908 (No. 1219ᵃ; bush 2–3 m., flowers greenish yellow); without locality, ravine, alt. 1300 m., July and October 1903 (Veitch Exped. No. 3233; tree 8 m. tall); Mt. Omei, June 1904 (Veitch Exped. No. 4745).

This handsome species is closely related to *P. napaulensis* Rehder & Wilson [1] which differs in its thinner leaves, its umbellate-corymbose inflorescence and in its smaller fruits. We have not seen specimens from Formosa, but Hayata says that Wilson's No. 3233 in Herb. Kew is exactly like his new species.

In woods and thickets and rocky places at low altitudes in western Szech'uan *P. daphniphylloides* is rather common as a bush or small tree. The leaf without the petiole is sometimes 25 cm. long and 9 cm. wide.

Pittosporum glabratum Lindley in *Jour. Hort. Soc. London,* I. 230 (1846). — Walpers, *Ann.* I. 77 (1848). — Bentham, *Fl. Hongk.* 19 (1861). — Hooker f. & Thomson in Hooker f., *Fl. Brit. Ind.* I. 198 (1872). — Hemsley in *Jour. Linn. Soc.* XXIII. 58 (1886). — E.

---

[1] **Pittosporum napaulensis** Rehder & Wilson, n. comb.

Celastrus verticillatus Roxburgh, *Hort. Beng.* 18 (nomen nudum) (1814); *Fl. Ind.* I. 391 (1824).

Senacia Napaulensis De Candolle, *Prodr.* I. 347 (1824).

Pittosporum Wallich, *Cat.* No. 8127 (1832).

Pittosporum floribundum Putterlick, *Syn. Pittospor.* 9 (pro parte, non Wight & Arnott) (1839). — Hooker f. & Thomson in Hooker f., *Fl. Brit. Ind.* I. 199 (1872).

INDIA. East Himalaya: *W. Griffith* (No. 270 in Herb. Gray). Sikkim: "alt. 5–7000 ped." *J. D. Hooker* (in Herb. Gray). Assam: Khasia, "alt. 4–6000 ped.," *J. D. Hooker & T. Thomson* (in Herb. Gray).

The *P. floribundum* Wight & Arnott with which Putterlick and others have confused this plant is well distinguished by its inflorescence, which is an erect much branched paniculate-corymb. In the *Flora of British India* the description is based upon the two species.

Pritzel in *Bot. Jahrb.* XXIX. 378 (1900). — Gagnepain in Lecomte, *Fl. Gén. Indo-Chine,* I. 239 (1909). — Dunn & Tutcher in *Kew Bull. Misc. Inform.* add. ser. X. 37 (*Fl. Kwangtung & Hongk.*) (1912).

Pittosporum Fortunei Turczaninow in *Bull. Soc. Nat. Mosc.* XXXVI. pt. 1, 562 (1863).
Pittosporum glabratum, var. *angustifolium* E. Pritzel in *Bot. Jahrb.* XXIX. 378 (nomen nudum) (1900).
Pittosporum pauciflorum, var. *brevicalyx* E. Pritzel l. c. (non Oliver) (1900).
Pittosporum glabratum, var. *chinense* Pampanini in *Nuov. Giorn. Bot. Ital.* n. ser. XVII. 285 (1910).

Kiangsi: Kuling, side of streams, alt. 1300 m., July 28, 1907 (No. 1674; bush 1–2 m. high). Western Hupeh: north and south of Ichang, woods, thickets and cliffs, alt. 30–1000 m., May 11, November 1907 (No. 657; bush 1.5–2.5 m. high, flowers yellow); same locality, *A. Henry* (Nos. 3387ᵃ, 3414); Patung Hsien, thickets, alt. 600–1000 m., May, October 1907 (Nos. 3180, 456, bush 1–2.5 m. high, flowers yellow); same locality, April 1900 (Veitch Exped. No. 279); Chienshi Hsien, woods, June, October 1900 (Veitch Exped. No. 886); without locality, November 1900 (Veitch Exped. Nos. 279ᵃ, 583ᵃ); "lungo il fiume Jang-se-kiang," alt. 700 m., June 20–30, 1907, *C. Silvestri* (No. 874, co-type of *P. glabratum,* var. *chinense* Pampanini); without locality, *A. Henry* (Nos. 4369, 7850ᵃ ex Herb. Kew as *P. pauciflorum forma,* 5999 ex Herb. Kew as *P. glabratum,* var. *angustifolium*). Shensi: "Thui-kio-Tsuen," September 25, 1897, *G. Giraldi.* Eastern Szech'uan: Wushan Hsien, May 1900 (Veitch Exped. No. 583); same locality, *A. Henry* (No. 7042). Wan Hsien, banks of Yangtsze River, alt. 300 m., April 1908 (No. 3178; bush 1–2 m. high, flowers yellow); Fu-chou, banks of Yangtsze River, alt. 300 m., April 1908 (No. 3177; bush 5 m. high, flowers deep yellow, fragrant); Nanch'uan, *A. von Rosthorn* (Nos. 1933, 2074); Nanch'uan, Pên-sha-ai, August 1891, *A. von Rosthorn* (No. 674 received as *P. pauciflorum,* var. *brevicalyx*). Western Szech'uan: Mt. Omei, June 1904 (Veitch Exped. No. 4744); same locality, 1899, *Hugh Scallan.* Yunnan: Mengtsze, mountains to north, alt. 1800–2300 m., *A. Henry* (Nos. 10551, 10553, 10191ᵃ); Yuanchiang, alt. 1600–2000 m., *A. Henry* (Nos. 13299, 13403). Chekiang: Changhua Hsien, woods, among rocks, alt. 300 m., July 11, 1915, *F. N. Meyer* (No. 1540). Fokien: Dunn's Exped., April to June 1905 (Hongk. Herb. No. 2369). Kwangtung: Lao-fou-shan, *C. Ford* (No. 592). Hongkong: *C. Wright* (No. 26).

INDIA. Manipur: 1881–82, *G. Watts* (No. 6538). East Bengal: *W. Griffith* (No. 271 in part). Assam: Khasia Hills, alt. 1600 m., *Hooker f. & Thomson.*

This is the common Pittosporum of China and is frequent in woods, thickets and shady places throughout the warmer parts from the sea-coast to the extreme west. The leaves vary much in size and in shape from obovate and obovate-oblong to oblanceolate-oblong and occasionally lanceolate-oblong. The pedicels are glabrous or furnished with few to many crisped and glandular hairs. The capsule is large and relatively thick and woody and is usually three-valved, but may vary from 2 to 4 on the same branch. The flowers vary in color from pale to deep yellow and are sweetly scented. The characters on which Pampanini bases his var. *chinense* are inconstant.

Pittosporum glabratum, var. **neriifolium** Rehder & Wilson n. var.

A typo recedit foliis anguste oblongo-lanceolatis v. anguste lanceolatis 10–15 cm. longis et 1–2 cm. latis, interdum ad 20 cm. longis et 2.6 cm. latis.

Western Hupeh: Patung Hsien, cliffs, alt. 600–1000 m., May, November 1907 (No. **3181**, type; bush 1.5–2 m. high, flowers yellow); without locality, May 1900 (Veitch Exped. No. 361); without locality, *A. Henry* (Nos. 5359, 5422). Western Szech'uan: without locality, May 1904 (Veitch Exped. Nos. 3231, 3231ᵃ); Mt. Omei, 1899, *Hugh Scallan.* Yunnan: Mentgsze, forests, alt. 1800–2300 m., *A. Henry* (Nos. 10191ᵇ, 10545) Hongkong: Little Hongkong, woods, April 12, 1909, *W. J. Tutcher* (ex Herb. Hongkong No. 9501).

INDIA. Assam: Khasia Hills, alt. 1600 m., *Hooker f. & T. Thomson.* East Bengal, *W. Griffith* (No. 271 pro parte).

In its typical form with its long and narrow oblong-lanceolate leaves, this plant looks sufficiently distinct to be regarded as a species, but the numbers Veitch Exped. 361, 3231, 3231ᵃ; Henry Nos. 5359, 10191ᵇ, and the Mt. Omei specimen collected by Father Hugh Scallan are more or less intermediate, and we do not think that it is entitled to rank higher than a variety. Its distribution is similar to that of the type with which it grows.

Pittosporum truncatum E. Pritzel in *Bot. Jahrb.* XXIX. 378 (1900).— Pampanini in *Nuov. Giorn. Bot. Ital.* n. ser. XVII. 286 (1910).

Pittosporum pauciflorum, var. *brevicalyx* Oliver in *Hooker's Icon.* XVI. t. 1579 (1887).

Pittosporum brevicalyx Gagnepain in *Bull. Soc. Bot. France,* LV. 545 (1908).

Western Hupeh: Ichang, cliffs of glens and ravines, alt. 30–600 m., May 3, November 1907 (No. **650**; bush 1–2.5 m., flowers yellow); same locality, April and November 1900 (Veitch Exped. No. 51); same locality *A. Henry* (No. 1524); without locality *A. Henry* (Nos. 1081, 3414, 7306). Eastern Szech'uan: Wushan Hsien, *A. Henry* (No. 5513 in Herb. Gray); Nanch'uan, *A. von Rosthorn* (No. 2071, co-type).

This is a very common shrub on the cliffs at low altitudes in western Hupeh and eastern Szech'uan, where it is known colloquially as Ai-hua-tzŭ. The typical form, with rhombic-obovate leaves attenuate at the base and often abruptly caudate-acuminate, is very well marked, but some specimens approach *P. heterophyllum* Franchet, and it may be that these species are not really distinct. Henry's specimens were distributed by the Kew herbarium as doubtful forms of *P. pauciflorum* Hooker & Arnott, but they are not this species, which is characterized by its few-flowered umbellate inflorescence, its oblong acute sepals and by its fruit, which is three times the size of that of *P. truncatum*. On E. Pritzel's type the leaves are somewhat malformed, but such leaves may often be found on specimens, as on Henry No. 5513. Gagnepain clearly defines the difference between this and *P. pauciflorum* Hooker & Arnott, but in taking up Oliver's varietal name he unfortunately overlooked that given by E. Pritzel.

Pittosporum heterophyllum Franchet in *Bull. Soc. Bot. France,* XXXIII. 415 (1886); *Pl. Delavay.* 76, t. 18 (1889). — Diels in *Not. Bot. Gard. Edinburgh,* VII. 30, 95 (*Pl. Chin. Forrest.*) (1912).

Western Szech'uan: Wên-ch'uan Hsien, valley of Min River, alt. 1000–1300 m., July and October 1908 (No. **1146**; bush 1–1.5 m., flowers yellow); Yachou Fu, thickets, alt. 330–600 m., November 1908 (No. **3179**; bush 1 m. high); without precise locality, cliffs, alt. 2000–3500 m., May, June 1904 (Veitch Exped. Nos. 3229, 3232). Yunnan: Lichiang Range, alt. 2800 m., May 1906, *G. Forrest* (No. 2084).

This pretty species is common on cliffs and in rocky places generally in the warm and arid valleys of western Szech'uan. The leaves are heteromorphic, but on some specimens they are rather similar in shape to those of *P. truncatum* Pritzel.

Pittosporum saxicola Rehder & Wilson, n. sp.

Frutex 0.3–2 m. altus, saepe prostratus; ramuli graciles, teretes, flavo- v. brunneo-cinerei, hornotini puberuli. Folia persistentia, coriacea, oblonga v. anguste elliptico-oblonga, obtusa, mucronata, basi cuneata v. attenuata, margine revoluta interdum leviter crenulata, 1–5, pleraque 2–3.5 cm. longa, et 0.3–0.9 cm. lata, glabra, supra nitidula, intense viridia, subtus pallidiora, nervis utrinsecus 4–8 supra ut costa leviter impressa subtus leviter elevatis; petioli 1–5 mm. longi, marginati. Flores lutei, fragrantes, solitarii v. plures in cymis umbelliformibus ramulos breves terminantibus; pedicelli 3–8 mm. longi, puberuli; bracteae membranaceae, lanceolatae, 1–2 mm. longae, caducae; sepala libera v. fere libera, ovata, acuta v. obtusa, 2–2.5 mm. longa, glanduloso-ciliolata; corolla tubuloso-campanulata, 7–8 mm. longa, petalis oblongis 2–2.5 mm. latis coherentibus et tubum formantibus, parte libera patente tubum dimidium subaequante; stamina petalis breviora sed tubum superantia, filamentis subulatis 4–4.5

mm. longis glabris, antheris luteis sagittatis; pistillum staminibus brevius, ovario leviter compresso glabro v. sparse pubescente, stylo crasso, stigmate simplice. Capsula subglobosa, compressa, 6–8 mm. alta, leviter verruculosa, stylo persistente coronata 2-valvis, 4-sperma.

Western Szech'uan: near Mao-chou, valley of Min River, alt. 1300–1800 m., May 26, 1908 (No. 3182, type; flowers); same locality, alt. 1000 m., August 1904 (Veitch Exped. No. 3230; co-type fruit).

This new species is most closely related to *P heterophyllum* Franchet, which is distinguished by its thinner, heteromorphic and larger leaves. It may also be compared with *P. humile* Hooker f. & Thomson, which has larger leaves, a racemose sub-umbellate inflorescence, shorter pedicels with longer bracts, and smaller flowers with sepals half as long as the corolla. *P. saxicola* is common in rocky places in arid parts of the valley of the Min River. Often it forms low scrub with stems prostrate on the ground, but under favorable conditions it is a bush 2 m. high and as much through. We are not sure that the fruit always contains four seeds only. Here may be added a note on a species from Formosa.

Pittosporum formosanum Hayata in *Jour. Coll. Sci. Tokyo*, XXII. 32, t. 4 (*Enum Pl. Formosan.*) (1906); in XXX. art. 1, 34 (1911).

Pittosporum sp. nova Henry in *Trans. Asiat. Soc. Japan*, XXIV. Suppl. 18 (*List Pl. Formosa*) (1896).

Formosa: Takow, Ape's Hill, *A. Henry* (Nos. 1058, 1070ª, 1888; shrub or small tree 1–3 m. high); same locality, *G. M. Playfair* (No. 52); Bankinsing, *A. Henry* (Nos. 48, 822); South Cape, *A. Henry* (Nos. 256, 971, 977).

We do not think that this species has any affinity with *P. pauciflorum* Hooker & Arnott as stated by Hayata. To us it appears most closely related to *P. Fernandezii* Vidal.

APOCYNACEAE.

Determined by Camillo Schneider.

MELODINUS Först.

Melodinus Hemsleyanus Diels in *Bot. Jahrb.* XXIX. 539 (1900).

Melodinus? sp. nov. Hemsley in *Jour. Linn. Soc.* XXVI. 94 (1889).

Eastern Szech'uan: Taning Hsien, alt. 650 m., July 1910 (No. 4563; climber over rocks, 1.5–3 m. long, flowers white); Mt. Omei, July 1904 (Veitch Exped. No. 5070; climber). Southern Szech'uan: "Nanch'uan, 'Kung chia p'ing,' Felswände, August," *A. von Rosthorn* (No. 511, type).

This species may be distinguished by the small and sometimes indistinct scales at the mouth of the corolla. It is closely related to the following species.

Here may be added the description of another species not collected during the Arnold Arboretum Expeditions.

Melodinus Seguini Léveillé in Fedde, *Rep. Spec. Nov.* II. 114 (descriptio satis incorrecta) (1906), fide specim. originalis; *Fl. Kouy-Tchéou*, 30 (1914).

Frutex ad 1 m. altus; ramuli hornotini distincte, annotini laxius breviter villosuli v. fere tomentosi, cinereo-brunnescentes. Folia subcoriacea, ovato-oblonga v. elliptico-oblonga, basi rotundata, brevissime acutata, apicem versus sensim v. abruptius acuminata, integra, 5–11 cm. longa, 2.5–5.5 cm. lata, supra viridia, lucida, in nervis et costa impressa densius, in facie sparse griseo-villosula, subtus pallidiora, molliter pubescenti-villosula, nervis lateralibus utrinsecus circiter 10 patentibus paulo prominulis costa distincte prominente; petioli superne sulcati, dense breviter villosuli, 6–8 mm. longi. Inflorescentia pro sectione normalis, axibus calycibus alabastris modo ramulorum pilosis; pedicelli versus apicem bracteolis 2 ovato-lanceolatis instructi, 1–1.3 m. longi; calycis segmenta ovato-lanceolata, acuta, sublibera, subimbricata, plus minusve appressa, circiter 6 mm. longa; corolla alba, pro genere normalis; tubus extus breviter tomentellus, circiter 1.6–1.9 cm. longus, 1.5 mm. crassus, sub faucem paulo angustatus, intus sparse puberulus; lobi circiter 1.4 cm. longi et 5–6 mm. lati, oblanceolati, obtusi, margine interiore paulo supra basim plus minusve irregulariter cristato-undulato, extus in partibus in alabastro non inflatis nudi, intus glabri v. sub microscopio minime puberuli, ad faucem lobulis lineari-bus puberulis circiter 2 mm. longis paracorollam irregularem formantibus instructi; stamina 5 versus basim tubo inserta gynaeceumque glabra, pro genere normalia; stigmata apice breviter bifida; stylus calycem subaequans; fructus ignoti.

Kweichou: "cascade de Hoangko-Chou sur les rochers, 9 juin 1898," *J. Seguin* (No. 2390, type). Kwangsi: Lungchou, *H. B. Morse* (No. 171 — shrub 1 m. tall, flowers white; ex collectione A. Henry).

This species seems to be most closely related to *M. Hemsleyanus* Diels, judging by the shape of the calyx. Otherwise it has a much more distinct paracorolla, but not so well developed as that of *M. suaveolens* Champion, which may be at once distinguished by its shorter, obtuser calyx and by the peculiar lobulation of the dolabriform shorter lobes. I have seen only fragments of Léveillé's type, and it seems to be a little different from Morse's plant before me. The leaves of the type are narrower, the lobes of the corolla are a little longer and not so distinctly undulate-cristate, the corolla tube is more hairy within, and the branches of the stigma are very slender.

TRACHELOSPERMUM Lemaire.

Trachelospermum gracilipes Hooker f., var. Cavaleriei Schneider, n. var.

Melodinus Cavaleriei Léveillé in Fedde, *Rep. Spec. Nov.* II. 113 (descriptio incorrecta) (1912), fide specim. originalis.
Trachelospermum rubrinerve Léveillé, *Fl. Kouy-Tchéou*, 32 (1914).

Frutex scandens; ramuli hornotini brunnescentes, flavo-hirto-villosuli, annotini plus minusve glabrescentes. Folia crassa, obovato-oblonga, basi cuneata, apice satis subito acutata, leviter anguste emarginata (an normaliter apiculata?), margine integra, 5–6.5 cm. longa, 1.8–2.8 cm. lata; supra viridia, ut videtur paulo nitentia, costa impressa, glabra, subtus ut videtur laete viridia, in costa nervisque elevatis et in facie plus minusve ut ramuli villosula, nervis lateralibus utrinsecus circiter 10; petioli supra sulcati, utrinque villosuli, 4–6 mm. longi; stipulae non visae; glandulae inter- et intrapetiolares in annulum dispositae, nigro-purpureae. Inflorescentiae laxe cymoso-paniculatae, terminales axillaresque, satis longe pedunculatae, folia superantes; pedunculus communis 3–5 cm. longus, compressus, plus minusve pilosulus; pedunculi laterales 3–5, breviores, similes, bracteis ut videtur deciduis lanceolatis ad 4 mm. longis ciliatis in dorso pilosulis suffulti; pedicelli 4–13 mm. longi, interdum subglabri, bracteolis minutis ciliatis praediti. Flores albi; sepala oblonga, 2–2.5 mm. longa, subacuta v. obtusiuscula, ciliata (interdum calyx in basi extus pilosus), apice plus minusve conniventia intus basi squamulis 10 subrotundis denticulatis instructa; corolla hypocrateriformis; tubus circiter 9 mm. longus, cylindricus, sed apice dilatatus, intus post stamina pilosus, ad faucem glaber; lobi oblique obovati, tubo aequilongi, apice ad 6 mm. lati, subretusi, glabri; stamina gynaeceumque pro genere normalia; ovarium glabrum bipartitum disco quinquelobo cinctum. Fructus ignotus.

Western Hupeh: Hsing-shan Hsien, on rocks, alt. 900–1200 m., July 1907 (No. **2341**, type; flowers white); Patung Hsien, alt. 900–

1600 m., July 1907 (No. 2342ᵃ). Kweichou: "environs de Kouy-yang, mont du Collège, rare, 25 avril 1898," *J. Chaffaanjon;* "environs de Lo-Pie, avril 1898," *J. Seguin;* "environs de Tou-chan, 3 juin 1899," *J. Cavalerie* (ex Léveillé).

On account of the glabrous mouth of the corolla and of the exserted anthertips I refer Wilson's plant as a variety to *T. gracilipes* Hooker f. So far as I can judge by fragments kindly sent to me by Mgr. Léveillé from one of his type specimens the plant from Kweichou is the same as that from Hupeh. The stamens are exserted and not included as stated by the author in his description. I am not sure if all three specimens cited by Léveillé really belong to our plant.

Trachelospermum cathayanum Schneider n. sp.

Trachelospermum jasminoides Hemsley in *Jour. Linn. Soc.* XXVI. 99 (pro parte, non Lemaire) (1889). — Diels in *Bot. Jahrb.* XXIX. 540 (pro parte) (1900); in *Not. Bot. Gard. Edinburgh,* VII. 407 (*Pl. Chin. Forrest.*) (1913).

Frutex scandens, ad 6.5 m. altus; ramuli hornotini brunnescentes, flavo-puberuli v. interdum breviter subvillosuli, rarius subglabri, annotini glabrescentes v. glabri, purpurascentes. Folia crasse chartacea, oblonga v. oblongo-elliptica v. obovato-oblonga, basi acuta, versus apicem sensim v. subito acuminata, imo apice minime mucronulata, margine integra, minimis exceptis pleraque 5–10 cm. longa et 1.8–4 cm. lata, superne satis viridia, lucida, costa plana v. subimpressa, glabra, subtus pallide viridia, subnitida, costa elevata, nervis nervillisque prominentibus graciliter reticulata, glabra; petioli 3–7 mm. longi, glabri v. ut ramuli puberuli superne sulcati; glandulae normales. Inflorescentiae laxe cymoso-paniculatae, terminales axillaresque, foliis breviores v. subaequilongae; pedunculus communis 2–4 cm. longus, glaber v. minute puberulus, compressus, pedunculi laterales 2–3 v. nulli, illo similes sed breviores, basi bracteis 2–3 mm. longis lanceolatis ciliolatis suffulti; pedicelli 3–6 mm. longi, plerique glabri, bracteolati; flores albi, fragrantes; sepala oblonga v. oblongo-lanceolata, acuta v. subobtusa, glabra margine ciliolato excepto, adpressa v. apice paulo patula, 2–2.5 v. fere 3 mm. longa, intus basi squamulis 10 denticulato-incisis praedita; corolla hypocraterimorpha; tubus circiter 7–8 mm. longus, supra medium dilatatus, intus post stamina et ad faucem pilosus; lobi oblique obovato-oblongi, tubo plerique subaequilongi v. fere sublongiores, apice circiter 5 mm. lati subretusi, glabri; stamina inclusa v. interdum (in No. 2347) paulo e fauce exserta gynaeceumque normalia; ovarium glabrum, bipartitum, disco quinquepartito subduplo breviore v. aequilongo cinctum. Fructus unus tantum visus; folliculi paralleli, 12 cm. longi, circiter 5 mm. crassi, utrinque paulo attenuati, glabri; semina pro subgenere normalia, circiter 16 mm. longa et 1.5 mm. crassa, apice coma 18–20 mm. longa instructa.

Western Hupeh: Changlo Hsien, rocks, etc., alt. 900–1300 m., June 1907 (No. 2344; climber, flowers white); without precise locality, *A. Henry* (No. 5976). Eastern Szech'uan: Nanch'uan, *A. von Rosthorn* (No. 2345). Western Szech'uan: Ta-p'ao-shan, northeast of Tachien-lu, alt. 1800–2400 m., July 3, 1908 (No. 2345; on rocks and trees, flowers white); west of Kuan Hsien, Niu-tou-shan, rocks, alt. 1300 m., June 18, 1908 (No. 2346; climber, flowers white, fragrant); Mupin, on rocks, alt. 900–2200 m., June 1908 (No. 2347;

climber 2.5–5 m., flowers white); Wa-shan, cliffs, alt. 1300 m., July 1908 (No. **2348**, type; flowers white, fragrant); Mt. Omei, June 1904 (Veitch Exped. No. 5072); Yunnan: Tali Range, alt. 2100–2300 m., July 1906, *G. Forrest* (No. 4719); Fêng-chen-lin, mountains south of Red River, alt. 1900 m., *A. Henry* (No. 10651; large climber, flowers white); Mengtsze, mountains to southeast, woods, alt. 1600 m., *A. Henry* (No. 10651ᵃ; climber, white flowers).

According to the short-pointed apex of its flower-buds this species shows affinity to *T. jasminoides* Lemaire, but it seems to be most closely related to *T. lucidum* Kanitz, differing in the characters given in the key on p. 336. There are slight differences between some of the specimens mentioned above in the size of the flowers, the length of the lobes, and the shape of the sepals and of the flower-buds. The leaves are glabrous, but in Henry's No. 10651 there are a few hairs on the midrib of the under surface.

Trachelospermum jasminoides Lemaire in *Gard. Fleur.* I. t. 61 (1851). — Wallis in *Gard. Chron.* n. ser. VII. 116, fig. 19 (1877). — Franchet in *Nouv. Arch. Mus. Paris*, sér. 2, VI. 86 (*Pl. David.* I. 206) (1883). — Hemsley in *Jour. Linn. Soc.* XXVI. 99 (pro parte) (1889). — Bean, *Trees & Shrubs Brit. Isl.* II. 599 (1914), exclud. synon. pro parte.

Rhynchospermum jasminoides Lindley in *Jour. Hort. Soc. Lond.* I. 74, fig. (1846). — A. Henfrey in *Gard. Mag. Bot.* II. 113, t. (1850). — Lindley & Paxton, *Flow. Gard.* II. 26, fig. 147 (1851). — Hooker in *Bot. Mag.* LXXIX. t. 4737 (1853).
Parechites adnascens Hance in *Jour. Bot.* VI. 299 (1868).
Trachelospermum divaricatum K. Schumann in Engler & Prantl, *Nat. Pflanzen-fam.* IV. Abt. 2. 173, fig. 58 J–L (1895), quoad syn. Lemaire et icon. — Pampanini in *Nuov. Giorn. Bot. Ital.* n. ser. XVII. 695 (1910). — Schneider, *Ill. Handb. Laubholzk.* II. 851, fig. 534 a–e (1912), quoad syn. Lemaire et icon.

Western Hupeh: south of Ichang, on rocks and trees, abundant, alt. 300–900 m., June 1907 (No. **578ᵃ**; flowers white, fragrant); same locality, October 1907 (No. **578**; fruits); same locality, *A. Henry* (Nos. 504, 792); Patung Hsien, rocks, alt. 900–1600 m., July 1907 (No. **2342**; climber, flowers white); same locality, on trees, etc., alt. 900 m., May 1907 (No. **2343**; climber, flowers white); "Kao-kien-scian," alt. 800 m., May–June 1907, *C. Silvestri* (No. 1869). Fokien: "in ins. Taitan, prope Amoy, in rupibus lichenosis, Majo 1866," *T. Sampson* (No. 11071, Herb. Hance, type of *P. adnascens*). Chekiang: vicinity of Ningpo, 1908, *D. Macgregor* (with flowers and fruits); Changhua Hsien, on rocks and tree trunks in shady places, alt. 250 m., July 12, 1915, *F. N. Meyer* (flowers white, fragrant). Kiangsu: Shanghai, *A. Fortune* (type of *Rhynchospermum jasminoides* ex Lindley). Formosa:

without locality, *A. Henry's Collector*. Hongkong: top of Ridge, May 9, 1859, *Colonel Urquhart* (1861 ex Herb. Kew; forma incerta).

The true *T. jasminoides* is a Chinese plant and is distinct from the Japanese *T. divaricatum* K. Schumann, with which it has been united by previous authors. The latter may be distinguished at once even from glabrous forms of *T. jasminoides* by the exserted stamens and by the longer-pointed flower-buds. Most of the cultivated plants of *T. jasminoides* and apparently also Lindley's type are much more glabrous than the wild forms mentioned above, but there is in the herbarium of the Arnold Arboretum a specimen from the Villa Thuret at Antibes, France, which shows the same pubescence on the inflorescence, the sepals and on the under surfaces of the leaves that we find in the Chinese specimens. Even the apparently wholly glabrous cultivated plants are more or less sparsely hairy on the under surface of the leaves, but the mouth of the corolla is mostly without any hairs. The stamens are distinctly included, and the sepals are spreading or reflexed and comparatively longer than those of *T. cathayanum*. The pubescence of the branchlets is variable and we find in almost every species quite glabrous and rather pubescent forms. See also under *T. divaricatum* Kanitz p. 337.

Trachelospermum axillare Hooker f., *Fl. Brit. Ind.* III. 668 (1882), ut videtur tantum pro parte. — Diels in *Bot. Jahrb.* XXIX. 540 (1900). — Brandis, *Ind. Trees*, 464 (1906). — Léveillé, *Fl. Kouy-Tchéou*, 31 (1914), fide Léveillé.

> *Melodinus Chaffanjoni* Léveillé in Fedde, *Rep. Spec. Nov.* II. 114 (1906), fide Léveillé.
> *Periploca astacus* Léveillé, *Fl. Kouy-Tchéou*, 43 (1914), secund. specim. originale.

Western Szech'uan: Mupin, thickets, alt. 900–1300 m., June 1908 (No. **1939**; climber, 2.5 m. tall, flowers orange); Mt. Omei, July 1904 (Veitch Exped. No. 5073); without precise locality, July 1904 (Veitch Exped. 4107; climber, flowers orange). Western Hupeh: Hsing-shan Hsien, side of streams, alt. 900 m., July 1907 (No. **1940**; scandent shrub, 2.5 m., flowers reddish); same locality, June 1901 (Veitch Exped. No. 856; climber, flowers orange); without precise locality, *A. Henry* (No. 3618). Fokien: Dunn's Exped., April to June 1905 (Hongkong Hcrb. No. 2918). Yunnan: Mengtsze, mountains to southwest, alt. 1900 m., *A. Henry* (No. 9854; large climber; with ripe fruits); same locality, alt. 1600 m., *A. Henry* (No. 9854ᵇ; large climber, yellow flowers); same locality, mountains to southeast, alt. 1600 m., *A. Henry* (No. 9854ᵃ; large climber); Szemao, mountains to east, alt. 1600 m., *A. Henry* (No. 12050; reddish flowers); same locality, mountains to northwest, alt. 1600 m., *A. Henry* (No. 13115; climber, red flowers).

INDIA. Sikkim: "reg. subtrop. alt. 4–6000 ped.," *J. D. Hooker* (type; "*Rhynchospermum* 5, Herb. Ind. Or. Hook. f. et T.").

The co-type of this species in the Gray Herbarium agrees well with the Chinese specimens mentioned above, but it is strange that Hooker in his description says: " ovary pubescent, follicles tomentose," while in all the specimens before me there is not the slightest indication of any pubescence on the ovaries or fruits. It might be possible that there are two different plants mixed under the type specimens, or it was a mistake of Hooker's, who also describes the mouth and throat of the corolla as glabrous, while in the specimens before me both are distinctly pubescent. I suggest that the fruits described by Hooker do not belong to a *Trachelospermum* at all. See also my remarks under subgen. *Axillanthus*.

CLAVIS SPECIERUM GENERIS TRACHELOSPERMI.[1]

Ovarium fructusque glaberrimus. Squamae calycis distincti, 10 (v. plures?), 2 ante segmentum quodque insertae.

Tubus corollae apice dilatatus; stamina in parte dilatata inserta; styli elongati. Valvae fructuum (an semper?) tenues.

Subgen. 1. EUTRACHELOSPERMUM.

Alabastra satis longe acuminata, parte loborum convolutorum quam pars superior dilatata tubi plus minusve duplo longiore. Antherae e fauce glabra paulo exsertae.

Sepala circiter 3.5 mm. longa, apice plus minusve patula.

Corollae tubus circiter 8–9 mm. longus. Folia vix ad 9 cm. longa et 3.5 cm. lata . . . . 1. *T. divaricatum.*

---

[1] As is the case with many other genera of the *Apocynaceae* it is very difficult to limit properly the genus *Trachelospermum.* After a careful examination of all the material at my disposition and of the description given by authors I think it best not to include the following plants referred to this genus:

1. *T. difforme* Gray, *Syn. Fl. N. Am.* II. pt. 1, 85 (1878), and the apparently very different *T. stans* Gray in *Proc. Am. Acad.* XXI. 394 (1886) (see the specimens collected by *E. Palmer*, No. 724, at Chapala in 1886, and of *C. G. Pringle*, No. 8743, at Monterey, Mexico, 1903). Both may be kept in the genus *Secondatia* A. DC.

2. *T. verrucosa* Boerlage, *Handl. Fl. Nederl. Ind.* II. 400 (1899), which is the same as *Triadenia verrucosa* Miquel, *Fl. Ind. Bat.* II. 459 (1856) and referred to our genus by Bentham & Hooker. I have not seen a specimen.

3. *T. Bowringii* Hemsley in *Jour. Linn. Soc.* XXVI. 99˙(1889), which was first described as *Parechites Bowringii* Hance in *Jour. Bot.* VI. 299 (1868). I know this plant only from Hance's description, according to which, if correct, I do not believe that the plant is a Trachelospermum in our sense.

4. *T. philippinense* Elmer in *Leafl. Philipp. Bot.* II. 488 (1908), of which I have seen the type (*A. D. E. Elmer*, No. 9135, Lukban, 1907). The author says " near to *T. fragrans* Hook. f.," but, as he states himself, the very short corolla-lobes are " slightly twisted to the right in the bud, but reflexed to the left in the anthesis." Besides this there are only five scales of the calyx and the inflorescence is different. It certainly has no relationship with *T. fragrans* or any other Trachelospermum.

5. *T. Cavaleriei* Léveillé, *Fl. Kouy-Tchéou*, 31 (1912), of which I saw some leaves, is no Trachelospermum. *T. Esquirolii* Léveillé, l. c. 32, is a Melodinus, near to *M. monogynus. T. Navillei* Léveillé, l. c., is the same as *Aganosma cymosum*, var. *elegans* Hooker f.

Corollae tubus circiter 1.4 cm. longus. Folia ad 13.5 cm. longa et 4.5 cm.
lata (fide Craib) . . . . 2. *T. siamense.*
Sepala 2–2.5 mm. longa, plus minusve adpressa; corollae tubus 8–10 mm.
longus (si tubus tantum 6–7 mm. longus, confer 1. *T. divaricatum,* var.
brevisepalum) . . . . 3. *T. gracilipes.*
Alabastra acuta v. subobtusa, lobis parte dilatata tubi vix v. paulo longioribus,
interdum subbrevioribus v. faux distincte pilosa; antherae fere semper
inclusae.
Sepala brevia, parte inferiore haud dilatata tubi duplo v. ultra breviora,
plus minusve adpressa, vix 2.5 (–3) mm. longa (confer etiam 1. *T. di-*
varicatum, var. *brevisepalum*).
Folliculi visi ad 13 cm. longi et 0.5 cm. lati; comae seminales seminibus
aequilongae . . . . 4. *T. cathayanum.*
Folliculi ad 35 cm. longi et 4 mm. lati; comae seminales seminibus duplo
longiores . . . . 6. *T. lucidum.*
Sepala longiora, parte inferiore tubi tantum ⅓ breviora v. aequilonga v.
3.5–5 mm. longa, apice distincte patula v. reflexa.
Corollae tubus circiter 9–10 mm. longus; sepala fere semper apice reflexa.
Folliculi ad 21 cm. longi, angulo 180° divergentes. 5. *T. jasminoides.*
Corollae tubus c. 13–14 mm. longus; sepala apice paulo v. vix patula.
Folliculi ad 40 cm. longi, paralleli . . . . 7. *T. tetanocarpum.*
Tubus corollae brevis, tantum basi paulo dilatatus; stamina in parte inferiore
inserta; styli breves. Valvae fructuum satis crassae.
Subgen. 2. AXILLANTHUS.
8. T. axillare.
Ovarium fructusque pilosus.
Squamae calycis ut in subgen. 1 et 2, florum structura ut in subgen. *Axillanthus,*
sed sepala medio tubi paulo longiora. Subgen. 3. PSEUDAXILLANTHUS.
9. T. Dunnii.
Squamae calycis minores, ut videtur decem v. plures (v. tantum 5, ex King &
Gamble); tubus corollae apice attenuatus, infra medium ad insertionem sta-
minum dilatatus sed basi constrictus. Folliculi maturi angulo recto divergentes.
Subgen. 4. LACHNOCARPUS.
Folliculi ad 23 cm. longi et 7 mm. lati . . . . 10. *T. Curtisii.*
Folliculi ad 14 cm. longi et 12 mm. lati . . . . 11. *T. auritum.*

ENUMERATIO SPECIERUM.

Subgen. 1. EUTRACHELOSPERMUM K. Schumann in Engler & Prantl,
Natürl. Pflanzenfam. IV. Abt. 2, 173 (pro sectione) (1895).
The principal differences of the four subgenera are briefly stated in the key.
They are found in the structure of the flowers and of the fruits, while in the inflorescences there are only slight differences. The seeds of *Eutrachelospermum* are
more or less angular-cylindric with a furrow, while they are broader and flatter in
the two other subgenera. The angle in which the follicles diverge seems to be
rather variable in the first subgenus, and the fruits and seeds of species No. 1–7 need
further investigation.

1. **Trachelospermum divaricatum** Kanitz in *Termész. Füzet.* II. 46 (*Anthoph.*
Jap.) (1878). — K. Schumann in Engler & Prantl, *Nat. Pflanzenfam.* IV. Abt. 2,
173 (1895), exclud. icone et syn. Lemaire. — Schneider, *Ill. Handb. Laubholzk.*
II. 851 (pro parte) (1912). — Matsumura, *Ind. Pl. Jap.* II. pt. 2, 507 (1912).

Nerium divaricatum Thunberg, *Fl. Jap.* 110 (non Linnaeus) (1794).
Malouetia asiatica Siebold & Zuccarini in *Abh. Akad. Münch.* IV. pt. 3, 163
(*Fl. Jap. Fam. Nat.* II. 39) (1846).
Parechites Thunbergii A. Gray in *Mem. Am. Acad.* ser. 2, VI. 403 (*Bot. Jap.*)
(1859). — Oliver in *Jour. Linn. Soc.* IX. 166 (1867).
Trachelospermum jasminoides Franchet & Savatier, *Enum. Pl. Jap.* II. 438
(non Lemaire) (1879). — Maximowicz in *Bot. Jahrb.* VI. 65 (1885). —
Matsumura & Hayata in *Jour. Coll. Sci. Tokyo*, XXII. 252 (*Enum. Pl.
Formos.*) (1906). — Nakai in *Jour. Coll. Sci. Tokyo* XXXI. 91 (*Fl. Kor.* II.)
(1911). — Makino in *Tokyo Bot. Mag.* XXVI. 122 (1912).
Trachelospermum jasminoides, var. *pubescens* Makino, l. c. (1912).
Trachelospermum crocostomum Stapf in *Kew Bull. Misc. Inform.* 1906, 74. —
Bean, *Trees & Shrubs Brit. Isl.* II. 599 (1914).

EASTERN ASIA. Korea: without precise locality "in arboribus et rupibus
serpens," June 1907, *U. Faurie* (No. 1899); Quelpaert, in dumosis, October 1906,
U. Faurie (No. 739); same locality "scandens in sepibus torrentium Hogno," May
12, 1908, *Taquet* (No. 1122); in dumosis Hogno," October 1907, *Taquet* (No.
279).
JAPAN. Hondo: prov. Musashi, Mitsumine-san, on rocks, common, alt.
300 m., June 8, 1914, *E. H. Wilson* (No. 6956; flowers white, fragrant; throat of
the flowers somewhat hairy); prov. Shinano, Nakatsu-gawa to Nojiri, September 6,
1905, *J. G. Jack* (climbing high on trunk of trees); prov. Mino, near Gifu, October
20, 1892, *C. S. Sargent* (climbing high on Pine trees); prov. Sagami, Yokohoma,
1862, *C. Maximowicz;* prov. Idzu: Simoda, *C. Wright;* prov. Harima, vicinity
of Sayo, July 23, 1903, *S. Arimoto*. Shikoku: prov. Tosa, Nanokawa, June 9,
1889, *K. Watanabe;* same locality, October 24, 1889, *K. Watanabe*. Kyushu:
prov. Osumi, base of Higashi-Kirishima, common on trees, alt. 500–800 m., June
3, 1914, *E. H. Wilson;* prov. Hizen, Nagasaki, 1862, *R. Oldham* (No. 562); prov.
Bungo, Beppu, June 6, 1910 (ex Herb. Sakurai). Tsusima Island: 1859, *C.
Wilford*. Bonin Islands: *C. Wright* (No. 192).
This Japanese species is different from *T. jasminoides* Lemaire, as I have already pointed out (p. 335), and seems to be most nearly related to *T. gracilipes*
Hooker f. and *T. siamense* Craib. Unfortunately these two species are very little
known and I have not been able to examine the type specimens of them. Their
fruits have not yet been described. The follicles of *T. divaricatum* are rather parallel
or spreading in an acute angle not widely divergent as in *T. jasminoides*. In Wilson's No. 6956 the mouth of the flowers is a little hairy, and Arimoto's specimen
resembles in its pubescence the wild forms of *T. jasminoides*, but the buds are
distinctly long-pointed, while in the latter species they are often very acute but
with shorter points. Through the kindness of Mr. Bean we have received flowering branchlets of the type-plant of *T. crocostomum* Stapf, which cannot be separated
from the true *T. divaricatum*. Stapf compared his plant only with *T. jasminoides*
Lemaire, which hitherto has been regarded as being identical with the Japanese
species.

Trachelospermum divaricatum, var. brevisepalum Schneider, n. var.
A typo praecipue recedit sepalis 2–2.5 mm. tantum longis appressis v. vix apice
patulis, floribus paulo minoribus, alabastris apice interdum minus longe acuminatis.
JAPAN. Liuku Islands (ex Herb. Yokohama Nursery Co.).
CHINA. Formosa: South Cape, *A. Henry* (No. 1370, fruiting co-type; No.
1301); "ubique frequens in dumosis montium," May 1913, *U. Faurie* (No. 249,
type; with flowers); without precise locality, 1864, *R. Oldham* (No. 329; forma
ut videtur inter typum et varietatem intermedia).

This variety seems to be connected with the type by forms like Oldham's No. 329, and it needs further observation. In Henry's No. 1370 the coma of the seed is twice as long as the seed, while in the type, as far as I have seen, it seems to be scarcely longer than the seed.

2. **Trachelospermum siamense** Craib in *Kew Bull. Misc. Inform.* 1911, 414.
SIAM. Chiangmai: Doi Sootep, alt. 600 m., *A. F. G. Kerr* (No. 1133, type, ex Craib).
I know this species only from the author's description according to which it seems to differ from *T. gracilipes* Hooker f. only in its longer sepals. I believe the Siamese plant is very near *T. borneanum* Boerlage, *Handl. Fl. Nederl. Ind.* II. 400 (*Borneana*) (1899), the type of which is *Parechites borneana* Miquel in *Versl. Med. Kon. Akad. Weten. Afdel. Nat.* VI. 193 (*Nov. Gen. Apoc. Ind.* 3) (1857) from Borneo. Miquel's description agrees well with that of Craib.

3. **Trachelospermum gracilipes** Hooker f., *Fl. Brit. Ind.* III. 668. (1882).
INDIA. East Bengal: "Khasia and Jynthea Mts.," *T. Lobb* (Type, ex Hooker f.).
According to Hooker's short description this species is characterized by "throat glabrous, anther-tips exserted." It belongs to the group formed by *T. divaricatum* K. Schumann, *T. siamense* Craib and *T. borneanum* Boerlage mentioned above. The following variety is only a provisional one.

Trachelospermum gracilipes, var. **Cavaleriei** Schneider. See p. 332.

4. **Trachelospermum cathayanum** Stapf. See p. 333.

5. **Trachelospermum jasminoides** Lemaire. See p. 334.

6. **Trachelospermum lucidum** K. Schumann in Engler & Prantl, *Nat. Pflanzenfam.* IV. Abt. 2, 173 (1895).
Alstonia lucida D. Don, *Prodr. Fl. Nepal.* 131 (1825), fide *J. D. Hooker.*
Echites triangularis Hamilton apud D. Don, l. c. (pro synon.) (1825).
Blaperopus lucidus A. De Candolle, *Prodr.* VIII. 411 (1844).
Ichnocarpus fragrans Wallich apud A. De Candolle, l. c. 435 (1844).
Trachelospermum fragrans Hooker f., *Fl. Brit. Ind.* III. 667 (1782). — Brandis, *Ind. Trees,* 464 (1906).

INDIA. Kumaon: Pyura, alt. 1800 m., *R. Strachey & J. E. Winterbottom* (No. 1); "Him. Bor. Occ. Reg. trop. alt. 2000 ped.," *T. Thomson.* Assam: Goalpara, Chrang Duar, plains, December 1890, *G. Mann;* near Jingali Bam Garden, Joboca, October 1898, *Dr. Prain's Collector* (No. 155); Jingali Bam Jungle, March 1899, *Dr. Prain's Collector* (No. 810); without precise locality, *Colonel Jenkins.*
A well-marked species which apparently is closely related to *T. tetanocarpum* Schneider, but this species has larger flowers, longer leaves and longer seeds, with a coma only a little longer than the seed. The type of *T. lucidum* was collected by Wallich in Nepal and seems to be No. 1675 of his Catalogue.

7. **Trachelospermum tetanocarpum** Schneider, n. sp.
Frutex alte scandens; ramuli annotini hornotinique plus minusve villoso-pubescentes, cinereo-purpurascentes. Folia coriacea, elliptica, integra utrinque acuta v. apice breviter acuminata, utrinque lucida et glabra, supra intensius viridia, costa fere plana, subtus pallide viridia, 5–13 cm. longa, 2.3–4.5 cm. lata, costa elevata, nervis lateralibus utrinsecus 10–13 prominulis; petioli superne sulcati, interdum pilosi, subtus convexi, plerique glabri, 5–10 mm. longi; glandulae inter- et intrapetiolares normales. Inflorescentia pro sectione normalis, ad 8 cm. longa, glabra, pedunculo communi 2–3.5 cm. longo, pedunculis secundariis 3 circiter 2–2.5 cm. longis,

bracteis lanceolato-linearibus 5–10 mm. longis instructa; pedicelli glabri, basi normaliter bracteati, calycibus vix v. ad duplo longiores; sepala lanceolata, acuta, tenuiter ciliata, 4–5 mm. longa, apice plus minusve patula, intus basi squamulis 10 instructa; flores albi, normales; tubus 13–14 mm. longus, parte superiore dilatata quam pars inferior cylindrica vix v. paulo breviore; lobi oblique obovato-oblongi, circiter 10 mm. longi et apice 5 mm. lati, in fauce puberuli; stamina inclusa gynaeceaque normalia; discus plus minusve 5-lobatus, ovario brevior. Folliculi subparalleli v. angulo satis acuto divergentes ut videtur versus apicem convergentes, sed ima apice subito paulo recurvati, 20–41 cm. longi, 4–8 mm. lati, valvis satis tenuibus; semina normalia, circiter 2.8 cm. longa, comis ad 4 cm. longis apice instructa.

CHINA. Yunnan: Szemao, forests, alt. 1300 m., A. Henry (No. 11949, type; large climber, white flowers); same locality, forests to southeast, A. Henry (No. 12800; fruiting co-type).

A well-marked species, related to the last as indicated above.

Subgen. 2. AXILLANTHUS K. Schumann in Engler & Prantl, Nat. Pflanzenfam. IV. Abt. 2, 173 (pro sectione) (1895).

Trachelospermum, sect. (vel genus nov.) Pycnanthes Bentham & Hooker, Gen. II. 720 (1876), in textu, ut videtur tantum pro parte.

Bentham & Hooker say that there are two species — one from Sikkim, the other from Khasia — with dense axillary subsessile inflorescences, and they propose the name Pycnanthes for a section of Trachelospermum or for a new genus. The Sikkim species, first mentioned, I believe, is identical with T. axillare Hooker f., and therefore one might accept the name Pycnanthes. With regard to the fruits they say: "folliculi similes nisi rigidiores et magis divaricati." This does not agree with what I have seen. The ripe fruit on the specimens before me of T. axillare (Henry's Nos. 9854, 9854[b]) may be described as follows: folliculi 13 cm. longi, medio ad 1.8 cm. lati, fere paralleli, apicem versus angustati, paulo incurvi, ima apice cohaerentes (vide juniores in No. 9854[b],) glabri, cinereo-fusci; valvae satis tenues, sed firmae, intus laeves, flavo-brunneae. Semina lapidea, satis magna, oblonga v. obovato-oblonga, circiter 21 mm. longa, versus basim obtusum circiter 7 mm. lata, apice setifera circiter 2–3 mm. lata, circiter 0.5–1 mm. crassa, sed plus minusve irregulariter curvata et carinata, obscure purpurascentia, infra comam parte paulo flavescente instructa; comae densae, basi flavescentes, ad 5 cm. longae.

8. **Trachelospermum axillare** Hooker f. See p. 335.

Subgen. 3. PSEUDAXILLANTHUS Schneider, n. subgen. Descriptio in clavi.

The species referred to this new subgenus has the appearance of an Eutrachelospermum, while the structure of the flowers is the same as in Axillanthus. The sepals are comparatively long, and the ovary is hairy as in Lachnocarpus. The fruits are not yet known.

9. **Trachelospermum Dunnii** Léveillé, Fl. Kouy-Tchéou, 31 (1912).

Melodium Dunnii Léveillé in Fedde, Rep. Spec. Nov. IX. 453 (1911), secund. specim. originale.

CHINA. Kweichou: "environs de Kouy-yang, mont du Collège, Avril 1898," J. Chaffanjon; "Pin-fa, bois de Si-Tchéou-Goi, Auguste–Septembre 1902," J. Cavalerie (Nos. 344, 558, ex Léveillé).

By the kindness of Mgr. Léveillé this Arboretum has received a fragment of one of his type specimens. According to this specimen this species represents a very distinct new subgenus well distinguished by the characters given in the key on p. 337 and above.

Subgen. 4. LACHNOCARPUS Schneider, n. subgen. Descriptio in clavi. As stated before by King & Gamble, the two species referred to this subgenus may be found to form a new genus,[1] but without having made a thorough study of all the genera of the subfamily *Echitoideae* I am afraid to propose one. The leaves and the shape of the flowers look very different from those of the other species of Trachelospermum, but the flat (possibly not yet fully ripe) follicles I have seen resemble to a certain degree those of the subgen. *Axillanthus*. See the description of *T. auritum* Schneider.

10. **Trachelospermum Curtisii** King & Gamble in *Jour. As. Soc. Bengal*, LXXIV. pt. 2, extra number, 499 (*Mat. Fl. Mal. Penins.* No. 19, 709) (1907).

MALAYA. Penang: at Balick Pulan, *C. Curtis* (No. 838, type ex King & Gamble). Perak: *King's Collector* (No. 5194, co-type ex King & Gamble).

King & Gamble mention the following species as a new unnamed one " which approaches very near to this." According to their description *T. Curtisii* differs mainly in its longer fruits (from 8 to 9 in. long); the corolla-lobes are described as " obliquely triangular-spathulate, dentate at apex," and the tube is apparently a little shorter than that of the following species, which has at least ten small scales at the base of the sepals or sometimes more (or the scales are deeply divided), while in *T. Curtisii* according to the author there are small subulate scales between the lobes, that is, only five.

11. **Trachelospermum auritum** Schneider, n. sp.

Frutex scandens; ramuli hornotini rotundi, fistulosi, laxe (ad nodos densius) subhirtello-villosi, sordide rubro-brunnescentes. Folia membranacea, late elliptica v. paulo obovata, basi aurito-cordata, apice valde subito in acumen angustum caudatum contracta, margine integra, strigoso-ciliata, 8–15 cm. longa (apice 8–15 mm. longo incluso), 4–10 cm. lata, supra intense viridia, in costa nervisque tenuiter impressis v. etiam in facie pilis strigosis satis longis sparse pilosa, interdum subscabra, subtus paulo (in sicco) discoloria, similiter sed brevius pilosa, costa nervisque lateralibus utrinque circiter 10 angulo 50–80° ab illa divergentibus, distincte transversaliter reticulata; petioli breves, supra sulcati, dense strigosi, 0.6–1 cm. longi; glandulae inter- et intrapetiolares, pilis strigosis occultae, purpureae. Inflorescentiae pro genere normales, terminales, sessiles, modo ramulorum strigoso-pilosae, pedunculis primariis tribus 2–3 cm. longis; bracteae lanceolatae, satis breves, pilosae; pedicelli 4–8 mm. longi, bracteolis parvis suffulti; segmenta calycis lineari-lanceolata, $\frac{2}{3}$–$\frac{3}{5}$ libera, 3–4 mm. longa, apice patula, extus plus minusve pilosa, ciliata, intus basi squamulis ut videtur 10 v. pluribus satis parvis oblongis instructa; flores albi, hypocraterimorphi; tubus circiter 15 mm. longus, ima basi constrictus, infra medium ad insertionem staminum dilatatus, circiter 2 mm. crassus, a medio ad apicem angustatus, extus lineis 5 pilosis (in alabastro multo distinctius) praeditus, ceterum sparse tenuiter pilosus, intus ab apice staminum ad fauces sparse v. vix pilosus, ceterum glaber; lobi sinistrorsum convoluti, oblongo-spathulati, subaequilongi, apice leviter emarginati et 8–9 mm. lati, in alabastro extus pilosuli; stamina pro genere normalia; gynaeceum fere typicum; ovarium apice villosulum, disco 5-lobato fere aequilongo cinctum; stylus gracilis, stigmate conico oblongo apice

---

[1] The authors say that in the Kew Herbarium *T. Curtisii* was placed with *Chonemorpha* G. Don, and that in some respects it comes nearest to *Epigynum* Wight. The type of *Chonemorpha* (*Ch. macrophylla* G. Don) is widely different in the structure of the flowers, while that of *Epigynum* (*E. Griffithianum* Wight, *Icon. Pl. Ind. Orient.* IV. 4, t. 1308 [1850]), which I know only from Wight's plate, in the structure of the flowers seems less different, but I have been unable to find a good description of the fruits of this genus.

apiculato paulo longior. Fructus pro genere normalis; folliculae angulo recto divergentes, ad 16 cm. longae et 1.4 cm. latae, apicem versus paulo angustatae imo apice breviter recurvatae, modo ramulorum pilosae, cinereo-purpurascentes; valvae satis crassae, intus flavae; semina (an satis matura?) ad 20 mm. longa et versus apicem 4 mm. lata, basi breviter acutata, apice paulo v. vix angustata, coma ad 3–3.5 cm. longa densa detersili coronata, valde compressa, dura, purpureo-brunnea.

CHINA. Yunnan: Szemao, mountains to south, alt. 1300 m., A. Henry (No. 12136, type; large climber, flowers white); same locality, A. Henry (No. 12852; fruiting, co-type).

According to King & Gamble there are specimens in the Kew Herbarium from the Khasia Hills, Hooker f. & Thomson, and from Manipur, G. Watt (No. 7272) which are identical with Henry's No. 12136. See my remarks under the subgen. Lachnocarpus and T. Curtisii King & Gamble.

SINDECHITES Oliv.

Sindechites Henryi Oliver in *Hooker's Icon.* XVIII. t. 1772 (1888). — Hemsley in *Jour. Linn. Soc.* XXVI. 100 (1889).

Western Hupeh: Fang Hsien, thickets, alt. 900–1300 m., June 1910 (No. **4562**; climber 1.5–3 m., flowers yellow); Ichang, A. Henry (No. 3636, type).

A very interesting plant of which the fruits are not yet known. In Wilson's specimen and in the co-type before me the anthers have longer spurs than appear on Oliver's plate.

Here may be added notes on two genera not collected during the Arnold Arboretum Expeditions.

VALLARIS Burm.

Vallaris grandiflora Hemsley & Wilson in *Kew Bull. Misc. Inform.* 1906, 162.

Western Szech'uan: Tung River valley, alt. 700 m., June 1904 (Veitch Exped. No. 4158, type; flowers pale yellow).

A very distinct large-flowered species which apparently represents a separate section of the genus.

ECDYSANTHERA Hook. & Arn.

Ecdysanthera rosea Hooker & Arnott, *Bot. Voy. Beechey,* 198, t. 42 (1836). — A. De Candolle, *Prodr.* VIII. 442 (1844). — Bentham, *Fl. Hongk.* 222 (1861). — Hemsley in *Jour. Linn. Soc.* XXVI. 98 (1889). — Matsumura & Hayata in *Jour. Coll. Sci. Tokyo,* XXII. 251 (*Enum. Pl. Formos.*) (1906). — Dunn & Tutcher in *Kew Bull. Misc. Inform.* add. ser. X. 169 (*Fl. Kwangtung & Hongk.*) (1912).

CHINA. Eastern Szech'uan: Min River valley, June 1903 (Veitch Exped. No. 4092; climber, flowers red). Fokien: Dunn's Exped., April to June 1905 (Hongk. Herb. No. 2914). Formosa: Bankinsing, A. Henry (Nos. 388, 838). Hainan: C. Ford. Hongkong: May 20, 1899, E. Faber (ex Herb. Hongk. No. 9057).

According to the authors the type was collected at Canton and in Macao and on the adjacent islands.

ASCLEPIADACEAE.

Determined by CAMILLO SCHNEIDER.

PERIPLOCA L.

Periploca calophylla Falconer in *Proc. Linn. Soc.* I. 115 (1842); in *Ann. Mag. Nat. Hist.* VIII. 449 (1842). — Decaisne in De Candolle, *Prodr.* VIII. 498 (1844). — Hooker f., *Fl. Brit. Ind.* IV. 12 (1883). — Hemsley in *Jour. Linn. Soc.* XXVI. 101 (1889). — Diels and Schlechter in *Bot. Jahrb.* XXIX. 541 (1900). — Léveillé, *Fl. Kouy-Tchéou*, 43 (1914); *Cat. Pl. Yun-Nan*, 14 (1915).

Streptocaulon calophyllum Wight, *Contrib. Bot. Ind.* 65 (1834); *Ill. Ind. Bot.* II. 230, t. 182, fig. 1 (1850).

Western Hupeh: north and south of Ichang, thickets, common, alt. 3–600 m., July 5, 1907 (No. **1942**; climber 1.5–3 m. tall, flowers white with pink strips down petals); same locality, *A. Henry* (No. 3419); Changyang Hsien, thickets, alt. 600 m., May and October 1907 (No. **1942ª**); without exact locality, *A. Henry* (No. 4119). Yunnan: Mengtsze, woods, alt. 1800 m., *A. Henry* (No. 11311; climber on trees, greenish flowers).

INDIA. Assam: without precise locality, *Dr. King's Collector.*

I have been able to see a specimen of Herb. Falconer distributed from Kew in 1869. The flowers of this specimen agree fully with those of Wilson's. Wight's figure shows a flower with glabrous corolla-lobes which are always hairy within in the specimens before me, with the exception of those of a specimen in Herb. Gray. from Sikkim " Regio temp. alt. 6000 ped.," *J. D. Hooker.* *Forrest* has collected a species with glabrous corolla-segments near Yunnan-fu, and it was described by Schlechter (in *Not. Bot. Gard. Edinburgh*, VIII. 15 [*Pl. Chin. Forrest.*] [1913]), as *P. Forrestii*, which I know only from the description.

Periploca sepium Bunge in *Mém. Sav. Étr. Acad. Sci. St. Pétersbourg*, II. 117 (*Enum. Pl. Chin. Bor.* 43) (1835). — Decaisne in De Candolle, *Prodr.* VIII. 498 (1844). — Hance in *Jour. Linn. Soc.* XIII. 134 (1875). — Maximowicz in *Bull. Acad. Sci. St. Pétersbourg*, XXIII. 352 (1877); in *Mél. Biol.* IX. 774 (1877). — Hemsley in *Jour. Linn. Soc.* XXVI. 101 (1889). — Schlechter & Diels in *Bot. Jahrb.* XXIX. 541 (1900). — Léveillé, *Fl. Kouy-Tchéou*, 43 (1914).

Western Szech'uan: Mao-chou, arid rocky places, alt. 1300–1900 m., May 24 and October 1908 (No. 2251; semiscandent bush 0.25–1.5 m. tall, flowers greenish-yellow with chocolate markings). N o r t h - ern Shensi: "Lao-y-san," June 4, 1897, *G. Giraldi.* Chili: Pagoda Hill near summer palace near Peking, 1903, *C. S. Sargent;* hills near Nankow, October 5, 1905, *J. G. Jack.*

This species is in cultivation in the Arnold Arboretum, from seeds collected by Jack in 1905 and by Purdom.

HOLOSTEMMA R. Br.

Holostemma sinense Hemsley in *Jour. Linn. Soc.* XXVI. 103 (1889).

Ad descriptionem addenda: Frutex scandens ad 2.5 m. altus. Folia late ovata v. ovato-oblonga, basi profunde cordata, apice sensim acuta et abrupte breviterque acuminata, 5–10 cm. longa et 2.5–5.5 v. 8 cm. lata, superne intense viridia, praesertim ad costam pilosula, ceterum fere glabra, subtus cinerascentia, sub microscopio dense tenuissime papillosa, ad costam sparsissime pilosa, ceterum glabra; petioli 2–5.5 cm. longi. Cymae racemoso-paniculatae, breviter pilosae, ad 5 cm. longae et 3 cm. crassae, 6–20-florae, pedunculis 2–5 cm. longis nudis suffultae. Flores ad 12 mm. diametientes, albi; pedicelli puberuli circiter 5 mm. longi, basi bracteis parvis lanceolatis pilosis instructi; calycis segmenta fere ad basim libera, lanceolata, acuta, extus sparse pilosa, margine ciliata, circiter 2.5 mm. longa; corollae lobi late oblongi, obtusi, circiter 5 mm. longi, glabri; corona brevis, 5-lobulata, infra antheras affixa, lobis ut videtur patentibus obtusiusculis; antherarum membrana ovato-triangularis; stigma ultra caput planum 5-lobum subito breviter filamentose productum imo apice paulo bifidum; pollinia oblongo-elliptica, paulo curvata, pendula, translatoribus brevibus versus pollinia dilatatis, retinaculo oblongo ad insertionem translatorum angustato polliniis vix duplo breviore. Folliculi oblongi, laxe puberuli, ut videtur valde divaricati (angulo 180°?), basi subito paulo contracti, a medio ad apicem sensim attenuati, circiter 7 cm. longi et 1.8 cm. infra medium crassi, in dorso paulo rugoso-angulati; semina late oblonga, circiter 6 mm. longa et 4 mm. lata, margine subalata et basi denticulata, apice coma sericea iis circiter 4-plo longiore instructa, rubro-brunnea.

Western Hupeh: north and south of Ichang, thickets, abundant, alt. 300–900 m., August 1907 (No. 648; climber 1.25–2.5 m., flowers white); same locality, November 1907 (No. 648ᵃ; with fruits); same locality, *A. Henry* (No. 2755, co-type; and No. 3992, type ex Hemsley); Hsing-shan-Hsien, thickets, alt. 600-900 m., September 1907 (No. **2250**; climber 1.2–1.8 m., flowers white).

In Henry's No. 2755 in the Gray Herbarium the inflorescences are very few flowered; otherwise there is no difference between this co-type and Wilson's specimens. I doubt if our plant belongs to *Holostemma* at all, the type of the genus being *H. annularis* K. Schumann (in Engler & Prantl, *Nat. Pflanzenfam.* IV. Abt. 2, 250, fig. 71 J–K [1895]. — *Asclepias annularia* Roxburgh, *Fl. Ind.* ed. 2, II. 37 [1832]). Roxburgh mentions as synonym "Ada-kodien *Rheed. mal.* 9, t. 7," which is also cited by R. Brown in describing his new genus. There is a plate in Wight, *Icon. Ind. Orient.* II. t. 597, under the name of *Holostemma Rheedii* Sprengel.

CYNANCHUM L.

Cynanchum decipiens Schneider, n. sp.

Suffrutex scandens ad 1.2–1.8 m. altus; ramuli teretes, flavobrunnescentes, unifariam puberuli, ceterum fere glabri, remote foliati. Folia membranacea, ovato-lanceolata, basi subauriculato-cordata, apicem versus sensim angustata, acuminatissima, utrinque sparse puberula, subtus paulo pallidiora, majora 5–8 cm. longa, ad basim 2–4 cm. lata; petioli ad 2 cm. longi, superne sulcati, laxe puberuli, ad basim folioli glandulosi. Cymae umbelliformes, ad 25-florae, pedunculo puberulo 4–9.5 cm. longo suffultae; flores purpurascentes, aperti, circiter 7 mm. diametientes; pedicelli filiformes, plus minusve unifariam puberuli, basi bracteis parvis lineari-lanceolatis acutis puberulis instructi, ad 18 mm. longi; calycis segmenta fere libera, lanceolata, acuta, intus glabra, extus sparse pilosa, margine ciliata, circiter 2 mm. longa, intus basi inter se squamulis circiter 3 (toti circiter 15) minimis instructa; corolla subrotata; lobi patentes (v. reflexi?), oblongi, subacuti, circiter 4 mm. longi, extus glabri, intus margine excepto albo-puberuli; corona gynostegio brevior, basin appendicis antherarum attingens, lobis 5 obtusis distinctioribus intus appendice ligulato-triangulato iis simili et paulo breviore medio ornatis, inter lobos iterum brevius lobulato-plicata; antherae rhomboideae; subquadratae, marginibus cartilagineis, apice appendice hyalino albo subacuto in caput stigmatis paulo incurvo coronatae; pollinia pendentia, retinaculo paulo longiora, translatoribus versus pollinia dilatatis; stigma basi 5-angulum, convexum, obscure bilobum. Folliculi singuli (an semper?), oblongi, basi brevius et apicem versus longe attenuati, subglabri, ad 11 cm. longi et versus medium 1.2 cm. crassi, olivacei, purpurascenti-striati; semina oblongo-spathulata, basi dilatata, rotundata, denticulata, versus apicem angustata, apice coma sericea fere 4-plo longiore coronata, circiter 7 mm. longa, plano-compressa.

Western Szech'uan: Wên-chuan Hsien, dry and rocky valleys, alt. 1300–2300 m., June and November 1908 (No. **1239**, type; climber 1.2–1.8 m., flowers purple; with flowers and ripe fruits).

According to the description and figures given by Maximowicz in *Bull. Acad. Sci. St. Pétersbourg*, XXIII. 372 (1877), this species seems to be closely related to *C. acutum* Linnaeus, var. *longifolium* Ledebour, of which I have not been able to see good material. It also resembles the Chinese species of the sect. *Endotropis*, but these have a different corona, which in our species looks somewhat like that represented in Maximowicz's fig. 10.

Cynanchum (§ Endotropis) **auriculatum** Royle in herb. apud Wight, *Contrib. Bot. Ind.* 58 (1834). — Hooker f., *Fl. Brit. Ind.* IV. 25 (1883). — Hemsley in *Jour. Linn. Soc.* XXVI. 105 (1889). — Diels & Schlechter in *Bot. Jahrb.* XXIX. 541 (1900).

Endotropis auriculata Decaisne in De Candolle, *Prodr.* VIII. 546 (1844).

Western Hupeh: Changyang Hsien, thickets, alt. 650–900 m., August 1907 (No. **2249**; climber 3 m. tall, flowers white); Ichang, *A. Henry* (No. 2481); without precise locality, *A. Henry* (Nos. 2196, 4271, 6546). Kiangsi: Kuling, common, alt. 1100 m., July 30, 1907 (No. **1533**; climber 1.8 m. tall, perianth green, corona waxy-white).

In the plant from Kuling the flowers are a little smaller than in the other specimens.

Cynanchum (§ Endotropis) **amphibolum** Schneider, n. sp.

Suffrutex scandens ad 3 m. altus; ramuli teretes, unifariam puberuli. Folia ovato-lanceolata, basi auriculato-cordata, apicem versus satis sensim acuminata, fere caudata, utrinque sparse puberula v. subglabra, subtus pallidiora, 5–10 cm. longa et 3–5.5 cm. lata, ad insertionem petioli superne glandulosa; petioli 2–6 cm. longi, superne sulcati et sparse pilosi. Cymae 10–20-florae, umbelliformes, pedunculis 2.5–4.5 cm. longis ut pedicelli ad 2 cm. longi filiformes unifariam pilosulis suffulti. Flores purpurascentes, aperti ad 0.8–1 cm. diametientes; calycis segmenta ovato-lanceolata, acuminata, margine ciliolata, 2–2.5 mm. longa; corolla subrotata; lobi oblongi, subacuti, 3.5–4.5 mm. longi, intus puberuli, plerique reflexi; corona gynostegio aequilonga, corollae lobis circiter duplo brevior, laciniis oblongis obtusis intus medio appendice ligulato-triangulari foliolum aequante instructis extus versus basim ut in *C. auriculato* Royle constructis; antherae rhomboideo-subquadratae, marginibus cartilagineis, apice appendice albo hyalino subobtuso in caput stigmatis planum 5-lobatum medio bilobulatum incurvo praeditae; pollinia retinaculo oblongo-rhomboideo satis magno vix longiora, ovato-elliptica, translatoribus versus pollinia satis dilatatis iis subduplo brevioribus. Folliculi ignoti.

Western Hupeh: Patung Hsien, thickets, alt. 600–1300 m., August 1907 (No. **2247**, type; climber 3 m. tall, flowers purple); without precise locality, *A. Henry* (No. 6030).

In its flowers this species is closely related to *C. auriculatum* Royle, which has larger whitish flowers with a corolla longer than the gynostegium and much broader leaves. It seems to be also similar to *C. Giraldii* Schlechter (in *Bot. Jahrb.* XXXVI. Beibl. 82, 92 [1905]), which I know only from the description. The author does not mention the color of the flowers and describes the corolla-lobes as glabrous on both surfaces and the peduncles as about as long as the pedicels. The type was collected by Father Hugh Scallan on Mt. Omei (Giraldi No. 2272).

Cynanchum (§ Cynoctonum) **otophyllum** Schneider, n. sp.

Suffrutex scandens ad 1.8 m. altus; ramuli teretes, flavescentes, unifariam pilosuli, subdense foliati. Folia membranacea, ovato-lanceolata, basi profunde auriculato-cordata, supra basim attenuata et apicem versus sensim acuminata, utrinque sparse pilosula v. subglabrescentia, subtus paulo pallidiora, 4–9 cm. longa, 2.5–7 cm. ad basim lata; petioli superne sulcati et puberuli, 1.5–4.5 cm. longi. Cymae elongatae, pseudo-racemoso-paniculatae, ad 4 cm. longae, pedunculo nudo ad 3.5 cm. longo excluso, puberulae v. subglabrae. Flores albescentes, satis parvae, ut videtur vix 4 mm. diametientes; pedicelli filiformes, subpuberuli, 3–5 mm. longi, basi bracteis minimis lanceolatis subglabris instructa; calycis segmenta fere libera, ovato-lanceolata, acuta, corolla 2½-plo breviora, fere glabra, intus inter se squamula 1 minima instructa; corolla subrotata; lobi alte fissi, oblongi, obtusiusculi, circiter 2.5–3 mm. longi, extus glabri, intus margine excepto puberuli; corona quam corolla paulo brevior et gynostegio sublongior, laciniis 5 aequalibus satis alte fissis oblongis obtusiusculis simplicibus; antherae et cetera ut in specie praecedente descripta. Folliculi ignoti.

Western Szech'uan: Tung River valley, near Wa-shan, alt. 500–900 m., September 1908 (No. **2252**, type; climber 1.8 m. tall, flowers white).

I do not know a species of eastern Asia with an elongated inflorescence and small flowers to which I can refer Wilson's plant. It apparently belongs to the group *Cynoctonum* and may be related to the following species, which widely differs in its denser and shorter inflorescence, its larger flowers, the corona of which is scarcely half as long as the corolla-lobes, and in other characters.

Cynanchum (§ Cynoctonum) **Wilfordii** Franchet et Savatier. *Enum. Pl. Jap.* II. 445 (1879). — Hemsley in *Jour. Linn. Soc.* XXVI. 109

(1889). — Komarov in *Act. Hort. Petrop.* XXV. 294 (*Fl. Mansh.* III.)
(1907). — Nakai in *Jour. Coll. Sci. Tokyo,* XXXI. 94 (*Fl. Kor.* II.)
(1911).

Endotropis auriculata Franchet & Savatier, *Enum. Pl. Jap.* I. 319 (non Decaisne)(1875).
Cynoctonum Wilfordii Maximowicz in *Bull. Ac. Sci. St. Pétersbourg,* XXIII.
370 (1877); in *Mél. Biol.* IX. 799 (1877).

Western Hupeh: Patung Hsien, thickets, alt. 900–1300 m., July
1907 (No. **2248**; climber 1.8 m. tall, flowers yellowish).

This species is well distinguished from those of the sect. *Endotropis* by the
corolla-lobes, which are not spreading or reflexed, and by the simple lobes of the
petaloid corona which are almost orbicular. K. Schumann (in Engler & Prantl, *Nat.
Pflanzenfam.* IV. Abt. 2, 252 [1895]) apparently mistook this species for *C. auriculatum* Wight (see above), as he stated that it had a reflexed corolla.

SECAMONE R. Br.

Secamone emetica R. Brown [1] apud Roemer & Schultes, *Syst.*
VI. 124 (1820). — R. Wight, *Contrib. Bot. Ind.* 60 (1834); *Icon. Pl. Ind.
Orient.* IV. t. 1283 (1850); *Ill. Ind. Bot.* II. 169, t. 155ᵇ, fig. d (1850). —
Decaisne in De Candolle, *Prodr.* VIII. 501 (1844). — Hooker f., *Fl.
Brit. Ind.* IV. 13 (1883). — K. Schumann in Engler & Prantl, *Nat.
Pflanzenfam.* III. Abt. 2, 262, fig. 76 b–e (1895). —Costantin in Le-
comte, *Fl. Gén. Indo-China,* IV. 43 (1912).

Periploca emetica Retzius, *Obs. Bot.* II. 14 (1781).

Western Szech'uan: Tung River valley, near Wa-shan, thickets,
alt. 1300 m., June 1908 (No. **2246**; climber, flowers yellow), without
precise locality, alt. 1100 m., May 1904 (Veitch Exped. No. 4090;
climber, flowers yellow). **Yunnan:** between Yung-peh-ting and Tai-
naoko, thickets, alt. about 1800 m., July 3, 1914, *C. Schneider* (No.
1732; little climbing shrub up to 3 m. tall).

INDIA. **South Deccan Peninsula:** "Maisor & Carnatic. Regio trop."
T. Thomson; without precise locality (No. 1895 Herb. Wight).

Wilson's specimens agree with Wight's picture and the Indian plants mentioned
above. My Yunnan specimen is a little different, the corona-lobes being somewhat
longer and extending to the tips of the anthers. In Schumann's figure these lobes
are intermediate between the typical form and that of my Yunnan specimen. I
have not seen any fruit.

---

[1] It is always quoted "Br. in Mem. Wern. Soc. I. 56," but Brown did not make
the combination; he only described the genus *Secamone* and mentions as belonging
to it *Periploca emetica* Retzius. The correct quotation for the genus seems to be
R. Brown in *Mem. Wern. Nat. Hist. Soc.* I. 55 (1810), fide C. G. Nees von Esenbeck,
R. Brown's Verm. Bot. Schrift. II. 392 (1826); *Prodr. Fl. Holl.* I. 464 (1810);
Misc. Bot. Works, II. 229 (1867).

TOXOCARPUS Wight & Arn.

Toxocarpus villosus Decaisne in De Candolle, *Prodr.* VIII. 506
(1844). — Miquel, *Fl. Ned. Ind.* II. 475 (1856). — Hooker f., *Fl. Brit.
Ind.* IV. 13 (1883). — Costantin in Lecomte, *Fl. Gén. Indo-Chine,* IV.
51, fig. 7 (12–16) (1912).

> *Secamone villosa* Blume, *Bijdr.* XVI. 1050 (1826). — Koorders, *Exkursionsfl.
> Java,* III. 87 (1912).
> *Toxocarpus Wightianus* Hemsley in *Jour. Linn. Soc.* XXVI. 101 (non Hooker
> & Arnott) (1889).
> *Secamone Wightiana* Diels & Schlechter in *Bot. Jahrb.* XXIX. 543 (non K.
> Schumann) (1900).
> *Toxocarpus Wightianus,* var. *hirta* A. Henry, Mss. in Herb.

Western Hupeh: Hsing-shan Hsien, ravines, rare, alt. 300–900 m.,
May 9, 1907 (No. **1943**; climber 2.5 m. tall, flowers yellow); Ichang,
A. Henry (Nos. 3281, 3515).

SIAM. Without precise locality, *Mrs. M. C. Bradley* (No. 34).

Wilson's and Henry's plants agree well with *T. villosus* as described and figured
by Costantin. I have not seen a type specimen, and a careful study of the whole
genus would be necessary to show if our plant is indeed identical with Blume's
Javanese type. The Hupeh plants certainly are very different from *T. Wightianus*
Hooker & Arnott (*Bot. Voy. Beechey,* 200 [1836]. — *Secamone Wightiana* K. Schumann in Engler & Prantl, *Nat. Pflanzenfam.* IV. Abt. 2, 263 [1895]), of which I
have seen specimens from Hainan and Hongkong. It has smaller, more glabrous
and shining leaves and a very short and dense inflorescence. The flowers are smaller
too, and the stigmas are shorter, undivided and enlarged, while in our species the
stigma (the part of style above the gynostegium) is enlarged only at the base, being
slender toward the end and divided at the apex. The length of the coma of the
seeds seems to be variable, as well as the pubescence at the base of the corolla-
lobes.

CEROPEGIA L.

Ceropegia (§ Phananthe) **driophila** Schneider, n. sp.

Suffrutex 0.75–1.25 m. altus, scandens; caules filiformes, teretes v.
leviter striato-angulati, flavescentes, glabri, laxe foliati. Folia tenuiter
membranacea, oblonga v. ovato-oblonga, basi subcordata, apicem versus sensim distincte acuminata, superne sparse breviter pilosa, subtus
vix pallidiora, glabra v. glabriuscula, margine ciliata, 2.5–6.5 cm. longa,
basim versus 1–2.4 cm. lata (in specimine Henryano ad 5: 2.8 cm.
magna); petioli tenues, superne ut videtur paulo sulcati, glabri v. ad
apicem superne sparse pilosuli, 1–2.2 cm. longi. Cymae pedunculatae,
2–8-florae; pedunculi 1–1.5 cm. longi, glabri; pedicelli graciles, glabri,
1–1.5 cm. longi, basi bracteis linearibus glabris v. vix pilosis 1–1.5 mm.

longis suffulti. Flores obscure rubri; calycis laciniae lineares, acuminatae, 2–3 mm. longae, glabrae, basi intus squamulis 5 parvis sed satis latis inter se instructae; corolla tota circiter 2.5 cm. longa, ad basim ellipsoideo-inflatam paulo obliquam circiter 4 mm. crassa, deinde in tubum brevem contracta, ad medium tubi circiter 2 mm. crassa, versus faucem iterum ampliata, ad faucem fere 5 mm. crassa; lobi circiter 12 mm. longi, apice cohaerentes, in alabastro sub apicem dilatatam constricti, oblongo-ligulati, apice fere cochleari v. spathulato, intus carinati, glabri, margine sparse longe ciliato; corona cyathiformis, foliolis altius connatis in dentes 2 triangulares acutos hirsuto-ciliatos ligulis interioribus linearibus subobtusis glabris 4–5-plo breviores productis; pollinia oblique oblongo-elliptica, apice uno apiculata, translatoribus ut videtur dilatatis brevibus; retinacula polliniis vix breviora, anguste oblonga sed ut videtur apice uno dilatata.

Western Hupeh: Patung Hsien, thickets, alt. 600–900 m., July 1907 (No. 2316; climber 0.75–1.25 m., flowers dark red); without precise locality, A. Henry (No. 6222).

I am unable to refer Wilson's plant to any described species from eastern Asia as far as I know them from descriptions, pictures and herbarium specimens. Hemsley does not mention Henry's number, the poor specimen of which before me does not differ from Wilson's plant and probably came from the same region. I do not know the real relationship of our plant. The shape of the corolla seems to be a very characteristic one, the head (the lobes) being much contracted below the apex, which is a little impressed, at least in the dried flowers.

Ceropegia (§ Phananthe) **stenophylla** Schneider, n. sp.

Suffrutex 1–1.5 m. altus, scandens; caules filiformes, teretes v. leviter striati, viridi-flavescentes, glabri, v. sparse indistincte pilosuli, laxe foliati. Folia tenuiter membranacea, linearia v. lineari-lanceolata, basim versus in petiolum brevem ad 6 mm. longum pilosulum attenuata, apice sensim acuta, ad 7 cm. longa et infra medium 0.7 cm. lata, superne sparse breviterque pilosiuscula, subtus subglabra v. glabra, vix pallidiora. Cymae breviter pedunculatae v. subsessiles, ut videtur tantum uniflorae; pedunculi ad 6 mm. longi, plerique glabri; pedicelli ad 1 cm. longi, glabri v. vix pilosi, basi bracteis linearibus glabris 2–4 mm. longis suffulti; flores colore obscuro incerto, omnino circiter 4–4.3 cm. longi; calycis laciniae lineares v. lineari-lanceolatae, acuminatae, 5–7 mm. longae, glabrae v. in margine sparse ciliatae, intus basi squamulis parvis compluribus (ut videtur 16–20) instructae; tubus corollae circiter 2.6–2.8 cm. longus, basi ellipsoideo-inflata obliqua circiter 6 mm. crassus, deinde cylindricus, circiter 3 mm. crassus, versus faucem iterum sensim dilatatus, fauce fere 1 cm. crassus; lobi 1.2–1.4 cm. longi, apice cohaerentes, e basi lata anguste elliptico-oblongi, medio circiter 3 mm. lati, intus ad carinam sparse pilosuli, ceterum ut videtur glabri, margine ciliato; corona cyathiformis, foliolis fere ad medium connatis, in dentes anguste triangulares acutos apice incisos hirsuto-ciliatos ligulis interioribus oblongo-linearibus subobtusis glabris circiter 3-plo breviores productis; pollinia oblonga, apice uno apiculata, translatoribus ut videtur angustis brevibus; retinacula anguste oblongoidea, apice uno leviter dilatata.

Western Szech'uan: Hsiao-chin Ho valley, near Monkong Ting, alt. 1900–2600 m., June 30, 1908 (No. 2313, type; climber 1–1.5 m., flowers dark).

Our plant seems most closely related to *C. sootepensis* Craib (in *Kew Bull. Misc. Inform.* 1911, 420. — Costantin in Lecomte, *Fl. Gén. Indo-Chine*, IV. 152 [1912]) from Siam, Chiengmai, in open deciduous jungle on Doi Sootep, alt. 450 m., *A. F. G. Kerr* (No. 695, type ex Craib) as far as I can judge by the author's description. In the Siamese species the leaves are described as subsessile, and I can hardly believe that the plant from western Szech'uan can be identical with this southern species.

TYLOPHORA R. Br.

Tylophora pseudotenerrima Costantin in Lecomte, *Fl. Gén. Indo-Chine*, IV. 108 (1912).

Western Szech'uan: west of Kuan Hsien, Pan-lan-shan valley, alt. 2100–2900 m., June 1908 (No. 2244; climber 0.75–1.5 m. tall, flowers green).

Our plant is very similar to *Tylophora tenuissima* Wight & Arnott according to the plate in Wight, *Icon.* II. t. 588, which differs in the dark color of the flowers and in its longer inflorescence, and is usually regarded as a form or even a synonym of *T. tenuis* Blume. Otherwise it agrees almost perfectly with Costantin's description which was drawn from specimens from Indo-China, the type having been collected by *Godefroy* in " Cambodga: mᵗ de Pursat." The pollinia seem to be erect and included in the upper part of the anthers.

Tylophora nana Schneider, n. sp.

Frutex 10–30 cm. altus, aspectu *Thesii;* ramuli gracillimi, teretes, striati, viridescentes, villoso-puberuli, basi glabrescentes, satis dense foliati. Folia lineari-lanceolata, margine revoluta, 1.2–2 cm. longa, 1.5–3 mm. lata, utrinque acuta, superne obscure viridia, puberula, costa impressa, subtus pallida, vix v. paulo pilosa, costa prominente, ceterum fere enervia; petioli 1–2 mm. longi, modo ramulorum puberuli. Cymae axillares, sessiles, pseudo-umbellatae, 4–7-florae; flores parvi, viridi-flavescentes, aperti circiter 8 mm. diametientes; pedicelli gracillimi, vix ad 5 mm. longi, plerique glabri, basi bracteis parvis linearibus fere glabris instructi; alabastra oblongo-subacuminata, circiter 5 mm. longa, lobis sinistrorsum convolutis; calycis segmenta fere libera, lineari-lanceolata, acuminata, pleraque glabra, 1–1.5 mm. longa, intus basi interdum inter se squamula una minima instructa; corollae lobi fere ad basim liberi, anguste lanceolati, acuti, 3–4 mm. longi, glabri, saepe paulo contorti; gynostegium calyci aequilongum, sessile; coronae foliola ut videtur minima, ad basim gynostegii affixa, breviter gibbosa sed in floribus siccis difficile certe recognoscenda; antherae breves, subquadratae, appendice brevi hyalino rotundato in stigma inflexo coronatae; pollinia ut videtur erecto-patentia, circuitu oblongo-subrotunda, translatoribus tenuibus curvatis patentibus brevibus; retinacula fere rotunda, polliniis paulo minora; stigmatis caput fere planum, medio paulo depressum. Folliculi ignoti.

Western Szech'uan: Wênch'uan Hsien, Min River valley, arid places, alt. 1300 m., May 3, 1908 (No. 2253; shrub 10–30 cm. tall).

This plant looks like a *Thesium* and also somewhat resembles *Cynanchum sibiricum* R. Brown, which has longer and broader leaves and shortly pedunculate inflorescence. On account of the very small polliniae which are included in the upper part of the anthers I refer this species to *Tylophora* without being able to indicate its relationship to any species hitherto described from eastern Asia or India. The shape of the gynaeceum and especially of the anthers is somewhat similar to those of *T. cuspidata* Zippel as represented by Decaisne on t. 10, fig. e in *Ann. Sci. Nat.* sér. 2, IX. (1838), but otherwise this species has very different leaves, inflorescence and corolla-lobes. The corona seems to be very small, and it is most difficult to ascertain its exact form in dried specimens.

DREGEA E. Mey.

Dregea sinensis Hemsley in *Jour. Linn. Soc.* XXVI. 115 (1889). — K. Schumann in Engler & Prantl, *Nat. Pflanzenfam.* IV. Abt. 2, 293 (1895).—Diels & Schlechter in *Bot. Jahrb.* XXIX. 544 (ut videtur pro parte) (1900). — Diels in *Not. Bot. Gard. Edinburgh*, VII. 357 (*Pl. Chin. Forrest.*) (1913).

Western Hupeh: north and south of Ichang, alt. 30–900 m., cliffs, etc., May 1907 (No. 682, climber 1.5–2.5 m. tall, flowers white or pink); same locality, etc., November 1907 (No. 682ᵃ; fruits); same locality, *A. Henry* (No. 3554; also probably No. 1767, type ex Hemsley); without exact locality, *A. Henry* (Nos. 6164, 6387; Hemsley also mentions as co-types Nos. 3005, 3125, 3868). Western Szech'-uan: Mupin, alt. 1300–1600 m., thickets, June 1908 (No. 682ᵇ; climber 1.5–2.5 m. tall, flowers pink); Mao-chou, arid places, alt. 1300–1600 m., May 24, 1908 (No. **2245**; climber 1.25–1.50 m. tall). Yunnan: between Yung-peh-ting and Tai-nao-ko, climbing on shrubs and creeping on the ground, alt. about 2400 m., July 3, 1914, *C. Schneider* (No. 1730; flowers white with violet markings, semiscandent shrub up to 2 m.).

The plant collected by me in Yunnan (No. 1730) is very similar to Wilson's No. 2245, both having small leaves from 2.5 to 3.5 cm. in length and 2 to 3.5 cm. in width, but there is scarcely any other difference on which a variety could be based. Possibly, however, the fruits are somewhat different. The ripe follicles before me of Wilson's No. 282ᵃ are up to 7 cm. long and 1.8 cm. broad at the base and are long-acuminate at the apex. The seeds are about 11 mm. long and bear a coma of about the same length (12–13 mm.). The shape of the calyx-lobes, the amount of the pubescence on the leaves and inflorescence and other details seem to be rather variable. I also collected specimens in southern Szech'uan in the district of Kuapie, of which I have only a small flowering piece before me which is very similar to the following species, but as far as I can see from my notes the color of the flowers is the same as that given for No. 1730.

Dregea corrugata Schneider, n. sp.

Dregea sinensis Schlechter & Diels in *Bot. Jahrb.* XXIX. 544 (pro parte, non Hemsley) (1900).

Frutex scandens, ad 3 m. altus; ramuli vetustiores glauco-cinerei, valde rimosi, subvillosi, hornotini dense v. laxius tomentoso-villosi. Folia membranacea v. adulta crassiuscula, late ovata v. ovato-oblonga, basi pleraque profunde cordata, rarius tantum subcordata, apice sensim acuta v. subacuminata, 5–9.5 cm. longa et 2.8–5.5 cm. lata, margine integra, supra intense viridia, sparse (in costa distinctius) villosula, subtus albo-(v. flavo-)villoso-tomentosa, mollia, in nervis lateralibus utrinsecus 4–5 conspicuis minus quam in facie pilosis; petioli villoso-tomentosi, 1.5–4 cm. longi. Inflorescentiae axillares, umbellato-cymosae, pedunculis breviter-villosulis ad 6 cm. longis suffultae, 8–15 florae; pedicelli 2.3–2.7 cm. longi, filiformes, villosuli, basi bracteis parvis linearibus villosulis praediti; flores flavescentes (an lilacino-maculati?), pro genere typici, aperti ad 1.2–1.5 cm. diametientes; calycis segmenta oblonga, plus minusve acuta, 3–4 mm. longa, extus medio villosula versus margines glabrescentia sed margine ipso ciliata, intus glabra, basi inter segmenta squamulis minimis 5 (v. interdum pluribus?) instructa; corollae lobi ovati, obtusi, glabri sed margine satis dense ciliati, circiter 5–6 mm. longi et basi subaequilati; corona pro genere typica, lobis ut videtur ovoideis inflato-saccatis apice distincte apiculatis; pollinia oblonga, leviter curvata, pendula, quam retinaculum anguste oblongum circiter duplo longiora, translatoribus satis distinctis curvatis retinaculo subaequilongis; ovaria minute pilosa, stigmatibus capitatis antheras vix v. paulo superantibus. Folliculi (an satis maturi?) angulo 180° divergentes, 6.5 cm. longi et basim-versus circiter 1.3 cm. crassi, versus apicem valde attenuati, leviter curvati, breviter et laxe villosuli, insigniter (apice excepto) subannulato-plicato-corrugati, in sicco cinereo-flavescentes; semina subobovato-oblonga, circiter 9–10 mm. longa, basi obtusa 4–5 mm. lata, valde compressa, apice coma sericea iis duplo longiore instructa.

Western Szech'uan: Mao-chou, arid regions, alt. 1300–1900 m., June 1, 1908 (No. 1190 type; climber 1.5–3 m. tall, flowers yellowish); same locality, etc., August 1908 (No. 1190ᵃ; with fruits). Southern Szech'uan: Nanch'uan, *A. von Rosthorn* (No. 2351; with young fruits; No. 2350; with flowers).

This species is very similar to *D. sinensis* Hemsley, but the fruits look distinct, as they have numerous annular crest-like excrescences, while those of *D. sinensis* have indistinct and irregular longitudinal excrescences. The pubescence of the new species is even more dense than that of *D. sinensis*, especially on the under surface of the leaves. The specimens from Nanch'uan mentioned above seem to belong to *D. corrugata* according to a young fruit on No. 2351. It is difficult to decide from dried specimens whether the differences in the structure of the flowers are sufficient to distinguish the two species and even some of the forms of *D. sinensis*. The color of the flowers is described differently by different collectors. I suppose the flowers are creamy white marked with purplish or lilac spots which more or less or wholly disappear in drying.

CONVOLVULACEAE.

Determined by CAMILLO SCHNEIDER.

CONVOLVULUS L.

Convolvulus Ammanii Desrousseaux in Lamarck, *Encycl. Méth.*
III. 549 (1791). — Choisy in De Candolle, *Prodr.* IX. 402 (1845). —
Ledebour, *Fl. Ross.* III. pt. 1, 90 (1847). — Hance in *Jour. Bot.* XX.
292 (1882). — Hemsley in *Jour. Linn. Soc.* XXVI. 165 (1890).

Western Szech'uan: valley of Hsao-chin Ho, near Monkong
Ting, rare, alt. 2400 m., June 1908 (No. **2288**; forming dense tufts,
flowers white, tinged pink).

PORANA Burm.

Porana sinensis Hemsley in *Jour. Linn. Soc.* XXVI. 167 (1890). —
Diels in *Bot. Jahrb.* XXIX. 544 (1900).

Western Hupeh: Changlo Hsien, side of streams, alt. 600 m.,
June 1907 (No. **3221**; climber 3–5 m., flowers purple). Western
Szech'uan: Shih-ch'uan Hsien, thickets, alt. 1000–1300 m., August
1910 (No. **4625**; climber 2.5–3 m., flowers purple). Southeastern
Szech'uan: Nanch'uan, "Kung chia p'ing," August 1891, *A. von
Rosthorn* (No. 509; forma affinis ex Diels). Yunnan: Szemao, mountains to west, alt. 1600 m., *A. Henry* (No. 11892; large climber, blue
flowers); Mengtsze, mountains, alt. 1600 m., *A. Henry* (Nos. 10715[a],
10715[b]; large climber, flowers blue). Kwangtung: without precise
locality, *C. Ford* (No. 290, type, ex Hemsley).

The specimens of Wilson and Henry before me seem to represent a form of this
species, the type of which I have not seen; the leaves are less pubescent and almost
glabrous when the fruits are ripe. The size of the flowers seems to be rather variable; Hemsley says they are about an inch broad, which is the size of the flowers of
Henry's No. 11892, while on Wilson's No. 3221 they are only 2 cm. broad and in
No. 4625 they seem to be even a little smaller. In the form mentioned by Diels
the flowers measure about 2 cm. in width. Otherwise I cannot detect any good
character on which to base even a variety. On Wilson's No. 4625 the corolla is
sparsely hairy within below the lobes, while on No. 3221 the flowers are glabrous
within except the top of the ovary and the lower part of the style. Hemsley did
not see the fruits, and it may happen that the fruits of the true *P. sinensis* have 3
of the sepals enlarged, as is the case in *P. spectabilis* Kurz, which has much larger

flowers (see key, p. 358). On Wilson's No. 4625 the young fruits show only two enlarged sepals, and they differ from the ripe fruits of Henry's No. 10715[b] only in the size and shape of the smaller three sepals, which are longer-pointed and about twice as long as the unripe fruit. Of the ripe fruits I add the following description:

Fructus maturi sepalis scariosis valde inaequalibus circumcincti, olivacei, obovato-oblongi, circiter 1–1.2 cm. longi et 0.8 cm. crassi, apice flavo-pilosi, stylo sicco aequilongo versus basim piloso et incrassato filiformi coronati. Sepala 2 (exteriora?) valde elongata oblonga, 6.5–7 cm. longa, fere tota longitudine 1.2–1.5 cm. lata, basi paulo contracta, apice rotundata, utrinque sparse puberula, nervis 5 distinctis parallelis longitudinaliter percursa, 3 (interiora?) fructu subaequilonga apice subito in acumen brevem pilosum contracta; subdenticulata; semen 1, flavobrunneum, a latere compressum, irregulariter subrotundum, circiter 8 mm. latum, 3–4 mm. crassum.

Porana triserialis Schneider, n. sp.

Frutex scandens, ad 5 m. altus; ramuli floriferi herbacei, ut videtur flavo-virides, in sicco flavo-brunnei, subteretes v. angulato-compressi et saepe contorti, glabri. Folia alterna, membranacea, rotundato-ovata, basi profunde lateque cordata, apice satis subito acuminata, brevicaudata, margine integra v. pleraque utrinque 1, varius 2 lobulis brevibus acutis instructa, supra ut videtur laete viridia, nervis primariis praecipue 5 e basi palmatim orientibus plus minusve puberulis ceterum glabra, subtus paulo pallidiora, tantum ad basim axillis nervorum sparse breviter pilosa, 6–10 cm. longa (acumine c. 1.5 cm. longo incluso), 5–7.5 cm. lata; petioli graciles, 2–4 cm. longi, a latere compressi, glabri. Inflorescentia axillaris, pedunculo 8–15 cm. longo parte superiore foliolis paucis cordato-oblongis longe acuminatis ad 3 cm. longis petiolatis instructo, cymoso-pseudoracemosa, circiter 15-flora, ad 10 cm. longa, glabra; flores pulchre coerulei, glabri, infundibuliformes, parte explanata corollae apertae circiter 3–3.5 cm. diametiente et circiter 1.8 cm. longa, lobis 5 circiter 9 mm. longis 13 mm. latis rotundis medio apiculatis apiculo piloso, tubo circiter 1.5–1.7 cm. longo vix 1.5–1.8 mm. crasso ima basi intus piloso; stamina 5, triserialia, in tubo inclusa, filamentis liberis filiformibus quam antherae oblongae subbrevioribus; gynaeceum glabrum, tubo triplo brevius, stylo quam ovarium subduplo longiore filiformi stamina seriei infimae vix attingente, stigmate obovato-oblongo subclavato apice emarginato; ovarium ut videtur biloculatum, ovariis 2 sessilibus instructum; sepala 5, anguste lanceolata, acuminata, inaequalia, longiora ad 8 mm. longa, glabra, nervis 3 subparallelis longitudinaliter percursa. Prophylla 2, sepalis proxima, brevioribus valde similia. Pedicelli graciles, 8–12 mm. longi, glabri, bracteis linearibus triente brevioribus instructi. Fructus adhuc ignoti; vide var. *lasiam*, p. 362.

Western Szech'uan: Tung River valley, near Wa-shan, alt. 100–
900 m., September, 1908 (No. 3220, type; climber 2.5–5 m. tall,
flowers beautiful blue). Eastern Szech'uan: without precise locality,
A. Henry (No. 8866; distributed as *Porana?* aff. *P. grandiflorae*
[= *Ipomaea* No. 26¹ of Ind. Fl. Sin.]). Yunnan: Mengtsze, woods,
alt. 1500–1600 m., *A. Henry* (Nos. 9229, 9229ᶜ partim; large climber,
lavender flowers).

Without having seen the fruits of var. *lasia* (p. 362) I should have taken this
species for an *Ipomaea*.¹ It is somewhat similar to *P. grandiflora* Wallich, from
which it differs as indicated in the key, p. 358, especially by the stamens inserted in
3 rows in the tube and not at the same height.

To describe this new species correctly I was obliged to look over all the species
of *Porana* that have been described, and have prepared the following short synopsis
of them, adding descriptions of three new species and a variety collected by Henry
in Yunnan.

CLAVIS SPECIERUM GENERIS PORANAE ²

Ovaria ovulis 4 instructa; stigmata plus minusve globoso-capitata, pleraque latiora
quam longa; stamina versus basim corollae fere eadem altitudine inserta
filamentis liberis aequilongis v. inaequalibus quam antheras semper longioribus. Corolla plus minusve campanulata . . . . Subgen. 1. EUPORANA.
Sepala omnia in flore aequalia.
Stamina inaequalia, tria longiora corollam aequilonga v. superantia, dua
breviora in tubo inclusa.
Stylus ovario 4–5-plo longior, apice bipartitus; discus nullus; sepala in
flore satis lata, oblongo-elliptica, in fructu omnia amplificata.
1. *P. volubilis*.
Stylus ovario subbrevior, simplex; discus satis distincte evolutus; sepala in
flore anguste lanceolata, in fructu ignota . . . . 2. *P. discifera*.
Stamina subaequalia, in tubo inclusa; stylus ovario brevior; sepala in fructu
tantum tria valde amplificata . . . . 3. *P. paniculata*.
Sepala in flore inaequalia; stamina subaequilonga, stylus iis sublongior.
Flores aperti vix ad 2.5 cm. lati tubo circiter 8 mm. longo; sepala tantum 2
in fructu valde amplificata, ad 7 cm. longa.   4. *P. sinensis*.

¹ This *Ipomaea* is said by Hemsley to be " *I. coccinea* simillima sed sepalis
puberulis exapiculatis "; it apparently has nothing to do with our plant.
² I do not know *P. Esquirolii* Léveillé (in Fedde, *Rep. Spec. Nov.* IX. 444 [1911],
of which the author gives only the following short description: " Affinis *P. paniculatae* Roxb. a quo tamen differt sequentibus notis; planta tota villosa subtus in
foliis et floribus tantum rufo-tomentosa; petioli 3–5 cm. longi, sepala lanceolata
obtusa; flores coerulei." Collected in Kweichou by *J. Esquirol* (No. 976.).
I have not been able to see the descriptions of *P. Delavayi*, *P. Duclouxii*,
and *P. Mairei* published by Gagnepain in Lecomte's *Notulae Systematicae*, III.
fasc. 5, as no parts of this publication have been received here since 1913. Also
fascicle 3 of Lecomte, *Fl. Gén. Indo-Chine*, vol. IV., has not been received; this
contains *Convolvulaceae* by Gagnepain & Courchet with 3 species of *Porana* of
which *P. spectabilis* Kurz is figured. I unfortunately did not know of these
publications until after this paper was in type.

Flores aperti 2.5–3 cm. lati, tubo ad 12–15 mm. longo; sepala (fide Kurz)
  tria in fructu amplificata, ad 4 cm. longa . . . . 5. *P. spectabilis.*
Ovaria ovulis 2 instructa.
Stigmata staminaque fere ut in Euporana, discus distinctus.
                                   Subgen. 2. PSEUDODINETUS.
                                             6. *P. dinetoides.*
Stigmata plus minusve claviformia, paulo longiora quam lata, apice emarginata;
  stamina 1–v. 3-seriatim in tubo inserta filamentis liberis quam antherae
  distincte brevioribus; discus nullus (an semper?).   Subgen. 3. DINETUS.
Sepala in fructu omnia amplificata, aequalia; flores parvi, 8–12 mm. longi.
Sepala in fructu ad 2.5 cm. longa et 9–11 mm. lata, elliptico-oblonga.
                                             7. *P. malabarica.*
Sepala in fructu ad 12–15 mm. longa, versus apicem 3–4 mm. lata, oblongo-
  spathulata . . . . 8. *P. racemosa.*
Sepala in fructu inaequalia 3 valde 2 minus aucta; flores majores, 2.5–3.5 cm.
  longi.
Sepala in flore subaequalia; stamina parte basali leviter dilata 1 seriatim
  inserta; tubus intus infra stamina circa gynaeceum pilosus.
Sepala fructifera majora circiter 1 cm. lata; bracteae inflorescentiae
  parvae, lanceolato-filiformes . . . . 9. *P. grandiflora.*
Sepala fructifera majora (fide Kurz) vix 4 mm. lata; bracteae inflorescentiae foliaceae v. lanceolatae, caudatae . . . 10. *P. stenoloba.*
Sepala in flore distincte inaequalia; stamina in tubo haud dilatato 3-
  seriatim inserta; tubus tantum ima basi intus circa ovarium pilosulus.
                                             11. *P. triserialis.*

ENUMERATIO SPECIERUM GENERIS PORANAE.[1]

Subgen. 1. EUPORANA Peter in Engler-Prantl, *Nat. Pflanzenfam.* IV. Abt.
3ᵃ, 24 (emend.) (1891).

The type of *Porana* is *P. volubilis* Burman, the ovary of which contains 4 ovula
and bears a two-cleft style. I think it best to refer to *Euporana* all the species with
ovaries containing 4 ovula. All those species have globular stigmas.

1. **Porana volubilis** Burman, *Fl. Ind.* 51, t. 21, fig. 1 (1768). — Roxburgh, *Fl.
Ind.* II. 40 (1824); ed. 2, 465 (1832). — Wight, *Icon. Pl. Ind. Orient.* II. t. 347
(1843); *Ill. Ind. Bot.* II. t. 168ᵇ, fig. 8 (1850). — Choisy in De Candolle, *Prodr.*
IX. 436 (1845). — Kurz in *Jour. Bot.* XI. 137 (1873). — Clarke in Hooker f., *Fl.
Brit. Ind.* IV. 222 (1883). — Koorders, *Excursionsfl. Java,* III. 114 (1912).

INDIA. Malaya: (No. 58741, Herb. Griffith); without locality (No. 2000,
Herb. Wight).

2. **Porana discifera** Schneider, n. sp.

Frutex alte scandens; ramuli biennes purpureo-brunnei, contorto-striati, lenticellosi, sparse pubescentes, floriferi dense flavo-tomentelli. Folia late cordato-
rotundata, basi profunde et subanguste cordata, apice satis subito late acutata,

---

[1] Bentham, F. v. Mueller and Peter refer to *Porana* as a distinct subgenus also
Duperreya sericea Gaudichaud, *Voy. Freyc. Bot.* I. 452, t. 63 (1826) (*Ipomoea modesta*
F. v. Mueller, *Fragm. Phys. Austral.* II. 22 [1860]. — *Porana sericea* F. v. Mueller,
l. c. VI. 100 [1867]) from Australia. But this plant is so distinct regarding the single
axillary flowers, the linear leaves with very short petioles, the shape of the flowers
and the silky pubescence that I think it best not to unite *Duperreya* with *Porana*.

membranacea, subtus viridia, sparse (praecipue in nervis) v. vix puberula, subtus plus minusve flavo-pubescentia, basi 5-nervata, margine obscure breviter denticulata v. subintegra, ad 11 cm. longa et 9 cm. lata, inflorescentiorum minora, subacuminata, supra laxe, subtus dense flavo-tomentella, 2–7 cm. longa, 1.2–4 cm. lata; petioli foliorum majorum circiter 1.8 cm. longa, plus minusve puberula, foliorum inflorescentiarum breviores, tomentelli. Inflorescentiae multiflorae, cymoso-paniculatae, axibus primariis foliatis ad 50 cm., secundariis ad 15 cm. longis, partibus omnibus flavo-tomentellae; pedunculi (axes tertiarii) multiflori, bracteis lanceolatis parvis suffulti; flores flavo-albi, minimi, 4–5 mm. longi, campanulati, limbo aperto circiter 4 mm. diametiente; petioli floribus subaequilongi, medio et apice prophyllis 2 minimis instructi; sepala 5, aequalia, lanceolata, subacuta, corolla fere 3-plo breviora, extus ut corolla dense flavo-tomentella, fructifera ignota; lobi corollae lati, apiculati, ad margines extus apice excepto glabriusculi, tubo 3-plo breviores; stamina 5, inaequalia, omnia ad partem inferiorem sparse pilosum tubi insertae, 3 longiora corollam superantia, 2 tubo breviora; filamenta libera glabra, antheris versatilibus oblongis pro genere normalibus longiora; gynaeceum breve, vix medium tubi aequans; ovarium apice pilosum stylo simplici vix aequilongo stigmatibus 2 subglobosis cohaerentibus coronato, basi disco satis distincto circumcinctum, uniloculare, ovulis 4 sessilibus instructum. Fructus ignotus.

CHINA. Yunnan: Szemao, forests to south, alt. 1600 m., *A. Henry* (No. 12694, type; large climber, flowers yellowish white); same locality, eastern forests, alt. 1600 m., *A. Henry* (No. 12622; large climber).

This species is most closely related to *P. paniculata* Roxburgh, but differs in its even smaller flowers, its unequal stamens, its pilose ovaries and in its distinct disk. The fruits may show some other distinguishing character.

3. **Porana paniculata** Roxburgh, *Pl. Coromand.* III. 31, t. 235 (1819); *Fl. Ind.* II. 39 (1824); ed. 2, I. 464 (1831). — Choisy in De Candolle, *Prodr.* IX. 436 (1845). — Kurz in *Jour. Bot.* XI. 137 (1873). — Clarke in Hooker f., *Fl. Brit. Ind.* IV. 222 (1883). — Hooker f. in *Bot. Mag.* XLVIII. t. 7240 (1892). — Koorders, *Excursionsfl. Java,* III. 114 (1912).

Dinetus paniculatus Sweet, *Hort. Brit.* 289 (1827); ed. 2, 373 (1830).

CHINA. Yunnan: Mengtsze, alt. 1500 m., *A. Henry* (No. 9489; large climber, covering rocks, etc., white flowers in masses).

INDIA. Kumaon: "The Bhabar," alt. 300 m., *R. Strachey & J. E. Winterbottom* (No. 1). Punjab: "reg. trop.," *T. Thomson*. Central India: Malva, plateau of Amarkántak, January 21/24, 1856, *Schlagintweit* (No. 11859). Sikkim: "reg. temp. alt. 1–5000 ped.," *J. D. Hooker*. East Bengal: "Mont. Khasia, reg. trop., alt. 3–5000 ped.," *J. D. Hooker & Thomson*.

See the note under *P. discifera* and the key on p. 357.

4. **Porana sinensis** Hemsley. See p. 355.

5. **Porana spectabilis** Kurz in *Jour. Bot.* XI. 136 (1873). — Clarke in Hooker f., *Fl. Brit. Ind.* IV. 221 (1883).

Porana speciosa Bentham in Bentham & Hooker, *Gen. pl.* II. 876 (sphalm.) (1876).

INDIA. Andaman Islands: "Baratang Island," December 30, 1903, *Dr. Prain's Collector* (No. 8). Burma: in the tropical evergreen forests of Martaban, in the Toukyeghat Valley, east of Tongu (type, ex Kurz).

See the note under *P. sinensis* Hemsley and the key, p. 357.

Subgen. 2. PSEUDODINETUS Schneider, n. subgen. — Descriptio in clavi et subtus.

6. **Porana dinetoides** Schneider, n. spec.

Frutex scandens; ramuli floriferi flavo-brunnei, angulato-rotundi, contorti, flavo-hirsuto-pubescentes. Folia profunde cordato-ovata v. cordato-rotundata apice subito acuminata, saepe caudata, supra laxius subtus densius pilis flavidis molliter hirsuto-pubescentia, basi 5-nervia, margine integra, ciliata, majora 7–10 cm. longa, 4.5–9 cm. lata, minora (inflorescentiarum) 2.5–5.5 cm. longa, 1.2–4 cm. lata; petioli foliorum majorum 3–6 cm., minorum 0.3–1.5 cm. longi, flavo-hirsutuli. Inflorescentiae pseudopaniculatae, axillares; pedunculi circiter 15-flori, inferne nudi, foliis plus minusve longiores, ut pedicelli flavo-hirsutuli v. glabri; pedicelli 4–7 mm. longi, bracteis triplo brevioribus lanceolatis hirsutis suffulti; flores albi, campanulati, 7–8 mm. longi, corolla aperta circiter 7 mm. lata; prophylla 2 minima; sepala satis aequalia, anguste lanceolata, glabra v. sparse flavo-hirsutula, tubo corollae vix dimidio breviora; corolla extus glabra, tantum in apicibus loborum satis brevium hirsutulia, intus in basi infra insertionem staminum pilosa; stamina 5, inclusa, eodem altitudine inserta, filamentis liberis inaequalibus brevissimis quam antherae oblongae deinde paulo contortae vix longioribus, ceteris iis 1½–2-plo longioribus; gynaeceum sepalis aequilongum, glabrum, disco distincto basi instructum; ovarium ovulis 2 sessilibus praeditum, stylo simplici paulo longiore stigma simplex ovato-globosum gerente coronatum. Fructus pro genere normales, sepalis scariosis 3 valde acutis late oblongis circiter 1 cm. longis 4 mm. latis apiculati 2 paulo brevioribus angustioribus omnibus nervis 5 longitudinaliter percursis glabris circumcincti, elliptico-oblongi, circiter 7 mm. longi, 4 mm. crassi, styli basi coronati, glabri, flavo-brunnei; semen 1, obovato-rotundum, a latere compressum, rubro-brunneum, circiter 6 mm. longum.

CHINA. Yunnan: Mengtsze, alt. 1700 m., *A. Henry* (No. 9340, type; climber, flowers white; No. 9340ᵃ; with fruits); same locality, alt. 1500 m., *A. Henry* (No. 9340ᵇ; hairy form with younger inflorescences).

At first sight this species resembles *P. racemosa* Roxburgh, but the structure of the flowers is very different. It is somewhat intermediate between the subgenera *Euporana* and *Dinetus*, and I think it best to refer it to a separate subgenus, *Pseudodinetus*. The stamens are of unequal length but are inserted in one row; the stigma seems to be different from those of *Euporana* and of *Dinetus*, being globular as in the first but also slightly emarginate as in the last subgenus. The inflorescence of the type is glabrous or nearly so, except the small bracts, while in No. 9340ᵃ the peduncles, petioles and sepals are more or less hirsute at least when young.

Subgenus 3. DINETUS Peter in Engler & Prantl, *Nat. Pflanzenfam.* IV. Abt. 3 a, 25 (1891), exclud. *P. paniculata*, Roxburgh.

To this subgenus, the type of which is *Porana racemosa* Roxburgh, I refer all the species with the exception of *P. dinetoides*, the ovaries of which contain only two ovules. They have a more distinct narrower tube, and the stamens have very short filaments. According to the size of the flowers there are two distinct groups in this subgenus, the first representing *Dinetus* proper, the flowers of which are small and all the sepals are equally enlarged on the fruit; and the second with larger flowers and the sepals unequally enlarged on the fruit. (See also the key, p. 358.)

7. **Porana malabarica** Clarke in Hooker f., *Fl. Brit. Ind.* IV. 223 (1883).

Porana racemosa Dalzell & Gibson, *Bombay Fl.* 162 (non Roxburgh, fide Clarke) (1861).

INDIA. Malabar and Concan: without precise locality; *Stocks & Law Porana*, No. 3, Herb. Ind. Or.).

This species is closely related to *P. racemosa* Roxburgh, differing from it apparently only in the larger size of the sepals on the fruit.

According to C. B. Clarke *P. truncata* Kurz (in *Jour. Bot.* XI. 136 [1873]) " is intermediate between that species and *P. malabarica.*" It was collected in Burma (Pegu and Martaban), and I have not seen a specimen. Kurz mentions Griffith's No. 5876 from East Bengal as related to his species. I have a specimen of this before me, bearing only fruits which are very similar to those of *P. malabarica*, but the sepals are more ovate, rather acute at the apex, and more contracted at the base; the fruits appear to be more globular.

8. **Porana racemosa** Roxburgh, *Hort. Bengal.* 13 (nomen nudum) (1814); *Fl. Ind.* II. 41 (1824); ed. 2, I. 466 (1832). — D. Don, *Prodr. Fl. Nepal.* 98 (1825). — Choisy in De Candolle, *Prodr.* IX. 436 (1825). — Wight, *Ill. Ind. Bot.* II. t. 168[b], fig. 9 (1850); *Icon. Pl. Ind. Orient.* IV. t. 1376 (1850). — Kurz in *Jour. Bot.* XI. 137 (1873). — Clarke in Hooker f., *Fl. Brit. Ind.* IV. 222 (1883). — Hemsley in *Jour. Linn. Soc.* XXVI. 166 (1890). — Diels in *Bot. Jahrb.* XXIX. 544 (1900); in *Not. Bot. Gard. Edinburgh*, VII. 387 (*Pl. Chin. Forrest.*) (1913). — Koorders, *Excursionsfl. Java*, III. 114 (1912).

> *Porana cordifolia* Ledebour, *Ind. Hort. Dorp.* 1824, suppl. 6, fide Choisy.
> *Porana dichotoma* Hamilton ex D. Don, *Prodr. Fl. Nepal.* 99 (pro synon.) (1825).
> *Dinetus racemosus* Hamilton apud Sweet, *Brit. Fl. Gard.* II. t. 127 (1825).
> *Porana elegans* Zollinger & Moritzi in *Natuur. en Geneesk. Archief*, II. 571 (fide Koorders) (1845).

CHINA. Western Hupeh: near Ichang, *A. Henry* (No. 2595); Nanto and mountains to northward, *A. Henry* (No. 3062); without precise locality, *A. Henry* (Nos. 2905, 4357, 7004). Yunnan: Mi-lê, *A. Henry* (No. 9954; large climber, white flowers). INDIA. East Bengal: " Mont. Khasia, reg. trop., alt. 2–5000 ped.," *J. D. Hooker & T. Thomson;* without locality (No. 5877[1], Herb. Griffith.). Sikkim: " reg. trop., alt. 2–4000 ped.," *J. D. Hooker.* Sunda Islands: Timor, Futunaba Hills, 1881–2, *H. O. Forbes* (No. 4104).

This is a widely distributed half-shrubby or almost herbaceous climber, the " Snow-creeper " of the English, according to Clarke, " one of the most beautiful of Indian plants." The type was collected in Nepal.

9. **Porana grandiflora** Wallich in Roxburgh, *Fl. Ind.* II. 41 (1824). — Choisy in De Candolle, *Prodr.* IX. 436 (1845). — Kurz in *Jour. Bot.* XI. 138 (1873).— C. B. Clarke in Hooker f., *Fl. Brit. Ind.* IV. 221 (1883).

INDIA. Sikkim: " reg. trop. 5–7000 ped.," *J. D. Hooker.* Nepal: Katumanda, mountains of Shivapoor (Sheopore) (type, ex Roxburgh, which seems to be the same as Wallich's No. 1324).

A very distinct species, the tube of the flowers of which is a little enlarged at the point of insertion of the stamens, which are in one row.

10. **Porana stenoloba** Kurz in *Jour. Bot.* XI. 136 (1873). — C. B. Clarke in Hooker f., *Fl. Brit. Ind.* IV. 221 (1883).

INDIA. Sikkim: " not unfrequent along the post-road from Kersiang to Darjeeling at 5–6000 feet elevation, on metamorphic rocks," October 1868, *S. Kurz* (type, ex Kurz).

According to Clarke " this differs from *P. grandiflora* in the shallowly cordate leaves, the remarkable bracts, the color of the flowers, and the exceedingly narrow fruit-sepals." The author and Clarke do not say anything about the stamens and the gynaeceum, and I have not seen a specimen. This species needs further observation.

11. **Porana triserialis** Schneider. See p. 356.

Porana triserialis, var. lasia Schneider, n. var.

A typo recedit pedunculis, petiolis calycibusque satis dense subglanduloso-puberulis. Fructus maturi sepalis inaequalibus 3 majoribus late elliptico-oblongis circiter 2.5 cm. longis et medio 1 cm. latis nervis 5–7 longitudinaliter percursis utrinque laxe puberulis 2 minoribus acutioribus fructu paulo longioribus, prophyllis 2 parvis satis dilatis pilosis circumcincti, flavescentes, venosi, glabri, subglobosi, circiter 9 mm. crassi, styli basi coronati; semen 1, flavo-brunneum, laeve, obovato-rotundum, a latere compressum, circiter 7 mm. crassum.

CHINA. Yunnan: probably at Mengtsze, *A. Henry* (No. 9229ᶜ partly, flowering type; No. 9229ᵇ; with fruits).

The flowers and leaves agree well with those of the type; see my remarks on p. 357.

BORAGINACEAE.

Determined by E. H. WILSON.

EHRETIA L.

Ehretia acuminata R. Brown, *Prodr. Fl. Nov. Holland.* I. 497 (1810). — De Candolle, *Prodr.* IX. 503 (1845). — Bentham, *Fl. Austral.* IV. 387 (1869). — Clarke in Hooker f., *Fl. Brit. Ind.* IV. 141 (1883). — Hemsley in *Jour. Linn. Soc.* XXVI. 143 (1890). — Henry in *Trans. Asiat. Soc. Jap.* XXIV. suppl. 62 (*List Pl. Formosa*) (1896). — Matsumura in *Tokyo Bot. Mag.* XII. 83 (1898); *Ind. Pl. Jap.* II. pt. 2, 524 (1912). — Diels in *Bot. Jahrb.* XXIX. 545 (1900). — Matsumura & Hayata in *Jour. Coll. Sci. Tokyo*, XXII. 253 (*Enum. Pl. Formos.*) (1906). — Shirasawa, *Icon. Ess. For. Jap.* II, t. 69, fig. 12–24 (1908). — Pampanini in *Nuov. Giorn. Bot. Ital.* n. ser. XVII. 699 (1910). — Dunn & Tutcher in *Kew Bull. Misc. Inform.* add. ser. X. 177 (*Fl. Kwangtung & Hongk.*) (1912). — Bean, *Trees & Shrubs Brit. Isl.* I. 503 (1914).

Ehretia serrata Roxburgh, *Hort. Beng.* 17 (nomen nudum) (1814); *Fl. Ind.* II. 34 (1824). — Lindley in *Bot. Reg.* XIII. t. 1097 (1827). — De Candolle, *Prodr.* IX. 503 (1845). — Franchet & Savatier, *Enum. Pl. Jap.* I. 333 (1875).
Ehretia pyrifolia D. Don, *Prodr. Fl. Nepal.* 102 (1825).
Ehretia serrata, β *obovata* Lindley in *Bot. Reg.* XIII. sub t. 1097 (1827). — De Candolle, *Prodr.* IX. 503 (1845).
Ehretia ovalifolia Hasskarl in *Cat. Hort. Bogor.* 137 (non Wight) (1844).
Ehretia serrata, β *pyrifolia* De Candolle, *Prodr.* 503 (1845).
Cordia thyrsiflora Siebold & Zuccarini in *Abh. Akad. Münch.* IV. pt. 3, 150 (*Fl. Jap. Fam. Nat.* II. 26) (1846).
Ehretia acuminata, var. *grandifolia* Pampanini in *Nuov. Giorn. Bot. Ital.* n. ser. XVII. 699 (1910).

Western Hupeh: north and south of Ichang, woods, alt. 300–1150 m., June and August 1907 (No. 74; tree 6–15 m. tall, girth 0.3–2 m., flowers white, fruit yellowish); Hsing-shan Hsien, woodlands, alt. 300–1150 m., June 10 and August 1907 (No. 74ᵃ; tree 6–15 m. tall); Changyang Hsien, woods, alt. 1000–1500 m., June and September 1900 (Veitch Exped. Nos. 819, 1103, Seed No. 1025); Nanto and mountains to northward, *A. Henry* (No. 1941); without locality, *A. Henry* (Nos. 4556, 6538); "Ma-pan-scian," alt. 1000 in May 1907, *C. Silvestri*

(No. 1919). Yunnan: Szemao, alt. 1500 m., *A. Henry* (No. 10454; tree 10 m.; flowers white). Kwangtung: Hainan, 1889, *A. Henry* (No. 8274). Hongkong: *C. Wright* (No. 415). Formosa: Takow, *A. Henry* (Nos. 1135, 1874, 1778, 1778[a]); Bankinsing, *A. Henry* (Nos. 443, 506); South Cape, *A. Henry* (Nos. 952, 922); Maruyama, woods, April 30, 1903, *U. Faurie* (No. 303); " Paehiran," June 13, 1903, *U. Faurie* (No. 302); Tamsui, 1864, *R. Oldham* (No. 348).

This tree is common in woods throughout western Hupeh, where it is known as the Tsu-kang, and the wood is similar to and used for the same purposes as that of *E. macrophylla* Wallich.

The only Australian specimens I have seen are one collected on Richmond River by C. Moore and another from Hastings River collected by Dr. Beckler and both referred by Bentham to his var. *laxiflora*. The Chinese specimens agree well with Bentham's full description of Brown's species, and in the absence of Australian material of the typical plant there is nothing to be done but to follow Clarke and others in referring the Asiatic plant to Brown's species. Nevertheless, I am sceptical as to the two being conspecific, since such distribution is contrary to all we know of Chinese woody plants.

Ehretia macrophylla Wallich in Roxburgh, *Fl. Ind.* II. 343 (1824); *Cat.* No. 901 (1828). — De Candolle, *Prodr.* IX. 503 (1845). — Clarke in Hooker f., *Fl. Brit. Ind.* IV. 141 (1883). — Hemsley in *Jour. Linn. Soc.* XXVI. 145 (1890). — Henry in *Trans. Asiat. Soc. Jap.* XXIV. suppl. 62 (*List Pl. Formosa*) (1896). — Matsumura in *Tokyo Bot. Mag.* XII. 83 (1898); *Ind. Pl. Jap.* II. pt. 2, 524 (1912). — Diels in *Bot. Jahrb.* XXIX. 545 (1900). — Matsumura & Hayata in *Jour. Coll. Sci. Tokyo,* XXII. 254 (*Enum. Pl. Formos.*) (1906). — Shirasawa, *Icon. Ess. For. Jap.* II. t. 69, fig. 1–11 (1908). — Pampanini in *Nuov. Giorn. Bot. Ital.* n. ser. XVII. 699 (1910). — Bean, *Trees & Shrubs Brit. Isl.* I. 504 (1914).

Ehretia Dicksoni Hance in *Ann. Sci. Nat.* sér. 4, XVIII. 224 (1862).
Ehretia corylifolia Wright in *Kew Bull. Misc. Inform.* 1896, 25. — Diels in *Not. Bot. Gard. Edinburgh,* VII. 357 (*Pl. Chin. Forrest.*) (1913).

Kiangsi: near Kuling, alt. 1000 m., common August 1, 1907 (No. 1571; tree 4 m. tall). Western Hupeh: north and south of Ichang, open country, alt. 30–1000 m., abundant, June and September 1907 (Nos. 3554, 3554[a]; tree 6–13 m. tall, girth 0.6–2 m.); same locality, April 1900 (Veitch Exped. No. 84); same locality, *A. Henry* (Nos. 1885, 7171); " Fansien," alt. 800 m., May 20 to June 3, 1906, *C. Silvestri* (No. |1923); " Ou-tan-scien," alt. 2090 m., July 1907, *C. Silvestri* (No. 1925). Western Szech'uan: near Mao-chou, valley of Min River, May 26, 1908 (No. 3554[b]; tree 5–10 m., flowers white, ragrant); without locality, *A. von Rosthorn* (No. 2364). Yunnan:

Mengtsze, woods, alt. 1500–1800 m., *A. Henry* (Nos. 10515, 10548, 10548ᵃ, 10548ᵇ; tree 6 m. tall, flowers white, fragrant, fruit yellow); Tali Range, lat. 25° 40′, alt. 2600–3000 m., on margins of thickets, *G. Forrest* (No. 4769). Kwangtung: Hainan, near Hoihow, alt. 100 m., March 25, 1905 (Herb. Hongk. Bot. Gard. No. 2183). Kianghuai: Nanking, June 3, 1915, *F. N. Meyer* (No. 1415); Formosa, without locality, 1864, *R. Oldham* (No. 347); Takow, *A. Henry* (No. 313; shrub 1–2 m.); Bankinsing, *A. Henry* (Nos. 190, 456; tree 6 m. tall); South Cape, *A. Henry* (No. 323; tree 4 m., fruit yellow).

This tree is very common in warm, rocky valleys throughout western Hupeh and in Szech'uan from river level up to 1500 m. altitude. It would also appear to be plentiful in Yunnan and other warm parts of China and in Formosa. In Japan it has been reported from the island of Shikoku, but I did not meet with it there. At its best it is a tree some 15 m. tall, with a trunk 2 m. in girth, but usually it is only about half that size. The bark is pale gray and fissured; the branches, which are relatively stout, often somewhat tortuous, spread to form a rather broad flattened crown. The leaves are always scabrid on the upper surface, but often this character is marked by the presence of a covering of soft, appressed hairs, which however more or less completely disappear toward autumn. The flowers vary somewhat in size and are white and rather overpoweringly fragrant; the fruit when ripe is yellowish, globose, apiculate and often as much as 1.5 cm. in diameter. This plant varies greatly in degree of pubescence. In Forrest's No. 4769 both surfaces of the leaves, the peduncle and branches of the inflorescence, the calyx and the outside of the corolla are clothed with a soft pale gray pubescence, while my Nos. 3554 and 3554ᵇ are nearly glabrous. These extremes look very different, but in the series of specimens before me is every conceivable intermediate condition, and I do not think that any useful purpose would be served by attempting to separate the extreme forms. The inflorescence varies in size and may be loosely or densely branched and as a character is of no importance in this species. I suspect that this plant is more or less polygamo-diœcious, and it is worthy of note that all the fruiting specimens I have seen have the leaves pubescent on the under side and many of them slightly so above. Wright, when establishing his *E. corylifolia*, overlooked the fact that the scabrid character of the leaves was merely hidden by a soft tomentum.

The colloquial name for this tree in Hupeh is Tzu-kang, and the wood, which is light, tough and strong, makes the best carrying-poles, which are commonly used by the people for carrying loads everywhere in central and western China.

A picture of this tree will be found under No. 0345 of the collection of my photographs.

VERBENACEAE.

Determined by ALFRED REHDER.

CALLICARPA.

Callicarpa Giraldiana Hesse in *Mitt. Deutsch. Dendr. Ges.* XXI. 366, 2 fig. (sine descriptione) (1912).

Callicarpa longifolia Hemsley in *Jour. Linn. Soc.* XXVI. 253 (pro parte, non Lamarck) (1890). — Diels in *Bot. Jahrb.* XXIX. 548 (pro parte) (1900).
Callicarpa Giraldiana Hort. Hesse apud Schneider, *Ill. Handb. Laubholzk.* II. 1048 (nomen nudum) (1912).
Callicarpa Giraldii Hesse apud Rehder in Bailey, *Stand. Cycl. Hort.* II. 629 (1914).

Frutex 1–2-metralis (ex Wilson); ramuli teretes, initio pube stellata farinosa sparsa mox evanescente v. densa et diu persistente vestiti, hornotini saepius purpurascentes, annotini pallide brunnei v. cinereobrunnei. Folia decidua, membranacea, elliptica v. ovato-elliptica v. oblongo-elliptica, rarius obovato-elliptica, acuminata, basi late cuneata, satis dense denticulata v. irregulariter dentata, 6–10 cm. longa et 2.5–6 cm. lata, supra intense viridia, sparse breviter fasciculato-pilosa, demum glabra v. fere glabra, subtus pallide viridia et initio praecipue in nervis et venulis sparse fasciculato-pilosa, demum plus minusve glabrescentia sed vix perfecte glabra, insuper utrinque subtus densius resinoso-punctata, nervis utrinsecus circiter 10 supra ut costa leviter impressis subtus elevatis et trabeculis elevatis conjunctis; petioli 1–1.5 cm. longi, graciles, initio farinoso-stellato-pilosi, demum fere glabri. Cymae multiflorae, densae, 2–3 cm. diam., pedunculatae pedunculo dimidium petiolum subaequante, fasciculato-pilosae; pedicelli circiter 1 mm. longi ut calyx sparse fasciculato-pilosi; calyx 1.5–2 mm. longus, 4-dentatus dentibus ovato-triangularibus v. interdum late triangularibus; corolla calyce duplo longior, purpureo-rosea, lobis brevibus late ovatis; stamina corollam duplo superantes, 5–6 mm. longa, antheris ovalibus basi vix v. leviter emarginatis per totam longitudinem dehiscentibus, circiter 1 mm. longis; stylus stamina superans. Fructus globosus, 3–4 mm. diam., violaceo-purpureus, in cymis densis.

Kiangsi: Kuling, thickets, common, alt. 1200 m., August 1, 1907 (No. **1531**; bush 1.3 m., flowers rose-purple). Western Hupeh:

Hsing-shan Hsien, thickets, alt. 600–900 m., June and October 1907 (No. 439; bush 2–2.6 m. tall, flowers purple, fruit purple); same locality, July 1907 (No. 2193; bush 2–3 m. tall, flowers purple); Changyang Hsien, thickets, alt. 1200 m., June and November 1907 (No. 439ª; bush 2 m., flowers rose-purple, fruits violet-purple); Fang Hsien, thickets, alt. 900–1200 m., June and November 1907 (No. 2196; bush 1–2 m. tall, flowers purple, fruits purple); "Kin-ki-ken," alt. 800 m., June–July 1906, *C. Silvestri* (No. 1948); "Ma-pan-scian," alt. 1000 m., May 1907, *C. Silvestri* (No. 1950); "monti di Nan-tcian, On-scengau," October 1907, *C. Silvestri* (No. 1952); "monti di Nan-tcian, Ta p'in," November 1907, *C. Silvestri* (No. 1953); "catena di Outan-scian," alt. 2000 m., August 1909, *C. Silvestri* (No. 3158); without precise locality, *A. Henry* (Nos. 3107, 5864, 5992, 7312, 7497). Eastern Szech'uan: Nanch'uan, *A. von Rosthorn* (No. 1896). Western Szech'uan: west and near Wên-ch'uan Hsien, August 1908 (No. 2194; bush 2 m., flowers purple); Mt. Omei, July 1904 (Veitch Exped. No. 5100; bush 3 m. tall). Shensi: Tu-kia-po, October 1897, *G. Giraldi;* without precise locality, *G. Giraldi* (Nos. 1360, 1361, 5896). Kwangtung: Hainan, *A. Henry* (No. 6127). Yunnan: Szemao, alt. 1400 m., *A. Henry* (Nos. 12119, 12119ª, 12119ᶜ; shrub 1.5 m., flowers purple, fruits purple). Cultivated: Hort. H. A. Hesse, Weener, Germany, August and October 1910, *H. A. Hesse;* same locality, July 1911, *A. Rehder,* and July and October 1912, *P. Kache.*

This species seems to be most closely related to *C. formosana* Rolfe and *C. japonica* Thunberg; the first is easily distinguished by the denser pubescence, the longer peduncles which exceed the petioles, and by the rounded or obtuse base of the leaves; *C. japonica* differs in its glabrous narrower leaves, smaller glabrous inflorescence on a peduncle about as long as the petiole and in its longer anthers opening at the apex with a pore. It is also near the Indian *C. psilocalyx* Clarke, which differs chiefly in its densely pubescent branchlets, in its long-acuminate leaves usually obtuse or rounded at the base, shorter petioles, smaller inflorescence, and in the filaments scarcely exceeding the corolla-lobes. Wilson's Nos. 439, 439ᵉ, 2193 and 2496, Giraldi's Nos. 1360, 1361 and 5896 and Henry's No. 5992 agree very well with the specimens from Hesse's nursery, which must be considered the type of this species. Hesse's plants were raised from seed sent by Giraldi from Shensi. Most of the other specimens are more glabrescent, and some fruiting specimens with rather narrower leaves may belong to *C. japonica*, but without flowers their position remains uncertain. Von Rosthorn's No. 1896 is more pubescent and approaches var. *subcanescens.*

Here may be added a variety not collected by Wilson.

Callicarpa Giraldiana, var. Rosthornii Rehder, n. comb.

Callicarpa longifolia, var. *Rosthornii* Diels in *Bot. Jahrb.* XXIX. 548 (1900).

Eastern Szech'uan: "Shou-tzu-p'ing," *A. von Rosthorn*, August 1891 (No. 293).

This variety differs chiefly in its smaller and narrower leaves very gradually narrowed into the petiole, in the denser pubescence and in the somewhat longer antlers.

Callicarpa Giraldiana, var. subcanescens Rehder, n. var.

A typo recedit indumento omnium partium densiore, ramulis junioribus pube farinoso fasciculato-piloso obtectis, foliis supra sparse et minute fasciculato-pilosis subtus densius praecipue in nervis et venulis fasciculato-pilosis, cymis et calycibus densius fasciculato-pilosis.

Western Hupeh: without precise locality, *A. Henry* (No. 5864, type). Kiangsi: Kuling, thickets, not common, alt. 1200 m., July 31, 1907 (No. **1528**; bush 1.6–2 m. tall, with immature fruits).

This variety differs from the type in the denser pubescence of all its parts; the leaves are loosely covered on the whole under surface with rather long fascicled hairs. The leaves of Henry's specimens are comparatively narrow, elliptic-oblong to elliptic-lanceolate, the larger ones not exceeding 3.5 cm. in width, while those of Wilson's No. 1528 are mostly elliptic or obovate and up to 6 cm. broad.

Callicarpa japonica Thunberg, *Fl. Jap.* 60 (1784). — Siebold & Zuccarini in *Abh. Akad. Münch.* IV. pt. 3, 154 (*Fl. Jap. Fam. Nat.* II. 30) (1846). — Siebold in *Jaarb. Nederl. Maatsch. Anmoed. Tuinb.* 1845, 71, t. 5, 6. — Miquel in *Ann. Mus. Lugd.-Bat.* II. 98 (1866); *Prol. Fl. Jap.* 30 (1866). — Franchet & Savatier, *Enum. Pl. Jap.* I. 358 (1875). — Maximowicz in *Bull. Acad. Sci. St. Pétersbourg*, XXXI. 77 (1886); in *Mél. Biol.* XII. 508 (1886). — Hemsley in *Jour. Linn. Soc.* XXVI. 253 (1890). — Gilg & Loesener, in *Bot. Jahrb.* XXXIV. Beibl. LXXV. 61 (1904). — Shirasawa, *Icon. Ess. For. Jap.* II. t. 70, fig. 11–19 (1908).

Callicarpa mimurasaki Hasskarl, *Cat. Hort. Bogor. Alt.* 136 (nomen nudum) (1844).
Callicarpa Murasaki Siebold in *Jaarb. Nederl. Maatsch. Anmoed. Tuinb.* 1844 25 (nomen nudum) (1844).
Callicarpa longifolia, a subglabrata Schauer in De Candolle, *Prodr.* XI. 645 (pro parte) (1847).

Shantung: Tsingtau, 1911, *R. Zimmermann* (No. 210). Widely distributed in Japan and Korea.

The specimen cited above is the only one referable to the typical form of this species I have seen from China, and on account of its rather small leaves it resembles somewhat *C. dichotoma* K. Koch. It presents, however, the chief characters by which *C. japonica* can be distinguished from that species, as the perfectly terete branchlets, the long-acuminate leaves with spreading teeth, the strictly axillary inflorescence, larger flowers and the longer anthers opening at the apex by a pore.

Callicarpa japonica, var. angustata Rehder, n. var.

Callicarpa longifolia Hemsley in *Jour. Linn. Soc.* XXVI. 253 (pro parte, non Lamarck) (1890). — Diels in *Bot. Jahrb.* XXIX. 548 (pro parte) (1900).

A typo recedit foliis angustioribus, oblongo-lanceolatis v. oblanceolatis rarius elliptico-lanceolatis, 5–12 cm. longis et 1.2–3.5 cm. latis.

W e s t e r n H u p e h : Hsing-shan Hsien, thickets, common, alt. 1500 m., July 1907 (No. **2195**, type; bush 1.6 m., flowers pink); without precise locality, July 1900 (Veitch Exped. 1342); without precise locality, 1907, *C. Silvestri* (No. 1954); without precise locality, *A. Henry* (No. 6679). K i a n g s i : Kuling, thickets, not common, alt. 1000–1200 m., July 30, 1907 (Nos. **1530**, **1530**ᵃ; bush 1–1.5 m., fruits rosy purple). S h e n s i : "monte Kin-qua-san July 10, 1897, *G. Giraldi.*

This variety seems to differ from the type only in its narrower leaves and might be considered a mere form of *C. japonica* if it did not occupy a different geographical area. From narrow-leaved glabrescent forms of *C. Giraldiana* it is easily distinguished by the oblong anthers opening by a pore at the apex, and by the smaller longer-stalked inflorescence. It has little relation to the true *C. longifolia* Lamarck (*Encycl. Méth.* I. 563 [1785]; *Tab. Encycl. Méth.* I. 293, t. 69, fig. 2 [1791]), which is a southern species and has according to Lamarck's description and figure narrow-lanceolate glabrous leaves about 25 cm. long and a longer-stalked larger inflorescence. Hemsley's *C. longifolia*, var. *? longissima* (in *Jour. Linn. Soc.* XXVI. 253 [1890]. — Hayata, *Icon. Fl. Formos.* II. 125, t. 36 [1912]) represents apparently this species; Hance's No. 4956, which I have before me, agrees perfectly with Lamarck's figure, and has the leaves perfectly glabrous except a short pubescence on the midrib and on the veins on the upper surface.

Here may be added the description of a new variety from Korea and remarks on three species not collected during the Arnold Arboretum Expeditions.

Callicarpa japonica, var. luxurians Rehder, n. var.

A typo recedit ramulis plus minusve pallide lenticellatis, foliis majoribus ovato-ellipticis v. ovatis acuminatis basi late cuneatis v. e basi rotundata subito in petiolum productis dentatis dentibus triangularibus v. latissime triangularibus 10–18 cm. longis et 4.5–10 cm. latis, petiolis 1–2.5 cm. longis, pedunculis petiolis paullo brevioribus.

K o r e a n A r c h i p e l a g o : Quelpaert "in insula Septlum," May 28, 1908 and July 31, 1910, *Taquet* (Nos. 912, 4092, type), "in insula Pemtlum," June 7, 1910, *Taquet* (No. 4097), "in sylvis Hallaisan," September 1909, *Taquet* (No. 3083). L i u - k i u I s l a n d s : Oshima, July 1900, *U. Faurie* (No. 4047). K y u s h u : without precise locality, *C. Wright.*

This variety with large membranous leaves up to 18 cm. long is apparently only a luxuriant form owing its origin to the moist subtropical climate of the islands where it grows.

Callicarpa rubella Lindley in *Bot. Reg.* XI. t. 883 (1825). — Schauer in De Candolle, *Prodr.* XI. 645 (1847). — Bentham, *Fl. Hongk.* 270 (1861). — C. B. Clarke in Hooker f., *Fl. Brit. Ind.* IV. 569 (1885). — Maximowicz in *Bull. Acad. Sci. St. Pétersbourg*, XXXI. 75 (1886); in *Mél. Biol.* XII. 506 (1886). — Hemsley in *Jour. Linn. Soc.* XXVI. 255 (1890). — Dunn & Tutcher in *Kew Bull. Misc. Inform.* add. ser. X. 203 (*Fl. Kwangtung & Hongk.*) (1912).

Callicarpa tenuifolia Champion in *Hooker's Jour. Bot. & Kew Gard. Misc.* V.
135 (1853). — Walpers, *Ann.* V. 709 (1858).

Callicarpa purpurea? Van Houtte in *Fl. des Serr.* XIII. 127, fig., t. 1359 (non
Jussieu) (1858). — Lemaire in *Ill. Hort.* VI. t. 202 (1859). — Groenland in
Rev. Hort. 1859, 106, fig. 24–25. — *Gard. Chron.* 1859, 56, fig. — Hérincq in
Hort. Franç. 1861, 11, t. 4. — *Garden,* XXIII. 540, t. 392 (1883).

Hongkong: *C. Wright* (No. 380); *C. Ford.* Yunnan: Mengtsze, woods, *A.
Henry* (Nos. 9412, 9412ᵃ; shrub 1.25 m., flowers purple, fruits purple); Salwin
River valley, near Tali, alt. 2000 m., October 1914, *C. Schneider* (No. 3161; shrub
2 m. tall, fruit red).

The Yunnan specimens agree with those from Hongkong except that the pubescence on the upper surface of the leaves is shorter and denser.

Callicarpa rubella, var. **Hemsleyana** Diels in *Bot. Jahrb.* XXIX. 548 (1900).

Szech'uan: Nanch'uan, Ta'u-chia-wan, August 1891, *A. von Rosthorn* (No. 390);
without precise locality, alt. 900 m., June 1903 (Veitch Exped. No. 4318; bush
2 m., flowers lilac).

This variety differs from the type of the species only in the slighter pubescence of
the leaves and in their somewhat coarser serration. In Wilson's specimen, however, the under surface of the leaves is much more densely pubescent than in Rosthorn's specimen, but their dentation is much coarser and very unequal, the largest
teeth being 6–8 mm. long and bearing from 1 to 3 small teeth near the base; the
leaves measure up to 18 cm. in length and to 10 cm. in width.

Callicarpa dichotoma Raeuschel, *Nomencl. Bot.* ed. 3, 37 (nomen nudum)
(1797). — K. Koch, *Dendr.* II. 336 (1872).

Porphyra dichotoma Loureiro, *Fl. Cochin.* 70 (1790).

Callicarpa purpurea Jussieu in *Ann. Mus. Paris,* VII. 69 (1806). — Schauer in
De Candolle, *Prodr.* XI. 645 (1847). — Miquel in *Ann. Mus. Lugd.-Bat.*
II. 98 (1866); *Prol. Fl. Jap.* 30 (1866). — Franchet & Savatier, *Enum. Fl.
Jap.* I. 358 (1875). — Franchet in *Mém. Soc. Sci. Nat. Cherbourg,* XXIV. 240
(*Cat. Pl. Tché-fou*) (1882). — Maximowicz in *Bull. Acad. Sci. St. Pétersbourg,*
XXXI. 78 (1886); in *Mél. Biol.* XII. 509 (1886). — Hemsley in *Jour. Linn.
Soc.* XXVI. 254 (pro parte) (1890). — Schneider, *Ill. Handb. Laubholzk.* II.
593, fig. 385 m. (1911). — Dunn & Tutcher in *Kew Bull. Misc. Inform.* add.
ser. X. 202 (*Fl. Kwangtung & Hongk.*) (1912).

Callicarpa gracilis Siebold & Zuccarini in *Abh. Akad. Münch.* IV. pt. 3, 154
(*Fl. Jap. Fam. Nat.* II. 30) (1846).

Callicarpa japonica, var. *angustifolia* Savatier, *Livres Kwawi,* 78 (1873).

Kwangtung: "secus fl. West River, prov. Canton," July 1872, *H. F. Hance* (No.
335). Fokien: Dunn's Exped. to central Fokien, April to June 1905 (Hongkong
Herb. No. 3383). Chekiang: Ningpo, 1908, *D. Macgregor;* near Hanchou, June
1907, *F. N. Meyer* (No. 425).

JAPAN. Hondo: prov. Musashi, Yokohama, 1862, C. Maximowicz (Iter II.);
Omiya, June 18, 1911 (Herb. K. Sakurai). Kyushu: Nagasaki, 1862, *R. Oldham*
(No. 628).

This species is closely related to *C. japonica* Thunberg, but can be distinguished
from it by the young branchlets being slightly grooved or obscurely angled, by the
smaller, less acuminate leaves with fewer teeth pointing more or less forward, by
the distinctly supra-axillary slender-stalked inflorescence, and by the smaller
flowers with the anthers dehiscent to the base. It is of more southern distribution
and seems to be confined to southeastern China and southern Japan, where it reaches
Yokohama.

Callicarpa gracilipes Rehder, n. sp.

Frutex, ut videtur; ramuli teretes annotini tomento farinoso-stellato flavido obtecti. Folia chartacea, ut videtur persistentia, elliptico-ovata v. ovato-oblonga, manifeste acuminata, basi cuneata, triente inferiore excepto sinuato-dentata v. remote denticulata, interdum fere integra, 4–7.5 cm. longa et 1.7–3.2 cm. lata, supra initio sparse farinose stellato-pilosa, demum glabra v. fere glabra, nitidula, atro-viridia, subtus cana venis flavescentibus, pilis stellatis multo-radiatis dense ob-tecta et glandulis a tomento absconditis conspersa, nervis utrinsecus 6–7, supra leviter impressis subtus elevatis et colore flavescente distinctis; petioli 6–8 mm. longi, flavescenti-tomentosi. Cymae parvae, pauciflorae, 1–1.5 cm. diam., graciliter pedicellatae pedicello petiolum superante circiter 1 cm. longo, farinose flavescenti-tomentosae; calyx in fructu sparse stellato-pilosus, glabrescens, obsolete 4–denti-culatus; flores non visi. Fructus ut videtur purpureus, subglobosus, vix 3 mm. diam.

Western Hupeh: without precise locality, *A. Henry* (No. 7690, type in Herb. Gray).

This species does not seem to be closely related to any other Chinese species; it is characterized by its comparatively small, generally elliptic-ovate leaves, dark green above, whitish tomentose beneath, and by its small inflorescence and slender petioles and peduncles, the latter slightly exceeding the former.

PREMNA L.

Premna puberula Pampanini in *Nuov. Giorn. Bot. Ital.* n. ser. XVII. 701 (1910).

> *Premna microphylla* Hemsley in *Jour. Linn. Soc.* XXVI. 256 (pro parte, non Turczaninow) (1890), quoad plantam e Hupeh et Szech'uan. — Diels in *Bot. Jahrb.* XXIX. 548 (1900).

Western Hupeh: north and south of Ichang, thickets, alt. 300–1000 m., August 6, 1907 (No. **3217**, in part; semiscandent bush 2–4 m., flowers yellow); same locality, May 1900 (Veitch Exped. No. 763); same locality, *A. Henry* (No. 3582); without precise locality, *A. Henry* (Nos. 4099, 6123ᴬ). Southeastern Szech'uan: Nanch'uan, *R. von Rosthorn* (Nos. 2367, 2368). Western Szech'uan: Yachou Fu, alt. 600 m., June 1908 (No. **3217**, in part; bush 2–3 m., flowers yellow).

The leaves of this species are almost always entire; only a few leaves of Wilson's No. 3217 from Ichang and of Rosthorn's No. 2367 show a slight dentation. The inflorescence on the specimens before me is mostly 7–11 cm. long, while Pampanini gives their size as 5–7 cm.

Premna microphylla Turczaninow in *Bull. Soc. Nat. Mosc.* XXXVI. pt. 2, 217 (1863). — Maximowicz in *Bull. Acad. Sci. St. Pétersbourg*, XXXI. 79 (1886); in *Mél. Biol.* XII. 510 (1886). — Hemsley in *Jour. Linn. Soc.* XXVI. 256 (pro parte) (1890), exclud. planta e Hupeh et Szech'uan.

Premna japonica Miquel in *Ann. Mus. Lugd.-Bat.* II. 97 (1865); *Prol. Fl. Jap.*
29 (1866). — Hance in *Jour. Bot.* XVI. 111 (1878.) — Maximowicz in *Bull.*
Soc. Nat. Mosc. LIV. pt. 1, 40 (1879). — Shirasawa, *Icon. Ess. For. Jap.* II.
t. 70, fig. 1–10 (1908).
Priva spec. Siebold & Zuccarini in *Abh. Akad. Münch.* IV. pt. 3, 152 (*Fl. Jap.*
Fam. Nat. II. 28) (nomen nudum) (1846).

Kiangsi: Kuling, thickets, rare, alt. 1200 m., July 27, 1907 (No.
1670; bush 0.6–1 m., fruit blue). Chekiang: vicinity of Ningpo,
1908, *D. Macgregor;* Mokan-shan, August 5, 1915, *F. N. Meyer* (No.
1618). Fokien: Dunn's Exped. to central Fokien, April to June
1905 (Hongkong Herb. No. 3392).

The Kiangsi and Fokien specimens differ from those from Chekiang in their
smaller dentate and more pubescent leaves and in their more compact inflorescence,
the peduncles of the lower lateral cymes of the panicle scarcely exceeding 1 cm. in
length, while the Chekiang specimens have the looser panicle of the Japanese
specimens.

Premna ligustroides Hemsley in *Jour. Linn. Soc.* XXVI. 256 (1890).

Western Szech'uan: Kiating Fu, roadsides, alt. 300–600 m.,
May and June 1908 (No. 3216; bush 1–2 m., tall, flowers yellow, fruit
dark red); Mt. Omei, June 1904 (Veitch Exped. No. 5101; bush 1.3–
1.6 m. tall); banks of Yangtsze River, May 1903 (Veitch Exped. No.
4306; bush 0.6–1.3 m. tall). Eastern Szech'uan: Kweichou Fu,
E. Faber (No. 590).

VITEX L.

Vitex Negundo, Linnaeus, *Spec.* 638 (1753). — Roxburgh, *Fl. Ind.*
ed. 2, III. 72 (1832). — Schauer in De Candolle, *Prodr.* XI. 684
(1847). — Wight, *Icon. Pl. Ind. Orient.* II. 519 (1843). — Brandis,
Forest Fl. Ind. 369 (1874); *Ind. Trees*, 503 (1906). — C. B. Clarke
in Hooker f., *Fl. Brit. Ind.* IV. 583 (1885). — Hemsley in *Jour. Linn.*
Soc. XXVI. 258 (1890). — Diels in *Bot. Jahrb.* XXIX. 549 (1900). —
Pampanini in *Nuov. Giorn. Bot. Ital.* n. ser. XVII. 702 (1910). — Diels
in *Not. Bot. Gard. Edinburgh*, VII. 79 (*Pl. Chin. Forrest.*) (1912).

Vitex bicolor Willdenow, *Enum. Hort. Berol.* 660 (1809). — Schauer in De
Candolle, *Prodr.* XI. 683 (1847).
Vitex arborea Fischer apud Desfontaines, *Cat. Hort. Paris*, ed. 3, 391 (1829).
Vitex paniculata Lamarck, *Encycl. Méth.* II. 612 (1788). — Roxburgh, *Fl.*
Ind. ed. 2, III. 71 (1832).

Kiangsi: Kuling, thickets, alt. 750 m., August 1, 1907 (No. 1697;
bush 1–2 m., flowers lavender). Western Hupeh: Ichang, hillsides,
alt. 300–600 m., June, August and December 1907 (No. 2702, bush

0.6–2.5 m., flowers lavender); Patung Hsien, thickets, side of streams, alt. 600–900 m., July and December 1907 (No. **790**; bush 1–2.5 m. tall, flowers lavender); Changlo Hsien, roadsides, etc., alt. 600 m., June 1907 (No. **2701**; bush 2–2.5 m.; flowers lavender); without precise locality, *A. Henry* (No. 92). Southern Szech'uan: between Telipu and Lomapu, alt. 1500–2000 m., May 9, 1914, *C. Schneider* (No. 1139; bush 1 m.; fruits). Yunnan: Yuan-chiang, alt. 750 m., *A. Henry* (No. 13210; shrub 2 m. tall, purple flowers); Yunnan Fu, *A. Henry* (No. 9750); Yangtsze Valley, between Likiang and Chungtien, moist places, alt. 1500 m., August 6, 1914, *C. Schneider* (No. 2196; bush 1–1.25 m.; flowers blue). Formosa: South Cape, *A. Henry* (No. 905); Bankinsing, *A. Henry* (without No.).

A picture of this plant will be found under No. 0202 of the collection of Wilson's photographs.

Vitex cannabifolia Siebold & Zuccarini is often referred as a synonym to this species, but, though undoubtedly very closely related to *V. Negundo*, it is easily distinguished by the greenish and glabrescent under surface of the leaves with the midrib and veins covered with short more or less spreading hairs. To *V. cannabifolia* I refer the specimens from Chekiang, Fokien, Hongkong and Kwangtung which I have seen.

Here may be added a note on a variety not collected during the Arnold Arboretum Expeditions.

Vitex Negundo, var. **incisa** Clarke in Hooker f., *Fl. Brit. Ind.* IV. 584 (1885).

Vitex chinensis Miller, *Dict.* No. 5 (1768).[1]

Vitex incisa Lamarck, *Encycl. Méth.* II. 612 (1788). — Poiret, *Tab. Encycl. Méth.* III. 92, t. 541, fig. 2 (1823). — Schauer in De Candolle, *Prodr.* XI. 684 (1847). — Bunge in *Mém. Sav. Étr. Acad. Sci. St. Pétersbourg,* II. 126 (*Enum. Pl. Chin. Bor.* 52) (1833). — Debeaux in *Act. Soc. Linn. Bordeaux,* XXXI. 346 (*Fl. Tché-fou,* 113) (1876); XXXIII. 59 (*Fl. Tien-tsin*) (1879). — Franchet, *Mém. Soc. Nat. Cherbourg,* XXIV. 241 (*Cat. Pl. Tché-fou*) (1882); in *Nouv. Arch. Mus. Paris,* sér. 2, VI. 112 (*Pl. David.* I. 232) (1883). — Maximowicz in *Bull. Acad. Sci. St. Pétersbourg,* XXXI. 82 (1886); in *Mél. Biol.* XII. 516 (1886). — Hemsley in *Jour. Linn. Soc.* XXVI. 257 (1890). — Diels in *Bot. Jahrb.* XXIX. 549 (1900). — Gilg & Loesener in *Bot. Jahrb.* XXXIV. Beibl. LXXV. 62 (1904). — Pavolini in *Nuov. Giorn. Bot. Ital.* n. ser. XV. 432 (1908). — Schneider, *Ill. Handb. Laubholzk.* II. 594, fig. 384 m–n, fig. 385 r–t (1911).

Vitex Negundo Curtis in *Bot. Mag.* XI. t. 364 (non Linnaeus) (1797).

Vitex laciniatus Hort. ex Schauer in De Candolle, *Prodr.* XI. 684 (pro synon.) (1847).

Agnus castus incisa Carrière in *Rev. Hort.* 1871, 415.

Western Szech'uan: Tung River valley, July 1903 (Veitch Exped. No. 4308ᵃ; bush 2 m.); same locality, alt. 1400 m., July 1903 (Veitch Exped. No. 4308; bush 1–2 m.). Northern Shensi: without precise locality, 1897, *G. Giraldi.* Chili: near Peking, August 1865, *S. W. Williams* (Herb. Hance, No. 11435): same

[1] Figured in Miller, *Fig. Pl.* II. 183, t. 275 (1760).

locality, August 1876, *S. W. Williams;* Peking, Temple of Heaven, September, 15, 1903, *C. S. Sargent;* same locality, October 3, 1905, *J. G. Jack;* Peking, city wall, July 5, 1915, *F. N. Meyer* (No. 1008); Peking-Kolgan road, hills near Great Wall, October 5, 1905, *J. G. Jack.* Shantung: Chifu, September 22, 1903, *C. S. Sargent;* Tsingtau, 1901, *R. Zimmermann* (No. 442).

This variety has been by most authors considered a distinct species, but it differs only in its more deeply and incisely serrate or even pinnatifid leaflets; it represents the northern form of the species and seems well separated geographically, ranging from northeastern Szech'uan through Shensi and northwestern Hupeh to northern Chili and eastern Shantung, while the typical form ranges southward to India and Malaya. The latter has also been recorded from the Philippine Islands, but the specimens I have seen differ considerably in the character of the inflorescence. The var. *incisa* shows a great range of variation in the shape of the leaflets; some specimens approach the type by their broad only deeply serrate leaflets, as Zimmermann's No. 442 from Tsingtau, others have serrate or sometimes entire and pinnatisect leaflets often on the same branch like Wilson's No. 4308[a], and this form has been distinguished by Franchet as *V. incisa,* var. *heterophylla* (in *Nouv. Arch. Mus. Paris,* sér. 2, VI. 112 [*Pl. David.* I. 232] [1883]), while the most extreme forms have deeply pinnatifid leaflets with comparatively narrow and often remote segments as in Wilson's No. 4308; this form has been named *V. incisa,* var. *multifida* Schneider (*Ill. Handb. Laubholzk.* II. 594 [1911]. — *Agnus castus incisa,* var. *multifida* Carrière in *Rev. Hort.* 1871, 416).

Vitex quinata F. N. Williams in *Bull. Herb. Boissier,* sér. 2, V. 431 (1905).

Cornutia quinata Loureiro, *Fl. Cochin.* 387 (1790).
Vitex heterophylla Roxburgh, *Hort. Beng.* 46 (nomen nudum) (1814); *Fl. Ind.* ed. 2, III. 75 (1832). — Wallich, *Pl. As. Rar.* III. 15, t. 226 (1832). — Schauer in De Candolle, *Prodr.* XI. 686 (1847). — Clarke in Hooker f., *Fl. Brit. Ind.* IV. 585 (1885). — Hemsley in *Jour. Linn. Soc.* XXVI. 257 (1890).
Vitex Loureiri Hooker & Arnott, *Bot. Voy. Beechey,* 206, t. 48 (1841). — Bentham, *Fl. Hongk.* 273 (1861). — Hance in *Jour. Linn. Soc.* XIII. 117 (1873). — Maximowicz in *Bull. Acad. Sci. St. Pétersbourg,* XXXI. 81 (1886); in *Mél. Biol.* XII. 514 (1886).

Eastern Szech'uan: Wushan Hsien, thickets, roadsides, etc., alt. 300–750 m., June and October 1907 (No. **408**; bush 2–4 m. tall, flowers white). Yunnan: Szemao, alt. 1500 m., *A. Henry* (Nos. 9787, 12638, 12638[a]; tree 7–13 m.). Hongkong: *C. Ford.* Kwangtung: Tai-shek near Canton, *T. Sampson* (Herb. Hance, No. 7494). Formosa: Takow, *A. Henry* (Nos. 1182, 1182[c]; Bankinsing, *A. Henry* (No. 1182[a, b]).

This species does not seem to have been recorded before from western China. Wilson's No. 408 has very large leaflets, broadly cuneate or nearly rounded at the base, the terminal one attaining 14 cm. in length, and large inflorescences from 16 to 24 cm. long; it resembles thus more the Indian plant, but differs from it in the finely pubescent veins of the under side of the leaves, while the specimens from eastern China which represent the typical *V. quinata* have a smaller inflorescence and smaller leaflets narrower at the base. Henry's specimens from Yunnan have longer and narrower glabrous or nearly glabrous leaflets.

CLERODENDRON L.

Clerodendron foetidum Bunge in *Mém. Sav. Étr. Acad. Sci. St. Pétersbourg*, II. 126 (*Enum. Pl. Chin. Bor.* 52) (non D. Don) (1833). — Schauer in De Candolle, *Prodr.* XI. 672 (1847). — Hooker in *Bot. Mag.* LXXXI. t. 4880 (1855). — Maximowicz in *Bull. Acad. Sci. St. Pétersbourg*, XXXI. 84 (1886); in *Mél. Biol.* XII. 518 (1886). — Hemsley in *Jour. Linn. Soc.* XXVI. 259 (1890). — Diels in *Bot. Jahrb.* XXIX. 549 (1900); in *Not. Bot. Gard. Edinburgh*, VII. 29, 72, 184 (*Pl. Chin. Forrest.*) (1912). — Schneider, *Ill. Handb. Laubholzk.* II. 595, fig. 384 h–i, 386 g–i (1911).

> *Clerodendron Bungei* Steudel, *Nomencl. Bot.* ed. 2, I. 382 (1840). — Planchon in *Fl. des Serres*, IX. 17, t. 863–64 (1853).

Western Hupeh: Patung Hsien, glens at low altitudes, alt. 300 m., June and August 1907 (No. 2224; sub-shrub 1 m., flowers pink); without precise locality, *A. Henry* (Nos. 262, 2307). Yunnan: near Tengyueh, woods, alt. 1800 m., October 1914, *C. Schneider* No. 2573; bush 0.5–1.5 m.); near Yangpi (Tali Fu), in hedges, alt. 2300 m., October 1914, *C. Schneider* (No. 3272).

Clerodendron mandarinorum Diels in *Bot. Jahrb.* XXIX. 549 (1900).

Western Hupeh: Hsing-shan Hsien, cliffs, alt. 600–1000 m., July and October 1907 (No. 425; small tree 5–7 m. tall, flowers white, fruits blue-black). Yunnan: Mengtze, mountains to southeast, alt. 1500 m., *A. Henry* (Nos. 9026, 9026ª; flowers white).

The inflorescence of Wilson's specimens is from 8 to 12 cm. high, and from 10 to 20 cm. broad, while Diels gives 25 by 30 cm.; it is distinctly cymose with from 4 to 6 subumbellate primary branches, the lowest pair more strongly developed than the others. The fruit, which is not described by Diels, is bluish black, from 8 to 10 mm. high, and according to the number of seeds it contains ovoid to subglobose and slightly 2–4-lobed; at the base it is surrounded by the enlarged, irregularly split and finally reflexed calyx; seeds 1–4, light grayish-brown, oblong-ovoid, about 8 mm. long and 5 mm. broad, plain on the ventral sides and with a deep curved furrow on one side, strongly reticulate on the dorsal side. Henry's No. 9026ª differs in the more densely tomentose undersurface of the leaves and in the densely tomentose inflorescence and calyx. Young plants raised from seeds of Wilson's No. 425 and growing in the Arnold Arboretum have the leaves coarsely crenate-dentate.

Clerodendron trichotomum Thunberg, *Fl. Jap.* 256 (1784). — Schauer in De Candolle, *Prodr.* XI. 668 (1847). — Debeaux in *Act. Soc. Linn. Bordeaux*, XXX. 100 (*Fl. Shanghai*, 48) (1875). — Hooker f. in *Bot. Mag.* CVII. t. 6561 (1881). — Maximowicz in *Bull. Acad.*

Sci. St. Pétersbourg, XXXI. 85 (1886); in *Mél. Biol.* XII. 519 (1886). — Hemsley in *Jour. Linn. Soc.* XXVI. 262 (1890). — Shirasawa, *Icon. Ess. For. Jap.* II. t. 70, fig. 28–36 (1908). — Nash in *Addisonia*, I. 29, t. 15 (1916).

   Clerodendron serotinum Carrière in *Rev. Hort.* 1867, 351, fig.

Western Hupeh: common around Ichang, alt. 40 m., July 1907 (No. 2225; bush 2–3 m., flowers white). Also in Korea and Japan. Shantung: Chifu, cultivated, September 22, 1903; *C. S. Sargent*, Lan-shan, August 1907, *F. N. Meyer* (No. 279).

Clerodendron trichotomum, var. Fargesii Rehder, n. var.

   Clerodendron trichotomum Hemsley in *Jour. Linn. Soc.* XXVI. 262 (pro parte, non Thunberg) (1890), quoad plantam e Hupeh et Szech'uan. — Diels in *Bot. Jahrb.* XXIX. 550 (non Thunberg) (1900). — Pavolini in *Nuov. Giorn. Bot. Ital.* n. ser. XV. 432 (1908). — Pampanini in *Nuov. Giorn. Bot. Ital.* n. ser. XVII. 701 (1910).
   Clerodendron Fargesii Dode in *Bull. Soc. Dendr. France*, 1907, 207, figs. — Pinelle in *Rev. Hort.* 1911, 522, fig. 206–207.
   Clerodendron trichotomum, var. *Fargesii* Hort. ex Bailey, *Stand. Cycl. Hort.* II. 800 (pro synon.) (1914).

Western Hupeh: north and south of Ichang, common in thickets, alt. 1000–1200 m., July and September 1907 (No. 216; bush 2–4 m., flowers white); without precise locality, July 1900 (Veitch Exped. No. 1489); "monte Cia-inen-ku presso Siang-yong," July 1904, *C. Silvestri* (No. 1964); "catena di Ou-tan-scian," August to November 1909, *C. Silvestri* (No. 3158); without precise locality, *A. Henry* (Nos. 4520, 4646, 6422, 6422ª). Western Szech'uan: Wa-shan, thickets, alt. 1500–2000 m., July 1908 (No. 2223; tree 8 m. tall, girth 0.6 m., flowers white); west and near Wên-ch'uan Hsien, thickets, alt. 2000–2300 m., July and September 1908 (No. 1040; bush 2–4 m., flowers white); near Tachien-lu, alt. 2000–2300 m., October 1908 (No. 1040ª; bush 2 m., flowers white). Northern Shensi: "Thui-kio-tsen," September 1897, *G. Giraldi;* "Lao-y-san," September 6, 1897, *G. Giraldi*. Yunnan: between Yungpating and Taowang, thickets, alt., 1800–2300 m., July 2, 1914, *C. Schneider* (No. 1717); tree 6 m. tall, girth 0.3 m., flowers white).

This variety differs from the type chiefly in its glabrous or glabrescent smaller leaves from 6 to 12 cm. long and from 2 to 6 cm. broad, in its nearly glabrous branches, glabrous or glabrescent inflorescence and in the narrower more acute or acuminate greenish sepals, but all these characters are very variable and I am unable to find a constant morphological character to separate these two forms specifically. Wilson's No. 1040 has the leaves rather villose on the veins beneath, and his No. 216 has pubescent veins, while some specimens of typical *C. trichotomum* from Korea and those from Shantung are pubescent only on the veins beneath and are otherwise nearly glabrous. Var. *Fargesi* is apparently a geographical form of the mountains of western China and extends east to Ichang, where it is found at an altitude of about 1000 m., while the typical *C. trichotomum* which seems to reach there its western limit was collected by Wilson at river-level at about 40 m. altitude.

Clerodendron cyrtophyllum Turczaninow in *Bull. Soc. Nat. Mosc.* XXXVI. pt. 3, 222 (1863). — Maximowicz in *Bull. Acad. Sci. St. Pétersbourg*, XXXI. 86 (1886); in *Mél. Biol.* XII. 520 (1886). — Hemsley in *Jour. Linn. Soc.* XXVI. 259 (1890). — Dunn & Tutcher in *Kew Bull. Misc. Inform.* add. ser. X. 205 (*Fl. Kwangtung & Hongk.*) (1912).

Clerodendron amplius, Hance in *Ann. Sci. Nat.* sér. 5, V. 233 (1866). — Franchet in *Nouv. Arch. Mus. Paris*, sér. 2, VI. 111 (*Pl. David.* I. 231) (1883).
Clerodendron formosanum Maximowicz in *Bull. Acad. Sci. St. Pétersbourg*, XXXI. 85 (1886); in *Mél. Biol.* XII. 519 (1886).

Kiangsi: Kuling, thickets abundant, alt. 750 m., July 27, 1907 (No. **1525**, bush 1–1.6 m. tall, flowers white). Kiangsu: Shanghai, 1915, *D. Macgregor.* Chekiang: near Hanchou, June 1907, *F. N. Meyer* (No. 436); Mokan-shan, August 3, 1915, *F. N. Meyer* (No. 1611); Ningpo Mts., 1888, *E. Faber* (No. 645). Fokien: Amoy, *H. F. Hance* (No. 397). Hainan: *C. Ford, A. Henry.* Formosa: Bankinsing, *A. Henry* (Nos. 1, 562); South Cape, *A. Henry* (No. 1295); Takow, *A. Henry* (No. 1873).

Clerodendron japonicum Sweet, *Hort. Brit.* 322 (1826). — Makino in *Tokyo Bot. Mag.* XVII. 91 (1903). — Matsumura, *Ind. Fl. Jap.* II. 2, 532 (1912).

Volkameria japonica Thunberg, *Fl. Jap.* 255 (1784).
Volkameria Kaempferi Jacquin, *Icon. Pl. Rar.* III. 7, t. 500 (*V. Kaempferiana* in tab.) (1786); *Coll.* III. 207 (1789).
Clerodendron squamatum Vahl, *Symb.* II. 74 (1791). — Lindley in *Bot. Reg.* VIII. t. 649 (1822). — Schauer in De Candolle, *Prodr.* XI. 669 (1847). — Hance in *Jour. Bot.* XVII. 13 (1879). — Clarke in Hooker f., *Fl. Brit. Ind.* IV. 593 (1885). — Maximowicz in *Bull. Acad. Sci. St. Pétersbourg*, XXXI. 86 (1886); in *Mél. Biol.* XII. 521 (1886). — Hemsley in *Jour. Linn. Soc.* XXVI. 262 (1890). — Dunn & Tutcher in *Kew Bull. Misc. Inform.* add. ser. X. 205 (*Fl. Kwangtung & Hongk.*) (1912).
Volkameria coccinea Loiseleur-Deslongchamps, *Herb. Amat.* VIII. t. 519 (1827).
Clerodendron dentatum Wallich, *Cat.* No. 1799 (nomen nudum) (1828).
Clerodendron Kaempferi Siebold, herb. ex Miquel in *Ann. Mus. Lugd.-Bat.* II. 299 (pro synon., non Siebold 1830) (1865–66); *Prol. Fl. Jap.* 31 (1866).
Volkameria dentata Roxburgh, *Hort. Bengal.* 46 (nomen nudum) (1814); *Fl. Ind.* ed. 2, III. 61 (1832).
Clerodendron coccineum D. Dietrich, *Syn. Pl.* III. 616 (1843).

Western Szech'uan: cultivated as an ornamental plant in Szech'uan generally, alt. 30–600 m., July 1910 (No. **4555**). Yunnan: Szemao, alt. 1200 m., *A. Henry* (No. 12060). Fokien: Dunn's Exped. to central Fokien, April to June 1905 (Hongk. Herb. No. 3395).

A picture of this plant will be found under No. 0216 of the collection of Wilson's photographs.

According to the *Index Kewensis Clerodendron speciosissimum* Paxton (in *Mag. Bot.* III. 217, t. 271, fig., t. [1837]; in *Hort. Belg.* III. 322, t. 68 [1837].—Loiseleur-Deslongchamps in *Herb. Amat.* sér. 2, I. t. 92 [1839]) belongs here, but that plant differs apparently in its much smaller calyx and in the pubescent under surface of the leaves, though in its general appearance it strongly resembles *C. japonicum,* and may after all be a form of that species.

CARYOPTERIS Bge.

Caryopteris incana Miquel in *Ann. Mus. Lugd.-Bat.* II. 97 (1866); *Prol. Fl. Jap.* 29 (1866). — Maximowicz in *Bull. Acad. Sci. St. Péters-bourg,* XXXI. 87 (1886); in *Mél. Biol.* XII. 523 (1886). — Diels in *Bot. Jahrb.* XXIX. 550 (1900). — Schneider, *Ill. Handb. Laubholzk.* II. 595, fig. 382 m–o, 386 k–p (1911).

> *Nepeta incana* Thunberg, *Fl. Jap.* 244 (1784).
> *Barbula sinensis* Loureiro, *Fl. Cochin.* 367 (1790).
> *Nepeta japonica* Willdenow, *Spec.* III. 52 (1800).
> *Mastacanthus sinensis* Endlicher in Walpers, *Rep.* IV. 2 (1844). — Lindley in *Bot. Reg.* XXXII. t. 2 (1846).
> *Caryopteris Mastacanthus* Schauer in De Candolle *Prodr.* XI. 625 (1847). — Bentham, *Fl. Hongk.* 268 (1861). — Hance in *Jour. Linn. Soc.* XIII. 116 (1873). — Henriques in *Bot. Soc. Brot.* III. 144 (1885). — Hooker f. in *Bot. Mag.* CIII. t. 6799 (1885). — Hemsley in *Jour. Linn. Soc.* XXVI. 263 (1890).
> *Caryopteris sinensis* Dippel, *Handb. Laubholzk.* I. 59 (1889).

Western Szech'uan: Arid parts of Min River valley, Wên-ch'uan Hsien, alt. 1200–1800 m., August 1908 (No. **2221**; bush 0.3–6 m., flowers blue); without precise locality, alt. 1200 m., July 1903 (Veitch Exped. No. 4312); Wei-kuan, Ta-chai-tzu, August 1891, *A. von Rosthorn* (No. 2519ᵃ). Western Hupeh: Ichang, *A. Henry* (No. 2782); without precise locality, *A. Henry* (Nos. 4581, 7771); without precise locality, June 1901 (Veitch Exped. No. 2153). Western Kansu: Lao-chow district, alt. 3000 m., *W. Purdom* (No. 792; bush 0.45 m., flowers blue). Chekiang: Ningpo, 1908, *D. Macgregor.*

Here may be added the description of a new species not collected during the Arnold Arboretum Expeditions.

Caryopteris glutinosa Rehder, n. sp.

Frutex ½-metralis, ramis erectis; ramuli hornotini teretes, crispulo-villosuli, vetustiores cinereo-brunnei. Folia crassiuscula, lanceolata, obtusiuscula, basi in petiolum brevem attenuata, margine revoluta integra, 1.5–2.5 cm. longa et 3–6 mm. lata, supra glabra, atroviridia, glutinosa, subtus venis glabrescentibus initio glutinosis exceptis dense albido- v. cano-tomentosa, nervis utrinsecus 6–8 adscendentibus supra obsoletis subtus colore fusco ab tomento albido bene distinctis et conspicuis; petioli 1–2 mm. longi, minute villosuli, flavescentes.   Flores pallide coerulei, in cymis congestis plurifloris pedunculatis ex axillis foliorum superiorum orientibus et in apice ramulorum saepius corymbum 2–3 cm. diam. formantibus; pedunculi 0.5–1 cm. longi, minute villosuli; pedicelli breves ad 1 mm. longi v. fere nulli; calyx 3 mm. longus, circiter ad medium 5-partitus, lobis anguste ovatis acutiusculis cinereo-tomentosis sed apice et ad costam saepe plus minusve glabrescentibus et glutinosis; corolla circiter 6 mm. longa, extus tubo et margine labii inferioris glabris exceptis dense cano-villosula, intus ad faucem annulo denso pilorum albidorum instructa, labio inferiore simpliciter fimbriato-dentata tubum subaequante, lobis ceteris dimidio brevioribus ovatis obtusiusculis; stamina 4, labium inferius subaequantia, loculis antherarum parallelis; stylus staminibus subaequilongus, apice bifido, ramis longiusculis.   Calyx fructiferus 4–5 mm. longus; fructus non visus.

Western Szech'uan: Min River valley, alt. 1600 m., August 1903 (Veitch Exped. No. 4309).

This species differs from the other species of the genus in the presence of a ring of long white hairs closing the mouth of the corolla-tube and in the glutinous leaves which are entire and whitish tomentose beneath and marked with dark veins.

LABIATAE.

Determined by ALFRED REHDER.

COLQUHOUNIA.

Colquhounia Seguinii Vaniot in *Bull. Acad. Intern. Geog. Bot.* XIV. 165 (1904).

> *Colquhounia coccinea* Hemsley in *Jour. Linn. Soc.* XXVI. 299 (non Wallich) (1890).
> *Colquhounia elegans*, var. *pauciflora* Prain in *Jour. As. Soc. Beng.* LXII. 38 (1893). — Dunn in *Not. Bot. Gard. Edinburgh*, VI. 179 (1915).
> *Colquhounia pauciflora* Prain in *Jour. As. Soc. Bengal*, LXII. 38 (pro synon.) (1893).
> *Colquhounia decora* Diels in *Not. Bot. Gard. Edinburgh*, V. 240 (1912).

Western Hupeh: Ichang and immediate neighborhood, *A. Henry* (No. 3334). Prain and Dunn consider this a variety of *C. elegans* Wallich, but it differs considerably from all other species of the genus in the shape of the corolla, which has a short wide tube scarcely twice as long as the long upper lip, while in *C. elegans* and the other species the tube is at least three times as long as the comparatively short upper lip. *C. Seguinii* differs further from the other species in its glabrescent stems and leaves. Vaniot describes the calyx also as glabrous, but in Henry's specimens it is pubescent, as it is described by Diels in his *C. decora*. Vaniot possibly overlooked the very fine and close pubescence. I have no doubt that Dunn, who had seen the types of Vaniot's, Prain's and Diels's plants, is right in uniting them. A more pubescent form occurs in western Szech'uan and this may be separated as a distinct variety.

Colquhounia Seguinii, var. **pilosa** Rehder, n. var.

A typo recedit ramis dense crispulo-pilosis et foliis supra satis dense et subtus praecipue ad costam nervosque longe pilosis.

Western Szech'uan: Wa-shan, thickets, alt. 1200–1500 m., October 1908 (No. **3530**; sub-shrub 1.5 m., flowers reddish).

Wilson's No. 3580 differs from Henry's No. 3334 chiefly in the pilose pubescence of the stem and of the leaves; the lateral flowering branchlets are shorter, measuring from 3 to 8 cm. in length; the corolla is from 20 to 23 mm. long, with the tube from 12 to 13 and the upper lip from 8 to 10 mm. long. These measurements agree well with those given by Diels for his *C. decora*.

In Yunnan the following two species have been collected.

Colquhounia elegans Wallich, var. **tenuiflora** Prain in *Jour. As. Soc. Bengal*, LXII. 38 (1893). — Dunn in *Not. Bot. Gard. Edinburgh*, VI. 179 (1915).

> *C. elegans* Kurz, *Forest Fl. Brit. Burma*, II. 278 (non Wallich) (1877).
> *C. tenuiflora* Hooker f., *Fl. Brit. Ind.* IV. 674 (1885).
> *C. martabanica* Kurz mss. ex Prain in *Jour. As. Soc. Bengal*, LXII. 38 (pro synon.) (1893).

Yunnan: Szemao, mountains to east, alt. 1200 m., *A. Henry* (No. 12607; climber, red flowers).
Though Prain says that if *C. elegans*, var. *pauciflora* is considered specifically distinct this would necessitate also the recognition of var. *tenuiflora* as a distinct species, I am inclined to keep it as a variety of *C. elegans*. Its flowers have the same short upper lip and long though slenderer corolla, and the plant is pubescent throughout, though not quite as densely as in typical *C. elegans*. The inflorescence may vary on the same plant; particularly on the long scandent stems the inflorescences in the lower axils elongate into shorter or longer racemes, while toward the end of the stems they change gradually into dense, nearly sessile heads.

Colquhounia coccinea Wallich, var. **mollis** Prain in *Jour. As. Soc. Bengal*, LXII. 37 (1893). — Diels in *Not. Bot. Gard. Edinburgh*, VII. 46, 47, 245 (1912).

Colquhounia vestita Bentham in De Candolle, *Prodr.* XII. 457 (non Wallich) (1848), quoad plantam Assamicam. — Hooker f., *Fl. Brit. Ind.* IV. 674 (1885), exclud. pl. Kumaonica. — Collett & Hemsley in *Jour. Linn. Soc.* XXVII. 116 (1890).
Colquhounia mollis Schlechtendal in *Linnaea*, VIII. 681 (1851). — Walpers, *Ann.* V. 689 (1858).
Colquhounia tomentosa Jacques in *Jour. Soc. Hort. France*, VII. 47, 179 (1861). — Houllet in *Rev. Hort.* 1873, 131.
Colquhounia vestita, var. *rugosa*, C. B. Clarke mss. ex Prain in *Jour. As. Soc. Bengal*, LXII. 37 (pro synon.) (1893).

Yunnan: "inter Yangpi et Tengyueh, in dumetis mont. ad vias," alt. 2600 m., October 1914, *C. Schneider* (No. 2630); "ad lat. mont. prope Tali in dumetis apertis," alt. 3000 m., September 1914, *C. Schneider* (No. 3886).
This variety differs from the type chiefly in its rusty pubescence and in the crenulate leaves. Schneider's specimens have capitate and racemose inflorescences on the same branch.

ELSHOLTZIA.

Elsholtzia fruticosa Rehder, n. comb.

Colebrookia oppositifolia Loddiges, *Bot. Cab.* V. t. 487 (non Smith) (1820).
Perilla fruticosa D. Don, *Prodr. Fl. Nepal.* 115 (1825).
Mentha fruticosa Roxburgh mss. ex D. Don, l. c. (pro synon.) (1825).
Aphanochilus polystachyus Bentham in Wallich, *Cat.* No. 1554 (nomen nudum) (1828); in Wallich, *Pl. As. Rar.* I. 28, t. 33 (1830).
Elsholtzia polystachya Bentham, *Labiat.* 116 (1832); in De Candolle, *Prodr.* XII. 160 (1848). — Gamble, *Man. Indian Timbers*, 301 (1881). — Hooker f., *Fl. Brit. Ind.* IV. 643 (1885). — Hemsley in *Jour. Linn. Soc.* XXVI. 268 (1890). — Diels in *Bot. Jahrb.* XXIX. 561 (1900). — Collett, *Fl. Siml.* 388, fig. 123 (1902). — Brandis, *Ind. Trees*, 516 (1906). — Diels in *Not. Bot. Gard. Edinburgh*, VII. 47 (*Pl. Chin. Forrest.*) (1912). — Dunn in *Not. Bot. Gard. Edinburgh*, V. 150 (1915).
Elsholtzia tristis Léveillé in Fedde, *Rep. Spec. Nov.* VIII. 424 (1910).
Elsholtzia Dielsii Léveillé, l. c. IX. 441 (1911).
Elsholtzia Soulei Léveillé, l. c. 248 (1911).

Western Szech'uan: Min River valley, north of Kuan Hsien, alt. 600–1200 m., August and November 1908 (No. **3532**; sub-shrub

1–1.75 m., flowers white); near Yachou Fu, alt. 600–800 m., September 1908 (No. 3733; sub-shrub 1–2 m., flowers pale pink or white); Tung River valley, near Wa-shan, thickets, September 1908 (No. 3533; sub-shrub 1.25 m., flowers white); Wa-shan, October 1903 (Veitch Exped. No. 4307; sub-shrub 1.75 m., flowers white); Wên-ch'uan, August 1891, *A. von Rosthorn* (No. 3004). Western Hupeh: without precise locality, *A. Henry* (No. 6755). Yunnan: Mengtsze, grass mountains, alt. 1500 m., *A. Henry* (No. 11305; shrub 0-6–1 m., yellowish flowers); "in dumetis collium orient. prope pagum Ngu-leh-keh in reg. Lichiang," alt. 3200 m., August 14, 1914. *C. Schneider* (No. 2255; flowers white); Tali Range, *G. Forrest* (No. 4759).

This species seems very variable in the pubescence of the leaves; in Wilson's No. 3532 and in Rosthorn's No. 3004 the under surface of the leaves is densely villose, while in Wilson's No. 3733 and in Henry's No. 6755 the pubescence is shorter and much less dense; Wilson's Nos. 3533 and 4307 have the under side covered by a close white tomentum and represent apparently a distinct variety.

Elsholtzia fruticosa, var. **tomentella** Rehder, n. var.

A typo recedit foliis subtus tomento minuto albido v. canescente dense obtectis et reticulatis margine minus argute serratis plerumque in triente v. dimidio inferiore integra et supra minute et remote serrulatis v. grossius serratis.

Western Szech'uan: Wa-shan, October 1903 (Veitch Exped. No. 4307, type; sub-shrub 1.6 m., flowers white); near Wa-shan, Tung River valley, thickets, alt. 900–1200 m., September 1908 (No. 3533ᵃ; sub-shrub 1.3 m., flowers white).

In the close whitish tomentum of the under side of the leaves this variety differs markedly from the type, but as there is apparently no difference in the flowers, it does not seem advisable to treat it as a distinct species.

Elsholtzia flava Bentham, *Labiat.* 161 (1832); in De Candolle, *Prodr.* XII. 160 (1848). — Hooker f., *Fl. Brit. Ind.* IV. 642 (1885). — Diels in *Not. Bot. Gard. Edinburgh*, VII. 40, 260 (*Pl. Chin. Forrest.*) (1912). — Dunn in *Not. Bot. Gard. Edinburgh*, VI. 149 (1915).

Aphanochilus flavus Bentham in Wallich, *Cat.* 1553 (nomen nudum) (1828); in Wallich, *Pl. As. Rar.* I. 28, t. 34 (1830).

Western Szech'uan: Hungya Hsien, foot of Wa-wu-shan, alt. 1200 m., September 13, 1908 (No. 3535, sub-shrub 1 m., flowers bronzy-red). Yunnan: "in dumetis ad lat. orient. mont. niveorum prope Lichiang," alt. 3200 m., October 1914, *C. Schneider* (No. 3907).

Wilson's specimen differs in the broadly cuneate base of the leaves from the Himalayan and Yunnan specimens before me, which have the leaves subcordate at the base.

Elsholtzia dependens Rehder, n. sp.

Frutex 1–2-metralis, ramis sarmentosis apice pendentibus; ramuli teretes, initio farinaceo-tomentosi, mox glabrescentes, demum pallide brunnei, medulla ampla alba repleti. Folia membranacea, oblongo-elliptica v. obovato-elliptica, acuta v. breviter acuminata, basi cuneata v. interdum superiora infra inflorescentiam basi plus minusve sagittata, triente inferiore plerumque integro excepto inaequaliter serrata, 4–8 cm. longa et 1.7–3.2 cm. lata, supra glabra, subtus initio pube laxo praecipue ad costam nervosque farinaceostellato conspersa, mox glabrescentia, viridia, utrinsecus nervis 5–6 adscendentibus et ante marginem anastomosantibus; petioli ut costa subtus pube farinaceo-stellato demum evanescente obtecti, 2–4 mm. longi. Spicae terminales pendentes v. nutantes, plerumque plures in apice ramulorum, dense albido-tomentosae, cylindricae, 10–20 cm. longae et 1–1.5 cm. diam., verticillastris sessilibus approximatis; verticillastri 6–10-flori ante anthesin bracteis late triangulari-ovatis acuminulatis extus albido-tomentosis intus glabris 4–7 mm. longis decussatis caducis suffulti et bracteolis anguste ellipticis brevioribus instructi; calyx 10-nervosus, campanulatus, subaequaliter 5-dentatus, dentibus triangulari-ovatis acutis, anticis paulo latioribus, extus dense stellato-tomentosus, intus glaber; corolla rosea, tubo vix exserto glabro, limbo bilabiato extus sparse stellato-piloso et glanduloso, labio antico patente v. erecto-patente 3-lobato lobo medio majore circiter 2 mm. longo concavo, labio postico recto rotundato-ovato integro circiter 1 mm. longo lobis lateralibus subaequilongis angustiore; stamina valde exserta, longiora corollam fere duplo superantia, filamentis glabris basi in excrescentias disciformes subito dilatatis ut crista semi-circularis infra basin labii inferioris dense papilloso-pilosis, antheris suborbicularibus; discus cupularis, margine irregulariter sinuato-dentatus, fere dimidium ovarium aequans; stylus glaber, apice bifidus, stamina longiora paullo superans. Nuculi trigono-oblongi, circiter 3 mm. longi, apice rostrati rostro curvato circiter 0.75 mm. longo, brunnei, opaci.

Western Szech'uan: near Wa-shan, thickets, alt. 600–1200 m., September and November 1908 (No. **3534**, type); without precise locality, ravine, alt. 800 m., August 1903 (Veitch Exped. No. 4313).

This species seems not closely related to any other species of the genus. According to its broadly bracted spikes it ought to be placed in the group *Platyelasmeae* Briquet of the section *Aphanochilus* Bentham, but it differs from the species of this group as from those of the other groups in the entire upper lip of the corolla, in the irregular ring of hairs at the mouth of the corolla formed by hairy disk-like excrescences at the base of the filaments and in a hairy crescent-shaped crest below the base of the lower lip, and in the rostrate nutlets. The drooping habit of the long and slender spikes is also very peculiar and, as far as I know, does not occur in any other Elsholtzia.

PLECTRANTHUS L'Hérit.

Plectranthus discolor Dunn in *Not. Bot. Gard. Edinburgh*, VIII. 155 (1913); VI. 140 (1915).

Western Szech'uan: valley of Hsao-chin Ho, near Romi-chango, alt. 2100–2700 m., June 1908 (No. **3529**; shrub 1–1.3 m., flowers lavender); Min River valley, alt. 1200–2700 m., August 1903 (Veitch Exped. No. 4322; shrub 0.3–1.m., flowers blue); without precise locality, alt. 1200 m., July 1903 (Veitch Exped. No. 4321; 0.6 m., flowers blue). Kansu: without precise locality, *W. Purdom* (No. 804***, seeds only; plants growing in the Arnold Arboretum).

Here may be added a note on another species not collected during the Arnold Arboretum Expeditions.

P. leucophyllus Dunn in *Not. Bot. Gard. Edinburgh*, VIII. 157 (1913); VI. 140 (1915). Western Szech'uan: Tung River valley, alt. 1200 m., July 1903 (Veitch Exped. No. 4319; shrub 1.3 m., flowers light purple).

This species is closely related to the preceding species, but is easily distinguished by its rugose leaves.

SOLANACEAE.

Determined by CAMILLO SCHNEIDER.

LYCIUM L.

Lycium chinense Miller, *Gard. Dict.* ed. 8, No. 5 (1768). — Lamarck, *Enc. Méth.* III. 509 (1791). — Poiret[1] in *Nouv. Duhamel,* I. 116, t. 30 (1801?). — Watson, *Dendr. Brit.* I. 8, t. 8 (1825). — Loudon, *Arb. Brit.* III. 1271, fig. 1110, 1111 (1838). — G. Don, *Gen. Syst.* IV. 458 (1838). — Dunal in De Candolle, *Prodr.* XIII. pt. 1, 510 (1852). — K. Koch, *Dendr.* II. pt. 1, 348 (1872). — Dippel, *Handb. Laubholzk.* I. 25, fig. 11 (1889). — Hemsley in *Jour. Linn. Soc.* XXVI. 175 (1890). — Koehne, *Deutsche Dendr.* 518 (1893). — Diels in *Bot. Jahrb.* XXIX. 563 (1900); in *Not. Bot. Gard. Edinburgh,* VII. 374 (*Pl. Chin. Forrest.*) (1913). — Schneider, *Ill. Handb. Laubholzk.* II. 611, fig. 394 f–g, 395 f–k (1911). — Bean, *Trees & Shrubs Brit. Isl.* II. 61 (1914). — Rehder in Bailey, *Stand. Cycl. Hort.* IV. 1930, fig. 2229 (1916).

> Lycium barbarum, var. *chinense* Aiton, *Hort. Kew.* I. 257 (1789).
> Lycium Trewianum[2] Roemer & Schultes, *Syst.* IV. 693 (1820). — G. Don, *Gen. Syst.* IV. 458 (1838). — Loudon, *Arb. Brit.* III. 1271 (1838).

Western Hupeh: Changlo Hsien, roadsides, etc., alt. 600–1300 m., July 1907 (No. **3536**; shrub 0.5–1 m. tall, flowers purple); Ichang, *A. Henry* (without No.). Western Szech'uan: Min River valley, Mao-chou, alt. 1100–2000 m., September 1908 (No. **3536ᵃ**; shrub 0.5–1 m., fruits scarlet, edible). Chili: Peking, on city wall, July 15, 1913, *F. N. Meyer* (No. 1019); same locality, yellow temple, September 16, 1903, *C. S. Sargent.* Yunnan: " Rocky moist situations below high water level, Salwin-Irrawadi divide. Only at one place near Chongwa," alt. 1000 m., November 1904, *G. Forrest* (No. 952; spreading shrub 0.5–1 m., flowers heliotrope); eastern flank of the Lichiang Range, open situations in pine forests, alt. about 3000 m., September 1914, *C. Schneider* (No. 2222; shrub up to 1.25 m. tall). Chekiang: vicinity of Ningpo, 1908, *D. Macgregor.*

[1] According to Dunal, who has seen the specimens of Lamarck, Poiret is the author of the descriptions in the *Nouveau Duhamel,* which appeared in eighty parts between 1801 and 1819.

[2] The species is founded on *Lycium* foliis oblongo-lanceolatis sine ordine, ramorum spinis rarioribus of Trew, *Pl. Select. Ehret.* t. 68 (1750). It seems to be in no way different from typical *L. chinense.*

As far as I can judge by the specimens before me they seem to belong to what I have called *L. chinense*, var. *typicum* Schneider (*Ill. Handb. Laubholzk*. II. 611 [1911]). There is also a var. *ovatum* Schneider, l. c., the synonymy of which may be stated as follows: *?Jasminoides rhombifolium* Moench, *Meth*. 470 (1794).— *Lycium ovatum* Poiret in *Nouv. Duhamel*, I. 117 (1801?). — *Lycium barbarum* Watson, *Dendr. Brit.* I. 9, t. 9 (non Linnaeus) (1825). — *Lycium megistocarpum*, var. *ovatum* Dunal in De Candolle, *Prodr.* XIII. 510 (1852). — *Lycium rhombifolium* Dippel in Dosch & Scriba, *Excursionsfl. Grossh. Hess.*, ed. 3, 218 (1888); *Handb. Laubholzk*. I. 24, fig. 10 (1889). Of this variety the leaves are more rhombic, broader and larger, and the fruits are somewhat larger and very obtuse at the apex.

L. chinense is closely related to *L. halimifolium* Miller (*L. vulgare* Dunal), which may be distinguished by the somewhat narrower, longer and therefore more distinct tube of the corolla, the lobes of which usually are a little shorter than those of *L. chinense*. The leaves of this species are brighter green and fall later in the autumn. *L. halimifolium* is often said to be a native of China, but as far as I know it is entirely absent from central and eastern Asia. Its native country seems to be the Mediterranean region (Spain) and probably also Hungary.

GESNERACEAE.

Determined by ALFRED REHDER.

LYSIONOTUS D. Don.

Lysionotus pauciflorus Maximowicz in *Bull. Acad. Sci. St. Péters-bourg*, XIX. 534 (1874); in *Mél. Biol.* IX. 366 (1874). — Franchet & Savatier, *Enum. Pl. Jap.* I. 327 (1875). — S. Moore in *Jour. Bot.* XIII. 231 (1875). — C. B. Clarke in De Candolle, *Monog. Phaner.* V. 599 (1883). — Hemsley in *Jour. Linn. Soc.* XXVI. 225 (1890).

Lysionotus pauciflorus, var. ? Henry in *Trans. As. Soc. Japan*, XXIV. suppl. 68 (1896). — Matsumura & Hayata, in *Jour. Coll. Sci. Tokyo*, XXII. 288 (*Enum. Pl. Formos.*) (1906).
Lysionotus warleyensis Willmott in *Gard. Chron.* ser. 3, LIV. 125 (1913).

Western Hupeh: Hsing-shan Hsien, cliffs, alt. 1200–1500 m., July and October 1907 (No. 370; flowers creamy white, spotted purple); Changyang Hsien, cliffs, alt. 600–900 m., July 1907 (No. 3519; flowers purple); without precise locality, *A. Henry* (No. 6139).

I have not been able to compare flowering specimens of the Japanese plant, but a co-type in fruit in the Gray Herbarium of Maximowicz's species agrees very well with the Hupeh plant; the capsules are identical with those of Wilson's No. 370, and the leaves do not differ from those of Wilson's No. 3519, while the leaves of No. 370 are smaller and narrower, mostly linear-oblong in outline and up to 3.5 cm. long and to 7 mm. wide; the leaves of Henry's No. 6139 are similar, but somewhat shorter and broader. The color of the flowers varies apparently from white and spotted or striped purple to purple, as appears from Wilson's field notes.

Lysionotus brachycarpus Rehder, n. sp.

Suffrutex, saepe epiptyticus, circiter 0.25 cm. altus; ramuli teretes, carnosuli, glabri. Folia plerumque in apice ramulorum 5–7 subverticillata, crassiuscula, elliptico-ovata v. ovato-lanceolata, acuta, basi late cuneata v. subrotundata, serrulata v. interdum serrata dentibus parvis glandula fusco terminatis, in triente v. dimidio inferiore integra 2–4 cm. longa et 0.7–1.5 cm. lata, glabra, supra laete viridia, subtus albescentia, costa media supra impressa subtus elevata, nervis utrinsecus 4–5, supra leviter impressis, subtus obsoletis v. interdum rubescentibus et magis conspicuis; petioli 2–6 mm. longi, supra canaliculati et sparse pilosi. Flores albi, in cymis axillaribus 1–3-floris in

387

apice ramulorum aggregatis graciliter pedunculatis pedunculo 2–3.5 cm.
longo; pedicelli tenues, 3–8 mm. longi; calycis lobi lineari-lanceolati
v. lineares, 5–6 mm. longi, plerumque purpurascentes, glabri; corolla
alba, bilabiata, 3.5 mm. longa, extus infra medium villosa, ceterum
glabra, tubo 2.5 cm. longo e basi satis lata sensim et paulo ampliato,
labio postico 3–4 mm. longo bilobato lobis rotundatis multo latioribus
quam longis, labio antico 9–10 mm. longo 3-lobato lobis rotundatis
margine crenulatis lateralibus quam medius brevioribus; stamina
trienti inferiori tubi inserta, 2 fertilia tubo paulo brevior filamentis
valde complanatis apice conniventibus, antheris rotundatis coherentibus, 2 sterilia valde rudimentaria circiter 1 mm. longa apice interdum leviter incurva antheris destitutis; discus cylindrico-cupularis,
obsolete sinuato-dentatus, 1.5–2 mm. longus; stylus ovario fere aequilongus.   Capsula lineari-oblanceolata, compressa, 3–4 cm. longa et
3–4 mm. lata; semina oblonga v. fere fusiformia, vix 1 mm. longa, pallide brunnea, funiculo et seta apicali paulo brevioribus instructa.

W e s t e r n   S z e c h ' u a n : Wa-shan, on rocks and trees, alt. 1200 m.,
July 1908 (No. 2254, type); same locality, on trees, alt. 1500–2200 m.,
October 1908 (No. 1111ᵃ); without precise locality, *A. Henry* (No.
8997).

This species seems most closely related to *L. pauciflorus* Maximowicz, which is
easily distinguished by the narrower, coarsely dentate leaves on shorter and stouter
petioles, by the shorter peduncles, the shorter lanceolate calyx-teeth, the glabrous corolla with a tube more abruptly enlarged from a narrow base, by the sterile
filaments similar to the fertile ones, but only half as long and with rudimentary co-
herent anthers, by the longer disk, and by the linear capsule from 6 to 7 cm. long.

Lysionotus Wilsonii Rehder, n. sp.

Suffrutex sarmentosus v. repens, ramis carnosulis angulatis de-
bilibus, glaberrimus.   Folia opposita, crasse coriacea, elliptica v. oblonga, breviter obtuse acuminata, basi cuneata saepe inaequaliter v.
fere rotundata, margine leviter revoluta, integra v. minute remote
denticulata, denticulis saepe ad glandulam reductis, 4.5–7 longa et
2.3–3.7 lata v. 4–7.5 longa et 1–2.3 lata, supra laete viridia, subtus
pallida, utrinque nervis 5–6 angulo angusto a costa divergentibus
supra inconspicuis vix impressis subtus invisibilibus; petioli crassi,
0.5–1 cm. longi.   Flores albi, plerumque bini in pedunculis axillaribus 1.5–4 cm. longis gracilibus; pedicelli circiter 5 mm. longi; calycis
dentes anguste lanceolati, 5–6 mm. longi, purpurascentes; corolla
alba, bilabiata, 4.5 cm. longa, tubo in triente inferiore satis angusto
supra ampliato et leviter ventricoso; labium posticum bilobatum cir-

citer 4 mm. longum lobis rotundatis latissimis, labium anticum trilobatum, 8–10 mm. longum, lobis rotundatis subaequalibus leviter crenulatis; stamina 4, trienti inferiori tubi inserta, 2 fertilia antica tubo paulo breviore, filamentis crassiusculis leviter compressis, antheris rotundatis coherentibus, 2 sterilia duplo minora, filamentis fere filiformibus, antheris minutis rudimentariis coherentibus; discus tubuloso-cupularis, 2 mm. longus, sinuato-dentatus; ovarium circiter 2 cm. longus; stylus vix 1 cm. longus. Capsula linearis, compressus, 13–14 cm. longa; semina oblonga, flavo-brunnea, 0.75 mm. longa, funiculo 1.5 mm. longo et seta apicali subaequilonga.

Western Szech'uan: Wa-shan, on rocks, alt. 1200–1500 m., July and October 1908 (No. IIII).

This species seems most nearly related to *L. conferta* C. B. Clarke, which differs in its straight, undivided branches, in its larger, ternate leaves with the veins diverging at right angles, and in its many-flowered cymes on peduncles about 10 cm. long. The flowers of *C. conferta* are not known and may show further differences.

RUBIACEAE.

Determined by J. HUTCHINSON.

ADINA Salisb.

Adina rubella Hance in *Jour. Bot.* VI. 114 (1868). — Maximowicz in *Mél. Biol.* IX. 270 (1873); in *Bull. Acad. Sci. St. Pétersbourg,* XIX. 286 (1874). — Hemsley in *Jour. Bot.* XIV. 208 (1876); in *Jour. Linn. Soc.* XXIII. 371 (1888). — E. Pritzel in *Bot. Jahrb.* XXIX. 580 (1901). — Pampanini in *Nuov. Giorn. Bot. Ital.* n. ser. XVII. 718 (1910).

Western Hupeh: Ichang, by the sides of streams, abundant, alt. 32–330 m., July 1907 (No. **1949**; bush 0.65–1.65 m. tall; flowers white); banks of Yangtsze River, Nanto, June 1900 (Veitch Exped. No. 1255; bush 1 m. tall; flowers blush). Kwangsi: Lungchou, "Caobung River banks," *H. B. Morse* (No. 599; shrub up to 1.25 m. tall; flowers brownish).

Adina racemosa Miquel, *Cat. Mus. Lugd.-Bat.* I. 44, *Fl. Jap.* (1870). — Maximowicz in *Mél. Biol.* IX. 270 (1873); in *Bull. Acad. Sci. St. Pétersbourg,* XIX. 286 (1874). — Franchet et Savatier, *Enum. Pl. Jap.* I. 206 (1875). — Hemsley in *Jour. Linn. Soc.* XXIII. 370 (1888). — E. Pritzel in *Bot. Jahrb.* XXIX. 580 (1901).

Nauclea racemosa Siebold et Zuccarini in *Abh. Akad. Muench.* IV. pt. 3, 178 (*Fl. Jap. Fam. Nat.* II. 54) (1846). — Miquel in *Ann. Mus. Lugd.-Bat.* III. 108 (1867).

Western Hupeh: Hsing-shan Hsien, woodlands, alt. 330–950 m., June 1907 (No. **1951**; slender tree 5–15 m. tall; flowers white, anthers pink); Ichang, hills near Shih-tu-ya, June 1900 (Veitch Exped. No. 1268; tree 3 m. tall, flowers dirty white); Nanto, mountains, June 1901 (Veitch Exped. No. 1181, 1181ᵃ).

Here may be added notes on two species not collected during the Arnold Arboretum Expeditions.

Adina cordifolia Hooker f. in Bentham & Hooker, *Gen.* II. 30 (1873); *Fl. Brit. Ind.* III. 24 (1880). — Brandis, *Forest Fl. Ind.* 263, t. 33 (1874).

Nauclea cordifolia Roxburgh, *Pl. Coromand.* I. t. 53 (1798); *Hort. Beng.* 14 (1814); *Fl. Ind.* ed. 2, I. 514 (1832). — Wallich, *Cat.* 692 (1828). — Wight

& Arnott, *Prodr. Fl. Ind.* 391 (1834). — Kurz, *Forest Fl. Brit. Burma*, II. 66 (1877). — Dalziel & Gibson, *Bombay Fl.* 118 (1861).

Yunnan: Red River, Hsu-kai, alt. 330 m., *A. Henry* (No. 9570; tree 10–13 m. tall); Manpan, Red River valley, alt. 330–660 m., *A. Henry* (No. 9570ª; trees 5–8 m. tall).

Adina mollifolia Hutchinson, n. sp.

Arbor 3–6.5 m. alta; ramuli juniores breves, glabri v. fere glabri. Folia in speciminibus floriferis leviter oblique ovata, abrupte acuminata, acuta (acumine 1–1.2 cm. longo), 6–12 cm. longa, 4–6.5 cm. lata, in speciminibus fructiferis multo majora, usque ad 25 cm. longa et 18 cm. lata, tenuiter chartacea, supra minutissime puberula, infra praecipue in nervis lateralibus et tertiariis molliter pubescentia; costa infra prominens; nervi laterales utrinsecus 7–8, utrinque distincti, a costa sub angulo 45° abeuntes, prope marginem arcuati et conjuncti; nervi tertiarii primum subparalleli, demum subflexuosi; petioli 1–1.5 cm., demum 5 cm. longi, puberuli; stipulae mox caducae, in vaginam connatae, circiter 0.8 cm. longae, rigide subcoriaceae, brunneae, glabrae. Capitula globosa, 2.5–3 cm. diametro, in racemum terminalem laxum disposita; pedunculi patuli, usque ad 4 cm. longi, circiter 1.25 mm. crassi, minute puberuli v. tomentelli; bracteae lineares, usque ad 0.8 cm. longae, acutae, puberulae; calycis lobi 5, triangulares, minuti, circiter 0.5 mm. longi, dense lanati; corollae tubus inferne cylindricus, superne breviter expansus, 0.5 cm. longus, minute puberulus; lobi 5, oblongo-ovati, obtusissimi, 1 mm. longi, 0.75 mm. lati, marginibus breviter pubescentibus; antherae inclusae, utrinque acutae, 1.75 mm. longae, apice minute pubescentes; stylus longe exsertus, 1.3 cm. longus, glaber, stigmate ellipsoideo 0.35 mm. longo coronatus. Infructescentia globosa, 2–2.5 cm. diametro.

Yunnan: Szemao, mountains to west, alt. 1640 m., *A. Henry* (No. 11888, type; tree 5 m tall; flowers); same locality, alt. 1320 m., *A. Henry* (No. 12853; tree 6 m. tall; fruits); Talang, alt. 1640 m., *A. Henry* (No. 13265; shrub 3 m. tall; fruits).

This is a very distinct species with leaves softly hairy as in *A. cordifolia* Hooker f., but not at all cordate, very different stipules, and a racemose inflorescence as in *A. racemosa* Miquel; the hairy leaves and larger fruiting capitula serve to distinguish it from the latter species.

EMMENOPTERYS Oliv.

Emmenopterys Henryi Oliver in *Hooker's Icon.* XIX. t. 1823 (1889). — E. Pritzel in *Bot. Jahrb.* XXIX. 580 (1901). — Pampanini in *Nuov. Giorn. Bot. Ital.* n. ser. XVII. 718 (1910).

Western Hupeh: woodlands north and south of Ichang, alt. 670–1340 m., July and November 1907 (No. **622**; tree 10–30 m. tall; girth 1.25–4 m.; flowers white; fruits); without definite locality, June 1900 (Veitch Exped. No. 1253).

This is one of the most strikingly beautiful trees of the Chinese forests, with its flattish to pyramidate corymbs of pure white, rather large flowers and still larger white bracts. In July, when in full bloom, the trees are conspicuous splashes of white. The bracts persist and become pink as the capsular fruit ripens. The tree grows from 15 to 26 m. tall with a trunk from 2 to 3 m. in girth and has gray rough bark which is scaly on young trees. The branches on adult trees are stout, flat and spreading, and the leafage is abundant. The tree is common in moist forests on the mountains of western Hupeh and of Szech'uan between 600 and 1300 m. altitude, where it is colloquially known as the Hsiang-kuo-shu.

Pictures will be found under Nos. 067, 283, 318, 319 and 518 of the collection of Wilson's photographs and in his *Vegetation of Western China*, Nos. 211, 212.

E. H. W.

WENDLANDIA Bartl.

Wendlandia longidens Hutchinson, n. comb.

Hedyotis longidens Hance in *Jour. Bot.* XX. 289 (1882).
Wendlandia Henryi Oliver in *Hooker's Icon.* XVIII. t. 1712 (1887). — Hemsley in *Jour. Linn. Soc.* XXIII. 372 (1888). — E. Pritzel in *Bot. Jahrb.* XXIX. 580 (1901).

Hupeh: Ichang, glens, alt. 30–300 m. (No. **2359**; bush up to 1.5 m. tall; flowers white). Szech'uan: Yangtsze River banks, May 1903 (Veitch Exped. No. 3756; shrub 0.75–1.25 m. tall; flowers white, fragrant).

Here may be added notes on species collected in Yunnan by A. Henry.

Wendlandia tinctoria De Candolle, *Prodr.* IV. 411 (1830). — Brandis, *Forest Fl. Ind.* 269 (1874). — Kurz, *Forest Fl. Brit. Burma*, II. 74 (1877). — Hooker f., *Fl. Brit. Ind.* III. 38 (1882).

Yunnan: Red River valley, near Manpan, grass hills, alt. 820 m., *A. Henry* (No. 10176; shrub 2.5 m. tall; flowers white); Mi-lê district, *A. Henry* (No. 10568; tree 3 m. tall; flowers white).

Wendlandia paniculata De Candolle, *Prodr.* IV. 411 (1830). — Hooker f., *Fl. Brit. Ind.* III. 39 (partim, excl. syn. *W. luzoniensis* De Candolle) (1882). — Hemsley in *Jour. Linn. Soc.* XXIII. 372 (1888).

Wendlandia uvariifolia Hance in *Jour. Bot.* 73 (1870). — Maximowicz in *Bull. Acad. Sci. St. Pétersbourg*, XXIX. 158 (1883); in *Mél. Biol.* XI. 776 (1883).

Yunnan: Mengtsze, mountains to southeast, forests, alt. 1640 m., *A. Henry* (No. 10953; shrub 2.5 m. tall; No. 11479; tree 3 m. tall; flowers white; No. 11479); Szemao, mountains to southwest, alt. 1330 m., *A. Henry* (No. 12982; shrub 3 m. tall; flowers white). Kwangsi: Lungchou, *H. B. Morse* (No. 372; tree 5–6.5 m. tall; flowers white).

Wendlandia glabrata De Candolle, *Prodr.* IV. 411 (1830). — Kurz, *Forest Fl. Brit. Burma*, II. 74 (1877). — Hooker f., *Fl. Brit. Ind.* III. 39 (1882). — Maximowicz in *Bull. Acad. Sci. St. Pétersbourg*, XXIX. 158 (1883); in *Mél. Biol.* XI. 777 (1883). — Engler & Maximowicz in *Bot. Jahrb.* VI. 67 (1885). — Hemsley in *Jour. Linn. Soc.* XXIII. 371 (1888).

Yunnan: Red River valley near Manpan, grass hills, alt. 820 m., *A. Henry* (No. 10176; shrub 2.5 m. tall, flowers white); Szemao, *A. Henry* (No. 13014; small tree); near Kouang-yu, Feb. 13, 1891, *J. Delavay*.

Wendlandia floribunda Craib in *Kew Bull. Misc. Inform.* 1913, 200.

Wendlandia glabrata, var. *floribunda* Craib, l. c. 1911, 386; in *Aberdeen Univ. Studies*, No. 57, 100 (*Contrib. Fl. Siam*) (1912).

Yunnan: Mengtsze, mountains to southeast, forests, alt. 1640 m., *A. Henry* (No. 10176ᵃ; tree 3 m. tall; flowers).

Wendlandia bouvardioides Hutchinson, n. sp.

Frutex 2 m. altus; ramuli annotini pubescentes v. fere glabri, cortice cinereo obtecti, hornotini breves, appresse hirsuti. Folia elliptica v. lanceolata, utrinque attenuata, apice longe et subobtuse acuminata, 8–16 cm. longa, 2–5 cm. lata, tenuiter chartacea, utrinque infra costa et nervis puberulis exceptis glabra; nervi laterales utrinsecus 6–7, a costa sub angulo 65–75° abeuntes, supra distincti, infra prominuli, graciles, intra marginem conjuncti et crenato-undulati; venae gracil, limae, laxae; petioli 4–8 mm. longi, puberuli; stipulae suborbiculares, foliaceaei 4–5 mm. latae, extra puberulae, intra glabrae. Inflorescentia thyrsoidea, densiflora, 4–5 cm. longa, circiter 3.5 cm. diametro; bracteae subulatae v. lineares, parvae; pedicelli brevissimi; receptaculum puberulum; calycis lobi subulato-lanceolati, subacuti, 1.25 mm. longi, purpurascentes, fere glabri; corolla rubra; tubus cylindricus, 1 cm. longus, glaber; lobi 5, oblongo-ovati, obtusi, 1 mm. longi, glabri; antherae semiexsertae, 1 mm. longae; stylus gracillimus, glaber, corollae tubo aequilongus.

Yunnan: Mengtsze, mountains to southeast, forests, alt. 1640 m., *A. Henry* (No. 10956; shrub 2 m. tall; flowers red).

The corolla of this species is somewhat similar to that of *W. longidens* Hutchinson, but in our plant the anthers are scarcely exserted, the style is shorter, and the leaves are much longer and long-acuminate. The species reminds one of certain cultivated Bouvardias.

MUSSAENDA L.

Mussaenda Wilsonii Hutchinson, n. sp.

Frutex 1.3–2 m. altus; ramuli teretes, circiter 4 mm. crassi, breviter et parce pubescentes, internodiis 5–8 cm. longis. Folia late elliptica v. elliptico-orbicularia, basi breviter acuminata, apice subabrupte acuminata, 10–20 cm. longa, 5–12 cm. lata, membranaceo-chartacea, supra costa et nervis lateralibus parce puberulis exceptis glabra, infra in costis et nervis minute pubescentia; costa supra plana, infra prominens; nervi laterales utrinsecus circiter 9, a costa sub angulo 55° abeuntes, arcuati, intra marginem conjuncti et flexuosi; nervi tertiarii subparalleli, undulati; petioli 1.5–3.5 cm. longi, puberuli; stipulae ovato-lanceolatae, apice plerumque bifidae v. bilobatae, acutae, 0.8–1 cm. longae, tenuiter chartaceae, extra parce appresse pubescentes, subpersistentes. Flores in cymas laxas terminales dispositi; bracteae stipulis similes sed leviter minores; bracteolae lineari-lanceolatae, acute acuminatae, 0.5–1 cm. longae, extra breviter pubescentes; receptaculum breviter appresse pubescens; calycis lobi subfoliacei, lanceolati, longe acuminati, acuti, fere 1 cm. longi, 2 mm. lati, extra breviter pubescentes, interdum uno lobo petaloideo obovato breviter acuminato stipitato 3–4 cm. longo 1.5–2 cm. lato albo fere glabro; corolla flava; tubus 1.4 cm. longus, inferne cylindricus, superne leviter ampliatus, extra dense appresse pubescens, intra in parte ampliata dense villosus, infra glaber; lobi 5, ovati, breviter apiculati, 2 mm. longi, basi 3 mm. lati, utrinque breviter pubescentes; stamina ad tubi medium inserta; antherae inclusae, 3 mm. longae; filamenta corollae adnata, glabra; stylus glaber, stigmate complanato bilobato leviter exserto coronatus.

Hupeh: Changlo Hsien, ravines, up to 650 m. alt. (No. 3265, type; bush 1.5–2 m. tall; calyx-lobe white; flowers yellow); Nanto, on low mountains (Veitch Exped. No. 1248); Changyang Hsien, *A. Henry* (No. 6233; large climber; petaloid calyx-lobe white; flowers yellow; fruit green).

Mussaenda divaricata Hutchinson, n. sp.

Frutex erectus v. scandens, ramuli appresse pubescentes, demum subglabri et parce lenticellati. Folia elliptica v. ovato-elliptica, basi cuneata v. acuta, apice abrupte acuminata, 7–12 cm. longa, 5–7 cm. lata, tenuiter chartacea v. submembranacea, supra parce setulosa v. fere glabra, infra praesertim in nervis brevissime pubescentia; costa infra prominens; nervi laterales utrinsecus 9–11, arcuati, a costa sub angulo 65°–75° abeuntes, intra marginem elongati, utrinque conspicui; nervi tertiarii conferti, subparalleli; petioli 0.5–1 cm. longi, appresse setulosi; stipulae profunde bilobatae, 1 cm. longae, lobis subulato-acuminatis acutissimis parce setulosis. Inflorescentia plerumque divaricato-cymosa, laxiflora, calycis lobis petaloideis plerumque 2–4 ornata; bracteae lineares v. subulatae, usque ad 1 cm. longae, breviter ciliatae; receptaculum parce setulosum, 2.5–3 mm. longum; calycis lobi normales subulati, acuti, 4.5–5 mm. longi, brevissime ciliati; lobus petaloideus late ellipticus v. ovato-rotundatus, breviter acute acuminatus, unguiculatus, 4–6 cm. longus, 3–5 cm. latus, plerumque 7-nervus, albus, utrinque solum in nervis puberulus; corolla flava, extra subdense appresse setoso-pubescens, intra superne dense villosa; tubus 2–2.5 cm. longus, superne sensim ampliatus, fauce dense villosus; lobi ovati, acuti, vix appendiculati, 3.5 mm. longi, intra glabri; antherae 5 mm. longae; stylus brevissimus, bilobatus, 3 mm. longus, glaber.

Hupeh: glens around Ichang, alt. 33–475 m., June 1907 (No. 3266, type; bush 0.7–1.7 m. tall; flowers yellow; petaloid calyx-lobe white); same locality, *A. Henry* (No. 3118, 4120); ravines, Chang-yang Hsien, alt. 320–640 m., June 1907 (No. 3264; shrub 1.2 m. tall; flowers yellow; petaloid calyx-lobe white); Nanto, *A. Henry* (No. 6366, flowers; No. 2703, fruits); side of rice fields, Yangtsze River valley, above Lu Chou, June 8, 1903 (Veitch Exped. No. 3759; bush 1 m. tall; flowers yellow; petaloid calyx-lobe white). Szech'uan: Min River valley, *E. Faber* (No. 645, flowers; No. 263, young fruits); Mt. Omei, alt. 1160 m., *E. Faber* (No. 642; fruits). Yunnan: Mengtsze, mountains to south, alt. 1640 m., *A. Henry* (No. 9650; fruits); mountains to southwest, alt. 1330 m., *A. Henry* (No. 9650ᵃ; climber; flowers yellow); same locality, *A. Henry* (No. 9884); same locality, *A. Henry* (No. 11391; fruits).

Mussaenda elliptica Hutchinson, n. sp.

Frutex 1–2 m. altus; ramuli breviter et parce pubescentes, sicco nigrescentes, prominenter lenticellati. Folia elliptica, basi rotundata v. leviter angustata, apice ? (non visa) usque ad 20 cm. longa et 10 cm. lata, tenuiter chartacea, supra fere glabra, infra solum in nervis et venis brevissime pubescentia; costa infra leviter elevata, basi 2 mm. lata; nervi laterales utrinsecus 6–7, a costa sub angulo latissimo abeuntes, arcuati, utrinque conspicui; nervi tertiarii tenuissimi, laxi; petioli 0.7–1 cm. longi, appresse setulosi; stipulae ovato-triangulares, acuminatae, circiter 0.8 cm. longae, extra parce puberulae. Cymae trichotome ramosae, laxae, calycis lobis petaloideis circiter 5 ornatae; cymulae longe pedunculatae, subdensiflorae; bracteae lineari-subulatae, acutae, usque ad 8 mm. longae, appresse pubescentes; receptaculum parce appresse pubescens; calycis lobi normales receptaculo longiores, subulati, acuti, 6–7 mm. longi, extra pubescentes; lobus petaloideus breviter unguiculatus, late ovatus, acute breviter acuminatus, 5–7.5 cm. longus, 3.5–5 cm. latus, 5–6-nervus, solum in nervis minute puberulus; corolla flava; tubus 2 cm. longus, superne sensim ampliatus, extra appresse setuloso-pubescens, intra superne villosus; lobi 5, triangulari-ovati, longe apiculati, 1 mm. longi; antherae inclusae, 5 mm. longae; stylus brevissimus, glaber.

Central Szech'uan: Suiting Fu, ravines, alt. 660–980 m. (No. 4604; bush 1–2 m. tall; flowers yellow; calyx-lobe white).

CLAVIS SPECIERUM CHINENSIUM MUSSAENDAE.

Calycis lobi omnes aequales, parvi . . . . 1. *M.* (?) *Bodinieri.*
Calycis lobus unus plerumque aliis major, petaloideus.
  * Calycis lobi normales lati et foliacei, lanceolati, 1.5–3 mm. lati.
    Caules, calyx corollaque pilis minutis adpressis induta . . 2. *M. Wilsonii.*
    Caules, calyx corollaque pilis longis debilibus patulis v. subreflexis induta.
                                  3. *M. macrophylla.*
  ** Calycis lobi normales filiformes v. lineares, haud foliacei, 1 mm. lati v. angustiores.

† Folia fere sessilia, basi rotundata v. subcordata. Ramuli et folia infra pilis
   rubris dense et molliter villosa.
  Folia abrupte acuminata. Corollae tubus indumento denso et rubro in-
   dutus . . . . 4. *M. sessilifolia.*
  . Folia sensim acuminata. Corollae tubus indumento sparso cinereo brevi
   indutus . . . . 5. *M. Rehderiana.*
†† Folia distincte petiolata v. basi cuneata v. acuta.
  ‡ Calycis lobi normales sub anthesi receptaculo aequilongi v. breviores.
    Folia basi rotundata v. obtusissima, haud cuneata.
      Folia 13–15 cm. longa, parce pubescentia. Corollae lobi longe appen-
       diculati . . . . 6. *M. Henryi.*
      Folia usque ad 10 cm. longa, dense pubescentia. Corollae lobi vix
       v. brevissime appendiculati . . . . 7. *M. breviloba.*
    Folia basi plerumque longe cuneata.
      Folia magna, 18–25 cm. longa, utrinque attenuata. . 8. *M. elongata.*
      Folia usque ad 15 cm. longa.
       Ramuli pubescentes. Nervi tertiarii conferti et paralleli.
                                  9. *M. divaricata.*
       Ramuli glabri v. fere glabri. Nervi tertiarii laxi et reticulati.
                                10. *M. erosa.*
  ‡‡ Calycis lobi normales sub anthesi receptaculo multo longiores.
    Corollae loborum appendices brevissimae v. fere nullae. 11. *M. elliptica.*
    Corollae loborum appendices conspicuae, circiter 1 mm. longae.
      Inflorescentia laxe cymosa, ramosa. Folia plus minusve obovata, infra
       dense pubescentia. Calycis lobi 4–5 mm. longi . . 12. *M. laxiflora.*
      Inflorescentia arcte cymosa, vix ramosa. Folia infra conspicue pube-
       scentia, elliptica v. obovata. Calycis lobi 8 mm. longi
                              13. *M. hirsutula.*
      Inflorescentia arcte cymosa, pauciflora. Folia minute pubescentia,
       ovato-lanceolata. Calycis lobi circiter 3 mm. longi.
                              14. *M. pubescens.*
Species imperfecte descripta, non visa . . . . 15. *M. Cavaleriei.*

ENUMERATIO SPECIERUM CHINENSIUM.

1. **Mussaenda (?) Bodinieri** Léveillé & Vaniot in *Bull. Soc. Bot. France*, **LV.**
59 (1908).
  Yunnan: foot of "Cay-mo-chan," May 6, 1895, *E. Bodinier* (No. 1159).
  A specimen of this species has not been seen by the writer.

2. **Mussaenda Wilsonii** Hutchinson. See p. 393.

3. **Mussaenda macrophylla** Wallich in *Roxburgh, Fl. Ind.* II. 228 (1824). —
Wallich, *Pl. As. Rar.* II. 77, t. 180 (1832.) — Wallich, *Cat.* 6295 (1830). — De
Candolle, *Prodr.* IV. 371 (1830). — Lindley in *Bot. Reg.* XXXII. t. 24 (1846.) —
Hooker f., *Fl. Brit. Ind.* III. 89 (1880). — King & Gamble, *Mat. Fl. Malay Penins.*
No. 14, 73 (1904).

  Mussaenda hispida, D. Don, *Prodr. Fl. Nepal.* 139 (1825). — De Candolle,
    Prodr. IV. 371 (1830).
  Mussaenda calycina Wallich, *Cat.* 6253 (nomen nudum) (1830). — Kurz, *Forest*
    Fl. Brit. Burma, II. 58 (1877).
  Mussaenda frondosa Wallich, *Cat.* 6250ᵃ (non Linnaeus) (nomen nudum)
    (1830).

Yunnan: Szemao, forests, alt. 1300 m., *A. Henry* (No. 12265; large climbing shrub; flowers yellow).

4. **Mussaenda sessilifolia** Hutchinson, n. sp.

Frutex scandens; ramuli elongati, subteretes, 3–4 mm. crassi, pilis rufis v. rubescentibus dense villosi, internodiis 3–4.5 cm. longis. Folia sessilia, oblonga v. oblongo-elliptica, basi rotundata, apice subacuta, sensim acuminata, 6–10 cm. longa, 3–4.5 cm. lata, chartacea, utrinque infra praecipue in nervis dense rubro-villosa; costa utrinque prominens; nervi laterales utrinsecus 9–10, a costa sub angulo 65° abeuntes, arcuati; stipulae profunde bilobatae, 0.8 cm. longae, lobis lineari-subulatis dense villosis. Inflorescentia terminalis, subdense cymosa, calycis lobis magnis petaloideis circiter 5 ornata; bracteae subulatae, dense villosae; receptaculum, calycis lobi et corolla dense rubro-villosa; calycis lobi normales subulati, 2 mm. longi; lobus petaloideus unguiculatus, late ovatus, obtusus, basi rotundatus, 5–6 cm. longus, 3–4.5 cm. latus, circiter 7-nervus, breviter pubescens; corolla flava; tubus 1.7 cm. longus, superne sensim ampliatus et infra dense villosus; lobi ovati, prominenter appendiculati, appendiculis 0.75 mm. longis; antherae inclusae, 0.6 cm. longae; filamenta fere ad basin corollae tubi adnata, glabra; stylus brevis, 6 mm. longus, glaber, lobis 2 mm. longis.

Yunnan: Szemao, forests south, alt. 1300 m., *A. Henry* (No. 12774; climber; flowers yellow).

Very similar to *M. Rehderiana* Hutchinson, but with more elliptic leaves, a denser and longer indumentum on the leaves and on the much stouter corolla tube.

5. **Mussaenda Rehderiana** Hutchinson, n. sp.

Frutex scandens; ramuli subteretes, 3–4 mm. crassi, pilis brevibus subreflexis brunneis hirsuti, internodiis 3.5–7 cm. longis. Folia brevissime petiolata, subelongato-oblonga, obliqua, basi subcordata v. rotundata, apice sensim acute acuminata, 8–14 cm. longa, 3–4.5 cm. lata, submembranacea, utrinque pilis brevibus rubro-cinereis molliter pubescentia; costa supra rigide tomentosa, infra patule pubescentia; nervi laterales utrinsecus circiter 9, a costa sub angulo 75° abeuntes, arcuati, utrinque conspicui; petioli 3–6 mm. longi, dense brunneo-hirsuti; stipulae persistentes, profunde bilobatae, circiter 7 mm. longae, lobis subulatis acutis dense pubescentibus. Inflorescentia laxe paniculato-cymosa, multiflora, calycis lobis magnis petaloideis 4–5 ornata; bracteae lineares, usque ad 7 mm. longae, hirsutae; receptaculum breviter pubescens; calycis lobi normales subulati, 1.5–2 mm. longi, pubescentes; lobus petaloideus ovato-orbicularis, brevissime obtuse acuminatus, 5–7 cm. longus, 3–5 cm. latus, 7–9-nervus, utrinque breviter pubescens; corolla flava, gracilis; tubus 2.5 cm. longus, apicem versus breviter ampliatus, extra parce et breviter cinereo-pubescens, superne intra villosus; lobi ovato-orbiculares, circiter 2 mm. lati, conspicue mucronati; antherae inclusae, 5–6 mm. longae; filamenta glabra; stylus brevis, 5 mm. longus, glaber, lobis 1 mm. longis.

Yunnan: Szemao, mountains to south, forests, alt. 1600 m., *A. Henry* (No. 11790; climbing shrub; flowers yellow).

A very distinct species with subsessile elongate-oblong leaves, rounded or subcordate at the base.

6. **Mussaenda Henryi** Hutchinson, n. sp.

Arbor 6 m. alta; ramuli subteretes, subdense appresse pubescentes, internodiis 2–6 cm. longis. Folia elliptica v. oblongo-elliptica, basi rotundata, apice acute acuminata, 10–15 cm. longa, 4–6 cm. lata, fere membranacea, utrinque appresse setuloso-pubescentia; costa utrinque prominens, pubescens; nervi laterales utrinsecus 9–11, suboppositi, a costa sub angulo 65° abeuntes, graciles, arcuati, utrinque prominentes; venae inconspicuae; petioli 0.7–1 cm. longi, appresse pubescentes;

stipulae profunde bilobatae, lobis lineari-subulatis acutis 1 cm. longis pilosis. Inflorescentia terminalis, laxe cymoso-corymbosa, calycis lobis magnis petaloideis plerumque 4 ornata; bracteae lineares, usque ad 1.4 cm. longae, parce pubescentes; receptaculum puberulum; calycis lobi normales lineari-subulati, acuti, sub anthesi 2.5 mm. longi, puberuli; lobus petaloideus magnus, unguiculatus, ovatus, sensim acute acuminatus, 7–10 cm. longus, 4.5–7 cm. latus, prominenter 7-nervus, solum in nervis parce puberulus, ungue 2–2.5 cm. longo; corollae tubus cylindricus, parte inferiore angusta 1.5–2 cm. longa, parte superiore abrupte ampliata 4–6 mm. longa, extra minute puberulus v. fere glaber, superne villosus; lobi 5, triangulares, prominenter appendiculati, 3 mm. longi, appendicibus 1.75 mm. longis, marginibus alabastro inflexis glabrescentibus; antherae ad basin partis superioris positae, 3 mm. longae; stylus longus, gracilis, glaber, stigmatibus 2 complanatis leviter exsertis circiter 2 mm. longis coronatus.

Yunnan: south of the Red River, A. Henry (No. 13660; tree 6 m. tall).

7. **Mussaenda breviloba** S. Moore in Jour. Bot. XLIII. 137 (1905).

Yunnan: Szemao, alt. 1450 m., A. Henry (No. 11931; 11931ª; climbing shrub; flowers yellow).

8. **Mussaenda elongata** Hutchinson, n. sp.

Frutex 3 m. altus; ramuli sulcati, circiter 4 mm. crassi, breviter appresse pubescentes, internodiis 6–7 cm. longis. Folia elongato-elliptica, utrinque attenuata, apice sensim acute acuminata, 18–24 cm. longa, 5–7 cm. lata, membranacea, utrinque solum in costa nervis et venis parce pubescentia; costa supra angusta, infra crassa, prominens, basi circiter 1.5 mm. lata; nervi laterales circiter 9, ascendentes, a costa sub angulo 45° abeuntes, intra marginem elongati et flexuosi, supra conspicui, infra prominentes; nervi tertiarii pauci, laxi, undulati, graciles, infra conspicui; petioli 1–1.3 cm. longi, breviter appresse pubescentes; stipulae deciduae ? lineares, circiter 0.8 cm. longae, ciliatae. Inflorescentia aperte cymosa, calycis lobis petaloideis circiter 4 ornata; bracteae lineari-lanceolatae, acutae, usque ad 1 cm. longae; receptaculum 3 mm. longum, parce breviter pubescens; calycis lobi normales brevissimi, sub anthesi 1 mm. longi, pubescentes, recurvati; lobus petaloideus ovato-lanceolatus, sensim subacute acuminatus, longe unguiculatus, 8–12 cm. longus, 4–5 cm. latus, prominenter 3- v. subquinquenervus, utrinque solum in nervis parce puberula, ungue 3–3.5 cm. longo pubescente; corolla flava; tubus gracilis, 2.5 cm. longus, superne breviter (circiter 3 mm.) ampliatus, extra breviter pubescens, fauce dense villosus; lobi conspicue apiculati; antherae in parte superiore ampliata positae, 3 mm. longae; stylus gracilis, corolla aequilongus, glaber, breviter bilobus.

Yunnan: Szemao, forests south, alt. 1500 m., A. Henry (No. 12363; shrub 3 m. tall; flowers yellow; in Herb. Kew).

A very striking species with the leaves long-attenuated to the ends, and with long-stalked petaloid calyx-lobes. It resembles M. Kerrii Craib, from Siam, but has broader leaves, a stouter corolla-tube and much larger petaloid calyx-lobes.

9. **Mussaenda divaricata** Hutchinson. See p. 394.

Mussaenda divaricata, var. **mollis** Hutchinson, n. var.

A typo ramulis et foliis infra molliter dense pubescentibus differt.

Tonkin: Yenbay, Red River, A. Henry (No. 9519; straggling shrub; in Herb. Kew).

10. **Mussaenda erosa** Champion in Hooker's Jour. Bot. & Kew Gard. Misc. IV. 193 (1852). — Walpers, Ann. V. 136 (1858). — Bentham, Fl. Hongk. 153 (1861). — Henriques in Bol. Soc. Brot. III. 148 (1884). — Hemsley in Jour. Linn. Soc. XXIII. 378 (1888).

Mussaenda glabra Hooker & Arnott, *Bot. Voy. Beechey,* 264 (1836–40). — Hemsley in *Jour. Linn. Soc.* XXIII. 379 (non Vahl) (1888).

Hongkong: *C. Wilford* (No. 160); *C. Wright* (No. 236); *Col. Urquhart* (No. 163). Kwangtung: Tingu-shan, West River, May 6, 1882, *C. Ford.* Yunnan: Man-kuo, south of Red River, alt. 1340 m., *A. Henry* (No. 10646; shrub 2.5 m. tall; flowers yellow); valley near Manpan, alt. 660 m., *A. Henry* (No. 10646[a]; climbing shrub); Feng-chen-lin, alt. 1960 m., *A. Henry* (No. 13648); Mengtsze, *A. Henry* (No. 13694).

Also in Formosa and on the Liukiu Islands.

11. **Mussaenda elliptica** Hutchinson. See p. 395.

12. **Mussaenda laxiflora** Hutchinson, n. sp.

Frutex 3 m. altus; ramuli juniores graciles, appresse tomentosi. Folia obovata v. obovato-oblanceolata, apice acute acuminata acumine 1–1.5 cm. longo, basi cuneata, 7–14 cm. longa, 2.5–5 cm. lata, chartacea, supra parce setulosa, infra praecipue in nervis et venulis molliter puberula; costa infra conspicua, basi circiter 1.5 mm. lata; nervi laterales utrinsecus 8–9, a costa sub angulo 60°–75° abeuntes, supra prominuli, infra prominentes, prope marginem arcuati et elongati; nervi tertiarii subparalleli, leviter conferti; petioli 0.5–1 cm. longi, appresse tomentelli; stipulae subulatae, profunde bilobatae, acutae, 0.8 cm. longae, hirsutae. Cymae terminales, laxiflorae, circiter 8 cm. latae, calycis lobis petaloideis 2–3 ornatae; bracteae plerumque deciduae, lineares, acutae, usque ad 1 cm. longae, tenuiter pubescentes; calycis lobi normales filiformes, acuti, receptaculo appresse pubescente longiores; sub anthesi 6–7 mm. longi, parce puberuli; lobus petaloideus lanceolatus, subacutus, unguiculatus, 3.5–7 cm. longus, 1.5–2.5 cm. latus, 5–6-nervatus, supra glaber, infra praecipue in nervis parce puberulis, reticulatus; corolla flava; tubus gracilis, superne vix ampliatus, 2.5–3 cm. longus, extra appresse pubescens, intra superne pilosus; lobi triangulari-lanceolati, appendice acuta 1–1.5 mm. longa, intra glabri; antherae paullo supra medium tubi insertae, 4 mm. longae; stylus corollae tubo aequilongus, glaber, bilobatus, lobis 4 mm. longis. Fructus sicco niger, late ellipsoideus, parce pubescens, 1 cm. longus.

Yunnan: Mengtsze, mountains to southeast, forests, alt. 1600 m., *A. Henry* (No. 9650[b]; shrub 3 m. tall; flowers yellow; No. 11493; flowers; in Herb. Kew).

13. **Mussaenda hirsutula** Miquel in *Jour. Bot. Néerl.* I. 109 (1861). — Hemsley in *Jour. Linn. Soc.* XXIII. 379 (1888).

Mussaenda frondosa, Hemsley l. c. (non Linnaeus) (1888).

Kwangtung: West River, May 1882, *C. Ford; Luheang, Krone* (ex Miquel). Hainan: *R. Swinhoe; A. Henry* (No. 8080; flowers); *A. Henry* (No. 8124; 8548; fruits); one mile south of Hoihow, in the midst of dense Pandanus thickets, April 18, 1879, *W. Hancock* (No. 13; flowers yellow).

The above specimens have been identified from Miquel's description; they are certainly not *M. frondosa* Linnaeus, as determined by Hemsley.

14. **Mussaenda pubescens** Aiton f., *Hort. Kew.* ed. 2, I. 372 (1810). — De Candolle, *Prodr.* IV. 371 (1830). — Bentham, *Fl. Hongk.* 153 (1861). — Henriques in *Bol. Soc. Brot.* III. 148 (1884). — Hemsley in *Jour. Linn. Soc.* XXIII. 379 (partim) (1888).

Hongkong: *J. G. Champion* (No. 102); *R. B. Hinds; G. R. Tate; Col. Urquhart; R. Swinhoe; C. Wright; C. Wilford* (No. 160); without locality, *R. Fortune* (No. 80). Fokien: Amoy, *de la Touche* ex *A. Henry* (No. 1855). Yunnan: Szemao, forests, alt. 1500 m., *A. Henry* (No. 12157; small climbing shrub; flowers yellow).

15. **Mussaenda Cavaleriei** Léveillé in Fedde, *Rep. Sp. Nov.* XIII. 178 (1914). Kweichou; on rocks at Touan-Po, near Pin-ue, August 1905, *J. Cavalerie* (No. 2481).

I am quite unable to obtain an idea of the appearance of this species from Léveillé's inadequate description. It may not be a *Mussaenda*.

RANDIA Houst.

Randia Henryi E. Pritzel in *Bot. Jahrb.* XXIX. 581 (1901).

Randia acutidens Hemsley & Wilson in *Kew Bull. Misc. Inform.* 1906, 160.
Diplospora sp.? Hemsley in *Jour. Linn. Soc.* XXIII. 384 (1888).

Western Szech'uan: Kiating Fu, thickets, alt. 320 m., May 1908 (No. **2957**; bush 2–2.5 m. tall; flowers white).

Here may be added the description of two new species from Yunnan collected by A. Henry.

Randia yunnanensis Hutchinson, n. sp.
Frutex inermis, 2.5–3 m. altus; ramuli glabri. Folia elliptico-lanceolata, acute longissime acuminata, basi cuneata, 10–15 cm. longa, 2.5–5 cm. lata, tenuiter chartacea, utrinque costa infra minute puberula excepta glabra; nervi laterales utrinsecus circiter 7, arcuati, supra conspicui, infra prominentes, prope marginem conjuncti; venae laxae, infra prominulae; petioli 3–4 mm. longi, glabri; stipulae sublato-lanceolatae, acutae, 5–7 mm. longae, glabrae. Flores axillares, fasciculati, breviter pedicellati; pedicelli 2–3 mm. longi, puberuli; receptaculum campanulatum, puberulum; calyx cupularis, 1 mm. longus, puberulus, dentatus, dentibus ovato-triangularibus 0.75 mm. longis; corollae tubus 3 mm. longus, extra glaber, fauce villosus; lobi mox reflexi, oblongi, obtusi, 4–5 mm. longi, 2 mm. lati, glabri; antherae exsertae, 0.5 cm. longae; stylus exsertus, 1 cm. longus, anguste clavatus. Fructus globosus, 5–6 mm. diametro, parce puberulus v. fere glaber.

Yunnan: Szemao, mountains to south, alt. 1640 m., *A. Henry* (Nos. 11750, type; shrub 3 m. tall; flowers white; 11750[a]; 11750[b]; fruit red; 11750[c]; 11750[d]).

Very closely related to *Randia densiflora* Bentham, but without the line of dense hairs on the back of the corolla lobes characteristic of that species, and with fasciculate not cymose flowers.

Randia evenosa Hutchinson, n. sp.
Frutex 3 m. altus; ramuli spinosi, cortice cinereo obtecti, juniores minute puberuli, sicco nigrescentes. Folia obovata, basi breviter cuneata, apice rotundata et minute mucronata, 3–6 cm. longa, 2–3 cm. lata, rigide chartacea, nervorum axillis puberulis exceptis glabra, evenosa; nervi laterales utrinsecus circiter 5, plerumque suboppositi, a costa sub angulo 45° abeuntes, utrinque prominuli; petioli 5 mm. longi; stipulae coriaceae, caducae, triangulari-ovatae, 2.5–3 mm. longae, glabrae; spinae rectae 1 cm. longae. Flores flavi, ad apicem ramulorum brevium fasciculati, circiter 8-nati; pedicelli 0.5–1 cm. longi, glabri, basi bracteati, bracteis stipulis subsimilibus; receptaculum obconicum, 1.5 mm. longum, glabrum; calycis tubus cupularis, 2 mm. longus, glaber, 5-dentatus; corollae tubus cylindricus, 5 mm. longus, extra glaber, fauce breviter pubescens; lobi 5, oblongo-elliptici, obtusi, 6 mm. longi, 3–3.5 mm. lati, glabri; antherae inter lobos corollae insertae, exsertae, 5 mm. longae; ovarium biloculare; stylus 1 cm. longus, superne clavato-bilobatus, glaber. Fructus non visus.

Yunnan: Mengtsze, woods, alt. 1520 m., *A. Henry* (No. 10363; shrub 2.3. m. tall; fruits black; No. 10363ª, type; shrub 3 m. tall; flowers yellow).

GARDENIA L.

Gardenia florida Linnaeus, *Spec.* ed. 2, 1679; *flore pleno.*

For synonymy of the species see Hemsley in *Jour. Linn. Soc.* XXIII. 382 (1888).

Western Hupeh: wild and cultivated round Ichang up to 330 m. alt. (No. **2438**; bush 2–3 ft. tall; flowers white).

DIPLOSPORA DC.

Diplospora fruticosa Hemsley in *Jour. Linn. Soc.* XXIII. 383 (1888).

Tricalysia fruticosa K. Schumann apud E. Pritzel in *Bot. Jahrb.* XXIX. 582 (1901).

Western Hupeh: Ichang, thickets, alt. 320–460 m., May 1907 (No. **2968**; bush 2–2.5 m. tall; fruits red).

Here may be added a new species from Yunnan collected by A. Henry.

Diplospora mollissima Hutchinson, n. sp.

Arbor 3–6.5 m. alta; ramuli elongati, subteretes, minute tomentelli v. puberuli, internodiis 2–4.5 cm. longis. Folia oblonga v. oblongo-elliptica, basi leviter cuneata v. rotundata, apice obtuse acuminata acumine 1–1.5 cm. longo, 10–20 cm. longa, 3–7 cm. lata, rigide chartacea, supra costa puberula excepta glabra et leviter nitidula, infra tomentella; costa supra leviter impressa, infra prominens, subteres, basi 2 mm. lata; nervi laterales utrinsecus 7–9, graciles, arcuati, a costa sub angulo 70° abeuntes; nervi tertiarii subparalleli, gracillimi; petioli 0.8–1 cm. longi, tomentelli; stipulae inferne connatae, 0.8–1 cm. longae, parte libera triangulari acute acuminata, extra puberulae. Flores subsessiles, ad axillas foliorum delapsorum dense glomerati, glomerulis 1.5–2 cm. longis; bracteae minutae; receptaculum puberulum; calyx cupularis, 1.25 mm. altus, dentato-undulatus, extra appresse puberulus; corollae tubus 3 mm. longus, glaber; lobi 4, oblongo-elliptici, obtusi, 3.5 mm. longi, 2 mm. lati; stamina inter corollae lobos inserta, exserta; filamenta 1.5 mm. longa, glabra; antherae 3 mm. longae, 1.25 mm. latae; stylus glaber, robustus, 2-lobus, 0.5 cm. longus. Fructus ruber, subglobosus, verruculosus, circiter 7 mm. diametro, parce puberulus, calyce persistente parvo coronatus.

Yunnan: Szemao, forests to east, alt. 1460 m., *A. Henry* (No. 12246; tree 3–6 m. tall); Szemao, forests to southeast, alt. 1320 m., *A. Henry* (No. 12246ª; tree 6 m. tall; fruits red); Yulo forests to the south, alt. 1320 m., *A. Henry* (No. 12928).

Related to *D. siamica* Craib, but with much more densely hairy leaves and hairy calyx.

LASIANTHUS Jack.

Lasianthus Henryi Hutchinson, n. sp.

Frutex 2 m. altus; ramuli elongati, subteretes, dense appresse tomentelli, internodiis 2.5–5.5 cm. longis. Folia oblongo- v. elliptico-lanceo-

lata, longe sensim acuminata, basi obtusa, 10–15 cm. longa, 2.5–
4.5 cm. lata, chartacea, supra glabra, infra in nervis puberula; nervi
laterales utrinsecus 6–8, arcuati, ascendentes, intra marginem elongati,
infra prominentes; nervi tertiarii patuli, paralleli, numerosi, infra
prominentes; petioli 5–7 mm. longi, puberuli; stipulae subulato-tri-
angulares, acutae, 1.5–2 mm. longae, puberulae.   Flores axillares,
subsessiles; bracteae parvae, subulatae; receptaculum turbinatum,
appresse puberulum; calycis lobi subulati, acuti, 2 mm. longi, puberuli;
corolla non visa.   Fructus maturus leviter carnosus, subglobosus,
circiter 6 mm. diametro, glaber.

Western Szech'uan: Kiating Fu, alt. 330–660 m., May 1908
(No. **3344**; bush 2 m. tall); same locality (Veitch Exped. No. 3762;
shrub 2 m. tall; fruits blue).   Yunnan: Mengtsze, forests to southeast, alt. 1640 m., *A. Henry* (No. 11253, type; shrub 1.7 m. tall).

Very closely related to *Lasianthus micranthus* Hooker f., but with densely hairy
branchlets and longer calyx-lobes.

Here may be added notes on some species from Yunnan collected by A. Henry.

Lasianthus Hookeri C. B. Clarke apud Hooker f., *Fl. Brit. Ind.* III. 184 (1880).

> *Mephitidia* sp., Hooker f. & Thomson, *Herb. Ind. Or.* No. 18 ex C. B. Clarke,
> l. c. (1880).

Yunnan: Szemao, mountains to east, alt. 1350 m., *A. Henry* (No. 12031; shrub
2 m. tall; flowers white); forests, to south, alt. 1640 m., *A. Henry* (No. 12031ᵃ;
shrub 2 m. tall); Szemao, forests, alt. 1450 m., *A. Henry* (No. 13570).

Lasianthus inconspicuus Hooker f., *Fl. Brit. Ind.* III. 187 (1880).

> *Hyptianthera stricta* Wallich, *Cat.* No. 8313[1] (nomen nudum) (non Wight &
> Arnott) (1847).

Yunnan: Szemao, mountains to east, alt. 1300 m., *A. Henry* (No. 12608; shrub
1 m. tall; flowers white); Szemao, forests, alt. 1450 m., *A. Henry* (No. 12608ᵃ; in
Herb. Kew).

Lasianthus inconspicuus var. hirtus Hutchinson, n. var.
A typo ramulis breviter tomentellis differt.
Yunnan: Mengtsze, mountains to south, alt. 1970 m., *A. Henry* (No. 9775).

Lasianthus Kerrii Craib in *Kew Bull. Misc. Inform.* 1911, 396.
Yunnan: Szemao, forests, alt. 1320 m., *A. Henry* (No. 12571; shrub 1 m. tall;
fruits blue; in Herb. Kew).

Lasianthus longicauda Hooker f., *Fl. Brit. Ind.* III. 190 (1880).
Yunnan: Mengtsze, mountains to southeast, alt. 1950 m., *A. Henry* (No.
9035; slender tree 3 m. tall; fruit blue; No. 10633; flowers purple).

Lasianthus micranthus Hooker f., *Fl. Brit. Ind.* III. 190 (1880).
Yunnan: south of Red River, *A. Henry* (No. 13670; shrub 3 m. tall).

Lasianthus Biermanni King ex Hooker f., *Fl. Brit. Ind.* III. 190 (1880).
Yunnan: Mengtsze, mountains to southeast, forests, alt. 1640 m., *A. Henry*
(No. 11148; climber with purple flowers).

PAEDERIA L.

Paederia tomentosa Blume, *Bijdr.* 968 (1826). — De Candolle, *Prodr.* IV. 471 (1830). — Hooker f., *Fl. Brit. Ind.* III. 197 (1880). — Maximowicz in *Bull. Acad. Sci. St. Pétersbourg*, XXIX. 173 (1883); in *Mél. Biol.* XI. 798 (1883). — Hemsley in *Jour. Linn. Soc.* XXIII. 389 (1888). — Diels in *Bot. Jahrb.* XXIX. 582 (1900). — Pavolini in *Nuov. Giorn. Bot. Ital.* n. ser. XV. 420 (1915). — Pampanini in *Nuov. Giorn. Bot. Ital.* n. ser. XVII. 720 (1910); XVIII. 139 (1911). — Ito, *Icon. Pl. Jap.* I. t. 6 (1911). — Dunn & Tutcher in *Kew Bull. Misc. Inform.* add. ser. X. 134 (*Fl. Kwangtung & Hongk.*) (1912).

> *Paederia foetida* Thunberg, *Fl. Jap.* 106 (1784). — Hooker & Arnott, *Bot. Voy. Beechey*, 194 (1836). — Bentham, *Fl. Hongk.* 162 (1861). — Hance in *Jour. Bot.* XII. 261 (1874).
> *Paederia chinensis* Hance in *Jour. Bot.* XVI. 228 (1878); XVII. 12 (1879). — Franchet in *Nouv. Arch. Mus. Paris*, sér. 2, VI. 35 (*Pl. David.* I. 155) (1883).
> *Paederia Wilsonii* Hesse in *Mitt. Deutsch. Dendr. Ges.* XXII. 268 (1913).

Kiangsi: Kuling, hedgerows, very abundant, alt. 1200 m., July 27, 1907 (No. **1663**; climber 3–5 m., flowers dark red, silvery without). Western Hupeh: north and south of Ichang, thickets, alt. 300–1000 m., June and October 1907 (No. **435**; climber 3–5 m., flowers dark red); Patung Hsien, thickets, alt. 900 m., July 1907 (Nos. **3209, 3211**; climber 3–4 m., flowers dark red). Hsing-shan Hsien, thickets, alt. 600–900 m., July 1907 (No. **3210**; climber 2–4 m., flowers red); without precise locality, July 1901 (Veitch Exped. No. 1240). Szech'uan: Nanch'uan, 1891, *A. von Rosthorn* (Nos. 510, 2273). Yunnan: Tengyueh, alt. 1700 m., August 1904, *G. Forrest* (No. 34). Chekiang: Ningpo, 1908, *D. Macgregor*. Fokien: Dunn's Exped., April to June 1905 (*Hongk. Herb.* No. 2804).

LEPTODERMIS Wall.

Leptodermis oblonga Bunge in *Mém. Sav. Étr. Acad. Sci. St. Pétersbourg*, II. 108 (*Enum. Pl. Chin. Bor.* 34) (1833). — Walpers, *Rep.* II. 488 (1843). — Hance in *Jour. Bot.* XX. 290 (1882). — Hemsley in *Jour. Linn. Soc.* XXIII. 390 (1888). — E. Pritzel in *Bot. Jahrb.* XXIX. 582 (1901).

> *Hamiltonia oblonga* Franchet in *Nouv. Arch. Mus. Paris*, sér. 2, VI. 34 (*Pl. David.* I. 154) (1883).

Western Hupeh: rocky places at Ichang, alt. 330–660 m., June 1907 (No. 3511; bush 1 m. tall; flowers pink) without precise locality, June 1901 (Veitch Exped. No. 1957). Western Szech'uan: without precise locality, July 1903 (Veitch Exped. No. 3757); shrub 0.75 m. tall; flowers white; cliffs, August 1904 (Veitch Exped. No. 3763); near Monkong Ting, alt. 2300–3000 m., June 1908 (No. 3517; bush 0.75–1.25 m. tall; flowers mauve-purple). North China: without precise locality, *W. Purdom* (seeds only; plants in *Hort. Sargent*, October 1911).

Leptodermis parvifolia Hutchinson, n. sp.

Frutex prostratus v. 0.75–1 m. altus; rami graciles, cortice puberulo demum glabro obtecti; ramuli laterales brevissimi, foliati. Folia parva, spatulato-obovata v. oblanceolata, obtusa v. subacuta, basi in petiolum brevem attenuata, 0.5–1 cm. longa, 2–3.5 mm. lata, infra reticulata, nervis lateralibus inconspicuis; stipulae subulato-lanceolatae, 1 mm. longae. Flores ad ramulorum apices solitarii v. subsolitarii, sessiles; bracteae oppositae, ovatae, abrupte acute acuminatae, 1.5 mm. longae, membranaceae, glabrae; receptaculum costatum, glabrum; calycis segmenta 5, triangulari-lanceolata, acuta, 1.25 mm. longa, breviter ciliolata; corollae tubus superne sensim ampliatus, 1.2 cm. longus, extra puberulus, intra parce pubescens; lobi 5, ovato-lanceolati, acuti, 2.5 mm. longi, marginibus inflexis; antherae leviter exsertae, 2.5 mm. longae; stylus corollae tubo aequilongus, glaber, 5-lobatus, lobis 1.75 mm. longis gracilibus. Fructus non visus.

Western Szech'uan: Tung River valley, near Wa-shan, prostrate over rocks, June 1908 (No. 3518, type; flowers white or pink); without precise locality, cliffs, alt. 330–660 m., June 1903 (Veitch Exped. No. 3760, partly).

A distinct species with slender puberulous branchlets, small leaves and terminal solitary flowers.

Leptodermis nervosa Hutchinson, n. sp.

Frutex usque ad 1 m. altus; rami satis robusti, cortice cinereo breviter puberulo v. demum fere glabro obtecti; ramuli juniores puberuli. Folia plerumque fasciculata, obovata v. oblanceolata, acuta v. subacuta, basi in petiolum brevem attenuata, 1.5–4 cm. longa, 0.7–1.3 cm. lata, tenuiter chartacea, utrinque conspicue nervosa, utrinque infra nervis parce puberulis exceptis glabra, sicco cinereo-viridia; nervi laterales utrinsecus 3–4, ascendentes, utrinque prominentes; venae laxae, inconspicuae; stipulae in segmentis subulatis 2 mm. longis divisae, basi latae et submembranaceae, parce pubescentes. Flores ad apicem ramulorum glomerati, sessiles; bracteae rhomboideo-ellipticae, sensim longe acuminatae, membranaceae, 6 mm. longae, parce ciliolatae; receptaculum glabrum; calycis segmenta 5, rigida, elongato-lanceolato-subulata, acutissima, 4 mm. longa, ciliata; corollae tubus 4.5 mm. longus, extra glaber, fauce pilosus; lobi 5, oblongo-lanceolati, subacuti, 2.5 mm. longi; antherae inclusae, 1.25 mm. longae; stylus gracilis, 7 mm. longus, bilobus, lobis 1.5 mm. longis. Fructus non visus.

Western Hupeh: Changlo Hsien, roadsides and dry places, alt. 330–1000 m., June 1907 (No. 3513, type; bush 0.35–1 m. tall). Kiangsi: Kiukiang, abundant on the plain, alt. 100 m., July 27, 1907 (No. 1622; shrub 1 m. tall; flowers white).

A very distinct species with tripartite stipules, leaves strongly nerved on the upper surface, very long bracts and calyx segments and a bifid style.

Leptodermis Wilsonii Hort. Kew. apud Diels in *Not. Bot. Gard. Edinburgh*, V. 275 (1912).

Western Hupeh: Hsing-shan Hsien, roadsides, alt. 330–920, June 1907 (No. 3510; bush 0.3–1 m. tall; flowers pink); Changlo Hsien, roadsides and dry places, alt. 330–920, June 1907 (No. 3512; flowers white). Western Szech'uan: dry places near Monkong Ting, alt. 2600 m., July 4, 1908 (No. 3514; bush 0.7–1.3 m. tall; flowers white); west and near Wen-ch'uan Hsien, alt. 1640–2000 m., July 1908 (No. 3515; flowers white); cliffs, Niu-tou-shan, west of Kuan Hsien, alt. 2300–2620 m., June 21, 1908 (No. 3516; flowers white).

Here may be added notes on three species not collected during the Arnold Arboretum Expeditions.

Leptodermis pilosa Diels in *Not. Bot. Gard. Edinburgh*, V. 275 (1912).

Hamiltonia pilosa Franchet in sched.

Yunnan: Mengtsze, grass mountains, alt. 1640 m., *A. Henry* (No. 10974; shrubby plant 0.35 m. tall; flowers pink). North-central China: "Miowansan," August 1897, *Hugh Scallan* (Nos. 2, 49, 50).

Leptodermis Purdomii Hutchinson, n. sp.

Frutex 1.5 m. altus; ramuli gracillimi, elongati, juniores minute puberuli, mox glabri. Folia plerumque fasciculata, lineari-oblanceolata, obtusa, basi angustata, 0.5–1 cm. longa, 1.5–3.5 mm. lata, chartacea, glabra, enervosa, marginibus conspicue recurvata; stipulae ovatae, cartilagineo-carinatae, circiter 1.25 mm. longae. Flores ramulorum apices versus aggregati, sessiles v. subsessiles; bracteae oppositae, basi connatae, ovatae, abrupte acutae, 1.5 mm. longae, leviter carinatae, membranaceae, fere glabrae; receptaculum glabrum; calycis segmenta 5, oblongo-ovata, apice triangulari-acuta, 1.5 mm. longa, 0.75 mm. lata, coriacea, ciliolata; corollae tubus superne leviter expansus, 1 cm. longus, fauce 2 mm. diametro, extra dense puberulus; lobi 5, ovato-lanceolati, subobtusi, 2 mm. longi, utrinque intra parce puberuli; stamina fauci inserta; antherae inclusae, 1.5 mm. longae; stylus glaber, breviter exsertus. Fructus non visus.

Western Kansu: Lotoni, on the road to Siku, *W. Purdom* (No. 804, type). Eastern Kansu: Wei River valley, July 20, 1885, *G. N. Potanin*.

Very similar to *Leptodermis virgata* Edgeworth, from northwestern India, but with long slender branchlets at length quite glabrous and not scabrid-puberulous as in that species, and with shorter and broader stipules.

Leptodermis glomerata Hutchinson, n. sp.

Herba basi lignosa, usque ad 60 cm. alta; caulis basi ramosa, ramis erectis gracilibus, pilorum lineis parce puberula, internodiis 3–4 cm. longis. Folia opposita, lanceolata v. ovato-lanceolata, sensim acute acuminata, 2.5–5 cm. longa, 1–2 cm. lata, tenuiter chartacea, supra marginem versus scabridula, infra glabra, reticulata; nervi laterales utrinsecus circiter 5–6, arcuati, inferne prominuli; petioli 1–2 mm. longi; stipulae e basi lato subulatae, acutae, 4 mm. longae, extra basi parce puberulae. Flores ad ramulorum apice glomerati, sessiles; bracteae oppositae, basi connatae, subulato-acuminatae; 2.5 mm. longae, membranaceae, glabrae; receptaculum glabrum; calycis segmenta 5, lanceolato-subulata, acutissima, rigida, 2 mm. longa, glabra, eciliata; corollae tubus leviter curvatus, 1 cm. longus, sensim ampliatus, extra minute puberulus, fauce pilosus; lobi 5, ovati, apice obtuse mucronati, 3.5 mm. longi, 2.5 mm. lati; antherae subexsertae, 1.75 mm. longae; stylus glaber 8 mm. longus, trilobus, lobis 3 mm. longis gracilibus. Fructus non visus.

Yunnan: Mi-lê district, mountain forest, *A. Henry* (No. 9949; flowers dark red).

Very similar in general appearance to *Leptodermis lanceolata* Wallich, from India, but with almost glabrous leaves, much smaller bracts and a trilobed style.

Here may be added notes on several genera not collected during the Arnold Arboretum Expeditions.

NAUCLEA L.

Nauclea Griffithii Haviland in *Jour. Linn. Soc.* XXXIII. 51 (1897).

Adina Griffithii Hooker f., *Fl. Brit. Ind.* III. 24 (1880).

Yunnan: Szemao; forests to south, 1330 m., *A. Henry* (No. 12676; tree 6.5 m. tall; flowers red); mountains to the westward, 1600 m., *A. Henry* (No. 12880; tree 5 m. tall).

UNCARIA Schreb.

Uncaria sessilifructus Roxburgh, *Fl. Ind.* ed. 2, I. 520 (1832). — Wallich, *Cat.* No. 6109 (1830). — Kurz, *Forest Fl. Brit. Burma*, II. 71 (1877). — Hooker f., *Fl. Brit. Ind.* III. 30 (1880).

Nauclea scandens Roxburgh mss. Ic. pict. No. 1218 ex Hooker f., l. c. (pro synon.) (1880).

Yunnan: Chienhung, *M. Bons d'Anty* in Herb. A. Henry (No. 11995); Szemao, forests to west, alt. 1450 m., *A. Henry* (No. 11995[a]; large climber; flowers yellowish white; known as "Kou-teng"); Szemao, alt. 1320 m., *A. Henry* (No. 13449; climber; flowers).

Uncaria scandens Hutchinson, n. comb.

Nauclea scandens Smith in *Rees Cyclop.* XXIV. No. 9 (1819).

Uncaria pilosa Roxburgh, *Fl. Ind.* ed. 2, I. 520 (1832). — Wallich, *Cat.* 6108[a, b] (1828). — Hooker f., *Fl. Brit. Ind.* III. 32 (1880).

Yunnan: Szemao, mountains to west, alt. 1640 m., *A. Henry* (No. 11868; large climber; flowers white; in Herb. Kew). This is the first record of this species from China.

Uncaria macrophylla Wallich in Roxburgh, *Fl. Ind.* II. 132 (1824). — Wallich, *Cat.* 6107, partim (1828). — Hooker f., *Fl. Brit. Ind.* III. 32 (1880).

Uncaria sessilifolia Roxburgh mss. in Wallich, *Cat.* ex Hooker f., l. c. (pro synon.) (1880).

Yunnan: Szemao, near Nantan River, alt. 1490 m., *A. Henry* (No. 12820; large climber;[1] flowers pale yellow).

Uncaria lancifolia Hutchinson, n. sp.

Frutex scandens magnus; ramuli quadrangulares, leviter alati, glabri, superne armati; spinae axillares geniculatae, recurvatae, usque ad 1.5 cm. longae, glabrae. Folia oblongo-lanceolata v. ovato-lanceolata, subacute acuminata, basi rotundata v. leviter cordata, 10–12 cm. longa, 3–5 cm. lata, submembranacea, glabra, supra leviter nitida, infra pallide viridia, tenuiter reticulata; costa utrinque prominens; nervi laterales utrinsecus 7–10, a costa sub angulo 65° abeuntes, arcuati, intra marginem conjuncti, tenues, utrinque conspicui; petioli brevissimi, 2–3 mm. longi; stipulae persistentes, late ovatae, profunde bilobatae, 7–8 mm. longae, 0.7–1 cm. latae, chartaceae, venosae, glabrae, lobis triangularibus acuminatis. Flores non visi. Infructescentiae longe pedunculatae, subglobosae, 3–4 cm. diametro; pedunculi robusti, 4–7 cm. longi, medio bracteati, bracteis deciduis; fructus obovoideo-ellipsoideus, 1.2 cm. longus, 4–5 mm. diametro, parce hispidus, calyce parvo coronatus; semina suborbicularia, 0.75 mm. diametro, utrinque longe alata.

Yunnan: Mengtsze, forests to east, alt. 1640 m., *A. Henry* (No. 11389; large climber; fruits).

Closely related to *Uncaria homomalla* Miquel, from the Jyntea Hills, northeastern India (*Wallich* 6108[c]), but that species has pubescent branches and leaves, and much smaller and narrower stipules. Another close ally is *U. laevigatus* Wallich, from Manipur, Khasia Hills and Tenasserim, with broader leaves, only 5 pairs of side nerves and very early caducous stipules.

HYMENOPOGON Wall.

Hymenopogon parasiticus Wallich in Roxburgh, *Fl. Ind.* II. 157 (1824). — Wallich, *Pl. As. Rar.* III. t. 227 (1832); *Cat.* No. 6113 (1828). — Kurz, *Forest Fl. Brit. Burma*, II. 73 (1877). — Hooker f., *Fl. Brit. Ind.* III. 34 (1880).

Yunnan: Mengtsze, mountains to east, alt. 1930 m., *A. Henry* (No. 9813; shrub 0.75 m. tall; fruits); same locality, on rocks, alt. 2160 m., *A. Henry* (No. 9813; plant 0.35 m. tall; flowers white); same locality, mountains to north, alt. 2300 m., *A. Henry* (No. 9813[a]; shrub up to 1 m. tall; fruits); same locality, mountains to east, cliffs, alt. 2300 m., *A. Henry* (No. 9813[b]; flowers white); same locality, mountains to northwest, cliffs, alt. 2000 m., *A. Henry* (No. 9813[c]; flowers white); Puerh Fu, cliffs, *A. Henry* (No. 9813[d]).

LUCULIA Sweet.

Luculia gratissima Sweet, *Brit. Fl. Gard.* II. t. 145 (1826). — Hooker in *Bot. Mag.* LXVIII. t. 3946 (1842). — Kurz, *Forest Fl. Brit. Burma*, II. 71 (1877). — Hooker f., *Fl. Brit. Ind.* III. 36 (1880).

---

[1] The specimens in Herb. Kew and in Herb. Arnold Arboretum are both in fruit.

Cinchona gratissima Wallich in Roxburgh, *Fl. Ind.* II. 154 (1824). — Wallich, *Tent. Fl. Nap.* t. 21 (1826); *Cat.* No. 6177ᵃ (1828).
Mussaenda Luculia Hamilton in D. Don, *Prodr. Fl. Nepal.* 139 (1825).

Yunnan: Szemao, mountains to west, 1500 m., *A. Henry* (No. 12489; shrub 4 m. tall; flowers pinkish); same locality, 1660 m., *A. Henry* (No. 12489ᵇ; fruits).

Luculia intermedia Hutchinson, n. sp.

Frutex usque ad 3.5 m. altus; ramuli subteretes, prominenter lenticellati, ceterum glabri, internodiis 3–4 cm. longis. Folia oblonga v. oblongo-oblanceolata, acute acuminata, ad basin sensim angustata, 10–15 cm. longa, 2.5–5.5 cm. lata, integra, tenuiter chartacea, supra glabra, infra solum in nervis lateralibus et costa breviter pubescentia; costa supra plana, infra conspicua, apicem versus tenuissima, basi circiter 1.25 mm. crassa; nervi laterales utrinsecus 9–12, a costa sub angulo 45° abeuntes, intra marginem leviter flexuosi et arcuati, utrinque distincti, infra prominuli; nervi tertiarii et venae vix evidentes; petioli 1–1.5 cm. longi, minute pubescentes; stipulae caducae, triangulares, acuminatae, circiter 1 cm. longae et 0.5 cm. latae, subchartaceae, glabrae. Inflorescentia terminalis, laxe cymosa, cymulis circiter 3 plerumque 3–4-floris; bracteae foliaceae, lineari-lanceolatae, acutae, glabrae, usque ad 1.5 cm. longae; pedicelli usque ad 0.5 cm. longi, glabri; receptaculum glabrum; calycis lobi foliacei, lanceolati, acuti, 1 cm. longi, 2–2.5 cm. lati, 1-nervi, glabri; corolla rubescens; tubus cylindricus, 3 cm. longus, extra striatus, glaber; lobi 5, elliptici v. oblongo-ellipitici, 1.5–2 cm. longi, 1.5 cm. lati, basin versus in utroque latere cristis descendentibus intra sinum confluentibus ornati, marginibus minute crenulatis; stamina fauci inserta; antherae inclusae, 4 mm. longae; ovarium 2-loculare, ovulis numerosissimis; stylus breviter exsertus, breviter bilobus, glaber, lobis obtusis 2 mm. longis. Fructus turbinatus, 2–2.5 cm. longus, circiter 0.5 cm. diametro, 10-costatus, minute lenticellatus, glaber.

Yunnan: Mengtsze, mountain forests, 2000–2300 m., December 1894, *W. Hancock* (No. 261; flowers); same locality, 2600 m., *A. Henry* (Nos. 9023, 9023ᵇ; shrub 1.25 m. tall; fruits); Feng-chen-lin Mt., 2000 m., *A. Henry* (No. 9023ᶜ; shrub 3.5 m. tall; flowers reddish).

A very remarkable species distinguished from *Luculia Pinceana* Hooker by the striking flap-like processes upon and continuous between the lobes of the corolla. This species occurs also in Burma as follows: Kachin Hills, Bhamo, alt. 2000 m., September, *G. E. S. Cubitt* (No. 3 and 4 in Herb. *Lace; Vern. Panlawng*); near Laukhang, alt. 1640 m., September, *Kyaw* (No. 34 in Herb. *Lace*).

HEDYOTIS Linn.

Hedyotis capitellata Wallich, *Cat.* No. 837 (nomen nudum) (1828). — G. Don, *Gen. Syst.* III. 527 (1834). — Kurz in *Jour. As. Soc. Bengal*, XLVl, pt. 2, 135 (1877) (excl. var.). — Hooker f., *Fl. Brit. Ind.* III. 57 (1888).

Hedyotis Finlaysoniana Wallich, *Cat.* 6189 (nomen nudum) (1828).
Oldenlandia rubioides Miquel, *Fl. Ind. Bat.* II. 353 (1856).

Yunnan: south of the Red River from Manmei, *A. Henry* (No. 9807; flowers white); Szemao, woods, alt. 1550 m., *A. Henry* (No. 12264); shrub or climber; flowers white).

Hedyotis macrostemon Hooker & Arnott, *Bot. Voy. Beechey*, 192 (1841). — Walpers, *Rep.* II. 493 (1843). — Hemsley in *Jour. Linn. Soc.* XXIII. 374 (1888).

Hedyotis recurva, Bentham in *Lond. Jour. Bot.* I. 486 (1842); in *Hooker's Jour. Bot. & Kew Gard. Misc.* IV. 170 (1852). — Walpers, *Rep.* II. 494

(1843). — Seemann, *Bot. Voy. Herald*, 382, t. 84 (1857). — Miquel in *Jour. Bot. Néerl.* I. 108 (1861). — Bentham, *Fl. Hongk.* 148 (1861). — Hance in *Ann. Sci. Nat.* sér. 4, XVIII. 222 (1862). — Maximowicz in *Bull. Acad. Sci. St. Pétersbourg*, XXIX. 159 (1883); in *Mél. Biol.* XI. 778 (1883). — Henriques in *Bol. Soc. Brot.* III. 148 (1885).

Yunnan: Mengtsze, mountain forests to southeast, alt. 1640 m., *A. Henry* (No. 9555[a]; climber).

Hedyotis scandens Roxburgh, *Hort. Beng.* 10 (nomen nudum) (1814). — *Fl. Ind.* I. 364 (1820). — D. Don, *Prodr. Fl. Nepal.* 134 (1825). — Wallich, *Cat.* No. 839 (1828). — De Candolle, *Prodr.* IV. 422 (1830). — Hooker f., *Fl. Brit. Ind.* III. 57 (1880).

Hedyotis polycarpa R. Brown in Wallich, *Cat.* No. 838 (nomen nudum) (1828). *Hedyotis volubilis* Wallich, *Cat.* No. 838, 840 (nomen nudum) (1828).

Yunnan: Szemao, woods, alt. 1520 m., *A. Henry* (No. 11700; climbing shrub); Szemao, mountains to south, alt. 1480 m., *A. Henry* (No. 11700[a]; climber with purple flowers); Szemao, mountains, alt. 1950 m., *A. Henry* (No. 11700[b]; climber over shrubs; flowers white); Szemao, alt. 1640 m., *A. Henry* (No. 13484; climber with white flowers).

MYCETIA Reinw.

Mycetia gracilis Craib in *Kew Bull. Misc. Inf.* 1914, 125.

Mycetia cauliflora Craib l. c. 1911, 390 (non Reinwardt); in *Aberdeen Univ. Studies*, No. 57, 104 (*Contrib. Fl. Siam*) (1912).

Yunnan: Szemao, forests to east, alt. 1420 m., *A. Henry* (No. 12248; shrub 0.7 m. tall; flowers yellow); Szemao, forests, alt. 1400 m., *A. Henry* (No. 12248[a]; shrub 0.65 m. tall; flowers yellow); *A. Henry* (No. 12248[c]; shrub 1 m. tall; fruits white); *A. Henry* (No. 12248[d]; slender shrub 0.75–1 m. tall; flowers yellow).

Mycetia bracteata Hutchinson, n. sp.

Frutex 1.25 m. altus; ramuli satis robusti, puberuli, demum glabri, cortice cinereo laevi obtecti. Folia elongato-oblanceolata v. lanceolata, sensim acuminata, subacuta, basi attenuata, 9–16 cm. longa, 1.5–4 cm. lata, fere membranacea, utrinque costa et nervis lateralibus minute puberulis exceptis glabra; costa infra prominens, basi circiter 1 mm. lata; nervi laterales utrinsecus 16–20, graciles, a costa sub angulo 45°–65° abeuntes, fere recti, prope marginem prominenter conjuncti, supra conspicui, infra prominentes; nervi tertiarii subparalleli, infra prominuli; petioli 0.5–1 cm. longi, puberuli; stipulae densae, persistentes, late triangulari-ovatae, acutae, rigide chartaceae, 0.7–1 cm. longae, extra nervosae, minute puberulae. Inflorescentia terminalis, laxe cymosa, usque ad 7 cm. longa; bracteae conspicuae, persistentes, interdum subfoliaceae, plus minusve lanceolatae, acutae, usque ad 1.5 cm. longae, puberulae; pedicelli graciles, 1–1.5 cm. longi, puberuli; receptaculum 2 mm. longum, fere glabrum; calycis lobi 5–6, filiformes, 3–6 mm. longi, fere glabri; corolla flava, matura non visa, alabastro 1 cm. longa, subclavata, glabra. Fructus non visus.

Yunnan: Szemao, mountains to south, alt. 1300 m., *A. Henry* (No. 11930[a]; shrub 1.25 m. tall; flowers yellow).

Related to *Mycetia gracilis* Craib, but with relatively longer and broader leaves, with more numerous lateral nerves, a more congested conspicuously bracteate inflorescence and shorter pedicels.

Mycetia glandulosa Craib in *Kew Bull. Misc. Inf.* 1914, 125.

Mycetia longifolia Craib, l. c. 390 (non K. Schumann); in *Aberdeen Univ. Studies*, No. 57, 104 (*Contrib. Fl. Siam*) (1912).

Yunnan: Mengtsze, forests, alt. 1650 m., *A. Henry* (No. 11930; shrubby plant 0.3–0.7 m. tall; flowers yellow). Szemao, forests, alt. 1300 m., *A. Henry* (No. 12248[b]; shrub 0.7 m. tall; fruits white); same locality, *A. Henry* (No. 12700; fruits white).

Mycetia hirta Hutchinson, n. sp.

Frutex 1–2 m. altus; ramuli superne parce ioliati, pilis crispis laxe tomentosi. Folia elongato-elliptica v. late lanceolata, apice acuta, longe sensim acuminata, basi triangulari-cuneata v. sensim angustata, 8–25 cm. longa, 3.5–9 cm. lata, fere membranacea, supra setuloso-pilosa, infra breviter crispato-pubescentia; costa supra insculpta, pilosa, infra prominens; nervi laterales utrinsecus 18–23, suboppositi, a costa sub angulo lato abeuntes, arcuati, prope marginem conjuncti, utrinque prominentes; nervi tertiarii gracillimi, laxi, undulati; petioli 1–3 cm. longi, pubescentes; stipulae oblongae, obtusae, 1–2 cm. longae, extra medio parce pubescentes. Cymae breviter pedunculatae, usque ad 8 cm. expansae, hirtae; bracteae ovatae, stipitato-glandulosae; pedicelli ad 4 mm. longi, pubescentes; receptaculum globoso-campanulatum, setuloso-pubescens; calycis lobi 5, triangulares, subacuti, glanduloso-lobulati, 2.5 mm. longi, parce ciliati; corolla flava; tubus cylindricus, 6 mm. longus, 1.75 mm. diametro, superne longe pilosus; lobi 5, triangulares, subobtusi, 1.75 mm. longi, extra longe pilosi; antherae corollae fauci insertae, 2 mm. longae; stylus 1.5 mm. longus, glaber, ramis 1.25 mm. longis lanceolatis. Capsula subglobosa, circiter 3.5 mm. diametro, hirta, calyce persistente coronata.

Yunnan: Szemao, mountain forests to east, alt. 1600 m., *A. Henry* (No. 11633; shrub 1 m. tall; fruit); same locality, *A. Henry* (No. 12299, type; shrub 2 m. tall; flowers yellow); same locality, 1300 m., *A. Henry* (No. 12299[a]).

Related to and closely resembling the Indian and Malayan *M. longifolia* K. Schumann (*Adenosacme longifolia* Wallich), but distinguished from it by the hairy receptacle and pilose corollas.

MYRIONEURON R. Br.

Myrioneuron Faberi Hemsley in *Jour. Linn. Soc.* XXIII. 380 (1888). — E. Pritzel in *Bot. Jahrb.* XXIX. 581 (1901).

Western Szech'uan: Mt. Omei, August 1904 (Veitch Exped. No. 5043); without definite locality, dripping rocks, alt. 660 m., July 1903 (Veitch Exped. No. 4265; flowers yellow). Yunnan: Mengtsze, mountains to the south, *A. Henry* (No. 9376; shrub 0.35–0.75 m. tall); mountains to southeast, ravines, alt. 1320 m., *A. Henry* (No. 9376[a]; flowers yellow); same locality, forests, alt. 1640 m., *A. Henry* (No. 9376[b]; shrub 2 m. tall; flowers yellow; No. 9376[c]; fruits); Szemao, mountains to west, alt. 1460 m., *A. Henry* (No. 9376[e]; shrub 1 m. tall; fruit white); Szemao, mountains to south, alt. 1320 m., *A. Henry* (No. 12168; shrub 1.65 m. tall; flowers yellow).

TARENNA Schreb.

Tarenna pallida Hutchinson, n. comb.

Webera pallida Franchet apud Brandis, *Ind. Trees*, 378 (1906).

Yunnan: near Manpan, Red River valley, alt. 820 m., *A. Henry* (No. 10686; tree 3–6.5 m. tall; flowers yellow); Szemao, mountains to west, alt. 1640 m., *A. Henry* (No. 11923; tree 6 m. tall; flowers yellow); Szemao, forests, alt. 1320 m.,

A. Henry (No. 11923ᵉ; tree 6 m. tall); mountains to east, alt. 1640, *A. Henry* (No. 11923ᶠ; tree 9 m. tall).

Tarenna sylvestris Hutchinson, n. sp.

Frutex 2–3 m. altus; ramuli glabri. Folia oblanceolato-elliptica v. elliptica, acute acuminata, basi subacuta, 8–15 cm. longa, 3–6.5 cm. lata, chartacea, supra nitidula, reticulata, subtus opaca, evenosa, glabra; nervi laterales utrinsecus 9–13, a costa sub angulo lato abeuntes, utrinque prominuli, arcuati; petioli 1–2 cm. longi, glabri; stipulae lanceolato-ovatae, acute caudato-acuminatae, 1 cm. longae, coriaceae, glabrae. Inflorescentia ·ultiflora, terminalis, corymbosa, 4–5 cm. longa, circiter 6 cm. diametro; bracteae lanceolato-triangulares, acutissimae, usque ad 7 mm. longae, basi 3.5 mm. latae, glabrae; pedicelli laterales 2 mm. longi, glabri v. minute puberuli; bracteolae receptaculum subtendentes oppositae, bracteis minores; receptaculum subcampanulatum, basin versus minute puberulum; calyx 5-partitus; segmenta ovata, acuta, 2 mm. longa, 1.5 mm. lata, coriacea, glabra; corolla fere ad basin 5-partita; tubus 1.5 mm. longus, extra glaber, fauce dense villosus; lobi oblongo-lanceolati, subobtusi, 8–9 mm. longi, 3 mm. lati, marginibus submembranacei, eciliati; antherae exsertae, 6 mm. longae; ovarium 2-loculare, loculis uniovulatis; stylus anguste clavatus, profunde bilobatus, glaber, 1 cm. longus. Fructus viridis, globosus, 8 mm. diametro, reticulato-verrucosus.

Yunnan: Mengtsze woods, alt. 1520 m., *A. Henry* (No. 10006; shrub 3 m. tall; No. 10006ᵃ, type; flowers yellow; No. 10006ᶜ; shrub 3 m. tall; fruit green).

Related to *T. attenuata* Hutchinson, n. comb. (*Webera attenuata* Hooker f.), but the pedicels much shorter, and with larger bracts and calyx-lobes.

Tarenna depauperata Hutchinson, n. sp.

Frutex 2–2.5 m. altus; rami cortice cinereo glabro obtecti; ramuli foliati, glabri. Folia elliptico-obovata, breviter subacute acuminata, basi cuneata, 6–12 cm. longa, 3–6.5 cm. lata, chartacea, supra nitidula, infra glabra; costa utrinque prominens, straminea; nervi laterales utrinsecus 7–11, a costa sub angulo lato abeuntes, utrinque prominuli; venae infra prominulae; petioli 0.5–1 cm. longi; stipulae caducae, triangulari-ovatae, acutae, 4–5 mm. longae, glabrae. Inflorescentia depauperata, terminalis, breviter pedunculata; pedunculus circiter 1 cm. longus, glaber; pedicelli ultimi nulli v. usque ad 3 mm. longi; bracteae triangulares, acutae, 1–1.5 mm. longae, intra puberulae; receptaculum glabrum; calycis lobi ovato-rotundati, 0.75 mm. longi, extra parce pubescentes, ciliolati; corollae tubus lobis aequilongus, 4 mm. longus, extra glaber, intra pilosus; lobi 5, oblongi, apice rotundati, striati, basi intra pubescentes; antherae exsertae, 4 mm. longae; stylus clavatus, 1.3 cm. longus, inferne puberulus. Fructus globosus, 0.8 cm. longus, nitidus, 1-spermus.

Yunnan: Mengtsze woods, alt. 1640 m., *A. Henry* (No. 10816, type; shrub 2.5 m. tall; flowers white; No. 10816ᵃ; fruits black).

A very distinct species with the midrib prominent on each surface of the somewhat shining leaves, deciduous stipules, depauperate inflorescence and shining globose fruits.

Tarenna pubinervis Hutchinson, n. sp.

Frutex circiter 2 m. altus; ramuli juniores nigrescentes, puberuli. Folia oblongo-elliptica v. elongato-lanceolata, longe acute acuminata, ad basin attenuata, 9–15 cm. longa, 2.5–4.5 cm. lata, membranaceo-chartacea, utrinque costa nervisque puberulis exceptis glabra, sicco nigra; costa utrinque prominula, arcte puberula; nervi laterales utrinsecus 7–10, intra marginem ramosi et conjuncti, utrinque prominuli; venae laxae, prominulae; petioli 0.5–1.3 cm. longi, scabrido-puberuli; stipulae paleaceae, rigidae, triangulares, acute acuminatae, 0.5–0.9 cm. longae, glabrae. Inflorescentia subsessilis, pauciflora, circiter 3 cm. diametro; bracteae lineares, 1.5–2 mm. longae, breviter pilosae; pedicelli breviter pubescentes; receptaculum fere glabrum; calycis lobi 5, lanceolati, subacuti, 1.75 mm. longi, parce ciliati, sicco apice straminei; corollae tubus lobis paullo brevior, 4 mm. longus, extra glaber, intra villosus; lobi 5, oblongi, obtusi; antherae exsertae, acutae, 0.5 cm. longae; stylus anguste clavatus, 0.8 cm. longus, inferne pubescens. Fructus 3-spermus, anguste ellipsoideus, sicco niger, 1 cm. longus, glaber, apice annulo stramineo coronatus.

Yunnan: Feng-chen-lin Mt., south of Red River, forests, alt. 2640 m., *A. Henry* (No. 10678, type; shrub 2 m. tall); Mengtsze, forests to southeast, alt. 1640 m., *A. Henry* (No. 10059; shrub 1.6 m. tall; fruits black).

This species is distinguished by its elongated leaves puberulous on the petioles and nerves, paleaceous stipules, a small almost sessile inflorescence and a pubescent style.

IXORA Linn.

Ixora yunnanensis Hutchinson, n. sp.

Frutex usque ad 1 m. altus; caulis parce ramosus (v. subsimplex?), glaber, internodiis 2–4 cm. longis. Folia geminata v. ternata, oblique et anguste oblanceolata, sensim acute acuminata, ad basin longe attenuata, 7–22 cm. longa, 1–4 cm. lata, chartacea, glabra, sicco cinereo-viridia; costa supra angustissima, prominula, infra prominens, sicco longitudinaliter sulcata, basi 1.5 mm. lata; nervi laterales utrinsecus 12–15, a costa sub angulo lato abeuntes, intra marginem circiter 0.5 cm. distanter prominenter furcati, supra prominuli, infra prominentes; nervi tertiarii et venae laxae, vix prominulae; petioli 0.5–1.5 cm. longi, demum transverse verrucosi; stipulae e basi late ovato longe acuminatae, 6–7 mm. longae, coriaceae, glabrae. Flores albi, dense cymosi; cymae subsessiles, circiter 7 cm. diametro, glabrae; bracteae parvae, lanceolatae, acutae; pedicelli vix 1 mm. longi v. nulli; receptaculum turbinatum, 1.5–2 mm. longum; calycis lobi lineari-lanceolati, subacuti, 3–4 mm. longi; corollae tubus gracilis, 3 cm. longus, vix 1 mm. diametro; lobi oblongo-elliptici, obtusi, 6 mm. longi, 2.5 mm. lati; antherae exsertae, 2.5 mm. longae; stylus brevissime exsertus, ramis 1.25 mm. longis subacutis. Fructus 1–2-lobus, 1 cm. longus, oblongo-ellipsoideus.

Yunnan: Hsin-kei, Red River, *A. Henry* (No. 9584, type; shrub 1 m. tall; flowers white); Red River banks, Manpan, *A. Henry* (No. 10370; shrub 0.65 m. tall; fruits red).

This is a very distinct species with extremely narrow leaves and a congested inflorescence. It differs from *Ixora subsessilis* Wallich from eastern India in those characters and in its larger calyx-lobes.

PAVETTA Linn.

Pavetta indica Linnaeus, var. **polyantha**, Hooker f., *Fl. Brit. Ind.* III. 150 (1880). — King & Gamble, *Jour. As. Soc. Bengal*, LXIII. pt. 3, 84 (*Mat. Fl. Malay. Penins.* No. 15) (1904).

   Pavetta polyantha Wallich, *Cat.* 6176 (nomen nudum) (1828).
   Pavetta indica Wallich, l. c. 6175ᶠ (nomen nudum) (1828). — Ker in *Bot. Reg.* III. t. 198 (1817).
   Pavetta Rothiana De Candolle, *Prodr.* IV. 491 (1830). — Wight & Arnott, *Prodr. Fl. Ind.* 431 (1834).
   Pavetta villosa Heyne in Roth, *Nov. Pl. Spec.* 88 (non Vahl) (1821).
   Ixora tomentosa, var. *glabrescens* Kurz, *Forest Fl. Brit. Burma*, II. 19 (1877).

Yunnan: Mengtsze, forests to southeast, alt. 1640 m., *A. Henry* (No. 10777; tree 5 m. tall; flowers white); Szemao, east on the Tea Hills, *A. Henry* (No. 11934); same locality, mountains to south, alt. 1480 m., *A. Henry* (No. 11934ª; shrub 2 m. tall).

MORINDA Linn.

Morinda umbellata Linnaeus, *Spec.* 176 (1763). — De Candolle, *Prodr.* IV. 449 (1830). — Wight & Arnott, *Prodr. Fl. Ind.* 420 (1834). — Wallich, *Cat.* No. 8431 (1847). — Bentham, *Fl. Hongk.* 159 (1861). — Kurz, *Forest Fl. Brit. Burma*, II. 62 (1877). — Hooker f., *Fl. Brit. Ind.* III. 157 (1880). — Maximowicz in *Bull. Acad. Sci. St. Pétersbourg*, XXIX. 170 (1883); in *Mél. Biol.* XI. 795 (1883). — Hemsley in *Jour. Linn. Soc.* XXIII. 386 (1888).

Morinda Padavara A. L. de Jussieu, *Gen.* 206 (1789).
Morinda scandens Roxburgh, *Fl. Ind.* I. 548 (1820). — De Candolle, *Prodr.* IV. 449 (1830).
Morinda tetrandra Jack in *Malay. Misc.* I. 13 (1820). — Wallich, *Cat.* 8432 (1847). — Roxburgh, *Fl. Ind.* II. 203 (1824). — De Candolle, *Prodr.* IV. 449 (1830).

Yunnan: Mengtsze, forests to southeast, alt. 1640 m., *A. Henry* (No. 11246; large climber on tree; fruits reddish).

PRISMATOMERIS Thwaites.

Prismatomeris tetrandra K. Schumann in Engl. & Prantl, *Nat. Pflanzenfam.* IV. abt. 4, 138 (1891). — Craib in *Kew Bull. Misc. Inform.* 1911, 395.

Coffea tetrandra Roxburgh, *Fl. Ind.* I. 538 (1820). — Wallich, *Cat.* 6242 (1830). — De Candolle, *Prodr.* IV. 499 (1830). — Kurz, *Forest Fl. Brit. Burma*, II. 28 (1877).
Prismatomeris albiflora Thwaites in *Hooker's Jour. Bot. & Kew Gard. Misc.* VIII. 268, t. VII. (1856); *Enum. Pl. Zeylan.* 154, 421 (1859). — Hooker f., *Fl. Brit. Ind.* III. 159 (1880).

Yunnan: I-wu, *A. Henry* (No. 13573; in Herb. Kew).

Prismatomeris brevipes Hutchinson, n. sp.
Frutex 2–3 m. altus; ramuli graciles, flexuosi, sulcati, cortice cinereo-brunneo obtecti, glabri. Folia lanceolata v. oblongo-lanceolata, basi acute cuneata, apice longe acute acuminata acumine 1.5–2 cm. longo, 7–12 cm. longa, 1.7–3.3 cm. lata, tenuiter chartacea, utrinque glabra; nervi laterales utrinsecus circiter 6, intra marginem conspicue conjuncti, graciles, infra prominuli; nervi tertiarii et venae vix evoluta; petioli 3–5 mm. longi, glabri; stipulae minutae. Flores axillares et terminales, brevissime pedicellati, plerumque geminati; pedicelli 1–2 mm. longi; receptaculum turbinatum, 1.5 mm. longum; calyx brevissimus, minutus, denticulatus; corolla alba; tubus superne sensim leviter ampliatus, 1 cm. longus, extra glaber, intra tenuiter pubescens; lobi ovato-lanceolati, subobtusi, 2 mm. longi; stamina fauci inserta, inclusa; antherae 2 mm. longae; stylus gracillimus, corollae tubo aequilongus, glaber; ovarium 2-loculare. Fructus leviter 2-lobus, globosus, 5 mm. diametro, glaucescens, calycis lobis suberectis persistentibus coronatus; pedicelli 2.5–4 mm. longi.
Yunnan: Mengtsze, mountains to southeast, alt. 1950 m., *A. Henry* (No. 9040ª; shrub 3 m. tall; fruit reddish; No. 9040ᵈ; No. 9040ᵉ, type; No. 9040ᶠ).
This species is very similar in general appearance to *Prismatomeris tetrandra*

K. Schumann (*P. albiflora* Thwaites), but with narrower, more attenuated leaves, much shorter pedicels, and flowers in pairs and not several in a fascicle as in that species.

Prismatomeris linearis Hutchinson, n. sp.

Fruticulus gracilis 0.75 m. altus; ramuli graciles, angulares, sulcati, cortice pallide brunneo glabro obtecti. Folia linearia, basi acuta, apice acute acuminata, 5.5–13 cm. longa, 0.6–2 cm. lata, tenuiter chartacea, glabra; nervi laterales numerosi, circiter 14, patuli, infra prominuli; petioli, stipulae, flores et fructus quam in *P. brevipes* Hutchinson.

Yunnan: Mengtsze, mountain woods to southeast, *A. Henry* (No. 9040; small shrub 0.75 m. tall; flowers and fruit whitish; in Herb. Kew).

PSYCHOTRIA Linn.

Psychotria yunnanensis Hutchinson, n. sp.

Frutex usque ad 2 m. altus; ramuli glabri, internodiis longis. Folia obovata v. elongato-oblanceolata, sensim acute acuminata, basi longe cuneata, 12–20 cm. longa, 3–7 cm. lata, membranaceo-chartacea, utrinque opaca et glabra; nervi laterales utrinsecus 10–15, fere recti v. arcuati, graciles, utrinque prominuli; venae vix evolutae; petioli usque ad 3.5 cm. longi, glabri; stipulae oblongae, longe caudato-acuminatae, 1.5 cm. longae, submembranaceae, basi intra hirsutae. Cymae laxiflorae, terminales, thyrsoideae, pedunculatae, usque ad 10 cm. longae; pedunculus 4–6 cm. longus, glaber v. parce puberulus; bracteae lanceolatae, acute acuminatae, 5 mm. longae, fere glabrae; receptaculum glabrum; calyx campanulatus glaber, tubo 1 mm. longo, lobis 5 subulato-lanceolatis acutis 1.25 mm. longis; corollae tubus cylindricus, apice leviter ampliatus, 0.5 cm. longus, extra glaber, fauce villosus; lobi obtusi; antherae leviter exsertae; stylus 3.5 mm. longus, gracilis, glaber, breviter bilobus. Fructus oblongo-ellipsoideus, extra subcarnosus, 1–1.2 cm. longus; semina dorso leviter 4-carinata, facie ventrali plana.

Yunnan: Mengtsze, forests to southwest, alt. 1640 m., *A. Henry* (No. 11447; 11447[a]); Szemao, mountains to east, alt. 1450 m., *A. Henry* (No. 12032; 12032[b]); Szemao, mountains to south, *A. Henry* (No. 12032[a], type; flowers white); forests to southeast, *A. Henry* (No. 12806; fruit red; No. 12806[a]).

Psychotria morindoides Hutchinson, n. sp.

Frutex 1 m. altus; ramuli compressi, crispato-pubescentes v. subtomentosi, internodiis 1–3 cm. longis. Folia elongato-obovato-oblanceolata, subsensim acute acuminata, ad basin longe attenuata, 12–20 cm. longa, 4–8 cm. lata, tenuiter chartacea, supra glabra, infra solum in nervis puberula; costa supra leviter elevata, infra prominens; nervi laterales utrinsecus 10–14, a costa sub angulo 50°–60° abeuntes gracillimi, intra marginem elongati et vix conjuncti; venae vix evolutae; petioli 2–4 cm. longi, breviter pubescentes; stipulae oblongo-ovatae, bilobatae, 1 cm. longae, longissime acuminatae, submembranaceae, mox caducae, puberulae. Inflorescentia terminalis, dense capitata, breviter pedunculata, circiter 2 cm. diametro; pedunculi 0.5–1 cm. longi, pubescentes; bracteae lineares, acutae, 5 mm. longae, parce ciliatae; receptaculum costatum, glabrum; calycis tubus 1.25 mm. longus, glaber; lobi 5, lanceolato-lineares, acuti, 2 mm. longi, parce ciliolati. Corollae tubus cylindricus, apice leviter ampliatus, 4 mm. longus, extra glaber, fauce villosus; lobi 5, apice cucullati, lanceolati, 1.5 mm. longi, intra appresse villosi; stamina fauci inserta; antherae apice paulum exsertae, 1.75 mm. longae; discus epigynus magnus, glaber; stylus bifidus, 2 mm. longus, glaber. Fructus ellipsoideus, 6–7 mm. longus, 4 mm. latus, costatus, glaber, calycis lobis persistentibus coronatus; semina dorso 4-costata, convexa, facie ventrali plana.

Yunnan: Szemao, mountains to east, alt. 1300 m., *A. Henry* (No. 12069; 12069ᵃ, type; slender shrub, 1 m. tall; flowers white); Szemao, forests, *A. Henry* (No. 12069ᵇ; 12069ᵈ; fruits red).

Very similar to the East Indian *P. fulva* Hamilton, but with usually longer petioles, a shorter peduncle, a denser inflorescence with the flowers in a single cluster, and shorter and relatively broader fruits.

Psychotria siamica Hutchinson, n. comb.

Cephaelis siamica Craib in *Kew Bull. Misc. Inform.* 1911, 395.

Yunnan: Yuanchiang, ravine, alt. 1470 m., *A. Henry* (No. 11589; shrub 0.75 m. tall; flowers; No. 11589ᵃ; fruits); Szemao, forests to south, alt. 1640 m., *A. Henry* (No. 12368ᵃ). Kwangsi: Lungchou, *H. B. Morse* (No. 704; shrub with green flowers).

Psychotria Henryi Léveillé in Fedde, *Rep. Sp. Nov.* XIII. 179 (1914).

Yunnan: Szemao, forests, alt. 1320 m., *A. Henry* (No. 12146; spreading shrub 1 m. tall; flowers white; No. 12146ᶜ; fruits red; No. 12146ᵈ).

Very similar to *P. sulcata* Wallich from eastern India, but the midrib of the leaves glabrous below and not closely hairy as in that species, and the stipules are very small and caducous.

Psychotria symplocifolia Kurz, *Forest Fl. Brit. Burma*, II. 11 (1877). — Hooker f., *Fl. Brit. Ind.* III. 172 (1880).

Psychotria sp. Wallich, *Cat.* No. 8357 (1828). — Hooker f. & Thomson, *Herb. Ind. Or.* No. 27 ex Hooker f., *Fl. Brit. Ind.* III. 172 (pro synon.) (1880).

Morinda sp.? Wallich, *Cat.* No. 8428 (nomen nudum) (1828).

Yunnan: Szemao, mountains to east, alt. 1460 m., *A. Henry* (No. 12065; shrub 1.7 m. tall; flowers white; No. 12065ᵃ; No. 12065ᵇ; fruits red).

Psychotria pilifera Hutchinson, n. sp.

Suffrutex 0.75–1.75 m. altus; rami parce foliati, circiter 4 mm. crassi, pilis articulatis longissimis dense induti. Folia longe petiolata, elliptica v. elliptico-obovata, basi breviter v. longe cuneata, apice acuta v. breviter acuminata, 12–17 cm. longa, 4.5–9.5 cm. lata, membranaceo-chartacea, utrinque pilis longis appressis articulatis induta; nervi laterales utrinsecus 8–10, a costa sub angulo 45° abeuntes, prope marginem elongati et crenulati, utrinque prominuli; petioli 3–5 cm. longi, complanati, circiter 2.5 mm. lati, dense hirsuti; stipulae subdeciduae, magnae, latae, longe caudato-acuminatae, pilosae, usque ad 2 cm. longae. Flores non visi. Infructescentia pedunculata, axillaris, sublaxe cymoso-paniculata; pedunculus 3–5 cm. longus, dense hirsutus; bracteae lineari-lanceolatae, acuminatae, usque ad 1 cm. longae, parce ciliatae; fructus ruber, oblongo-ellipsoideus, calycis lobis lineari-lanceolatis superne ciliatis 1.5 mm. longis persistentibus coronatus, costatus, glaber, circiter 1 cm. longus, 4–5 mm. latus; semina nigra, dorso leviter bisulcata, facie ventrali levitor concava, 5–6 mm. longa.

Yunnan: Mengtsze, forests to southeast, alt. 1640 m., *A. Henry* (No. 11409; shrub 0.75 m. tall; fruits red); *A. Henry* (No. 13307, type; shrub 1.75 m. tall; No. 13308).

Related to *P. Helferiana* Kurz, from Tenasserim, but with much longer petioles, longer cuneate leaves clothed with remarkably long articulated hairs.

Psychotria calocarpa Kurz in *Jour. As. Soc. Bengal*, XLI. pt. 2, 315 (1872); *Forest Fl. Brit. Burma*, II. 9 (1877). — Hooker f., *Fl. Brit. Ind.* III. 173 (1880).

Psychotria asiatica Wallich in Roxburgh, *Fl. Ind.* II. 160 (non Linnaeus, nec Roxburgh) (1824). — Wallich, *Cat.* No. 8331 (1828).

Psychotria viridiflora, var. *2 undulata* Kurz, *Forest Fl. Brit. Burma*, II. 13 (1877).
Psychotria picta Wallich, *Cat.* No. 8353 (nomen nudum) (1847).

Yunnan: Szemao forests, alt. 1640 m., *A. Henry* (No. 12283[b]; shrub 0.5 m. tall).

Psychotria straminea Hutchinson, n. sp.

Frutex usque ad 2.5 m. altus; ramuli breves, complanati, glabri. Folia ovato-lanceolata v. elliptico-lanceolata, basi longe cuneata, apice sensim v. subabrupte acuminata, 10–25 cm. longa, 4–10 cm. lata, membranacea, glabra; costa utrinque fere plana; nervi laterales utrinsecus circiter 8, a costa sub angulo 45°–70° abeuntes, graciles, utrinque prominuli, straminei; nervi tertiarii venaeque vix evoluta; petioli 1–2 cm. longi; stipulae parvae, coriaceae. Cymae terminales, pauciflorae, breviter pedunculatae, 3 cm. longae, glabrae; bracteae minutae, recurvatae; receptaculum 1.5 mm. longum, glabrum; calyx brevissimus, denticulatus, submembranaceus; corollae tubus late cylindricus, 3 mm. longus, striatus, fauce villosus; lobi 5, ovati, subacuti, 2.5 mm. longi; stamina inter corollae lobos inserta, exserta; filamenta 1.5 mm. longa, glabra; antherae 1 mm. longae; discus crassus, glaber; stylus 2 mm. longus, ad medium bilobus. Fructus viridis (*A. Henry*), sicco niger, obovoideo-ellipsoideus, basi leviter stipitatus, circiter 1.3 cm. longus, 6–7 mm. diametro; semina dorso ecostata, convexa, facie ventrali concava.

Yunnan: Mengtsze, mountains to southeast, forests, alt. 1640 m., *A. Henry* (No. 11138, type; shrub 2 m. tall; flowers white); *A. Henry* (No. 13461; shrub 1.25 m. tall; fruits); *A. Henry* (No. 11428; fruit green).

This species bears considerable resemblance to *Psychotria montana* Blume, from eastern India, but with a smaller and denser inflorescence, subturgid not ridged fruits, and seeds deeply concave on their ventral face.

CHASALIA Comms.

Chasalia curviflora Thwaites, *Enum. Pl. Zeylan.* 150, 421 (1859). — Kurz, *Forest Fl. Brit. Burma*, II. 14 (1877). — Hooker f., *Fl. Brit. Ind.* III. 176 (1880).

Psychotria curviflora and *P. ophioxyloides* Wallich in Roxburgh, *Fl. Ind.* II. 167, 168 (1824). — De Candolle, *Prodr.* IV. 520 (1830).
Chasalia lurida, *C. curviflora*, *C. tetrandra* (excl. syn. *C. rostrata*) and *C. Sangiana* Miquel, *Fl. Ind. Bat.* II. 282 (1856); in *Ann. Mus. Lugd.-Bat.* IV. 202, 203 (1869); *Fl. Ind. Bat. Suppl.* 546 (1860).
Psychotria lurida Blume, *Bijdr.* 959 (1825). — De Candolle, *Prodr.* IV. 521 (1830).
Psychotria ambigua Wight & Arnott, *Prodr. Fl. Ind.* 433 (1834). — Wight, *Ill.* t. 127 (1850).
Psychotria tetrandra Blume, *Bijdr.* 959 (1825). — De Candolle, *Prodr.* IV. 521 (1830).
Zwardekronia lurida Korthals in *Ned. Kruidk. Arch.* II. 252 (1849).
Ixora attenuata Wallich, *Cat.* No. 6164 (1828).

Yunnan: near Manpan, Red River valley, alt. 830 m., *A. Henry* (No. 10691; shrub 1.7 m. tall; flowers white); Szemao, *A. Henry* (Nos. 11966, 11966[b], 11966.[c] 11966[d], 11966[e], 11966[f], 12283); Mengtsze, mountains to southeast, woods, alt, 1640 m., *A. Henry* (No. 10758; shrub 2.5 m. tall; flowers pink).

SAPROSMA Blume.

Saprosma ternatum, Hooker f. in Bentham & Hooker f., *Gen. Pl.* II. 131 (1873); *Fl. Brit. Ind.* III. 193 (1881). — Kurz, *Forest Fl. Brit. Burma*, II. 29 (1877).

Paederia ternata Wallich in Roxburgh, *Fl. Ind.* II. 520 (1824). — Wallich, *Cat.* No. 6248 (1828). — De Candolle, *Prodr.* IV. 471 (1830). *Serissa ternata* Kurz, *Rep. Veget. Andaman Isl.* App. A. 40 (1870). *Mephitidia* sp. Griffith, *Notul.* IV. 267 (1854); *Icon. Pl. Asiat.* t. 476 (1854).

Yunnan: Szemao, ravine, alt. 1320 m., *A. Henry* (No. 11965; shrub 2 m. tall; flowers white); same locality, mountains to south, alt. 1320 m., *A. Henry* (No. 11965[a]; shrub 2.25 m. tall; flowers white); same locality, alt. 1640 m., *A. Henry* (No. 11965[b]; shrub 2.65 m. tall; fruits red).

Saprosma Henryi Hutchinson, n. sp.
Frutex 3 m. altus; ramuli cortice cinereo glabro obtecti. Folia opposita, elliptica v. oblongo-elliptica, breviter acute acuminata, basi paullo cuneata, 6–9 cm. longa, 3–5 cm. lata, tenuiter chartacea, glabra; costa supra leviter impressa, infra prominens; nervi laterales utrinsecus 4–6, a costa sub angulo fere 90° abeuntes, arcuati, intra marginem dupliciter conjuncti, utrinque prominuli; venae infra laxae, prominulae; petioli 3–7 mm. longi, glabri; stipulae parvae; puberulae. Cymae axillares, pedunculatae, circiter 4-florae; pedunculi 1–2 cm. longi, costati; flores subsessiles, albi; receptaculum glabrum; calycis lobi 4, triangulares, subacuti, circiter 1.25 mm. longi, fere glabri; corollae tubus superne sensim leviter expansus, 1 cm. longus, extra glaber, intra pubescens; lobi 4, ovato-triangulares, obtusi, 2 mm. longi; antherae 2.25 mm. longae; ovarium 2-loculare, ovulis solitariis erectis basi angustatis; stylus bifidus, 2 mm. longus, glaber. Fructus oblique ellipsoideus, 0.8 cm. longus, verrucosus.

Yunnan: Szemao, mountains to south, alt. 1320 m., *A. Henry* (No. 12145, type; shrub 3 m. tall; flowers white); Szemao, forests, alt. 1640 m., *A. Henry* (No. 12646; small climber; fruits).

COMPOSITAE.

Determined by Camillo Schneider.

PLUCHEA Cass.

Pluchea rubicunda C. Schneider, n. sp.

Frutex ut videtur erectus et laxe ramosus, 1.5–5 m. altus; ramuli floriferi teretes, dense cano-tomentelli pubescentes, partim glandulosi. Folia alterna, membranacea, elliptica v. elliptico-lanceolata v. sub-ovato-elliptica, basi acuta, vix in petiolum decurrentia, apice plus minusve breviter acuminata acumine interdum satis subito producto, margine indistincte et distanter subrepando-brevidenticulata v. fere integra, supra viridia, omnino (in nervis densius) brevissime glanduloso-puberula, subtus canescentia, satis dense ut ramuli tomentella, nervis lateralibus utrinsecus circiter 7–8, etiam reticulo paulo prominulo, majora ad 16 cm. longa et paulo infra medium ad 7 cm. lata, minora basi inflorescentiae 5–12 cm. longa, 2–4 cm. lata; petioli cano-tomentelli, 8–11 mm. longi. Inflorescentiae paniculato-corymbosae, circiter 6 cm. altae et ad 20 cm. latae, ut ramuli cano-tomentellae; capitula numerosa, discoidea, ut videtur homogama, 5–7 mm. crassa; involucri bracteae pluriseriatae, adpressae, subscariosae, exteriores ovato-ellipticae, interiores elongatae, omnes obtusae, in dorso sericeo-tomentellae; receptacula plana, nuda. Flores visi omnes hermaphroditi, numerosi, circiter 10–12 mm. longi, rubicundi; pappi setae albae, biseriatae, exteriores interioribus 3–4-plo breviores, achaenio vix aequilongae, interiores corollam subaequantes; corolla extus pilis glandulosis paucis obsita, lobis lanceolatis tubo 3-plo brevioribus; antherae basi caudatae; stylus pro genere normalis, apice bifidus et pilosus; achaenia immatura tetragona, sparse pilosa.

W e s t e r n   S z e c h ' u a n : Kiating Fu, thickets, alt. 100–800 m., September 1908 (No. 3403; bush 1.50–5 m. tall, flowers pink).

Judging by the bracts of the involucre, the style and the anthers this plant belongs to the genus *Pluchea* as fixed by Hoffmann in Engler & Prantl, *Nat. Pflanzenfam.* IV. 5. Abt. 176 (1890), but, so far as I know, all the species hitherto described have a 1-serrate pappus. It looks different from all the species of this genus or of *Blumea* known from Asia. Further observations are needed to determine if our plant belongs to *Pluchea*, but I know of no other genus to which it could be referred.

PERTYA Schultz Bip.

Pertya sinensis Oliver in *Hooker's Icon.* XXIII. t. 2214 (1892). — Bean in *Kew Bull. Misc. Inform.* 174 (*Portya*) (1910).

Western Hupeh: Fang Hsien, thickets, alt. 1800 m., August 1907 (No. **3222**; shrub 0.75–1 m. tall, flowers pink); Hsing-shan Hsien, alt. 2600 m., *A. Henry* (No. 6982, type; ex Oliver).

Here may be added notes on another species not collected during the Arnold Arboretum Expeditions.

Pertya phylicoides J. F. Jeffrey in *Not. Bot. Gard. Edinburgh,* V. 200 (1912).

Yunnan: "Dry limestone country on the descent from Chung-tien plateau to Yangtze, near Chiao-tou. Elevation 10,000 ft., September 1904," *G. Forrest* (No. 112, type; ex Jeffrey); same locality, September 1914, *C. Schneider* (No. 2175; shrub 0.6–1 m. tall).

SENECIO L.

Senecio scandens Hamilton in D. Don, *Prodr. Fl. Nepal.* 178 (1825). — Hooker f., *Fl. Brit. Ind.* III. 352 (1881). — Hemsley in *Jour. Linn. Soc.* XXVI. 457 (1888). — Diels in *Bot. Jahrb.* XXIX. 620 (1900); in *Not. Bot. Gard. Edinburgh,* VII. 401 (*Pl. Chin. Forrest.*) (1913).

Western Hupeh: Ichang, glens, alt. 100–300 m., March 1907 (No. **2497**; scandent sub-shrub, flowers yellow).

CORRECTIONS AND ADDITIONS TO VOLUMES I–III

By ALFRED REHDER and E. H. WILSON, with contributions by CAMILLO SCHNEIDER.

VOLUME I

Philadelphus brachybotrys, var. **purpurascens** (p. 6). This variety has been separated from *P. brachybotrys* Koehne as a distinct species:

PHILADELPHUS PURPURASCENS Rehder in *Mitt. Deutsch. Dendr. Ges.* XXIV. (1915). (*Philadelphus Delavayi* Hutchinson in *Bot. Mag.* CXXXVI. t. 8324 [non L. Henry] [1910]).

Deutzia coriacea (p. 9). The number of the type specimen which is given as No. **4481** should read No. **4487**.

Deutzia discolor (p. 12). Add the following number:
Eastern Szech'uan: Wushan Hsien, alt. 1300 m., 1907 (No. **777**; plants only).

Deutzia Vilmorinae (p. 20). Add the following number:
Western Hupeh: Changlo Hsien, alt. 1300–1600 m., June 1907 (No. **4467**).
Like Wilson's Nos. 940 and 1998 this specimen differs from the type in the sparser pubescence of the under side of the leaves, the stellate hairs being more distant, not forming a dense tomentum.

Hydrangea Davidii (p. 25). Add the following number:
Western Szech'uan: Mupin, thickets, 1908 (No. **1159ᵃ**, seeds only).

Hydrangea xanthoneura, var. **setchuenensis** (p. 28, 579). Add the following numbers:
Western Hupeh: Fang Hsien, 1907 (No. **573**; shrub 1 m. tall, flowers white; seeds only). Western Szech'uan: west of Kuan Hsien, woodlands, alt. 2600–3000 m., October 1910 (No. **4303**; bush 3–5 m. tall or small tree, flowers white).

Itea (p. 44). Add the following new species:

ITEA OMEIENSIS Schneider, n. sp.
Frutex altus; ramuli floriferi flavo-cinerascentes, teretiusculi, glabri v. parce minutissime pilosi. Folia alterna, sempervirentia, tenuiter coriacea, longe elliptica, basi subrotundata, apice satis subito breviter acuminata, 11–12.5 cm. longa, 4–5 cm. lata, glabra, superne ut videtur intense sed obscure viridia, costa incisa, nervis lateralibus distinctis sed vix prominulis, subtus pallidiora, ut videtur viridescentia, costa venisque lateralibus utrinque 7–8 distincte elevatis, reta nervorum tenui paulo distincto, margine distanter breviter subglanduloso-serrata versus basim plus minusve integra; petioli crassi, glabri; superne anguste sulcati, 10–13 mm. longi. Inflorescentiae spiciformes, pseudo-racemosae, axillares, singulae

421

binaeve, subdensiflorae, 5–7 mm. longae, rhachi striato-angulato brevissime piloso; flores singuli v. plerique ad 2–4 aggregati, ut videtur albi; pedicelli 2–3.5 mm. longi, filiformes, minus quam rhachis pilosi, basi bracteis iis paulo brevioribus subulatis v. lanceolato-subulatis subpilosis suffulti; calyx 5-partitus, laciniis anguste lanceolatis acuminatis tantum basi paulo pilosis quam petala circiter triente v. fere duplo brevioribus; petala 5, lanceolata, acuminata, 2.5–3 mm. longa, erecta, glabra; stamina 5, normalia, filamentis glabris quam petala vix longioribus; ovarium oblongum, apice àttenuatum, stylo simplici, stigmate capitato subbilobo incluso circiter 4 mm. longum, glabrum, basi disco annulari cinctum.   Fructus ignotus; receptaculum fructiferum ut videtur paulo concavum.

Western Szech'uan: Mt. Omei, May 1904 (Veitch Exped. No. 4904; shrub up to 6 m. tall).

This species seems to be most nearly related to *I. chinensis* Hooker & Arnott, the type of which was collected near Hongkong. It differs from the plant of central China in its somewhat broader elliptic or obovate-elliptic leaves and in its distinctly puberulent ovaries. The Chinese and Indian species of the genus need a thorough investigation. As I have already pointed out in my *Ill. Handb. Laubholzk.* I. 396 (1905), *I. chinensis* does not occur in Assam, where it is represented by *I. khasiana* Schneider, nor in Formosa, where *I. Oldhamii* Schneider (*I. chinensis,* var. *subserrata* Maximowicz) takes its place.

It may be mentioned that, according to the author's description, *I. yunnanensis* Franchet (in *Jour. de Bot.* X. 268 [1896]) seems to be only a synonym of *I. ilicifolia* Oliver.   C. S.

Ribes himalayense Decaisne, var. *α* **glandulosum** (p. 44).   In the enumeration of specimens No. **799** should be changed to No. **1799**.

Ribes longeracemosum, *α* **Davidii** (p. 45).   Add the following number:
Western Szech'uan: Mupin, alt. 2000 m., 1908 (No. **898ᵃ**); Ma-ngan-shan, thickets, 1908 (No. **1389**; seeds only).

Ribes longeracemosum, *β* **Wilsonii** (p. 45).   Add the following number:
Western Hupeh: Fang Hsien, woodlands, 2000–2600 m., 1910 (No. **4421**; bush 1–2 m. tall, fruit black; seeds only).

Ribes tenue (p. 45).   Add the following numbers:
Western Hupeh: Fang Hsien, among rocks in woods, rare, alt. 1000 m., July 1907 (No. **38ᵃ**; bush 1 m. tall, fruit red); same locality, alt. 2300 m., May 19, 1907 (No. **38ᵇ**; bush 2–2.5 m. tall, flowers green); Hsing-shan Hsien, cliffs, alt. 2600 m., May 1907 (No. **38ᶜ**; bush 1.5 m. tall, flowers green).

Ribes (p. 46).   Add at the end of the genus:
Ribes spec.
Western Szech'uan: Mupin, 1908 (No. **1074**; shrub 1.25–2.6 m. tall, leaves bristly above, racemes 7.5 cm. long; seeds only).

Sorbaria arborea, var. **subtomentosa** (p. 47).   Add the following number:
Western Szech'uan: west of Kuan Hsien, Pan-lan-shan, woodlands, alt. 3000 m., 1910 (No. **4296**; bush 6.5 m. tall, panicles 0.3 m. long, leaves pubescent; seeds only).

Rubus xanthocarpus (p. 49).   Add the following number:
Western Szech'uan: roadsides around Tachien-lu, alt. 2600–3000 m., October 1910 (No. **4137**, 1–1.25 m.).

Rubus (p. 49).   Insert after *R. Playfairianus:*

Rubus tephrodes Hance in *Jour. Bot.* XII. 260 (1874). — Hemsley in *Jour. Linn. Soc.* XXIII. 238 (1887). — Focke in *Bibl. Bot.* LXXII. 50 (1911).

Kiangsi: Kiukiang Plain, in hedge-rows, alt. 100 m., July 27, 1907 (No. **1683**; bush 4 m., flowers pink).

Rubus (p. 50).   Insert after *R. fusco-rubens:*

Rubus sepalanthus Focke in *Bot. Jahrb.* XXIX. 391 (1900); in *Bibl. Bot.* LXXII. 53 (1911).

Western Szech'uan: near Wa-shan, thickets, alt. 1000–2300 m., September 20, 1908 (No. **935**; climber, 2–3 m., fruit black); banks of Min River, June 1903 (Veitch Exped. No. 3481).

Rubus ichangensis (p. 50).   Add as a synonym:

Rubus eugenius Focke in *Bot. Jahrb.* XXIX. 393 (1900).

Rubus Lambertianus (p. 51).   Instead of " (No. **482**) " read " (Nos. **482**; fruit red; **482ᵃ**; fruit yellow)."

Rubus clemens Focke (p. 51).   This species is to be referred as a synonym to:

Rubus setchuenensis Bureau & Franchet in *Jour. de Bot.* V. 46 (1891). — Focke in *Bibl. Bot.* LXXXIII. 32 (*Spec. Rub.* III. 256) (1914).

Rubus clemens Focke in *Bibl. Bot.* LXXII. 105, fig. 46 (*Spec. Rub.*) (1910); in Sargent, *Pl. Wilson,* I. 51 (1913).

Rubus (p. 51).   After *R. clemens* insert:

Rubus flagelliflorus Focke in *Bot. Jahrb.* XXIX. 393 (1900); in *Bibl. Bot.* LXXII. 111 (1910).

Western Hupeh: Ichang, woods and shady places, alt. 1000–1600 m., 1907 (No. **750**; plants only); without precise locality, June 1900 (Veitch Exped. No. 1215).

Rubus irenaeus (p. 51).   Add the following number:

Western Hupeh: north and south of Ichang, woods, alt. 1300–2300 m., 1907 (No. **751**; plants only).

Rubus corchorifolius (p. 51).   Add the following number:

Western Hupeh: Hsing-shan Hsien, alt. 1000 m., 1907 (No. **15ᵃ**; bush 2–2.6 m., fruit red, of good flavor).

Rubus trianthus (p. 51).   Add to this species the following synonyms and number:

Rubus conduplicatus Duthie in *Kew Bull. Misc. Inform.* 1912, 36.
Rubus incisus Thunberg, var. *conduplicatus* Koidzumi in *Jour. Coll. Sci. Tokyo,* XXXIV. art. 2, 122 (1913).

Kiangsi: Kuling, abundant, alt. 1200 m., August 1, 1907 (No. **1693**; bush 6 ft.).

Rubus Thunbergii, var. **glabellus** (p. 52).   No. 2 from Ichang does not seem to belong here, it is probably *R. eustephanos* Focke apud Diels in *Bot. Jahrb.* XXXVI. Beibl. LXXXII. 54 (1905).

Rubus pungens (p. 52).   Add the following number:

Western Szech'uan: Niu-tou-shan, thickets, alt. 1000–2600 m., 1907 (No. **901**; bush 2 m. tall, fruit red, good eating; seeds only).

Rubus coreanus (p. 54). No. 152 enumerated by Focke under this species belongs to the following species:

RUBUS TELEDAPOS Focke in *Bot. Jahrb.* XXIX. 398 (1900).

Western Hupeh: Changlo Hsien, alt. 1200 m., September 1907 (No. **152**; bush 2.5 m.).

Rubus Kuntzeanus (p. 54). This name must be referred as a synonym to *R. innominatus* S. Moore.

Rubus (p. 54). After *R. coreanus* insert:

RUBUS TRIPHYLLUS Thunberg, *Fl. Jap.* 215 (1784). — Focke in *Bibl. Bot.* LXXII. 187 (*Spec. Rub.*) (1910).

Rubus parvifolius Linnaeus, *Spec.* ed. 2, 707 (non Linnaeus, *Amoen. Acad. IV.* 129 [1754]) (1762), quoad plantam Chinensem.

Rubus macropodus Seringe in De Candolle, *Prodr.* II. 557 (1825).

Rubus Thunbergii Blume, *Bijdr.* 1109 (1825).

Rubus Zahlbrucknerianus Endlicher, *Atact.* t. 35 (sine descript.) (1833).

Rubus innominatus, subsp. *plebejus* Focke in Sargent, *Pl. Wilson.* I. 55 (1911).

Kiangsi: Kuling, roadsides, alt. 1300 m., July 30, 1907 (No. **1679**; bush 1.3–2 m. tall, flowers rose-pink, fruit red). Western Hupeh: Ichang, roadsides, alt. 300–1000 m., June and July 1907 (No. **42**; prostrate, flower pink, fruit red); without precise locality, *A. Henry* (No. 6375).

Wilson's No. 42 described as *R. innominatus*, subsp. *plebejus* by Focke differs from the typical plant in the presence on the branches and the inflorescence of numerous prickles and glandular bristles. Henry's No. 6375 has many of the leaves quinate.

Rubus innominatus (p. 55). Add to the literature cited: Wilson in *Gard. Chron.* ser. 3, XXXVIII. 290, fig. 112 (1905); add the following synonym and specimen and refer the subsp. *plebejus* Focke to *R. triphyllus*, which see above.

Rubus Kuntzeanus Hemsley in *Jour. Linn. Soc.* XXIII. 232 (1887). — Focke in *Bibl. Bot.* LXXII. 195 (*Spec. Rub.*) (1911); in Sargent, *Pl. Wilson.* I. 54 (1911).

Kiangsi: Kuling, roadsides, common, alt. 1300 m., July 28, 1907 (No. **1685**; straggling bush). Western Hupeh: north and south of Ichang, thickets, alt. 300–1500 m., June and August 1907 (No. **92**; bush 1.3–2.6 m. tall, flowers rose-purple, fruit red). Yunnan: Mengtsze, alt. 1600–2000 m., *A. Henry* (No. 10922).

The form described as *R. Kuntzeanus*, represented by Wilson's No. 92, differs from typical *R. innominatus* only in the absence of the glandular pubescence and in the always ternate leaves.

Rubus (p. 56). At the end of the genus add:

RUBUS spec.

Western Hupeh: Ichang, alt. 1000–1300 m., 1907 (No. **80**; stems long, arching, leaves cordate, hairy; seeds only).

RUBUS spec.

Western Hupeh: Changlo Hsien, woods, alt. 1300 m., 1907 (No. **155**; bush 1 m. tall, leaves 3-foliolate, green, fruit red; seeds only).

Prunus brachypoda, var. pseudossiori (p. 65). Add the following number:

Western Szech'uan: Tachien-lu, woods, alt. 2000 m., September 1908 (No. **986**; tree 8–10 m. tall).

Prunus pubigera, var. **obovata** (p. 68). Add the following number:
Kiangsi: Kuling, thickets, rare, alt. 1300 m., July 31, 1907 (No. **1675**; bush 2 m.
tall; sterile).

Prunus (p. 71). After *P. laxiflora* insert:

PRUNUS MACROPHYLLA Siebold & Zuccarini in *Abh. Akad. Münch.* IV. part 2, 120
(*Fl. Jap. Fam. Nat.* I. 14) (1845). — Hance in *Jour. Bot.* XVI. 87 (1878). — Maximowicz in *Bull. Acad. Sci. St. Pétersbourg*, XXIX. 111 (1883); in *Mél. Biol.* XI.
710 (1883). — Hemsley in *Jour. Linn. Soc.* XXIII. 219 (1887). — Shirasawa,
Icon. Ess. For. Jap. II. t. 29, fig. 1–9 (1908). — Dunn & Tutcher in *Kew Bull.
Misc. Inform.* Add. ser. X. 93 (*Fl. Kwangtung & Hongk.*) (1912).

> *Pygeum oxycarpum* Hance in *Jour. Bot.* VIII. 242 (1870).
> *Prunus oxycarpa* Maximowicz in *Bull. Acad. Sci. St. Pétersbourg*, XXIX. 111
> (1883); in *Mél. Biol.* XI. 710 (1883).
> *Laurocerasus macrophylla* C. Schneider, *Ill. Handb. Laubholzk.* I. 647, fig. 355 e
> (1906).

Western Szech'uan: Hung-yah Hsien, foot of Wawu-shan, alt. 600–1000 m.,
September 1908 (No. **2541**; tree 15 m. tall, girth 1 m., bark brown and scaly, flowers
white).

PRUNUS HYPOTRICHA Rehder, n. sp.
Arbor 7-metralis; ramuli flavo-cinerei, obsolete lenticellati, crassiusculi. Folia
persistentia, elliptica, breviter acuminata, basi rotundata v. late cuneata, serrulata
dentibus minutis glandula conica adusta terminatis plerumque 4–8 pro 1 cm., 8–11
cm. longa et 4–5.5 cm. lata, supra pallide glauco-viridia, glabra, subtus vix pallidiora,
tota facie breviter molliter pilosa costa subglabra excepta, costa media supra incisa,
nervis utrinsecus circiter 10 subtus elevatis supra fere obsoletis, nervis tertiariis
leviter v. vix elevatis; petioli supra canaliculati, crassiusculi, 5–8 mm. longi, apice
glandulis duobus instructi. Racemus fructifer brevis, glaber, 2 cm. longus; drupa
unica ovoidea, exocarpio ut videtur livido et vix carnoso; putamen oblongo-ovoideum, leviter compressum 2 cm. longum, laeve, pariete tenui fragili.
Western Szech'uan: Kuan Hsien, thickets, alt. 800 m., November 1908 (No.
2540; tree 7 m., girth 0.6 m.).
The pubescent under side of the leaves distinguishes *P. hypotricha* easily from
the other Asiatic evergreen species of the subgen. *Padus*. It is apparently most
closely related to *P. macrophylla* Siebold & Zuccarini which differs in the longer and
narrower, more remotely and coarsely serrate leaves glabrous beneath and of thicker
texture.   A. R.

Ilex yunnanensis (p. 76). Add to this species the following numbers, note and
variety:
Western Szech'uan: south-east of Tachien-lu, woodlands, alt. 2600 m., October 1910 (No. **4135**; bush 2–2.5 m. tall, fruit red); west and near Wên-ch'uan Hsien,
woodlands, alt. 2300–2600 m., October 1910 (No. **4159**; bush 2–3 m. tall, fruit
red).
Very common in thickets and woodlands throughout western Szech'uan. The
flowers are white and fragrant.

ILEX YUNNANENSIS, β GENTILIS Loesener in *Bot. Jahrb.* XXIX. 435 (1900); in
Nov. Act. Leop.-Carol. LXXVIII. 132 (1901).

> *Ilex gentilis* Franchet apud Loesener in *Nov. Act. Leop.-Carol.* LXXVIII. 133
> (1901).

Western Hupeh: Fang Hsien, thickets, alt. 1300–1600 m., November 1910 (No. **4458**; bush 1–2 m. tall, fruit red); without precise locality, July 1901 (Veitch Exped. No. 2344, Seed No. 1190).

Ilex Fargesii, var. megalophylla (p. 77). Add to this variety the following numbers:
Western Szech'uan: west and near Wên-ch'uan Hsien, woodlands, alt. 2600–3000 m., November 1910 (No. **4094**, in part; bush 4 m. tall); west of Kuan Hsien, Pan-lan-shan, thickets, alt. 2300–2800 m., October 1910 (No. **4094**, in part; bush 3–5 m. tall, fruit red); Tachien-lu, thickets, alt. 2600–3000 m., October 1910 (No. **4094**, in part; bush 5 m. tall, fruit red).

Ilex Franchetiana (p. 77). Add to this species the following numbers and note:
Western Szech'uan: west of Kuan Hsien, Pan-lan-shan, woodlands, alt. 3000 m., October 1910 (No. **4316**; bush 3–4 m. tall, fruit red); Mt. Omei, June 1904 (Veitch Exped. Nos. 4794, 4796); without locality, May 1904 (Veitch Exped. No. 3318).
This handsome species is common in the woods and thickets on Mt. Omei and elsewhere in western Szech'uan. The flowers are white and fragrant.

Ilex corallina (p. 80). Add the following number and note:
Western Szech'uan: Mupin, alt. 1000 m., October 1910 (No. **4222**; bush 1.5–3 m. tall, fruit coral-red).
A common lowland species throughout western Hupeh and Szech'uan.

Ilex fragilis, β Kingii (p. 82). Add to this variety the following numbers:
Western Szech'uan: Lungan Fu, Tu-ti-liang mountains, woodlands, alt. 2300 m., August 1910 (No. **4540**; tree 6 m. tall, fruit dark red); Wa-shan, alt. 2800 m., July 1903 (Veitch Exped. No. 3341).

Acer robustum (p. 89). In the enumeration of specimens insert No. **1893** before **1899** and change No. 540 of Veitch Exped. to No. 590.

Acer Davidii (p. 92). Add the following number:
Western Szech'uan: southeast of Tachien-lu, woods, alt. 2100–2400 m., October 1908 (No. **1916**; tree 8 m. tall).

Acer laxiflorum (p. 93). Add the following synonym and numbers:

Acer Forrestii Diels in *Not. Bot. Gard. Edinburgh*, V. 165 (*Pl. Chin. Forrest.*) (1912).

Yunnan: Lichiang Range, alt. 3000 m., May 1906, *G. Forrest* (No. 2106); same locality, alt. 3000 m., October 1914, *C. Schneider* (No. 3281); Lichiang, toward Taku, alt. 3200 m., August 27, 1914, *C. Schneider* (No. 3338). Southern Szech'uan: "in reg. Yen-yuan Hsien, inter Ouentin et Kalapa in silvis, alt. circ. 2800 m.," June 4, 1914, *C. Schneider* (No. 1462); "inter viculos Huan-ka et Wo-lo-ho, in dumetis apertis, alt. circ. 3200 m.," June 13, 1914, *C. Schneider* (No. 1499).

Acer Henryi (p. 97). This species is apparently much more closely related to *Acer Negundo* Linnaeus than to the other species of the section *Trifoliata*, as its flowers are borne in lateral pendulous racemes and the winter-buds have two valvate scales. Therefore *A. Henryi* as well as *A. cissifolium* K. Koch are better referred to the section *Negundo*, as I proposed in 1914 (in Bailey, *Stand. Cycl. Hort.* I. 204). By Nieuwland all the species of the section *Trifoliata* in the sense of

Pax have been separated from *Acer* as a distinct genus under the name of *Crula* (in *Am. Midland Nat.* II. 140 [1911]).   A. R.

Acer griseum (p. 97). Add the following number:
Western Hupeh: north of Ichang, margin of woods, alt. 1500–2000 m., 1907 (No. 719; trees 8–16 m. tall; plants only).

Acer (p. 98). At the end of the genus insert:
ACER spec.
Western Hupeh: north and south of Ichang, woods, alt. 1500–2000 m., 1907 (No. 684; tree 10 m., girth 0.6 m., cymes large; seeds only).

ACER spec.
Western Szech'uan: west and near Wên-ch'uan Hsien, alt. 2600–2800 m., 1910 (No. 4109; tree 8–12 m. tall, girth 1–1.6 m.).

Vitaceae

Ampelopsis aconitifolia (p. 100). Insert the following variety:
AMPELOPSIS ACONITIFOLIA, var. PALMILOBA Rehder in *Mitt. Deutsch. Dendr. Ges.* XXI. 190 (1912); in Bailey, *Stand. Cycl. Hort.* I. 278 (1914).

Ampelopsis palmiloba Carrière in *Rev. Hort.* 1867, 451, fig. 40.
Ampelopsis tripartita, Carrière, l. c. 1868, 39.
Ampelopsis aconitifolia, var. *typica* Koehne, *Deutsch. Dendr.* 400 (1893).
Vitis aconitifolia J. H. Veitch in *Jour. Roy. Hort. Soc.* XXVIII. 392, fig. 87 (non Hance) (1904).
Ampelopsis rubricaulis Schneider, *Ill. Handb. Laubholzk.* II. 321, fig. 213 h-h' (non Carrière) (1909).

Western Szech'uan: Mao-chou, valley of Min River, common on cliffs, alt. 2300 m., August 1910 (No. 4024).
This variety was first known only from cultivated plants. The original description is based on plants introduced by David from China, probably from the neighborhood of Peking, and it is possible that the plant is distributed from the mountains west of Peking through Mongolia to northwestern Szech'uan.

Ampelopsis (p. 101). After *A. megalophylla* insert:
AMPELOPSIS WATSONIANA Wilson in *Jour. Roy. Hort. Soc.* XLII. (1916).

Vitis leeoides J. H. Veitch in *Jour. Roy. Hort. Soc.* XXVIII. 395, fig. 95, 96 (non Maximowicz) (1904); *Hortus Veitch.* 384 (1906).

Western Hupeh: without precise locality, 1900 (Veitch Exped. Seed No. 629).
This species is easily distinguished from *Ampelopsis leeoides* Planchon by its always simply pinnate leaves.

Parthenocissus Thomsonii (p. 101). Add the following numbers:
Western Szech'uan Wên-ch'uan Hsien, on rocks, alt. 2300–2800 m., September 1910 (No. 4189; 3–4 m., fruit black). Yunnan: Mengtsze, ravines, alt. 2600–2700 m., *A. Henry* (Nos. 10754, 10754ᵃ; large climber, yellow flowers).

Parthenocissus himalayana (p. 101). Add the following numbers:
Eastern Szech'uan: Nanch'uan, " Ma tzu ai," August 1891, *A. von Rosthorn* (No. 409). Western Szech'uan: Ta-p'ao-shan, northeast of Tachien-lu, climb-

ing over rocks, alt. 2300–2600 m., July 3, 1908 (No. 2730); Tachien-lu, rocks, alt. 2600–3000 m., October 1910 (No. 4171; fruit blue-black); west and near Wên-ch'uan Hsien, covering rocks, alt. 2000–2300 m., October 1910 (No. 4184; 3–7 m., fruit black).

Parthenocissus landuk (p. 102). Wilson's No. 2730 enumerated by Gagnepain under this species we refer to *P. himalayana*.

Parthenocissus (p. 102). At the end of the genus insert:

PARTHENOCISSUS spec.

Western Szech'uan: northeast of Sungpan Ting, common on rocks, alt. 2700–3000 m., August 1910 (No. 4480).

This number is represented by a single specimen with immature fruit. It belongs apparently to a new species, but the material is too incomplete for a sufficient description. It seems most nearly related to *P. landuk* Planchon and to *P. tricuspidata* Planchon; from the former it differs in its much smaller coarsely dentate leaflets, the middle one obovate, and from the latter in the small, constantly 3-foliolate leaves.

PARTHENOCISSUS spec.

Western Hupeh: Chanyang Hsien, alt. 1000 m., 1907 (No. 469; climber, pedicels bright red and fruit blue-black; seeds only).

Vitis pentagona (p. 103). Add as synonyms:

Vitis Coignetiae Diels in *Bot. Jahrb.* XXIX. 461 (non Planchon) (1900).
Vitis ficifolia Bge., var. *pentagona* Pampanini in *Nuov. Giorn. Bot. Ital.* XVII. 116 (1910).

Transfer No. 1046ª to *Vitis betulifolia* Diels & Gilg and insert after the note:

VITIS PENTAGONA var. BELLULA Rehder, n. var.

A typo recedit ramulis petiolisque glabris, foliis minoribus, 3–6 cm. longis et 2–4.5 cm. latis remote minute denticulatis, inflorescentiis gracilioribus basi tantum breviter ramulosis fere glabris, floribus et fructibus seminibusque minoribus.

Western Hupeh: Changlo Hsien, climbing over rocks, alt. 600–1000 m., July 1907 (No. 77, type; 2–3 m., fruit black), without precise locality, May 1900 (Veitch Exped. No. 777).

This variety differs from the type chiefly in the much smaller leaves and bears the same relation to the typical form as *V. flexuosa*, var. *parvifolia* Gagnepain bears to its type. Gagnepain had enumerated it under typical *V. pentagona*, but remarks that it is a small-leaved form resembling *V. flexuosa*, var. *parvifolia*. As young plants raised from seed of No. 77 and growing in the Arnold Arboretum have preserved the characters of the parent plant, it seems advisable to distinguish this form by a varietal name.

Vitis reticulata (p. 103). Change the author citation as follows and add the following synonym and specimens: Pampanini in *Nuov. Giorn. Bot. Ital.* XVII. 118, fig. 13 (1910). — Gagnepain in Lecomte, *Not. Syst.* II. 12 (1911).

Vitis Wilsonae Hort. Veitch apud *Gard. Chron.* ser. 3, XLVI. 236, fig. 101 (nomen nudum) (1909).

Western Hupeh: Hsing-shan Hsien, thickets, alt. 1300–2000 m., September 1907 (No. 242, 3–4 m.); without precise locality, June 1902 (Veitch Exped. No. 1151). Western Szech'uan: Mt. Omei, June 1904 (Veitch Exped. No. 4825).

Vitis reticulata Pampanini and *V. reticulata* Gagnepain refer to the same plant; there is no apparent discrepancy in the descriptions and a leaf kindly sent by Dr. Pampanini from his type specimen matches exactly the leaves of Gagnepain's type. It is a rather peculiar coincidence that both authors chose independently the same name for this plant.

Vitis betulifolia (p. 103). Add the following specimens:
Western Hupeh: Fang Hsien, alt. 1600 m., September 1907 (No. **291**; 5 m., fruit black); without precise locality, April and May 1900 (Veitch Exped. Nos. 2704, 792). Western Szech'uan: near Tachien-lu, thickets, alt. 1300–2000 m., October 1908 (No. **1046ª**; 3–5 m., fruit black); west and near Wên-ch'uan Hsien, alt. 2–2500 m., October 1910 (No. **4188**; 3–7 m., fruit blue-black); Mt. Omei, June 1904 (Veitch Exped. No. 4822).

Vitis armata, var. **cyanocarpa** (p. 104). Transfer No. 291 to *Vitis betulifolia* Diels & Gilg.

Vitis (p. 105). After V. ROMANETII insert:
VITIS spec.
Kiangsi: Kuling, amidst rocks, rare, alt. 1000 m., August 1, 1907 (No. **1698**; climber 2–3 m.).
This number is represented by sterile specimens only; it resembles *V. pentagona*, var. *bellula* Rehder, but differs markedly in the dense short villose pubescence of the upper surface of the leaves and of the petioles and branchlets.   A. R.

Caprifoliaceae

Viburnum sympodiale (p. 109). Add the following number:
Western Hupeh: north and south of Ichang, woods, alt. 1600 m., autumn 1907 (No. **714**; bush 2–4 m. tall; plants only).

Viburnum buddleifolium (p. 109). Add the following number:
Western Hupeh: north and south of Ichang, thickets, alt. 600–1300 m., 1907 (No. **716**; plants only).

Viburnum Davidii (p. 111). Add the following number:
Western Szech'uan: Mupin, woodlands, alt. 2000–2600 m., 1908 (No. **963ª**; seeds only.)

Triosteum himalayanum, var. **chinense** (p. 117). Add the following number:
Western Szech'uan: Tachien-lu, uplands, 1908 (No. **998**; herb 0.5–0.8 m. tall, fruits white; seeds only).

Abelia Graebneriana (p. 118). Add the following number:
Western Hupeh: north and south of Ichang, thickets, alt. 1000–2000 m. (No. **717**; plants only).

Abelia Zanderi (p. 121). Add the following synonym and numbers:
Linnaea brachystemon Diels in *Not. Bot. Gard. Edinburgh*, V. 178 (*Pl. Chin. Forrest.*) (1912).
Western Szech'uan: Mupin, woodland, alt. 2300–3000 m., October 1910 (No· **4210**; bush 3–5 m. tall, flowers white, small). Yunnan: Lichiang Range, alt. 3000 m., May 1906, *G. Forrest* (No. 2155, type of *Linnaea brachystemon*).
Linnaea brachystemon is apparently only a glabrescent form of the variable *A.*

Zanderi. In Forrest's specimens before me some of the leaves are sparingly pilose on the midrib beneath, very much as in Purdom's No. 320.

Lonicera Henryi (p. 141). Add the following number:

Western Szech'uan: near Tachien-lu, thickets, alt. 2000 m., October 1908 (No. 3481; climber 3 m., flowers reddish and yellow).

Lonicera subaequalis (p. 142). Add after No. 940: 940ᵃ; seeds only.

Lonicera (p. 143). Add after *L. tragophylla:*

LONICERA spec.

Western Hupeh: Hsing-shan Hsien, thickets, alt. 1300 m., 1907 (No. 196; seeds only).

LONICERA spec.

Western Szech'uan: Mupin, thickets, 1908 (No. 1091; seeds only).

No plants of these two Loniceras are growing at the Arnold Arboretum.

Diervilla japonica (p. 144). Change the entry under Diervilla as follows:

DIERVILLA JAPONICA, var. SINICA Rehder in *Mitt. Deutsch. Dendr. Ges.* XXII. 264 (1913).

> *Diervilla floribunda* Hemsley in *Jour. Linn. Soc.* XXIII. 369 (non Siebold & Zuccarini) (1888).
> *Diervilla japonica* Rehder in Sargent, *Pl. Wilson.* I. 144 (non Siebold & Zuccarini) (1911).

Western Hupeh: north and south of Ichang, thickets, alt. 1000–2500 m., May, June and December 1907 (No. 762; bush 1.5–6.5 m. tall, flowers vinous color). Hsing-shan Hsien, roadsides, thickets, alt. 1000–2000 m., May and June 1907 (Nos. 2916, 2917, 2918; bush 1–4 m. tall, flowers pale rose pink); without precise locality, May 1900 (Veitch Exped. No. 266). Eastern Szech'uan: Wushan Hsien, *A. Henry* (No. 5485).

DIERVILLA CORAEENSIS De Candolle, *Prodr.* IV. 330 (1830).

> *Weigela coraeensis* Thunberg in *Trans. Linn. Soc.* II. 331 (1794).
> *Diervilla grandiflora* Siebold & Zuccarini, *Fl. Jap.* I. 71, t. 31 (1838).

Kiangsi: Kuling, thickets, abundant, alt. 1300 m., July 28, 1907 (No. 1568; bush 1.5–3 m. tall). A. R.

Cotoneaster foveolata (p. 162). Add the following number:

Western Hupeh: Hsing-shan Hsien, thickets, alt. 1300–2000 m., 1907 (No. 187; bush 2–3 m.; seeds only).

Cotoneaster Zabelii (p. 166). Transfer Henry's No. 7918 to *C. gracilis* Rehder & Wilson and add the following variety:

COTONEASTER ZABELII, var. MINIATA Rehder & Wilson, n. var.

A typo recidit praecipue fructibus minoribus miniatis, pyrenis 5 mm. longis.

Western Hupeh: Changlo Hsien, alt. 1300–2000 m., autumn 1907 (No 153; bush 2.5 m. tall, seeds only; plants growing at the Arnold Arboretum, specimens collected July 3, 1914 and October 14, 1916, type).

The fruits of this variety are smaller and light orange-scarlet, much lighter than those of the type; the habit of the plant is more branched and denser and the leaves are lighter colored and turn yellow in autumn, while the leaves of typical *C. Zabelii* do not change much before falling.

Cotoneaster gracilis (p. 167). Add the following:
Western Hupeh: without precise locality, *A. Henry* (No. 7918).
Henry's number had been referred by mistake to *C. Zabelii*, but it certainly belongs to *C. gracilis;* it has red fruits 6 or 7 mm. long. *C. gracilis* is apparently closely related to *C. Zabelii* and there are intermediate forms, as Wilson's No. 2174, which has the leaves glabrous above like *C. gracilis*, but a pubescent calyx.

Cotoneaster racemiflora (p. 168). Add the following variety:
COTONEASTER RACEMIFLORA, var. VEITCHII Rehder & Wilson, n. var.

Cotoneaster acutifolia Veitch, *New Hardy Pl. China*, 1910, p. 5 (non Turczaninow).

Varietati *soongaricae* Schneider affinis sed differt praecipue habitu virgato, foliis acutis v. rarius acutiusculis mucronulatis ovalibus v. ellipticis 2–3.5 cm. longis subtus densius cinereo-villosis, floribus majoribus 13–15 mm. diam., calyce densius villoso, fructu majore subgloboso 8–12 mm. diam. purpureo, pyrena plerumque una, 5–6 mm. diam. longitudinaliter irregulariter sulcato, hypostylio trientem pyrenam obtegente.

Western Hupeh: Hsing-shan Hsien, 1900 (Veitch Exped. Seed No. 1079; plants growing in the Arnold Arboretum, specimens collected June 5 and November 7, 1916, type).

This variety is closely related to var. *soongarica* Schneider, which differs in its more divaricate irregular habit, obtuse leaves more thinly pubescent beneath, smaller flowers and in its smaller bright red fruits with usually 2 stones. As *C. racemiflora*, var. *Veitchii* grows in this Arboretum it forms an upright shrub with rather slender virgate branches covered in June along their whole length with white or slightly pinkish flowers followed in autumn by dark red fruits.

Osteomeles Schwerinae (p. 184). To this species add the following variety:
OSTEOMELES SCHWERINAE, var. MICROPHYLLA Rehder & Wilson, n. var.

A typo recedit foliis circiter 1.5 cm. longis, foliolis plerumque 11–13 obovatis v. ellipticis obtusiusculis fere glabris v. subtus sparse tantum pubescentibus 2–5 mm. longis, calyce glabro, sepalis triangulari-ovatis saepe obtusiusculis, et habitu magis dumoso.

Western Szech'uan: valley of Min River, around Mao-chou, alt. 1300–2000 m., May and September 1908 (No. 1016, fruiting [type], and flowering; bush 1–3 m. tall, flowers white, fruits with purple bloom); valley of Tung River, July 1903 (Veitch. Exped. No. 3518ᵃ).

This variety differs from the type chiefly in the much smaller glabrescent leaves, the glabrous calyx with shorter lobes and in its denser more compact habit. Wilson's No. 3518ᵃ is intermediate between the variety and the type, but seems closer to the variety.

Photinia (p. 190). After *Photinia amphidoxa* insert:
PHOTINIA spec.
Western Szech'uan: Taning Hsien, thickets, alt. 1000–1300 m., July 1910 (No. 4619; bush 1.3–2 m.).
This specimen, which consists of leafy shoots only, may be a juvenile form of an evergreen Photinia; the short-petioled elliptic to oblong leaves are coarsely and irregularly spinose-dentate and 5–7 cm. long.

Prunus discadenia (p. 200). On p. 201, line 4, after No. 62 instead of " as to flowering branches . . ." read: as to fruiting branches, the flowering branches belong to *P. laxiflora* Koehne);

Prunus pilosiuscula, var. **media** (p. 204). Add the following specimen:
Western Hupeh: Hsing-shan Hsien, side of streams, common, alt. 1000 m.,
June 1907 (No. 16; tree 5–8 m. tall).

Prunus involucrata (p. 206). This becomes:
PRUNUS PSEUDO-CERASUS Lindley in *Trans. Hort. Soc. Lond.* VI. 90 (1827). —
Wilson, *Cherries Jap.* 3 (1916), which see for further literature and synonyms.
Chili: purchased on Peking-Henkow Railroad, 1910 (No. 4002, tree 5–8 m.
tall; seeds only).

Prunus (p. 225). After *P. tomentosa*, var. *endotricha* add:
PRUNUS spec.
Western Szech'uan: near Sungpan, woodlands, alt. 3000–3300 m., 1908
(No. 4008; tree 8 m. tall, very bushy; seeds only).

PRUNUS spec.
Western Szech'uan: west of Kuan Hsien, Pan-lan-shan, woodlands, 2600–
3000 m., 1908 (No. 4034; tree 3–4 m. tall, fruits black, leaves white below; seeds
only).
There are no plants of this and the preceding number growing at the Arnold
Arboretum.

Prunus (p. 237). For changes in the nomenclature of the Japanese species mentioned by Koehne in the Enumeratio specierum omnium subgen. Cerasi (pp. 237–
271), see Wilson, *Cherries of Japan* (1916).

Prunus stipulacea (p. 258). In the citation between " *St. Pétersbourg* " and
" XI." insert: XXIX. 97 (1883); *Mél. Biol.* In the enumeration of specimens
for " Prewalski " read " Przewalski."

Prunus dehiscens (p. 271). On p. 272 instead of No. 4028 read 4029.
Prunus dehiscens is probably not different from *P. tangutica* Koehne.

Prunus Persica (p. 273). Add the following numbers:
Western Hupeh: mountains south of Ichang, naturalized, 800–1600 m., 1907
(No. 125, bush 1–2 m. tall, flowers dark; seeds only); Patung Hsien, side of streams,
alt. 800 m., 1907 (No. 744; bush 3 m. tall, flowers double white, only one bush seen;
scions for grafting only.

Prunus triflora (p. 276). Add the following numbers to this species, the oldest
name of which is *P. salicina* Lindley, as stated on p. 580 of vol. I:
Western Hupeh: cultivated around Ichang, 1907 (No. 27; bush 3–5 ft. tall,
flowers white, fruit small, round, yellow; seeds only); same locality, 1907 (Nos. 46,
61; bushy tree 3–7 m. tall, fruit yellow; seeds only); Hsing-shan Hsien, thickets,
alt. 1000–1500 m., September 1907 (No. 243; bushy tree 5 m. tall, fruit yellow;
seeds only).

Styrax japonicus (p. 291). In the enumeration of specimens on p. 292 for
" *E. J. Taquet* (Nos. 725, 727, 1109, 1876, 3033, 3034) " read " *U. Faurie* (Nos.
725, 727, 1876), *Taquet* (Nos. 1109, 3033, 3034)."

Syringa

Syringa Sargentiana (p. 298). Judging by living plants, this species must be
regarded as a variety only of *S. Komarowii* Schneider. The insertion of the anthers
in the corolla-tube is about the same in both species; it is possible to find slight
variations in different flowers, the apex of the stamens sometimes only reaching the

mouth of the corolla, sometimes being more or less slightly exserted. Therefore I suggest the following combination:

Syringa Komarowii, var. Sargentiana Schneider, n. var.
Add to the enumeration of specimens: Western Szech'uan: Wa-shan, alt. 2100–2400 m., July 1903 (Veitch Exped. No. 4081; shrub 1.8–4.5 m. tall, flowers rose).

Syringa tetanoloba (p. 299). This species described from a meagre specimen proves identical with *S. Sweginzowii* Koehne & Lingelsheim (see p. 301). It comes from the Sungpan region from which the seeds of the type probably had been introduced to Petrograd by Russian collectors.

Syringa tomentella (p. 300). Add as a synonym: *S. alborosea* N. E. Brown in *Kew Bull. Misc. Inform.* 1914, 187. The type of *S. alborosea* is Wilson's No. 1739 (Veitch Exped.), the seeds of which were collected near Tachien-lu in western Szech'uan.

Syringa Wilsonii (p. 300). According to observations on living plants this species is scarcely different from *S. tomentella* Bureau & Franchet, of which even *S. Rehderiana* Schneider may represent only a variety.

Syringa microphylla, var. glabriuscula (p. 301). This variety is scarcely distinct and is connected by many intermediate forms with the type, the synonymy of which follows:

Syringa microphylla Diels in *Bot. Jahrb.* XXIX. 531 (1900). — Schneider in *Bot. Jahrb.* XXXVI. Beibl. LXXXII. 87 (1905); *Ill. Handb. Laubholzk.* II. 778, fig. 487 n–p, 486 z–z² (1911).

Syringa Dielsiana Schneider in *Bot. Jahrb.* XXXVI. Beibl. 82, 88 (1905); *Ill. Handb. Laubholzk.* II. 778, fig. 487 g–k, 488 a–d (1911).

To this species belongs Wilson's specimen from western Hupeh, June 1901 (Veitch Exped. No. 2024). *S. microphylla* is in cultivation in this Arboretum, where plants were raised from seeds collected by Purdom (No. 583). It flowered well in the first days of June 1915 and 1916, and in 1916 a second time in August.

Syringa Meyeri (p. 301). This species is not yet known in a wild state. F. N. Meyer (according to his note in *Bull. U. S. Dept. Agric. Bur. Pl. Industry*, CXLII. 57, No. 23032 [1909]) sent cuttings from Fengtai, near Peking, Chili. " (No. 694, Mar. 31, 1908). A small-leaved Lilac, bearing many panicles of purple flowers, grafted upon a small-leaved Privet. Used much in forcing; quite rare and expensive; not hardy. Chinese name Shan ting hsien." In this Arboretum *S. Meyeri* has proved quite hardy. It is apparently a slow-growing species, forming a densely branched small bush.   C. S.

Catalpa Duclouxii (p. 304). Add the following number:
Eastern Szech'uan: Wushan Hsien, A. *Henry* (No. 5856), distributed as *Aleurites* (*Elaeococca cordata* Muell. Arg.).

Sambucus Schweriniana (p. 306). This species flowered at the Arnold Arboretum for the first time in August 1914. This enables us to add the following description of the flowers which were unknown when the species was described: Corymbus concavus, glaber v. fere glaber, 9 cm. diam., 5-radiatus, radius medius lateralibus multo brevior: flores 5–7-meri, plerumque 6-meri, sessiles v. pedicellati pedicellis circiter 1 mm. longis; sepala subulata, 0.5 mm. longa, albida v. leviter purpurascentia; corolla rotata, 3 mm. diam., alba, extus leviter rosea, lobis suborbicularibus fere ad basin fissis; stamina corolla paulo breviora, antheris suborbicularibus flavo-albis demum ochraceis; stylus conicus, brevissimus, stigmatibus 3.

A. R.

Clematis nutans, var. thyrsoidea (p. 324). This variety has been raised to specific rank and Wilson's No. 1422 separated as a distinct species:

CLEMATIS REHDERIANA Craib in *Kew Bull. Misc. Inform.* 1914, 150.

Here belong all the synonyms cited under *C. nutans*, var. *thyrsoidea* and the numbers quoted except Wilson's seed number 1422.

CLEMATIS VEITCHIANA Craib, l. c. 151.

Here belongs Wilson's seed number 1422, represented at Kew and at the Arnold Arboretum by plants received from Veitch and raised from that seed.

Clematis Pavoliniana (p. 328). Add the following numbers:

Western Hupeh: Ichang, alt. 300–1500 m., ravines, 1907 (No. **402**; climber, flowers white, fruit tails yellowish; seeds only); Changyang Hsien, thickets, 1907 (No. **668**; flowers small, white, fruit tails brown).

Akebia lobata, var. australis (p. 348). Add the following numbers:

Western Szech'uan: without precise locality, alt. 2000 m., 1908 (Nos. **826, 4181**; large climber, fruits 10 cm. long, purplish, edible; seeds only).

Berberis

Berberis dictyophylla, var. epruinosa (p. 353). Add as a synonym *B. Ambrozyana* Schneider, l. c., p. 356, which was based on a rather poor specimen representing only an insignificant form of the Szech'uan variety of *B. dictyophylla* Franchet. To the enumeration of specimens under var. *epruinosa* should be added Wilson's No. 4170 which had been referred to *B. diaphana* Maximowicz.

Here it may be also stated that *B. approximata* Sprague seems to represent a good species the type of which was raised from seeds collected by Farges in eastern Szech'uan and sent to Mr. M. L. de Vilmorin. It is the same as *B. dictyophylla* Vilmorin & Bois, *Frutic. Vilmorin.* 1904, 19, fig. (non Franchet) (1905), exclud. descript. lat. Franchetii.

Berberis diaphana (p. 353). After a careful examination of the dried and living specimens of this Arboretum I believe that Wilson's No. 930 referred to this species represents the type of the following new species:

BERBERIS AEMULANS Schneider, n. sp.

Berberis diaphana Schneider in Sargent, *Pl. Wils.* I. 353 (pro parte, non Maximowicz) (1913), quoad specim. Wilsonii No. 930.

Frutex 0.9–1.8 m. altus, glaberrimus; ramuli annotini intense purpurascentes, glabri, plus minusve sulcati, vetustiores cinereo-nigrescentes (hornotini plantae juvenilis cultae intense rubescentes et leviter pruinosae); internodia 2–4 cm. longa; spinae normales, 3-fidae, debiles, mediae vix ad 1.2 cm. longae. Folia ramulorum floriferorum ad 6 fasciculata, membranacea, oblonga, oblongo-elliptica v. obovato-oblonga, apice rotundata, mucronulata, basim versus in petiolum brevissimum v. ad 5 mm. longum attenuata, margine utrinque breviter spinuloso-serrata v. interdum integra, minimis exceptis 2.4–5 cm. longa et 1–1.8 cm. supra medium lata, ramulorum fructiferorum firmiora, chartacea, usque ad 5 cm. longa et 2.3 cm. lata, omnia serrata, superne viridia, subtus albescentia, pruinosa, haud distincte papillosa, utrinque distincte elevato-reticulata. Inflorescentiae 1-3-florae, pedunculi nulli v. brevissimi; flores ab iis *B. diaphanae* vix diversae, petalis paulo angustioribus, ovulis 7-11; pedicelli fructiferi 2-3 cm. longi. Fructus aurantiaci, ut videtur leviter pruinosi, ovato-elliptici, 15-18 mm. longi et 8-10 mm. crassi, apice paulo attenuati (an semper?) et stylo brevissimo coronati; semina matura 3-5.

Western Szech'uan: near Was-han (No. 930, type; see p. 353); without exact locality (Veitch Exped. No. 3145, in part, flowering specimens; see p. 354).

On account of the dark-colored branchlets which are bright red on the cultivated specimens this species has to be kept distinct from *B. diaphana*. Young plants of *B. aemulans* cannot be distinguished from those of *B. Tischleri* Schneider, but this species is very distinct in its mostly elongate subracemose inflorescence and in its elongated almost rostrate fruits.

Berberis diaphana, var. circumserrata (p. 354). This variety differs in several respects from the true *B. diaphana* Maximowicz, especially in its ovaries containing only 4 ovules (not 6-12) and in its habit. It seems to represent a good species which inhabits a different geographical area. It may be described as follows:

BERBERIS CIRCUMSERRATA Schneider, n. sp.

Frutex erectus, satis dense ramosus; ramuli hornotini annotinique flavescentes, flavo-brunnescentes v. paulo rubescentes, glabri, striato-angulati, vetustiores cinerascentes; internodia 1.5-3 cm. longa; spinae normales 3-fidae, flavescentes v. brunnescentes, interdum ad 5-fidae v. reductae 1-fidae, dorso sulcatae, mediae surculorum ad 2.5 cm. longae. Folia ramulorum fertilium ad 5 fasciculata, matura chartacea, obovato-oblonga v. late obovata, apice rotunda v. leviter emarginata, basim versus in petiolum ad 0.5 cm. longum sensim v. subito contracta, 1.5-3.5 cm. longa et 0.5-2.5 cm. lata, superne intense viridia, subtus albescentia, pruinosa v. in planta vetustiori pallide viridescentia, utrinque distincte satis anguste v. laxius ut in *B. diaphana* graciliter elevato-reticulata, margine circumcirca dense spinuloso-serrulata, dentibus 0.5-2 mm. longis gracilibus 3-8 pro 5 mm., folia surculorum (juvenilia longe petiolata excepta) majora v. latiora, ovato- v. elliptico-rotunda v. late elliptica v. obovata, minus dense grossius spinuloso-serrata v. etiam fere integra, 1.5 : 1 cm. ad 4 : 3 cm. v. ad 3 : 3 cm. magna, subtus distinctius albescentia, saepe laxius reticulata. Inflorescentiae 1-3 florae, pedunculo ad 2.5 cm. longo; flores lutei, circiter 10 mm. diametientes, ab iis *B. diaphanae* v. *B. Tischleri* vix diversi, sed staminibus apice vix apiculatis; ovula pleraque 4, interdum 3 v. ad 7. Fructus (plantae cultae) anguste elliptici, apice plus minusve ut in *B. Tischleri* in stylum elongati, flavo-rubri, partim leviter pruinosi, stylo incluso circiter 13-15 mm. longi et 5-6 mm. crassi; semina 1-3, aurantiaca, laevia; pedicelli 10-17 mm. longi, bracteis acuminatis ad 2-2.5 mm. longis suffulti.

Shensi: Tai-pei-shan, 1910, *W. Purdom* (No. 4, type; with flowers); same locality, February 8, 1911, *W. Purdom* (Seed Nos. 604, 604ᵃ, 608); young plants in the Arnold Arboretum, fruiting co-type).

There are also young plants in this Arboretum under No. 182 of Purdom of which a record is wanting. These plants have not yet flowered and may possibly represent a somewhat different form. *B. circumserrata* seems to be a good species, resembling in its fruits *B. Tischleri* but in its inflorescences and in its growth much more *B. diaphana*. It does not form such regular dense bushes as the latter species and it grows much less vigorously than *B. Tischleri*.

Berberis yunnanensis (p. 355). This species does not occur in western Szech'uan. The plants mentioned under *B. yunnanensis* should be referred to *B. Tischleri*

Schneider (see below). Of the typical *B. yunnanensis* Franchet may be mentioned the following specimens:

Yunnan: "Tali Range, alt. 9–10000 ft., June–July 1906," *G. Forrest* (No. 4344); "in dumetis ad pedem montium niveorum prope Lichiang, alt. circiter 3000 m.," October 1914, *C. Schneider* (No. 2860); "in declivibus montium prope Lichiang versus Ta ku, alt. circiter 3400 m.," August 27, 1914, *C. Schneider* (No. 3343).

Berberis Tischleri (p. 355). Besides the specimens from western Szech'uan mentioned under *B. yunnanensis* Franchet, add the following specimens:

Western Szech'uan: west of Kuan Hsien, Pan-lan-shan, woodlands, alt. 3300–3600 m., 1910 (No. **4313**; bush 2 m. tall, fruits ovoid, solitary; seeds only); Tachien-lu, 1904 (Veitch Exped. Seed No. 1719; bush 6 ft.; fruits paniculate, salmon; Seed No. 1731; fruits solitary, scarlet).

There is before me a specimen collected by *E. Faber* on Mt. Omei (No. 229), the leaves of which resemble, as I mentioned in *Bull. Herb. Boiss.* sér. 2, VIII. 201 (1908), those of *B. pallens* Franchet. The ovaries contain 6 ovules, of which two are more or less stalked. It belongs to the affinity of *B. Tischleri* or *B. diaphana* and may represent a new species.

Berberis Ambrozyana (p. 356). See my remarks above under *B. dictyophylla,* var. *epruinosa* Schneider. In the note under this species I mentioned the following species:

BERBERIS PARVIFOLIA Sprague in *Kew Bull. Misc. Inform.* 1908, 445. The type of this species is Wilson's No. 3154ᵃ Veitch Exped. which probably had been collected in 1904 in the valley of the Min River in western Szech'uan. Here belongs probably also Veitch Exped. Seed No. 1682 from the same region. In his note to *B. Wilsonae* Hemsley in *Bot. Mag.* CXXXVIII. t. 8414 (1912) W. J. Bean says that *B. parvifolia* "has been in cultivation at Kew since 1896, when seeds were received from St. Petersburg." According to this statement seeds of this species must have been collected by Potanin or Przewalski in western Kansu, but there seems to be no herbarium specimen of it in Petrograd. But there are living plants in this Arboretum raised from seeds collected by W. Purdom (No. 826) in western Kansu: "Chone and Touchow district, alt. 8–9000', April 25, 1912, shrub 2½ ft." The plants in the Arboretum have not yet flowered, but others sent from here to the Golden Gate Park in San Francisco have flowered and I have before me branches of them with young fruits collected in August 1916. Purdom's plant well agrees with a fragment of the type kindly sent to this Arboretum by the Keeper of the Kew Herbarium. The inflorescence of this is more or less compound and is even more so in the specimens from San Francisco. They are almost identical with those of the typical *B. aggregata* Schneider, the flowers being only less numerous (1–5 in the type, up to 12 in Purdom's plants). The fruits are very much like those of *B. aggregata*, but the leaves are smaller. The largest I have seen are up to 2 cm. long and 1 cm. wide, obovate-oblong, with a spinose apex and two spreading pungent teeth on each side of the margin. The young shoots are somewhat downy or almost glabrescent and of a yellowish-brown color. In the fruits of Purdom's form I have found only 2 yellow seeds and no trace of any immature ovule, while the flowers of the type contain 4 ovules. This fact needs further investigation. It is a very pretty and apparently quite hardy species. The name *parvifolia* has been used before by Lindley for Berberis (in *Jour. Hort. Soc.* II. 243, fig. [1847]), of which I saw a specimen in 1905. Lindley's species being a mere synonym of *B. ruscifolia* Lamarck, Sprague's name must be kept according to the International rules.

Berberis triacanthophora (p. 359). Strike out Veitch Exped. No. 951.

Berberis levis (p. 360). This species needs further investigation. I believe that there are several different species mixed under what I have referred to *B. levis* and even in the specimens mentioned by Franchet. Wilson's Nos. 1284 and 4287 with jet-black fruits from western Szech'uan represent a new species different from the true *B. levis* as well as from *B. Soulieana* Schneider (see below). It may be described as follows:

BERBERIS ATROCARPA Schneider, n. sp.

Frutex 1–1.5 m. altus; ramuli hornotini (plantae in Arboreto Arnoldiano olim cultae) laeves, striato-angulato-sulcati, flavescentes, annotini vetustioresque grisei v. cinerascentes, sulcati; internodia 2.5–4 cm. longa; spinae 3-fidae, flavescentes, dorso applanatae, mediae 1–4.5 cm. longae. Folia sempervirentia, firma sed satis flexilia, coriacea, lanceolata v. rarius anguste elliptico-lanceolata, basi in petiolum vix ad 5 mm. longum contracta, apice acuta, mucronata, plana, margine spinoso-serrata dentibus porrecto-divergentibus 0.5–1.5 mm. longis 2–3 pro 1 cm., latiora ad 5.5 cm. longa et 1.5 cm. lata, angustiora ad 7 cm. longa et 1.4 cm. lata, superne saturate viridia, subnitentia, nervis paulo v. vix visibilibus, subtus plus minusve discoloria, flavo-viridia, fere enervia. Flores ignoti. Fructus in inflorescentia unica ad 8 fasciculati (ramulis fructiferis brevissimis saepe pluribus congestis), ovato-globosi, stylo distincto excepto circiter 5 mm. longi et 4 mm. crassi, atri, nituli; pedicelli vix ultra 8 mm. longi, saepe rubescentes; ovula 2–3 subsessilia, semina pleraque 2.

Western Szech'uan: Mupin, Nos. **1284**, type, and **4287** enumerated on p. 360 of vol. I.

As far as I know there is no other species of this section which has such jet-black, almost globose fruits. According to Franchet the fruits of *B. levis* are " ovatae nigro-coeruleae." As I said on p. 360 the fruits I have seen of the type of *B. levis* are without bloom, but it is often difficult to tell the real color of fruits from dried specimens only. Apparently *B. atrocarpa* and *B. levis* are closely related, but a very careful examination of the Chinese Berberis shows that there are many more different species, each confined to a certain region.

The typical *B. levis* Franchet of which I have a cotype (Delavay's No. 993 from He-chan-men) from the Kew Herbarium before me, is also represented by a specimen collected by *G. Forrest* in Yunnan: " shady situations amongst scrub on the ascent of the Sung-kwei pass from the Lang-kong valley, alt. 9–10000 ft., April 1906 " (No. 2012; spreading shrub of 3–6 ft., flowers bright-yellow) which is named *B. levis* by Diels (in *Not. Bot. Gard. Edinburgh*, VII. 341 [1913]).

BERBERIS SOULIEANA Schneider is, as I now believe, a distinct species and *B. stenophylla* Hance is a synonym of it. The type was raised in Hort. Vilmorin from seeds collected by R. Farges in the region of " Tchen keou tin " in eastern Szech'uan. In the same region probably, E. H. Parker collected the type specimen of *B. stenophylla* and not at Chungking as stated by Hance because there is no evergreen Berberis at all at this locality according to Wilson's observations. *B. Soulieana* has very firm leaves, rather patent spinose teeth (1–2 mm. long and three or four or even five in 1 cm. of the length of the margin) and distinctly glaucous fruits with mostly three seeds and otherwise similar to those of *B. atrocarpa*.

To *B. Soulieana* apparently belong the following specimens: Shensi: Tai-pei-shan, 1910, *W. Purdom* (No. 7), which I doubtfully referred to *B. Julianae* on p. 361 and southwestern Kansu: near Kua-tsa, on decomposed rock slope, alt. 1400 m., November 9, 1914, *F. N. Meyer* (No. 1823; shrub). Both are exactly alike.

Berberis Julianae (p. 360). As stated above, Purdom's No. 7 should be removed from the enumeration of specimens to which should be added Wilson's No. 2878, mentioned on p. 362 under *B. Bergmanniae* Schneider (see below), and A. Henry's No. 1458 from Hupeh. *B. Julianae* seems to be the hardiest of the evergreen Chinese Barberries growing in the Arnold Arboretum. The remark under *B. Sargentiana* Schneider on p. 360 should be transferred to *B. Julianae*. The true *B. Sargentiana* is also cultivated in this Arboretum and may prove as hardy and useful as the other species.

Berberis Bergmanniae (p. 362). As already stated, Wilson's specimen No. 2878 from Hupeh has to be transferred to *B. Julianae* Schneider. The differences mentioned by me regarding the glands of the petals seem to be due to a condition caused by the process of drying. I have made a new investigation and I find that otherwise No. 2878 is very much like the type-number 417 of *B. Julianae*. *B. Bergmanniae*, of which No. 2877 should be regarded as the type, may be distinguished by its coarser more distant serration and the glossy under surface of its leaves and by its somewhat shorter and thicker fruits, which are without the style about 8–9 mm. long and 6–7 mm. thick, while in *B. Julianae* they are more elliptic-oblong and about half as thick as long. This species is confined to western Szech'uan. *B. Bergmanniae*, var. *acanthophylla* Schneider may perhaps better be referred as a synonym to the type. The leaves of young plants of this form have a strong distant and sinuate serrature. *B. Bergmanniae* is closely related to *B. Julianae* and the other species with thick leaves and with pruinose fruits bearing a short style, and has apparently little connection with *B. pruinosa* Franchet, as I remarked on p. 362.

Berberis Veitchii (p. 363). The quotation should read "*acuminata* J. H. Veitch, *Hortus Veitchii*, 391 (non Franchet) (1906)." To the description add: Flores extus rubescentes; ovula pleraque 3–4. Fructus late elliptici, nigri sed pruinosi, ad 9 mm. longi et 7 mm. crassi, stigmatibus sessilibus coronati. This species may be most closely related to *B. Gagnepainii* Schneider, from which it is easily distinguished by its bronzy yellow flowers, its reddish and smooth young branchlets and by its different habit, forming a loose shrub with gracefully arching branches, while *B. Gagnepainii* forms a densely branched shrub.

Berberis Griffithiana (p. 364). Remove from enumeration of specimens Wilson's No. 535 in Herb. Hofmuseum, Vienna, which is the same as the specimen in Herb. New York Bot. Garden and belongs to *B. Julianae* Schneider, *B. Griffithiana* being entirely absent from China.

Berberis Delavayi (p. 364). Wilson's No. 2879 which I doubtfully referred to this species represents a new species which may be described as follows:

BERBERIS SILVICOLA Schneider, n. sp.

Frutex semiprostratus, 0.3–0.6 m. altus; ramuli juniores ignoti, vetustiores cinerei, striato-angulati v. teretes; internodia 3.5–5 cm. longa; spinae valde reductae v. trifidae, flavae, graciles, mediae ad 1 cm. longae, subteretes. Folia sempervirentia, tenuiter coriacea, 3–5 fasciculata, elliptica v. elliptico-oblonga, basi in petiolum vix ad 3 mm. longum satis subito attenuata, apice breviter acutata, mucronulata, margine distincte et satis anguste graciliter spinoso-serrata dentibus subpatentibus saepe inaequilongis 0.5–1.5 mm. longis 4–5 pro 1 cm., 2–4.5 cm. longa, 1–1.8 cm. lata, utrinque fere concoloria, viridia, subnitentia (?), superne fere enervata, subtus subdistincte laxe reticulata. Flores 1–5 fasciculata, lutei (an extus pubescentes?), circiter 1 cm. diam. pedicellis 5–7 mm. longis gracilibus rubescentibus, bracteis obtusis rubescentibus; prophylla flori adjuncta parva, lanceolata, subacuta; sepala exteriora elliptico-oblonga, interioribus obovatis apice rotundis basi paulo attenuatis circiter 7 mm. longis vix breviora; petala circiter 5 mm. longa, obovata, apice incisa, basi subunguiculata, glandulis 2 distinctis oblongis; stamina normalia connectivo obtuso vix producto, ovariis oblongo-ellipticis subaequilonga; stylus ut videtur brevis; ovula 2, subsessilia. Fructus ignotus.

Western Hupeh: Hsing-shan Hsien, woods, alt. 1900–2400 m., May 31, 1907 (No. 2879, type; semiprostrate shrub, 0.3–0.6 m. tall, flowers yellow).

In the texture and serrature of the leaves this species resembles *B. arguta* Schneider, which has larger leaves and an even closer serration (see p. 366). Of both species the ripe fruits are unknown.

Berberis Francisci-Ferdinandi (p. 368). Add to the description: Fructus . . . magni, stylo brevissimo coronati.

After this species may be inserted:

BERBERIS BEANIANA Schneider, n. sp.

Berberis Veitchii Hort. Veitch, non Schneider.

Frutex; ramuli annotini flavo-brunnescentes v. subrubescentes, vix v. obtuse angulati, laeves, glabri, vetustiores cinerascentes; internodia 2–3 cm. longa; spinae 3-fidae, graciles, flavescentes, subtus vix sulcatae, mediae ad 2 cm. longae. Folia ad 6 fasciculata, inaequalia, crasse chartacea, elliptico-lanceolata, apice acuta, mucronulata, basi in petiolum subnullum sensim attenuata, minimis exceptis 2–4 cm. longa et 0.8–1.4 cm. lata, superne intense viridia (an subnitentia?), distinctius quam in pagina inferiore discolori albescente pruinosa laxe reticulata, margine utrinque dentibus 5–12 plus minusve divaricatis 0.5–3 mm. longis graciliter spinoso-serrata. Inflorescentiae fasciculato-cymoso-paniculatae, 10–20-florae, ad 4 cm. longae; flores ut videtur flavi, circiter 6 mm. diam.; pedicelli 6–15 mm. longi, bracteis parvis circiter 1.5 mm. longis ovato-lanceolatis acutis margine saepe denticulatis glabris suffulti; prophylla non visa, sepala externa parva, ovata, subacutata, media majora subrotunda, interna maxima circiter 5 mm. longa, fere orbiculata, concava; petala circiter 4 mm. longa, obovata, apice incisa, basi contracta, glandulis 2 distinctis praedita; stamina petalis breviora, normalia, apice obtusa; ovaria oblongo-elliptica, estylosa, ovultis 3–4 fere sessilibus instructa. Fructus elliptici v. obovato-elliptici ut videtur purpurei et leviter pruinosi, 9–10 mm. longi, estylosi; semina 2 (ovulis 1–2 rudimentariis additis), laevia, dilute purpureo-rubescentia.

Western Szech'uan: without precise locality, 1904 (Veitch Exped. Seed No. 1930; plants cultivated in Hort. Kew; flowers June 18, 1914; fruits October 9, 1914).

This is a very distinct species of which the taxonomic position is yet unknown. The mature leaves of vigorous shoots somewhat resemble those of *B. Veitchii* Schneider, but this is an evergreen and otherwise totally different species. Unfortunately there is no record of the locality where the seeds were collected by E. H. Wilson. At Mr. Wilson's request I have with pleasure named this Berberis in compliment to the well-known dendrologist, Mr. W. J. Bean, assistant curator of the Royal Gardens, Kew.

Berberis Caroli, var. **hoanghensis** (p. 368). This variety has to be referred as a synonym to *B. Vernae* Schneider.

Berberis Wilsonae (p. 368). In the citation from *Bot. Mag.* read "t. 8414."

440 CORRECTIONS AND ADDITIONS TO VOLUME I

Berberis subcaulialata (p. 369). Add as a synonym: *Berberis Coryi* **Veitch,** *New Hardy Pl. China,* 1913, 7.

Berberis Vernae (p. 372). This species belongs to sect. *Integerrimae* **Schneider** and as already said *B. Caroli,* var. *hoanghoensis* Schneider is a synonym of it.

Berberis (p. 374). After *B. Lecomtei* may be added at the end of sect. *Sinenses* the following new and still imperfectly known species:

BERBERIS VIRGETORUM Schneider, n. spec.

Frutex ad 1.8 m. altus, glaber; ramuli juniores ignoti, vetustiores cinerei, paulo striato-angulati; spinae simplices, ad 2.5 cm. longae, rubro-brunneae, subtus sulcatae. Folia ad 6 fasciculata, inaequalia, membranacea, oblongo-rhomboidea, utrinque sensim acutata, basi in petiolum ad 2 cm. longum producta, minimis exceptis petiolo incluso 3.5–10 cm. longa et 1–2.8 cm. lata, supra in medio laminae lata, superne intense viridia, subtus pallida, pruinosa, haud papillosa, nervis paucis lateralibus plus minusve elevatis nervillis invisibilibus, margine integra, interdum paulo undulata. Inflorescentiae racemosae v. subumbellatae, 3–10-florae, 2–2.5 cm. longae. Flores ignoti. Fructus immaturi oblongo-elliptici, circiter 8 mm. longi et 3 mm. crassi, stigmate sessili; pedicelli 5–8 mm. longi, bracteis parvis lanceolatis acuminatis 1–1.5 mm. longis suffulti; semen immaturum 1 (an ovula interdum 2), sessile.

Kiangsi: Kuling, thickets, alt. 1400 m., July 29, 1907 (No. 1517, type; shrub 1.2 –1.8 m. tall).

This species is well marked by its rather large leaves and short racemes. Unfortunately there are no young branchlets, flowers or ripe fruits. It may be most closely related to *B. Lecomtei* Schneider, but in its leaves it resembles somewhat *B. Tschonoskyana* Regel from Japan. It differs from both in the shape of the leaves, which are much smaller and obtuse in *B. Lecomtei* and are not distinctly acute at the apex in *B. Tschonoskyana,* which has longer pedicels and inflorescences.

Berberis brachypoda (p. 375). This species has to be regarded as the type of the following new section:

Sect. BRACHYPODAE Schneider, n. sect.

Berberis, sect. *Vulgares,* subsect. *Dasystachyae* Schneider in *Mitt. Deutsch. Dendr. Ges.* XIV. 1905, 115 (1906), exclud. *B. dasystachya.*

Folia et plerumque inflorescentiae ramulique pubescentes. Inflorescentiae spiciformes, densiflorae.

To this section also belongs *B. Gilgiana* Fedde (p. 375), of which *B. pubescens* Pampanini in *Nuov. Giorn. Bot. Ital.* n. s. XVII. 273 (1910), is a synonym.

Berberis (p. 375). After *B. amurensis* the following three species have to be inserted which were not enumerated in 1913 for lack of sufficient material at that time:

BERBERIS HENRYANA Schneider in *Bull. Herb. Boiss.* sér. 2, V. 664 (1905); l. c. VIII. 261 (1908), exclud. specim. Giraldii Nos. 2303, 2310 et Wilsonii No. 3151.

Ad descriptionem addenda v. emendanda: Frutex ad 2 metralis v. ultra; ramuli annotini flavo- v. atrobrunnei, vix purpurascentes, paulo sulcato-angulati, glabri; spinae flavescentes, subtus sulcatae, 1–3-fidae, mediae ad 3 cm. longae. Folia ramulorum sterilium surculorumque satis variabilia, matura crasse membranacea v. fere chartacea, elliptica, elliptico-oblonga, late obovato-oblonga v. interdum ovato-elliptica, apice pleraque satis obtusa v. subrotunda, basim versus satis subito in petiolum 3–10 (–13) mm. longum contracta v. attenuata, minora 1.5–3 cm. longa, 0.8–2.8 cm. lata, majora ad 5 : 2.5 cm. et maxima longe petiolata ad 6 : 3 cm. magna (petiolo excluso), supra intense viridia, subtus discoloria, cinerascentia, pruinosa, utraque pagina satis laxe subdistincte elevato-reticulata, margine integra v. paucidentata (minora) v. (majora) satis dense graciliter spinuloso-serrata serraturis divaricatis 0.5–1 mm. longis, ramulorum sterilium praecedentibus similia v. subtus subviridescentia et saepe densius breviter serrulata. Inflorescentiae breviter v. longius racemosae, basi haud paniculatae, circiter 10–20-florae, 2.5–6 cm. longae (pedunculo nudo interdum ad 2 cm. longo incluso), pendentes; flores flavi, 9–10 mm. diametientes; pedicelli 5–9 mm. longi, basi bracteis lanceolatis acuminatis 1–1.5 mm. longis suffulti; prophylla bracteis similia sed latiora; sepala externa elliptica v. ovato-elliptica, internis obovatis 5–7 mm. longis plus minusve v. fere duplo breviora; petala late ovato-oblonga, circiter 5–6 mm. longa, apice incisa, basi contracta, glandulis 2 ellipticis separatis aurantiacis instructa; stamina normalia, apice vix v. haud producta; ovarium elliptico-oblongum, versus apicem leviter attenuatum, ovulis 2 sessilibus oblongis instructum. Fructus (in vivo) elliptici, circiter 9 mm. longi et 6 mm. crassi, obscure sanguinei, praesertim basi leviter pruinosi, stigmate subsessili coronati; semina 2, flavo-brunnea, laevia.

Western Hupeh: Hsing-shan Hsien, summit of Wan-tiao-shan, alt. 2400 m., June 5, 1907 (No. 2864; shrub 1.5–2.4 m. tall, flowers yellow); Changyang Hsien, mountains, May 1900 (Veitch Exped. No. 645; bush 1.2 m. high, flowers yellow); without precise locality, A. Henry (Nos. 5470, 5470ᴬ, 5470ᴮ, type, in Herb. Hofmus., Vienna). Eastern Szech'uan: Wushan Hsien, 1907 (No. 391; bush 1.5 m. tall, fruits red; seeds only; young plants in this Arboretum from which the above description is partly drawn); without precise locality, A. Henry (No. 5470ᴰ).

This Barberry represents apparently B. vulgaris in Hupeh and eastern Szech'uan, but it is very different from the European species and its nearest relatives especially in its brownish, sometimes almost purplish branches which are yellowish gray in B. vulgaris Linnaeus and B. amurensis Ruprecht. On the other hand the plant of Hupeh is closely related to the following species, but living plants of both look rather distinct. See also my remarks under B. Dielsiana Fedde.

Berberis Dielsiana Fedde in Bot. Jahrb. XXXVI. Beibl. 82, 41 (1905). — Schneider in Bull. Herb. Boiss. sér. 2, VIII. 261 (1908), quoad specim. Giraldii No. 2295.

Berberis Henryana Schneider, l. c. (1908), quoad specim. Giraldiana.

A B. Henryana praecipue recedit: foliis ramulorum fertilium anguste ellipticis v. elliptico-lanceolatis utrinque acutioribus 3 : 0.8 cm. ad 7 : 2.8–3 cm. magnis, petiolis longioribus instructis subtus plerisque distinctius viridescentibus, inflorescentiis majoribus saepe infra paniculatis ad 7 cm. longis, floribus in vivo luteis, pedicellis infimis ad 13 mm. longis, fructibus 9–10 mm. longis ellipticis, sed basi paulo angustatis splendidius coloratis.

Western Szech'uan: Wa-shan, thickets, alt. 1600–1800 m., May 1908 (No. 2863; shrub 1.2–3 m. tall, flowers yellow, foliage often bronzy). Shensi: "Sciu jan shan, am Kan y huo, blühend am 15. Mai," G. Giraldi (No. 2298, type, ex Fedde); south of Yenan Fu, 1910, W. Purdom (No. 341); same locality, alt. 1300 m., February 8, 1911, W. Purdom (Seed No. 549; bush 1.8 m. high, red fruits; plants in this Arboretum); at Chempu, February 1911, W. Purdom (Seed No. 543; plants in this Arboretum); at Chung-pu, February 8, 1911, W. Purdom (Seed No. 605ᵃ; plants in this Arboretum).

Not having seen the type of Fedde's species I am not entirely convinced that

Wilson's and Purdom's plants belong to the same species, but Fedde's description agrees well enough with the specimens before me. It was drawn from a flowering plant only with probably rather young flowers because the author says: " flores circiter 3 mm. diamet." The living plants in this Arboretum look somewhat different from those of the preceding species, forming a more spreading, loosely branched shrub. The leaves are mostly narrowly elliptic and acute, but those of the vigorous young shoots are much more obtuse and broader, and more distinctly whitish beneath. The inflorescences are variable, being shorter and almost fasciculate-racemose at the top of the branches and longer and more or less paniculate at the base. The fruits have a very short, sometimes very indistinct style. It needs further investigation to determine if B. Dielsiana represents a good species or only a variety of B. Henryana. Wilson's plant from western Szech'uan when fully known may even represent another distinct variety.

BERBERIS FEDDEANA Schneider in Bull. Herb. Boiss. sér. 2, V. 665 (1905).

?Berberis vulgaris Fedde in Bot. Jahrb. XXIX. 341 (1900).
?Berberis Henryana Schneider, l. c. VIII. 262 (1908), quoad specim. Wilsonii No. 3151.

Western Szech'uan: northeast of Tachien-lu, thickets, alt. 2800–3100 m., July 9, 1908 (No. 2862; bush 1.8–2.4 m. tall; flowers yellow).
Wilson's specimen consists of old branches with very straight, apparently upright, 4–13 cm. long inflorescences which in the somewhat verticillate arrangement of the rather small flowers are similar to those of B. dasystachya Maximowicz. The leaves are almost entire or very indistinctly serrate. The type, A. von Rosthorn's No. 2044, was collected in Nanchuan, southeastern Szech'uan, and has the same kind of inflorescences the flowers of which had fallen, and the fruits are unknown. The leaves are similar in shape but distinctly serrulate. I described the racemes as pendent, but probably they are upright. The flowers of Wilson's specimen are about 5–7 mm. wide and similar to those of B. dasystachya, and I suppose that B. Feddeana may rather belong to the following than to the section Vulgares.
Here may be added the following species representing in my opinion a distinct section:

Sect. DASYSTACHYAE Schneider, sect. nov.

Berberis, sect. Vulgares, subsect. Dasystachyae Schneider in Mitt. Deutsch. Dendr. Ges. XIV. 1905, 118 (1906), sensu stricto.

Plantae glaberrimae. Inflorescentiae iis sectionis Brachypodae similes, sed erectae, magis laxiflorae, floribus minoribus, pedicellis longioribus.

BERBERIS DASYSTACHYA Maximowicz, in Bull. Acad. Sci. St. Pétersbourg, XXIII. 308 (1877); in Mél. Biol. IX. 711 (1877); Fl. Tangut. 30, t. 7, fig. 1–7 (1889). — Hemsley in Jour. Linn. Soc. XXXIII. 3 (1886). — Schneider in Bull. Herb. Boiss. sér. 2, V. 664 (1905); l. c. VIII. 262 (1908). — Fedde in Bot. Jahrb. XXXVI. Beibl. 82, 43 (1905).

?Berberis dolichobotrys Fedde, l. c. 41 (1905).

Western Hupeh: Hsing-shan Hsien, woods, alt. 2100 m., June and September 1907 (No. 307; bush 0.9–1.5 m. tall, flowers yellow, fruits coral red); without exact locality, A. Henry (No. 6816). Shensi: Tai-pei-shan, 1910, W. Purdom (Nos. 1, 5 and 9).
The type of B. dolichobotrys was collected on the Tai-pei-shan, and I cannot detect in Fedde's description any character sufficient to separate his species except

the long (10–12 cm.) inflorescences which are said to be " infra subpaniculatae." Such long inflorescences I know only in *B. Feddeana* Schneider. Not having before me Fedde's type I cannot decide the question of its identity with any of the species known to me. Purdom's specimens are not quite identical with the typical *B. dasystachya*, especially the flowers of No. 5 are somewhat larger. The typical form is in cultivation in this Arboretum; the plants came from Vilmorin and the exact region where the seeds were collected is not known.

Berberis Prattii (p. 376). This species is connected with *B. aggregata* Schneider by many intermediate forms according to the living plants in this Arboretum. I therefore reduce it to the following variety:

BERBERIS AGGREGATA, var. PRATTII Schneider, n. var.

B. brevipaniculata Bean, *Trees & Shrubs Gr. Brit.* I. 236 (non Schneider) (1914).
B. Geraldii Veitch, *New Hardy Pl. China (Cat.)*, 1913, 7 (nomen nudum).

Bean's plant is entirely different from *B. brevipaniculata* Schneider which is a Hupeh plant and not yet in cultivation. The cultivated form was introduced by Wilson in 1904 from near Tachien-lu, and it is growing in this Arboretum under No. 1320 from Hort. Veitch. It is one of those intermediate forms, and very similar to the plant figured in *Bot. Mag.* CXL. t. 8549 (1914). Bean mentions, and it is also mentioned by Sprague in the *Botanical Magazine*, that the leaves of *B. Prattii* are pale green beneath and not glaucous, as in those of what he calls *B. brevipaniculata*. But according to my observations in this Arboretum these differences in the color of the leaves are apparently not constant and are due to local conditions or the individual behavior of the plants.

There is also the following variety:

BERBERIS AGGREGATA, var. RECURVATA Schneider, n. comb.
Berberis Prattii, var. *recurvata* Schneider (p. 377).
As the cultivated plants in this Arboretum have not flowered, it is impossible to decide whether this is a good variety or a mere form.

Berberis (p. 377). After *B. Liechtensteinii* insert the following numbers:

BERBERIS spec.
Western Szech'uan: southeast of Tachien-lu, alt. 2500 m., 1910 (No. **4130**; shrub 1–1.5 m., flowers yellow, fruits black; seeds only).

BERBERIS spec.
Western Szech'uan: west and near Wên-ch'uan Hsien, thickets, alt. 2300–2800 m., 1910 (No. **4153**; bush 2–2.5 m. tall, fruits scarlet in cymose racemes).

BERBERIS spec.
Western Szech'uan: Sungpan, alt. 3000 m., 1910 (No. **4203**; bush 3–4 m. tall, leaves b oadly obovate, fruits black in cymose racemes).

This and the preceding number are not available at present for determination, as they have been loaned to a European collection.   C. S.

Mahonia

Mahonia Zemanii (p. 378, 382). This name becomes a synonym of the older

MAHONIA CONFUSA Sprague in *Kew Bull. Misc. Inf.* 1912, 339; 1914, 232. — T[urrill] in *Kew Bull. Misc. Inform.* 1915, 128.

Mahonia Fortunei Fedde in *Bot. Jahrb.* XXXI. 130, fig. 3 E (pro parte, non Mouillefert) (1901).

When I described *M. Zemanii* I overlooked Sprague's species, of which Henry's No. 3117 is a type. This number had been referred to *M. Fortunei* (Lindley) Mouillefert by Fedde, whom I followed, not having seen that specimen.

Mahonia decipiens (p. 379). I now regard this name as a mere synonym of *M. Sheridaniana* Schneider. The ovary apparently has no style, and there is a pair of very small leaflets present at the base of the petiole on another specimen of Wilson's in the Herbarium of this Arboretum.

Mahonia Fortunei (p. 380). The author of the combination is Mouillefert, *Traité Arb. Arbriss.* I. 165 (1892).

Mahonia japonica (p. 382). The citation of the second synonym should read: *Berberis japonica* Sprengel, *Syst.* II. 119 (1825). R. Brown in Tuckey, *Narr. Exp. Congo*, App. V. 441, in adnot. (reprint p. 22) (1818 [not 1816]) only mentions that *Ilex japonica* Thunberg has to be referred to Berberis, but he does not make the combination. See also Nees v. Esenbeck, *R. Brown's Verm. Bot. Schrift.* I. 230 (1825).

Mahonia nepalensis (p. 382). Read "MAHONIA NAPAULENSIS."

Mahonia Veitchiorum (p. 383). This species has to be referred as a synonym to *Mahonia polyodonta* Fedde. The distinguishing characters given by me in the key on p. 381 have proved unreliable.

Mahonia Sheridaniana (p. 384). See note above under *M. decipiens*.

C. S.

Liriodendron chinense (p. 410). Add the following number:
Western Hupeh: Ichang Fu, rare, alt. 1000–1500 m., 1907 (No. **701**; tree 8–20 m. tall; seeds only).

Corylopsis Willmottiae (p. 425). This species flowered profusely in the spring of 1916 in this Arboretum, which enables us to add the description of the flowers:
Flores pallide flavi in racemis densis multifloris cum pedunculo villoso v. fere glabro 5–7 cm. longis, rhachi villosa; bracteae suborbiculares v. late ovatae, rotundatae, margine villosae, extus glabrae, intus plus minusve villosae, basales magnae, florales minores circiter 5 mm. latae et 4 mm. longae, interdum margine irregulariter et sparse erosae; bracteolae obovatae v. ovales, circiter 3 mm. longae; sepala late ovata, 1–1.5 mm. longa, glabra, tubum subaequantia, petale irregulariter orbicularia, e basi truncata in unguiculum brevem contracta, lamina plerumque paulo latiora quam longa, 4–5 mm. diam.; nectaria sepalis paullo breviora v. subaequilonga, plerumque ultra medium bifida, apice truncata v. irregulariter lobulati; stamina petalis paulo breviora, circiter 3.5 mm. longa, filamentis basin versus dilatatis; styli erecti petala fere subaequantes.

Corylopsis platypetala, var. levis (p. 427). This variety has also flowered in the spring of 1916 and we add the following description of the flowers:
Flores pallide flavi, in racemis cum pedunculo 3–5 cm. longis, rhachi villosa; bracteae basales obovatae v. obovato-oblongae, acuminulati, extus glabrae, intus sericeo-villosae, florales oblongo-ovatae concavae, circiter 5 longae et 2.5 latae, extus sparse villosulae et minute glandulosae, intus villosae; bracteolae oblongae; sepala ovata v. triangulari-ovata, interdum acutiuscula, 1–1.5 mm. longa, glabra; petala suborbicularia, e basi truncata v. late cuneata in unguiculum brevem attenuata, lamina paulo latiora quam longa v. aeque ac longa quam lata, 3–4 mm. diam.; nectaria apice tantum biloba et irregulariter lobulata, interdum bifida,

sepalis paulo breviora v. subaequilonga; stamina petalis fere triente breviora, circiter 2.5 mm. longa, filamentis basin versus manifeste dilatatis; styli recti, staminibus paulo breviores.

Hamamelis mollis (p. 431). Add the following number:
Western Hupeh: Changyang Hsien, woods, alt. 1300–2000 m., 1907 (No. **725**; bush 2–7 m. tall; plants only).

Exochorda racemosa, var. Wilsonii (p. 456). This variety has been referred to *E. Giraldii* Hesse (*E. racemosa*, var. *Giraldii* Rehder) which we now consider a distinct species. The variety becomes:

ExocHorda Giraldii Hesse, var. Wilsonii Rehder in Bailey, *Stand. Cycl. Hort.* II. 1194 (1914); in *Mitt. Deutsch. Dendr. Ges.* XXIII. 258 (1914).

Aesculus Wilsonii (p. 498). Add the following number:
Western Hupeh: north and south of Ichang, woods, rare, alt. 1200–2000 m., 1907 (No. **702**, tree 10–20 ft. tall; seeds only).

Rhododendron Wasonii (p. 532). In the enumeration of specimens for "3965" read "3956."

Rhododendron (p. 548). Three lines from below read "*R. quinquefolium*" instead of "*R. quinqueloculare.*"

Rhododendron (p. 549). Insert at the end of the genus:
Rododendron spec.
Western Szech'uan: Wa-shan, thickets, 1908 (No. [**1384**; shrub 3 m. tall, flowers red, fruit russet; seeds only).

Enkianthus deflexus (p. 550). Add the following number:
Western Szech'uan: Mupin, alt. 3300 m., October 1910 (No. **4336**; bush 3–6 m. tall).

Cassiope selaginoides (p. 551). In the enumeration of specimens for "Veitch Exped. No. **3912**" read "Veitch Exped. No. 3910."

Brandisia glabrescens (p. 574). For No. 9176ª read 9716ª.

Paulownia Fortunei (p. 578). Add as a synonym:
Paulownia Mikado Ito, *Icon. Pl. Jap.* I. t. 9–12 (1912).

Ampelopsis micans (p. 579). Add Wilson's No. **2724** enumerated under *Ampelopsis heterophylla*, var. *amurensis* on p. 100 and strike out No. **129**, which is *Cornus controversa* Hemsley.

Ampelopsis micans, var. cinerea (p. 579). Strike out No. **168**, which is *Sargentodoxa cuneata* Rehder & Wilson.

VOLUME II

Picea asperata (p. 22). Add the following number:
Western Szech'uan: west of Kuan Hsien, Pan-lan-shan, alt. 2600–1200 m., 1910 (No. **4399**; tree 18–24 m. tall, girth 2 m., leaves square, shoots shining; plants only).

Picea Watsoniana (p. 26).   Add the following number:
Western Hupeh: Fang Hsien, alt. 2300–3000 m., 1910 (No. 4452; tree 10–24 m. tall, girth 0.6–2 m.; plants only).

Tsuga chinensis (p. 37).   Add the following number:
Western Hupeh: Fang Hsien, forests, alt. 2300–3000 m., 1910 (No. 4453; tree 14–28 m. tall, girth 1.5–4 m.; plant only).

Abies Delavayi (p. 41).   Add the following number:
Western Szech'uan: Mupin, forming forests above 2300 m. alt., 1910 (No. 4396; tree 20–40 m. tall; plants only).

Abies Faxoniana (p. 42).   Add the following number:
Western Szech'uan: west of Kuan Hsien, Pan-lan-shan, forming forests, alt. 2600–3300 m., 1910 (No. 4400; tree 26–32 m. tall, girth 2.3–4 m.; plants only).

Litsea (p. 78).   After *L. Wilsonii* insert:

LITSEA spec.
Western Hupeh: Fang Hsien, thickets, alt. 1000–1600 m., 1907 (No. 4446; bush 2–2.6 m. tall, flowers yellow; plants only).

Lindera strychnifolia, var. **Hemsleyana** (p. 82).   In the enumeration of specimens for "Veitch Exped. No. 4439" read "Veitch Exped. No. 4429."

Leguminosae (p. 87).   Insert before *Albizzia:*

ACACIA Willd.

ACACIA INTSIA Willdenow, var. OXYPHYLLA Baker in Hooker f., *Fl. Brit. Ind.* II. 297 (1878).
Western Szech'uan: near Wa-shan, valley of Tung River, arid regions, alt. 600–1300 m., November 1908 (No. 3527; large scandent bush, to 7 m. tall, flowers yellow); dry valleys, May 1904 (Veitch Exped. No. 3406).

Albizzia (p. 87).   After *A. julibrissin* insert:

ALBIZZIA KALKORA Prain in *Jour. As. Soc. Bengal*, LXVI. 511 (1897); *Novic. Ind.* 345 (1905). — Gagnepain in Lecomte, *Fl. Gén. Indo-Chine* II. 92 (1913).

Mimosa Kalkora Roxburgh, *Hort. Beng.* 40 (nomen nudum) (1814); *Fl. Ind.* ed. 2, II. 547 (1832), fide Prain.
Acacia macrophylla Bunge in *Mém. Sav. Étr. Acad. Sci. St. Pétersbourg*, II. 135 (*Enum. Pl. Chin. Bor.* 61) (1833).
Albizzia Lebbek Hemsley in *Jour. Linn. Soc.* XXIII. 216 (non Bentham) (1897), exclud. synon. pro parte. — Harms in *Bot. Jahrb.* XXIX. 408 (1900).

Western Hupeh: north and south of Ichang, woods, alt. 300–900 m., July 6, 1907 and September 1907 (No. 2033; tree 7–24 m. tall, girth 0.9–3 m., flat-topped, flowers white); same locality, May 1900 (Veitch Exped. No. 511; tree 7–10 m. tall, flowers yellowish); Patung Hsien, woods, alt. 900–1400 m., November 1907 (No. 792; tree 10–13 m. tall, girth 1.2–1.5 m., wood good); Chienshi Hsien, July 1900 (Veitch Exped. No. 1316; tree 8 m. tall). Kiangsi: common on plain and hills around Kiukiang, alt. 8000 m., August 2, 1907 (No. 1506; tree 8 m. tall). Kianghuai: Spirit Valley near Nanking, on open rocky hill slopes, also in open woods, June 4, 1915, F. N. Meyer (No. 1448; tree or tall shrub; wood used in turnery and for furniture). Shantung: near Boshan, September 1907, F. N. Meyer (No. 257;

same as Seed No. 768ª; see also *Bull. U. S. Dep. Agr. Bur. Pl. Industr.* No. 127, 29 [1909] under No. 21969; this number has been distributed as *Albizzia chinensis* Meyer or sp. cf. *stipulacea* [1]); Chifu, garden, September 22, 1903, *C. S. Sargent.*

The specimens before me agree well with the descriptions given by Prain and Gagnepain. *A. lebbek* (Linnaeus) Bentham seems to be entirely absent from central and northern China.   C. S.

Cercis chinensis (p. 87). Add the following number:

Western Hupeh: Ichang, alt. 1000 m., 1907 (No. **730**; tree 7–15 m. tall; plants only).

Bauhinia densiflora (p. 88). Add the following number:

Western Szech'uan: southeast of Tachien-lu, valley of Tung River, alt. 600–1300 m., September 1908 (No. **3376**; climber 3 m. tall).

Bauhinia yunnanensis (p. 89). For "Veitch Exped. No. 3378" read "Veitch Exped. No. 3401."

Bauhinia (p. 90). Insert after *B. hupehana,* var. *grandis:*

BAUHINIA spec.

Western Szech'uan: valley of Min River and regions around Mao-chou, alt. 1300–2000 m., 1908 (No. **3375**; bush 0.3–2 m., flowers white).

Gleditsia sinensis (p. 91). Add the following numbers:

Western Szech'uan: near town of Mupin, alt. 1300 m., November 1908 (No. **1364**; tree 10–26 m. tall, girth 1.25–4 m.); southeast of Tachien-lu, valley of Tung River, alt. 1300–1500 m., November 1908 (No. **1364**ª; tree 16 m. tall, girth 2.6 m.).

These specimens differ from the others in their larger fruits which are 25–30 cm. long and 2.5–3 cm. wide, while those of the other numbers are about 15 cm. long and 2 cm. wide.

Sophora (p. 96). After *S. flavescens* insert:

SOPHORA GLAUCA Leschenault, var. ALBESCENS Rehder, var. nov.

A typo recedit foliolis obtusis v. leviter emarginatis rarius acutiusculis et mucronulatis, floribus albidis.

Western Szech'uan: near Wa-shan, side of streams, alt. 600–2000 m., June and October 1908 (No. **1179**, type; bush 6 ft. tall, flowers white); southeast of Tachien-lu, stony places, alt. 1300 m., May 1908 (No. **2562**; bush 1–1.3 m. tall, flowers white); without precise locality, alt. 2000 m., May 1904 (Veitch Exped. No. 3389; bush 1.6 m. tall); without precise locality, 1300–1600 m., July 1903 (Veitch Exped. No. 3391; decumbent shrub, 0.6 m. tall).

Wilson's No. 3391 differs in the acutish mucronulate leaflets and in the less dense pubescence of the whole plant. Whether Henry's No. 10720 from Yunnan and the specimens from the same province enumerated by Franchet (*Pl. Delavay.* 188 [1889]) belong to this variety or to the typical species, I am unable to state, as the color of the flowers is unknown to me.   A. R.

Piptanthus nepalensis (p. 99). The Chinese plant has been separated from typical *P. nepalensis* under the following name:

PIPTANTHUS CONCOLOR Harrow apud Craib in *Gard. Chron.*, ser. 3, LX. 289 (1916).

---

[1] This apparently refers to *A. stipulata* Boivin, a tropical species absent from China.

Indigofera (p. 100). After *I. pseudotinctoria* insert:
INDIGOFERA SOULIEI Craib in *Not. Bot. Gard. Edinburgh*, VIII. 61 (1913).
Western Szech'uan: northeast of Tachien-lu, roadsides, alt. 1600 m., October 1910 (No. 4332; bush 1–2 m. tall; flowers rosy-pink).

Indigofera (p. 100). After *I. myosurus* insert:
INDIGOFERA POTANINII Craib in *Not. Bot. Gard. Edinburgh*, VIII. 60 (1913).
Kansu: without precise locality, *W. Purdom* (No. 539ᵃ; seeds only; plants raised have flowered and fruited in the Arnold Arboretum).

Caragana Boisii (p. 102). Add the following number:
Western Szech'uan: valley of Min River, Mao-chou to Sungpan, alt. 2000–3000 m., October 1910 (No. 4375; bush 2–3 m. tall, flowers bronzy yellow).

Desmodium laburnifolium (p. 103). Add the following number:
Kiangsi: Kuling, not common, alt. 1000 m., August 1, 1907 (No. 1627; bush 1–1.3 m., flowers purple).

Desmodium podocarpum, var. szechuenense (p. 104). Before this variety insert:
DESMODIUM PODOCARPUM De Candolle in *Ann. Sci. Nat.* IV. 102 (1825); *Prodr.* II. 336 (1825). — Baker in Hooker f., *Fl. Brit. Ind.* II. 165 (1876). — Hemsley in *Jour. Linn. Soc.* XXIII. 174 (1887).
Kiangsi: Kuling, thickets, common, alt. 1300 m., July 28, 1907 (No. 1623; 0.6–1 m., flowers red).

Lespedeza formosa (p. 107). Strike out Veitch Exped. No. 1304.

Lespedeza (p. 107). After *L. formosa* insert:
LESPEDEZA spec.
Kiangsi: Kuling, grassy places, alt. 1500 m., 1907 (No. 1619; sub-shrub 0.6 m. tall, flowers purple).

Lespedeza Friebeana (p. 111). This species had been previously described under the name:
LESPEDEZA MAXIMOWICZII Schneider, *Ill. Handb. Laubholzk.* II. 113 (1907).

Dalbergia hupeana (p. 115). Instead of No. 2094 read 2904.

Dalbergia (p. 116). After this genus insert:

VICIA L.

VICIA UNIJUGA A. Braun in *Ind. Sem. Hort. Berol.* 1853, 12. — Hemsley in *Jour. Linn. Soc.* XXIII. 186 (1887).
Western Hupeh: Hsing-shan Hsien, thickets, 600–1300 m., 1907 (No. 3486; bush 1–1.6 m., flowers purple). Western Szech'uan: Kuan Hsien, roadsides, alt. 1000–1300 m., 1908 (No. 4589; bush 0.5–0.6 m., flowers red-purple).

Rhynchosia (p. 119). After *R. volubilis* insert:
RHYNCHOSIA spec.
Western Szech'uan: near Tachien-lu, alt. 1300–1600 m., October 1910 (No. 4588; twining shrub, 1.6–2 m., flowers red).

The specimen has immature fruits only. As the pods often have 3 seeds, it may not belong to the genus at all, but otherwise it bears a strong resemblance to *R. Dielsii* in the size and shape of its leaflets and the character of its inflorescence except that it is appressed-pubescent, not villose.

Evodia hupehensis (p. 133). Change No. 2693ª to " Veitch Exped. No. 2693ª " and add the following number:
Western Hupeh: Hsing-shan Hsien, woods, alt. 1300 m., 1907 (No. 387, tree 7 m.).

Orixa japonica (p. 135). In the enumeration of specimens for No. 3887 read 3847.

Phellodendron chinense, var. **glabriusculum** (p. 137). In the enumeration of specimens for No. 4603 read 4003.

Citrus (p. 145). After *C. ichangensis* add the following numbers:
Citrus spec.
Western Hupeh: Ichang, cultivated, November 1907 (No. 4735); without locality (No. 4733).

Ailanthus cacodendron (p. 153). Instead of *Ailanthus cacodendron* Schinz & Thellung insert the following name and transfer *Ailanthus cacodendron* to the synonyms:
Ailanthus altissima Swingle in *Jour. Washington Acad. Sci.* VI. 495 (1916).

Toxicodendron Altissimum Miller, *Gard. Dict.* ed. VIII. (1768).
Rhus sinense Ellis apud Houttuyn, *Natuurl. Hist.* II. 2, 212 (1774).

Ailanthus cacodendron, var. **sutchuenense** (p. 153). This combination must become a synonym of the following:
Ailanthus altissima, var. sutchuenensis Rehder & Wilson, n. comb.

Polygala arillata (p. 160). Instead of " Veitch Exped. No. 3226ª " read " Veitch Exped. No. 3236ª."

Spondias axillaris, var. **pubinervis** (p. 173). For Veitch Exped. No. 3363 read "3365."

Pistacia chinensis (p. 173). Strike out "Veitch Exped. No. 1965."

Staphylea holocarpa (p. 185). On p. 186 instead of "*W. Purdom* (Nos. 31661, 316)" read " *W. Purdom* (Nos. 316, 316***)."

Sageretia Cavaleriei (p. 228). Add the following numbers to this species which becomes Sageretia Henryi Drummond & Sprague as stated on p. 623 of vol. II.
Szech'uan : without precise locality, *A. von Rosthorn* (Nos. 2141, 2142, 2143, distributed as *Maesa* spec.).

Rhamnus heterophyllus (p. 232). Add the following number:
Western Hupeh: Fang Hsien, thickets, alt. 1000 m., September 1907 (No. 2911, bush 0.5–1 m., flowers yellowish).

Fraxinus Griffithii (p. 258, 623). Strike out note on p. 623 and add the following specimen:
Western Hupeh: Changlo Hsien, alt. 600 m., May 1907 (No. 2775ª).

Fraxinus Paxiana (p. 259). Strike out "Veitch Exped. No. 4085" which belongs to *Fraxinus chinensis*.

Fraxinus chinensis (p. 260). Add the following number:
Western Hupeh: north and south of Ichang, alt. 300–1000 m., 1907 (No. 423; tree 7 m. tall; seeds only).

Fraxinus (p. 262). At the end of the genus insert:
FRAXINUS spec.
Western Hupeh: Changyang Hsien, alt. 1500 m., 1907 (No. 1392; tree 8 m tall, girth 0.6 m.; seeds only).

FRAXINUS spec.
Western Hupeh: Fang Hsien, woods, alt. 2000–2600 m., 1910 (No. 4431; tree 5–12 m. tall; plants only).

Pyrus serotina (p. 263). Add the following number:
Western Hupeh: north and south of Ichang, alt. 1000 m., 1907 (No. 395; tree 7 m. tall, flowers white, fruit globose; seeds only).

Pyrus Calleryana (p. 264). Instead of "No. 2775" read "No. 2975," and add the following number:
Western Hupeh: common around Ichang, alt. 1000 m., 1907 (No. 365; bush or small tree; seeds only).

Malus Sieboldii, var. **arborescens** (p. 294). For "Wilson No. 7428" read "Wilson No. 7423," and add as synonyms:

Pirus subcrataegifolia Léveillé in Fedde, *Rep. Spec. Nov.* VII. 199 (1909).
Crataegus Taquetii Léveillé, l. c. X. 377 (1912).

Chaenomeles japonica (p. 298). Add as a synonym:

Cydonia japonica, var. *alpina* Ito, *Icon. Pl. Jap.* I. t. 3 (1911).

Rosa caudata (p. 321). To this species probably belong the following synonyms:

Rosa setipoda Baker in Willmott, *Gen. Rosa*, I. fig. 55 (non Hemsley & Wilson) (1911), quoad figuram fructus tantum.
Rosa jaluana Baker in Willmott, *Gen. Rosa*, II. fig. 164, 499 (non Komarov) (1914), quoad fructus descriptionem figuramque et plantam Chinae occidentalis.

The figures of the fruits of *R. jaluana* and of *R. setipoda* represent apparently a form of *R. caudata* with naked or nearly naked pedicels and fruits. The immature fruits of *R. jaluana* are described by Komarov as sphaeroid or sphaeroid-pyriform which is an entirely different shape from that of the fruit figured and described by Baker. It is, moreover, very unlikely that a species of northeastern Asia should appear also in western China.

Rosa setipoda (p. 323). In the citation "Willmott, *Gen. Rosa*," strike out "fig. 55" and insert " (exclud. fig. 55)," also add the following synonym:

Rosa caudata Baker in Willmott, *Gen. Rosa*, II. fig. 163 (1914), quoad figuram fructus tantum.

The figures of *R. caudata* and *R. setipoda* have been probably interchanged by mistake; they are easily distinguished at the first glance by the shape of the sepals which are correctly described by Baker.

Pygeum (p. 344).   Add to this genus:

PYGEUM spec.

Western Szech'uan: valley of Tung River, east of Tachien-lu, thickets, alt. 1000–1200 m., November 1908 (No. 2542; tree 10 m. tall).

The specimen resembles *P. latifolium* Miquel, which we know only from the description, but for geographical reasons it is unlikely that it is identical with that species, which is a native of Java, as it does not seem to occur in the regions between Java and western China. The Chinese localities given for *P. latifolium* and cited on p. 344 are at least doubtful. Our specimen is too incomplete for definite determination, as it is in fruit only.

Thea (p. 394).   Insert after *T. Grijsii*:

THEA spec. v. gen. novum Theeacearum.

Western Szech'uan: Kiating Fu, roadside thickets, alt. 500 m., May 11, 1908 (No. 2586; bush 5 m. tall).

This is either an aberrant species of Thea or a new genus. It differs from Thea chiefly in the perfectly distinct styles and in the fruit thickly covered with warty excrescences; in its leaves it strongly resembles *Thea sinensis* Linnaeus. The specimens are unfortunately too incomplete for a sufficient description; they consist of leafy branches with immature fruit. The young fruits are very short-stalked, with a persistent calyx, 3-celled and crowned by 3 slender, nearly 1.5 cm. long, perfectly distinct styles.

Eurya japonica, var. **aurescens** (p. 399).   For " No. 22 " read " Veitch Exped. No. 22."

Hypericum

Hypericum (p. 402).   After *H. Ascyron* insert the following two new species:

HYPERICUM MACROSEPALUM Rehder, n. sp.

Herba semi-metralis, ut videtur adscendens, glabra; caules teretes. Folia papyracea, oblongo-ovata, obtusa, basi semiamplexicaulia, subcordata, 2–7 cm. longa et 1.5–3 cm. lata, punctulis striisque pallucidis instructa, subtus glaucescentia. Inflorescentia 3–5-flora; flores longe pedunculati pedunculo 2–4 cm. longo, circiter 5 cm. diam.; sepala foliacea, valde inaequalia, oblonga v. ovato-oblonga, acutiuscula, majora circiter 2.5 cm., minora vix 1.5 cm. longa, 5-nervata, ut folia pellucido- punctata et striata; petala lutea, obliqua, oblonga, sepalis majoribus breviora, circiter 2 cm. longa; stamina pentadelpha, circiter 15 in fasciculo quoque; ovarium conicum, 5-loculatum, apice uniloculatum; styli 5, liberi, leviter patentes, circiter 5 mm. longi ovario breviores, stigmate capitato. Fructus ignotus.

Western Szech'uan: Tai-pei-shan, northeast of Tachien-lu, roadsides, alt. 3300 m., July 5, 1908 (No. 2426).

This new species seems to be most nearly related to *H. Ascyron* Linnaeus and particularly its var. *punctato-striatum* Keller which is readily distinguished by the less unequal sepals which are several times shorter than the petals. *H. macrosepalum* also bears a strong general resemblance to *H. Androsaemum* Linnaeus, but that species has only 3 styles, much smaller flowers and a berry-like fruit, while the fruit of *H. macrosepalum*, which is not known, is apparently capsular as far as one can judge from the appearance of the ovary.

HYPERICUM URALOIDES Rehder, n. sp.

Suffrutex semi-metralis, ramosus, ut videtur erectus, glaber; caules bilineati et superne plus minusve quadrangulati et leviter bialati. Folia membranacea, punctulis striisque pellucidis instructa, caulinia rhombico-ovata, obtusa et mucronulata, basi in petiolum brevissimum attenuata, circiter 2 cm. longa et 1 cm. lata, ea ramulorum lateralium oblonga v. rhombico-oblonga, 1–1.5 cm. longa et 0.4–0.6 cm. lata, acutiuscula, basi attenuata, omnia margine cartilaginea, recurva et plus minusve crispula, subtus glaucescentia, nervo medio prominente. Inflorescentia plerumque 3-flora, rarius floribus 1–2 tantum: flores breviter pedicellati pedicellis 3–5 mm. longis bibracteolatis bracteis subulatis; sepala ovalia v. anguste ovalia, obtusiuscula, 4–5 mm. longa et 1.75–2.5 lata; petala rhombico-obovata, circiter 1 cm. longa et 0.6–0.8 cm. lata; stamina pentadelpha, circiter 35 in quoque fasciculo; styli 5, liberi, suberecti, apice tantum recurvi, 2 mm. longi, ovario duplo breviores, stigmate capitato et papilloso. Capsula non visa.

Western Szech'uan: banks of Min River, June 1903 (Veitch Exped. No. 3258; sub-shrub, 0.5 m.).

This new species seems to be most closely related to *H. patulum* Thunberg and particularly to its var. *uralum* Koehne, which differs chiefly in its more shrubby habit, in the only slightly two-edged, never four-angled branches, larger and broader leaves, much larger flowers with suborbicular or broadly ovate rounded sepals.

Hypericum (p. 404). After *H. longistylum* insert:

HYPERICUM PERFORATUM Linnaeus, *Spec.* 785 (1753). — De Candolle, *Prodr.* I. 549 (1824). — Reichenbach, *Icon. Fl. Germ. Helv.* VI. 68, t. 343, fig. 5177 (1844). — Maximowicz in *Bull. Acad. Sci. St. Pétersbourg*, XXVII. 432 (1882); in *Mél. Biol.* XI. 166 (1881). — Hemsley in *Jour. Linn. Soc.* XXIII. 74 (1886). — Diels in *Bot. Jahrb.* XXIX. 476 (1900). — R. Keller in *Bot. Jahrb.* XXXIII. 554 (1904). — Pampanini in *Nuov. Giorn. Bot. Ital.* XVII. 672 (1910).

Western Szech'uan: Wa-shan, grasslands, alt. 2300–2600 m., 1908 (No 2425; herb 1 m. tall). Western Hupeh: Ichang, *A. Henry* (No. 1569).

Hypericum (p. 405). At the end of the genus insert:

HYPERICUM spec.

Eastern Szech'uan: Wushan Hsien, cliffs and thickets, alt. 1000 m., 1907 (No. 256; shrubby, 0.6 m. tall, flowers yellow; seeds only). A. R.

Melastomataceae (p. 422). At the end of the family insert:

PLAGIOPETALUM Rehder, n. gen.

Flores tetrameri; calycis tubus campanulatus, 4-angulatus, 8-nervius, limbus 4-dentatus dentibus subulatis brevibus, saepe sparsissime serrulatis, inaequalibus persistentibus; petala 4, ovalia v. obovata, inaequilateralia, apice fere truncata et uno latere acuminulata altero rotundata; stamina 8, interiora exterioribus paulo breviora, ceterum vix diversa petala paulo superantia, antheris lineari-lanceolatis leviter curvatis basi bilobis apice uniporosis, connectivo basi postice tuberculis 2 instructo antice non producto; ovarium 4-loculare, loculis multi-ovulatis, totum inferum, calycis tubo duplo longiori adnatum, vertice exsculptum et margine lobis 4 apice laceratis basi cum tubo calycis trabeculis conjunctis; stylus rectus filiformis, stigmate punctiformi. Capsula campanulata, 4-gona, vertice late exsculpta, calycis tubo marginata; semina recta, oblonga, hilo basilari, minute punctulata. — Suffrutex gracilis ramis quadrangulatis, foliis supra sparse setosis exceptis glaber, foliis petiolatis ellipticis v. oblongis trinerviis saepe inaequalibus, floribus terminalibus solitariis v. binis majusculis roseis. Species unica Chinae occidentalis incola. Genus novum Melastomacearum ad tribum Sonerileas referendum. Proximum videtur generibus *Fordiophyto* Stapf et *Sonerilae* Roxburgh sed antheris sub-aequalibus a priori et floribus tetrameris ab altero facile distinguitur.

This new genus seems to be most closely related to *Fordiophytum* which differs chiefly in its unequal stamens, deciduous rather large sepals, in the ovary only partly adnate to the calyx, in the symmetrical petals and the more herbaceous habit. In its floral structure it seems nearest to *Sonerila*, but that genus is easily distinguished by its trimerous flowers and symmetrical petals. In its habit and general appearance the new genus resembles certain species of *Monochaetum*, like *M. alpestre* Naudin and *M. Pringlei* Rose, but that genus is otherwise entirely different and belongs to a different tribe. In the shape of the petals *Plagiopetalum* resembles *Carionia* as they are figured by Naudin (in *Ann. Sci. Nat. Bot.* sér. 3, XV. t. 15, fig. 6·[1851]) except that the petals are broader at the base.

PLAGIOPETALUM QUADRANGULUM Rehder, n. sp.

Suffrutex 30–45 cm. altus, dichotome patenti-ramosus, glaber; ramuli et rami graciles, anguste quadri-alati, alis in ramulis vetustioribus magis prominentibus et suberosis. Folia membranacea, graciliter petiolata, anguste elliptica v. elliptico-oblonga, obtusiuscula v. obtuse breviter acuminata, basi angustata, adpresse crenulato-serrulata dentibus seta albida 0.5–1 mm. longa terminatis, 1.8–4.5 cm. longa et 0.7–1.5 cm. lata, opposita saepe valde inaequalia, supra obscure coeruleo-viridia, inter nervos primarios utrinque serie setarum albidarum distantium in-structa, subtus paulo pallidiora glabra, trinervia, nervis primariis supra inconspicuis subtus elevatis trabeculis 6–10 leviter elevatis conjunctis; petioli graciles, 3–8 mm. longi, supra canaliculati, marginati. Flores rosei, terminales, solitarii v. bini, circiter 2 cm. diam.; pedicelli angulati, 3–5 mm. longi; calyx campanulatus, 4-angulatus, 8-nervius, 5 mm. longus, lobis 4 subulatis saepe inaequalibus 0.5–2 mm. longis longioribus pauci-serrulatis; petala quadrangulo-ovalia, inaequilateralia, 8–10 mm. longa et circiter 6 mm. lata; stamina petala paulo superantia, antheris filamentis subaequilongis paulo inaequalia filamentis filiformibus basin versus leviter dilatatis, rectis, antheris lineari-lanceolatis leviter curvatis eis staminum longiorum 5.5 mm. longis, staminum breviorum 4 mm. longis, connectivo stami-num longiorum postice paulo supra basin antherarum tuberculis 2 levibus con-fluentibus notato, staminum breviorum ad basin antherarum tuberculis 2 levibus distinctis notato; stylus stamina vix superans. Capsula campanulata, circiter 8 mm. longa; semina (immatura tantum visa) oblonga, basin versus paulo an-gustata, minute punctulata, 1–1.2 mm. longa.

Western Szech'uan: Hungya Hsien, sandstone cliffs, alt. 600–1000 m., June 9, 1908 (No. **3261**; sub-shrub, 0.3–0.45 m. tall, flowers carmine red).

A. R.

Carpinus cordata, var. **chinensis** (p. 425). Strike out "Veitch Exped. No. **420**" and add the following numbers:

Western Hupeh: Changyang Hsien, woods, alt. 1300–2000 m., 1907 (No. **1440**; plants only); Fang Hsien, woodlands, alt. 2000–2500 m., 1907 (No. **4444**; plants only).

Carpinus Henryana (p. 429). Add the following numbers:

Western Hupeh: Changyang Hsien, woods, alt. 1300–2000 m., 1907 (No. **1436**; plants only); Fang Hsien, woodlands, alt. 2000–2500 m., 1907 (No. **4443**; plants only).

Corylus (p. 447). After *C. heterophylla* insert:
CORYLUS spec.
Western Szech'uan: Fang Hsien, woods, alt. 1600–2000 m., 1907 (No.
729; bush or tree 3–10 m. tall; plants only).

Betula utilis, var. **Prattii** (p. 457). Add the following number:
Western Szech'uan: Tachien-lu, alt. 2500–4000 ft., 1908 (No. 1415; trees
13–33 m. tall, girth 2–6.5 m.; plants only).

Betula albo-sinensis (p. 457). Add the following numbers:
Western Hupeh: Fang Hsien, alt. 2000–3000 m., 1907 (No. 705; tree 7–20
ft. tall, bark white and rough; plants only); same locality, 1907 (No. 705ᵃ; tree
7–12 m. tall; plants only).

Betula japonica, var. **mandshurica** (p. 461). The true var. *mandshurica* from
Mandshuria remains somewhat doubtful and, as I have already stated, the Chinese
White Birch of western Szech'uan seems geographically well separated from the
other forms of *B. japonica* Siebold. The living plants, too, in this Arboretum
look very different from those of typical *B. japonica* and of the var. *kamtschatica.*
I think it best, therefore, to describe the western form as a distinct variety:

BETULA JAPONICA, var. SZECHUANICA Schneider, nov. var.
Betula alba, var. *vulgaris* Franchet, and also Burkill, see p. 461.

Arbor ad 20 m. alta, trunco ambitu ad 2.4 m. cortice albo secedente tecto;
ramuli hornotini (et biennes) plantae maturae glabri, vix v. parce glandulosi,
atrobrunnei, vetustiores cinerascentes epidermide secedenti, plantae juvenilis et
surculi distincte glanduliferi, interdum parce pilosi. Folia ramulorum fructiferorum
late ovato-triangularia, ovato-rotunda v. ovato-rhomboidea, basi rotunda, late
truncata v. plus minusve cuneata, versus apicem satis subito breviter acuminata v.
in foliis rhomboideis sensim acuminata, 4–8 cm. longa, 3–6 cm. lata, superne
intense et obscure viridia, glabra, satis dense glanduloso-punctata, subtus pallide
viridia, glabra v. ad nervos sparsissime pilosa, satis dense glandulosa, margine
paulo inaequaliter acute aperte glanduloso-dentato-serrata; folia ramulorum
sterilium surculorumque latiora, late ovato-triangularia, basi truncata v. leviter
subcordata, apice subsensim breviter acuminata, 7–13 cm. longa, 6–8.5 cm. lata
(in planta juvenili folia sunt plus minusve pilosa) margine magis inaequaliter dentata, pleraque sublobulata; petioli plerique 1.5–2.3 cm. longi, glabri, glandulosi,
surculorum in comparatione laminae breviores. Amenta fructifera solitaria, pendula,
cylindrica, plantae chinensis 3–4 cm. longa et 6–6.5 mm. crassa, plantae juvenilis
cultae ad 5 cm. longa et 1 cm. crassa, pedunculis tenuibus glabris glandulosis 10–
14 mm. longis; bracteae satis variabiles, pilosulae, ab iis varietatum alterarum vix
distincte diversae, 4.5–5.5. mm. longae, lobis lateralibus plerisque patentibus
paulo recurvis obtusis lobo mediano lanceolato acuto circiter subduplo longiore
quam lato; nuculae anguste ellipticae circiter 3 mm. longae, alis latis nuculis vix
v. ½-plo latioribus.
Western Szech'uan: see p. 461. The type is Wilson's No. 983. — Southern Szech'uan: "in silvis inter Liuku et Kuapie, alt. circiter 2800 m., Maj.
19, 1914," *C. Schneider* (No. 1320; tall bush or tree, up to 5 m. high).
Young vigorous trees in this Arboretum raised from seeds of Wilson's Nos. 983
and 4088 are very different in their widely spreading branches from those of
typical *B. japonica* and also from var. *kamtschatica.* The color of the leaves is
a dull dark, sometimes rather bluish green and not as bright as it is in the other

varieties. I saw in southern Szech'uan only small trees, mostly shoots from old trees which had been cut down, and I was able to collect only sterile shoots.

C. S.

Wistaria venusta (p. 514). Add the following form:

WISTARIA VENUSTA, f. PLENA, Rehder & Wilson, n. f.

Wistaria chinensis, var. *alba plena* Bean, *Trees & Shrubs Brit. Isl.* II. 681 (1914).

A typo recedit floribus plenis.

Cult. in Hort. Bot. Kew, " 333–13 Yokohama," June 8, 1916, *W. J. Bean.*

A specimen sent by Mr. Bean has enabled us to refer this form, which we have already mentioned on p. 513, definitively to *W. venusta.* This form is, as far as we know, the only white double-flowered Wistaria in cultivation.

Bischofia javanica (p. 521). Add the following synonym and the citation of its type specimen:

Celtis polycarpa Léveillé in Fedde, *Rep. Spec. Nov.* XI. 296 (1912).

Kweichou: " Gan-Chouen (Choin-Tang-Tchiai)," May and October 1910, *J. Cavalerie* (No. 3790).

Wikstroemia angustifolia (p. 535). For Henry's "No. 3313 " read "3313 b."

Wikstroemia gracilis (p. 536). For "No. 6440 " read "6540."

Wikstroemia Bodinieri (p. 538). The flowers of Léveillé's species belong to *Melodinus.*

Brassaiopsis fatsioides (p. 556). For "No. 3967 " read "No. 3697."

Symplocos botryantha (p. 596). In the enumeration of specimens for "No. 5272^{b}" read "No. 5242^{b}."

VOLUME III

Smilax longipes (p. 5). Instead of "No. 667" read "No. 677."

Smilax (p. 7). After *Smilax cocculoides* insert:

SMILAX spec.

Western Hupeh: south of Ichang, thickets, alt. 1000 m., 1907 (No. 474; climber, fruit blue-black; seeds only).

SMILAX spec.

Western Hupeh: Changyang Hsien, thickets, alt. 1000–1300 m., 1907 (No. 675; climber, fruit black, leaves large, ovate; seeds only).

SMILAX spec.

Western Szech'uan: Wa-shan, 1908 (No. 1119; bush 1.3–2 m., fruit black, on filiform peduncles; seeds only).

Chloranthus brachystachyus (p. 15). Instead of " brachystachyus " read " brachystachys."

Populus szechuanica (p. 20). Add the following number:

Western Szech'uan: without precise locality, alt. 1300–2000 m., 1908 (No. 1416; trees 8–12 m. tall, girth 1–1.6 m.; plants only).

Populus tremula, var. Davidiana (p. 24). Add the following number: Western Hupeh: Fang Hsien, uplands, alt. 2000–3000 m., 1907 (No. 4448; trees 10–12 m. tall, young leaves bronzy; plants only).

Salix

Salix hypoleuca (p. 53). Line 5 from below for "*S. glandulosa*" read "*S. Wilsonii.*"

Salix macroblasta (p. 58). Change No. 1402 to No. 1402ᵃ.

Salix etosia (p. 73). Add after this species:

SALIX spec.
Western Szech'uan: west and near Wên-ch'uan Hsien, alt. 1300–2000 m., November 1908 (No. 1419; bush 3–4 m. tall).

SALIX spec.
Western Szech'uan: Pan-lan-shan, west of Kuan Hsien, woodlands, alt. 2600–3000 m., October 1910 (No. 4349; bush 3–7 m. tall).

SALIX spec.
Western Hupeh: Fang Hsien, upland thickets, alt. 2000–3000 m., 1907 (No. 4433; shrub; plants only).

SALIX spec.
Western Hupeh: Fang Hsien, upland thickets, alt. 2000–3000 m., 1907 (No. 4435; shrub; plants only).

Salix fruticulosa (p. 119). This name must be changed on account of the earlier *S. fruticulosa* de Lacroix in *Bull. Soc. Bot. France*, VI. 566 (1859). I therefore propose the following new name:

SALIX HYLEMATICA Schneider, n. nom.

Salix fruticulosa Andersson in *Jour. Linn. Soc.* IV. 53 (non de Lacroix) (1860).

The specific name is derived from ὑληματικός, shrubby.   C. S.

Pterocarya Paliurus (p. 182). Add the following number: Western Hupeh: Fang Hsien, moist woodlands, rare, alt. 2000–2300 m., 1910 (No. 4412; tree 20–23 m. tall, girth 2–2.6 m.; seeds only).

Juglans regia (p. 184). Add the following numbers: Western Szech'uan: near Wa-shan, cultivated, alt. 1000–2500 m., 1908 (No. 969; large tree, 20–26 m. tall, girth 3.3–4 m., fruit round; seeds only; No. 969ᵃ; fruit elongated; seeds only).

Castanea mollissima (p. 192). Add the following number: Western Hupeh: Changyang Hsien, woodlands, alt. 1000–1300 m., 1907 (No. 1452; tree 10 m. tall, girth 1 m.; plants only).

Castanea Seguinii (p. 194). Add the following number: Western Hupeh: bare hills around Ichang, alt. 100–600 m., 1907 (No. 1474; bush 0.3–1.6 m., flowers white; plants only).

Quercus glandulifera (p. 212). Add the following number: Western Szech'uan: west and near Wên-ch'uan Hsien, woods, 1907 (No. 1092; tree 7–14 m. tall; seeds only).

Quercus variabilis (p. 219).   Add the following number:
Western Hupeh: Ichang, alt. 300–1000 m., 1907 (No. **351**; tree 7–18 m.; seeds only).

Quercus (p. 230).   Insert after *Q. oxyodon*, var. *Fargesii:*
QUERCUS spec.
Western Hupeh: Changyang Hsien, woods, 1907 (No. **784**; tree 7–10 m. tall, leaves obovate, membranous, pubescent).

Ulmus Bergmanniana (p. 240).   Add the following number:
Western Hupeh: Hsing-shan Hsien, woods, alt. 2000–2600 m., 1907 (No. **1451**; thin tree, 3–10 m. tall; plants only).

Celtis Vandervoetiana (p. 267).   In the enumeration of specimens for "Veitch Exped. No. 4467" read "Veitch Exped. No. 4469."

Morus cathayana (p. 292).   In the enumeration of specimens for Henry's No. "5543" read "5548."

Morus acidosa (p. 297).   Add the following number:
Western Hupeh: south of Ichang, cliffs, alt. 1000 m., 1907 (No. **84**; large bush, 5 m. tall, fruit red and black; seeds only).

Ficus (p. 311).   Insert at the end of the genus:
FICUS spec.
Western Szech'uan: Kiating Fu, thickets, alt. 300 m., 1908 (No. **2792**; bush 2 m. tall).

Polygonum multiflorum (p. 325).   In the note under this species I mentioned *P. Aubertii* L. Henry, but I did not know that it had been collected by Wilson.
POLYGONUM AUBERTII L. Henry in *Rev. Hort.* 1907, 82, figs. 23, 24. — Schneider, *Ill. Handb. Laubholzk.* II. 907 (1912).
Western Szech'uan: thickets around Tachien-lu, alt. 1800–2400 m., July 1908 (No. **3183**; climber 1.8–2.4 m. high; flowers white).   C. S.

Trachelospermum (p. 336).   Insert after *T. axillare:*
TRACHELOSPERMUM spec.
Western Hupeh: Fang Hsien, alt. 1300 m., 1907 (No. **776**; climber on rocks and trees; seeds only).

Cynanchum (p. 348).   Add at the end of this genus:
CYNANCHUM spec.
Western Hupeh: Ichang, thickets, alt. 500–1000 m., 1907 (No. **514**; climber, flowers white; seeds only).

Dregea sinensis (p. 352).   In the enumeration of specimens for Henry's No. "6164" read "4164."

Ehretia acuminata (p. 363).   In the enumeration of specimens for Henry's No. "6538" read "6358."

Ehretia macrophylla (p. 364).   In the enumeration of specimens for Henry's No. "1885" read "1885 bis."

Callicarpa Giraldiana (p. 366).   Add the following number:
Western Hupeh: Fang Hsien, thickets, alt. 1200 m., 1907 (No. **633**; fruit purple; seeds only).

Premna (p. 372). At the end of the genus insert the following two new species:

Premna subcapitata Rehder, n. sp.

Frutex metralis; ramuli teretes, hornotini dense patenti-villosi, annotini tarde glabrescentes, fusci, vetustiores cinereo-fusci. Folia membranacea, ovato-oblonga, acuminata, basi rotundata, basi apiceque exceptis dentato-serrata, 3.5–6 cm. longa et 1.5–2.3 cm. lata, supra obscure luteo-viridia, subaccumbenti-pilosa, subtus molliter dense villosa, nervis utrinsecus 4–5 angulo angusto divergentibus et ad-scendentibus ut costa supra impressis et subtus elevatis fuscescentibus; petioli dense patenti-villosi, 0.5–1 cm. longi, oppositi saepe inaequales. Inflorescentia terminalia in apice ramulorum lateralium patentium, capitato-corymbosa, subsessilis v. pedunculo ad 1 cm. longo suffulta, villosa, circiter 2 cm. diam.; pedicelli brevissimi; flores virescentes; calyx turbinatus, circiter 4 mm. longus, fere ad medium 5-partitus lobis lanceolatis, villosus; corolla bilabiata et 4-lobata, circiter 6 mm. longa, extus pilosula et glandulosa, labium posticum rotundato-ovatum tubo paulo brevius ut videtur concavum (flores nondum perfecte aperti tantum exstant), anticum trilobum, postico brevius lobo medio late ovato, lateralibus similibus, sed duplo minoribus, tubus fauce villosus; stamina 4, exserti; stylus staminibus longior; ovarium glandulosum. Fructus desideratur.

Western Szech'uan: without precise locality, cliffs, alt. 1200 m., June 1903 (Veitch Exped. No. 3761).

This species seems to be most closely related to *P. pinguis* Clarke and *P. procumbens* Moon, but is easily distinguished from both by the smaller and narrower densely pubescent leaves and by the deeply divided calyx.

Premna urticifolia Rehder, n. sp.

Frutex metralis; ramuli teretes, pallide fusci, hornotini sparse villosuli. Folia membranacea, ovata, acuminata, basi cordata, dentato-serrata, 5–8.5 cm. longa et 3–6 cm. lata, supra obscure viridia, sparse scabrido-pilosula et punctulis impressis minutis notata, secus costam et nervos villosula, subtus paulo pallidiora, initio secus nervos crispulo-villosula, demum fere glabra, minute impresse reticulata et punctu-lata, utrinsecus nervis 4–5 adscendentibus, ut costa et nervi tertiarii supra impressis subtus elevatis; petioli 1–3.5 cm. longi, oppositi saepe inaequales, supra dense crispo-villosuli, ceterum sparse villosuli. Inflorescentia corymbosa, densa, 2.5–3.5 cm. diam., brevior quam lata, pedunculata pedunculo 1–1.5 cm. longo ut ramuli et pedicelli breves v. brevissimi villosulo; calyx fructifer cupularis, breviter sinuato-dentatus, circiter 4 mm. longus et latus, fere glaber; flores non visi. Drupa obo-voidea, 4–5 mm. longus, purpureo-nigra, putamine osseo, 3–4-loculari.

Yunnan: Szemao, alt. 1600 m., *A. Henry* (No. 13389).

This species belongs to the same group as the preceding species, but differs from it in the glabrescent leaves and the sinuate-dentate calyx; from *P. pinguis* Clarke it is easily distinguished by the glabrescent stems, the smaller leaves and the pe-duncled corymb, and from *P. procumbens* Moon by the broader ovate leaves and the peduncled corymb. The minute impressed dots on the under side of the leaves I have not seen in any other species except in the African *P. Schimperi* Engler. Both new species, *P. subcapitata* and *P. urticifolia*, resemble in their general appearance certain species of Viburnum of the section Odontotinus.

Clerodendron japonicum (p. 377). According to Makino (in *Tokyo Bot. Mag.* XVII. 91 [1903]) *Clerodendron Kaempferi* Siebold in *Verh. Batav. Genoot.* XII. 51 (*Syn. Pl. Oec. Jap.*) (1830) is a synonym of *Firmiana simplex* W. F. Wight, but we do not think that this view is correct; Siebold apparently merely confused the Japanese names.

Caryopteris incana (p. 378).   Strike out Henry's No. 7771.

Compositae

Compositae (p. 418).   Before *Pluchea* insert the following two genera:

ASTER Linnaeus

ASTER HERSILEOIDES Schneider, n. sp.
Suffrutex v. frutex parvus, ramis ut videtur ex parte procumbentibus, ad 0.6 m. altus; ramuli hornotini floriferi praesertim ad apicem pilosi annotinique subglabri, flavo-brunnescentes v. brunnei, subalato-striato-angulati, vetustiores cinerascentes, glabri, satis lignescentes.   Folia ramulorum juvenilium tantum visa, membranacea, anguste oblanceolata, sessilia, apice subrotundata v. subobtusa, apiculata, versus basim sensim attenuata, 0.7–2 cm. longa et versus apicem 0.2–0.5 cm. lata, superne ad costam parce breviter pilosa v. glabra, subtus glabra, concoloria, margine plus minusve dense crispo-ciliata, integra.   Capitula in apice ramulorum 3–6 cm. longorum superiore parte nudorum solitaria, floribus marginalibus explanatis 2.5–3 cm. lata; involucri phylla triserialia, anguste lanceolata, plus minusve acuminata, margine scariosa subfimbriato-ciliolata, dorso glabra v. subpuberula, interiora ad 5 mm. longa, exteriora breviora, flores radii anguste lineares, ad 15 mm. longi, rubicundi; flores disci circiter 8 mm. longi; pappi setae albo-rubicundae tubum vix superantes, ut videtur biseriales exterioribus valde breviores; ovarium anguste ovato-ellipticum pilosum.   Fructus ignoti.
Western Szech'uan: Mao-chou, arid places, alt. 1600–1800 m., May 24, 1908 (No. 2234, type; sub-shrub 0.6–0.9 m. tall; flowers light purple); same locality, cliffs, etc., alt. 1300 m., July 1908 (No. 2243; sub-shrub 0.6 m. tall, flowers pink).
I cannot refer this plant to any other genus than *Aster* (sect. *Calimeris*), and it does not resemble any other described Chinese species.   It is apparently a distinctly shrubby plant, and the flowering branches most resemble the figure given by Klotzsch of his *Hersilea ramosa* in Klotzsch & Garcke, *Bot. Ergeb. Reise Prinz. Wald.* 76, t. 83 (1862).   The receptacle is flat.

ASTER POLIA Schneider, n. sp.
Suffrutex erectus, 0.6–1.2 m. altus; ramuli hornotini annotinique subteretes, purpurascentes, laxe araneoso-pubescentes, dein glabrescentes, vetustiores non visi.   Folia ramulorum floriferarum crasse membranacea, anguste lanceolata, apice obtusiuscula, basi acuta, petiolo 1–2 mm. longo satis distincto suffulta, 7–12 mm. longa et 2–3.5 mm. lata, superne viridia, breviter tuberculato-pilosa, subtus albido-araneoso-pubescentia, integra.   Inflorescentiae, apice ramulorum hornotinorum lateralium ad 7 cm. longorum corymbosae, capitulis 3–7; capitula explanata vix ad 1.5 cm. diametientia; involucri phylla triserialia, anguste lanceolata, acuta, margine scariosa, subfimbriato-ciliolata, dorso laxe puberula, exteriora circiter 2 mm., interiora circiter 5 mm. longa; receptaculum planum, alveolatum; flores radii 9–10 mm. longi, albi (ex Wilson) sed ut videtur initio rubicundi; flores disci circiter 7 mm. longi; pappi setae uniseriales, albae (v. levissime rubicundae), tubum vix superantes; ovarium anguste ellipticum, pilosum.   Fructus ignoti.
Western Szech'uan: near Monkong Ting, valley of Hsiao-chin Ho, alt. 2200–2700 m., June 1908 (No. 2233, type; sub-shrub 0.6–1.2 m. tall, flowers white).
This species differs from the preceding in its smaller corymbose flower-heads, its white pubescence and its single series of pappus-hairs, and approaches *Microglossa* in the details of its flowers.   Nevertheless I cannot refer it to any other genus with

more certainty, and according to O. Hoffmann and other authors it belongs to sect. *Calimeris*.

The specific name is derived from πολιός, with white hair.

MICROGLOSSA De Candolle

MICROGLOSSA SALICIFOLIA Diels in *Bot. Jahrb.* XXIX. 612 (1900).

Western Szech'uan: Wên-chuan Hsien, thickets, alt. 1300–2000 m., July 1908 (No. 2235; sub-shrub 1.2–1.8 m. tall, flowers white).

C. S.

NUMERICAL LISTS

GENERAL INDEX

NUMERICAL LISTS

NUMBERS AND NAMES
OF SPECIMENS COLLECTED DURING THE TWO
ARNOLD ARBORETUM EXPEDITIONS[1]

1. Prunus involucrata, I. 206.
2. Rubus eustephanus, I. 51; III. 423.
3. Prunus malifolia, I. 207.
3. Prunus Conradinae, I. 211.
3a. Prunus tenuiflora, I. 209.
3b. Prunus Conradinae, I. 211.
4, 4a. Rubus Playfairianus, I. 49.
5. Prunus Conradinae, I. 211.
6. Ilex corallina, I. 80.
7. Prunus Conradinae, I. 211.
8, 8a, b, c, d, e, f. Morus mongolica, III. 269.
9. Prunus cyclamina, I. 207.
10, 10a, b, c, d, e, f. Morus cathayana, III. 292.
11. Prunus Conradinae, I. 211.
12, 12a. Coriaria sinica, II. 170.
13. Prunus tenuiflora, I. 209.
14. Lonicera Standishii, var. lancifolia, I. 135.
15. Rubus corchorifolius, I. 51.
15a. Rubus corchorifolius, III. 423.
16. Prunus pilosiuscula, var. media, III. 432.
16a. Prunus pilosiuscula, var. media, I. 204.
17. Betula luminifera, II. 455.
18, 18a. Prunus pilosiuscula, var. barbata, I. 203.
19. Daphne genkwa, II. 538.
20. Prunus tenuiflora, I. 209.
21, 21a. Debregeasia longifolia, III. 313.
26. Prunus Armeniaca, I. 278.
27. Prunus salicina, III. 432.
29, 29a. Polygala Wattersii, II. 161.

30. Arundinaria dumetosa, II. 63.
31. Rubus coreanus, I. 54.
32. Lonicera saccata, I. 133.
33, 33a, b, c, d, e, f, g, h. Morus acidosa, III. 297.
34, 34a. Broussonetia Kaempferi, III. 304.
35. Cayratia oligocarpa, I. 99.
36. Meratia praecox, I. 419.
36a. Meratia praecox, var. grandiflora, I. 420.
37, 37a. Prunus Dielsiana, var. laxa, I. 208.
38. Ribes tenue, I. 45.
38a. Ribes tenue, III. 423.
39. Prunus pilosiuscula, var. barbata, I. 203.
39a. Prunus pilosiuscula, var. media, I. 204.
40. Wikstroemia capitata, II. 530.
41. Prunus pilosiuscula, var. subvestita, I. 204.
42. Rubus triphyllus, I. 55; III. 424.
43, 43a. Prunus glandulosa, var. trichostyla, f. Faberi, I. 224.
44. Rubus Parkeri, var. longisetosus, I. 50.
45. Prunus Zappeyana, I. 221.
45a. Prunus Zappeyana, var. subsimplex, I. 222.
46. Prunus salicina, III. 432.
47. Prunus polytricha, I. 204.
48. Rubus Henryi, I. 49.
49. Prunus tomentosa, var. endotricha, I. 225.
50, 50a. Elaeagnus magna, II. 411.

[1] This is a complete enumeration of all the numbers referring to woody plants; numbers omitted refer to herbaceous plants. For the numbers and names of Ferns see H. Christ, *Filices Wilsonianae* (in *Bot. Gaz.* LI. 345–359, 2 fig. [1911]).

51, 51a. Prunus tenuiflora, I. 209.
52, 52a. Rubus mesogaeus, I. 56.
53. Prunus venusta, I. 230.
54. Prunus Persica, I. 273.
55. Prunus Persica, I. 273.
56. Morus acidosa, III. 297.
57. Rubus trullissatus, I. 53.
58, 58a, b. Schoepfia jasminodora, III. 321.
59. Elaeagnus cuprea, II. 414.
60. Malus prunifolia, var. rinki, II. 279.
61. Prunus salicina, III. 432.
62. { Prunus laxiflora, I. 70.
      { Prunus discadenia, I. 200; III. 431.
63. Maddenia Wilsonii, I. 58.
64. Prunus variabilis, I. 201.
65. Prunus canescens, I. 215.
66. Prunus salicina, I. 276, 580.
68. Prunus Dielsiana, var. laxa, I. 208.
69. Prunus tenuiflora, I. 209.
70. { Prunus pilosiuscula, var. barbata, I. 203.
      { Prunus Zappeyana, I. 221.
71. Rubus mesogaeus, I. 56.
72. Rubus pungens, I. 52.
73. Ribes Franchetii, I. 46.
74, 74a. Ehretia acuminata, III. 363.
75. Prunus mume, I. 278.
76. Rubus Henryi, I. 49.
77. Vitis pentagona, var. bellula, III. 428.
78. Rubus trianthus, I. 51.
79. Rubus chroosepalus, I. 49.
80. Rubus spec., III. 424.
81. Rubus adenophorus, I. 55.
82. Prunus salicina, I. 276, 580.
83. Cornus controversa, II. 573.
84. Morus acidosa, III. 456.
85, 85a, b. Betula luminifera, II. 455.
86. Neillia sinensis, I. 436.
87. Cotinus coggygria, var. pubescens, II. 175.
88. Cornus chinensis, II. 577.
89. Stachyurus chinensis, I. 287.
90. Ribes tenue, I. 45.
91, 91a. Prunus venosa, I. 60.
92. Rubus innominatus, I. 54; III. 424.
93, 93a. Lonicera Koehneana, I. 140.
94, 94a. Betula luminifera, II. 455.
95. Clematis Armandi, I. 326.
95a. Clematis Armandi, f. Farquhariana, I. 327.

97. Rubus inopertus, I. 54.
98. Lonicera Koehneana, I. 140.
99. Rubus hupehensis, I. 49.
100. Ribes luridum, I. 46.
104. Prunus salicina, I. 276, 580.
105. Paliurus orientalis, II. 209.
106. Caesalpinia sepiaria, II. 92.
107. Bauhinia hupehana, II. 89.
108, 108a. Tapiscia sinensis, II. 188.
109, 109a. Grewia parviflora, var. glabrescens, II. 371.
110. Clematis grata, var. grandidentata, I. 338.
111. Staphylea holocarpa, II. 185.
112. Kerria japonica, II. 301.
113. Zanthoxylum dimorphophyllum, II. 126.
114. Euscaphis japonica, II. 187.
115. Prunus brachypoda, var. pseudossiori, I. 65.
116. Prunus salicina, I. 276, 580.
117. Viburnum tomentosum, I. 111.
118. Prunus venosa, I. 60.
119. Ilex Pernyi, I. 78.
120. Prunus salicina, I. 276, 580.
121. Ampelopsis heterophylla, var. amurensis, I. 100, 579.
122. Staphylea holocarpa, var. rosea, II. 186.
122a, b, c. Staphylea holocarpa, II. 185.
123. Rhus verniciflua, II. 181.
124. Ampelopsis heterophylla, var. Delavayana, I. 100.
125. Prunus Persica, III. 432.
125a, b. Prunus Persica, I. 273.
126. Vitis betulifolia, I. 103.
126a. Vitis Piasezkii, I. 103.
127. { Prunus Wilsonii, var. leiobotrys, I. 63.
       { Prunus sericea, var. brevifolia, I. 64.
128. Aralia chinensis, II. 566.
128a. Aralia chinensis, var. glabrescens, II. 567.
129. Cornus controversa, II. 573.
130. Ampelopsis heterophylla, var. Delavayana, I. 100.
131. Ficus impressa, III. 311.
132. Helwingia japonica, II. 570.
132a. Helwingia japonica, var. hypoleuca, II. 570.
133. Cornus controversa, II. 573.

134. Vitis pentagona, I. 103.
135. Viburnum tomentosum, I. 111.
136. Cornus paucinervis, II. 576.
137. Rhamnus crenatus, II. 232.
138. Zanthoxylum alatum, var. plani-spinum, II. 125.
139, 139a. Euptelea Franchetii, I. 314.
140. Cornus macrophylla, II. 575.
141. Rubus irenæus, I. 51.
142. Alangium platanifolium, II. 554.
143. Ampelopsis megalophylla, I. 101.
144. Akebia lobata, var. australis, I. 348.
145. Rubus flosculosus, f. parvifolius, I. 54.
145a. Rubus flosculosus, I. 54.
146. Holboellia coriacea, I. 345.
147. Cotoneaster foveolata, I. 162.
148. Ilex Franchetiana, I. 77.
149. Zanthoxylum dimorphophyllum, II. 126.
150. Vitis betulifolia, I. 103.
151. { Ilex macrocarpa, var. *a* genuina, I. 81. / Ilex macrocarpa, var. β tricho-phylla, I. 81.
152. Rubus teledapos, I. 54; III. 424.
153. Cotoneaster Zabelii, var. miniata, III. 430.
153a. Cotoneaster divaricata, I. 157.
154. Meliosma Beaniana, II. 204.
155. Rubus spec., III. 424.
156. { Cotoneaster foveolata, I. 162. / Cotoneaster acutifolia, var. villo-sula, I. 158.
157. Ampelopsis micans, I. 579.
158. Schisandra pubescens, I. 413.
159. Ampelopsis micans, I. 579.
160, 160a. Rhus sylvestris, II. 180.
161. Phellodendron chinense, II. 136.
162, 162a. Betula luminifera, II. 455.
163. Cephalotaxus drupacea, var. si-nensis, f. globosa, II. 4.
164, 164a. Schisandra glaucescens, I. 413.
165. Alangium chinense, II. 552.
166. Holboellia Fargesii, I. 346.
167, 167a. Cephalotaxus drupacea, var. sinensis, II. 3.
168. Sargentodoxa cuneata, I. 351.
169. Prunus salicina, I. 276, 580.
170. Vitis flexuosa, I. 102.
171. Akebia lobata, I. 348.

172. Helwingia chinensis, var. macro-carpa, II. 571.
172a, b. Helwingia chinensis, II. 571.
173. Corylopsis Veitchiana, I. 425.
173a. Corylopsis sinensis, I. 424.
174. Prunus discadenia, I. 200.
175. Prunus salicina, I. 276, 580.
176. Morus acidosa, III. 297.
177. { Prunus venosa, I. 60. / Prunus stellipila, I. 61.
178. Prunus gracilifolia, I. 223.
179. Rosa omeiensis, II. 331.
180. Ribes glaciale, I. 46.
181. Prunus pubigera, var. Prattii, I. 68.
182. { Prunus litigiosa, var. abbreviata, I. 205. / Prunus Rossiana, I. 223.
183. Lonicera longa, I. 134.
184. Corylopsis platypetala, I. 426.
185, 185a. Staphylea holocarpa, var. rosea, II. 186.
185b. Staphylea holocarpa, II. 185.
186. Prunus pubigera, var. obovata, I. 68.
187. Cotoneaster foveolata, III. 430.
188. Rubus lasiostylus, I. 52.
189. Neillia sinensis, I. 436.
190. Prunus brachypoda, var. pseudo-ssiori, I. 65.
191. Ribes Franchetii, I. 46.
192. Stachyurus chinensis, I. 287.
192a. Stachyurus himalaicus, I. 287.
193. Smilax discotis, var. concolor, III. 6.
194. Lonicera Maackii, var. podo-carpa, I. 140.
195. Spiraea Miyabei, var. glabrata, I. 454.
196. Lonicera spec., III. 430.
198. Lonicera Koehneana, I. 140.
199. Triosteum Fargesii, I. 116.
200. Aesculus Wilsonii, I. 498.
203. Juglans regia, III. 184.
204. Rosa banksiopsis, II. 322.
205. Parthenocissus himalayana, var. rubrifolia, I. 101.
206. Juglans regia, III. 184.
215. Vitis Piasezkii, I. 103.
216. Clerodendron trichotomum, var. Fargesii, III. 376.
217. Cotoneaster acutifolia, var. villo-sula, I. 158.
218. Viburnum theiferum, I. 112.

219. Euptelea Franchetii, I. 314.
220, 220a. Viburnum rhytidophyllum, I. 110.
220b. Viburnum Rosthornii, I. 110.
221. Viburnum ichangense, I. 115.
222. Prunus Wilsonii, var. leiobotrys, I. 63.
222b. Prunus sericea, var. Batalinii, I. 64.
223, 223a, b. Cornus kousa, II. 577.
224. Viburnum ovatifolium, I. 113.
225. Acer Davidii, I. 92.
226, 226a. Sinofranchetia chinensis, I. 349.
227. Cotoneaster horizontalis, I. 154.
228. Viburnum ichangense, I. 115.
229. Acer Maximowiczii, I. 94.
230. Viburnum ovatifolium, I. 113.
231. Ilex Fargesii, I. 77.
232. Cotoneaster divaricata, I. 157.
233. Acer Wilsonii, I. 90.
234, 234a. Viburnum tomentosum, I. 111.
235. Parthenocissus Thomsonii, I. 101.
236. Viburnum theiferum, I. 112.
237. Viburnum hupehense, I. 115.
238. Viburnum lobophyllum, I. 114.
238a. Viburnum betulifolium, I. 114.
239. Viburnum ichangense, I. 115.
240. Viburnum ovatifolium, I. 113.
241. Prunus salicina, I. 276, 580.
242. Vitis reticulata, III. 428.
243. Prunus salicina, III. 432.
243a. Prunus salicina, I. 276, 580.
244. Vaccinium japonicum, I. 562.
245. Schisandra sphenanthera, I. 414.
246. Vitis betulifolia, I. 103.
247. Evodia glauca, II. 129.
248. Vitis Piasezkii, I. 103.
249. Viburnum betulifolium, I. 114.
250. Vitis reticulata, I. 103.
251. Evodia officinalis, II. 130.
252, 252a. Cudrania tricuspidata, III. 306.
253. Evodia glauca, II. 129.
254. Lonicera Henryi, I. 141.
255, 255a, b. Pteroceltis Tatarinowii, III. 284.
256. Hypericum spec., III. 452.
257. Acer oblongum, I. 92.
258, 258a. Meliosma Beaniana, II. 204.
259. Rhus verniciflua, II. 181.

261. Malus baccata, var. mandshurica, II. 291.
263. { Schisandra sphenanthera, I. 414. / Schisandra grandiflora, I. 411.
264. Malus kansuensis, II. 286.
266. Lonicera gynochlamydea, I. 134.
267. Cephalotaxus drupacea, var. sinensis, II. 3.
268, 268a. Pteroceltis Tatarinowii, III. 284.
270. Viburnum Henryi, I. 106.
271. Cotoneaster foveolata, I. 162.
272. Rosa setipoda, II. 323.
273. Cotoneaster foveolata, I. 162.
274. Acer tetramerum, I. 94.
275, 275a, b. Rhus punjabensis, var. sinica, II. 176.
276. Acanthopanax Giraldii, var. inermis, II. 560.
277. Ribes alpestre, a commune, I. 45.
278. Magnolia denudata, var. purpurascens, I. 401.
279. Rubus lasiostylus, var. dizygos, I. 53.
280. Ribes longeracemosum, β Wilsonii, I. 45.
281. Viburnum Sargentii, I. 116.
282. Rubus simplex, I. 48, 579.
283. Vitis betulifolia, I. 103.
284, 284a. Rhus orientalis, II. 179.
285. Crataegus Wilsonii, I. 180.
286. Vitis betulifolia, I. 103.
287, 287a. Rosa banksiopsis, II. 322.
288. Acanthopanax leucorrhizus, var. fulvescens, II. 558.
289. Abelia Engleriana, I. 120.
290. Staphylea Bumalda, II. 185.
291. Vitis betulifolia, III. 429.
292. Stachyurus chinensis, I. 287.
293. Amelanchier asiatica, var. sinica, I. 195.
294, 294a. Viburnum sympodiale, I. 109.
295. Litsea pungens, II. 76.
296. Lindera communis, II. 79.
297. Litsea ichangensis, II. 77.
298. Litsea ichangensis, II. 77.
299. Lindera cercidifolia, II. 85.
300. Lindera obtusiloba, II. 85.
301. Lindera obtusiloba, II. 85.
302. Lindera megaphylla, II. 80.
303. Pachysandra terminalis, II. 164.
304. Lonicera lanceolata, I. 140.

305. Viburnum erubescens, var. gracilipes, I. 107.
306. Rosa caudata, II. 321.
307. Berberis dasystachya, III. 442.
308. Styrax Veitchiorum, I. 290.
309. Acer caudatum, var. multiserratum, I. 91.
310. Acer pictum, var. parviflorum, I. 83.
311. Ficus heteromorpha, III. 311.
313. Schisandra sphenanthera, I. 414.
314. Cercis racemosa, II. 88.
315, 315a. Ligustrum acutissimum, II. 600.
316. Rosa saturata, II. 324.
317. Ribes himalayense, γ urceolatum, I. 44.
318. Schisandra grandiflora, I. 411.
319. { Cotoneaster foveolata, I. 162.
Cotoneaster acutifolia, var. villosula, I. 158.
320. Sorbus hupehensis, var. syncarpa, I. 467.
320a. Sorbus Koehneana, I. 471.
321. Zanthoxylum dissitum, II. 128.
322. Meliosma Veitchiorum, II. 204.
323, 323a, 323b. Acanthopanax leucorrhizus, var. scaberulus, II. 558.
324. Evodia Henryi, II. 133.
325. Itea ilicifolia, I. 44.
326. Meliosma cuneifolia, II. 199.
326a. Meliosma pendens, II. 200.
327. Cotoneaster acutifolia, var. villosula, I. 158.
329. Viburnum ichangense, I. 115.
330, 330a. Decaisnea Fargesii, I. 344.
331. Cotoneaster Zabelii, I. 166.
332. Nothopanax Davidii, II. 556.
333. Photinia villosa, var. sinica, I. 186.
334. Cotoneaster hupehensis, I. 169.
335. Cotoneaster salicifolia, var. rugosa, I. 172.
330. Sinomenium acutum, var. cinereum, I. 387.
337. Acer Franchetii, I. 97.
338. Clematis grata, var. grandidentata, I. 338.
339. Acer robustum, I. 89.
340. Acer griseum, I. 97.
341. Acer Davidii, I. 92.
342. Cayratia oligocarpa, I. 99.
343. Celtis Bungeana, III. 269.

344. Castanea mollissima, III. 192.
344a. Castanea mollissima, III. 192.
345. Magnolia denudata, var. elongata, I. 402.
346. Lonicera tragophylla, I. 143.
347. Actinidia chinensis, II. 385.
348. Sorbus caloneura, II. 269.
350. Quercus glandulifera, III. 212.
351. Quercus variabilis, III. 457.
352. Castanea Henryi, III. 196.
353. Evonymus lanceifolia, I. 491.
353, 353a. Evonymus yedoensis, var. Koehneana, I. 491.
354. Evonymus alata, I. 493.
354a. Evonymus alata, I. 493.
355. Acer Maximowiczii, I. 94.
356. { Evonymus Giraldii, var. ciliata, I. 495.
Evonymus Giraldii, var. angustialata, I. 495.
Evonymus porphyrea, I. 495.
357. Celastrus rugosa, II. 349.
357a. Celastrus Loesneri, II. 350.
358. Actinodaphne confertifolia, II. 74.
360. Symplocos paniculata, II. 593.
361, 361a, b. Magnolia aulacosperma, I. 396.
362. Celastrus hypoleuca, II. 346.
363. Celastrus gemmata, II. 352.
364, 364a, b. Celastrus angulata, II. 346.
365. Pyrus Calleryana, III. 450.
366, 366a. Evodia officinalis, II. 120.
367. Aristolochia heterophylla, III. 323.
370. Lysionotus pauciflorus, III. 387.
371, 371a, b, c, d, e, f, g. Juglans cathayensis, III. 185.
372. Styrax dasyanthus, I. 289.
372a. Styrax dasyanthus, var. cinerascens, I. 289.
373. Magnolia denudata, var. purpurascens, I. 401.
374, 374a. Castanea Seguinii, III. 194.
375. Cornus Walteri, II. 576.
376. Acer oblongum, var. latialatum, I. 92.
377. Chaenomeles lagenaria, var. cathayensis, II. 297.
378. Vitis reticulata, I. 103.
379. Acanthopanax Henryi, II. 557.
379a. Acanthopanax villosulus, II. 562.
380, 380a. Pistacia chinensis, II. 173.
381. Evodia glauca, II. 129.

382. Stranvaesia Davidiana, var. undulata, I. 192.
383, 383a. Eucommia ulmoides, I. 433.
385. Spiraea Miyabei, var. glabrata, I. 454.
386. Zanthoxylum pteracanthum, II. 123.
387, 387a, b. Evodia hupehensis, II. 133; III. 449.
388. Ailanthus Vilmoriniana, II. 154.
389. Juglans regia, III. 184.
390. Juglans regia, III. 184.
391. Berberis Henryana, III. 440.
392, 392a. Viburnum ichangense, I. 115.
393. Viburnum lobophyllum, I. 114.
394. Viburnum ovatifolium, I. 113.
395. Pyrus serotina, III. 450.
396. Evonymus japonica, var. radicans, I. 485.
397. Exochorda Giraldii, var. Wilsonii, I. 456; III. 445.
398. Photinia subumbellata, I. 189.
399, 399a. Acanthopanax trifoliatus, II. 563.
400. Viburnum betulifolium, I. 114.
401. Desmodium floribundum, II. 103.
402. Clematis Pavoliniana, III. 434.
403. Clematis uncinata, I. 327.
404. Pterocarya hupehensis, III. 182.
405. Photinia amphidoxa, I. 190.
406. Sorbus expansa, I. 457.
407. Grewia parviflora, var. glabrescens, II. 371.
408. Vitex quinata, III. 374.
409. Vitis Davidii, var. cyanocarpa, I. 104.
410. Rhamnus leptophyllus, II. 239.
411. Viburnum lobophyllum, I. 114.
412. Lonicera Maackii, var. podocarpa, I. 140.
413. Liriodendron chinense, I. 410.
414. Tilia Henryana, II. 367.
415. Pyrus serotina, II. 263.
415a. Pyrus Calleryana, II. 264.
416. Clematis Pavoliniana, I. 328.
417. Berberis Julianae, I. 360.
418. Cephalotaxus Oliveri, II. 6.
419. Ampelopsis heterophylla, var. Delavayana, I. 100.
420. Smilax discotis, var. concolor, III. 6.
421. Rhamnus leptophyllus, II. 239.
422. Rhamnus paniculiflorus, II. 233.
423. Fraxinus chinensis, III. 450.

424, 424a. Acer Henryi, I. 97.
425. Clerodendron mandarinorum, III. 375.
426. Clematis quinquefoliolata, I. 328.
427, 427a. Clematis grata, var. grandidentata, I. 338.
427b. Clematis apiifolia, var. obtusidentata, I. 336.
429. Diospyros kaki, var. silvestris, II. 590.
429a. Diospyros Lotus, II. 587.
430. Acer tetramerum, I. 94.
431. Rosa Helenae, II. 310.
431b. Rosa Helenae, II. 310.
431c. Rosa Rubus, II. 308.
432. Rhamnus utilis, II. 240.
433. Rhamnus leptophyllus, II. 239.
434. Acer longipes, I. 88.
435. Paederia tomentosa, III. 403.
436. Acer Davidii, I. 92.
437. Polygonum multiflorum, III. 325.
438. Lespedeza formosa, II. 107.
439, 439a. Callicarpa Giraldiana, III. 366.
440. Parthenocissus laetevirens, I. 580.
441. Diospyros Lotus, II. 587.
442. Celtis cerasifera, III. 271.
444, 444a. Celtis labilis, III. 267.
445. Photinia parvifolia, I. 189.
446. Crataegus hupehensis, I. 178, 181
447. { Viburnum brevipes, I. 113.
{ Viburnum corylifolium, I. 112.
447a. Viburnum brevipes, I. 310.
448. Smilax china, III. 4.
449. Photinia serrulata, I. 184.
450. Evonymus lanceifolia, I. 491.
451. Malus theifera, II. 283.
452. Pterocarya Paliurus, III. 182.
453. Rhus succedanea, II. 182.
453b. Spondias axillaris, II. 172.
454. Parthenocissus Henryana, I. 101.
455. Smilax scobinicaulis, III. 2.
456. Pittosporum glabratum, III. 326.
457. Lonicera Maackii, var. podocarpa, I. 140.
458. Ardisia Henryi, II. 582.
459. Evonymus aculeatus, I. 490.
460. Fagus longipetiolata, III. 190.
461. Ilex szechwanensis, I. 80.
462. Phoebe neurantha, II. 72.
463. Viburnum dasyanthum, I. 115.
464. Parthenocissus tricuspidata, I. 102.
465. Photinia amphidoxa, I. 190.

466. Cotoneaster Dielsiana, I. 166.
467. Viburnum dasyanthum, I. 115.
468. Photinia Beauverdiana, var. notabilis, I. 188.
469. Parthenocissus spec., III. 428.
470. Sorbus caloneura, II. 269, 272, 274.
470a, b. Sorbus Keissleri, II. 269, 272, 274.
471. Ligustrum myrianthum, II. 607.
472. Diospyros kaki, II. 588.
473. Decumaria sinensis, I. 152.
474. Smilax spec., III. 455.
475. Rhamnus leptophyllus, II. 239.
476. Photinia Schneideriana, I. 188.
477. Ilex pedunculosa, f. β continentalis, I. 76.
478. Evonymus japonica, var. acuta, I. 485.
479. Pyrus serrulata, II. 263.
479a, b. Pyrus serotina, II. 263.
480. Spondias axillaris, II. 172.
480a. Spondias axillaris, var. pubinervis, II. 173.
481. Cotoneaster Dammeri, I. 176.
482. Rubus Lambertianus, I. 51.
482a. Rubus Lambertianus, III. 423.
483. Smilax longipes, III. 5.
484. Photinia Davidsoniae, I. 185.
485. Schisandra propinqua, var. sinensis, I. 416.
486. Tilia tuan, var. chinensis, II. 369.
487. Idesia polycarpa, II. 284.
488. Photinia subumbellata, I. 189.
489. Smilax longipes, III. 5.
490. Spiraea Henryi, I. 447.
491. Hydrangea anomala, I. 34, 36, 40.
492, 492a. Pieris ovalifolia, var. elliptica, I. 552.
493. Fagus longipetiolata, III. 190.
494, 494a. Dalbergia hupeana, II. 115.
495. Platycarya strobilacea, III. 180.
496. Cotoneaster horizontalis, var. perpusilla, I. 155.
497. Pterostyrax hispidus, I. 295, 580.
498. Viburnum propinquum, I. 111.
499. Sorbaria arborea, var. glabrata, I. 48.
499a. Sorbaria arborea, I. 48.
500, 500a. Poliothyrsis sinensis, I. 285.
501. Cercis chinensis, II. 87.
502. Celastrus gemmata, II. 352.
503. Celastrus Loeseneri, II. 350.

504. Evonymus acanthocarpa, var. sutchuenensis, I. 490.
505. Evonymus japonica, var. acuta, I. 485.
506. Millettia Dielsiana, II. 101.
507. Castanea Seguinii, III. 194.
507a. Castanea mollissima, III. 192.
508, 508a, b, c, d. Carpinus cordata, var. chinensis, II. 437.
509. Rhododendron sutchuenense, I. 544.
510. Davidia involucrata, II. 255.
511. Diospyros kaki, var. silvestris, II. 590.
512. Actinidia callosa, var. Henryi, II. 382.
513. Liquidambar formosana, I. 421.
514. Cynanchum spec., III. 547.
515. Quercus aliena, var. acuteserrata, f. calvescens, III. 215.
516. Quercus aliena, var. acuteserrata, III. 215.
516a. Quercus Fabri, III. 216.
517. Quercus aliena, var. acuteserrata, III. 215.
518. Quercus glandulifera, III. 212.
519. Quercus glandulifera, III. 212.
520. Quercus glandulifera, III. 212.
521. Quercus glandulifera, III. 212.
522. Quercus Fabri, III. 216.
523. Quercus Fabri, III. 216.
524. Quercus glandulifera, III. 212.
525. Quercus glandulifera, III. 212.
526. Quercus glandulifera, III. 212.
527. Quercus aliena, var. acuteserrata, III. 215.
527a. Quercus Fabri, III. 216.
527b. Quercus aliena, III. 214.
528. Quercus glandulifera, III. 212.
529. Quercus aliena, var. acuteserrata, III. 215.
530. Quercus glandulifera, III. 212.
531. Quercus aliena, var. acuteserrata, f. calvescens, III. 215.
532. Quercus variabilis, III. 219.
533. Quercus variabilis, III. 219.
534. Quercus variabilis, III. 219.
535. Quercus variabilis, III. 219.
536. Quercus variabilis, III. 219.
536a. Quercus serrata, III. 217.
537. Quercus variabilis, III. 219.
538. Quercus variabilis, III. 219.
539. Quercus variabilis, III. 219.

540. Quercus spathulata, III. 226.
541. Quercus glauca, III. 226.
541a. Quercus glauca, III. 226.
542. Castanopsis sclerophylla, III. 201.
542a. Quercus glauca, III. 226.
543, 543a, b, c. Lithocarpus Henryi, III. 209.
544. Quercus oxyodon, III. 228.
545. Quercus glauca, III. 226.
546. Quercus glandulifera, III. 212.
547. Castanea mollissima, III. 192.
548. Castanea mollissima, III. 192.
549. Castanea mollissima, III. 192.
550. Castanea mollissima, III. 192.
551. Castanea Henryi, III. 196.
551a. Castanea Henryi, III. 196.
551b. Castanea Henryi, III. 196.
551c. Castanea Henryi, III. 196.
552. Castanea Henryi, III. 196.
553. { Sorbus Wilsoniana, I. 458. Sorbus alnifolia, II. 270.
554. Berberis brachypoda, I. 375.
555. Berberis Sargentiana, I. 359.
556, 556a. Pyrus Calleryana, II. 264.
556b, c. Pyrus serotina, II. 263.
557. { Evonymus kiautschovica, var. patens, I. 486. Evonymus myriantha, I. 487.
558, 558a. Evonymus cornuta, I. 489.
559. Evonymus venosa, I. 488.
560. Celastrus Loeseneri, II. 350.
561, 561a. Celastrus Hindsii, var. Henryi, II. 353.
562. { Evonymus japonica, var. acuta, I. 485. Evonymus kiautschovica, var. patens, I. 486.
562a. Evonymus japonica, var. acuta, I. 485.
563. Lindera glauca, II. 80.
564. Berberis Sargentiana, I. 359.
565. Fortunearia sinensis, I. 428.
567. Rhododendron stamineum, I. 546.
568. Spiraea Veitchii, I. 449.
569. Rhododendron indicum, var. ignescens, I. 547.
570. Deutzia discolor, I. 12.
571. Hydrangea strigosa, var. macrophylla, I. 32.
572. Tsuga chinensis, II. 37.
573. Hydrangea xanthoneura, var. setchuenensis, III. 421.
574. Philadelphus incanus, I. 5.

575. Lespedeza formosa, II. 107.
576. Campylotropis macrocarpa, II. 113.
577. Spiraea Miyabei, var. glabrata, I. 454.
578, 578a. Trachelospermum jasminoides, III. 334.
579. Spiraea japonica, var. acuminata, I. 452.
580. Hydrangea longipes, I. 33.
581. Philadelphus Wilsonii, I. 4.
582. Hypericum Ascyron, II. 402.
583. Philadelphus incanus, I. 5.
584. Sinowilsonia Henryi, I. 429.
585, 585a. Cedrela sinensis, II. 156.
586. Rhododendron discolor, I. 542.
587. Clematis montana, var. rubens, I. 333.
588. Euptelea Franchetii, I. 314.
589. Lonicera similis, var. Delavayi, I. 142.
590. Viburnum betulifolium, I. 114.
591. Tilia Oliveri, II. 366.
592. Cornus Walteri, II. 576.
593. Celtis cerasifera, III. 271.
594. Tilia paucicostata, II. 363.
595, 595a. Celtis Bungeana, III. 269.
596. Alangium chinense, II. 552.
597, 597a. Tilia Henryana, II. 367.
597c. Tilia Oliveri, var. cinerascens, II. 367.
598. Jasminum Giraldii, II. 614.
599. Ligustrum sinense, var. nitidum, II. 606.
601. Viburnum hupehense, I. 115.
602. Kalopanax ricinifolius, II. 564.
603. Vitis Davidii, var. cyanocarpa, I. 104.
604. Ampelopsis heterophylla, var. Delavayana, I. 100.
606. Rhododendron Mariesii, I. 548.
607. Abelia umbellata, I. 122.
608. Rhododendron Augustinii, I. 524.
609, 609a. Rosa Gentiliana, II. 312.
610. Photinia villosa, var. sinica, I. 186.
611. Prunus Persica, I. 273.
612. Rhamnus hupehensis, II. 236.
613. { Buddleia Davidii, var. magnifica, I. 567. Buddleia Davidii, var. superba, I. 568.
613a. Buddleia Davidii, I. 567.

614. Nothopanax Davidii, II. 556.
615. Tilia Oliveri, II. 366.
616. Rosa multiflora, var. cathayensis, II. 304.
617. Sinomenium acutum, var. cinereum, I. 387.
617a. Sinomenium acutum, I. 387.
618. Rhamnus leptophyllus, II. 239.
619. Rosa Banksiae, var. normalis, II. 317.
619a. Rosa Banksiae, f. lutescens, II. 317.
619b, c, d. Rosa Banksiae, var. normalis, II. 317.
620. Acanthopanax setchuenensis, II. 559.
621. Diospyros Lotus, II. 587.
622. Emmenopterys Henryi, III. 391.
623. Rhamnus utilis, II. 240.
624, 624a. Hamamelis mollis, I. 431.
625. Rosa corymbulosa, II. 323.
626. Cedrela microcarpa, II. 157.
627. Smilax scobinicaulis, III. 2.
628. Rosa Giraldii, var. venulosa, II. 328.
629. Photinia Beauverdiana, var. notabilis, I. 188.
630, 630a. Rosa corymbulosa, II. 323.
631. Ligustrum Quihoui, II. 607.
632. Dalbergia stenophylla, II. 116.
633. Callicarpa Giraldiana, III. 457.
634. Tilia Oliveri, II. 366.
635, 635a, b, c, d, e. Celtis Julianae, III. 265.
636. Catalpa Fargesii, I. 305.
637. { Forsythia suspensa, f. atrocaulis, I. 580.
Forsythia suspensa, f. pubescens, I. 302.
638. Acer nikoense, var. megalocarpum, I. 98.
639. Pterocarya hupehensis, III. 182.
040. Catalpa Duclouxii, I. 304.
641. Orixa japonica, II. 135.
642. Acer pictum, var. parviflorum, I. 83.
643. Lespedeza formosa, II. 107.
644. Lespedeza Buergeri, II. 106.
645, 645a. Carpinus Henryana, II. 429.
646. Carpinus Seemeniana, II. 430.
647. Abies chensiensis, II. 44.
648. Holostemma sinense, III. 344.
648a. Holostemma sinense, III. 344.

649. Acer Davidii, I. 92.
650. Pittosporum truncatum, III. 328.
651. Sophora japonica, II. 96.
652. Magnolia officinalis, I. 391.
653. Juglans regia, III. 184.
654, 654a. Juglans regia, III. 184.
657. Pittosporum glabratum, III. 326.
659. Tetracentron sinense, I. 417.
659a. Tetracentron sinense, I. 417.
660. Rhododendron micranthum, I. 513.
661. Smilax megalantha, III. 4.
662. Pyracantha crenulata, I. 177.
663. Rubus ichangensis, I. 50.
664. Ilex micrococca, I. 82.
665. Clematis grata, var. lobulata, I. 337.
666. Rosa Helenae, II. 310.
666a. Rosa Rubus, II. 308.
667. Firmiana simplex, II. 377.
668. Clematis Pavoliniana, III. 434.
669. Viburnum betulifolium, I. 114.
670. Ardisia japonica, II. 582.
671. Smilax scobinicaulis, III. 2.
672, 672a. Clematis Gouriana, var. Finetii, I. 339.
673. Clematis lasiandra, I. 322.
674. Heterosmilax Gaudichaudiana, III. 13.
675. Smilax spec., III. 455.
676. Viburnum brevipes, I. 113, 310.
677. Smilax longipes, III. 5, 455.
678. Smilax micropoda, var. reflexa, III. 6.
679, 679a. Clematis lasiandra, I. 322.
680. Smilax scobinicaulis, III. 2.
681. Sarcococca ruscifolia, II. 163.
682, 682a, b. Dregea sinensis, III. 352.
683. { Acer tetramerum, I. 94.
Acer flabellatum, I. 91.
684. Acer spec., III. 427.
685. Photinia Davidsoniæ, I. 185.
686. Castanopsis Fargesii, III. 198.
687. Quercus glauca, f. gracilis, III. 228.
688. Celastrus Loeseneri, II. 350.
689. Ilex purpurea, var. a Oldhamii, I. 76.
696, 696a. Juniperus formosana, II. 56.
697. Viburnum cylindricum, I. 110.
698. Berberis triacanthophora, I. 358.
699. Ligustrum Henryi, II. 601.

700, 700a, b. Fagus longipetiolata, III. 190.
701. Liriodendron chinense, III. 444.
702. Æsculus Wilsonii, III. 444.
703. Fagus Engleriana, III. 191.
704. Fagus Engleriana, III. 191.
705, 705a. Betula albo-sinensis, III. 454.
706. Populus lasiocarpa, III. 17.
706a, b. Populus Wilsonii, III. 16.
707. Cercis racemosa, II. 88.
708. Fraxinus platypoda, II. 623.
709. Maackia hupehensis, II. 98.
710. Sassafras tzumu, II. 74.
711. Tilia Oliveri, II. 366.
712. Corylus tibetica, II. 443.
714. Viburnum sympodiale, III. 429.
715. Fagus lucida, III. 191.
716. Viburnum buddleifolium, III. 429.
717. Abelia Graebneriana, III. 429.
718. Symphoricarpos sinensis, I. 117.
719. Acer griseum, III. 427.
720. Salix Fargesii, III. 47.
720a, b, c. Salix Fargesii, III. 47.
722. Populus tremula, var. Davidiana, III. 24.
724. Populus adenopoda, III. 23.
725. Hamamelis mollis, III. 445.
726. Ulmus parvifolia, III. 244.
729. Corylus spec., III. 454.
730. Cercis chinensis, III. 447.
738. Rhamnus Leveilleanus, II. 237.
739. Rhamnus leptacanthus, II. 236.
742, 742a, b. Cercidiphyllum japonicum, var. sinense, I. 316.
743. Ulmus Bergmanniana, III. 240.
744. Prunus Persica, III. 432.
745. Ulmus Wilsoniana, III. 238.
746. Ulmus Wilsoniana, III. 238.
747. Abelia parvifolia, I. 121, 124.
748. Catalpa Fargesi, I. 305.
749. Morus cathayana, III. 292.
750. Rubus flagelliflorus, III. 423.
751. Rubus irenaeus, III. 423.
752. Parthenocissus Thomsonii, I. 101.
753, 753a. Celastrus angulata, II. 346.
754. Ligustrum sinense, var. Stauntonii, II. 606.
754a. Ligustrum sinense, var nitidum, II. 606.
754b. Ligustrum sinense, var. Stauntonii, II. 606.
755. Clausena suffruticosa, II. 140.

756. Ilex metabaptista, I. 76.
757. Hydrangea strigosa, var. macrophylla, I. 32.
758. Spiraea chinensis, I. 444.
759. Sapindus mukorossi, II. 191.
760. Gymnocladus chinensis, II. 91.
761. Cercis chinensis, II. 87.
762. Diervilla japonica, var. sinica, III. 430.
763. Clematis heracleaefolia, var. ichangensis, I. 321.
764. Maesa hupehensis, II. 583.
765. Hydrangea strigosa, I. 31.
766. Campylotropis macrocarpa, II. 113.
767. Deutzia Schneideriana, var. laxiflora, I. 7.
768. Dalbergia hupeana, II. 115.
769. Paulownia tomentosa, var. lanata, I. 574.
769a. Paulownia recurva, I. 577.
770. Enkianthus quinqueflorus, var. serrulatus, I. 550.
771. Spiraea chinensis, I. 444.
772. Hydrangea Sargentiana, I. 29.
773. Hydrangea strigosa, I. 31.
774. Lespedeza formosa, II. 107.
775. Gymnosporia variabilis, II. 359.
776. Trachelospermum spec., III. 457.
777. Deutzia discolor, III. 421.
778. Ligustrum Quihoui, II. 607.
779. Pyrus serrulata, II. 263.
780. Ligustrum Quihoui, II. 607.
781. Jasminum lanceolarium, II. 612.
781a. Jasminum lanceolarium, var. puberulum, II. 612.
782. Quercus aliena, III. 214.
783. Quercus Fabri, III. 216.
784. Quercus spec., III. 457.
785. Quercus glandulifera, III. 212.
786. { Indigofera amblyantha, II. 99. Indigofera pseudotinctoria, II. 100.
787. Lespedeza formosa, II. 107.
788. Pterocarya stenoptera, III. 181.
789. Jasminum floridum, II. 614.
790. Vitex Negundo, III. 372.
791. Evodia Bodinieri, II. 130.
792. Albizzia kalkora, III. 446.
794, 794a, b. Cunninghamia lanceolata, II. 50.
795, 795a. Liquidambar formosana, var. monticola, I. 422.

796. Viscum album, III. 318.
797, 797a. Keteleeria Davidiana, II. 39.
798. Cupressus funebris, II. 55.
798a. Cupressus torulosa, II. 54.
799. Gleditsia macracantha, II. 90.
800. Rhododendron sinense, I. 549.
801. Lonicera mucronata, I. 136.
802. Daphne Wilsonii, II. 540.
803. Ribes Meyeri, *a* tanguticum, I. 44.
804. Rubus thibetanus, I. 54.
805. Viburnum erubescens, var. Prattii, I. 107.
805a. Viburnum oliganthum, I. 108.
806. Rubus xanthocarpus, I. 49.
808, 808a. Lonicera deflexicalyx, I. 140.
809. Cornus poliophylla, II. 574.
810. { Prunus Twymaniana, I. 211.
      { Prunus lobulata, I. 220.
811. Prunus perulata, I. 61.
813. Rubus pileatus, I. 52.
815. Rubus Giraldianus, I. 55.
816. Betula luminifera, II. 455.
817. Ribes laurifolium, I. 46.
818. Sabia latifolia, II. 195.
820. Ulmus Bergmanniana, var. lasiophylla, III. 241.
821. Litsea citrata, II. 75.
822. Ribes moupinense, γ laxiflorum, I. 44.
823, 823a. Ribes tenue, I. 45.
824. Prunus micromeloides, I. 218.
825. Cornus chinensis, II. 577.
826. Akebia lobata, var. australis, III. 434.
827. Lonicera thibetica, I. 130.
828. Rubus tricolor, I. 49.
829. Gaultheria Veitchiana, I. 554.
829a, 829b. Gaultheria Veitchiana, I. 554.
830. Rubus amabilis, I. 52.
831. Lonicera tangutica, I. 132.
831a. Lonicera Schneideriana, I. 133.
831b. { Lonicera saccata, I. 133.
       { Lonicera tangutica, I. 132.
831c. Lonicera szechuanica, I. 132.
832. Rubus biflorus, var. quinqueflorus, I. 53.
833. Lonicera nitida, I. 580.
834. Rubus pungens, I. 52.
835. Eleagnus umbellata, II. 410.
836. Ribes alpestre, β giganteum, I. 45.

838. Magnolia Nicholsoniana, I. 394, 404, 407.
841, 841a. Sophora viciifolia, II. 95.
844. Buddleia officinalis, I. 565.
850. Rubus macilentus, I. 53.
854. Actinidia spec., II. 385.
854a. Actinidia kolomikta, II. 380.
855. Clematis Armandi, I. 326.
856. { Lonicera trichosantha, I. 141.
     { Lonicera deflexicalyx, I. 140.
856a, 856b. Lonicera trichosantha, I. 141.
857. Cotoneaster moupinensis, I. 163.
858, 858a, b. Rubus pileatus, var., I. 52.
859. Viburnum kansuense, I. 116.
860. Sabia Schumanniana, I. 196.
861. Lonicera mupinensis, I. 138.
862, 862a. Rhamnus Sargentianus, II. 235.
864. Sorbus unguiculata, I. 473.
864, 864a. Sorbus multijuga, var. microdonta, I. 473.
864b. Sorbus setschwanensis, I. 475.
865. Acanthopanax setchuenensis, II. 559.
866. Schisandra sphenanthera, I. 414.
867. Litsea pungens, II. 76.
868. Clematis Spooneri, I. 334.
868a. Clematis montana, var. Wilsonii, I. 333.
868b. Clematis Spooneri, I. 334.
869, 869a. Schisandra sphenanthera, I. 414.
870. Ribes Maximowiczii, I. 46.
871. Rubus setchuenensis, I. 51; III. 423.
872, 872a. Rhamnus Rosthornii, II. 236.
873. Cotoneaster bullata, var. macrophylla, I. 164.
874. Sorbus setschwanensis, I. 475.
874a. Sorbus unguiculata, I. 473.
875. Nothopanax Davidii, II. 556.
876. Triosteum himalayanum, var. chinense, I. 117.
877. Lonicera pileata, I. 135.
878. Tetrastigma obtectum, var. pilosum, I. 99.
879. Rhododendron moupinense, I. 525.
882, 882a. Rhododendron Hanceanum, I. 517.
883. Dipteronia sinensis, I. 83.
884. Styrax Wilsonii, I. 293, 580.

885. Piptanthus concolor, II. 99; III. 447.
886, 886a, b, c. Clematoclethra lasioclada, var. grandis, II. 386.
887. Sorbus Sargentiana, I. 461.
888. Actinidia venosa, II. 383.
889. Holboellia grandiflora, I. 346.
890. Clematoclethra actinidioides, II. 386.
890a. Actinidia tetramera, II. 381.
891. Actinidia venosa, II. 383.
892. Ilex fragilis, α genuina, I. 81.
892a. Ilex fragilis, β Kingii, I. 82.
893. Actinidia purpurea, II. 378.
894. Stauntonia sp. nov.? I. 345.
895, 895a, b. Clematoclethra scandens, II. 387.
896. Ribes Meyeri, atanguticum, I. 44.
897. Schisandra sphenanthera, I. 414.
898, 898a. Ribes longeracemosum, Davidii, I. 45; III. 423.
899. Prunus brachypoda, var. pseudossiori, I. 65.
900, 900a. Betula albo-sinensis, var. septentrionalis, II. 458.
901. Rubus pungens, III. 423.
902. Lonicera subdentata, I. 136.
902a. Lonicera setifera, var. trullifera, I. 136.
903. Larix Potaninii, II. 18.
904. Prunus conadenia, I. 197.
904a. Prunus pleiocerasus, I. 198.
905. Pinus sinensis, var. densata, II. 17.
906. Larix Mastersiana, II. 19.
906a. Larix Potaninii, II. 18.
907. Prunus plurinervis, I. 208.
909. { Maddenia hypoxantha, I. 57.
       { Maddenia Wilsonii, I. 58.
910. Larix Potaninii, II. 18.
911, 911a. Prunus tomentosa, var. endrotricha, I. 225.
912. Prunus lobulata, I. 220.
913. Ribes Vilmorinii, I. 45.
914. Magnolia Sargentiana, I. 398.
915. Betula luminifera, II. 455.
916. Neillia longiracemosa, I. 434.
916a. Neillia affinis, I. 434.
917, 917a, b. Sloanea Hemsleyana, II. 361.
918. Viburnum cordifolium, I. 109.
919. Morus notabilis, III. 293.
920, 920a, b. Gaultheria pyroloides, var. cuneata, I. 554.

921, 921a, b. Schisandra rubiflora, I. 412.
922. Sorbus scalaris, I. 462.
923. Magnolia Sargentiana, I. 398.
923a. Magnolia Sargentiana, var. robusta, I. 399.
924. Gaultheria nummularioides, I. 555.
925. { Akebia lobata, I. 348.
      { Akebia lobata, var. australis, I. 348.
927, 927a. Lonicera lanceolata, I. 140.
928. Hippophæ rhamnoides, var. procera, II. 409.
928a. Hippophæ rhamnoides, II. 409.
929. Ribes longiracemosum, α Davidii, I. 45.
930. Berberis aemulans, I. 353; III. 434.
931. Rosa Moyesii, f. rosea, II. 325.
932. Actinidia coriacea, II. 384.
933. Celtis Biondii, var. Cavaleriei, III. 273.
934, 934a. Actinidia polygama, II. 380.
935. Rubus sepalanthus, III. 423.
936. Lonicera similis, var. Delavayi, I. 142.
937. Alangium platanifolium, II. 554.
938. Lonicera alseuosmoides, I. 141.
939, 939a. Perrottetia racemosa, II. 359.
940. Lonicera subaequalis, I. 142.
940a. Lonicera subaequalis, III. 430.
941. Sorbus unguiculata, I. 473.
942. Lonicera chaetocarpa, I. 137.
944. Clematoclethra Franchetii, II. 388.
946. Rubus inopertus, I. 54.
948. Rubus vicarius, I. 56.
950. Lonicera tangutica, I. 132.
952, 952a. Celastrus glaucophylla, II. 347.
955. Berberis Silva-Taroucana, I. 370.
956. Sorbus megalocarpa, II. 266.
958. Ribes Maximowiczii, I. 46.
959, 959a, b. Rosa omeiensis, II. 331.
960, 960a. Hosiea sinensis, II. 190.
961. Zanthoxylum Bungei, II. 121.
962. Idesia polycarpa, var. vestita, I. 285.
963. Viburnum Davidii, I. 111.
963a. Viburnum Davidii, III. 429.
964. Zanthoxylum stenophyllum, II. 127.
965, 965a, b. Spiraea Rosthornii, I. 451.

967. Evonymus porphyrea, I. 495.

967a. { Evonymus cornuta, I. 489.
{ Evonymus porphyrea, I. 495.

968. Evonymus sanguinea, var. β camptoneura, I. 494.

968. Evonymus porphyrea, I. 495.

969, 969a. Juglans regia, III. 456.

971. Vaccinium japonicum, I. 562.

972. Celtis Biondii, var. Cavaleriei, III. 273.

973. Picea asperata, var. notabilis, II. 23.

974. Neillia longiracemosa, I. 434.

975. Malus prunifolia, var. rinki, II. 279.

976, 976a. Betula utilis, var. Prattii, II. 457.

977. { Prunus obtusata, I. 66.
{ Prunus bicolor, I. 69.

978. Prunus lobulata, I. 220.

979. Acer laevigatum, I. 92.

980. Prunus pubigera, var. Potaninii, I. 68.

981. Prunus pleuroptera, I. 221.

983, 983a, b. Betula japonica, var. szechuanica, II. 461; III. 454.

984. Prunus pleuroptera, I. 221.

985. Juniperus squamata, II. 57.

986. Prunus brachypoda, var. pseudossiori, III. 424.

987. Lindera cercidifolia, II. 85.

988. Prunus serrula, var. tibetica, I. 213.

989. Spiraea myrtilloides, I. 440.

990. Betula utilis, var. Prattii, II. 457.

991. Sorbus munda, f. a tatsienensis, I. 469.

992. Rubus aurantiacus, I. 56.

993. Prunus salicina, I. 276, 580.

994. Evodia velutina, II. 134.

995. Cotoneaster Franchetii, I. 165.

996, 996a. Ilex ciliospina, I. 78.

997. Sorbus caloneura, II. 269.

998. Triosteum himalayanum, var. chinense, III. 429.

999. Ribes himalayense, a glandulosum, I. 44.

1000, 1000a. Malus yunnanensis, II. 287.

1001. Acanthopanax leucorrhizus, var. fulvescens, I. 558.

1002. Rubus Fockeanus, I. 48.

1003. Clematis montana, var. Wilsonii, f. platysepala, I. 334.

1004. Acer fulvescens, I. 84.

1005, 1005a. Acer Davidii, I. 92.

1006. Acer flabellatum, I. 91.

1007, 1007a. Acer laxiflorum, I. 93.

1008, 1008a. Acer Davidii, I. 92.

1009. Acer cappadocicum, var. sinicum, I. 85.

1011. Stellera chamaejasme, II. 551.

1012, 1012a. Berberis Silva-Taroucana, I. 370.

1013. Rhamnus heterophyllus, II. 232.

1014. Acanthopanax Giraldii, II. 560.

1015. Sorbus Conradinae, I. 460.

1016. Osteomeles Schwerinae, var. microphylla, III. 431.

1017. Cornus Walteri, II. 576.

1018. Prunus mume, I. 278.

1019. Smilax vaginata, III. 2.

1020. Corylopsis platypetala, var. levis, I. 427.

1021, 1021a. Juniperus squamata, var. Fargesii, II. 59.

1022. Evonymus grandiflora, I. 484.

1023. Acanthopanax leucorrhizus, var. fulvescens, II. 558.

1024. Ilex yunnanensis, I. 76.

1025. Viburnum lobophyllum, I. 114.

1025a. Viburnum Wilsonii, I. 112.

1026. Prunus glyptocarya, I. 219.

1027. Prunus salicina, I. 276, 580.

1029, 1029a. Actinidia venosa, II. 383.

1030. Clematoclethra lasioclada, var. grandis, II. 386.

1031. Viburnum oliganthum, I. 108.

1034. Ilex Fargesii, β megalophylla, I. 77.

1035. { Sorbus Rehderiana, var. grosseserrata, I. 465.
{ Sorbus multijuga, I. 472.

1036, 1036a. Dipelta ventricosa, I. 118.

1038. Berberis yunnanensis, I. 355.

1039. Berberis Mouillacana, I. 371.

1040, 1040a. Clerodendron trichotomum, var. Fargesii, III. 376.

1041. Berberis Mouillacana, I. 371.

1042. Rubus mesogaeus, f., I. 56.

1043. Viburnum betulifolium, I. 114.

1044. Acanthopanax setchuenensis, II. 559.

1045. Prunus pubigera, var. obovata, I. 68.

1046. Vitis betulifolia, I. 103.
1046a. Vitis betulifolia, III. 429.
1047. Evonymus cornuta, I. 489.
1048. Berberis polyantha, I. 376.
1049. Evonymus cornuta, I. 489.
1050. Berberis aggregata, I. 375.
1050a. Berberis aggregata, var. Prattii, I. 376; III. 443.
1051. Smilax trachypoda, III. 3.
1052. Rubus ichangensis, I. 50.
1053. Rosa multibracteata, II. 328.
1054. Skimmia melanocarpa, I. 138.
1055. Rosa multibracteata, II. 328.
1056. Rosa Moyesii, II. 325.
1057. Chionanthus retusus, II. 611.
1058, 1058a. Actinidia kolomikta, II. 380.
1059. Berberis Silva-Taroucana, I. 370.
1060. Rosa Davidii, II. 322.
1061. Cotoneaster multiflora, I. 170.
1062. Rosa Moyesii, f. rosea, II. 325.
1063. Rosa Davidii, II. 322.
1064. Stranvaesia Davidiana, I. 192.
1065. Smilax trachypoda, III. 3.
1067. Sophora Wilsonii, II. 94.
1068. Schizophragma integrifolium, I. 41.
1069. Acer laxiflorum, I. 93.
1070. Schisandra propinqua, var. sinensis, I. 416.
1071. Cotoneaster Dammeri, var. radicans, I. 176.
1072. Vaccinium urceolatum, I. 560.
1073. Berberis aggregata, var. recurvata, I. 377; III. 443
1074. Ribes spec., III. 423.
1075, 1075a. Ligustrum Delavayanum, II. 601.
1076. Eleagnus multiflora, f. angustata, II. 413.
1077, 1077a. Lonicera chaetocarpa, I. 137.
1078. Prunus rufomicans, I. 65.
1079. Vaccinium fragile, I. 559.
1079a. Desmodium floribundum, II. 103.
1083. Berberis verruculosa, I. 357.
1089. Evonymus acanthocarpa, I. 490.
1091. Lonicera spec., III. 430.
1092. Quercus glandulifera, III. 456.
1093. Quercus Fabri, III. 216.
1094. Quercus aliena, var. acuteserrata, III. 215.

1095. Quercus aliena, var. acuteserrata, III. 215.
1095a. Quercus glandulifera, III. 212.
1096, 1096a, b. Castanopsis platyacantha, III. 200.
1097. Pinus sinensis, II. 15.
1098, 1098a. Rosa longicuspis, II. 313.
1099. Rosa Davidii, var. elongata, II. 323.
1100. Clematis grata, ' ar andidentata, I. 338.
1102. Cladrastis Wilsonii, II. 97.
1102a. Cladrastis sinensis, II. 97.
1104. Rosa Moyesii, f. rosea, II. 325.
1105, 1105a. Evonymus lanceifolia, I. 491.
1106. Celastrus rugosa, II. 349.
1107, 1107a. Malus Prattii, II. 281.
1108. Viburnum cinnamomifolium, I. 111.
1109. Parthenocissus himalayana, var. rubrifolia, I. 101.
1110. Acer caudatum, var. multiserratum, I. 91.
1111. Lysionotus Wilsonii, III. 388.
1111a. Lysionotus brachycarpus, III. 387.
1113. Acanthopanax setchuenensis, II. 559.
1114. Rosa Davidii, var. elongata, II. 323.
1115. Cephalotaxus drupacea, var. sinensis, II. 3.
1116. Kadsura spec., I. 411.
1117. Actinodaphne confertifolia, II. 74.
1119. Smilax spec., III. 455.
1120. Viburnum Wilsonii, I. 112.
1121. Prunus salicina, I. 276, 580.
1122. Jasminum urophyllum, II. 613.
1123, 1123a. Rosa Moyesii, f. rosea, II. 325.
1124. Chloranthus brachystachys, III. 15.
1125. Rosa Brunonii, II. 306.
1126. Rosa Davidii, var. elongata, II. 323.
1127. Rubus Gentilianus, I. 55.
1128. Sorbus Esserteauiana, I. 459.
1129. Malus theifera, II. 283.
1130. Sarcococca Hookeriana, var. digyna, II. 164.

1131. Viburnum foetidum, var. rectangulum, I. 112.
1132. Evonymus Rehderiana, I. 488.
1133. Cotoneaster salicifolia, I. 172.
1133a. Cotoneaster salicifolia, var. floccosa, I. 173.
1134. Rosa Murielae, II. 326.
1135. { Lonicera ligustrina, I. 134.
{ Lonicera pileata, I. 135.
1136. Nothopanax Davidii, II. 557.
1137, 1137a. Berberis Gagnepainii, I. 358.
1138. Betula utilis, var. Prattii, II. 457.
1140. Betula Potaninii, II. 459.
1141, 1141a. Castanea mollissima, III. 192.
1142. Castanopsis ceratacantha, III. 199.
1143. Quercus glandulifera, III. 212.
1144. Juglans regia, III. 184.
1145, 1145a. Juglans regia, III. 184.
1146. Pittosporum heterophyllum, III. 329.
1147. Celastrus Rosthorniana, II. 351.
1148. Celastrus rugosa, II. 349.
1149. Corylus heterophylla, var. sutchuenensis, II. 445.
1150. Rosa Prattii, II. 329.
1151, 1151a, b. Pinus Armandi, II. 12.
1153. Spiraea japonica, var. ovalifolia, I. 452.
1154. Acer laxiflorum, I. 93.
1155. Enkianthus deflexus, I. 550.
1156. Hydrangea Rosthornii, I. 33.
1157. Pieris ovalifolia, var. elliptica, I. 552.
1158. Spiraea mollifolia, I. 441.
1159. Hydrangea Davidii, I. 25.
1159a. Hydrangea Davidii, III. 421.
1160. Spiraea Veitchii, I. 449.
1161. Acer caudatum, var. multiserratum, I. 91.
1162. Acer fulvescens, I. 84.
1164. Evodia Baberi, II. 131.
1166. Berberis Boschanii, I. 369.
1168. Pterostyrax hispidus, I. 295.
1172. Spiraea Henryi, I. 447.
1174. Zanthoxylum Piasezkii, II. 122.
1175. Celastrus Rosthorniana, II. 351.
1176. Celastrus spiciformis, var. laevis, II. 349.
1177. Berberis Tischleri, I. 355.

1178. Rosa Davidii, var. elongata, II. 323.
1179. Sophora glauca, var. albescens, III. 447.
1180. Berberis Francisci-Ferdinandi, I. 367.
1182. Cassiope selaginoides, I. 551.
1183. Hydrangea xanthoneura, var. Wilsonii, I. 27, 150.
1184. Celastrus Hookeri, II. 352.
1186. Deutzia longifolia, I. 13.
1187. Evonymus Sargentiana, I. 487.
1188. Deutzia glomeruliflora, I. 10.
1188a. Deutzia subsessilis, I. 11.
1189. Zanthoxylum Bungei, II. 121.
1190, 1190a. Dregea corrugata, III. 353.
1192. Clethra monostachya, I. 501.
1193. Spiraea japonica, var. ovalifolia, I. 452.
1194. Schizophragma integrifolium, I. 41.
1195. Rhododendron lutescens, I. 516.
1196. Rhododendron yanthinum, I. 518.
1196a. Rhododendron yanthinum, I. 518.
1197. Rhododendron Augustinii, I. 524.
1197a. Rhododendron lutescens, I. 516.
1198. Rhododendron Hunnewellianum, I. 535.
1199. Rhododendron lutescens, I. 516.
1200. Rhododendron micranthum, I. 513.
1201. Rhododendron yanthinum, I. 518.
1202. Rhododendron flavidum, I. 512.
1203. Rhododendron pachytrichum, I. 530.
1204. Rhododendron longistylum, I. 514.
1205. Rhododendron polylepis, I. 521.
1206. Rhododendron Watsonii, I. 545.
1207. Rhododendron Augustinii, I. 524.
1207a. Rhododendron polylepis, I. 521.
1208. Rhododendron Sargentianum, I. 504.
1209, 1209a. Rhododendron decorum, I. 541.
1210. Rhododendron argyrophyllum, I. 525.
1211. Rhododendron oreodoxa, I. 540
1212. Carrierea calycina, I. 284.

1213, 1213a. Potentilla fruticosa, II. 301.

1213a. Potentilla fruticosa, var. albicans, II. 302.

1214. Gleditsia sinensis, II. 91.

1215. Evonymus subsessilis, I. 489.

1216. Evonymus subsessilis, var. latifolia, I. 489.

1217. Syringa Komarowii, I. 301.

1218, 1218a. Quercus oxyodon, var. Fargesii, III. 229.

1219, 1219a. Pittosporum daphniphylloides, III. 326.

1220, 1220a. Rhododendron villosum, I. 524.

1221, 1221a, b. Rhododendron polylepis, I. 521.

1222. Rhododendron Souliei, I. 537.

1223. Rhododendron Davidsonianum, I. 515.

1224. Rhododendron calophytum, I. 544.

1225. Rhododendron Websterianum, I. 511.

1226. Carpinus Henryana, II. 429.

1227. Hydrangea villosa, I. 29.

1228. Rosa filipes, II. 311.

1229. Clematis Gouriana, I. 339.

1230. Abelia Schumannii, I. 121.

1231. Smilax pekingensis, III. 2.

1232. Clematis Delavayi, I. 325.

1233. Clematis grata, var. grandidentata, I. 338.

1233a. Clematis apiifolia, var. obtusidentata, I. 336.

1234. Acer laxiflorum, I. 93.

1235. { Sorbaria arborea, I. 47. Sorbaria arborea, var. subtomentosa, I. 47.

1236, 1236a. Desmodium serriferum, I. 104.

1237. Rhododendron Augustinii, I. 524.

1238. Rosa Davidii, II. 322.

1239. Cynanchum decipiens, III. 345.

1240. Pieris ovalifolia, var. lanceolata, I. 552.

1240a. Pieris ovalifolia, var. elliptica, I. 552.

1241. Magnolia Dawsoniana, I. 397.

1242. Rhamnus heterophyllus, II. 232.

1244. Rosa Prattii, II. 329.

1245. Zanthoxylum stenophyllum, II. 127.

1246. Ribes pulchellum, I. 45.

1246a. Rubus flosculosus, f. laxiflorus, I. 55.

1249. Hydrangea Davidii, I. 25.

1250. Hydrangea villosa, I. 29.

1251. { Schizophragma integrifolium, var. molle, I. 42. Schizophragma integrifolium,′ I. 41.

1252. Malus Prattii, II. 281.

1253. Xylosma racemosum, var. pubescens, I. 283.

1254. Rosa Prattii, II. 239.

1255. Sorbus pallescens, II. 266.

1256. Smilax microphylla, III. 2.

1257. Ilex Franchetiana, I. 77.

1258. Smilax menispermoidea, III. 5.

1259. Vaccinium moupinense, I. 560.

1260. Rhus Delavayi, var. quinquejuga, II. 184.

1261. Berberis aggregata, var. Prattii, I. 376; III. 443.

1262. Viburnum betulifolium, I. 114.

1263, 1263a. Viburnum betulifolium, I. 114.

1264. Rubus flosculosus, I. 54.

1265. Taxus cuspidata, var. chinensis, II. 8.

1266. Sorbus Rehderiana, I. 464.

1267. Berberis subcaulialata, I. 369.

1268, 1268a. Schisandra sphenanthera, var. lancifolia, I. 415.

1268b. Schisandra propinqua, var. sinensis, I. 416.

1269. Ilex corallina, I. 80.

1270. Cotoneaster ambigua, I. 159.

1271. Smilax pekingensis, III. 2.

1272. Thuja orientalis, II. 53.

1273. Syringa Wilsonii, I. 300.

1273a. Syringa Rehderiana, I. 299.

1273b. Syringa tomentella, I. 300.

1274. Rhododendron Davidsonianum, I. 515.

1275. Rhododendron Davidsonianum, I. 515.

1276. Rhododendron Davidsonianum, I. 515.

1278. Rhododendron longesquamatum, I. 529.

1280. Caragana erinacea, II. 622.

1281. Sorbus Rehderiana, I. 464.

1282. Berberis thibetica, I. 369.

1283. Berberis Mouillacana, I. 371.

1284. Berberis atrocarpa, III. 437.
1285. Malus transitoria, var. toringoides, II. 286.
1286. Phellodendron sachalinense, II. 136.
1287. Cotoneaster Dielsiana, var. elegans, I. 166.
1288. Viburnum Veitchii, I. 109.
1288a. Viburnum Veitchii, I. 109.
1289. Rosa Moyesii, II. 325.
1290. Ligustrum Delavayanum, II. 601.
1291. Symplocos paniculata, II. 593.
1292. Sinofranchetia chinensis, I. 349.
1293. Pyrus serotina, II. 263.
1294. Quercus glandulifera, III. 212.
1295. Quercus variabilis, III. 219.
1296. Quercus variabilis, III. 219.
1297. Quercus dentata, III. 210.
1298. Quercus glauca, III. 226.
1300. Berberis aggregata, var. Prattii, I. 376; III. 443.
1301, 1301a. Corylus tibetica, II. 443.
1302. Celastrus gemmata, II. 352.
1303. Clematis montana, var. Wilsonii, I. 333.
1305. Celtis Julianae, var. calvescens, III. 266.
1306. Rosa glomerata, II. 309.
1307. Vitis betulifolia, I. 103.
1308. Evonymus sanguinea, var. brevipedunculata, I. 495.
1309. Acer laxiflorum, I. 93.
1311. Pentapanax Henryi, II. 565.
1313. Acanthopanax lasiogyne, II. 563.
1314, 1314a. Actinidia purpurea, II. 378.
1315. Clematis Rehderiana, I. 324; III. 434.
3115a. Clematis lasiandra, I. 322.
1316. Corylopsis Willmottiæ, I. 425.
1317. Cotoneaster racemiflora, var. soongorica, I. 168.
1318. Spiraea Henryi, I. 447.
1318a. Spiraea Sargentiana, I. 447.
1319, 1319a. Rhododendron Edgarianum, I. 508.
1320. Rhododendron micranthum, I. 513.
1321. Deutzia longifolia, I. 13.
1322. Deutzia longifolia, I. 13.
1323. Hydrangea xanthoneura, var. setchuenensis, I. 579.

1324, 1324a, b, c. Rhododendron ambiguum, I. 518.
1325. Rhododendron taliense, I. 533.
1326. Rhododendron pachytrichum, I. 530.
1327. Hydrangea xanthoneura, var. Wilsonii, I. 27, 150.
1328. Rhododendron trichostomum, I. 505.
1329. Rhododendron longistylum, I. 514.
1330, 1330a. Rhododendron ambiguum, I. 518.
1331. Clematis Spooneri, I. 334.
1332. Quercus serrata, III. 217.
1333. Paeonia Delavayi, var. angustiloba, I. 318.
1334. Rosa longicuspis, II. 313.
1334a. Rosa glomerata, II. 309.
1335. Pyrus pashia, II. 264.
1336. Sarcococca ruscifolia, var. chinensis, II. 163.
1337. Fraxinus chinensis, var. typica, II. 260.
1339. Rhododendron insigne, I. 528.
1340. Deutzia longifolia, I. 13.
1341. Rhododendron strigillosum, I. 530.
1342. Rhododendron villosum, I. 524.
1343. Rhododendron Searsiae, I. 522.
1344. Berberis Gagnepainii, I. 358.
1345. Rhododendron lutescens, I. 516.
1346. Philadelphus purpurascens, I. 6; III. 421.
1347. Hydrangea xanthoneura, var. Wilsonii, I. 27, 150.
1348. Hydrangea Rosthornii, I. 33.
1349. Rhododendron pachytrichum, I. 530.
1350. Rhododendron Williamsianum, I. 538.
1351. Buddleia stenostachya, I. 565.
1351a. Buddleia nivea, var. yunnanensis, I. 570.
1352. Rhododendron Davidsonianum, I. 515.
1353. Rhododendron Wiltonii, I. 531.
1354. Hydrangea xanthoneura, I. 26.
1355. Hypericum patulum, var. Henryi, II. 403.
1355a. Hypericum Hookerianum, II. 403.
1356. Berberis Wilsonae, I. 368.

1357. Clematis chinensis, I. 329.
1358. Acer cappadocicum, f. tricaudatum, I. 86.
1359. Acer catalpifolium, I. 84.
1360. Viburnum foetidum, I. 111.
1361. Rhododendron longesquamatum, I. 529.
1362. Chaenomeles lagenaria, var. Wilsonii, II. 298.
1363. Gleditsia sinensis, II. 91.
1364, 1364a. Gleditsia sinensis, III. 447.
1365. Hydrangea Rosthornii, I. 33.
1366. Evonymus Aquifolium, I. 484.
1367. Rhododendron calophytum, I. 544.
1368. Pinus sinensis, var. densata, II. 17.
1369. Pinus sinensis, II. 15.
1370. Pinus sinensis, II. 15.
1372. Hydrangea Rosthornii, I. 33.
1373. Ceratostigma Willmottianum, II. 586.
1374. Magnolia Wilsonii, I. 395.
1375. Buddleia Lindleyana, var. sinuato-dentata, I. 564.
1375a. Buddleia Lindleyana, I. 564.
1376. Pinus sinensis, II. 15.
1377, 1377a, b, c. Alnus lanata, II. 488.
1378. Pinus Massoniana, II. 14.
1384. Rhododendron spec., III. 445.
1385. Pilostegia viburnoides, I. 151.
1386. Cephalotaxus Fortunei, II. 4.
1387. Pinus Armandi, II. 12.
1388, 1388a, b, d. Alnus cremastogyne, II. 488.
1389. Ribes longeracemosum, a Davidii, III. 423.
1390. Pinus sinensis, II. 15.
1391. Rhododendron ovatum, I. 546.
1392. Fraxinus spec., III. 450.
1393. Pinus sinensis, var. yunnanensis, II. 17.
1394. Pinus sinensis, var. yunnanensis, II. 17.
1395. Pinus sinensis, var. yunnanensis, II. 17.
1396. Pinus sinensis, var. yunnanensis, II. 17.
1397. Pinus sinensis, var. densata, II. 17.
1398. Pinus sinensis, var. densata, II. 17.

1399. Pinus sinensis, var. yunnanensis, II. 17.
1400, 1400a. Populus adenopoda, III. 23.
1401, 1401a, b, c. Salix magnifica, III. 44.
1401d. Salix ulotricha, III. 44.
1402. Salix cathayana, III. 57.
1402a. Salix macroblasta, III. 58, 456.
1403. Salix Rehderiana, III. 66.
1404. Salix hypoleuca, III. 53.
1405, 1405a. Salix cathayana, III. 57.
1406. Salix Ernesti, III. 47.
1407, 1407a. Salix plocotricha, III. 49.
1408, 1408a. Salix cathayana, III. 57.
1409, 1409a. Salix phaidima, III. 51.
1410. Salix allochroa, III. 72.
1411, 1411a. Salix hypoleuca, III. 53.
1412. Salix moupinensis, III. 46.
1413. Populus szechuanica, III. 20.
1414, 1414a. Salix Bockii, III. 71.
1415. Betula utilis, var. Prattii, III. 454.
1416. Populus szechuanica, III. 455.
1417, 1417a. Salix moupinensis, III. 46.
1418, 1418a. Salix hypoleuca, var. platyphylla, III. 54.
1419. Salix spec., III. 456.
1420. Populus Simonii, III. 21.
1421. Salix Rehderiana, III. 66.
1422. Magnolia globosa, var. sinensis, I. 393.
1423. Ulmus Bergmanniana, var. lasiophylla, III. 241.
1424. Salix Rehderiana, III. 66.
1427. Betula insignis, II. 459.
1428. Salix cathayana, III. 57.
1429. { Salix dyscrita, III. 53.
Salix Rehderiana, var. brevisericea, III. 67.
1430. Populus adenopoda, III. 23.
1431. Populus suaveolens, III. 18.
1432. { Populus suaveolens, III. 18.
Populus rotundifolia, var. Duclouxiana, III. 25.
1434. Populus szechuanica, III. 20.
1435, 1435a. Salix babylonica, III. 42.
1436. Carpinus Henryana, III. 453.
1437. Populus adenopoda, III. 23.
1440. Carpinus cordata, var. chinensis, III. 453.
1441. Salix paraplesia, III. 40.
1451. Ulmus Bergmanniana, III. 457.

1452. Castanea mollissima, III. 456.
1453. Corylus chinensis, II. 444.
1454. Populus Simonii, III. 21.
1459. Populus tremula, var. Davidiana, III. 24.
1460. Populus adenopoda, III. 23.
1461. Salix heterochroma, III. 61.
1462. Arundinaria spec., II. 64.
1464. Pinus sinensis, var. yunnanensis, II. 17.
1465. Pinus sinensis, var. densata, II. 17.
1466. Pinus sinensis, var. densata, II. 17.
1467. Pinus sinensis, var. densata, II. 17.
1468. Pinus Massoniana, II. 14.
1469. Pinus Massoniana, II. 14.
1470. Pinus Armandi, II. 12.
1471. Pinus sinensis, II. 15.
1472. Pinus sinensis, II. 15.
1473. Pinus Massoniana, II. 14.
1474. Castanea Seguinii, III. 456.
1475. Pinus sinensis, II. 15.
1476. Pinus Massoniana, II. 14.
1477. Pinus sinensis, II. 15.
1478. Pinus sinensis, var. densata, II. 17.
1479. Pinus sinensis, var. densata, II. 17.
1480. Pinus Massoniana, II. 14.
1481. Pinus Massoniana, II. 14.
1482. Pinus Massoniana, II. 14.
1483. Pinus Massoniana, II. 14.
1484. Pinus sinensis, II. 15.
1485. Pinus sinensis, II. 15.
1486. Pinus sinensis, II. 15.
1487. Pinus sinensis, II. 15.
1488. Pinus sinensis, II. 15.
1489. Pinus sinensis, II. 15.
1490. Pinus sinensis, II. 15.
1491. Pinus sinensis, II. 15.
1492. Pinus sinensis, II. 15.
1493. Pinus sinensis, II. 15.
1494. Pinus sinensis, II. 15.
1495. Pinus sinensis, II. 15.
1496. Pinus sinensis, II. 15.
1497. Pinus sinensis, II. 15.
1498. Pinus sinensis, II. 15.
1499. Pinus sinensis, II. 15.
1500. Pinus sinensis, var. densata, II. 17.
1501. Acer Davidii, I. 92.

1502. Acer amplum, var. tientaiense, I. 87.
1503. Acer trifidum, var. ningpoense, I. 92.
1504. Acer palmatum, I. 88.
1505. Acer palmatum, I. 88.
1506. Albizzia kalkora, III. 446.
1507. Aristolochia debilis, III. 323.
1508. Aralia chinensis, II. 566.
1509. Akebia lobata, var. australis, I. 348.
1510. Aleurites Fordii, II. 528.
1512. Actinidia purpurea, II. 378.
1513. Berchemia pycnantha, II. 215.
1514. Berchemia floribunda, var. megalophylla, II. 213.
1515. Rhamnella obovalis, II. 223.
1516. Buddleia Lindleyana, I. 564.
1517. Berberis virgetorum, III. 440.
1518. Boehmeria nivea, III. 312.
1519. Celastrus gemmata, II. 352.
1520. Celtis Biondii, III. 272.
1521. Castanea Henryi, III. 196.
1522. Castanopsis caudata, III. 201.
1523. Aphananthe aspera, III. 290.
1524. Crataegus cuneata, I. 179.
1525. Clerodendron cyrtophyllum, III. 377.
1526. Crataegus kulingensis, I. 179.
1527. Thea oleifera, II. 393.
1528. Callicarpa Giraldiana, var. subcanescens, III. 368.
1529. Castanea Seguinii, III. 194.
1530, 1530a. Callicarpa japonica, var. angustata, III. 369.
1531. Callicarpa Giraldiana, III. 366.
1532. Thea oleifera, II. 393.
1533. Cynanchum auriculatum, III. 346.
1534. Carpinus laxiflora, var. Davidii, II. 426.
1535. Cladrastis Wilsonii, II. 97.
1536. Cornus controversa, II. 573.
1537. Ternstroemia japonica, II. 397.
1538. Corylus heterophylla, var. sutchuenensis, II. 445, 451.
1539. Cocculus trilobus, I. 388.
1540. Cinnamomum Camphora, II. 68.
1541. Stephania Delavayi, I. 389.
1542. Castanea Seguinii, III. 194.
1543. Cedrela sinensis, II. 156.
1544. Camptotheca acuminata, II. 254.

1545. Cardiandra sinensis, I. 24.
1546. Eurya ochnacea, II. 399.
1548. Clematis paniculata, I. 330.
1549. Clematis uncinata, I. 327.
1550. Clematis chinensis, I. 329.
1551. Clematis brevicaudata, var. lissocarpa, I. 340.
1552. Clematis brevicaudata, var. lissocarpa, I. 340.
1553. Clematis brevicaudata, var. lissocarpa, I. 340.
1554. Camptotheca acuminata, II. 254.
1555. Corylopsis sinensis, var. glandulifera, I. 424.
1556. Corylopsis sinensis, I. 424.
1557, 1557a. Nyssa sinensis, II. 254.
1558. Quercus glauca, III. 226.
1559. Quercus glauca, III. 226.
1560. Quercus glandulifera, III. 212.
1561. Tilia Henryana, II. 367.
1562. Tilia spec., II. 369.
1563. Campsis chinensis, I. 303.
1565. Ulmus pumila, III. 242.
1566. Zanthoxylum setosum, II. 124.
1567. Dalbergia hupeana, II. 115.
1568. Diervilla japonica, var. sinica,
III. 430.
1569. Deutzia scabra, I. 6.
1571. Ehretia macrophylla, III. 364.
1572. Eurya japonica, var. nitida, II.
398.
1574. Sapium japonicum, II. 527.
1575. Phyllanthus flexuosus, II. 519.
1576. Eugenia microphylla, II. 420.
1577. Euscaphis japonica, II. 187.
1578. Elaeagnus multiflora, II. 412.
1579. Eurya japonica, var. nitida, II.
398.
1580. Evonymus alata, I. 493.
1581. Evonymus grandiflora, I. 484.
1582. Edgeworthia chrysantha, II. 550.
1583. Evodia glauca, II. 129.
1584. Evodia glauca, II. 129.
1585. Evodia glauca, II. 129.
1586. Trema spec., III. 289.
1587. Ficus impressa, III. 311.
1588. Ficus heteromorpha, III. 311.
1590. Fraxinus Mariesii, II. 260.
1591. Fraxinus Mariesii, II. 260.
1592. Fraxinus Mariesii, II. 260.
1593. Ligustrum acutissimum, II. 600.
1594. Fraxinus chinensis, var. rhynchophylla, II. 261.

1595, 1595a. Fraxinus chinensis, var.
rhynchophylla, II. 261.
1596. Ficus pumila, III. 311.
1597. Grewia parviflora, II. 371.
1598. Gymnocladus chinensis, II. 91.
1599. Glochidion puberum, II. 518.
1600. Glochidion Wilsonii, II. 518.
1601. Hydrangea paniculata, I. 25.
1602. Hamamelis mollis, I. 431.
1603. Pterostyrax corymbosus, I. 295.
1604. Hypericum Prattii, II. 404.
1605. Hydrangea umbellata, I. 25.
1606. Hypericum chinensis, II. 404.
1607. Helwingia japonica, II. 570.
1608. Ilex cornuta, I. 78.
1609. Ilex pedunculosa, f. β continentalis, I. 76.
1610. Ilex Wilsonii, I. 80.
1611. Ilex rotunda, I. 76.
1612. Idesia polycarpa, var. vestita, I.
285.
1616. Lespedeza Davidii, II. 107.
1617. Lespedeza formosa, II. 107.
1618. Phoebe neurantha, II. 72.
1619. Lespedeza spec., III. 448.
1620. Lindera glauca, II. 80.
1621. Lindera reflexa, II. 82.
1622. Leptodermis nervosa, III. 404.
1623. Desmodium podocarpum, III.
448.
1624. Lindera rubronervia, II. 84.
1625. Litsea citrata, II. 75.
1626. Loropetalum chinense, I. 430.
1627. Desmodium laburnifolium, III.
448.
1628. Liquidambar formosana, I. 421.
1629. Liquidambar formosana, I. 421.
1630. Liriodendron chinense, I. 410.
1631. Lespedeza Buergeri, II. 106.
1632. Lindera reflexa, II. 82.
1633. Ligustrum acutissimum, II. 600.
1634. Lindera umbellata, II. 81.
1635. Lindera strychnifolia, II. 82.
1636. Litsea fruticosa, II. 77.
1637. Lindera rubronervia, II. 84.
1638. Ligustrum acutissimum, II. 600.
1639. Lindera reflexa, II. 82.
1640. Ligustrum acutissimum, II. 600.
1641. Apios Fortunei, II. 117.
1642. Lindera cercidifolia, II. 85.
1643. Lespedeza formosa, II. 107.
1644. Mallotus apelta, II. 525.
1645. Mallotus apelta, II. 525.

1646. Alangium chinense, II. 552.
1647. Meliosma pendens, II. 200.
1648. Millettia reticulata, II. 102.
1649. Magnolia officinalis, var. biloba, I. 392, 406.
1650. Meliosma Oldhamii, II. 206.
1651. Alangium platanifolium, II. 554.
1652. Myrica rubra, III. 189.
1653. Morus alba, III. 294.
1654, 1654a. Magnolia denudata, I. 399, 400, 405, 409.
1655. Lonicera modesta, var. lushanensis, I. 139.
1656. Lonicera modesta, var. lushanensis, I. 139.
1657. Lonicera modesta, var. lushanensis, I. 139.
1658, 1658a. Lonicera modesta, I. 139.
1659. Litsea fruticosa, II. 77.
1660. Pterocarya stenoptera, III. 181.
1661. Pueraria hirsuta, II. 118.
1662. Pyrus Calleryana, II. 264.
1663. Paederia tomentosa, III. 403.
1664. Photinia subumbellata, I. 189.
1665. Platycarya strobilacea, III. 180.
1666. Photinia villosa, var. sinica, I. 186.
1668. Malus theifera, II. 283, 288, 291.
1669. Philadelphus incanus, var. Sargentianus, f. kulingensis, I. 145.
1670. Premna microphylla, III. 371.
1671. Picrasma quassioides, II. 152.
1672. Photinia Beauverdiana, I. 187.
1673. Photinia subumbellata, I. 189.
1674. Pittosporum glabratum, III. 326.
1675. Prunus pubigera, var. obovata, III. 425.
1676. Photinia serrulata, I. 184.
1677. Sorbus Folgneri, II. 271, 272, 276.
1678. Rosa microcarpa, II. 314, 337, 624.
1679. Rubus triphyllus, III. 424.
1680. Rhamnus Wilsonii, II. 240, 244, 252.
1681. Rhododendron Mariesii, I. 548.
1682. Rhododendron indicum, var. ignescens, I. 547.
1683. Rubus tephrodes, III. 423.
1684. Rhamnus crenatus, II. 232, 241, 244.
1685. Rubus innominatus, III. 424.

1686. Rhododendron Fortunei, I. 541.
1687. Rhus silvestris, II. 180.
1689. Ribes luridum, I. 46, 153.
1690. Rhododendron ovatum, I. 546.
1691. Rhus trichocarpa, II. 180.
1692. Rosa laevigata, II. 318, 337.
1693. Rubus trianthus, III. 423.
1694. Rhus javanica, II. 178.
1695. Vitis Davidii, I. 104.
1695a. Vitis Davidii, I. 104.
1696. { Parthenocissus Landuk, I. 102. Parthenocissus tricuspidata, I. 102.
1697. Vitex Negundo, III. 372.
1698. Vitis spec., III. 429.
1699. Vitis pentagona, I. 103.
1700. Vaccinium Donianum, I. 557.
1701. Vaccinium Donianum, I. 557.
1702. Vaccinium bracteatum, I. 558.
1703. Ampelopsis heterophylla, var., I. 580.
1704. Vaccinium Donianum, I. 557.
1705. Cayratia tenuifolia, I. 99.
1706. Parthenocissus tricuspidata, I. 102.
1708. Viburnum sympodiale, I. 109.
1709. Viburnum hirtulum, I. 112.
1710. Viburnum ichangense, I. 115.
1711. Viburnum theiferum, I. 112.
1712. Viburnum hirtulum, I. 112.
1713. Sapium sebiferum, II. 527.
1714. Sassafras tzumu, II. 74.
1715. Symplocos stellaris, II. 597.
1716. Maackia hupehensis, II. 98.
1717. Spiraea japonica, var. Fortunei, I. 451.
1718. Spiraea chinensis, I. 444.
1719. Smilax megalantha, III. 4.
1720. Symplocos paniculata, II. 593.
1721. Spiraea Blumei, I. 446.
1722. Stewartia sinensis, II. 395.
1723. Smilax glabra, III. 1.
1724. Stephanandra chinensis, I. 437.
1725. Stachyurus chinensis, I. 287.
1726. Schisandra sphenanthera, I. 414.
1727. Schizophragma integrifolium, var. denticulatum, I. 42.
1728. Sabia emarginata, II. 196.
1729. Smilax herbacea, var. oblonga, III. 1.
1730. Sambucus javanica, I. 307.
1731. Smilax glauco-china, III. 5.
1732. Styrax philadelphoides, I. 289.

1733. Styrax calvescens, I. 290.
1734. Styrax Veitchiorum, I. 290.
1735. Kadsura peltigera, I. 410.
1736. Kadsura peltigera, I. 410.
1737. Kadsura peltigera, I. 410.
1738. Populus adenopoda, III. 23.
1740. Cunninghamia lanceolata, II. 50.
1741. Cephalotaxus Fortunei, var. concolor, II. 6.
1741a. Cephalotaxus Fortunei, II. 4.
1742. Pseudolarix Kaempferi, II. 21.
1743. Ginkgo biloba, II. 1.
1744. Pinus Massoniana, II. 14.
1745. Pinus sinensis, II. 15.
1746. Cryptomeria japonica, II. 52.
1747. Pinus sinensis, II. 15.
1788. Ribes humile, I. 45.
1789. Ribes pulchellum, I. 45.
1790. Ribes glaciale, I. 46.
1791. Ribes glaciale, I. 46.
1792. Ribes glaciale, β glandulosum, I. 46.
1793. Ribes glaciale, β glandulosum, I. 46.
1794. Ribes tenue, I. 45.
1795. Ribes tenue, I. 45.
1796. Ribes acuminatum, I. 46.
1797. Ribes moupinense, γ laxiflorum, I. 44.
1798. Ribes longeracemosum, a Davidii, I. 45.
1799. Ribes himalayense, a glandulosum, I. 44; III. 423.
1800. Ribes himalayense, a glandulosum, I. 44.
1801. Ribes moupinense, γ laxiflorum, I. 44.
1802. Ribes moupinense, β tripartitum, I. 44.
1803. Ribes moupinense, β tripartitum, I. 44.
1804. Viburnum ichangense, I. 115.
1805. Viburnum dasyanthum, I. 115.
1806. Viburnum ichangense, I. 115.
1807. Viburnum lobophyllum, I. 114.
1808. Viburnum lobophyllum, var. flocculosum, I. 114.
1809. Viburnum betulifolium, I. 114.
1810. Viburnum dasyanthum, I. 115.
1811. Viburnum lobophyllum, var. flocculosum, I. 114.
1812. Viburnum lobophyllum, var. flocculosum, I. 114.

1813. Viburnum Wilsonii, I. 112.
1814. Viburnum Henryi, γ erubescens, I. 106. Viburnum erubescens, var. gracilipes, I. 106.
1815. Viburnum Henryi, γ erubescens, I. 106.
1816. Viburnum betulifolium, I. 114.
1817. Viburnum ovatifolium, I. 113.
1818. Viburnum dasyanthum, I. 115.
1819. Viburnum ovatifolium, I. 113.
1820. Viburnum dasyanthum, I. 115.
1821. Viburnum dasyanthum, I. 115.
1822. Viburnum ovatifolium, I. 113.
1823. Viburnum hupehense, I. 115.
1824. Viburnum erubescens, var. Prattii, I. 107.
1825. Viburnum erubescens, var. Prattii, I. 107.
1826. Viburnum erubescens, var. Prattii, I. 107.
1827. Viburnum erubescens, var. Prattii, I. 107.
1828. Viburnum erubescens, var. gracilipes, I. 107.
1829. Viburnum Henryi, I. 106.
1830. Viburnum propinquum, I. 111.
1831. Viburnum propinquum, I. 111.
1832. Viburnum tomentosum, I. 111.
1833. Viburnum foetidum, var. rectangulum, I. 112.
1834. Viburnum macrocephalum, I. 110.
1835. Viburnum macrocephalum, I. 110.
1836. Viburnum hypoleucum, I. 110.
1837. Viburnum utile, I. 110.
1838. Viburnum buddleifolium, I. 109.
1839. Viburnum shensianum, I. 109.
1840. Viburnum brachybotryum, I. 108.
1841. Cornus ulotricha, II. 574.
1842. Cornus Walteri, II. 576.
1843. Cornus Hemsleyi, II. 574.
1844. Cornus Hemsleyi, II. 574.
1845. Cornus Hemsleyi, II. 574.
1846. Cornus poliophylla, II. 574.
1847. Cornus Hemsleyi, II. 574.
1848. Cornus controversa, II. 573.
1849. Cornus capitata, II. 578.
1850. Cornus capitata, var. mollis, II. 579.
1851. Cornus capitata, II. 578.
1852. Lonicera deflexicalyx, I. 140.
1853. Lonicera hispida, I. 137.

1854. Lonicera chaetocarpa, I. 137.
1855. Lonicera hispida, I. 137.
1856. Lonicera praecox, I. 138.
1857. Lonicera chaetocarpa, I. 137.
1858. Lonicera pileata, I. 135.
1859. Lonicera Schneideriana, I. 133.
1860. Lonicera Schneideriana, I. 133.
1861. Lonicera saccata, I. 133.
1862. Lonicera saccata, I. 133.
1863. Lonicera saccata, I. 133.
1864. Lonicera saccata, I. 133.
1865. Lonicera saccata, f. Wilsonii, I. 134.
1866. Lonicera trichogyne, I. 131.
1867. Lonicera shensiensis, I. 131.
1868. Lonicera flavipes, I. 132.
1869. Lonicera similis, var. Delavayi, I. 142.
1870. Lonicera tatsienensis, I. 139.
1871. Lonicera Hemsleyana, I. 139.
1872. Lonicera syringantha, I. 130.
1873. Lonicera syringantha, I. 130.
1874. Lonicera Ferdinandi, var. leycesterioides, I. 135.
1875. Lonicera japonica, I. 142.
1876. Lonicera Ferdinandi, var. leycesterioides, I. 135.
1877. Lonicera retusa, I. 139.
1878. Lonicera crassifolia, I. 141.
1879. { Lonicera Henryi, I. 141.
Lonicera Henryi, var. subcoriacea, I. 142.
1880. Lonicera nervosa, I. 140.
1881. Lonicera lanceolata, I. 140.
1882. Lonicera alseuosmoides, I. 141.
1883. Lonicera tubuliflora, I. 129.
1884. Acer cappadocicum, var. sinicum, I. 85.
1885. Acer sinense, I. 90.
1886. Acer sutchuenense, I. 97.
1887. Acer sutchuenense, I. 97.
1888. Acer Franchetii, I. 97.
1889. Acer pictum, var. parviflorum, I. 83.
1890. Acer robustum, I. 89.
1891. Acer flabellatum, I. 91.
1892. Acer cappadocicum, f. tricaudatum, I. 86.
1893. Acer robustum, III. 426.
1894. Acer tetramerum, var. elobulatum, I. 95.
1895. Acer tetramerum, var. elobulatum, I. 95.

1896. Acer tetramerum, var. elobulatum, f. longeracemosum, I. 96.
1897. Acer pictum, var. parviflorum, I. 83.
1898. Acer tetramerum, var. elobulatum, I. 95.
1899. Acer robustum, I. 89.
1900. Acer robustum, I. 89.
1901. Acer tetramerum, var. betulifolium, I. 95.
1902. Acer flabellatum, I. 91.
1903. Acer cappadocicum, var. sinicum, I. 85.
1904. Acer laxiflorum, I. 93.
1905. Acer pictum, var. parviflorum, I. 83.
1906. Acer amplum, I. 86.
1907. Acer fulvescens, I. 84.
1908. Acer flabellatum, I. 91.
1909. Acer longipes, I. 88.
1910. Acer flabellatum, I. 91.
1911. Acer flabellatum, I. 91.
1912. Acer flabellatum, I. 91.
1913. Acer robustum, I. 89.
1914. Acer Maximowiczii, I. 94.
1915. Acer pictum, var. parviflorum, I. 83.
1916. Acer Davidii, III. 426.
1917. Acer Davidii, I. 92.
1918. Acer Davidii, I. 92.
1919. Acer pictum, var. parviflorum, I. 83.
1920. Acer robustum, I. 89.
1921. Acer pictum, var. parviflorum, I. 83.
1922. Acer pictum, var. parviflorum, I. 83.
1923. Acer pictum, var. parviflorum, I. 83.
1924. Acer laevigatum, I. 92.
1925. Acer cappadocicum, var. sinicum, I. 85.
1926. Acer pictum, var. parviflorum, I. 83.
1927. Acer laxiflorum, var. longilobum, I. 94.
1928. Acer caudatum, var. multiserratum, I. 91.
1929. Acer oblongum, I. 92.
1930. Acer caudatum, var. multiserratum, I. 91.
1931. Acer erianthum, I. 90.
1932. Acer robustum, I. 89.

1933. Acer tataricum, I. 91.
1934. Acer ceriferum, I. 89.
1935. Acer pictum, var. parviflorum, I. 83.
1936. Acer Oliverianum, I. 90.
1937. Acer Fargesii, I. 92.
1938. Acer amplum, I. 86.
1939. Trachelospermum axillare, III. 335.
1940. Trachelospermum axillare, III. 335.
1941. Gardneria multiflora, I. 563.
1942. 1942a. Periploca calophylla, III. 343.
1943. Toxocarpus villosus, III. 349.
1949. Adina rubella, III. 390.
1951. Adina racemosa, III. 390.
1952. Nothopanax Davidii, II. 556.
1953. Nothopanax Davidii, II. 556.
1954. Nothopanax Davidii, II. 556.
1955. Nothopanax Davidii, II. 556.
1956. Nothopanax Davidii, II. 556.
1957. Nothopanax Davidii, II. 556.
1958. Nothopanax Davidii, II. 556.
1959. Nothopanax Davidii, II. 556.
1960. Nothopanax Davidii, II. 556.
1961. Nothopanax Rosthornii, II. 557.
1962. Kalopanax ricinifolius, II. 564.
1963. Kalopanax ricinifolius, II. 564.
1964. Aralia Wilsonii, II. 567.
1965. Acanthopanax leucorrhizus, II. 557.
1966. Acanthopanax leucorrhizus, var. scaberulus, II. 558.
1967. Acanthopanax leucorrhizus, II. 557.
1968. Acanthopanax setchuenensis, II. 559.
1969. Acanthopanax Giraldii, II. 560.
1970. Acanthopanax Giraldii, II. 560.
1971. Acanthopanax Giraldii, II. 560.
1972. Acanthopanax Wilsonii, II. 560.
1973. Acanthopanax spinosus, II. 562.
1974. Acanthopanax Rehderianus, II. 561.
1975. Acanthopanax leucorrhizus, var. fulvescens, II. 558.
1976. Acanthopanax Giraldii, var. inermis, II. 560.
1977. Acanthopanax Henryi, II. 557.
1980. Ardisia Henryi, II. 582.
1981. Ardisia crispa, II. 581.

1982. Aristolochia moupinensis, III. 324.
1983. Aristolochia moupinensis, III. 324.
1984. Aristolochia mollissima, III. 324.
1985. Aristolochia thibetica, III. 323.
1986. Aristolochia debilis, III. 323.
1987. Acalypha acmophylla, II. 523.
1988. Acalypha szechuanensis, II. 524.
1989. Alchornea rufescens, II. 524.
1990. Alchornea rufescens, II. 524.
1991. Alchornea rufescens, II. 524.
1992. Alchornea rufescens, II. 524.
1993. Alchornea Davidii, II. 524.
2005. Actinidia kolomikta, II. 380.
2006. Actinidia tetramera, II. 381.
2007. Clematoclethra integrifolia, II. 386.
2008. Actinidia kolomikta, II. 380.
2009. Actinidia kolomikta, II. 380.
2010. Actinidia polygama, II. 380.
2011. Actinidia tetramera, II. 381.
2012. Actinidia callosa, var. Henryi, II. 382.
2013. Actinidia polygama, II. 380.
2014. Clematoclethra lanosa, II. 388.
2015. Clematoclethra Hemsleyi, II. 389.
2016. Actinidia callosa, var. Henryi, II. 382.
2017. Abelia Graebneriana, II. 118.
2018. Abelia Graebneriana, II. 118.
2019. Abelia Schumannii, II. 121.
2020. Abelia Graebneriana, II. 118.
2021. Abelia Zanderi, II. 121.
2022. Abelia Zanderi, II. 121.
2023. Abutilon sinense, II. 373.
2024. Abelia chinensis, II. 121.
2030. Alnus cremastogyne, II. 488.
2031, 2031a. Aleurites Fordii, II. 528.
2032. Albizzia julibrissin, II. 87.
2033. Albizzia kalkora, III. 446.
2034. Ailanthus altissima, var. sutchuenensis, II. 153; III. 449.
2042. Amphicome arguta, II. 303.
2047. Aucuba chinensis, II. 572.
2048. Alniphyllum Fortunei, I. 294.
2049. Acer pictum, var. parviflorum, I. 83.
2050. Acer robustum, I. 89.
2051. Picea Watsoniana, II. 27.
2052. Picea brachytyla, II. 33.
2053. Picea Wilsonii, II. 27.
2054. Picea Neoveitchii, II. 26.

2055. Picea Balfouriana, II. 30.
2056. Picea montigena, II. 33.
2057. Picea likiangensis, var. rubescens, II. 31.
2058. Picea aurantiaca, II. 26.
2059. Picea Balfouriana, II. 30.
2060. Picea montigena, II. 33.
2061. Picea likiangensis, II. 31.
2062. Picea montigena, II. 33.
2063. Picea likiangensis, II. 31.
2064. Picea likiangensis, var. rubescens, II. 31.
2065. Picea retroflexa, II. 25.
2066. Picea likiangensis, var. rubescens, II. 31.
2067. Picea gemmata, II. 24.
2068. Picea asperata, var. notabilis, II. 23.
2069. Picea aurantiaca, II. 26.
2070. Picea Sargentiana, II. 35.
2071. Picea ascendens, II. 34.
2072. Picea ascendens, II. 34.
2073. Picea ascendens, II. 34.
2074. Picea retroflexa, II. 25.
2075. Picea complanata, II. 35.
2076. Picea complanata, II. 35.
2077. Picea complanata, II. 35.
2078. Picea Sargentiana, II. 35.
2079. Picea complanata, II. 35.
2080. Picea asperata, II. 22.
2081. Picea complanata, II. 35.
2082. Picea hirtella, II. 32.
2083. Picea complanata, II. 35.
2084. Picea hirtella, II. 32.
2085. Picea Sargentiana, II. 35.
2086. Picea complanata, II. 35.
2087. Picea Wilsonii, II. 27.
2088. Abies Fargesii, II. 48.
2089. Abies Delavayi, II. 41.
2090. Abies Beissneriana, II. 46, 621.
2091. Abies Beissneriana, II. 46, 621.
2092. Abies Faxoniana, II. 42.
2093. Abies Delavayi, II. 41.
2094. Abies Delavayi, II. 41.
2095. Abies Beissneriana, II. 46.
2096. Tsuga chinensis, II. 37.
2097. Tsuga chinensis, II. 37.
2098. Tsuga yunnanensis, II. 36.
2099. Tsuga yunnanensis, II. 36.
2100. Tsuga chinensis, III. 37.
2101. Juniperus squamata, var. Fargesii, II. 59.
2102. Thuja orientalis, II. 53.

2103. Juniperus chinensis, II. 60.
2104. Juniperus chinensis, II. 60.
2105. Cupressus torulosa, II. 54.
2106. Cupressus torulosa, II. 54.
2107. Cephalotaxus argotaenia, II. 6.
2108. Torreya grandis, II. 7.
2109. Ginkgo biloba, II. 1.
2110. Cephalotaxus Fortunei, II. 4.
2111, 2111a. Salix Wallichiana, III. 64.
2112. Salix etosia, III. 73.
2113, 2113a, b. Salix Wallichiana, III. 64.
2114, 2114a, b. Salix Wallichiana, III. 64.
2115, 2115a, b. Salix Wallichiana, III. 64.
2116. Salix polyclona, III. 55.
2117, 2117a. Salix rhoophila, III. 54.
2118, 2118a. Salix mictotricha, III. 56.
2119, 2119a, b. Salix heterochroma, III. 61.
2120, 2120a. Salix variegata, III. 70.
2121. Salix Wilsonii, II. 40.
2122, 2122a. Salix babylonica, III. 42.
2123. Salix driophila, III. 59.
2124. Salix amphiloba, III. 60.
2125. Salix Rehderiana ?, III. 66.
2126. Salix moupinensis, III. 46.
2127. Salix cathayana, III. 57.
2128. Salix Wallichiana, III. 64.
2129. Salix driophila?, III. 59.
2130. Salix cathayana, III. 57.
2131. Salix cathayana, III. 57.
2132, 2132a. Salix Ernesti, III. 47.
2133, 2133a. Salix dolia, III. 65.
2134, 2134a. Salix atopantha, III. 43.
2135, 2135a. Salix cathayana, III. 57.
2136, 2136a. Salix dissa, III. 52.
2137, 2137a. Salix atopantha, III. 43.
2138, 2138a. Salix hylonoma, III. 68.
2139, 2139a. Salix opsimantha, III. 63.
2140. Salix Wilsonii, III. 40.
2141. Salix microphyta, III. 62.
2142, 2142a. Salix microphyta, III. 62.
2143. Salix Souliei, III. 62.
2144. Salix hylonoma, III. 68.
2145, 2145a. Salix dissa, III. 52.
2146, 2146a. Salix cheilophila, III. 69.
2147, 2147a. Salix plocotricha, III. 49.
2148. Salix Wallichiana, III. 64.
2149, 2149a. Salix Ernesti, III. 47.
2150. Salix driophila, III. 59.
2151, 2151a. Salix Ernesti, III. 47.

2152. Salix Ernesti, III. 47.
2153. Salix Ernesti, III. 47.
2154. Salix argyrophegga, III. 49.
2155. Salix phanera, III. 50.
2156. Salix Fargesii, III. 47.
2157. Salix plocotricha?, III. 49.
2158. Salix plocotricha?, III. 49.
2159. Salix Ernesti, III. 47.
2160. Salix dissa, III. 52.
2161. Salix myrtillacea, III. 71.
2162. Populus suaveolens, III. 18.
2163. Populus szechuanica, III. 20.
2164. Populus suaveolens, III. 18.
2165. Populus szechuanica, III. 20.
2166. Cotoneaster racemiflora, var. so-ongorica, I. 168.
2167. Cotoneaster divaricata, I. 157.
2168. Cotoneaster racemiflora, var. so-ongorica, I. 168.
2169. Cotoneaster gracilis, I. 167.
2170. Cotoneaster Dielsiana, var. elegans, I. 166.
2171. Cotoneaster Franchetii, I. 165.
2172. Cotoneaster Zabelii, I. 166.
2173. Cotoneaster Zabelii, I. 166.
2174. Cotoneaster Zabelii, I. 166.
2175. Cotoneaster foveolata, I. 162.
2176. Cotoneaster gracilis, I. 167.
2177. Cotoneaster acutifolia, var. laete-virens, I. 159.
2178. Cotoneaster ambigua, I. 159.
2179. Cotoneaster ambigua, I. 159.
2180. { Cotoneaster moupinensis, I. 163. Cotoneaster bullata, var. macro-phylla, I. 164.
2181. Cotoneaster bullata, var. macro-phylla, I. 164.
2182. Cotoneaster Henryana, I. 174.
2183. Cotoneaster Henryana, I. 174.
2184. Cotoneaster rhytidophylla, I. 175.
2185. Cotoneaster glabrata, I. 171.
2186. Cotoneaster disticha, var. tongo-lensis, I. 154.
2187. Cotoneaster adpressa, I. 155.
2188. Cotoneaster microphylla, var. vellaea, I. 176.
2189. Cotoneaster microphylla, var. cochleata, I. 176.
2190. Cotoneaster Zabelii, I. 166.
2193. Callicarpa Giraldiana, III. 366.
2194. Callicarpa Giraldiana, III. 366.
2195. Callicarpa japonica, var. angus-tata, III. 369.

2196. Callicarpa Giraldiana, III. 366.
2197. Cassia Leschenaultiana, II. 90.
2198. Catalpa ovata, I. 303.
2199. Caragana Boisii, II. 102.
2200. Caragana bicolor, II. 102.
2201. Caragana jubata, II. 103.
2202. Caragana jubata, II. 103.
2203. Caragana chamlagu, II. 102.
2204. Thea Grijsii, II. 394.
2205. Thea oleifera, II. 393.
2206. Thea Grijsii, II. 394.
2207. Thea fraterna, II. 390.
2208. Thea cuspidata, II. 390.
2209. Carpinus cordata, var. chinensis, II. 437.
2210. Carpinus Henryana, II. 429.
2211. Carpinus Fargesiana, II. 428.
2212. Carpinus Seemeniana, II. 430.
2213. Carpinus Seemeniana, II. 430.
2214. Carpinus polyneura, II. 430.
2215. Ostrya japonica, II. 424.
2216. Carpinus polyneura, II. 430.
2217. Carpinus laxiflora, var. macro-stachya, II. 425.
2218. Carpinus Seemeniana, II. 430.
2219. Ternstroemia japonica, var. Wightii, II. 397.
2220. Ostrya japonica, II. 424.
2221. Caryopteris incana, III. 378.
2222. Clethra Fargesii, I. 502.
2223. Clerodendron trichotomum, var. Fargesii, III. 376.
2224. Clerodendron foetidum, III. 375.
2225. Clerodendron trichotomum, III. 375.
2226. Cinnamomum hupehanum, II. 69.
2227. Cinnamomum Wilsonii, II. 66.
2228. Citrus nobilis, var. deliciosa, II. 143.
2229. Citrus grandis, II. 144.
2230, 2230a, b. Citrus ichangensis, II. 144.
2231. Poncirus trifoliata, II. 149.
2232. Heptacodium miconioides, II. 618.
2233. Aster polia, III. 459.
2234. Aster hersileoides, III. 459.
2235. Microglossa salicifolia, III. 460.
2240. Prunus hypotricha, III. 425.
2241. Prunus macrophylla, III. 425.
2243. Aster hersileoides, III. 459.

2244. Tylophora pseudotenerrima, III. 351.
2245. Dregea sinensis, III. 352.
2246. Secamone emetica, III. 348.
2247. Cynanchum amphibolum, III. 346.
2248. Cynanchum Wilfordii, III. 347.
2249. Cynanchum auriculatum, III. 346.
2250. Holostemma sinense, III. 345.
2251. Periploca sepium, III. 343.
2252. Cynanchum otophyllum, III. 347.
2253. Tylophora nana, III. 351.
2254. Lysionotus brachycarpus, III. 387.
2272. Corylopsis platypetala, var. levis, I. 427.
2273. Corylopsis platypetala, var. levis, I. 427.
2274. Corylopsis platypetala, var. levis, I. 427.
2275. Corylopsis Willmottiae, I. 425.
2276. Corylus heterophylla, var. sutchuenensis, II. 445.
2277, 2277a. Corylus heterophylla, var. sutchuenensis, II. 445.
2278, 2278a. Corylus heterophylla, var. sutchuenensis, II. 445.
2279. Corylus heterophylla, var. sutchuenensis, II. 445.
2280. Corylus chinensis, II. 444.
2281, 2281a. Corylus chinensis, II. 444.
2282. Corylus chinensis, II. 444.
2283. Corylus heterophylla, var. sutchuenensis, II. 445.
2284. Cocculus trilobus, I. 388.
2285. Cyclea racemosa, I. 390.
2286. Diploclisia affinis, I. 389.
2287. Stephania japonica, I. 389.
2288. Convolvulus Ammanii, III. 355.
2304. Coriaria terminalis, II. 170.
2305, 2305a. Celastrus gemmata, II. 352.
2306. Celastrus spiciformis, II. 348.
2307. Celastrus Hindsii, var. Henryi, II. 353.
2308. Celastrus articulata, var. cuneata, II. 350.
2309. Celastrus Loeseneri, II. 350.
2310. Celastrus rugosa, II. 349.
2311. Celastrus Rosthorniana, II. 351.
2312. Celastrus spiciformis, II. 348.
2313. Ceropegia stenophylla, III. 350.

2314. Ceratostigma Willmottianum, II. 586.
2315. Ceratostigma minus, II. 586.
2316. Ceropegia driophila, III. 349.
2317. Celtis Bungeana, III. 269.
2318. Celtis labilis, III. 267.
2319. Celtis Bungeana, III. 267.
2320. Celtis Vandervoetiana, III. 267.
2321. Celtis Biondii, III. 272.
2323. Tilia chinensis, II. 364.
2324. Tilia intonsa, II. 365.
2325. Tilia intonsa, II. 365.
2326. Tilia nobilis, II. 363.
2327. Tilia nobilis, II. 363.
2328. Tilia intonsa, II. 365.
2329. Tilia intonsa, II. 365.
2330. Tilia intonsa, II. 365.
2331. Tilia tuan, var. chinensis, II. 369.
2332. Tilia Oliveri, II. 366.
2333. Tilia Oliveri, II. 366.
2334. Tilia tuan, II. 368.
2335. Tilia Oliveri, II. 366.
2336. Tilia Oliveri, II. 366.
2337. Tilia Oliveri, II. 366.
2338. Tilia Oliveri, var. cinerascens, II. 367.
2339. Triosteum Rosthornii, I. 117.
2340. Triosteum Rosthornii, I. 117.
2341. Trachelospermum gracilipes, var. Cavaleriei, III. 332.
2342. Trachelospermum jasminoides, III. 334.
2343. Trachelospermum jasminoides, III. 334.
2344. Trachelospermum cathayanum, III. 333.
2345. Trachelospermum cathayanum, III. 333.
2346. Trachelospermum cathayanum, III. 333.
2347. Trachelospermum cathayanum, III. 333.
2348. Trachelospermum cathayanum, III. 333.
2349. Torricellia angulata, II. 569.
2350. Tamarix chinensis, II. 406.
2351. Turpinia nepalensis, II. 187.
2354. Urena lobata, II. 373.
2359. Wendlandia longidens, III. 392.
2360. Wistaria sinensis, II. 509.
2361. Daphne modesta, II. 541.
2362. Daphne gemmata, II. 543.

2363. Wikstroemia ligustrina, II. 531.
2364. Wikstroemia micrantha, II. 530.
2365. Wikstroemia stenophylla, II. 530.
2366. Wikstroemia stenophylla, II. 530.
2367. Daphne penicillata, II. 542.
2368. Daphne gemmata, II. 543.
2370. Koelreuteria apiculata, II. 191.
2371. Osteomeles Schwerinae, I. 184.
2372. Ormosia Hosiei, II. 94.
2373. Ormosia Henryi, II. 93.
2374. Osmanthus fragrans, II. 609.
2374a. Osmanthus venosus, II. 611.
2379. Nandina domestica, I. 386.
2380. Neillia affinis, I. 434.
2381. Neillia longiracemosa, I. 434.
2382. Neillia ribesioides, I. 435.
2383. Holboellia Fargesii, I. 346.
2384. Holboellia Fargesii, I. 346.
2385. Holboellia coriacea, I. 345.
2386. Akebia lobata, var. australis, I. 348.
2387. Akebia lobata, var. australis, I. 348.
2388. Akebia quinata, I. 347.
2389. Stauntonia Duclouxii, I. 344.
2390. Hydrangea strigosa, I. 31.
2391. Hydrangea glabripes, I. 30.
2392. Hydrangea strigosa, I. 31.
2393. Hydrangea strigosa, var. angustifolia, I. 32.
2394. Hydrangea strigosa, I. 31.
2395. Hydrangea strigosa, I. 31.
2396. Hydrangea strigosa, var. angustifolia, I. 32.
2397. Hydrangea hypoglauca, I. 26.
2398. Hydrangea xanthoneura, var. Wilsonii, I. 27, 150.
2399. Hydrangea xanthoneura, var. setchuenensis, I. 28, 579.
2400. Hydrangea longipes, I. 33.
2401. Hydrangea longipes, I. 33.
2402. Hydrangea longipes, I. 33.
2403. Hydrangea aspera, var. velutina, I. 30.
2404. Hydrangea aspera, var. velutina, I. 30.
2405. Hydrangea aspera, var. velutina, I. 30.
2406. Hydrangea longipes, I. 33.
2407. Hydrangea xanthoneura, var. Wilsonii, I. 27, 150.

2408. Hydrangea xanthoneura, var. Wilsonii, I. 27, 150.
2409. Hydrangea xanthoneura, I. 26.
2410. Hydrangea xanthoneura, var. Wilsonii, I. 27, 150.
2411. Hydrangea pubinervis, I. 27.
2412. Hydrangea xanthoneura, var. lancifolia, I. 579.
2413. Hydrangea longipes, I. 33.
2414. Hydrangea Rosthornii, I. 33.
2415. Helwingia himalaica, II. 571.
2416. Hibiscus Manihot, II. 374.
2417. Hibiscus Manihot, II. 374.
2418. Hypericum patulum, var. Henryi, II. 403.
2419. Hypericum patulum, var. Henryi, II. 403.
2420. Hypericum Prattii, II. 404.
2421. Hypericum chinense, II. 404.
2421a. Hypericum Prattii, II. 404.
2422. Hypericum chinense, II. 404.
2423. Hypericum longistylum, II. 404.
2424. Hypericum longistylum, II. 404.
2425. Hypericum perforatum, III. 452.
2426. Hypericum macrosepalum, III. 451.
2427. Hedera himalaica, II. 555.
2429, 2429a. Hovenia dulcis, II. 252.
2430. Hibiscus syriacus, II. 374.
2434. Glochidion puberum, II. 518.
2435. Glochidion Wilsonii, II. 518.
2436. Sauropus albicans, II. 518.
2437. Grewia parviflora, var. glabrescens, II. 371.
2438. Gardenia florida, III. 401.
2444. Gleditsia macracantha, II. 90.
2445. Gleditsia macracantha, II. 90.
2446. Gleditsia officinalis, II. 91.
2447. Gordonia sinensis, II. 395.
2448. Gordonia axillaris, II. 394.
2458. Sterculia lanceæfolia, II. 376.
2459. Nitraria Schoberi, II. 120.
2460. Clematis montana, I. 332.
2461. Clematis montana, var. Wilsonii, f. platysepala, I. 334.
2462. Clematis montana, var. grandiflora, I. 333.
2463. Clematis montana, var. rubens, I. 333.
2464. Clematis montana, I. 332.
2465. Clematis montana, var. rubens, I. 333.

2466. Clematis montana, var. rubens, I. 333.
2467. Clematis Fargesii, var. Souliei, I. 336.
2468. Clematis Armandi, I. 326.
2469. Clematis pogonandra, var. pilosula, I. 320.
2470. Clematis Prattii, I. 320.
2471. Clematis obscura, I. 329.
2472. Clematis obscura, I. 329.
2473. Clematis chinensis, I. 329.
2474. Clematis chinensis, f. vestita, I. 330.
2475. Clematis Benthamiana, I. 330.
2476. Clematis Benthamiana, I. 330.
2477. Clematis chinensis, I. 329.
2478. Clematis obscura, I. 329.
2479. Clematis brevicaudata, var. subsericea, I. 341.
2480. Clematis gracilifolia, I. 331.
2481. Clematis pogonandra, I. 319.
2482. Clematis Faberi, I. 320.
2483. Clematis fruticosa, I. 326.
2484. Clematis trullifera, I. 324.
2485. Clematis Henryi, I. 342.
2486. Clematis Henryi, I. 342.
2487. Clematis tangutica, var. obtusiuscula, I. 343.
2488. Clematis pterantha, var. grossedentata, I. 322.
2489. Clematis pogonandra, I. 319.
2497. Senecio scandens, III. 419.
2500. Pinus sinensis, II. 15.
2501. Pinus sinensis, II. 15.
2502. Pinus sinensis, var. densata, II. 17.
2503. Pinus Massoniana, II. 14.
2504. Pinus sinensis, var. densata, II. 17.
2505. Pinus Armandi, II. 12.
2506. Pinus Armandi, II. 12.
2509. Pinus Armandi, II. 12.
2510. Pinus Armandi, II. 12.
2511. Pinus Armandi, II. 12.
2512. Pinus Bungeana, II. 13.
2513. Pinus sinensis, II. 15.
2520. Sambucus javanica, I. 106.
2521. Sapium sebiferum, II. 527.
2522. Sapium discolor, II. 527.
2528. Sambucus Sieboldiana, I. 106.
2529. Sabia Schumanniana, var. longipes, II. 197.
2530. Sabia latifolia, II. 195.
2531. Sabia latifolia, II. 195.

2532. Sabia gracilis, II. 198.
2533. Sabia Ritchieae, II. 195.
2534. Sabia Schumanniana, var. pluriflora, II. 197.
2534a, b. Sabia puberula, II. 197.
2535. Symplocos paniculata, II. 593.
2536. Symplocos paniculata, II. 593.
2536a. Symplocos paniculata, II. 593.
2537. Symplocos laurina, II. 594.
2537a. Symplocos javanica, II. 597.
2538. Symplocos laurina, II. 594.
2539. Symplocos javanica, II. 597.
2540. Prunus hypotricha, III. 425.
2541. Prunus macrophylla, III. 425.
2542. Pygeum spec., III. 451.
2543. Symplocos stellaris, II. 597.
2544. Symplocos stellaris, II. 597.
2545. Symplocos anomala, II. 596.
2546. Symplocos caudata, II. 595.
2547. Symplocos anomala, II. 596.
2548. Symplocos caudata, II. 595.
2549. Symplocos botryantha, II. 596.
2550. Symplocos myriantha, II. 596.
2551. Schisandra sphenanthera, **var.** pubinervis, I. 415.
2552. Schisandra sphenanthera, **var.** lancifolia, I. 415.
2553. Schisandra sphenanthera, I. 414.
2554. Schisandra sphenanthera, I. 414.
2555. Stachyurus yunnanensis, I. 288.
2556. Stachyurus chinensis, I. 287.
2557. Sophora japonica, II. 96.
2558. Sophora japonica, II. 96.
2559. Sophora japonica, II. 96.
2560. Sophora viciifolia, II. 95.
2561. Sophora flavescens, II. 96.
2562. Sophora glauca, var. albescens, III. 447.
2563. Schizophragma integrifolium, **var.** denticulatum, I. 42.
2564. Schizophragma integrifolium, **var.** denticulatum, I. 42.
2565. Schizophragma integrifolium, **var.** denticulatum, I. 42.
2566. Schizophragma integrifolium, **var.** denticulatum, I. 42.
2567. Schizophragma integrifolium, **var.** denticulatum, I. 42.
2568. Schizophragma integrifolium, I. 41.
2569. Schizophragma integrifolium, **var.** glaucescens, I. 42.
2570. Styrax suberifolius, I. 290.

2571. Styrax dasyanthus, var. cinerascens, I. 289.
2572. Styrax dasyanthus, I. 289.
2573. Styrax japonicus, I. 291.
2574. Styrax Hemsleyanus, I. 291.
2574a. Styrax Hemsleyanus, var. griseus, I. 291.
2575. Styrax roseus, I. 291.
2576. Styrax Perkinsiae, I. 292.
2577. Styrax roseus, I. 291.
2578. Styrax Hemsleyanus, I. 291.
2579. Syringa verrucosa, I. 298.
2580. Syringa Komarowii, I. 301.
2581. Syringa Sargentiana, I. 298.
2582. Syringa reflexa, I. 297.
2583. Syringa Potaninii, I. 297.
2584. Syringa tomentella, I. 300.
2585. Syringa pinnatifolia, I. 297.
2586. Thea spec., III. 451.
2587. Firmiana simplex, II. 377.
2588. Sageretia rugosa, II. 227.
2589. Sageretia Henryi, II. 623.
2590. Sarcococca ruscifolia, II. 163.
2591. Sarcococca ruscifolia, II. 163.
2592. Sarcococca ruscifolia, II. 163.
2595. Skimmia melanocarpa, II. 138.
2596. Skimmia melanocarpa, II. 138.
2597. Skimmia melanocarpa, II. 138.
2683. Toddalia asiatica, II. 137.
2684. Zanthoxylum dissitum, II. 128.
2685. Zanthoxylum stenophyllum, II. 127.
2686. Zanthoxylum stenophyllum, II. 127.
2687. Zanthoxylum pilosulum, II. 123.
2688. Zanthoxylum setosum, II. 124.
2689. Zanthoxylum undulatifolium, II. 124.
2690. Zanthoxylum setosum, II. 124.
2691. Zanthoxylum setosum, II. 124.
2692. Zanthoxylum Bungei, II. 121.
2693. Zanthoxylum alatum, var. planispinum, II. 125.
Zanthoxylum alatum, var. planispinum, f. ferrugineum, II. 125.
2694. Evodia Bungei, II. 121.
2695. Evodia Bungei, II. 121.
2696. Evodia dimorphyllum, var. spinifolium, II. 126.
2697. Evodia micranthum, II. 127.
2697a. Evodia micranthum, II. 127.
2698. Evodia micranthum, II. 127.
2699, 2699a. Zelkova sinica, III. 286.

2700. Ziziphus sativa, var. inermis, II. 212.
2701. Vitex Negundo, III. 372.
2702. Vitex Negundo, III. 372.
2703. Vaccinium Henryi, I. 561.
2704. Vaccinium iteophyllum, var. fragrans, I. 558.
2705. Vaccinium Donianum, I. 557.
2706. Vaccinium Donianum, I. 557.
2707. Vaccinium Donianum, var. laetum, I. 558.
2708. Gaultheria nummularioides, var. elliptica, I. 555.
2709. Viscum articulatum, III. 318.
2710. Vaccinium Donianum, I. 557.
2712. Gaultheria Veitchiana, I. 554.
2713. Vitis betulifolia, I. 103.
2714. Vitis flexuosa, I. 102.
2715. Vitis betulifolia, I. 103.
2716. Vitis betulifolia, I. 103.
2717. Vitis Piasezkii, I. 103.
2718. Ampelopsis micans, var. cinerea, I. 100, 579.
2719. Ampelopsis micans, var. cinerea, I. 579.
2720. Ampelopsis heterophylla, var. vestita, I. 579.
2721. Ampelopsis micans, I. 100, 579.
2722. Ampelopsis micans, var. cinerea, I. 100, 579.
2723. Ampelopsis micans, I. 100, 579.
2724. Ampelopsis micans, I. 100; III. 445.
2725. Vitis flexuosa, I. 102.
2726. Vitis flexuosa, var. parvifolia, I. 103.
2727. Vitis flexuosa, var. parvifolia, I. 103.
2728. Vitis Thunbergii, var. cinerea, I. 105.
2729. Vitis Thunbergii, var. adstricta, I. 105.
2730. Parthenocissus himalayana, III. 427.
2731. Parthenocissus Landuk, I. 102.
2732. Vitis Davidii, var. cyanocarpa, I. 104.
2733. Vitis Romanetii, I. 105.
2734. Ampelopsis heterophylla, var. Gentiliana, I. 100.
2735. Ampelopsis aconitifolia, I. 100.
2736. Ampelopsis micans, var. cinerea, I. 101, 579.

2737. Cayratia oligocarpa, I. 99.
2738. Tetrastigma obtectum, var. pilosum, I. 99.
2739. Tetrastigma serrulatum, I. 99.
2741. Sorbaria arborea, I. 47.
2742. Sorbaria arborea, I. 47.
2743, 2743a. Sorbaria arborea, I. 47.
2744. Sorbaria arborea, var. subtomentosa, I. 47.
2745. Sorbaria arborea, I. 47.
2746. Spiraea japonica, var. ovalifolia, I. 452.
2747. Spiraea laeta, I. 442.
2748. Spiraea Rosthornii, I. 451.
2749. Spiraea Fritschiana, var. angulata, I. 453.
2750. Spiraea Fritschiana, I. 453.
2751. Spiraea Miyabei, var. glabrata, I. 454.
2752. Spiraea japonica, var. Fortunei, I. 451.
2753. Spiraea japonica, var. acuminata, I. 452.
2754. Spiraea hypericifolia, var. hupehensis, I. 439.
2755. Spiraea prunifolia, var. plena, I. 438.
2756. Spiraea Miyabei, var. pilosula, I. 455.
2757. Spiraea Miyabei, var. pilosula, I. 455.
2758. Spiraea Rosthornii, I. 451.
2759. Spiraea Rosthornii, I. 451.
2760. Spiraea myrtilloides, I. 440.
2761. Spiraea myrtilloides, I. 440.
2762. Spiraea canescens, var. oblanceolata, I. 450.
2763. Spiraea Schneideriana, var. amphidoxa, I. 450.
2764. Spiraea tortuosa, I. 445.
2765. Spiraea Henryi, I. 447.
2766. Spiraea Sargentiana, I. 447.
2767. Spiraea laeta, I. 442.
2768. Spiraea hirsuta, I. 444.
2769. Spiraea Blumei, I. 446.
2770. Spiraea hirsuta, var. rotundifolia, I. 445.
2771. Spiraea alpina, I. 440.
2772. Spiraea laeta, var. tenuis, I. 443.
2773. Sibiraea laevigata, var. angustata, I. 445.
2774. Fraxinus Griffithii, II. 258.

2775. Fraxinus Griffithii, II. 258.
2775a. Fraxinus Griffithii, III. 449.
2776. Fraxinus platypoda, II. 262.
2777. Fraxinus Sargentiana, II. 261.
2778. Fraxinus platypoda, II. 623.
2779. Fraxinus inopinata, II. 262.
2780. Fraxinus platypoda, II. 623.
2781. Fraxinus retusa, var. Henryana, II. 258.
2782. Fraxinus chinensis, II. 623.
2783. Fraxinus retusa, var. Henryana, II. 258.
2784. Fraxinus retusa, var. Henryana, II. 258.
2785. Fraxinus retusa, var. Henryana, II. 258.
2786. Fraxinus retusa, var. Henryana, II. 258.
2787. Fraxinus retusa, var. Henryana, II. 258.
2788. Fraxinus retusa, var. Henryana, II. 258.
2789. Fraxinus retusa, var. Henryana, II. 258.
2790. Fraxinus chinensis, var. rhynchophylla, II. 261.
2791, 2791a. Fagus longipetiolata, III. 190.
2792. Ficus spec., III. 457.
2793. Ficus heteromorpha, III. 311.
2794. Ficus heteromorpha, III. 311.
2795. Ficus heteromorpha, III. 311.
2796. Ficus lacor, III. 309.
2797. Ficus foveolata, var. Henryi, III. 310.
2798. Ficus pumila, III. 311.
2799. Ficus clavata, III. 310.
2801. Ulmus parvifolia, III. 244.
2802. Ulmus Wilsoniana, var. psilophylla, III. 239.
2803. Ulmus Wilsoniana, var. psilophylla, III. 239.
2804, 2804a. Ulmus Bergmanniana, III. 240.
2805, 2805a. Ulmus Bergmanniana, III. 240.
2806. Ulmus Wilsoniana, III. 238.
2807. Jasminum officinale, II. 613.
2808. Jasminum sinense, II. 612.
2809. Jasminum humile, II. 614.
2810. Jasminum humile, II. 614.
2811. Jasminum humile, II. 614.
2812. Trema virgata, III. 289.

2813. Prunus platysepala, I. 277.
2814. Prunus Armeniaca, I. 278.
2815. Amelanchier asiatica, var. sinica, I. 195.
2816. Amelanchier asiatica, var. sinica, I. 195.
2817. Prunus trichostoma, I. 216.
2818. Prunus hirtifolia, I. 209.
2819. Prunus latidentata, I. 217.
2820. Prunus latidentata, I. 217.
2821. Prunus droseracea, I. 215.
2822. Prunus oxyodonta, I. 218.
2823. Prunus conadenia, I. 197.
2824. Prunus saltuum, I. 213.
2825. Prunus concinna, I. 210.
2826. Prunus Helenae, I. 212.
2827. Prunus pulchella, I. 197.
2828. Prunus variabilis, I. 201.
2829. Prunus discadenia, I. 200.
2830. Prunus variabilis, I. 201.
2831. Prunus Rehderiana, I. 205.
2832. Prunus discadenia, I. 200.
2833. Prunus Herincquiana, I. 214.
2834. Prunus Buergeriana, var. nudiuscula, I. 60.
2835. Prunus Wilsonii, var. leiobotrys, I. 63.
2836. Prunus brachypoda, var. microdonta, I. 66.
2837. Prunus pubigera, var. Prattii, I. 68.
2838. Prunus brachypoda, var. microdonta, I. 66.
2839. { Prunus venosa, I. 60.
        Prunus brachypoda, var. pseudossiori, I. 65.
2840. Prunus velutina, I. 69.
2841. Prunus Grayana, I. 69.
2842. Prunus perulata, I. 61.
2843. Prunus brachypoda, var. pseudossiori, I. 65.
2844. Prunus obtusata, I. 66.
2845. { Prunus obtusata, I. 66.
        Prunus pubigera, var. Prattii, I. 68.
2846. Prunus brachypoda, var. pseudossiori, I. 65.
2847. Prunus microbotrys, I. 62.
2848. Maddenia hypoleuca, I. 56.
2849. Maddenia hypoleuca, I. 56.
2850. Maddenia hypoleuca, I. 56.
2851. Maddenia Wilsonii, I. 58.
2852. Berberis yunnanensis, I. 355.

2853. Berberis Tischleri, I. 355.
2854. Berberis Tischleri, I. 355.
2855. Berberis yunnanensis, I. 355.
2856. Berberis Tischleri, I. 355.
2857. Berberis Silva-Taroucana, I. 370.
2858. Berberis Silva-Taroucana, I. 370.
2859. Berberis Tischleri, I. 355.
2860. Berberis Silva-Taroucana, I. 370.
2861. Berberis Silva-Taroucana, I. 370.
2862. Berberis Feddeana, III. 442.
2863. Berberis Dielsiana, III. 441.
2864. Berberis Henryana, III. 440.
2865. Berberis diaphana, I. 354.
2866. Berberis dictyophylla, var. epruinosa, I. 353.
2867. Berberis Silva-Taroucana, I. 370.
2868. Berberis polyantha, var. oblanceolata, I. 376.
2869. Berberis Francisci-Ferdinandi, I. 367.
2870. Berberis aggregata, I. 375.
2871. Berberis Liechtensteinii, I. 377.
2872. Berberis polyantha, I. 376.
2873. Berberis Asmyana, I. 357.
2874. Berberis Gagnepainii, I. 358.
2875. Berberis sanguinea, I. 359.
2876. Berberis Bergmanniae, I. 362.
2877. Berberis Bergmanniae, I. 362; III. 438.
2878. Berberis Julianae, III. 438.
2879. Berberis silvicola, III. 438.
2880. Berberis candidula, I. 357.
2881. Mahonia nitens, I. 379.
2882. Mahonia Fortunei, I. 380.
2883. Mahonia confusa, I. 378; III. 443.
2884. Mahonia Sheridaniana, I. 379; III. 444.
2885. Deutzia densiflora, I. 12.
2886. Deutzia discolor, I. 12.
2887. Deutzia discolor, I. 12.
2888. Deutzia discolor, I. 12.
2889. Deutzia Schneideriana, I. 7.
2890. Deutzia glomeruliflora × longifolia, I. 10.
2891. Deutzia glomeruliflora, I. 10.
2892. Deutzia longifolia, I. 13.
2893. Deutzia glomeruliflora longifolia, I. 10.
2894. Deutzia mollis, I. 13.
2895. Deutzia setchuenensis, var. longidentata, I. 8.
2896. Deutzia pilosa, I. 8.

2897. Deutzia rubens, I. 13.
2898. Deutzia rubens, I. 13.
2899. Deutzia glomeruliflora, I. 10.
2900. Deutzia glomeruliflora × longifolia, I. 10.
2901. Deutzia glomeruliflora, I. 10.
2902. Deutzia rubens, I. 13.
2903. Dalbergia Dyeriana, II. 115.
2904. Dalbergia hupeana, II. 115.
2905. Dalbergia spec., II. 116.
2906. Dalbergia hupeana, II. 115.
2907. Dalbergia hupeana, II. 115.
2908. Dalbergia Dyeriana, II. 115.
2909. Dalbergia Dyeriana, II. 115.
2910. Diospyros sinensis, II. 591.
2911. Rhamnus heterophyllus, II. 623.
2912. Diospyros armata, II. 591.
2913. Diospyros kaki, var. silvestris, II. 590.
2914. Diospyros kaki, var. silvestris, II. 590.
2915. Diospyros Lotus, II. 587.
2916. Diervilla japonica, var. sinica, III. 430.
2917. Diervilla japonica, var. sinica, III. 430.
2918. Diervilla japonica, var. sinica, III. 430.
2919. Davidia involucrata, var. Vilmoriniana, II. 256.
2920. Davidia involucrata, var. Vilmoriniana, II. 256.
2921. Dioscorea zingiberensis, III. 14.
2922. Dioscorea japonica, III. 14.
2923. Dioscorea acerifolia, III. 14.
2924. Lespedeza formosa, II. 107.
2925. Lespedeza formosa, II. 107.
2926. Campylotropis macrocarpa, II. 113.
2927. Desmodium serriferum, II. 104.
2928. Desmodium serriferum, II. 104.
2929. Desmodium laxiflorum, II. 103.
2930. Flemingia fluminalis, II. 119.
2931. Lespedeza sericea, II. 105.
2932. Flemingia congesta, var. viridis, II. 119.
2933. Desmodium podocarpum, var. szechuenense, II. 104.
2934. Rhynchosia Craibiana, II. 118.
2935. Uraria hamosa, var. sinensis, II. 104.
2936. Desmodium tiliaefolium, II. 104.
2937. Desmodium tiliaefolium, II. 104.

2938. Campylotropis trigonoclada, II. 114.
2939. Desmodium laburnifolium, II. 103.
2940. Desmodium tiliaefolium, II. 104.
2941. Desmodium sinuatum, II. 104.
2942. Daphne Wilsonii, II. 540.
2943. Daphne Wilsonii, II. 540.
2944. Daphne acutiloba, II. 539.
2945. Daphne retusa, II. 541.
2946. Daphne acutiloba, II. 539.
2947. Daphne Wilsonii, II. 540.
2948. Daphne Wilsonii, II. 540.
2949. Daphne acutiloba, II. 539.
2950. Dipelta ventricosa, I. 118.
2951. Dipelta ventricosa, I. 118.
2952. Dipelta floribunda, I. 118.
2953. Dipelta floribunda, I. 118.
2954. Dipelta floribunda, I. 118.
2955. Dipelta floribunda, I. 118.
2956. Dichroa febrifuga, I. 43.
2957. Randia Henryi, III. 400.
2958. Gardneria multiflora, I. 563.
2959. Daphniphyllum angustifolium, II. 521.
2960. Daphniphyllum glaucescens, II. 522.
2961. Distylium chinense, I. 423.
2968. Diplospora fruticosa, III. 401.
2969. Photinia Beauverdiana, var. notabilis, I. 188.
2970. Photinia Beauverdiana, I. 187.
2971. Photinia Beauverdiana, I. 187.
2972. Photinia villosa, var. sinica, I. 186.
2973. { Photinia Beauverdiana, var. notabilis, I. 188.
Photinia Schneideriana, I. 188.
2974. Photinia Beauverdiana, I. 187.
2975. Pyrus Calleryana, II. 264; III. 450.
2976. Pyrus Calleryana, II. 264.
2077. Pyrus serotina, II. 263.
2978. Malus prunifolia, var. rinki, II. 279.
2979. Malus baccata, var. mandshurica, II. 291.
2980. Malus theifera, f. rosea, II. 284.
2981. Malus theifera, II. 283.
2982. Malus theifera, II. 283.
2983. Malus theifera, II. 283.
2984. Pyracantha crenulata, I. 177.
2985. Pyracantha crenulata, I. 177.

2986. Pyracantha crenulata, I. 177.
2987. Crataegus sanguinea, I. 180.
2988. Crataegus hupehensis, I. 178.
2989. Crataegus cuneata, I. 179.
2990. Chaenomeles lagenaria, II. 296.
2991. Chaenomeles lagenaria, var. cathayensis, II. 297.
2992. Sorbus caloneura, II. 269.
2992a. Sorbus alnifolia, II. 270.
2993. Sorbus aronioides, II. 268.
2994. Malus yunnanensis, II. 287.
2995. Malus kansuensis, II. 286.
2996. Malus kansuensis, II. 286.
2997. Sorbus Folgneri, II. 271.
2998. Docynia Delavayi, II. 296.
2999. Eriobotrya grandiflora, I. 193.
3000. Eriobotrya japonica, I. 193.
3001. Sorbus glomerulata, I. 470.
3002. Sorbus aestivalis, I. 469.
3003. Sorbus pogonopetala, I. 473.
3004. Sorbus unguiculata, I. 473.
3005. Sorbus Rehderiana, I. 464.
3006. Sorbus Prattii, f. laevis, I. 468.
3007. Sorbus setschwanensis, I. 475.
3008. Sorbus laxiflora, I. 466.
3009. Sorbus Helenae, f. subglabra, I. 463.
3010. Sorbus Helenae, f. rufidula, I. 463.
3011. Sorbus Sargentiana, I. 461.
3012. Sorbus Esserteauiana, I. 459.
3013. Rubus mesogaeus, f., I. 56.
3014. Rubus Buergeri, I. 51.
3015. Rubus mesogaeus, f., I. 56.
3016. Rubus eucalyptus, I. 53.
3017. Rubus eucalyptus, I. 53.
3018. Rubus pileatus, I. 52.
3019. Rubus Thunbergii, I. 51.
3020. Rubus chiliadenus, I. 55.
3021. Rubus pungens, I. 52.
3022. Rubus irenaeus, I. 51.
3023. Rubus Parkeri, var. brevisetosus, I. 50.
3024. { Rubus pinfaensis, I. 55.
        { Rubus chiliadenus, I. 55.
3025. Rubus fusco-rubens, I. 50.
3026. Rubus lutescens, I. 53.
3032. Meliosma cuneifolia, II. 199.
3033. Meliosma cuneifolia, II. 199.
3034. Meliosma cuneifolia, II. 199.
3035. Meliosma cuneifolia, II. 199.
3036. Meliosma parviflora, II. 201.
3037. Meliosma Kirkii, II. 207.

3037a. Meliosma Kirkii, II. 207.
3038. Meliosma Oldhamii, II. 206.
3039. Philadelphus subcanus, I. 4.
3040. Philadelphus subcanus, I. 4.
3041. Philadelphus Wilsonii, I. 4.
3042. Philadelphus subcanus, I. 4.
3043. Philadelphus subcanus, I. 4.
3044. Philadelphus subcanus, var. dubius, I. 4.
3045. Philadelphus subcanus, var. dubius, I. 4.
3046. Philadelphus purpurascens, I. 6; III. 421.
3047. Philadelphus incanus, I. 5.
3048. Philadelphus incanus, I. 5.
3049. Philadelphus incanus, I. 5.
3050. Philadelphus incanus, I. 5.
3051. Philadelphus incanus, I. 5.
3052. Philadelphus incanus, I. 5.
3053. Philadelphus incanus, I. 5.
3054. Philadelphus incanus, I. 5.
3055. Philadelphus incanus, I. 5.
3057. Itoa orientalis, I. 286.
3074. Indigofera Wilsonii, II. 101.
3075. Indigofera Monbeigii, II. 100.
3075a. Indigofera szechuensis, II. 101.
3076. Indigofera pseudotinctoria, II. 100.
3077. Indigofera amblyantha, II. 99.
3078, 3078a. Indigofera amblyantha, II. 99.
3079. Indigofera amblyantha, II. 99.
3080. Indigofera Monbeigii, II. 100.
3081. Indigofera ichangensis, f. leptantha, II. 100.
3082. Indigofera ichangensis, f. calvescens, II. 100.
3083. Indigofera ichangensis, f. rigida, II. 100.
3084. Indigofera dichroa, II. 100.
3085. Illicium Henryi, I. 417.
3086, 3086a. Illicium Henryi, I. 417.
3087. Illicium Henryi, I. 417.
3088. Ilex Henryi, I. 81.
3089. { Ilex macrocarpa, var. a genuina, I. 81.
        { Ilex macrocarpa, var. β trichophylla, I. 81.
3090. Ilex dubia, var. e pseudomacropoda, I. 82.
3091. Myrsine semiserrata, II. 580.
3092. Ilex yunnanensis, I. 76.
3093. Ilex yunnanensis, I. 76.

3094. Ilex yunnanensis, I. 76.
3095. Ilex pedunculosa, f. β continentalis, I. 76.
3096. Ilex purpurea, var. α Oldhamii, I. 76.
3097. Ilex purpurea, var. α Oldhamii, I. 76.
3098. Ilex Fargesii, I. 77.
3099. Ilex subrugosa, I. 80.
3100. Ilex Aquifolium, var. c chinensis, I. 78.
3101. Ilex cornuta, I. 78.
3102. Evonymus alata, var. operta, I. 494.
3103. Evonymus alata, I. 493.
3104. Evonymus cornuta, I. 489.
3105. Evonymus cornuta, I. 489.
3106. Evonymus cornuta, I. 489.
3107. Evonymus cornuta, I. 489.
3108. Evonymus porphyrea, I. 495.
3109. Evonymus porphyrea, I. 495.
3110. { Evonymus porphyrea, I. 495.
       { Evonymus dasydictyon, I. 496.
3111. Evonymus sanguinea, var. β camptoneura, I. 494.
3112. Evonymus yedoensis, var. Koehneana, I. 491.
3113. Evonymus verrucosoides, var. viridiflora, I. 493.
3114. Evonymus elegantissima, I. 496.
3115. Evonymus mupinensis, I. 489.
3116. Evonymus acanthocarpa, var. sutchuenensis, I. 490.
3117. Evonymus acanthocarpa, I. 490.
3118. Evonymus myriantha, I. 487.
3119. Evonymus Dielsiana, I. 488.
3120. Evonymus acanthocarpa, var. sutchuenensis, I. 490.
3121. Evonymus acanthocarpa, var. sutchuenensis, I. 490.
3122. Evonymus myriantha, I. 487.
3123. Evonymus myriantha, I. 487.
3124. Evonymus verrucosoides, I. 493.
3125. Evonymus oblongifolia, I. 486.
3126. Evonymus Semenovii, I. 492.
3127. Evonymus verrucosoides, I. 493.
3128. Evonymus grandiflora, I. 484.
3129. Evonymus verrucosoides, I. 493.
3170. { Paulownia Fargesii, I. 575.
       { Paulownia Duclouxii, I. 577.
3171. Paulownia thyrsoidea, I. 576.
3172. Potentilla fruticosa, var. mandshurica, II. 303.

3173. Potentilla fruticosa, var. rigida, II. 302.
3174. Potentilla fruticosa, II. 301.
3175. Potentilla fruticosa, II. 301.
3176. Potentilla fruticosa, var. Veitchii, II. 303.
3177. Pittosporum glabratum, III. 326.
3178. Pittosporum glabratum, III. 326.
3179. Pittosporum heterophyllum, III. 329.
3180. Pittosporum glabratum, III. 326.
3181. Pittosporum glabratum, var. neriifolium, III. 328.
3182. Pittosporum saxicola, III. 329.
3183. Polygonum Aubertii, III. 457.
3188. Pieris ovalifolia, var. elliptica, I. 552.
3189. Pieris ovalifolia, var. elliptica, I. 552.
3190. Pieris ovalifolia, var. elliptica, I. 552.
3191. Pieris villosa, var. pubescens, I. 554.
3192. Pieris villosa, I. 553.
3193. Pieris villosa, I. 553.
3194. Pieris villosa, I. 553.
3196. Polygala arillata, II. 160.
3197. Polygala arillata, II. 160.
3198. Polygala arillata, II. 160.
3199. Polygala arillata, II. 160.
3200. Polygala arillata, II. 160.
3201. Polygala arillata, II. 160.
3204. Andrachne hirsuta, II. 516.
3205. Andrachne capillipes, II. 516.
3206. Securinega fluggeoides, II. 520.
3207. Osyris Wightiana, III. 320.
3208. Phyllanthus flexuosus, II. 519.
3209. Paederia tomentosa, III. 403.
3210. Paederia tomentosa, III. 403.
3211. Paederia tomentosa, III. 403.
3212. Pterocarya insignis, III. 183.
3213. Pterocarya Paliurus, III. 182.
3214. Pterocarya stenoptera, III. 181.
3215. Pteroceltis Tatarinowii, III. 284.
3216. Premna ligustroides, III. 372.
3217. Premna puberula, III. 371.
3218. Punica Granatum, II. 419.
3220. Porana triserialis, III. 356.
3221. Porana sinensis, III. 355.
3222. Pertya sinensis, III. 419.
3223. Picrasma quassioides, II. 152.
3224. Paliurus ramosissimus, II. 210.
3225. Pterolobium punctatum, II. 92.

3226. Pueraria hirsuta, II. 118.
3227. Phellodendron chinense, var. glabriusculum, II. 137.
3228. Phellodendron chinense, var. glabriusculum, II. 137.
3231. Smilax discotis, III. 6.
3232. Smilax discotis, III. 6.
3233. Smilax scobinicaulis, III. 2.
3234. Smilax china, III. 4.
3235. Smilax discotis, III. 6.
3236. Smilax discotis, III. 6.
3237. Smilax discotis, III. 6.
3238. Smilax discotis, III. 6.
3239. Smilax discotis, III. 6.
3240. Smilax discotis, III. 6.
3241. Smilax glauco-china, III. 5.
3242. Smilax trachypoda, III. 3.
3243. Smilax discotis, III. 6.
3244. Smilax discotis, III. 6.
3245. Smilax herbacea, var. oblonga, III. 1.
3246. Smilax herbacea, var. acuminata, III. 1.
3247. Smilax microphylla, III. 2.
3248. Smilax menispermoidea, III. 5.
3249. Smilax vaginata, III. 2.
3250. Smilax trachypoda, III. 3.
3251. Smilax trachypoda, III. 3.
3252. Smilax cocculoides, III. 7.
3253. Smilax megalantha, III. 4.
3254. Smilax megalantha, III. 4.
3255. Caesalpinia szechuenensis, II. 92.
3256. Mezoneurum sinense, II. 93.
3257. Maesa castaneifolia, II. 583.
3258. Maesa Wilsonii, II. 584.
3259. Melastoma normale, II. 421.
3260. Osbeckia crinita, II. 421.
3261. Plagiopetalum quadrangulum, III. 453.
3262. Viscum articulatum, III. 318.
3263. Viscum articulatum, III. 318.
3264. Mussaenda divaricata, III. 394.
3265. Mussaenda Wilsonii, III. 393.
3266. Mussaenda divaricata, III. 394.
3267. Melia Azedarach, II. 157.
3268. Myricaria dahurica, II. 407.
3270. Rhamnus paniculiflorus, II. 233.
3271. Mucuna sempervirens, II. 117.
3272. Alangium platanifolium, II. 554.
3273. Alangium chinense, II. 552.
3274. Alangium chinense, II. 552.
3275. Alangium chinense, II. 552.
3276. Alangium chinense, II. 552.

3277. Alangium chinense, II. 552.
3278. Millettia Dielsiana, II. 101.
3279. Millettia Dielsiana, II. 101.
3280. Millettia Dielsiana, II. 101.
3281. Millettia Dielsiana, II. 101.
3282. Millettia reticulata, II. 102.
3283. Mallotus apelta, II. 525.
3284. Mallotus apelta, II. 525.
3285. Mallotus apelta, II. 525.
3286. Mallotus apelta, II. 525.
3287. Mallotus apelta, II. 525.
3288, 3288a. Mallotus repandus, II. 526.
3289. Mallotus tenuifolius, II. 525.
3290. Mallotus tenuifolius, II. 525.
3291. Mallotus tenuifolius, II. 525.
3292, 3292a. Mallotus tenuifolius, II. 525.
3293. Mallotus tenuifolius, II. 525.
3294. Mallotus philippinensis, II. 526.
3295. Mappia pittosporoides, II. 190.
3296. Mappia pittosporoides, II. 190.
3297. Myrsine semiserrata, II. 580.
3298. Myrsine africana, II. 580.
3299. Morus acidosa, III. 297.
3300. Morus alba, III. 294.
3301, 3301a. Morus acidosa, III. 297.
3302. Morus acidosa, III. 297.
3303, 3303a. Morus alba, III. 294.
3304. Morus alba, III. 294.
3305. Morus acidosa, III. 297.
3306. Morus acidosa, III. 297.
3307, 3307a. Morus acidosa, III. 297.
3308. Morus alba, III. 294.
3309, 3309a. Morus acidosa, III. 297.
3310. Morus cathayana, III. 292.
3314. Rhus verniciflua, II. 181.
3315. Rhus verniciflua, II. 181.
3316. Rhus trichocarpa, II. 180.
3317. Rhus punjabensis, var. sinica, II. 176.
3317a. Rhus Potaninii, II. 177.
3318. Rhus punjabensis, var. sinica, II. 176.
3318a. Rhus Potaninii, II. 177.
3319. Rhus Potaninii, II. 177.
3320. Rhus punjabensis, var. sinica, II. 176.
3321, 3321a, b. Rhus javanica, II. 178.
3322. Rhamnus Rosthornii, II. 236.
3323. Rhamnus Rosthornii, II. 236.
3324. Rhamnus Hemsleyanus, II. 234.
3325. Rhamnus iteinophyllus, II. 239.

3326. Rhamnus rugulosus, II. 238.
3327. Rhamnus crenatus, II. 232.
3328. Rhamnus crenatus, II. 232.
3329, 3329a. Rhamnus crenatus, II. 232.
3330. Rhamnus crenatus, II. 232.
3331. Rhamnus utilis, II. 240.
3332. Rhamnus utilis, II. 240.
3333, 3333a. Rhamnus utilis, II. 240.
3334. Rhamnus Esquirolii, II. 233.
3335. Fluggea leucopyrus, II. 520.
3336. Fluggea leucopyrus, II. 520.
3337. Rhamnella Wilsonii, II. 222.
3338. Rhamnella Julianae, II. 223.
3339. Sageretia pycnophylla, II. 226.
3340. Sageretia perpusilla, II. 226.
3341. Sageretia theezans, II. 227.
3342. Sageretia Henryi, II. 623.
3343. Sageretia subcaudata, II. 228.
3344. Lasianthus Henryi, III. 401.
3345. Rhodotypus kerrioides, II. 300.
3346. Buddleia Davidii, var. magnifica, I. 567.
3347. Buddleia Davidii, I. 567.
3348. Buddleia Davidii, var. Wilsonii, I. 568.
3349. Buddleia Davidii, I. 567.
3350. Buddleia Davidii, I. 567.
3351. Buddleia nivea, var. yunnanensis, I. 570.
3352. Buddleia Davidii, var. superba, I. 568.
3353. Buddleia nivea, var. yunnanensis, I. 570.
3354. Buddleia nivea, var. yunnanensis, I. 570.
3355. Buddleia Davidii, I. 567.
3356. Buddleia nivea, I. 570.
3357. Buddleia nivea, var. yunnanensis, I. 570.
3358. Buddleia nivea, I. 570.
3359. Buddleia nivea, var. yunnanensis, I. 570.
3300. Buddleia albiflora, I. 569.
3361, 3361a, b, c. Buddleia albiflora, I. 569.
3362. Buddleia asiatica, I. 566.
3363. Betula albo-sinensis, II. 457.
3364. Betula luminifera, II. 455.
3365. Betula insignis, II. 459.
3366. Betula luminifera, II. 455.
3367. Betula luminifera, II. 455.
3368. Betula luminifera, II. 455.

3369. Betula luminifera, II. 455.
3370, 3370a. Betula luminifera, II. 455.
3371, 3371a, b. Betula luminifera, II. 455.
3372. Bauhinia hupehana, var. grandis, II. 90.
3373. Bauhinia hupehana, II. 89.
3374. Bauhinia Faberi, var. microphylla, II. 89.
3375. Bauhinia spec., III. 447.
3376. Bauhinia densiflora, III. 447.
3377. { Bauhinia densiflora, II. 88. { Bauhinia Faberi, II. 88.
3378. Bauhinia yunnanensis, II. 89.
3379. Berchemia pycnantha, II. 215.
3380. Berchemia sinica, II. 215.
3381. Berchemia pycnantha, II. 215.
3382. Berchemia hypochrysa, II. 214.
3383, 3383a. Berchemia hypochrysa, II. 214.
3384. Berchemia yunnanensis, II. 216.
3385. Berchemia yunnanensis, II. 216.
3386. Berchemia sinica, II. 215.
3387. Berchemia Giraldiana, II. 213.
3388. Chaydaia Wilsonii, II. 221.
3389. Rhamnella obovalis, II. 223.
3390. Berchemia Giraldiana, II. 213.
3396, 3396a. Buxus microphylla, var. sinica, II. 165.
3397. Buxus microphylla, var. sinica, II. 165.
3398, 3398a. Buxus microphylla, var. sinica, II. 165.
3399. Buxus Harlandii, II. 166.
3400, 3400a. Broussonetia Kaempferi, III. 304.
3401, 3401a. Broussonetia papyrifera, III. 303.
3402. Bischoffia javanica, II. 521.
3403. Pluchea rubicunda, III. 418.
3404. Brandisia Hancei, II. 573.
3405. Bambusa spec., II. 65.
3406. Phyllostachys nidularia, II. 65.
3407. Phyllostachys nidularia, II. 65.
3408. Arundinaria szechuanensis, II. 64.
3409. Phyllostachys puberula, II. 65.
3410. Bambusa Beechyana, II. 65.
3411. Rhododendron yanthinum, I. 518.
3412. Rhododendron maculiferum, I. 531.
3413. Rhododendron Hanceanum, I. 517.

3414. Rhododendron Openshawianum, I. 543.
3415. Rhododendron Davidii, I. 543.
3416. Rhododendron Fargesii, I. 540.
3417. Rhododendron Fargesii, I. 540.
3418. Rhododendron orbiculare, I. 540.
3419. Rhododendron yanthinum, var. lepidanthum, I. 519.
3420. Rhododendron polylepis, I. 521.
3421. Rhododendron bracteatum, I. 519.
3422. Rhododendron apiculatum, I. 520.
3423. Rhododendron Wasonii, I. 532.
3424. Rhododendron longipes, I. 528.
3425. Rhododendron ochraceum, I. 534.
3426. Rhododendron Davidsonianum, I. 515.
3427. Rhododendron auriculatum, I. 544.
3428. Rhododendron racemosum, I. 516.
3429. Rhododendron strigillosum, I. 530.
3430. Rhododendron strigillosum, I. 530.
3431. Rhododendron lacteum, I. 545.
3432. Rhododendron Faberi, I. 533.
3433. Rhododendron Przewalskii, I. 534.
3434. Rhododendron Przewalskii, I. 534.
3435. Rhododendron Przewalskii, I. 534.
3436. Rhododendron Faberi, I. 533.
3437. Rhododendron Faberi, I. 533.
3438. Rhododendron longesquamatum, I. 529.
3439. Rhododendron longesquamatum, I. 529.
3440. Rhododendron pachytrichum, I. 530.
3441. Rhododendron argyrophyllum, var. cupulare, I. 526.
3442. Rhododendron argyrophyllum, var. cupulare, I. 526.
3443. Rhododendron hypoglaucum, I. 527.
3444. Rhododendron Amesiae, I. 523.
3445. Rhododendron villosum, I. 524.
3446. Rhododendron concinnum, I. 522.
3447. Rhododendron yanthinum, I. 518.
3448. Rhododendron concinnum, I. 522.
3449. Rhododendron Searsiae, I. 522.
3450. Rhododendron polylepis, I. 521.
3451. Rhododendron polylepis, I. 521.
3452. Rhododendron flavidum, var. psilostylum, I. 513.
3453. Rhododendron cephalanthum, I. 503.

3454. Rhododendron Sargentianum, I. 504.
3455. Rhododendron rufescens, I. 503.
3456, 3456a. Rhododendron yanthinum, I. 518.
3457. Rhododendron Augustinii, I. 524.
3458. Rhododendron nitidulum, I. 509.
3459. Rhododendron Edgarianum, I. 508.
3460. Rhododendron violaceum, I. 511.
3461. Rhododendron nitidulum, var. nubigenum, I. 510.
3462. Rhododendron Websterianum, I. 511.
3463. Rhododendron violaceum, I. 511.
3464. Rhododendron verruculosum, I. 507.
3465. Rhododendron alpicola, I. 506.
3466. Rhododendron intricatum, I. 505.
3467. Rhododendron Edgarianum, I. 508.
3467a. Rhododendron alpicola, var. strictum, I. 506.
3468. Rhododendron ramosissimum, I. 507.
3469. Rhododendron ramosissimum, I. 507.
3470. Rhododendron stamineum, I. 546.
3471. Rhododendron dendrocharis, I. 525.
3472. Rhododendron indicum, var. ignescens, I. 547.
3473. Rhododendron indicum, var. ignescens, I. 547.
3474. Rhododendron indicum, var. ignescens, I. 547.
3475. Rhododendron indicum, var. ignescens, I. 547.
3476. Leycesteria formosa, var. stenosepala, I. 312.
3477. Leycesteria formosa, var. stenosepala, I. 312.
3478. Leycesteria formosa, var. stenosepala, I. 312.
3479. Leycesteria formosa, var. stenosepala, I. 312.
3480. Lonicera tragophylla, I. 143.
3481. Lonicera Henryi, III. 430.
3483. Dumasia hirsuta, II. 116.
3484. Rhynchosia Dielsii, II. 118.
3485. Rhynchosia Dielsii, II. 118.
3486. Vicia unijuga, III. 448.
3487. Rhynchosia volubilis, II. 119.

3488. Apios Fortunei, II. 117.
3489. Indigofera pseudotinctoria, II. 100.
3490. Indigofera pseudotinctoria, II. 100.
3491. Lespedeza floribunda, II. 105.
3492. Campylotropis Sargentiana, II. 113.
3493. Lespedeza formosa, II. 107.
3494. Lespedeza Buergeri, II. 106.
3495. Ligustrum sinense, var. Stauntonii, II. 606.
3496. Ligustrum sinense, II. 605.
3497. Ligustrum sinense, var. nitidum, II. 606.
3498. Ligustrum gracile, II. 602.
3499. Ligustrum sinense, var. Stauntonii, II. 606.
3500. Ligustrum thibeticum, II. 604.
3501. Ligustrum strongylophyllum, II. 605.
3502. Ligustrum Quihoui, II. 607.
3503. Ligustrum acutissimum, II. 600.
3504. Ligustrum expansum, II. 600.
3505. Ligustrum compactum, II. 604.
3506. Ligustrum lucidum, II. 603.
3507. Ligustrum lucidum, II. 603.
3508. Ligustrum lucidum, II. 603.
3509. Ligustrum compactum, II. 604.
3510. Leptodermis Wilsonii, III. 405.
3511. Leptodermis oblonga, III. 403.
3512. Leptodermis Wilsonii, III. 405.
3513. Leptodermis nervosa, III. 404.
3514. Leptodermis Wilsonii, III. 405.
3515. Leptodermis Wilsonii, III. 405.
3516. Leptodermis Wilsonii, III. 405.
3517. Leptodermis oblonga, III. 403.
3518. Leptodermis parvifolia, III. 404.
3519. Lysionotus pauciflorus, III. 387.
3520, 3520a. Loropetalum chinense, I. 430.
3521. Loranthus yadoriki, var. hupehanus, III. 315.
3522. Loranthus sutchuenensis, III. 316.
3523. Loranthus caloreas, var. oblongifolius, III. 315.
3524. Loranthus Delavayi, III. 316.
3525. Elytranthe ampullacea, var. tonkinensis, III. 317.
3526. Loranthus Balfourianus, III. 315.
3527. Acacia intsia, var. oxyphylla, III. 446.

3528. Tinospora sagittata, I. 390.
3529. Plectranthus discolor, III. 384.
3530. Colquhounia Seguinii, III. 380.
3531. Lagerstroemia indica, II. 418.
3532. Elsholtzia fruticosa, III. 381.
3533. Elsholtzia fruticosa, III. 381.
3533a. Elsholtzia fruticosa, var. tomentella, III. 382.
3534. Elsholtzia dependens, III. 383.
3535. Elsholtzia flava, III. 382.
3536, 3536a. Lycium chinense, III. 385.
3537. Distylium chinense, I. 423.
3538. Evonymus alata, I. 493.
3539. Andrachne capillipes, var. pubescens, II. 516.
3540. Phyllanthus flexuosus, II. 519.
3541. Excoecaria acerifolia, II. 528.
3542. Mallotus repandus, II. 526.
3543. Sapium japonicum, II. 527.
3544. Eurya japonica, var. nitida, II. 398.
3545. Eurya japonica, var. aurescens, II. 399.
3546, 3546a. Euptelea pleiosperma, I. 313, 315.
3547. Enkianthus chinensis, I. 551.
3548. Enkianthus chinensis, I. 551.
3549. Enkianthus chinensis, I. 551.
3550. Enkianthus deflexus, I. 550.
3551. Enkianthus deflexus, I. 550.
3552. Daphniphyllum macropodum, II. 522.
3553. Elaeocarpus japonicus, II. 360.
3554, 3554a, b. Ehretia macrophylla, III. 364.
3555. Edgeworthia chrysantha, II. 550.
3556. Elaeagnus umbellata, II. 410.
3557. Elaeagnus viridis, II. 414.
3558. Elaeagnus multiflora, f. angustata, II. 413.
3559. Elaeagnus magna, II. 411.
3560. Elaeagnus multiflora, II. 412.
3561. Elaeagnus umbellata, II. 410.
3561a. Elaeagnus magna, II. 411.
3562. Elaeagnus lanceolata, II. 413.
3563. Elaeagnus lanceolata, II. 413.
3564. Elaeagnus lanceolata, II. 413.
3565. Elaeagnus Henryi, II. 414.
3565, 3565a. Elaeagnus cuprea, II. 414.
3566. Phellodendron chinense, var. glabriusculum, II. 137.
3567. Phellodendron chinense, var. glabriusculum, II. 137.

3568. Elaeagnus magna, II. 411.
3569. Evodia Baberi, II. 131.
3570. Evodia Baberi, II. 131.
3571. Evodia Henryi, var. villicarpa, II. 134.
3572. Evodia Bodinieri, II. 130.
3573. Evodia Bodinieri, II. 130.
3574. Evodia hupehensis, II. 133.
3575. Evodia hupehensis, II. 133.
3576. Evodia hupehensis, II. 133.
3577. Evodia officinalis, II. 130.
3578. Zanthoxylum micranthum, II.127.
3579, 3579a. Evodia glauca, II. 129.
3580. Rosa Murielae, II. 326.
3581. Rosa Prattii, II. 329.
3582. Rosa Murielae, II. 326.
3583, 3583a. Rosa graciliflora, II. 330.
3584. Rosa Sweginzowii, II. 324.
3585. Rosa Davidii, II. 322.
3586. Rosa Moyesii, f. rosea, II. 325.
3587. Rosa Moyesii, II. 325.
3588. Rosa Sweginzowii, II. 324.
3589. Rosa Moyesii, f. rosea, II. 325.
3590. Rosa Moyesii, f. rosea, II. 325.
3591. Rosa banksiopsis, II. 322.
3592. Rosa banksiopsis, II. 322.
3593. Rosa sertata, II. 327.
3594. Rosa omeiensis, f. pteracantha, II. 332.
3595. Rosa omeiensis, II. 331.
3596. Rosa omeiensis, II. 331.
3597. Rosa omeiensis, f. pteracantha, II. 332.
3598. Rosa Gentiliana, II. 312.
3599. Rosa Gentiliana, II. 312.
3600. Rosa Gentiliana, II. 312.
3601. Rosa Gentiliana, II. 312.
3602. Rosa Willmottiae, II. 329.
3603. Rosa microcarpa, II. 314.
3604. Rosa microcarpa, II. 314.
3605. Rosa Hugonis, II. 330.
3606. Rosa multiflora, var. cathayensis, II. 304.
3607. Rosa multiflora, var. cathayensis, II. 304.
3607a. Rosa multiflora, var. carnea, II. 305.
3608. Rosa multiflora, var. cathayensis, II. 304.
3609. Rosa multiflora, var. cathayensis, II. 304.
3610. Rosa multiflora, var. carnea, f. platyphylla, II. 306.

3611. Rosa chinensis, II. 320.
3612. Rosa Roxburghii, f. normalis, II. 319.
3613. Rosa Soulieana, II. 314.
3614. Rosa laevigata, II. 318.
3615. Castanea mollissima, III. 192.
3616. Castanea Henryi, III. 196.
3617. Castanea Henryi, III. 196.
3618. Castanea Seguinii, III. 194.
3619. Castanopsis ceratacantha, III. 199.
3620. Castanopsis Fargesii, III. 198.
3621, 3621a. Castanopsis ceratacantha, III. 199.
3622. Castanopsis platyacantha, III. 200.
3623. Castanopsis hystrix, III. 197.
3624. Lithocarpus cleistocarpa, III. 205.
3625. Quercus spathulata, III. 226.
3626. Quercus aquifolioides, var. rufescens, III. 223.
3627. Quercus glandulifera, III. 212.
3628. Quercus Baronii, III. 226.
3629. Lithocarpus spicata, var. mupinensis, III. 207.
3630. Lithocarpus viridis, III. 210.
3631. Lithocarpus megalophylla, III. 208.
3632. Quercus Engleriana, III. 220.
3632a. Quercus aquifolioides, III. 222.
3633. Quercus Engleriana, III. 220.
3634. Quercus glauca, III. 226.
3635, 3635a. Lithocarpus cleistocarpa, III. 205.
3636, 3636a, b, c, d. Lithocarpus cleistocarpa, III. 205.
3637, 3637a. Lithocarpus Henryi, III. 209.
3638. Lithocarpus spec., III. 209.
3639. Quercus glauca, III. 226.
3640. Quercus spathulata, III. 226.
3641. Castanopsis Fargesii, III. 198.
3642. Castanopsis sclerophylla, III. 201.
3643. Quercus spinosa, III. 224.
3644. Quercus acrodonta, III. 225.
3645. Quercus serrata, III. 217.
3646. Quercus glauca, III. 226.
3647. Quercus variabilis, III. 219.
3648. Quercus variabilis, III. 219.
3649. Quercus glandulifera, III. 212.
3650. Quercus glandulifera, III. 212.
3651. Quercus Fabri, III. 216.

3652. Quercus glandulifera, III. 212.
3653. Quercus Fabri, III. 216.
3654. Quercus glandulifera, III. 212.
3655, 3655a. Quercus Fabri, III. 216.
3656. Quercus glandulifera, III. 212.
3657. Quercus Fabri, III. 216.
3658. Quercus aliena, var. acuteserrata, III. 215.
3658a. Quercus aliena, var. acuteserrata, f. calvescens, III. 215.
3659. Castanea Seguinii, III. 194.
3660. Castanea Seguinii, III. 194.
3661. Castanea mollissima, III. 192.
3662. Castanea mollissima, III. 192.
3663. Castanea mollissima, III. 192.
3664. Castanea Seguinii, III. 194.
3665. Castanea Seguinii, III. 194.
3666. Castanea Seguinii, III. 194.
3667. Castanea Seguinii, III. 194.
3668. Castanea mollissima, III. 192.
3669. Lindera cercidifolia, II. 85.
3670. Litsea fruticosa, II. 77.
3671. Lindera membranacea, II. 81.
3672. Litsea Veitchiana, II. 76.
3673. Lindera umbellata, II. 81.
3674. Litsea sericea, II. 75.
3675. Lindera umbellata, II. 81.
3676. Litsea pungens, II. 76.
3677. Lindera glauca, II. 80.
3678. { Lindera membranacea, II. 81.
       { Lindera spec., II. 86.
3679. Lindera spec., II. 86.
3680. Lindera umbellata, II. 81.
3681. Lindera membranacea, II. 81.
3682. Litsea pungens, II. 76.
3683. Litsea pungens, II. 76.
3684. Lindera umbellata, II. 81.
3685. Lindera spec., II. 86.
3686. Lindera spec., II. 86.
3687. Litsea sericea, II. 75.
3688. Litsea sericea, II. 75.
3689. Litsea citrata, II. 75.
3690. Lindera obtusiloba, II. 85.
3691. Lindera obtusiloba, II. 85.
3692. Lindera obtusiloba, II. 85.
3693. Litsea populifolia, II. 77.
3694. Litsea Wilsonii, II. 78.
3695. Phoebe Sheareri, II. 621.
3696. Phoebe Sheareri, II. 72.
3697. Phoebe Sheareri, II. 72.
3698. Phoebe Sheareri, II. 72.
3699. Machilus Bournei, II. 73.
3700. Machilus ichangensis, II. 621.

3701. Machilus Bournei, II. 73.
3702. Alseodaphne omeiensis, II. 70.
3703. Phoebe neurantha, II. 72.
3704. Machilus Bournei, II. 73.
3705. Phoebe macrophylla, II. 71.
3706. Lindera megaphylla, II. 80.
3707. Neolitsea lanuginosa, var. chinensis, II. 79.
3708. Cinnamomum argenteum, II. 67.
3709. Cinnamomum hupehanum, II. 69.
3710. Cinnamomum inunctum, var. longepaniculatum, II. 69.
3711. Cinnamomum Camphora, II. 68.
3712. Cinnamomum Wilsonii, var. multiflorum, II. 67.
3713. Cinnamomum inunctum, var. albosericeum, II. 69.
3714. Lindera Prattii, II. 83.
3715. Lindera Prattii, II. 83.
3716. Lindera strychnifolia, var. Hemsleyana, II. 82.
3717. Lindera strychnifolia, var. Hemsleyana, II. 82.
3718. Lindera strychnifolia, var. Hemsleyana, II. 82.
3719, 3719a. Lindera strychnifolia, var. Hemsleyana, II. 82.
3720. Lindera fragrans, II. 83.
3721. Lindera strychnifolia, var. Hemsleyana, II. 82.
3722. Lindera strychnifolia, var. Hemsleyana, II. 82.
3723. Lindera fragrans, II. 83.
3724. Lindera fragrans, II. 83.
3725. Lindera strychnifolia, var. Hemsleyana, II. 82.
3726. Lindera communis, II. 79.
3727. Daphniphyllum glaucescens, II. 522.
3727a. Daphniphyllum angustifolium, II. 521.
3728, 3728a. Daphne Wilsonii, II. 540.
3729. Dipteronia sinensis, I. 83.
3730. Diospyros armata, II. 591.
3731. Perrottetia racemosa, II. 359.
3732. Ligustrum Delavayanum, II. 601.
3733. Elsholtzia fruticosa, III. 381.
3758. Stizolobium hassjoo, II. 117.
3759. Boehmeria nivea, III. 312.
3767. Stizolobium hassjoo, II. 117.
3816. Canarium album, II. 155.
3817. Picea Wilsonii, II. 27.

4000. Ulmus pumila, III. 242.
4002. Prunus pseudocerasus, III. 432.
4005. Buddleia officinalis, I. 565.
4006. Lonicera praecox, I. 138.
4007. Lonicera saccata, I. 133.
4008. Prunus spec., III. 432.
4009. Rosa Willmottiae, II. 329.
4010. Lonicera hispida, I. 137.
4011. Lonicera nervosa, I. 140.
4012. Rosa omeiensis, II. 331.
4013. Prunus pleiocerasus, I. 198.
4014. Cotoneaster racemiflora, var. microcarpa, I. 169.
4015. Cotoneaster multiflora, var. calacarpa, I. 170.
4015a. Cotoneaster racemiflora, var. soongorica, I. 168.
4016. Prunus macradenia, I. 199.
4017. Cornus Hemsleyi, II. 574.
4018, 4018a. Acanthopanax Giraldii, II. 560.
4019. Daphne retusa, II. 541.
4020. Sambucus Schweriniana, I. 306.
4021. Cotoneaster nitens, I. 156.
4022. Berberis Vernae, I. 368; III. 439.
4023. Lonicera Schneideriana, I. 133.
4024. Ampelopsis aconitifolia, var. palmiloba, III. 427.
4025. Arctous alpinus, var. ruber, I. 556.
4026. Rosa multibracteata, II. 328.
4026a. Rosa Willmottiae, II. 329.
4027. Clematoclethra integrifolia, II. 386.
4027a. Clematoclethra actinidioides, II. 386.
4028. Rosa Sweginzowii, II. 324.
4028x. Rosa Moyesii, f. rosea, II. 325.
4029. Prunus dehiscens, I. 271; III. 432.
4030. Viburnum Veitchii, I. 109.
4031. Viburnum erubescens, var. Prattii, I. 107.
4032. Dipelta ventricosa, I. 118.
4033. Ribes tenue, I. 153.
4034. Prunus spec., III. 432.
4035. Betula utilis, var. Prattii, II. 457.
4036. Prunus pubigera, var. obovata, I. 196.
4037. Lonicera tatsienensis, I. 139.
4038. Lonicera saccata, I. 133.
4039. Prunus tatsienensis, var. stenadenia, I. 201.

4040. Prunus glyptocarya, I. 219.
4041. Rhododendron yanthinum, var. lepidanthum, I. 519.
4043. Sambucus Sieboldiana, I. 106.
4044. Lonicera prostrata, I. 141.
4045. Clematis gracilifolia, I. 331.
4046. Picea asperata, II. 22.
4047. Picea asperata, II. 22.
4048. Picea Sargentiana, II. 35.
4048a. Picea ascendens, II. 34.
4049. Abies Delavayi, II. 41.
4050. Picea ascendens, II. 34.
4051. Abies recurvata, II. 44.
4052, 4052a. Abies Faxoniana, II. 42.
4053. Taxus cuspidata, var. chinensis, II. 8.
4054. Cephalotaxus Fortunei, II. 4.
4055. Pinus sinensis, var. densata, II. 17.
4056. Pinus sinensis, II. 15.
4057, 4057a. Abies recurvata, II. 44.
4058. Picea ascendens, II. 34.
4059. Picea purpurea, II. 29.
4060. Abies Faxoniana, II. 42.
4061. Picea asperata, II. 22.
4062. Picea purpurea, II. 29.
4063. Picea purpurea, II. 29.
4064. Picea heterolepis, II. 24.
4065. Picea Balfouriana, II. 30.
4066. Picea asperata, II. 22.
4067. Picea asperata, var. notabilis, II. 23.
4068. Picea asperata, var. ponderosa, II. 23.
4069. Abies Faxoniana, II. 42.
4070. Abies Faxoniana, II. 42.
4071. Larix Potaninii, II. 18.
4072. Tsuga chinensis, II. 37.
4073. Pinus sinensis, II. 15.
4074. Pinus sinensis, II. 15.
4076. Cunninghamia lanceolata, II. 50.
4077. Cryptomeria japonica, II. 52.
4078. Abies Delavayi, II. 41.
4079. Abies squamata, II. 48.
4080. Picea Balfouriana, II. 30.
4081. Picea aurantiaca, II. 26.
4082. Abies Delavayi, II. 41.
4083. Picea retroflexa, II. 25.
4084. Picea montigena, II. 33.
4085. Juniperus squamata, II. 57.
4086. Abies Delavayi, II. 41.
4087. Betula utilis, var. Prattii, II. 457.

4088. Betula japonica, var. mandshurica, II. 461.
4089. Betula utilis, var. Prattii, II. 457.
4090. Cotoneaster obscura, I. 161.
4091. Viburnum Veitchii, I. 109.
4092. Sorbus Rehderiana, I. 464.
4093. Rosa Prattii, II. 329.
4094. Ilex Fargesii, var. megalophylla, III. 426.
4095. Rosa omeiensis, f. pteracantha, II. 332.
4096. Rhamnus dumetorum, var. crenoserratus, II. 238.
4097. Lonicera Henryi, var. subcoriacea, I. 142.
4098. Rosa Moyesii, II. 325.
4099. Acer laxiflorum, I. 93.
4100. Acer laxiflorum, I. 93.
4101. Acer flabellatum, I. 91.
4102. Acer tetramerum, var. betulifolium, I. 95.
4103. Acer caudatum, var. Prattii, I. 91.
4104. { Acer tetramerum, var. betulifolium, f. latialatum, I. 95. Acer tetramerum, var. elobulatum, f. longeracemosum, I. 96.
4105. Carpinus Turczaninovii, var. ovalifolia, II. 427.
4106. Betula albo-sinensis, II. 457.
4107. Acer tetramerum, var. tiliifolium, I. 96.
4108. Acer laxiflorum, var. longilobum, I. 94.
4109. Acer spec., III. 427.
4110. Clematis montana, var. Wilsonii, I. 333.
4111. Rosa Moyesii, II. 325.
4112. Malus Prattii, II. 281.
4113. Rhamnus dumetorum, II. 237.
4114. Sorbus Rehderiana, I. 464.
4115. Malus kansuensis, II. 286.
4116. Magnolia Dawsoniana, II. 397.
4117. Celastrus rugosa, II. 349.
4118. Rosa omeiensis, f. pteracantha, II. 332.
4118a. Rosa banksiopsis, II. 322.
4119. Sorbus unguiculata, I. 473.
4120. Chaenomeles lagenaria, var. Wilsonii, II. 298.
4121. Viburnum cordifolium, I. 112.
4122. Celastrus glaucophylla, II. 347.
4123. Berberis Mouillacana, I. 371.

4124. Gaultheria nummularioides, I. 555.
4125. Viburnum lobophyllum, var. flocculosum, I. 114.
4126. Ribes Meyeri, α tanguticum, I. 152.
4127. Rosa longicuspis, II. 313.
4128. Viburnum betulifolium, I. 114.
4129. Skimmia melanocarpa, II. 138.
4130. Berberis spec., III. 443.
4131. Cotoneaster multiflora, I. 170.
4132. Pyrus pashia, II. 264.
4133. Prunus pubigera, var. Potaninii, I. 196.
4134. Berberis Tischleri, I. 355.
4135. Ilex yunnanensis, III. 425.
4136. Cotoneaster adpressa, I. 155.
4137. Rubus xanthocarpus, III. 423.
4138, 4138a. Osyris Wightiana, III. 320.
4139. Lonicera mitis, I. 136.
4140. Lonicera serpyllifolia, I. 131.
4141. Pterocarya hupehensis, III. 182.
4142. Acer laxiflorum, I. 93.
4143. Acer caudatum, var. multiserratum, I. 91.
4144. Clematis Fargesii, I. 335.
4145. Clematis montana, I. 332.
4146. Prunus mume, I. 278.
4147. Viburnum lobophyllum, I. 114.
4148. Zanthoxylum stenophyllum, II. 127.
4149. Berberis Bergmanniae, var. acanthophylla, I. 362.
4150. Viburnum ichangense, I. 115.
4152. Evonymus alata, var. aperta, I. 494.
4153. Berberis spec., III. 443.
4154. Berberis Liechtensteinii, I. 377.
4155. Sorbus aperta, I. 465.
4156. Sorbus Conradinae, I. 460.
4157. Celastrus rugosa, II. 349.
4159. Ilex yunnanensis, III. 425
4160. Cotoneaster Franchetii, I. 165.
4161. Triosteum himalayanum, var. chinense, I. 117.
4162. Elaeagnus multiflora, f. angustata, II. 413.
4163. Rosa omeiensis, II. 331.
4164. Rosa Soulieana, II. 314.
4165. Evonymus microcarpa, I. 487.
4166. Ribes himalayense, γ glandulosum, I. 44, 152.

4167. Acanthopanax lasiogyne, II. 563.
4168. Clematoclethra lasioclada, var. grandis, II. 386.
4170. Berberis diaphana, I. 354.
4171. Parthenocissus himalayana, III. 427.
4172. Malus Halliana, II. 285.
4173. Berberis aggregata, var. Prattii, I. 376; III. 443.
4174. Rosa Rubus, II. 308.
4175. Rosa glomerata, II. 309.
4176. Caragana erinacea, II. 622.
4177. Evonymus sanguinea, var. β camptoneura, I. 494.
4178. Evonymus alata, var. aperta, I. 494.
4179. Lonicera deflexicalyx, I. 140.
4180. Evonymus lanceifolia, I. 491.
4181. Akebia lobata, var. australis, III. 434.
4182. Smilax trachypoda, III. 3.
4183. Evonymus cornuta, I. 489.
4184. Parthenocissus himalayana, III. 427.
4185. Prunus pubigera, var. obovata, I. 196.
4186. Rhamnus heterophyllus, II. 232.
4187. Celastrus Rosthorniana, II. 351.
4188. Vitis betulifolia, III. 429.
4189. Ampelopsis Thomsonii, III. 427.
4190. Berberis diaphana, I. 354.
4191. Cotoneaster reticulata, I. 160.
4192. Rhamnella Julianae, II. 223.
4193. Clematoclethra lasioclada, var. grandis, II. 386.
4194. Viburnum lobophyllum, I. 114.
4195. Celastrus Rosthorniana, II. 351.
4196. Viburnum Wilsonii, I. 112.
4197. Rosa multibracteata, II. 328.
4198. Sorbus multijuga, I. 472.
4199. Cotoneaster salicifolia, var. floccosa, I. 173.
4200. Rosa filipes, II. 311.
4201. Evonymus grandiflora, I. 484.
4202. Prunus salicina, I. 276, 580.
4203. Berberis spec., III. 443.
4204. Acanthopanax evodiaefolius, II. 563.
4205. Prunus mira, I. 272.
4206. Clematoclethra scandens, II. 387.
4207. Sorbus Sargentiana, I. 461.
4208. Acer catalpifolium, I. 87.
4209. Dipelta ventricosa, I. 118.

4210. Abelia Zanderi, III. 429.
4211. Acer caudatum, var. Prattii, I. 91.
4212. Ribes moupinense, γ laxiflorum, I. 152.
4213. Zanthoxylum stenophyllum, II. 127.
4214. Lonicera Schneideriana, I. 133.
4215. Sorbus megalocarpa, var. cuneata, II. 267.
4216. Osmanthus serrulatus, II. 610.
4217. Phellodendron sachalinense, II. 136.
4218. Cornus chinensis, II. 577.
4219. Prunus sericea, II. 63.
4220, 4220a. Neillia thibetica, I. 435.
4221. Sorbus meliosmifolia, II. 270.
4222. Ilex corallina, III. 426.
4223. Rosa Davidii, II. 322.
4224. Corylopsis Willmottiae, I. 425.
4225. Lonicera mupinensis, I. 138.
4226. Ribes himalayense, a glandulosum, I. 152.
4227. Lonicera prostrata, I. 141.
4228. Viburnum cinnamomifolium, I. 111.
4229. Ribes Maximowiczii, I. 153.
4230. Lonicera chaetocarpa, I. 137.
4231. Rhododendron Przewalskii, I. 534.
4232. Rhododendron taliense, I. 533.
4233. Rhododendron Amesiae, I. 523.
4234. Rhododendron Faberi, I. 533.
4235. Rhododendron Weldianum, I. 532.
4236. Rhododendron yanthinum, I. 518.
4237. Rhododendron Sargentianum, I. 504.
4238. Rhododendron Augustinii, I. 524.
4239. Rhododendron Davidsonianum, I. 515.
4240. Rhododendron ambiguum, I. 518.
4241. Rhododendron yanthinum, var. lepidanthum, I. 519.
4242. Rhododendron villosum, I. 524.
4243. Rhododendron Przewalskii, I. 534.
4244. Rhododendron Watsonii, I. 545.
4245. Rhododendron oreodoxa, I. 540.
4246. Rhododendron pachytrichum, I. 530.
4247. Rhododendron oreodoxa, I. 540.
4248. Rhododendron Hunnewellianum, I. 535.

4249. Rhododendron Wasonii, I. 532.
4250. Rhododendron Weldianum, I. 532.
4251. Rhododendron Watsonii, I. 545.
4252. Rhododendron ambiguum, I. 518.
4253. Rhododendron bracteatum, I. 519.
4254. Rhododendron lacteum, I. 545.
4255. Rhododendron Hanceanum, I. 517.
4256. Rhododendron moupinense, I. 525.
4257. Rhododendron decorum, I. 541.
4258. Rhododendron strigillosum, I. 530.
4259. Rhododendron Watsonii, I. 545.
4260. Rhododendron oreodoxa, I. 540.
4261. Rhododendron Davidii, I. 543.
4262. Rhododendron micranthum, I. 513.
4263. Rhododendron longesquamatum, I. 529.
4264. Rhododendron Wiltonii, I. 531.
4265. Rhododendron ambiguum, I. 518.
4266. Rhododendron floribundum, I. 535.
4267. Rhododendron strigillosum, I. 530.
4268. Rhododendron stamineum, I. 546.
4269. Rhododendron violaceum, I. 511.
4270. Rhododendron pachytrichum, I. 530.
4271. Rhododendron oreodoxa, I. 540.
4272. Rhododendron Faberi, I. 533.
4273. Rhododendron Thayerianum, I. 529.
4274. Rhododendron Souliei, I. 537.
4275. Rhododendron argyrophyllum, var. cupulare, I. 526.
4275a. Rhododendron argyrophyllum, var. cupulare, I. 526.
4276. Rhododendron argyrophyllum, I. 525.
4277. Rhododendron lutescens, I. 516.
4278. Rhododendron polylepis, I. 521.
4279. Rhododendron calophytum, I. 544.
4280. Rhododendron Davidsonianum, I. 515.
4281. Quercus Gilliana, III. 223.
4281a. Quercus semicarpifolia, III. 221.
4282. Quercus aliena, III. 214.

4283. Corylus tibetica, II. 443.
4284. Pentapanax Henryi, II. 565.
4286. Berberis aggregata, I. 375.
4287. Berberis atrocarpa, III. 437.
4288. Berberis Silva-Taroucana, I. 370.
4289. Schisandra rubriflora, I. 412.
4290. Clematoclethra actinidioides, II. 386.
4291. Evonymus cornuta, I. 489.
4292. Clematoclethra Faberi, II. 387.
4293. Lonicera deflexicalyx, I. 140.
4294. Vaccinium Dunalianum, I. 560.
4295. Tilia chinensis, II. 364.
4296. Sorbaria arborea, var. subtomentosa, III. 423.
4297. Philadelphus Wilsonii, I. 145.
4298. Deutzia longifolia, I. 13.
4299, 4299a, b. Betula Potaninii, II. **459**.
4300. Deutzia longifolia, I. 13.
4301. Cercidiphyllum japonicum, var. sinense, I. 316.
4302. Hydrangea villosa, I. 29.
4303. Hydrangea xanthoneura, var. sutchuenensis, III. 421.
4304. Syringa Sargentiana, I. 298.
4305. Spiraea Rosthornii, I. 451.
4306. Cotoneaster obscura, I. 161.
4307. Berberis Tischleri, I. 355.
4308. Evonymus sanguinea, var. β camptoneura, I. 494.
4309. Rosa Moyesii, II. 325.
4310. Viburnum lobophyllum, var. flocculosum, I. 114.
4311. Cotoneaster apiculata, I. 156.
4312. Rosa Murielae, II. 326.
4313. Berberis Tischleri, III. 436.
4315. Viburnum lobophyllum, var. flocculosum, I. 114.
4316. Ilex Franchetiana, III. 426.
4317. Celastrus glaucophylla, II. 347.
4318. Clematis montana, var. Wilsonii, I. 333.
4319. Lonicera tatsienensis, I. 139.
4320. Vaccinium moupinense, I. 560.
4321. Sorbus Conradinae, I. 460.
4322. Actinidia tetramera, II. 381.
4323. Sorbus munda, f. b. subarachnoidea, I. 469.
4325. Philadelphus sericanthus, var. Rehderianus, I. 145.
4326. Deutzia longifolia, I. 13.
4327. Spiraea Henryi, I. 447.
4328. Tetracentron sinense, I. 417.

4329. Spiraea gemmata, I. 441.
4330. Hydrangea longipes, I. 33, 36, 40.
4331. Tilia intonsa, II. 365.
4332. Indigofera Souliei, III. 448.
4333. Tilia intonsa, II. 365.
4334. Philadelphus Magdalenae, I. 145.
4335. Acer caudatum, var. multiserratum, I. 91.
4336. Enkianthus deflexus, I. 550.
4336. Salix hypoleuca, III. 53.
4337. Hydrangea Rosthornii, I. 33.
4338. Hypericum Hookerianum, II. 403.
4339. Schizophragma integrifolium, I. 41.
4340. Sorbaria arborea, var. glabrata, I. 48.
4341. Hydrangea Rosthornii, I. 33.
4342. Hydrangea longipes, I. 33.
4343. Hydrangea Davidii, I. 25.
4344. Myricaria brateata, II. 407.
4345. Clematis gracilifolia, I. 331.
4346. Populus szechuanica, III. 20.
4347. Populus tremula, var. Davidiana, III. 24.
4348. Populus szechuanica, III. 20
4349. Salix spec., III. 456.
4350. Salix pella, III. 45.
4351. Salix Bockii, III. 71.
4352. { Salix phanera, III. 50.
       { Salix Wallichiana, III. 64.
4353. Salix Wallichiana, III. 64.
4354. Salix macroblasta, III. 58.
4355. Populus szechuanica, III. 20.
4356. Salix macroblasta, III. 58.
4357. Salix Rehderiana, III. 66.
4358. Populus tremula, var. Davidiana, III. 24.
4359. Populus tremula, var. Davidiana, f. tomentella, III. 25.
4360. Ulmus Bergmanniana, var. lasiophylla, III. 241.
4361. Populus szechuanica, III. 20.
4362. Larix Potaninii, II. 18.
4363. Salix magnifica, III. 44.
4364. Salix Rehderiana, III. 66.
4365. Salix cathayana, III. 57.
4366. Enkianthus deflexus, III. 445.
4367. Salix hypoleuca, III. 53.
4368. { Salix cathayana, III. 57.
       { Salix Rehderiana, III. 66.
4369. Salix Bockii, III. 71.
4370. Salix moupinensis, III. 46.
4371. Salix Ernesti, III. 47.

4372. Quercus aquifolioides, var. rufescens, III. 223.
4372a. Quercus aquifolioides, III. 222.
4373. Clematis Spooneri, I. 334.
4374. Caragana jubata, II. 103.
4375. Caragana Boisii, III. 448.
4376. Caragana jubata, II. 103.
4377. Cassiope selaginoides, I. 551.
4378. Evonymus saxicola, I. 491.
4379. Tilia intonsa, II. 365.
4380. Abelia Graebneriana, I. 118.
4381. Berberis verruculosa, I. 357.
4382. Betula Delavayi, II. 460.
4383. Deutzia glomeruliflora, I. 10.
4384. Philadelphus Wilsonii, I. 145.
4385. Berberis thibetica, I. 369.
4386. Aralia chinensis, var. glabrescens, II. 567.
4387. Philadelphus Magdalenae, I. 145.
4388. Cladrastis sinensis, II. 97.
4389. Buddleia nivea, var. yunnanensis, I. 570.
4390. Zanthoxylum Piasezkii, II. 122.
4391. Hydrangea xanthoneura, I. 26.
4392. Syringa pinnatifolia, I. 297.
4393, 4393a. Davidia involucrata, II. 255.
4395. Reevesia pubescens, II. 376.
4396. Abies Delavayi, III. 446.
4399. Picea spec., III. 446.
4400. Abies Faxoniana, III. 446.
4402. Spiraea mollifolia, I. 441.
4403. Buddleia nivea, var. yunnanensis, I. 570.
4404. Spiraea Henryi, I. 447.
4405. Camptotheca acuminata, II. 254.
4406. Corylopsis Willmottiae, I. 425.
4407. Syringa Komarowii, I. 301.
4408. Syringa tomentella, I. 300.
4409. Tilia tuan, II. 368.
4410. Tilia Oliveri, II. 366.
4411. Tilia Oliveri, II. 366.
4412. Pterocarya Paliurus, III. 456.
4413. Ribes Maximowiczii, I. 153.
4414. Ribes himalayense, γ urceolatum, I. 152.
4415. Lonicera longa, I. 134.
4416. Berberis brachypoda, I. 375.
4417. Rosa sertata, I. 327.
4418. Rosa caudata, II. 321.
4418a. Rosa banksiopsis, II. 322.
4421. Ribes longeracemosum, var. Wilsonii, III. 423.

4422. Abelia Graebneriana, I. 118.
4423. Fraxinus platypoda, II. 262, 623.
4424. Dipelta floribunda, I. 118.
4425. Rhus succedanea, II. 182.
4427. Acer Maximowiczii, I. 94.
4428. Acer erianthum, I. 90.
4429. Pterocarya hupehensis, III. 182.
4430. Pinus sinensis, II. 15.
4431. Fraxinus spec., III. 450.
4432. Salix Wallichiana, III. 64.
4433. Salix spec., III. 456.
4434. Salix hypoleuca, III. 53.
4435. Salix spec., III. 456.
4436. Salix Wallichiana, III. 64.
4437. Salix hypoleuca, III. 53.
4438. Betula albo-sinensis, II. 457.
4439. Salix Fargesii, III. 47.
4440. Populus adenopoda, III. 23.
4441. Fraxinus platypoda, II. 623.
4442. Ulmus Wilsoniana, III. 238.
4442a. Ulmus Bergmanniana, III. 240.
4443. Carpinus Henryana, III. 453.
4444. Carpinus cordata, var. chinensis, III. 453.
4445. Tilia Oliveri, II. 366.
4446. Litsea spec., III. 446.
4447. Salix Wallichiana, III. 64.
4448. Populus tremula, var. Davidiana, III. 456.
4449. Tilia tuan, II. 368.
4450, 4450a. Populus Wilsonii, III. 16.
4451. Abies Fargesii, II. 48.
4452. Picea Watsoniana, III. 446.
4453. Tsuga chinensis, III. 446.
4454. Daphne Wilsonii, II. 540.
4455. Clematoclethra Hemsleyi, II. 389.
4456. Triosteum Fargesii, I. 116.
4457. Meliosma Veitchiorum, II. 204.
4458. Ilex yunnanensis, β gentilis, III. 425.
4459. Actinidia melanandra, II. 378.
4460. Syringa reflexa, I. 297.
4461. Spiraea Veitchii, I. 449.
4462. Schizophragma integrifolium, var. denticulatum, I. 42.
4467. Deutzia Vilmorinae, III. 421.
4470. Ostryopsis Davidiana, II. 423.
4471. Helwingia chinensis, II. 571.
4472. Helwingia chinensis, II. 571.
4473. Helwingia chinensis, II. 571.
4474. Sorbaria arborea, var. subtomentosa, I. 47.

4475. Sorbaria arborea, var. glabrata, I. 48.
4476. Sorbaria arborea, var. glabrata, I. 48.
4477. Sorbaria arborea, var. glabrata, I. 48.
4478. Lonicera Koehneana, I. 140.
4479. Lonicera Ferdinandii, var. leycesterioides, I. 135.
4480. Parthenocissus spec., III. 428.
4481. Hydrangea longipes, I. 33.
4482. Hydrangea aspera, var. velutina, I. 30.
4483. Hydrangea villosa, I. 29.
4484. Hydrangea xanthoneura, I. 26.
4485. Hydrangea aspera, var. scabra, I. 31.
4486. Deutzia setchuenensis, var. corymbiflora, I. 9.
4487. Deutzia coriacea, I. 9; III. 421.
4488. Deutzia Fargesii, I. 10.
4489. Triosteum Fargesii, I. 116.
4490. Sambucus Wightiana, I. 306.
4491. Abelia Engleriana, I. 120.
4492. Abelia Zanderi, I. 121.
4493. Abelia parvifolia, I. 121.
4494. Abelia Schumannii, I. 121.
4495. Abelia myrtilloides, I. 120.
4496. Philadelphus sericanthus, I. 145.
4497. Viburnum Rosthornii, I. 110.
4498. Viburnum Veitchii, I. 109.
4499. Viburnum Henryi, I. 106.
4500. Viburnum Rosthornii, I. 110.
4501. Ribes moupinense, γ laxiflorum, I. 152.
4502. Ribes moupinense, γ laxiflorum, I. 152.
4503. Ribes Vilmorinii, I. 152.
4504. Ribes humile, I. 153.
4505. Alangium Faberi, II. 552.
4506. Acer Giraldii, I. 90.
4507. Acer pictum, var. parviflorum, I. 83.
4508. Acer erianthum, I. 90.
4509. Acer laxiflorum, var. longilobum, I. 94.
4510. Acer caudatum, var. multiserratum, I. 91.
4511. Acer tetramerum, var. betulifolium, I. 95.
4512. Acer tetramerum, var. betulifolium, I. 95.
4513. Acer laxiflorum, I. 93.

4540. Ilex fragilis, β Kingii, III. 426.
4541. Stachyurus yunnanensis., var. pedicellatus, I. 288.
4542. Cotoneaster Zabelii, I. 166.
4543. Cotoneaster obscura, var. cornifolia, I. 162.
4544. Cotoneaster tenuipes, I. 171.
4545. Clematoclethra lasioclada, II. 386.
4546. Clematis Rehderiana, I. 324; III. 434.
4547. Clematis brevicaudata, var. tenuisepala, I. 340.
4548. Clematis glauca, var. akebioides, I. 342.
4549. Clematis obscura, I. 329.
4550. Clematis pogonandra, I. 319.
4551. Corylus heterophylla, var. sutchuenensis, II. 445.
4552. Corylus heterophylla, var. sutchuenensis, II. 445.
4553. Corylus chinensis, I. 444.
4554. Thea oleifera, II. 393.
4555. Clerodendron japonicum, III. 377.
4556. Catalpa Fargesii, I. 305.
4557. Actinidia tetramera, I. 381.
4558. Acanthopanax leucorrhizus, var. fulvescens, II. 558.
4559. Schefflera Delavayi, II. 555.
4560. Aralia chinensis, var. glabrescens, II. 567.
4561. Acanthopanax Wilsonii, II. 560.
4562. Sindechites Henryi, III. 342.
4563. Melodinus Hemsleyanus, III. 331.
4564. Aristolochia heterophylla, III. 323.
4565. Erythina arborescens, II. 117.
4566. Evonymus Giraldii, var. angustialata, I. 495.
4567. Evonymus nanoides, I. 492.
4568. Evodia Bodinieri, II. 130.
4569. Syringa Sweginzowii, I. 299; III. 433.
4570. Spiraea Blumei, I. 446.
4571. Spiraea aemulans, I. 448.
4572. Spiraea Veitchii, I. 449.
4573. Spiraea ovalis, I. 446.
4574. Schisandra grandiflora, I. 411.
4575. Salix apatela, III. 46.
4576. Salix cheilophila, III. 69.
4577. Populus suaveolens, III. 18.
4578. Populus szechuanica, III. 20.
4579. Quercus semicarpifolia, III. 221.

4580. Quercus aquifolioides, III. 222.
4581. Castanea mollissima, III. 192.
4582. Castanea Henryi, III. 196.
4583. Quercus Gilliana, III. 223.
4584. Quercus dentata, III. 210.
4585. Quercus Fabri, III. 216.
4586. Lindera szechuenensis, II. 82.
4587. Litsea fruticosa, II. 77.
4588. Rhynchosia spec., III. 448.
4589. Vicia unijuga, III. 448.
4590. Phoebe nanmu, II. 72.
4591. Phoebe nanmu, II. 72.
4592. Ligustrum Delavayanum, II. 601.
4593. Lindera strychnifolia, II. 82.
4594. Lagerstroemia indica, II. 418.
4596. Ligustrum strongylophyllum, II. 605.
4597. Campylotropis macrocarpa, II. 113.
4598. Michelia spec., I. 409.
4599. Maddenia hypoleuca, I. 56, 59.
4600. Meliosma subverticillaris, II. 201.
4601. Magnolia denudata, var. purpurascens, I. 401.
4602. Meliosma Oldhami, II. 206.
4603. Millettia Dielsiana, II. 101.
4604. Mussaenda elliptica, III. 395.
4605. Myrsine semiserrata, II. 580.
4606. Meliosma Beaniana, II. 204.
4607. Meliosma Beaniana, II. 204.
4608. Millettia pachycarpa, II. 102.
4609. Fraxinus Paxiana, II. 623.
4610. Fraxinus Paxiana, II. 259.
4611. Torricellia angulata, II. 569.
4612. Ulmus pumila, III. 242.
4613. Wikstroemia brevipaniculata, II. 532.
4614. Daphne gemmata, II. 543.
4615. Gleditsia officinalis, II. 91.
4616. Diospyros sinensis, II. 591.
4617. Diospyros sinensis, II. 591.
4618. Campylotropis Wilsonii, II. 114.
4619. Photinia spec., III. 431.
4620. Zanthoxylum echinocarpum, II. 128.
4621. Zanthoxylum Bungei, II. 121.
4622. Zanthoxylum Bungei, II. 121.
4623. Zanthoxylum Bungei, II. 121.
4624. Zanthoxylum dimorphophyllum, II. 126.
4625. Porana sinensis, III. 355.
4626. Passiflora cupiformis, II. 408.

4627. Potentilla fruticosa, var. Veitchii, II. 303.
4628. Evodia Henryi, II. 133.
4629. { Potentilla fruticosa, II. 301.
Potentilla fruticosa, var. tangutica, II. 303.
4630. Palurius orientalis, II. 209.
4631. Spondias axillaris, var. pubinervis, II. 173.
4632. Sorbus aronioides, II. 268.
4633. Berberis dictyoneura, I. 374.
4634. Berberis polyantha, I. 376.
4635. Berberis Wilsonae, I. 368.
4636. Betula japonica, var. szechuanica, II. 461; III. 453.
4637. Berberis sanguinea, I. 359.
4638. Buddleia Davidii, var. alba, I. 568.
4639. Buddleia Davidii, var. magnifica, I. 567.
4640. Buddleia albiflora, var. Giraldii, I. 569.
4641. Buddleia albiflora, var. Giraldii, I. 569.
4642. Rosa multibracteata, II. 328.
4643. Rosa sertata, II. 327.
4644. Rosa Giraldii, f. glabriuscula, II. 328.
4645. Rosa sertata, II. 327.
4646. Rosa saturata, II. 324.
4647. Rhus orientalis, II. 179.

4648. Rhus Delavayi, II. 183.
4649, 4649a. Rosa chinensis, f. spontanea, II. 320.
4650. Torreya grandis, II. 7.
4651. Cunninghamia lanceolata, II. 50.
4723. Thea sinensis, II. 391.
4726. Rhododendron longistylum, I. 514.
4727. Rhododendron Hunnewellianum, I. 535.
4728. Rosa rugosa, var. Chamissoniana, II. 321.
4729. Larix Potaninii, II. 18.
4730. Larix Mastersiana, II. 19.
4731. Picea asperata, II. 22.
4732. Citrus nobilis, var. deliciosa, II. 143.
4733. Citrus spec., III. 449.
4734. Citrus nobilis, II. 142.
4735. Citrus spec., III. 449.
4736. Citrus ichangensis, II. 144.
4737. Citrus ichangensis, II. 144.
4738. Citrus Medica, II. 141.
4739. Citrus Medica, var. sarcodactylis, II. 141.
4740. Citrus Medica, var. sarcodactylis, II. 141.
4741. Citrus Medica, II. 141.
4742. Citrus Medica, II. 141.
4743. Citrus grandis, II. 144.
4744. Cornus Walteri, II. 576.

NUMBERS AND NAMES OF OTHER COLLECTIONS CITED IN THIS WORK[1]

Aitchison, James Edward Tierney.
111. Salix Wilhelmsiana, III. 169.
161. Populus afghanica, III. 36.
341. Abelia corymbosa, I. 128.
389. Salix Wallichiana, III. 64
403. Ulmus spec., III. 265.
413. Salix denticulata, III. 117.
574. Salix insignis, III. 152.
677. Ulmus villosa, III. 251.
711. Ulmus campestris, var., III. 262.

719. Betula Jacquemontii, II. 473.
1112. Ulmus spec., III. 265.
1116. Salix sericocarpa, III. 112.
1118. Salix alba, III. 112.
1207. Salix sericocarpa, III. 112.

Anderson, Thomas.
466. Prunus trichantha, I. 254.

Bailie, Joseph.
6. Celtis sinensis, III. 277.

[1] Some numbers, chiefly of specimens collected by Wilson on the Veitch Expeditions and by Henry, which are not cited in this work, have been added as far as seen and determined; the names of these additional numbers are followed by the author citation only. Numbers cited but not seen have not been enumerated.

Balansa, A.
3787. Salix tonkinensis, III. 97.
3788. Salix tonkinensis, III. 97.
4753. Salix Balansaei, III. 96.
4999. Salix Balansaei, III. 96.

Barchet, William.
229. Exochorda racemosa, I. 456.

Bodinier, Émile.
54d. Prunus Davidiana, I. 275.
65. Salix Cavaleriei, III. 101.
1413. Cudrania Bodinieri, III. 309.
1540. Deutzia Bodinieri, I. 147.
1569. Deutzia cinerascens, I. 146.
1604. Rosa microcarpa, II. 314.
1620. Rhamnus pseudofrangula, II. 244.
1633. Celtis Bodinieri, III. 276.
2070. Salix Wallichiana, III. 64.
2102. Salix Wallichiana, III. 64.
2134. Salix Camusii, III. 119.
2143. Berberis Griffithiana, I. 364.
2216. Deutzia pilosa, var. ochrophloeos,
    I. 146.
2223. Deutzia lancifolia, I. 147.
2465. Mahonia Bodinieri, I. 384.
2469. Mahonia Leveilleana, I. 385.
2587. Celtis Bodinieri, III. 276.
2697. Rhamnus Bodinieri, II. 246.

Bons d'Anty.
437. Buddleia yunnanensis, I. 564.

Bornmüller, Joseph.
3430. Sageretia Brandrethiana, var.
    Bornmuelleri, II. 230.

Carles, William Richard.
88. Loropetalum chinense, I. 430.
202. Cephalotaxus drupacea, II. 3.
439. Styrax philadelphoides, I. 289.
470. Lespedeza sericea, II. 105.
473. Lespedeza virgata, II. 110.

Cavalerie, Julien.
117. Rosa microcarpa, II. 314.
394. Celtis Biondii, var. Cavaleriei,
    III. 273.
725. Sageretia Henryi, II. 623.
728. Salix kouytchensis, III. 171.
974. Populus rotundifolia, var. Du-
    clouxiana, III. 25.
992. Rhamnus hamatidens, II. 252.
1317. Salix dodecandra, III. 101.
2069. Salix angiolepis, III. 104.
2212. Prunus Davidiana, I. 275.

2225. Prunus Davidiana, I. 275.
2477. Rhamnus heterophyllus, II. 232.
3283. Morus acidosa, III. 299.
3348. Rhamnus Leveilleanus, II. 237.
3784. Pteroceltis Tatarinowii, III. 285.
3790. Bischofia javanica, III. 454.

Chaffanjon, J.
2076. Daphne Feddei, II. 547.
2292. Rosa microcarpa, II. 314.

Champion, John George.
74. Turpinia nepalensis, II. 187.
102. Mussaenda pubescens, III. 399.

Chanet, L.
232. Sageretia Chanetii, II. 228.

Clarke, Charles Baron.
5911a, b. Salix tetrasperma, III. 94.
6780. Salix tetrasperma, III. 94.
13693. Alnus nepalensis, II. 502.
14580. Salix tetrasperma, III. 94.
25151d, h. Alnus nepalensis, II. 502.
25479. Alnus nepalensis, II. 502.
28928. Betula Jacquemontii, II. 473.
34176a. Salix tetrasperma, III. 94.
36465b. Alnus nepalensis, II. 502.
46331b. Alnus nepalensis, II. 502.

Cubitt, J. E. S.
3. Luculia intermedia, III. 408.
4. Luculia intermedia, III. 408.

David, Armand.
750. Carpinus laxiflora, var. Davidii,
    II. 426.
1681. Salix purpurea, var. stipularis,
    III. 168.
1687. Populus tremula, var. Davidiana,
    III. 24.
1694. Ostryopsis Davidiana, II. 423.
1704. Quercus dentata, III. 210.
1728. Malus baccata, II. 291.
1783. Celastrus articulata, II. 356.
2676. Salix microstachya, III. 170.
2719. Salix microstachya, III. 170.

Delavay, Jean Marie.
37. Ostryopsis Davidiana, var. cineras-
    cens, II. 423.
128. Corylus heterophylla, var. sutchu-
    enensis, II. 445.
197. Corylus chinensis, II. 444.
212. Corylus chinensis, II. 444.
485. Berberis Delavayi, I. 364.
555. Corylus heterophylla, var. yunna-
    nensis, II. 451.

DELAVAY, J. M. (*continued*)
822. Stachyurus yunnanensis, I. 288.
993. Berberis levis, III. 437.
1031. Picea likiangensis, II. 31.
1047. Berberis Lecomtei, I. 373.
1049. Prunus yunnanensis, I. 239.
1124. Berberis Griffithiana, I. 364.
1148. Hydrangea heteromalla, var. mollis, I. 151.
1210. Abies Delavayi, II. 41.
2200. Salix floccosa, III. 148.
2353. Mahonia Duclouxiana, I. 384.
2447. Berberis Lecomtei, I. 373.
2658. Prunus caudata, I. 259.
3105. Salix floccosa, III. 148.
3543. Deutzia calycosa, I. 149.
3718. Helwingia japonica, II. 570.
3725. Betula Delavayi, II. 460.
3773. Prunus serrula, I. 253.
3790. Prunus serrula, I. 253.
4323. Salix floccosa, III. 148.
4618. Tsuga yunnanensis, II. 36.
4678. Salix floccosa, III. 148.
5024. Mahonia gracilipes, I. 385.
6830. Mahonia setosa, I. 385.

Ducloux, F.
77. Prunus Duclouxii, I. 242.
653. Salix Cavaleriei, III. 101.
658. Salix Cavaleriei, III. 101.
669. Salix Cavaleriei, III. 101.
670. Salix Duclouxii, III. 170.
2323. Mahonia Duclouxiana, I. 384.
3055. Mahonia Duclouxiana, I. 384.

Dunn, Stephen Troyte. *See* HONG-KONG HERB.

Elmer, Adolph Daniel Edward.
8779. Pistacia chinensis, II. 173.
11317. Berchemia Elmeri, II. 220.

Esquirol, Joseph.
327. Salix andropogon, III. 170.
368. Salix pachyclada, III. 150.
392. Rhamnus Esquirolii, II. 233.
567. Salix erioclada, III. 118.
775. Daphne Feddei, II. 547.
1517. Rosa microcarpa, II. 314.
2100. Rosa Gentiliana, II. 312.
3618. Celastrus Esquiroliana, II. 357.

Faber, Ernst.
3. Salix Bockii, III. 71.
4. Andrachne montana, II. 517.
26. Evodia officinalis, II. 130.

57. Elaeocarpus japonicus, II. 360.
74. Salix Bockii, III. 71.
87. Symplocos botryantha, II. 596.
92. Staphylea Bumalda, II. 185.
101. Quercus dentata, III. 210.
103. Salix ?cathayana, III. 57.
114. Cotinus coggygria, var. pubescens, II. 175.
116. Salix Wilsonii, III. 40.
202. Acer amplum, var. tientaiense, I. 87.
207. Diospyros sinensis, II. 591.
210. Deutzia Faberi, I. 18.
211. Salix Bockii, III. 71.
212. Betula luminifera, II. 455.
215. Quercus glandulifera, III. 212.
216. Quercus serrata, III. 217.
229. Berberis Tischleri, III. 436.
274. Smilax herbacea, var. oblonga, III. 1.
495. Melastoma normale, II. 421.
528. Rosa omeiensis, II. 331.
543. Spiraea japonica, var. acuminata, I. 452.
572. Melia Azedarach, II. 157.
590. Premna ligustroides, III. 372.
637. Cornus paucinervis, II. 576.
642. Mussaenda divaricata, III. 394.
645. (Herb. Arnold Arb.) Clerodendron cyrtophyllum, III. 377.
645. Mussaenda divaricata, III. 394.
669. Cipadessa baccifera, var. sinensis, II. 159.
736. Clematis Pavoliniana, I. 328.
901. Salix Wilsonii, III. 40.
902. Buxus microphylla, var. sinica, II. 165.
999. Elaeagnus multiflora, II. 412.
1714. Malus theifera, II. 283.

Falconer, Hugh.
955. Alnus nitida, II. 501.
961. Salix denticulata, III. 117.

Farges, Paul.
14. Carpinus cordata, var. chinensis, II. 437.
83. Betula insignis, II. 459.
128. Torreya grandis, II. 7.
153. Juniperus squamata, var. Fargesii, II. 59.
681. Celtis Biondii, III. 272.
699. Carpinus laxiflora, var. macrostachya, II. 425.

FARGES, P. (*continued*)
758. Helwingia japonica, var. hypoleuca, II. 570.
795. Salix Fargesii, III. 47.
806. Picea brachytyla, II. 33.
808. Tsuga chinensis, II. 37.
809. Corylus heterophylla, II. 450.
998. Prunus Herincquiana, var. biloba, I. 254.
1010. Betula luminifera, II. 455.
1012. Betula Fargesii, II. 478.
1043. Deutzia Fargesii, I. 10.
1273. Carpinus Turczaninovii, var. ovalifolia, II. 427.
1292. Keteleeria Davidiana, II. 39.

Faurie, Urbain (Formosan plants).
4. Gordonia axillaris, II. 394.
22. Zanthoxylum alatum, var. planispinum, II. 125.
28. Euphoria longana, II. 193.
32. Sapindus mukorossi, II. 191.
33. Rhus succedanea, II. 182.
45. Liquidambar formosana, I. 421.
49. Eurya chinensis, II. 400.
129. Rosa laevigata, II. 318.
150. Lespedeza sericea, II. 105.
249. Trachelospermum divaricatum, var. brevisepalum, III. 338.
302. Ehretia acuminata, III. 363.
303. Ehretia acuminata, III. 363.
314. Symplocos paniculata, II. 593.
465. Buddleia asiatica, I. 566.
516. Alnus formosana, II. 501.
519. Salix glandulosa, var. Warburgii, III. 99.

Faurie, Urbain (Japanese plants).
10. Salix Buergeriana, III. 162.
11. Salix Buergeriana, III. 162.
100. Prunus subhirtella, I. 255.
251. Salix Urbaniana, III. 103.
252. Salix Urbaniana, III. 103.
253. Salix cardiophylla, III. 103.
254. Salix Caprea, III. 149.
255. Salix viminalis, var. yezoensis, III. 158.
256. Salix hondoensis, III. 110.
257. Salix ?sapporoensis, III. 166.
258. Salix vulpina, III. 130.
258bis. Salix vulpina, III. 130.
260. Salix opaca, III. 159.
263. Salix hamatidens, III. 108.
265. Salix sachalinensis, III. 158.

266. Salix sapporoensis, III. 166.
268. Salix sapporoensis, III. 166.
270. Salix sachalinensis, III. 158.
376. Prunus Grayana, I. 69.
507. Salix Urbaniana, III. 103.
513. Rhamnus oreigenes, II. 244.
783. Alnus Fauriei, II. 495.
2093. Prunus parvifolia, f. aomoriensis, I. 251.
2096. Prunus nikkoensis, I. 260.
2099. Prunus incisa, I. 258.
3156. Prunus tomentosa, var. insularis, I. 269.
3157. Prunus tomentosa, var. insularis, I. 269.
3158. Prunus japonica, var. eujaponica, f. Fauriei, I. 266.
3160. Prunus incisa, I. 258.
3211. Prunus glandulosa, var. glabra, f. Sieboldiana, I. 263.
3653. Celtis sinensis, III. 277.
3654. Aphananthe aspera, III. 290.
4047. Callicarpa japonica, var. luxurians, III. 369.
4841. Aphananthe aspera, III. 290.
5009. Prunus Grayana, I. 69.
5068. Alnus pendula, II. 507.
5085. Populus Sieboldii, III. 38.
5087. Populus Sieboldii, III. 38.
5088. Populus Sieboldii, III. 38.
5362. Alnus Matsumurae, II. 500.
5380. Malus Sieboldii, var. arborescens, II. 294.
5421. Quercus mongolica, III. 230.
5535. Broussonetia Kaempferi, III. 305.
5751. Salix vulpina, III. 130.
5752. Salix kakista, III. 128.
5757. Salix vulpina, III. 130.
5758. Salix ampherista, III. 175.
5759. Salix Caprea, III. 149.
5760. Salix amygdalina, var. nipponica, III. 106.
5766. Salix amygdalina, var. nipponica, III. 106.
5767. Salix gracilistyla, III. 164.
5772. Carpinus laxiflora, II. 438.
5773. Carpinus laxiflora, II. 438.
5774. Carpinus japonica, II. 433.
5775. Carpinus cordata, II. 434.
5776. Carpinus cordata, II. 434.
5777. Carpinus cordata, II. 434.
5778. Carpinus Tschonoskii, II. 441.

FAURIE, U. (*continued*)
5779. Carpinus japonica, II. 433.
5780. Carpinus japonica, II. 433.
5781. Betula Maximowicziana, II. 465.
5782. Betula japonica, II. 485.
5783. Betula Ermanii, var. subcordata, II. 471.
5784. Betula Ermanii, var. subcordata, II. 471.
5785. Betula Ermanii, var. subcordata, II. 471.
5787. Alnus Maximowiczii, II. 505.
5791. Alnus firma, subsp. hirtella, II. 506.
5792. Alnus firma, subsp. hirtella, II. 506.
5802. Corylus heterophylla, II. 450.
5803. Corylus Sieboldiana, var. brevirostris, II. 453.
5877. Ulmus japonica, III. 258.
5878. Zelkova serrata, III. 288.
5879. Zelkova serrata, III. 288.
5880. Morus acidosa, III. 300.
6044. Malus Sieboldii, var. arborescens, II. 294.
6051. Malus Halliana, II. 285.
6052. Prunus nipponica, I. 259.
6054. Prunus Grayana, I. 69.
6595. Salix vulpina, III. 130.
6599. Betula Maximowicziana, II. 465.
6600. Salix sachalinensis, var. Pilgeriana, III. 159.
6601. Salix hondoensis, III. 110.
6602. Salix hondoensis, III. 110.
6605. Salix vulpina, III. 130.
6606. Salix vulpina, III. 130.
6608. Salix vulpina, III. 130.
6609. Salix sachalinensis, III. 158.
6610. Salix sachalinensis, III. 158.
6611. Salix Caprea, III. 149.
6612. Salix Caprea, III. 149.
6613. Salix ?Reinii, III. 127.
6615. Salix Makinoana, III. 110.
6621. Salix vulpina, III. 130.
6622. Salix hondoensis, III. 110.
6623. Salix eriocarpa, III. 108.
6624. Salix vulpina, III. 130.
6626. Salix ?Reinii, III. 127.
6627. Betula grossa, II. 477.
6628. Betula Ermanii, var. subcordata, II. 471.
6629. Betula Ermanii, var. subcordata, II. 471.

6630. Betula japonica, var. kamtschatica, II. 486.
6631. Betula japonica, var. kamtschatica, II. 486.
6632. Betula Maximowicziana, II. 465.
6633. Betula Ermanii, var. subcordata, II. 471.
6634. Betula japonica, var. kamtschatica, II. 486.
6635. Betula japonica, var. kamtschatica, II. 486.
6636. Betula Ermanii, var. subcordata, II. 471.
6637. Betula Ermanii, var. subcordata, II. 471.
6638. Betula grossa, II. 477.
6639. Betula Ermanii, var. subcordata, II. 471.
6641. Carpinus japonica, var. caudata, II. 434.
6642. Populus Maximowiczii, III. 32.
6643. Salix cardiophylla, III. 103.
6644. Populus Sieboldii, III. 38.
6645. Alnus Maximowiczii, II. 505.
6646. Alnus hirsuta, var. sibirica, II. 498.
6647. Alnus Maximowiczii, II. 505.
6648. Corylus Sieboldiana, II. 452.
6649. Corylus Sieboldiana, II. 452.
6698. Prunus Herincquiana, I. 214.
6699. Prunus iwagiensis, I. 259.
6700. Prunus subhirtella, I. 255.
6816. Corylus Sieboldiana, II. 452.
6837. Abelia spathulata, I. 125.
6838. Abelia spathulata, I. 125.
6848. Corylus Sieboldiana, II. 452.
6873. Rhamnus japonicus, II. 251.
6923. Wistaria floribunda, II. 510.
7588. Alnus Maximowiczii, II. 505.
10026. Salix vulpina, III. 130.
10027. Salix vulpina, III. 130.

Faurie, Urbain (Korean plants).
51. Lespedeza Maximowiczii, II, 111; III. 448.
77. Prunus Nakaii, I. 267.
82. Sorbus alnifolia, var. lobulata, II. 275.
174. Salix ?Caprea, III. 149.
175. Salix koreensis, III. 111.
176. Salix koreensis, III. 111.
177. Salix Maximowiczii, III. 100.

FAURIE, U. (*continued*)
179. Salix purpurea, III. 167.
180. Salix purpurea, III. 167.
181. Salix koreensis, III. 111.
184. Castanopsis cuspidata, III. 204.
185. Quercus glauca, III. 226.
186. Quercus dentata, III. 210.
187. Quercus dentata, III. 210.
188. Quercus dentata, III. 210.
189. Quercus mongolica, III. 230.
190. Quercus mongolica, III. 230.
191. Quercus mongolica, III. 230.
192. Quercus aliena, var. acuteserrata, III. 215.
193. Quercus dentata, III. 210.
194. Quercus variabilis, III. 219.
195. Quercus aliena, var. acuteserrata, III. 215.
197. Quercus aliena, III. 214.
198. Quercus glandulifera, III. 212.
202. Carpinus Turczaninovii, II. 439.
203. Betula chinensis, II. 479.
204. Betula chinensis, II. 479.
205. Carpinus cordata, II. 434.
206. Alnus japonica, var. arguta, II. 494.
207. Corylus Sieboldiana, var. mandshurica, II. 454.
233. Rhamnus koraiensis, II. 249.
234. Rhamnus Schneideri, II. 250.
235. Rhamnus koraiensis, II. 249.
236. Rhamnus koraiensis, II. 249.
305. Sorbus alnifolia, var. lobulata, II. 275.
306. Sorbus alnifolia, var. lobulata, II. 275.
312. Spiraea Fritschiana, var. angulata, I. 453.
324. Salix ?Caprea, III. 149.
325. Rosa multiflora, var. quelpaertensis, II. 335.
334. Prunus Nakaii, I. 267.
335. Prunus tomentosa, var. insularis, I. 269.
358. Hydrangea petiolaris, I. 41.
360. Deutzia glabrata, I. 24.
361. Deutzia parviflora, I. 23.
362. Deutzia glabrata, I. 24.
364. Deutzia coreana, I. 22.
411. Lespedeza Maximowiczii, II. 111; III. 448.
412. Lespedeza Maximowiczii, II. 111; III. 448.

413. Lespedeza sericea, II. 105.
414. Lespedeza inschanica, II. 108.
415. Lespedeza trichocarpa, II. 109.
416. Lespedeza formosa, II. 107.
419. Lespedeza tomentosa, II. 110.
447. Euscaphis japonica, II. 187.
459. Orixa japonica, II. 135.
463. Zanthoxylum Bungei, var. Zimmermannii, II. 122.
465. Zanthoxylum alatum, var. planispinum, II. 125.
468. Evodia Daniellii, II. 135.
469. Cedrela sinensis, II. 156.
478. Staphylea Bumalda, II. 185.
479. Citrus nobilis, var. deliciosa, II. 143.
480. Citrus sinensis, II. 148.
481. Citrus nobilis, var. deliciosa, II. 143.
482. Citrus nobilis, var. deliciosa, II. 143.
483. Poncirus trifoliata, II. 149.
484. Rhus trichocarpa, II. 180.
485. Rhus sylvestris, II. 180.
486. Rhus javanica, II. 178.
487. Rhus javanica, II. 178.
490. Tilia mandshurica, II. 370.
495. Eurya ochnacea, II. 399.
500. Rhamnus davuricus, II. 251.
502. Rhamnus koraiensis, II. 249.
503. Rhamnus davuricus, II. 251.
504. Rhamnella franguloides, II. 225.
508. Zizyphus sativa, var. inermis, II. 212.
509. Zizyphus sativa, var. inermis, II. 212.
512. Rhamnus koraiensis, II. 249.
583. Zelkova serrata, III. 288.
601. Betula davurica, II. 483.
605. Betula chinensis, II. 479.
608. Quercus mongolica, III. 230.
624. Carpinus laxiflora, II. 438.
639. Rhamnella franguloides, II. 225.
640. Rhamnella franguloides, II. 225.
660. Vaccinium bracteatum, I. 558.
725. Styrax japonicus, I. 291; III. 432.
727. Styrax japonicus, I. 291; III. 432.
739. Trachelospermum divaricatum, III. 338.
903. Hemiptelea Davidii, III. 288.
904. Celtis sinensis, III. 277.
1447. Salix koreensis, III. 111.
1502. Salix ?Caprea, III. 149.

FAURIE, U. (*continued*)
1503. Salix Caprea, III. 149.
1524. Castanopsis cuspidata, III. 204.
1525. Quercus serrata, III. 217.
1527. Quercus glandulifera, III. 212.
1529. Quercus salicina, III. 234.
1530. Carpinus cordata, II. 434.
1531. Carpinus cordata, II. 434.
1532. Carpinus laxiflora, II. 438.
1535. Carpinus Fauriei, II. 442.
1536. Carpinus laxiflora, II. 438.
1538. Betula Ermanii, var. genuina, subvar. Saitôana, II. 470.
1539. Betula Ermanii, var. genuina, subvar. Saitôana, II. 470.
1540. Betula Ermanii, var. genuina, subvar. Saitôana, II. 470.
1541. Myrica rubra, III. 189.
1546. Prunus Herincquiana, I. 214.
1552. Sorbus alnifolia, var. lobulata, II. 275.
1554. Sorbus alnifolia, var. lobulata, II. 275.
1555. Malus Sieboldii, var. arborescens, II. 294.
1566. Rosa Wichuraiana, II. 335.
1568. Rosa Wichuraiana, II. 335.
1604. Euscaphis japonica, II. 187.
1606. Sapindus mukorossi, II. 191.
1607. Melia Azedarach, II. 157.
1608. Picrasma quassioides, II. 152.
1623. Rhamnella franguloides, II. 225.
1627. Zanthoxylum alatum, var. planispinum, II. 125.
1628. Zanthoxylum Bungei, var. Zimmermannii, II. 122.
1654. Hydrangea petiolaris, I. 41.
1689. Lespedeza bicolor, II. 112.
1690. Lespedeza bicolor, II. 112.
1843. Lonicera Tatarinovii, I. 144.
1844. Lonicera Tatarinovii, I. 144.
1867. Meliosma Oldhamii, II. 206.
1876. Styrax japonicus, I. 291; III. 432.
1889. Symplocos caudata, II. 595.
1890. Rhamnus Taquetii, II. 248.
1899. Trachelospermum divaricatum, III. 338.
2008. Ulmus parvifolia, III. 244.
3234. Betula Ermanii, var. genuina, subvar. Saitôana, II. 470.
3240. Salix koreensis, III. 111.
3241. Salix koreensis, III. 111.

3242. Salix koreensis, III. 111.
3243. Salix koreensis, III. 111.
3244. Salix koreensis, III. 111.
3245. Salix Blinii, III. 161.
3246. Salix sibirica, var. subopposita, III. 154.
3247. Salix sibirica, var. subopposita, III. 154.
3248. Salix Blinii, III. 161.
3249. Salix Blinii, III. 161.
4706. Salix koreensis, III. 111.
4707. Salix koreensis, III. 111.
4708. Salix koreensis, III. 111.

Faurie, Urbain (Saghalien plants).
5. Betula Ermanii, var. genuina, II. 470.
14. Malus baccata, var. mandshurica, II. 281.
272, 272bis. Salix Caprea, III. 149.
273. Salix viminalis, III. 157.
273ter. Salix Caprea, III. 149.
274. Salix sachalinensis, III. 158.
275. Salix sachalinensis, III. 158.
276. Salix sachalinensis, III. 158.
278. Salix rorida, III. 155.
280, 280bis. Populus Maximowiczii, III. 32.
286. Alnus hirsuta, II. 496.
288. Alnus Maximowiczii, II. 505.
289. Betula Ermanii, var. genuina, II. 470.
291. Betula Ermanii, var. genuina, II. 470.
292. Betula Ermanii, var. subcordata, II. 471.
293. Betula Ermanii, var. genuina, II. 470.
555. Malus baccata, var. mandshurica, II. 281.

Fernandes, W. S.
60. Cudrania javanensis, III. 308.

Forbes, Henry Ogg.
4104. Porana racemosa, III. 361.

Ford, Charles.
17. Mahonia Fordii, I. 383.
68. Pyrus Calleryana, II. 264.
94. Abelia chinensis, I. 121.
280. Paliurus hirsutus, II. 211.
327. Aphananthe aspera, III. 290.
357. Quercus Fabri, III. 216.
386. Keteleeria Fortunei, II. 40.

FORD, C. (*continued*)
592. Pittosporum glabratum, III. 326.
612. Spondias axillaris, II. 172.
1795. Abelia chinensis, I. 121.

Forrest, George.
13. Wikstroemia holosericea, II. 537.
34. Paederia tomentosa, III. 403.
44. Wikstroemia scytophylla, II. 538.
133. Wikstroemia dolichantha, II. 537.
146. Prinsepia utilis, II. 335.
522. Osyris Wightiana, III. 320.
540. Loranthus Delavayi, III. 316.
952. Lycium chinense, III. 385.
2012. Berberis levis, III. 437.
2022. Rosa omeiensis, II. 331.
2033. Salix Cavaleriei, III. 101.
2039. Tamarix juniperina, II. 407.
2049. Rosa odorata, var. gigantea, f. erubescens, II. 339.
2054. Schoepfia jasminodora, III. 321.
2084. Pittosporum heterophyllum, III. 329.
2106. Acer laxiflorum, III. 427.
2111. Malus yunnanensis, II. 287.
2115. Daphne aurantiaca, II. 547.
2155. Abelia Zanderi, III. 430.
2180. Smilax discotis, III. 6.
2256. Rosa omeiensis, II. 331.
2318. Salix floccosa, III. 148.
2370. Rosa Soulieana, II. 314.
2402. Rosa Moyesii, II. 325.
2425. Hypericum Hookerianum, II. 403.
2600. Loranthus caloreas, III. 315.
2723. Potentilla fruticosa, II. 301.
4300. Hypericum Hookerianum, II. 403.
4344. Berberis yunnanensis, III. 435.
4442. Rosa Moyesii, II. 325.
4443. Rosa irridens, II. 337.
4444. Rosa multiflora, var. brachyacantha, II. 334.
4445. Rosa sertata, II. 327.
4447. Rosa sertata, II. 327.
4448. Rosa omeiensis, II. 331.
4449. { Rosa multiflora, var. brachyacantha, II. 334. Rosa multiflora, var. carnea, II. 305.
4450. Rosa Roxburghii, f. normalis, II. 319.
4451. Rosa omeiensis, II. 331.

4452. Rosa odorata, var. gigantea, f. erubescens, II. 339.
4453. Rosa sertata, II. 327.
4454. Rosa longicuspis, II. 313.
4600. Salix cathayana, III. 57.
4602. Salix resecta, III. 121.
4719. Trachelospermum cathayanum, III. 334.
4745. Osyris Wightiana, III. 320.
4759. Elsholtzia fruticosa, III. 382.
4769. Ehretia macrophylla, III. 364.
4779. Smilax vaginata, III. 2.
4782. Boehmeria nivea, III. 312.
4785. Smilax rigida, III. 11.
4793. Smilax menispermoidea, III. 5.
4796. Smilax discotis, III. 6.
4967. Salix psilostigma, III. 116.
4974. Prinsepia utilis, II. 335.
4997. Hedera himalaica, II. 555.
5024. Daphne papyracea, II. 546.
5026. Abutilon sinense, II. 373.
5546. Betula Delavayi, var. Forrestii, II. 479.
5835. Betula Delavayi, var. calcicola, II. 479.
10343. Ostryopsis nobilis, II. 423.

Fortune, Robert.
80. Mussaenda pubescens, III. 399.
103. Tamarix chinensis, II. 406.

Gamble, James Sykes.
2646a. Populus Gamblei, III. 34.
6707. Populus Gamblei, III. 34.

Gammie, George Alexander.
312. Malus theifera, II. 283.

Giraldi, Guiseppe.
26. Berchemia Giraldiana, II. 213.
771. Deutzia discolor, I. 12.
931. Rhamnus hypochrysus, II. 252.
932. Rhamnus hypochrysus, II. 252.
940. Rhamnus hypochrysus, II. 252.
944. Berchemia Giraldiana, II. 213.
949. Berchemia Giraldiana, II. 213.
1082. Sorbus tapashana, I. 482.
1083. Sorbus Koehneana, I. 471.
1084. Sorbus Koehneana, I. 471.
1085. Sorbus Koehneana, I. 471.
1086. Sorbus Koehneana, I. 471.
1134. Prunus dictyoneura, I. 262.
1135. Prunus dictyoneura, I. 262.
1137. Prunus glandulosa, var. trichostyla, f. sinensis, I. 265.

GIRALDI, G. (*continued*)

1137. Prunus triloba, var. plena, I. 274.
1138. Prunus tomentosa, var. endotricha, I. 225.
1141. Prunus brachypoda, var. pseudossiori, I. 65.
1160. Salix Caprea, III. 149.
1167. Hydrangea Bretschneideri, var. Giraldii, I. 39.
1168. Hydrangea Bretschneideri, var. Giraldii, I. 39.
1169. Hydrangea Bretschneideri, var. Giraldii, I. 39.
1171. Hydrangea Bretschneideri, var. Giraldii, I. 39.
1172. Hydrangea Bretschneideri, var. Giraldii, I. 39.
1360. Callicarpa Giraldiana, III. 366.
1361. Callicarpa Giraldiana, III. 366.
1656. Deutzia grandiflora, var. Baroniana, I. 21.
1696. Prunus dictyoneura, I. 262.
1766. Abelia onkocarpa, I. 128.
1796. Berchemia Giraldiana, II. 213.
1815. Abelia Dielsii, I. 128.
2115. Acer Giraldii, I. 90.
2118. Acer tetramerum, var. betulifolium, I. 95.
2119. Acer tetramerum, var. betulifolium, I. 95.
2136. Acer Giraldii, I. 90.
2231. Acanthopanax stenophyllus, II. 564.
2526. Deutzia micrantha, I. 23.
3287. Cornus macrophylla, II. 575.
3290. Cornus Hemsleyi, II. 574.
3789. Prunus Giraldiana, I. 257.
4522. Deutzia grandiflora, var. Baroniana, I. 21.
4924. Berchemia Giraldiana, II. 213.
4928. Berchemia Giraldiana, II. 213.
4931. Prunus velutina, I. 69.
5118. Sorbus Koehneana, I. 171.
5119. Sorbus Koehneana, I. 471.
5120. Sorbus Koehneana, I. 471.
5123. Sorbus Koehneana, I. 471.
5124. Sorbus Koehneana, I. 471.
5126. Sorbus tapashana, I. 482.
5127. Sorbus tapashana, I. 482.
5129. Sorbus aperta, I. 465.
5195. Prunus dictyoneura, I. 262.
5200. Prunus brachypoda, var. pseudossiori, I. 65.

5253. Sorbus Koehneana, I. 471.
5290. Prunus tomentosa, var. endotricha, I. 225.
5291. Prunus tomentosa, var. endotricha, I. 225.
5295. { Prunus tomentosa, var. breviflora, I. 270. Prunus tomentosa, var. tsuluensis, I. 270.
5359. Salix spathulifolia, III. 114.
5362. Salix hypoleuca, III. 53.
5896. Callicarpa Giraldiana, III. 366.
6081. Prunus sericea, var. septentrionalis, I. 64.
7137. Acer tetramerum, var. betulifolium, I. 95.
7148. Prunus tomentosa, var. endotricha, I. 225.
7173. Sorbus Koehneana, I. 471.
7179. Deutzia micrantha, I. 23.
7186. Prunus tomentosa, var. endotricha, I. 225.
7267. Carpinus Turczaninovii, II. 439.
9293. Prunus tomentosa, var. endotricha, I. 225.

Griffith, William.

270. Pittosporum napaulensis, III. 326.
271. { Pittosporum glabratum, III. 328. Pittosporum glabratum, var. neriifolium, III. 328.
957. Populus ciliata, III. 31.
2029. Rhamnus Bodinieri, II. 246.
2032. Rhamnus nipalensis, II. 245.
2074. Sorbus ferruginea, II. 277.
2076. Sorbus Griffithii, II. 277.
2078. Sorbus khasiana, II. 278.
2140. Rosa longicuspis, II. 313.
2559. Populus ciliata, III. 31.
4486. Betula alnoides, II. 467.
4488. Carpinus viminea, II. 437.
4489. Carpinus faginea, II. 442.
4490. Alnus nepalensis, II. 501.
4495. Populus rotundifolia, III. 39.
4498. Salix Daltoniana, III. 115.
4500. Salix psilostigma, III. 116.
4501. Salix Wallichiana, III. 64.
4503. Salix tetrasperma, III. 94.
4505. Salix dealbata, III. 105.
4678. Ulmus lanceaefolia, III. 263.
4986. Juniperus squamata, II. 57.

Hance, Henry Fletcher.
322. Buxus Harlandii, II. 166.
335. Callicarpa dichotoma, III. 370.
367. Castanopsis sclerophylla, III. 201.
379. Chloranthus brachystachys, III. 15.
397. Clerodendron cyrtophyllum, III. 377.
497. Eurya chinensis, II. 400.
663. Alangium chinense, II. 552.
787. Quercus bambusifolia, III. 235.
1068. Celtis sinensis, III. 277.
1414. Ulmus parvifolia, III. 244.
4956. Callicarpa longifolia, III. 369.
6683. Buxus stenophylla, II. 169.
7046. Prunus campanulata, I. 253.
7494. Vitex quinata, III. 374.
10130. Prunus pogonostyla, var. globosa, I. 265.
11071. Trachelospermum jasminoides, III. 334.
11209. Liquidambar formosana, I. 421.
11416. Quercus variabilis, III. 219.
11435. Vitex Negundo, var. incisa, III. 373.
12008. Cotinus coggygria, var. cinerea, II. 176.
12681. Carpinus Turczaninovii, II. 439.
13757. Salix babylonica, III. 42.
14113. Rhamnus oreigenes, II. 244.
16446. Salix Mesnyi, III. 95.
17623. Pterocarya stenoptera, III. 181.
19216. Cudrania javanensis, III. 308.
22009. Salix microstachya, III. 170.

Hancock, William.
13. Mussaenda hirsutula, III. 399.

Henry, Augustine (Chinese plants).
5. Juniperus formosana, II. 56.
22. Myricaria bracteata, II. 407.
35. Abelia chinensis, I. 121.
46. Salix variegata, III. 70.
48. Salix variegata, III. 70.
68a. Quercus aliena, var. acuteserrata, III. 215.
82. Polygonum multiflorum, III. 325.
92. Vitex Negundo, III. 372.
98. Ardisia japonica, II. 582.
130. Gymnosporia variabilis, II. 359.
150. Dioscorea acerifolia, III. 14.
156a. Buddleia albiflora, I. 569.
157. Quercus glandulifera, III. 212.

167. Desmodium podocarpum, var. szechuenensis, II. 104.
185. Hydrangea strigosa, var. sinica, I. 32.
246. Salix Wilsonii, III. 40.
254. Loropetalum chinense, I. 430.
262. Clerodendron foetidum, III. 375.
268. Rhododendron sinense, I. 549.
270. Symplocos paniculata, II. 593.
300 (Herb. Arnold Arb.). Cornus Wilsoniana, II. 579.
309. Clematis paniculata, I. 330.
314. Rhus verniciflua, II. 181.
449a. Gymnosporia variabilis, II. 359.
504. Trachelospermum jasminoides, III. 334.
706. Bauhinia hupehana, II. 89.
714a. Clematis uncinata, I. 327.
718. Hypericum longistylum, II. 404.
767. Actinidia polygama, II. 380.
791. Clematis florida, I. 325.
792. Trachelospermum jasminoides, III. 334.
843. Myricaria bracteata, II. 407.
844. Euscaphis japonica, II. 187.
957. Salix variegata, III. 70.
1032. Sapindus mukorossi, II. 191.
1062. Clematis Benthamiana, I. 330.
1073. Ligustrum lucidum, II. 603.
1081. Pittosporum truncatum, III. 328.
1083. Hydrangea strigosa, I. 31.
1104. Polygala Wattersii, II. 161.
1105. Elaeagnus viridis, II. 414.
1117. Buddleia officinalis, I. 565.
1143. Rosa laevigata, I. 318.
1149. Symplocos paniculata, II. 593.
1150. Brandisia Hancei, I. 573.
1151. Rosa chinensis, f. spontanea, II. 320.
1172. Evonymus alata, I. 493.
1173. Clematis uncinata, I. 327.
1181. Spiraea Blumei, I. 446.
1222. Smilax longipes, III. 5.
1224. Acanthopanax spinosus, II. 562.
1241. Staphlea Bumalda, II. 185.
1277. Salix Wilsonii, III. 40.
1287a. Stachyurus himalaicus, I. 287.
1289. Coriaria sinica, II. 170.
1306. Toddalia asiatica, II. 137.
1309. Prunus consociiflora, I. 279.
1328. Salix babylonica, III. 42.
1332. Pterocarya stenoptera, III. 181.
1339. Morus alba, III. 294.

HENRY, A. (*continued*)

1356. Abelia longituba, I. 126.
1397. Evonymus microcarpa, I. 487.
1399. Alangium chinense, II. 552.
1409. Morus cathayana, III. 292.
1415. Akebia lobata, var. australis, I. 348.
1449. Corylus heterophylla, var. sutchuenensis, II. 445.
1458. Berberis Julianae, III. 437.
1466. Salix Wallichiana, III. 64.
1468. Clematis Armandi, I. 326.
1493. Rhamnus leptophyllus, II. 239.
1497. Clematis Benthamiana, I. 330.
1500. Pistacia chinensis, II. 173.
1502. Diospyros kaki, var. silvestris, II. 590.
1503. Alangium chinense, II. 552.
1504. Zanthoxylum setosum, II. 124.
1518. Clematis Benthamiana, I. 330.
1519. Wikstroemia angustifolia, II. 535.
1524. Pittosporum truncatum, III. 328.
1535. Campsis chinensis, I. 303.
1541. Daphne genkwa, II. 538.
1544. Schisandra propinqua, var. sinensis, I. 416.
1553. Clematis uncinata, I. 327.
1555. Grewia parviflora, var. glabrescens, II. 371.
1556. Clematis apiifolia, var. obtusidentata, I. 336.
1559. Ligustrum strongylophyllum, II. 605.
1561. Quercus glauca, III. 226.
1569. Hypericum perforatum, III. 452.
1570. Castanea mollissima, III. 192.
1571. Zanthoxylum setosum, II. 124.
1586. Broussonetia Kaempferi, III. 304.
1601. Clematis chinensis, I. 329.
1621. Dioscorea zingiberensis, III. 14.
1627. Cotinus coggygria, var. pubescens, II. 175.
1634. Loropetalum chinense, I. 430.
1637. Elaeagnus magna, II. 411.
1649. Photinia Davidsoniae, I. 185.
1650. Evonymus microcarpa, I. 487.
1652. Sapindus mukorossi, II. 191.
1664. Stephania Delavayi, I. 389.
1676. { Evodia officinalis, II. 130.
{ Evodia rutaecarpa, II. 132.

1685. Acer oblongum Wallich.
1689. Rhamnus rugulosus, II. 238.
1694. Sambucus javanica, I. 307.
1702. Ilex pedunculosa, f. β continentalis, III.
1706. Helwingia japonica, var. hypoleuca, II. 570.
1731. Alangium platanifolium, II. 554.
1737. Abelia Engleriana, I. 120.
1762. Symplocos paniculata, II. 593.
1763. Ligustrum sinense, var. Stauntonii, II. 606.
1767. Dregea sinensis, III. 352.
1768. Abelia Graebneriana, I. 118.
1786. Hydrangea longipes, var. lanceolata, I. 40.
1885[bis] (Gray Herb.). Ehretia macrophylla, III. 364, 457.
1893. Abelia macrotera, I. 126.
1907. Eurya japonica, var. nitida, II. 398.
1912. Euscaphis japonica, II. 187.
1914. Diospyros Lotus, II. 587.
1922. Ligustrum sinense, var. nitidum, II. 606.
1929. Grewia parviflora, var. glabrescens, II. 371.
1941. Ehretia acuminata, III. 363.
1968. Deutzia Schneideriana, I. 7.
1973. Polygala Wattersii, II. 161.
1985. Orixa japonica, II. 135.
1999. Hypericum longistylum, II. 404.
2002. Stachyurus chinensis, I. 287.
2007. Sambucus javanica, I. 307.
2015. Clematis grata, var. grandidentata, I. 338.
2021. Zizyphus sativa, var. inermis, II. 212.
2076. Actinidia chinensis, II. 385.
2077. Evodia officinalis, II. 130.
2083. Hydrangea strigosa, var. macrophylla, I. 32.
2084. Celastrus angulata, II. 346.
2095. Zanthoxylum micranthum, II. 127.
2096. Evonymus alata, I. 493.
2126a. Hibiscus syriacus, II. 374.
2196. Cynanchum auriculatum, III. 346.
2206. Hydrangea strigosa, var. sinica, I. 32.
2214. Thea sinensis, II. 391.
2246a. Kalopanax ricinifolius, II. 564.

HENRY, A. (*continued*)
2253. Acanthopanax trifoliatus, II. 563.
2262. Celtis cercidifolia, III. 276.
2270. Wikstroemia angustifolia, II. 535.
2275. Ardisia japonica, II. 582.
2288. Osmanthus fragrans, II. 609.
2291, 2291a. Quercus variabilis, III. 219.
2293. Quercus aliena, var. acutesser-rata, III. 215.
2294. Quercus Fabri, III. 216.
2301a. Symplocos paniculata, II. 593.
2307. Clerodendron foetidum, III. 375.
2322a. Cudrania tricuspidata, III. 306.
2343. Polygonum multiflorum, III. 325.
2344. Eurya japonica, var. nitida, II. 398.
2351. Buddleia albiflora, I. 569.
2388. Sambucus javanica, I. 307.
2473. Hydrangea villosa, var. strigosior, I. 39.
2481. Cynanchum auriculatum, III. 346.
2488. Polygonum multiflorum, III. 325.
2497. Osmanthus fragrans, II. 609.
2499. Thea sinensis, II. 391.
2507. Lagerstroemia indica, II. 418.
2537. Sambucus javanica, I. 106, 307.
2588. Sarcococca ruscifolia, II. 163.
2589. Sarcococca ruscifolia, II. 163.
2595. Porana racemosa, III. 361.
2636. Quercus Fabri, III. 216.
2639. Acanthopanax trifoliatus, II. 563.
2683. Ligustrum sinense, var. nitidum, II. 606.
2688. Abelia chinensis, I. 121.
2690. Zanthoxylum micranthum, II. 127.
2695. Grewia parviflora, var. glabres-cens, II. 371.
2700. Jasminum floridum, II. 614.
2702. Acanthopanax trifoliatus, II. 563.
2703. Mussaenda divaricata, III. 394.
2705. Nandina domestica, I. 386.
2721. Clematis grata, var. lobulata, I. 337.
2729. Jasminum lanceolarium, var. puberulum, II. 612.
2755. Holostemma sinense, III. 344.
2760. Quercus glandulifera, III. 212.
2773. Clematis Benthamiana, I. 330.
2782. Caryopteris incana, III. 378.

2815. Styrax japonicus, I. 291.
2826. Vaccinium japonicum, I. 562.
2837. Celastrus hypoleuca, II. 346.
2838. Clethra Fargesii, I. 502.
2853. Symplocos paniculata, II. 593.
2854. Corylus heterophylla, var. sutch-uenensis, II. 445.
2858. Cotoneaster horizontalis, var. perpusilla, I. 155.
2866. Corylus heterophylla, var. sutch-uenensis, II. 445.
2867. Castanea Seguinii, III. 194.
2876. Juniperus formosana, II. 56.
2878. Castanea Henryi, III. 196.
2905. Porana racemosa, III. 361.
2912. Polygala Wattersii, II. 161.
2913. Stachyurus himalaicus, I. 287.
2938, 2938a. Bauhinia hupehana, II. 89.
2946. Clematis Gouriana, I. 339.
2953. Elaeagnus viridis, II. 414.
2954, 2954b. Quercus acrodonta, III. 225.
2969. Nothopanax Davidii, II. 556.
2981. Ilex macrocarpa, III.
2993. Sarcococca ruscifolia, II. 163.
2998. Symplocos paniculata, II. 593.
3000. Jasminum lanceolarium, var. puberulum, II. 612.
3005. Dregea sinensis, III. 352.
3007. Brandisia Hancei, I. 573.
3014. Diospyros Lotus, II. 587.
3024. Thea cuspidata, II. 390.
3040. Stachyurus himalaicus, I. 287.
3053. Clematis heracleaefolia, var. ichangensis, I. 321.
3056. Gymnosporia variabilis, II. 359.
3062. Porana racemosa, III. 361.
3067. Vaccinium bracteatum, I. 558.
3072. Ligustrum sinense, var. nitidum, II. 606.
3073. Evonymus microcarpa, I. 487.
3077, 3077a. Sarcococca ruscifolia, II. 163.
3080. Quercus aliena, III. 214.
3087. Evonymus alata, I. 493.
3088. Quercus glauca, f. gracilis, III. 228.
3090. Clematis Gouriana, I. 339.
3098. Quercus glauca, III. 226.
3099. Evonymus microcarpa, I. 487.
3100. Celtis Biondii, III. 272.
3101. Kalopanax ricinifolius, II. 564.

HENRY, A. (*continued*)
3104. Ligustrum strongylophyllum, II. 605.
3106. Rosa microcarpa, II. 314.
3107. Callicarpa Giraldiana, III. 366.
3110. Buddleia officinalis, I. 565.
3115. Celastrus Rosthorniana, II. 351.
3117. Mahonia confusa, I. 380; III. 443.
3118. Mussaenda divaricata, III. 394.
3121. Zanthoxylum dissitum, II. 128.
3124. Ligustrum Henryi, II. 601.
3125. Dregea sinensis, III. 352.
3127. Clausena suffruticosa, II. 140.
3129. Rosa multiflora, var. cathayensis, II. 304.
3150. Celtis Biondii, III. 272.
3157. Rhus punjabensis, var. sinica, II. 176.
3170. Berberis Ferdinandi-Coburgii, I. 364.
3187. Eurya japonica, var. nitida, II. 398.
3190. Ligustrum sinense, var. nitidum, II. 606.
3198. Rosa Banksiae, var. normalis, II. 317.
3206. Viscum articulatum, III. 318.
3216. Thea cuspidata, II. 390.
3218, 3218a. Castanopsis sclerophylla, III. 201.
3219. Ulmus parvifolia, III. 244.
3233. Hypericum chinense, II. 404.
3234. Hypericum chinense, II. 404.
3261. Hedera himalaica, II. 555.
3265. Ardisia crispa, II. 581.
3277, 3277b, d. Myrsine semiserrata, II. 580.
3280, 3280a, b. Clematis Henryi, I. 342.
3281. Toxocarpus villosus, III. 349.
3285. Buddleia Davidii, I. 567.
3287. Sarcococca ruscifolia, II. 163.
3293a. Buxus microphylla, var. aemulans, II. 160.
3299. Ilex Aquifolium, var. chinensis Loesener.
3301. Ternstroemia japonica, var. Wightii, II. 397.
3307, 3307a. Elaeagnus Henryi, II. 414.
3310. Ligustrum Henryi, II. 601.
3313. Buxus Harlandii, II. 166.
3313b. Wikstroemia angustifolia, II. 535; III. 455.
3314. Distylium chinense, I. 423.

3320. Ligustrum sinense, var. nitidum, II. 606.
3325. Zanthoxylum dissitum, II. 128.
3327. Smilax micropoda, var. reflexa, III. 6.
3334. Colquhounia Seguinii, III. 380.
3335. Thea Grijsii, II. 394.
3344. Ilex corallina, III.
3353, 3353a. Aucuba chinensis, f. obcordata, II. 572.
3355. Salix babylonica, III. 42.
3362. Stachyurus himalaicus, I. 287.
3374, 3374a. Thea fraterna, II. 390.
3376. Ilex corallina Franchet.
3377. Clematis Armandi, I. 326.
3382. Akebia lobata, var. australis, I. 348.
3387. Buxus Henryi, II. 168.
3387a. Pittosporum glabratum, III. 326.
3388. Illicium Henryi, I. 417.
3394. Evonymus alata, I. 493.
3399. Salix Wallichiana, III. 64.
3404. Celtis labilis, III. 267.
3405. Celastrus angulata, II. 346.
3406, 3406a. Acanthopanax spinosus, II. 562.
3407, 3407a. Rhamnus leptophyllus, II. 239.
3414. { Pittosporum glabratum, III. 326.
       { Pittosporum truncatum, III. 328.
3416b. Zanthoxylum echinocarpum, II. 128.
3417, 3417a. Rubus Playfairianus, I. 49.
3419. Periploca calophylla, III. 343.
3425. Quercus acrodonta, III. 225.
3426. Quercus glauca, III. 226.
3431. Tinospora sagittata, I. 390.
3436. Abelia Graebneriana, I. 118.
3441. Diospyros kaki, var. silvestris, II. 590.
3442. Salix Wilsonu, III. 40.
3446. Schisandra sphenanthera, I. 414.
3449a. Stachyurus himalaicus, I. 287.
3455. Ardisia Henryi, II. 582.
3456. Buddleia asiatica, I. 566.
3469. Schisandra sphenanthera, I. 414.
3480. Deutzia setchuenensis, I. 18.
3485. Diospyros kaki, var. silvestris, II. 590.
3494. Actinidia callosa, var. Henryi, II. 382.

HENRY, A. (*continued*)

3495, 3495a. Celastrus Hindsii, var. Henryi, II. 353.

3496. Toddalia asiatica, II. 137.

3498. Morus alba, III. 294.

3506a. Spiraea hirsuta, var. rotundifolia, I. 445.

3511, 3511a, b. Evonymus subsessilis, I. 489.

3515. Toxocarpus villosus, III. 349.

3516, 3516a. Clematis florida, I. 325.

3527. Perrottetia racemosa, II. 359.

3529. Clematis Pavoliniana, I. 328.

3536. Mappia pittosporoides, II. 190.

3537a. Mappia pittosporoides, II. 190.

3538. Salix Wilsonii, III.

3539. Tetrastigma obtectum Planch., var. pilosum Gagnepain.

3549. Ligustrum sinense, var. nitidum, II. 606.

3551, 3551a. Bauhinia hupehana, II. 89.

3552a. Ficus foveolata, var. Henryi, III. 310.

3554. Dregea sinensis, III. 352.

3561. Ligustrum sinense, var. Stauntonii, II. 606.

3564. Actinidia callosa, var. Henryi, II. 382.

3565. Meratia praecox, I. 419.

3570. { Spiraea Blumei, I. 446.
Spiraea hirsuta, var. rotundifolia, I. 445.

3571. Deutzia Schneideriana, I. 7.

3575. Ligustrum Henryi, II. 601.

3580. Evonymus microcarpa, I. 487.

3582. Premna puberula, III. 371.

3585. Deutzia setchuenensis, I. 18.

3593a. Rhamnus rugulosus, II. 238.

3594. Wikstroemia angustifolia, II. 535.

3597. Rosa microcarpa, II. 314.

3598a. Prunus japonica, var. eujaponica, f. Oldhamii, I. 266.

3606. Rhamnus utilis, II. 240.

3609. Rhamnus rugulosus, II. 238.

3611, 3611a. Rhamnus rugulosus, II. 238.

3618. Trachelospermum axillare, III. 335.

3619. Ligustrum sinense, II. 605.

3622a. Clematis uncinata, I. 327.

3629. Grewia parviflora, var. glabrescens, II. 371.

3636. Sindechites Henryi, III. 342.

3640. Cocculus trilobus, I. 388.

3657. Cedrela sinensis, II. 156.

3666. Broussonetia papyrifera, III. 303.

3668a. Broussonetia Kaempferi, III. 304.

3669. Jasminum lanceolarium, var. puberulum, II. 612.

3673. Eurya japonica, var. aurescens, II. 399.

3690. Evonymus kiautschovica, var. patens, I. 486.

3694. Clematis lasiandra, I. 322.

3699. Schisandra propinqua, var. sinensis, I. 416.

3713. Zanthoxylum dissitum, II. 128.

3770. Stachyurus chinensis, I. 287.

3774. Pterostyrax hispidus, I. 295.

3807. Akebia quinata, I. 347.

3808. Myrsine africana, II. 580.

3822. Abutilon sinense, II. 373.

3826. Distylium chinense, I. 423.

3829. Rhododendron Mariesii, I. 548.

3832. Sarcococca ruscifolia, II. 163.

3834. Juglans cathayensis, III. 185.

3847. Orixa japonica, II. 135; III. 449.

3848. Illicium Henryi, I. 417.

3856. Celastrus Hindsii, var. Henryi, II. 353.

3868. Dregea sinensis, III. 352.

3870. Castanopsis sclerophylla, III. 201.

3876. Styrax japonicus, I. 291.

3878. Keteleeria Davidiana, II. 39.

3882. Melia Azedarach, II. 157.

3883. Celastrus angulata, II. 346.

3886. Ailanthus altissima, var. sutchuenensis, II. 153; III. 449.

3891. Cornus Walteri, II. 576.

3906. Quercus glandulifera, III. 212.

3918. Vaccinium Donianum, I. 557.

3938. Zanthoxylum undulatifolium, II. 124.

3955. Actinidia callosa, var. Henryi, II. 382.

3956. Zanthoxylum Bungei, II. 121.

3961. Schisandra propinqua, var. sinensis, I. 416.

3969. Philadelphus incanus, III.

3972. Ardisia Henryi, II. 582.

3973. Ligustrum Quihoui, II. 607.

3974. Salix variegata, III.

3979. Buddleia Lindleyana, var. sinuato-dentata, I. 564.

HENRY, A. (*continued*)
3992. Holostemma sinense, III. 344.
4003. Phellodendron chinense, var. glabriusculum, II. 137; III. 449.
4013. Populus lasiocarpa, III. 317.
4031. Rhododendron stamineum, I. 546.
4045. Sabia puberula, II. 197.
4059. Schisandra sphenanthera, I. 414.
4064. Photinia subumbellata, I. 189.
4073. Cornus kousa, II. 577.
4099. Premna puberula, III. 371.
4115. Spiraea hirsuta, I. 444.
4118. Mappia pittosporoides, II. 190.
4119. Periploca calophylla, III. 343.
4120. Mussaenda divaricata, III. 394.
4122. Clausena suffruticosa, II. 140.
4127, 4127a. Zanthoxylum micranthum, II. 127.
4139. Deutzia setchuenensis, I. 18.
4144a. Aristolochia mollissima, III. 324.
4164. Dregea sinensis, III. 352, 457.
4166a. Buddleia Davidii, I. 567.
4171. Ligustrum sinense, II. 605.
4173. Wikstroemia micrantha, II. 530.
4179. Ilex macrocarpa, III.
4183. Hibiscus Manihot, II. 374.
4184. Wikstroemia angustifolia, II. 535.
4185. Clematis quinquefoliolata, I. 328.
4214. Celtis Bungeana, III. 269.
4225. Abelia parvifolia, I. 121.
4271. Cynanchum auriculatum, III. 346.
4280. Distylium chinense, I. 423.
4314. Ardisia Henryi, II. 582.
4328. Clematis chinensis, I. 329.
4329. Clematis Gouriana, I. 339.
4330. Clematis grata, var. lobulata, I. 337.
4332. Clematis quinquefoliolata, I. 328.
4341. Clematis brevicaudata, I. 340.
4348. Clematis chinensis, I. 329.
4357. Porana racemosa, III. 361.
4359. Clematis heracleaefolia, var. ichangensis, I. 321.
4361. Clematis brevicaudata, I. 340.
4368. Clematis Benthamiana, I. 330.
4369. Pittosporum glabratum, III. 326.
4377a. Actinidia callosa, var. Henryi, II. 382.
4385. Clematis uncinata, I. 327.
4440. Ilex purpurea, var. a Oldhamii, III.

4464. Jasminum sinense, II. 612.
4478. Clematis heracleaefolia, var. ichangensis, I. 321.
4482. Evodia hupehensis, II. 133.
4520. Clerodendron trichotomum, var. Fargesii, III. 376.
4526. Vaccinium Donianum, I. 557.
4536. Staphylea holocarpa, II. 185.
4556. Ehretia acuminata, III. 363.
4560. Rhamnus rugulosus, II. 238.
4562. Jasminum lanceolarium, var. puberulum, II. 612.
4573. Kalopanax ricinifolius, II. 564.
4577. Evodia glauca, II. 129.
4581. Caryopteris incana, III. 378.
4583. Clematis brevicaudata, I. 340.
4585. Diervilla japonica, var. sinica Rehder.
4646. Clerodendron trichotomum, var. Fargesii, III. 376.
4689. Buddleia albiflora, I. 569.
4703. Vaccinium Henryi, I. 561.
4770. Alangium platanifolium, II. 554.
4826. Vaccinium Henryi, I. 561.
4832. Acanthopanax Henryi, II. 557.
4890. Rhus javanica, II. 178.
5030. Cephalotaxus drupacea, II. 3.
5151. Thea cuspidata, II. 390.
5154. Eleagnus cuprea, II. 414.
5162. Eurya japonica, var. aurescens, II. 399.
5165. Thea cuspidata, II. 390.
5167. Eurya japonica, var. aurescens, II. 399.
5170. Eurya japonica, var. aurescens, II. 399.
5218, 5218b. Liquidambar formosana, I. 421.
5219a, 5219b. Decumaria sinensis, I. 152.
5223a, 5223b, c. Clematis Armandi, I. 526.
5225. Holboellia coriacea, I. 345.
5233a. Juglans cathayensis, III. 185.
5242b. Symplocos botryantha, II. 596; III. 455.
5247. Orixa japonica, II. 135.
5249. Chaenomeles lagenaria, II. 296.
5254. Spiraea prunifolia, var. plena, I. 438.
5256, 5256a. Holboellia Fargesii, I. 346.
5273. Sorbus Folgneri, II. 271.
5274. Rhododendron Mariesii, I. 548.

HENRY, A. (*continued*)

5276. Celtis Biondii, III. 272.
5277a. Quercus oxyodon, III. 228.
5277b. Quercus glauca, III. 226.
5278. Rhododendron ovatum, I. 546.
5281, 5281a. Populus adenopoda, III. 23.
5282b. Helwingia chinensis, II. 571.
5282c, 5282e. Helwingia japonica, II. 570.
5283. Berchemia yunnanensis, II. 216.
5285. Rhododendron sutchuenense, I. 544.
5289. Rosa multiflora, var. cathayensis, II. 304.
5292. Symplocos caudata, II. 595.
5295. Prunus litigiosa, I. 239.
5298a. Ilex Pernyi Franchet.
5299. Pyrus serotina, II. 263.
5307. Cornus capitata, II. 578.
5308. Prunus Wildeniana, I. 249.
5314. Sabia emarginata, II. 196.
5319. Symplocos caudata, II. 595.
5334, 5334a. Fagus longipetiolata, III. 190.
5335. Evonymus myriantha, I. 487.
5335a. Evonymus aculeatus, I. 490.
5343. Eriobotrya japonica, I. 193.
5349. Salix heterochroma, III. 61.
5354. Rhododendron Fortunei, var. Houlstonii, I. 541.
5356. Acer Davidii, I. 92.
5357. Symplocos botryantha, II. 596.
5359. Pittosporum glabratum, var. neriifolium, III. 328.
5360. Thea cuspidata, II. 390.
5383, 5383a. Aucuba chinensis, f. angustifolia, II. 573.
5389, 5389a. Magnolia officinalis, I. 391.
5405, 5405a. Decaisnea Fargesii, I. 344.
5421, 5421b. Sabia Schumanniana, var. pluriflora, II. 197.
5422. Pittosporum glabratum, var. neriifolium, III. 328.
5423, 5423a. Populus lasiocarpa, III. 17.
5437, 5437a. Clematis montana, var. rubens, I. 333.
5445. Evonymus sanguinea, var. β camptoneura, I. 494.
5453. Morus acidosa, III. 297.
5463. Cotoneaster Zabelii, I. 166.

5468. Staphylea holocarpa, II. 185.
5469a, 5469b. Rubus pungens, I. 52.
5470, 5470a, b, d. Berberis Henryana, III. 440.
5473. Elaeagnus cuprea, II. 414.
5475. Enkianthus quinqueflorus, var. serrulatus, I. 550.
5477. Prunus Helenae, I. 212.
5478. Schisandra glaucescens, I. 413.
5483. Elaeagnus lanceolata, II. 413.
5484b. Elaeagnus multiflora, f. angustata, II. 413.
5485. Diervilla japonica, var. sinica, III. 430.
5487. Morus cathayana, III. 292.
5490. Photinia serrulata, I. 184 (by error as No. 1490).
5492. Symplocos paniculata, II. 593.
5494. Zanthoxylum dimorphophyllum, var. spinifolium, II. 126.
5495. Styrax japonicus, I. 291.
5498. Ulmus castaneifolia, III. 256.
5502. Daphne odora, var. atrocaulis, II. 545.
5506. Cornus macrophylla, II. 575.
5512. Zanthoxylum dimorphophyllum, II. 126.
5513. Pittosporum truncatum, III. 328.
5517. Photinia parvifolia, I. 189.
5518. Photinia subumbellata, I. 189.
5520, 5520a. Carpinus polyneura, II. 430.
5521. Amelanchier asiatica, var. sinica, I. 195.
5524. Torricellia angulata, II. 569.
5525. Cotoneaster hupehensis, I. 169.
5527a. Schisandra sphenanthera, I. 414.
5529a. Rhus Potaninii, II. 177.
5529b. Rhus punjabensis, var. sinica, II. 176.
5529c. Rhus Potaninii, II. 177.
5532. Sambucus Sieboldiana, I. 106.
5533. Sambucus Sieboldiana, I. 106.
5534, 5534a. Picrasma quassioides, II. 152.
5535. Spondias axillaris, var. pubenervis, II. 173.
5538. Acer laevigatum, III.
5540. Evonymus myriantha, I. 487.
5548. Morus cathayana, III. 292, 457.
5550. Rosa Rubus, II. 308.
5551. Castanopsis Fargesii, III. 198.

HENRY, A. (*continued*)
5552. Rosa Banksiae, var. normalis, II. 317.
5554a. Neillia sinensis, I. 436.
5557. Hydrangea anomala, I. 34.
5559. Celastrus Hookeri, II. 352.
5560. Zanthoxylum stenophyllum, II. 127.
5562. Evonymus sanguinea, var. β camptoneura, I. 494.
5563. Abelia Engleriana, I. 120.
5565, 5565a. Photinia amphidoxa, I. 190.
5573. Abelia Engleriana, I. 120.
5574. Actinidia chinensis, II. 385.
5577, 5577b. Davidia involucrata, var. Vilmoriniana, II. 256.
5578. Clematis grata, var. grandidentata, I. 338.
5595, 5595b, c, d. Pterostyrax hispidus, I. 295.
5597, 5597b. Schoepfia jasminodora, III. 321.
5598, 5598b, c. Hosiea sinensis, II. 190.
5599a. Photinia Beauverdiana, I. 187.
5600f. Smilax herbacea, var. angustata, III. 9.
5604. Prunus pilosiuscula, var. media, I. 204.
5608. Skimmia melanocarpa, II. 138.
5616. Eurya japonica, var. aurescens, II. 399.
5620. Aristolochia heterophylla, III. 323.
5622. Actinidia purpurea, II. 378.
5623c. Rubus mesogaeus, I. 56.
5628. Spiraea Miyabei, var. pilosula, I. 455.
5632, 5632a, d. Quercus oxyodon, III. 228.
5632e. Quercus glauca, III. 226.
5636. Akebia lobata, var. australis, I. 348.
5638. Malus yunnanensis, II. 287.
5639a. Styrax japonicus, I. 291.
5640. Celastrus Loeseneri, II. 350.
5641. Acer sinense Pax.
5643. Acer Davidii Franchet.
5645a. Spiraea Henryi, I. 447.
5646. Zanthoxylum undulatifolium, II. 124.
5647, 5647a. Clematis grata, var. grandidentata, I. 338.

5649. Picrasma quassioides, II. 152.
5651. Magnolia denudata, var. purpurascens, I. 401.
5653. Juniperus formosana, II. 56.
5658. Hydrangea anomala, I. 34.
5665, 5665a. Rubus Henryi, III.
5667. Betula luminifera, II. 455.
5668. Morus acidosa, III. 297.
5669. Morus acidosa, III. 297.
5671. Salix heterochroma, III.
5672, 5672b, c. Cornus kousa, II. 577.
5676. Styrax Hemsleyanus, I. 291.
5677. Rhamnus Hemsleyanus, II. 234.
5678. Salix Fargesii, III.
5679. Rosa sertata, II. 327.
5681. Berberis triacanthophora, I. 358.
5682, 5682a. Quercus Engleriana, III. 220.
5683. Staphylea holocarpa, II. 185.
5690. Ulmus Bergmanniana, III. 241.
5691. Viburnum erubescens, var. gracilipes, I. 107.
5695. Neillia sinensis, I. 436.
5698. Stranvaesia Davidiana, var. undulata, I. 192.
5701. Cotoneaster divaricata, I. 157.
5715. Sorbus Keissleri, II. 269.
5715a. Sorbus Keissleri, II. 269.
5725. Schisandra glaucescens, I. 413.
5733. Cornus chinensis, II. 577.
5734. Celastrus Loeseneri, II. 350.
5735. Celtis Biondii, III. 272.
5739. Prunus brachypoda, var. pseudossiori, I. 65.
5741. Morus acidosa, III. 297.
5742b. Mallotus tenuifolius, II. 525.
5744. Stachyurus chinensis, I. 287.
5745. Morus acidosa, III. 297.
5746. Rosa banksiopsis, II. 322.
5750. Spiraea Henryi, I. 447.
5751, 5751a. Staphylea holocarpa, II. 185.
5752. Cotoneaster Henryana, I. 174.
5755, 5755b, c. Symplocos paniculata, II. 593.
5760. Symplocos paniculata, II. 593.
5763. Prunus brachypoda, var. pseudossiori, I. 65.
5764. Actinidia polygama, II. 380.
5767, 5767c. Euscaphis japonica, II. 187.
5772. Rubus inopertus, I. 54.
5773. Rosa Gentiliana, II. 312.

HENRY, A. (*continued*)
5774. Prunus velutina, I. 69.
5778. Evonymus venosa, I. 488.
5779. Styrax japonicus, I. 291.
5780. Prunus rufoides, I. 244.
5781. Zanthoxylum Bungei, II. 121.
5783, 5783b. Polygala arillata, I. 160.
5786. Rosa multiflora, var. cathayensis, II. 304.
5787. Rhododendron stamineum, I. 546.
5788. Rubus lasiostylus, I. 52.
5793, 5793a. Castanea Henryi, III. 196.
5796. Vitis Piasezkii Maxim.
5797. Actinidia callosa, var. Henryi, II. 382.
5800. Castanea Henryi, III. 196.
5800a. Castanea Seguinii, III. 194.
5801. Rubus corchorifolius Linnaeus.
5804. Sorbus Folgneri, II. 271.
5805, 5805a. Lithocarpus Henryi, III. 209.
5806a. Pieris ovalifolia, var. elliptica, I. 552.
5807b. Vaccinium Donianum, I. 557.
5812. Prunus Dielsiana, I. 243.
5813, 5813a. Alangium platanifolium, II. 554.
5818. Clethra Fargesii, I. 502.
5820a. Diospyros Lotus, II. 587.
5827. Acer Davidii Franchet.
5833. Prunus tenuiflora, I. 209.
5834, 5834a. Actinidia chinensis, II. 385.
5836, 5836a. Liriodendron chinense, I. 410.
5839a, 5839b. Hydrangea longipes, I. 33.
5843. Salix heterochroma, III. 61.
5849, 5849a. Meliosma cuneifolia, II. 199.
5849e. Meliosma pendens, II. 200.
5850, 5850a, b. Ampelopsis megalophylla Diels & Gilg.
5853. Rubus flosculosus, III.
5854. Corylopsis sinensis, I. 424.
5855, 5855a, b. Alangium chinense, II. 552.
5856. Catalpa Duclouxii, III. 433.
5856a. Catalpa Fargesii, I. 305.
5860. Morus cathayana, III. 292.
5863. Meliosma Oldhamii, II. 206.
5864. Callicarpa Giraldiana, III. 366.

5864. Callicarpa Giraldiana, var. subcanescens, III. 368.
5870, 5870b. Dioscorea acerifolia, III. 14.
5872. Rubus eucalyptus, III.
5875. Pyrus serotina, II. 263.
5881. Ligustrum acutissimum, II. 600.
5886, 5886a. Carpinus cordata, var. chinensis, II. 437.
5887. Celastrus hypoleuca, II. 346.
5890. Acanthopanax villosulus, II. 562.
5892. Aesculus Wilsonii, I. 498.
5899, 5899a. Rhus verniciflua, II. 181.
5903. Rhus Potaninii, II. 177.
5908, 5908a (Herb. Arnold Arb.). Ligustrum expansum, II. 600.
5909. Celastrus Loeseneri, II. 350.
5911. Vaccinium Henryi, I. 561.
5915c. Rhamnus iteinophyllus, II. 239.
5922, 5922a, b. Actinidia polygama, II. 380.
5925a. Celastrus angulata, II. 346.
5930. Acanthopanax Rehderianus, II. 561.
5931. Schisandra glaucescens, I. 413.
5935. Celastrus spiciformis, II. 348.
5938a. Actinidia melanandra, II. 378.
5944. Jasminum urophyllum, var. Henryi, II. 613.
5945. Evonymus myriantha, I. 487.
5948, 5948b. Staphlea Bumalda, II. 185.
5949. Hydrangea fulvescens, I. 39.
5950b. Acanthopanax setchuenensis, II. 559.
5950c. Acanthopanax leucorrhizus, var. scaberulus, II. 558.
5965. Schizophragma integrifolium, var. minus, I. 43.
5968. Rhamnus leptophyllus Schneider.
5972. Broussonetia Kaempferi, III. 304.
5973. Rosa Helenae, II. 310.
5975. Schoepfia jasminodora, III. 321.
5976. Trachelospermum cathayanum, III. 333.
5977 (Herb. Gray). Styrax dasyanthus, I. 289.
5977 (Herb. Arn. Arb.). Styrax Hemsleyanus, I. 291.
5981a, 5981c. Quercus spinosa, III. 224.
5982a, b. Rubus simplex Focke.
5986, 5986a. Celastrus Loesneri, II. 350.

HENRY, A. (*continued*)

5987. Ternstroemia japonica, var. Wrightii, II. 397.
5988. Prunus brachypoda, var. pseudossiori, I. 65.
5992. Callicarpa Giraldiana, III. 366.
5993. Spiraea japonica, var. Fortunei, I. 451.
5998. Clematis Armandi, I. 326.
5999. Pittosporum glabratum, III. 326.
6000. Meliosma pendens, II. 200.
6002. Lithocarpus cleistocarpa, III. 205.
6008. Clematis uncinata, I. 327.
6010. Actinidia callosa, var. Henryi, II. 382.
6014. Platycarya strobilacea, III. 180.
6015. Philadelphus sericanthus Koehne.
6016. Gardneria multiflora, I. 563.
6018. Rhamnus utilis, II. 240.
6019. Sycopsis sinensis, I. 431.
6021. Vaccinium japonicum, I. 562.
6022. Sabia puberula, II. 197.
6023. Lithocarpus Henryi, III. 209.
6030. Cynanchum amphibolum, III. 346.
6035a. Rhus verniciflua, II. 181.
6039. Evonymus sanguinea, var. β camptoneura, I. 494.
6046. Castanea Seguinii, III. 194.
6056. Hydrangea hypoglauca, I. 26.
6058. Rhamnus crenatus, II. 232.
6071a. Rosa banksiopsis, II. 322.
6093. Philadelphus incanus Koehne.
6094. Morus acidosa, III. 297.
6096. Rubus innominatus S. Moore.
6099. Hypericum patulum, var. Henryi, II. 403.
6114. Sabia Schumanniana, var. pluriflora, II. 197.
6115. Dumasia hirsuta, II. 116.
6120. Styrax japonicus, I. 291.
6121. Machilus ichangensis, II. 621.
6123a. Premna puberula, III. 371.
6126. Evonymus myriantha, I. 487.
6127. Callicarpa Giraldiana, III. 366.
6128. Pieris ovalifolia, var. elliptica, I. 552.
6129. Vaccinium Donianum, I. 557.
6136a. Evodia officinalis, II. 130.
6139. Lysionotus pauciflorus, III. 387.
6158. Pterocarya hupehensis, III. 182.

6166. Stewartia sinensis, II. 395.
6167. Quercus Engleriana, III. 220.
6193. Buddleia albiflora, I. 569.
6194. Vitis betulifolia Diels & Gilg.
6199, 6199a. Evodia officinalis, II. 130.
6210. Trema spec., III. 289.
6211. Ilex purpurea, var. a Oldhamii Loesener.
6212. Clematis uncinata, I. 327.
6217. Ligustrum lucidum, II. 603.
6219. Schisandra propinqua, var. sinensis, I. 416.
6220. Clematis chinensis, I. 329.
6222. Ceropegia driophila, III. 349.
6226. Schisandra Henryi, I. 413.
6233. Mussaenda Wilsonii, III. 393.
6235. Diospyros Lotus, II. 587.
6245. Sorbaria arborea, var. glabrata, I. 48.
6249a. Morus acidosa, III. 297.
6268. Photinia Beauverdiana, I. 187.
6269. Sorbus Hemsleyi, II. 276.
6274. Salix psilostigma, III. 116.
6277. Enkianthus chinensis, I. 551.
6279. Symplocos paniculata, II. 593.
6288. Jasminum floridum, II. 614.
6299. Ligustrum strongylophyllum, II. 605.
6300a. Cornus macrophylla, II. 575.
6323. Thea cuspidata, II. 390.
6325. Salix babylonica, III. 42.
6327. Prunus Grayana, I. 69.
6328. Cotoneaster foveolata, I. 162.
6333. Spiraea Miyabei, var. pilosula, I. 455.
6346. Torreya grandis, II. 7.
6349. Rhus sylvestris, II. 180.
6352. Ligustrum Quihoui, II. 607.
6358. Ehretia acuminata, III. 363, 457.
6359. Quercus spathulata, III. 226.
6366. Mussaenda divaricata, III. 394.
6367. Sapindus mukorossi, II. 191.
6370. Photinia subumbellata, I. 189.
6375. Rubus triphyllus, III. 424.
6378. Morus cathayana, III. 292.
6379. Paliurus orientalis, II. 209.
6382. Cornus Walteri, II. 576.
6383. Schisandra Henryi, I. 413.
6387. Dregea sinensis, III. 352.
6392. Acer oblongum Wallich.
6398. Abelia macrotera, I. 126.
6407. Clethra Fargesii, I. 502.
6412. Hamamelis mollis, I. 431.

HENRY, A. (continued)

6413. Corylus heterophylla, var. sutchuenensis, II. 445.
6416. Alangium chinense, II. 552.
6417. Aristolochia heterophylla, III. 323.
6422, 6422a. Clerodendron trichotomum, var. Fargesii, III. 376.
6431, 6431a. Vaccinium japonicum, I. 562.
6432. Rhododendron stamineum, I. 546.
6447. Schisandra sphenanthera, I. 414.
6448. Rhus orientalis, II. 179.
6455. Euptelea Franchetii, II. 314.
6456. Acer Franchetii Pax.
6461. Clematis Gouriana, var. Finetii, I. 339.
6462. Clematis brevicaudata, var. lissocarpa, I. 340.
6466. Zanthoxylum stenophyllum, II. 127.
6474. Tilia tuan, var. chinensis, II. 369.
6475. Orixa japonica, II. 135.
6477. Hydrangea strigosa, var. macrophylla, I. 32.
6478. Torreya grandis, II. 7.
6479. Vitis Piasezkii Maxim.
6480. Sinofranchetia chinensis, I. 349.
6483. Celtis Bungeana, III. 269.
6491. Rosa corymbulosa, II. 323.
6503a. Acanthopanax Simonii, II. 559.
6503b. Acanthopanax leucorrhizus, var. scaberulus, II. 558.
6504. Rhamnus lamprophyllus, II. 252.
6507. Evonymus sanguinea, var. β camptoneura, I. 494.
6509. Malus theifera, II. 283.
6511. Hydrangea anomala, I. 34.
6512. Acer Oliverianum Pax.
6521. Acanthopanax setchuenensis, II. 559.
6524. Lithocarpus cleistocarpa, III. 205.
6538. Lithocarpus cleistocarpa, III. 205.
6540. Wikstroemia gracilis, II. 536; III. 455.
6543. Viburnum erubescens, var. gracilipes, I. 107.
6546. Cynanchum auriculatum, III. 346.
6555. Zanthoxylum stenophyllum, II. 127.

6556. Evonymus sanguinea, var. β camptoneura, I. 494.
6559. Sinowilsonia Henryi, I. 429.
6560. Cornus chinensis, II. 577.
6569. Evodia officinalis, II. 130.
6576. Juniperus chinensis, II. 60.
6581. Ostrya japonica, II. 424.
6583. Ligustrum acutissimum, II. 600.
6584. Evonymus elegantissima, I. 496.
6608b. Nothopanax Davidii, II. 556.
6612. Enkianthus chinensis, I. 551.
6623. Vaccinium Henryi, I. 561.
6630. Acanthopanax setchuenensis, II. 559.
6644. Actinidia polygama, II. 380.
6648. Evonymus yedoensis, var. Koehneana, I. 491.
6652. Elaeagnus lanceolata, II. 413.
6653. Zanthoxylum Bungei, II. 121.
6679. Callicarpa japonica, var. angustata, III. 369.
6690. Tetracentron sinense, I. 417.
6691. Symplocos anomala, II. 596.
6693. Eurya japonica, var. aurescens, II. 399.
6707. Cornus chinensis, II. 577.
6708. Evonymus alata, I. 493.
6712. Evodia Henryi, II. 133.
6713. Clematis lasiandra, I. 322.
6714. Rosa corymbulosa, II. 323.
6715. Lithocarpus cleistocarpa, III. 206.
6719. Helwingia chinensis, var. macrocarpa, II. 571.
6730. Osmanthus armatus, II. 611.
6747. Rosa saturata, II. 324.
6754. Malus kansuensis, II. 286.
6755. Elsholtzia fruticosa, III. 381.
6760. Ilex Fargesii Franchet.
6763. Picea Watsoniana, II. 27.
6766. Sorbus Koehneana, I. 471.
6778, 6778a. Corylus tibetica, II. 443.
6782. Rosa omeiensis, II. 331.
6783. Acer Maximowiczii Pax.
6791. Sorbus alnifolia, II. 270.
6793. Fagus lucida, III. 191.
6797. Fagus Engleriana, III. 191.
6798, 6798a. Betula albo-sinensis, II. 457.
6811. Celastrus hypoleuca, II. 346.
6816. Berberis dasystachya, III. 442.
6817. Clematis pogonandra, I. 319.
6818. Clematoclethra Hemsleyi, II. 389.

HENRY, A. (*continued*)
6821. Actinidia tetramera, II. 381.
6823. Picea Watsoniana, II. 27.
6829. Vaccinium Henryi, I. 561.
6830. Sorbus xanthoneura, II. 272.
6830a. Sorbus Hemsleyi, II. 272.
6838. Daphne Wilsonii, II. 540.
6849. Rubus pileatus Focke.
6856. Rubus amabilis Focke.
6879. Betula Fargesii, II. 478.
6881. Abies Fargesii, II. 48.
6885. Clematoclethra Hemsleyi, II. 389.
6886. Buxus microphylla, var. sinica, II. 165.
6887. Clematis montana, I. 332.
6888. Skimmia melanocarpa, II. 138.
6891. Acanthopanax Giraldii, II. 560.
6892. Rubus mesogaeus Focke.
6895. Styrax Hemsleyanus, I. 291.
6896. Juniperus squamata, II. 57.
6900. Acer flabellatum Rehder.
6903. Zanthoxylum Bungei, II. 121.
6907. Tsuga chinensis, II. 37.
6908. Picea brachytyla, II. 33.
6909. Pinus sinensis, II. 15.
6910. Acer Davidii Franchet.
6913. Taxus cuspidata, var. chinensis, II. 8.
6914. Rhododendron sutchuenense, I. 544.
6917. Stachyurus chinensis, I. 287.
6935. Juniperus squamata, II. 57.
6937. Acer caudatum, var. multiserratum Rehder.
6967. Potentilla fruticosa, var. mandshurica, II. 303.
6968. Spiraea myrtilloides, I. 440.
6982. Pertya sinensis, III. 419.
6997. Rosa sertata, II. 327.
7003. Zanthoxylum dimorphophyllum, II. 126.
7004. Porana racemosa, III. 361.
7007. Rosa Rubus, II. 308.
7008. Buddleia Davidii, var. magnifica, I. 567.
7013. Carpinus laxiflora, var. macrostachya, II. 425.
7016. Evonymus myriantha, I. 487.
7018. Cephalotaxus Fortunei, II. 4.
7019. Evonymus venosa, I. 488.
7020. Carpinus Turczaninovii, var. ovalifolia, II. 427.

7020a. Carpinus polyneura, II. 430.
7021. Sorbus Zahlbruckneri, II. 272.
7026. Thea cuspidata, II. 390.
7027. Sorbus caloneura, II. 269.
7030, 7030a. Lithocarpus Henryi, III. 209.
7031. Cornus oblonga, II. 579.
7032. Rubus Lambertianus Seringe.
7042. Pittosporum glabratum, III. 326.
7044. Diospyros Lotus, II. 587.
7048a. Rhamnus leptophyllus Schneider.
7054. Evonymus alata, I. 493.
7063. Carpinus Henryana, II. 429.
7065. Sarcococca Hookeriana, var. humilis, II. 164.
7074a. Picea brachytyla, II. 33.
7075. Sorbus Folgneri, II. 271.
7076. Quercus glauca, f. gracilis, III. 228.
7083. Abelia umbellata, I. 122.
7089. Tilia Oliveri, II. 366.
7095. Photinia Beauverdiana, I. 187.
7096. Torreya grandis, II. 7.
7097. Taxus cuspidata, var. chinensis, II. 8.
7098. Keteleeria Davidiana, II. 39.
7099. Eurya japonica, var. nitida, II. 398.
7100. Rosa Helenae, II. 310.
7111. Corylus chinensis, II. 444.
7117. Ligustrum pedunculare, II. 609.
7118. Sageretia Henryi, II. 623.
7119. Daphne odora, var. atrocaulis, II. 545.
7135. Actinidia trichogyna, II. 384.
7137. Zanthoxylum dissitum, II. 128.
7139. Stachyurus himalaicus, I. 287.
7155. Taxus cuspidata, var. chinensis, II. 8.
7157. Picea brachytyla, II. 33.
7158. Ligustrum acutissimum, II. 600.
7159. Buxus microphylla, var. sinica, II. 165.
7166. Sorbus Keissleri, II. 269.
7168. Evonymus alata, I. 493.
7170. Trema virgata, III. 289.
7171. Ehretia macrophylla, III. 364.
7175. Salix variegata, III. 70.
7179. Bauhinia Faberi, var. microphylla, II. 89.
7180b. Urena lobata, II. 373.
7182. Salix variegata, III. 70.

HENRY, A. (*continued*)

7183. Rosa multiflora, var. cathayensis, II. 304.
7186. Cephalotaxus Fortunei, II. 4.
7189. Perrottetia racemosa, II. 359.
7203, 7203a. Aesculus Wilsonii, I. 498.
7205. Paliurus orientalis, II. 209.
7219. Carpinus Turczaninovii, var. ovalifolia, II. 427.
7230. Clematis grata, var. lobulata, I. 337.
7240. Sabia puberula, II. 197.
7243. Actinidia callosa, var. Henryi, II. 382.
7254. Evonymus sanguinea, var. β camptoneura, I. 494.
7267. Clematis grata, var. grandidentata, I. 338.
7270. Clethra Fargesii, I. 502.
7306. Pittosporum truncatum, III. 328.
7312. Callicarpa Giraldiana, III. 366.
7333. Rubus simplex Focke.
7335. Spiraea Henryi, I. 447.
7337. Euptelea Franchetii, I. 314.
7358. Dioscorea acerifolia, III. 14.
7382. Ilex macrocarpa Oliver.
7385. Hydrangea opuloides, var. Hortensia, I. 37.
7389. Photinia amphidoxa, I. 190.
7392, 7392a. Stewartia sinensis, II. 295.
7402. Betula luminifera, II. 455.
7412. Evonymus alata, I. 493.
7424. Elaeagnus lanceolata, II. 413.
7432. Pieris ovalifolia, var. elliptica, I. 552.
7440. Symplocos anomala, II. 596.
7442. Thea cuspidata, II. 390.
7444. Fagus longipetiolata, III. 190.
7447. Ligustrum acutissimum, II. 600.
7452. Tilia tuan, II. 368.
7452a. Tilia Henryana, II. 367.
7452b. Tilia Oliveri, II. 366.
7462. Rhamnus Hemsleyanus, II. 234.
7479. Cephalotaxus Oliveri, II. 6.
7497. Callicarpa Giraldiana, III. 366.
7518. Ulmus parvifolia, III. 244.
7522. Crataegus hupehensis, I. 178.
7533. Corylus chinensis, II. 444.
7550. Meliosma cuneifolia, II. 199.
7570, 7570b. Picrasma quassioides, II. 152.
7574. Sycopsis sinensis, I. 431.
7576. Keteleeria Davidiana, II. 39.

7578. Rhus succedanea, II. 182.
7589. Rosa microcarpa, II. 314.
7591. Koelreuteria bipinnata, II. 193.
7596. Castanopsis sclerophylla, III. 201.
7601. Grewia parviflora, var. glabrescens, II. 371.
7604. Photinia Davidsoniae, I. 185.
7606. Camptotheca acuminata, II. 254.
7609. Acanthopanax Henryi, II. 557.
7614. Celastrus gemmata, II. 352.
7619. Quercus acrodonta, III. 225.
7620. Melia Azedarach, II. 157.
7630. Liquidambar formosana, I. 421 (by error as No. 1630).
7639. Rosa multiflora, var. cathayensis, II. 304.
7651, 7651a. Staphylea holocarpa, II. 185.
7653. Salix Wallichiana, III. 64.
7655. Akebia lobata, I. 348.
7660. Vaccinium Donianum, I. 557.
7664. Photinia subumbellata, I. 189.
7666. Alangium chinense, II. 552.
7683. Ardisia japonica, II. 582.
7685. Myrsine semiserrata, II. 580.
7687. Zanthoxylum alatum, var. planispinum, II. 125.
7690. Callicarpa gracilipes, III. 371.
7694. Pterocarya Paliurus, III. 182.
7702. Pistacia chinensis, II. 173.
7704. Styrax suberifolius, I. 290.
7707. Castanopsis sclerophylla, III. 201.
7714. Polygala caudata, II. 161.
7717. Diospyros armata, II. 591.
7721. Acer caudatum Wallich.
7722. Osmanthus fragrans, II. 609.
7723. Symplocos paniculata, II. 593.
7724. Photinia villosa, var. sinica, I. 186.
7744. Ligustrum Quihoui, II. 607.
7747. Ardisia Henryi, II. 582.
7766. Xylosma racemosum, var. pubescens, I. 283.
7773. Orixa japonica, II. 135.
7775. Elaeagnus magna, II. 411.
7780. Ulmus castaneifolia, III. 256.
7784b. Clematis Armandi, I. 326.
7788. Holboellia coriacea, I. 345.
7805. Distylium chinense, I. 423.
7807. Buxus microphylla, var. aemulans, II. 169.
7822. Thea sinensis, II. 391.

HENRY, A. (*continued*)
7823. Evonymus myriantha, I. 487.
7830. Eurya japonica, var. aurescens, II. 399.
7831. Cephalotaxus drupacea, II. 3.
7832a. Cephalotaxus Oliveri, II. 6.
7834. Sarcococca Hookeriana, var. humilis, II. 164.
7849. Loranthus Delavayi, III. 316.
7850a. Pittosporum glabratum, III. 326.
7856a. Akebia lobata, var. australis, I. 348.
7864. Thea cuspidata, II. 390.
7866. Celtis ?labilis, III. 268.
7873. Daphne genkwa, II. 538.
7883. Viscum album, III. 318.
7884. Buddleia officinalis, I. 565.
7904. Clematis Armandi, I. 326.
7909. Acanthopanax leucorrhizus, II. 557.
7917. Thea cuspidata, II. 390.
7918. Cotoneaster gracilis, III. 431.
7921. Thea Grijsii, II. 394.
7937. Vaccinium Henryi, I. 561.
7942. Corylus heterophylla, var. sutchuenensis, II. 445.
7945. Quercus aliena, var. acuteserrata, III. 215.
7946. Eurya japonica, var. nitida, II. 398.
7958. Sapindus mukorossi, II. 191.
7960. Euphoria longana, II. 193.
7976. Alangium chinense, II. 552.
8080. Mussaenda hirsutula, III. 399.
8124. Mussaenda hirsutula, III. 399.
8274. Ehretia acuminata, III. 363.
8417. Ardisia crispa, II. 581.
8548. Mussaenda hirsutula, III. 399.
8559. Trema virgata, III. 289.
8713. Symplocos javanica, II. 597.
8799. Acer tetramerum, var. elobulatum, I. 95.
8806. Actinidia kolomikta, II. 380.
8808. Hypericum Prattii, II. 404.
8862. Rhododendron lutescens, I. 516.
8866. Porana triserialis, III. 356.
8871. Cassiope selaginoides, I. 551.
8883. Pterostyrax hispidus, I. 295.
8890. Alnus cremastogyne, II. 488.
8891. Salix moupinensis, III. 46.
8944. Rosa Davidii, II. 322.
8947. Rosa omeiensis, II. 331.
8955. Potentilla fruticosa, II. 301.

8957. Sorbus Henryi, II. 274.
8960. Sorbus pluripinnata, I. 481.
8961. Rosa omeiensis, II. 331.
8968. Neillia affinis, I. 434.
8975. Sorbus setschwanensis, I. 475.
8976. Melastoma normale, II. 421.
8990. Tapiscia sinensis, II. 188.
8997. Lysionotus brachycarpus, III. 387.
9000a. Pistacia weinmannifolia, II. 174.
9013. Brandisia Hancei, I. 573.
9015, 9015a, e. Myrica esculenta, III. 189.
9021. Eurya acuminata, var. multiflora, II. 401.
9023, 9023b, c. Luculia intermedia, III. 408.
9025, 9025b. Buddleia Henryi, I. 571.
9026, 9026a. Clerodendron mandarinorum, III. 375.
9027. Fagus longipetiolata, III. 190.
9032, 9032a, b, c. Helwingia himalaica, II. 571.
9035. Lasianthus longicauda, III. 402.
9039, 9039a. Eurya japonica, var. nitida, II. 398.
9039d. Eurya acuminata, II. 400.
9040. Prismatomeris linearis, III. 414.
9040a, 9040d, e, f. Prismatomeris brevipes, III. 413.
9070a, 9070b. Castanopsis calathiformis, III. 204.
9091, 9091b, e, f. Pieris ovalifolia, I. 552.
9091, 9091d. Pieris ovalifolia, var. elliptica, I. 552.
9098a, 9098c. Rosa odorata, var. gigantea, II. 338.
9099. Koelreuteria bipinnata, II. 193.
9100. Cephalotaxus Fortunei, II. 4.
9103. Koelreuteria bipinnata, II. 193.
9104. Clematis Gouriana, I. 339.
9112. Loranthus Delavayi, III. 316.
9134. Campylotropis yunnanensis, II. 114.
9135, 9135a. Campylotropis trigonoclada, II. 114.
9148a. Cunninghamia lanceolata, II. 50.
9155, 9155a. Rhododendron decorum, I. 541.
9170, 9170b, e. Vaccinium Dunalianum, I. 560.

HENRY, A. (*continued*)
9170c. Vaccinium Dunalianum, var. urophyllum, I. 560.
9173b, 9173c, d, e. Myrsine semiserrata, II. 580.
9176, 9176a, b. Cornus capitata, II. 578.
9185. Rhamnus Henryi, II. 44.
9201. Quercus dentata, III. 210.
9202a. Castanopsis Fargesii, III. 198.
9208. Hydrangea aspera, I. 40.
9214. Schefflera Delavayi, II. 555.
9223. Alnus nepalensis, II. 502.
9225. Smilax ovalifolia, III. 12.
9229, 9229c. Porana triserialis, III. 356.
9229b, 9229c. Porana triserialis, var. lasia, III. 362.
9236a. Rosa longicuspis, II. 313.
9243a, 9243b. Campylotropis diversifolia, II. 115.
9263. Hibiscus Manihot, II. 374.
9280. Spiraea japonica, var. stellaris, I. 452.
9281. Prinsepia utilis, II. 345.
9299a. Quercus glauca, III. 226.
9323. { Celtis amphibola, III. 279.
{ Celtis Bungeana, III. 269.
9323a. Celtis yunnanensis, III. 279.
9330, 9330a. Smilax trigona, III. 10.
9338, 9338c, d. Salix araeostachya, III. 96.
9338b. Salix psilostigma, III. 116.
9340, 9340a, b. Porana dinetoides, III. 360.
9341. Diospyros kaki, var. silvestris, II. 590.
9361a, 9361c. Chloranthus brachystachys, III. 15.
9364. Polygala congesta, II. 162.
9366, 9366c, d. Zanthoxylum alatum, II. 125.
9366a, 9366e, f. Zanthoxylum alatum, var. planispinum, II. 125.
9367a. Dichotomanthes tristaniaecarpa, II. 344.
9376, 9376a, b, c, e. Myrioneuron Faberi, III. 410.
9377. Clematis fulvicoma, I. 327.
9394, 9394a. Quercus aliena, var. acuteserrata, III. 215.
9395. Polygala Wattersii, II. 161.
9408. Itoa orientalis, I. 286.
9411, 9411a, b. Prunus majestica, I. 252.

9412, 9412a. Callicarpa rubella, III. 369.
9415a. Smilax ovalifolia, III. 12.
9426, 9426a, b. Crataegus Henryi, I. 181.
9427, 9427b. Paliurus sinicus, II. 211.
9431a, 9431b, c, d. Clematis uncinata, I. 327.
9433, 9433a, b. Jasminum nintooides, II. 615.
9456, 9456a, b. Vitis pentagona Diels & Gilg.
9461, 9461b, c. Cipadessa baccifera, var. sinensis, II. 159.
9464, 9464a, b. Maesa castaneifolia, II. 583.
9465. Hovenia dulcis, II. 252.
9466, 9466b. Symplocos paniculata, II. 593.
9475. Deutzia aspera, I. 149.
9475a. Deutzia purpurascens, var. pauciflora, I. 19.
9489. Porana paniculata, III. 359.
9519. Mussaenda divaricata, var. mollis, III. 398.
9555a. Hedyotis macrostemon, III. 408.
9570, 9570a. Adina cordifolia, III. 390.
9581a. Gardneria multiflora, I. 563.
9584. Ixora yunnanensis, III. 412.
9586. Ceratostigma Griffithii, II. 586.
9600, 9600b. Pistacia weinmannifolia, II. 174.
9622. Campylotropis trigonoclada, II. 114.
9623a. Pieris ovalifolia, var. lanceolata, I. 552.
9626a, 9626b. Campylotropis polyantha, II. 114.
9650, 9650a. Mussaenda divaricata, III. 394.
9650b. Mussaenda laxiflora, III. 399.
9657, 9657a, b. Jasminum sinense, II. 612.
9667, 9667a. Cryptomeria japonica, II. 52.
9669. Neillia sinensis, I. 436.
9679a. Celastrus gemmata, II. 352.
9682. Corylus heterophylla, var. yunnanensis, II. 451.
9689. Campylotropis hirtella, II. 115.
9692, 9692b. Leycesteria formosa, I. 311.

HENRY, A. (*continued*)

9695. Tetrastigma serrulatum Planchon.

9702, 9702a. Campylotropis yunnanensis, II. 114.

9716, 9716a. Brandisia glabrescens, I. 574; III. 445.

9722. Thea sinensis, II. 391.

9744, 9744a. Tetracentron sinense, I. 417.

9750. Vitex Negundo, III. 372.

9775. Lasianthus inconspicuus, var. hirtus, III. 402.

9782a. Celastrus gemmata, II. 352.

9787. Vitex quinata, III. 374.

9791, 9791a. Ardisia crispa, II. 581.

9795. Photinia serrulata, I. 184.

9803, 9803b. Campylotropis Prainii, II. 115.

9807. Hedyotis capitellata, III. 408.

9813, 9813a, b, c, d. Hymenopogon parasiticus, III. 407.

9815. Lespedeza tomentosa, II. 110.

9854, 9854a, b. Trachelospermum axillare, III. 335.

9856, 9856a. Hedera himalaica, II. 555.

9857. Elaeagnus Loureirii, II. 416.

9858. Elaeagnus Loureirii, II. 416.

9859, 9859a. Sarcococca Hookeriana, var. humilis, II. 164.

9859b. Sarcococca ruscifolia, var. chinensis, II. 163.

9864, 9864a. Clematis Henryi, I. 342.

9867. Smilax megalantha, III. 4.

9868. Pinus Armandi, II. 12.

9877. Tetrastigma serrulatum Planchon.

9878. Eriobotrya prinoides, I. 194.

9881, 9881a, c, d. Tetrastigma yunnanense, var. triphyllum Gagnepain.

9884. Mussaenda divaricata, III. 394.

9889. Campylotropis latifolia, II. 115.

9894. Corylus heterophylla, var. yunnanensis, II. 451.

9898, 9898d, e. Diospyros Lotus, II. 587.

9898c. Diospyros kaki, var. silvestris, II. 590.

9899, 9899b. Photinia serrulata, I. 184.

9900a, b, c, d. Rhododendron indicum, var. ignescens, I. 547.

9901, 9901a, b. Xylosma longifolium, I. 284.

9906, 9906a. Osyris Wightiana, III. 320.

9913, 9913a. Quercus variabilis, III. 219.

9924. Sambucus javanica, I. 307.

9928, 9928d. Toddalia asiatica, II. 137.

9929. Carpinus pubescens, II. 442.

9930. Cornus oblonga, II. 579.

9937. Platycarya strobilacea, III. 180.

9938. Celtis Biondii, var. Cavaleriei, III. 273.

9942. Viscum articulatum, III. 318.

9947. Loranthus ferrugineus, III. 316.

9948, 9948a. Symplocos paniculata, II. 593.

9949. Leptodermis glomerata, III. 406.

9954. Porana racemosa, III. 361.

9959a. Pachysandra axillaris, II. 164.

9968. Ligustrum compactum, II. 604.

9972, 9972b. Myrsine africana, II. 580.

9973. Brandisia glabrescens, I. 574.

9986. Hypericum patulum, var. Henryi, II. 403.

9987. Cudrania tricuspidata, III. 306.

9987a. Cudrania javanensis, III. 308.

9991. Loranthus philippensis, III. 317.

9992, 9992a, b. Tetrastigma Henryi Gagnepain.

9998, 9998a, b, c. Zanthoxylum multijugum, II. 129.

9999, 9999c. Polygala arillata, II. 160.

10006, 10006a, c. Tarenna sylvestris, III. 411.

10011. Trema virgata, III. 289.

10021, 10021a. Rhamnus leptophyllus, var. milensis, II. 250.

10035, 10035b, c. Pyrus pashia, II. 264.

10035a. Pyrus pashia, var. kumaoni, II. 265.

10036, 10036a. Docynia Delavayi, II. 296.

10050b. Actinidia callosa, II. 382.

10057. Loranthus estipitatus, var. longiflorus, III. 316.

10059. Tarenna pubinervis, III. 411.

10065, 10065a. Viburnum brachybotryum, I. 309.

10114, 10114a. Clematis fasciculiflora, I. 331.

10118, 10118a. Daphne papyracea, II. 546.

10122. Aucuba chinensis, II. 572.

10123, 10123a. Aucuba chinensis, II. 572.

HENRY, A. (*continued*)
10123b. Aucuba chinensis, f. angustifolia, II. 573.
10126. Stellera chamaejasme, II. 551.
10136. Sorbus granulosa, II. 274.
10138, 10138a. Stachyurus himalaicus,
I. 287.
10142a, b. Alangium chinense, II. 552.
10144, 10144a. Sageretia gracilis, II. 623.
10148. Clematis Gouriana, I. 339.
10153. Maesa castaneifolia, II. 583.
10176. { Wendlandia glabrata, III. 392.
{ Wendlandia tinctoria, III. 392.
10176a. Wendlandia floribunda, III.
392.
10180. Mahonia flavida, I. 382.
10191a. Pittosporum glabratum, III.
326.
10191b. Pittosporum glabratum, var.
neriifolium, III. 328.
10209. Salix psilostigma, III. 116.
10217. Quercus spinosa, III. 224.
10220a. Staphylea holocarpa, var.
rosea, II. 186.
10231, 10231a. Neillia pauciflora, I.
437.
10235. Hydrangea xanthoneura, var.
Wilsonii, I. 27, 150.
10236, 10236b. Hydrangea yunnanensis, I. 37.
10245. Evodia rugosa, II. 132.
10251, 10251a, c, d. Buddleia macrostachya, I. 572.
10254, 10254a. Pistacia chinensis, II.
173.
10255, 10255a. Dichotomanthes tristaniaecarpa, II. 344.
10257. Berberis Ferdinandi-Coburgii, I.
366.
10277. Wikstroemia dolichantha, II.
537.
10283. Rhus Delavayi, II. 183.
10306. Clematis Gouriana, I. 339.
10308. Ilex macrocarpa Oliver.
10315. Loranthus sootepensis, III. 317.
10363, 10363a. Randia evenosa, III.
400.
10369. Castanopsis concolor, III. 203.
10370. Ixora yunnanensis, III. 412.
10376a. Berchemia yunnanensis, II.
216.
10377, 10377a. Thea sinensis, II. 391.
10395. Altingia yunnanensis, I. 422.

10397. Smilax lanceaefolia, III. 11.
10398. Gordonia axillaris, II. 394.
10425. Elaeagnus Henryi, II. 414.
10429. Zanthoxylum dissitum, II. 128.
10433. Ilex rotunda Thunberg.
10437, 10437a. Betula alnoides, II.
467.
10443, 10443a, b, c, d. Buddleia asiatica, I. 566.
10445. Citrus Limonia, II. 146.
10445a. Citrus Medica, var. sarcodactylis, II. 141.
10446. Celastrus monosperma, II. 357.
10454. Ehretia acuminata, III. 363.
10464, 10464a. Rhamnus Bodinieri, II.
246.
10469. Skimmia melanocarpa, II. 138.
10493, 10493a. Salix psilostigma, III.116.
10498, 10498a. Juglans cathayensis, III.
185.
10504, 10504a. Quercus Delavayi, III.
236.
10506. Ribes moupinense Franchet.
10507. Juglans regia, III. 184.
10510, 10510a, c. Pieris ovalifolia, var.
lanceolata, I. 552.
10511, 10511a. Polygala congesta, II.
162.
10515. Ehretia macrophylla, III. 364.
10519. Pinus Armandi, II. 12.
10520. Lithocarpus viridis, III. 210.
10522. Celastrus Hookeri, II. 352.
10530, 10530b. Tetrastigma yunnanense, var. triphyllum Gagnepain.
10531. Celastrus gemmata, II. 352.
10532, 10532c. Chionanthus retusus,
II. 611.
10535, 10535b. Morus acidosa, III. 297.
10536, 10536b. Debregeasia longifolia,
III. 313.
10543. Stachyurus himalaicus, I. 287.
10545. Pittosporum glabratum, var.
neriifolium, III. 328.
10548, 10548a, b. Ehretia macrophylla,
III. 364.
10549, 10549a. Vitis flexuosa, var. parvifolia Gagnepain.
10551. Pittosporum glabratum, III.
326.
10553. Pittosporum glabratum, III.
326.
10554. Symplocos paniculata, II. 593.

Henry, A. (continued)
10559. Celastrus Hindsii, II. 357.
10564. Viburnum calvum, I. 310.
10566. Smilax discotis, III. 6.
10568. Wendlandia tinctoria, III. 392.
10569, 10569a, b. Broussonetia papyrifera, III. 303.
10571. Ulmus lanceaefolia, III. 263.
10573. Pterocarya stenoptera, III. 181.
10576. Photinia serrulata, I. 184.
10587. Aleurites Fordii, II. 528.
10593. Alniphyllum Fortunei, I. 294.
10605. Schoepfia jasminodora, III. 321.
10610. Castanopsis platyacantha, III. 200.
10618. Berberis Julianae, I. 360.
10625. Pyracantha crenulata, I. 177.
10629. Prunus Henryi, I. 240.
10629b. Prunus neglecta, I. 241.
10633. Lasianthus longicauda, III. 402.
10639. Acanthopanax spinosus, II. 562.
10646, 10646a. Mussaenda erosa, III. 398.
10647. Alangium chinense, II. 552.
10651, 10651a. Trachelospermum cathayanum, III. 333.
10662. Spiraea fulvescens, I. 439.
10678. Tarenna pubinervis, III. 411.
10679. Akebia lobata, var. australis, I. 348.
10686. Tarenna pallida, III. 410.
10691. Chasalia curviflora, III. 416.
10694. Turpinia nepalensis, II. 187.
10697. Schisandra sphenanthera, I. 414.
10698. Symplocos pilosa, II. 598.
10701. Castanea mollissima, III. 192.
10702. Debregeasia longifolia, III. 313.
10715a, 10715b. Porana sinensis, III. 355.
10719. Schisandra propinqua, I. 416.
10746. Euptelea pleiosperma, I. 313.
10747. Cornus macrophylla, II. 575.
10747a. Cornus controversa, II. 573.
10748. Clematis montana, var. Wilsonii, I. 333.
10754, 10754a. Parthenocissus Thomsonii, III. 427.
10758. Chasalia curviflora, III. 416.
10772. Sambucus adnata, I. 308.
10777. Pavetta indica, III. 412.
10784. Cladrastis sinensis, II. 97.
10785. Cotoneaster Harroviana, I. 173.
10786. Deutzia Henryi, I. 148.

10800, 10800a. Cornus paucinervis, II. 576.
10803. Berchemia Giraldiana, II. 213.
10804. Xylosma racemosum, var. pubescens, I. 283.
10814, 10814a. Rhamnus Bodinieri, II. 246.
10816, 10816a. Tarenna depauperata, III. 411.
10821. Cudrania javanensis, III. 308.
10824. Actinidia callosa, II. 382.
10828. Rosa odorata, II. 338.
10835. Meliosma Wallichii, II. 207.
10847. Castanopsis tribuloides, var. echidnocarpa, III. 203.
10848. Celtis Salvatiana, III. 283.
10859. Daphne papyracea, var. crassiuscula, II. 546.
10864. Melia Azedarach, II. 157.
10872. Acer cappadocicum, var. sinicum Rehder.
10874. Symplocos longipetiolata, II. 599.
10889. Ventilago calyculata, II. 253.
10901, 10901a, b. Polygala caudata, II. 161.
10904. Vaccinium fragile, I. 559.
10911. Smilax lanceaefolia, III. 11.
10914b. Eurya acuminata, var. multiflora, II. 401.
10922. Rubus innominatus, III. 424.
10929. Rhamnella Martinii, II. 225.
10932. Zanthoxylum echinocarpum, II. 128.
10938. Diospyros yunnanensis, II. 592.
10951. Evodia trichotoma, II. 132.
10953. Wendlandia paniculata, III. 392.
10954. Melastoma normale, II. 421.
10956. Wendlandia bouvardioides, III. 393.
10957. Acer oblongum Wallich.
10959, 10959a. Broussonetia Kaempferi, III. 304.
10974. Leptodermis pilosa, III. 405.
10978. Deutzia crassifolia, I. 148.
11006. Celastrus spiciformis, var. laevis, II. 349.
11008. Actinidia purpurea, II. 378.
11009. Enkianthus quinqueflorus, var. serrulatus, II. 550.
11034. Rhus javanica, II. 178.
11069. Skimmia melanocarpa, II. 138.
11079. Polygala congesta, II. 162.

HENRY, A. (*continued*)
11082. Altingia yunnanensis, I. 422.
11131. Cedrela sinensis, II. 156.
11138. Psychotria straminea, III. 416.
11148. Lasianthus Biermanni, III. 402.
11157. Buxus microphylla, var. sinica, II. 165.
11158a. Acanthopanax trifoliatus, II. 563.
11161. Cornus oblonga, II. 579.
11162. Gordonia axillaris, II. 394.
11165. Evonymus lanceifolia, I. 491.
11171. Eurya acuminata, II. 400.
11188. Gouania javanica, II. 253.
11192. Passiflora cupiformis, II. 408.
11200. Skimmia melanocarpa, II. 138.
11211. Schisandra sphenanthera, I. 414.
11237. Smilax ocreata, III. 11.
11238. Smilax indica, III. 12.
11239. Smilax cocculoides, III. 7.
11240. Sageretia Henryi, II. 623.
11246. Morinda umbellata, III. 413.
11250. Salix araeostachya, III. 96.
11253. Lasianthus Henryi, III. 401.
11267. Celastrus Hookeri, II. 352.
11268. Pieris ovalifolia, var. lanceolata, I. 552.
11272. Rosa chinensis, II. 320.
11296. Sorbus polycarpa, II. 274.
11298. Quercus aliena, var. acuteserrata, III. 215.
11305. Elsholtzia fruticosa, III. 381.
11311. Periploca calophylla, III. 343.
11313. Castanopsis hystrix, III. 197.
11321. Daphne acutiloba, II. 539.
11341. Cotoneaster amoena, I. 165.
11343. Prinsepia utilis, II. 345.
11347. Clematis urophylla, I. 323.
11355. Keteleeria Davidiana, II. 39.
11362. Viburnum corylifolium, I. 112.
11363. Daphne papyracea, II. 546.
11387. Betula alnoides, II. 467.
11389. Uncaria lancifolia, III. 407.
11391. Mussaenda divaricata, III. 394.
11392. Elaeagnus Henryi, II. 414.
11397. Cornus oblonga, II. 579.
11399. Celastrus monosperma, II. 357.
11409. Psychotria pilifera, III. 415.
11413. Zanthoxylum alatum, II. 125.
11414a. Eurya acuminata, II. 400.
11416. Polygala congesta, II. 162.
11428. Psychotria straminea, III. 416.

11434, 11434a. Lithocarpus viridis, III. 210.
11437. Zanthoxylum dissitum, II. 128.
11439. Elaeagnus sarmentosa, II. 417.
11447, 11447a. Psychotria yunnanensis, III. 414.
11449. Elaeagnus Henryi, II. 414.
11457. Elaeagnus Loureirii, II. 416.
11469. Prunus majestica, I. 252.
11471. Celastrus gemmata, II. 352.
11472. Polygala congesta, II. 162.
11473. Elytranthe ampullacea, III. 317.
11479. Wendlandia paniculata, III. 392.
11489. Smilax lanceaefolia, III. 11.
11493. Mussaenda laxiflora, III. 399.
11501. Sloanea sterculiacea, II. 362.
11510. Reevesia pubescens, II. 376.
11564, 11564b, c. Eriosolena involucrata, II. 550.
11565, 11565a, c, d. Castanopsis tribuloides, var. echidnocarpa, III. 203.
11574. Elaeagnus Loureirii, II. 416.
11575. Clematis fasciculiflora, I. 331.
11578, 11578a. Rhus paniculata, II. 184.
11589, 11589a. Psychotria siamica, III. 415.
11603. Docynia Delavayi, II. 296.
11604. Elytranthe Henryi, III. 317.
11605. Citrus Aurantium, II. 147.
11608. Alniphyllum Fortunei, I. 294.
11612, 11612a, b, d, e, f, g, h, i. Turpinia nepalensis, II. 187.
11614, 11614a, b, c, d. Lithocarpus spicata, var. brevipetiolata, III. 208.
11615, 11615a, b, e, f. Stranvaesia nussia, var. oblanceolata, I. 193.
11617, 11617a. Berberis Ferdinandi-Coburgii, I. 364.
11618, 11618b, c, d. Diospyros kaki, var. silvestris, II. 590.
11625, 11625a. Maesa castaneifolia, II. 583.
11626. Vaccinium Donianum, I. 557.
11633. Mycetia hirta, III. 410.
11640, 11640a, b. Carpinus Londoniana, II. 438.
11647, 11647b. Tetrastigma yunnanense, var. triphyllum Gagnepain.
11648, 11648a. Vaccinium iteophyllum, I. 558.

HENRY, A. (*continued*)
11664. Loranthus adpressus, III. 317.
11671. Symplocos paniculata, II. 593.
11675, 11675b, c, d, e. Quercus vestita, III. 236.
11679. Buddleia asiatica, I. 566.
11687. Quercus Griffithii, III. 233.
11690. Spondias axillaris, II. 172.
11690a. Rhus succedanea, II. 182.
11696a, 11696b, c. Castanopsis calathiformis, III. 204.
11698. Quercus myrsinaefolia, III. 236.
11698a, 11698b. Quercus Schottkyana, III. 237.
11700, 11700a, b. Hedyotis scandens, III. 409.
11713. Jasminum lanceolarium, II. 612.
11716, 11716a. Photinia glomerata, I. 190.
11726. Zizyphus spec., II. 213.
11736a. Tetrastigma dubium Planchon.
11737, 11737a, b. Meliosma glomerulata, II. 203.
11738, 11738a, b. Castanopsis hystrix, III. 197.
11745, 11745b. Sloanea tomentosa, II. 362.
11747a. Berchemia floribunda, var. megalophylla, II. 213.
11750, 11750a, b, c, d. Randia yunnanensis, III. 400.
11755, 11755b. Elytranthe ampullacea, var. tonkinensis, III. 317.
11756, 11756a, b, c, d. Tetrastigma Henryi Gagnepain.
11770, 11770a. Quercus Schottkyana, III. 237.
11790. Mussaenda Rehderiana, III. 397.
11800. Lonicera pileata, var. linearis, I. 143.
11868. Uncaria scandens, III. 406.
11870. Vitis pentagona Diels & Gilg.
11884c. Xylosma controversum, I. 284.
11885, 11885a, b. Styrax suberifolius, I. 290.
11888. Adina mollifolia, III. 391.
11892. Porana sinensis, III. 355.
11893. Schisandra propinqua, I. 416.
11906, 11906a, b. Cudrania pubescens, III. 307.
11908. Zanthoxylum alatum, II. 125.
11909. Torricellia tiliifolia, II. 569.
11913. Pistacia weinmannifolia, II. 174.

11917. Vaccinium Donianum, I. 557.
11923, 11923c, f. Tarenna pallida, III. 410.
11930. Mycetia glandulosa, III. 410.
11930a. Mycetia bracteata, III. 409.
11931, 11931a. Mussaenda breviloba, III. 398.
11934, 11934a. Pavetta indica, III. 412.
11937, 11937a, b. Cudrania javanensis, III. 308.
11941, 11941a. Rubus asper Wallich.
11949. Trachelospermum tetanocarpum, III. 339.
11955, 11955a. Rhus succedanea, II. 182.
11957, 11957a. Alniphyllum Fortunei, I. 294.
11959, 11959a. Dichotomanthes tristaniaecarpa, var. glabrata, II. 344.
11965, 11965a, b. Saprosma ternatum, III. 416.
11966, 11966b, c, d, e, f. Chasalia curviflora, III. 416.
11972, 11972b. Celastrus monosperma, II. 357.
11975. Morus laevigata, III. 301.
11992, 11992b. Helwingia himalaica, II. 571.
11993. Celastrus paniculata, II. 355.
11995, 11995a. Uncaria sessilifructus, III. 406.
12016, 12016a, b. Meliosma Wallichii, II. 207.
12019. Broussonetia Kaempferi, III. 304.
12019a. Morus laevigata, III. 301.
12022, 12022a, b. Schisandra sphenanthera, I. 414.
12031, 12031a. Lasianthus Hookeri, III. 402.
12032, 12032a, b. Psychotria yunnanensis, III. 414.
12034, 12034a. Hovenia dulcis, II. 252.
12039, 12039b, c. Turpinia nepalensis, II. 187.
12040, 12040a. Rhamnus paniculiflorus, II. 233.
12050. Trachelospermum axillare, III. 335.
12060. Clerodendron japonicum, III. 377.

HENRY, A. (*continued*)

12065, 12065a, b. Psychotria symplocifolia, III. 415.

12069, 12069a, b, d. Psychotria morindoides, III. 414.

12086a, b. Zizyphus yunnanensis, II. 212.

12092a. Evodia simplicifolia, II. 135.

12108b, c, d, e, f. Ternstroemia japonica, var. Wightii, II. 397.

12110. Sloanea tomentosa, II. 362.

12114. Meliosma velutina, II. 202.

12115, 12115a. Smilax hypoglauca, III. 10.

12119, 12119a, c. Callicarpa Giraldiana, III. 366.

12122, 12122a, b. Celastrus dependens, II. 355.

12136. Trachelospermum auritum, III. 341.

12137, 12137a. Evodia trichotoma, II. 132.

12141. Cayratia tenuifolia Gagnepain.

12141a. Tetrastigma serrulatum Planchon.

12145. Saprosma Henryi, III. 417.

12146, 12146c, d. Psychotria Henryi, III. 415.

12157. Mussaenda pubescens, III. 399.

12168. Myrioneuron Faberi, III. 410.

12192. Schisandra propinqua, I. 416.

12214. Buddleia yunnanensis, I. 564.

12237, 12237a, b. Evodia trichotoma, II. 132.

12246, 12246a. Diplospora mollissima, III. 401.

12248, 12248a, c, d. Mycetia gracilis, III. 409.

12248b. Mycetia glandulosa, III. 410.

12249. Zanthoxylum alatum, II. 125.

12264. Hedyotis capitellata, III. 408.

12265. Mussaenda macrophylla, III. 396.

12272, 12272b, d. Polygala congesta, II. 162.

12274, 12274a, b, c. Schoepfia fragrans, III. 322.

12283. Chasalia curviflora, III. 416.

12283b. Psychotria calocarpa, III. 415.

12299, 12299a. Mycetia hirta, III. 410.

12312, 12312a. Kadsura peltigera, I. 410.

12313, 12313a, b. Pygeum Henryi, II. 344.

12328a. Castanopsis tribuloides, III. 203.

12329, 12329a, b, c, d, e. Lithocarpus viridis, III. 210.

12340. Sambucus javanica, I. 307.

12341, 12341b, c. Chloranthus brachystachys, III. 15.

12363. Mussaenda elongata, III. 398.

12368a. Psychotria siamica, III. 415.

12385, 12385b. Cudrania javanensis, III. 308.

12387, 12387a. Debregeasia longifolia, III. 313.

12435, 12435a. Quercus Griffithii, III. 233.

12458. Osbeckia crinita, II. 421.

12489, 12489b. Luculia gratissima, III. 407.

12490, 12490a. Loropetalum chinense, I. 430.

12561. Acanthopanax trifoliatus, II. 563.

12571. Lasianthus Kerrii, III. 402.

12572, 12572b. Celastrus paniculata, II. 355.

12577. Smilax cocculoides, var. lanceolata, III. 11.

12605. Brandisia laetevirens, I. 573.

12605b. Brandisia discolor, I. 573.

12607. Colquhounia elegans, var. tenuiflora, III. 380.

12608, 12608a. Lasianthus inconspicuus, III. 402.

12621, 12621a. Polygala arillata, II. 160.

12622. Porana discifera, III. 358.

12631. Elytranthe bibracteolata, var. sinensis, III. 317.

12635. Xylosma longifolium, I. 284.

12638, 12638a. Vitex quinata, III. 374.

12646. Saprosma Henryi, III. 417.

12660. Schoepfia fragrans, III. 322.

12676. Nauclea Griffithii, III. 406.

12680. Elaeagnus conferta, II. 417.

12694. Porana discifera, III. 358.

12700. Mycetia glandulosa, III. 410.

12708. Pygeum Henryi, II. 344.

12712. Campylotropis parviflora, II. 115.

12719. Smilax ovalifolia, III. 12.

12734. Keteleeria Davidiana, II. 39.

12745. Vaccinium Donianum, I. 557.

12757. Xylosma controversum, I. 284.

Henry, A. (*continued*)

12758. Viscum stipitatum, III. 319.
12770. Acanthopanax trifoliatus, II. 563.
12774. Mussaenda sessilifolia, III. 397.
12787, 12787a, b. Elaeagnus Loureirii, II. 416.
12790, 12790a. Viburnum brachybotryum, I. 309.
12799. Smilax lanceaefolia, III. 11.
12800. Trachelospermum tetanocarpum, III. 339.
12806, 12806a. Psychotria yunnanensis, III. 414.
12809. Sambucus javanica, I. 307.
12818. Elaeagnus macrantha, II. 416.
12820. Uncaria macrophylla, III. 407.
12831. Castanopsis ceratacantha, III. 199.
12831a. Castanopsis tribuloides, III. 203.
12833. Photinia lancifolia, I. 191.
12834. Castanopsis indica, III. 202.
12852. Trachelospermum auritum, III. 341.
12853. Adina mollifolia, III. 391.
12855. Keteleeria Davidiana, II. 39.
12864. Ulmus lanceaefolia, III. 263.
12869, 12869a, b. Betula alnoides, II. 467.
12880. Nauclea Griffithii, III. 406.
12881. Celtis Bodinieri, III. 276.
12889. Melia Azedarach, II. 157.
12892. Loranthus Delavayi, var. latifolius, III. 316.
12902, 12902a, b. Smilax lanceaefolia, III. 11.
12903, 12903a, c. Smilax ocreata, III. 11.
12928. Diplospora mollissima, III. 401.
12980. Morus acidosa, III. 297.
12982. Wendlandia paniculata, III. 392.
12984, 12984a. Diospyros yunnanensis, II. 592.
13014. Wendlandia glabrata, III. 392.
13115. Trachelospermum axillare, III. 335.
13030. Alangium chinense, II. 552.
13049. Engelhardtia chrysolepis, III. 186.
13091. Camptotheca acuminata, II. 254.
13151. Tapiscia sinensis, II. 188.

13156. Cudrania fruticosa, III. 307.
13183. Thea sinensis, II. 391.
13210. Vitex Negundo, III. 372.
13213. Osmanthus fragrans, II. 609.
13238. Lithocarpus spicata, III. 207.
13247. Quercus variabilis, III. 219.
13265. Adina mollifolia, III. 391.
13267. Berberis subacuminata, I. 363.
13270. Myrsine semiserrata, II. 580.
13273. Symplocos javanica, II. 597.
13288. Diospyros yunnanensis, II. 592.
13293. Daphne papyracea, II. 546.
13297. Torricellia tiliifolia, II. 569.
13299. Pittosporum glabratum, III. 326.
13300. Aucuba chinensis, f. angustifolia, II. 573.
13303. Diospyros mollifolia, II. 591.
13307. Psychotria pilifera, III. 415.
13308. Psychotria pilifera, III. 415.
13311. Myrsine semiserrata, II. 580.
13326, 13326a. Zanthoxylum dissitum, II. 128.
13328. Skimmia melanocarpa, II. 138.
13334. Xylosma longifolium, I. 284.
13342. Castanopsis concolor, III. 203.
13353. Thuja orientalis, II. 53.
13367. Wikstroemia effusa, II. 538.
13389. Premna urticifolia, III. 458.
13403. Pittosporum glabratum, III. 326.
13404. Vaccinium Dunalianum, I. 560.
13411. Evonymus lanceifolia, I. 491.
13412. Photinia lancifolia, I. 191.
13432. Clematis Gouriana, I. 339.
13433. Camptotheca acuminata, II. 254.
13449. Uncaria sessilifructus, III. 406.
13461. Psychotria straminea, III. 416.
13484. Hedyotis scandens, III. 409.
13509. Celtis Biondii, var. Cavaleriei, III. 273.
13519. Polygala congesta, II. 162.
13570. Lasianthus Hookeri, III. 402.
13573. Prismatomeris tetrandra, III. 413.
13627. Clematis fasciculiflora, I. 331.
13644. Broussonetia Kaempferi, III. 304.
13648. Mussaenda erosa, III. 398.
13649. Smilax herbacea, var. angustata, III. 9.
13650. Toddalia asiatica, II. 137.

HENRY, A. (*continued*)
13654. Sloanea assamica, II. 362.
13660. Mussaenda Henryi, III. 397.
13670. Lasianthus micranthus, III. 402.
13682. Castanopsis calathiformis, III. 204.
13689. Vitis pentagona Diels & Gilg.
13692. Meliosma Wallichii, II. 207.
13694. Mussaenda erosa, III. 398.

Henry, Augustine (Formosan plants).
1. Clerodendron cyrtophyllum, III. 377.
35. Stachyurus himalaicus, I. 287.
38. Deutzia pulchra, I. 18.
42. Clausena punctata, II. 140.
48. Pittosporum formosanum, III. 330.
52. Smilax stenopetala, III. 12.
55. Smilax stenopetala, III. 12.
62. Rhus succedanea, II. 182.
74. Heterosmilax japonica, III. 13.
98. Hydrangea chinensis, I. 37.
100. Morus acidosa, III. 297.
115. Smilax stenopetala, III. 12.
134. Morus acidosa, III. 297.
135. Cudrania javanensis, III. 308.
140. Aucuba chinensis, II. 572.
144. Smilax stenopetala, III. 12.
150. Citrus sinensis, II. 148.
190. Ehretia macrophylla, III. 364.
196. Eurya acuminata, II. 400.
200. Buddleia asiatica, I. 566.
214. Sambucus javanica, I. 307.
215. Gordonia axillaris, II. 394.
256. Pittosporum formosanum, III. 330.
284. Smilax stenopetala, III. 12.
313. Ehretia macrophylla, III. 364.
318. Citrus grandis, II. 144.
319. Akebia quinata, var. longeracemosa, I. 349.
323. Ehretia macrophylla, III. 364.
331. Ligustrum formosanum, II. 608.
348. Rhus javanica, var. Roxburghii, II. 179.
375. Eurya chinensis, II. 400.
388. Ecdysanthera rosea, III. 342.
425. Liquidambar formosana, I. 421.
428. Quercus glauca, III. 226.
429. Idesia polycarpa, I. 284.
443. Ehretia acuminata, III. 363.
456. Ehretia macrophylla, III. 364.

477. Deutzia pulchra, I. 18.
488. Pistacia chinensis, II. 173.
506. Ehretia acuminata, III. 363.
535. Eurya acuminata, II. 400.
536. Styrax suberifolius, I. 290.
552. Sapindus mukorossi, II. 191.
553. Sambucus javanica, I. 307.
562. Clerodendron cyrtophyllum, III. 377.
590. Hydrangea chinensis, I. 37.
591. Vaccinium bracteatum, **var.** Wrightii, I. 559.
592. Styrax suberifolius, I. 290.
635. Symplocos paniculata, II. 593.
636. Vaccinium bracteatum, **var.** Wrightii, I. 559.
690. Melia Azedarach, II. 157.
720. Cudrania javanensis, III. 308.
744. Morus acidosa, III. 297.
822. Pittosporum formosanum, III. 330.
838. Ecdysanthera rosea, III. 342.
876. Grewia parviflora, var. glabrescens, II. 371.
904, 904a. Clematis grata, var. lobulata, I. 337.
905. Vitex Negundo, III. 372.
921. Sambucus javanica, I. 307.
922. Ehretia acuminata, III. 363.
947. Vaccinium bracteatum, **var.** Wrightii, I. 559.
952. Ehretia acuminata, III. 363.
971. Pittosporum formosanum, III. 330.
977. Pittosporum formosanum, III. 330.
979. Loranthus chinensis, var. formosanus, III. 316.
1058. Pittosporum formosanum, III. 330.
1068, 1068a, b. Salix Kusanoi, III. 100.
1070a. Pittosporum formosanum, III. 330.
1092. Hibiscus syriacus, II. 374.
1131. Wikstroemia retusa, II. 534.
1135. Ehretia acuminata, III. 363.
1177. Buxus microphylla, var. sinica, II. 165.
1182, 1182a, b, c. Vitex quinata, III. 374.
1197, 1197a. Wikstroemia indica, II. 534.
1281. Melia Azedarach, II. 157.

HENRY, A. (*continued*)

1295. Clerodendron cyrtophyllum, III. 377.
1301. Trachelospermum divaricatum, var. brevisepalum, III. 338.
1302. Heterosmilax japonica, II. 13.
1313. Cestanopsis cuspidata, III. 204.
1369. Styrax suberifolius, I. 290.
1370. Trachelospermum divaricatum, var. brevisepalum, III. 338.
1382. Paliurus ramosissimus, II. 210.
1387. Firmiana simplex, II. 377.
1389. Alnus Henryi, II. 495.
1394. Alnus formosana, II. 501.
1404. Salix Kusanoi, III. 100.
1473. Salix Kusanoi, III. 100.
1529. Ulmus parvifolia, III. 244.
1616. Celtis formosana, III. 280.
1716. Hydrangea chinensis, I. 37.
1732. Alnus formosana, II. 501.
1747. Sambucus javanica, I. 307.
1778, 1778a. Ehretia acuminata, III. 363.
1784. Morus acidosa, III. 297.
1829. Akebia quinata, var. longeracemosa, I. 349.
1873. Clerodendron cyrtophyllum, III. 377.
1874. Ehretia acuminata, III. 363.
1883. Wikstroemia indica, II. 534.
1888. Pittosporum formosanum, III. 330.
1893. Celastrus Kusanoi, II. 356.
2035. Celtis nervosa, III. 280.

Hilgendorf.
192. Prunus Grayana, I. 69.

Hongkong; Herb. Bot. Garden.
435. Rosa microcarpa, II. 314.
987. Eurya chinensis, II. 400.
1067. Quercus bambusifolia, III. 235.
1112. Chionanthus retusus, II. 611.
1256. Wikstroemia monnula, II. 535.
1704. Quercus myrsinaefolia, III. 236.
1766. Daphne Championii, II. 544.
1782. Osbeckia crinita, II. 421.
2183. Ehretia macrophylla, III. 364.
2336. Diploclisia affinis, I. 389.
2344. Nandina domestica, I. 386.
2369. Pittosporum glabratum, III. 326.
2396. Eurya japonica, var. nitida, II. 398.

2412. Hibiscus syriacus, II. 374.
2419. Grewia parviflora, var. glabrescens, II. 371.
2433. Toddalia asiatica, II. 137.
2440. Illicium Henryi, I. 417.
2460. Ailanthus altissima, II. 153; III. 449.
2461. Canarium album, II. 155.
2479, 2479b. Celastrus Hookeri, II. 352.
2483. Celastrus articulata, var. punctata, II. 356.
2488. Rhamnus crenatus, II. 232.
2498. Wikstroemia monnula, II. 535.
2525. Rhus sylvestris, II. 180.
2528. Rhus sylvestris, II. 180.
2534. Sabia coriacea, II. 198.
2541. Acer cordatum, I. 92.
2545. Acer Oliverianum, var. serrulatum, I. 90.
2560. Lespedeza Dunnii, II. 111.
2597. Sorbus Dunnii, II. 273.
2599. Crataegus cuneata, I. 179.
2636. Rosa bracteata, II. 337.
2637. Rosa microcarpa, II. 314.
2638. Rosa anemoneflora, II. 336.
2639. Rosa anemoneflora, II. 336.
2641. Rosa Gentiliana, var. australis, II. 336.
2643. Rosa multiflora, var. cathayensis, II. 104.
2647. Rosa Gentiliana, II. 312.
2657. Spiraea chinensis, I. 444.
2664. Hydrangea chinensis, I. 37.
2676. Deutzia setchuenensis, I. 18.
2683. Loropetalum chinense, I. 430.
2703. Eugenia microphylla, II. 420.
2729. Lagerstroemia indica, II. 418.
2760. Hedera himalaica, II. 555.
2763. Viburnum luzonicum, var. formosanum, I. 116.
2771. Viburnum laterale, I. 311.
2773. Euscaphis japonica, II. 187.
2777. Lonicera affinis, I. 144.
2778. Lonicera affinis, var. pubescens, I. 141.
2804. Paederia tomentosa, III. 403.
2825. Rhus trichocarpa, II. 180.
2877. Vaccinium iteophyllum, I. 558.
2880. Rhododendron ovatum, I. 546.
2882. Rhododendron Mariesii, I. 548.
2891. Ligustrum sinense, var. Stauntonii, II. 606.

HONGKONG (*continued*)
2892. Ligustrum lucidum, II. 603.
2896. Styrax suberifolius, I. 290.
2899. Styrax philadelphoides, I. 289.
2903. Symplocos botryantha, II. 596.
2906. Symplocos stellaris, II. 597.
2909. Symplocos paniculata, II. 593.
2912. Diospyros kaki, II. 588.
2914. Ecdysanthera rosea, III. 342.
2918. Trachelospermum axillare, III. 335.
2930. Buddleia Lindleyana, I. 564.
3337. Chloranthus brachystachys, III. 15.
3369. Paulownia Fortunei, I. 578.
3369 (or 3368?). Paulownia thyrsoidea, I. 576.
3383. Callicarpa dichotoma, III. 370.
3392. Premna microphylla, III. 371.
3395. Clerodendron japonicum, III. 378.
3420. Celastrus Hookeri, II. 352.
3433. Celtis Biondii, var. heterophylla, III. 282.
3462. Myrica rubra, III. 189.
3479. Cudrania javanensis, III. 308.
3482. Quercus phillyraeoides, III. 233.
3491. Castanopsis sclerophylla, III. 201.
3494. Quercus glauca, III. 226.
3498. Castanopsis hystrix, III. 197.
3500. Castanopsis fissa, III. 203.
3501. Castanea mollissima, III. 192.
3509. Salix Dunnii, III. 97.
3515. Buxus Harlandii, II. 166.
3570. Smilax riparia, III. 10.
3576. Smilax riparia, III. 10.
3808. Juniperus formosana, II. 56.
3809. Juniperus formosana, II. 56.
3863. Pterocarya stenoptera, III. 181.
7447. Eurya chinensis, II. 400.
9057. Ecdysanthera rosea, III. 342.
9501. Pittosporum glabratum, var. neriifolium, III. 328.
10218. Ternstroemia japonica, var. Wightii, II. 397.

Jyer Subramaina, P. K.
86. Populus ciliata, III. 31.

King, Sir George.
100. Ulmus lanceaefolia, III. 263.
173. Debregeasia longifolia, III. 313.
244. Sorbus khasiana, II. 278.

Komarov, Vladimir.
485. Betula costata, II. 476.
487. Betula Ermanii, var. genuina, II. 470.
488. Betula fruticosa, II. 482.
492. Alnus viridis, II. 503.
836. Deutzia parviflora, I. 23.
963. Lespedeza juncea, II. 109.

Maack, Richard.
5. Salix rorida, III. 155.
6. Salix rorida, III. 155.
10. Populus tremula, var. Davidiana, III. 24.
11. Populus tremula, var. Davidiana, III. 24.
103. Ulmus pumila, III. 242.
113. Ulmus japonica, var. levigata, III. 260.
126. Salix Starkeana, var. cinerascens, III. 151.
207. Salix pentandra, III. 101.
379. Ulmus japonica, var. levigata, III. 260.
611. Alnus hirsuta, var. sibirica, II. 498.

Macgregor, D.
44. Castanea mollissima, III. 192.

Mackenzie, William.
165. Keteleeria Fortunei, II. 40.

Macoun, James Melville.
18884. Salix arctica, III. 136.

Maingay, Alexander Caroll.
192. Quercus glandulifera, III. 212.

Makino, Tomitarō.
37a. Salix Sieboldiana, III. 162.
277. Carpinus Tanakaeana, II. 440.

Manickam, R.
87. Betula cylindrostachya, II. 466.

Martin, L.
2071. Prunus Davidiana, I. 275.
2076. Daphne Feddei, II. 547.
2299. Rhamnella Martinii, II. 225.
2603. Rosa Rubus, II. 308.

Matsumura, Jinzo.
54. Salix gracilistyla, III. 164.
70. Salix japonica, III. 132.

Maximowicz, Carl.
1. Salix sachalinensis, III. 158.
2. Salix Urbaniana, III. 103.

MAXIMOWICZ, C. (*continued*)
5. Salix opaca, III. 159.
6. Salix lepidostachys, III. 166.
527. Salix gracilistyla, III. 164.
719. Salix gracilistyla, III. 164.

Meyer, Frank N.
17. See No. 1254.
22. Populus Maximowiczii, III. 32.
24. Evonymus alata, I. 493.
38. Quercus aliena, var. acuteserrata, III. 215.
38. Quercus dentata, III. 210.
39. Quercus serrata, III. 217.
40. Quercus liaotungensis, III. 233.
42. Lonicera Tatarinovii, I. 144.
43. Populus tremula, var. Davidiana, III. 24.
46. Diospyros Lotus, II. 587.
54. Salix Matsudana, III. 107.
68. Corylus Sieboldiana, var. mandshurica, II. 454.
81. Malus baccata, II. 291.
81. See No. 1300.
89. Cornus Bretschneideri, II. 579.
90. Betula chinensis, II. 479.
90. Cornus Bretschneideri, II. 579.
92. Spiraea Fritschiana, var. angulata, I. 453.
94. Celtis Bungeana, III. 269.
100. Tilia mandshurica, II. 370.
101. Populus tremula, var. Davidiana, III. 24.
104. Malus baccata, II. 291.
105. See No. 1239.
108. Potentilla fruticosa, var. Veitchii, II. 303.
109. Ulmus japonica, III. 258.
121. Berberis Poiretii, I. 372.
123. Tilia mandshurica, II. 370.
124. Ulmus laciniata, III. 255.
129. See No. 991.
130. Tilia mandshurica, II. 370.
156. Betula Ermanii, var. genuina, II. 470.
181. Pistacia chinensis, II. 173.
183. Diospyros Lotus, II. 587.
187. Ulmus parvifolia, III. 244.
197. Quercus serrata, III. 217.
198. Juglans regia, III. 184.
201. Ulmus Davidiana, III. 261.
208. Ulmus macrocarpa, III. 251.
214. Quercus dentata, III. 210.

216. Meratia praecox, I. 419.
225. Quercus dentata, III. 210.
227. Castanopsis sclerophylla, III. 201.
228. Myrica rubra, III. 189.
231. Ulmus macrocarpa, III. 251.
247. Pteroceltis Tatarinowii, III. 285.
249. Rhamnus Meyeri, II. 249.
250. Salix Matsudana, III. 107.
255. Cotinus coggygria, var. cinerea, II. 176.
256. Chionanthus retusus, II. 611.
257. Albizzia kalkora, III. 446.
258. Carpinus Turczaninovii, II. 439.
262. Evodia Daniellii, II. 135.
263. Forsythia suspensa, var. latifolia, I. 302.
265. Grewia parviflora, II. 371.
266. Chionanthus retusus, II. 611.
272. Celtis Bungeana, III. 269.
275. Ulmus macrocarpa, III. 251.
277. Styrax japonicus, var. calycothrix, I. 292.
278. Evodia Daniellii, II. 135.
279. Clerodendron trichotomum, III. 376.
280. Hovenia dulcis, II. 252.
285. Crataegus dsungarica, I. 183.
296. Catalpa ovata, I. 303.
307. Quercus variabilis, III. 219.
312. Picrasma quassioides, II. 152.
315. Pistacia chinensis, II. 173.
320. Symplocos paniculata, II. 593.
321. Quercus glandulifera, III. 212.
326. Quercus variabilis, III. 219.
327. Pinus Bungeana, II. 13.
342. Zelkova ?sinica, III. 287.
393. Thuja orientalis, II. 53.
396. Pieris ovalifolia, var. elliptica, I. 552.
399. Tamarix chinensis, II. 406.
413. Elaeagnus umbellata, II. 410.
414. Rosa xanthina, f. normalis, II. 342.
416. Ulmus ?japonica, III. 261.
425. Callicarpa dichotoma, III. 370.
427. Platycarya strobilacea, III. 180.
432. Cephalotaxus Fortunei, II. 4.
433. Taxus cuspidata, var. chinensis, II. 8.
436. Clerodendron cyrtophyllum, III. 377.
440. Styrax philadelphoides, I. 289.
441. Castanea Seguinii, III. 194.
442. Quercus aliena, III. 214.

MEYER, F. N. (*continued*)
443. Rhus sylvestris, II. 180.
733. Salix polia, III. 174.
783. Populus laurifolia, III. 35.
820. Salix herbacea, III. 143.
923. Ulmus pumila, III. 242.
928. Ulmus pumila, III. 242.
929. Broussonetia papyrifera, III. 303 (by error as No. 928).
955. Quercus aliena, var. acuteserrata, f. calvescens, III. 215.
965. Celtis Bungeana, III. 269.
973. Rosa davurica, II. 340.
974. Populus Simonii, III. 21.
975. Populus Simonii, III. 21.
976. Quercus dentata, III. 210.
977. Quercus serrata, III. 217.
989. Corylus heterophylla, II. 450.
990. Hemiptelea Davidii, III. 288.
991. Tilia mandshurica, II. 370 (by error also as No. 129).
993. Ulmus japonica, III. 258.
998. Malus baccata, II. 291.
1008. Vitex Negundo, var. incisa, III. 374.
1019. Lycium chinense, III. 385.
1050. Wikstroemia chamaedaphne, II. 536.
1069. Ulmus ?pumila, by error under U. parvifolia, III. 244.
1088. Ulmus macrocarpa, III. 251.
1113. Populus tremula, var. Davidiana, III. 24.
1125. Ostryopsis Davidiana, II. 423.
1160. Salix characta, III. 125.
1163. Betula japonica, II. 485.
1176. Potentilla fruticosa, var. Veitchii, II. 303.
1204. Salix phylicifolia, III. 123.
1207. Betula davurica, II. 483.
1211. Quercus liaotungensis, III. 233.
1227. Corylus Sieboldiana, var. mandshurica, II. 454.
1239. Rosa davurica, II. 340 (by error as No. 105).
1240. Rosa davurica, II. 340.
1245. Tilia mandshurica, II. 370.
1248. Salix Caprea, III. 149.
1254. Malus baccata, II. 291 (by error as No. 17).
1258. Tilia mongolica, II. 369.
1288. Castanea mollissima, III. 192.
1297. Diospyros Lotus, II. 587.

1299. Wistaria venusta, II. 514.
1300. Grewia parviflora, II. 371 (by error as No. 81).
1311. Populus suaveolens, III. 18.
1323. Hippophae rhamnoides, II. 409.
1328. Ulmus japonica, III. 258.
1385. Ulmus ?pumila, III. 242 (by error also under U. parvifolia).
1390. Smilax vaginata, III. 2.
1391. Diospyros Lotus, II. 587.
1400a. Castanea mollissima, III. 192.
1407. Hemiptelea Davidii, III. 288.
1409. Castanea mollissima, III. 192.
1415. Ehretia macrophylla, III. 364.
1419. Morus cathayana, III. 292.
1422. Cudrania tricuspidata, III. 306.
1425. Celtis Biondii, III. 272.
1444. Zelkova ?sinica, III. 286.
1448. Albizzia kalkora, III. 446.
1453. Juglans cathayensis, III. 185.
1465. Castanea mollissima, III. 192.
1478. Schoepfia jasminodora, III. 321.
1479. Castanea Seguinii, III. 194.
1485. Quercus Fabri, III. 216.
1486. Castanea Seguinii, III. 194.
1501. Hemiptelea Davidii, III. 288.
1508. Ulmus japonica, III. 258.
1521. Carya cathayensis, III. 187.
1540. Pittosporum glabratum, III. 326.
1542. Carpinus Tschonoskii, II. 441.
1546. Quercus serrata, III. 217.
1587. Broussonetia Kaempferi, III. 305.
1588. Morus cathayana, III. 292.
1589. Castanea mollissima, III. 192.
1611. Clerodendron cyrtophyllum, III. 377.
1618. Premna microphylla, III. 371.
1692. Quercus Baronii, III. 226.
1786a. Zelkova serrata, III. 287.
1823. Berberis Soulieana, III. 437.
2100a. Smilax vaginata, III. 2.
21620. Rosa xanthina, II. 342.
22671. Picea Wilsonii, II. 27.
22672. Picea Meyeri, II. 28.
22673. Pinus sinensis, II. 15.
22674. Larix dahurica, var. Principis Rupprechtii, II. 21.
22680. Pinus sinensis, II. 15.
23032. Syringa Meyeri, I. 301.
23913. Pinus sinensis, II. 15.
35611. Juglans regia, III. 184.
35612. Juglans regia, III. 184.
35613. Juglans regia, III. 184.

Monbeig, T.
2. Rhododendron Monbeigii, I. 536.
3. Rhododendron foveolatum, I. 537.
4. Rhododendron gymnanthum, I. 539.
6. Rhododendron polylepis, I. 521.
16. Rhododendron Monbeigii, I. 536.

Morse, Hosea Ballou.
171. Melodinus Seguinii, III. 331.
328. Wikstroemia indica, II. 534.
372. Wendlandia paniculata, III. 392.
599. Adina rubella, III. 390.
704. Psychotria siamica, III. 415.

Nelson, E. W.
1053. Abelia floribunda, I. 127.

Nicholson, George.
1677. Castanopsis cuspidata, f. latifolia, III. 205.
2291. Castanopsis cuspidata, f. variegata, III. 205.

Niederlein, Gustav.
98. Salix variegata, III. 70.

Oldham, Richard.
70. Berchemia formosana, II. 220.
87. Rhus javanica, var. Roxburghii, II. 179.
97. Rosa Wichuraiana, II. 335.
105. { Prunus pogonostyla, var. globosa, I. 265. Prunus pogonostyla, var. obovata, I. 265.
107. Deutzia taiwanensis, I. 18.
183. Meliosma Oldhamii, II. 206.
187. Rhus sylvestris, II. 180.
190. Prunus glandulosa, var. glabra, subf. rosea, I. 263.
200. Prunus japonica, var. eujaponica, f. Oldhamii, I. 266.
202. Sambucus javanica, I. 307.
293. Styrax suberifolius, I. 290.
329. Trachelospermum divaricatum, var. brevisepalum, III. 338.
347. Ehretia macrophylla, III. 364.
348. Ehretia acuminata, III. 363.
436. Polygonum multiflorum, III. 325.
508. Alnus formosana, II. 501.
509. Salix glandulosa, var. Warburgii, III. 99.
512. Celtis sinensis, III. 277.
513. Celtis sinensis, III. 277.

530. Morus acidosa, III. 299.
551. Salix Harmsiana, III. 163.
562. Trachelospermum divaricatum, III. 338.
628. Callicarpa dichotoma, III. 370.
720. Salix japonica, var. Oldhamiana, III. 132.
721. Alnus japonica, var. arguta, II. 494.
722. Celtis sinensis, III. 277.
723. Aphananthe aspera, III. 290.
724. Aphananthe aspera, III. 290.
746. Corylus heterophylla, II. 450.
778. Morus alba, III. 294.
881. Liquidambar formosana, I. 421.

Parry, C. C., & **Palmer,** E.
299. Abelia coriacea, I. 127.

Perrottet, George Samuel.
454. Morus alba, III. 294.

Playfair, George Macdonald.
52. Pittosporum formosanum, III. 330.
297. Melia Azedarach, II. 157.
458. Celtis nervosa, III. 280.

Prain, Sir Daniel.
8. Porana spectabilis, III. 359.
61. Sorbus Griffithii, II. 277.
155. Trachelospermum lucidum, III. 339.
274. Cudrania javanensis, III. 308.
508. Rhamnus nipalensis, II. 245.
810. Trachelospermum lucidum, III. 339.

Pratt, A. E.
3. Aristolochia moupinensis, III. 324.
5. Corylus tibetica, II. 443.
8. Enkianthus deflexus, I. 550.
58. Rhododendron Faberi, I. 533.
78. Clematis grata, var. grandidentata, I. 338.
93. Malus Prattii, II. 281.
99. Helwingia japonica, II. 570.
108. Polygala arillata, II. 160.
116. Rosa Prattii, II. 329.
122. Sambucus adnata, I. 308.
125. Clematis montana, var. grandiflora, I. 333.
136. Abelia tereticalyx, I. 127.
189. Pieris villosa, I. 553.
234. Sorbus Prattii, f. striata, I. 468.
236. Betula utilis, var. Prattii, II. 457.
237. Clematis tangutica, I. 343.

PRATT, A. E. (*continued*)
271. Abelia Schumannii, I. 121.
278. Rosa Brunonii, II. 306.
285. Hydrangea xanthoneura, var. Wilsonii, I. 27.
292. Hypericum Hookerianum, II. 403.
335. Salix moupinensis, III. 46.
424. Alnus cremastogyne, II. 488.
475. Pieris villosa, I. 553.
592. Clematis Rehderiana, I. 324; III. 434.
677. Deutzia longifolia, I. 13.
751. Salix myrtillacea, III. 71.
799. Actinodaphne cupularis, II. 75.
806. Litsea populifolia, II. 77.
809. Lindera Prattii, II. 83.
824. Malus Prattii, II. 281.
833. Gaultheria trichophylla, I. 556.
835. Salix myrtillacea, III. 71.
848. Rhododendron ambiguum, I. 518.

Pringle, Cyrus Guernsey.
2546. Abelia coriacea, I. 127.
4649. Abelia floribunda, I. 127.

Przewalski, Nicolai Mikhailovich.
19. Prunus stipulacea, I. 258.
187. Prunus stipulacea, I. 258.
338. Prunus stipulacea, I. 258.

Purdom, William.
1. Aralia chinensis, var. glabrescens, II. 567.
1. Berberis dasystachya, III. 442.
1. Clematis grata, var. grandidentata, I. 338.
1. Exochorda Giraldii, I. 457.
1. Neillia sinensis, I. 436.
1. Rhododendron yanthinum, I. 518.
1. Spiraea Fritschiana, var. angulata, I. 453.
2. Acanthopanax setchuenensis, II. 559.
2. Berberis Poiretii, f. weichangensis, I. 372.
2. Populus suaveolens, III. 18.
2. Prunus dictyoneura, I. 262.
2. Rhododendron micranthum, I. 513.
2. Spiraea gemmata, I. 441.
3. Acanthopanax Giraldii, II. 560.
3. Berberis Purdomii, I. 372.
3. Prunus tomentosa, I. 268.
3. Spiraea Fritschiana, var. angulata, I. 453.

4. Acanthopanax stenophyllus, II. **764.**
4. Berberis circumserrata, I. 354; III. 435.
4. Rhododendron Purdomii, I. 538.
4. Sorbus Koehneana, I. 471.
4. Spiraea hirsuta, I. 444.
5. Berberis dasystachya, III. 442.
5. Spiraea alpina, I. 440.
6. Berberis Gilgiana, I. 375.
6. Evonymus cornuta, I. 489.
7. Berberis Soulieana, III. 437.
7. Spiraea Miyabei, var. glabrata, I. 454.
8. Berberis Gilgiana, I. 375.
8. Evonymus yedoensis, var. Koehneana, I. 491.
9. Berberis dasystachya, III. 442.
9. Spiraea hirsuta, I. 444.
11. Populus tremula, var. **Davidiana,** III. 24.
12. Evonymus microcarpa, I. 487.
12. Prunus glandulosa, var. Purdomii, I. 264.
15. Prunus Padus, var. pubescens, f. Purdomii, I. 196.
16. Deutzia grandiflora, I. 21.
21. Larix dahurica, var. Principis Rupprechtii, II. 21.
29. Rhamnus davuricus, II. 251.
30. Evonymus alata, var. pilosa, I. 494.
32. Rhamnus argutus, II. 250.
34. Berberis amurensis, I. 375.
35. Berberis Poiretii, f. weichangensis, I. 372.
40. Deutzia parviflora, I. 23.
42. Quercus mongolica, III. 230.
43. Quercus mongolica, III. 230.
49. Rosa davurica, II. 340.
50. Rhododendron micranthum, I. 513.
53, 53a. Tilia mandshurica, II. 370.
55. Rhododendron micranthum, I. 513.
61. Ulmus pumila, III. 242.
63. Betula davurica, II. 483.
67. Tilia mongolica, II. 369.
69. Ostryopsis Davidiana, II. 423.
71. Acanthopanax Giraldii, II. 560.
82. Lonicera Tatarinovii, I. 144.
85. Betula chinensis, II. 479.
86. Tilia mongolica, II. 369.
87. Ulmus laciniata, III. 255.
92. Corylus heterophylla, II. 450.
95. Quercus mongolica, III. 230.
96. Ulmus pumila, III. 242.

PURDOM, W. (continued)
97. Quercus mongolica, III. 230.
100. Picea Schrenkiana, II. 29.
103. Betula japonica, II. 485.
104. Betula fruticosa, II. 482.
105. Quercus mongolica, III. 230.
107. Rosa davurica, II. 340.
109. Betula davurica, II. 483.
114. Quercus mongolica, III. 230.
124. Betula japonica, II. 485.
127. Potentilla fruticosa, II. 301.
143. Abies sibirica, var. nephrolepis, II. 49.
144. Picea Meyeri, II. 28.
145. Picea Wilsonii, II. 27.
161, 161a, b. Larix dahurica, var. Principis Rupprechtii, II. 21.
182. Berberis circumserrata, III. 435.
189. Hippophae rhamnoides, II. 409.
202. Picea Schrenkiana, II. 29.
203. Picea Schrenkiana, II. 29.
204. Larix dahurica, var. Principis Rupprechtii, II. 21.
205. Picea Schrenkiana, II. 29.
246. Larix dahurica, var. Principis Rupprechtii, II. 21.
261. Ulmus Davidiana, III. 261.
262. Ulmus Davidiana, III. 261.
273. Quercus dentata, III. 210.
274. Quercus dentata, III. 210.
281. Salix Matsudana, III. 107.
296. Hemiptelea Davidii, III. 288.
297. Abelia biflora, I. 128.
298. Rhamnus parvifolius, II. 250.
299. Rhamnus parvifolius, II. 250.
314. Rosa bella, II. 341.
316. Staphylea holocarpa, II. 185.
316ˣˣˣ. Staphylea holocarpa, III. 449.
320. Abelia Zanderi, I. 121.
321. Melia Azedarach, II. 157.
324. Prinsepia uniflora, II. 345.
329. { Malus baccata, II. 291.
{ Malus theifera, II. 283.
341. Berberis Dielsiana, III. 441.
342. Evonymus alata, I. 493.
344. Prunus dictyoneura, I. 262.
345. Berberis Purdomii, I. 372.
346. Spiraea hypericifolia, var. hupehensis, I. 439.
347. Prunus Davidiana, I. 275.
350. Acanthopanax spinosus, II. 562.
355. Thuja orientalis, II. 53.
360. Cotoneaster Zabelii, I. 166.

361. Exochorda Giraldii, I. 457; III. 444.
362. Picrasma quassioides, II. 152.
364. Cotinus coggygria, var. pubescens, II. 175.
365. Staphylea holocarpa, II. 185.
367. Cotoneaster acutifolia, var. villosula, I. 158.
375. Evonymus alata, I. 493.
381. Clematis grata, var. grandidentata, I. 338.
382. Clematis obscura, I. 329.
396, 396a, 396ˣˣˣ. Potentilla fruticosa, var. Veitchii, II. 303.
400. Betula albo-sinensis, II. 457.
404. Larix Potaninii, II. 18.
405. Abies sutchuenensis, II. 48.
427. Spiraea Miyabei, var. glabrata, I. 454.
432. Evonymus cornuta, I. 489.
433. Sorbus Koehneana, I. 471.
440. Rhododendron fastigiatum, I. 507.
446. Staphylea holocarpa, II. 185.
467. Neillia sinensis, I. 436.
501. Decaisnea Fargesii, I. 344.
536. Rosa multiflora, var. cathayensis, II. 304.
539a. Indigofera Potaninii, III. 448.
540. Clematis Fargesii, var. Souliei, I. 336.
543. Berberis Dielsiana, III. 441.
549. Berberis Dielsiana, III. 441.
583. Syringa microphylla, III. 433.
604, 604a. Berberis circumserrata, III. 435.
605a. Berberis Dielsiana, III. 441.
608. Berberis circumserrata, III. 435.
668. Tsuga chinensis, II. 37.
669. Tetracentron sinense, I. 417.
670. Tetracentron sinense, I. 417.
673. Picea Watsoniana, II. 27.
681. Caragana Maximowicziana, II. 622.
752. Betula albo-sinensis, var. septentrionalis, II. 458.
760. Larix Potaninii, II. 458.
772. Wikstroemia chamaedaphne, II. 536.
787. Carpinus Turczaninovii, var. ovalifolia, II. 427.
790. { Picea Meyeri, II. 28.
{ Picea Schrenkiana, II. 29.

Purdom, W. (*continued*)
791. Sibiraea laevigata, var. angustata, I. 455.
792. Caryopteris incana, III. 378.
804. Leptodermis Purdomii, III. 405.
804ˣˣˣ. Plectranthus discolor, III. 384.
805. Abies sutchuenensis, II. 48.
806. Picea Schrenkiana, II. 29.
812. Betula Delavayi, II. 460.
812. Pinus sinensis, II. 15.
813. Picea Meyeri, II. 28.
813. Picea Schrenkiana, II. 29.
813. Pinus sinensis, II. 15.
814. Pinus sinensis, II. 15.
815. Picea purpurea, II. 29.
817. Larix Potaninii, II. 18.
819. Potentilla fruticosa, var. Veitchii, II. 303.
821, 821ˣˣˣ. Potentilla fruticosa, var. parvifolia, II. 304.
823. Abies sutchuenensis, II. 48.
826. Berberis parvifolia, III. 436.
864. Tilia mongolica, II. 369.
874. Aesculus chinensis, I. 499.

Rosthorn, A. von.
11. Pilostegia viburnoides, I. 151.
32. Corylopsis sinensis, I. 424.
34. Abelia Engleriana, I. 120.
51. Hydrangea anomala, I. 34.
75. Hydrangea opuloides, var. Hortensia, I. 37.
135. Symplocos stellaris, II. 597.
144. Mahonia Fortunei, I. 380.
147. Gordonia axillaris, II. 394.
149. Prunus pleiocerasus, I. 198.
152. Lagerstroemia indica, II. 418.
158. Prunus ampla, I. 243.
163. Jasminum floridum, II. 614.
164. Loropetalum chinense, I. 430.
168. Quercus variabilis, III. 219.
174. Mahonia japonica, I. 382.
211. Photinia parvifolia, I. 189.
278. Toddalia asiatica, II. 137.
291. Elaeagnus Henryi, II. 414.
293. Callicarpa Giraldiana, var. Rosthornii, III. 367.
294. Carpinus Seemeniana, II. 430.
299. Hydrangea strigosa, I. 31.
301. Celtis Biondii, III. 272.
303. Cotoneaster horizontalis, var. perpusilla, I. 155.
354. Hydrangea xanthoneura, I. 26.

390. Callicarpa rubella, var. Hemsleyana, III. 370.
409. Parthenocissus himalayana, III. 428.
430. Rhododendron stamineum, I. 546.
438. Buddleia Davidii, I. 567.
454. Picrasma quassioides, II. 152.
456. Aucuba chinensis, II. 572.
459. Abelia Schumannii, I. 121.
460. Alangium platanifolium, II. 554.
471. Hydrangea Rosthornii, I. 33.
510. Paederia tomentosa, III. 403.
511. Melodinus Hemsleyanus, III. 331.
512. Rhamnus Rosthornii, II. 236.
540. Prunus pleiocerasus, I. 198.
602. Hydrangea opuloides, var. Hortensia, I. 37.
621. Quercus phillyraeoides, III. 233.
622. Prunus pleiocerasus, I. 198.
628. Osmanthus fragrans, II. 609.
629. Hydrangea strigosa, var. sinica, I. 32.
674. Pittosporum glabratum, III. 326.
790. Celastrus angulata, II. 346.
798. Symplocos anomala, II. 596.
804. Celtis Bungeana, III. 269.
806. Ligustrum lucidum, II. 603.
842. Tilia tuan, II. 368.
855. Osmanthus fragrans, II. 609.
953. Hydrangea strigosa, var. macrophylla, I. 32.
966. Corylopsis sinensis, var. glandulifera, I. 424.
1052. Ligustrum Delavayanum, II. 601.
1053. Enkianthus chinensis, I. 551.
1104. Elaeagnus lanceolata, II. 413.
1125. Cornus controversa, II. 573.
1213. Mahonia japonica, I. 382.
1251. Mahonia eurybracteata, I. 384.
1509. Salix Bockii, III. 71.
1511. Salix dictyoneura, III. 98.
1525. Fagus Engleriana, III. 191.
1533. Betula ?luminifera, II. 455.
1534. Betula luminifera, II. 455.
1542. Torricellia angulata, II. 569.
1557. Ampelopsis micans, var. cinerea, I. 579.
1585. Rhamnus hypochrysus, II. 252.
1652. Lespedeza Buergeri, II. 106.
1672. Helwingia chinensis, II. 571.
1684. Helwingia chinensis, II. 571.
1685. Helwingia japonica, II. 570.
1689. Alangium chinense, II. 552.

Rosthorn, A. von (*continued*)
1697. Photinia serrulata, I. 184.
1698. Helwingia chinensis, II. 571.
1805. Cotoneaster acutifolia, var. villosula, I. 158.
1806. Cotoneaster acutifolia, var. laetevirens, I. 159.
1820. Symplocos paniculata, II. 593.
1824. Actinidia callosa, var. Henryi, II. 382.
1837. Spiraea Rosthornii, I. 451.
1839. Prunus tomentosa, var. endotricha, I. 225.
1842. Prunus tomentosa, var. endotricha, I. 225.
1843. Abelia Koehneana, I. 126.
1884. Evodia hupehensis, II. 133.
1896. Callicarpa Giraldiana, III. 366.
1898. Hovenia dulcis, II. 252.
1931. Hydrangea longipes, I. 33.
1932. Hydrangea mandarinorum, I. 39.
1933. Pittosporum glabratum, III. 326.
1969. Rhus punjabensis, var. sinica, II. 176.
1971. Rhus punjabensis, var. sinica, II. 176.
1997. Actinidia chinensis, II. 385.
2000. Stachyurus chinensis, I. 287.
2006. Daphne gracilis, II. 548.
2023. Clematis grata, var. grandidentata, I. 338.
2040. Osmanthus armatus, II. 611.
2043. Mahonia polyodonta, I. 383.
2071. Pittosporum truncatum, III. 328.
2074. Pittosporum glabratum, III. 326.
2075. Pterostyrax hispidus, I. 295.
2078. Styrax Hemsleyanus, I. 291.
2080. Enkianthus chinensis, I. 551.
2092. Diospyros Lotus, II. 587.
2099. Ligustrum myrianthum, II. 607.
2138a. Pieris ovalifolia, var. elliptica, I. 552.
2141. Sageretia Cavaleriei, III. 449.
2142. Sageretia Cavaleriei, III. 449.
2143. Sageretia Cavaleriei, III. 449.
2145. Rhododendron micranthum, I. 513 (by error as No. 2545).
2146. Rhododendron micranthum, I. 513.
2147. Rhododendron micranthum, I. 513.
2148. Rhododendron indicum, var. ignescens, I. 547.

2156. Rhododendron discolor, I. 542.
2162. Rhododendron micranthum, I. 513.
2165. Eurya japonica, var. aurescens, II. 399.
2167. Symplocos anomala, II. 596.
2169. Myrsine africana, II. 580.
2250. Andrachne capillipes, II. 516.
2253. Andrachne capillipes, II. 516.
2273. Paederia tomentosa, III. 403.
2283. Rhus verniciflua, II. 181.
2293. Rhamnus rugulosus, II. 238.
2300. Rhamnus rugulosus, II. 238.
2345. Trachelospermum cathayanum, III. 333.
2350. Dregea corrugata, III. 354.
2351. Dregea corrugata, III. 354.
2364. Ehretia macrophylla, III. 364.
2367. Premna puberula, III. 371.
2368. Premna puberula, III. 371.
2415. Hibiscus syriacus, II. 374.
2420. Prunus malifolia, var. Rosthornii, I. 243.
2423. Styrax japonicus, I. 291.
2443. Loranthus yadoriki, var. hupehanus, III. 315.
2517. Euptelea pleiosperma, I. 313.
2519a. Caryopteris incana, III. 378.
2523. Koelreuteria apiculata, II. 191.
2527. Berberis polyantha, I. 376.
2533. Quercus Baronii, III. 226.
2538. Clematis grata, var. grandidentata, I. 338.
2545. Daphne gemmata, II. 543.
2546. Hydrangea villosa, var. strigosior, I. 39.
2547. Corylus heterophylla, var. sutchuenensis, II. 445.
2549. Cotoneaster horizontalis, I. 154.
2554. Rhus Potaninii, II. 177.
2555. Spiraea Sargentiana, I. 447.
2556. Rhododendron micranthum, I. 513.
2573. Acanthopanax setchuenensis, II. 559.
3001. Ceratostigma minus, II. 586.
3004. Elsholtzia fruticosa, III. 382.
3007. Buddleia Davidii, I. 567.
3112. Hypericum patulum, var. Henryi, II. 403.
3113. Wikstroemia stenophylla, II. 530.
3144. Elaeagnus Bockii, II. 416.

Savatier, Louis.
1139. Salix amygdalina, var. nipponica, III. 106.
1145. Salix purpurea, subsp. amplexicaulis, var. multinervis, III. 168.
1152. Alnus firma, subsp. hirtella, II. 506.
1172. Carpinus Tschonoskii, II. 441.
2717. Salix amygdalina, III. 106.
2718. Salix eriocarpa, III. 108.

Scallan, Hugh.
2. Leptodermis pilosa, III. 405.
49. Leptodermis pilosa, III. 405.
50. Leptodermis pilosa, III. 405.
75. Evodia hupehensis, II. 133.
119. Cudrania tricuspidata, III. 306.
2334. Alnus cremastogyne, II. 488.
2335. Alnus cremastogyne, II. 488.
5361. Salix Biondiana, III. 118.

Schlagintweit, Robert von.
209. Rosa longicuspis, II. 313.
258. Sorbus verrucosa, II. 278.
1162. Populus szechuanica, var. tibetica, III. 33.
1358. Morus serrata, III. 302.
1563. Populus szechuanica, var. tibetica, III. 33.
4970. Salix alba, III. 112.
4976. Populus szechuanica, var. tibetica, III. 33.
5261. Salix oxycarpa, III. 171.
5327. Populus szechuanica, var. tibetica, III. 33.
5329. Salix hastata, var. himalayensis, III. 134.
5458. Salix dealbata, III. 105.
6096. Salix hastata, var. himalayensis, III. 134.
6531. Salix pycnostachya, var. alpina, III. 171.
6568. Betula Jacquemontii, II. 473.
6592. Salix hastata, var. himalayensis, III. 134.
6744. Salix oxycarpa, III. 171.
9372. Alnus nepalensis, II. 502.
9626. Salix Lindleyana, III. 145.
11859. Porana paniculata, III. 359.
12319. Ulmus lanceaefolia, III. 263.

Schneider, Camillo.
1113. Ulmus Wilsoniana, var. subhirsuta, III. 257.

1139. Vitex Negundo, III. 372.
1148. Osyris Wightiana, III. 320.
1320. Betula japonica, var. szechuanica, III. 453.
1462. Acer laxiflorum, III. 427.
1499. Acer laxiflorum, III. 427.
1667. Celtis ?labilis, III. 269.
1717. Clerodendron trichotomum, var. Fargesii, III. 376.
1730. Dregea sinensis, III. 352.
1732. Secamone emetica, III. 348.
2173. Schoepfia jasminodora, III. 321.
2196. Vitex Negundo, III. 372.
2222. Lycium chinense, III. 385.
2255. Elsholtzia fruticosa, III. 382.
2482. Rumex hastatus, III. 325.
2573. Clerodendron foetidum, III. 375.
2597. Osyris Wightiana, III. 320.
2630. Colquhounia coccinea, III. 381.
2709. Celtis ?yunnanensis, III. 279.
2860. Berberis yunnanensis, III. 435.
2880. Rumex hastatus, III. 325.
3161. Callicarpa rubella, III. 370.
3272. Clerodendron foetidum, III. 375.
3281. Acer laxiflorum, III. 427.
3338. Acer laxiflorum, III. 427.
3343. Berberis circumserrata, III. 435.
3886. Colquhounia coccinea, III. 381.
3907. Elsholtzia flava, III. 382.

Schrenk, Alexander Gustav.
118. Populus suaveolens, III. 18.

Seguin, J.
2390. Melodinus Seguinii, III. 331.

Shirai, Mitsutarō.
33. Salix Urbaniana, III. 103.
42. Salix Shiraii, III. 135.
43. Salix Shiraii, III. 135.

Silvestri, Cipriano.
94. Cephalotaxus Fortunei, II. 4.
100. Pinus Massoniana, II. 14.
104. Cunninghamia lanceolata, II. 50.
105. Cryptomeria japonica, II. 52.
106. Cryptomeria japonica, II. 52.
107. Thuja orientalis, II. 53.
299. Salix Wallichiana, III. 64.
309. Platycarya strobilacea, III. 180.
311. Platycarya strobilacea, III. 180.
316. Pterocarya stenoptera, III. 181.
319. Carpinus cordata, II. 435.
324. Corylus heterophylla, var. sutchuenensis, II. 445.

SILVESTRI, C. (*continued*)

328. Corylus heterophylla, var. sutchuenensis, II. 445.
329. Corylus heterophylla, var. sutchuenensis, II. 445.
331. Castanea Henryi, III. 196.
339. Quercus aliena, var. acuteserrata,
f. calvescens, III. 215.
344. Quercus dentata, III. 210.
348. Quercus glandulifera, III. 212.
350. Quercus glandulifera, III. 212.
357. Celtis Biondii, III. 272.
358, 358a. Celtis Biondii, III. 272.
359. Celtis Bungeana, III. 269.
360. Ulmus Wilsoniana, var. ?psilo-
phylla, III. 240.
629. Clematis Armandi, I. 326.
630. Clematis grata, var. grandidentata, I. 338.
631. Clematis montana, var. Wilsonii,
I. 333.
632. Clematis heracleaefolia, var.
ichangensis, I. 321.
633. Clematis heracleaefolia, var.
ichangensis, I. 321.
637. Clematis lasiandra, I. 322.
671. Deutzia Silvestrii, I. 19.
672. Clematis uncinata, I. 327.
868. Deutzia Silvestrii, I. 19.
869. Deutzia Silvestrii, I. 19.
870. Deutzia Silvestrii, I. 19.
872. Deutzia Silvestrii, I. 19.
874. Pittosporum glabratum, III. 326.
908. Chaenomeles sinensis, II. 299.
928. Kerria japonica, II. 301.
974. Prunus glabra, I. 241.
986. Rosa Roxburghii, f. normalis, II.
319.
1026. Spiraea Blumei, I. 446.
1043. Stephanandra chinensis, I. 437.
1218. Evodia glauca, II. 129.
1239. Picrasma quassioides, II. 152.
1302. Mallotus tenuifolius, II. 525.
1316. Coriaria sinica, II. 170.
1318. Pistacia chinensis, II. 173.
1321. Rhus javanica, II. 178.
1323. Rhus javanica, II. 178.
1336. Celastrus angulata, II. 346.
1368. Staphylea Bumalda, II. 185.
1388. Aesculus Wilsonii, I. 498.
1389. Aesculus Wilsonii, I. 498.
1449. Grewia parviflora, var. glabrescens, II. 371.

1451. Tilia Henryana, II. 367.
1452. Tilia Henryana, II. 367.
1467. Actinidia chinensis, II. 385.
1534. Daphne genkwa, II. 538.
1540. Wikstroemia Pampaninii, II. 537.
1541. Wikstroemia Pampaninii, II.
537.
1548. Elaeagnus umbellata, II. 410.
1564. Punica Granatum, II. 419.
1601. Kalopanax ricinifolius, II. 564.
1609. Aralia chinensis, var. glabrescens, II. 567.
1613. Hedera himalaica, II. 555.
1671. Alangium chinense, II. 552.
1672. Alangium chinense, II. 552.
1679. Cornus controversa, II. 577.
1684. Cornus kousa, II. 577.
1701. Rhododendron indicum, var.
ignescens, I. 547.
1703. Rhododendron micranthum, I.
513.
1706. Myrsine africana, II. 580.
1762. Symplocos paniculata, II. 593.
1763. Symplocos paniculata, II. 593.
1794. Ligustrum Quihoui, II. 607.
1802. Ligustrum lucidum, II. 603.
1869. Trachelospermum jasminoides,
III. 334.
1919. Ehretia acuminata, III. 363.
1923. Ehretia macrophylla, III. 364.
1925. Ehretia macrophylla, III. 364.
1948. Callicarpa Giraldiana, III. 366.
1950. Callicarpa Giraldiana, III. 366.
1952. Callicarpa Giraldiana, III. 366.
1953. Callicarpa Giraldiana, III. 366.
1954. Callicarpa japonica, var. angustata, III. 369.
1964. Clerodendron trichotomum, var.
Fargesii, III. 376.
2923. Platycarya strobilacea, III. 180.
2925. Carpinus cordata, II. 435.
2931. Quercus aliena, var. acuteserrata,
III. 215.
2939. Celtis Bungeana, III. 269.
2960. Euptelea Franchetii, I. 314.
3001, 3001a. Deutzia sessilifolia, I.
150.
3025. Prunus glabra, I. 241.
3028, 3028a. Prunus Sprengeri, I. 252.
3073. Polygala Wattersii, II. 161.
3097. Aesculus Wilsonii, I. 498.
3133. Myrsine africana, II. 580.
3158. Callicarpa Giraldiana, III. 366.

Silvestri, C. (*continued*)
3159. Clerodendron trichotomum, var.
Fargesii, III. 376 (by error as
No. 3158).
3331. Loropetalum chinense, I. 430.
3332. Loropetalum chinense, I. 430.
3339. Evodia officinalis, II. 130.
3348. Sarcococca ruscifolia, II. 163.
3349. Sarcococca ruscifolia, II. 163.
3353. Rhus verniciflua, II. 181.
3355. Meliosma Oldhamii, II. 206.

Singh, Sulakhan.
98. Betula Jacquemontii, II. 473.

Siuzev, P. V.
34. Salix Siuzevii, III. 160.

Soulié, Jean André.
293. Juniperus squamata, II. 57.
509. Salix myrtillacea, III. 71.
564. Juniperus formosana, II. 56.
566. Betula japonica, var. szechuanica,
II. 461; III. 453.
2289. Salix Souliei, III. 62.

Strachey, Sir Richard, & Winterbottom, J. E.
1. Betula Jacquemontii, II. 473.
1. Populus ciliata, II. 31.
1. Porana paniculata, III. 359.
1. Trachelospermum lucidum, III.
339.
2. Betula Jacquemontii, II. 473.
2. Carpinus faginea, II. 442.
2. Ulmus spec., III. 264.
3. Betula cylindrostachya, II. 466.
3. Salix Wallichiana, III. 64.
3. Ulmus Brandisiana, III. 252.
4. Betula cylindrostachya, II. 466.
5. Salix Wallichiana, III. 64.
6. Salix denticulata, III. 117.
7. Salix denticulata, III. 117.
8. Salix sclerophylla, III. 112.
9. Salix denticulata, III. 117.
10. Salix sclerophylla, III. 112.
11. Salix Wallichiana, III. 64.
12. Salix Lindleyana, III. 145.
13. Salix hylematica, III. 119, 455.

Taquet, *Père*.
4. Sapindus mukorossi, II. 191.
77. Lespedeza bicolor, II. 112.
79. Lespedeza sericea, II. 105.
82. Lespedeza formosa, II. 107.

279. Trachelospermum divaricatum,
III. 338.
284. Symplocos caudata, II. 595.
326. Quercus mongolica, III. 230.
329. Castanopsis cuspidata, III. 204.
333. Corylus hallaisanensis, II. 451.
344. Celtis sinensis, III. 277.
346. Aphananthe aspera, III. 290.
542. Cocculus trilobus, I. 388.
543. Cocculus trilobus, I. 388.
567. Xylosma racemosum, I. 283.
587. Carpinus Fauriei, II. 442.
591. Eurya ochnacea, II. 399.
595. Hibiscus syriacus, II. 374.
598. Melia Azedarach, II. 157.
618. Orixa japonica, II. 135.
620. Zanthoxylum alatum, var. planispinum, II. 125.
621. Zanthoxylum Bungei, var. Zimmermannii, II. 122.
633. Celastrus articulata, II. 356.
643. Sageretia theezans, var. tomentosa, II. 228.
660. Sapindus mukorossi, II. 191.
664. Euscaphis japonica, II. 187.
666. Rhus javanica, II. 178.
667. Rhus sylvestris, II. 180.
693. Lespedeza formosa, II. 107.
695. Lespedeza bicolor, II. 112.
696. Lespedeza bicolor, II. 112.
748. Sorbus alnifolia, var. lobulata, II.
275.
758. Sorbus alnifolia, var. lobulata, II.
275.
770. Rosa multiflora, var. quelpaertensis, II. 335.
770. Rosa Wichuraiana, II. 335.
772. Rosa Wichuraiana, II. 335.
778. Rosa multiflora, var. quelpaertensis, II. 335.
786. Prunus Herincquiana, I. 214.
790. Prunus tomentosa, var. insularis,
I. 269.
807. Schizophragma hydrangeoides, I.
43.
808. Schizophragma hydrangeoides, I.
43.
809. Hydrangea petiolaris, I. 41.
823. Lagerstroemia indica, II. 418.
912. Callicarpa japonica, var. luxurians, III. 369.
1083. Vaccinium bracteatum, I. 558.
1084. Vaccinium bracteatum, I. 558.

TAQUET, *Père* (*continued*)
1107. Symplocos caudata, II. 595.
1109. Styrax japonicus, I. 291.
1122. Trachelospermum divaricatum, III. 338.
1210. Rhamnus Taquetii, II. 248.
1334. Ulmus parvifolia, III. 244.
1376. Celtis jessoensis, III. 281.
1387. Aphananthe aspera, III. 290.
1392. Morus acidosa, III. 300.
1393. Morus acidosa, III. 300.
1394. Morus acidosa, III. 300.
1396. Morus acidosa, III. 300.
1402. Myrica rubra, III. 189.
1427. Quercus glauca, III. 226.
1428. Castanopsis cuspidata, III. 204.
1429. Castanopsis cuspidata, III. 204.
1432. Quercus glauca, III. 226.
1434. Quercus dentata, III. 210.
1435. Quercus serrata, III. 217.
1436. Quercus aliena, var. acuteserrata, f. calvescens, III. 216.
1438. Castanopsis cuspidata, III. 204.
1439. Betula Ermanii, var. genuina, subvar. Saitôana, II. 470.
1440. Betula Ermanii, var. genuina, subvar. Saitôana, II. 470.
1442. Salix ?Caprea, III. 149.
1443. Salix ?Caprea, III. 149.
1444. Salix ?Caprea, III. 149.
1466. Idesia polycarpa, I. 284.
1509. Cudrania tricuspidata, III. 306.
1510. Cudrania tricuspidata, III. 306.
1511. Cudrania tricuspidata, III. 306.
1512. Cudrania tricuspidata, III. 306.
1513. Cudrania tricuspidata, III. 306.
1514. Staphylea Bumalda, II. 185.
1515. Chionanthus retusus, II. 611.
1598. Smilax herbacea, var. nipponica, III. 10.
1601. Smilax herbacea, var. nipponica, III. 10.
1602. Smilax Sieboldii, III. 10.
1603. Smilax Sieboldii, III. 10.
2433. Quercus glandulifera, III. 212.
2440. Carpinus Tschonoskii, II. 441.
2517. Prunus Leveilleana, I. 250.
2519. Prunus Leveilleana, I. 250.
2525. Malus baccata, var. mandshurica, II. 281.
2526. Prunus tomentosa, var. insularis, I. 269.
2536. Hemiptelea Davidii, III. 288.

2537. Zelkova serrata, III. 288.
2542. Celtis Biondii, var. heterophylla, III. 282.
2544. Quercus serrata, III. 217.
2546. Quercus aliena, var. acuteserrata, f. calvescens, III. 216.
2547. Quercus mongolica, III. 230.
2548. Quercus serrata, III. 217.
2549. Quercus dentata, III. 210.
2550. Quercus mongolica, III. 230.
2551. Quercus glandulifera, III. 212.
2554. Castanopsis cuspidata, III. 204.
2556. Castanopsis cuspidata, III. 204.
2557. Castanopsis cuspidata, III. 204.
2558. Quercus salicina, III. 234.
2559. Quercus salicina, III. 234.
2560. Castanopsis cuspidata, III. 204.
2561. Quercus salicina, III. 234.
2562. Castanopsis cuspidata, III. 204.
2563. Castanopsis cuspidata, III. 204.
2597. Cocculus trilobus, I. 388.
2598. Cocculus trilobus, I. 388.
2599. Sinomenium acutum, I. 387.
2600. Cocculus trilobus, I. 388.
2601. Akebia quinata, I. 347.
2692. Eurya ochnacea, II. 399.
2693. Eurya ochnacea, II. 399.
2700. Melia Azedarach, II. 157.
2707. Zanthoxylum alatum, var. planispinum, II. 125.
2708. Zanthoxylum alatum, var. planispinum, II. 125.
2709. Zanthoxylum Bungei, var. Zimmermannii, II. 122.
2715. Poncirus trifoliata, II. 149.
2722. Celastrus articulata, II. 356.
2746. Sapindus mukorossi, II. 191.
2759. Rhus javanica, II. 178.
2824. Sorbus alnifolia, var. lobulata, II. 275.
2825. Sorbus alnifolia, var. lobulata, II. 275.
2828. Malus Sieboldii, var. arborescens, II. 294.
2864. Rosa Wichuraiana, II. 335.
2865. Rosa Wichuraiana, II. 335.
2866. Rosa Wichuraiana, II. 335.
2867. Rosa Wichuraiana, II. 335.
2870. Rosa multiflora, var. quelpaertensis, II. 335.
2871. Rosa multiflora, var. quelpaertensis, II. 335.
2875. Prunus Herincquiana, I. 214.

TAQUET, *Père* (*continued*)
2876. Prunus Herincquiana, I. 214.
2884. Hydrangea petiolaris, I. 41.
2885. Schizophragma hydrangeoides, I. 43.
2896. Lagerstroemia indica, II. 418.
2931. Ternstroemia japonica, II. 397.
2932. Euscaphis japonica, II. 187.
2933. Diospyros kaki, var. silvestris, II. 590.
2934. Staphylea Bumalda, II. 185.
2978. Diospyros kaki, var. silvestris, II. 590.
2980. Diospyros kaki, var. silvestris, II. 590.
3033. Styrax japonicus, I. 291.
3034. Styrax japonicus, I. 291.
3055. Chionanthus retusus, II. 611.
3081. Rhamnus Taquetii, II. 248.
3083. Callicarpa japonica, var. luxurians, III. 369.
3112. Orixa japonica, II. 135.
3116. Paliurus ramosissimus, II. 210.
3120. Rhus sylvestris, II. 180.
3211. Celtis sinensis, III. 277.
3212. Celtis sinensis, III. 277.
3213. Celtis Biondii, var. heterophylla, III. 282.
3214. Rhamnella franguloides, II. 222.
3215. Zelkova serrata, III. 288.
3216. Morus acidosa, III. 300.
3217. Morus acidosa, III. 300.
3218. Morus acidosa, III. 300.
3220. Broussonetia kazinoki, III. 306.
3221. Broussonetia Kaempferi, III. 305.
3237. Carpinus laxiflora, II. 438.
3238. Corylus hallaisanensis, II. 451.
3239. Corylus hallaisanensis, II. 451.
3251. Salix ?Caprea, III. 149.
3252. Salix ?Caprea, III. 149.
3253. Salix ?Caprea, III. 149.
3255. Salix ?Caprea, III. 149.
3256. Salix ?Caprea, III. 149.
3257. Salix ?Caprea, III. 149.
3258. Salix ?Caprea, III. 149.
3259. Salix ?Caprea, III. 149.
3260. Salix ?Caprea, III. 149.
3267. Zelkova serrata, III. 288.
3268. Ulmus parvifolia, III. 244.
3303. Smilax Sieboldii, III. 10.
3304. Smilax Sieboldii, III. 10.
3305. Smilax Sieboldii, III. 10.
3306. Smilax china, III. 4.

3307. Smilax china, III. 4.
3308. Smilax china, III. 4.
3335. Carpinus cordata, II. 434.
3336. Carpinus cordata, II. 434.
4059. Smilax Oldhamii, III. 9.
4060. Smilax Oldhamii, III. 9.
4061. Smilax Sieboldii, III. 10.
4090. Picrasma quassioides, II. 152.
4092. Callicarpa japonica, var. luxurians, III. 369.
4096. Idesia polycarpa, I. 284.
4097. Callicarpa japonica, var. luxurians, III. 369.
4136. Eurya ochnacea, II. 399.
4149. Orixa japonica, II. 135.
4172. Euscaphis japonica, II. 187.
4173. Rhus sylvestris, II. 180.
4219. Malus Sieboldii, var. arborescens, II. 294.
4220. Malus Sieboldii, var. arborescens, II. 294.
4235. Ternstroemia japonica, II. 397.
4281. Viburnum erosum, var. Taquetii, I. 311.
4304. Vaccinium bracteatum, I. 558.
4344. Meliosma Oldhamii, II. 206.
4345. Meliosma Oldhamii, II. 206.
4351. Staphylea Bumalda, II. 185.
4417. Celtis sinensis, III. 277.
4418. Celtis jessoensis, III. 281.
4419. Aphananthe aspera, III. 290.
4420. Cudrania tricuspidata, III. 306.
4421. Morus acidosa, III. 300.
4426. Myrica rubra, III. 189.
4439. Ostrya japonica, II. 424.
4441. Carpinus cordata, II. 434.
4442. Quercus glandulifera, III. 212.
4443. Quercus glandulifera, III. 212.
4444. Quercus glandulifera, III. 212.
4446. Quercus serrata, III. 217.
4447. Quercus mongolica, III. 230.
4448. Quercus mongolica, III. 230.
4458. Citrus sinensis, II. 148.
4459. Citrus Aurantium, II. 147.
4531. Citrus nobilis, var. deliciosa, II. 143.
4612. Akebia quinata, I. 347.
4616. Orixa japonica, II. 135.
4617. Poncirus trifoliata, II. 149.
4701. Morus acidosa, III. 300.
4702. Celtis sinensis, III. 277.
4703. Myrica rubra, III. 189.
4704. Myrica rubra, III. 189.

TAQUET, *Père* (*continued*)
4705. Carpinus laxiflora, II. 438.
4821. Myrica rubra, III. 189.

Tokubuchi, Y.
17. Salix viminalis, var. yezoensis, III. 158.
31. Salix lepidostachys, III. 166.

Tokyo, Herb. Bot. Gard.
11. Salix Caprea, III. 149.
21. Salix vulpina, var. discolor, III. 131.
33. Salix vulpina, III. 130.
51. Salix eriocarpa, III. 108.
53. Salix sibirica, var. subopposita, III. 154.
63. Salix lasiogyne, III. 111.
66. Salix Reinii, III. 127.
88. Salix opaca, III. 159.
89. Salix opaca, III. 159.
106. Salix kakista, III. 128.
109. Salix vulpina, III. 130.
114. Salix japonica, var. padifolia, III. 133.

Tschonoski.
4. Salix Urbaniana, III. 103.

Urquhart, *Colonel.*
163. Mussaenda erosa, III. 399.

Wallich, Nathaniel.
2793. Betula alnoides, II. 467.
2796, 2796b. Populus ciliata, III. 31.
2798. Corylus Jacquemontii, II. 449.
2799. Alnus nepalensis, II. 502.
3699a. Salix denticulata, III. 117.
3699b. { Salix denticulata, III. 117.
{ Salix denticulata, var. himalensis, III. 118.
3703a, b. Salix apiculata, III. 176.
3704. Salix eriostachya, III. 114.
3705. Salix pyrina, III. 95.
3707, 3707b, c. Salix tetrasperma, III. 94.
3708. Salix dealbata, III. 105.

Watt, Sir George.
6538. Pittosporum glabratum, III. 328.

Wichura, Max.
1147b. Prunus autumnalis, I. 259.

Wight, Robert.
1895. Secamone emetica, III. 348.
2000. Porana volubilis, III. 358.

Wilford, Charles.
2. Quercus dentata, III. 210.
160. { Mussaenda erosa, III. 399.
{ Mussaenda pubescens, III. 399.

Wilson, Ernest Henry (Arnold Arboretum Expedition to Japan).
6029. Alnus firma, II. 506.
6033. Quercus salicina, III. 234.
6099. Chaenomeles lagenaria, II. 296.
6136. Chaenomeles lagenaria, II. 296.
6155, 6155a. Alnus japonica, II. 493.
6165, 6165a. Salix Sieboldiana, III. 162.
6179, 6179a. Salix Sieboldiana, III. 162.
6180, 6180a. Salix purpurea, III. 167.
6185. Alnus firma, II. 506.
6195. Ulmus japonica, III. 258.
6215. Salix gracilistyla, III. 164.
6248, 6248a. Salix Buergeriana, III. 162.
6256. Quercus mongolica, var. grosseserrata, III. 231.
6277, 6277a. Salix Sieboldiana, III. 162.
6279. Ulmus japonica, III. 258.
6301, 6301a. Salix Sieboldiana, III. 162.
6330, 6330a. Salix purpurea, subsp. amplexicaulis, var. multinervis, III. 168.
6331. Salix gracilistyla, III. 164.
6332. Chaenomeles japonica, II. 298.
6340. Chaenomeles lagenaria, II. 296.
6341. Chaenomeles lagenaria, II. 296.
6360. Malus Halliana, II. 285.
6365. Chaenomeles lagenaria, II. 296.
6366. Chaenomeles lagenaria, II. 296.
6367. Chaenomeles lagenaria, II. 296.
6377. Zelkova serrata, III. 287.
6387. Chaenomeles japonica, II. 298.
6389. Carpinus laxiflora, II. 438.
6402. Salix japonica, III. 132.
6413. Carpinus cordata, II. 434.
6419. Salix viminalis, var. yezoensis, III. 158.
6423. Salix Caprea, III. 149.
6427. Alnus firma, subsp. hirtella, II. 506.
6435. Alnus hirsuta, var. sibirica, II. 498.
6447. Salix japonica, III. 132.
6456. Salix opaca, III. 159.
6477. Celtis sinensis, III. 277.
6572. Chaenomeles sinensis, II. 299.
6578. Wistaria floribunda, f. rosea, II. 512.

Wilson, E. H. (*continued*)
6579. Wistaria sinensis, II. 509.
6580. Wistaria venusta, II. 514.
6580a. Wistaria venusta, II. 514.
6581. Wistaria floribunda, f. alba, II. 512.
6582. Wistaria floribunda, f. rosea, II. 512.
6597. Aphananthe aspera, III. 290.
6602. Salix japonica, var. Oldhamiana, III. 132.
6616. Chaenomeles japonica, II. 298.
6620, 6620a. Wistaria floribunda, II. 510.
6637, 6637a. Salix algista, III. 174.
6638. Salix ?gracilistyla, III. 165.
6639. Salix purpurea, subsp. amplexicaulis, var. multinervis, III. 168.
6642, 6642a. Salix Reinii, III. 127.
6645. Alnus firma, subsp. hirtella, II. 506.
6646. Alnus hirsuta, var. sibirica, II. 498.
6656. Chaenomeles japonica, II. 298.
6658. Malus Sieboldii, var. arborescens, II. 294.
6659. Malus zumi, II. 292.
6660. Malus zumi, II. 292.
6664. Malus Halliana, II. 285.
6671. Wistaria floribunda, f. macrobotrys, II. 513.
6672. Wistaria floribunda, f. alba, II. 512.
6677. Wistaria floribunda, II. 510.
6686. Salix ?Shiraii, III. 135.
6691. Wistaria floribunda, f. macrobotrys, II. 513.
6710. Betula grossa, II. 477.
6714. Wistaria venusta, II. 514.
6732. Salix Caprea, III. 149.
6735. Betula grossa, II. 477.
6738. Carpinus japonica, II. 433.
6740. Carpinus cordata, II. 434.
6741. Betula Schmidtii, II. 475.
6748. Salix Gilgiana, III. 169.
6749. Salix purpurea, subsp. amplexicaulis, var. multinervis, III. 168.
6751. Betula Ermanii, var. subcordata, II. 471.
6752. Alnus Maximowiczii, II. 505.
6758. Carpinus japonica, II. 433.
6765, 6765a. Salix purpurea, III. 167.

6767. Betula japonica, var. kamtschatica, II. 486.
6769. Carpinus cordata, II. 434.
6774. Malus Sieboldii, var. arborescens, II. 294.
6775. Salix Urbaniana, III. 103.
6777. Ulmus japonica, III. 258.
6779. Salix Caprea, III. 149.
6788. Alnus Maximowiczii, II. 505.
6789. Betula Ermanii, var. subcordata, II. 471.
6797. Malus zumi, II. 292.
6799. Malus Sieboldii, var. arborescens, II. 294.
6800. Malus Sieboldii, var. arborescens, II. 294.
6803. Chaenomeles lagenaria, II. 296.
6836. Alnus Matsumurae, II. 500.
6842. Sorbus japonica, var. calocarpa, II. 276.
6844, 6844a. Salix Reinii, III. 127.
6847. Betula corylifolia, II. 467.
6860. Betula Ermanii, var. subcordata, II. 471.
6864. Betula Ermanii, var. subcordata, II. 471.
6865. Alnus Maximowiczii, II. 505.
6874. Sorbus alnifolia, var. submollis, II. 275.
6915. Carpinus Tschonoskii, II. 441.
6956. Trachelospermum divaricatum, III. 338.
6994. Carpinus japonica, II. 433.
6997. Carpinus laxiflora, II. 438.
7001. Carpinus cordata, II. 434.
7019. Alnus Maximowiczii, II. 505.
7024. Betula corylifolia, II. 476.
7070. Betula Maximowicziana, II. 465.
7080. Salix Reinii, III. 127.
7094. Betula Ermanii, var. subcordata, II. 471.
7098. Alnus Maximowiczii, II. 505.
7103. Salix kakista, III. 128.
7114. Alnus pendula, II. 507.
7161. Carpinus japonica, II. 433.
7163. Betula grossa, II. 477.
7172. Betula Ermanii, var. subcordata, II. 471.
7175. Alnus Maximowiczii, II. 505.
7177. Quercus mongolica, var. grosseserrata, III. 231.
7178. Alnus japonica, II. 493.
7215. Corylus Sieboldiana, II. 452.

Wilson, E. H. (*continued*)
7228. Betula corylifolia, II. 476.
7254. Alnus japonica, var. arguta, II. 494.
7260. Alnus Maximowiczii, II. 505.
7261. Alnus ?borealis, II. 500.
7262. Alnus hirsuta, II. 496.
7267. Salix Caprea, III. 149.
7268. Corylus Sieboldiana, var. brevirostris, II. 453.
7279. Betula Ermanii, var. genuina, II. 470.
7304. Betula Ermanii, var. genuina, II. 470.
7311. Corylus Sieboldiana, II. 452.
7325. Ulmus japonica, var. levigata, III. 260.
7326. Alnus hirsuta, II. 496.
7329. Ulmus japonica, III. 258.
7332. Salix cardiophylla, III. 103.
7334. Salix opaca, III. 159.
7338. Betula Ermanii, var. subcordata, II. 471.
7341. Salix viminalis, III. 157.
7342. Salix Caprea, III. 149.
7343. Salix rorida, III. 155.
7344. Salix opaca, III. 159.
7355. Betula japonica, var. kamtschatica, II. 486.
7366. Salix Caprea, III. 149.
7369. Malus baccata, var. mandshurica, II. 291.
7372. Quercus mongolica, III. 230.
7375. Malus baccata, var. mandshurica, II. 291.
7377. Alnus hirsuta, II. 496.
7381. Alnus japonica, var. arguta, II. 494.
7383. Carpinus cordata, II. 434.
7384. Quercus mongolica, III. 230.
7403. Salix rorida, III. 155.
7411. Malus Sicboldii, var. arborescens, II. 294.
7414. Salix jessoensis, III. 110.
7416. Quercus mongolica, var. grosseserrata, III. 231.
7426. Corylus Sieboldiana, II. 452.
7428. Alnus hirsuta, var. sibirica, II. 498.
7448. Quercus mongolica, var. grosseserrata, III. 231.
7449. Alnus firma, subsp. hirtella, II. 506.

7452. Alnus firma, subsp. hirtella, II. 506.
7454. Carpinus japonica, II. 433.
7456. Betula japonica, II. 485.
7460. Corylus Sieboldiana, II. 452.
7463. Malus Sieboldii, II. 293.
7468. Alnus japonica, II. 493.
7483. Carpinus japonica, II. 433.
7485. Betula grossa, II. 477.
7499. Salix opaca, III. 159.
7500. Salix Caprea, III. 149.
7501. Alnus Matsumurae, II. 500.
7502. Betula corylifolia, II. 476.
7505. Alnus Maximowiczii, II. 505.
7511. Betula Ermanii, var. subcordata, II. 471.
7527. Chaenomeles japonica, II. 298.
7528. Salix gracilistyla, III. 164.
7532. Wistaria floribunda, II. 510.
7534. Malus baccata, var. mandshurica, II. 291.
7540. Malus Sieboldii, II. 293.
7548. Betula davurica, II. 483.
7553. Alnus hirsuta, var. sibirica, II. 498.
7556. Alnus Maximowiczii, II. 505.
7562. Quercus aliena, var. acuteserrata, III. 215.
7577. Betula Ermanii, var. genuina, subvar. brevidentata, II. 471.
7580. Quercus mongolica, var. grosseserrata, III. 231.
7586. Carpinus cordata, II. 434.
7587. Carpinus cordata, II. 434.
7596. Malus baccata, var. mandshurica, II. 291.
7608. Alnus hirsuta, var. sibirica, II. 498.
7619. Malus prunifolia, var. rinki, II. 279.
7643. Sorbus japonica, var. calocarpa, II. 276.
7651. Betula corylifolia, II. 476.
7655. Malus Sieboldii, var. arborescens, II. 294.
7662. Alnus firma, subsp. hirtella, II. 506.
7669. Betula japonica, var. kamtschatica, II. 486.
7674. Alnus firma, subsp. hirtella, II. 506.
7675. Carpinus japonica, II. 433.
7680. Betula grossa, II. 477.

WILSON, E. H. (*continued*)
7687. Betula Schmidtii, II. 475.
7692. Carpinus japonica, II. 433.
7701. Betula grossa, II. 477.
7708. Betula globispica, II. 479.
7711. Alnus hirsuta, var. sibirica, II. 498.
7738. Malus micromalus, II. 290.
7745. Carpinus laxiflora, II. 438.
7750. Carpinus laxiflora, II. 438.
7751. Alnus hirsuta, var. sibirica, II. 498.
7752. Alnus firma, subsp. hirtella, II. 506.
7753. Malus Tschonoskii, II. 295.
7796. Wistaria japonica, II. 515.
7846. Ostrya japonica, II. 424.
7852. Wistaria venusta, II. 514.
7856. Ulmus parvifolia, III. 244.
7868. Quercus salicina, III. 234.

Wilson, Ernest Henry (Veitch expeditions to China).
1. Caragana chamlagu, II. 102.
2. Clematis Armandi, I. 326.
3. Prunus glandulosa, var. trichostyla, f. paokangensis, I. 264.
6. Symplocos caudata, II. 595.
7. Populus adenopoda, III. 23.
8. Thea fraterna, II. 390.
13. Salix Wallichiana, III. 64.
14. Staphylea holocarpa, II. 185.
16. Acer laevigatum Wallich.
17. Rhododendron sutchuenense, I. 544.
19. Caesalpinia sepiaria, II. 92.
20. Daphniphyllum macropodum, II. 522.
21, 21a. Magnolia denudata, var. purpurascens, I. 401.
22. { Eurya japonica, var. nitida, II. 398.
    Eurya japonica, var. aurescens, II. 399; III. 451.
23. Rubus illecebrosus, III.
26. Clematis Armandi, I. 326.
27. Prunus glandulosa, var. trichostyla, f. paokangensis, I. 264.
28. Prunus tomentosa, var. endotricha, I. 225.
29. Rhododendron Mariesii, I. 548.
33. Daphne genkwa, II. 538.
34. { Lindera umbellata, II. 81.
    Litsea ichangensis, II. 77.
36, 36b. Evonymus alata, I. 493.
38. Osmanthus fragrans, II. 609.

40. Carpinus laxiflora, var. macrostachya, II. 425.
41. Lindera fragrans, II. 83.
42. Sageretia subcaudata, II. 228.
46. Ilex corallina Franchet.
48, 48a. Betula luminifera, II. 455.
49. Lindera glauca, II. 80.
51. Pittosporum truncatum, III. 328.
56. Edgeworthia chrysantha, II. 550.
57. Corylus tibetica, II. 443.
59. Lindera megaphylla, II. 80.
61. Hamamelis mollis, I. 431.
64. Sassafras tzumu, II. 74.
65. Corylopsis sinensis, I. 424.
66. Prunus Veitchii, I. 257.
66a. Prunus tenuiflora, I. 209.
67. Spiraea chinensis, I. 444.
70. Prunus Herincquiana, I. 214.
72. Cephalotaxus Oliveri, II. 6.
75. Rosa laevigata, II. 318.
76. Symplocos paniculata, II. 593.
77. Fraxinus chinensis, var. typica, II. 260.
81. Sophora viciifolia, II. 95.
83. Jasminum floridum, II. 614.
84. Ehretia macrophylla, III. 364.
85. Cinnamomum hupehanum, II. 69.
89. Chionanthus retusus, II. 611.
90. Ribes tenue Janczewski.
91. Acer Franchetii Pax.
92. Eukianthus quinqueflorus, var. serrulatus, I. 550.
93. Prunus Herincquiana, I. 214.
97. Spiraea laeta, var. subpubescens, I. 444.
99. Machilus ichangensis, II. 622.
99a. Phoebe macrophylla, II. 71.
100. Cephalotaxus Fortunei, II. 4.
101. Aucuba chinensis, f. obcordata, II. 572.
102. Orixa japonica, II. 135.
105. Akebia quinata, I. 347.
105a. Akebia lobata, var. australis, I. 348.
106. Elaeagnus magna, II. 411.
107. Ilex corallina Franchet.
108. Ilex Aquifolium, var. chinensis Loesener.
110. Loropetalum chinense, I. 430.
115. Distylium chinense, I. 423.
117 (Herb. Arnold Arb.). Clematis Henryi, I. 342.
117. Pterocarya stenoptera, III. 181.
119. Machilus ichangensis, II. 621.

WILSON, E. H. (*continued*)
124. Prunus salicina, I. 580.
125. Stachyurus chinensis, I. 287.
127. Sarcococca ruscifolia, II. 163.
129. Chaenomeles lagenaria, II. 296.
134, 134a. Rubus corchorifolius, III.
136. Pteroceltis Tatarinowii, III. 284.
138. Ilex metabaptista Loesener.
141. Tinospora sagittata, I. 390.
144. Itea ilicifolia, I. 44.
146. Prunus scopulorum, I. 241.
148. Zanthoxylum dissitum, II. 128.
150. Coriaria sinica, II. 170.
152. Prunus Conradinae, I. 211.
155. Buddleia officinalis, I. 565.
162. Polygala Wattersii, II. 161.
166. Clematis florida, I. 325.
167. Photinia serrulata, I. 184.
167a. Photinia Davidsoniae, I. 185.
172. Spondias axillaris, II. 172.
175. Rubus Playfairianus Hemsley.
178. Rosa multiflora, var. cathayensis, II. 304.
179a. Schisandra sphenanthera, I. 414.
181. Celastrus articulata, var. cuneata, II. 350.
183. Rosa Banksiae, var. normalis, II. 317.
185. Actinidia chinensis, II. 385.
190. Deutzia discolor, I. 12.
192. Magnolia liliflora, I. 402.
194. Rhus trichocarpa, II. 180.
196. Crataegus hupehensis, I. 178.
202 (Herb. Arnold Arb.). Citrus ichangensis, II. 144.
217a. Rhamnus utilis, II. 240.
218. Liquidambar formosana, I. 421.
221. Cotinus coggygria, var. pubescens, II. 175.
232. Gleditsia macracantha, II. 90.
233. Cornus controversa, II. 573.
234. Acer pictum, var. parviflorum Schneider.
234a. Acer cappadocicum, var. sinicum Rehder.
236. Castanopsis sclerophylla, III. 201.
246. Lindera communis, II. 79.
248. Mezoneurum sinense, II. 93.
249. Celtis Biondii, III. 272.
250. Ilex macrocarpa Oliver.
256. Fagus Engleriana, III. 191.
266. Diervilla japonica, var. sinica, III. 430.

267, 267a. Abelia Graebneriana, I. 118.
271. Ilex Pernyi Franchet.
277. See No. 2077.
279, 279a. Pittosporum glabratum, III. 326.
287. Acer amplum, I. 86.
289. Lindera fragrans, II. 83.
295. Carpinus laxiflora, var. macrostachya, II. 425.
298. Acer tetramerum, f. lobulatum Rehder.
299, 299a. Acer Davidii Franchet.
301. Sorbus caloneura, II. 269.
302. Rhododendron Augustinii, I. 524.
303. Acer Wilsonii Rehder.
308. Prunus Dielsiana, var. conferta, I. 244.
311. Rhododendron hypoglaucum, I. 527.
312. Rhododendron Fortunei, var. Houlstonii, I. 541.
314. Aucuba chinensis, II. 572.
315. Ribes Franchetii Janczewski.
316. Prunus pilosiuscula, var. media, I. 204.
320. Ulmus Bergmanniana, III. 240.
326. Staphylea Bumalda, II. 185.
327. Acer longipes Rehder.
330. Juglans regia, III. 184.
334. { Salix hypoleuca, III. 53. / Salix Wilsonii, III. 40.
337. Decumaria sinensis, I. 152.
348. Actinidia callosa, var. Henryi, II. 382.
349. Pyracantha crenulata, I. 177.
352. Sorbus Folgneri, II. 271.
356. Rhamnus leptophyllus, II. 239.
359. Photinia Beauverdiana, var. notabilis, I. 188.
361. Pittosporum glabratum, var. neriifolium, III. 328.
363. Quercus phillyraeoides, III. 233.
365. Morus cathayana, III. 292.
366. { Machilus ichangensis, II. 621. / Phoebe macrophylla, II. 71.
368. Acer nikoense, var. megalocarpum, I. 98.
370. Juniperus formosana, II. 56.
371. Magnolia officinalis, I. 391.
375c. See No. 3754c.
384. { Populus lasiocarpa, III. 17. / Populus Wilsonii, III. 16.
391. Photinia serrulata, I. 184.

WILSON, E. H. (*continued*)
391a. Photinia Davidsoniae, I. 185.
394. Rhamnus utilis, II. 240.
397, 397a. Acer oblongum Wallich.
407. Pistacia chinensis, II. 173.
408. Indigofera pseudotinctoria, II. 100.
410. Chaenomeles lagenaria, var. cathayensis, II. 297.
411. Cercis chinensis, II. 87.
412a. Fluggea leucopyrus, II. 520.
414. Ardisia Henryi, II. 582.
415a. Salix Wilsonii, III. 40.
418. Malus prunifolia, var. rinki, II. 279.
420. Keteleeria Davidiana, II. 39.
421. Actinodaphne cupularis, II. 75.
421a. Actinodaphne confertifolia, II. 74.
424. Litsea sericea, II. 75.
424a. Lindera spec., II. 86.
426. Mahonia Sheridaniana, I. 384.
428. Juniperus formosana, II. 56.
431. Illicium Henryi, I. 417.
433. Buxus microphylla, var. aemulans, II. 167.
433a. Buxus microphylla, var. sinica, II. 165.
444. Magnolia denudata, var. elongata, I. 402.
446. Prunus venusta, I. 239.
447. Fagus Engleriana, III. 191.
450. Celastrus Hindsii, var. Henryi, II. 353.
451. Platycarya strobilacea, III. 180.
453. Euscaphis japonica, II. 187.
457. Clematis uncinata, I. 327.
459. Ligustrum myrianthum, II. 607.
462. Photinia Davidsoniae, I. 185.
462a. Photinia serrulata, I. 184.
463. Meliosma Oldhamii, II. 206.
464. Cinnamomum hupehanum, II. 69.
466. Indigofera ichangensis, II. 100.
467. Celastrus angulata, II. 346.
471. Lonicera ligustrina, I. 134.
473. Hypericum longistylum, II. 404.
474. Prunus Macgregoriana, I. 240.
477. Carpinus laxiflora, var. macrostachya, II. 425.
478. Lindera pulcherrima, II. 85.
480. Vitis flexuosa, var. parvifolia Gagnepain.
495. Cornus paucinervis, II. 576.
497, 497a. Rhamnus rugulosus, II. 238.

510. Alangium chinense, II. 552.
511. Albizzia kalkora, III. 446.
512. Rubus Parkeri Hance.
515. Amelanchier asiatica, var. sinica, I. 195.
527. Carpinus cordata, var. chinensis, II. 437.
535. Berberis Julianae, I. 360; III. 438.
537. Fagus longipetiolata, III. 190.
538. (Herb. Arnold Arb.). Acer robustum, I. 89.
545. Ulmus castaneifolia, III. 256.
548. Acer Henryi, Pax.
550. Acer longipes, Rehder.
551. Acer cappadocicum, var. sinicum Rehder.
552. Cornus chinensis, II. 577.
562. Quercus aliena, var. acuteserrata, III. 215.
564. Cotoneaster horizontalis, var. perpusilla, I. 115.
568. Acer oblongum Wallich.
572. Tsuga chinensis, II. 37.
573. Quercus myrsinaefolia, III. 236.
574, 574a. Acer laevigatum Wallich.
575. Castanopsis sclerophylla, III. 201.
576. Evonymus myriantha, I. 487.
583, 583a. Pittosporum glabratum, III. 326.
587. Carpinus cordata, var. chinensis, II. 437.
589, 589a. Evonymus sanguinea, var. β camptoneura, I. 494.
590. Acer robustum, I. 89.
591. Acer pictum, var. parviflorum Schneider.
593. Sorbus Keissleri, II. 269.
594. Litsea pungens, II. 76.
596. Thea cuspidata, II. 390.
600. Lindera communis, II. 79.
601. Zanthoxylum dimorphophyllum, II. 126.
603. Illicium Henryi, I. 417.
604. Prunus Grayana, I. 69.
605. Acer amplum, I. 86.
607. Cercis racemosa, II. 88.
608. Fagus longipetiolata, III. 190.
609. Rhododendron Fortunei, var. Houlstonii, I. 541.
610. Lindera cercidifolia, II. 85.
610a. Lindera umbellata, var. latifolia, II. 81.
616. Acer sinense, var. concolor Pax.

WILSON, E. H. (*continued*)
623. Litsea sericea, II. 75.
624. Taxus cuspidata, var. chinensis, II. 8.
627. Meliosma Beaniana, II. 204.
629. Eucommia ulmoides, I. 433.
633. Clematis montana, var. rubens, I. 333.
634. Paliurus orientalis, II. 209.
637. Daphne Wilsonii, II. 540.
639. Acer sutchuenense Franchet.
642. Davidia involucrata, var. Vilmoriniana, II. 256.
645. Berberis Henryana, III. 440.
647. Quercus oxyodon, III. 228.
648, 648a. Holboellia Fargesii, I. 346.
651. Quercus myrsinaefolia, III. 236.
662. Pinus Armandi, II. 12.
663. Ailanthus altissima, var. sutchuenensis, II. 153; III. 449.
669a, b. Cotoneaster acutifolia, var. villosula, I. 158.
670. Malus yunnanensis, II. 287.
671. Pterostyrax hispidus, I. 295.
672. Rubus mesogaeus, III.
675. Schisandra sphenanthera, I. 414.
677. { Ulmus castaneifolia, III. 256. Ulmus Wilsoniana, III. 238.
678. Quercus Engleriana, III. 220.
679. Salix Fargesii Burkill.
681. Cornus kousa, II. 577.
682. Illicium Henryi, I. 417.
683. Prunus consociiflora, I. 279.
685. Salix heterochroma, III. 61.
701. Neillia sinensis, I. 436.
703. Aucuba chinensis, f. angustifolia, II. 573.
708. Acer flabellatum Rehder.
710. Deutzia discolor, I. 12.
714. { Photinia amphidoxa, I. 190. Photinia villosa, var. sinica, I. 186.
716. Acer Oliverianum Pax.
719. Rhododendron ovatum, I. 546.
723. Prunus tenuiflora, I. 209.
724. Acer Maximowiczii Pax.
726. Sargentodoxa cuneata, I. 351.
731. Spiraea prunifolia, var. plena, I. 438.
747. Fagus Engleriana, III. 191.
749, 749a. Malus theifera, II. 283.
758. Rhododendron stamineum, I. 546.
759. Castanea Henryi, III. 196.
760. Sapium japonicum, II. 527.

763. Premna puberula, III. 371.
764. { Cornus Walteri, II. 576. Cornus Wilsoniana, II. 579.
765. Evonymus lanceifolia, I. 491.
765a. Evonymus yedoensis, var. Koehneana, I. 491.
777. Vitis pentagona, var. bellula, III. 428.
779. Schisandra sphenanthera, I. 414.
782. Dalbergia Dyeriana, II. 115.
786. Rubus Henryi Hemsley & Kuntze.
788. Cayratia oligocarpa Gagnepain.
794. Photinia Beauverdiana, I. 187.
808. Ormosia Hosiei, II. 94.
811. Cornus kousa, II. 577.
812. Castanea mollissima, III. 192.
819. Ehretia acuminata, III. 363.
822. Diploclisia affinis, I. 389.
824. Millettia Dielsiana, II. 101.
828. Rosa microcarpa, II. 314.
833. Philadelphus incanus, var. Sargentianus, f. hupehensis, I. 145.
834. Bauhinia hupehana, II. 89.
836. Cinnamomum hupehanum, II. 69.
841, 841a. Betula luminifera, II. 455.
848. Zizyphus sativa, var. inermis, II. 212.
851. Vitis pentagona Diels & Gilg.
856. Trachelospermum axillare, III. 335.
864. Dalbergia hupeana, II. 115.
866a. Ilex metabaptista Loesener.
867. Rubus Swinhoei Hance.
868. Ilex purpurea, var. Oldhamii Loesener.
877. Cotoneaster divaricata, I. 157.
883. Schisandra Henryi, I. 413.
885. Acer cappadocicum, var. sinicum Rehder.
886. Pittosporum glabratum, III. 326.
889. Clematis montana, var. rubens, I. 333.
891. Evonymus lanceifolia, I. 491.
891a. Evonymus yedoensis, var. Koehneana, I. 491.
894. Hydrangea anomala, I. 34.
895². Prunus glandulosa, var. trichostyla, f. paokangensis, I. 264.
901. Pterocarya Paliurus, III. 182.
903. Picrasma quassioides, II. 152.
905. Pterocarya hupehensis, III. 182.
915. Styrax Hemsleyanus, I. 291.
917. Vitis flexuosa Thunberg.
919. Celastrus gemmata, II. 352.

WILSON, E. H. (*continued*)
921. Spiraea Henryi, I. 447.
922. Abelia umbellata, I. 122.
925. Aristolochia heterophylla, III. 323.
930. Idesia polycarpa, I. 284.
938. Ligustrum acutissimum, II. 600.
940. Deutzia Vilmorinae, I. 16.
944. Viburnum brevipes, I. 113.
945. Rosa Helenae, II. 310.
948. Carpinus laxiflora, var. macrostachya, II. 425.
950. Rubus eucalyptus Focke.
951. Sorbus Folgneri, II. 271.
952. Berberis triacanthophora, I. 358.
956. Symplocos paniculata, II. 593.
957. Acanthopanax villosulus, II. 562.
959. Meliosma cuneifolia, II. 199.
964. Photinia Beauverdiana, I. 187.
972. Celastrus gemmata, II. 352.
974. Clematis grata, var. grandidentata, I. 338.
976. Catalpa Duclouxii, I. 304.
978a. Viburnum tomentosum Thunberg.
984. Cornus macrophylla, II. 575.
985. Sorbus Wilsoniana, I. 458.
986. Rubus lasiostylus Focke.
987. Spiraea Fritschiana, I. 453.
993. Actinidia chinensis, II. 385.
994. Rhus verniciflua, II. 181.
995. Evonymus myriantha, I. 487.
996. Rubus Henryi Hemsley & Kuntze.
997. $\left\{ \begin{array}{l} \text{Spiraea japonica, var. acuminata,} \\ \text{I. 452.} \\ \text{Spiraea Miyabei, var. pilosula, I.} \\ \text{455.} \end{array} \right.$
1000. Photinia Beauverdiana, I. 187.
1001. Photinia parvifolia, I. 189.
1002. Enkianthus chinensis, I. 551.
1010. Vaccinium Donianum, I. 557.
1016. Grewia parviflora, var. glabrescens, II. 371.
1017. Millettia reticulata, II. 102.
1018. Jasminum lanceolarium, var. puberulum, II. 612.
1020. $\left\{ \begin{array}{l} \text{Acanthopanax Giraldii, var. inermis, II. 560.} \\ \text{Acanthopanax Rehderianus, II.} \\ \text{561.} \end{array} \right.$
1028. Ilex dipyrena, c. paucispinosa, III.
1036. Schisandra sphenanthera, I. 414.
1041. Pieris ovalifolia, var. elliptica, I. 552.

1045. Rosa banksiopsis, II. 322.
1046. Meliosma Veitchiorum, II. 204.
1048. Euptelea Franchetii, I. 314.
1049. Liriodendron chinense, I. 410.
1062. Picrasma quassioides, II. 152.
1063. Celastrus hypoleuca, II. 346.
1065. $\left\{ \begin{array}{l} \text{Schizophragma integrifolium,} \\ \text{var. denticulatum, I. 42.} \\ \text{Schizophragma integrifolium,} \\ \text{var. minus, I. 43.} \end{array} \right.$
1067. Stranvaesia Davidiana, var. undulata, I. 192.
1068, 1068a. Actinidia melanandra, II. 378.
1077. $\left\{ \begin{array}{l} \text{Rhododendron discolor, I. 542.} \\ \text{Rhododendron Fortunei, var.} \\ \text{Houlstonii, I. 541.} \end{array} \right.$
1079. Cotoneaster racemiflora, var. Veitchii, III. 431.
1089. Millettia Dielsiana, II. 101.
1093. Pieris ovalifolia, var. elliptica, I. 552.
1102. Glochidion Wilsonii, II. 518.
1103. Ehretia acuminata, III. 363.
1104. Carrieria calycina, I. 284.
1106. Styrax dasyanthus, I. 289.
1125. Vitis Piasezkii Maximowicz.
1126. Meliosma platypoda, II. 201.
1127. Cotoneaster Dielsiana, I. 166.
1128. Rhus orientalis, II. 179.
1130. 1130a. Betula insignis, II. 459.
1134. Alangium chinense, II. 552.
1135, 1135a. Rhamnus crenatus, II. 232.
1138. Berberis Veitchii, I. 363.
1141, 1141a. Ampelopsis micans, I. 579.
1142. Acanthopanax evodiaefolius, II. 563.
1150. Hydrangea longipes, var. lanceolata, I. 40.
1151. Vitis reticulata, III. 428.
1152. Cornus macrophylla, II. 575.
1155. Rubus irenaeus Focke.
1158. Schizophragma integrifolium, var. denticulatum, I. 42.
1160. Evonymus sanguinea, var. β camptoneura, I. 494.
1163. Phoebe Sheareri, II. 72.
1165. Actinidia purpurea, II. 378.
1167. Cotoneaster Zabelii, I. 166.
1169. Acer pictum, var. parviflorum Schneider.
1170. Carpinus Seemeniana, II. 430.

Wilson, E. H. (*continued*)
1174. Dichroa febrifuga, I. 43.
1177. { Rhus Potaninii, II. 177.
Rhus punjabensis, var. sinica, II. 176.
1179. Berchemia yunnanensis, II. 216.
1181, 1181a. Adina racemosa, III. 390.
1183, 1183a. Spiraea japonica, var. Fortunei, I. 451.
1191. Hovenia dulcis, II. 252.
1193. Ligustrum compactum, II. 604.
1199. Spiraea Miyabe, var. glabrata, I. 454.
1201. Cocculus trilobus, I. 388.
1202. Styrax suberifolius, I. 290.
1203. Sinomenium acutum, var. cinereum, I. 387.
1204. Evonymus Dielsiana, I. 488.
1213. Hydrangea longipes, I. 33.
1218, 1218a. Acer Davidii Franchet.
1222. Lespedeza Buergeri, II. 106.
1227. Evonymus japonica, var. acuta, I. 485.
1230. Indigofera chalara, II. 101.
1231. Rubus inopertus Focke.
1232. Acer flabellatum Rehder.
1233. Acer tetramerum, var. elobulatum, I. 95.
1238. Lespedeza formosa, II. 107.
1240. Paederia tomentosa, III. 403.
1242. Tilia tuan, var. chinensis, II. 369.
1246. Wikstroemia angustifolia, II. 535.
1248. Mussaenda Wilsonii, III. 393.
1249. Buddleia Davidii, var. magnifica, I. 567.
1253. Emmenopterys Henryi, III. 391.
1254. Firmiana simplex, II. 377.
1255. Adina rubella, III. 390.
1261a. Sambucus javanica, I. 307.
1265a. Grewia parviflora, var. glabrescens, II. 371.
1268. Adina racemosa, III. 390.
1271. Hydrangea hypoglauca, I. 26.
1274. Polygala arillata, II. 160.
1275. { Styrax dasyanthus, I. 289.
Styrax dasyanthus, var. cinerascens, I. 289.
1282. Philadelphus sericanthus Koehne.
1291. Cotoneaster foveolata, I. 162.
1294. Rosa rugosa, var. Chamissoniana, II. 321.
1298. Rosa multiflora, var. cathayensis, II. 304.

1303. Aesculus Wilsonii, I. 498.
1304. Schisandra propinqua, var. sinensis, I. 416.
1305. Ilex pedunculosa, var. continentalis, III.
1306. Clematis chinensis, I. 329.
1308d. Clematis grata, var. grandidentata, I. 338.
1309. Evodia officinalis, II. 130.
1316. Albizzia kalkora, III. 446.
1320. Schizophragma integrifolium, var. denticulatum, I. 42.
1325. Ilex pedunculosa, var. continentalis Loesener.
1326. Clethra Fargesii, I. 502.
1329. Phoebe Sheareri, II. 72.
1330. Dumasia hirsuta, II. 116.
1333. Ilex szechwanensis Loesener.
1342. Callicarpa japonica, var. angustata, III. 369.
1347. Buddleia Davidii, I. 567.
1350. Dalbergia stenophylla, II. 116.
1363, 1363a. Actinidia polygama, II. 380.
1371. Sophora flavescens, II. 96.
1373. Hydrangea fulvescens, I. 39.
1382. Viburnum erubescens, var. gracilipes, I. 107.
1383. Celastrus spiciformis, II. 348.
1384. Cotoneaster Henryana, I. 174.
1385. Cornus Hemsleyi, II. 574.
1391. Lespedeza formosa, II. 107.
1392. Tilia Henryana, II. 367.
1393. See No. 1373.
1401. Sorbaria arborea, I. 47.
1406. Pueraria hirsuta, II. 118.
1410, 1410a. Parthenocissus laetevirens, I. 580.
1414. Evonymus Dielsiana, I. 488.
1417. Lespedeza virgata, II. 110.
1418. Ampelopsis heterophylla, var. Delavayana Gagnepain.
1420. Abelia chinensis, I. 121.
1424. Rubus ichangensis Hemsley & Kuntze.
1429. Cocculus trilobus, I. 388.
1438. Rosa corymbulosa, II. 323.
1442. Clematis quinquefoliolata, I. 328.
1446. Evonymus yedoensis, var. Koehneana, I. 491.
1456. Acanthopanax Simonii, II. 559.
1458, 1458a, b. Meliosma cuneifolia, II. 199.

WILSON, E. H. (*continued*)

1467. Rhododendron auriculatum, I. 544.
1471, 1471a. Clematis lasiandra, I. 322.
1473. Hydrangea strigosa, I. 31.
1473a. Hydrangea villosa, var. strigosior, I. 39.
1475. Poliothyrsis sinensis, I. 285.
1480. Rubus inopertus Focke.
1483. Sinomenium acutum, var. cinereum, I. 387.
1489. Clerodendron trichotomum, var. Fargesii, III. 376.
1491. Vaccinium Henryi, I. 561.
1493. Quercus spinosa, III. 224.
1499. Jasminum urophyllum, var. Henryi, II. 613.
1503. Tilia Henryana, II. 367.
1504. Ligustrum strongylophyllum, II. 605.
1510. Spiraea japonica, var. acuminata, I. 452.
1516. Maackia hupehensis, II. 98.
1520. Evonymus japonica, var. radicans, I. 485.
1521. Lithocarpus cleistocarpa, III. 205.
1526. Rhododendron micranthum, I. 513.
1532. Tilia paucicostata, II. 363.
1557. Rubus Lambertianus Seringe.
1558. Rubus innominatus S. Moore.
1564. Nothopanax Davidii, II. 556.
1573. Lespedeza formosa, II. 107.
1582. Maackia hupehensis, II. 98.
1590. Clematis uncinata, I. 327.
1609. Koelreuteria bipinnata, II. 193.
1614. Rhamnus paniculiflorus, II. 233.
1621. Vaccinium japonicum, I. 562.
1630. Lespedeza formosa, II. 107.
1631. Catalpa ovata, I. 303.
1672. Symplocos anomala, II. 596.
1679. Clematis chinensis, I. 329.
1681. Camptotheca acuminata, II. 254.
1688. Quercus glandulifera, III. 212.
1694. Lespedeza tomentosa, II. 110.
1700. Corchoropsis crenata Siebold & Zuccarini.
1706. Clematis Gouriana, I. 339.
1710. Rhus javanica, II. 178.
1716. Lespedeza sericea, II. 105.
1723. Clematis brevicaudata, I. 340.
1751. Eurya japonica, var. nitida, II. 398.

1753. { Meratia praecox, I. 419.
Meratia praecox, var. grandiflora, I. 420.
1754. Spiraea hypericifolia, var. hupehensis, I. 439.
1757. Abutilon sinense, II. 373.
1761. Celtis Biondii, III. 272.
1761a. Celtis cercidifolia, III. 276.
1770. Litsea pungens, II. 76.
1781. Prunus salicina, I. 580.
1784. Litsea elongata, II. 78.
1787. Lindera strychnifolia, var. Hemsleyana, II. 82.
1788. Lindera glauca, II. 80.
1789. Prunus velutina, I. 69.
1795. Litsea citrata, II. 75.
1795a. Lindera membranacea, II. 81.
1802. Symplocos botryantha, II. 596.
1807. Lindera fragrans, II. 83.
1809. Carpinus polyneura, II. 430.
1819. Acer caudatum Wallich.
1820. Salix Fargesii, III. 47.
1821. Viburnum tomentosum Thunberg.
1823. Picrasma quassioides, II. 152.
1824. Fraxinus retusa, var. Henryana, II. 258.
1825. Sycopsis sinensis, I. 431.
1826. Machilus microcarpa, II. 74.
1827. Ilex Fargesii Franchet.
1855. Ulmus Bergmanniana, III. 240.
1856. Lindera obtusiloba, II. 85.
1860. Staphylea holocarpa, var. rosea, II. 186.
1862. Prunus canescens, I. 215.
1864. Prunus tomentosa, var. Kashkarovii, I. 269.
1868 (Herb. Arnold Arb.). Cotoneaster salicifolia, I. 172.
1870. Deutzia grandiflora, I. 16.
1873. Litsea pungens, II. 76.
1875. Pteroceltis Tatarinowii, III. 284.
1877. Rhododendron Fargesii, I. 540.
1878. Rhododendron maculiferum, I. 531.
1882. Acer Davidii Franchet.
1883. Acer griseum Pax.
1886. Populus tremula, var. Davidiana, III. 24.
1889. Betula albo-sinensis, II. 457.
1889a. Betula luminifera, II. 455.
1890. Betula albo-sinensis, II. 457.
1891. Acer Maximowiczii Pax.

Wilson, E. H. (*continued*)

1895. Abies Fargesii, II. 48.
1896. Picea brachytyla, II. 33.
1897, 1897a. Picea Wilsonii, II. 27.
1900. Tamarix parviflora, II. 406.
1902. Diospyros sinensis, II. 591.
1915. Berberis brachypoda, I. 375.
1916a. Deutzia Wilsonii, I. 20.
1916b. Deutzia discolor, I. 12.
1917. Deutzia discolor × mollis, I. 20.
1919. Deutzia rubens, I. 13.
1919a. Deutzia hypoglauca, I. 24.
1921a. Ternstroemia japonica, var. Wightii, II. 397.
1925. Ilex purpurea, var. Oldhamii Loesener.
1926. Fraxinus Griffithii, II. 258.
1929. Schoepfia jasminodora, III. 321.
1930, 1930a. Evodia glauca, II. 129.
1933. Machilus ichangensis, II. 621.
1935. Cornus macrophylla, II. 575.
1936. Rubus triphyllus Thunberg.
1941. Celtis Julianae, III. 265.
1944. Cinnamomum hupehanum, II. 69.
1946. Litsea fruticosa, II. 77.
1950. Fraxinus retusa, var. integra, II. 259.
1953. Rosa Giraldii, var. venulosa, II. 328.
1957. Leptodermis oblonga, III. 403.
1964. Ilex szechwanensis Loesener.
1965. Picrasma quassioides, II. 152.
1966. Cotoneaster Dammeri, I. 176.
1968. Schisandra sphenanthera, I. 414.
1969. Rhododendron yanthinum, I. 518.
1972. Evodia rutaecarpa, II. 132.
1976. Ilex pedunculosa, var. continentalis Loesener.
1980. Sorbus Zahlbruckneri, II. 274.
1984. Grewia parviflora, var. glabrescens, II. 371.
1990. Rhamnella Martinii, II. 225.
1993. Acer Fargesii Franchet.
1994. Ormosia Hosiei, II. 94.
1998. Deutzia Vilmorinae, I. 20.
2000. Bauhinia Faberi, II. 88.
2003. Cinnamomum Wilsonii, II. 66.
2006. { Machilus Bournei, II. 73.
       { Phoebe neurantha, II. 72.
2007. Phoebe Sheareri, II. 72.
2015. Styrax Veitchiorum, I. 290.
2017. Indigofera amblyantha, II. 99.

2024. Syringa microphylla, III. 433.
2030a. Cotoneaster hupehensis, I. 169.
2034. Indigofera ichangensis, I. 100.
2041. Berchemia hypochrysa, II. 214.
2045. Salix Biondiana, III. 118.
2053. Evonymus sanguinea, var. β camptoneura, I. 494.
2075. Prunus discadenia, I. 200.
2077. { Prunus Wilsonii Koehne.
       { Prunus sericea, var. Batalinii,
       {   I. 64 (by error as No. 277).
2078. Syringa reflexa, I. 297.
2082. Sorbus hupehensis, I. 467.
2085. Schisandra grandiflora, I. 411.
2086. Rosa saturata, II. 324.
2094. Rosa omeiensis, II. 331.
2096. Actinidia tetramera, II. 381.
2098. Cinnamomum Wilsonii, II. 66.
2101. Ilex Wilsonii Loesener.
2102. Carpinus polyneura, II. 430.
2103. Paliurus orientalis, II. 209.
2104. Ribes longeracemosum Franchet.
2105. Ribes moupinense Franchet.
2106. Evonymus myriantha, I. 487.
2114. Cephalotaxus drupacea, II. 3.
2121. Phoebe neurantha, II. 72.
2122. Tetrastigma obtectum Planchon.
2123. Betula albo-sinensis, II. 457.
2126, 2126a. Fraxinus Paxiana, II. 259.
2143. Fraxinus Griffithii, II. 258.
2145. Lespedeza Buergeri, II. 106.
2148. Stewartia sinensis, II. 395.
2152, 2152a. Deutzia Schneideriana, I. 7.
2153. Caryopteris incana, III. 378.
2154. Rhododendron discolor, I. 542.
2156. Tetracentron sinense, I. 417.
2167. Cornus poliophylla, II. 574.
2181. Clematoclethra lanosa, II. 388.
2183, 2183a. Evodia Henryi, II. 133.
2184. Hydrangea xanthoneura, var. Wilsonii, I. 27, 150.
2187. Potentilla fruticosa, var. Veitchii, II. 303.
2194. Tapiscia sinensis, II. 188.
2197. Tamarix chinensis, II. 406.
2204. Actinidia trichogyna, II. 384.
2210, 2210a. Evodia glauca, II. 129.
2211. Zanthoxylum micranthum, II. 127.
2215. Celastrus spiciformis, II. 348.
2215a. Celastrus gemmata, II. 352.
2216. Hypericum patulum, var. Henryi, II. 403.

WILSON, E. H. (*continued*)

2217. Carpinus polyneura, II. 430.
2228. Lithocarpus Henryi, III. 209.
2229. Acanthopanax leucorrhizus, II. 557.
2234. Schisandra pubescens, I. 413.
2247, 2247a. Buddleia albiflora, I. 569.
2255. Prunus brachypoda, var. microdonta, I. 66.
2257. Ilex pedunculosa, var. continentalis Loesener.
2265. Acer Fargesii Franchet.
2266. Neolitsea lanuginosa, var. chinenrsis, II. 79.
2267. Sinomenium acutum, var. cinereum, I. 387.
2274. Tilia Oliveri, II. 366.
2276. Spiraea Veitchii, I. 449.
2310. Clematis quinquefoliolata, I. 328.
2312. Sophora flavescens, II. 96.
2316. Tilia tuan, II. 368.
2325. Rosa sertata, II. 327.
2325a. Rosa Giraldii, f. glabriuscula, II. 328.
2333. Tilia chinensis, II. 364.
2335, 2335a. Deutzia discolor, I. 12.
2341a. Cornus ulotricha, II. 574.
2343. Acer Maximowiczii Pax.
2344. Ilex yunnanensis, β gentilis, III. 426.
2371 (Herb. Arnold Arb.). Meliosma Kirkii, II. 207.
2397. Clematis Gouriana, I. 339.
2398. Cladrastis sinensis, II. 97.
2409, 2409a. Rosa setipoda, II. 323.
2410. Rubus flosculosus Focke.
2411. Picea Watsoniana, II. 27.
2415. Evodia hupehensis, II. 133.
2422. Tilia paucicostata, II. 363.
2431. Bischofia javanica, II. 521.
2435. Pterolobium punctatum, II. 92.
2446. Hydrangea strigosa, I. 31.
2450. Corchoropsis crenata Siebold & Zuccarini.
2485. Sophora japonica, II. 96.
2497. Nothopanax Rosthornii, II. 557.
2514. Hydrangea longipes, I. 33.
2527. Hydrangea strigosa, I. 31.
2537. Rhododendron sutchuenense, I. 544.
2558. Osbeckia crinita, II. 421.
2561. Cocculus trilobus, I. 388.
2580. Symplocos anomala, II. 596.

2587. Ormosia Henryi, II. 93.
2591. Litsea elongata, II. 78.
2596a. Clematis heracleaefolia, **var.** ichangensis, I. 321.
2618. Bauhinia Faberi, II. 88.
2637. Litsea citrata, II. 75.
2637a. Actinodaphne cupularis, II. 75.
2645. Osmanthus armatus, II. 611.
2650. Ulmus parvifolia, III. 244.
2687. Sloanea Hemsleyana, II. 361.
2688. Eurya ochnacea, II. 399.
2693, 2693a. Evodia hupehensis, II. 133; III. 449.
2695. Ilex macrocarpa Oliver.
2697. Ilex purpurea, var. Oldhamii Loesener.
2698. Ilex pedunculosa, var. continentalis Loesener.
2706. Acer cappadocicum, var. sinicum Rehder.
2708. Cedrela sinensis, II. 156.
2711. Indigofera pseudotinctoria, II. 100.
2722a. Rosa Gentiliana, II. 312.
2756. Styrax japonicus, I. 291.
2783. Machilus microcarpa, II. 74.
2784. Lindera membranacea, II. 81.
2785. Elaeagnus Henryi, II. 414.
2786. Elaeagnus lanceolata, II. 413.
2800. Betula luminifera, II. 455.
3000. Pinus sinensis, var. yunnanensis, II. 17.
3001. Pinus sinensis, II. 15.
3005. Cephalotaxus argotaenia, II. 6.
3007. Podocarpus neriifolius, II. 9.
3008. Cryptomeria japonica, II. 52.
3009. Larix Potaninii, II. 18.
3010. Juniperus convallium, II. 62.
3011. Thuja orientalis, II. 53.
3012. Cupressus torulosa, II. 54.
3013. Juniperus saltuaria, II. 61.
3015. Pinus sinensis, var. densata, II. 17.
3016. Pinus sinensis, var. densata, II. 17.
3017. Pinus Armandi, II. 12.
3019. Abies squamata, II. 48.
3020. Abies Beissneriana, II. 46.
3021. { Abies Delavayi, II. 41. Abies recurvata, II. 44.
3022. Abies Delavayi, II. 41.
3023. Picea Watsoniana, II. 27.
3024. Picea ascendens, II. 34.
3025. Picea asperata, II. 22.

Wilson, E. H. (*continued*)

3026. Picea purpurea, II. 29.
3027. Picea montigena, II. 33.
3028. Picea likiangensis, II. 31.
3029. Picea aurantiaca, II. 26.
3029a. Picea retroflexa, II. 25.
3030. Picea complanata, II. 35.
3031. Picea complanata, II. 35.
3032. Picea complanata, II. 35.
3033. Paeonia Delavayi, var. angustiloba, I. 318.
3114. Clematis Spooneri, I. 334.
3114a. Clematis montana, var. Wilsonii, I. 333.
3115. Clematis Delavayi, I. 325.
3116. Clematis Armandi, I. 326.
3117. Clematis Fargesii, var. Souliei, I. 336.
3118. Clematis trullifera, I. 324.
3119. Clematis lasiandra, I. 322.
3120, 3120a, b. Clematis Rehderiana, I. 324; III. 434.
3121. Clematis urophylla, I. 323.
3122. Clematis repens, I. 320.
3123. Clematis pseudo-pogonandra, I. 323.
3124. Clematis chinensis, I. 329.
3125. Clematis Faberi, I. 320.
3126. Clematis Prattii, I. 320.
3127. Clematis fasciculiflora, I. 331.
3131, 3131a. Clematis glauca, var. akebioides, I. 342.
3132, 3132a. Clematis glauca, var. akebioides, I. 342.
3133. Euptelea pleiosperma, I. 313.
3134. Schisandra sphenanthera, var. lancifolia, I. 415.
3137. Magnolia Wilsonii, I. 391.
3139. Holboellia grandiflora, I. 346.
3140. Sinofranchetia chinensis, I. 349.
3142. Mahonia polyodonta, I. 383; III. 444.
3145. { Berberis dictyophylla, var. epruinosa, I. 353.
B. aemulans, III. 434.
3146. Berberis dictyophylla, var. epruinosa, I. 353.
3146a. Berberis Ambrozyana, I. 356.
3147. Berberis Wilsonae, I. 368.
3148, 3148a. Berberis Gagnepainii, I. 358.
3150, 3150a. Berberis verruculosa, I. 357.

3151. Berberis Francisci-Ferdinandi, I. 367.
3151a. Berberis Silva-Taroucana, I. 370.
3152. Berberis polyantha, I. 376.
3152a. Berberis aggregata, var. Prattii, I. 376; III. 442.
3154. Berberis Wilsonae, I. 368.
3154a. Berberis parvifolia, III. 436.
3156. Berberis Boschanii, I. 369.
3157. Berberis Vernae, I. 368; III. 439.
3191. Myricaria bracteata, II. 407.
3192. Myricaria bracteata, II. 407.
3193. Myricaria dahurica, II. 407.
3226. Idesia polycarpa, var. vestita, I. 285.
3227. Carrierea calycina, I. 284.
3228. Itoa orientalis, I. 286.
3229. Pittosporum heterophyllum, III. 329.
3230. Pittosporum saxicola, III. 329.
3231, 3231a. Pittosporum glabratum, var. neriifolium, III. 328.
3232. Pittosporum heterophyllum, III. 329.
3233. Pittosporum daphniphylloides, III. 326.
3234. Mappia pittosporoides, II. 190.
3236, 3236a. Polygala arillata, II. 160; III. 449.
3237. Polygala Wattersii, II. 161.
3258. Hypericum uraloides, III. 452.
3261. Hypericum patulum, II. 402.
3262. Hypericum chinense, II. 404.
3263. Hypericum patulum, II. 402.
3264. Gordonia axillaris, III. 394.
3265. Thea oleifera, II. 393.
3266. Thea elongata, II. 392.
3269. Actinidia purpurea, II. 378.
3271. Actinidia tetramera, II. 381.
3272, 3272a. Actinidia coriacea, II. 384.
3273. Actinidia callosa, var. Henryi, II. 382.
3274. Clematoclethra tiliacea, II. 389.
3275. Actinidia venosa, II. 383.
3278. Eurya acuminata, var. multiflora, II. 401.
3279, 3279a. Eurya acuminata, II. 400.
3280, 3280a. Eurya japonica, var. nitida, II. 398.
3280b. Eurya chinensis, II. 400.
3285. Tilia nobilis, II. 363.
3286. Tilia chinensis, II. 364.

WILSON, E. H. (*continued*)

3287. Tilia intonsa, II. 365.
3306. Coriaria terminalis, II. 170.
3307. Citrus ichangensis, II. 144.
3308. Zanthoxylum Bungei, II. 121.
3309. Zanthoxylum Piasezkii, II. 122.
3310. Zanthoxylum Piasezkii, II. 122.
3311. Zanthoxylum dimorphophyllum, II. 126.
3312. Zanthoxylum stenophyllum, II. 127.
3314. Meliosma parviflora, II. 201.
3316. Ilex ciliospina Loesener.
3318. Ilex Franchetiana, III. 426.
3319. Ilex ciliospina Loesener.
3320. Ilex corallina Franchet.
3324, 3324a. Celastrus Hindsii, var. Henryi, II. 353.
3325. Celastrus rugosa, II. 349.
3329. Evonymus microcarpa, I. 487.
3331. Evonymus sanguinea, var. β camptoneura, I. 494.
3332. Evonymus microcarpa, I. 487.
3333. Evonymus cornuta, I. 489.
3334. Evonymus Giraldii, var. angustialata, I. 495.
3336. Sageretia Henryi, II. 623.
3337. Berchemia sinica, II. 215.
3338. Berchemia hypochrysa, II. 214.
3339, 3339a. Rhamnus Sargentianus, II. 235.
3340. Paliurus ramosissimus, II. 210.
3341. Ilex fragilis, β Kingii, III. 426.
3342. Rhamnus Hemsleyanus, II. 234.
3343. Sageretia pycnophylla Schneider.
3344. Fluggea leucopyrus, II. 520.
3345. Acer sinense Pax.
3346. Acer laevigatum Wallich.
3347. Acer caudatum, var. multiserratum Rehder.
3347a. Acer erianthum Schwerin.
3348. Acer tetramerum, var. elobulatum, I. 95.
3349a. Acer laxiflorum Pax.
3350. Acer catalpifolium, I. 87.
3352a. Acer Oliverianum Pax.
3353. Ampelopsis heterophylla, var. Gentiliana Gagnepain.
3354. Tetrastigma obtectum Planchon.
3357. Euscaphis japonica, II. 187.
3358. Staphylea holocarpa, var. rosea, II. 186.
3359. Turpinia nepalensis, II. 187.

3360. Euphoria longana, II. 193.
3361. Picrasma quassioides, II. 152.
3362. Tapiscia sinensis, II. 188.
3363. Pistacia chinensis, II. 173.
3364. Koelreuteria apiculata, II. 191.
3365. Spondias axillaris, var. pubinervis, II. 173; III. 440.
3366. Cipadessa baccifera, var. sinensis, II. 159.
3366a. Cipadessa baccifera, II. 159.
3367. Pistacia weinmannifolia, II. 174.
3368. Spondias axillaris, var. pubinervis, II. 173.
3369. Rhus punjabensis, var. sinica, II. 176.
3372. Rhus Delavayi, var. quinquejuga, II. 184.
3373. Spondias axillaris, var. pubinervis, II. 173.
3374. Clausena punctata, II. 140.
3375. Canarium album, II. 155.
3377. Campylotropis trigonoclada, II. 114.
3378. Campylotropis polyantha, II. 114.
3382. Indigofera myosurus, II. 101.
3383. Campylotropis Muehleana, II. 114.
3385. Indigofera scabrida, II. 101.
3386. Indigofera lenticellata, II. 101.
3387, 3387a. Campylotropis Wilsonii, II. 114.
3388. Sophora viciifolia, II. 95.
3389. Sophora glauca, var. albescens, III. 447.
3390. Sophora Wilsonii, II. 94.
3391. Sophora glauca, var. albescens, III. 447.
3392, 3392a. Cladrastis sinensis, II. 97.
3393, 3393a. Sophora japonica, II. 96.
3395. Caragana erinacea, II. 103, 622.
3396. Caragana jubata, II. 103.
3398. Caragana erinacea, II. 103, 622.
3400. Bauhinia hupehana, II. 89.
3401. Bauhinia yunnanensis, II. 89; III. 447.
3403. Piptanthus concolor, II. 99; III. 447.
3406. Acacia intsia, var. oxyphylla, III. 446.
3407. Ormosia Hosiei, II. 94.
3411. Mucuna sempervirens, II. 117.
3412, 3412a. Millettia Dielsiana, II. 101.

WILSON, E. H. (*continued*)

3415. Zanthoxylum stenophyllum, II. 127.
3450. Mezoneurum sinense, II. 93.
3466. Potentilla fruticosa, II. 301.
3467. Potentilla fruticosa, var. parvifolia, II. 304.
3468. Potentilla fruticosa, var. Veitchii, II. 303.
3468a. Potentilla fruticosa, var. parvifolia, II. 304.
3469. Rubus amabilis Focke.
3470. Rubus macilentus Cambessedes.
3471. Rubus tricolor Focke.
3477. Rubus triphyllus Thunberg.
3478. Rubus pileatus Focke.
3479. Rubus mallodes Focke.
3480. Rubus malifolius Focke.
3481. Rubus sepalanthus Focke.
3482. Rubus fusco-rubens Focke.
3483. Rubus thibetanus Franchet.
3486. Rubus ichangensis Hemsley & Kuntze.
3487. Rubus ellipticus Smith, var. obcordatus Focke.
3488. Rubus biflorus, var. quinqueflorus Focke.
3491. Malus yunnanensis, II. 287.
3493. Docynia Delavayi, II. 296.
3494. Malus transitoria, var. toringoides, II. 286.
3495. Malus kansuensis, II. 286.
3496. Sorbus caloneura, II. 269.
3498. Malus Prattii, II. 281.
3505. Stranvaesia Davidiana, I. 192.
3506. Eriobotrya grandiflora, I. 193.
3507. Eriobotrya prinoides, I. 194.
3508. Photinia berberidifolia, I. 191.
3509. Cotoneaster hupehensis, I. 169.
3510. Cotoneaster Dammeri, var. radicans, I. 176.
3513. Cotoneaster breviramea, I. 177.
3514. Cotoneaster racemiflora, var. soongorica, I. 168.
3515a. Cotoneaster bullata, var. macrophylla, I. 164.
3518. Osteomeles Schwerinae, I. 184.
3518a. Osteomeles Schwerinae, var. microphylla, III. 431.
3523. Prunus serrula, var. tibetica, I. 213.
3524. Prunus latidentata, I. 217.
3524a. Prunus trichostoma, I. 216.

3525. Prunus oxyodonta, I. 218.
3525a. Prunus podadenia, I. 258.
3526. Prunus Zappeyana?, var. subsimplex, I. 222.
3527. Prunus trichostoma, I. 216.
3528. Prunus latidentata, I. 217.
3529. Rosa Roxburghii, f. normalis, II. 309.
3530. Rosa multiflora, var. cornea, II. 305.
3532. Rosa Soulieana, II. 314.
3533. Rosa Giraldii, var. venulosa, II. 328.
3534. Rosa Willmottiae, II. 329.
3535, 3535a. Rosa Murielae, II. 326.
3536. Rosa Rubus, II. 308.
3537. Rosa longicuspis, II. 313.
3538. Rosa Banksiae, var. normalis, II. 317.
3539a. Rosa microcarpa, II. 314.
3541. Rosa multiflora, var. cathayensis, II. 304.
3542. Rosa multiflora, var. cathayensis, II. 304.
3542a. Rosa multibracteata, II. 328.
3543. Rosa Moyesii, II. 325.
3544. Rosa Moyesii, f. rosea, II. 325.
3545. Rosa Davidii, II. 322.
3546. Rosa omeiensis, II. 331.
3547. Sibiraea laevigata, var. angustata, I. 455.
3548. Spiraea alpina, I. 440.
3549. Spiraea Veitchii, I. 449.
3550. Spiraea papillosa, I. 443.
3553. Spiraea mollifolia, I. 441.
3554. Spiraea mollifolia, I. 441.
3555. Spiraea japonica, var. ovalifolia, I. 452.
3555a. Spiraea Rosthornii, I. 451.
3556, 3556a. Spiraea myrtilloides, I. 440.
3557. Spiraea Schneideriana, I. 449.
3557a. Spiraea Schneideriana, var. amphidoxa, I. 450.
3558. Neillia longiracemosa, I. 434.
3559. Neillia affinis, I. 434.
3560. Neillia affinis, I. 434.
3562. Schizophragma integrifolium, I. 41.
3563. Hydrangea Davidii, I. 25.
3566. Deutzia rubens, I. 13.
3567, 3567a. Deutzia longifolia, I. 13.
3568. Deutzia glomeruliflora, I. 10.
3569. Ribes tenue Janczewski.

WILSON, E. H. (*continued*)
3570. Ribes Davidii Franchet.
3571. Ribes luridum Hooker f. & Thomson.
3572. Ribes glaciale Wallich.
3573. Ribes luridum Hooker f. & Thomson.
3573a. Ribes moupinense Franchet.
3574. Ribes alpestre, β giganteum Janczewski.
3575. Ribes longeracemosum Franchet.
3576, 3576a. Ribes Meyeri Maximowicz.
3577. Ribes Meyeri Maximowicz.
3578. Ribes himalayense Decaisne.
3579, 3579a. Ribes Maximowiczii Batalin.
3648. Melastoma normale, II. 421.
3690. Acanthopanax Wilsonii, II. 560.
3692. Aralia chinensis, var. glabrescens, II. 567.
3693. Acanthopanax setchuenensis, II. 559.
3693a. Acanthopanax leucorrhizus, var. fulvescens, II. 558.
3696. Helwingia chinensis, II. 571.
3697. Brassaiopsis fatsioides, II. 556; III. 455.
3699. Alangium Faberi, II. 552.
3700. Camptotheca acuminata, II. 254.
3701. Aucuba chinensis, II. 572.
3702. Davidia involucrata, II. 255.
3717. Aralia Wilsonii, II. 567.
3719. Leycesteria formosa, var. stenosepala, I. 312.
3720. Abelia Graebneriana, I. 118.
3721. Abelia Zanderi, I. 121.
3722. Abelia myrtilloides, I. 120.
3741. Lonicera mupinensis, I. 138.
3754. Lonicera chaetocarpa, I. 137.
3754c. Lonicera montigena, I. 143 (by error as No. 375c).
3756. Wendlandia longidens, III. 392.
3757. Leptodermis oblonga, III. 403.
3759. Mussaenda divaricata, III. 394.
3760. Leptodermis parvifolia, III. 404.
3761. Premna subcapitata, III. 458.
3762. Lasianthus Henryi, III. 401.
3763. Leptodermis oblonga, III. 403.
3910. Cassiope selaginoides, I. 551; III. 445.
3911. Gaultheria nummularioides, I. 555.

3912. Enkianthus deflexus, I. 550.
3914. Vaccinium moupinense, I. 560.
3915. Gaultheria trichophylla, I. 556.
3916. Gaultheria Veitchiana, I. 554.
3917, 3917a. Vaccinium fragile, I. 559.
3918, 3918a. Vaccinium Donianum, var. laetum, I. 558.
3920. Pieris ovalifolia, var. lanceolata, I. 552.
3921a. Pieris ovalifolia, var. lanceolata, I. 552.
3922. Pieris villosa, I. 553.
3923. Vaccinium viburnoides, I. 561.
3924. Vaccinium urceolatum, I. 560.
3927. Clethra monostachya, I. 501.
3929. Rhododendron trichostomum, I. 505.
3930. Rhododendron rufescens, I. 503.
3931. Rhododendron Sargentianum, I. 504.
3932. Rhododendron flavidum, I. 512.
3933. Rhododendron Sargentianum, I. 504.
3934. Rhododendron intricatum, I. 505.
3935. Rhododendron nitidulum, var. nubigenum, I. 510.
3936. Rhododendron fastigiatum, I. 507.
3937. Rhododendron moupinense, I. 525.
3938. Rhododendron dendrocharis, I. 525.
3939. Rhododendron lutescens, I. 516.
3942. Rhododendron yanthinum, I. 518.
3943. Rhododendron ambiguum, I. 518.
3944. Rhododendron villosum, I. 524.
3945. Rhododendron villosum, I. 524.
3946. Rhododendron villosum, I. 524.
3947. Rhododendron Davidsonianum, I. 515.
3949. Rhododendron polylepis, I. 521.
3950. Rhododendron Davidsonianum I. 515.
3951. Rhododendron obiculare, I. 540.
3952. Rhododendron Wiltonii, I. 531.
3953. Rhododendron taliense, I. 533.
3955. Rhododendron Wasonii, I. 532.
3956. Rhododendron Wasonii, I. 532; III. 444.
3957. Rhododendron Przewalskii, I. 534.

Wilson, E. H. (continued)
3958, 3958a. Rhododendron Faberi, I. 533.
3959. Rhododendron Faberi, I. 533.
3961. Rhododendron Faberi, I. 533.
3962. Rhododendron argyrophyllum, I. 525.
3962a. Rhododendron argyrophyllum, var. omeiense, I. 527.
3963. Rhododendron argyrophyllum, var. cupulare, I. 526.
3964. Rhododendron Watsonii, I. 545.
3965. Rhododendron insigne, I. 528.
3966. Rhododendron longipes, I. 528.
3967. Rhododendron floribundum, I. 535.
3968. Rhododendron Przewalskii, I. 534.
3969. Rhododendron Wasonii, I. 532.
3970. Rhododendron taliense, I. 533.
3971. Rhododendron Souliei, I. 537.
3972. Rhododendron oreodoxa, I. 540.
3973. Rhododendron longesquamatum, I. 529.
3975. Rhododendron decorum, I. 541.
3976. Rhododendron pachytrichum, I. 530.
3978. Rhododendron Davidii, I. 543.
3979. Rhododendron calophytum, I. 544.
4056. Ardisia crispa, II. 581.
4057. Maesa Wilsonii, II. 584.
4058. Antidesma delicatulum, II. 522.
4059. Maesa castaneifolia, II. 583.
4060. Diospyros kaki, var. silvestris, II. 590.
4061. Diospyros sinensis, II. 591.
4062. Diospyros sinensis, II. 591.
4063. Diospyros mollifolia, II. 591.
4064. Skimmia melanocarpa, II. 138.
4065. Styrax roseus, I. 291.
4066. Styrax dasyanthus, I. 289.
4067. Symplocos stellaris, II. 597.
4068. Symplocos caudata, II. 595.
4069. Symplocos caudata, II. 595.
4070. Symplocos laurina, II. 594.
4072. ?Antidesma Seguinii Léveillé.
4074. Jasminum Beesianum, II. 615.
4074a. Jasminum officinale, II. 613.
4075. Jasminum urophyllum, var. Wilsonii, II. 613.
4077. Jasminum humile, I. 614.
4078. Jasminum humile, I. 614.

4079, 4079a. Osmanthus serrulatus, II. 610.
4080. Syringa Sweginzowii, I. 301.
4080a. Syringa Potaninii, I. 297.
4081. Syringa Komarowii, var. Sargentiana, III. 433.
4082. Syringa pinnatifolia, I. 297.
4083. Ligustrum Delavayanum, II. 601.
4085. Fraxinus chinensis, var. typica, II. 260; III. 450.
4086. Ligustrum lucidum, II. 603.
4087. Fraxinus Paxiana, II. 623.
4088. Fraxinus chinensis, var. acuminata, II. 261.
4088a. Fraxinus chinensis, II. 623.
4090. Secamone emetica, III. 348.
4092. Ecdysanthera rosea, III. 342.
4107. Trachelospermum axillare, III. 335.
4108. Vallaris grandiflora, III. 342 (by error as No. 4158).
4116. Buddleia Lindleyana, I. 564.
4117. Buddleia albiflora, var. Giraldii, I. 569.
4118. Buddleia alata, I. 570.
4119. Buddleia nivea, var. yunnanensis, I. 570.
4120. Buddleia Davidii, I. 567.
4121. Buddleia nivea, I. 570.
4170. Berberis dictyophylla, var. epruinosa, III. 434.
4207. Paulownia Fargesii, I. 575.
4265. Myrioneuron Faberi, III. 410.
4289. Catalpa Duclouxii, I. 577.
4306. Premna ligustroides, III. 372.
4307. { Elsholtzia fruticosa, III. 381.
Elsholtzia fruticosa, var. tomentella, III. 382.
4308, 4308a. Vitex Negundo, var. incisa, III. 373.
4309. Caryopteris glutinosa, III. 379.
4312. Caryopteris incana, III. 378.
4313. Elsholtzia dependens, III. 383.
4318. Callicarpa rubella, var. Hemsleyana, III. 370.
4319. Plectranthus leucophyllus, III. 384.
4321. Plectranthus discolor, III. 384.
4322. Plectranthus discolor, III. 384.
4418. Elaeagnus stellipila, II. 415.
4419. Elaeagnus umbellata, II. 410.
4420. Elaeagnus Bockii, II. 416.
4421, 4421a. Hippophae rhamnoides, var. procera, II. 409.

WILSON, E. H. (*continued*)

4422, 4422a. Litsea Wilsonii, II. 78.
4424. Actinodaphne reticulata, II. 75.
4425. Actinodaphne confertifolia, II. 74.
4426. {Litsea Veitchiana, II. 76.
{Machilus Bournei, II. 73.
4427. Lindera umbellata, II. 81.
4428. Lindera Prattii, II. 83.
4429. Lindera strychnifolia, var. Hemsleyana, II. 82; III. 446.
4431. Daphne leuconeura, II. 548.
4432. Wikstroemia micrantha, II. 530.
4433. Daphne angustiloba, II. 547.
4434. Daphne rosmarinifolia, II. 549.
4435. Wikstroemia stenophylla, II. 530.
4436. Wikstroemia effusa, II. 538.
4438. Daphne acutiloba, II. 539.
4439. Daphne retusa, II. 541.
4440. Mappia pittosporoides, II. 190.
4441. Loranthus caloreas, var. Fargesii, III. 315.
4443. Osyris Wightiana, III. 320.
4444. Mallotus philippinensis, II. 526.
4446. Glochidion puberum, II. 518.
4447. Croton Tiglium, II. 523.
4448. Excoecaria acerifolia, II. 528.
4449. Antidesma delicatulum, II. 522.
4450. Schoepfia jasminodora, III. 321.
4453. Andrachne hirsuta, II. 516.
4467. Morus acidosa, III. 297.
4468. Morus cathayana, III. 292.
4469. Celtis Vandervoetiana, III. 267, 457.
4472. Ficus clavata?, III. 310.
4480. Antidesma delicatulum, II. 522.
4481. Aristolochia moupinensis, III. 324.
4482. Viscum articulatum, III. 318.
4484. Viscum articulatum, III. 318.
4485. Loranthus Balfourianus, III. 315.
4486. Loranthus chinensis, III. 316.
4487. Engelhardtia chrysolepis, III. 186.
4488. Carpinus Henryana, II. 429.
4489. Carpinus Turczaninovii, var. ovalifolia, II. 427.
4490. Betula Potaninii, II. 459.
4491. Betula Delavayi, II. 460.
4492, 4492a. Betula utilis, var. Prattii, II. 457.
4493. Betula Delavayi, II. 460.

4494. Betula japonica, var. szechuanica, II. 461; III. 453.
4495. Betula utilis, var. Prattii, II. 457.
4496. Betula utilis, var. Prattii, II. 457.
4497. Alnus cremastogyne, II. 488.
4498. Alnus lanata, II. 488.
4499. Fagus longipetiolata, III. 190.
4500. Juglans cathayensis, III. 185.
4501. Quercus aliena, III. 214.
4502. Castanopsis hystrix, III. 197.
4503, 4503a. Quercus semicarpifolia, III. 221.
4504. Quercus spathulata, III. 226.
4508. Castanopsis platyacantha, III. 200.
4509. Corylus Sieboldiana, var. mandshurica, II. 454.
4510. Salix Bockii, III. 71.
4511. Salix brachista, III. 145.
4512. Salix Souliei, III. 62.
4513. Salix microphyta, III. 62.
4515. Salix oreinoma, III. 138.
4516. Salix cathayana, III. 57.
4517. Salix cathayana, III. 57.
4518. Salix paraplesia, III. 40.
4518a. Salix paraplesia, III. 40.
4519. Salix opsimantha, III. 63.
4519a. Salix opsimantha, III. 63.
4520. Salix Ernesti, III. 47.
4521. Salix Ernesti, III. 47.
4522. Salix moupinensis, III. 46.
4524. Salix isochroma, III. 122.
4525. Salix oritrepha, III. 113.
4526. Salix magnifica, III. 44.
4527. Populus szechuanica, var. tibetica, III. 33.
4530. Betula utilis, var. Prattii, II. 457.
4538. Populus lasiocarpa, III. 17.
4539. Populus suaveolens, III. 18.
4693. Clematis montana, var. grandiflora, I. 333.
4694. Clematis Gouriana, I. 339.
4695. Clematis Armandi, I. 326.
4696. Clematis lasiandra, I. 322.
4697. Clematis grata, var. grandidentata, I. 338.
4698. Clematis uncinata, I. 327.
4718. Sinomenium acutum, I. 387.
4721. Tetracentron sinense, I. 417.
4722. Schisandra Henryi, I. 413.
4726. Berberis Silva-Taroucana, I. 370.
4741. Polygala arillata, II. 160.
4742. Polygala Wattersii, II. 161.

Wilson, E. H. (*continued*)

4744. Pittosporum glabratum, III. 326.
4745. Pittosporum daphniphylloides, III. 326.
4752. Carrierea calycina, I. 284.
4753. Elaeocarpus japonicus, II. 360.
4754. Urena lobata, II. 373.
4755. Thea elongata, II. 392.
4756. Thea oleifera, II. 393.
4757. Thea sinensis, II. 391.
4758. Gordonia axillaris, II. 394.
4760. Actinidia coriacea, II. 384.
4761. Actinidia kolomikta, II. 380.
4762. Actinidia callosa, var. Henryi, II. 382.
4763. Clematoclethra scandens, II. 387.
4764. Actinidia tetramera, II. 381.
4765. Actinidia venosa, II. 383.
4766. Clematoclethra Faberi, II. 387.
4767. Eurya acuminata, II. 400.
4768. Eurya acuminata, var. multiflora, II. 401.
4769. Zanthoxylum alatum, var. planispinum, f. ferrugineum, II. 125.
4770a. Zanthoxylum dimorphophyllum, II. 126.
4772. Evodia Baberi, II. 131.
4773. Zanthoxylum echinocarpum, II. 128.
4774. Cipadessa baccifera, var. sinensis, II. 159.
4776. Berchemia hypochrysa, II. 214.
4777. Berchemia yunnanensis, var. trichoclada, II. 217.
4778. Paliurus ramosissimus, II. 210.
4779. Rhamnus leptophyllus, II. 239.
4780. Rhamnus Hemsleyanus, II. 234.
4781. Celastrus Rosthorniana, II. 351.
4782. Celastrus glaucophylla, II. 347.
4783. Ilex szechwanensis Loesener.
4784. Evonymus subsessilis, var. latifolia, I. 489.
4784a. Evonymus acanthocarpa, var. sutchuenensis, I. 490.
4785. Evonymus subsessilis, I. 489.
4786. Evonymus myriantha, I. 487.
4787. Evonymus myriantha, I. 487.
4794. Ilex Franchetiana Loesener.
4795. Ilex macrocarpa Oliver.
4796. Ilex Franchetiana Loesener.
4800. Rhamnus Esquirolii, II. 233.
4802, 4802a. Perrottetia racemosa, II. 359.

4803. Turpinia nepalensis, II. 187.
4804. Euscaphis japonica, II. 187.
4805. Gordonia sinensis, II. 395.
4806. Sabia gracilis, II. 198.
4807. Sageretia Henryi, II. 623.
4808. Stachyurus salicifolius Franchet.
4809. Gardneria multiflora, I. 563.
4810. Gardneria multiflora, I. 563.
4811. Rhus succedanea, II. 182.
4812. Rhus javanica, II. 178.
4813. Rhus punjabensis, var. sinica, II. 176.
4814. Meliosma cuneifolia, II. 199.
4816. Sageretia omeiensis, II. 230.
4817. Meliosma Fischeriana, II. 203.
4818. Acer laevigatum Wallich.
4821. Acer Oliverianum Pax.
4822. Vitis betulifolia Diels & Gilg.
4823. Ampelopsis heterophylla, var. Delavayana Gagnepain.
4824. Tetrastigma obtectum Planchon.
4825. Vitis reticulata Pampanini.
4826. Acer Davidii Franchet.
4828. Bauhinia hupehana, II. 89.
4830. Desmodium podocarpum, var. szechuenense, II. 104.
4831. Caesalpinia sepiaria, II. 92.
4832. Cladrastis sinensis, II. 97.
4833. Dalbergia Dyeriana, II. 115.
4834. Millettia Dielsiana, II. 101.
4835. Millettia Dielsiana, II. 101.
4840. Spiraea Miyabei, var. tenuifolia, I. 455.
4841. Spiraea japonica, var. ovalifolia, I. 452.
4842. Spiraea japonica, var. acuminata, I. 452.
4844. Rubus xanthoneurus Hemsley.
4845. Rubus Playfairianus Hemsley.
4846. Rubus pectinaris Focke.
4847. Rubus gracilis Roxburgh.
4848. Rubus malifolius Focke.
4849. Rubus eucalyptus Focke.
4849a. Rubus pileatus Focke.
4850. Rubus inopertus Focke.
4851. Rubus pacatus Focke.
4852. Rubus pungens Cambessedes.
4859. Prunus cyclamina, var. biflora, I. 243.
4860. Prunus trichostoma, I. 216.
4864. Sorbus caloneura, II. 269.
4867. Sorbus aronioides, II. 268.

WILSON, E. H. (*continued*)
4871, 4871a. Pyracantha crenulata, I. 177.
4872. Stranvaesia Davidiana, I. 192.
4873. Photinia serrulata, I. 184.
4874. Rosa Roxburghii, II. 319.
4876. Rosa longicuspis, II. 313.
4877. Rosa microcarpa, II. 314.
4879. Rosa omeiensis, II. 331.
4880. Rosa Rubus, II. 308.
4881, 4881a, b. Rosa multiflora, var. cathayensis, II. 304.
4882. Ribes tenue Janczewski.
4883. Philadelphus subcanus Koehne.
4884. Deutzia pilosa, I. 8.
4885. Schizophragma hypoglaucum, I. 43.
4886. Neillia affinis, I. 434.
4890. Dichroa febrifuga, I. 43.
4899. Hydrangea xanthoneura, I. 26.
4900. Hydrangea xanthoneura, I. 26.
4901. Hydrangea anomala, I. 34.
4902. ⎰ Hydrangea strigosa, I. 31.
⎱ Hydrangea strigosa, var. macrophylla, I. 32.
4903. Hydrangea Rosthornii, I. 33.
4904. Itea omeiensis, III. 421.
4905. Maesa Wilsonii, II. 584.
4907. Melastoma normale, II. 421.
4936. Acanthopanax leucorrhizus, II. 557.
4937. Acanthopanax leucorrhizus, II. 557.
4938. Nothopanax Rosthornii, II. 557.
4942. Helwingia himalaica, II. 571.
4945. Mappia pittosporoides, II. 190.
4946. Alangium chinense, II. 552.
4947. Alangium Faberi, II. 552.
4949. Camptotheca acuminata, II. 254.
4950. Cornus paucinervis, II. 576.
4951. Cornus controversa, II. 573.
4951a. Cornus macrophylla, II. 575.
4952. Cornus Hemsleyi, II. 574.
4953. Cornus chinensis, II. 577.
4954. Cornus capitata, II. 578.
4955. Aucuba chinensis, II. 572.
4956. Skimmia Fortunei, II. 139.
5015. Ligustrum sinense, var. nitidum, II. 606.
5016. Ligustrum pedunculare, II. 609.
5017, 5017a. Ligustrum sinense, var. nitidum, II. 606.
5031. Abelia Graebneriana, I. 118.
5038. Buddleia Davidii, I. 567.

5040. Osmanthus fragrans, II. 609.
5042. Jasminum lanceolarium, II. 612.
5043. Myrioneuron Faberi, III. 410.
5062. Ardisia Henryi, II. 582.
5063. Ardisia japonica, II. 582.
5066. Symplocos laurina, II. 594.
5067. Symplocos caudata, II. 595.
5069, 5069a. Symplocos paniculata, II. 593.
5070. Melodinus Hemsleyanus, III. 331.
5072. Trachelospermum cathayanum, III. 333.
5073. Trachelospermum axillare, III. 335.
5100. Callicarpa Giraldiana, III. 366.
5101. Premna ligustroides, III. 372.
5134. Vaccinium Donianum, I. 557.
5135. Elaeocarpus omeiensis, II. 360.
5136. Vaccinium urceolatum, I. 560.
5137. Pieris ovalifolia, var. elliptica, I. 552.
5137. Rhododendron argyrophyllum, I. 525.
5137a. Rhododendron argyrophyllum, var. omeiense, I. 527.
5138. Vaccinium Dunalianum, I. 560.
5140. Rhododendron stamineum, I. 546.
5142. Rhododendron Faberi, I. 533.
5143. Rhododendron indicum, var. ignescens, I. 547.
5144. Rhododendron ambiguum, I. 518.
5168. Mallotus repandus, II. 526.
5169. Mallotus tenuifolius, II. 525.
5172. Sauropus albicans, II. 518.
5173. Andrachne montana, II. 517.
5174. Daphniphyllum macropodum, II. 522.
5175. Litsea Faberi, II. 79.
5176. Litsea sericea, II. 75.
5177. Actinodaphne reticulata, II. 75.
5178. Cinnamomum inunctum, II. 68.
5179. Actinodaphne cupularis, II. 75.
5180. Neolitsea umbrosa, II. 79.
5181. Lindera strychnifolia, var. Hemsleyana, II. 82.
5182. Litsea populifolia, II. 77.
5183. Cinnamomum Wilsonii, II. 66.
5184. Machilus Bournei, II. 73.
5184a. Phoebe Sheareri, II. 621.
5185. Alseodaphne omeiensis, II. 70.
5186. Castanea Henryi, III. 196.

WILSON, E. H. (*continued*)
5187. Castanopsis hystrix, III. 197.
5188. Castanopsis platyacantha, III. 200.
5189. Quercus serrata, III. 217.
5190. Quercus oxyodon, var. Fargesii, III. 229.
5191. Carpinus polyneura, var. Wilsoniana, II. 443.
5192. Betula luminifera, II. 455.
5193. Salix omeiensis, III. 122.
5194. Salix heterochroma, III. 61.

Wilson, Ernest Henry (Veitch Expedition Seed numbers).
98. Cotoneaster amoena, I. 165.
99a (in Hort. Veitch). Ribes moupinense Franchet.
118a/2. See No. 1118a.
219. Rosa odorata, var. gigantea, II. 338.
305. Ulmus Bergmanniana, III. 240.
357. Malus prunifolia, var. rinki, II. 279.
448. Corylopsis sinensis, I. 424.
512 (in Hort. Veitch). Rosa banksiopsis, II. 322.
519. Cotoneaster Dielsiana, I. 166.
538. Aristolochia heterophylla, III. 323.
573 (in Hort. Veitch). Ilex purpurea, var. Oldhamii Loesener.
596. Cotoneaster acutifolia, var. villosula, I. 158.
623. Cotoneaster Zabelii, I. 166.
628. Sorbus Zahlbruckneri, II. 272.
629. Ampelopsis Watsoniana, III. 427.
680. Acer tetramerum, f. elobulatum Rehder.
688. Magnolia denudata, var. purpurascens, I. 401.
728. Acer cappadocicum, f. tricaudatum Rehder.
737, 737a. Carpinus laxiflora, var. macrostachya, II. 425.
766. Malus theifera, var. rosea, II. 284.
766a. Malus theifera, II. 283.
775. Lithocarpus Henryi, III. 209.
777. Castanea Henryi, III. 196.
800. Quercus oxyodon, III. 228.
803. Ilex Pernyi, var. Veitchii Rehder.
804. Ilex Fargesii Franchet.
831. Acer Oliverianum Pax.
835. Acer oblongum Wallich.

839. Malus theifera, II. 283.
862 (in Hort. Veitch). Acer longipes Rehder.
863. Acer pictum, var. parviflorum, f. tricuspis Rehder.
880. Clematis lasiandra, I. 322.
889, 889a. Acer amplum Rehder.
900. Sarcococca Hookeriana, var. humilis, II. 164.
926. Tilia tuan, var. chinensis, II. 369.
927. Sarcococca ruscifolia, II. 163.
946. Ilex Fargesii Franchet.
952. Tsuga chinensis, II. 37.
1025. Ehretia acuminata, III. 363.
1028 (in Hort. Veitch). Ilex Aquifolium, var. chinensis Loesener.
1046. Ribes moupinense Franchet.
1047. Rosa setipoda, II. 323.
1053. Ribes moupinense Franchet.
1083. Prunus Wilsonii Koehne.
1087. Potentilla fruticosa, var. Veitchii, II. 303.
1099. Thea cuspidata, II. 390.
1118a. Deutzia globosa, I. 20.
1156. Ilex Fargesii Franchet.
1157. Betula albo-sinensis, II. 457.
1160a. Cotoneaster acutifolia, var. villosula, I. 158.
1185. Bischofia javanica Blume.
1190. Ilex yunnanensis, β gentilis, III. 425.
1196. Acanthopanax setchuenensis, II. 559.
1204. Lithocarpus cleistocarpa, III. 205.
1223. Quercus Engleriana, III. 220.
1246. Philadelphus incanus Koehne.
1253. Deutzia reflexa, I. 20.
1282. Picea brachytyla, II. 33.
1285. Stewartia sinensis, II. 395.
1288. Cotoneaster Dielsiana, I. 166.
1309. Picea Watsoniana, II. 27.
1315. Cotoneaster Harroviana, I. 173.
1320. Berberis aggregata, var. Prattii, III. 443.
1322. Ilex dipyrena, c. paucispinosa, Loesener.
1422 (in Hort. Veitch). Clematis Veitchiana, I. 324; III. 434.
1423. Aristolochia moupinensis, III. 324.
1428 (in Hort. Veitch). Buddleia nivea, I. 570.

Wilson, E. H. (*continued*)
1440. Rosa Davidii, II. 322.
1447. Rosa Sweginzowii, II. 324.
1492. Rosa sertata, II. 327.
1507. Cotoneaster ambigua, I. 159.
1515. Abies Delavayi, II. 41.
1530. Picea complanata, II. 35.
1535. Rhododendron Davidsonianum, I. 515.
1541. Rhododendron oreodoxa, I. 540.
1563. Rosa Prattii, II. 329.
1569. Tilia intonsa, II. 365.
1570. Abies Beissneriana, II. 46.
1682. Berberis parvifolia, III. 436.
1700. Clematis glauca, var. akebioides, I. 342.
1718. Cotoneaster obscura, I. 161.
1719. Berberis Tischleri, III. 436.
1720. Malus Halliana, II. 285.
1723. Cotoneaster ambigua, I. 159.
1727. Rosa Prattii, II. 329.
1730. Malus transitoria, var. toringoides, II. 286.
1731. Berberis Tischleri, III. 436.
1750. Carpinus laxiflora, var. macrostachya, II. 425.
1766. Rhododendron yanthinum, I. 518.
1768. Philadelphus purpurascens Rehder.
1772. Philadelphus Magdalenae Koehne.
1773. Rhododendron flavidum, I. 512.
1778a (in Hort. Veitch). Philadelphus purpurascens Rehder.
1782. Rhododendron decorum, I. 541.
1797. Clematis trullifera, I. 324.
1834. Picea likiangensis, II. 31.
1836. Picea likiangensis, II. 31.
1837. Acer pictum, var. parviflorum Schneider.
1838. Ilex szechwanensis, III.
1841. Rosa Murielae, II. 326.
1857. Rhododendron polylepis, I. 521.
1862. Rhododendron villosum, I. 524.
1879. Rhododendron yanthinum, I. 518.
1888. Rhododendron Sargentianum, I. 504.
1889. See No. 98.
1930. Berberis Beaniana, III. 439.

Wright, Charles.
22. Polygala arillata, II. 160.
26. Pittosporum glabratum, III. 326.

58. Gordonia axillaris, II. 394.
90. Turpinia nepalensis, II. 187.
107. Spondias axillaris, II. 172.
170. Vaccinium bracteatum, var. Wrightii, III. 559.
192. Trachelospermum divaricatum, III. 338.
236. Mussaenda erosa, III. 399.
284. Buxus liukiuensis, II. 168.
318. Ligustrum myrianthum, II. 607.
380. Callicarpa rubella, III. 370.
397. Elaeagnus Loureirii, II. 416.
415. Ehretia acuminata, III. 363.
422. Buxus microphylla, var. sinica, II. 165.
457. Celtis sinensis, III. 277.
466. Quercus bambusifolia, III. 235.
467. Castanopsis fissa, III. 203.
469. Chloranthus brachystachys, III. 15.
473. Spondias axillaris, II. 172.
608. Alangium chinense, II. 552.

Zimmerman, R.
184. Evonymus alata, I. 493.
210. Callicarpa japonica, III. 368.
216. Cocculus trilobus, I. 388.
223. Grewia parviflora, II. 371.
243. Lespedeza formosa, II. 107.
261. Lagerstroemia indica, II. 418.
271. Lagerstroemia indica, II. 418.
276. Lespedeza formosa, II. 107.
335. Deutzia hamata, I. 22.
344. Spiraea Fritschiana, var. angulata, I. 453.
348. Deutzia hamata, I. 22.
349. Deutzia glabrata, I. 24.
352. Castanea mollissima, III. 192.
361. Picrasma quassioides, II. 152.
370. Elaeagnus umbellata, II. 410.
374. Ailanthus altissima, II. 153; III. 449.
382. Spiraea Fritschiana, I. 453.
422. Styrax japonicus, I. 291.
442. Vitex Negundo, var. incisa, III. 374.
460. Zanthoxylum Bungei, var. Zimmermannii, II. 122.
463. Firmiana simplex, II. 377.
529. Buxus microphylla, var. sinica, II. 165.
535. Daphne genkwa, II. 538.

GENERAL INDEX

Names of families and subfamilies are in small capitals; names of admitted genera, subgenera, sections, species and varieties are in roman type; synonyms in *italics*.

Abelia, I. 118.
Abelia adenotricha, I. 129.
Abelia angustifolia, I. 124, 128.
Abelia Aschersoniana, I. 124, 127.
Abelia biflora, I. 125, 128.
Abelia Buchwaldii, I. 123, 125.
Abelia chinensis, I. 121, 124, 127.
Abelia chinensis × uniflora, I. 128.
Abelia coriacea, I. 124, 127.
Abelia corymbosa, I. 124, 128.
Abelia Davidii, I. 128.
Abelia Dielsii, I. 125, 128.
Abelia Engleriana, I. 120, 123, 126.
Abelia floribunda, I. 124, 127.
Abelia floribunda hybrida, I. 128.
Abelia Graebneriana, I. 118, 123, 126; III. 429.
Abelia grandiflora, I. 128.
Abelia gymnocarpa, I. 123, 125.
Abelia Hanceana, I. 121.
Abelia hirsuta, I. 127.
Abelia Koehneana, I. 124, 126.
Abelia longituba, I. 123, 126.
Abelia macrotera, I. 123, 126.
Abelia multiflora hybrida, I. 128.
Abelia myrtilloides, I. 120, 124, 127.
Abelia onkocarpa, I. 125, 128.
Abelia parvifolia, I. 121, 124.
Abelia rupestris, I. 124, 127.
Abelia rupestris, I. 128.
Abelia rupestris alba, I. 128.
Abelia rupestris, var. *grandiflora*, I. 128.
Abelia rupestris hybrida, I. 128.
Abelia Schumannii, I. 121, 124, 127.
Abelia serrata, I. 123, 125.
Abelia serrata, I. 118, 125.
Abelia shikokiana, I. 128.
Abelia spathulata, I. 123, 125.
Abelia speciosa, I. 127.
Abelia splendens, I. 129.
Abelia tereticalyx, I. 124, 127.
Abelia triflora, I. 124, 127.
Abelia triflora, var. parvifolia, I. 128.

Abelia umbellata, I. 123, 125, 128.
Abelia uniflora, I. 123, 126.
Abelia uniflora, I. 118.
Abelia Zanderi, I. 121, 125, 128; III. 429.
Abelicea acuminata, III. 287.
Abelicea hirta, III. 287.
Abelicea Keaki, III. 287.
Abies, II. 41.
Abies Beissneriana, II. 46, 621.
Abies brachytyla, II. 33.
Abies chensiensis, II. 44.
Abies chinensis, II. 37.
Abies Davidiana, II. 39.
Abies Delavayi, II. 41; III. 446.
Abies dumosa, var. *chinensis*, II. 36, 37.
Abies Fargesii, II. 48.
Abies Fargesii, II. 41.
Abies Fargesii, var. *sutchuensis*, II. 48.
Abies Faxoniana, II. 42; III. 446.
Abies firma, II. 47.
Abies firma, II. 44, 46.
Abies Fortunei, II. 41.
Abies Fortuni, II. 40.
Abies Jezoensis, II. 40.
Abies Kaempferi, II. 21.
Abies lanceolata, II. 50.
Abies likiangensis, II. 31.
Abies Mariesii, II. 49.
Abies nephrolepis, II. 49.
Abies recurvata, II. 44.
Abies sacra, II. 39.
Abies Schrenkiana, II. 29.
Abies sibirica, II. 50.
Abies sibirica, var. nephrolepis, II. 49.
Abies Smithiana, II. 29.
Abies sp., II. 44.
Abies squamata, II. 48.
Abies sutchuenensis, II. 48.
Abies thié-sha, II. 37.
Abies Tsuga, II. 37.
Abies Veitchii, II. 49.
Abies Veitchii, II. 50.

579